Birkhäuser

Yuming Qin

Integral and Discrete Inequalities and Their Applications

Volume II: Nonlinear Inequalities

Birkhäuser

Yuming Qin
Department of Applied Mathematics
College of Science
Donghua University
Shanghai, China

ISBN 978-3-319-33303-8 ISBN 978-3-319-33304-5 (eBook)
DOI 10.1007/978-3-319-33304-5

Library of Congress Control Number: 2016950915

Mathematics Subject Classification (2010): 26D20, 34A40, 35A23

Printed on acid-free paper

This book is published under the trade name Birkhäuser, www.birkhauser-science.com
The registered company is Springer International Publishing AG

To my Parents Zhenrong Qin and Xilan Xia
and to my wife and son Yu Yin and Jia Qin

Preface

This book is Part II of a two-volume work on inequalities. It concentrates on one- and multidimensional nonlinear integral and discrete Gronwall-Bellman-type inequalities and complements the book on linear inequalities.

Integral and discrete inequalities are very important tools in classical analysis and play a crucial role in establishing the well-posedness of the related equations, i.e. differential, difference and integral equations.

Chapters 1–3 and 5–7 provide introductions to one-dimensional and multidimensional, respectively, nonlinear continuous integral inequalities, discrete (difference) inequalities and discontinuous integral inequalities. Chapters 4 and 8 study relevant applications of these inequalities.

I am thankful for the generous help that I received in the process of writing this book and for the support I experienced during my visits at the University of Potsdam (Germany), TU Bergakademie Freiberg (Germany), Georg-August-Universität Göttingen (Germany), Chinese University of Hong Kong, the Institute of Mathematics of the Czech Academy of Sciences of the Czech Republic, the University of São Paulo (Brazil), National Laboratory for Scientific Computing (LNCC, Brazil) and the University of Minnesota at Duluth. I would like to take this opportunity to thank these universities for their hospitality. I also greatly appreciate the help of Professors Bert-Wolfgang Schulze, Eduard Feireisl, Michael Reissig, Ingo Witt, Zhouping Xin, Tofu Ma, Jaime Muñoz Rivera and Zhuangyi Liu. Moreover, I would like to thank my students Lan Huang, Xinguang Yang, Shuxian Deng, Xin Liu, Zhiyong Ma, Taige Wang, Guili Hu, Xiaoke Su, Yaodong Yu, Lili Xu, Dongjie Ge, Xiaona Yu, Songtao Li, Tao Li, Xiaozhen Peng, Baowei Feng, Ming Zhang, Wei Wang, Haiyan Li, Jianlin Zhang, Xing Su, Yang Wang, Jie Cao, Tianhui Wei, Jia Ren, Jianpeng Zhang, Linlin Sun, Pengda Wang, Cheng Chen, Ying Wang, Mei Feng, Xiuzhen Zheng, Jiaolong Li, Wensong Hu, Xiaolei Dong, Ying Liu and Qiujuan Cai for their hard work in typewriting and checking the galley proof of the book manuscript.

I also acknowledge the NNSF of China for its support. Currently, this book project is being supported by the NNSF of China with contract nos. 11031003, 11271066 and 11671075 and by a grant from the Shanghai Municipal Education Commission with contract no. 13ZZ048.

Last but not least, I would like to express my deepest thanks to my parents Zhenrong Qin and Xilan Xia, sisters Yujuan Qin and Yuzhou Qin, brother Yuxing Qin, my wife Yu Yin and my son Jia Qin for their great help, constant concern and advice.

Shanghai, China Yuming Qin

Contents

Volume II: Nonlinear Integral and Discrete Inequalities

Chapter 1
Nonlinear One-Dimensional Continuous Integral Inequalities

1.1 Nonlinear One-Dimensional Bellman-Gronwall Inequality, Reid Inequality, Bihari Inequality, Langenhop Inequality and LaSalle Inequality

1.1.1 The One-Dimensional Bihari Inequality, Langenhop Inequality and LaSalle Inequality

The attractive Gronwall-Bellman inequality [259] plays a vital role in studying stability and asymptotic behavior of solution of differential equations (see, e.g., [50, 69]). Many linear and nonlinear generalizations have appeared in the literature [137, 560]. Bihari's inequality [82] is the most important generalization of the Gronwall-Bellman inequality. Several integral inequalities similar to Bellman-Bihari's inequality are introduced in this section.

There can be found a lot of its generalizations in various cases from literature (see, e.g. [42, 207, 307, 355, 396, 507]). An important contribution was made by Bihari [82] for the integral inequality (1.1.1).

Theorem 1.1.1 (The Bihari Inequality [82]) *Assume that $x(t)$ and $v(t)$ are non-negative continuous functions on $[0, \tau)$, and $f(u)$ is a positive non-decreasing continuous function for all $u \in (0, +\infty)$ such that for all $t \in [0, \tau)$,*

$$x(t) \le \eta + \int_0^t v(s)f(x(s))ds, \tag{1.1.1}$$

where $\eta > 0$ is a constant, then for all $t \in [0, \tau_1)$,

$$x(t) \le \Phi^{-1}(\Phi(\eta) + \int_0^t v(s)ds), \tag{1.1.2}$$

Y. Qin, *Integral and Discrete Inequalities and Their Applications*,
DOI 10.1007/978-3-319-33304-5_1

where

$$\Phi(u) = \int_{u_0}^{u} \frac{1}{f(t)} dt, \quad u \geq u_0 > 0,$$

and Φ^{-1} *is the inverse function of* Φ *and*

$$\tau_1 = \sup \left\{ t \in [0, \tau) : \Phi(+\infty) \geq \Phi(\eta) + \int_0^t v(s)ds \right\}.$$

Proof Putting, for all $t \in [0, \tau)$,

$$y(t) := \int_0^t f(x(s))v(s)ds,$$

we have $y(0) = 0$, and from the relation (1.1.1), we derive, for all $t \in [0, \tau)$,

$$y'(t) \leq f(\eta + y(t))v(t).$$

By integrating on $[0, t]$, we thus conclude, for all $t \in [0, \tau)$,

$$\int_0^{y(t)} \frac{ds}{f(\eta + s)} \leq \int_0^t v(s)ds + \Phi(\eta),$$

that is, for all $t \in [0, \tau)$,

$$\Phi(y(t) + \eta) \leq \int_0^t v(s)ds + \Phi(\eta),$$

which gives us the desired estimate (1.1.2). □

The following similar conclusions are obtained by Dragomir and Kim in [207].

Theorem 1.1.2 (The Dragomir-Kim Inequality [207]) *Let* g *be a monotone continuous function in an interval* I*, containing a point* u_0 *which vanishes in* I*. Let* u *and* k *be continuous functions in an interval* $J = [\alpha, \beta]$ *such that* $u(J) \subset I$*, and suppose that* k *is fixed sign in* J*. Let* $a \in I$*.*

(i) Assume that g *is non-decreasing and* k *is non-negative. If for all* $t \in J$*,*

$$u(t) \leq a + \int_\alpha^t k(s)g(u(s))ds,$$

then for all $\alpha \leq t \leq \beta_1$*,*

$$u(t) \leq G^{-1}\left(G(a) + \int_\alpha^t k(s)ds\right),$$

where $G(u) = \int_{u_0}^{u} \frac{dx}{g(x)}$, $u \in I$, *and* $\beta_1 = \min(v_1, v_2)$, *with for all* $\alpha \le t \le v$,

$$\begin{cases} v_1 = \sup\left\{ v \in J : a + \int_\alpha^t k(s)g(u(s))ds \right\} \in I, \\ v_2 = \sup\left\{ v \in J : G(a) + \int_\alpha^t k(s)ds \right\}. \end{cases}$$

(ii) Assume that $J = (\alpha, \beta]$. *If for all* $t \in J$,

$$u(t) \le a + \int_t^\beta k(s)g(u(s))ds,$$

then for all $\alpha_1 \le t \le \beta$,

$$u(t) \le G^{-1}\left(G(a) + \int_t^\beta k(s)ds \right),$$

where $\alpha_1 = \max(\mu_1, \mu_2)$, *with for all* $\mu \le t \le \beta$,

$$\begin{cases} \mu_1 = \sup\left\{ \mu_1 \in J : a + \int_t^\beta k(s)g(u(s))ds \right\} \in I, \\ \mu_2 = \sup\left\{ \mu_2 \in J : G(a) + \int_t^\beta k(s)ds \right\}. \end{cases}$$

The proof of the inequalities in (i), (ii) is similar to that of Theorem 1.1.1, and hence we omit the details.

Theorem 1.1.3 (The Yang Inequality [695]) *Let* $\phi \in C(\mathbb{R}_+, \mathbb{R}_+)$ *be strictly increasing function with* $\phi(+\infty) = +\infty$ *and* $\psi \in C(\mathbb{R}_+, \mathbb{R}_+)$ *be non-decreasing. Let* $c \ge 0$ *be a real constant. Then the following conclusions are true. If* $u, F \in C(\mathbb{R}_+, \mathbb{R}_+)$, *and the integral inequality holds, for all* $t \in \mathbb{R}_+$,

$$\phi[u(t)] \le c + \int_0^t F(s)\psi[u(s)]ds, \tag{1.1.3}$$

then for all $t \in [0, T)$,

$$u(t) \le \phi^{-1}\left(G^{-1}[G(c) + \int_0^t F(s)ds] \right), \tag{1.1.4}$$

where ϕ^{-1}, G^{-1} *are the inverse functions of* ϕ, G *respectively, and*

$$G(z) := \int_{z_0}^{z} \frac{ds}{\psi[\phi^{-1}(s)]}, \; z \ge z_0 > 0 \tag{1.1.5}$$

and $T > 0$ is chosen so that, for all $t \in [0, T)$,

$$G(c) + \int_0^t F(s)ds \in Dom\,(G^{-1}). \tag{1.1.6}$$

Proof We define on $\mathbb{R}_+$ a positive, non-decreasing and differentiable with function Z by, for all $t \in \mathbb{R}_+$,

$$Z(t) := c + \epsilon + \int_0^t F(s)\psi[u(s)]ds, \tag{1.1.7}$$

where ϵ is an arbitrary positive number, since ϕ is strictly increasing and $\phi(+\infty) = +\infty$, from inequality (1.1.3), it follows that, for all $t \in \mathbb{R}_+$,

$$u(t) \le \phi^{-1}(Z(t)). \tag{1.1.8}$$

Differentiating both sides of (1.1.7), we obtain for all $t \in \mathbb{R}_+$,

$$Z'(t) = F(t)\psi\big[u(t)\big] \le F(t)\psi\big\{\varphi^{-1}\big[Z(t)\big]\big\}. \tag{1.1.9}$$

Here we have used (1.1.8) and the monotonicity of ψ and ψ^{-1}, Because, for all $t \in \mathbb{R}_+$,

$$\psi\{\varphi^{-1}[Z(t)]\} \ge \psi\{\varphi^{-1}[Z(0)]\} > 0,$$

from (1.1.9) we may derive that, for all $t \in \mathbb{R}_+$,

$$\frac{d}{dt}G[Z(t)] = \frac{Z'(t)}{\psi\{\varphi^{-1}[Z(t)]\}} \le F(t),$$

where the function G is defined by

$$G(z) := \int_{z_0}^{z} \frac{ds}{\psi[\phi^{-1}(s)]}, \quad z \ge z_0 > 0.$$

Letting $t = s$ in the last relation and then integrating with respect to s from $s = 0$ to t, we can obtain (after letting $\varepsilon \to 0^+$): for all $t \in \mathbb{R}_+$,

$$G[Z(t)] \le G(c) + \int_0^t F(s)ds. \tag{1.1.10}$$

According to condition (1.1.6), the right-hand side of (1.1.10) is in the definition domain of G^{-1} as long as $t \in [0, T)$. Hence, the desired estimate in (1.1.4) follows from (1.1.8) and (1.1.10). $\square$

In 1949, LaSalle [330] established the following remarkable result, which has not received the extensive attention. LaSalle's lemma can be restated as follows:

Theorem 1.1.4 (The LaSalle Inequality [330]) *Let $f(t), g(t)$ be non-negative continuous functions on $[0, T]$, and assume that $K > 0$ is a constant, $F(t)$ is a non-negative, non-decreasing continuous functions for all $0 \leq u < K$, $G(u) = \int_K^u 1/F(s)ds$ and the following inequality holds, for all $0 \leq t \leq T$,*

$$g(t) \leq K + \int_0^t f(s)F(g(s))ds. \tag{1.1.11}$$

Then we have for all $0 \leq t \leq T$,

$$G(g(t)) \leq \int_0^t f(s)ds.$$

Moreover, if F is an identity mapping, i.e., $F(u) = u$, then the LaSalle inequality (1.1.11) reduces to the classical Bellman-Gronwall inequality.

Proof The proof is similar to that of Theorem 1.1.1. □

Obviously, the above LaSalle inequality furnishes an upper bound.

Theorem 1.1.5 (The LaSalle Inequality [330]) *If $x(t)$, $v_0(t)$ and $v(t)$ are positive continuous functions on $[0, \tau)$, and $f(u)$ is a positive, non-decreasing, sub-additive, continuous function for all $u \in (0, +\infty)$ such that, for all $t \in [0, \tau)$,*

$$x(t) \leq v_0(t) + \int_0^t v(s)f(x(s))ds,$$

then for all $t \in [0, \tau_3)$,

$$x(t) \leq v_0(t) + \Phi^{-1}\left[\Phi(\int_0^t v(s)f(v_0)(s))ds + \int_0^t v(s)ds\right],$$

where Φ and Φ^{-1} are given as in Theorem 1.1.1 and

$$\tau_3 = \sup\left\{t \in [0, \tau) : \Phi(+\infty) \geq \Phi\left(\left(\int_0^t v(s)f(v_0(s))ds\right) + \int_0^t v(s)ds\right\}.$$

Proof The proof of the above theorem is similar to that of Theorem 1.1.1, and hence will be omitted. □

We shall say that the function $h(x, r)$ possesses the property I if $h(x, r) \geq 0$ for the specified range of values of x and r, if it is measurable in x for fixed $r \geq 0$, continuous in r for fixed x, $x_0 \leq x < +\infty$, $r \geq 0$, and if $r(x)$ is the maximal solution of the differential equation $r' = h(x, r)$ passing through the point $(x_0, 0)$. We have the following lemma.

Lemma 1.1.1 (The Lakshmikantham Inequality [320]) *Suppose that $h(x, r)$ has property I. Let $y(x)$ be continuous on $x_0 \leq x < +\infty$ and satisfy the inequality for all $\Delta x \in [0, +\infty)$,*

$$|\Delta y(x)| \leq \int_x^{x+\Delta x} h(t, y(t))dt,$$

then for all $x \in [x_0, +\infty)$,

$$y(x) \leq r(x).$$

Proof The inequality shows that $y(x)$ is absolutely continuous in $[x_0, +\infty)$, which implies that $y'(x)$ exists almost everywhere in $[x_0, +\infty)$. Furthermore, it is clear from the assumed inequality that

$$|y'(x)| \leq h(x, y(x)), \tag{1.1.12}$$

almost everywhere.

Suppose that $b(x, \epsilon)$ is a solution of $r' = h(x, r) + \epsilon, r(x_0) = 0$, where ϵ is an arbitrary small quantity. It is easy to show that for all $x \in [x_0, +\infty)$,

$$y(x) \leq b(x, \epsilon). \tag{1.1.13}$$

For suppose that this relation does not hold. Then, without loss of generality, let $[x_0, x_1]$ be an interval where $y(x) \geq b(x, \epsilon)$. At x_0, we have $y(x_0) \geq b(x_0, \epsilon)$. Hence taking right-hand derivatives at x_0, we obtain

$$y'(x_0) \geq b'(x_0, \epsilon). \tag{1.1.14}$$

From this we obtain the further inequality

$$h(x_0, y(x_0)) \geq h(x_0, b(x_0, \epsilon)) + \epsilon, \tag{1.1.15}$$

which leads to a contradiction. Hence (1.1.13) holds.

Since we know that $\lim_{\epsilon \to 0} b(x, \epsilon) = r(x)$, the lemma is proved. □

Now let y and $f(x, y)$ be vectors with real components, $(y_1, y_2, \cdots, y_n)$ and $(f_1(x, y), f_2(x, y), \cdots, f_n(x, y))$ respectively. Consider the system

$$y' = f(x, y), \qquad y(x_0) = 0, \tag{1.1.16}$$

where $f(x, y)$ is continuous on $x_0 \leq x < +\infty$, $\|y\| < +\infty$. We can obtain the following conclusion.

Lemma 1.1.2 ([320]) *Suppose that $h(x, r)$ has property I and that*

$$\|f(x, y)\| \leq h(x, \|y\|). \tag{1.1.17}$$

Then if $r(x) = O(1)$ as $x \to +\infty$, the norm of every solution (1.1.16) tends to a finite limit as $x \to +\infty$. If, in particular, $r(x) = o(1)$ then each component of every solution of (1.1.16) tends to zero as $x \to +\infty$.

Proof Let a solution of (1.1.16) be $y(x) = \int_0^x f(t, y(t))dt$, and let $\Delta y(x) = y(x + \Delta x) - y(x)$, for $\Delta x > 0$. It follows that

$$\|\Delta y(x)\| \le \int_x^{x+\Delta x} \|f(t, y(t))\|dt \tag{1.1.18}$$

and hence that

$$\|\Delta y(x)\| \le \int_x^{x+\Delta x} h(t, \|y(t)\|)dt. \tag{1.1.19}$$

Using Lemma 1.1.1, we obtain for all $x \in [0, +\infty)$,

$$\|y(x)\| \le r(x). \tag{1.1.20}$$

This together with the assumptions of the theorem yield the stated results. □

Lemma 1.1.3 (The Lakshmikantham Inequality [322]) *Let the function $g(x, u) \ge 0$ be continuous in the region $a \le x \le b, u \ge 0$. Let the function $f(x, y)$ of (1.1.16) satisfy the condition*

$$|f(x, y)| \le g(x, |y|).$$

Let $y(x)$ satisfy $|y(x)| > 0$ and be a solution of (1.1.16) in the region $a \le x \le b$. Then we have for all $x \in [a, b]$,

$$|y(x)| \le M(x) \tag{1.1.21}$$

and

$$|y(x)| \le m(x) \tag{1.1.22}$$

where $M(x)$ and $m(x)$ are the maximal and minimal solutions of $u'(x) = \pm g(x, u), u(a) = |y(a)|$, respectively.

Proof The inequality (1.1.21) follows from Lemma 1.1.2. To prove (1.1.22), we have to use essentially the same argument as in Lemma 1.1.2 but now we have to consider the minimal solution of $u'(x) = -g(x, u), u(a) = |y(a)|$ instead of the maximal solution of $u'(x) = g(x, u), u(a) = |y(a)|$. This completes the proof. □

Theorem 1.1.6 (The Willett-Wong Inequality [673]) *Let the functions $v(t)$, $w(t)$, $v(t)u(t)$, and $w(t)u^p(t)$, be locally integrable non-negative functions on $\mathbb{R}_+$. If $u_0 >$*

0 and $p \geq 0$, $p \neq 1$, *and the following inequality holds for all* $t \in \mathbb{R}_+$,

$$u(t) \leq u_0 + \int_0^t v(s)u(s)ds + \int_0^t w(s)u^p(s)ds \tag{1.1.23}$$

then for all $t \in \mathbb{R}_+$,

$$u(t)\exp\Big(-\int_0^t v(s)ds\Big) \leq \Big(u_0^q + q\int_0^t w(s)\exp\Big(-q\int_0^s v(r)dr\Big)ds\Big)^{\frac{1}{q}}, \quad q = 1-p. \tag{1.1.24}$$

Proof Let $\varphi(t)$ be defined as the right-hand side of (1.1.23); so for all $t \in \mathbb{R}_+$,

$$\varphi'(t) \leq v(t)\varphi(t) + w(t)\varphi^p(t), \tag{1.1.25}$$

since $p \geq 0$. By Lemma 1.1.3, we know that $\varphi(t)$ is bounded by the maximal solution $r(t)$ of

$$r'(t) = v(t)r(t) + w(t)r^p(t), \quad r(0) = u_0; \tag{1.1.26}$$

and we can solve (1.1.26) explicitly as a Bernoulli equation. However, we need not refer to Lemma 1.1.3 at all for this special case, but can obtain directly from (1.1.25) that

$$\theta'(t) \leq w(t)\theta^p(t)\exp\Big(-q\int_0^t v(s)ds\Big), \quad q = 1-p, \tag{1.1.27}$$

where

$$\theta(t) = \varphi(t)\exp\Big(-\int_0^t v(s)ds\Big). \tag{1.1.28}$$

Since $\theta(t) > 0$ on $\mathbb{R}_+$, we can divide (1.1.27) by $\theta^p(t)$ and integrate to obtain equation (1.1.24) for all $p \geq 0, p \neq 0 (q = 1-p)$.

If $u_0 = 0$, then equation (1.1.23) is valid for all positive constants u_* in place of u_0. By letting $u_* \to 0$ in the corresponding equation (1.1.24), we get that equation (1.1.24) as it now stands is also valid when $u_0 = 0$, if we agree to first write the right hand side with a factor u_0 when $q < 0$. □

Corollary 1.1.1 (The Li Inequality [348]) *If* $x(t)$, $v(t)$, *and* $w(t)$ *are non-negative continuous functions on* $[0, \tau)$ *and* η *and* α *are constants,* $\eta > 0$, $\alpha < 1$ *such that, for all* $t \in [0, \tau)$,

$$x(t) \leq \eta + \int_0^t v(s)x(s)ds + \int_0^t w(s)x^{1-\alpha}(s)ds,$$

then for all $t \in [0, \tau_5)$,

$$x(t) \leq E(t)W_\alpha(t),$$

where

$$E(t) = \exp\left(\int_0^t v(s)ds\right), \quad W_0(t) = \eta \exp\left(\int_0^t w(s)ds\right),$$

$$W_\alpha(t) = \left(\eta^\alpha + \alpha \int_0^t W(s)E^{-\alpha}(s)ds\right)^{1/\alpha}, \quad \alpha \neq 0,$$

and

$$\tau_5 = \sup\left\{t \in [0, \tau) : \eta^\alpha + \alpha \int_0^t W(s)E^{-\alpha}(s)ds \geq 0\right\},$$

(e.g., $\tau_5 = \tau$ *when* $0 \leq \alpha \leq 1$*).*

The next result called the Langenhop inequality (see, e.g., Langenhop [328]) will give us a lower bound.

Theorem 1.1.7 (The Langenhop Inequality [328]) *Let*

(1) x *be a real variable and* z *and* F *be finite-dimensional complex vectors with* n *components* z_i *and* F_i *respectively;*
(2) F *be continuous in* (x, z) *for all* z *and all* $x \in [a, b]$ *with* $a < b$*;*
(3) for some norm, say $\|z\| = \sum_{i=1}^n |z_i|$, F *satisfies*

$$\|F(x, z)\| \leq v(x)g(\|z\|)$$

where $v(x)$ *is continuous,* $v(x) \geq 0$ *for all* $x \in [a, b]$, $g(u)$ *is continuous and non-decreasing for all* $u \geq 0$*, and* $g(u) > 0$ *for all* $u > 0$.

If $z(x)$ *is continuous, and is a solution of* $\frac{dz}{dx} = F(x, z)$ *for all* $x \in [a, b]$*, where* F *satisfies the conditions above, then for all* $x \in [a, b]$, $z(x)$ *satisfies the inequality*

$$|z(x)| \geq G^{-1}\left(G(|z(a)|) - \int_a^x v(s)ds\right)$$

where

$$G(u) = \int_{u_0}^u [g(s)]^{-1}ds, \quad u \geq u_0 \geq 0$$

and for all $x \in [a, b]$ *for which* $G(|z(a)|) - \int_a^x v(s)ds$ *is in the domain of* G^{-1}.

Proof The proof is left to the reader an exercise. □

In 1971, Györi [260] extended the Bihari inequality and the LaSalle inequality.

Theorem 1.1.8 (The Györi Inequality [260]) *Let $u(t), v(t)$ be non-negative continuous on $[t_0, +\infty)$, $a(t), b(t), g(u)$ be differentiable, $a(t) \geq 0$ and $g > 0$ be increasing, and $b(t) \geq 0$ be decreasing. If for any $t \geq t_0$, there holds*

$$u(t) \leq a(t) + b(t) \int_{t_0}^{t} v(s)g(u(s))ds,$$

and for all non-negative continuous function ϕ, there holds

$$a'(t)[\frac{1}{g(\phi(t))} - 1] \leq 0,$$

then

$$u(t) \leq G^{-1}\left(a(t_0) + \int_{t_0}^{t} [b(s)v(s) + a'(s)]ds\right)$$

where $G(u) = \int_{u_0}^{u} \frac{1}{g(s)}ds,\ u \geq u_0 > 0$.

Proof We leave the proof to the reader as an exercise. □

Theorem 1.1.9 (The Pachpatte Inequality [518]) *Let $a, b \in C(I, \mathbb{R}_+)$ be non-decreasing with $\alpha(t) \leq t$ on I, and $k \leq 0$, $c \leq 1$, and $p > 1$ are constants.*

(a_1) *If $u \in C(I, \mathbb{R}_+)$ and for all $t \in I$,*

$$u(t) \leq k + \int_{t_0}^{t} a(s)u(s)ds + \int_{\alpha(t_0)}^{\alpha(t)} b(s)u(s)ds, \tag{1.1.29}$$

then for all $t \in I$,

$$u(t) \leq k\exp(A(t) + B(t)), \tag{1.1.30}$$

where for all $t \in I$,

$$A(t) = \int_{t_0}^{t} a(s)ds, \tag{1.1.31}$$

$$B(t) = \int_{\alpha(t_0)}^{\alpha(t)} b(s)ds. \tag{1.1.32}$$

(a_2) *If $u \in C(I, \mathbb{R}_+)$ and for all $t \in I$,*

$$u(t) \leq c + \int_{t_0}^{t} a(s)u(s)\log u(s)ds + \int_{\alpha(t_0)}^{\alpha(t)} b(s)u(s)\log u(s)ds, \tag{1.1.33}$$

then for all $t \in I$,

$$u(t) \leq c^{\exp(A(t)+B(t))}, \tag{1.1.34}$$

where $A(t)$ *and* $B(t)$ *are defined by (1.1.31) and (1.1.32).*

(a_3) *If* $u \in C(I, \mathbb{R}_+)$ *and for all* $t \in I$,

$$u^p(t) \leq k + \int_{t_0}^{t} a(s)u(s)ds + \int_{\alpha(t_0)}^{\alpha(t)} b(s)u(s)ds, \tag{1.1.35}$$

then for all $t \in I$,

$$u(t) \leq \left[k^{\frac{p-1}{p}} + \left(\frac{p-1}{p}\right)[A(t) + B(t)]\right]^{\frac{1}{p-1}}, \tag{1.1.36}$$

where $A(t)$ *and* $B(t)$ *are defined by (1.1.31) and (1.1.32).*

Proof (a_1) Let $k > 0$ and defined a function $z(t)$ by the right-hand side of (1.1.29). Then, $z(t) > 0, z(t_0) = k, u(t) \leq z(t)$, and

$$\begin{aligned} z'(t) &= a(t)u(t) + b(\alpha(t))u(\alpha(t))\alpha'(t) \\ &\leq a(t)z(t) + b(\alpha(t))z(\alpha(t))\alpha'(t) \\ &\leq a(t)z(t) + b(\alpha(t))z(t)\alpha'(t), \end{aligned}$$

i.e.,

$$\frac{z'(t)}{z(t)} \leq a(t) + b(\alpha(t))\alpha'(t). \tag{1.1.37}$$

Integrating (1.1.37) from t_0 to $t, t \in I$, and the change of variable yield for all $t \in I$,

$$z(t) \leq k \exp(A(t) + B(t)). \tag{1.1.38}$$

Using (1.1.38) in $u(t) \leq z(t)$, we get the inequality in (1.1.30). If $k \geq 0$, we carry out the above procedure with $k + \epsilon$ instead of k, where $\epsilon > 0$ is an arbitrary small constant, and subsequently pass the limit as $\epsilon \to 0$ to obtain (1.1.30).

(a_2) Define a function $z(t)$ by the right-hand side of (1.1.33). Then $z(t) > 0, z(t_0) = c$, and $u(t) \leq z(t)$, and as in the proof of (a_1), we may get

$$\frac{z'(t)}{z(t)} \leq a(t) \log z(t) + b(\alpha(t)) \log z(\alpha(t))\alpha'(t). \tag{1.1.39}$$

Integrating (1.1.39) from t_0 to $t, t \in I$, and the change of variable yield

$$\log z(t) \leq \log c + \int_{t_0}^{t} a(s) \log z(s) ds + \int_{\alpha(t_0)}^{\alpha(t)} b(s) \log z(s) ds. \tag{1.1.40}$$

Now by a suitable application of the inequality given in (a_1) to (1.1.40), we get

$$\begin{aligned} \log z(t) &\leq (\log c) \exp(A(t) + B(t)) \\ &= \log c^{\exp(A(t)+B(t))}. \end{aligned} \tag{1.1.41}$$

Thus from (1.1.41), we may conclude that

$$z(t) \leq c^{\exp(A(t)+B(t))}. \tag{1.1.42}$$

Now by using (1.1.42) in $u(t) \leq z(t)$, we may get the required inequality in (1.1.34).

(a_3) Let $k > 0$ and define a function $z(t)$ by the right-hand side of (1.1.35). Then $z(t) > 0, z(t_0) = k, u(t) \leq \{z(t)\}^{1/p}$, and as in the proof of (a_1), we have

$$\{z(t)\}^{-1/p} z'(t) \leq a(t) + b(\alpha(t))\alpha'(t). \tag{1.1.43}$$

Integrating (1.1.43) from t_0 to $t, t \in I$, and the change of variable, we get

$$z(t) \leq \left[k^{(p-1)/p} + (\frac{p-1}{p})[A(t) + B(t)] \right]^{1/(p-1)}. \tag{1.1.44}$$

The desired inequality in (1.1.36) follows by using (1.1.44) in $u(t) \leq \{z(t)\}^{1/p}$. The case $k \geq 0$ can be completed as mentioned in the proof of (a_1). □

Theorem 1.1.10 (The Pachpatte Inequality [518]) *Let a, b, α, k, c, p be as in Theorem 1.1.9. For $i = 1, 2$, let $g_i \in C(\mathbb{R}_+, \mathbb{R}_+)$ be non-decreasing functions with $g_i(u) > 0$ for all $u > 0$.*

(b_1) *If $u \in C(I, \mathbb{R}_+)$ and for all $t \in I = [t_0, T)$, there holds that*

$$u(t) \leq k + \int_{t_0}^{t} a(s) g_1(u(s)) ds + \int_{\alpha(t_0)}^{\alpha(t)} b(s) g_2(u(s)) ds, \tag{1.1.45}$$

then for all $t_0 \leq t \leq t_1$,

(i) in case $g_2(u) \leq g_1(u)$, we have

$$u(t) \leq G_1^{-1}[G_1(t) + A(t) + B(t)], \tag{1.1.46}$$

(ii) in case $g_1(u) \leq g_2(u)$, we have

$$u(t) \leq G_2^{-1}[G_2(t) + A(t) + B(t)], \tag{1.1.47}$$

where $A(t)$ and $B(t)$ are defined by (1.1.31) and (1.1.32) and for $i = 1, 2$; G_i^{-1} are inverse functions of

$$G_i(r) = \int_{r_0}^{r} \frac{ds}{g_i(s)}, \qquad r \geq r_0 > 0, \tag{1.1.48}$$

and $t_1 \in I$ is chosen so that

$$G_i(k) + A(t) + B(t) \in Dom\,(G_i^{-1}),$$

respectively, for all $t \in [t_0, t_1]$.

(b_2) *If $u \in C(I, \mathbb{R}_1)$, $\mathbb{R}_1 = [1, +\infty)$, and for all $t \in I$,*

$$u(t) \leq c + \int_{t_0}^{t} a(s)u(s)g_1(\log u(s))ds + \int_{\alpha(t_0)}^{\alpha(t)} b(s)u(s)g_2(\log u(s))ds, \tag{1.1.49}$$

then for all $t_0 \leq t \leq t_2$,

(i) in case $g_2(u) \leq g_1(u)$, we have

$$u(t) \leq \exp\left(G_1^{-1}[G_1(\log c) + A(t) + B(t)]\right); \tag{1.1.50}$$

(ii) in case $g_1(u) \leq g_2(u)$, we have

$$u(t) \leq \exp\left(G_2^{-1}[G_2(\log c) + A(t) + B(t)]\right), \tag{1.1.51}$$

where $G_i, G_i^{-1}, A(t), B(t)$ are as in (b_1) and t is chosen so that for $i = 1, 2$,

$$G_i(\log c) + A(t) + B(t) \in Dom\,(G_i^{-1}),$$

respectively, for all $t \in [t_0, t_2]$.

(b_3) *If $u \in C(I, \mathbb{R}_+)$ and for all $t \in I$,*

$$u^p(t) \leq k + \int_{t_0}^{t} a(s)g_1(u(s))ds + \int_{\alpha(t_0)}^{\alpha(t)} b(s)g_2(u(s))ds, \tag{1.1.52}$$

then for all $t_0 \leq t \leq t_3$,

(i) in case $g_2(u) \leq g_1(u)$, we have

$$u(t) \leq \left[H_1^{-1}[H_1(k) + A(t) + B(t)]\right]^{1/p}, \tag{1.1.53}$$

(ii) in case $g_1(u) \leq g_2(u)$, we have

$$u(t) \leq \left[H_2^{-1}[H_2(k) + A(t) + B(t)]\right]^{1/p}, \tag{1.1.54}$$

where $A(t)$ *and* $B(t)$ *are defined by (1.1.31) and (1.1.32) and for* $i = 1, 2, H_i^{-1}$ *are the inverse functions of*

$$H_i(r) = \int_{r_0}^{r} \frac{ds}{g_i(s^{1/p})}, \qquad r \geq r_0 > 0, \tag{1.1.55}$$

and $t_3 \in I$ *is chosen so that*

$$H_i(k) + A(t) + B(t) \in Dom\ (H_i^{-1}),$$

respectively, for all $t \in [t_0, t_3]$.

Proof Since the proofs resemble one another, we give the details for (b_1) only; the proofs of the remaining inequalities can be completed by following the proofs of the above mentioned inequalities.

From the hypotheses, we observe that $\alpha'(t) \geq 0$ for all $t \in I, \alpha'(x) \geq 0$ for all $x \in J_1 = [x_0, X), \beta'(y) \geq 0$ for $y \in J_2 = [y_0, Y)$.

(b_1) Let $k > 0$ and define a function $z(t)$ by the right-hand side of (1.1.45). Then $z(t) > 0, z(t_0) = k$, and $u(t) \leq z(t)$, we get

$$z'(t) \leq a(t)g_1(z(t)) + b(\alpha(t))g_2(z(\alpha(t)))\alpha'(t). \tag{1.1.56}$$

(i) when $g_2(u) \leq g_1(u)$, then from (1.1.56), we derive

$$z'(t) \leq g_1(z(t))[a(t) + b(\alpha(t))\alpha'(t)]. \tag{1.1.57}$$

Thus from (1.1.48) and (1.1.57) it follows that

$$\frac{d}{dt}G_1(z(t)) = \frac{z'(t)}{g_1(z(t))} \leq a(t) + b(\alpha(t))\alpha'(t). \tag{1.1.58}$$

Integrating (1.1.58) from t_0 to $t, t \in I$, and making the change of variable, we have

$$G_1(z(t)) \leq G_1(k) + A(t) + B(t). \tag{1.1.59}$$

Since $G_1^{-1}(z)$ is increasing, from (1.1.59) it follows

$$z(t) \leq G_1^{-1}[G_1(k) + A(t) + B(t)]. \tag{1.1.60}$$

Using (1.1.60) in $u(t) \leq z(t)$, this gives the required inequality in (1.1.46). The case $k \geq 0$ can be completed as mentioned in the proof of (a_1) in Theorem 1.1.9. The proof of the case when $g_1(u) \leq g_2(u)$ can be completed similarly. The sub-interval $t_0 \leq t \leq t_1$ is obvious. □

Next, we shall introduce some generalizations of the Bellman-Bihari inequality [69], where the nonlinear functions appearing on the right-hand side belong to certain classes of functions.

Definition 1.1.1 A function $\omega : [0, +\infty) \to [0, +\infty)$ is said to belong to the class H, if

(H_1) $\omega(u)$ is non-decreasing and continuous for all $u \geq 0$ and positive for all $u > 0$.

(H_2) There exists a function ϕ, continuous on $[0, +\infty)$ with $\omega(\alpha u) \leq \phi(\alpha)\omega(u)$ for all $\alpha > 0, u \geq 0$.

Several examples and properties of the class H have been obtained in [181]. In particular, H includes all functions $w \in F$, with corresponding function ϕ defined by $\phi(\alpha) = 1$ $(0 \leq \alpha \leq 1)$, $\phi(\alpha) = \alpha$ $(\alpha \geq 1)$. Also H includes all sub-multiplicative functions which satisfy (H_1), with corresponding function $\phi = w$.

Example 1.1.1 Every function ω which is continuous and non-decreasing on $[0, +\infty)$ with $\omega(u) > 0$ for all $u > 0$ and which is sub-multiplicative is of class H with $\phi = \omega$.

Example 1.1.2 Any function ω for which (H_1) holds and (H_2) with $\phi(\alpha) = \alpha, \alpha \geq 1$ and $\phi(\alpha) \equiv 1, 0 < \alpha \leq 1$ belongs to H. In fact, this type of function has been used in [197, 455, 456] and modified by Beesack [54]. The modification in [54] is essential to avoid trivialities.

Example 1.1.3 Every super-multiplicative function ω which satisfies (H_1) is of class H with $\phi(\alpha) = 1/\omega(1/(\alpha + 1))$. In fact, if $\omega(ku) \geq \omega(k)\omega(u)$, then $\omega(u) \leq (1/\omega(k))\omega(ku)$ for all $k > 0$. Let $k = 1/(1 + \alpha)$ and $u = (\alpha + 1)z$. Then

$$\omega(\alpha z) \leq \omega((\alpha + 1)z) \leq \omega(z)\left[1/\omega\left(\frac{1}{\alpha + 1}\right)\right].$$

Example 1.1.4 The function $u^2/(1 + u)$ belongs to H with $\phi(\alpha) = \alpha^2, \alpha \geq 1$ and $\phi(\alpha) \equiv 1, 0 < \alpha \leq 1$. Note that this function is not sub-multiplicative.

Now we note some properties of function $\phi(\alpha)$.

(a) $\phi(\alpha) > 0$ for all $\alpha > 0$.
(b) $\phi(\alpha) \geq 1$ for all $\alpha \geq 1$. This follows from $\omega(u) \leq \omega(\alpha u) \leq \phi(\alpha)\omega(u)$.
(c) If $\omega(0) = 0$, then $\phi(+\infty) = +\infty$ must hold. This follows from $0 < \omega(1) \leq \phi(\alpha)\omega(\alpha^{-1})$ for all $\alpha > 0$.
(d) $\phi(\alpha)\phi(\alpha^{-1}) \geq 1$ for all $\alpha > 0$. In fact, for all $\alpha > 0, u > 0$, we have $\omega(1) \leq \phi(1/u)\omega(u)$ and $\omega(\alpha) \leq \phi(\alpha)\omega(1)$. Hence

$$\omega(1) \leq \phi(\alpha^{-1})\omega(\alpha) \leq \phi(\alpha^{-1})\phi(\alpha)\omega(1),$$

so $\phi(\alpha)\phi(\alpha^{-1}) \geq 1$.

In what follows, we shall give some properties of class H.

Lemma 1.1.4 ([181]) *Let $f(u)$ and $g(u)$ be of class H with corresponding multiplier functions $\phi(\alpha)$ and $\psi(\alpha)$, respectively. Then*

(i) $f(u)+g(u), f(u)g(u)$ *and* $f(g(u))$ *are of class H.*
(ii) $h(u)=\int_0^u f(s)ds$ *belongs to H.*

Proof Obviously, the assertion (i) can be proved easily. To prove assertion (ii), we note first that $h(u)$ satisfies (H_1). Also, we have

$$h(\alpha u)=\int_0^{\alpha u} f(s)ds=\alpha\int_0^u f(\alpha z)dz\le\alpha\phi(\alpha)\int_0^u f(z)dz=\alpha\phi(\alpha)h(u),$$

which gives us that $h(u)$ satisfies (H_2). □

Lemma 1.1.5 ([181]) *Let* $F(u)$ *be a convex continuous function satisfying* (H_2) *with corresponding multiplier function* $\phi(\alpha)$*. Furthermore, we assume that* $F(u)>0$ *for all* $u>0$ *and* $F(0)=0$*. Let* $G(u)$ *be a concave continuous function such that* $G(u)>0$ *for all* $u>0$, $G(0)=0$ *and one of the following is satisfied:*

(i) There exists a function ψ*, continuous and positive on* $[0,+\infty)$ *with* $G(\alpha u)\ge\psi(\beta)G(u)$.
(ii) $0<l\le G(x)\le m$*, where l and m are constants.*

Then $T(u)=(F(u)/G(u))\in H$.

Proof It follows from [87] that $T(u)$ satisfies (H_1) in Definition 1.1.1. If $G(u)$ satisfies (i), then we have

$$T(\alpha u)=\frac{F(\alpha u)}{G(\alpha u)}\le\frac{\phi(\alpha)}{\psi(\alpha)}\frac{F(u)}{G(u)}=\frac{\phi(\alpha)}{\psi(\alpha)}T(u).$$

This proves that (H_2) holds for $T(u)$. Now if (ii) holds, then

$$T(\alpha u)=\frac{F(\alpha u)}{G(\alpha u)}\le\frac{\phi(\alpha)F(u)G(u)}{lG(u)}\le\frac{m\phi(\alpha)}{l}T(u)$$

and $T(u)$ satisfies (H_2). □

Theorem 1.1.11 (The Dannan Inequality [181]) *Assume that* $x(t)$ *and* $f(t)$ *are positive continuous functions on* $I=[0,+\infty), \omega(u)\in H$ *with corresponding multiplier function* ϕ *and* $h(t)>0$ *is a monotonic, non-decreasing and continuous function on* $[0,+\infty)$*. If for all* $t\in I$,

$$x(t)\le h(t)+\int_0^t f(s)\omega(x(s))ds, \tag{1.1.61}$$

then for all $0<t\le b$,

$$x(t)\le h(t)W^{-1}\left[W(1)+\int_0^t f(s)\frac{\phi(h(s))}{h(s)}ds\right], \tag{1.1.62}$$

where

$$W(u) = \int_{u_0}^{u} \frac{ds}{\omega(s)}, \quad u \geq u_0 > 0, \tag{1.1.63}$$

and W^{-1} *is the inverse of* W *and* $(0, b]$ *is the sub-interval for which, for all* $t \in (0, b]$,

$$W(1) + \int_0^t f(s)\frac{\phi(h(s))}{h(s)} ds \in Dom\ (W^{-1}). \tag{1.1.64}$$

Proof From (1.1.61), we derive, for all $t \in I$,

$$\frac{x(t)}{h(t)} \leq 1 + \int_0^t \frac{f(s)\omega(x(s))}{h(s)} ds \leq 1 + \int_0^t f(s)\frac{\phi(h(s))}{h(s)}\omega(\frac{x(s)}{h(s)})ds$$

since $\omega(u) \in H$ and $h(t)$ is monotonic non-decreasing. Considering $x(t)/h(t)$ as a function, using Bihari's inequality (i.e., Theorem 1.1.1), the results (1.1.62) follows. □

Corollary 1.1.2 (The Dannan Inequality [181]) *Let* x, f, ω, h, W *all be as in Theorem 1.1.11 and suppose* $b(t)$ *is non-negative, continuous and non-decreasing on* $I = [0, +\infty)$. *If, for all* $t \in I$,

$$x(t) \leq h(t) + b(t) \int_0^t f(s)\omega(x(s))ds, \tag{1.1.65}$$

then, for all $0 \leq t \leq t_0$,

$$x(t) \leq h(t)W^{-1}\left[W(1) + b(t)\int_0^t f(s)\frac{\phi(h(s))}{h(s)}ds\right], \tag{1.1.66}$$

where $W(u) = \int_{u_0}^{u} ds/\omega(s)$ *for all* $u \leq u_0 > 0$, W^{-1} *is the inverse of* W *and* $[0, t_0]$ *is the sub-interval for which*

$$W(1) + b(t_0)\int_0^{t_0} f(s)\frac{\phi(h(s))}{h(s)} \in Dom\ (W^{-1}).$$

Proof Fix any $T > 0$. Then, for all $0 \leq t \leq T$,

$$x(t) \leq h(t) + \int_0^t b(T)f(s)\omega(x(s))ds.$$

Hence by Theorem 1.1.11,

$$x(t) \leq h(t)W^{-1}\left[W(1) + b(t)\int_0^t f(s)\frac{\phi(h(s))}{h(s)}ds\right] \tag{1.1.67}$$

holds for all $0 \le t \le T$ provided, that for all $0 \le t \le T$,

$$W(1) + b(T)\int_0^t f(s)\frac{\phi(h(s))}{h(s)} \in Dom\ (W^{-1}).$$

This will be the case provided that

$$W(1) + b(T)\int_0^T f(s)\frac{\phi(h(s))}{h(s)} \in Dom\ (W^{-1}). \tag{1.1.68}$$

Hence for all $T > 0$ such that (1.1.68) holds it follows that (1.1.67) holds for all $0 \le t \le T$. In particular, taking $t = T$ in (1.1.67), we have

$$x(T) \le h(T)W^{-1}\left[W(1) + b(t)\int_0^T f(s)\frac{\phi(h(s))}{h(s)}ds\right]. \tag{1.1.69}$$

Now replace T by t in (1.1.68), (1.1.69) and we obtain the result stated valid for all $0 \le t \le t_0$, provided that

$$W(1) + b(t_0)\int_0^{t_0} f(s)\frac{\phi(h(s))}{h(s)} \in Dom\ (W^{-1}).$$

□

Theorem 1.1.12 (The Dannan Inequality [181]) *Let $x(t)$, $f(t)$ and $g(t)$ be positive continuous functions on $I = [0, +\infty)$ and $\omega \in H$ with corresponding multiplier function ϕ, for which the inequality holds, for all $t \in I$,*

$$x(t) \le x_0 + \int_0^t f(s)\omega(x(s))ds + \int_0^t g(s)(\int_0^s f(\tau)\omega(x(\tau))ds, \tag{1.1.70}$$

where $x_0 > 0$ is a constant. Then, for all $0 \le t \le t_0$,

$$x(t) \le x_0A(t)E(t)W^{-1}\left[W(1) + x_0^{-1}E(t)\int_0^t f(s)\frac{\phi(x_0A(s)E(s))}{A(s)E(s)}ds\right] \tag{1.1.71}$$

where

$$E(t) \equiv \exp\left(\int_0^t g(s)ds\right),\quad A(t) \equiv \int_0^t g(s)ds/E(t),\quad W(r) = \int_{r_0}^r \frac{ds}{\omega(s)},\quad r \ge r_0 > 0,$$

and W^{-1} is the inverse of W and t is in the sub-interval $[0, t_0]$ so that, for all $t \in [0, t_0]$,

$$W(1) + x_0^{-1}E(t)\int_0^t f(s)\frac{\phi(x_0A(s)E(s))}{A(s)E(s)}ds \in Dom\ (W^{-1}).$$

Proof Let $z(t) = \int_0^t f(s)\omega(x(s))ds$ and

$$u(t) = x_0 + z(t) + \int_0^t g(s)z(s)ds.$$

Then from (1.1.70), it follows

$$x(t) \leq u(t) \tag{1.1.72}$$

and

$$u'(t) = f(t)\omega(x(t)) + g(t)z(t) \leq f(t)\omega(u(t)) + g(t)[u(t) - x_0].$$

The integration the above inequality from 0 to t gives us

$$u(t) \leq p_1(t) + \int_0^t g(s)u(s)ds, \tag{1.1.73}$$

where

$$p_1(t) \equiv x_0 - x_0 \int_0^t g(s)ds + \int_0^t f(s)\omega(u(s))ds.$$

From (1.1.73) and the most general linear Gronwall inequality (see, e.g., Qin [557], Theorem 1.1.4), it follows that

$$\begin{aligned} u(t) &\leq p_1(t) + \int_0^t p_1(s)g(s)\exp\left(\int_0^t g(\tau)d\tau\right)ds \\ &\leq p_1(t) + p(t)\int_0^t g(s)\exp\left(\int_s^t g(\tau)d\tau\right)ds, \end{aligned} \tag{1.1.74}$$

where for all $t \in I$,

$$p(t) = x_0 + \int_0^t f(s)\omega(u(s))ds.$$

After evaluating the integral in (1.1.74), we may get, for all $t \geq 0$,

$$u(t) \leq x_0\left[E(t) - \int_0^t g(s)ds\right] + E(t)\int_0^t f(s)\omega(u(s))ds. \tag{1.1.75}$$

Since $E(t) - \int_0^t d(s)ds = A(t)E(t) > 0$ is non-decreasing and $x(t) \leq u(t)$, Corollary 1.1.2 can be applied to (1.1.75) to give us (1.1.71). □

Several integral inequalities similar to (1.1.70) have been obtained by Pachpatte [441, 445, 446, 451, 455, 456, 458], where the nonlinear terms were assumed to be sub-additive or sub-multiplicative or both. In Theorems 1.1.11 and 1.1.12, the nonlinear function $\omega(u)$ is assumed to belong to certain class of functions. In what follows, we assume that $\omega(u)$ satisfies a Lipschitz condition.

Theorem 1.1.13 (The Dannan Inequality [181]) *Let the functions $x(t), f(t), g(t)$, $h(t)$ be non-negative continuous on $I = [0, +\infty)$, and let $\omega(u) \geq 0$ be monotonic non-decreasing function and satisfy a Lipschitz condition: for all $u, v \geq 0$,*

$$|\omega(u+v) - \omega(u)| \leq kv$$

where k is a positive constant. Suppose, for all $t \in I$,

$$x(t) \leq x_0(t) + h(t)\left[\int_0^t f(s)\omega(x(s))ds + \int_0^t g(s)\left(\int_0^s f(\tau)\omega(x(\tau))d\tau\right)ds\right]. \quad (1.1.76)$$

Then, for all $t \in I$,

$$x(t) \leq x_0(t) + h(t)\int_0^t f(s)\omega(x_0(s))\exp\left(\int_0^t [kh(\tau)f(\tau) + g(\tau)]d\tau\right)ds. \quad (1.1.77)$$

Proof Let $u(t) = \int_0^t f(s)\omega(x(s))ds$. Then $u'(t) = f(t)\omega(x(t))$ and

$$x(t) \leq x_0(t) + h(t)v(t), \quad (1.1.78)$$

where

$$v(t) = u(t) + \int_0^t g(s)u(s)ds, \quad v(0) = 0.$$

Now

$$v'(t) = u'(t) + g(t)u(t) \leq f(t)\omega(x(t)) + g(t)v(t). \quad (1.1.79)$$

Since $\omega(u)$ is a non-decreasing function, then by (1.1.78), and the Lipschitz condition,

$$\begin{aligned} v'(t) &\leq f(t)\omega(x_0(t) + h(t)v(t)) + g(t)v(t) \\ &\leq f(t)[\omega(x_0(t)) + kh(t)v(t)] + g(t)v(t). \end{aligned}$$

Hence

$$v'(t) - [g(t) + kh(t)f(t)]v(t) \leq f(t)\omega(x_0(t)),$$

which implies

$$v(t) \leq \int_0^t f(s)\omega(x_0(s)) \exp\left(\int_s^t [g(\tau) + kh(\tau)f(\tau)]d\tau\right)ds. \tag{1.1.80}$$

Thus (1.1.78) and (1.1.80) give us (1.1.77). □

Putting $g(t) \equiv 0$ in Theorem 1.1.13, we may obtain the following corollary.

Corollary 1.1.3 (The Dannan Inequality [181]) *Let $x(t), h(t), f(t)$ and $\omega(u)$ all be as in Theorem 1.1.13 and suppose that, for all $t \geq 0$,*

$$x(t) \leq x_0(t) + h(t) \int_0^t f(s)\omega(x(s))ds.$$

Then, for all $t \geq 0$,

$$x(t) \leq x_0(t) + h(t) \int_0^t f(s)\omega(x_0(s)) \exp\left(\int_s^t kh(\tau)f(\tau)d\tau\right)ds.$$

We introduce some of the nonlinear generalizations of Gronwall-Bellman inequalities. Due to [182], where the nonlinear functions appearing on the right-hand side belong to the classes of functions defined. Also, we obtain several integral inequalities similar to the Bellman-Bihari inequality in [82], which is devoted to the nonlinear versions of the main inequality of Pachpatte [441] and its extension by Agarwal [5].

Definition 1.1.2 A function $w : [0, +\infty) \to [0, +\infty)$ is said to belong to the class M if

(M_1) $w(u)$ is non-decreasing and continuous for all $u \geq 0$ and positive for all $u > 0$.
(M_2) There exists a function ψ, continuous on $[0, +\infty)$ with $w(\alpha + u) \leq \psi(\alpha) + w(u)$ for all $\alpha > 0,\ u \geq 0$.

Example 1.1.5 Every function w which is continuous and non-decreasing on $[0, +\infty)$ with $w(u) > 0$ for all $u > 0$ which is sub-additive is of class M with $\psi = w$.

Example 1.1.6 Any non-decreasing continuous function w on $[0, +\infty)$ with $w(u) > 0$ for all $u > 0$ which satisfies a Lipschitz condition of order $n > 0$,

$$w(\alpha + u) - w(u) \leq K\alpha^n,$$

is of class M with $\psi = K\alpha^n$, where K is a non-negative constant.

Example 1.1.7 The function $w(u) = \ln(\cosh u)$ belong to M with $\psi(\alpha) = \ln(2\cosh\alpha)$.

Example 1.1.8 The function $w(u) = u^3/(u^2+1)$ belong to M with $\psi(\alpha) = \alpha^3/(\alpha^2+1) + k\alpha$, for any $k \geq \frac{3}{2}$.

Now we note some properties of the function $\psi(\alpha)$.

(a) $\psi(\alpha) \geq 0$ for all $\alpha \geq 0$. This follows from

$$w(u) \leq w(u+\alpha) \leq w(u)\psi(\alpha).$$

(b) If $w(0) = 0$, then $w(\alpha) \leq \psi(\alpha)$ for all $\alpha \geq 0$.

In what follows, we shall give some properties of the class M.

Lemma 1.1.6 ([182]) *Let $w(u) \in M$ with corresponding function $\psi(\alpha)$. Then $T(u) = (1/u)\int_0^u w(s)ds$ for all $u > 0$ with $T(0) = w(0)$ is of class M.*

Proof It follows from [87] that $T(u)$ satisfies (M_1). Now we note that

$$\begin{aligned} T(\alpha+u) &= \frac{1}{\alpha+u}\left[\alpha T(\alpha) + \int_a^u w(\alpha+\theta)d\theta\right] \\ &\leq \frac{1}{\alpha+u}\left[\alpha T(\alpha) + \int_a^u (\psi(\alpha) + w(\theta))d\theta\right] \\ &\leq \frac{\alpha}{\alpha+u}T(\alpha) + \frac{u\psi(\alpha) + uT(u)}{\alpha+u} \\ &= T(u) + [\psi(\alpha) + T(\alpha)]. \end{aligned}$$

Thus T satisfies (M_2) with corresponding function $T + \psi$. □

Lemma 1.1.7 ([182]) *Let $F(u)$ be a convex continuous function on $[0, +\infty)$ with $F(0) = 0$ and $F(u) > 0$ for all $u > 0$, which satisfies (M_2) with corresponding function $\psi(\alpha)$. Assume also that $G(u)$ is a concave continuous function on $[0, +\infty)$ with $G(0) = 0$ for which there exists a function χ defined on $[0, +\infty)$ such that $G(u+\alpha) \geq \chi(\alpha) + G(u)$ for all $u \geq 0$, $\alpha > 0$. If, in addition, $\lim_{\alpha\to 0_+} F(u)/G(u) + A$ exists (finite), then $F(u)/G(u)$ is of class M.*

Proof Observe that $G(\alpha) \geq \chi(\alpha) > 0$ for all $\alpha > 0$. So $F(u)/G(u)$ is defined positive, and continuous for all $u > 0$. By [87], $F(u)/G(u)$ is also non-decreasing on $(0, +\infty)$, and since $F(u)/G(u) > 0$ for all $u > 0$, it follows that $B = \lim_{\alpha\to 0_+} F(u)/G(u)$ exist $(B \geq 0)$. Now, we show that if the function $F(u)/G(u)$ is defined to have the value B for $u = 0$, then it is of class M. For, as now proved, it satisfies (M_1). For all $\alpha > 0$, $u \geq 0$,

$$\frac{F(\alpha+u)}{G(\alpha+u)} \leq \frac{\psi(\alpha) + F(u)}{\chi(\alpha) + G(u)} \leq \frac{\psi(\alpha)}{\chi(\alpha)} + \frac{F(u)}{G(u)}.$$

A corresponding function for $F(u)/G(u)$ is therefore the function $\psi(\alpha)$ defined by $\psi(0) = A$, $\psi(\alpha) = \psi(\alpha)/\chi(\alpha)$ for all $\alpha > 0$; ψ is thus continuous on $[0, +\infty)$ as required by (M_2). □

Lemma 1.1.8 (The Dannan Inequality [182]) *Let $f(x) \in M$ with corresponding function $\psi(\alpha)$. Then, for all $\alpha \geq 0,\ x \geq 0$,*

$$f(\alpha x) \leq ([\alpha] + 1)\psi(x) + f(0)$$

where $[\alpha]$ is the largest integer less than or equal to α.

Proof Since $f(x) \in M$, then

$$f(x + y) \leq \psi(x) + f(y).$$

Putting $y = 0$, we may obtain

$$f(x) \leq \psi(x) + f(0).$$

Therefore

$$f(x + y) \leq \psi(x) + \psi(y) + f(0),$$

and

$$f(2x) \leq 2\psi(x) + f(0).$$

It is easy to prove by induction that

$$f(x\alpha) \leq \alpha\psi(x) + f(0),$$

for a natural number $\alpha \in \mathbb{N}$. If α does not belong to $\mathbb{N}$, then $m < \alpha < m + 1$, where $m \in \mathbb{N}$. Hence

$$f(x\alpha) \leq f((m + 1)x) \leq (m + 1)\psi(x) + f(0).$$

The proof is thus complete. □

The next generalization of the Bihari inequality [82] will be obtained, by considering two nonlinear terms on the right-hand side. For this purpose, one of the nonlinear functions must be sub-additive; whereas, for the second nonlinearity, a class of functions $\mathcal{F}$ (see, [2, 3]) has been defined below.

Dohongade and Deo [199] were the first who defined a class $\mathcal{F}$ of functions $w(u)$, which are continuous, positive and non-decreasing on $[0, x)$, and satisfy the condition: for all $u \geq 0,\ v > 0$,

$$\frac{1}{v}g(u) \leq g(\frac{u}{v}).$$

In fact, the previous condition implies that $g(u) \equiv g(1)u$ for all $u > 0$. To avoid this triviality, an essential modification has been given by Beesack [54], namely to require the above inequality to hold only for all $u \geq 0,\ v > 1$.

Definition 1.1.3 A function $w : [0, +\infty) \to [0, +\infty)$ is said to the class $\mathcal{F}$ if

(i) $w(u) > 0$ is non-decreasing and continuous for all $u \geq 0$,
(ii) $\frac{1}{n}w(u) \leq w(\frac{u}{n})$, *for all* $u \geq 0,\ n > 0$.

But actually, just as proved in [56], the function $w(u)$ satisfying (ii) must be a linear function, so all results on nonlinear inequalities of those papers are not of any meaning.

In [200], condition (ii) is changed to, for all $u \geq 0, v \geq 1$,

$$\frac{1}{v}w_1(u) \leq w_1\left(\frac{u}{v}\right)$$

where $w_1 > 0$ is non-decreasing function. However, just as indicated in [686], its proof is wrong because $f(x)g(x) \geq 1$ does not hold.

In what follows, we shall prove a similar result to that of Pachpatte [455], where the nonlinear function belongs to H rather than just to $\mathcal{F}$.

Theorem 1.1.14 (The Dannan Inequality [182]) *Let $x(t)$, $f(t)$, $g(t)$, $p(t)$, and $k(t)$ be real-valued positive functions defined on $I = [0, +\infty)$, let $w(u) \in H$ with corresponding multiplier function ϕ and let $k(t)$ also be a monotonic, non-decreasing function, for which the inequality*

$$x(t) \leq k(t) + p(t)\int_0^t f(s)x(s)ds + \int_0^t g(s)w(x(s))ds \tag{1.1.81}$$

holds for all $t \in I$. Then, for all $t \in [0, b]$,

$$x(t) \leq k(t)r(t)W^{-1}[W(1) + \int_0^t \frac{1}{k(s)}g(s)\phi(k(s))\phi(r(s))ds], \tag{1.1.82}$$

where, for all $t \in I$,

$$\begin{cases} r(t) = 1 + p(t)\displaystyle\int_0^t f(s)\exp\left(\int_s^t p(\theta)f(\theta)d\theta\right)ds, & (1.1.83)\\ W(r) = \displaystyle\int_{r_0}^r \frac{ds}{w(s)}, \quad r \geq r_0 > 0 & (1.1.84)\end{cases}$$

and W^{-1} is the inverse function of W, and $t \in [0, b]$ so that

$$W(1) + \int_0^t \frac{1}{k(s)}g(s)\phi(k(s))\phi(r(s))ds \in Dom\ (W^{-1}).$$

Proof Since $k(t)$ is positive, monotonic, and non-decreasing, we derive from (1.1.81) that

$$\frac{x(t)}{k(t)} \le 1 + p(t)\int_0^t f(s)\frac{x(s)}{k(s)}ds + \int_0^t w(x(s))\frac{g(s)}{k(s)}ds. \tag{1.1.85}$$

Let $z(t) = x(t)/k(t)$ and use the fact that $w \in H$. Then from (1.1.85), it follows

$$z(t) \le 1 + p(t)\int_0^t f(s)z(s)ds + \int_0^t w(x(s))h(s)ds, \tag{1.1.86}$$

where $h(s) = g(s)\phi(k(s))/k(s)$.
Now define

$$n(t) = 1 + \int_0^t w(x(s))h(s)ds, \quad n(0) = 1 \tag{1.1.87}$$

and observe that $n(t)$ is positive monotonic non-decreasing. We obtain from Theorem 1.2.7 in Qin [557] and (1.1.86) that

$$z(t) \le n(t)r(t). \tag{1.1.88}$$

Furthermore,

$$w(z(t)) \le \phi(r(t))w(n(t))$$

since $w \in H$. Hence

$$\frac{w(z(t))h(t)}{w(n(t))} \le \phi(r(t))h(t).$$

Because of (1.1.84) and (1.1.87), this reduces to

$$\frac{d}{dt}W(n(t)) \le \phi(r(t))h(t).$$

Now integrating from 0 to t, we may obtain

$$W(n(t)) - W(1) \le \int_0^t \phi(r(s))h(s)ds. \tag{1.1.89}$$

Thus the desired bound in (1.1.82) follows from (1.1.88) and (1.1.89). □

Now we establish an extension of Theorem 5.6 in [54], where the nonlinear function under the integral sign belongs to M and is not just sub-additive as in [54].

We note that, somewhat earlier, Deo and Murdeshwar [196] had obtained the same estimate as that given in ([54], Theorem 5.6), but the proof in [196] is unfortunately incorrect. See also Beesack [55], Theorem 1.

In next theorem, $h \in\uparrow$ (respectively $h \in\downarrow$) denotes that h is an increasing (respectively, decreasing) function.

Theorem 1.1.15 (The Beesack Inequality [56]) *Let x, a, k be continuous functions such that k does not change sign on $J = [\alpha, \beta)$. Let g be continuous, monotonic, and never zero on an interval I_0 such that $x(J) \subset I_0$ and $ax(J) \subset I_0$. Suppose also that the function h is continuous and monotonic on an interval I such that $0 \in I$, $h(I) \subset I_0$, and that any one of the conditions*

(i) $h \in\uparrow$, $g \in\uparrow$, g is sub-additive, $k \geq 0$, $g > 0$,
(ii) $h \in\downarrow$, $g \in\uparrow$, g is sub-additive, $k \leq 0$, $g > 0$,
(iii) $h \in\uparrow$, $g \in\downarrow$, g is sub-additive, $k \leq 0$, $g < 0$,
(iv) $h \in\downarrow$, $g \in\downarrow$, g is sub-additive, $k \geq 0$, $g < 0$,

is satisfied. Then for all $t \in J$,

$$x(t) \leq a(t) + h\left(\int_\alpha^t k(s)g(x(s))ds\right), \tag{1.1.90}$$

implies for all $\alpha \leq t < \beta_1$,

$$x(t) \leq a(t) + h\left\{G^{-1}\left[\int_\alpha^t kds + G\left(\int_\alpha^t k(s)g(a(s))ds\right)\right]\right\} \tag{1.1.91}$$

where for all $u \in I(u_0 \in I)$, $G(u) = \int_{u_0}^u dy/g(h(y))$ and $\beta_1 = \min_{1\leq i\leq 3} u_i$, with

$$u_1 = \sup\left\{u \in J : a(t) + h\left(\int_\alpha^t k(s)g(x(s))ds\right) \in I_0,\ \alpha \leq t \leq u\right\},$$

$$u_2 = \sup\left\{u \in J : \int_\alpha^t k(s)\left\{g\left(a(s) + g \circ h\left(\int_\alpha^t k(r)g(x(r))dr\right)\right\}ds \in I\right\},$$

$$u_3 = \sup\left\{u \in J : \int_\alpha^t kds + G\left(\int_\alpha^T k(s)g(a(s))ds\right) \in G(I),\ \alpha \leq t \leq T \leq u\right\},$$

The result is valid if $\leq$ is replaced by $\geq$ in both (1.1.90) and (1.1.91) provided the conditions (i)–(iv) are replaced by

(i') $h \in\downarrow$, $g \in\downarrow$, g is sub-additive, $k \geq 0$, $g > 0$,
(ii') $h \in\uparrow$, $g \in\downarrow$, g is sub-additive, $k \leq 0$, $g > 0$,
(iii') $h \in\downarrow$, $g \in\uparrow$, g is sub-additive, $k \leq 0$, $g < 0$,
(iv') $h \in\uparrow$, $g \in\uparrow$, g is sub-additive, $k \geq 0$, $g < 0$.

Finally, both results remain valid if $[\alpha, \beta)$, $[\alpha, \beta_1)$ *and* $\int_\alpha^t$ *are replaced by* $(\alpha, \beta]$, $(\alpha_1, \beta]$ *and* $\int_t^\beta$ *respectively, where now* $\alpha_1 = \max_{1\leq i\leq 3} v_i$, *with*

$$v_1 = \inf\left\{v \in J : a(t) + h\left(\int_t^\beta k(s)g(x(s))ds\right) \in I_0,\ v \leq t \leq \beta\right\},$$

$$v_2 = \inf\left\{v \in J : \int_v^\beta k(s)\left\{g\left(a(s) + g\circ h\left(\int_s^\beta k(r)g(x(r))dr\right)\right\}ds \in I\right\},$$

$$v_3 = \inf\left\{v \in J : \int_t^\beta kds + G\left(\int_T^\beta k(s)g(a(s))ds\right) \in G(I),\ v \leq t \leq T \leq \beta\right\}.$$

Proof Define $U(t) \equiv \int_\alpha^t k(s)g(x(s))ds$ and note that (1.1.90) implies that $U(J) \subset I$, and that

$$g(x(s)) \gtreqless g[a(s0 + h(U(s))] \gtreqless g(a(s)) + g\circ h(U(s))$$

where $\leq$ or $\geq$ holds according as g is non-decreasing and sub-additive, or non-increasing and super-additive. Therefore

$$U'(x(s)) \gtreqless k(s)g(a(s)) + k(s)g\circ h(U(s))$$

where $\leq$ or $\geq$ holds according as (a_1) : $(g \in\uparrow,\ g\,\text{sub}-\text{additive, k} \geq 0) \vee (g \in \downarrow,$ $g\,\text{super}-\text{additive, k} \leq 0)$, or (b_1) : $(g \in\uparrow,\ g\,\text{sub}-\text{additive, k} \leq 0) \vee (g \in\downarrow,$ $g\,\text{super}-\text{additive, k} \geq 0)$.
Integrating from α to t, this reduces to

$$U(t) \gtreqless \int_\alpha^t k(s)g(a(s))ds + \int_\alpha^t k(s)g\circ h(U(s))ds \tag{1.1.92}$$

where $\leq$ or $\geq$ holds according as (a_1) or (b_1) holds.

Now, fix $T \in (\alpha, \beta_1)$. Then by (1.1.92), if we set $A(t) \equiv \int_\alpha^t k(s)g(a(s))ds$, it follows that for all $\alpha \leq t \leq T$,

$$U(t) \gtreqless A(T) + \int_\alpha^t k(s)g\circ h(U(s))ds \tag{1.1.93}$$

where $\leq$ or $\geq$ holds according as (a_1) holds and k, g have the same sign, or (b_1) holds and k, g have the opposite sign. Since $U([\alpha, T]) \subset I$, it follows from Theorem 1.1.1 that for all $\alpha \leq t \leq T$,

$$U(t) \gtreqless G^{-1}\left[\int_\alpha^t kds + G(A(T))\right]. \tag{1.1.94}$$

(Observe that $\alpha \le t \le T << \beta_1$ implies that $A(T) + \int_\alpha^t k(s)g \circ h(U(s))ds$ lies between 0 and $A(T) + \int_\alpha^T k(s)g \circ h(U(s))ds$ in all cases (i)–(iv), (i')–(iv'), and hence $A(T)+\int_\alpha^t k(s)g\circ h(U(s))ds \in I$ follows.) By Theorem 1.1.1, $\le$ or $\ge$ holds in (1.1.94) according as $(a_2) : ((a_1),\ g\circ h \in\uparrow,\ k \ge 0,\ g \ge 0) \vee ((a_1),\ g\circ h \in\downarrow,\ k \le 0,\ g \le 0)$, or $(b_2) : ((b_1),\ g \circ h \in\uparrow,\ k \ge 0,\ g \le 0) \vee ((b_1),\ g \circ h \in\downarrow,\ k \le 0,\ g \ge 0)$. On analysis, these conditions reduce to (a_2) : (i) or (iii) hold, or (b_2) : (ii) or (iv) hold. From (1.1.94), with $t = T$ and a change of notation,

$$h(U(t)) \le h\left\{G^{-1}\left[\int_\alpha^t kds + G(At))\right]\right\} \tag{1.1.95}$$

holds for either (a_2) or (b_2). Hence, (1.1.91) follows in all four cases (i)–(iv), as asserted.

If $\le$ is replaces by $\ge$ in (1.1.90), the only change in the analysis preceding conditions (a_2), (b_2) is that the roles of "$g \in\uparrow$" and "$g \in\downarrow$" are interchanged. The new (a_2) reduces o conditions (i') or (ii'), and the new (b_2) reduces to conditions (ii') or (iv'). Hence we obtain (1.1.95) and (1.1.91) with $\le$ replaced by $\ge$ in all four cases (i')–(iv'). The final part of the theorem follows precisely as in the proof of Theorem 1.1.1 with suitable changes. □

Theorem 1.1.16 (The Dannan Inequality [182]) *Let $x(t)$, $a(t)$, $k(t)$, and $h(t)$ be real-valued positive functions defined on $J = [0, \beta)$, let $g(u) \in M$ with corresponding function ψ on an internal I such that $x(J) \subset J$ and $a(J) \subset I$. Suppose also that the function h be a monotonic, non-decreasing function on an internal K such that $0 \in K$, $h(K) \in I$. If the following inequality holds for all $t \in J$,*

$$x(t) \le a(t) + h(t)\left[\int_0^t k(s)g(x(s))ds\right], \tag{1.1.96}$$

then for all $0 \le t < \beta_1$,

$$x(t) \le a(t) + h(t)\left(G^{-1}[\int_0^t k(s)ds + G(\int_0^t k(s)\psi(a(s))ds)]\right) \tag{1.1.97}$$

where $G(u) = \int_{u_0}^u dy/g(h(y))$ for all $u \ge u_0 \in K$ and $\beta_1 = \min(u_1, u_2, u_3)$ with

$$\begin{cases} u_1 = \sup\{u \in J : a(t) + h(t)\big(\int_0^t k(s)g(x(s))ds\big) \in I, \quad 0 \le t \le u\}, \\ u_2 = \sup\{u \in J : \int_0^u k(s)\big[\phi(a(s)) + g \circ h\big(\int_0^s k(\theta)g(x(\theta))d\theta\big)\big]ds \in K\}, \\ u_3 = \sup\{u \in J : \int_0^t k(s)ds + G\big(\int_0^T k(s)\psi(a(s))ds\big) \in G(K), \quad 0 \le t \le T \le u\}. \end{cases}$$

Proof The proof can be finished in a similar way to that of Theorem 1.1.15. □

Remark 1.1.1 It is not difficult to show that the same estimate for $x(t)$ can be obtained when $k(t)$ is non-positive and $h(t)$ is non-increasing.

Remark 1.1.2 When $h(u) = u$, Theorem 1.1.16 reduces to a generalization of Lemma 2 by Muldowney and Wong [402].

The case when the nonlinear function h in (1.1.96) is multiplied by $b(t)$ has been considered in detail by Beesack ([56], Theorem 5.4, or [55], Theorems 2 and 3). Under several sets of conditions on (x, a, b, h, k, g), different incomparable estimates for $x(t)$ have been obtained.

Corollary 1.1.4 (The Dannan Inequality [182]) *Let x, a, k, g all be as in Theorem 1.1.16 and suppose $b(t)$ is non-negative, continuous, and non-decreasing on $I = [0, \beta]$. If, for all $t \in I$,*

$$x(t) \leq a(t) + b(t) \int_0^t k(s)g(x(s))ds, \tag{1.1.98}$$

then, for all $0 \leq t < t_0$,

$$x(t) \leq a(t) + G^{-1} \left[b(t) \int_0^t k(s)ds + G(b(t) \int_0^k k(s)\psi(a(s))ds) \right]$$

where G, G^{-1} are defined in Theorem 1.1.16, but with $h(u) \equiv u$ there,

$$b(t) \int_0^t k(s)ds + G(b(t) \int_0^k k(s)\phi(a(s))ds) \in Dom\ (G^{-1}).$$

Proof The proof of this corollary follows by an argument similar to that in the proof of Corollary 1.1.2. □

In what follows, we shall give an estimate for $x(t)$ under different set of conditions on (x, a, b, h, k, g).

Theorem 1.1.17 (The Dannan Inequality [182]) *Let $x(t)$, $a(t)$, $k(t)$, $b(t)$ be continuous and non-negative on $J = [0, \beta]$, with $b(t) > 0$ and $a(t)/b(t) \leq \gamma$ for some positive constant γ. Let $g(u)$ be of class H with corresponding function ϕ. Suppose that the function h is a continuous, non-negative and non-decreasing function on $[0, +\infty)$. If, for all $t \in J$,*

$$x(t) \leq a(t) + b(t)h[\int_0^t k(s)g(x(s))ds], \tag{1.1.99}$$

then, for all $0 \leq t < \beta_1$,

$$x(t) \leq a(t) + b(t)h \circ L^{-1}(\int_0^t k(s)\phi(b(s))ds), \tag{1.1.100}$$

where, for all $u \geq 0$,

$$L(u) = \int_0^u \frac{dz}{g(\gamma + h(z))},$$

and

$$\beta_1 = \sup\left\{t \in J : \int_0^t k(s)g(x(s))ds \in L(\mathbb{R}_+)\right\}.$$

Proof Let

$$z(t) = \int_0^t k(s)g(x(s))ds.$$

Then from (1.1.99) and the hypotheses on g and $a,\ b$, it follows that

$$\begin{aligned}\frac{dz}{dt} &= k(t)g(x(t)) \leq k(t)g[a(t) + b(t)h(z(t))] \\ &\leq k(t)\phi(b(t))g[\frac{a(t)}{b(t)} + h(z(t))]\end{aligned}$$

and

$$\frac{dz}{\gamma + h(z)} \leq k(t)\phi(b(t))dt. \tag{1.1.101}$$

Integrating both sides of (1.1.101) from 0 to t, we may obtain

$$L(z) \leq \int_0^t k(s)\phi(b(s))ds$$

and, for all $0 \leq t < \beta_1$,

$$z \leq L^{-1}[\int_0^t k(s)\phi(b(s))ds]. \tag{1.1.102}$$

The substitution of (1.1.102) in (1.1.99) implies (1.1.100). □

Remark 1.1.3 In Theorem 1.1.17, it is clear that hypotheses $b > 0$ and $a/b \leq \gamma$ can be replaced by $a > 0$ and $b/a \leq \gamma$, Therefore $\phi(b(s))$ in (1.1.100) will be replaced by $\phi(a(s))$ and

$$L(u) = \int_0^u dz/g(1 + \gamma h(z)).$$

Remark 1.1.4 Let $g(u) = u^2/(1+u)$. Then $g(u)$ is not sub-multiplicative and does not satisfy the condition $g(u)/v \leq g(u/v)$ for all $u > 0$ and all $v \geq 1$. Therefore, all theorems in [54–56, 195] are not applicable. Theorem 1.1.17 can be applied, since $u^2/(1+u)$ is of class H with corresponding function ϕ defined by $\phi(\alpha) = \alpha$ $(0 \leq \alpha \leq 1)$, $\phi(\alpha) = \alpha^2$ $(\alpha \geq 1)$.

Remark 1.1.5 In the case when g is strictly increasing and $h \equiv g^{-1}$, we may obtain from Theorem 1.1.17 and following estimate for $x(t)$,

$$x(t) \leq a(t) + b(t)g^{-1} \circ L^{-1}(\int_0^t k(s)\phi(a(s))ds)$$

where now $L(u) = \int_0^u dz/g(\gamma + g^{-1}(z))$.

This estimate is not comparable with a result obtained by Gollwitzer ([250], Theorem 1).

There is another upper bound for $x(t)$ when g satisfies different, but general conditions. The following result essentially is the variation of Grollwitzer's Theorem 1 of [250] in which the conditions: g convex and sub-multiplicative, are replaced by: $g \in H \cap M$.

Theorem 1.1.18 (The Dannan Inequality [182]) *Let $a(t)$, $k(t)$, $b(t)$ be continuous, non-negative functions on $J = [0, \beta)$, with $b(t) > 0$ and $g \in H$ and M with corresponding function ϕ and ψ, respectively. Assume also that the function g is strictly increasing. If the inequality holds for all $t \in J$,*

$$x(t) \leq a(t) + b(t)g^{-1}(\int_0^t k(s)g(x(s))ds), \tag{1.1.103}$$

then for all $t \in [0, \beta_1]$,

$$x(t) \leq b(t)g^{-1}(B(t)), \tag{1.1.104}$$

where

$$B(t) = \psi(\frac{a(t)}{b(t)}) + \int_0^t k(s)\phi(b(s))\psi(\frac{a(s)}{b(s)})\exp(\int_s^t k(\theta)\phi(b(\theta))d\theta)ds \tag{1.1.105}$$

and

$$\beta_1 = \sup\{t \in J : \int_0^t k(s)\phi(b(s))B(s)ds \in L(\mathbb{R}_+)\}.$$

Proof From (1.1.103) it follows that

$$\frac{x(t)}{b(t)} \leq \frac{a(t)}{b(t)} + g^{-1}\left(\int_0^t k(s)g(\frac{x(s)}{b(s)})\phi(b(s))ds\right). \tag{1.1.106}$$

Let $x(t)/b(t) = z(t)$ and use the hypotheses on g to obtain

$$g(z) \leq \psi(\frac{a}{b}) + \int_0^t k\phi(b)g(z)ds. \tag{1.1.107}$$

Considering $g(z)$ as a function, using the most general linear Bellman-Gronwall inequality (see, e.g., Beesack [54], or Theorem 1.1.2 in Qin [557]), it follows that $g(z(t)) \leq B(t)$ so $z(t) \leq g^{-1}(B(t))$, but since $x(t) = b(t)z(t)$, (1.1.195) follows. □

Several integral inequalities similar to Bellman-Bihari type have been obtained by Pachpatte [441, 445, 446, 451, 455, 456, 458, 463, 472, 475, 476, 478]. Most of these inequalities are based on a main inequality in [441], in which an estimate for $x(t)$ has been obtained, when

$$x(t) \leq x_0 + \int_0^t f(s)x(s)ds + \int_0^t f(s)\left(\int_0^s g(\theta)x(\theta)d\theta\right)ds,$$

where $f(t)$, $g(t)$ and $x(t)$ are supposed to be non-negative with x_0 being a positive constant and $t \in [0, +\infty)$.

Later on, Agarwal [5] proved a general version of Pachpatte inequality, when $x(t)$ satisfies the inequality, for all $t \geq 0$,

$$x(t) \leq p(t) + \int_0^t f_1(s)x(s)ds + \int_0^t f_2(s)\int_0^s f_3(\theta)x(\theta)d\theta. \tag{1.1.108}$$

Several linear and nonlinear generalizations have been obtained by Agarwal and Thandapani in [21]. In the following two theorems, we consider nonlinear versions of (1.1.108).

These two theorems are related to the special case $m = 2$ of Theorem 11 and Theorem 13 of [21], which dealt with $g, h \in \mathcal{F}$ (see, below Definition 1.1.3) rather than $g, h \in H$ or $g, h \in M$. See also the case $k = 2$ of Theorem 1 of Beesack [58] for related results.

Theorem 1.1.19 (The Dannan Inequality [182]) *Let $x(t)$, $a(t)$, $k(t)$, $l(t)$ and $m(t)$ be real-valued non-negative, continuous functions defined on $I = [0, +\infty)$ with $a(t)$ positive, non-decreasing. Assume that $g(u)$ and $h(u)$ belong to H with corresponding multiplier function ϕ and ψ, respectively, with $\phi(u) \leq cu$ for all $u \geq 1$, where c is a positive constant. If the inequality holds for all $t \in I$,*

$$x(t) \leq a(t) + \int_0^t k(s)g(x(s))ds + \int_0^t l(s)\int_0^s m(\theta)h(x(\theta))d\theta ds, \tag{1.1.109}$$

then for all $0 \leq t \leq \beta$,

$$x(t) \leq a(t)F(t)H^{-1}\left\{H(1) + \int_0^t k_2(s)\psi(F(s))ds\right\}, \tag{1.1.110}$$

where

$$\begin{cases} F(t) = G^{-1}[G(1) + c\int_0^t k_1(s)ds], \\ k_1(t) = k(t)\phi(a(t))/a(t), \\ k_2(t) = l(t)\int_0^t \dfrac{m(\theta)\phi(a(\theta))}{a(\theta)}d\theta, \\ H(t) = \int_{u_0}^u ds/h(s), \quad G(u) = \int_{u_0}^u ds/g(s), \quad u \geq u_0 > 0, \end{cases} \tag{1.1.111}$$

and H^{-1} *and* G^{-1} *are the inverse function of* H *and* G, *respectively,* $\beta = \min(b_1, b_2)$,

$$b_1 = \sup\left\{t \in I : G(1) + c\int_0^t k_1(s)ds \in Dom\ (G^{-1})\right\}$$

and

$$b_2 = \sup\left\{t \in I : H(1) + c\int_0^t k_2(s)\psi(F(s))ds \in Dom\ (H^{-1})\right\}.$$

Proof Let $x(t)/a(t) \equiv y(t)$. Since g and h belong to H, from (1.1.109) it follows that

$$y(t) \leq R(t), \tag{1.1.112}$$

where, for all $t \in I$,

$$R(t) = 1 + \int_0^t \frac{h(s)\phi(a(s))}{a(s)}g(y(s))ds + \int_0^t l(t)\int_0^t \frac{m(\theta)\phi(a(\theta))}{a(\theta)}h(y(\theta))d\theta ds. \tag{1.1.113}$$

Noting (1.1.113) and the non-decreasing property of g and h, we arrive

$$R'(t) \leq k_1(t)g(R) + k_2(t)h(R), \quad R(0) = 1. \tag{1.1.114}$$

Integrating (1.1.114) from 0 to t, we may obtain

$$R(t) \leq 1 + \int_0^t k_1(s)g(R)ds + \int_0^t k_2(t)h(R)ds, \quad R(0) = 1. \tag{1.1.115}$$

Using Theorem 1.1.11,

$$n(t) \le 1 + \int_0^t k_2(t)h(s)(R(s))ds \tag{1.1.116}$$

we obtain for all $0 \le t \le b_1$,

$$R(t) \le n(t)G^{-1}\left[G(1) + \int_0^t k_1(t)\frac{\phi(n(s))}{n(s)}ds\right]. \tag{1.1.117}$$

Since $\phi(n)/n \le c$, from (1.1.117) it follows that

$$R(t) \le n(t)F(t), \tag{1.1.118}$$

where $F(t)$ is defined by (1.1.111). Furthermore,

$$h(R(t)) \le \psi(F(t))h(n(t)),$$

since $h \in H$. Hence,

$$\frac{k_2(t)h(R(t))}{h(n(t))} \le k_2(t)\psi(F(t)).$$

Because of (1.1.113) and (1.1.116), this reduces to

$$\frac{d}{dt}H(n(t)) \le k_2(t)\psi(F(t)).$$

Now integrating from 0 to t, we obtain

$$H(n(t)) \le H(1) + \int_0^t k_2(s)\psi(F(s))ds. \tag{1.1.119}$$

Thus the desired bound in (1.1.110) follows from (1.1.118), (1.1.119), and (1.1.116). □

We point out that the conditions $g(\alpha u) \le \phi(\alpha)g(u) \le c\alpha g(u)$ for all $u \ge 0,\ \alpha \ge 1$, imply that $g(u) \ge g(1)u/c$ for $0 < u \le 1$, and $g(u) \le cg(u)u$ for all $u \ge 1$ (and that $c \ge 1$).

Remark 1.1.6 We may get a similar bound for $x(t)$, when the condition $\phi(u) \le cu$ is replaced by $\psi(u) \le cu$ for all $u \ge 1$.

Theorem 1.1.20 (The Dannan Inequality [182]) *Let $x(t)$, $a(t)$, $k(t)$, $l(t)$ and $m(t)$ be real-valued non-negative, continuous functions defined on $I = [0, +\infty)$, let $g(u)$ and $h(u)$ be of class M with corresponding function ϕ and ψ, respectively* *and let (i) $g \in H$ or (ii) $h \in H$ with corresponding multiplier function χ such that $\chi(u) \le cu$, where c is a positive constant. If the inequality holds for all $t \in I$,*

$$x(t) \le a(t) + \int_0^t k(s)g(x(s))ds + \int_0^t l(s)\int_0^s m(\theta)h(x(\theta))d\theta ds, \tag{1.1.120}$$

then, for case (i), that

$$x(t) \le a(t) + r(t)N(t) + H^{-1}\Big\{r(t)\int_0^t p(s)ds + H\big[r(t)\int_0^t p(s)\psi(r(s)N(s))ds\big]\Big\}, \tag{1.1.121}$$

while in case (ii), we have, for all $t \in [0, \beta]$,

$$x(t) \le a(t) + r_1(t)N(t) + G^{-1}\Big\{r_1(t)\int_0^t k(s)ds + G\big[r_1(t)\int_0^t k(s)\psi(r_1(s)N(s))ds\big]\Big\}$$

where

$$r_1(t) = H^{-1}\big[H(1) + c\int_0^t p(s)ds\big], \tag{1.1.122}$$

H and G are as defined in Theorem 1.1.19, H^{-1} and G^{-1} are the inverse functions of H and G, respectively,

$$\begin{cases} N(t) = \displaystyle\int_0^t \Big[k(s)\phi(a(s)) + l(s)\int_0^s m(\theta)\phi(a(\theta))d\theta\Big]ds, & (1.1.123) \\ p(t) = l(t)\displaystyle\int_0^t m(s)ds,\ r_1(t) = G^{-1}\Big[G(1) + c\int_0^t k(s)ds\Big],\ \beta = \min(\beta_1, \beta_2), & (1.1.124) \\ \beta_1 = \sup\{u \in I : G(1) + c\displaystyle\int_0^t k(s)ds \in Dom\ (G^{-1}),\quad 0 \le t \le u\}, & (1.1.125) \\ \beta_2 = \sup\{u \in I : r(t)\displaystyle\int_0^t p(s)ds & \\ \qquad +H\Big[r(t)\displaystyle\int_0^t p(s)\psi(r(s))N(s)ds\Big] \in Dom\ (H^{-1}),\quad 0 \le t \le u\}. & (1.1.126) \end{cases}$$

Proof It suffices to consider case (i), since case (ii) can be treated in a similar way. Let for all $t \in I$,

$$R(t) = \int_0^t k(s)g(x(s))ds + \int_0^r l(s)\int_0^s m(\theta)h(x(\theta))d\theta ds.$$

Since g and $h \in M$, we obtain for all $t \in I$,

$$R'(t) \le k(t)\phi(a(t)) + k(t)g(R(t)) + \Big[l(t)\int_0^t m(s)ds\Big]h(R(t))$$

and

$$R(t) \le N(t) + \int_0^t k(s)g(R(s))ds + \int_0^t p(s)h(R(s))ds. \tag{1.1.127}$$

If we put

$$M(t) = N(t) + \int_0^t p(s)h(R(s))ds, \tag{1.1.128}$$

then from Theorem 1.1.11 it follows that, for all $t \in [0, \beta)$,

$$R(t) \le M(t)G^{-1}\left[G(1) + \int_0^t \frac{k(s)\chi(M(s))}{M(s)}ds\right] \tag{1.1.129}$$

where β is defined by (1.1.126). Since G and G^{-1} are strictly increasing and $\chi(M) \le cM$, then from (1.1.129), we conclude, for all $t \in [0, \beta)$,

$$R(t) \le r(t)N(t) + r(t)\int_0^t p(s)h(R(s))ds. \tag{1.1.130}$$

Thus, applying Corollary 1.1.2 to (1.1.130) completes the proof. □

Remark 1.1.7 When $g(u) = h(u) = u$, Theorem 1.1.20 reduces to Lemma 1 of Agarwal [5].

For the Gronwall-Bellman inequality like (1.1.131), [198, 672] have given some estimates of the upper bounds of its solutions; but to obtain the results, not only do they have to impose several restrictions on its functions such that the useful scopes are reduced, but also the estimates are not sharp.

Now we also give a definition of a function class $\mathcal{F}_1$ given in [56].

Definition 1.1.4 A function $w : \mathbb{R}_+ \to \mathbb{R}_+$ is said to belong to a function class $\mathcal{F}_1$ if it satisfies the following conditions:

(i) $w(u) \ge 0$ is non-decreasing and continuous on $\mathbb{R}_+$,
(ii) $\frac{1}{v}w(u) \le w\left(\frac{u}{v}\right)$ for all $u \ge 0, v \ge 1$.

Lemma 1.1.9 (The Kong-Zhang Inequality [311]) *Suppose that* $f(x) \ge 0, g(x) > 0, h(x) \ge 0$ *and* $y(x)$ *are continuous on* $\mathbb{R}_+, w(u) \in \mathcal{F}_1$. *If for all* $x \in \mathbb{R}_+$,

$$y(x) \le f(x) + g(x)\int_0^x h(s)w(y(s))ds, \tag{1.1.131}$$

then, for all $x \in \mathbb{R}_+$,

$$y(x) \leq f(x) + g(x)G^{-1}\left[G\left(\int_0^x h(s)\bar{g}(s)w\left(\frac{f(s)}{g(s)}\right)ds\right) + \int_0^x h(s)\bar{g}(s)ds\right],$$

where

$$G(u) = \int_{u_0}^{u} \frac{ds}{w(s)}, \quad u_0 \geq u \geq 0,$$

Proof Assume $y(x)$ is a solution of (1.1.131), denote $z(x) = \max\{y(x), 0\}, x \in \mathbb{R}_+$, then $z(x)$ is also a solution of (1.1.131), i.e.,

$$z(x) \leq f(x) + g(x)\int_0^x h(s)w(z(s))ds. \tag{1.1.132}$$

At first, consider the case for $g(x) \equiv 1$. Define for all $x \in \mathbb{R}_+$,

$$R(x) = \int_0^x h(s)w(z(s))ds.$$

Then

$$R'(x) = h(x)w(z(s)).$$

From $w(u) \in \mathcal{F}_1$, we know that

$$R'(x) \leq h(x)[w(f(x)) + w(R(x))].$$

Integrating both sides from 0 to x, we get

$$R(x) \leq \int_0^x h(s)w(f(s))ds + \int_0^x h(s)w(R(s))ds.$$

For any $X \geq 0$, when $x \leq X$,

$$R(x) \leq \int_0^X h(s)w(f(s))ds + \int_0^x h(s)w(R(s))ds.$$

Using the Bihari inequality (i.e., Theorem 1.1.1), we have

$$R(x) \leq G^{-1}\left[G\left(\int_0^X h(s)w(f(s))ds\right) + \int_0^x h(s)ds\right].$$

Letting $x = X$ and replacing X by x, we see that

$$R(x) \leq G^{-1}\left[G\left(\int_0^x h(s)w(f(s))ds\right) + \int_0^x h(s)ds\right].$$

Therefore

$$z(x) \leq f(x) + G^{-1}\left[G\left(\int_0^x h(s)w(f(s))ds\right) + \int_0^x h(s)ds\right]. \quad (1.1.133)$$

When $g(x) \neq 1$, from (1.1.132), we derive

$$\begin{aligned}\frac{z(x)}{g(x)} &\leq \frac{f(x)}{g(x)} + \int_0^x h(s)\bar{g}(s)w\left(\frac{z(s)}{\bar{g}(s)}\right)ds \\ &\leq \frac{f(x)}{g(x)} + \int_0^x h(s)\bar{g}(s)w\left(\frac{z(s)}{g(s)}\right)ds. \quad (1.1.134)\end{aligned}$$

Substituting $\frac{z(x)}{g(x)}$, $\frac{f(x)}{g(x)}h(x)\bar{g}(x)$ for $z(x), f(x), h(x)$ in (1.1.133) respectively, we have

$$\frac{z(x)}{g(x)} \leq \frac{f(x)}{g(x)} + G^{-1}\left[G\left(\int_0^x h(s)\bar{g}(s)w\left(\frac{f(s)}{g(s)}\right)ds\right) + \int_0^x h(s)\bar{g}(s)ds\right]$$

i.e.,

$$z(x) \leq f(x) + g(x)G^{-1}\left[G\left(\int_0^x h(s)\bar{g}(s)w\left(\frac{f(s)}{g(s)}\right)ds\right) + \int_0^x h(s)\bar{g}(s)ds\right].$$

Noting that $y(x) \leq z(x)$, the result inequality becomes true. However, $w(u) \in \mathcal{F}_1$ implies that

$$\int_{u_0}^{u} \frac{ds}{w(s)} = +\infty.$$

In fact, for all $u \geq 1$,

$$\frac{w(u)}{u} \leq w(1),$$

i.e., $w(u) \leq w(1)u$, which follows

$$\int_1^{+\infty} \frac{ds}{w(s)} \geq \int_1^{+\infty} \frac{ds}{w(1)s} = +\infty.$$

So, for all $x \in \mathbb{R}_+$,

$$G\Big(\int_0^x h(s)\overline{g}(s)w(\frac{f(s)}{g(s)})ds\Big) + \int_0^x h(s)\overline{g}(s)ds \in Dom\,(G^{-1}).$$

The proof is thus complete. □

Theorem 1.1.21 (The Kong-Zhang Inequality [311]) *Suppose* $f(x)$, $g(x)$, $\overline{g}(x)$, $h(x)$, $y(x)$ *and* $w(x)$ *are defined as Lemma 1.1.9,* $\psi(u) \geq 0$ *is non-decreasing, continuous on* $\mathbb{R}_+$. *If*

$$y(x) \leq f(x) + g(x)\psi\Big(\int_0^x h(s)w\big(y(s)\big)ds\Big),$$

then, for all $x \in [0, b)$,

$$y(x) \leq f(x) + g(x)\psi\Big\{F^{-1}\Big[F\Big(\int_0^x h(s)\overline{g}(s)w(\frac{f(s)}{g(s)})ds\Big) + + \int_0^x h(s)\overline{g}(s)ds\Big]\Big\},$$

where

$$F(u) = \int_{u_0}^u \frac{ds}{w(\psi(s))}, u_0 \geq u \geq 0,$$

$$b = \sup_{x\in\mathbb{R}_+}\left\{x : F\left(\int_0^x h(s)\bar{g}(s)w\left(\frac{f(s)}{g(s)}\right)ds\right) + \int_0^x h(s)\bar{g}(s)ds \in Dom\,(F^{-1})\right\}.$$

Proof The proof is similar to that of Lemma 1.1.9, we omit it here. □

Theorem 1.1.22 (The Kong-Zhang Inequality [311]) *Suppose* $f(x), g_i(x)$ *and* $h_i(x)$ $(i = 1, 2, \ldots, m)$ *are non-negative and continuous on* $\mathbb{R}_+$, $w(u), \psi(u)$ *as in Theorem 1.1.21,* $g_{n+1}(x) > 0, h_{n+1}(x) \geq 0$ *are continuous on* $\mathbb{R}_+$. *If for all* $x \in \mathbb{R}_+$,

$$y(x) \leq f(x) + \sum_{i=1}^n g_i(x)\int_0^x h_i(s)y(s)ds + g_{n+1}(x)\psi\left(\int_0^x h_{n+1}(s)w(y(s))ds\right), \quad (1.1.135)$$

then, for all $x \in [0, b)$,

$$y(x) \leq A_n(f) + A_n(g_{n+1})\psi\left\{F^{-1}\left[F\left(\int_0^x h_{n+1}\bar{A}_n(g_{n+1})w\left(\frac{A_n(f)}{A_n(g_{n+1})}\right)ds\right)\right.\right.$$
$$\left.\left.+ \int_0^x h_{n+1}\bar{A}_n(g_{n+1})ds\right]\right\},$$

where $F(s) = \int_{u_0}^{u} \frac{ds}{w(\psi(s))}$, $u \geq u_0 \geq 0$, $A_{n+1}(u)$ *is defined as in Theorem 1.2.12 in Qin [557],* $\bar{A}_n(g_{n+1}) = \max\{A_n(g_{n+1}), 1\}, x \in \mathbb{R}_+$, *the determination of* b *must make* F^{-1} *be well-defined.*

Proof To simplify the notation, let

$$\psi(\cdot) = \psi\left(\int_0^x h_{n+1} w(y) ds\right).$$

Therefore

$$y(x) \leq (f + g_{n+1}\psi(\cdot)) + \sum_{i=1}^{n} g_i \int_0^x h_i y ds.$$

According to Theorem 1.2.12 in Qin [557],

$$y(x) \leq A_n(f + g_{n+1}\psi(\cdot)).$$

Noting that

$$\psi\left(\int_0^x h_{n+1} w(y) ds\right)$$

is non-decreasing in x, from the inequality (2) of Lemma 1.2.1 in Qin [557], we derive

$$y(x) \leq A_n(f) + A_n(g_{n+1})\psi\left(\int_0^x h_{n+1} w(y) ds\right).$$

By Theorem 1.1.21, the conclusion holds. □

When $w(u)$ is a concave function, we have the following lemma.

Lemma 1.1.10 (The Kong-Zhang Inequality [311]) *Suppose that* $w(u)$ *is non-negative and concave on* $\mathbb{R}_+$, $w(0) = 0$. *Then for all* $u \geq 0$, $0 < v \leq 1$,

$$\frac{1}{v} w(u) \leq w\left(\frac{u}{v}\right).$$

Proof By definition, $w(u)$ is concave, which implies that, for all $\alpha \geq 0, \beta \geq 0, \alpha + \beta = 1$,

$$w(\alpha u_1 + \beta u_2) \leq \alpha w(u_1) + \beta w(u_2).$$

Letting

$$\alpha = v,\ u_1 = \frac{u}{v},\ u_2 = 0,$$

and considering $w(0) = 0$, we have

$$w(u) \leq vw\left(\frac{u}{v}\right),$$

i.e.,

$$\frac{1}{v}w(u) \leq w\left(\frac{u}{v}\right).$$

□

Lemma 1.1.11 (The Kong-Zhang Inequality [311]) *Suppose that* $f(x) \geq 0, h(x) \geq 0, 0 < g(x) \leq 1$, *and* $y(x)$ *are continuous on* $\mathbb{R}_+, w(0) = 0, w$ *is non-negative and concave on* $\mathbb{R}_+$. *If, for all* $x \in \mathbb{R}_+$,

$$y(x) \leq f(x) + g(x)\int_0^x h(s)w(y(s))ds, \tag{1.1.136}$$

then, for all $x \in [0, b)$,

$$y(x) \leq f(x) + g(x)H^{-1}\left[H\left(\alpha\int_0^x h(s)g(s)w\left(\frac{f(s)}{\alpha g(s)}\right)ds\right) + \beta\int_0^x h(s)g(s)ds\right],$$

with all $\alpha > 0, \beta > 0, \alpha + \beta = 1$, *where*

$$H(u) = \int_{u_0}^u \frac{ds}{w(s/\beta)}, \quad u \geq u_0 > 0,$$

and determination of b must make H^{-1} *have meaning.*

Proof Assume that $y(x)$ is a solution of (1.1.136). Denote $z(x) = \max\{y(x), 0\}$. Then $z(x)$ is also a solution of (1.1.136), i.e.,

$$z(x) \leq f(x) + g(x)\int_0^x h(s)w(z(s))ds. \tag{1.1.137}$$

At first, consider the case for $g(x) \equiv 1$. Define

$$R(x) = \int_0^x h(s)w(z(s))ds.$$

Then

$$R'(x) = h(x)w(z(x)) \leq h(x)w[f(x) + R(x)].$$

Since $w(u)$ is concave, we have for all $\alpha > 0, \beta > 0, \alpha + \beta = 1$,

$$R'(x) = h(x)w\left[\alpha\left(\frac{f(x)}{\alpha}\right) + \beta\left(\frac{R(x)}{\beta}\right)\right]$$
$$\leq \alpha h(x)w\left(\frac{f(x)}{\alpha}\right) + \beta h(x)w\left(\frac{R(x)}{\beta}\right).$$

Integrating both sides from 0 to x, we conclude

$$R(x) \leq \alpha\int_0^x h(s)w\left(\frac{f(x)}{\alpha}\right)ds + \beta\int_0^x h(s)w\left(\frac{R(x)}{\beta}\right)ds.$$

For any $X > 0$, when $x \leq X$,

$$R(x) \leq \alpha\int_0^X h(s)w\left(\frac{f(s)}{\alpha}\right)ds + \beta\int_0^x h(s)w\left(\frac{R(s)}{\beta}\right)ds.$$

Using the Bihari inequality, we have

$$R(x) \leq H^{-1}\left[H\left(\alpha\int_0^X h(s)w\left(\frac{f(s)}{\alpha}\right)ds\right) + \beta\int_0^x h(s)ds\right].$$

Letting $x = X$ and replacing X by x, we see that

$$R(x) \leq H^{-1}\left[H\left(\alpha\int_0^x h(s)w\left(\frac{f(s)}{\alpha}\right)ds\right) + \beta\int_0^x h(s)ds\right].$$

Hence

$$z(x) \leq f(x) + H^{-1}\left[H\left(\alpha\int_0^x h(s)w\left(\frac{f(s)}{\alpha}\right)ds\right) + \beta\int_0^x h(s)ds\right]. \quad (1.1.138)$$

When $g(x) \neq 1$, since $0 < g(x) \leq 1$, from (1.1.134) and (1.1.137) it follows

$$\frac{z(x)}{g(x)} \geq \frac{f(x)}{g(x)} + \int_0^x h(s)g(s)w\left(\frac{z(s)}{g(s)}\right)ds.$$

Substituting $\frac{z(x)}{g(x)}, \frac{f(x)}{g(x)}, h(x)g(x)$ for $z(x), f(x), h(x)$ in (1.1.138) respectively, we may get

$$\frac{z(x)}{g(x)} \leq \frac{f(x)}{g(x)} + H^{-1}\left[H\left(\alpha\int_0^x h(s)g(s)w\left(\frac{f(s)}{\alpha g(s)}\right)ds\right) + \beta\int_0^x h(s)g(s)ds\right].$$

Noting that $y(x) \leq z(x)$, we can obtain the result. □

Theorem 1.1.23 (The Kong-Zhang Inequality [311]) *Suppose $f(x), g(x), h(x)$, $y(x)$ and $w(u)$ are defined as in Lemma 1.1.11, $\psi(u) \geq 0$ is non-decreasing and continuous on $\mathbb{R}_+$, If, for all $x \in \mathbb{R}_+$,*

$$y(x) \leq f(x) + g(x)\psi\left(\int_0^x h(s)w(y(s))ds\right), \tag{1.1.139}$$

then, for all $x \in [0, b)$,

$$y(x) \leq f(x) + g(x)\psi\left\{I^{-1}\left[I\left(\alpha\int_0^x h(s)g(s)w\left(\frac{f(s)}{\alpha g(s)}\right)ds\right) + \beta\int_0^x h(s)g(s)ds\right]\right\}, \tag{1.1.140}$$

with all $\alpha > 0, \beta > 0, \alpha + \beta = 1$, where

$$I(u) = \int_{u_0}^u \frac{ds}{w(\psi(s)/\beta)}, \quad u \geq u_0 > 0,$$

and the determination of b must make I^{-1} be well-defined.

Proof The proof is similar to that of Lemma 1.1.11. □

In [55], results of the inequality (1.1.139) and its inverse inequality are obtained. But the requirement that $w(u)$ should be sub-additive and sub-multiplicative is so exacting that even the function $w(u) = k + u^\alpha$ $(0 < k < 1, \alpha \in \mathbb{R})$ is not applicable.

Theorem 1.1.24 (The Kong-Zhang Inequality [311]) *Suppose that $f(x), g(x)$, $h_i(x)$ $(i = 1, 2, \ldots, n)$ and $A_{n+1}(u)$ are defined as in Theorem 1.2.12 in Qin [557], $w(u), \psi(u)$ as in Theorem 1.1.23, $h_{n+1}(x) \geq 1, g_{n+1}(x) > 0$ are continuous on $\mathbb{R}_+$ and $A_n(g_{n+1}(x)) \leq 1$. If, for all $x \in \mathbb{R}_+$,*

$$y(x) \leq f(x) + \sum_{i=1}^n g_i(x)\int_0^x h_i(s)y(s)ds + g_{n+1}(x)\psi\left(\int_0^x h_{n+1}(s)w(y(s))ds\right), \tag{1.1.141}$$

then for all $x \in [0, b)$,

$$\begin{aligned} y \leq A_n(f) + A_n(g_{n+1})\psi\left\{I^{-1}\left[I\left(\alpha\int_0^x h_{n+1}A_n(g_{n+1})w\left(\frac{A_n(f)}{\alpha A_n(g_{n+1})}\right)ds\right)\right.\right. \\ \left.\left. +\beta\int_0^x h_{n+1}A_n(g_{n+1})ds\right]\right\}, \end{aligned} \tag{1.1.142}$$

with $\alpha > 0, \beta > 0, \alpha + \beta = 1$, where $I(u)$ and b are the same as in Theorem 1.1.23.

Proof The proof is similar to that of Theorem 1.1.22. □

Remark 1.1.8 The result of Lemma 1.1.11, Theorems 1.1.23 and 1.1.24 can be easily generalized to the case that $g(x)$ or $A_n(g_{n+1}(x))$ is non-negative and bounded on $\mathbb{R}_+$, we omit them here.

The next result is due to Pachpatte [519].

Theorem 1.1.25 (The Pachpatte Inequality [519]) *Let $u(t), f(t) \in C(\mathbb{R}_+, \mathbb{R}_+)$, $h(t,s) \in C(\mathbb{R}_+^2, \mathbb{R}_+)$, for $0 \le s \le t < +\infty$ and $c \ge 0$, $p > 1$ are real constants. Let $g \in C(\mathbb{R}_+, \mathbb{R}_+)$ be a non-decreasing function, $g(u) > 0$ for all $u > 0$ and, for all $t \in \mathbb{R}_+$,*

$$u^p(t) \le c + \int_0^t \left[f(s)g(u(s)) + \int_0^s h(s,\sigma)g(u(\sigma))d\sigma \right] ds, \tag{1.1.143}$$

then for all $0 \le t \le t_1$,

$$u(t) \le [G^{-1}[G(c) + A(t)]]^{1/p}, \tag{1.1.144}$$

where

$$A(t) = \int_0^t \left[f(s) + \int_0^s h(s,\sigma)d\sigma \right] ds, \;\; G(r) = \int_{r_0}^r \frac{ds}{g(s^{1/p})}, \;\; r \ge r_0 > 0, \tag{1.1.145}$$

and G^{-1} is the inverse function of G and $t_1 \in \mathbb{R}_+$ is chosen so that, for all $t \in [0, t_1]$,

$$G(c) + A(t) \in Dom\,(G^{-1}).$$

Proof We first assume that $c > 0$ and define a function $z(t)$ by the right hand side of (1.1.143). Then $z(t) > 0$, $z(0) = c$, $u(t) \le (z(t))^{1/p}$ and

$$\begin{aligned} z'(t) &= f(t)g(u(t)) + \int_0^t h(t,\sigma)g(u(\sigma))d\sigma \\ &\le f(t)g(z(t))^{1/p} + \int_0^t h(t,\sigma)g((z(\sigma))^{1/p})d\sigma \\ &\le g((z(t))^{1/p}) \left[f(t) + \int_0^t h(t,\sigma)d\sigma \right]. \end{aligned} \tag{1.1.146}$$

From (1.1.145) and (1.1.146), it follows that

$$\begin{aligned} \frac{d}{dt}G(z(t)) &= \frac{z'(t)}{g((z(t))^{1/p})} \\ &\le f(t) + \int_0^t h(t,\sigma)d\sigma. \end{aligned} \tag{1.1.147}$$

By setting $t = s$ in (1.1.147) and integrating it from 0 to t, we can get

$$G(z(t)) \le G(c) + A(t). \tag{1.1.148}$$

Since G^{-1} is increasing, from (1.1.148) we can derive

$$z(t) \le G^{-1}[G(c) + A(t)]. \tag{1.1.149}$$

Using (1.1.149) in $u(t) \le (z(t))^{1/p}$, we conclude the required inequality in (1.1.144). If c is non-negative, we carry out the above procedure with $c + \varepsilon$ instead of c, where $\varepsilon > 0$ is an arbitrary small constant, and by letting $\varepsilon \to 0^+$, we can obtain (1.1.144). The interval $0 \le t \le t_1$ is obvious. □

Remark 1.1.9 We note that the definition of the function G in (1.1.145) is motivated from the work of Medved' [387]. If $\int_{r_0}^{+\infty} \frac{ds}{g(s^{1/p})} = +\infty$, then $G(+\infty) = +\infty$ and the inequality in (1.1.144) is true for all $t \in \mathbb{R}_+$.

The next result is a special version of Theorem 1.1.25.

Corollary 1.1.5 (The Pachpatte Inequality [519]) *Let u, f, h, c, p be as in Theorem 1.1.25. If, for all $t \in \mathbb{R}_+$,*

$$u^p(t) \le c + \int_0^t \left[f(s)u(s) + \int_0^s h(s,\sigma)u(\sigma)d\sigma \right] ds, \tag{1.1.150}$$

then, for all $t \in \mathbb{R}_+$,

$$u(t) \le \left[c^{(p-1)/p} + \frac{p-1}{p} A(t) \right]^{1/(p-1)}, \tag{1.1.151}$$

where $A(t)$ is defined by (1.1.145).

Proof Let $g(u) = u$ in Theorem 1.1.25. Then (1.1.143) reduces to (1.1.150) and

$$G(r) = \frac{p}{p-1}[r^{(p-1)/p} - r_0^{(p-1)/p}],\ G^{-1}(r) = [\frac{p-1}{p} r + r_0^{(p-1)/p}]^{p/(p-1)}$$

and consequently the bound in (1.1.144) reduces to the bound in (1.1.151).

Remark 1.1.10 In the special case when $p = 2$, then inequality given in Corollary 1.1.5 reduces to a variant of the inequality given in [507]. For an example, a bound on a different version of the inequality (1.1.150), see also, Willett and Wong [673].

Willett [671] (see also Willett and Wong [673]) generalized the Gronwall inequality to functions in $L^p(J)$ $(1 \le p < +\infty)$.

Theorem 1.1.26 (The Willett Inequality [671]) *Let v, g, h be non-negative functions of class of $L^p(J)$ $(1 \le p < +\infty)$, $J = [0, T]$, and for all $t \in J$,*

$$v(t) \le g(t) + h(t) \left(\int_0^t v^p(\tau)d\tau \right)^{1/p}. \tag{1.1.152}$$

Then for all $t \in J$,

$$\left(\int_0^t v^p(\tau)d\tau\right)^{1/p} \le \left(\int_0^t g^p(\tau)\varepsilon(\tau)d\tau\right)^{1/p} /\{1-[1-\varepsilon(t)]^{1/p}\} \tag{1.1.153}$$

where

$$\varepsilon(t) = \exp\left(-\int_0^t h^p(\tau)d\tau\right).$$

Proof Define a function $\vartheta(t)$ on I by

$$\vartheta(t) = \varepsilon(t)\int_a^t \varphi^p(s)ds. \tag{1.1.154}$$

It follows from inequality (1.1.152) that

$$\vartheta'(t) = \varepsilon(t)\varphi^p(t) - \psi^p(t)\vartheta'(t) \le (\zeta\varepsilon^{1/p} + \psi\vartheta^{1/p})^p - \psi^p\vartheta.$$

Since $\vartheta(a) = 0$, we obtain next by integration

$$\vartheta(t) \le \int_a^t (\zeta\varepsilon^{1/p} + \psi\vartheta^{1/p})^p ds - \int_a^t \psi^p\vartheta ds.$$

But by the triangle inequality;

$$\left(\int_a^t (\zeta\varepsilon^{1/p} + \psi\vartheta^{1/p})^p ds\right)^{1/p} \le \left(\int_a^t \zeta^p\varepsilon ds\right)^{1/p} + \left(\int_a^t \psi^p\vartheta ds\right)^{1/p}$$

hence,

$$\left(\vartheta(t) + \int_a^t \psi^p\vartheta\right)^{1/p} - \left(\int_a^t \psi^p\vartheta\right)^{1/p} \le \left(\int_a^t \zeta^p\varepsilon\right)^{1/p}. \tag{1.1.155}$$

The left-hand side of inequality (1.1.155) is a function of the form $m(x) = (\alpha + x)^{1/p} - x^{1/p}$. For any $p \ge 1$ and $\alpha \ge 0$, $m(x)$ is a non-increasing function of x for all $x \ge 0$. Thus, we may replace $\int_a^t \psi^p\vartheta ds$ in inequality (1.1.155) by a larger quantity and still have a valid inequality.

It is easy to see from the definition of $\vartheta(t)$, given by equation (1.1.154), that

$$\int_a^t \psi^p\vartheta ds \le \left(\int_a^t \varphi^p ds\right)\left(\int_a^t \varepsilon\psi^p ds\right) = (1-\varepsilon(t))\int_a^t \varphi^p ds. \tag{1.1.156}$$

The conclusion (1.1.153) follows by substituting from equations (1.1.154) and (1.1.156) into equation (1.1.155). □

Theorem 1.1.27 (The Willett-Wong Inequality [673]) *Let the functions* $v(t)u^p(t)$, $v(t)w^p(t)$ *and* $v(t)u_0^p(t)$ *be locally integrable non-negative functions on* I. *If the following inequality holds for* $1 \leq p < +\infty$, *and for all* $t \in I$,

$$u(t) \leq u_0(t) + w(t)\left(\int_0^t v(s)u^p(s)ds\right)^{1/p}, \tag{1.1.157}$$

then for all $t \in I$,

$$\left(\int_0^t v(s)u^p(s)ds\right)^{1/p} \leq \frac{(\int_0^t v(s)u_0^p(s)e(s)ds)^{1/p}}{1-(1-e(t))^{1/p}}, \tag{1.1.158}$$

where

$$e(t) = \exp\left(-\int_0^t v(s)w^p(s)ds\right). \tag{1.1.159}$$

Proof Theorem 1.1.27 with $v(t) = 1$ is proved as Theorem 1.1.26. The case for general $v(t)$ follows easily from this case by multiplying inequality (1.1.157) by $v^{1/p}(t)$ and identifying $v^{1/p}(t)u(t)$ with $u(t)$. A bound on $u(t)$, which is independent of $u(t)$, can be obtained now by substituting for $(\int_0^t v(s)u^p(s)ds)^{1/p}$ in equation (1.1.158). □

The next result, due to Deo and Murdenshwar [196], which generalizes the Gollwitzer inequality.

Theorem 1.1.28 (The Deo-Murdeshwar Inequality [196]) *If*

(i) x, η *and* F *are positive continuous functions on* $[0, +\infty)$,
(ii) Ω *is a positive, continuous, sub-additive and non-decreasing function on* $[0, +\infty)$,
(iii) $h : (0, +\infty) \to (0, +\infty)$ *is a non-decreasing continuous function and*
(iv) for all $t \in (0, +\infty)$,

$$x(t) \leq \eta(t) + h\left(\int_0^t F(s)\Omega(x(s))ds\right),$$

then for all $t \in I$,

$$x(t) \leq \eta(t) + h\left\{G^{-1}\left[G\left(\int_0^t F(s)\Omega(\eta(s))ds\right) + \int_0^t F(s)ds\right]\right\} \tag{1.1.160}$$

where

$$G(u) = \int_\varepsilon^u \frac{dy}{\Omega(h(y))}, \quad u \geq \varepsilon > 0, \tag{1.1.161}$$

and G^{-1} is the inverse function of G and

$$I = \left\{ t \in (0, +\infty) | G(+\infty) \geq G\left(\int_0^t F(s)\Omega(\eta(s))ds\right) + \int_0^t F(s)ds \right\}.$$

Proof Without loss of generality, we may assume that $x(t) \geq \eta(t)$. Let $T \in I$ be any arbitrary number. From the hypothesis (iv), we have, for all $0 < t \leq T$, using the sub-additivity of Ω and the monotonicity of h,

$$x(t) - \eta(t) \leq h\left(\int_0^t F(s)\Omega(x(s) - \eta(s))ds + \int_0^T F(s)\Omega(\eta(s))ds\right). \tag{1.1.162}$$

Denote the expression in the parentheses by $v(t)$. Then for all $t \in (0, T]$,

$$F(t)\Omega(x(t) - \eta(t)) \leq F(t)\Omega[h(v(t))],$$

which implies for all $t \in (0, T]$,

$$\frac{v'(t)}{\Omega[h(v(t))]} \leq F(t).$$

Using (1.1.161), this further reduces to, for all $t \in (0, T]$,

$$\frac{d}{dt}G(v(t)) \leq F(t).$$

Now integrating from 0 to T, we may obtain

$$G(v(T)) - G(v(0)) \leq \int_0^T f(t)dt.$$

which, together with (1.1.162), gives us

$$\begin{aligned} x(T) - \eta(T) &\leq h[v(T)] \\ &\leq h\left[G^{-1}\left\{G\left(\int_0^T F(s)\Omega(\eta(s))ds\right) + \int_0^T F(s)ds\right\}\right]. \end{aligned}$$

This thus completes the proof. □

Theorem 1.1.29 (The Deo-Murdeshwar Inequality [196]) *If, in addition to the assumptions (i), (ii) and (iii) in Theorem 1.1.28, Ω is an even function on $(-\infty, +\infty)$ and*
(iv) for all $t \in (0, +\infty)$,

$$x(t) \geq \eta(t) - h\left(\int_0^t F(s)\Omega(x(s))ds\right),$$

then for all $t \in I$,

$$x(t) \geq \eta(t) + h\left\{G^{-1}\left[G\left(\int_0^t F(s)\Omega(\eta(s))ds\right) + \int_0^t F(s)ds\right]\right\} \tag{1.1.163}$$

where the function G is the same as (1.1.161).

Proof As in Theorem 1.1.28, we derive from (iv), for all $t \in (0, T]$,

$$x(t) - \eta(t) \geq -h(v(t)).$$

Hence for all $t \in (0, T]$,

$$F(t)\Omega(x(t) - \eta(t)) \geq F(t)\Omega(h(v(t))).$$

Now we can complete the proof by following the argument as in Theorem 1.1.28. The details are omitted. □

Remark 1.1.11 The inequality (iv) has been studied, when (i) $h(u) = u$, $\eta(t) =$ *constant* in [82], (ii) $h(u) = u$ in [402], and (iii) $h = \Omega^{-1}$ in [250]. A lower estimate for $x(t)$ has been obtained by considering the inequality (iv) when (i) $h(u) = u$ in [328], and (ii) $h = \Omega^{-1}$ in [250].

The following inequality, considered in (1.1.164), combines the features of the inequalities in [50, 82], since on the right-hand side we consider two integrals, one containing a linear term and the other a nonlinear term. The bound obtained in (1.1.165) contains several special cases from the existing literature.

Theorem 1.1.30 (The Dhongade-Deo Inequality [197]) *Suppose*

(i) $y(x), f(x), g(x) : (0, +\infty) \to (0, +\infty)$ *and continuous on* $(0, +\infty)$,
(ii) $\Omega(u)$ *be a non-negative, monotonic, non-decreasing, continuous, sub-multiplicative for all* $u > 0$.

If, for all $0 < x < +\infty$,

$$y(x) \leq k + \int_0^x f(s)y(s)ds + \int_0^x g(s)\Omega(y(s))ds, \tag{1.1.164}$$

where $k > 0$ is a constant, then for all $0 < x \leq b$,

$$y(x)\exp\left(-\int_0^x f(s)ds\right) \leq G^{-1}\left[G(k) + \int_0^x g(s)\Omega\left(\exp\int_0^s f(t)dt\right)ds\right], \tag{1.1.165}$$

where

$$G(u) = \int_{u_0}^u \frac{ds}{\Omega(s)}, \quad 0 < u_0 \leq u \tag{1.1.166}$$

and G^{-1} *is the inverse of* G *and* x *is in the sub-interval* $(0, b]$ *of* $(0, +\infty)$ *so that*

$$G(k) + \int_0^x g(s)\Omega\left(\exp\int_0^s f(t)dt\right)ds \in Dom\,(G^{-1}).$$

Proof Define, for all $0 \leq x < +\infty$,

$$n(x) = k + \int_0^x g(s)\Omega(y(s))ds.$$

Then (1.1.164) can be written as, for all $0 < x < +\infty$,

$$y(x) \leq n(x) + \int_0^x f(s)y(s)ds.$$

Since $n(x)$ is monotonic, non-decreasing on $[0, +\infty)$, we may derive from Theorem 1.1.4 in Qin [557], for all $0 < x < +\infty$,

$$y(x) \leq n(x)\exp\left(\int_0^x f(s)ds\right), \tag{1.1.167}$$

which implies, since Ω is sub-multiplicative,

$$\Omega(y(x)) \leq \Omega(m(x))\Omega\left(\exp\left(\int_0^x f(s)ds\right)\right).$$

Hence,

$$\frac{\Omega(y(x))g(x)}{\Omega(n(x))} \leq g(x)\Omega\left(\exp\left(\int_0^x f(s)ds\right)\right).$$

This, because of (1.1.166), reduces to, for all $0 < x < +\infty$,

$$\frac{d}{dx}G(m(x)) \leq g(x)\Omega\left(\exp\left(\int_0^x f(s)ds\right)\right).$$

Now integrating from 0 to x, we can get, for all $0 < x \leq b$,

$$G(n(x)) - G(n(0)) \leq \int_0^x g(s)\Omega\left(\exp\left(\int_0^x f(t)dt\right)\right)ds. \tag{1.1.168}$$

Thus the desired result (1.1.165) follows from (1.1.167) and (1.1.168). The sub-interval $(0, b]$ is obvious. □

Remark 1.1.12

(i) The result in (1.1.161) is due to Bellman [50], when $\Omega(y) = 0$.

(ii) In inequality (1.1.164), when $\Omega(y) = y$, the inequality (1.1.165) reduces to, for all $0 < x < +\infty$,

$$y(x)\exp\left(-\int_0^x f(s)ds\right) \leq k\exp\left[\int_0^x g(s)\exp\left(\int_0^s f(t)dt\right)ds\right].$$

This is a linear generalization of the integral inequality resulting from Bellman [50].

(iii) When $\Omega(y) = y^p, p \neq 1 > 0$, Theorem 1.1.30 reduces to Theorem 1.1.6, proved by Willett and Wong [673].

If the constant $k > 0$ is replaced by a continuous function $p(x)$ in (1.1.164), the function $p(x)$ in (1.1.164) must be sub-additive. The sub-additivity property was first employed by Muldowney and Wong [402].

Theorem 1.1.31 (The Willett-Wong Inequality [673]) *Let, in addition to assumptions (i), (ii) of Theorem 1.1.30, the function Ω be sub-additive, the functions $p(x) > 0, \Psi(x) \geq 0$ be non-decreasing in x and continuous on $(0, +\infty)$ for all $x > 0$.*
If, for all $x > 0$,

$$y(x) \leq p(x) + \int_0^x f(s)y(s)ds + \Psi\left(\int_0^x g(s)\Omega(y(s))ds\right), \tag{1.1.169}$$

then, for all $0 < x \leq b$,

$$\begin{aligned} y(x)\exp\left(-\int_0^x f(s)ds\right) \leq p(x) + \Psi\Bigg[G^{-1}\Bigg\{G\Bigg[\int_0^x g(s)\Omega(p(s)\exp\int_0^s f(t)dt)ds\Bigg] \\ + \int_0^x g(s)\Omega(\exp\int_0^s f(t)dt)ds\Bigg\}\Bigg], \end{aligned} \tag{1.1.170}$$

where $G(u)$ is defined as

$$G(u) = \int_{u_0}^u \frac{ds}{\Omega(\Psi(s))}, \quad 0 < u_0 \leq u, \tag{1.1.171}$$

and G^{-1} is the inverse of G and x is in the sub-interval $(0, b]$ so that

$$\begin{aligned} G\left[\int_0^x g(s)\Omega\left(p(s)\exp\int_0^s f(t)dt\right)ds\right] \\ + \int_0^x g(s)\Omega\left(\exp\int_0^s f(t)dt\right)ds \in Dom\,(G^{-1}). \end{aligned}$$

Proof We can complete the proof by following the same argument as in Theorem 1.1.30, together with Theorem 1.1.28. The details are omitted. □

The inequality (1.1.169) was studied in [402] when $\Psi(u) = u$ and the linear term in (1.1.169) is absent. Further, when $p(x)$ is constant, $\Psi = \Omega^{-1}$, and Ω is a concave function, (1.1.169) reduces to the inequality in [250]. The case where the linear term on the right-hand side of (1.1.169) is absent, was first studied in [196]. Vidyasagar [656] has also studied the inequality (1.1.169), without assuming monotonicity of $p(x)$. However, the estimates there are different from (1.1.170).

Now we give the following lemma concerning some properties of class $\mathcal{F}_1$.

Lemma 1.1.12 ([311])

(1) If $w(u) \in \mathcal{F}_1$, then $w(u)$ is sub-additive;
(2) If $w(u)$ satisfies (i) and is convex on $\mathbb{R}_+$, then $w(u) \in \mathcal{F}_1$.

Proof (1) For any $u, v \in \mathbb{R}_+$, without loss of generality, we assume $v \le u$. Then $v = \lambda u, 0 \le \lambda \le 1$. Because $w(u) \in \mathcal{F}_1, w(u)$ satisfies (ii). Hence

$$\frac{1}{1+\lambda} w[(1+\lambda)u] \le w(u),$$

i.e.,

$$w(u + \lambda u) \le w(u) + \lambda w(u), \tag{1.1.172}$$

which, using (ii) again, yields

$$w(u + v) \le w(u) + w(v). \tag{1.1.173}$$

So $w(u)$ is sub-additive.

(2) Let $w(u)$ is convex on $\mathbb{R}_+$, i.e., for all $\alpha \ge 0, \beta \ge 0, \alpha + \beta = 1$,

$$w(\alpha u_1 + \beta u_2) \ge \alpha w(u_1) + \beta w(u_2). \tag{1.1.174}$$

Let $u_1 = u, u_2 = 0$ in (1.1.174). Considering $w(0) \ge 0$, we get for all $0 \le \alpha \le 1$,

$$\alpha w(u) \le w(\alpha u).$$

For fixed $v \ge 1$, let

$$\alpha = \frac{1}{v}.$$

Hence

$$\frac{1}{v} w(u) \le w\left(\frac{u}{v}\right).$$

Therefore $w(u) \in \mathcal{F}_1$. □

Lemma 1.1.12 explains that all non-negative, non-decreasing and convex functions belong to $\mathcal{F}_1$. In fact, besides those functions, there are many functions in $\mathcal{F}_1$. For instance,

$$w(u) = u\left(1 + \sin\frac{\pi}{2(u+1)}\right)$$

is not a convex function, but we can prove that it is in $\mathcal{F}_1$.

In [686], condition (2) of Lemma 1.1.12 is modified as to

$$\frac{1}{v}\omega(u) \leq \mu(v)w\left(\frac{u}{v}\right)$$

for all $u \geq 0, v > 0$, where $w > 0$ is non-decreasing, $\mu \geq 0$, and under that condition, the following inequality

$$u(t) \leq \alpha(t) + \sum_{i=1}^{n}\int_0^t f_i(t,s)\Omega(u(s))ds$$

is discussed. But except for power functions, verification of the above condition and selection of $\mu(v)$ are rather difficult. In fact, Corollary 2 in [686] is not true because the $\mu(v)$ is unsuitable. Furthermore, with an example, we shall see, the exactness of its estimate is also not satisfactory.

Theorem 1.1.32 (The Kong-Zhang Inequality [311]) *Let*

(i) $y(x), f(x) : (0, +\infty) \to (0, +\infty)$ *and be continuous on* $(0, +\infty)$,
(ii) $w \in \mathcal{F}$,
(iii) $n > 0$ *be monotonic, non-decreasing and continuous on* $[0, +\infty)$.

If, for all $x > 0$,

$$y(x) \leq n(x) + \int_0^x f(s)w(y(s))ds, \tag{1.1.175}$$

then for all $0 < x \leq b$,

$$y(x) \leq n(x)G^{-1}\left[G(1) + \int_0^x f(s)ds\right], \tag{1.1.176}$$

where $(0, b] \subset (0, +\infty)$,

$$G(u) = \int_{u_0}^{u}\frac{ds}{w(s)}, \quad 0 < u_0 \leq u \tag{1.1.177}$$

and G^{-1} *is the inverse of* G *and the sub-interval* $(0, b]$ *is so chosen that*

$$G(1) + \int_0^x f(s)ds \in Dom\ (G^{-1}).$$

Proof Since $n(x)$ is monotonic, non-decreasing and $w \in \mathcal{F}$, we derive from (1.1.175) that

$$\frac{y(x)}{n(x)} \leq 1 + \int_0^x \frac{f(s)w(y(s))}{n(s)}ds \leq 1 + \int_0^x f(s)w(\frac{y(s)}{n(s)})ds, \quad 0 < x < +\infty.$$

Now, considering $y(x)/n(x)$ as a function, by Bihari's inequality [82], i.e., Theorem 1.1.1, the result (1.1.176) follows; the existence of the sub-interval $(0, b] \subset (0, +\infty)$ is obvious. □

Note that the above theorem provides a nonlinear generalization of the lemma stated above. The next two theorems depend heavily on this result.

Theorem 1.1.33 (The Kong-Zhang Inequality [311]) *Let the conditions (i), (ii) of Theorem 1.1.32 hold, and* $w \in \mathcal{F}$.
If, for all $x > 0$,

$$y(x) \leq k + \int_0^x f(s)w(y(s))ds + \int_0^x g(s)\Omega(y(s))ds, \tag{1.1.178}$$

where $k > 0$ *is a constant, then, for all* $0 < x \leq b$,

$$y(x)\left[G^{-1}\left(G(1) + \int_0^x f(s)ds\right)^{-1}\right] \leq F^{-1}\left\{F(k) + \int_0^x g(s)\Omega\left[G^{-1}\left(G(1) + \int_0^s f(t)dt\right)\right]ds\right\}, \tag{1.1.179}$$

where $G(u)$ *is defined as in Theorem 1.1.32,* F *is defined as*

$$F(u) = \int_{u_0}^u \frac{ds}{\Omega(s)}, \quad 0 < u_0 \leq u, \tag{1.1.180}$$

and G^{-1}, F^{-1} *are the inverse of* G, F, *respectively and* x *is in the sub-interval* $(0, b]$ *of* $(0, +\infty)$ *such that*

$$G(1) + \int_0^x f(s)ds \in Dom\ (G^{-1}),$$

and

$$F(k) + \int_0^x g(s)\Omega\left[G^{-1}\left(G(1) + \int_0^s f(t)dt\right)\right]ds \in Dom\ (F^{-1}).$$

Proof Define, for all $x > 0$,

$$n(x) = k + \int_0^x g(s)\Omega(y(s))ds.$$

Then (1.1.178) can be written as, for all $x > 0$,

$$y(x) = n(x) + \int_0^x f(s)w(s)ds.$$

Since $n > 0$ is monotonic, non-decreasing, and $w \in \mathcal{F}$, we obtain, in by Theorem 1.1.32, for all $0 < x \leq b'$,

$$y(x) \leq n(x)G^{-1}\left[G(1) + \int_0^s f(s)ds\right]. \tag{1.1.181}$$

Further,

$$\Omega(y(x)) \leq \Omega(n(x))\Omega\left[G^{-1}\left(G(1) + \int_0^x f(s)ds\right)\right],$$

since Ω is sub-multiplicative. Hence, for all $0 < x \leq b'$,

$$\frac{\Omega(y(x))g(x)}{\Omega(n(x))} \leq g(x)\Omega\left[G^{-1}\left(G(1) + \int_0^x f(s)ds\right)\right].$$

Because of (1.1.180), this reduces to

$$\frac{d}{dx}F(n(x)) \leq g(x)\Omega\left\{G^{-1}\left(G(1) + \int_0^x f(s)ds\right)\right\}, \quad 0 < x \leq b'.$$

Now, integrating from 0 to x, we obtain

$$F(n(x)) - F(n(0)) \leq \int_0^x g(s)\Omega\left[G^{-1}\left(G(1) + \int_0^s f(t)dt\right)\right]ds. \tag{1.1.182}$$

The result (1.1.179) now follows from (1.1.181) and (1.1.182) on the sub-interval $(0, b] \subset (0, b']$.

If the constant $k > 0$ in (1.1.178) is replaced by a monotonically non-decreasing function $p(x)$, we require Ω to be sub-additive. □

Theorem 1.1.34 (The Kong-Zhang Inequality [311]) *If, in addition to assumptions of Theorem 1.1.33, Ω is sub-additive, and the functions $p(x) > 0, \Psi(x) \geq 0$, be non-decreasing, continuous on $(0, +\infty)$, and if, for all $x > 0$,*

$$y(x) \leq p(x) + \int_0^x f(s)w(y(s))ds + \Psi\left(\int_0^x g(s)\Omega(y(s))ds\right), \tag{1.1.183}$$

then, for all $0 < x \le b$,

$$y(x)\left[G^{-1}\left(G(1)+\int_0^x f(s)ds\right)^{-1}\right]$$
$$\le p(x)+\Psi[F^{-1}\{F[\int_0^x g(s)\Omega[p(s)\cdot G^{-1}(G(1)+\int_0^x f(t)dt)ds]$$
$$+\int_0^x g(s)\Omega[G^{-1}(G(1)+\int_0^x f(t)dt)]ds\}], \tag{1.1.184}$$

where G is defined as in Theorem 1.1.30, and F is defined as

$$F(u)=\int_{u_0}^u \frac{ds}{\Omega(\Psi(s))}, \quad 0<u_0\le u, \tag{1.1.185}$$

and G^{-1}, F^{-1} *have the same meaning as in Theorem 1.1.33, and x is in the subinterval* $(0,b)$ *so that*

$$G(1)+\int_0^x f(s)ds \in Dom\ (G^{-1})$$

and

$$F\left\{\int_0^x g(s)\Omega\left[p(s)G^{-1}\left(G(1)+\int_0^x f(s)ds\right)\right]ds\right\}$$
$$+\int_0^x g(s)\Omega\left[G^{-1}\left(G(1)+\int_0^x f(s)ds\right)\right]ds \in Dom\ (G^{-1}).$$

Proof We can complete the proof of Theorem 1.1.32 by following the argument of Theorem 1.1.33. The details are omitted. □

Corollary 1.1.6 ([311]) *In Theorem 1.1.33, let*

$$w(y)=y^\beta,\quad 0<\beta<1,\quad \Omega(y)=y^\theta,\quad 0<\theta<1,$$

and

$$\Psi(u)=u^r,\quad r\ge 1.$$

If, for all $x>0$,

$$y(x)\le p(x)+\int_0^x f(s)y^\theta(s)ds+\left(\int_0^x g(s)y^\beta(s)\right)^r, \tag{1.1.186}$$

then for all $x > 0$,

$$y(x)\left[1+\alpha\int_0^x f(s)ds\right]^{-1/\alpha} \le p(x) + \left[\left(\int_0^x g(s)\left[p^\alpha(s)\left(1+\alpha\int_0^\alpha f(t)dt\right)\right]^{\frac{\theta}{\alpha}} ds\right)^\delta \right.$$

$$\left. +\delta\int_0^x g(s)\left[1+\alpha\int_0^\alpha f(t)dt\right]^{\frac{\theta}{\alpha}} ds\right]^{\frac{\theta}{\delta}},$$

where $\delta = 1 - r\theta$ *and* $\alpha = 1 - \beta$.

Note that in Corollary 1.1.6 $w \in \mathcal{F}$ and Ω are sub-additive, and sub-multiplicative.

We now apply Corollary 1.2.4 and Corollary 1.2.16 in Qin [557] to establish the following interesting and useful integral inequalities.

Theorem 1.1.35 (The Pachpatte Inequality [441]) *Let* $x(t), f_1(t), f_2(t), f_3(t)$ *and* $f_4(t)$ *be real-valued non-negative continuous on* $I = [0, +\infty)$*, and* $w(u)$ *be a positive, continuous, non-decreasing, sub-multiplicative function for all* $u > 0$*, with* $w(0) = 0$*. Furthermore, assume the following inequality holds for all* $t \in I$,

$$x(t) \le x_0 + \int_0^t f_1(s)x(s)ds + \int_0^t f_2(s)[\int_0^s f_3(\tau)x(\tau)d\tau]ds + \int_0^t f_4(s)w(x(s))ds \quad (1.1.187)$$

where x_0 *is a positive constant. Then*

$$x(t) \le G^{-1}[G(x_0) + \int_0^t f_4(s)w(\phi(s))ds]\phi(t), \quad (1.1.188)$$

where $G(r) = \int_{r_0}^r (ds/w(s))$, $r \ge r_0 > 0$; G^{-1} *is the inverse of* G*; and* t *is in the sub-interval* $[0, b]$ *of* I *so that*

$$G(x_0) + \int_0^t f_4(s)w(\phi(s))ds \in Dom\,(G^{-1}).$$

Proof Since in (1.1.187) the term $x_0 + \int_0^t f_4(s)w(x(s))ds$ is non-decreasing. From Corollary 1.2.16 in Qin [557] we find

$$x(t) \le \phi(t)[x_0 + \int_0^t f_4(s)w(x(s))ds].$$

Let

$$R(t) = x_0 + \int_0^t f_4(s)w(x(s))ds, \quad R(0) = x_0;$$

then, on using the assumption on w, it follows that

$$R'(t)/w(R(t)) \leq f_3(t)w(\phi(t)).$$

Integrating the above inequality, the result (1.1.188) follows. □

Corollary 1.1.7 (The Pachpatte Inequality [456]) *Let $x(t)$, $f(t)$, $g(t)$ and $h(t)$ be real-valued positive continuous functions defined on I; $W(u)$ be a positive, continuous, monotonic, non-decreasing and sub-multiplicative function for all $u > 0$, $W(0) = 0$, and suppose further that the following inequality holds for all $t \in I$,*

$$x(t) \leq x_0 + \int_0^t f(s)x(s)\,ds + \int_0^t f(s)\left(\int_0^s g(\tau)x(\tau)\,d\tau\right)ds + \int_0^t h(s)W(x(s))\,ds, \quad (1.1.189)$$

where x_0 is a positive constant. Then, for all $0 \leq t \leq b$,

$$\begin{aligned} x(t) \leq G^{-1}&\left[G(x_0) + \int_0^t h(s)W\left(1 + \int_0^s f(\tau)\exp\left(\int_0^\tau (f(k) + g(k))\,dk\right)d\tau\right)ds\right] \\ &\times\left[1 + \int_0^t f(s)\exp\left(\int_0^t (f(\tau) + g(\tau))\,d\tau\right)ds\right], \end{aligned} \quad (1.1.190)$$

where

$$G(r) = \int_{r_0}^r \frac{ds}{W(s)}, \quad r \geq r_0 > 0, \quad (1.1.191)$$

and G^{-1} is the inverse function of G, and $t \in [0, b)$ of I so that

$$G(x_0) + \int_0^t h(s)W\left(1 + \int_0^s f(\tau)\exp\left(\int_0^\tau (f(k) + g(k))\,dk\right)d\tau\right)ds \in Dom\,(G^{-1}).$$

Proof Define

$$n(t) = x_0 + \int_0^t h(s)W(x(s))\,ds, \quad n(0) = x_0.$$

Then (1.1.189) can be rewritten as

$$x(t) \leq n(t) + \int_0^t f(s)x(s)\,ds + \int_0^t f(s)\left(\int_0^s g(\tau)x(\tau)\,d\tau\right)ds.$$

Since $n(t)$ is positive, monotonic, non-decreasing on I, we have from Corollary 1.2.5 in Qin [557]

$$x(t) \leq n(t)\left(1 + \int_0^t f(s)\exp\left(\int_0^s (f(\tau) + f(\tau))\,d\tau\right)ds\right). \quad (1.1.192)$$

Furthermore, since W is sub-multiplicative, we have

$$W(x(t)) \leq W(n(t))W\left(1+\int_0^t f(s)\exp\left(\int_0^s (f(\tau)+g(\tau))\,d\tau\right)ds\right).$$

Hence

$$\frac{h(t)W(x(t))}{W(n(t))} \leq h(t)W\left(1+\int_0^t f(s)\exp\left(\int_0^s (f(\tau)+g(\tau))\,d\tau\right)ds\right).$$

Because of (1.1.191), this reduces to

$$\frac{d}{dt}G(n(t)) \leq h(t)W\left(1+\int_0^t f(s)\exp\left(\int_0^s (f(\tau)+g(\tau))\,d\tau\right)ds\right).$$

Now integrating from 0 to t, we may obtain

$$G(n(t))-G(n(0)) \leq \int_0^t h(s)W\left(1+\int_0^t f(\tau)\exp\left(\int_0^\tau (f(k)+g(k))\,dk\right)d\tau\right)ds. \quad (1.1.193)$$

Thus the desired bound in (1.1.190) follows now from (1.1.192) and (1.1.193). The sub-interval [0,b] is obvious. □

Theorem 1.1.36 (The Pachpatte Inequality [456]) *Let $x(t)$, $f(t)$, $g(t)$ and $h(t)$ be real-valued positive continuous functions defined on I; $W(u)$ be a positive, continuous, monotonic, non-decreasing, sub-additive and sub-multiplicative function for all $u>0$, $W(0)=0$, the functions $p(t)>0$, $M(t)\geq 0$ be non-decreasing in t and continuous on I, $M(0)=0$, and suppose further that the inequality holds for all $0\leq t\leq b$, $t\in I$,*

$$\begin{aligned}x(t) \leq p(t)&+\int_0^t f(s)x(s)\,ds+\int_0^t f(s)\left(\int_0^s g(\tau)x(\tau)\,d\tau\right)ds\\&+M\left(\int_0^t h(s)W(x(s))\,ds\right).\end{aligned} \quad (1.1.194)$$

Then for all

$$\begin{aligned}x(t) \leq \Big[p(t)&+M\Big(G^{-1}\Big[G\Big(\int_0^t h(s)W\Big(p(s)\Big(1+\int_0^s f(\tau)\exp\Big(\int_0^\tau (f(k)+g(k))\,dk\Big)\,d\tau\Big)\Big)\,ds\Big)\\&+\int_0^t h(s)W\Big(1+\int_0^s f(\tau)\exp\Big(\int_0^\tau (f(k)+g(k))\,dk\Big)\,d\tau\Big)\,ds\Big]\Big)\Big]\\&\times\Big[1+\int_0^t f(s)\exp\Big(\int_0^s (f(\tau)+g(\tau))\,d\tau\Big)\,ds\Big],\end{aligned} \quad (1.1.195)$$

where

$$G(r) = \int_{r_0}^{r} \frac{ds}{W(M(s))}, \quad r \geq r_0 \geq 0, \tag{1.1.196}$$

and G^{-1} *is the inverse of* G, *and* $t \in [0, b)$ *of* I *so that*

$$G\left(\int_0^t h(s)W\left(p(s)\left(1+\int_0^s f(\tau)\exp\left(\int_0^\tau (f(k)+g(k))\,dk\right)d\tau\right)ds\right)\right)$$
$$+\int_0^t h(s)W\left(1+\int_0^s f(\tau)\exp\left(\int_0^\tau (f(k)+g(k))\,dk\right)d\tau\right)ds \in Dom\,(G^{-1}).$$

Proof The proof of Theorem 1.1.36 follows by the similar argument as in the proof of Corollary 1.1.7, together with Theorem 1.1.28. We omit the details. □

In [197], the authors have studied the integral inequalities in Corollary 1.1.7 and Theorem 1.1.36 when the second integral term on the right-hand side in (1.1.189) and (1.1.194) is absent. However, the bounds obtained in Corollary 1.1.7 and Theorem 1.1.36 are different from those given in [197].

Next, we introduce some integral inequalities, by considering one linear and two nonlinear terms on the right-hand side.

Before giving these results, we first introduce the following theorem which is useful in our further discussion.

Theorem 1.1.37 (The Pachpatte Inequality [456]) *Let* $x(t)$, $g(t)$ *be real-valued positive continuous functions defined on* I, $n(t)$ *be a positive, monotonic, non-decreasing continuous function defined on* I, *and* $H \in \mathcal{F}$, *for which the following inequality holds for all* $t \in I$,

$$x(t) \leq n(t) + \int_0^t g(s)\left(\int_0^s g(\tau)H(x(\tau))\,d\tau\right)ds. \tag{1.1.197}$$

Then for all $0 \leq t \leq b$,

$$x(t) \leq n(t)\left(1+\int_0^s g(s)G^{-1}\left(G(1)+\int_0^s g(\tau)\,d\tau\right)ds\right), \tag{1.1.198}$$

where

$$G(r) = \int_{r_0}^{r} \frac{ds}{s+H(s)}, \quad r \geq r_0 > 0, \tag{1.1.199}$$

and G^{-1} *is the inverse of* G, *and* $t \in [0, b)$ *of* I *so that*

$$G(1) + \int_0^t g(s)\,ds \in Dom\,(G^{-1}).$$

Proof Since $n(t)$ is positive, monotonic, non-decreasing and $H \in \mathcal{F}$, we have

$$\begin{aligned}\frac{x(t)}{n(t)} &\leq 1 + \int_0^t g(s) \left(\frac{x(s)}{n(t)} + \int_0^s \frac{g(\tau)H(x(\tau))}{n(t)} d\tau \right) ds \\ &\leq 1 + \int_0^t g(s) \left(\frac{x(s)}{n(t)} + \int_0^s g(\tau)H\left(\frac{x(\tau)}{n(\tau)}\right) d\tau \right) ds. \qquad (1.1.200)\end{aligned}$$

Define $v(t)$ by the right-hand side of (1.1.200). Then

$$v'(t) = g(t) \left(\frac{x(t)}{n(t)} + \int_0^t g(\tau)H\left(\frac{x(\tau)}{n(\tau)}\right) d\tau \right), \quad v(0) = 1,$$

which, in view of (1.1.200), implies

$$v'(t) \leq g(t) \left(v(t) + \int_0^t g(\tau)H(v(\tau))\, d\tau \right). \qquad (1.1.201)$$

If we put

$$m(t) = v(t) + \int_0^t g(\tau)H(v(\tau))\, d\tau, \quad m(0) = v(0) = 1, \qquad (1.1.202)$$

then, it follows from (1.1.201), (1.1.202) and the fact that $v(t) \leq m(t)$, and

$$m'(t) \leq g(t)(m(t) + H(m(t))). \qquad (1.1.203)$$

Dividing both sides of (1.1.203) by $(m(t)+H(m(t)))$, using (1.1.199) and integrating from 0 to t, we can obtain

$$G(m(t)) - G(m(0)) \leq \int_0^t g(s)\, ds. \qquad (1.1.204)$$

Then from (1.1.201) and (1.1.204), we derive

$$v'(t) \leq g(t)G^{-1} \left(G(1) + \int_0^t g(s)\, ds \right). \qquad (1.1.205)$$

Now, integrating both sides of (1.1.205) from 0 to t and substituting the value of $v(t)$ in (1.1.197), we can obtain the desired bound in (1.1.198). □

We note that the estimate for $x(t)$ in (1.1.197), when $n(t)$ is not monotonic non-decreasing and H is a positive, continuous, non-decreasing and sub-additive function for $u > 0$, $h(0) = 0$ is already obtained in [445].

We now apply Theorem 1.1.37 to establish the following more general integral inequalities.

Theorem 1.1.38 (The Pachpatte Inequality [456]) *Let $x(t)$, $g(t)$ and $h(t)$ be real valued positive continuous functions defined on I, $H \in \mathcal{F}$, and W is the same function as defined in Corollary 1.1.7, and suppose further that the following inequality holds for all $t \in I$,*

$$x(t) \leq x_0 + \int_0^t g(s)\left(x(s) + \int_0^s g(\tau)H(x(\tau))\,d\tau\right) ds + \int_0^t h(s)W(x(s))\,ds, \quad (1.1.206)$$

where x_0 is a positive constant. Then for all $0 \leq t \leq b$,

$$x(t) \leq \Omega^{-1}\left[\Omega(x_0) + \int_0^t h(s)W\left(1 + \int_0^s g(\tau)G^{-1}\left(G(1) + \int_0^\tau g(k)dk\right) d\tau\right) ds\right]$$
$$\times\left[1 + \int_0^t g(s)G^{-1}\left(G(1) + \int_0^s g(\tau)\right) ds\right] \quad (1.1.207)$$

where G, G^{-1} are as defined in Theorem 1.1.37, Ω is defined by

$$\Omega(r) = \int_{r_0}^r \frac{ds}{W(s)}, \quad r \geq r_0 > 0, \quad (1.1.208)$$

and Ω^{-1} is the inverse function of Ω, and $t \in [0, b)$ of I such that

$$G(1) + \int_0^t g(s)\,ds \in Dom\,(G^{-1})$$

and

$$\Omega(x_0) + \int_0^t h(s)W\left(1 + \int_0^s g(\tau)G^{-1}\left(G(1) + \int_0^\tau g(k)\,dk\right) d\tau\right) ds \in Dom\,(\Omega^{-1}).$$

Proof Define

$$n(t) = x_0 + \int_0^t h(s)W(x(s))\,ds, \quad n(0) = x_0.$$

Then (1.1.206) can be restated as

$$x(t) \leq n(t) + \int_0^t g(s)\left(x(s) + \int_0^s g(\tau)H(x(\tau))\,d\tau\right) ds.$$

Since $n(t)$ is positive, monotonic, non-decreasing, and $H \in \mathcal{F}$, we derive from Theorem 1.1.37,

$$x(t) \leq n(t) + \left(1 + \int_0^t g(s)G^{-1}\left(G(1) + \int_0^s g(\tau)\right) ds\right). \quad (1.1.209)$$

Further, since W is sub-multiplicative,

$$W(x(t)) \leq W(n(t))W\left(1+\int_0^t g(s)G^{-1}\left(G(1)+\int_0^s g(\tau)\,d\tau\right)ds\right).$$

Hence

$$\frac{h(t)W(x(t))}{W(n(t))} \leq h(t)W\left(1+\int_0^t g(s)G^{-1}\left(G(1)+\int_0^s g(\tau)\,d\tau\right)ds\right).$$

Because of (1.1.208), this reduces to, for all $0 \leq t \leq b'$,

$$\frac{d}{dt}\Omega(n(t)) \leq h(t)W\left(1+\int_0^t g(s)G^{-1}\left(G(1)+\int_0^s g(\tau)\,d\tau\right)ds\right).$$

Now, integrating from 0 to t, we may obtain

$$\Omega(n(t))-\Omega(n(0)) \leq \int_0^t h(s)W\left(1+\int_0^s g(\tau)G^{-1}\left(G(1)+\int_0^\tau g(k)\,dk\right)d\tau\right)ds, \tag{1.1.210}$$

Thus the desired bound in (1.1.207) follows from (1.1.209) and (1.1.210) on the sub-interval $[0,b) \subset [0,b']$ of I. □

Finally, we now introduce a more general form of Theorem 1.1.38 under some additional conditions.

Theorem 1.1.39 (The Pachpatte Inequality [456]) *Let $x(t)$, $g(t)$ and $h(t)$ be real-valued positive continuous functions defined on I, $H \in \mathcal{F}$, and W is the same function as defined in Theorem 1.1.36, the functions $p(t) > 0$, $M(t) \geq 0$ be non-decreasing in t and continuous on I, $M(0) = 0$, and suppose further that the following inequality holds for all $t \in I$,*

$$x(t) \leq p(t)+\int_0^t g(s)\left(x(s)+\int_0^s g(\tau)H(x(\tau))\,d\tau\right)ds+M\left(\int_0^t h(s)W(x(s))\,ds\right).$$

Then for all $0 \leq t \leq b$,

$$\begin{aligned}x(t) \leq \Big[p(t)+M\Big(\Omega^{-1}\Big[\Omega\Big(\int_0^t h(s)W\Big(p(s)\Big(1+\int_0^s g(\tau)\\ \times G^{-1}\Big(G(1)+\int_0^\tau g(k)\,dk\Big)d\tau\Big)\Big)ds\Big)\\ +\int_0^t h(s)W\Big(1+\int_0^s g(\tau)G^{-1}\Big(G(1)+\int_0^\tau g(k)\,dk\Big)d\tau\Big)ds\Big)\Big]\Big)\Big]\\ \times\left[1+\int_0^t g(s)G^{-1}\left(G(1)+\int_0^s g(\tau)\,d\tau\right)ds\right],\end{aligned}$$

where G is as defined in Theorem 1.1.37, and Ω is defined as

$$\Omega(r) = \int_{r_0}^{r} \frac{ds}{W(M(s))}, \quad r \geq r_0 > 0,$$

and G^{-1}, Ω^{-1} are the inverse functions of G and Ω, respectively, and $t \in [0, b]$ of I so that

$$G(1) + \int_0^t g(s)\, ds \in Dom\ (G^{-1}),$$

and

$$\Omega\left(h(s)W\left(p(s)\left(1 + \int_0^s g(\tau)G^{-1}\left(G(1) + \int_0^\tau g(k0\, dk\right) d\tau\right)\right) ds\right)$$
$$+ \int_0^t h(s)W\left(1 + \int_0^s g(\tau)G^{-1}\left(G(1) + \int_0^\tau g(k)\, dk\right) d\tau\right) ds \in Dom\ (\Omega^{-1}).$$

Proof The proof of this theorem follows by the similar arguments as in the proof of Theorem 1.1.38. We omit the details. □

Theorem 1.1.40 (The Denche-Khellaf Inequality [193]) *Let $u(t)$, $f(t)$ be non-negative continuous functions in a real interval $I = [a, b]$. Suppose that $k(t, s)$ and its partial derivatives $k_t(t, s)$ exist and are non-negative continuous functions for almost every $t, s \in I$. Let $\phi(u(t))$ be real-valued, positive, continuous, strictly non-decreasing, sub-additive, and sub-multiplicative function for $u(t) \geq 0$ and let $W(u(t))$ be real-valued, positive, continuous, and non-decreasing function defined for all $t \in I$. Assume that $a(t)$ is a positive continuous function and non-decreasing for all $t \in I$. If, for all $a \leq \tau \leq s \leq t \leq b$,*

$$u(t) \leq a(t) + \int_a^t f(s)u(s)ds + \int_a^t f(s)W\left(\int_a^s k(s, \tau)\phi(u(\tau))d\tau\right)ds, \tag{1.1.211}$$

then for all $a \leq t \leq t_1$,

$$u(t) \leq p(t)\left\{a(t) + \int_a^t f(s)\psi^{-1}(\psi(\zeta) + \int_a^s k(s, \tau)\phi\left(p(\tau)\right)\phi\left(\int_a^\tau f(\sigma)d\sigma\right)d\tau\right)ds\right\}, \tag{1.1.212}$$

where

$$\begin{cases} p(t) = 1 + \int_a^t f(s)\exp\left(\int_a^s g(\sigma)d\sigma\right)ds, & (1.1.213) \\ \zeta = \int_a^b k(b, s)\phi(p(s)a(s))ds, & (1.1.214) \\ \psi(x) = \int_{x_0}^{x} \frac{ds}{\phi(W(s))}, \quad x \geq x_0 > 0. & (1.1.215) \end{cases}$$

Here ψ^{-1} is the inverse of ψ and t_1 is chosen so that, for all $a \le s \le t_1$,

$$\psi(\zeta) + \int_a^s k(s,\tau)\phi(p(\tau))\phi\left(\int_a^\tau f(\sigma)d\sigma\right)d\tau \in Dom\ (\psi^{-1}).$$

Proof Define a function $z(t)$ by

$$z(t) = a(t) + \int_a^t f(s)W\left(\int_a^s k(s,\tau)\phi(u(\tau))d\tau\right)ds, \tag{1.1.216}$$

then (1.1.216) can be rewritten as

$$u(t) \le z(t) + \int_a^t f(s)u(s)ds. \tag{1.1.217}$$

Clearly, $z(t)$ is non-negative and continuous in $t \in I$, using Theorem 1.1.4 in Qin [557] to (1.1.217), we can get

$$u(t) \le z(t) + \int_a^t f(s)z(s)\exp\left(\left(\int_a^s f(\sigma)d\sigma\right)ds.\right. \tag{1.1.218}$$

Moreover, if $z(t)$ is non-decreasing in $t \in I$, we can obtain

$$u(t) \le z(t)p(t), \tag{1.1.219}$$

where $p(t)$ is defined by (1.1.213). Thus from (1.1.216) it follows that

$$z(t) \le a(t) + \int_a^t f(s)W(v(s))ds, \tag{1.1.220}$$

where

$$v(t) = \int_a^t k(t,s)\phi(u(s))ds. \tag{1.1.221}$$

From (1.1.219), we derive that

$$\begin{aligned} v(t) &\le \int_a^t k(t,s)\phi[p(s)(a(s) + \int_a^s f(\tau)W(v(\tau))d\tau)]ds \\ &\le \int_a^t k(t,s)\phi(p(s)(a(s))ds + \int_a^t k(t,s)\phi(p(s)\int_a^s f(\tau)W(v(\tau))d\tau)ds \\ &\le \int_a^t k(t,s)\phi(p(s)(a(s))ds + \int_a^t k(t,s)\phi(p(s)\int_a^s f(\tau)d\tau)\phi(W(v(s)))ds \\ &\le \zeta + \int_a^t k(t,s)\phi\left(p(s)\int_a^s f(\tau)d\tau\right)\phi(W(v(s)))ds \end{aligned} \tag{1.1.222}$$

where ζ is defined by (1.1.214).

Since ϕ is sub-additive and sub-multiplicative, W and $v(t)$ are non-decreasing. Define $r(t)$ as the right-hand side of (1.1.222), then $r(a) = \zeta$ and $v(t) \leq r(t)$, $r(t)$ is positive non-decreasing in $t \in I$ and

$$\begin{aligned} r'(t) &= k(t,t)\phi\left(p(t)\int_a^t f(\tau)d\tau\right)\phi(W(v(t))) \\ &\quad + \int_a^t k_t(t,s)\phi\left(p(s)\int_a^s f(\tau)d\tau\right)\phi(W(v(s)))ds \\ &\leq \phi(W(v(r)))\left[k(t,t)\phi\left(p(t)\int_a^t f(\tau)d\tau\right)\right. \\ &\quad \left. + \int_a^t k_t(t,s)\phi(p(s)\int_a^s f(\tau)d\tau)ds\right]. \end{aligned} \tag{1.1.223}$$

Dividing both sides of (1.1.223) by $\phi(W(r(t)))$, we can obtain

$$\frac{r'(t)}{\phi(W(r(t)))} \leq \left[\int_a^t k(t,s)\phi\left(p(s)\int_a^s f(\tau)d\tau\right)ds\right]'. \tag{1.1.224}$$

Note that for

$$\psi(x) = \int_{x_0}^x \frac{ds}{\phi(W(s))}, \quad x \geq x_0 > 0,$$

it follows that

$$[\psi(r(t))]' = \frac{r'(t)}{\phi(W(r(t)))}. \tag{1.1.225}$$

From (1.1.225) and (1.1.224), we have

$$[\psi(r(t))]' \leq \left[\int_a^t k(t,s)\phi(p(s)\int_a^s f(\tau)d\tau)ds\right]'. \tag{1.1.226}$$

Integrating (1.1.226) from a to t leads to

$$\psi(r(t)) \leq \psi(\zeta) + \int_a^t k(t,s)\phi\left(p(s)\int_a^s f(\tau)d\tau\right)ds,$$

then

$$r(t) \leq \psi^{-1}\left(\psi(\zeta) + \int_a^t k(t,s)\phi(p(s))\phi\left(\int_a^s f(\tau)d\tau\right)ds\right). \tag{1.1.227}$$

By (1.1.227), (1.1.221), (1.1.220) and (1.1.219), we have the desired result. □

The above theorem is a generalization of the result obtained in Theorem 2.1 by Pachpatte in [517].

Theorem 1.1.41 (The Denche-Khellaf Inequality [193]) *Let $u(t)$, $f(t)$, $b(t)$, $h(t)$ be non-negative continuous functions in a real interval $I = [a, b]$. Suppose that $h(t) \in C^1(I, \mathbb{R}_+)$ is non-decreasing. Let $\phi(u(t))$, $W(u(t))$ and $a(t)$ be as defined in Theorem 1.1.40. If for all $a \leq \tau \leq s \leq t \leq b$,*

$$u(t) \leq a(t) + \int_a^t f(s)u(s)ds + \int_a^t f(s)h(s)W\left(\int_a^s b(\tau)\phi(u(\tau))d\tau\right)ds,$$

then for all $a \leq t \leq t_2$,

$$u(t) \leq p(t)\left\{a(t) + \int_a^t f(s)h(s)\psi^{-1}(\psi(v)\right.$$
$$\left. + \int_a^s b(\tau)\phi(p(\tau))\phi\left(\int_a^\tau f(\sigma)h(\sigma)d\sigma)d\tau\right)ds\right\},$$

where $p(t)$ is defined by (1.1.213), ψ is defined by (1.1.215) and

$$v = \int_a^b b(s)\phi(p(s)a(s))ds,$$

the t_2 is chosen so that for all $a \leq s \leq t_2$,

$$\psi(v) + \int_a^s b(\tau)\phi\left(p(\tau))\phi(\int_a^\tau f(\sigma)h(\sigma)d\sigma\right)d\tau \in Dom\ (\psi^{-1}).$$

Proof The proof of this theorem follows from similar arguments as the proof of Theorem 1.1.40. Therefore we omit it. □

Theorem 1.1.42 (The Denche-Khellaf Inequality [193]) *Let $u(t)$, $f(t)$, $a(t)$, $k(t,s)$, ϕ and W be as defined in Theorem 1.1.40, let $g \in \mathcal{F}_1$. If, for all $a \leq \tau \leq s \leq t \leq b$,*

$$u(t) \leq a(t) + \int_a^t f(s)g(u(s))ds + \int_a^t f(s)W\left(\int_a^s k(s,\tau)\phi(u(\tau))d\tau\right)ds, \tag{1.1.228}$$

then for all $a \leq t \leq t_3$,

$$u(t) \leq \overline{p}(t)\left\{a(t) + \int_a^t f(s)\psi^{-1}(\psi(\overline{\zeta}) + \int_a^s k(s,\tau)\phi(\overline{p}(\tau))\phi\left(\int_a^\tau f(\sigma)d\sigma)d\tau\right)ds\right\}, \tag{1.1.229}$$

where

$$\begin{cases} \overline{p}(t) = \Omega^{-1}\left(\Omega(1) + \int_a^t f(s)ds\right), & (1.1.230) \\ \overline{\zeta} = \int_a^b k(b,s)\phi(\overline{p}(s)a(s))ds, & (1.1.231) \\ \Omega(\delta) = \int_\epsilon^\delta \frac{ds}{g(s)}, \quad \delta \geq \epsilon > 0. & (1.1.232) \end{cases}$$

Here Ω^{-1} *is the inverse of* Ω *and* ψ, ψ^{-1} *be as defined in Theorem 1.1.40,* t_3 *is chosen so that* $\Omega(1) + \int_a^t f(s)ds \in Dom\ (\Omega^{-1})$, *and*

$$\psi(\overline{\zeta}) + \int_a^s k(s,\tau)\phi(\overline{p}(\tau))\phi\left(\int_a^\tau f(\sigma)d\sigma\right)d\tau \in Dom\ (\psi^{-1}).$$

Proof Define the function

$$z(t) = a(t) + \int_a^t f(s)W\left(\int_a^s k(s,\tau)\phi(u(\tau))d\tau\right)ds. \tag{1.1.233}$$

Then (1.1.228) can be restated as

$$u(t) \leq z(t) + \int_a^t f(s)g(u(s))ds. \tag{1.1.234}$$

When $z(x)$ is a positive, continuous, non-decreasing in $x \in I$ and $g \in \mathcal{F}_1$, then it can be restated as

$$\frac{u(t)}{z(t)} \leq 1 + \int_a^t f(s)g\left(\frac{u(s)}{z(s)}\right)ds. \tag{1.1.235}$$

The inequality (1.1.235) may be treated as the one-dimensional Bihari-LaSalle inequality (see, e.g., [42]), i.e., Theorem 1.1.4, which implies

$$u(t) \leq \overline{p}(t)z(t), \tag{1.1.236}$$

where $\overline{p}(t)$ is defined by (1.1.230). By (1.1.233) and (1.1.236), we derive

$$u(t) \leq \overline{p}(t)\left[a(t) + \int_a^t f(s)W(v(s))ds\right]$$

where

$$v(s) = \int_a^s k(s,\tau)\phi(u(\tau))d\tau.$$

Now by following the argument as in the proof of Theorem 1.1.40, we can obtain the desired inequality in (1.1.229). □

Theorem 1.1.43 (The Denche-Khellaf Inequality [193]) *Let $u(t)$, $f(t)$, $b(t)$, $h(t)$, $\phi(u(t))$, $W(u(t))$, and $a(t)$ be as defined in Theorem 1.1.41, let $g \in \mathcal{F}_1$. If for all $a \le \tau \le s \le t \le b$,*

$$u(t) \le a(t) + \int_a^t f(s)g(u(s))ds + \int_a^t f(s)h(s)W\left(\int_a^s b(\tau)\phi(u(\tau))d\tau\right)ds, \tag{1.1.237}$$

then for all $a \le t \le t_4$,

$$u(t) \le \overline{p}(t)\left\{a(t) + \int_a^t f(s)h(s)\psi^{-1}\left(\psi(\overline{v}) + \int_a^s b(\tau)\phi(\overline{p}(\tau))\phi(\int_a^\tau f(\sigma)h(\sigma)d\sigma)d\tau\right)ds\right\}$$

where $\overline{p}(t)$ is defined by (1.1.230), ψ is defined by (1.1.215) and

$$\overline{v} = \int_a^b b(s)\phi(\overline{p}(s)a(s))ds,$$

the t_4 is chosen so that

$$\psi(\overline{v}) + \int_a^s b(\tau)\phi(\overline{p}(\tau))\phi\left(\int_a^\tau f(\sigma)h(\sigma)d\sigma\right)d\tau \in Dom\ (\psi^{-1}).$$

Proof The proof of the above theorem follows similar arguments as the proof of Theorem 1.1.42, we omit it. □

The next is a nonlinear Bihari inequality in [450].

Theorem 1.1.44 (The Pachpatte Inequality [450]) *Let $x(t), f(t), g(t), h(t)$ and $k(t)$ be real-valued non-negative continuous functions defined on $I = [a, b]$, $H(u)$ be a positive, continuous, strictly increasing, sub-multiplicative and sub-additive function for all $u > 0$; $H(0) = 0$, and suppose further that the following inequality holds for all $t \in I$,*

$$x(t) \le f(t) + g(t)\left[\int_a^t h(s)H\left(Hx(s) + g(s)\int_a^s k(\tau)H(x(\tau))\,d\tau\right)ds\right]. \tag{1.1.238}$$

Then for all $t \in I_0$,

$$\begin{aligned} x(t) &\le f(t) + g(t) \\ &\times \left[c + \int_a^t h(s)H(g(s)G^{-1}\left(G(c) + \int_a^s H(g(\tau))(h(\tau) + k(\tau))\,d\tau\right)ds\right], \end{aligned} \tag{1.1.239}$$

where

$$\begin{cases} c = \int_a^b h(s)H\Big(f(s) + g(s)\int_a^s k(\tau)H(f(\tau))\,d\tau\Big)ds, & (1.1.240) \\ G(r) = \int_{r_0}^r \frac{ds}{H(s)}, \quad r \geq r_0 > 0, & (1.1.241) \end{cases}$$

and G^{-1} *is the inverse function of* G, *and*

$$I_0 = \left\{t \in I : G(+\infty) \geq G(c) + \int_a^t H(g(\tau))(h(\tau) + k(\tau))\,d\tau\right\}.$$

Proof Without loss of generality, we may assume that $x(t) \geq f(t)$. Using the subadditivity of H, we have from (1.1.238) that

$$x(t) - f(t) \leq g(t)\Bigg[\int_a^t h(s)H(x(s) - f(s) + g(s)\int_a^s k(\tau)H(x(\tau) - f(\tau))\,d\tau)\,ds + \int_a^b h(s)H\left(f(s) + g(s)\int_a^s k(\tau)H(f(\tau))\,d\tau\right)ds\Bigg]. \quad (1.1.242)$$

Let $u(t) = x(t) - f(t)$ and define

$$v(t) = \int_a^t h(s)H\left(u(s) + g(s)\int_a^s k(\tau)H(u(\tau))\,d\tau\right)ds + c, \quad v(a) = c. \quad (1.1.243)$$

Then (1.1.242) can be restated as

$$u(t) \leq g(t)v(t). \quad (1.1.244)$$

Differentiating (1.1.243) and using (1.1.344) in view of the properties of H, we can get

$$v'(t) \leq h(t)H(g(t)(v(t) + \int_a^t k(\tau)H\Big(g(\tau))H(v(\tau))d\tau)\Big). \quad (1.1.245)$$

If we put

$$m(t) = v(t) + \int_a^t k(\tau)H(g(\tau))H(v(\tau))\,d\tau, \quad m(a) = v(a) = c, \quad (1.1.246)$$

then it follows from (1.1.245), (1.1.246) and the fact that $v(t) \leq m(t)$ that

$$m'(t) \leq H(g(t))(h(t) = k(t))H(m(t)). \quad (1.1.247)$$

Dividing both sides of (1.1.247) by $H(m(t))$, using (1.1.241) and integrating from a to t, we may obtain

$$G(m(t)) - G(c) \leq \int_a^t H(g(g(\tau))(h(\tau) + k(\tau))\,d\tau. \tag{1.1.248}$$

Then from (1.1.245) and (1.1.248), it follows

$$v'(t) \leq h(t)H(g(t)G^{-1}\left(G(c) + \int_a^t H(g(\tau))(h(\tau) + k(\tau))\,d\tau\right). \tag{1.1.249}$$

Now, integrating both sides of (1.1.249) from a to t and substituting the value of $v(t)$ in (1.1.244), we obtain the desired bound in (1.1.239). □

Theorem 1.1.45 (The Pachpatte Inequality [450]) *Let $x(t), f(t), g(t), h(t)$ and $k(t)$ be real-valued non-negative continuous functions defined on I, $H(u)$ be a positive, continuous, strictly increasing, sub-multiplicative and sub-additive function for all $u > 0$; $H(u) = 0$, and H^{-1} denotes the inverse function of H, for which the following inequality holds for all $t \in I$,*

$$\begin{aligned} x(t) \leq\, & f(t) + g(t)H^{-1}\left[\int_a^t h(s)H(x(s))\,ds \right.\\ & \left. + \int_a^t h(s)H(g(s))\left(\int_a^s k(\tau)H(x(\tau))\,d\tau\right)ds\right]. \end{aligned} \tag{1.1.250}$$

Then for all $t \in I$,

$$\begin{aligned} x(t) \leq\, & f(t) + g(t)H^{-1}\left[\int_a^t h(s)\left(H(f(s)) + H(g(s)) \right.\right.\\ & \times\left\{\exp\left(\int_a^s H(g(\tau))(h(\tau) + k(\tau))\,d\tau\right)\cdot\int_a^s H(f(\tau))(h(\tau) + k(\tau))\right.\\ & \left.\left.\left.\times\left(-\int_a^\tau H(g(n))(h(n) + k(n))dn\right)d\tau\right\}\right)ds\right]. \end{aligned} \tag{1.1.251}$$

Proof Since H is sub-additive, sub-multiplicative and monotonic, we may derive from (1.1.250)

$$\begin{aligned} H(x(t)) \leq\, & H(f(t)) + H(g(t)) \\ & \times\left[\int_a^t h(s)H(x(s))ds + \int_a^t h(s)H(g(s))\left(\int_a^s k(\tau)H(x(\tau))\,d\tau\right)ds\right]. \end{aligned} \tag{1.1.252}$$

Define a function $v(t)$ such that

$$v(t) = \int_a^t h(s)H(x(s))ds + \int_a^t h(s)H(g(s)) \left(\int_a^s k(\tau)H(x(\tau))d\tau \right) ds, \quad v(a) = 0,$$

then we may obtain

$$v'(t) = h(t) \left(H(x(t)) + H(g(t)) \int_a^t k(\tau)H(x(\tau))d\tau \right),$$

which, in view of (1.1.252), implies

$$v'(t) \le h(t) \left(H(f(t)) + H(g(t)) \left\{ v(t) + \int_a^t k(\tau)(H(f(\tau)) + H(g(\tau))v(\tau))d\tau \right\} \right). \tag{1.1.253}$$

If we put

$$m(t) = v(t) + \int_a^t k(\tau)(H(f(\tau)) + H(g(\tau))v(\tau))d\tau, \quad m(a) = v(a) = 0, \tag{1.1.254}$$

then it follows from (1.1.253)–(1.1.254) and the fact that $v(t) \le m(t)$ that

$$m'(t) \le H(g(t))(h(t) + k(t))m(t) + H(f(t))(h(t) + k(t)),$$

which further implies, noting that $m(a) = 0$,

$$\begin{aligned} m(t) \le \exp\left(\int_a^t H(g(\tau))(h(\tau) + k(\tau))\, d\tau \right) \int_a^t H(f(\tau))(h(\tau) + k(\tau)) \\ \times \exp\left(-\int_a^\tau H(g(n))(h(n) + k(n))dn \right) d\tau. \end{aligned} \tag{1.1.255}$$

Then from (1.1.253) and (1.1.255), it follows

$$\begin{aligned} v'(t) \le h(t) \Bigg(H(f(t)) + H(g(t)) \Bigg\{ \exp\left(H(g(\tau))(h(\tau) + k(\tau))\, d\tau \right) \\ \times \int_a^t H(f(\tau))(h(\tau) + k(\tau)) \cdot \exp\left(-\int_a^\tau H(g(n))(h(n) + k(n))dn \right) d\tau \Bigg\} \Bigg). \end{aligned} \tag{1.1.256}$$

Now integrating both sides of (1.1.256) from a to t and substituting the value of $v(t)$ in (1.1.252) and then applying H^{-1} to both sides of (1.1.252), we obtain the desired bound in (1.1.251). □

Finally, we establish the following integral inequality which can be used in obtaining the lower bounds on unknown function.

Theorem 1.1.46 (The Pachpatte Inequality [450]) *Let $x(t), g(t), h(t)$ and $k(t)$ be real-valued non-negative continuous functions defined on I, $H(u)$ is the same function as defined in Theorem 1.1.45; and for all $a \le s \le t \le b$,*

$$x(t) \ge x(s) - g(t)H^{-1}\left[\int_s^{\tau} h(\tau)H(x(\tau))d\tau + \int_s^t h(\tau)\left(\int_{\tau}^t k(n)H(x(n))dn\right)d\tau\right]. \tag{1.1.257}$$

Then, for all $a \le s \le t \le b$,

$$x(t) \ge x(s)\left\{H^{-1}\left[1+H(g(t))\int_s^t h(\tau)\exp\left(\int_{\tau}^t (h(n)H(g(t))+k(n))dn\right)d\tau\right]\right\}^{-1}. \tag{1.1.258}$$

Proof In fact, we may rewrite (1.1.257) as

$$x(s) \le x(t) + g(t)H^{-1}\left[\int_s^t h(\tau)H(x(\tau))\,d\tau + \int_s^t h(\tau)\left(\int_{\tau}^t k(n)H(x(n))dn\right)d\tau\right]. \tag{1.1.259}$$

Since H is sub-additive, sub-multiplicative and monotonic, we may derive from (1.1.259)

$$\begin{aligned} H(x(s) \le H(x(t)) + H(g(t))\bigg[&\int_s^t h(\tau)H(x(\tau))d\tau \\ &+ \int_s^t h(\tau)\left(\int_{\tau}^t k(n)H(x(n))dn\right)d\tau\bigg]. \end{aligned} \tag{1.1.260}$$

For fixed t in the interval I, we define for $a \le s \le t$,

$$\begin{aligned} v(s) = H(x(t)) + H(g(t))\bigg[&\int_s^t h(\tau)H(x(\tau))d\tau \\ &+ \int_s^t h(\tau)\left(\int_{\tau}^t k(n)H(x(n))dn\right)d\tau\bigg], \quad v(t) = H(x(t)). \end{aligned} \tag{1.1.261}$$

Thus from (1.1.261), it follows that,

$$v'(s) = -H(g(t))h(s)\left[H(x(s)) + \int_s^t k(n)H(x(n))dn\right], \tag{1.1.262}$$

which, in view of $H(x(s)) \le v(s)$, implies

$$v'(s) \ge -H(g(t))h(s)\left[v(s) + \int_s^t k(n)v(n)dn\right]. \tag{1.1.263}$$

If we put

$$m(s) = v(s) + \int_s^t k(n)v(n)dn, \quad m(t) = v(t), \tag{1.1.264}$$

then it follows from (1.1.263)–(1.1.264) and the fact that $v(s) \le m(s)$, that

$$m'(s) + (h(s)H(g(t)) + k(s))m(s) \ge 0 \tag{1.1.265}$$

which implies, noting that $m(t) = H(x(t))$,

$$m(s) \le H(x(t)) \exp\left(\int_s^t (h(\tau)H(g(t)) + K(\tau))\, d\tau\right). \tag{1.1.266}$$

Then from (1.1.263) and (1.1.266), it follows that

$$v'(s) \ge -H(g(t))H(x(t))h(s) \exp\left(\int_s^t (h(\tau)H(g(\tau)) + k(\tau))\, d\tau\right). \tag{1.1.267}$$

Now integrating both sides (1.1.267) from s to t and substituting the value of $v(s)$ in (1.1.260), we obtain the desired bound in (1.1.258). □

Note that in [258, 328], the authors have obtained the lower bounds on unknown functions. However, the bound obtained in Theorem 1.1.46 is different from those given in [258, 328].

Next, we introduce a retarded Bihari inequality, a generalization of the Gronwall-Bellman inequality.

Theorem 1.1.47 (The Lipovan Inequality [355]) *Let $u, f \in C([t_0, T), \mathbb{R}_+)$. Moreover, let $\omega \in C(\mathbb{R}_+, \mathbb{R}_+)$ be non-decreasing with $\omega(u) > 0$ on $(0, +\infty)$ and $\alpha \in C^1([t_0, T), [t_0, T))$ be non-decreasing with $\alpha(t) \le t$ on $[t_0, T)$.*

If, for all $t_0 \le t < T$,

$$u(t) \le k + \int_{\alpha(t_0)}^{\alpha(t)} f(s)\omega(u(s))ds, \tag{1.1.268}$$

where k is a non-negative constant, then for all $t_0 \le t < t_1$,

$$u(t) \le G^{-1}\left(G(k) + \int_{\alpha(t_0)}^{\alpha(t)} f(s)ds\right), \tag{1.1.269}$$

where $G(r) = \int_1^\gamma \frac{ds}{\omega(s)}$, $\gamma > 0$, *and* $t_1 \in (t_0, T)$ *is chosen so that, for all* $t \in [t_0, t_1)$,

$$G(k) + \int_{\alpha(t_0)}^{\alpha(t)} f(s)ds \in Dom\ (G^{-1}). \qquad (1.1.270)$$

Proof Assume first that $k > 0$ and let us denote by $U(t)$ the right-hand side of (1.1.268). Then $U(t_0) = k$ and

$$\begin{aligned} U'(t) &= f(\alpha(t))\omega(u(\alpha(t)))\alpha'(t) \\ &\leq f(\alpha(t))\omega(U(\alpha(t)))\alpha'(t). \end{aligned}$$

As $\alpha(t) \leq t$ on $[t_0, T)$, we deduce that

$$U'(t) \leq f(\alpha(t))\omega(U(t))\alpha'(t).$$

From the definition of G and the above relation, we may infer

$$\frac{d}{dt}G(U(t)) \leq f(\alpha(t))\alpha'(t).$$

An integration on $[t_0, t]$ shows now that

$$G(U(t)) \leq G(k) + \int_{\alpha(t_0)}^{\alpha(t)} f(s)ds.$$

Since G^{-1} is increasing on $Dom\ (G^{-1})$, the above inequality thus yields, for all $t_0 \leq t < t_1$,

$$U(t) \leq G^{-1}\left(G(k) + \int_{\alpha(t_0)}^{\alpha(t)} f(s)ds\right).$$

The required inequality (1.1.269) is obtained in view of the relation $u(t) \leq U(t)$ on $[t_0, T)$. If $k = 0$, we carry out the above procedure with $\epsilon \to 0$, instead of k and subsequently let $\epsilon \to 0^+$. □

Remark 1.1.13 (i) For $\alpha(t) \equiv t$ in Theorem 1.1.47, we obtain Bihari's inequality [82], i.e., Theorem 1.1.12.

(ii) Note that if $\int_1^{+\infty} \frac{ds}{\omega(s)} = +\infty$, then $G(+\infty) = +\infty$ and (1.1.269) is valid on $[t_0, T)$. Examples of such functions are $\omega(u) \equiv u$ and $\omega(u) \equiv u ln(1+u)$.

Setting $\omega(u) \equiv u$ in Theorem 1.1.47, we may obtain the following corollary.

Corollary 1.1.8 (The Lipovan Inequality [355]) *Let* $u, f \in C([t_0, T), \mathbb{R}_+)$. *Furthermore, let* $\alpha \in C^1([t_0, T), [t_0, T))$ *be non-decreasing with* $\alpha(t) \leq t$ *on* $[t_0, T)$, *and let k be a non-negative constant. If the inequality holds for all $t_0 \leq t < T$,*

$$u(t) \leq k + \int_{\alpha(t_0)}^{\alpha(t)} f(s)u(s)ds,$$

then for all $t_0 \leq t < T$,

$$u(t) \leq k \exp\left(\int_{\alpha(t_0)}^{\alpha(t)} f(s)ds\right).$$

Remark 1.1.14

(i) With $\alpha(t) \equiv t$ in Corollary 1.1.8, we may obtain the celebrated Gronwall-Bellman inequality, i.e., Theorem 1.1.2 in Qin [557], see [65, 82, 165, 175, 210, 259].
(ii) Let us assume that $t_0 = 0$ and $T = +\infty$. In this case, note that $\alpha(0) = 0$ and the hypothesis (1.1.268) of Theorem 1.1.47 implies that for all $t \geq 0$,

$$u(t) \leq k + \int_0^t f(s)\omega(u(s))ds.$$

Hence Bihari's result [82] could also be applied in order to obtain an upper estimate for $u(t)$. However, the estimate provided by Theorem 1.1.47 is sharper. To see this, take $\omega(u) \equiv u$, $\alpha(t) \equiv ln(t+1)$, and $f(t) \equiv \frac{1}{t+1}$. Bihari's inequality yields

$$u(t) \leq k(t+1), \quad t \geq 0,$$

while Theorem 1.1.47 gives the estimate

$$u(t) \leq k(ln(t+1)+1), \quad t \geq 0.$$

Similarly as the same of the proof of Theorem 1.1.47 we may obtain the next result.

Theorem 1.1.48 (The Lipovan Inequality [355]) *Let $u, f, g \in C([t_0, T), \mathbb{R}_+)$, and $\omega \in C(\mathbb{R}_+, \mathbb{R}_+)$ be non-decreasing with $\omega(u) > 0$ for all $u > 0$, and $\alpha \in C^1([t_0, T), [t_0, T))$ be non-decreasing with $\alpha(t) \leq t$ on $[t_0, T)$. If, for all $t_0 \leq t < T$,*

$$u(t) \leq k + \int_{t_0}^t f(s)\omega(u(s))ds + \int_{\alpha(t_0)}^{\alpha(t)} g(s)\omega(u(s))ds,$$

where k is a non-negative constant, then for all $t_0 \leq t < t_1$,

$$u(t) \leq G^{-1}(G(k) + \int_{t_0}^t f(s)ds + \int_{\alpha(t_0)}^{\alpha(t)} g(s)ds),$$

with G being in Theorem 1.1.47 and t_1 being chosen so that the right-hand side above is well-defined.

Theorem 1.1.49 (The Pachpatte Inequality [520]) *Let $a(t) \in C(\mathbb{R}_+, \mathbb{R}_+)$, $b(t,s) \in C(\mathbb{R}_+, \mathbb{R}_+)$ for all $t_0 \le s \le t \le T$ and $g \in C(\mathbb{R}_+, \mathbb{R}_+)$ be a non-decreasing function with $g(u) > 0$ for all $u > 0$. Let $\alpha \in C^1(\mathbb{R}_+, \mathbb{R}_+)$ be non-decreasing with $\alpha(t) \le t$ on $\mathbb{R}_+$ and $k \le 0$ be a constant. If for all $t \in I$,*

$$u(t) \le k + \int_{\alpha(t_0)}^{\alpha(t)} \left[a(s)g(u(s)) + \int_{\alpha(t_0)}^{s} b(s,\sigma)g(u(\sigma))d\sigma \right] ds, \qquad (1.1.271)$$

then for all $t_0 \le t \le t_1$,

$$u(t) \le G^{-1}[G(k) + A(t)], \qquad (1.1.272)$$

where $A(t)$ is defined by

$$A(t) = \int_{\alpha(t_0)}^{\alpha(t)} \left[a(s) + \int_{\alpha(t_0)}^{s} b(s,\sigma)d\sigma \right] ds,$$

G^{-1} is the inverse function of

$$G(r) = \int_{r_0}^{r} \frac{ds}{g(s)}, \quad r \ge r_0 > 0, \qquad (1.1.273)$$

and $t_1 \in I$ is chosen so that $G(k) + A(t) \in Dom\ (G^{-1})$ for all $t \in [t_0, t_1]$.

Proof Let $k > 0$ and define a function $z(t)$ by the right-hand side of (1.1.271). Then $z(t) > 0, z(t_0) = k, u(t) \le z(t)$, we get

$$\frac{z'(t)}{g(z(t))} \le \left[a(\alpha(t)) + \int_{\alpha(t_0)}^{\alpha(t)} b(\alpha(t),\sigma)d\sigma \right] \alpha'(t). \qquad (1.1.274)$$

From (1.1.273) and (1.1.274), we derive

$$\frac{d}{dt}G(z(t)) = \frac{z'(t)}{g(z(t))} \le \left[a(\alpha(t)) + \int_{\alpha(t_0)}^{\alpha(t)} b(\alpha(t),\sigma)d\sigma \right] \alpha'(t). \qquad (1.1.275)$$

Integrating (1.1.275) from t_0 to $t, t \in I$ and by making the change of variables, we have

$$G(z(t)) \le G(k) + A(t). \qquad (1.1.276)$$

Since $G^{-1}(z)$ is increasing, from (1.1.276) it follows

$$z(t) \leq G^{-1}[G(k) + A(t)]. \tag{1.1.277}$$

Using (1.1.277) in $u(t) \leq z(t)$, we get (1.1.272). The case $k \geq 0$ can be easily completed in the same way. The sub-interval $t_0 \leq t \leq t_1$ for t is obvious. □

Theorem 1.1.50 (The Pachpatte Inequality [520]) *Let a, α be as in Theorem 1.1.49. Assume $k, w \in C(\mathbb{R}_+, \mathbb{R}_+)$ are non-decreasing functions with $k(0) > 0, w(t) > 0$ for all $t > 0$ and $\int_1^{+\infty} \frac{dt}{w(t)} = +\infty$. If $u \in C(\mathbb{R}_+, \mathbb{R}_+)$ satisfies for all $t \geq 0$,*

$$u(t) \leq k(t) + \int_0^{\alpha(t)} a(t, s)w(u(s))ds,$$

then for all $t \geq 0$,

$$u(t) \leq G^{-1}\left(G(k(t)) + \int_0^{\alpha(t)} a(t, s)ds\right), \tag{1.1.278}$$

where $G(t) = \int_0^t \frac{ds}{w(s)} ds$, for all $t \geq 0$.

Proof Let $T \geq 0$ be fixed and denote $z(t) = \int_0^{\alpha(t)} a(t, s)w(u(s))ds,\ t \geq 0$. Our assumptions on b, α imply that z is non-decreasing on $\mathbb{R}_+$. Hence for all $t \in [0, T]$, we have

$$\begin{aligned}
z'(t) &= a(t, \alpha(t))w(u(\alpha(t)))a'(t) + \int_0^{\alpha(t)} \partial_t a(t, s)w(u(s))ds \\
&\leq a(t, \alpha(t))w[k(\alpha(t)) + z(\alpha(t))] + \int_0^{\alpha(t)} \partial_t a(t, s)w[(z(s) + k(s))]ds \\
&\leq a(t, \alpha(t))\alpha'(t)w[k(\alpha(T)) + z(t)] + w[k(\alpha(T)) + z(t)] \int_0^{\alpha(t)} \partial_t a(t, s)ds \\
&\leq \left(a(t, \alpha(t))\alpha'(t) + \int_0^{\alpha(t)} \partial_t a(t, s)ds\right) w[k(\alpha(T)) + z(t)],
\end{aligned}$$

and for all $t \in [0, T]$,

$$\frac{z'(t)}{w[k(T) + z(t)]} \leq \frac{d}{dt}\left(\int_0^{\alpha(t)} a(t, s)ds\right). \tag{1.1.279}$$

Integrating both sides of (1.1.279) on $[0, t]$, we get, for all $t \in [0, T]$,

$$G(k(T) + z(t)) \le G(k(T)) + \int_0^{\alpha(t)} a(t, s)ds.$$

or, equivalently, for all $t \in [0, T]$,

$$k(T) + z(t) \le G^{-1}[G(k(T)) + \int_0^{\alpha(t)} a(t, s)ds]. \qquad (1.1.280)$$

Note that the right-hand side of (1.1.280) is well defined as $G(+\infty)$. Letting $t = T$ in the above relation, we can obtain

$$u(T) \le k(T) + z(T) \le G^{-1}[G(k(T)) + \int_0^{\alpha(T)} a(T, s)ds],$$

and since $T \ge 0$ was arbitrarily chosen, we get (1.1.278). □

Corollary 1.1.9 (The Pachpatte Inequality [520]) *Assume a, w, k, α are as in Theorem 1.1.50. Suppose $u \in C(\mathbb{R}_+, \mathbb{R}_+)$ is a solution to the nonlinear Volterra integral equation, for all $t \ge 0$,*

$$u(t) = k(t) + \int_0^{\alpha(t)} a(t, s)w(u(s))ds. \qquad (1.1.281)$$

If k is bounded and $\lim_{t\to+\infty} \int_0^{\alpha(t)} a(t, s)ds < +\infty$, then u is bounded.

Theorem 1.1.51 (The Pachpatte Inequality [520]) *Let $a, b, k \in C(\mathbb{R}_+, \mathbb{R}_+), \alpha \in C^1(\mathbb{R}_+, \mathbb{R}_+)$ and assume that a, k, α are non-decreasing functions with $\alpha(t) \le t$ for all $t \ge 0$. Let also $w \in C(\mathbb{R}_+, \mathbb{R}_+)$ be a non-decreasing function such that $w(t) > 0$ for all $t > 0$ and $\int_0^{+\infty} \frac{ds}{w(t)} = +\infty$. If $u \in C(\mathbb{R}_+, \mathbb{R}_+)$ satisfies, for all $t \ge 0$,*

$$u(t) \le k(t) + a(t) \int_0^{\alpha(t)} b(s)w(u(s))ds, \qquad (1.1.282)$$

then, for all $t \ge 0$,

$$u(t) \le G^{-1}\left(G(k(t)) + a(t) \int_0^{\alpha(t)} b(s)ds\right), \qquad (1.1.283)$$

where $G(t) = \int_0^t \frac{ds}{w(s)}ds,\ t \ge 0$.

Proof Let $T \geq 0$ be fixed. Then for all $t \in [0, T]$, relation (1.1.282) together with the hypotheses on a, k implies

$$u(t) \leq k(T) + a(T) \int_0^{\alpha(t)} b(s)w(u(s))ds. \qquad (1.1.284)$$

By the retarded version of Bihari's inequality, i.e., Theorem 1.1.47, relation (1.1.284) implies for all $t \in [0, T]$,

$$u(t) \leq G^{-1}\left(G(k(T)) + a(T) \int_0^{\alpha(t)} b(s)ds \right).$$

Now let $t = T$ in the above relation to obtain

$$u(T) \leq G^{-1}\left(G(k(T)) + a(T) \int_0^{\alpha(T)} b(s)ds \right),$$

and since $T \geq 0$ was arbitrarily chosen, we can get (1.1.283). □

Corollary 1.1.10 (The Pachpatte Inequality [520]) *Assume a, b, w, k, α are as in Theorem 1.1.51. Suppose $u \in C(\mathbb{R}_+, \mathbb{R}_+)$ is a solution to the integral equation, for all $t \geq 0$,*

$$u(t) = k(t) + a(t) \int_0^{\alpha(t)} b(s)w(u(s))ds. \qquad (1.1.285)$$

If a, k are bounded and $\int_0^{\alpha(t)} b(s)ds < +\infty$, then u is bounded on $\mathbb{R}_+$.

Using the similar arguments to those in the proofs of Theorems 1.1.50–1.1.51, we obtain the next result.

Theorem 1.1.52 (The Pachpatte Inequality [520]) *Let $a, b, k \in C(\mathbb{R}_+, \mathbb{R}_+)$, $\alpha \in C^1(\mathbb{R}_+, \mathbb{R}_+)$ and assume that α is non-decreasing with $\alpha(t) \geq t$ for all $t \geq 0$. If $u \in C(\mathbb{R}_+, \mathbb{R}_+)$ satisfies for all $t \geq 0$,*

$$u(t) \geq k(t) + a(t) \int_0^{\alpha(t)} b(s)u(s)ds, \qquad (1.1.286)$$

then for all $t \geq 0$,

$$u(t) \geq k(t) + a(t) \int_0^{\alpha(t)} e^{\int_r^{\alpha(t)} a(s)b(s)ds} b(r)k(r)ds. \qquad (1.1.287)$$

Corollary 1.1.11 (The Pachpatte Inequality [520]) *Assume a, b, k, α are sa in Theorem 1.1.51. If $u \in C(\mathbb{R}_+, \mathbb{R}_+)$ is a solution to the integral equation, for all $t \geq 0$,*

$$u(t) = k(t) + a(t) \int_0^{\alpha(t)} b(s)u(s)ds.$$

Then each of the following conditions is sufficient for u to be unbounded:

(i) *a is unbounded and $b, k, \alpha \equiv 0$;*
(ii) $\limsup_{t \to +\infty} a(t) > 0$ *and* $\int_0^{+\infty} b(s)k(s)ds = +\infty$.

Theorem 1.1.53 (The Pachpatte Inequality [520]) *Let $a, b, k \in C(\mathbb{R}_+, \mathbb{R}_+), \alpha \in C^1(\mathbb{R}_+, \mathbb{R}_+)$ with a, k are non-increasing on $\mathbb{R}_+$ and assume that α is non-decreasing with $\alpha(t) \geq t$ for all $t \geq 0$. Let also $w \in C(\mathbb{R}_+, \mathbb{R}_+)$ be a non-decreasing function such that $w(t) > 0$ for all $t > 0$. If $u \in C(\mathbb{R}_+, \mathbb{R}_+)$ satisfies for all $t \geq 0$,*

$$u(t) \geq k(t) + a(t) \int_0^{\alpha(t)} b(s)w(u(s))ds.$$

then for all $t_1 \geq t \geq 0$,

$$u(t) \geq G^{-1}\left(G(k(\alpha(t))) + a(t) \int_0^{\alpha(t)} b(s)ds\right),$$

where $G(t) = \int_0^t \frac{ds}{w(s)}, t \geq 0$, and t_1 is chosen so that $G(k(\alpha(t))) + a(t) \int_0^{\alpha(t)} b(s)ds \in$ Dom (G^{-1}), for all $t \in [0, t_1]$.

Setting $a(t) \equiv 1, k(t) \equiv k > 0$ in Theorem 1.1.53, we obtain the following inequality, which may be regarded as a reverse version of Bihari's inequality [42].

Corollary 1.1.12 (The Pachpatte Inequality [520]) *Consider $k > 0, b \in C(\mathbb{R}_+, \mathbb{R}_+), \alpha \in C^1(\mathbb{R}_+, \mathbb{R}_+)$, assume that α is non-decreasing with $\alpha(t) \geq t$ for all $t \geq 0$. Let also $w \in C(\mathbb{R}_+, \mathbb{R}_+)$ be a non-decreasing function such that $w(t) > 0$ for all $t > 0$. If $u \in C(\mathbb{R}_+, \mathbb{R}_+)$ satisfies for all $t \geq 0$,*

$$u(t) \geq k + \int_0^{\alpha(t)} b(s)w(u(s))ds,$$

then for all $t_1 \geq t \geq 0$,

$$u(t) \geq G^{-1}\left(G(k) + \int_0^{\alpha(t)} b(s)ds\right),$$

where $G(t) = \int_0^t \frac{ds}{w(s)}ds$, for all $t \geq 0$, and t_1 is chosen so that $G(k) + \int_0^{\alpha(t)} b(s)ds \in Dom(G^{-1})$ for all $t \in [0, t_1]$.

Corollary 1.1.13 (The Pachpatte Inequality [520]) *Assume k, b, α, w, G are as in Corollary 1.1.12. Suppose in addition $G(+\infty) = \int_1^{+\infty} \frac{ds}{w(s)} ds = L < +\infty$. Let $u \in C([0, t_0), \mathbb{R}_+)$ is a solution the integral equation for all $t \geq 0$,*

$$u(t) = k + \int_0^{\alpha(t)} b(s)w(u(s))ds.$$

Suppose also that $[0, t_0)$ is the maximal interval of existence for u. If $T = \inf\Big\{t \geq 0 : G(k) + \int_0^{\alpha(t)} b(s)ds \geq L\Big\}$ exists and is finite, then $t_0 \leq T$.

Proof Suppose T exists and is finite and the maximal existence time t_0 satisfies $t_0 > T$. Take now $t < T$. Then $0 \leq G(k) + \int_0^{\alpha(t)} b(s)ds < L$ and hence $G(k) + \int_0^{\alpha(t)} b(s)ds \in Dom\ (G^{-1})$. By Corollary 1.1.12, we can get, for all $0 \leq t < T$,

$$u(t) \geq G^{-1}\bigg(G(k) + \int_0^{\alpha(t)} b(s)ds\bigg).$$

Letting $t \to T$ in the above relation, we can deduce $\lim_{t \to T} u(t) \geq G^{-1}(L) = +\infty$, which contradicts our assumption $t_0 > T$. □

Theorem 1.1.54 (The Zhao-Meng Inequality [722]) *Let $\varphi \in C(\mathbb{R}_+, \mathbb{R}_+)$ be an increasing function with $\varphi(+\infty) = +\infty$. Let $\psi \in C(\mathbb{R}_+, \mathbb{R}_+)$ be a non-decreasing function and let c be a non-negative constant. Let $\alpha \in C^1(\mathbb{R}_+, \mathbb{R}_+)$ be non-decreasing with $\alpha(t) \geq t$ on $\mathbb{R}_+$. If $u, f \in C(\mathbb{R}_+, \mathbb{R}_+)$ and for all $t \in \mathbb{R}_+$,*

$$\varphi(u(t)) \leq c + \int_{\alpha(t)}^{+\infty} f(s)\varphi(u(s))ds, \tag{1.1.288}$$

then for all $0 \leq T \leq t < +\infty$,

$$u(t) \leq \varphi^{-1}\bigg(G^{-1}\bigg[G(c) + \int_{\alpha(t)}^{+\infty} f(s)ds\bigg]\bigg), \tag{1.1.289}$$

where $G(z) = \int_{z_0}^{z} \frac{ds}{\psi[\varphi^{-1}(s)]}, z \geq z_0 > 0, \varphi^{-1}, G^{-1}$ are, respectively, the inverse of φ and G, $T \in \mathbb{R}_+$ is chosen so that for all $t \in [T, +\infty)$,

$$\begin{cases} G(c) + \displaystyle\int_{\alpha(t)}^{+\infty} f(s)ds \in Dom\ (G^{-1}), & (1.1.290) \\ G^{-1}\bigg[G(c) + \displaystyle\int_{\alpha(t)}^{+\infty} f(s)ds\bigg] \in Dom\ (\varphi^{-1}). & (1.1.291) \end{cases}$$

Proof Define a non-increasing positive function $z(t)$ by, for all $t \in \mathbb{R}_+$,

$$z(t) = c + \varepsilon + \int_{\alpha(t)}^{+\infty} f(s)\psi(u(s))ds, \tag{1.1.292}$$

where ε is an arbitrary small positive number. From inequality (1.1.288) it follows that

$$u(t) \le \varphi^{-1}[z(t)]. \tag{1.1.293}$$

Differentiating (1.1.292) and using (1.1.293) and the monotonicity of φ^{-1} and ψ, we can deduce that

$$\begin{aligned} z'(t) = -f(\alpha(t))\psi[u(\alpha(t))]\alpha'(t) &\ge -f(\alpha(t))\psi[\varphi^{-1}(z(\alpha(t)))]\alpha'(t) \\ &\ge -f(\alpha(t))\psi[\varphi^{-1}(z((t))]\alpha'(t). \end{aligned} \tag{1.1.294}$$

Noting that

$$\psi[\varphi^{-1}(z(t))] \ge \psi[\varphi^{-1}(z(\infty))] = \psi[\varphi^{-1}(c+\varepsilon)] > 0$$

from the definition of G, (1.1.294) thus gives us

$$\frac{d}{dt}G(z(t)) = \frac{z'(t)}{\psi[\varphi^{-1}(z(t))]} \ge -f(\alpha(t))\alpha'(t). \tag{1.1.295}$$

Setting $t = s$, and integrating (1.1.295) from t to $+\infty$ and letting $\varepsilon \to 0$, we conclude for all $t \in \mathbb{R}_+$,

$$G(z(t)) \le G(c) + \int_{\alpha(t)}^{+\infty} f(s)ds$$

Thus from (1.1.290), (1.1.291), (1.1.293) and the above relation, we obtain the inequality (1.1.289). □

In fact, we can regard Theorem 1.1.54 as a generalized form of the Ou-Yang inequality with advanced argument.

Theorem 1.1.55 (The Zhao-Meng Inequality [722]) *Let u, f and g be non-negative continuous functions defined on $\mathbb{R}_+$, and let $\varphi \in C(\mathbb{R}_+, \mathbb{R}_+)$ be an increasing function with $\varphi(+\infty) = +\infty$ and let c be a non-negative constant. Moreover, let $\omega_1, \omega_2 \in C(\mathbb{R}_+, \mathbb{R}_+)$ be non-decreasing functions with $\omega_i(u) > 0$ $(i = 1, 2)$ on $(0, +\infty)$, $\alpha \in C^1(\mathbb{R}_+, \mathbb{R}_+)$ be non-decreasing with $\alpha(t) \ge t$ on $\mathbb{R}_+$. If for all $t \in \mathbb{R}_+$,*

$$\varphi(u(t)) \le c + \int_{\alpha(t)}^{+\infty} f(s)\omega_1(u(s))ds + \int_{t}^{+\infty} g(s)\omega_2(u(s))ds, \tag{1.1.296}$$

then for all $0 \le T \le t < +\infty$,

(*i*) *for the case* $\omega_2(u) \leq \omega_1(u)$,

$$u(t) \leq \varphi^{-1}\left(G_1^{-1}\left[G_1(c) + \int_{\alpha(t)}^{+\infty} f(s)ds + \int_t^{+\infty} g(s)ds\right]\right), \tag{1.1.297}$$

(*ii*) *for the case* $\omega_1(u) \leq \omega_2(u)$,

$$u(t) \leq \varphi^{-1}\left(G_2^{-1}\left[G_2(c) + \int_{\alpha(t)}^{+\infty} f(s)ds + \int_t^{+\infty} g(s)ds\right]\right), \tag{1.1.298}$$

where

$$G_i(r) = \int_{r_0}^{r} \frac{ds}{\omega_i(\varphi^{-1}(s))}, \quad r \geq r_0 > 0, \ i = 1, 2,$$

and $\varphi^{-1}, G_i^{-1} (i = 1, 2)$ *are, respectively, the inverse of* $\varphi, G_i, T \in \mathbb{R}_+$ *is chosen so that for all* $t \in [T, +\infty)$,

$$G_i(c) + \int_{\alpha(t)}^{+\infty} f(s)ds + \int_t^{+\infty} g(s)ds \in Dom\,(G_i^{-1}), \ i = 1, 2. \tag{1.1.299}$$

Proof Define a non-increasing positive function $z(t)$ by, for all $0 \leq T \leq t < +\infty$,

$$z(t) = c + \varepsilon + \int_{\alpha(t)}^{+\infty} f(s)\omega_1(u(s))ds + \int_t^{+\infty} g(s)\omega_2(u(s))ds, \tag{1.1.300}$$

where ε is an arbitrary small positive number. Thus from inequality (1.1.296), it follows that for all $t \in \mathbb{R}_+$,

$$u(t) \leq \varphi^{-1}[z(t)]. \tag{1.1.301}$$

Differentiating (1.1.300) and using (1.1.301) and the monotonicity of $\varphi^{-1}, \omega_1, \omega_2$, we can deduce

$$\begin{aligned} z'(t) &= -f(\alpha(t))\omega_1[u(\alpha(t))]\alpha'(t) - g(t)\omega_2[u(t)] \\ &\geq -f(\alpha(t))\omega_1[\psi^{-1}(z(\alpha(t)))]\alpha'(t) - g(t)\omega_2[\varphi^{-1}(z(t))] \\ &\geq -f(\alpha(t))\omega_1[\psi^{-1}(z(t))]\alpha'(t) - g(t)\omega_2[\varphi^{-1}(z(t))]. \end{aligned} \tag{1.1.302}$$

(*i*) When $\omega_2(u) \leq \omega_1(u)$, we have, for all $t \in \mathbb{R}_+$,

$$z'(t) \geq -f(\alpha(t))\omega_1[\psi^{-1}(z(t))]\alpha'(t) - g(t)\omega_2[\varphi^{-1}(z(t))], \tag{1.1.303}$$

which, by noting that

$$\omega_1[\psi^{-1}(z(t))] \leq \omega_1[\varphi^{-1}(z(+\infty))] = \omega_1[\varphi^{-1}(c+\varepsilon)] > 0,$$

from the definition of $G_1(\gamma)$, gives us, for all $t \in \mathbb{R}_+$,

$$\frac{d}{dt}G_1(z(t)) = \frac{z'(t)}{\omega_1[\varphi^{-1}(z(t))]} \geq -f(\alpha(t))\alpha'(t) - g(t).$$

Setting $t = s$ and integrating it from t to $+\infty$ and let $\varepsilon \to 0$, we get, for all $t \in \mathbb{R}_+$,

$$G_1(z(t)) \leq G_1(c) + \int_{\alpha(t)}^{+\infty} f(s)ds + \int_t^{+\infty} g(s)ds,$$

whence, for all $0 \leq T \leq t < +\infty$,

$$z(t) \leq G_1^{-1}\left(G_1(c) + \int_{\alpha(t)}^{+\infty} f(s)ds + \int_t^{+\infty} g(s)ds\right).$$

Using (1.1.301), we can conclude for all $0 \leq T \leq t < +\infty$,

$$u(t) \leq \psi^{-1}\left(G_1^{-1}\left[G_1(c) + \int_{\alpha(t)}^{+\infty} f(s)ds + \int_t^{+\infty} g(s)ds\right]\right).$$

(*ii*) When $\omega_1(u) \leq \omega_2(u)$, the proof can be done similarly. □

Theorem 1.1.56 (The Zhao-Meng Inequality [722]) *Let u, f and g be non-negative continuous functions defined on $\mathbb{R}_+$ and let c be a non-negative constant. Moreover, let $\varphi \in C(\mathbb{R}_+, \mathbb{R}_+)$ be an increasing function with $\varphi(+\infty) = +\infty$, $\psi \in C(\mathbb{R}_+, \mathbb{R}_+)$ be a non-decreasing function with $\psi(u) > 0$ on $(0, +\infty)$ and $\alpha \in C^1(\mathbb{R}_+, \mathbb{R}_+)$ be non-decreasing with $\alpha(t) \geq t$ on $\mathbb{R}_+$. If for all $t \in \mathbb{R}_+$,*

$$\varphi(u(t)) \leq c + \int_{\alpha(t)}^{+\infty} [f(s)u(s)\psi(u(s))ds + g(s)u(s)]ds, \tag{1.1.304}$$

then for all $0 \leq T \leq t < +\infty$,

$$u(t) \leq \psi^{-1}\left(\Omega^{-1}\left[G^{-1}\left(G\left[\Omega(c) + \int_{\alpha(t)}^{+\infty} g(s)ds\right] + \int_{\alpha(t)}^{+\infty} f(s)ds\right)\right]\right), \tag{1.1.305}$$

where

$$\Omega(r)=\int_{r_0}^{r}\frac{ds}{\varphi^{-1}(s)},\quad r\geq r_0>0;\quad G(z)=\int_{z_0}^{z}\frac{ds}{\psi\{\varphi^{-1}[\Omega^{-1}(s)]\}},\quad z\geq z_0>0$$

and $\Omega^{-1},\varphi^{-1},G^{-1}$ *are respectively the inverse of* Ω,φ,G *and* $T\in\mathbb{R}_+$ *is chosen so that for all* $t\in[T,+\infty)$

$$G\left(\Omega(c)+\int_{\alpha(t)}^{+\infty}g(s)ds\right)+\int_{\alpha(t)}^{+\infty}f(s)ds\in Dom\,(G^{-1})$$

and

$$G^{-1}\left(G\left[\Omega(c)+\int_{\alpha(t)}^{+\infty}g(s)ds\right]+\int_{\alpha(t)}^{+\infty}f(s)ds\right)\in Dom\,(\Omega^{-1}).$$

Proof Let us first assume that $c>0$. Define the non-increasing positive function $z(t)$ by the right-hand side of (1.1.304). Then $z(+\infty)=c, u(t)\leq\varphi^{-1}[z(t)]$ and

$$\begin{aligned}z'(t)&=-[f(\alpha(t))u(\alpha(t))\psi[u(\alpha(t))]-g(\alpha(t))u(\alpha(t))]\alpha'(t)\\&\geq-[f(\alpha(t))\varphi^{-1}(z(\alpha(t)))\psi[\varphi^{-1}(z(\alpha(t)))]-g(\alpha(t))\varphi^{-1}(z(\alpha(t)))]\alpha'(t)\\&\geq-[f(\alpha(t))\varphi^{-1}(z(t))\psi[\varphi^{-1}(z(\alpha(t)))]-g(\alpha(t))\varphi^{-1}(z((t))]\alpha'(t).\quad(1.1.306)\end{aligned}$$

Since $\varphi^{-1}(z(t))\geq\varphi^{-1}(c)>0$, we have

$$\frac{z'(t)}{\varphi^{-1}(z(t))}\geq-\left(f(\alpha(t))\psi[\varphi^{-1}(z(\alpha(t)))]+g(\alpha(t))\right)\alpha'(t).$$

Setting $t=s$ and integrating it from t to $+\infty$, we arrive

$$\Omega(z(t))\leq\Omega(c)+\int_{\alpha(t)}^{+\infty}g(s)ds+\int_{\alpha(t)}^{+\infty}f(s)\psi[\varphi^{-1}z(s)]ds.$$

Let $T\leq T_1$ be an arbitrary number. We denote $p(t)=\Omega(c)+\int_{\alpha(t)}^{+\infty}g(s)ds$. From the above relation, we can deduce that for all $T_1\leq t<+\infty$,

$$\Omega(z(t))\leq p(T_1)+\int_{\alpha(t)}^{+\infty}f(s)\psi[\varphi^{-1}z(s)]ds.$$

Now applying Theorem 1.1.55 gives us, for all $T_1\leq t<+\infty$,

$$z(t)\leq\Omega^{-1}\left(G^{-1}\left[G(p(T_1))+\int_{\alpha(t)}^{+\infty}f(s)ds\right]\right).$$

Therefore, for all $T_1 \leq t < +\infty$,

$$u(t) \leq \varphi^{-1}\left(\Omega^{-1}\left[G^{-1}(G(p(T_1)) + \int_{\alpha(t)}^{+\infty} f(s)ds)\right]\right).$$

Taking $t = T_1$ in the above inequality, since T_1 is arbitrary, we can prove the desired inequality (1.1.305).

If $c = 0$, then we carry out the above procedure with $\varepsilon > 0$ instead of c and subsequently let $\varepsilon \to 0$. □

Theorem 1.1.57 (The Zhao-Meng Inequality [722]) *Let u,f and g be non-negative continuous functions defined on $\mathbb{R}_+$, and let $\varphi \in C(\mathbb{R}_+, \mathbb{R}_+)$ be an increasing function with $\varphi(+\infty) = +\infty$ and let c be a non-negative constant. Moreover, let $\omega_1, \omega_2 \in C(\mathbb{R}_+, \mathbb{R}_+)$ be non-decreasing functions with $\omega_i(u) > 0$ ($i = 1, 2$) on $(0, +\infty)$ and $\alpha \in C^1(\mathbb{R}_+, \mathbb{R}_+)$ be non-decreasing with $\alpha(t) \geq t$ on $\mathbb{R}_+$. If for all $t \in \mathbb{R}_+$,*

$$\varphi(u(t)) \leq c + \int_{\alpha(t)}^{+\infty} f(s)u(s)\omega_1(u(s))ds + \int_t^{+\infty} g(s)\omega_2(u(s))ds, \tag{1.1.307}$$

then for all $0 \leq T \leq t < +\infty$,

(i) for the case $\omega_2(u) \leq \omega_1(u)$,

$$u(t) \leq \varphi^{-1}\left(\Omega^{-1}[G_1^{-1}(G_1(\Omega(c)) + \int_{\alpha(t)}^{+\infty} f(s)ds + \int_t^{+\infty} g(s)ds)]\right), \tag{1.1.308}$$

(ii) for the case $\omega_1(u) \leq \omega_2(u)$,

$$u(t) \leq \varphi^{-1}\left(\Omega^{-1}\left[G_2^{-1}[G_2(\Omega(c)) + \int_{\alpha(t)}^{+\infty} f(s)ds + \int_t^{+\infty} g(s)ds)\right]\right), \tag{1.1.309}$$

where

$$\begin{cases} \Omega(r) = \displaystyle\int_{r_0}^{r} \frac{ds}{\varphi^{-1}(s)}, & r \geq r_0 > 0, \\ G_i(z) = \displaystyle\int_{z_0}^{z} \frac{ds}{\omega_i\{\varphi^{-1}[\Omega^{-1}(s)]\}}, & z \geq z_0 > 0,\ i = 1, 2, \end{cases}$$

and $\Omega^{-1}, \varphi^{-1}, G^{-1}$ *are, respectively, the inverse of* Ω, φ, G, *and* $T \in \mathbb{R}_+$ *is chosen so that for all* $t \in [T, +\infty)$,

$$G_i\left(\Omega(c) + \int_{\alpha(t)}^{+\infty} f(s)ds + \int_t^{+\infty} g(s)ds\right) \in Dom\,(G_i^{-1}),$$

$$G_i^{-1}\left(G_i((c) + \int_{\alpha(t)}^{+\infty} f(s)ds + \int_t^{+\infty} g(s)ds)\right) \in Dom(\Omega^{-1}).$$

Proof Let $c > 0$, define the non-increasing positive function $z(t)$ as

$$z(t) = c + \int_{\alpha(t)}^{+\infty} f(s)\omega_1(u(s))ds + \int_t^{+\infty} g(s)\omega_2(u(s))ds. \tag{1.1.310}$$

From inequality (1.1.307) it follows

$$u(t) \leq \varphi^{-1}[z(t)]. \tag{1.1.311}$$

Differentiating (1.1.310) and using (1.1.311) and the monotonicity of $\varphi^{-1}, \omega_1, \omega_2$, we can deduce

$$\begin{aligned} z'(t) &= -f(\alpha(t))u(\alpha(t))\omega_1[u(\alpha(t))]\alpha'(t) - g(t)u(t)\omega_2[u(t)] \\ &\geq -f(\alpha(t))\varphi^{-1}(z(\alpha(t)))\omega_1[\psi^{-1}(z(\alpha(t)))]\alpha'(t) - g(t)\varphi^{-1}(z(t))\omega_2[\varphi^{-1}(z(t))] \\ &\geq -f(\alpha(t))\varphi^{-1}(z(t))\omega_1[\psi^{-1}(z(t))]\alpha'(t) - g(t)\varphi^{-1}(z(t))\omega_2[\varphi^{-1}(z(t))]. \end{aligned} \tag{1.1.312}$$

(*i*) When $\omega_2(u) \leq \omega_1(u)$,

$$\frac{z'(t)}{\varphi^{-1}(z(t))} \geq -f(\alpha(t))\omega_1[\psi^{-1}(z(t))]\alpha'(t) - g(t)\omega_2[\varphi^{-1}(z(t))], \tag{1.1.313}$$

setting$t = s$ and integrating from t to $+\infty$, we have

$$\Omega(z(t)) \leq \Omega(c) + \int_{\alpha(t)}^{+\infty} f(s)\omega_1[\varphi^{-1}(z(t))]ds + \int_t^{+\infty} g(s)\omega_1[\varphi^{-1}(z(t))]ds.$$

From Theorem 1.1.55, we can conclude, for all $0 \leq T \leq t < +\infty$,

$$z(t) \leq \Omega^{-1}\left(G_1^{-1}\left[G_1\left(\Omega(c) + \int_{\alpha(t)}^{+\infty} f(s)ds + \int_t^{+\infty} g(s)ds\right)\right]\right).$$

Using $u(t) \leq \varphi^{-1}[z(t)]$, we can get the inequality in (1.1.308). If $c = 0$,we can carry out the procedure with $\varepsilon > 0$ instead of c and subsequently let $\varepsilon \to 0$.

(*ii*) When $\omega_1(u) \leq \omega_2(u)$, the proof can be completed similarly. □

Lipovan [355] improved Bihari's results by investigating the following so-called retarded Bellman-Gronwall-like inequalities

$$u(t) \leq a + \int_{b(t_0)}^{b(t)} f(s)w(u(s))ds, \quad t_0 < t < t_1, \tag{1.1.314}$$

and

$$u(t) \leq a + \int_{t_0}^{t} f(s)w(u(s))ds + \int_{b(t_0)}^{b(t)} g(s)w(u(s))ds, \quad t_0 \leq t < t_1. \tag{1.1.315}$$

Their results were further generalized by Agarwal et al. [13] to the inequality

$$v(t) \leq a(t) + \sum_{i=1}^{n} \int_{b_i(t_0)}^{b_i(t)} g_i(t,s)w_i(u(s))ds, \quad t_0 \leq t < t_1, \tag{1.1.316}$$

where the constant a is replaced with a function $a(t)$ and $w_i's$ are continuous and non-decreasing positive functions such that each ratio w_{i+1}/w_i is also non-decreasing.

The following result is to establish some nonlinear retarded inequalities, which extend the results in [306].

First we introduce some notation, $J = [\alpha, \beta]$ is the given subset of $\mathbb{R}$. Denote by $C^i(M,N)$ the class of all i-times continuously differential functions defined on the set M to the set N for $i = 1, 2, \ldots$, and $C^0(M,N) = C(M,N)$.

Theorem 1.1.58 (The Agarwal-Ryoo-Kim Inequality [17]) *Let $u(t)$ and $a(t)$ be non-negative continuous function in $J = [\alpha, \beta]$ and let $f_i(t,s)$, $i = 1, \ldots, n$, be non-negative continuous functions for all $\alpha \leq s \leq t \leq \beta$ which are non-decreasing in t for fixed $s \in J$. Suppose that $\phi \in C^1(J,J)$ is non-decreasing with $\phi(t) \leq t$ on J, $g(u)$ is a non-decreasing continuous function for all $u \in \mathbb{R}_+$ with $g(u) > 0$ for all $u > 0$, and $\varphi \in C(\mathbb{R}_+, \mathbb{R}_+)$ is an increasing function with $\varphi(+\infty) = +\infty$. If, for all $t \in [\alpha, \beta]$,*

$$\varphi(u(t)) \leq a(t) + \int_{\phi(\alpha)}^{\phi(t)} f_1(t,t_1)\Big(\int_{\phi(\alpha)}^{\phi(t_1)} f_2(t_1,t_2)\cdots\Big(\int_{\phi(\alpha)}^{\phi(t_{n-1})} f_n(t_{n-1},t_n)g(u(t_n))dt_n\Big)\cdots\Big)dt_1, \tag{1.1.317}$$

then for all $t \in [\alpha, T_1]$,

$$u(t) \leq \varphi^{-1}\left[G^{-1}\left(G(a(t)) + \sum_{i=1}^{n} \int_{\phi(\alpha)}^{\phi(t)} f_i(t,s)ds\right)\right], \tag{1.1.318}$$

where

$$G(r) = \int_{r_0}^{r} \frac{ds}{s + g(s)}, \; r \geq r_0 > 0, \tag{1.1.319}$$

and G^{-1} denotes the inverse function of G, and $T_1 \in J$ is chosen so that $G(a(t)) + \sum_{i=1}^{n} \int_{\phi(\alpha)}^{\phi(t)} f_i(t,s)ds \in Dom\ (G^{-1})$.

Proof Let us first assume that $a(t) > 0$. Fix $T \in (\alpha, \beta]$. For all $\alpha \le t \le T$, we derive from (1.1.317)

$$\phi(u(t)) \le a(T) + \int_{\phi(\alpha)}^{\phi(t)} f_1(T,t_1)\Big(\int_{\phi(\alpha)}^{\phi(t_1)} f_2(T,t_2)\cdots\Big(\int_{\phi(\alpha)}^{\phi(t_{n-1})} f_n(T,t_n)g(u(t_n))dt_n\Big)\cdots\Big)dt_1. \tag{1.1.320}$$

Now we introduce the functions

$$\begin{aligned} m_1(t) &= a(T) + \int_{\phi(\alpha)}^{\phi(t)} f_1(T,t_1)\Big(\int_{\phi(\alpha)}^{\phi(t_1)} f_2(T,t_2)\cdots\Big(\int_{\phi(\alpha)}^{\phi(t_{n-1})} f_n(T,t_n)g(u(t_n))dt_n\Big)\cdots\Big)dt_1,\\ m_k(t) &= m_{n-1}(t) + \int_{\phi(\alpha)}^{\phi(t)} f_k(T,t_k)\Big(\int_{\phi(\alpha)}^{\phi(t_1)} f_{k+1}(T,t_{k+1})\cdots\\ &\quad\times\Big(\int_{\phi(\alpha)}^{\phi(t_{n-1})} f_n(T,t_n)g(m_{k-1}(t_n))dt_n\Big)\cdots\Big)dt_k, \end{aligned} \tag{1.1.321}$$

for all $t \in [\alpha, T]$ and $k = 2, \ldots, n$. Then we have $m_k(\alpha) = a(T)$ for $k = 1, \ldots, n$, and $m_1(t) \le m_2(t) \le \cdots \le m_n(t)$, $t \in [\alpha, T]$. From the inequality (1.1.320), we obtain $u(t) \le \varphi^{-1}(m_1(t))$, or $u(t) \le \varphi^{-1}(m_n(t))$ for all $t \in [\alpha, T]$. Moreover, the function $m_1(t)$ is non-decreasing. Differentiating $m_1(t)$, we may get

$$\begin{aligned} m_1'(t) &= f_1(T,\phi(t))\left[\int_{\phi(\alpha)}^{\phi(t)} f_2(T,t_2)\cdots\Big(\int_{\phi(\alpha)}^{\phi(t_{n-1})} f_n(T,t_n)g(u(t_n))dt_n)\cdots\Big)dt_2\right]\phi'(t)\\ &\le [-f_1(T,\phi(t))\phi'(t)m_1(t) + f_1(T,\phi(t))\phi'(t)m_2(t)]. \end{aligned} \tag{1.1.322}$$

Thus induction with respect to k gives us for all $t \in [\alpha, T]$, $k = 1, 2, \ldots, n-1$,

$$m_k'(t) \le \left(\sum_{i=1}^{k-1} f_i(T,\phi(t)) - f_k(T,\phi(t)))\phi'(t)m_k(t) + f_k(T,\phi(t)\right)\phi'(t)m_{k+1}(t). \tag{1.1.323}$$

From the definition of the function $m_n(t)$ and inequality (1.1.323), we have

$$\begin{aligned} m_n'(t) &= m_{n-1}'(t) + f_n(T,\phi(t))g(m_{n-1}(\phi(t)))\phi'(t)\\ &\le \left[\left(\sum_{i=1}^{n-2} f_i(T,\phi(t))\right)m_{n-1}(t) + f_{n-1}(T,\phi(t))m_n(t) + f_n(T,\phi(t))g(m_n(t))\right]\phi'(t)\\ &\le \left[\left(\sum_{i=1}^{n-1} f_i(T,\phi(t))\right)m_n(t) + f_n(T,\phi(t))g(m_n(t))\right]\phi'(t)\\ &\le \sum_{i=1}^{n} f_i(T,\phi(t))\phi'(t)\left(m_n(t) + g(m_n(t))\right), \end{aligned} \tag{1.1.324}$$

that is,

$$\frac{m_n'(t)}{m_n(t) + g(m_n(t))} \leq \sum_{i=1}^{n} f_i(T, \phi(t))\phi'(t). \tag{1.1.325}$$

Taking $t = s$ in (1.1.325) and then integrating it from α to any $t \in [\alpha, \beta]$, changing the variable and using the definition of the function G, we conclude for all $\alpha \leq t \leq T \leq \beta$,

$$G(m_n(t)) \leq G(m_n(\alpha)) + \sum_{i=1}^{n} \int_{\phi(\alpha)}^{\phi(t)} f_i(T, s)ds, \tag{1.1.326}$$

or

$$m_n(t) \leq G^{-1}\left(G(m_n(\alpha)) + \sum_{i=1}^{n} \int_{\phi(\alpha)}^{\phi(t)} f_i(T, s)ds\right). \tag{1.1.327}$$

Now a combination of $u(t) \leq \phi^{-1}(m_n(t))$ and the last inequality gives us the required inequality in (1.1.318) for $T = t$. If $a(t) = 0$, we replace $a(t)$ by some $\epsilon > 0$ and subsequently let $\epsilon \to 0$. This completes the proof. □

For the special case $g(u) = u^p$ ($p > 0$ is a constant), Theorem 1.1.58 gives us the following retarded integral inequality for iterated integrals.

Corollary 1.1.14 (The Agarwal-Ryoo-Kim Inequality [17]) *Let $u(t)$, $a(t)f_i(t)$, $\phi(t)$ and $\varphi(u)$ be as in Theorem 1.1.58, and let $p > 0$ be a constant. Suppose that, for all $t \in [\alpha, \beta]$,*

$$\varphi(u(t)) \leq a(t) + \int_{\phi(\alpha)}^{\phi(t)} f_1(t, t_1)\left(\int_{\phi(\alpha)}^{\phi(t_1)} f_2(t_1, t_2) \cdots \left(\int_{\phi(\alpha)}^{\phi(t_{n-1})} f_n(t_{n-1}, t_n)u^p(t_n)dt_n\right)\cdots\right)dt_1. \tag{1.1.328}$$

Then for any $t \in [\alpha, T_1]$,

$$u(t) \leq \varphi^{-1}\left[G_1^{-1}\left(G_1(a(t)) + \sum_{i=1}^{n} \int_{\phi(\alpha)}^{\phi(t)} f_i(t, s)ds\right)\right], \tag{1.1.329}$$

where

$$G_1(r) = \int_{r_0}^{r} \frac{ds}{s + s^p}, \quad r \geq r_0 > 0, \tag{1.1.330}$$

and G_1^{-1} denotes the inverse function of G_1, and $T_1 \in J$ is chosen so that $G_1(a(t)) + \sum_{i=1}^{n} \int_{\phi(\alpha)}^{\phi(t)} f_i(t, s)ds \in Dom\ (G_1^{-1})$.

Remark 1.1.15

(i) When $\varphi(u) = u$ and $g(u) = u$, from Theorem 1.1.58, we may deduce the following retarded integral inequality:

$$u(t) \leq a(t) \exp\left(2\sum_{i=1}^{n}\int_{\phi(\alpha)}^{\phi(t)} f_i(t,s)ds\right). \tag{1.1.331}$$

(ii) When $\varphi(u) = u$ in Theorem 1.1.58, we may obtain the following retarded integral inequality:

$$u(t) \leq G^{-1}\left(G(a(t)) + \sum_{i=1}^{n}\int_{\phi(\alpha)}^{\phi(t)} f_i(t,s)ds\right). \tag{1.1.332}$$

(iii) When $\varphi(u) = u^p$ ($p > 0$ is a constant) in Theorem 1.1.58. we may derive the following retarded integral inequality:

$$u(t) \leq \left[G^{-1}\left(G(a(t)) + \sum_{i=1}^{n}\int_{\phi(\alpha)}^{\phi(t)} f_i(t,s)ds\right)\right]^{\frac{1}{p}}. \tag{1.1.333}$$

Now we introduce the following notation. For $\alpha < \beta$, let $J_i = \{(t_1, t_2, \ldots, t_i) \in \mathbb{R}^i : \alpha \leq t_i \leq \ldots \leq t_1 \leq \beta\}$ for $i = 1, \ldots, n$.

Theorem 1.1.59 (The Agarwal-Ryoo-Kim Inequality [17]) *Let $u(t)$ and $a(t)$ be non-negative continuous functions in $J = [\alpha, \beta]$ with $a(t)$ non-decreasing in J, and let $p_i(t)$, $i = 1, \ldots, n$, be non-negative continuous functions for all $\alpha \leq t \leq \beta$. Suppose that $\phi \in C^1(J, J)$ is non-decreasing with $\phi(t) \leq t$ on J, $g(u)$ is a non-decreasing continuous function for all $u \in \mathbb{R}_+$ with $g(u) > 0$ for all $u > 0$, and $\varphi \in C(\mathbb{R}_+, \mathbb{R}_+)$ is an increasing function with $\varphi(+\infty) = +\infty$. If, for any $t \in J$,*

$$\begin{aligned}\varphi(u(t)) \leq\ & a(t) + \int_{\phi(\alpha)}^{\phi(t)} p_1(t_1)g(u(t_1))dt_1 \\ & + \sum_{i=2}^{n}\int_{\phi(\alpha)}^{\phi(t)} p_1(t_1)\Big(\int_{\phi(\alpha)}^{\phi(t_1)} p_2(t_2)\Big(\cdots\Big(\int_{\phi(\alpha)}^{\phi(t_{i-2})} p_{i-1}(t_{i-1}) \\ & \times\Big(\int_{\phi(\alpha)}^{\phi(t_{i-1})} p_i(t_i)g(u(t_i))dt_i\Big)dt_{i-1}\Big)\cdots\Big)dt_2\Big)dt_1,\end{aligned} \tag{1.1.334}$$

then for all $t \in [\alpha, T_2]$,

$$u(t) \leq \varphi^{-1}[G^{-1}(G(a(t)) + F(t))] \tag{1.1.335}$$

where $T_2 \in I$ *is chosen so that* $G(a(t)) + F(t) \in Dom\ (G^{-1})$,

$$G(r) = \int_{r_0}^{r} \frac{ds}{g(\varphi^{-1}(s))}, \quad r \geq r_0 > 0, \tag{1.1.336}$$

and G^{-1} *denotes the inverse function of* G, *and for all* $t \in I$,

$$F(t) = \int_{\phi(\alpha)}^{\phi(t)} p_1(t_1)dt_1 + \sum_{i=2}^{n} \int_{\phi(\alpha)}^{\phi(t)} p_1(t_1)\Big(\int_{\phi(\alpha)}^{\phi(t_1)} p_2(t_2)\Big(\cdots\Big(\int_{\phi(\alpha)}^{\phi(t_{i-2})} p_{i-1}(t_{i-1}) \times\Big(\int_{\phi(\alpha)}^{\phi(t_{i-1})} p_i(t_i)dt_i\Big)dt_{i-1}\Big)\cdots\Big)dt_2\Big)dt_1. \tag{1.1.337}$$

Proof Assume the function $a(t)$ is positive. Define a function $v(t)$ by the right-hand side of (1.1.334). Clearly, $v(t)$ is non-decreasing continuous, $u(t) \leq \phi^{-1}(v(t))$ for all $t \in I$ and $v(\alpha) = a(\alpha)$. Differentiating $v(t)$ and rewriting, we arrive at

$$\frac{v'(t) - a'(t)}{\phi'(t)p_1(\phi(t))} - g(u(\phi(t))) \leq v_1(t), \tag{1.1.338}$$

where

$$v_1(t) = \int_{\phi(\alpha)}^{\phi(t)} p_2(t_2)g(u(t_2))dt_2 + \sum_{i=3}^{n} \int_{\phi(\alpha)}^{\phi(t)} p_2(t_2)\Big(\int_{\phi(\alpha)}^{\phi(t_2)} p_3(t_3)\Big(\cdots\Big(\int_{\phi(\alpha)}^{\phi(t_{i-2})} p_{i-1}(t_{i-1}) \times\Big(\int_{\phi(\alpha)}^{\phi(t_{i-1})} p_i(t_i)g(u(t_i))dt_i\Big)dt_{i-1}\Big)\cdots\Big)dt_3\Big)dt_2. \tag{1.1.339}$$

Now differentiating $v_1(t)$ and rewriting, we may get

$$\frac{v_1'(t)}{\phi'(t)p_2(\phi(t))} - g(u(\phi(t))) \leq v_2(t), \tag{1.1.340}$$

$$v_2(t) = \int_{\phi(\alpha)}^{\phi(t)} p_3(t_3)g(u(t_3))dt_3 + \sum_{i=4}^{n} \int_{\phi(\alpha)}^{\phi(t)} p_3(t_3)\Big(\int_{\phi(\alpha)}^{\phi(t_3)} p_4(t_4)\Big(\cdots\Big(\int_{\phi(\alpha)}^{\phi(t_{i-2})} p_{i-1}(t_{i-1}) \times\Big(\int_{\phi(\alpha)}^{\phi(t_{i-1})} p_i(t_i)g(u(t_i))dt_i\Big)dt_{i-1}\Big)\cdots\Big)dt_4\Big)dt_3. \tag{1.1.341}$$

Continuing in the same way, we may obtain

$$\frac{v_{n-2}'(t)}{\phi'(t)p_{n-1}(\phi(t))} - g(u(\phi(t))) \leq v_{n-1}(t), \tag{1.1.342}$$

where

$$v_{n-1}(t) = \int_{\phi(\alpha)}^{\phi(t)} p_n(t_n)g(u(t_n))dt_n. \tag{1.1.343}$$

From the definition of $v_{n-1}(t)$ and the inequality $u(t) \leq \varphi^{-1}(v(t))$, we may find

$$\frac{v'_{n-1}(t)}{g(\varphi^{-1}(v(t)))} \leq \phi'(t)p_n(\phi(t)). \tag{1.1.344}$$

Integrating the inequality (1.1.344), we get

$$\int_\alpha^t \frac{v'_{n-1}(s)}{g(\varphi^{-1}(v(s)))}ds \leq \int_{\phi(\alpha)}^{\phi(t)} p_n(s)ds. \tag{1.1.345}$$

Now integrating by parts the left-hand side of (1.1.345), we may obtain

$$\begin{aligned}\int_\alpha^t \frac{v'_{n-1}(s)}{g(\varphi^{-1}(v(s)))}ds &= \frac{v_{n-1}(t)}{g(\varphi^{-1}(v(t)))} + \int_\alpha^t \frac{v_{n-1}g'(\phi^{-1}(v))}{g^2(\varphi^{-1}(v))} \times \frac{v'}{\varphi'[\varphi^{-1}(v)]}ds \\ &\geq \frac{v_{n-1}(t)}{g(\varphi^{-1}(v(t)))}.\end{aligned} \tag{1.1.346}$$

Thus from the inequality (1.1.345) and (1.1.346), we may derive

$$\frac{v_{n-1}(t)}{g(\varphi^{-1}(v(t)))} \leq \int_{\phi(\alpha)}^{\phi(t)} p_n(s)ds. \tag{1.1.347}$$

Next, from the inequality (1.1.342), we observe that

$$v'_{n-1}(t) \leq \phi'(t)p_{n-1}(\phi(t))g(u(\phi(t))) + \phi'(t)p_{n-1}(\phi(t))v_{n-1}(t). \tag{1.1.348}$$

Thus it follows that

$$\begin{aligned}\frac{v'_{n-2}(t)}{g(\varphi^{-1}(v(t)))} &\leq \phi'(t)p_{n-1}(\phi(t))\frac{g(u(\phi(t)))}{g(\varphi^{-1}(v(t)))} + \phi'(t)p_{n-1}(\phi(t))\frac{v_{n-1}(t)}{g(\varphi^{-1}(v(t)))} \\ &\leq \phi'(t)p_{n-1}(\phi(t)) + \phi'(t)p_{n-1}(\phi(t))\frac{v_{n-1}(t)}{g(\phi^{-1}(v(t)))}.\end{aligned} \tag{1.1.349}$$

Using the same procedure from (1.1.345) to (1.1.347) to the inequality (1.1.349), we may get

$$\frac{v_{n-2}(t)}{g(\phi^{-1}(v(t)))} \leq \int_{\phi(\alpha)}^{\phi(t)} p_{n-1}(t_1)dt_1 + \int_{\phi(\alpha)}^{\phi(t)} p_{n-1}(t_1)\frac{v_{n-1}(t_1)}{g(\phi^{-1}(v(t_1)))}dt_1. \tag{1.1.350}$$

Now combining the inequalities (1.1.347) and (1.1.350), we may conclude

$$\frac{v_{n-2}(t)}{g(\varphi^{-1}(v(t)))} \le \int_{\phi(\alpha)}^{\phi(t)} p_{n-1}(t_1)dt_1 + \int_{\phi(\alpha)}^{\phi(t)} p_{n-1}(t_1) \int_{\phi(\alpha)}^{\phi(t_1)} p_n(t_2)dt_2dt_1. \quad (1.1.351)$$

Proceeding in this way, we may conclude

$$\frac{v_1(t)}{g(\varphi^{-1}(v(t)))} \le \int_{\phi(\alpha)}^{\phi(t)} p_2(t_1)dt_1 + \cdots$$
$$+ \int_{\phi(\alpha)}^{\phi(t)} p_2(t_1)\Big(\int_{\phi(\alpha)}^{\phi(t_1)} p_3(t_2)\Big(\cdots \int_{\phi(\alpha)}^{\phi(t_{n-2}} p_n(t_{n-1})dt_{n-1}\cdots\Big)dt_2\Big)dt_1. \quad (1.1.352)$$

On the other hand, from the inequality (1.1.338), we may get

$$v'(t) - a'(t) \le \phi'(t)p_1(\phi(t))g(u(\phi(t))) + \phi'(t)p_1(\phi(t))v_1(t), \quad (1.1.353)$$

or

$$\frac{v'(t) - a'(t)}{g(\varphi^{-1}(v(t)))} \le \phi'(t)p_1(\phi(t))\frac{g(u(\phi(t)))}{g(\phi^{-1}(v(t)))} + \phi'(t)p_1(\phi(t))\frac{v_1(t)}{g(\varphi^{-1}(v(t)))}$$
$$\le \phi'(t)p_1(\phi(t)) + \phi'(t)p_1(\phi(t))\frac{v_1(t)}{g(\varphi^{-1}(v(t)))} \quad (1.1.354)$$

that is,

$$\frac{v'(t)}{g(\varphi^{-1}(v(t)))} - \frac{a'(t)}{g(\varphi^{-1}(a(t)))} \le \phi'(t)p_1(\phi(t)) + \phi'(t)p_1(\varphi(t))\frac{v_1(t)}{g(\varphi^{-1}(v(t)))}. \quad (1.1.355)$$

Setting $t = t_1$, and integrating from α to t, and using the definition of G, we may obtain

$$G(v(t)) \le G\left(a(t) + \int_{\phi(\alpha)}^{\phi(t)} p_1(t_1)dt_1 + \int_{\phi(\alpha)}^{\phi(t)} p_1(t_1)\frac{v_1(t_1)}{g(\phi^{-1}(v(t_1)))}dt_1\right). \quad (1.1.356)$$

Consequently, using (1.1.352) to the inequality (1.1.356), we may conclude

$$v(t) \le G^{-1}[G(a(t)) + F(t)], \quad (1.1.357)$$

where the function $F(t)$ is defined in (1.1.337). Now, the desired inequality in (1.1.340) follows by the inequality $u(t) \le \phi^{-1}(v(t))$. If $a(t) = 0$, we replace $a(t)$ by some $\epsilon > 0$ and subsequently let $\epsilon \to 0^+$. This completes the proof. □

For the special case $\phi(u) = u^p$ ($p > 1$ is a constant), Theorem 1.1.59 gives us the following retarded integral inequality for iterated integrals.

Corollary 1.1.15 (The Agarwal-Ryoo-Kim Inequality [17]) *Let $u(t)$, $a(t)\,p_i(t)$, $\phi(t)$ and $g(u)$ be as in Theorem 1.1.59, and let $p > 0$ be a constant. If, for any $t \in J$,*

$$\begin{aligned} u^p(t) \le a(t) &+ \int_{\phi(\alpha)}^{\phi(t)} p_1(t_1)g(u(t_1))dt_1 \\ &+ \sum_{i=2}^{n} \int_{\phi(\alpha)}^{\phi(t)} p_1(t_1)\bigg(\int_{\phi(\alpha)}^{\phi(t_1)} p_2(t_2)\bigg(\cdots\bigg(\int_{\phi(\alpha)}^{\phi(t_{i-2})} p_{i-1}(t_{i-1}) \\ &\times\bigg(\int_{\phi(\alpha)}^{\phi(t_{i-1})} p_i(t_i)g(u(t_i))dt_i\bigg)dt_{i-1}\bigg)\cdots\bigg)dt_2\bigg)dt_1, \end{aligned} \tag{1.1.358}$$

then, for all $t \in [\alpha, T_3]$,

$$u(t) \le [G^{-1}(G(a(t)) + F(t))]^{\frac{1}{p}} \tag{1.1.359}$$

where $T_3 \in I$ is chosen so that $G_1(a(t)) + F(t) \in Dom\,(G_1^{-1})$, $G_1(r) = \int_{r_0}^{r} \frac{ds}{g(v^{\frac{1}{p}}(s))}$, $r \ge r_0 > 0$, and G^{-1} denotes the inverse function of G, and the function $F(t)$ is defined in (1.1.336) for any $t \in I$.

Theorem 1.1.60 (The Agarwal-Ryoo-Kim Inequality [17]) *Let $u(t)$ and $a(t)$ be non-negative continuous functions in $J = [\alpha, \beta]$ with $a(t)$ non-decreasing in J, and let $f_i(t)$, $p_i(t)$, $i = 1, \ldots, n$, be non-negative continuous functions for all $\alpha \le t \le \beta$. Suppose that $\phi \in C^1(J, J)$ is non-decreasing with $\phi(t) \le t$ on J, $g(u)$ is a non-decreasing continuous function for all $u \in \mathbb{R}_+$ with $g(u) > 0$ for all $u > 0$, and $\varphi \in C(\mathbb{R}_+, \mathbb{R}_+)$ is an increasing function with $\varphi(+\infty) = +\infty$. If, for all $t \in J$,*

$$\begin{aligned} \varphi(u(t)) \le a(t) &+ \int_{\phi(\alpha)}^{\phi(t)} p_1(t_1)f_1(t_1)u(t_1)g(u(t_1))dt_1 \\ &+ \sum_{i=2}^{n} \int_{\phi(\alpha)}^{\phi(t)} p_1(t_1)\bigg(\int_{\phi(\alpha)}^{\phi(t_1)} p_2(t_2)\bigg(\cdots\bigg(\int_{\phi(\alpha)}^{\phi(t_{i-2})} p_{i-1}(t_{i-1}) \\ &\times\bigg(\int_{\phi(\alpha)}^{\phi(t_{i-1})} p_i(t_i)f_i(t_i)u(t_i)g(u(t_i))dt_i\bigg)dt_{i-1}\bigg)\cdots\bigg)dt_2\bigg)dt_1, \end{aligned} \tag{1.1.360}$$

then for all $t \in [\alpha, T_4]$,

$$u(t) \le \varphi^{-1}\left(\Phi^{-1}[G_2^{-1}(G_2[\Phi(a(t))] + F(t))]\right) \tag{1.1.361}$$

where $T_4 \in I$ *is chosen so that* $G_2[\Phi(a(t))]+F_1(t)\in Dom\ (G_2^{-1})$, $G_2^{-1}(G_2[\Phi(a(t))]+F_1(t)) \in Dom\ (\Phi^{-1})$,

$$G_2(r) = \int_{r_0}^{r} \frac{ds}{g(\varphi^{-1}(\Phi^{-1}(s)))}, \quad \Phi(r) = \int_{r_0}^{r} \frac{ds}{\varphi^{-1}(s)}, \ r \geq r_0 > 0, \quad (1.1.362)$$

and G_2^{-1} *denotes the inverse function of* G_2, *for any* $t \in I$,

$$F_1(t) = \int_{\phi(\alpha)}^{\phi(t)} p_1(t_1)F_1(t_1)dt_1 + \sum_{i=2}^{n} \int_{\phi(\alpha)}^{\phi(t)} p_1(t_1)\bigg(\int_{\phi(\alpha)}^{\phi(t_1)} p_2(t_2)\bigg(\cdots\bigg(\int_{\phi(\alpha)}^{\phi(t_{i-2})} p_{i-1}(t_{i-1})$$
$$\times\bigg(\int_{\phi(\alpha)}^{\phi(t_{i-1})} p_i(t_i)F_i(t_i)dt_i\bigg)dt_{i-1}\bigg)\cdots\bigg)dt_2\bigg)dt_1. \quad (1.1.363)$$

Proof Let the function $a(t)$ be positive. Define a function $w(t)$ by the right-hand side of (1.1.360). Clearly, $w(t)$ is non-decreasing continuous, $u(t) \leq \phi^{-1}(w(t))$ for all $t \in I$ and $w(\alpha) = a(\alpha)$. Differentiating $w(t)$ and rewriting, we may have

$$\frac{w'(t) - a'(t)}{\phi'(t)p_1(\phi(t))} - f_1(\phi(t))u(\phi(t))g(u(\phi(t))) \leq w_1(t), \quad (1.1.364)$$

where

$$w_1(t) = \int_{\phi(\alpha)}^{\phi(t)} p_2(t_2)f_2(t_2)u(t_2)g(u(t_2))dt_2$$
$$+\sum_{i=3}^{n} \int_{\phi(\alpha)}^{\phi(t)} p_2(t_2)\bigg(\int_{\phi(\alpha)}^{\phi(t_2)} p_3(t_3)\bigg(\cdots\bigg(\int_{\phi(\alpha)}^{\phi(t_{i-2})} p_{i-1}(t_{i-1})$$
$$\times\bigg(\int_{\phi(\alpha)}^{\phi(t_{i-1})} p_i(t_i)f_i(t_i)u(t_i)g(u(t_i))dt_i\bigg)dt_{i-1}\bigg)\cdots\bigg)dt_3\bigg)dt_2. \quad (1.1.365)$$

Now differentiating the $w_1(t)$ and rewriting, we get

$$\frac{w_1'(t)}{\phi'(t)p_2(\phi(t))} - f_2(\phi(t))u(\phi(t))g(u(\phi(t))) \leq w_2(t), \quad (1.1.366)$$

where

$$w_2(t) = \int_{\phi(\alpha)}^{\phi(t)} p_3(t_3)g(u(t_3))dt_3$$
$$+\sum_{i=4}^{n} \int_{\phi(\alpha)}^{\phi(t)} p_3(t_3)\bigg(\int_{\phi(\alpha)}^{\phi(t_3)} p_4(t_4)\bigg(\cdots\bigg(\int_{\phi(\alpha)}^{\phi(t_{i-2})} p_{i-1}(t_{i-1})$$
$$\times\bigg(\int_{\phi(\alpha)}^{\phi(t_{i-1})} p_i(t_i)f_i(t_i)u(t_i)g(u(t_i))dt_i\bigg)dt_{i-1}\bigg)\cdots\bigg)dt_4\bigg)dt_3.$$
$$(1.1.367)$$

Continuing in this way, we may obtain

$$\frac{w'_{n-2}(t)}{\phi'(t)p_{n-1}(\phi(t))} - f_{n-1}(\phi(t))u(\phi(t))g(u(\phi(t))) \leq w_{n-1}(t), \tag{1.1.368}$$

where

$$w_{n-1}(t) = \int_{\phi(\alpha)}^{\phi(t)} p_n(t_n)f_n(t_n)u(t_n)g(u(t_n))dt_n. \tag{1.1.369}$$

From the definition of $w_{n-1}(t)$ and the inequality $u(t) \leq \phi^{-1}(w(t))$, we may find

$$\frac{w'_{n-1}(t)}{\varphi^{-1}(w(t))} \leq \phi'(t)p_n(\phi(t))f_n(\phi(t))g(\phi^{-1}(w(\phi(t)))). \tag{1.1.370}$$

Integrating the inequality (1.1.370), we can get

$$\int_{\alpha}^{t} \frac{w'_{n-1}(s)}{\varphi^{-1}(w(s))}ds \leq \int_{\phi(\alpha)}^{\phi(t)} p_n(s)f_n(s)g(\varphi^{-1}(w(s)))ds. \tag{1.1.371}$$

Now integrating by parts on the left-hand side of (1.1.371), we can obtain

$$\begin{aligned}\int_{\alpha}^{t} \frac{w'_{n-1}(s)}{\varphi^{-1}(w(s))}ds &= \frac{w_{n-1}(t)}{\varphi^{-1}(v(t))} + \int_{\alpha}^{t} \frac{w_{n-1}w'}{(\varphi^{-1}(w))^2\varphi'(\varphi^{-1}(s))}ds \\ &\geq \frac{w_{n-1}(t)}{\phi^{-1}((t))}.\end{aligned} \tag{1.1.372}$$

From the inequality (1.1.371) and (1.1.372), we have

$$\frac{w_{n-1}(t)}{\varphi^{-1}(v(t))} \leq \int_{\phi(\alpha)}^{\phi(t)} p_n(s)f_n(s)g(\varphi^{-1}(w(s)))ds. \tag{1.1.373}$$

Next, from the inequality (1.1.368), we may derive

$$w'_{n-2}(t) \leq \phi'(t)p_{n-1}(\phi(t))w_{n-1}(t) + \phi'(t)p_{n-1}(\phi(t))f_{n-1}(\phi(t))\phi^{-1}(w(t))g(\phi^{-1}(w(t))). \tag{1.1.374}$$

Also from the inequality (1.1.374), it follows that

$$\frac{w'_{n-2}(t)}{\varphi^{-1}(w(t))} \leq \phi'(t)p_{n-1}(\phi(t))\frac{w_{n-1}(t)}{\varphi^{-1}(w(t))} + \phi'(t)p_{n-1}(\phi(t))f_{n-1}(\phi(t))g(\varphi^{-1}(w(t)). \tag{1.1.375}$$

Using the same procedure from (1.1.371)–(1.1.373) to the inequality (1.1.375), we may conclude

$$\frac{w_{n-2}(t)}{\varphi^{-1}(w(t))} \leq \int_{\phi(\alpha)}^{\phi(t)} p_{n-1}(t_1)\frac{w_{n-1}(t_1)}{\phi^{-1}(w(t_1))}dt_1 + \int_{\phi(\alpha)}^{\phi(t)} p_{n-1}(t_1)f_{n-1}(t_1)g(\varphi^{-1}(w(t_1)))dt_1. \tag{1.1.376}$$

Next, using (1.1.374) in the inequality(1.1.376), we can get

$$\frac{w_{n-2}(t)}{\phi^{-1}(w(t))} \leq \int_{\phi(\alpha)}^{\phi(t)} p_{n-1}(t_1)\int_{\phi(\alpha)}^{\phi(t_1)} p_n(s)f_n(s)g(\varphi^{-1}(w(s)))dsdt_1 + \int_{\phi(\alpha)}^{\phi(t)} p_{n-1}(t_1)f_{n-1}(t_1)g(\varphi^{-1}(w(t_1)))dt_1. \tag{1.1.377}$$

Proceeding in this way, we arrive at

$$\frac{w_1(t)}{\varphi^{-1}(w(t))} \leq \int_{\phi(\alpha)}^{\phi(t)} p_2(t_1)f_2(t_1)g(\varphi^{-1}(w(t_1)))dt_1 + \cdots + \int_{\phi(\alpha)}^{\phi(t)} p_2(t_1)\Big(\cdots\int_{\phi(\alpha)}^{\phi(t_{n-2})} p_n(t_s)f_n(t_s)g(\varphi^{-1}(w(t_s)))ds\cdots\Big)dt_1. \tag{1.1.378}$$

On the other hand, from the inequality (1.1.364), we may derive

$$w'(t) - a'(t) \leq \phi'(t)p_1(\phi(t))f_1(\phi(t))\varphi^{-1}(w(t))g(\varphi^{-1}w(t)) + \phi'(t)p_1(\phi(t))w_1(t), \tag{1.1.379}$$

or

$$\frac{w'(t) - a'(t)}{\varphi^{-1}(w(t))} \leq \phi'(t)p_1(\phi(t))\frac{w_1(t)}{\varphi^{-1}(w(t))} + \phi'(t)p_1(\phi(t))f_1(\phi(t))g(\varphi^{-1}(w(t))). \tag{1.1.380}$$

Now the left-hand side of the inequality (1.1.380) implies that

$$\frac{w'(t)}{\varphi^{-1}(w(t))} - \frac{a'(t)}{\varphi^{-1}(a(t))} \leq \frac{w'(t) - a'(t)}{\varphi^{-1}(w(t))}. \tag{1.1.381}$$

In the inequalities (1.1.380) and (1.1.381), setting $t = t_1$, integrating from α to t, and using the definition of Φ, we can obtain

$$\Phi(w(t)) \leq \Phi(a(t)) + \int_{\phi(\alpha)}^{\phi(t)} p_1(t_1)\frac{w_1(t_1)}{\varphi^{-1}(w_1(t_1))}dt_1 + \int_{\phi(\alpha)}^{\phi(t)} p_1(t_1)f_1(t_1)g(\varphi^{-1}(w(t_1)))dt_1. \tag{1.1.382}$$

Consequently, from the inequality (1.1.378) and (1.1.382), we can conclude

$$w(t) \leq \Phi^{-1}[k(t)], \tag{1.1.383}$$

where the function $k(t)$ is defined in by, for some fixed T, $t \leq T \leq \beta$,

$$\begin{aligned} k(t) &= \Phi(a(T)) + \int_{\phi(\alpha)}^{\phi(t)} p_1(t_1)f_1(t_1)g(\varphi^{-1}(w(t_1))))dt_1 \\ &\quad + \sum_{i=2}^{n} \int_{\phi(\alpha)}^{\phi(t)} p_1(t_1)\Bigg(\int_{\phi(\alpha)}^{\phi(t_1)} p_2(t_2) \\ &\quad \times \Bigg(\cdots \Bigg(\int_{\phi(\alpha)}^{\phi(t_{i-1})} p_i(t_i)f_i(t_i)g(\varphi^{-1}(w(t_i)))dt_i \Bigg) \cdots \Bigg) dt_2 \Bigg) dt_1. \end{aligned} \tag{1.1.384}$$

Clearly, $k(t)$ is a non-decreasing continuous function and $k(\alpha) = \Phi(a(T))$. Differentiating $k(t)$ and rewriting, we can get

$$\frac{k'(t)}{\phi'(t)p_1(\phi(t))} - f_1(\phi(t))g(\varphi^{-1}(w(\phi(t)))) \leq k_1(t), \tag{1.1.385}$$

where

$$\begin{aligned} k_1(t) &= \int_{\phi(\alpha)}^{\phi(t)} p_2(t_2)f_2(t_2)g(\varphi^{-1}(w(t_2)))dt_2] \\ &\quad + \sum_{i=3}^{n} \int_{\phi(\alpha)}^{\phi(t)} p_2(t_2)\Bigg(\int_{\phi(\alpha)}^{\phi(t_2)} p_3(t_3)\Bigg(\cdots \Bigg(\int_{\phi(\alpha)}^{\phi(t_{i-2})} p_{i-1}(t_{i-1}) \\ &\quad \times \Bigg(\int_{\phi(\alpha)}^{\phi(t_{i-1})} p_i(t_i)f_i(t_i)g(\varphi^{-1}(w(t_i)))dt_i \Bigg) dt_{i-1} \Bigg) \cdots \Bigg) dt_3 \Bigg) dt_2. \end{aligned} \tag{1.1.386}$$

Using the same procedure from (1.1.366)–(1.1.380) to the equality (1.1.386), we can conclude

$$\begin{aligned} \frac{k_1(t)}{g(\varphi^{-1}(w(t)))} &\leq \int_{\phi(\alpha)}^{\phi(t)} p_2(t_2)f_2(t_2)dt_2 \\ &\quad + \sum_{i=3}^{n} \int_{\phi(\alpha)}^{\phi(t)} p_2(t_2)\Bigg(\int_{\phi(\alpha)}^{\phi(t_2)} p_3(t_3)\Bigg(\cdots \Bigg(\int_{\phi(\alpha)}^{\phi(t_{i-1})} p_i(t_i)f_i(t_i)dt_i \Bigg) \cdots \Bigg) dt_3 \Bigg) dt_2, \end{aligned} \tag{1.1.387}$$

$$\frac{k'(t)}{g(\varphi^{-1}(\Phi^{-1}(k(t))))} \leq \phi'(t)p_1(\phi(t))\frac{k_1(t)}{\varphi^{-1}(w(t))} + \phi'(t)p_1(\phi(t))f_1(\phi(t)). \tag{1.1.388}$$

In the inequality (1.1.388), setting $t = s$ and integrating from α to t, using the definition of G_2, we can obtain

$$G_2(k(t)) \le G_2(k(\alpha)) + \int_{\phi(\alpha)}^{\phi(t)} p_1(s)\frac{k_1(s)}{g(\varphi^{-1}(w(s)))}ds + \int_{\phi(\alpha)}^{\phi(t)} p_1(s)f_1(s)ds. \quad (1.1.389)$$

Finally from the inequalities (1.1.387) and (1.1.389), we may get

$$k(t) \le G_2^{-1}[G_2(\Phi(a(t))) + F_1(t)], \quad (1.1.390)$$

where the function $F_1(t)$ is defined in (1.1.363). In particular, for $T = t$, we can conclude that the desired inequality (1.1.363) follows from by the inequalities $u(t) \le \phi^{-1}(w(t))$ and $w(t) \le \Phi^{-1}(k(t))$. This thus completes the proof. □

1.1.2 Nonlinear One-Dimensional Bellman-Gronwall Inequalities

Now we introduce the Viswanatham inequality which is a generalization of Bellman's Lemma. We shall first prove the following result.

Theorem 1.1.61 (The Viswanatham Inequality [657]) *If*

$$\phi(x) \le \eta + \int_{x_0}^{x} f(s, \phi(s))ds \quad (1.1.391)$$

where $f(x,t)$ is continuous and monotonic increasing in y in the region R defined by $|x - x_0| \le a$; $|y - \eta| \le b$, where a and b are positive real numbers; and $\phi(x)$ is continuous in the interval $|x - x_0| \le a$, then

$$\phi(x) \le \chi(x) \quad (1.1.392)$$

where $\chi(x)$ is the maximal solution of the differential equation $z' = f(x,z)$ through (x_0, η) for all $x \ge x_0$ (We shall call this differential equation the associated differential equation of the above integral inequality).

Proof Take $\phi(x)$ as the zero approximation of the solution of the differential equation $z' = f(x,z)$ through (x_0, η) and set up the successive approximations recursively by

$$\phi_{k+1}(x) = \eta + \int_{x_0}^{x} f(s, \phi_k(s))ds.$$

These successive approximations exist at least on the interval $|x - x_0| \le \alpha$ where $\alpha = \min(a, b/M)$ where M is a positive number such that $|f(x,y)| \le M$. Furthermore, this sequence of successive approximations is equicontinuous in this interval, since

$$|\phi_n(x_1) - \phi_n(x_2)| = |\int_{x_1}^{x_2} f(s, \phi_{n-1}(s))ds| \leq \int_{x_1}^{x_2} |f(s, \phi_{n-1}(s))|ds$$
$$\leq |x_1 - x_2|M \leq \varepsilon\pounds \quad \text{if} \quad |x_1 - x_2| \leq \epsilon/M = \delta.$$

It is further uniformly bounded because

$$|\phi_n(x)| \leq |\eta| + M|x_2 - x_1| \leq \eta + M\alpha.$$

We can show by induction that these successive approximations form a monotonic increasing sequence, for, suppose that $\phi_k(x) \geq \phi_{k-1}(x)$. Then

$$\phi_{k+1}(x) - \phi_k(x) = \int_{x_0}^{x} \Big(f(s, \phi_k(s)) - f(s, \phi_{k-1}(s))\Big)ds \geq 0$$

since $f(x, z)$ is monotonically increasing in z. Therefore $\phi_{k+1}(x) \geq \phi_k(x)$. But the basic hypothesis on our theorem is that the zero approximation $\leq$ first approximation. So the successive approximations form a monotonically increasing, equicontinuous, and uniformly bounded function sequence in the interval $|x - x_0| \leq \alpha$, and therefore must converge uniformly to a function $\psi(x)$. Further, it is clear that $\psi(x)$ is a solution of the associated differential equation through (x_0, η) and for all $x_0 \leq x \leq x_0 + \alpha$,

$$\phi(x) \leq \psi(x).$$

Therefore, for all $x_0 \leq x \leq x_0 + \alpha$,

$$\phi(x) \leq \chi(x),$$

when $\chi(x)$ is the maximal solution through (x_0, η).

As a counterpart to Theorem 1.1.61, we can similarly prove the following theorem.

Theorem 1.1.62 (The Viswanatham Inequality [657]) *Under the same conditions as in Theorem 1.1.61 if, for all $x \in [x_0, x_0 + \alpha]$,*

$$\phi(x) \geq \eta + \int_{x_0}^{x} f(s, \phi(s))ds, \tag{1.1.393}$$

then $\phi(x) \geq$ minimal solution of the associated differential equation through (x_0, η) for all $x_0 \leq x \leq x_0 + \alpha$.

Proof The proof is the same except that in this case the successive approximations form a monotonically decreasing sequence converging to a solution of the associated equation. We leave the proof to the reader as an exercise. □

The following may be obtained as corollaries to the above Theorems 1.1.61–1.1.62.

Corollary 1.1.16 (The Viswanatham Inequality [657]) *Under the condition of Theorem 1.1.61, if, for all $x \geq x_0$,*

$$\phi(x) \leq \psi(x) + \int_{x_0}^{x} f(s, \phi(s))ds, \tag{1.1.394}$$

then for all $x \geq x_0$,

$$\phi(x) \leq \psi(x) + \chi(x), \tag{1.1.395}$$

where $\chi(x)$ is the maximal solution of $z' = f(x, z + \psi(x))$ through $(x_0, 0)$ as far as this maximal solution exists.

Proof Put $r(x) = \phi(x) - \psi(x)$ and the inequality becomes

$$r(x) \leq \int_{x_0}^{x} f(s, r(s) + \psi(s))ds.$$

Applying Theorem 1.1.61 to the above inequality, we can obtain $r(x) \leq \chi(x)$. Therefore $\phi(x) \leq \psi(x) + \chi(x)$. □

The counterpart to this may be stated as the next corollary.

Corollary 1.1.17 ([The Viswanatham Inequality [657]) *Under the condition of Theorem 1.1.61 if, for all $x \geq x_0$,*

$$\phi(x) \geq \psi(x) + \int_{x_0}^{x} f(s, \phi(s))ds \tag{1.1.396}$$

then for all $x \geq x_0$,

$$\phi(x) \geq \psi(x) + \chi(x), \tag{1.1.397}$$

where $\chi(x)$ is the minimal solution of the associated equation in Corollary 1.1.16.

Similar theorems may also be proved for intervals with x_0 as the right end point.

A very special case of Theorem 1.1.61 is what is known as Bellman's Lemma [69], i.e., Theorem 1.1.2 in Qin [557] which is as follows:

$$|y(x)| \leq M + \int_{0}^{x} |f(s)||y(s)|ds$$

then

$$|y(x)| \leq M \exp\left(\int_0^x |f(t)|dt\right).$$

This is obtained by putting $f(x, y) = |f(x)|y, x_0 = 0$ and $\eta = M$ in Theorem 1.1.61.

Another special case of the same theorem is obtained by putting $f(x, y) = v(x) \cdot g(y)$ where $v(x)$ is non-negative and $g(y)$ is monotonic increasing in y. This case was considered in [82] and also in [328]. It is not necessary to work through the details to prove results of [82] and [328] from Theorem 1.1.61.

It is clear that in many situations in ordinary differential equations where we use a Lipschitz condition or Lipschitz-like condition, we may obtain more general results by applying the above theorems. For example, [156] contains the following proposition on approximate solutions.

Suppose f satisfies a Lipschitz condition with Lipschitz constant k; ϕ_1 and ϕ_2 are ϵ_1 and ϵ_2 approximate solutions of the differential equation $x' = f(t, x)$ and for some τ, we have $|\phi_1(\tau) - \phi_2(\tau)| \leq \delta$. Then for all $t \geq \tau$, we have

$$|\phi_1(t) - \phi_2(t)| \leq \delta e^{k(t-\tau)} + \frac{\epsilon}{k}(e^{k(t-\tau)} - 1).$$

If $\epsilon = \epsilon_1 + \epsilon_2$, then f satisfies the more general condition

$$|f(x, y_1) - f(x, y_2)| \leq \omega(x, |y_1 - y_2|)$$

where $\omega(x, z)$ satisfies conditions of Theorem 1.1.61, we may show easily by applying Theorem 1.1.61, that $|\phi_1(t) - \phi_2(t)| \leq \chi(t)$ for all $t \geq \tau$ where $\chi(t)$ is the maximal solution of $z' \equiv \omega(t, z) + \epsilon$ through (t, δ). Further, the first few examples of page 37 of [156] can all be solved by the application of Corollary 1.1.17. Similarly Theorems 1.1.61–1.1.62 can be used in a natural way to extend the results of [328] concerning bounds on the norm of a solution of a differential equation.

The next result is concerned with a system of integral inequalities

$$\varphi_i(t) \leq c_i + \int_{t_0}^t f_i(\tau, \varphi_1(\tau), \ldots, \varphi_n(\tau))d\tau, \quad i = 1, \ldots, n, \tag{1.1.398}$$

for all $t \in [t_0, t_1] = J, t_1 \leq +\infty$.

For simplicity, we introduce the relation "$\leq$" in $\mathbb{R}^n$, namely, we set for any two points of $\mathbb{R}^n, x = (x_1, \ldots, x_n)$ and $y = (y_1, \ldots, y_n)$, for each $i = 1, \ldots, n$,

$$x \leq y \ \text{ iff } \ x_i \leq y_i. \tag{1.1.399}$$

Relation (1.1.399) is a semi-order in $\mathbb{R}^n$ and it is easy to see that for any bounded set $A \subset \mathbb{R}^n$, there exists the $\sup A$ with respect to relation (1.1.399),

$$\sup\ A = \min\{z \in \mathbb{R}^n : x \leq z \, for \, each \, x \leq A\}. \tag{1.1.400}$$

We shall need (1.1.400) only for two point sets. In that we have

$$\sup\{x, y\} = z = (z_1, \dots, z_n) \tag{1.1.401}$$

where $z_i = \max(x_i, y_i)$, (x_i and y_i are coordinates of x and y, respectively).

We can now write (1.1.398) in a shorter form, namely, for all $t \in J$,

$$\varphi(t) \leq c + \int_{t_0}^{t} f(\tau, \varphi(\tau))d\tau, \tag{1.1.402}$$

where φ, c and f are now n-vectors.

The next result is to present a simple and very short proof of a result due to Opial [435] concerning inequality (1.1.402) (Theorem 1.1.63 below). The proof is based on an idea used by Cafiero [130, 131] to prove an analogous result on differential inequality (cf. also [433]).

Theorem 1.1.63 (The Opial Inequality [435]) *Let a map $f : J \times \mathbb{R}^n \to \mathbb{R}^n$ be continuous and assume that for any $x \leq y$,*

$$f(t, x) \leq f(t, y). \tag{1.1.403}$$

If a continuous map $\varphi : J \to \mathbb{R}^n$ satisfies inequality (1.1.402) and $\psi : [t_0, t_2] \to \mathbb{R}^n$ ($t_2 \leq t_1$) is the maximal solution of

$$x(t) = c + \int_{t_0}^{t} f(\tau, x(\tau))d\tau, \tag{1.1.404}$$

then for all $t_0 \leq t < t_2$,

$$\varphi(t) \leq \psi(t). \tag{1.1.405}$$

A solution ψ is the maximal solution of (1.1.404) if for any other solution $x(t)$ of (1.1.404), the inequality $x(t) \leq \psi(t)$ holds on the common interval of existence; if (1.1.403) holds, then the maximal solution of (1.1.404) exists, cf. [666].

Proof Put

$$F(t, x) = f(t, \sup\{x, \varphi(t)\}). \tag{1.1.406}$$

By (1.1.401), $\varphi(t) \leq \sup\{x, \varphi(t)\}$; therefore from (1.1.403) and (1.1.406) we derive for each x,

$$F(t, x) \geq f(t, \varphi(t)). \tag{1.1.407}$$

Let $\mathcal{X} : [t_0, t_2] \to \mathbb{R}^n$ be the maximal solution of

$$x(t) = c + \int_{t_0}^{t} F(\tau, x(\tau))d\tau. \tag{1.1.408}$$

Then, using (1.1.407) and (1.1.402), we conclude

$$\mathcal{X} = c + \int_{t_0}^{t} F(\tau, \mathcal{X}(\tau))d\tau \geq c + \int_{t_0}^{t} f(\tau, \varphi(\tau))d\tau \geq \varphi(t). \tag{1.1.409}$$

It follows from (1.1.409) and (1.1.401) that $\sup\{\mathcal{X}, \varphi(t)\} = \mathcal{X}$. Therefore by (1.1.406) we have $F(t, \mathcal{X}(t)) = f(t, \mathcal{X})$, whence $\mathcal{X}$ is also the maximal solution of (1.1.404). Thus (1.1.409) proves (1.1.405) and completes the proof. □

Remark 1.1.16 The monotonicity assumption (1.1.403) is essential for Theorem 1.1.63 (cf. [435]).

The corresponding result concerning a system of differential inequalities in [666] requires that $f_i(t, x_1, \ldots, x_n)$ is non-decreasing with respect to x_i only for $j \neq i$. In that sense differential inequalities theory is more general. In fact, we can reduce a proof of Theorem 1.1.63 to a corresponding result on differential inequalities (cf. [667]). A special case of Theorem 1.1.63, namely $n = 1$ and f linear in x, is the celebrated Gronwall's inequality.

There exist various generalizations of the Bellman-Gronwall-Reid inequalities. The next result is a sufficiently general result in this direction which includes and unifies several works [126, 196, 250, 325].

We wish to develop a variation of constants formula for the scalar differential equation

$$u' = \lambda(t)g(u) + R(t, u), \; u(t_0) = v_0 \geq 0 \tag{1.1.410}$$

where $\lambda \in C(I, \mathbb{R}), g \in C(\mathbb{R}_+, \mathbb{R}_+), g(0) = 0, g(u) > 0$ for all $u > 0, R \in C(I \times \mathbb{R}_+, \mathbb{R})$ and $I = [t_0, T_0]$. It is well-known that a solution $v(t)$ of

$$v' = \lambda(t)g(v), \; v(t_0) = v_0 \geq 0 \tag{1.1.411}$$

may be expressed in the form, for all $t \in I_0$,

$$G(v(t)) = \int_{t_0}^{t} \lambda(s)ds + G(v_0), \tag{1.1.412}$$

where

$$G(u) = \int_{u_0}^{u} \frac{ds}{g(s)}, \; u \geq u_0 > 0$$

and

$$I_0 = \left\{ t \in I : G(v_0) + \int_{t_0}^{t} \lambda(s)ds \in Dom\,(G^{-1}) \right\}.$$

Let us now apply the variation of constants method. To this end, we need to determine v_0 as a function t such that $v(t)$ is a solution of (1.1.410). Thus, we find

$$G_u(v(t))v'(t) = \lambda(t) + G_u(v_0)v_0',$$

which, in view of definition (1.1.412) of G and (1.1.410), reduces to

$$v_0' = \frac{g(v_0)}{g(v(t))} R(t, v(t)).$$

Using now (1.1.412) to eliminate $v(t)$, we obtain the differential equation

$$v_0' = \omega(t, v_0), \ v_0(t_0) = v_0, \tag{1.1.413}$$

where

$$\omega(t, v_0) = \frac{g(v_0)R(t, G^{-1}(G(v_0) + \int_{t_0}^{t} \lambda(s)ds))}{g[G^{-1}(G(v_0) + \int_{t_0}^{t} \lambda(s)ds)]}.$$

Let $v_0(t)$ be a solution of (1.1.413) existing on I_0. Then (1.1.412) gives us the integral equation satisfied by a solution $v(t)$ of (1.1.410) in the form, for all $t \in I_0$,

$$v(t) = G^{-1}\left(\int_{t_0}^{t} \lambda(s)ds + G\left(v_0 + \int_{t_0}^{t} \frac{g\left(G^{-1}(G(v(s)) - \int_{t_0}^{s} \lambda(\xi)d\xi)\right)}{g(v(s))} R(s, v(s))ds \right) \right). \tag{1.1.414}$$

We thus have proved the following result.

Theorem 1.1.64 (The Bellman-Gronwall-Reid Inequality [323]) *Let λ, g, r and I be as given above. Then a solution $v(t)$ of (1.1.410) can be exhibited in the form (1.1.414) on I_0, which can be obtained by the method of variation of constants. We note, in passing, that the linear case is covered by (1.1.414).*

Corollary 1.1.18 (The Bellman-Gronwall-Reid Inequality [323]) *If, in addition to the assumptions of Theorem 1.1.61, we assume that $\lambda \geq 0, R(t, u) = \sigma(t) \geq 0$ and g is non-decreasing in u, then the following bound is true for $u(t)$: for all $t \in I_0$,*

$$u(t) \leq G^{-1}\left[\int_{t_0}^{t} \lambda(s)ds + G(v_0 + \int_{t_0}^{t} \sigma(s)ds) \right]. \tag{1.1.415}$$

Let $\mathbb{R}^n$ denote the real n-dimensional, euclidean space of elements $u = (u_1, u_2, \cdots, u_n)$. Sometimes, we shall denote also the $(t+1)$-tuple $(t, u_1, u_2, \cdots, u_n)$ as an element, and $\mathbb{R}^{n+1}$ shall denote the space of elements $(t, u_1, u_2, \cdots, u_n)$ or (t, u). Let $\|u\|$ be any convenient norm. As usual, we shall mean by $C(E, \mathbb{R}^n)$ the class of continuous mappings from E into $\mathbb{R}^n$. If f is a member of this class, one writes $f \in C(E, \mathbb{R}^n)$. Let us consider a system of first-order differential equations with an initial condition

$$u' = g(t, u), \quad u(t_0) = u_0, \tag{1.1.416}$$

where $u' = \frac{du}{dt}$, $u_0 = (u_{10}, u_{20}, \cdots, u_{n0})$, and $g \in C(E, \mathbb{R}^n)$. A solution of the initial value problem (1.1.416) is a differentiable function of t such that $u(t_0) = u_0$, $(t, u(t) \in E$, and $u'(t) = g(t, u(t))$ for a t-interval J containing t_0. This means that $u(t)$ has a continuous derivative. From these requirements on the continuous function $u(t)$, it follows that it satisfies the integral equation

$$u(t) = u_0 + \int_{t_0}^{t} g(s, u(s))ds, \quad t \in J.$$

The following theorem shows the fundamental property of such a family of functions, the proof of which will be omitted. To prove Theorem 1.1.65 below, we need the following lemmas.

Lemma 1.1.13 (Ascoli-Arzela Theorem) *Let $F = f$ be a sequence of functions defined on a compact u-set $E \subset \mathbb{R}^n$, which is equicontinuous and equibounded. Then, there exists a subsequence f_n, $n = 1, 2, \cdots$, which is uniformly convergent on E.*

Lemma 1.1.14 (Peano's Existence Theorem) *Let $g \in C(R_0, \mathbb{R}^n)$, where R_0 is the set $\big[(t, u) : t_0 \le t \le t_0 + a, \|u - u_0\| \le b\big]$; $\|g(t, u)\| \le M$ on R_0. Then, the initial value problem (1.1.416) possesses at least one solution $u(t)$ on $t_0 \le t \le t_0 + \alpha$, where $\alpha = min(a, \frac{b}{M})$.*

Proof Let $u_0(t)$ be a continuously differentiable function on $[t_0 - \delta, t_0]$, $\delta > 0$, such that $u_0(t_0) = u_0$, $\|u_0(t) - u_0\| \le b$, and $\|u_0'(t)\| \le M$. For $0 < \epsilon \le \delta$, we define a function $u_\epsilon = u_0(t)$ on $[t_0 - \delta, t_0]$ and

$$u_\epsilon = u_0 + \int_{t_0}^{t} g(s, u_\epsilon(s - \epsilon))ds \tag{1.1.417}$$

on $[t_0, t_0 + \alpha_1]$, where $\alpha_1 = \min(\alpha, \epsilon) > 0$. Observe that $u_\epsilon(t)$ is differentiable and

$$\|u_\epsilon(t) - u_0\| \le b \tag{1.1.418}$$

on $[t_0 - \delta, t_0 + \alpha_1]$. If $\alpha_1 < \alpha$, we can use (1.1.417) to extend $u_\epsilon(t)$ as a continuously differentiable function over $[t_0 - \delta, t_0 + \alpha_2]$, $\alpha_2 = min(\alpha, 2\epsilon)$, such that (1.1.418) holds. Continuing in this way, $u_\epsilon(t)$ can be defined over $[t_0 - \delta, t_0 + \alpha]$ so that it has a continuous derivative and satisfies (1.1.418) on the same interval. Furthermore, $\|u'_\epsilon(t)\| < M$, and therefore $u_\epsilon(t)$ forms a family of equicontinuous and uniformly bounded functions. An application of Lemma 1.1.13 shows the existence of a sequence $\{\epsilon_n\}$ such that $\epsilon_1 > \epsilon_2 > \cdots > \epsilon_n \to 0$ as $n \to \infty$, and $u(t) = \lim_{n\to\infty} u_{\epsilon_n}(t)$ exists uniformly on $[t_0 - \delta, t_0 + \alpha]$. Since g is uniformly continuous, we can obtain that $g(t, u_{\epsilon_n}(t - \epsilon_n))$ tends uniformly to $g(t, u(t))$ as $n \to \infty$, and, hence, term-by-term integration of (1.1.417) with $\epsilon = \epsilon_n$, $\alpha_1 = \alpha$ yields

$$u(t) = u_0 + \int_{t_0}^{t} g(s, u(s))ds.$$

This proves that $u(t)$ is a solution of (1.1.416). □

The following corollary of Peano's Theorem is useful in applications.

Corollary 1.1.19 (Peano's Existence Theorem) *Let E be an open (t, u)-set in $\mathbb{R}^{n+1}$ and E_0 be a compact subset of E. Suppose that $g \in C(E, \mathbb{R}^n)$ and $\|g(t, u)\| \leq M$ on E. Then, there exists an $\alpha = \alpha(E, E_0, M)$ such that, if $(t_0, u_0) \in E_0$, (1.1.26) has a solution, and every solution exists on $[t_0, t_0 + \alpha]$. In that case, when g is not bounded on E, we can replace the set E by an open subset E_1 having a compact closure in E and containing E_0.*

We adopt the following notation for Dini derivatives:

$$D^+u(t) = \limsup_{h\to 0^+} h^{-1}\big[u(t + h) - u(t)\big], \quad D_+u(t) = \liminf_{h\to 0^+} h^{-1}\big[u(t + h) - u(t)\big],$$

$$D^-u(t) = \limsup_{h\to 0^-} h^{-1}\big[u(t + h) - u(t)\big], \quad D_-u(t) = \liminf_{h\to 0^-} h^{-1}\big[u(t + h) - u(t)\big],$$

where $u \in C([t_0, t_0 + a), \mathbb{R})$. When $D^+u(t) = D_+u(t)$, the right derivative will be denoted by $u'_+(t)$. Similarly, $u'_-(t)$ denotes the left derivative.

Lemma 1.1.15 *Let E be an open (t, u)-set in $\mathbb{R}^2$ and $g \in C(E, \mathbb{R})$. Assume that $v\, w \in C\big([t_0, t_0 + a), \mathbb{R}\big)$ and $(t, v(t))$, $(t, w(t) \in E$, $t \in [t_0, t_0 + a)$. Suppose further that*

$$v(t_0) < w(t_0), \tag{1.1.419}$$

and, for all $t \in (t_0, t_0 + a)$, the inequalities

$$\begin{cases} D_v(t) \leq g(t, v(t)), & (1.1.420)\\ D_w(t) > g(t, w(t)), & (1.1.421)\end{cases}$$

hold. Then for all $t \in [t_0, t_0 + a]$,

$$v(t) < w(t). \tag{1.1.422}$$

Proof If assertion (1.1.422) is false, then the set

$$Z = \left[t \in [t_0, t_0 + a) : w(t) \leq v(t)\right]$$

is nonempty. Defining $t_1 = \inf Z$, it is clear from (1.1.419) that $t_0 < t_1$. Furthermore,

$$v(t_1) = w(t_1) \tag{1.1.423}$$

and for all $t \in [t_0, t_1)$,

$$v(t) < w(t). \tag{1.1.424}$$

Using (1.1.423) and (1.1.424), we can obtain, for small $h < 0$,

$$\frac{v(t_1 + h) - v(t_1)}{h} > \frac{w(t_1 + h) - w(t_1)}{h}, \tag{1.1.425}$$

which in its turn implies

$$D_v(t_1) \geq D_w(t_1). \tag{1.1.426}$$

The inequalities (1.1.420), (1.1.421) and (1.1.425) together with (1.1.420) lead us to the contradiction

$$g(t_1, v(t_1)) > g(t_1, w(t_1)).$$

Hence Z is empty, and the statement (1.1.422) follows. □

Remark 1.1.17 It is obvious from the proof that the inequalities (1.1.420) and (1.1.421) can also be replaced by

$$\begin{cases} D_v(t) < g(t, v(t)), & (1.1.427) \\ D_w(t) \leq g(t, w(t)), & (1.1.428) \end{cases}$$

respectively.

Lemma 1.1.16 (Zygmund) *Suppose that $u \in C\left([t_0, t_0 + a), \mathbb{R}\right)$ and the inequality $Dv(t) \leq 0$ for all $t \in [t_0, t_0 + a) - S$, D being a fixed Dini derivative. Then, $u(t)$ is non-increasing in t on $[t_0, t_0 + a)$.*

Lemma 1.1.17 *Suppose that $u \in C\left([t_0, t_0 + a), \mathbb{R}\right)$ and for some fixed Dini derivative $Dv(t) \leq w(t)$ for all $t \in [t_0, t_0 + a) - S$, Then, $D_v(t) \leq w(t)$ for all $t \in [t_0, t_0 + a)$.*

Proof Define the function

$$m(t) = v(t) - \int_{t_0}^{t} w(s)ds.$$

It then follows from the assumption that for all $t \in [t_0, t_0 + a) - S$,

$$Dm(t) = Dv(t) - w(t) \leq 0.$$

Hence, by Lemma 1.1.16, $m(t)$ is non-increasing in t on $[t_0, t_0 + a)$.

Consequently, for all $t \in [t_0, t_0 + a) - S$,

$$D_m(t) = D_v(t) - w(t) \leq 0,$$

and the lemma is proved. □

Remark 1.1.18 From Lemma 1.1.17, it is clear that Lemma 1.1.15 remains true when the inequalities (1.1.420) and (1.1.421) hold for all $t \in [t_0, t_0 + a) - S$, D being any fixed Dini derivative.

Definition 1.1.5 Let E be an open (t, u)-set in $\mathbb{R}^2$ and $g \in C(E, \mathbb{R})$. Consider the scalar differential equation with an initial condition

$$u' = g(t, u), \quad u(t_0) = u_0. \tag{1.1.429}$$

Suppose $v \in C\big([t_0, t_0 + a), \mathbb{R}\big)$, $v'_+(t)$ exists for all $t \in [t_0, t_0 + a)$, and $(t, v(t)) \in E$. If $v(t)$ satisfies the differential inequality for all $t \in [t_0, t_0 + a)$,

$$v'_+(t) < g(t, v(t)),$$

it is said to be an under-function with respect to the initial value problem (1.1.429). On the other hand, if for all $t \in [t_0, t_0 + a)$,

$$v'_+(t) > g(t, v(t)),$$

$v(t)$ is said to be an over-function.

Definition 1.1.6 Let $r(t)$ be a solution of the scalar differential equation (1.1.429) on $[t_0, t_0 + a)$. Then $r(t)$ is said to be a *maximal solution* of (1.1.429) if, for every solution $u(t)$ of (1.1.429) existing on $[t_0, t_0 + a)$, the inequality for all $t \in [t_0, t_0 + a)$,

$$u(t) \leq r(t) \tag{1.1.430}$$

holds. A minimal solution $\rho(t)$ may be defined similarly by reversing the inequality (1.1.430).

We shall now consider the existence of maximal and minimal solutions of problem (1.1.429) under the hypothesis of Peano's existence theorem.

Lemma 1.1.18 ([325]) *Let $g \in C[R_0, \mathbb{R}]$, where R_0 is the rectangle $t_0 \leq t \leq t_0 + a$, $|u - u_0| \leq b$, and $|g(t, u)| \leq M$ on R_0. Then there exist a maximal solution and a minimal solution of (1.1.429) on $[t_0, t_0 + \alpha]$, where $\alpha = \min\big(a, \frac{b}{2M+b}\big)$.*

Proof We shall prove the existence of the maximal solution only, since the case of the minimal solution is very similar. Let $0 < \epsilon \leq \frac{b}{2}$. Consider the differential equation with an initial condition

$$u' = g(t, u) + \epsilon, \quad u(t_0) = u_0 + \epsilon. \tag{1.1.431}$$

Observing that

$$g_\epsilon(t, u) = g(t, u) + \epsilon$$

is defined and continuous on

$$R_\epsilon : t_0 \leq t \leq t_0 + a, \quad |u - (u_0 + \epsilon)| \leq \frac{b}{2},$$

$R_\epsilon \subset R_0$ and $|g_\epsilon(t, u)| \leq M + \frac{b}{2}$ on R_ϵ, we deduce from Lemma 1.1.14 that the initial value problem (1.1.431) has a solution $u(t, \epsilon)$ on the interval $[t_0, t_0 + \alpha]$, where $\alpha = min\big(a, \frac{b}{2M+b}\big)$. For $0 < \epsilon_2 < \epsilon_1 \leq \epsilon$, we have for all $t \in [t_0, t_0 + \alpha]$,

$$\begin{aligned} u(t_0, \epsilon_2) &< u(t_0, \epsilon_1), \\ u'(t, \epsilon_2) &\leq g(t, u(t, \epsilon_2)) + \epsilon_2, \\ u'(t, \epsilon_1) &\leq g(t, u(t, \epsilon_1)) + \epsilon_2, \end{aligned}$$

We can apply Lemma 1.1.15 to get for all $t \in [t_0, t_0 + \alpha]$,

$$u(t, \epsilon_2) < u(t, \epsilon_1).$$

Since the family of functions $u(t, \epsilon)$ is equicontinuous and uniformly bounded on $[t_0, t_0 + \alpha]$, it follows from Lemma 1.1.13 that there exists a decreasing sequence ϵ_n such that $\epsilon_n \to 0$ as $n \to +\infty$, and the uniform limit

$$r(t) = \lim_{n \to \infty} u(t, \epsilon_n)$$

exists on $[t_0, t_0 + \alpha]$. Clearly, $r(t_0) = u_0$. The uniform continuity of g implies that $g(t, u(t, \epsilon_n))$ tends uniformly to $g(t, r(t))$ as $n \to +\infty$, and thus term-by-term integration is applied to

$$u(t, \epsilon_n) = u_0 + \epsilon_n + \int_{t_0}^{t} g(s, u(s, \epsilon_n))ds,$$

which in turn shows that the limit $r(t)$ is a solution of problem (1.1.429) on $[t_0, t_0 + \alpha]$.

We shall now show that $r(t)$ is the desired maximal solution of problem (1.1.429) on $[t_0, t_0 + \alpha]$ satisfying (1.1.430). Let $u(t)$ be any solution of problem (1.1.429) existing on $[t_0, t_0 + \alpha]$. Then, for all $t \in [t_0, t_0 + \alpha]$ and $\epsilon \leq \frac{b}{2}$,

$$u(t_0) = u_0 < u_0 + \epsilon = u(t_0, \epsilon),$$
$$u'(t) < g(t, u(t)) + \epsilon,$$
$$u'(t, \epsilon) \geq g(t, u(t, \epsilon)) + \epsilon.$$

By Remark 1.1.17, we can obtain that for all $t \in [t_0, t_0 + \alpha]$,

$$u(t) < u(t, \epsilon).$$

The uniqueness of the maximal solution shows that $u(t, \epsilon)$ tends uniformly to $r(t)$ on $[t_0, t_0 + \alpha]$ as $\epsilon \to 0$. This thus proves the lemma. □

Lemma 1.1.19 ([325]) *Let $g \in C(E, \mathbb{R}^n)$, where E is an open (t, u)-set in $\mathbb{R}^{n+1}$. Let $u(t)$ be a solution of problem (1.1.416) on an interval $t_0 \leq t < a$, $a < +\infty$. Assume that there exists a sequence $\{t_k\}$ such that $t_0 \leq t_k \to +\infty$ as $k \to +\infty$ and $u^0 = \lim_{k \to +\infty} u(t_k)$ exists. If $g(t, u)$ is bounded on the intersection of E and a neighborhood of (a, u^0), then*

$$\lim_{t \to a} u(t) = u^0. \tag{1.1.432}$$

If, in addition, $g(a, u^0)$ is defined such that $g(a, u)$ is continuous at (a, u^0), then $u(t)$ is continuously differentiable on $[t_0, a]$ and is a solution of problem (1.1.416) on $[t_0, a]$.

Proof Let $\epsilon > 0$ be sufficiently small. Consider the set $\hat{R} : 0 \leq a - t \leq \epsilon$, $\|u - u^0\| \leq \epsilon$. Let $M(\epsilon)$ be so large that $\|g(t, u)\| \leq M(\epsilon)$ for all $(t, u) \in E \bigcap \hat{R}$. If, for k sufficiently large, $0 < a - t_k \leq \frac{\epsilon}{2M(\epsilon)}$ and $\|u(t_k) - u^0\| \leq \frac{\epsilon}{2}$, then

$$\|u(t) - u(t_k)\| < M(\epsilon)(a - t_k) \leq \frac{\epsilon}{2} \tag{1.1.433}$$

for all $t_k \leq t < a$. If this is not true, there is a t_1 such that for all $t_k < t_1 < a$,

$$\|u(t_1) - u(t_k)\| = M(\epsilon)(a - t_k) \leq \frac{\epsilon}{2}.$$

It therefore follows that for all $t_k \leq t < t_1$,

$$\|u(t_1) - u^0\| \leq \frac{\epsilon}{2} + \|u(t_k)\| \leq \epsilon.$$

Consequently,

$$\|u(t_1) - u(t_k)\| \leq M(\epsilon)(t - t_k) < M(\epsilon)(a - t_k).$$

This proves (1.1.433), which, in turn, shows that (1.1.432) holds. The last part of the lemma follows from the fact that

$$u'(t) = g(t, u(t)) \to g(a, u^0), \quad \text{as } t \to a.$$

This lemma is thus proved. □

The next lemma deals with the problem of extending the solutions up to the boundary of E.

Lemma 1.1.20 ([325]) *Let E be an open (t, u)-set in $\mathbb{R}^{n+1}$, and let $g \in C(E, \mathbb{R}^n)$ and $u(t)$ be a solution of problem (1.1.416) on some interval $t_0 \le t \le a_0$. Then $u(t)$ can be extended as a solution to the boundary of E.*

Proof Let $E_1, E_2, \cdots$ be open subsets of E such that $E = \cup E_n$, the closures $\overline{E}_1, \overline{E}_2, \cdots$ are compact, and $\overline{E}_1 \subset E_{n+1}$. It then follows from Corollary 1.1.19 that there exists an $\epsilon_n > 0$ such that, if $(t_0, u_0) \in \overline{E}_n$, all solutions of problem (1.1.416) exist on $t_0 \le t \le t_0 + \epsilon_n$.

Chose n_1 so large that $(a_0, u(a_0)) \in \overline{E}_{n_1}$. Then, $u(t)$ can be extended over an interval $[a_0, a_0 + \epsilon_{n_1}]$, and, if $(a_0 + \epsilon_{n_1}, u(a_0 + \epsilon_{n_1})) \in \overline{E}_{n_1}$, $u(t)$ can be further extended over $[a_0 + \epsilon_{n_1}, a_0 + 2\epsilon_{n_1}]$. This argument can be repeated until we get the extension of $u(t)$ over the interval $t_0 \le t \le a_1$, where $a_1 = a_0 + N_1\epsilon_{n_1}$, N_1 is an integer ≥ 1, such that $(a_1, u(a_1)) \notin \overline{E}_{n_1}$.

Chose n_2 so large that $(a_0, u(a_0)) \in \overline{E}_{n_2}$. Arguing as before, we arrive at an integer $N_2 \ge 1$ such that $u(t)$ can be extended over $t_0 \le t \le a_2$, $a_2 = a_1 + N_2\epsilon_{n_2}$, and $(a_2, u(a_2)) \notin \overline{E}_{n_2}$.

Proceeding in this way, we are led to a sequence of integers $n_1 < n_2 < \cdots$ and numbers $a_0 < a_1 < a_2 < \cdots$ such that $u(t)$ has an extension over $[t_0, a)$, where $a = \lim_{k\to\infty} a_k$ and that $(a_k, u(a_k)) \notin \overline{E}_{n_k}$. Thus, the sequence $\{a_k, u(a_k)\}$ is either unbounded or has a cluster point on the boundary of E.

To show that $u(t)$ tends to the boundary of E as $t \to a$, we must show that no limit point of $\{t_k, u(t_k)\}$ is an interior point of E as $t_k \to a$. Since this follows from the Lemma 1.1.19, the lemma is proved. □

Lemma 1.1.18 together with the extension of Lemma 1.1.20, implies the following lemma.

Lemma 1.1.21 ([325]) *Let $g \in C(E, \mathbb{R})$, where E is an open (t, u)-set in $\mathbb{R}^2$ and $(t_0, u_0) \in E$. Then problem (1.1.429) has maximal and minimal solutions that can be extended to the boundary of E.*

Lemma 1.1.22 ([325]) *Let the hypothesis of Lemma 1.1.21 hold, and let $[t_0, t_0 + a)$ be the largest interval of existence of the maximal solution $r(t)$ of problem (1.1.429). Suppose $[t_0, t_1]$ is a compact sub-interval of $[t_0, t_0 + a)$. Then there is an $0 < \epsilon < \epsilon_0$, the maximal solution $r(t, \epsilon)$ of equation (1.1.431) exists over $[t_0, t_1]$, and*

$$\lim_{\epsilon\to 0} r(t, \epsilon) = r(t)$$

uniformly on $[t_0, t_1]$.

Proof Let E_0 be an open bounded set, $\overline{E}_0 \subset E$, and $(t, r(t)) \in E$ for all $t \in [t_0, t_1]$. We can choose a $b > 0$ such that, for all $t \in [t_0, t_1]$, the rectangle

$$R_t^\epsilon : [t, t+b], \quad |u - (r(t) + \epsilon)| \le b,$$

is included in E_0 for all $\epsilon \le \frac{b}{2}$. Let $|g(t,u)| \le M$ on E_0. Then it follows that

$$|g(t,u) + \epsilon| \le M + \frac{b}{2}$$

on R_t^ϵ, for all $t \in [t_0, t_1]$ and $0 < \epsilon \le \frac{b}{2}$. Consider the rectangle $R_{t_0}^\epsilon$. It follows from Lemma 1.1.18 that the maximal solution $r(t, \epsilon)$ of equation (1.1.431) exists on $[t_0, t_0 + \eta]$, $\eta = \min\left(b, \frac{b}{2M+b}\right)$. Note that η does not depend upon ϵ. Furthermore, proceeding as in Lemma 1.1.18, we can conclude, in view of the uniqueness of the maximal solution $r(t)$ of problem (1.1.429), that

$$\lim_{\epsilon \to 0} r(t_0 + \eta, \epsilon) = r(t_0 + \eta).$$

Consequently, there is an $\epsilon_1 \le \frac{b}{2}$ such that, for all $0 < \epsilon \le \epsilon_1$, we have

$$r(t_0 + \eta, \epsilon) \le r(t_0 + \eta) + \epsilon.$$

We can now repeat the foregoing argument with respect to the rectangle $R_{t_0+\eta}^\epsilon$, $\epsilon < \epsilon_1$, to show that there exists an $\epsilon_2 < \epsilon_1$ such that, for all $\epsilon < \epsilon_2$, the maximal solution $\hat{r}(t, \epsilon)$ of

$$u' = g(t,u) + \epsilon, \quad u(t_0 + \eta) = r(t_0 + \eta) + \epsilon$$

exists on $[t_0 + \eta, t_0 + 2\eta]$, and

$$\lim_{\epsilon \to 0} \hat{r}(t, \epsilon) = r(t)$$

uniformly on $[t_0 + \eta, t_0 + 2\eta]$. For all $\epsilon < \epsilon_2$, we can extend the function $r(t, \epsilon)$ by defining for all $t \in [t_0 + \eta, t_0 + 2\eta]$,

$$r(t, \epsilon) = \hat{r}(t, \epsilon).$$

It is clear that $r(t, \epsilon)$ is the maximal solution of equation (1.1.431) on $[t_0, t_0 + 2\eta]$, and

$$\lim_{\epsilon \to 0} r(t, \epsilon) = r(t)$$

uniformly on $[t_0, t_0 + 2\eta]$.

By induction, it can be shown that there is an $\epsilon_0 = \epsilon_n$ such that $[t_0, t_1] \subset [t_0, t_1 + n\eta]$, that the maximal solution $r(t, \epsilon)$ of equation (1.1.431) exists on $[t_0, t_1 + n\eta]$ for all $0 < \epsilon < \epsilon_0$, and that

$$\lim_{\epsilon \to 0} r(t, \epsilon) = r(t)$$

uniformly on $[t_0, t_1 + n\eta]$. The lemma is thus proved. □

Lemma 1.1.23 ([325]) *Let E be an open (t, u)-set in $\mathbb{R}^2$ and $g \in C(E, \mathbb{R})$. Suppose that $[t_0, t_0 + a)$ is the largest interval in which the maximal solution $r(t)$ of problem (1.1.429) exists. Let $m \in C((t_0, t_0 + a), \mathbb{R})$, $(t, m(t)) \in E$ for all $t \in [t_0, t_0 + a)$, $m(t_0) \le u_0$, and for a fixed Dini derivative, for all $t \in [t_0, t_0 + a) - S$,*

$$Dm(t) \le g(t, m(t)). \tag{1.1.434}$$

Then for all $t \in [t_0, t_0 + a)$,

$$m(t) \le r(t). \tag{1.1.435}$$

Proof From Lemma 1.1.17, it follows that (1.1.432) can be replaced by for all $t \in [t_0, t_0 + a)$,

$$D_m(t) \le g(t, m(t)). \tag{1.1.436}$$

Let $t_0 < \tau < t_0 + a$. By Lemma 1.1.22 the maximal solution $r(t, \epsilon)$ of problem (1.1.431) exists on $[t_0, \tau]$ for all $\epsilon > 0$ sufficiently small, and

$$r(t) = \lim_{\epsilon \to 0} r(t, \epsilon) \tag{1.1.437}$$

uniformly on $[t_0, \tau]$. Using (1.1.431) and (1.1.434) and applying Lemma 1.1.15, we derive that for all $t \in [t_0, \tau]$,

$$m(t) < r(t, \epsilon). \tag{1.1.438}$$

The last inequality, together with (1.1.435), proves the assertion of the theorem. □

Now we consider the functional inequality, for all $t \in I$,

$$f(x(t)) \le a(t) + b(t) + h[c(t) + \int_{t_0}^{t} k(t, s)\omega(s, x(s))ds], \tag{1.1.439}$$

under a variety of conditions on h.

Theorem 1.1.65 (The Bellman-Gronwall-Reid Inequality [323]) *Let $x, a, b, c \in C(I, \mathbb{R}_+)$, $f, h, \in C(\mathbb{R}_+, \mathbb{R}_+)$, f be strictly increasing, g be non-decreasing, $k \in$*

$C(I \times I, \mathbb{R}_+), \omega \in C(I + \mathbb{R}_+, \mathbb{R}_+)$ and $\omega(t, u)$ be non-decreasing in u for each $t > 0$. Define

$$A(t) = \max_{t_0 \le s \le t} a(s),\ B(t) = \max_{t_0 \le s \le t} b(s),\ C(t) = \max_{t_0 \le s \le t} c(s)$$

and

$$K(t, s) = \max_{s \le \sigma \le t} k(\sigma, s).$$

Then

(i) *for all $t \in I_{10}$,*

$$x(t) \le f^{-1}[a(t) + b(t)h(r_1(t, t_0, C(t)))], \tag{1.1.440}$$

where $r_1(T, t_0, r_{10})$ is the maximal solution of

$$r_1' = K(T, t)\omega[t, f^{-1}(a(t) + b(t)h(r_1))],\ r_1(t_0) = r_{10}, \tag{1.1.441}$$

existing on $I_1 \subset I$.

(ii) *If, in addition, $h_u(u)$ exists, is continuous and non-decreasing in u, then for all $t \in I_{20}$,*

$$x(t) \le f^{-1}[r_2(t, t_0, A(t) + B(t)h(C(t)))], \tag{1.1.442}$$

where $r_2(T, t_0, r_{20})$ is the maximal solution of

$$r_2' = B(T)h_u\left[h^{-1}\left(\frac{r_2 - A(T)}{B(T)}\right)\right]K(T, t)\omega(t, f^{-1}(r_2)),\ r_2(t_0) = r_{20}, \tag{1.1.443}$$

existing on $I_1 \subset I$.

(iii) *If, $h^{(-1)}(u)$ is convex, sub-multiplicative and $\alpha, \beta > 0$, continuous on I such that $\alpha(t) + \beta(t) = 1$, then the following two types of estimates are valid:*

(iiia) *for all $t \in I_{30}$,*

$$x(t) \le F^{-1}[r_3(t, t_0, C(t))], \tag{1.1.444}$$

where $r_3(T, t_0, r_{30})$ is the maximal solution of

$$r_3' = K(T, t)\omega[t, F^{-1}(m(t) + n(t)r_3)],\quad r_3(t_0) = r_{30}, \tag{1.1.445}$$

existing on $I_0 \subset I$. Here

$$F = h^{-1} \cdot f,\ m(t) = \alpha(t)h^{-1}[a(t)\alpha(t)^{-1}]$$

and

$$n(t) = \beta(t)h^{-1}[b(t)\beta(t)^{-1}];$$

(iiib) for all $t \in I_{40}$,

$$x(t) \leq F^{-1}[r_4(t, t_0, M(t) + N(t)C(t))], \tag{1.1.446}$$

where $r_4(T, t_0, r_{40})$ *is the maximal solution of*

$$r_4' = N(T)K(T, t)\omega(t, F^{-1}(r_4)), \ r_4(t_0) = r_{40}, \tag{1.1.447}$$

existing on $I_0 \subset I$. *Here* F *is as in (iiia),*

$$M(t) = \max_{t_0 \leq s \leq t} m(s) \text{ and } N(t) = \max_{t_0 \leq s \leq t} n(s).$$

In each of the above cases, I_{i0}, $i = 1, 2, 3, 4$, *is the appropriate interval contained in* I *subject to the domains of the inverse functions involved.*

Proof (i) We observe that for $t_0 \leq t \leq T \leq T_0$, we have from (1.1.439),

$$f(x(t)) \leq a(t) + b(t)h[v(t, T)], \tag{1.1.448}$$

where

$$v = v(t, T) = C(T) + \int_{t_0}^{t} K(T, t)\omega(s, x(s))ds. \tag{1.1.449}$$

Thus, it follows that

$$v' \leq K(T, t)\omega[t, f^{-1}\{a(t) + b(t)h(v)\}].$$

By Lemma 1.1.23, we readily get, for all $T \geq t_0$,

$$v(T, T) \leq r_1(T, t_0, v(t_0, T)),$$

where $r_1(T, t_0, r_{10})$ is the maximal solution of equation (1.1.441) with $r_{10} = v(t_0, T)$. By (1.1.448) and (1.1.449), we have, for all $T \geq t_0$,

$$f(x(T)) \leq a(T) + b(T)h[r_1(T, t_0, C(T))], \tag{1.1.450}$$

which implies the stated estimate (1.1.440).

(ii) In this case, we may write (1.1.439), for all $t_0 \leq t \leq T \leq T_0$, as

$$f(x(t)) \leq A(T) + B(T)h(v), \tag{1.1.451}$$

where $v = v(t, T)$ is the same function defined in (i). Setting

$$z = z(t, T) = A(T) + B(T)h(v),$$

we easily get

$$z' \leq B(T)h_u \left[h^{-1} \left(\frac{z - A(T)}{B(T)} \right) \right] K(T, t)\omega(t, f^{-1}(z)). \tag{1.1.452}$$

Again Lemma 1.1.23 yields, for all $T \geq t_0$,

$$z(T, T) \leq r_2(T, t_0, z(t_0, T)), \tag{1.1.453}$$

where $r_2(T, t_0, r_{20})$ is the maximal solution of equation (1.1.443) with

$$r_{20} = A(T) + B(T)h(C(T)).$$

By (1.1.441) and (1.1.443), the bound (1.1.442) follows because of the definition of z, arguing as before.

(iii) The inequality (1.1.439) can be written as, by using the convexity and submultiplicity of h^{-1},

$$h^{-1} \cdot f(x(t)) \leq \alpha(t)h^{-1}(a(t)\alpha(t)^{-1})$$
$$+\beta(t)h^{-1}(b(t)\beta(t)^{-1}) \left[c(t) + \int_{t_0}^{t} k(t, s)\omega(s, x(s))ds \right],$$

which reduces, because of the definition of f, m and n, to

$$F(x(t)) \leq m(t) + n(t) \left[c(t) + \int_{t_0}^{t} k(t, s)\omega(s, x(s))ds \right]. \tag{1.1.454}$$

If we treat (1.1.454) as a special case of (i) with $h(u) = u, F = f, a = m$ and $b = n$, we arrive at (1.1.444) and (1.1.445) from (1.1.440) and (1.1.441) under appropriate substitutions. If, on the other hand, we treat (1.1.454) as a particular case of (ii), we get (1.1.446) and (1.1.447) from (1.1.442) and (1.1.443), respectively. □

Employing the above theorems, we shall show that the results in [126, 196, 250, 325] can be derived as special cases. First, in order to get the result of Gollwutzer [250], we take $f(u) = u, c(t) = 0, k(t, s) = 1, \omega(t, u) = \lambda(t)g(u)$ and $h = g^{-1}$. Then it follows from Theorem 1.1.65 (iiia) that, for all $t \in I$,

$$x(t) \leq g^{-1}[m(t) + n(t)r_3(t, t_0, 0)],$$

where

$$r_3(t,t_0,0) = \int_{t_0}^{t} \exp\left(\int_{s}^{t} \lambda(\xi)n(\xi)d\xi\right)\lambda(s)m(s)ds.$$

Also, by Theorem 1.1.65 (iiib), we get for all $t \in I$,

$$x(t) \leq g^{-1}[M(t)\exp(\int_{t_0}^{t} N(t)\lambda(s)ds)].$$

Recall that the work of Butler and Rogers [126] follows from the above results on noting that $h(u) = u, c(t) = 0$ and $\omega(t,u) = g(u)$. We then get by Theorem 1.1.65 (ii),

$$x(t) \leq f^{-1}[r_2(t,t_0,A(t))],$$

where

$$r_2(T,t_0,A(T)) = \Omega^{-1}\left(\Omega(A(T)) + \int_{t_0}^{t} B(T)K(T,s)ds\right),$$

and Ω is given by

$$\Omega(u) = \int_{u_0}^{u} \frac{ds}{g \cdot f^{-1}(s)}, \quad u \geq u_0 > 0.$$

Next, we derive the result due to Deo and Murdeshwar [196]. Choosing $f(u) = u, b(t) = 1, k(t,s) = 1$ and $\omega(t,u) = \lambda(t)g(u), g(u)$ being sub-additive, then by Theorem 1.1.65 (i), we have

$$x(t) \leq a(t) + h(r_1(t,t_0,0)),$$

where $r_1(t,t_0,0)$ is the maximal solution of

$$r' = \lambda(t)H(r_1) + \sigma(t)$$

with $H = g \cdot h$ and $\sigma(t) = \lambda(t)g(a(t))$. This, by Corollary 1.1.18, nothing that H is non-decreasing, we arrive at

$$r_1(t,t_0,0) \leq G^{-1}\left[\int_{t_0}^{t} \lambda(s)ds + G\left(\int_{t_0}^{t} \sigma(s)ds\right)\right] \equiv r(t,t_0,0),$$

where

$$G(u) = \int_{u_0}^{u} \frac{ds}{H(s)}, \quad u \geq u_0 > 0.$$

We therefore obtain

$$x(t) \leq a(t) + h(r(t, t_0, 0)).$$

Finally, let $f(u) = u, b(t) = 1, h(u) = u, c(t) = 0$ and $k(t, s) = 1$, then we deduce Corollary 1.9.4 in [325] from (1.1.440) which is a generalization of the Bellman-Gronwall-Reid inequality in a rather general form. Clearly, for various choices of the functions involved in (1.1.439), we can derive from above results respective explicit bounds.

In 1981, Zahrout et al proved the following Bellman-Gronwall-Reid inequality (see, e.g., Kuang [315]).

Theorem 1.1.66 (The Bellman-Gronwall-Reid Inequality [714]) *Let f, g, u, v be continuous on $\mathbb{R}_+$. If there exist non-negative constants C, p $(0 \leq p < 1)$ such that for all $t \in \mathbb{R}_+$,*

$$u(t) \leq C + \int_0^t v(s)\left[u(s) + \int_0^s v(r)\left(\int_0^r [f(t)u(t) + g(t)u^p(t)]dt\right)dr\right]ds,$$

then we have

$$u(t) \leq C + \int_0^t v(s)\Bigg(C + \int_0^s v(r)\exp(\int_0^r [v(t) + f(t)]dt)$$

$$\times \left\{C^{1-p} + (1-p)\int_0^r g(t)\exp\left[-(1-p)\int_0^t (v(y) + f(y))dy\right]dt\right\}^{1/(1-p)} dr\Bigg)ds.$$

Proof The proof is left to the reader as an exercise. □

Theorem 1.1.67 (The Pachpatte-Pachpatte Inequality [525]) *Let $u(t), a(t), b(t)$ be real-valued non-negative continuous functions defined for all $t \in \mathbb{R}_+$ and $L : \mathbb{R}_+^2 \to \mathbb{R}_+$ be a continuous function which satisfies the condition for all $u \geq v \geq 0$,*

$$0 \leq L(t, u) - L(t, v) \leq M(t, v)(u - v),$$

where $M(t, v)$ is a real-valued non-negative continuous functions defined for all $t, v \in \mathbb{R}_+$. If for all $t \in \mathbb{R}_+$,

$$u(t) \leq a(t) + b(t)\int_t^{+\infty} L(s, u(s))ds, \tag{1.1.455}$$

then for all $t \in \mathbb{R}_+$,

$$u(t) \leq a(t) + b(t)e(t)\exp\Big(\int_t^{+\infty} M(s, a(s))b(s)ds\Big), \tag{1.1.456}$$

where for all $t \in \mathbb{R}_+$,

$$e(t) = \int_t^{+\infty} L(s, a(s))ds. \tag{1.1.457}$$

Proof Define a function $z(t)$ by

$$z(t) = \int_0^{+\infty} L(s, u(s))ds. \tag{1.1.458}$$

Then from (1.1.455), we have

$$u(t) \le a(t) + b(t)z(t). \tag{1.1.459}$$

From (1.1.458), (1.1.459) and the hypotheses on L, we derive that

$$\begin{aligned} z(t) \le & \int_t^{+\infty} [L(s, a(s) + b(s)z(s)) - L(s, a(s)(s)) + L(s, a(s))]ds \\ & \le e(t) + \int_t^{+\infty} M(s, a(s))b(s)z(s)ds \end{aligned} \tag{1.1.460}$$

where $e(t)$ is defined by (1.1.457). Clearly $e(t)$ is real-valued non-negative, continuous and non-increasing in $t \in \mathbb{R}_+$. An application to Theorem 1.1.5 in Qin [557] to (1.1.460) yields

$$z(t) \le e(t)\exp\left(\int_t^{+\infty} M(s, a(s))b(s)ds\right). \tag{1.1.461}$$

The desired inequality in (1.1.456) follows from (1.1.459) and (1.1.461). □

Theorem 1.1.68 (The Pachpatte-Pachpatte Inequality [525]) *Let* $u(t), a(t), b(t), M(t, v)$ *be as in Theorem 1.1.67 and* $L : \mathbb{R}_+^2 \to \mathbb{R}_+$ *be a continuous function which satisfies the condition for all* $u \ge v \ge 0$,

$$0 \le L(t, u) - L(t, v) \le M(t, v)\phi^{-1}(u - v),$$

where $\phi : \mathbb{R}_+ \to \mathbb{R}_+$ *be a continuous and strictly increasing function with* $\phi(0) = 0$, ϕ^{-1} *is the inverse function of* ϕ *and for all* $u, v \in \mathbb{R}_+$,

$$\phi^{-1}(uv) \le \phi^{-1}(u)\phi^{-1}(v).$$

If for all $t \in \mathbb{R}_+$,

$$u(t) \le a(t) + b(t)\phi\left(\int_t^{+\infty} L(s, u(s))ds\right), \tag{1.1.462}$$

then for all $t \in \mathbb{R}_+$,

$$u(t) \leq a(t) + b(t)\phi\Big(e(t)\exp\Big(\int_t^{+\infty} M(s, a(s))\phi^{-1}(b(s))ds\Big)\Big), \quad (1.1.463)$$

where $e(t)$ *is defined by (1.1.457).*

Proof Define a function $z(t)$ by (1.1.458). Then from (1.1.462), we have

$$u(t) \leq a(t) + b(t)\phi(z(t)). \quad (1.1.464)$$

From (1.1.458), (1.1.464) and the hypotheses on L and ϕ, we observe

$$\begin{aligned} z(t) &\leq \int_t^{+\infty} [L(s, a(s) + b(s)\phi(z(s))) - L(s, a(s)(s)) + L(s, a(s))]ds \\ &\leq e(t) + \int_t^{+\infty} M(s, a(s))\phi^{-1}(b(s))\phi(z(s))ds \\ &\leq e(t) + \int_t^{+\infty} M(s, a(s))\phi^{-1}(b(s))z(s)ds \end{aligned}$$

where $e(t)$ is defined by (1.1.457). Now by following the last arguments as in the proof of Theorem 1.1.67, we get the required inequality in (1.1.463). □

We next establish the following inequality which can be used in more general situation.

Theorem 1.1.69 (The Pachpatte-Pachpatte Inequality [525]) *Let* $u(t), a(t), b(t)$ *be as in Theorem 1.1.67 and* $L : \mathbb{R}_+^2 \to \mathbb{R}_+$ *be a continuous function which satisfies the condition for all* $u \geq v \geq 0$,

$$0 \leq L(t, u) - L(t, v) \leq M(t, v)(u - v),$$

where $M(t, v)$ *is a real-valued non-negative continuous functions defined for all* $t, v \in \mathbb{R}_+$. *If for all* $t \in \mathbb{R}_+$,

$$u(t) \leq a(t) + \int_t^{+\infty} b(s)u(s)ds + \int_t^{+\infty} L(s, u(s))ds, \quad (1.1.465)$$

then for all $t \in \mathbb{R}_+$,

$$u(t) \leq E(t)\Big[a(t) + A(t)\exp\Big(\int_t^{+\infty} M(s, E(s)a(s))E(s)ds\Big)\Big], \quad (1.1.466)$$

$$u(t) \leq a(t) + b(t)\exp\left\{\int_t^{+\infty} M[s, a(s)]b(s)ds\right\}\exp\left\{\int_t^{+\infty} L(s, a(s))ds\right\}, \quad (1.1.467)$$

where for all $t \in \mathbb{R}_+$,

$$E(t) = \exp\Big(\int_t^{+\infty} b(s)ds\Big),\ A(t) = \int_t^{+\infty} L(s, E(s)a(s))ds. \quad (1.1.468)$$

Proof Define a function $z(t)$ by (1.1.458). Then (1.1.465) can be restated as

$$u(t) \le a(t) + z(t) + \int_t^{+\infty} b(s)u(s)ds, \quad (1.1.469)$$

since $a(t) + z(t)$ is non-negative, continuous and non-increasing for all $t \in \mathbb{R}_+$, by applying Theorem 1.1.5 in Qin [557] to (1.1.469), we have

$$u(t) \le (a(t) + z(t))E(t). \quad (1.1.470)$$

From (1.1.458) and (1.1.470) and the hypotheses on L, we observe that

$$\begin{aligned} z(t) &\le \int_t^{+\infty} [L(s, E(s)a(s) + E(s)z(s)) - L(s, E(s)(s)) + L(s, E(s)a(s))]ds \\ &\le A(t) + \int_t^{+\infty} M(s, E(s)a(s))E(s)z(s)ds. \quad (1.1.471) \end{aligned}$$

Clearly, $A(t)$ is non-negative, continuous and non-increasing for all $t \in \mathbb{R}_+$. Now an application of Theorem 1.1.5 in Qin [557] to (1.1.471) yields

$$z(t) \le A(t)\exp\Big(\int_t^{+\infty} M(s, E(s)a(s))E(s)ds\Big). \quad (1.1.472)$$

Using (1.1.472) in (1.1.470), we get the required inequality in (1.1.466). The proof of (1.1.467) is similar. □

Recall that, Bihari [82] and Langenhop [328] have extended the classical integral inequalities of Gronwall [259] to cover certain non-linear situations. Since then, many works have appeared in generalizing the above mentioned results. These results, though quite similar in their conclusions, are quite different in their approaches. For instance, Antosiewicz [34] used the direct method of Lyapunov, Lakshmikantham [321] took the approach of maximal and minimal solutions; Stokes [625] applied the fixed point theorem of Tychonoff in certain abstract function spaces; Nohel [423] discussed his results from an integral equation point of view, and Viswanatham [657] based his proof on the method of successive approximations. Moreover, Bauer [46] has improved the results of [34] with simplified proofs; Waltman and Hanson [661] presented a lattice theoretical proof of the result given in [657], and Redheffer [572] gave a proof of all to the results of [82] and [328]. However, almost invariably in all the above mentioned works, the general scheme has been to compare two differential equations in order to conclude that the solution of one bounds the solution of the other if the derivative of one is bounded in some appropriate way by that of the other. The next result, due to [679], is to present a generalization of the result of Viswanatham which removes certain monotonity assumptions.

Consider the following first order equation:

$$u'(t) = f(t, u) \quad u(0) = c, \tag{1.1.473}$$

where $f(t, u)$ is continuous in the region R defined by $0 \leq t < +\infty$ and $-\infty < u < +\infty$, and its corresponding differential inequality:

$$v'(t) \leq f(t, v) \quad v(0) = c, \tag{1.1.474}$$

also defined over the region R.

The next result, due to [679], is as follows.

Theorem 1.1.70 ([679]) *Let $v(t)$ be a continuous solution of problem (1.1.474) and $u(t)$ be a maximal solution to equation (1.1.473). Then $v(t) \leq u(t)$ for all $t \geq 0$.*

Proof Consider the difference function $\delta(t) = u(t) - v(t)$. If $\delta(t) \geq 0$ for all $t \geq 0$, then there is nothing to prove. Otherwise, there exists an interval $[t_1, t_2], 0 \leq t_1 < t_2 < +\infty$ such that $\delta(t_1) = 0$ and $\delta(t) < 0$ for $t_1 < t < t_2$. Consider now the following one parameter family of differential equations, $0 < \varepsilon < 1$,

$$z'(t) = f(t, z) + \varepsilon, \quad z(t_1) = u(t_1), \tag{1.1.475}$$

restricted over the region defined by $t_1 \leq t \leq t_2$ and $|z| \leq M$, for some sufficiently large M. Since $u(t)$ is the maximal solution, there exists a family of solutions $\{u_\varepsilon(t) : 0 < \varepsilon < 1\}$, each $u_\varepsilon(t)$ is a solution to equation (1.1.475), and a number $t_3, t_1 < t_3 \leq t_2$ and t_3 independent of ε, such that $\lim_{\varepsilon \to 0} u_\varepsilon(t) = u(t)$ uniformly over $[t_1, t_3]$ (see, e.g., [156, 301]). Denote by $\delta_\varepsilon(t)$ the difference $u_\varepsilon(t) - v(t)$. For each $\varepsilon > 0$, we have $\delta_\varepsilon(t_1) = 0$ and $\delta_\varepsilon'(t_1) > 0$. Thus there exists, by continuity, a number t_4, (t_4 may depend on ε), such that $\delta_\varepsilon(t) > 0$ for $t_1 < t \leq t_4 \leq t_3$. If $t_4 < t_3$, then there must exist $t_5, t_4 \leq t_5 < t_3$ such that $\delta_\varepsilon(t_5) = 0$ and $\delta_\varepsilon'(t_5) \leq 0$; but we have on account of (1.1.474) and (1.1.475) that $\delta_\varepsilon'(t_5) \geq \varepsilon > 0$. Hence $t_4 = t_3$, and t_4 is independent of each ε. Finally, we have $\delta(t) = \lim_{\varepsilon \to 0} \delta_\varepsilon(t) \geq 0$ uniformly for all $t \in [t_1, t_3]$, which is the desired contradiction. □

Corollary 1.1.20 ([679]) *Let $w(t)$ be any solution of the following differential equation*

$$w'(t) = g(t, w) \quad w(0) = c,$$

where $g(t, w)$ is continuous over R and $g(t, x) \leq f(t, x)$. Then $w(t) \leq u(t)$, where $u(t)$ is the maximal solution of equation (1.1.473).

Corollary 1.1.21 ([657]) *If $w(t) \leq f(t,x)$. Then $w(t) \leq u(t)$ where $f(t,w)$ is continuous and monotonic increasing in w in the region defined by $t \geq 0$ and $-\infty < w < +\infty$, and $w(t)$ is continuous for all $t \geq 0$, then $w(t) \leq u(t)$, where $u(t)$ is the maximal solution to equation (1.1.473).*

Proof Let $v(t) = c + \int_0^t f(s, w(s))ds$. Observe that $v'(t) = f(t, w(t)) \leq f(t, v(t))$, by the monotonicity of f. Therefore, the conclusion follows readily from the main theorem. □

Remark 1.1.19 Let $f(t,u) = a(t)u$ where $a(t)$ is any integrable function over $[0, +\infty)$. Then Theorem 1.1.70 reduces to a result on differential operators as given in [34] (Theorem 1, p. 133–134).

Remark 1.1.20 Note that, Corollary 1.1.20 may be used in a similar way as that of Lakshmikantham [321] to prove results on systems of differential equations.

Remark 1.1.21 We note that the monotonicity condition of f in Corollary 1.1.5 cannot be waived, although it is certainly not needed in Theorem 1.1.65. Let $f(t, v) = \frac{5}{v+1}$, $v(t) = t^2$ and $v_0 = 1$. Clearly, $v(t) \leq 1 + \int_0^t f(s, v)ds$ for $0 \leq t \leq 2.5$. In this case, the maximal solution of equation (1.1.473) is easily computed to be $u(t) = \sqrt{10t + 3} - 1$; but $u(2) = \sqrt{23} - 1 < 4 = v(2)$.

As an application to Theorem 1.1.65, we present the following result on approximate solutions which also generalizes the application given in [657]. Let $\theta_1(t), \theta_2(t)$ be $\varepsilon_1(t), \varepsilon_2(t)$ approximate solutions of (1.1.473), and for some time τ, $|\theta_1(\tau) - \theta_2(\tau)| \leq \delta$. If $f(t,u)$ satisfies, in addition, the condition

$$|f(t, u_1) - f(t, u_2)| \leq g(t, |u_1 - u_2|)$$

where $g(t,w)$ satisfies the condition of Corollary 1.1.20, then we may easily show that, for all $t \geq \tau$,

$$|\theta_1(t) - \theta_2(t)| \leq \chi(t),$$

where $\chi(t)$ is the maximal solution of

$$w' = g(t,w) + \varepsilon(t), \quad w(\tau) = \delta$$

with $\varepsilon(t) = |\varepsilon_1(t)| + |\varepsilon_2(t)|$. In case $g(t,z) = k(t)z$, then we can easily verify that

$$\chi(t) = \delta \exp\left(\int_\tau^t k(s)ds\right) + \int_\tau^t \varepsilon(s) \exp\left(\int_s^t k(u)du\right) ds,$$

which is just the result of Rao [562].

Theorem 1.1.71 (The Opial Inequality [435]) *Let the mapping $f : [0, \tau] \times \mathbb{R}^n \to \mathbb{R}^n$ be continuous and satisfy that for any $x, y \in \mathbb{R}^n$,*

$$x \leq y \Longleftrightarrow f(t, x) \leq f(t, y).$$

Here the relation "$\leq$" between any two points $x = (x_1, \cdots, x_n), y = (y_1, \cdots, y_n)$ in $\mathbb{R}^n$ means that

$$x \leq y \ \text{ if and only if } \ x_i \leq y_i \ \text{ for } \ i = 1, 2, \cdots, n.$$

If the continuous $n-$vector function $u(t)$ mapping $[0, \tau]$ into $\mathbb{R}^n$ satisfies the inequality, for all $0 \leq t \leq \tau$,

$$u(t) \leq \eta + \int_0^t f(s, u(s))ds, \tag{1.1.476}$$

where η is an $n-$vector in $\mathbb{R}^n$, then for all $0 \leq t \leq \tau$,

$$u(t) \leq \phi(t), \tag{1.1.477}$$

where $\phi(t)$ is the maximal solution of, for all $0 \leq t \leq \tau$,

$$x(t) = \eta + \int_0^t f(s, x(s))ds. \tag{1.1.478}$$

In fact, the above inequality was established by Opial [435] in 1957. This theorem for the special case of $n = 1$ was firstly established by Viswanatham [657], and may easily be modified to include the case when η itself is a continuous map of $[0, \tau]$ into $\mathbb{R}^n$ and f depends on three arguments t, s, u (see, e.g., Viswanatham [657]). We can also obtain as a special case ($n = 1$ and $f(t, u) = g(t)\omega(u)$ were $g(t) \geq 0$ and $\omega(u)$ is non-decreasing in u) some useful nonlinear generalizations of Theorem 1.1.1 which is due to LaSalle and Bihari (see, e.g., Kuang [315]). However, we should point out that in Theorem 1.1.72 below, we have used the fact that nonlinear function f is non-decreasing in its second argument. In fact, the above result may not hold if f is non-increasing instead of non-decreasing. By considering the second iterate of the mapping defined by the right-hand side of (1.1.476), however, Ziebur [715] succeed in proving the following result in which f may be either non-decreasing or non-increasing.

Theorem 1.1.72 (The Ziebur Inequality [715]) *Define the operator P by*

$$Px(t) = \eta + \int_0^t f(s, x(s))ds, \quad 0 \leq t \leq \tau. \tag{1.1.479}$$

Let $f(t,u)$ be continuous and be either non-decreasing or non-increasing in its second argument. Suppose the integral equation

$$x(t) = P^2x(t) \tag{1.1.480}$$

has a maximal solution $\phi(t)$. If a continuous function $u(t)$ satisfies, for all $0 \le t \le \tau$,

$$u(t) \le P^2u(t), \tag{1.1.481}$$

then for all $0 \le t \le \tau$,

$$u(t) \le \phi(t). \tag{1.1.482}$$

Remark 1.1.22 If f is non-decreasing and continuous and if ϕ is the maximal solution of equation (1.1.478), then it was shown in Ziebur [715] that ϕ is also the maximal solution of equation (1.1.480). Further, (1.1.482) is satisfied whenever (1.1.481) holds. Hence Theorem 1.1.71 is contained in Theorem 1.1.72.

The following result, due to Agarwal, Deng and Zhang [13], generalizes a Lipovan's result of the Bellman-Gronwall inequalities [355] to a new type of retarded inequalities which includes both a nonconstant term outside the integrals and more than one distinct nonlinear integrals, from which, Bihari's result and Pinto's result reduce readily as some special cases.

In the next result, we shall consider such an inequality, for all $t_0 \le t < t_1$,

$$u(t) \le a(t) + \sum_{i=1}^{n} \int_{b_i(t_0)}^{b_i(t)} f_i(s)w_i(u(s))ds, \tag{1.1.483}$$

in a general form, where $a(t)$ is a function and w_i's may be distinct, and have improved Lipovan's result [355] (see, e.g., Theorem 1.1.12). Furthermore, we show that the results of [82, 537] can be deduced from this result some special cases.

As in [541], we say $w_1 \propto w_2$ for $w_1, w_2 : A \subset \mathbb{R} \to \mathbb{R} \setminus \{0\}$ if $w_2/w_1$is non-decreasing on A. This concept helps us compare monotonicity of different functions. Consider inequality (1.1.483) and suppose that

(H_1) all w_i $(i = 1, \ldots, n)$ are continuous and non-decreasing functions on $[0, +\infty)$ and are positive on $(0, +\infty)$ such that $w_1 \propto \ldots \propto w_n$,
(H_2) $a(t)$ is continuously differentiable in t and non-negative on $[t_0, t_1)$ where t_0, t_1 are constants and $t_0 < t_1$,
(H_3) all $b_i : [t_0, t_1) \to [t_0, t_1)$ $(i = 1, \ldots, n)$ are continuously differentiable and non-decreasing such that $b_i(t) \le t$ *on* $[t_0, t_1)$,
(H_4) all $f_i(t, s)$, $i = 1, \cdots, n$, are continuous and non-negative functions on $[t_0, t_1) \times [t_0, t_1)$.

We shall use the notation $W_i(u, u_i) := \int_{u_i}^{u} \frac{dz}{w_i(z)}$ for all $u > 0$, where $u_i > 0$ is given constant. It is denoted by $W_i(u)$ simply when there is no confusion. Clearly, W_i is strictly increasing, so its inverse W_i^{-1} is well defined, continuous and increasing in its corresponding domain.

Theorem 1.1.73 (The Agarwal-Deng-Zhang Inequality [13]) *Suppose* $(H_1)-(H_4)$ *hold and* $u(t)$ *is a continuous and non-negative function on* $[t_0, t_1)$ *satisfying (1.1.483). Then for all* $t_0 \le t \le T_1$,

$$u(t) \le W_n^{-1}\left[W_n(r_n(t)) + \int_{b_n(t_0)}^{b_n(t)} \max_{t_0 \le \tau \le t} f_n(\tau, s)ds\right], \tag{1.1.484}$$

where $r_n(t)$ *is determined recursively by*

$$\begin{cases} r_1(t) := a(t_0) + \displaystyle\int_{t_0}^{t} |a'(s)|ds, \\ r_{i+1} := W_i^{-1}\left[W_i(r_i(t)) + \displaystyle\int_{b_i(t_0)}^{b_i(t)} \max_{t_0 \le \tau \le t} f_i(\tau, s)ds\right], \quad i = 1, \cdots, n-1, \end{cases} \tag{1.1.485}$$

and $T_1 < t_1$ *is the largest number such that*

$$W_i(r_i(T_1)) + \int_{b_i(t_0)}^{b_i(T_1)} \max_{t_0 \le \tau \le t} f_i(\tau, s)ds \le \int_{u_i}^{+\infty} \frac{dz}{w_i(z)}, \quad i = 1, \cdots, n. \tag{1.1.486}$$

Proof Obviously, $\tilde{f}_i(t, s) := \max_{t_0 \le \tau \le t} f_i(\tau, s)$ is non-negative and non-decreasing in t for each fixed s and satisfies $\tilde{f}_i \ge f_i(t, s)$ for each $i = 1, \cdots, n$. We first discuss the case that $a(t) > 0$ for all $t \in [t_0, t_1)$. It means that $r_1(t) > 0$ for all $t \in [t_0, t_1)$. In such a case, $r_1(t)$ is positive, differentiable and non-decreasing on $[t_0, t_1)$ and $r_1(t) \ge a(t_0) + \int_{t_0}^{t} a'(s)ds = a(t)$.

Consider now the auxiliary inequality, for all $t_0 \le t < T$,

$$u(t) \le r_1(t) + \sum_{i=1}^{n} \int_{b_i(t_0)}^{b_i(t)} \tilde{f}_i(T, s)w_i(u(s))ds, \tag{1.1.487}$$

where T is chosen arbitrarily such that $t_0 \le T \le T_1$. Using (1.1.487), we can claim for all $t_0 \le t \le T \le T_2$,

$$u(t) \le W_n^{-1}\left[W_n(\tilde{r}_n(T, t)) + \int_{b_n(t_0)}^{b_n(t)} \tilde{f}_n(T, s)ds\right], \tag{1.1.488}$$

where

$$\tilde{r}_1(T, t) = r_1(t), \quad \tilde{r}_{i+1}(T, t) = W_i^{-1}\left[W_i(\tilde{r}_i(T, t)) + \int_{b_i(t_0)}^{b_i(t)} \tilde{f}_i(T, s)ds\right], \tag{1.1.489}$$

$i = 1, \cdots, n-1$, and $T_2 \le t_1$ is the largest number such that

$$W_i(\tilde{r}_i(T, T_2)) + \int_{b_i(t_0)}^{b_i(T_2)} \tilde{f}_i(T, s)ds \le \int_{u_i}^{+\infty} \frac{dz}{w_i(z)}, \quad i = 1, \cdots, n. \tag{1.1.490}$$

Notice that $T_1 \le T_2$. In fact, both $\tilde{r}_i(T, t)$ and $\tilde{f}_i(T, t)$ are non-decreasing in T. Thus, T_2 satisfying (1.1.490) gets smaller as T is chosen larger. In particular, T_2 satisfies the same (1.1.486) as T_1 when $T = T_1$.

To prove (1.1.488) for $n = 1$, we observe that (1.1.487) is equivalent to $u(t) \le r_1(t) + z(t)$ for all $t \in [t_0, T]$ where $z(t) := \int_{b_1(t_0)}^{b_1(t)} \tilde{f}_1(T, s) w_1(u(s))ds$ is a non-negative, non-decreasing and differentiable function on $[t_0, T]$. Moreover, $b_1(t)$ is differentiable and non-decreasing in t. So $b_1'(t) \ge 0$ for all $t \in [t_0, T]$. Since w_1 is non-decreasing, $z(t) + r_1(t) > 0$ and $b_1(t) \le t$ for all $t \in [t_0, T]$, we know

$$\begin{aligned}
\frac{z'(t) + r_1'(t)}{w_1(z(t) + r_1(t))} &\le \frac{b_1'(t)\tilde{f}_1(T, b_1(t)) w_1(u(b_1(t)))}{w_1(z(t) + r_1(t))} + \frac{r'(t)}{w_1(z(t) + r_1(t))} \\
&\le \frac{b_1'(t)\tilde{f}_1(T, b_1(t)) w_1(z(b_1(t)) + r_1(b_1(t)))}{w_1(z(t) + r_1(t))} + \frac{r'(t)}{w_1(r_1(t))} \\
&\le \frac{b_1'(t)\tilde{f}_1(T, b_1(t)) w_1(z(t) + r_1(t))}{w_1(z(t) + r_1(t))} + \frac{r'(t)}{w_1(r_1(t))} \\
&\le b_1'(t)\tilde{f}_1(T, b_1(t)) + \frac{r_1'(t)}{w_1(r_1(t))}.
\end{aligned}$$

Integrating both sides of the above inequality from t_0 to t, we can obtain for all $t_0 \le t \le T$,

$$\begin{aligned}
W_1(z(t) + r_1(t)) &\le W_1(r_1(t)) + \int_{t_0}^{t} b_1'(t)\tilde{f}_1(T, b_1(s))ds \\
&\le W_1(r_1(t)) + \int_{b_1(t_0)}^{b_1(t)} \tilde{f}_1(T, s)ds.
\end{aligned}$$

By (1.1.490), we may see that $W_1(r_1(t)) + \int_{b_1(t_0)}^{b_1(t)} \tilde{f}_1(T, s)ds$ is the domain of W_1^{-1} for all $t \in [t_0, T]$ for $n = 1$. Thus the monotonicity of W_1^{-1} implies, for all $t_0 \le t \le T \le T_2$,

$$u(t) \le r_1(t) + z(t) \le W_1^{-1}\left[W_1(r_1(t)) + \int_{b_1(t_0)}^{b_1(t)} \tilde{f}_1(T, s)ds \right],$$

i.e., (1.1.488) is true for $n = m$. Consider, for all $t_0 \leq t \leq T$,

$$u(t) \leq r_1(t) + \sum_{i=1}^{m+1} \int_{b_i(t_0)}^{b_i(t)} \tilde{f}_i(T,s)w_i(u(s))ds.$$

Let $z(t) = \sum_{i=1}^{m+1} \int_{b_i(t_0)}^{b_i(t)} \tilde{f}_i(T,s)w_i(u(s))ds$. Then $z(t)$ is differentiable, non-negative and non-decreasing on $[t_0, T]$ and satisfies $u(t) \leq r_1(t) + z(t)$ for all $t \in [t_0, T]$. Since w_i is non-decreasing, $z(t) + r_1(t) > 0$ and $b_i'(t) \geq 0$, we may get, for all $t_0 \leq t \leq T$,

$$\begin{aligned}
\frac{z'(t) + r_1'(t)}{w_1(z(t) + r_1(t))} &\leq \frac{\sum_{i=1}^{m+1} b_i'(t)\tilde{f}_i(T, b_i(t))w_i(u(b_i))}{w_1(z(t) + r_1(t))} + \frac{r_1'(t)}{w_1(z(t) + r_1(t))} \\
&\leq \sum_{i=1}^{m+1} b_i'(t)\tilde{f}_i(T, b_i(t))\frac{w_i(z(b_i(t))) + r_1(b_i(t))}{w_1(z(t) + r_1(t))} + \frac{r_1'(t)}{w_1(r_1(t))} \\
&\leq \sum_{i=1}^{m+1} b_i'(t)\tilde{f}_i(T, b_i(t))\frac{w_i(z(b_i(t)) + r_1(b_i(t)))}{w_1(z(t) + r_1(t))} + \frac{r'(t)}{w_1(r_1(t))} \\
&\leq b_1'(t)\tilde{f}_1(T, b_1(t)) + \sum_{i=2}^{m+1} b_i'(t)\tilde{f}_i(T, b_i(t))\phi_i(z(b_i(t)) + r_1(b_i(t))) \\
&\quad + \frac{r'(t)}{w_1(z(t) + r_1(t))} \\
&\leq b_1'(t)\tilde{f}_1(T, b_1(t)) + \sum_{i=1}^{m+1} b_{i+1}'(t)\tilde{f}_{i+1}(T, b_{i+1}(t))\phi_{i+1}(z(b_{i+1}(t)) + r_1(b_{i+1}(t))) \\
&\quad + \frac{r'(t)}{w_1(z(t) + r_1(t))}
\end{aligned}$$

where $\phi_{i+1}(u) := \frac{w_{i+1}(u)}{w_i(u)}$, $i = 1, \ldots, m$. Integrating the above inequality from t_0 to t, we may obtain for all $t_0 \leq t \leq T$,

$$\begin{aligned}
W_1(z(t) + r_1(t)) &\leq W_1(r_1(t)) + \int_{t_0}^{t} b_1'(s)\tilde{f}_1(T, b_1(s))ds \\
&\quad + \sum_{i=1}^{m} \int_{t_0}^{t} b_{i+1}'(s)\tilde{f}_{i+1}(T, b_{i+1}(s))\phi_{i+1}(z(b_{i+1}(s)) + r_1(b_{i+1}(s)))ds \\
&\leq W_1(r_1(t)) + \int_{b_1(t_0)}^{b_1(t)} \tilde{f}_1(T, s)ds \\
&\quad + \sum_{i=1}^{m} \int_{b_{i+1}(t_0)}^{b_{i+1}(t)} \tilde{f}_{i+1}(T, s)\phi_{i+1}(z(s) + r_1(s))ds
\end{aligned}$$

or equivalently, for all $t_0 \leq t \leq T$,

$$\xi(t) \leq c_1(t) + \sum_{i=1}^{m} \int_{b_{i+1}(t_0)}^{b_{i+1}(t)} \tilde{f}_{i+1}(T,s)\phi_{i+1}(W_1^{-1}(\xi(s)))ds,$$

the same form as (1.1.487) for $n = m$, where $\xi(t) = W_1(z(t) + r_1(t))$ and $c_1(t) = W_1(r_1(t)) + \int_{b_1(t_0)}^{b_1(t)} \tilde{f}_1(T,s)ds$. From the assumption (H_1), each $\phi_{i+1}(W_1^{-1}), i = 1, \cdots, m$, is continuous and non-decreasing on $[0, +\infty)$ and is positive on $(0, +\infty)$ since W_1^{-1} is continuous and non-decreasing on $[0, +\infty)$. Moreover, $\phi_2(W_1^{-1}) \propto \phi_3(W_1^{-1}) \propto \ldots \propto \phi_{m+1}(W_1^{-1})$. By the inductive assumption, we thus have, for all $t_0 \leq t \leq min\{T, T_3\}$,

$$\xi(t) \leq \Phi_{m+1}^{-1}[\Phi_{m+1}(c_m(t)) + \int_{b_{m+1}(t_0)}^{b_{m+1}(t)} \tilde{f}_{m+1}(T,s)ds], \qquad (1.1.491)$$

where $\Phi_{i+1}(u) = \int_{\tilde{u}_{i+1}}^{u} \frac{dz}{\phi_{i+1}(W_1^{-1}(z))}$, $u > 0$, $\tilde{u}_{i+1} = W_1(u_{i+1})$, Φ_{i+1}^{-1} is the inverse of Φ_{i+1}, $i = 1, \ldots, m$,

$$c_{i+1}(t) = \Phi_{i+1}^{-1}\left[\Phi_{i+1}(c_i(t)) + \int_{b_{i+1}(t_0)}^{b_{i+1}(t)} \tilde{f}_{i+1}(T,s)ds\right], \quad i = 1, .., m-1,$$

and $T_3 < t_1$ is the largest number such that

$$\Phi_{i+1}(c_i(T_3)) + \int_{b_{i+1}(t_0)}^{b_{i+1}(T_3)} \tilde{f}_{i+1}(T,s)ds \leq \int_{\tilde{u}_{i+1}}^{W_1(+\infty)} \frac{dz}{\phi_{i+1}(W_1^{-1}(z))}, \quad i = 1, \ldots, m. \qquad (1.1.492)$$

Note that

$$\begin{aligned}
\Phi_i &= \int_{\tilde{u}_i}^{u} \frac{dz}{\phi_i(W_1^{-1}(z))} = \int_{W_1(u_i)}^{u} \frac{w_1(W_1^{-1}(z))dz}{w_1(W_1^{-1}(z))} \\
&= \int_{u_i}^{W_1^{-1}(u)} \frac{dz}{w_i(z)} = W_i(W_1^{-1}(u)), \quad i = 2, \ldots, m+1.
\end{aligned}$$

Then from (1.1.491) it follows, for all $t_0 \leq t \leq min\{T, T_3\}$,

$$\begin{aligned}
u(t) &\leq r_1(t) + z(t) = W_1^{-1}(\xi(t)) \\
&\leq W_{m+1}^{-1}[W_{m+1}(W_1^{-1}(c_m(t))) + \int_{b_{m+1}(t_0)}^{b_{m+1}(t)} \tilde{f}_{m+1}(T,s)ds]. \qquad (1.1.493)
\end{aligned}$$

Let $\tilde{c}_i(t) = W_1^{-1}(c_i(t))$. Obviously, $\tilde{c}_1(t) = W_1^{-1}(c_1(t)) = W_1^{-1}[W_1(r_1(t)) + \int_{b_1(t_0)}^{b_1(t)} \tilde{f}_1(T,s)ds] = W_1^{-1}[W_1(\tilde{r}_1(t)) + \int_{b_1(t_0)}^{b_1(t)} \tilde{f}_1(T,s)ds]$. Moreover, with assumption that $\tilde{c}_m(t) = \tilde{r}_{m+1}(T, t)$, we can see that

$$\begin{aligned}\tilde{c}_{m+1} &= W_1^{-1}\left[\Phi_{m+1}^{-1}[\Phi_{m+1}(c_m(t)) + \int_{b_{m+1}(t_0)}^{b_{m+1}(t)} \tilde{f}_{m+1}(T, s)ds]\right] \\ &= W_{m+1}^{-1}\left[W_{m+1}(W_1^{-1}(c_m(t))) + \int_{b_{m+1}(t_0)}^{b_{m+1}(t)} \tilde{f}_{m+1}(T, s)ds\right] \\ &= W_{m+1}^{-1}\left[W_{m+1}(\tilde{c}_m(t)) + \int_{b_{m+1}(t_0)}^{b_{m+1}(t)} \tilde{f}_{m+1}(T, s)ds\right] \\ &= W_{m+1}^{-1}\left[W_{m+1}(\tilde{r}_{m+1}(T, t)) + \int_{b_{m+1}(t_0)}^{b_{m+1}(t)} \tilde{f}_{m+1}(T, s)ds\right] = \tilde{r}_{m+2}(T, t),\end{aligned}$$

which proves

$$\tilde{c}_i(t) = \tilde{r}_{i+1}(T, t), \quad i = 1, \ldots, m.$$

Therefore, (1.1.492) becomes

$$\begin{aligned}W_{i+1}(\tilde{r}_{i+1}(T, T_3)) + \int_{b_{i+1}(t_0)}^{b_{i+1}(T_3)} \tilde{f}_{i+1}(T, s)ds &\le \int_{\tilde{u}_{i+1}}^{W_1(\infty)} \frac{dz}{\phi_{i+1}(W_1^{-1}(z))} \\ &= \int_{u_{i+1}}^{+\infty} \frac{dz}{w_{i+1}(z)}, \quad i = 1, .., m,\end{aligned}$$

implying that $T_2 = T_3$ and $T \le T_3$. From (1.1.493) it thus follows, for all $t_0 \le t \le T \le T_2$,

$$u(t) \le W_{m+1}^{-1}[W_{m+1}(\tilde{r}_{m+1}(T, t)) + \int_{b_{m+1}(t_0)}^{b_{m+1}(t)} \tilde{f}_{m+1}(T, s)ds].$$

This proves (1.1.487) by induction. Finally, from (1.1.483) we derive

$$\begin{aligned}u(T) &\le a(T) + \sum_{i=1}^{n} \int_{b_i(t_0)}^{b_i(T)} \tilde{f}_i(T, s)w_i(u(s))ds \\ &\le r_1(T) + \sum_{i=1}^{n} \int_{b_i(t_0)}^{b_i(T)} \tilde{f}_i(T, s)w_i(u(s))ds,\end{aligned}$$

namely, the auxiliary inequality holds for $t = T$. By (1.1.487), we may get for all $t_0 \le T \le T_1$,

$$\begin{aligned}u(T) &\le W_n^{-1}\left[W_n(\tilde{r}_n(T, T)) + \int_{b_n(t_0)}^{b_n(T)} \tilde{f}_n(T, s)ds\right] \\ &\le W_n^{-1}\left[W_n(r_n(T)) + \int_{b_n(t_0)}^{b_n(T)} \tilde{f}_n(T, s)ds\right],\end{aligned}$$

where we have used the facts that $\tilde{r}_n(T,T) = r_n(T)$ and $T_2 = T_1$, which can be easily verified and found in the sentences after (1.1.489) respectively. This proves (1.1.484) because T is arbitrarily chosen.

In case $a(t) = 0$ for some $t \in [t_0, t_1)$. Let $r_{1,\epsilon}(t) := r_1(t) + \epsilon$ for all $t \in [t_0, t_1)$, where $\epsilon > 0$ is arbitrary. Then $r_{1,\epsilon}(t) > 0$ for all $t \in [t_0, t_1)$ since $r_1(t) \geq 0$. Using the same arguments as above, where $r_1(t)$ is replaced with the positive $r_{1,\epsilon}(t)$, we can derive, for all $t_0 \leq t \leq T_1$,

$$u(t) \leq W_n^{-1}\left[W_n(r_{n,\epsilon}) + \int_{b_n(t_0)}^{b_n(t)} \tilde{f}_n(t,s)ds\right].$$

Letting $\epsilon \to 0^+$, we obtain (1.1.484) because of continuity of $r_{i,\epsilon}$ in ϵ and continuity of W_i and W_i^{-1} for $i = 1, \cdots, n$. This completes the proof. □

Remark 1.1.23 T_1 is confined by (1.1.486). In particular, (1.1.484) is true for all $t \in [t_0, t_1)$ when all w_i $(i = 1, \cdots, n)$ satisfy $\int_{u_i}^{+\infty} \frac{dz}{w_i(z)}$.

Remark 1.1.24 Different choices of u_i in W_i do not affect the above result. In fact, for positive constants $v_i \neq u_i$, let $\tilde{W}_i(u) = \int_{v_i}^{u} \frac{dz}{w_i(z)}$. Then $\tilde{W_i}(u) = \tilde{W_i}(u_i) + W_i(u)$, and $\tilde{W}_i^{-1}(v) = W_i^{-1}(v - \tilde{W}_i(u_i))$. Let $\tilde{f}_i(\tilde{t},s) := \max_{t_0 \leq \tau \leq t} f_i(\tau, s)$. It thus follows that $\tilde{W}_i^{-1}[\tilde{W}_i(r_i(t)) + \int_{b_i(t_0)}^{b_i(t)} \tilde{f}_i(t,s)ds] = W_i^{-1}[W_i(r_i(t)) + \int_{b_i(t_0)}^{b_i(t)} f_i(t,s)ds]$ and

$$\begin{aligned}\tilde{W}_{r_i(T_1)} + \int_{b_i(t_0)}^{b_i(T_1)} \tilde{f}_i(T_1,s)ds &= \tilde{W}_i(u_i) + W_i(r_i(T_1)) + \int_{b_i(t_0)}^{b_i(T_1)} \tilde{f}_i(T_1,s)ds \\ &\leq \int_{v_i}^{+\infty} \frac{dz}{w_i(z)} = \int_{v_i}^{u_i} \frac{dz}{w_i(z)} = \int_{v_i}^{u_i} \frac{dz}{w_i(z)} + \int_{u_i}^{+\infty} \frac{dz}{w_i(z)} \\ &= \tilde{W}_i(u_i) + \int_{u_i}^{+\infty} \frac{dz}{w_i(z)}.\end{aligned}$$

That is, (1.1.484)–(1.1.486) are independent of the choice of $u_i > 0$.

Suppose that (1.1.483) and $(H_1) - (H_4)$ hold on the closed interval $[t_0, t_1]$ instead. By replacing the inequality $T_1 < t_1$ by that $T_1 \leq t_1$ in the line above (1.1.486) and using the same arguments in the proof of Theorem 1.1.73, we easily conclude that Theorem 1.1.73 also holds for all $t \in [t_0, t_1]$. In particular, when $a(t) \equiv a$ (a non-negative constant), $f_i(t,s) = \lambda_i(s)$, $b_i(t) = t$, for all $t \in [t_0, t_1]$, $i = 1, \cdots, n$, inequality (1.1.483) becomes

$$u(t) \leq a + \sum_{i=1}^{n} \int_{t_0}^{t} \lambda_i(s) w_i(u(s))ds, \tag{1.1.494}$$

which is the form of (5) in [541]. Thus Theorem 1.1.73 implies Theorem 1 of [541]. Even if sometimes $b_i(t)$ can be enlarged to t such that (1.1.483) is changed into the form of (1.1.494), Theorem 1.1.73 gives us a finer estimate. For example, the inequality, for all $0 \leq t \leq t_1$,

$$u(t) \leq 1 + 2\int_0^t (s+1)\sqrt{u(s)}ds + 2\int_0^{\sqrt{t}} su(s)ds, \tag{1.1.495}$$

implies, for all $0 \leq t \leq t_1$,

$$u(t) \leq 1 + 2\int_0^t (s+1)\sqrt{u(s)}ds + 2\int_0^{t} su(s)ds, \tag{1.1.496}$$

by enlarging $\sqrt{t}$ to t. Applying Theorem 1.1.73, we can obtain, for all $0 \leq t \leq t_1$,

$$u(t) \leq \frac{1+(1+t)^2}{2}e^t, \tag{1.1.497}$$

where $T_1 = t_1$ because $\int_{u_1}^{+\infty} \frac{dz}{\sqrt{z}} = +\infty$ and $\int_{u_2}^{+\infty} \frac{dz}{\sqrt{z}} = +\infty$. On the other hand, Theorem 1 of [541] gives us from (1.1.496) that, for all $0 \leq t \leq t_1$,

$$u(t) \leq \frac{1+(1+t_1)^2}{2}e^{t^2}. \tag{1.1.498}$$

Clearly, (1.1.497) is sharper than (1.1.498) for large t. In [182, 541], there are some examples of integral inequalities where the function term $a(t)$ also reduces to a constant by dividing $a(t)$ in (1.1.483). However, all functions w_i's are required to be given in the class $\mathcal{F}_1$ as defined in [181, 541] (see, Definitions 1.1.3 and 1.1.4). With Theorem 1.1.73, the restriction to the special class is not needed.

The next result, due to [306], is a new interesting Gronwall-like integral inequality involving iterated integrals.

Theorem 1.1.74 (The Kim Inequality [306]) *Let $u(t)$ be non-negative continuous function in $J = [\alpha, \beta]$ and let $a(t)$ be positive non-decreasing continuous function in J, and let $f_i(t,s)$, $i = 1, \ldots, n$, be non-negative continuous functions for $\alpha \leq s \leq t \leq \beta$ which are non-decreasing in t for fixed $s \in J$. If for all $t \in I$,*

$$u(t) \leq a(t) + \int_\alpha^t f_1(t,t_1)\bigg(\int_\alpha^{t_1} f_2(t_1,t_2)\cdots\bigg(\int_\alpha^{t_{n-1}} f_n(t_{n-1},t_n)u^p(t_n)dt_n\bigg)\cdots\bigg)dt_1 \tag{1.1.499}$$

where $p \geq 0$, $p \neq 1$, is a constant. Then $u(t) \leq Y_1(t,t)$, where $Y_1(T,t)$ can be successively determined from the formulas for all $t \in [\alpha, \beta_1)$,

$$Y_n(T,t) = \exp\left(\int_\alpha^t \sum_{i=1}^{n-1} f_i(T,s)ds\right)$$

$$\times \left[a^q(T) + q\int_\alpha^t f_n(T,s)\exp\left(-q\int_\alpha^s \sum_{i=1}^{n-1} f_i(T,\tau)d\tau\right)ds\right]^{\frac{1}{q}} \tag{1.1.500}$$

with $q = 1-p$ and β_1 is chosen so that the expression between $[\cdots]$ is positive in the sub-interval $[\alpha, \beta_1)$, and for $k = n-1, \cdots, 1$, $\alpha \leq t \leq t \leq T \leq \beta$,

$$\begin{cases} Y_k(T,t) = E_k(T,t)\left[a(T) + \displaystyle\int_\alpha^t f_k(T,s)\frac{Y_{k+1}(T,s)}{E_k(T,s)}ds\right], \\ E_k(T,t) = \exp\left(\displaystyle\int_\alpha^s \left[\sum_{i=1}^{k-1} f_i(T,\tau) - f_k(T,\tau)\right]d\tau\right). \end{cases} \tag{1.1.501}$$

Proof Fix $T \in (\alpha, \beta]$. For $\alpha \leq t \leq T$ we obtain from (1.1.499),

$$u(t) \leq a(T) + \int_\alpha^t f_1(T,t_1)$$

$$\times \left(\int_\alpha^{t_1} f_2(T,t_2)\cdots\left(\int_\alpha^{t_{n-1}} f_n(T,t_n)u^p(t_n)dt_n\right)\cdots\right)dt_1. \tag{1.1.502}$$

Now we introduce the functions

$$m_1(t) = a(T) + \int_\alpha^t f_1(T,t_1)$$

$$\times \left(\int_\alpha^{t_1}\cdots\left(\int_\alpha^{t_{n-1}} f_n(T,t_n)u^p(t_n)dt_n\right)\cdots\right)dt_1,$$

$$m_k(t) = m_{k-1}(t) + \int_\alpha^t f_k(T,t_k)$$

$$\times \left(\int_\alpha^{t_k}\cdots\left(\int_\alpha^{t_{n-1}} f_n(T,t_n)m_{k-1}^p(t_n)dt_n\right)\cdots\right)dt_k,$$

for all $t \in [\alpha, T]$ and $k = 2, \cdots, n$. Then the inequality (1.1.502) implies that $m_k(\alpha) = a(T), k = 1, \cdots, n$, and for all $t \in [\alpha, T]$,

$$u(t) \leq m_1(t) \leq \cdots \leq m_n(t).$$

Thus, induction with respect to k gives

$$m_k'(t) \leq \left(\sum_{i=1}^{k-1} f_i(T,t) - f_k(T,t)\right) m_k(t) + f_k(T,t)m_{k+1}(t), \qquad (1.1.503)$$

$$m_n'(t) \leq \left(\sum_{i=1}^{n-1} f_i(T,t)\right) m_n(t) + f_n(T,t)m_n^p(t) \qquad (1.1.504)$$

for all $t \in [\alpha, T], k = 1, 2, \cdots, n-1$. Lemma 1.2.1 and the inequality (1.1.504) imply that for $\alpha \leq t \leq T \leq \beta$,

$$\begin{aligned} m_n(t) &\leq \exp\left(\int_\alpha^t \sum_{i=1}^{n-1} f_i(T,s)ds\right) \\ &\quad \times \left(a^q(T) + q\int_\alpha^t f_n(T,s)\exp\left(-q\int_\alpha^s \sum_{i=1}^{n-1} f_i(T,\tau)d\tau\right)ds\right)^{\frac{1}{q}} \\ &= Y_n(T,t). \end{aligned}$$

Applying Lemma 1.1.1 in Qin [557] to inequality (1.1.503) for $k = n-1, \cdots, 2, 1$, we obtain

$$m_k(t) \leq E_k(T,t)\left(a(t) + \int_\alpha^t f_k(T,s)\frac{Y_{k+1}(T,s)}{E_k(T,s)}ds\right) = Y_k(T,t),$$

where the function $E_k(T,t)$ is defined by (1.1.501). Hence we get for $\alpha \leq t \leq T \leq \beta$,

$$u(t) \leq m_1(t) \leq Y_1(T,t),$$

which implies the result $u(t) \leq Y_1(t,t)$ for $T = t$. □

1.2 The One-Dimensional Ou-Yang Inequality and Its Generalization

In 1957, Ou-Yang Liang [438] established the Ou-Yang Inequality. Because of its fundamental importance, over the years, many its generalizations have been established. Such inequalities are in general known as the Bellman-Gronwall inequalities in the literature (see, e.g., [42, 47, 82, 140, 143, 144, 365, 395, 397, 507, 518]). Among various branches of Bellman-Gronwall inequalities, a very useful one is originated from Liang Ou-Yang. In his study of the boundedness of certain second order differential equations, Ou-Yang [438] established the following result which is generally known as Ou-Yang's inequality.

The importance of the Ou-Yang inequality stems from the fact that it is applicable in certain situations in which the other available inequalities do not apply directly. During the past few years, various investigators have discovered many useful and new integral inequalities centered around the celebrated Gronwall inequality, but the significance of the above inequality seems to have skipped any notice in the literature (see, e.g., [47, 54, 253, 476, 478, 663]).

Unlike many other types of integral inequalities, Ou-Yang-type inequalities or more generally, Gronwall-Bellman-Ou-Yang-type inequalities provide explicit bounds on the unknown function, and this special feature makes such inequalities especially important in many practical situations. In fact, over the years, such inequalities and their generalizations to various settings have proven to be very effective in the study of many qualitative as well as quantitative properties of solutions of differential equations. These include, among others, the global existence, boundedness, uniqueness, stability, and continuous dependence on initial data (see, e.g., [65, 82, 140–145, 147, 271, 345, 358, 360, 361, 365, 366, 388, 389, 397, 500, 507, 512, 520, 528, 695, 696]).

Theorem 1.2.1 (The Ou-Yang Inequality [438]) *Suppose that functions* $u(t) \geq 0,\ v(t) \geq 0$ *for all* $t \geq 0$ *satisfy, for all* $t \geq 0$,

$$u^2(t) \leq c + \int_0^t u(s)v(s)ds \tag{1.2.1}$$

with $c > 0$ *being a constant, then, for all* $t \geq 0$,

$$|u(t)| \leq \sqrt{c} + \frac{1}{2}\int_0^t v(s)ds \tag{1.2.2}$$

Proof Suppose that $w(t) = \int_0^t u(s)v(s)ds$, i.e., $\frac{dw}{dt} = uv$. Thus from (1.2.1) it follows

$$u^2v^2 \leq v^2(c+w),$$

i.e.,

$$(\frac{dw}{dt})^2 \leq v^2(c+w),$$

which implies

$$\frac{dw}{\sqrt{c+w}} \leq vdt.$$

Integrating the above inequality from 0 to t, we have

$$2(c+w)^{\frac{1}{2}} - 2c^{\frac{1}{2}} \le \int_0^t v(s)ds,$$

then

$$|u| \le (c+w)^{\frac{1}{2}} \le \sqrt{c} + \frac{1}{2}\int_0^t v(s)ds,$$

which completes the proof. □

While Ou-Yang's inequality is having a neat form and is interesting in its own right as an integral inequality, its importance lies equality heavily on its many beautiful applications in differential and integral equations (see, e.g., [507]).

Recent results in this direction include the works of Pachpatte [500], Pang and Agarwal [528], Ma [358], Meng and Li [388], Cheung [141], Cheung and Ren [147] and Ma and Cheung [360].

Note that, Ou-Yang's result [438] was generalized by Pachpatte [500] to the following inequalities

$$u^2(t) \le a^2 + 2\int_0^t (f(s)u(s) + g(s)u^2(s))ds, \quad \textit{for all } t \ge 0, \tag{1.2.3}$$

$$u^2(t) \le a^2 + 2\int_0^t (f(s)u(s) + g(s)u(s)w(u(s)))ds, \tag{1.2.4}$$

which actually includes Dafermos's generalization [180] (see below in Section 1.3).

The next result is a corollary of the Ou-Yang inequality, which can be found in Renardy, Hrusa and Nohel [573].

Corollary 1.2.1 *Let $f(t) \in L^1(0,T)$ such that $f(t) \ge 0$ a. e. on $[0,T]$, and let $a \ge 0$ be a constant. Assume that $w(t) \in C[0,T]$ satisfies*

$$\frac{1}{2}w^2(t) \le \frac{a^2}{2} + \int_0^t f(s)w(s)ds. \tag{1.2.5}$$

Then for any $t \in [0,T]$, we have

$$|w(t)| \le a + \int_0^t f(s)ds. \tag{1.2.6}$$

The next result is a generalization of the Ou-Yang inequality of a non-negative continuous function $y(t)$ (see, e.g., Caraballo, Rubin, and Valero [132]).

Theorem 1.2.2 (The Caraballo-Rubin-Valero Inequality [132]) *Suppose that $0 \le g(t) \in L^1(0,T)$ and $M \ge 0, 0 < \alpha \le 2$. Let $y(t)$ be a non-negative continuous* *function on* $[0, T]$ *such that, for all* $t \in [0, T]$,

$$y^2(t) \leq M^2 + 2\int_0^t g(\tau)y^\alpha(\tau)d\tau. \tag{1.2.7}$$

Then for all $t \in [0, T]$,

$$\begin{cases} y(t) \leq \left(M^{2-\alpha} + (2-\alpha)\int_0^t g(s)ds\right)^{1/(2-\alpha)}, \quad \text{if } \alpha < 2, & (1.2.8) \\ y(t) \leq M\exp\left(\int_0^t g(s)ds\right), \quad \text{if } \alpha = 2. & (1.2.9) \end{cases}$$

Proof Denote $U(s) = \sqrt{M^2 + 2\int_0^s g(\tau)y^\alpha(\tau)d\tau}$, which is a non-decreasing function. Differentiating $U^2(t)$, we have

$$2U(s)\frac{dU(s)}{ds} = 2g(s)y^\alpha(s) \leq 2g(s)U^\alpha(s). \tag{1.2.10}$$

Since $U(t)$ is non-decreasing, there exists $0 \leq \beta \leq T$ such that $U(t) = M$, for all $t \in [0, \beta]$, and $U(t) > M$, for all $\beta \in [\beta, T]$. Clearly, (1.2.10) is satisfied for $t \in [0, \beta]$. If $t > \beta$, then integrating over (β, t), we obtain

$$\frac{U^{2-\alpha}(t)}{2-\alpha} \leq \frac{M^{2-\alpha}}{2-\alpha} + \int_0^t g(s)ds \text{ if } \alpha < 2, \tag{1.2.11}$$

$$U(t) \leq M\exp\left(\int_0^t g(s)ds\right) \text{ if } \alpha = 2. \tag{1.2.12}$$

It follows that

$$y(t) \leq U(t) \leq \left[M^{2-\alpha} + (2-\alpha)\int_0^t g(s)ds\right]^{1/(2-\alpha)} \text{ if } \alpha < 2, \tag{1.2.13}$$

$$y(t) \leq U(t) \leq M\exp\left(\int_0^t g(s)ds\right) \text{ if } \alpha = 2. \tag{1.2.14}$$

The proof is complete. □

We easily prove the following result which may be viewed as variants of the Ou-Yang inequality.

Theorem 1.2.3 *Assume that* $T > 0, f(t) \in L^1(0, T), \phi(t) \in W^{1,1}(0, T), f(t) \geq 0$ *on* $[0, T]$, $\phi(t) \geq 0$ *on* $[0, T]$ *such that for all* $t \in [0, T]$,

$$\frac{d\phi}{dt} \leq 2f(t)\sqrt{\phi(t)}. \tag{1.2.15}$$

Then for any $t \in [0, T]$, we have

$$\sqrt{\phi(t)} \le \sqrt{\phi(0)} + \int_0^t f(s)ds. \tag{1.2.16}$$

Proof Let $h(t) = \sqrt{\phi(t)}$. It follows from (1.2.15) that

$$2h(t)h'(t) \le 2f(t)h(t),$$

i.e.,

$$h'(t) \le f(t). \tag{1.2.17}$$

Integrating (1.2.17) with respect to t implies (1.2.16). □

The next result, obtained by Kawashima, Nakao and Ono [304] in 1995, can be regarded as a generalization of Ou-Yang's inequality.

Theorem 1.2.4 (The Kawashima-Nakao-Ono Inequality [304]) *Let $y(t)$ be a non-negative function on $[0, T)$, $0 < T \le +\infty$, and satisfy the integral inequality for all $t \in [0, T)$,*

$$y(t) \le k_0(1+t)^{\alpha} + k_1 \int_0^t (1+t-s)^{-\beta}(1+s)^{-\gamma} y^{\mu}(s)ds \tag{1.2.18}$$

for some constants $k_0, k_1 > 0, \alpha, \beta, \gamma \ge 0$ and $0 \le \mu < 1$. Then

$$y(t) \le c(1+t)^{-\theta} \tag{1.2.19}$$

for some constant $c > 0$ and

$$\theta = \min\left\{\alpha, \beta, \frac{\gamma}{1-\mu}, \frac{\beta+\gamma-1}{1-\mu}\right\}, \tag{1.2.20}$$

with the following exceptional case: If $\alpha \ge \hat{\theta}$ and $(\beta+\gamma-1)/(1-\mu) = \hat{\theta} \le 1$ where

$$\hat{\theta} = \min\{\beta, \frac{\gamma}{1-\mu}\}, \tag{1.2.21}$$

then for all $t \in [0, T)$,

$$y(t) \le c(1+t)^{-\hat{\theta}}[\log(2+t)]^{1/(1-\mu)}. \tag{1.2.22}$$

Remark 1.2.1 Once we have known that $y(t)$ is a bounded function, we can also apply Theorem 1.2.4 to the case $\mu = 1$. In particular, if $\gamma > 0$ and $\beta + \gamma - 1 > 0$, we obtain (1.2.19) with

$$\theta = \min\{\alpha, \beta\}. \tag{1.2.23}$$

We note that even for the exceptional case, (1.2.19) is valid if θ is replaced $\theta_\varepsilon, 0 < \varepsilon << 1$.

Proof The case $\mu = 0$ is well-known. We define $M(t)$ by

$$M(t) = \sup_{0 \le s \le t} \{(1+s)^\theta y(s)\}. \tag{1.2.24}$$

Then we derive from (1.2.18) and (1.2.24) that

$$\begin{aligned} y(t) &\le k_0(1+t)^{-\alpha} + k_1 \int_0^t (1+t-s)^{-\beta}(1+s)^{-\gamma-\mu\theta} ds M^\mu(t) \\ &\le k_0(1+t)^{-\alpha} + c(1+t)^{-\theta *} M^\mu(t) \end{aligned}$$

with a constant $c > 0$ and $\theta * = \{\beta, \gamma + \mu\beta, \beta + \gamma + \mu\theta - 1\}$, where we have assumed that $\beta \neq 1$ and $\gamma + \mu\theta \neq 1$. Here it is easy to see that $\min\{\alpha, \theta^*\} = \theta$, and hence

$$(1+t)^\theta y(t) \le k_0 + cM(t)^\mu. \tag{1.2.25}$$

Since $0 < \mu < 1$, (1.2.25) implies $M(t) \le C < +\infty$, which is equivalent to (1.2.19). The exceptional case where $\beta = 1$ or $\gamma + \mu\theta = 1$ can be proved in the same manner. □

The following lemma is the differential form of Theorem 1.1.6, which is due to Oguntuase [428]. The proof presented here is differential from the one of Theorem 1.1.6.

Lemma 1.2.1 (The Oguntuase Inequality [428]) *Let $v(t)$ be a positive differentiable function satisfying the inequality, for all $t \in I = [a, b]$,*

$$v'(t) \le f(t)v(t) + g(t)v^p(t), \tag{1.2.26}$$

where the functions $f(t)$ and $g(t)$ are continuous in I, and $0 \le p \neq 1$, is a constant. Then for all $t \in [\alpha, \beta)$,

$$v(t) \le \exp\left(\int_a^t f(s)ds\right)\left[v^q(a) + q\int_a^t g(s)\exp(-q\int_a^s f(t)dt)ds\right]^{1/q}, \tag{1.2.27}$$

where $q = 1 - p$ and β is chosen so that the expression

$$\left[v^q(a) + q\int_a^t g(s)\exp(-q\int_a^s f(t)dt)ds\right]^{1/q}$$

is positive in the sub-interval $[a, \beta)$.

Proof We reduce (1.2.26) to a simpler differential inequality by the following substitution. Let $z(t) = v^q(t)/q$. Then by (1.2.26), noting that $q = 1 - p$, we have

$$\begin{aligned} z'(t) &= v^{q-1}(t)v'(t) \le v^{q-1}(t)(f(t)v(t) + g(t)v^p(t)) \\ &= qf(t)z(t) + g(t), \end{aligned}$$

which gives us

$$z(t) \le v^q/q\exp\left(\int_a^t qf(s)ds\right) + \int_a^t g(s)\exp\left(\int_s^t af(t)dt\right)ds.$$

That is,

$$z^q(t) \le \exp\left(\int_a^t qf(s)ds\right)\left[v^q(a) + \int_a^t g(s)exp\left(-\int_a^s qf(t)dt\right)ds\right],$$

which concludes

$$z(t) \le \exp\left(\int_a^t f(s)ds\right)\left[c^q(a) + q\int_a^t g(s)\exp\left(-q\int_a^s f(t)dt\right)ds\right]^{1/q}.$$

□

Theorem 1.2.5 (The Oguntuase Inequality [428]) *Let $u(t), f(t)$ be non-negative continuous functions in a real interval $I = [a, b]$. Suppose that the partial derivatives $k_t(t, s)$ exist and are non-negative continuous functions for almost every $t, s \in I$. If the following inequality holds for all $a \le \tau \le s \le t \le b$,*

$$u(t) \le c + \int_a^t f(s)u(s)ds + \int_a^t f(s)\left(\int_a^s k(s,\tau)u^p(\tau)d\tau\right)ds, \tag{1.2.28}$$

where $0 \le p < 1$, $q = 1 - p$ and $c > 0$ are constants, then for all $t \ge a$,

$$\begin{aligned} u(t) \le c + &\int_a^t f(s)\exp\left(\int_a^s f(t)dt\right) \\ &\times\left[c^{1-p} + (1-p)\int_a^s k(t,t)\exp\left(-(1-p)\int_a^t f(\delta)d\delta\right)dt\right]^{1/(1-p)} ds. \end{aligned} \tag{1.2.29}$$

Proof Define a function $v(t)$ by the right hand-side of (1.2.28) from which it follows that

$$u(t) \le v(t). \tag{1.2.30}$$

Then by (1.2.30)

$$\begin{aligned} v'(t) &= f(t)u(t) + f(t)\int_a^t k(t,\tau)u^p(\tau)d(\tau), \quad v(a) = c \\ &\le f(t)\left(v(t) + \int_a^t k(t,\tau)v^p(\tau)d\tau\right). \end{aligned} \tag{1.2.31}$$

If we put

$$m(t) = v(t) + \int_a^t k(t,\tau)v^p(\tau)d(\tau), \tag{1.2.32}$$

then

$$v(t) \le m(t). \tag{1.2.33}$$

Hence by (1.2.31)–(1.2.33), we obtain

$$m'(t) = v'(t) + k(t,t)v^p(t) + \int_a^t k_t(t,\tau)v^p(\tau)d\tau,\ m(a) = v(a) = c,$$

$$v'(t) + k(t,t)v^p(t) \le f(t)m(t) + k(t,t)v^p(t) \le f(t)m(t) + k(t,t)m^p(t).$$

By Lemma 1.2.1 or Theorem 1.1.6, we have

$$m(t) \le \exp\left(\int_a^t f(s)ds\right)\left[m^q + q\int_a^s k(s,s)\exp\left(-q\int_a^s f(\tau)d\tau\right)ds\right]^{1/q}. \tag{1.2.34}$$

Substituting (1.2.34) into (1.2.31), we arrive at

$$v'(t) \le f(t)\exp\left(\int_a^t f(s)ds\right)\left[m^q + q\int_a^s k(s,s)\exp\left(-q\int_a^s f(\tau)d\tau\right)ds\right]^{1/q}. \tag{1.2.35}$$

Integrating both sides of (1.2.35) from a to t and using (1.2.32), we can obtain

$$u(t) \le c + \int_a^t f(s)\exp\left(\int_a^s f(t)dt\right)\left[c^{(1-p)} + (1-p)\int_a^s k(t,t)\exp\left(-(1-p)\int_a^t f(\delta)d\delta\right)dt\right]^{\frac{1}{1-p}} ds.$$

This thus completes the proof. □

Remark 1.2.2 If, in Theorem 1.2.5, we put $k(t,s) = g(s)$, then Theorem 1.2.5 reduces to Theorem 2 in [441].

Theorem 1.2.6 (The Oguntuase Inequality [428]) *Let $u(t)$, $f(t)$, $h(t)$ and $g(t)$ be non-negative continuous functions in a real interval $I = [a,b]$. Suppose that $h'(t)$ exists and is a non-negative continuous function. If the following inequality holds for all $a \le t \le s \le t \le b$,*

$$u(t) \le c + \int_a^t f(s)u(s)ds + \int_a^t f(s)h(s)\left(\int_a^s g(t)u^p(t)dt\right)ds, \tag{1.2.36}$$

where $0 \le p < 1$, $q = 1-p$ and $c > 0$ are non-negative constants, then for all $t \ge a$,

$$u(t) \le c + \int_a^t f(s)\exp\left(\int_a^s f(t)dt\right)$$
$$\times\left[c^{(1-p)} + (1-p)\int_a^s \left(h(t)f(t) + h'(t)\int_a^t f(\delta)d\delta\right)\exp\left(-(1-p)\int_a^t f(\delta)d\delta\right)dt\right]^{\frac{1}{1-p}} ds.$$

Proof This follows by similar argument as in the proof of Theorem 1.2.5. We also omit the details. □

Remark 1.2.3 If, in Theorem 1.2.6, we set $h(t) = 1$, then Theorem 1.2.6 reduces to the estimate in Theorem 2 in [441].

Remark 1.2.4 If, in Theorem 1.2.6, $h'(t) = 0$, then Theorem 1.2.6 is more general than Theorem 2 in [441].

In 1994, Pachpatte [496] established some new generalizations of the Ou-Yang inequality. Next, we shall introduce these results. To this end, we first give some basic notation and definitions which are used in our subsequent discussion. Let define the differential operators L_i, $0 \le i \le n$, by

$$L_0x(t) = x(t), \qquad L_ix(t) = \frac{1}{r_i(t)}\frac{d}{dt}L_{i-1}x(t), \quad 1 \le i \le n,$$

with $r_n(t) = 1$, where $x(t)$ and $r_i(t) > 0$ are some functions defined for all $t \in \mathbb{R}_+$. For all $t \in \mathbb{R}_+$ and some functions $r_i(t) > 0$, $i = 1, \cdots, n-1$, and $r(t) \ge 0$ defined for all $t \in \mathbb{R}_+$, we set

$$A[t, r_1, \cdots, r_{n-1}, r] = \int_0^t r_i(t_1)\cdots\int_0^{t_{n-2}} r_{n-1}(t_{n-1})\int_0^{t_n} r(t_n)dt_ndt_{n-1}\cdots dt_1,$$

where $t_0 = t$.

Theorem 1.2.7 (The Pachpatte Inequality [496]) *Let $F(t) \ge 0$, $r(t) \ge 0$, and $r_i(t) > 0$ for $i = 1, \ldots, n-1$ be real-valued continuous functions defined on all*

$t \in \mathbb{R}_+$ and let $p > 1$ be a constant. If for all $t \in \mathbb{R}_+$,

$$F^p(t) \leq c + A[t, r_1, \ldots, r_{n-1}, rF], \tag{1.2.37}$$

where $c \geq 0$ is a constant, then for all $t \in \mathbb{R}_+$,

$$F(t) \leq \left(c^{\frac{(p-1)}{p}} + (\frac{p-1}{p})A[t, r_1, \ldots, r_{n-1}, r] \right)^{\frac{1}{p-1}}. \tag{1.2.38}$$

Proof In order to establish the inequality (1.2.38), we first assume that $c > 0$ and define a function $z(t)$ by

$$z(t) = c + A[t, r_1, \ldots, r_{n-1}, rF]. \tag{1.2.39}$$

From (1.2.39), it easily follows

$$L_n z(t) = r(t)F(t). \tag{1.2.40}$$

Using the fact that $F(t) \leq \sqrt[p]{z(t)}$ in (1.2.39), we have

$$L_n z(t) \leq r(t) \sqrt[p]{z(t)}. \tag{1.2.41}$$

From (1.2.41) and using the facts that $z(t)$ is positive, $(d/dt)[\sqrt[p]{z(t)}]$ and $L_{n-1}z(t)$ are non-negative for all $t \in \mathbb{R}_+$, we obtain

$$\frac{L_n(t)}{\sqrt[p]{z(t)}} \leq r(x) + \frac{(d/dt)[\sqrt[p]{z(t)}]L_{n-1}z(t)}{[\sqrt[p]{z(t)}]^2},$$

i.e.,

$$\frac{d}{dt}\left(\frac{L_{n-1}z(t)}{\sqrt[p]{z(t)}} \right) \leq r(t). \tag{1.2.42}$$

By setting $t = t_n$ in (1.2.42) and integrating with respect to t_n from 0 to t and using the fact that $L_{n-1}z(0) = 0$, we can obtain

$$\frac{L_{n-1}z(t)}{\sqrt[p]{z(t)}} \leq \int_0^t r(t_n)dt_n,$$

which implies

$$\frac{(d/dt)L_{n-2}z(t)}{\sqrt[p]{z(t)}} \leq r_{n-1}(t) \int_0^t r(t_n)dt_n. \tag{1.2.43}$$

Again as above, from (1.2.43), we may get

$$\frac{d}{dt}\left(\frac{L_{n-1}z(t)}{\sqrt[p]{z(t)}}\right) \leq r_{n-1}(t)\int_0^t r(t_n)dt_n. \tag{1.2.44}$$

By setting $t = t_{n-1}$ in (1.2.44) and integrating with respect to t_{n-1} from 0 to t and using the fact $L_{n-2}z(0) = 0$, we can conclude

$$\frac{L_{n-2}z(t)}{\sqrt[p]{z(t)}} \leq \int_0^t r_{n-1}(t_{n-1})\int_0^{t_{n-1}} r(t_n)dt_n.$$

Computing in this way, we may obtain

$$\frac{(d/dt)z(t)}{\sqrt[p]{z(t)}} \leq r_1(t)\int_0^t r_2(t_2)\cdots\int_0^{t_{n-2}} r_{n-1}(t_{n-1})\int_0^{t_{n-1}} r(t_n)dt_n dt_{n-1}\cdots dt_2. \tag{1.2.45}$$

By setting $t = t_1$ in (1.2.45) and integrating with respect to t_1 from 0 to t, we arrive at

$$[\sqrt[p]{z(t)}]^{p-1} - [\sqrt[p]{z(0)}]^{p-1} \leq (\frac{p-1}{p})A[t, r_1, \ldots, r_{n-1}, r]. \tag{1.2.46}$$

Thus from (1.2.46) and using the fact that $F(t) \leq \sqrt[p]{z(t)}$, we derive

$$F(t) \leq \left[c^{\frac{(p-1)}{p}} + (\frac{p-1}{p})A[t, r_1, \ldots, r_{n-1}, r\right]^{\frac{1}{p-1}}. \tag{1.2.47}$$

Now suppose that $c = 0$. Then from (1.2.37) we derive that the inequality

$$F^p(t) \leq \epsilon + A[t, r_1, \ldots, r_{n-1}, rF]$$

holds for every arbitrary positive small number ϵ and all $t \in \mathbb{R}_+$, which, by the above argument, yields

$$F(t) \leq \left[\epsilon^{\frac{(p-1)}{p}} + (\frac{p-1}{p})A[t, r_1, \ldots, r_{n-1}, r\right]^{\frac{1}{p-1}}. \tag{1.2.48}$$

Since $F(t) \geq 0$ and $\epsilon > 0$ is an arbitrary number independent of $t \in \mathbb{R}_+$, then as $\epsilon \to 0^+$, it follows from (1.2.48) that

$$F(t) \leq \left[(\frac{p-1}{p})A[t, r_1, \ldots, r_{n-1}, r]\right]^{\frac{1}{p-1}}.$$

This shows that (1.2.47) gives us the upper bound on $F(t)$ for all $c \geq 0$. This thus completes the proof. □

Another useful inequality is given in the following theorem which can also regarded as a variant of the Ou-Yang inequality.

Theorem 1.2.8 (The Pachpatte Inequality [496]) *Let $u(t) \geq 0, v(t) \geq 0, r_i(t) > 0$ for $i = 1, \ldots, n-1$ and $h_j(t) \geq 0$ for $j = 1,, 2, 3, 4$ be real-valued continuous functions defined for all $t \in \mathbb{R}_+$ and let $p > 1$ be a constant. If c_1, c_2 and μ are non-negative constants such that for all $t \in \mathbb{R}_+$,*

$$u^p(t) \leq c_1 + A[t, r_1, \ldots, r_{n-1}, h_1 u] + A[t, r_1, \ldots, r_{n-1}, h_2 \overline{v}], \tag{1.2.49}$$

$$v^p(t) \leq c_2 + A[t, r_1, \ldots, r_{n-1}, h_3 \overline{u}] + A[t, r_1, \ldots, r_{n-1}, h_4 v], \tag{1.2.50}$$

where $\overline{u}(t) = \exp(-p\mu t)u(t)$ and $\overline{v}(t) = \exp(p\mu t)v(t)$ for all $t \in \mathbb{R}_+$, then for all $t \in \mathbb{R}_+$,

$$u(t) \leq \exp(\mu t)\left[\{2^{p-1}(c_1 + c_2)\}^{(p-1)/p} + 2^{p-1}\left(\frac{p-1}{p}\right)A[t, r_1, \ldots, r_{n-1}, h]\right]^{\frac{1}{p-1}}, \tag{1.2.51}$$

$$v(t) \leq \left[\{2^{p-1}(c_1 + c_2)\}^{(p-1)/p} + 2^{p-1}\left(\frac{p-1}{p}\right)A[t, r_1, \ldots, r_{n-1}, h]\right]^{\frac{1}{p-1)}}, \tag{1.2.52}$$

where for all $t \in \mathbb{R}_+$,

$$h(t) = \max\{[h_1(t) + h_3(t)], [h_2(t) + h_4(t)]\}. \tag{1.2.53}$$

Proof Multiplying (1.2.49) by $\exp(-p\mu t)$ and observing that

$$\exp(-p\mu t)u^p(t) \leq c_1 + A[t, r_1, \ldots, r_{n-1}, h_1 \overline{u}] + A[t, r_1, \ldots, r_{n-1}, h_2 v]. \tag{1.2.54}$$

Define

$$F(t) = \exp(-\mu t)u(t) + v(t). \tag{1.2.55}$$

By taking the p-th power on both sides of (1.2.55) and using the elementary inequality

$$(d_1 + d_2)^q \leq 2^{q-1}(d_1^q + d_2^q), \tag{1.2.56}$$

where d_1, d_2 are non-negative reals and $q > 1$, and inequalities (1.2.54) and (1.2.50), we observe that

$$\begin{aligned} F^p(t) &\le 2^{p-1}\,(\exp(-p\mu t)u^p(t) + v^p(t)) \\ &\le 2^{p-1}(c_1 + c_2 + A[t, r_1 \ldots, r_{n-1}, [h_1 + h_3]\overline{u}]) \\ &\quad + A[t, r_1, \ldots, [h_2 + h_4]v]]. \end{aligned} \tag{1.2.57}$$

Now using the fact that $\exp(-p\mu t) \le \exp(\mu t)$ and (1.2.53) in (1.2.57), we conclude

$$F^p(t) \le 2^{p-1}(c_1 + c_2) + A[t, r_1 \ldots, r_{n-1}, 2^{p-1}hF]. \tag{1.2.58}$$

Thus the bounds in (1.2.51) and (1.2.52) follow from an application of Theorem 1.2.7 to (1.2.58) and splitting. The proof is thus complete. □

The next result was proved by Vaigant [654] (see also Kaliev and Podkuiko [300]), which can be regarded as a generalization of the Ou-Yang inequality.

Theorem 1.2.9 (The Vaigant Inequality [654]) *If $y(t)$ is continuous and non-negative function satisfying for all $t \in [0, T]$,*

$$y^n(t) \le a + b\int_0^t c(\tau)y^{n-1}(\tau)d\tau \tag{1.2.59}$$

where $a, b \ge 0$ and $n \ge 1$ are constants, $c(t) \in L^1[0, T]$, then there holds that for all $t \in [0, T]$,

$$y(t) \le a^{1/n} + \frac{b}{n}\int_0^t c(\tau)d\tau. \tag{1.2.60}$$

Proof Without loss of generality, we assume that $c(t)$ is a continuous and non-negative function. For the general case of $c(t)$, we may use the continuous functions from $L^1[0, T]$ to approximate $c(t)$.

Let $z(t) = \int_0^t c(\tau)y^{n-1}(\tau)d\tau$. Then it is easy to know that $z(t)$ is a continuous function verifying

$$z(0) = 0, \quad z'(t) = c(t)y^{n-1}(t). \tag{1.2.61}$$

From (1.2.59), we derive

$$y^n(t) \le a + bz(t) \tag{1.2.62}$$

which, along with (1.2.61), implies

$$\left(\frac{z'(t)}{c(t)}\right)^{1/(n-1)} \le a + bz(t), \quad z'(t) \le c(t)(a + bz(t))^{(n-1)/n}$$

or

$$\frac{d}{dt}(a+bz(t))^{1/n} \leq \frac{b}{n}c(t). \tag{1.2.63}$$

Integrating (1.2.63) with respect to t, we arrive at

$$[a+bz(t)]^{1/n} \leq a^{1/n} + \frac{b}{n}\int_0^t c(\tau)d\tau$$

which, together with (1.2.62), proves the theorem. □

The following theorems are some variants of the Ou-Yang inequality.

Theorem 1.2.10 (The Yang Inequality [695]) *Let $a > 0, b > 0$ and $c > 0$ be real constants. Then the following conclusions are true : If $u, E \in C(\mathbb{R}_+, \mathbb{R}_+)$, the following integral inequality holds for all $t \in \mathbb{R}_+$,*

$$u^2(t) \leq c + \int_0^t E(s)[au(s) + bu^2(s)]ds, \tag{1.2.64}$$

then for all $t \in \mathbb{R}_+$,

$$u(t) \leq \frac{a}{b}\Big\{(1+\frac{b}{a}\sqrt{c})\exp\Big[\frac{b}{2}\int_0^t E(s)ds\Big]-1\Big\}. \tag{1.2.65}$$

Remark 1.2.5 A slightly more general version of inequality $\varphi[u(t)] \leq c + \int_0^t F(s)\psi[u(s)]ds,\ \ t \in \mathbb{R}_+$ can be found in Butler and Rogers [129]. However, it was assumed in [129] that the function φ is strictly decreasing while function ψ is non-increasing, and it was treated on a finite closed interval.

Remark 1.2.6 We compare the bounds obtained by applying Theorem 1.2.10 to inequalities $u^2(t) \leq k^2 + 2\int_0^t [H(s)u(s) + F(s)u^2(s)]ds,\ \ t \in \mathbb{R}_+$, as follows: from inequality $u^2(t) \leq k^2 + 2\int_0^t [H(s)u(s) + F(s)u^2(s)]ds,\ \ t \in \mathbb{R}_+$, we have for all $t \in \mathbb{R}_+$,

$$u^2(t) \leq k^2 + 2\int_0^t E(s)[u(s) + u^2(s)]ds. \tag{1.2.66}$$

where $E(t) := max[F(t), H(t)]$. An application of Theorem 1.2.10 (with $a = b = 1$) to (1.2.66) yields for all $t \in \mathbb{R}_+$,

$$u(t) \leq \Big\{(1+k)\exp(\frac{1}{2}\int_0^t \max[F(s), H(s)]ds - 1)\Big\}. \tag{1.2.67}$$

Theorem 1.2.11 (The Pachpatte Inequality [500]) *Let u, f, g, h be real-valued non-negative continuous functions defined on $\mathbb{R}_+$ and c be a non-negative real constant.*

(a_1) *If for all $t \in \mathbb{R}_+$,*

$$u^2(t) \le c^2 + 2\int_0^t [f(s)u^2(s) + h(s)u(s)]ds, \tag{1.2.68}$$

then for all $t \in \mathbb{R}_+$,

$$u(t) \le p(t)\exp\left(\int_0^t f(s)ds\right), \tag{1.2.69}$$

where for all $t \in \mathbb{R}_+$,

$$p(t) = c + \int_0^t h(s)ds. \tag{1.2.70}$$

(a_2) *If for all $t \in \mathbb{R}_+$,*

$$u^2(t) \le c^2 + 2\int_0^t \left[f(s)u(s)\left(u(s) + \int_0^s g(\tau)u(\tau)d\tau\right) + h(s)u(s)\right]ds, \tag{1.2.71}$$

then for all $t \in \mathbb{R}_+$,

$$u(t) \le p(t)\left[1 + \int_0^t f(s)\exp\left(\int_0^s [f(\tau) + g(\tau)]d\tau\right)ds\right], \tag{1.2.72}$$

where $p(t)$ is defined by (1.2.70).

(a_3) *If for all $t \in \mathbb{R}_+$,*

$$u^2(t) \le c^2 + 2\int_0^t \left[f(s)u(s)\left(\int_0^s g(\tau)u(\tau)d\tau\right) + h(s)u(s)\right]ds, \tag{1.2.73}$$

then for all $0 \le t \le \bar{t}$, $\bar{t} \in \mathbb{R}_+$,

$$u(t) \le p(t)\exp\left(\int_0^t f(s)\left(\int_0^s g(\tau)d\tau\right)ds\right), \tag{1.2.74}$$

where $p(t)$ is defined by (1.2.70).

Proof Since the proofs resemble one another, we shall only give the details for (a_2).

(a_2) Define a function $z(t)$ by

$$z(t) = (c+\varepsilon)^2 + 2\int_0^t \left[f(s)u(s) + \left(u(s) + \int_0^s g(\tau)u(\tau)d\tau \right) + h(s)u(s) \right] ds, \tag{1.2.75}$$

where $\varepsilon > 0$ is an arbitrary small constant. Differentiating (1.2.75) and then using the fact that $u(t) \leq \sqrt{z(t)}$, we have

$$z'(t) \leq 2\sqrt{z(t)} \left[f(t) \left(\sqrt{z(t)} + \int_0^t g(\tau)\sqrt{z(\tau)}d\tau \right) + h(t) \right]. \tag{1.2.76}$$

Now differentiating $\sqrt{z(t)}$ and using (1.2.76), we can get

$$\frac{d}{dt}\left(\sqrt{z(t)}\right) = \frac{z'(t)}{\sqrt{z(t)}} \leq \left[f(t) \left(\sqrt{z(t)} + \int_0^t g(\tau)\sqrt{z(\tau)}d\tau \right) + h(t) \right]. \tag{1.2.77}$$

By taking $t = s$ in (1.2.77) and then integrating the above inequality from 0 to t, we know

$$\sqrt{z(t)} \leq p_\varepsilon(t) + \int_0^t f(s) \left(\sqrt{z(s)} + \int_0^s g(\tau)\sqrt{z(\tau)}d\tau \right) ds, \tag{1.2.78}$$

where $p_\varepsilon(t)$ is defined by (1.2.70) by replacing c by $c+\varepsilon$. Since $p_\varepsilon(t)$ is positive and monotone non-decreasing for all $t \in \mathbb{R}_+$, the inequality (1.2.78) implies the estimate (see, e.g. [456])

$$\sqrt{z(t)} \leq p_\varepsilon(t) + \left[1 + \int_0^t f(s) \exp\left(\int_0^s [f(\tau) + g(\tau)]d\tau \right) ds \right]. \tag{1.2.79}$$

Now using the fact that $u(t) \leq \sqrt{z(t)}$ in (1.2.79), and then letting $\varepsilon \to 0^+$, we get the desired inequality (1.2.72). □

Theorem 1.2.12 (The Pachpatte Inequality [500])

(1) If for all $t \in \mathbb{R}_+$,

$$u^2(t) \leq \left(c_1^2 + 2\int_0^t f(s)u(s)ds \right)\left(c_2^2 + 2\int_0^t h(s)u(s)ds \right), \tag{1.2.80}$$

then for $t \in \mathbb{R}_+$,

$$u(t) \leq p_0(t) \exp\left(\int_0^t \left[h(s)\left(\int_0^s f(\tau)d\tau \right) + f(s)\left(\int_0^s h(\tau)d\tau \right) \right] ds \right), \tag{1.2.81}$$

where for all $t \in \mathbb{R}_+$,

$$p_0(t) = c_1c_2 \int_0^t [c_1^2 h(s) + c_2^2 f(s)]ds. \tag{1.2.82}$$

(2) *If for all* $t \in \mathbb{R}_+$,

$$u^2(t) \le c_1 + \int_0^t [g_1(s)u^2(s) + h_1(s)u(s)]ds + \int_0^t [g_2(s)\overline{v}^2(s) + h_2(s)\overline{v}(s)]ds, \tag{1.2.83}$$

$$v^2(t) \le c_2 + \int_0^t [g_3(s)\overline{u}^2(s) + h_3(s)\overline{u}(s)]ds + \int_0^t [g_4(s)v^2(s) + h_4(s)v(s)]ds, \tag{1.2.84}$$

where $\overline{u}(t) = e^{-\mu t}u(t), \overline{v}(t) = e^{\mu t}v(t)$ *for all* $t \in \mathbb{R}_+$, *then for all* $t \in \mathbb{R}_+$,

$$u(t) \le e^{\mu t} p^*(t) \exp\left(\int_0^t G(s)ds\right), \tag{1.2.85}$$

$$v(t) \le p^*(t) \exp\left(\int_0^t G(s)ds\right), \tag{1.2.86}$$

where

$$p^*(t) = \sqrt{2(c_1 + c_2)} + \int_0^t H(s)ds, \tag{1.2.87}$$

where for all $t \in \mathbb{R}_+$,

$$\begin{cases} G(t) = \max\{[g_1(t) + g_3(t)], [g_2(t) + g_4(t)]\}, & (1.2.88)\\ H(t) = \max\{[h_1(t) + h_3(t)], [h_2(t) + h_4(t)]\}. & (1.2.89)\end{cases}$$

Proof Here we only give the proof of assertion in (2), the proof of assertion in (1) can be done similarly.

Multiplying both sides of (1.2.83) by $\exp(-2\mu t)$, we observe that

$$\begin{aligned}\overline{u}^2(t) \le c_1 &+ \int_0^t [g_1(s)\overline{u}^2(s) + h_1(s)\overline{u}(s)]ds\\ &+ \int_0^t [g_2(s)v^2(s) + h_2(s)v(s)]ds.\end{aligned} \tag{1.2.90}$$

Define

$$F(t) = \overline{u} + v(t). \tag{1.2.91}$$

Now by squaring both sides of (1.2.91) and using the elementary inequalities $(a + b)^2 \le 2(a^2 + b^2), (a^2 + b^2) \le (a + b)^2, a \ge 0, b \ge 0$ reals, and using (1.2.90),

(1.2.84), (1.2.88), (1.2.89), it is easy to observe that

$$F^2(t) \leq 2(c_1 + c_2) + 2\int_0^t [G(s)F^2(s) + H(s)F(s)]ds. \tag{1.2.92}$$

The bounds in (1.2.85)–(1.2.86) follow from an application of the inequality given in (1) to (1.2.92) and splitting. □

Theorem 1.2.13 (The Pachpatte Inequality [500]) *Let $u(t) \geq u_0 \geq 0$ be a real-valued continuous function defined on $\mathbb{R}_+$, u_0 is a real constant. Let f, g, h be real-valued non-negative continuous functions defined on $\mathbb{R}_+$ and c be a non-negative real constant. Let $W(u)$ be a continuous, non-decreasing real-valued function defined on $I = [u_0, +\infty)$ and $W(u) > 0$ on $(u_0, +\infty)$, $W(u_0) = 0$.*

(1) If for all $t \in \mathbb{R}_+$,

$$u^2(t) \leq c^2 + 2\int_0^t [f(s)u(s)W(u(s)) + h(s)u(s)]ds, \tag{1.2.93}$$

then for all $0 \leq t \leq t_1$,

$$u(t) \leq \Omega^{-1}\left[\Omega(p(t)) + \int_0^t f(s)ds\right], \tag{1.2.94}$$

where $p(t)$ is defined by (1.2.70) and

$$\Omega(r) = \int_{r_0}^r \frac{ds}{W(s)}, \quad r \geq r_0 \geq u_0, \tag{1.2.95}$$

and Ω^{-1} is the inverse of Ω and $t_1 \in \mathbb{R}_+$ is chosen so that for all $t \in [0, t_1]$,

$$\Omega(p(t)) + \int_0^t f(s)ds \in Dom\ (\Omega^{-1}).$$

(2) If for allt $\in \mathbb{R}_+$,

$$u^2(t) \leq c^2 + 2\int_0^t \left[f(s)u(s)\left(u(s) + \int_0^s g(\tau)W(u(\tau))d\tau\right) + h(s)u(s)\right]ds, \tag{1.2.96}$$

then for all $0 \leq t \leq t_2$,

$$u(t) \leq p(t) + \int_0^t f(s)E^{-1}\left[E(p(s)) + \int_0^s [f(\tau) + g(\tau)d\tau]\right]ds, \tag{1.2.97}$$

where $p(t)$ is defined by (1.2.70) and

$$E(r) = \int_{r_0}^{r} \frac{ds}{s + W(s)}, \quad r \geq r_0 \geq u_0, \tag{1.2.98}$$

E^{-1} is the inverse of E, and $t_1 \in \mathbb{R}_0$ is chosen so that for all $t \in [0, t_2]$,

$$E(p(t)) + \int_0^t (f(\tau) + g(\tau))d\tau \in Dom\,(E^{-1}).$$

(3) *If for all $t \in \mathbb{R}_+$,*

$$u^2(t) \leq c^2 + 2\int_0^t \left[f(s)u(s)\left(\int_0^s g(\tau)W(u(\tau))d\tau\right) + h(s)u(s)\right] ds, \tag{1.2.99}$$

then for all $0 \leq t \leq t_3$,

$$u(t) \leq \Omega^{-1}\left[\Omega(p(t)) + \int_0^t f(s)\left(\int_0^s g(\tau)d\tau\right) ds\right], \tag{1.2.100}$$

where $p(t)$ is defined by (1.2.70) and Ω, Ω^{-1} are as defined in (1) and $t_3 \in \mathbb{R}_+$ is chosen so that for all $t \in [0, t_3]$,

$$\Omega(p(t)) + \int_0^t [f(s)\left(\int_0^s g(\tau)d\tau\right) ds \in Dom\,(\Omega^{-1}).$$

Proof The proof is similar to that of the above two theorems, so we omit the details. □

Corollary 1.2.2 (The Cheung Inequality [142]) *Let $k \geq 0$ and $p > 1$ be constants. Let $a, b \in C(I, \mathbb{R}_+), \alpha, \gamma \in C^1(I, I)$, and $\varphi \in C(\mathbb{R}_+, \mathbb{R}_+)$ be functions satisfying*

(i) *α, γ, are non-decreasing with $\alpha, \gamma \leq id_I$; and*
(ii) *φ is non-decreasing with $\varphi(r) > 0$ for all $r > 0$.*

If $u \in C(\Delta, \mathbb{R}_+)$ satisfies for all $x \in I$,

$$u^p(x) \leq k + \frac{p}{p-1}\int_{\alpha(x_0)}^{\alpha(x)} a(s)u(s)ds + \frac{p}{p-1}\int_{\gamma(x_0)}^{\gamma(x)} b(s)u(s)\varphi(u(s))ds,$$

then for all $x \in [x_0, x_1]$,

$$u(x) \leq \left\{\Phi_{p-1}^{-1}\left[\Phi_{p-1}(k^{1-1/p} + A(x)) + B(x)\right]\right\}^{1/p-1} \tag{1.2.101}$$

where

$$A(x) := \int_{\alpha(x_0)}^{\alpha(x)} b(s)ds, \ \ B(x) := \int_{\gamma(x_0)}^{\gamma(x)} b(s)ds,$$

and $x_1 \in I$ *is chosen in such a way that* $\Phi_{p-1}(k^{1-1/p} + A(x)) + B(x) \in Dom\,(\Phi_{p-1}^{-1})$ *for all* $x \in [x_0, x_1]$.

Remark 1.2.7

(i) Same as before, in case $\Phi_{p-1}(+\infty) = +\infty$, inequality (1.2.101) holds for all $x \in I$.

(ii) Corollary 1.2.2 generalizes part (1) of Theorem 1.2.13. In fact, if we impose the conditions $p = 2, x_0 = 0$, and $\alpha(x) = \gamma(x) = x$ for all $x \in I$, Corollary 1.2.2 reduces to part (1) of Theorem 1.2.13.

Theorem 1.2.14 (The Yang Inequality [695]) *Let* $c \geq 0, p > 0$ *and* $q > 0$ *be real numbers. Suppose that* $u, F \in C(\mathbb{R}_+, \mathbb{R}_+)$, *and the inequality holds for all* $t \in \mathbb{R}_+$,

$$u^p(t) \leq c + \int_0^t F(s)u^q(s)ds. \tag{1.2.102}$$

If $p = q$, *then for all* $t \in \mathbb{R}_+$,

$$u(t) \leq c^{1/p} \exp\left(\frac{1}{p}\int_0^t F(s)ds\right), \tag{1.2.103}$$

while if $p > q$, *then for all* $t \in \mathbb{R}_+$,

$$u(t) \leq \left(c^{1-q/p} + (1-q/p)\int_0^t F(s)ds\right)^{1/(p-q)}. \tag{1.2.104}$$

Theorem 1.2.15 (The Yang Inequality [696]) *Let* $p \geq 1$ *be a constant,* $u(t), f(t)$ *be real-valued, non-negative and continuous functions defined on* $\mathbb{R}_+$. *Let further* $g(t,s)$, $h(t,s)$, $j(t,s)$, $k(t,s)$ *be real-valued, non-negative and continuous functions defined on* $\mathbb{R}_+ \times \mathbb{R}_+$ *and being non-decreasing in t for every s fixed. If the integral inequality holds for all* $t \in \mathbb{R}_+$,

$$u^p(t) \leq f(t) + \int_0^t pu^{p-1}(s)\left[h(t,s) + k(t,s)u(s) + j(t,s)\int_0^s g(s,m)u(m)dm\right]ds, \tag{1.2.105}$$

then for all $t \in \mathbb{R}_+$,

$$u(t) \leq K(t)Q(t)\left[f^{1/p}(t) + \int_0^t h(t,s)ds\right]\exp\left(-K(t)\int_0^t j(t,s)ds\right), \tag{1.2.106}$$

where

$$\begin{cases} K(t) := \exp\left(\int_0^t k(t,s)ds\right), & (1.2.107) \\ Q(t) := 1 + K(t)\int_0^t j(t,s)\left[\exp\left(\int_0^s (2K(s)j(s,m) + g(s,m)dm\right)\right]ds. & (1.2.108) \end{cases}$$

Proof Fix any positive number M and define on $[0, M)$ a function $\omega(t)$ by

$$\omega(t) := f(M) + \int_0^t pu^{p-1}(s)\left[h(M,s) + k(M,s)u(s) + j(M,s)\int_0^s g(M,m)u(m)dm\right]ds. \tag{1.2.109}$$

From (1.2.105), we have for all $t \in [0, M)$,

$$u(t) \leq \omega^{1/p}(t). \tag{1.2.110}$$

By differentiation with respect to t, we derive from (1.2.109) that

$$\frac{d}{dt}\omega(t) = pu^{p-1}(t)\left[h(M,t) + k(M,t)u(t) + j(M,t)\int_0^t g(M,m)u(m)dm\right],$$

whence, for all $t \in [0, M)$,

$$\begin{aligned} \frac{d}{dt}\omega^{1/p}(t) &= \frac{1}{p}[\omega(t)]^{(1-p)/p}\frac{d}{dt}\omega(t) \\ &= h(M,t) + k(M,t)u(t) + j(M,t)\int_o^t g(M,m)u(m)dm \\ &\leq h(M,t) + k(M,t)\omega^{1/p}(t) + j(M,m)\int_0^t g(M,m)\omega^{1/p}(m)dm, \end{aligned}$$

where we have used the relation (1.2.110). Letting $t = s$ in the last inequality and integrating its both sides from $s = 0$ to t, then we obtain

$$\begin{aligned} \omega^{1/p}(t) - \omega^{1/p}(0) \leq &\int_0^t h(M,s)ds + \int_0^t k(M,s)\omega^{1/p}(s)ds \\ &+ \int_0^t j(M,s)\left[\int_0^s g(M,m)\omega^{1/p}(m)dm\right]ds, \end{aligned}$$

i.e.,

$$\varphi(t) \leq N(t) + \int_0^t k(M,s)\varphi(s)ds + \int_0^t j(M,s)\left[\int_0^s g(M,s)\varphi(m)dm\right]ds, \tag{1.2.111}$$

where $\varphi(t) := \omega^{1/p}(t)$ and $N(t) := f^{1/p}(M) + \int_0^t h(M,s)ds$. Now applying Corollary 5.4.13 in Qin [557] to (1.2.111) yields, for all $t \in [0, M)$,

$$\omega^{1/p}(t) \le N(t)k^*(M,t)Q^*(M,t)\exp\left[-K^*(M,t)\int_0^t j(M,s)ds\right], \tag{1.2.112}$$

where $K^*(M,t) := \exp\left(\int_0^t k(M,s)ds\right)$ and

$$Q^*(M,t) := 1 + K^*(M,t)\int_0^t j(M,s)\left(\int_0^s [g(M,m) + 2K^*(M,s)j(M,m)]dm\right)ds.$$

Letting $t \to M$ in (1.2.112) and using (1.2.110), then we arrive at

$$u(M) \le K(M)Q(M)\left[f^{1/p}(M) + \int_0^M h(M,s)ds\right]\exp\left[-K(M)\int_0^M j(M,s)ds\right],$$

since $k^*(M,M) = K(M)$ and $Q^*(M,M) = Q(M)$. Hence the desired inequality (1.2.106) is valid for $t = M$, here M being any positive number. By (1.2.105), we see that (1.2.106) holds, then we observe that in the inequality, for all $t \in \mathbb{R}_+$,

$$u^p(t) \le f(t) + p\int_0^t h(t,s)u^{p-1}(s)ds,$$

implies, for all $t \in \mathbb{R}_+$,

$$u(t) \le f^{1/p}(t) + \int_0^t h(t,s)ds.$$

Obviously, the last result contains Ou-Yang's inequality as a particular case when $p = 2, f(t) = c$ and $h(t,s) = v(s)/2$. □

Remark 1.2.8 Obviously Pachpatte's inequalities (1.2.71) and (1.2.73) are special cases of inequality (1.2.105). By Theorem 1.2.15, a better bound than (1.2.72) can be derived from inequality (1.2.71). However, the bound on solutions of inequality (1.2.73) given by Theorem 1.2.15 is more complicated and it is not comparable with the bound (1.2.74) in Theorem 1.2.11. From the proof of Theorem 1.2.15, we also note that the function $u^{1/p}(t)$ on the right-hand side of inequality (1.2.105) can not be replaced by $u^q(t)$,with q being other than $p-1$.

The next result is also due to Pachpatte [454], which generalizes the above theorem.

Theorem 1.2.16 (The Pachpatte Inequality [454]) *Let $u_i(t), f_i(t)$ be real-valued non-negative continuous functions defined on $I = [0, +\infty)$, and $a_i(t,s), b_i(t,s)$ be continuous real-valued functions defined on $I \times I \to \mathbb{R}_+$, for which the following* *inequality holds for all $t \in I$,*

$$u_i(t) \le f_i(t) + \int_0^t a_i(t,s)u_i(s)ds + \int_0^t a_i(t,s)\Big(\int_0^s b_i(s,\tau)u_i(\tau)d\tau\Big)ds, \qquad (1.2.113)$$

where $i = 1, \cdots, n$. Define, for all $t \in I$,

$$\begin{cases} g(t)h(s) = \max\limits_{0\le s\le t} a_i(t,s), \quad i = 1,\cdots,n, & (1.2.114) \\ g(t)k(s) = \max\limits_{0\le s\le t} b_i(t,s), \quad i = 1,\cdots,n, & (1.2.115) \\ f(t) = \sum\limits_{i=1}^n f_i(t). & (1.2.116) \end{cases}$$

Then for all $t \in I$,

$$\sum_{i=1}^n u_i(t) \le f(t) + g(t)\Big(\int_0^t h(s)\Big\{f(s) + g(s)\exp\Big(\int_0^s g(\tau)(h(\tau) + k(\tau)d\tau\Big)$$
$$\times \int_0^s f(\tau)(h(\tau) + k(\tau))\exp\Big(-\int_0^\tau g(\eta)(h(\eta) + k(\eta))d\eta\Big)d\tau\Big\}ds\Big). \qquad (1.2.117)$$

Proof Substituting $i = 1, \cdots, n$ in (1.2.113), adding these inequalities and using (1.2.114)–(1.2.116), we obtain

$$\sum_{i=1}^n u_i(t) \le f(t) + g(t)\Big(\int_0^t h(s)\sum_{i=1}^n u_i(s)ds + \int_0^t h(s)g(s)\Big(\int_0^s k(\tau)\sum_{i=1}^n u_i(\tau)d\tau\Big)ds\Big).$$

Now applying Theorem 1.2.14 in Qin [557] yields the desired bound in (1.2.117). □

Now we shall study Henry's version of the Ou-Yang-Pachpatte inequality

$$u^2(t) \le a(t) + \int_0^t (t-s)^{\beta-1}F(s)\omega(u(s))ds, \qquad (1.2.118)$$

where $\beta > 0$ is a constant. Inequalities of such type with $\beta = 1$ and F continuous was studied by Pachpatte in [500].

We shall introduce the following theorem.

Theorem 1.2.17 (The Medved' Inequality [384]) *Let $a(t)$ be a non-decreasing, non-negative C^1 function on $[0, T)$ $(0 < T \le +\infty)$, $F(t)$ be a continuous, non-negative function, ω be as in Theorem 1.4.1 in Qin [557], and $u(t)$ be a continuous, non-negative function satisfying the inequality for all $t \in [0, T)$ and for a constant*

$\beta > 0$

$$u(t) \le a(t) + \int_0^t (t-s)^{\beta-1} F(s)\omega(u(s))ds.$$

Then the following assertion hold:

(i) Suppose $\beta > \frac{1}{2}$ and ω satisfies the condition (q) (see, Definition 1.4.1 in Qin [557]) with $q = 2$. Then for all $t \in [0, T_1)$,

$$u(t) \le e^t \left\{\Lambda^{-1}\left[\Lambda(2a^2(t)) + K\int_0^t F^2(s)R(s)ds\right]\right\}^{\frac{1}{4}}, \tag{1.2.119}$$

where

$$K = \frac{\Gamma(2\beta-1)}{4^{\beta-1}}, \quad \Lambda(v) = \int_{v_0}^{v} \frac{d\sigma}{\omega(\sqrt{\sigma})}, \; v \ge v_0 > 0, \tag{1.2.120}$$

and $T_1 \in \mathbb{R}_+$ is such that $\Lambda(2a^2(t)) + K\int_0^t F^2(s)R(s)ds \in Dom\,(\Lambda^{-1})$ for all $t \in [0, T_1)$.

(ii) Let $\beta \in (0, \frac{1}{2}]$ and ω satisfies the condition (q) (see, Definition 1.4.1 in Qin [557]) with $q = z + 2$, where $z = \frac{1-\beta}{\beta}$, i.e., $\beta = \frac{1}{z+1}$. Then for all $t \in [0, T_1]$,

$$u(t) \le e^t \left\{\Lambda^{-1}\left[\Lambda(2^{q-1}a^q(t)) + 2^{q-1}K_z^q\int_0^t F^q(s)R(s)ds\right]\right\}^{1/2q}, \tag{1.2.121}$$

where

$$K_z = \left[\frac{\Gamma(1-\beta p)}{p^{1-\beta p}}\right]^{\frac{1}{p}}, \qquad \beta = \frac{1}{z+1}, \qquad p = \frac{z+2}{z+1},$$

and $T_1 \in \mathbb{R}_+$ is such that $\Lambda(2^{q-1}a^q(t)) + 2^{q-1}K_z^q\int_0^t F^q(s)R(s)ds \in Dom\,(\Lambda^{-1})$ for all $t \in [0, T_1]$.

Proof First let us prove the assertion (i). Following the proof of Theorem 1.4.1 in Qin [557], we can show that

$$v^2(t) \le \alpha(t) + K\int_0^t F^2(s)R(s)\omega(v(s))ds, \tag{1.2.122}$$

with

$$v(t) = (e^{-t}u(t))^2, \quad \alpha(t) = 2a^2(t), \quad K = \frac{\Gamma(2\beta-1)}{4^{\beta-1}}. \tag{1.2.123}$$

Let $V(t)$ be the right-hand side of (1.2.122). Then $v(t) \leq \sqrt{V(t)}$. This yields $\omega(v(t)) \leq \omega(\sqrt{V(t)})$ and thus

$$\frac{V'(t)}{\omega(\sqrt{V(t)})} = \frac{\alpha'(t) + KF^2(t)R(t)\omega(v(t))}{\omega(\sqrt{V(t)})}$$
$$\leq \frac{\alpha'(t)}{\omega(\sqrt{\alpha(t)})} + KF^2(t)R(t)$$

which yields

$$\frac{d}{dt}\int_0^{V(t)} \frac{d\sigma}{\omega(\sqrt{V(\sigma)})} \leq \frac{d}{dt}\int_0^{\alpha(t)} \frac{d\sigma}{\omega(\sqrt{\alpha(\sigma)})} + KF^2(t)R(t). \tag{1.2.124}$$

Thus we have

$$\frac{d}{dt}\Lambda(V(t)) \leq \frac{d}{dt}\Lambda(\alpha(t)) + KF^2(t)R(t),$$

where Λ is defined by (1.2.120), which yields

$$V(t) \leq \Lambda^{-1}\left[\Lambda(\alpha(t)) + K\int_0^t F^2(s)R(s)ds\right]$$

whence

$$v(t) \leq \sqrt{V(t)} \leq \left\{\Lambda^{-1}\left[\Lambda(\alpha(t)) + K\int_0^t F^2(s)R(s)ds\right]\right\}^{\frac{1}{2}}.$$

Using (1.2.123), we can obtain (1.2.122).

Now we shall prove the assertion (ii). Following the proof of the assertion (ii) of Theorem 1.4.1 in Qin [557], we can show that

$$v^2(t) \leq \phi(t) + 2^{q-1}K_z^q \int_0^t F^q(s)R(s)\omega(v(s))ds, \tag{1.2.125}$$

where

$$v(t) = (e^{-t}u(t))^q, \quad \phi(t) = 2^{q-1}a^q(t). \tag{1.2.126}$$

Following the procedure from the proof of the assertion (i), we can obtain

$$v(t) \leq \left\{\Lambda^{-1}(\Lambda(\phi(t))) + 2^{q-1}K_z^q \int_0^t F^q(s)R(s)ds\right\}^{\frac{1}{2}}$$

and using (1.2.123), we can obtain (1.2.121). □

The next result is due to Lipovan [356] who introduced some retarded Ou-Yang-like integral inequalities.

Theorem 1.2.18 (The Lipovan Inequality [356]) *Let u, f, and g be non-negative continuous functions defined on $\mathbb{R}_+$ and let c be a non-negative constant. Moreover, let $w \in C(\mathbb{R}_+, \mathbb{R}_+)$ be non-decreasing with $w(u) > 0$ on $(0, +\infty)$ and $\alpha(t) \in C^1(\mathbb{R}_+, \mathbb{R}_+)$ be non-decreasing with $\alpha(t) \le t$ on $\mathbb{R}_+$. If for all $t \in \mathbb{R}_+$,*

$$u^2(t) \le c^2 + 2\int_0^{\alpha(t)} [f(s)u(s)w(u(s)) + g(s)u(s)]\, ds, \qquad (1.2.127)$$

then for all $0 \le t \le t_1$,

$$u(t) \le \Omega^{-1}\left[\Omega\left(c + \int_0^{\alpha(t)} g(s)ds\right) + \int_0^{\alpha(t)} f(s)ds\right], \qquad (1.2.128)$$

where

$$\Omega(r) = \int_1^r \frac{ds}{w(s)}, \quad r > 0,$$

and Ω^{-1} is the inverse of Ω, for all $t \in [0, t_1]$,

$$\Omega\left(c + \int_0^{\alpha(t)} g(s)ds\right) + \int_0^{\alpha(t)} f(s)ds \in Dom\,(\Omega^{-1}). \qquad (1.2.129)$$

Proof Let us first assume that $c > 0$. Define the non-decreasing positive function $z(t)$ by the right-hand side of (1.2.127) and let, for all $t \ge 0$,

$$p(t) = c + \int_0^{\alpha(t)} g(s)ds.$$

Then $z(0) = c^2, u(t) \le \sqrt{z(t)}$, and

$$\begin{aligned} z'(t) &= z[f(\alpha(t))u(\alpha(t))w(u(a(t))) + g(\alpha(t))u(\alpha(t))]\alpha'(t) \\ &\le 2\sqrt{z(\alpha(t))}\left[f(\alpha(t))w\left(\sqrt{z(\alpha(t))}\right)\right]\alpha'(t). \end{aligned}$$

Since $\alpha(t) \le t$ on $\mathbb{R}_+$, we deduce that

$$z'(t) \le 2\sqrt{z(t)}\left[f(\alpha(t))w\left(\sqrt{z(\alpha(t))}\right) + g(\alpha(t))\right]\alpha'(t)$$

which gives us

$$\frac{z'(t)}{2\sqrt{z(t)}} \leq \left[f(\alpha(t))w\left(\sqrt{z(\alpha(t))}\right) + g(\alpha(t))\right]\alpha'(t).$$

An integration on $[0, t]$ yields

$$\sqrt{z(t)} \leq p(t) + \int_0^{\alpha(t)} f(s)w\left(\sqrt{z(s)}\right)ds.$$

Let $T \leq t_1$ be an arbitrary number. From above relation, we deduce for all $0 \leq t \leq T$,

$$\sqrt{z(t)} \leq p(T) + \int_0^{\alpha} f(s)w\left(\sqrt{z(s)}\right)ds.$$

Now applying the retarded form of Bihari's inequality, i.e. Theorem 1.1.1 (see also, e.g., [355]) gives us for all $0 \leq t \leq T$,

$$\sqrt{z(t)} \leq \Omega^{-1}\left[\Omega(p(T)) + \int_0^{\alpha(t)} f(s)ds\right].$$

Taking $t = T$ in the above inequality and using the fact that $u(t) \leq \sqrt{z(t)}$ is true for $t = T$, we can obtain

$$u(T) \leq \Omega^{-1}\left[\Omega(p(T)) + \int_0^{\alpha(t)} f(s)ds\right].$$

Since $T \leq t_1$ is arbitrary, we prove the desired inequality (1.2.128). If $c = 0$, then we carry out the above procedure with $\varepsilon > 0$ instead of c and subsequently let $\varepsilon \to 0^+$. □

Remark 1.2.9

(i) Setting $\alpha(t) \equiv t$ in Theorem 1.2.18, we obtain Pachpatte's generalization [500] of Ou-Yang's inequality [438].
(ii) If $\int_1^{+\infty}(1/w(s))ds = +\infty$, then $\Omega(+\infty) = +\infty$ and (1.2.128) is valid on $\mathbb{R}_+$. Examples of such functions are $w(u) \equiv u$ and $w(u) \equiv u\ln(1+u)$.

Corollary 1.2.3 (The Lipovan Inequality [356]) *Let u and g be non-negative continuous functions defined on $\mathbb{R}_+$ and let c be a non-negative constant. If $\alpha \in C^1(\mathbb{R}_+, \mathbb{R}_+)$ is non-decreasing with $\alpha(t) \leq t$ on $\mathbb{R}_+$, and for all $t \geq 0$,*

$$u^2(t) \leq c^2 + 2\int_0^{\alpha(t)} g(s)u(s)ds, \tag{1.2.130}$$

then for all $t \geq 0$,

$$u(t) \leq c + \int_0^{\alpha(t)} g(s)ds. \tag{1.2.131}$$

Remark 1.2.10 For $\alpha(t) \equiv t$, Corollary 1.2.3 becomes Ou-Yang's inequality, i.e., Theorem 1.2.1 (see, e.g., [507]).

Corollary 1.2.4 (The Lipovan Inequality [356]) *Let u, f and g be non-negative continuous functions defined on $\mathbb{R}_+$ and let c be a non-negative constant. Moreover, let $\alpha \in C^1(\mathbb{R}_+, \mathbb{R}_+)$ be non-decreasing with $\alpha(t) \leq t$ on $\mathbb{R}_+$. If for all $t \geq 0$,*

$$u^2(t) \leq c^2 + 2\int_0^{\alpha(t)} \left(f(s)u^2(s) + g(s)u(s)\right) ds, \tag{1.2.132}$$

then for all $t \geq 0$,

$$u(t) \leq \left(c + \int_0^{\alpha(t)} g(s)ds\right) \exp\left[\int_0^{\alpha(t)} f(s)ds\right]. \tag{1.2.133}$$

Remark 1.2.11

(i) Corollary 1.2.4 is a retarded version of an inequality due to Pachpatte ([500], Theorem $1(a_1)$). Here we note that the hypotheses of Corollary 1.2.4 imply that for all $t \geq 0$,

$$u^2(t) \leq c^2 + 2\int_0^t [f(s)u^2(s) + g(s)u(s)]ds.$$

Hence Pachpatte's inequality [500] could be applied in order to obtain an upper estimate for $u(t)$. However, the estimate provided by Corollary 1.2.4 is sharper. To see this, let $c = 1, \alpha(t) \equiv \ln(t+1)$, and $f(t) \equiv g(t) \equiv 1/(t+1)$. Pachpatte's inequality yields, for all $t \geq 0$,

$$u(t) \leq (t+1)(1 + \ln(t+1)),$$

while Corollary 1.2.4 yields, for all $t \geq 0$,

$$u(t) \leq (1 + \ln(t+1))(1 + \ln(1 + \ln(t+1))).$$

(ii) For $g \equiv 0$, Corollary 1.2.4 becomes a retarded Gronwall-like inequality established in [355].

Theorem 1.2.19 (The Lipovan Inequality [356]) *Let u, f, and g be non-negative continuous functions defined on some interval $[0, T)$ and let c be a non-negative constant. Moreover, let $w \in C(\mathbb{R}_+, \mathbb{R}_+)$ be non-decreasing with $w(u) > 0$ on*

$(0,+\infty)$ and $\int_1^{+\infty}(1/w(s))ds = +\infty$. If $\alpha \in C^1(\mathbb{R}_+,\mathbb{R}_+)$ is non-decreasing with $\alpha(t) \le t$ on $\mathbb{R}_+$ and for all $0 \le t < T$,

$$u^2(t) \le c^2 + 2\int_0^{\alpha(t)} f(s)u(s)w(u(s)) + 2\int_0^{\alpha(t)} g(s)u(s)w(u(s))ds, \qquad (1.2.134)$$

then for all $0 \le t < T$,

$$u(t) \le \Omega^{-1}\left[\Omega(c) + \int_0^{\alpha(t)} (f(s)+g(s))\,ds\right], \qquad (1.2.135)$$

where $\Omega : (0,+\infty) \to (\Omega(0),+\infty)$ is the C^1-diffeomorphism defined by

$$\Omega(r) = \int_1^r \frac{ds}{w(s)}, \quad r > 0,$$

and Ω^{-1} is its inverse.

Proof Similarly to the proof of Theorem 1.2.18, let us first assume that $c > 0$. Denoting the right-hand side of (1.2.134) by $z(t)$, the same steps as in the case of Theorem 1.2.18 lead to

$$\frac{z'(t)}{2\sqrt{z(t)}} \le f(\alpha(t))w\left(\sqrt{z(\alpha(t))}\right)\alpha'(t) + g(\alpha(t))w\left(\sqrt{z(\alpha(t))}\right).$$

Integrating the above inequality on $[0,t]$, we can get for all $0 \le t \le T$,

$$\sqrt{z(t)} \le c + \int_0^{\alpha(t)} f(s)w\left(\sqrt{z(s)}\right)ds + \int_0^{\alpha(t)} g(s)w\left(\sqrt{z(s)}\right)ds,$$

An application of Theorem 1.1.47 yields, for all $0 \le t \le T$,

$$\sqrt{z(t)} \le \Omega^{-1}\left[\Omega(c) + \int_0^{\alpha(t)} f(s)ds + \int_0^{\alpha(t)} g(s)ds\right].$$

Since $u(t) \le \sqrt{z(t)}$ on $[0,T)$, inequality (1.2.135) follows immediately. The case $c = 0$ can be handled by repeating the above procedure with $\varepsilon > 0$ and subsequently letting $\varepsilon \to 0^+$. □

The next three corollaries are direct conclusions of Theorem 1.1.56.

Corollary 1.2.5 (The Zhao-Meng Inequality [722]) *Let u,f and g be non-negative continuous functions defined on $\mathbb{R}_+$ and let c be a non-negative constant. Moreover, let $\varphi \in C(\mathbb{R}_+,\mathbb{R}_+)$ be a non-decreasing function with $\psi(u) > 0$ on*

$(0, +\infty)$ *and* $\alpha \in C^1(\mathbb{R}_+, \mathbb{R}_+)$ *be non-decreasing with* $\alpha(t) \geq t$ *on* $\mathbb{R}_+$. *If for all* $t \in \mathbb{R}_+$,

$$u^2(t) \leq c^2 + \int_{\alpha(t)}^{+\infty} [f(s)u(s)\psi(u(s)) + g(s)u(s)]ds,$$

then for all $0 \leq T \leq t < +\infty$,

$$u(t) \leq \Omega^{-1}\left(\Omega\left(c + \frac{1}{2}\int_{\alpha(t)}^{+\infty} g(s)ds\right) + \frac{1}{2}\int_{\alpha(t)}^{+\infty} f(s)ds\right)$$

where

$$\Omega(r) = \int_1^r \frac{ds}{\omega(s)}, r \geq 1$$

and Ω^{-1} *is the inverse of* Ω *and* $T \in \mathbb{R}_+$ *is chosen so that, for all* $t \in [T, +\infty)$,

$$\Omega(c + \frac{1}{2}\int_{\alpha(t)}^{+\infty} g(s)ds) + \frac{1}{2}\int_{\alpha(t)}^{+\infty} f(s)ds \in Dom\,(\Omega^{-1}).$$

Corollary 1.2.6 (The Zhao-Meng Inequality [722]) *Let* u, f *and* g *be non-negative continuous functions defined on* $\mathbb{R}_+$ *and let* c *be a non-negative constant. Moreover, let* $\alpha \in C^1(\mathbb{R}_+, \mathbb{R}_+)$ *be non-decreasing with* $\alpha(t) \geq t$ *on* $\mathbb{R}_+$. *If for all* $t \in \mathbb{R}_+$,

$$u^2(t) \leq c^2 + \int_{\alpha(t)}^{+\infty} [f(s)u^2(s) + g(s)u(s)]ds,$$

then, for all $t \in \mathbb{R}_+$,

$$u(t) \leq \left(c + \frac{1}{2}\int_{\alpha(t)}^{+\infty} g(s)ds\right) \exp\left(\frac{1}{2}\int_{\alpha(t)}^{+\infty} f(s)ds\right).$$

Corollary 1.2.7 (The Zhao-Meng Inequality [722]) *Let* u, f *and* g *be non-negative continuous functions defined on* $\mathbb{R}_+$ *and let* c *be a non-negative constant. Moreover, let* $\alpha \in C^1(\mathbb{R}_+, \mathbb{R}_+)$ *be non-decreasing with* $\alpha(t) \geq t$ *on* $\mathbb{R}_+$. *If for all* $t \in \mathbb{R}_+$,

$$u^p(t) \leq c + \int_{\alpha(t)}^{+\infty} [f(s)u^q(s) + g(s)u(s)]ds,$$

then for all $t \in \mathbb{R}_+$,

$$\begin{cases} u(t) \le \left(c^{1-\frac{1}{p}} + \frac{p-1}{p}\int_{\alpha(t)}^{+\infty} g(s)ds\right)^{\frac{p}{p-1}} \exp\left(\frac{1}{p}\int_{\alpha(t)}^{+\infty} f(s)ds\right), & \text{when } p = q, \\ u(t) \le \left[\left(c^{1-\frac{1}{p}} + \frac{p-1}{p}\int_{\alpha(t)}^{+\infty} g(s)ds\right)^{\frac{p-q}{p-1}} + \frac{p-q}{p}\int_{\alpha(t)}^{+\infty} f(s)ds\right]^{\frac{1}{p-q}}, & \text{when } p > q. \end{cases}$$

The following results, due to Pachpatte [523], are retarded Ou-Yang inequality.

Theorem 1.2.20 (The Pachpatte Inequality [523]) *Let* $u, a_i, b_i \in C(I, \mathbb{R}_+)$, *and* $\alpha_i \in C^1(I, I)$ *be non-decreasing with* $\alpha_i(t) \le t$ *on* $I \equiv [t_0, T)$ *for* $i = 1, 2, \cdots, n$. *Let* $p > 1$ *and* $c \ge 0$ *be constants.*

(1) If for all $t \in I$,

$$u^p(t) \le c + p\sum_{i=1}^{n}\int_{\alpha_i(t_0)}^{\alpha_i(t)} [a_i(s)u^p(s) + b_i(s)u(s)]\, ds, \tag{1.2.136}$$

then for allt $\in I$,

$$u(t) \le \left\{A(t)\exp\left((p-1)\sum_{i=1}^{n}\int_{\alpha_i(t_0)}^{\alpha_i(t)} a_i(\sigma)d\sigma\right)\right\}^{1/(p-1)}, \tag{1.2.137}$$

where for allt $\in I$,

$$A(t) = c^{(p-1)/p} + (p-1)\sum_{i=1}^{n}\int_{\alpha_i(t_0)}^{\alpha_i(t)} b_i(\sigma)d\sigma. \tag{1.2.138}$$

(2) Let $w \in C(\mathbb{R}_+, \mathbb{R}_+)$ *be non-decreasing with* $w(u) > 0$ *on* $(0, +\infty)$. *If, for all* $t \in I$,

$$u^p(t) \le c + p\sum_{i=1}^{n}\int_{\alpha_i(t_0)}^{\alpha_i(t)} [a_i(s)u(s)w(u(s)) + b_i(s)u(s)]\, ds, \tag{1.2.139}$$

then for all $t_0 \le t \le t_1$,

$$u(t) \le \left\{G^{-1}\left[G(A(t)) + (p-1)\sum_{i=1}^{n}\int_{\alpha_i(t_0)}^{\alpha_i(t)} a_i(\sigma)d\sigma\right]\right\}^{1/(p-1)}, \tag{1.2.140}$$

where $A(t)$ is defined by (1.2.138), G^{-1} is the inverse function of

$$G(r) = \int_{r_0}^{r} \frac{ds}{w(s^{1/(p-1)})}, \quad r \geq r_0 > 0, \tag{1.2.141}$$

and $r_0 > 0$ is arbitrary and $t_1 \in I$ is chosen so that for all $t \in [0, t_1]$,

$$G(A(t)) + (p-1)\sum_{i=1}^{n} \int_{\alpha_i(t_0)}^{\alpha_i(t)} a_i(\sigma)d\sigma \in Dom\,(G^{-1}).$$

Proof We only give the details of the proof for (1); the proof of (2) is similar.

From the hypotheses, we observe that $\alpha_i'(t) \geq 0$ for all $t \in I, \alpha_i'(t) \geq 0$ for all $t \in I_1, \beta_i'(y) \geq 0$ for all $y \in I_2$.

(1) Let $c > 0$ and define a function $z(t)$ by the right-hand side of (1.2.136). Then $z(t) > 0, z(t_0) = c, z(t)$ is non-decreasing for all $t \in I, u(t) \leq \{z(t)\}^{1/p}$ and

$$\begin{aligned} z'(t) &= p\sum_{i=1}^{n} [a(\alpha_i(t))u^p(\alpha_i(t)) + b_i(\alpha_i(t))u(\alpha_i(t))]\,\alpha_i'(t) \\ &\leq p\sum_{i=1}^{n} \left[a_i(\alpha_i(t))z(\alpha_i(t)) + b_i(\alpha_i(t))\{z(\alpha_i(t))\}^{1/p}\right]\alpha_i'(t) \\ &= p\sum_{i=1}^{n} \left[a_i(\alpha_i(t))\{z(\alpha_i(t))\}^{(p-1)/p} + b_i(\alpha_i(t))\right]\{z(\alpha_i'(t))\}^{1/p}\alpha_i'(t) \\ &\leq p\sum_{i=1}^{n} \left[a_i(\alpha_i(t))\{z(\alpha_i(t))\}^{(p-1)/p} + b_i(\alpha_i(t))\right]\{z(t)\}^{1/p}\alpha_i'(t), \end{aligned} \tag{1.2.142}$$

i.e.,

$$\frac{z'(t)}{\{z(t)\}^{1/p}} \leq p\sum_{i=1}^{n} \left[a_i(\alpha_i(t))\{z(\alpha_i(t))\}^{(p-1)/p} + b_i(\alpha_i(t))\right]\alpha_i'(t). \tag{1.2.143}$$

By taking $t = s$ in (1.2.143) and integrating it with respect to s from t_0 to t, we can get

$$(z(t))^{\frac{p-1}{p}} \leq c^{(p-1)/p} + (p-1)\int_0^t \sum_{i=1}^{n} \left[a_i(\alpha_i(s))\{z(\alpha_i(t))\}^{(p-1)/p} + b_i(\alpha_i(s))\right] ds\alpha_i'(t). \tag{1.2.144}$$

Making the change of variables on the right-hand side in (1.2.144) and rewriting, we can get

$$(z(t))^{(p-1)/p} \leq A(t) + (p-1)\sum_{i=1}^{n}\int_{\alpha_i(t_0)}^{\alpha_i(t)} a_i(\sigma)(z(\sigma))^{(p-1)/p}d\sigma.$$

Clearly, $A(t)$ is a continuous, positive and non-decreasing function for all $t \in I$. Now following the idea used in the proof of Theorem 1.2.18 (see also [356] or [518]), we can obtain

$$(z(t))^{(p-1)/p} \leq A(t)\exp\left((p-1)\sum_{i=1}^{n}\int_{\alpha_i(t_0)}^{\alpha_i(t)} a_i(\sigma)d\sigma\right). \tag{1.2.145}$$

Using (1.2.145) in $u(t) \leq \{z(t)\}^{1/p}$, we can get the inequality in (1.2.137).

If $c \geq 0$, we carry out the above procedure with $c + \varepsilon$ instead of c, where $\varepsilon > 0$ is an arbitrary small constant, and subsequently pass to the limit $\varepsilon \to 0^+$ to obtain (1.2.137). □

Remark 1.2.12 If we take $p = 2, n = 1, \alpha_1 = \alpha, a_1 = f, b_1 = g$ in Theorem 1.2.20, then we recapture the inequalities given in [356] (see, Corollary 2 and Theorem 1).

Pachpatte [523] established further generalization (Theorem 1.2.21) of Theorem 1.2.18 as follows.

Theorem 1.2.21 (The Pachpatte Inequality [523]) *Let u, $a_i b_i \in C(I, \mathbb{R}_+)$ and let $\alpha_i \in C^1(I, I)$ be non-decreasing with $\alpha_i(t) \leq t$ on I for $i = 1, \cdots, n$. Let $p > 1$ and $c \geq 0$ be constants and $w \in C(\mathbb{R}_+, \mathbb{R}_+)$ be non-decreasing with $w(u) > 0$ on $(0, +\infty)$. If for all $t \in I$,*

$$u^p(t) \leq c + \sum_{i=1}^{n}\int_{\alpha_i(t_0)}^{\alpha_i(t)} u(s)(a_i(s)\psi(u(s)) + b_i(s))ds, \tag{1.2.146}$$

then for all $t_0 \leq t \leq t_1$,

$$u(t) \leq \left(G^{-1}(G(A(t))) + (p-1)\sum_{i=1}^{n}\int_{\alpha_i(t_0)}^{\alpha_i(t)} a_i(\sigma)d\sigma\right)^{1/(p-1)}, \tag{1.2.147}$$

where

$$\begin{cases} G(r) = \displaystyle\int_{r_0}^{r} \frac{ds}{w(s^{1/(p-1)})}, \ r \geq r_0 > 0, \\ A(t) = c^{(p-1)/p} + (p-1)\displaystyle\sum_{i=1}^{n}\int_{\alpha_i(t_0)}^{\alpha_i(t)} b_i(\sigma)d\sigma, \end{cases} \tag{1.2.148}$$

and $r_0 > 0$ is arbitrary, G^{-1} is the inverse function of G and $t_1 \in I$ is so chosen that, for all $t_0 \leq t \leq t_1$,

$$G(A(t)) + (p-1)\sum_{i=1}^{n}\int_{\alpha_i(t_0)}^{\alpha_i(t)} a_i(\sigma)d\sigma \in \mathrm{Dom}(\mathrm{G}^{-1}). \tag{1.2.149}$$

Proof The proof is similar to that of Theorem 1.2.20. □

In 2008, Agarwal, Kim and Sen [14] established some new retarded integral inequalities.

Theorem 1.2.22 (The Agarwal-Kim-Sen Inequality [14]) *Let $u, f_i, g_i \in C(I, \mathbb{R}_+)$ $i = 1, \ldots, n$, and let $\alpha_i \in C^1(I, I)$ be non-decreasing with $\alpha_i(t) \leq t$ on I for $i = 1, \cdots, n$. Let $q > 0$ and $c \geq 0$ be constants, $\psi \in C^1(\mathbb{R}_+, \mathbb{R}_+)$ is an increasing function with $\phi(+\infty) = +\infty$ on I, and $\psi(u)$ is a non-decreasing continuous function for all $u \in \mathbb{R}_+$ with $\psi(u) > 0$ for all $u > 0$. If for all $t \in I$,*

$$\phi(u(t)) \leq c + \sum_{i=1}^{n}\int_{\alpha_i(t_0)}^{\alpha_i(t)} u^q(s)[f_i(s)\psi(u(s)) + g_i(s)]ds, \tag{1.2.150}$$

then for all $t_0 \leq t \leq t_1$,

$$u(t) \leq \phi^{-1}\left(G^{-1}\left[\Psi^{-1}(\Psi(k(t_0)) + \sum_{i=1}^{n}\int_{\alpha_i(t_0)}^{\alpha_i(t)} f_i(s)ds)\right]\right) \tag{1.2.151}$$

where

$$\begin{cases} G(r) = \displaystyle\int_{r_0}^{r} \frac{ds}{[\varphi^{-1}(s)]^q}, \quad r \geq r_0 > 0, \\ \Psi(r) = \displaystyle\sum_{i=1}^{n}\int_{\alpha_i(t_0)}^{\alpha_i(t)} \frac{ds}{\psi[\varphi^{-1}(G^{-1}(s))]}, \quad r \geq r_0 > 0, \\ k(t_0) = G(c) + \displaystyle\sum_{i=1}^{n}\int_{\alpha_i(t_0)}^{\alpha_i(t)} g_i(s)ds, \end{cases} \tag{1.2.152}$$

and G^{-1} and Ψ^{-1} denote the inverse functions of G and Ψ, respectively, for all $t \in I$. $t_1 \in I$ is so chosen that for all $t_0 \leq t \leq t_1$,

$$\Psi(k(t_0)) + \sum_{i=1}^{n}\int_{\alpha_i(t_0)}^{\alpha_i(t)} f_i(s)ds \in Dom\,(\Psi^{-1}). \tag{1.2.153}$$

Proof Assume that $c > 0$. Define a function $z(t)$ by the right-hand side of (1.2.150). Clearly, $z(t)$ is non-decreasing, $u(t) \leq \phi^{-1}(z(t))$ for all $t \in I$ and $z(t_0) = c$.

Differentiating $z(t)$, we can get

$$z'(t) = \sum_{i=1}^{n}[u(\alpha_i(t))]^q[f_i(\alpha_i(t))\psi(u(\alpha_i(t))) + g_i(\alpha_i(t))]\alpha_i'(t)$$

$$\leq [\phi^{-1}(z(t))]^q \sum_{i=1}^{n}[f_i(\alpha_i(t))\psi(\phi^{-1}(z(\alpha_i(t)))) + g_i(\alpha_i(t))]\alpha_i'(t). \qquad (1.2.154)$$

Using the monotonicity of ϕ^{-1} and z, we may deduce

$$[\phi^{-1}(z(t))]^q = [\phi^{-1}(z(t_0))]^q \geq [\phi^{-1}(c)]^q > 0, \qquad (1.2.155)$$

that is,

$$\frac{z'(t)}{[(\phi^{-1}(z(t)))]^q} \leq \sum_{i=1}^{n}[f_i(\alpha_i(t))\psi(\phi^{-1}(z(\alpha_i(t)))) + g_i(\alpha_i(t))]\alpha_i'(t). \qquad (1.2.156)$$

Setting $t = s$ in the inequality (1.2.156), integrating it from t_0 to t, using the function G on the left-hand side, and changing variable on the right-hand side, we can obtain

$$G(z(t)) \leq G(c) + \sum_{i=1}^{n}\int_{\alpha_i(t_0)}^{\alpha_i(t)}[f_i(s)\psi(\phi^{-1}(z(s))) + g_i(s)]ds. \qquad (1.2.157)$$

From the inequality (1.2.157), we find

$$G(z(t)) \leq p(t) + \sum_{i=1}^{n}\int_{\alpha_i(t_0)}^{\alpha_i(t)}[f_i(s)\psi(\phi^{-1}(z(s))) + g_i(s)]ds, \qquad (1.2.158)$$

where

$$p(t) = G(c) + \sum_{i=1}^{n}\int_{\alpha_i(t_0)}^{\alpha_i(t)} g_i(s)ds. \qquad (1.2.159)$$

Thus from the inequality (1.2.158) it follows that for all $t \leq t_1$,

$$G(z(t)) \leq p(t_1) + \sum_{i=1}^{n}\int_{\alpha_i(t_0)}^{\alpha_i(t)} f_i(s)\psi(\phi^{-1}(z(s)))ds, \qquad (1.2.160)$$

Now, define a function $k(t)$ by the right-hand side of (1.2.160). Clearly, $k(t)$ is non-decreasing, $z(t) \le G^{-1}(k(t))$ for $t \in I$ and $k(t_0) = p(t_1)$. Differentiating $k(t)$, we can get

$$k'(t) = \sum_{i=1}^{n} [f_i(\alpha_i(t))\psi(\phi^{-1}z(\alpha_i(t)))]\alpha_i'(t) \le \psi(\phi^{-1}(G^{-1}(k(t)))) \sum_{i=1}^{n} [f_i(\alpha_i(t))]\alpha_i'(t). \tag{1.2.161}$$

Using the monotonicity of Ψ, ϕ^{-1}, G^{-1} and k, we deduce

$$\frac{k'(t)}{\Psi(\phi^{-1}(G^{-1}(k(t))))} \le \sum_{i=1}^{n} [f_i(\alpha_i(t))]\alpha_i'(t). \tag{1.2.162}$$

Setting $t = s$ in the inequality (1.2.162), integrating it from t_0 to t, using the function Ψ on the left-hand side, and changing variables on the right-hand side, we may obtain

$$\Psi(k(t)) \le \Psi(k(t_0)) + \sum_{i=1}^{n} \int_{\alpha_i(t_0)}^{\alpha_i(t)} f_i(s)ds. \tag{1.2.163}$$

From the inequalities (1.2.161) and (1.2.163), we conclude that for all $t_0 \le t \le t_1$,

$$z(t) \le G^{-1}[\Psi^{-1}(\Psi(p(t_1)) + \sum_{i=1}^{n} \int_{\alpha_i(t_0)}^{\alpha_i(t)} f_i(s)ds)]. \tag{1.2.164}$$

Now a combination of $u(t) \le \phi^{-1}(z(t))$ and the last inequality in (1.2.164) for $t_1 = t$ produces the required inequality. If $c = 0$, we carry out the above procedure with $\epsilon > 0$ instead of c and subsequently let $\epsilon \to 0^+$. This thus completes the proof. □

For the special case $\phi(u) = u^p$ ($p > q > 0$ is a constant), Theorem 1.2.22 gives us the following retarded integral inequality for nonlinear functions.

Corollary 1.2.8 (The Agarwal-Kim-Sen Inequality [14]) *Let u, f_i, g_i, and $\alpha_i \in C(I, \mathbb{R}_+)$, $I = 1, \ldots, n$, and let $\alpha_i \in C^1(I, I)$ be non-decreasing with $\alpha_i(t) \le t$ on I for $i = 1, \cdots, n$. Let $q > 0$ and $c \ge 0$ be constants, $\psi \in C^1(\mathbb{R}_+, \mathbb{R}_+)$ is an increasing function with $\phi(+\infty) = +\infty$ on I, and $\psi(u)$ is a non-decreasing continuous function for all $u \in \mathbb{R}_+$ with $\psi(u) > 0$ for all $u > 0$. If for all $t \in I$,*

$$u^p(t) \le c + \sum_{i=1}^{n} \int_{\alpha_i(t_0)}^{\alpha_i(t)} u^q(s)(f_i(s)\psi(u(s)) + g_i(s))ds, \tag{1.2.165}$$

then for all $t \in [t_0, \bar{t})$,

$$u(t) \le \left(\Psi_0^{-1}(\Psi_0(k_1(t_0)) + \frac{p-q}{p}\sum_{i=1}^{n}\int_{\alpha_i(t_0)}^{\alpha_i(t)} f_i(s)ds)\right)^{\frac{1}{p-q}} \quad (1.2.166)$$

where

$$\begin{cases} \Psi_0(r) = \displaystyle\int_{\alpha_i(t_0)}^{\alpha_i(t)} \frac{ds}{\psi(s^{1/(p-q)})}, \; r \ge r_0 > 0, \\ k_1(t_0) = c^{(p-q)/p} + \displaystyle\frac{p-q}{p}\sum_{i=1}^{n}\int_{\alpha_i(t_0)}^{\alpha_i(t)} g_i(s)ds, \end{cases} \quad (1.2.167)$$

and Ψ_0^{-1} *denote the inverse functions of* Ψ_0 *for all* $t \in I$ *and* $\bar{t} \in I$ *is so chosen that, for all* $t_0 \le t \le \bar{t}$,

$$\Psi_0(k_1(t_0)) + \frac{p-q}{p}\sum_{i=1}^{n}\int_{\alpha_i(t_0)}^{\alpha_i(t)} f_i(s)ds \in Dom\,(\Psi_0^{-1}). \quad (1.2.168)$$

Proof The proof follows by an argument similar to that in the proof of Theorem 1.2.22 with suitable modification. We omit the details here. □

Remark 1.2.13 When $q = 1$, from Corollary 1.2.8, we derive Theorem 1.2.22. When $p = 2$, $q = 1$, from Corollary 1.2.8, we derive Theorem 1.2.18.

Theorem 1.2.22 can easily be applied to generate other useful nonlinear integral inequalities in more general situations. For example, we have the following result.

Theorem 1.2.23 (The Agarwal-Kim-Sen Inequality [14]) *Let* $u \in C(I, \mathbb{R}_1)$, f_i, $g_i \in C(I, \mathbb{R}_+)$, $i = 1, \ldots, n$, *and let* $\alpha_i \in C^1(I, I)$ *be non-decreasing with* $\alpha_i(t) \le t$, $i = 1, \cdots, n$. *Suppose that* $c \ge 1$ *is a constant,* $\phi \in C^1(\mathbb{R}_+, \mathbb{R}_+)$ *is an increasing function with* $\phi(+\infty) = +\infty$ *on* I, *and* $\psi_j(u)$, $j = 1, 2$, *are non-decreasing continuous functions for all* $u \in \mathbb{R}_+$ *with* $\psi_j(u) > 0$ *for all* $u > 0$. *If for all* $t \in I$,

$$\phi(u(t)) \le c + \sum_{i=1}^{n}\int_{\alpha_i(t_0)}^{\alpha_i(t)} u^q(s)(f_i(s)\psi_1(u(s)) + g_i(s)\psi_2(\log(u(s))))ds, \quad (1.2.169)$$

then

(i) as the case $\psi_1(u) \ge \psi_2(\log(u))$, *we have for all* $t \in [t_0, t_1)$,

$$u(t) \le \phi^{-1}\left(G^{-1}\left[\Psi_1^{-1}(\Psi_1(G(c)) + \sum_{i=1}^{n}\int_{\alpha_i(t_0)}^{\alpha_i(t)} [f_i(s) + g_i(s)]ds)\right]\right), \quad (1.2.170)$$

(ii) as the case $\psi_1(u) < \psi_2(\log(u))$, we have for all $t \in [t_0, t_2)$,

$$u(t) \le \phi^{-1}\left(G^{-1}\left[\Psi_2^{-1}(\Psi_2(G(c)) + \sum_{i=1}^{n}\int_{\alpha_i(t_0)}^{\alpha_i(t)}[f_i(s) + g_i(s)]ds)\right]\right), \quad (1.2.171)$$

where

$$\Psi_j(r) = \sum_{i=1}^{n}\int_{\alpha_i(t_0)}^{\alpha_i(t)}\frac{ds}{\psi_j[\phi^{-1}(G^{-1}(s))]}, \quad r \ge r_0 > 0, \quad (1.2.172)$$

and $G^{-1}, \Psi_j^{-1}, j = 1, 2$, denote the inverse functions of G and $\Psi_j, j = 1, 2$, respectively, the function $G(t)$ is as defined in Theorem 1.2.22 for all $t \in I$ and $t_j \in I, j = 1, 2$, are so chosen that, for all $t_0 \le t \le t_j$,

$$\Psi_j(G(c)) + \sum_{i=1}^{n}\int_{\alpha_i(t_0)}^{\alpha_i(t)}[f_i(s) + g_i(s)]ds \in Dom\,(\Psi_j^{-1}). \quad (1.2.173)$$

Proof Let $c > 0$. Define a function $z(t)$ by the right-hand side of (1.2.169). Clearly, $z(t)$ is non-decreasing, $u(t) \le \phi^{-1}(z(t))$ for all $t \in I$ and $z(t_0) = c$. Differentiating $z(t)$, we can get

$$\begin{aligned} z'(t) &\le \sum_{i=1}^{n}[u(\alpha_i(t))]^q[f_i(\alpha_i(t))\psi_1(u(\alpha_i(t))) + g_i(\alpha_i(t))\psi_2(\log(u(\alpha_i(t))))]\alpha_i'(t) \\ &\le [\phi^{-1}(z(t))]^q\sum_{i=1}^{n}\Big[f_i(\alpha_i(t))\psi_1(\phi^{-1}(z(\alpha_i(t)))) \\ &\quad + g_i(\alpha_i(t))\psi_2(\log(\phi^{-1}(z(\alpha_i(t)))))\Big]\alpha_i'(t). \end{aligned} \quad (1.2.174)$$

Using the monotonicity of ϕ^{-1} and z, we can deduce

$$[\phi^{-1}(z(t))]^q \ge [\phi^{-1}(z(t_0))]^q \ge [\phi^{-1}(c)]^q > 0, \quad (1.2.175)$$

that is,

$$\frac{z'(t)}{[(\phi^{-1}(z(t)))]^q} \le \sum_{i=1}^{n}\left[f_i(\alpha_i(t))\psi_1(\phi^{-1}(z(\alpha_i(t)))) + g_i(\alpha_i(t))\psi_2(\log(\phi^{-1}(z(\alpha_i(t)))))\right]\alpha_i'(t). \quad (1.2.176)$$

Setting $t = s$ in the inequality (1.2.176), integrating it from t_0 to t, using the function G on the left-hand side, and changing variable on the right-hand side, we may obtain

$$G(z(t)) \leq G(c) + \sum_{i=1}^{n} \int_{\alpha_i(t_0)}^{\alpha_i(t)} [f_i(s)\psi_1(\phi^{-1}(z(s))) + g_i(s)\psi_2 log(\phi^{-1}(z(s)))]ds. \tag{1.2.177}$$

When $\psi_1(u) \geq \psi_2(\log(u))$, from the inequality (1.2.177), we derive

$$G(z(t)) \leq G(c) + \sum_{i=1}^{n} \int_{\alpha_i(t_0)}^{\alpha_i(t)} [f_i(s) + g_i(s)]\psi_1(\phi^{-1}(z(s)))ds. \tag{1.2.178}$$

Now, define a function $k(t)$ by the right-hand side of (1.2.178). Clearly, $k(t)$ is non-decreasing, $z(t) \leq G^{-1}(k(t))$ for all $t \in I$ and $k(t_0) = p(t_1)$. Differentiating $k(t)$, we may get

$$\begin{aligned} k'(t) &= \sum_{i=1}^{n} [f_i(\alpha_i(t)) + g_i(a_i(t))]\psi_1(\phi^{-1}z(\alpha_i(t)))\alpha_i'(t) \\ &\leq \psi_1(\phi^{-1}(G^{-1}(k(t)))) \sum_{i=1}^{n} [f_i(\alpha_i(t)) + g_i(a_i(t))]\alpha_i'(t). \end{aligned} \tag{1.2.179}$$

Using the monotonicity of ψ_1, ϕ^{-1}, G^{-1} and k, we may deduce

$$\frac{k'(t)}{\psi(\phi^{-1}(G^{-1}(k(t))))} \leq \sum_{i=1}^{n} [f_i(\alpha_i(t))]\alpha_i'(t). \tag{1.2.180}$$

Setting $t = s$ in the inequality (1.2.180), integrating it from t_0 to t, using the function Ψ_1 on the left-hand side, and changing variables on the right-hand side, we may obtain

$$\Psi_1(k(t)) \leq \Psi_1(k(t_0)) + \sum_{i=1}^{n} \int_{\alpha_i(t_0)}^{\alpha_i(t)} (f_i(s) + g_i(s))ds. \tag{1.2.181}$$

From the inequalities (1.2.181), we derive that for all $t \in I$,

$$z(t) \leq G^{-1}\left[\Psi_1^{-1}(\Psi_1(G(c)) + \sum_{i=1}^{n} \int_{\alpha_i(t_0)}^{\alpha_i(t)} [f_i(s) + g_i(s)]ds)\right]. \tag{1.2.182}$$

Now a combination of $u(t) \leq \phi^{-1}(z(t))$ and the last inequality produces the required inequality in (1.2.170).

When $\psi_1(u) < \psi_2(\log(u))$, from the inequality (1.2.177) it follows

$$G(z(t)) \leq G(c) + \sum_{i=1}^{n} \int_{\alpha_i(t_0)}^{\alpha_i(t)} (f_i(s) + g_i(s))\psi_1(\phi^{-1}(z(s)))ds. \quad (1.2.183)$$

Now, by the suitable application of the process from (1.2.178) to (1.2.181) in the inequality (1.2.183), we conclude that for all $t \in I$,

$$z(t) \leq G^{-1}\left[\Psi_2^{-1}(\Psi_2(G(c)) + \sum_{i=1}^{n} \int_{\alpha_i(t_0)}^{\alpha_i(t)} (f_i(s) + g_i(s))ds)\right]. \quad (1.2.184)$$

Now a combination of $u(t) \leq \phi^{-1}(z(t))$ and the last inequality produces the required inequality in (1.2.171). If $c = 0$, we carry out the above procedure with $\epsilon > 0$ instead of c and subsequently let $\epsilon \to 0$. This completes the proof. □

For special case $\phi(u) = u^p$ ($p > q > 0$ is a constant), Theorem 1.2.22 gives us the following retarded integral inequality for nonlinear functions.

Corollary 1.2.9 (The Agarwal-Kim-Sen Inequality [14]) *Let $u \in C(I, \mathbb{R}_1)$, f_i, $g_i \in C(I, \mathbb{R}_+)$, $i = 1, \ldots, n$, and let $\alpha_i \in C^1(I, I)$ be non-decreasing with $\alpha_i(t) \leq t$ $i = 1, \cdots, n$. Suppose that $c \geq 1$ and $p > q > 0$ are constants, and $\psi_j(u)$, $j = 1, 2$, are non-decreasing continuous functions for all $u \in \mathbb{R}_+$ with $\psi_j(u) > 0$ for all $u > 0$. If for all $t \in I$,*

$$u^p(t) \leq c + \sum_{i=1}^{n} \int_{\alpha_i(t_0)}^{\alpha_i(t)} u^q(s)\,(f_i(s)\psi_1(u(s)) + g_i(s)\psi_2(\log(u(s))))\,ds \quad (1.2.185)$$

then

(i) as the case $\psi_1(u) \geq \psi_2(\log(u))$, we have for all $t \in [t_0, t_1)$,

$$u(t) \leq \left(G_1^{-1}(G_1(c^{(p-q)/p}) + \frac{p-q}{p} \sum_{i=1}^{n} \int_{\alpha_i(t_0)}^{\alpha_i(t)} (f_i(s) + g_i(s))ds)\right)^{1/(p-q)} \quad (1.2.186)$$

(ii) as the case $\psi_1(u) < \psi_2(\log(u))$, we have for all $t \in [t_0, t_2)$,

$$u(t) \leq \left(G_2^{-1}(G_2(c^{(p-q)/p}) + \frac{p-q}{p} \sum_{i=1}^{n} \int_{\alpha_i(t_0)}^{\alpha_i(t)} (f_i(s) + g_i(s))ds)\right)^{1/(p-q)} \quad (1.2.187)$$

where $G_j^{-1}, j = 1, 2,$ *denotes the inverse functions of* $G_j, j = 1, 2,$ *for all* $t \in I$,

$$G_j(r) = \int_{r_0}^{r} \frac{ds}{\psi_j(s^{1/p-q})}, \quad r \geq r_0 > 0, \tag{1.2.188}$$

and $t_j \in I$, *are so chosen that for all* $t_0 \leq t \leq t_j$,

$$G_j(c^{(p-q)/p}) + \frac{p-q}{p} \sum_{i=1}^{n} \int_{\alpha_i(t_0)}^{\alpha_i(t)} (f_i(s) + g_i(s))ds \in Dom\ (G_j^{-1}). \tag{1.2.189}$$

Proof The proof is similar to that of Theorem 1.2.23. We omit the details here. □

Theorem 1.2.22 can easily be applied to generate another useful nonlinear inequalities in more general situations. For example, we have the following result.

Theorem 1.2.24 (The Agarwal-Kim-Sen Inequality [14]) *Let* $u, f_i, g_i \in C(I, \mathbb{R}_+)$, $i = 1, \ldots, n$, *and let* $\alpha_i \in C^1(I, I)$ *be non-decreasing with* $\alpha_i(t) \leq t$ *on* I *for* $i = 1, \cdots, n$. *Suppose that* $q > 0$ *and* $c \geq 0$ *are constants,* $\psi \in C^1(\mathbb{R}_+, \mathbb{R}_+)$ *is an increasing function with* $\phi(+\infty) = +\infty$ *on* I, *and* L, $M \in C(\mathbb{R}_+^2, \mathbb{R}_+)$ *satisfy, for all* t, v, $w \in \mathbb{R}_+$,

$$0 \leq L(t, v) - L(t, u) \leq M(t, w)(v - w), \tag{1.2.190}$$

with $v \geq w \geq 0$. *If for all* $t \in I$,

$$\phi(u(t)) \leq c + \sum_{i=1}^{n} \int_{\alpha_i(t_0)}^{\alpha_i(t)} u^q(s)(f_i(s)L(s, u(s)) + g_i(s)u(s))ds, \tag{1.2.191}$$

then for all $t \in [t_0, t_1)$,

$$u(t) \leq \phi^{-1}\left\{G^{-1}\left[\Omega^{-1}\left(\Omega(k_2(t_0)) + \sum_{i=1}^{n} \int_{\alpha_i(t_0)}^{\alpha_i(t)} (f_i(s)M(s) + g_i(s))ds\right)\right]\right\} \tag{1.2.192}$$

where

$$\begin{cases} \Omega(r) = \displaystyle\int_{r_0}^{r} \frac{ds}{\phi^{-1}(G^{-1}(s))}, \quad r \geq r_0 > 0, \\ k_2(t_0) = G(c) + \displaystyle\sum_{i=1}^{n} \int_{\alpha_i(t_0)}^{\alpha_i(t)} f_i(s)L(u(s))ds, \end{cases} \tag{1.2.193}$$

and G^{-1} *and* Ω^{-1} *denote the inverse functions of* G *and* Ω, *respectively, the function* G *is as defined in Theorem 1.2.22 for all* $t \in I$. $t_1 \in I$ *is so chosen that, for all*

$t_0 \le t \le t_1$,

$$\Omega(k_2(t_0)) + \sum_{i=1}^{n} \int_{\alpha_i(t_0)}^{\alpha_i(t)} f_i(s)ds \in Dom\ (\Omega^{-1}). \tag{1.2.194}$$

Proof Let $c > 0$. Define a function $z(t)$ by the right-hand side of (1.2.191). Clearly, $z(t)$ is non-decreasing, $u(t) \le \phi^{-1}(z(t))$ for all $t \in I$ and $z(t_0) = c$. Differentiating $z(t)$, we can get

$$\begin{aligned} z'(t) &= \sum_{i=1}^{n}[u(\alpha_i(t))]^q(f_i(\alpha_i(t))L(\alpha_i(t), u(\alpha_i(t)))+g_i(\alpha_i(t))u(\alpha_i(t)))\alpha_i'(t) \\ &\le [\phi^{-1}(z(t))]^q \sum_{i=1}^{n}\Big(f_i(\alpha_i(t))L(\alpha_i(t), \phi^{-1}(z(\alpha_i(t))))+g_i(\alpha_i(t))\varphi^{-1}(z(\alpha_i(t)))\Big)\alpha_i'(t). \end{aligned} \tag{1.2.195}$$

Using the monotonicity of ϕ^{-1} and z, we may deduce

$$\frac{z'(t)}{[(\phi^{-1}(z(t)))]^q} \le \sum_{i=1}^{n}\left(f_i(\alpha_i(t))L(\alpha_i(t), (\phi^{-1}(z(\alpha_i(t))))+g_i(\alpha_i(t))\varphi^{-1}(z(\alpha_i(t)))\right)\alpha_i'(t). \tag{1.2.196}$$

Setting $t = s$ in the inequality (1.2.196), integrating it from t_0 to t, using the function G on the left-hand side, and changing variable on the right-hand side, we may obtain

$$G(z(t)) \le G(c) + \sum_{i=1}^{n} \int_{\alpha_i(t_0)}^{\alpha_i(t)} [f_i(s)L(s, \phi^{-1}(z(s))) + g_i(s)\varphi^{-1}(z(s))]ds. \tag{1.2.197}$$

Thus from (1.2.190) and (1.2.197), we derive for all $t \le t_1$,

$$\begin{aligned} G(z(t)) \le\ & G(c) + \sum_{i=1}^{n} \int_{\alpha_i(t_0)}^{\alpha_i(t)} f_i(s)L(s)\varphi^{-1}(z(s))ds + \sum_{i=1}^{n} \int_{\alpha_i(t_0)}^{\alpha_i(t)} [f_i(s)M(s) \\ & + g_i(s)]\varphi^{-1}(z(s))ds. \end{aligned} \tag{1.2.198}$$

Now, define a function $k_2(t)$ by the right-hand side of (1.2.198). Clearly, $k_2(t)$ is non-decreasing, $z(t) \le G^{-1}(k_2(t))$ for $t \in I$. Differentiating $k_2(t)$, we get

$$\begin{aligned} k_2'(t) &= \sum_{i=1}^{n}(f_i(\alpha_i(t))M(\alpha_i(t)) + g_i(\alpha_i(t)))\phi^{-1}(z(s))\alpha_i'(t) \\ &\le \phi^{-1}(G^{-1}(k_2(t))) \sum_{i=1}^{n}(f_i(\alpha_i(t))M(\alpha_i(t)) + g_i(\alpha_i(t)))\alpha_i'(t). \end{aligned} \tag{1.2.199}$$

Using the monotonicity of ψ_1, ϕ^{-1}, G^{-1} and k_2, we deduce

$$\frac{k_2'(t)}{\psi(\phi^{-1}(G^{-1}(k_2(t))))} \le \sum_{i=1}^{n}[f_i(\alpha_i(t))M(\alpha_i(t)) + g_i(\alpha_i(t))]\alpha_i'(t). \quad (1.2.200)$$

Setting $t = s$ in the inequality (1.2.200), integrating it from t_0 to t, using the function Ω on the left-hand side, and changing variables on the right-hand side, we obtain

$$\Omega(k_2(t)) \le \Omega(k_2(t_0)) + \sum_{i=1}^{n}\int_{\alpha_i(t_0)}^{\alpha_i(t)}(f_i(s)M(s) + g_i(s))ds. \quad (1.2.201)$$

From (1.2.198) and (1.2.201), we conclude that

$$z(t) \le G^{-1}\left[\Omega^{-1}(\Omega(k_2(t_0)) + \sum_{i=1}^{n}\int_{\alpha_i(t_0)}^{\alpha_i(t)}(f_i(s)M(s) + g_i(s))ds)\right]. \quad (1.2.202)$$

for $t_0 \le t \le t_1$. Now a combination of $u(t) \le \phi^{-1}(z(t))$ and the last inequality produces the required inequality in (1.2.192) for $t_1 = t$. If $c = 0$, we carry out the above procedure with $\epsilon > 0$ instead of c and subsequently let $\epsilon \to 0$. This completes the proof. □

For the special case $\phi(u) = u^p$ ($p > q > 0$ is a constant), Theorem 1.2.24 gives us the following retarded integral inequality for nonlinear functions.

Corollary 1.2.10 (The Agarwal-Kim-Sen Inequality [14]) *Let* u, f_i, g_i, and α_i *be as defined in Theorem 1.2.24. Suppose that* $c \ge 0$ *and* $p > q > 0$ *are constants, and* $L, M \in C(\mathbb{R}_+^2, \mathbb{R}_+)$ *satisfies (1.2.190). If for all* $t \in I$,

$$u^p(t) \le c + \sum_{i=1}^{n}\int_{\alpha_i(t_0)}^{\alpha_i(t)} u^q(s)\,(f_i(s)L(s,u(s)) + g_i(s)u(s))\,ds, \quad (1.2.203)$$

then for all $t \in [t_0, t_1)$,

$$u(t) \le \left(\Omega_1^{-1}(\Omega_1(k_3(t_0)) + \frac{p-q}{p}\sum_{i=1}^{n}\int_{\alpha_i(t_0)}^{\alpha_i(t)}(f_i(s)M(s) + g_i(s))ds)\right)^{\frac{1}{p-q}} \quad (1.2.204)$$

where

$$\begin{cases} \Omega_1(r) = \displaystyle\int_{r_0}^{r}\frac{ds}{s^{1/(p-q)}}, \ r \ge r_0 > 0, \\ k_3(t_0) = c^{(p-q)/p} + \displaystyle\frac{p-q}{p}\sum_{i=1}^{n}\int_{\alpha_i(t_0)}^{\alpha_i(t)} f_i(s)L(s)ds, \end{cases} \quad (1.2.205)$$

and Ω_1^{-1} *denotes the inverse functions of* Ω_1 *for all* $t \in I$ *and* $t_1 \in I$ *is so chosen that, for all* $t_0 \le t \le t_1$,

$$\Omega_1(k_3(t_0)) + \frac{p-q}{p}\sum_{i=1}^{n}\int_{\alpha_i(t_0)}^{\alpha_i(t)}(f_i(s)M(s) + g_i(s))ds \in Dom\,(\Omega_1^{-1}). \tag{1.2.206}$$

Proof The proof is similar to that of Theorem 1.2.24 with suitable modification. We omit the details here. □

Pachpatte also showed the following inequalities (see, Kuang [315]).

Theorem 1.2.25 (The Pachpatte Inequality [499]) *Let* y, f, g *be real-valued non-negative continuous functions defined on* $\mathbb{R}_+$ *and* c_1, c_2 *be non-negative real constants. If for all* $t \in \mathbb{R}_+$,

$$y(t) \le \left(c_1 + \int_0^t f(s)y(s)ds\right)\left(c_2 + \int_0^t g(s)y(s)ds\right), \tag{1.2.207}$$

and $c_1c_2\int_0^t R(s)Q(s)ds < 1$, *for all* $t \in \mathbb{R}_+$, *then for all* $t \in \mathbb{R}_+$,

$$y(t) \le \frac{c_1c_2Q(t)}{1 - c_1c_2\int_0^t R(s)Q(s)ds}, \tag{1.2.208}$$

where for all $t \in \mathbb{R}_+$,

$$\begin{cases} R(t) = \left[g(t)\int_0^t f(\tau)d\tau + f(t)\int_0^t g(\tau)d\tau\right], & (1.2.209)\\ Q(t) = \exp\left(\int_0^t [c_1g(\tau) + c_2f(\tau)]d\tau\right). & (1.2.210)\end{cases}$$

Proof We first assume that c_1, c_2 are positive and define a function $z(t)$ by

$$z(t) = \left(c_1 + \int_0^t f(s)y(s)ds\right)\left(c_2 + \int_0^t g(s)y(s)ds\right). \tag{1.2.211}$$

Differentiating (1.2.211) and using the facts that $y(t) \le z(t)$ and $z(t)$ is monotone non-decreasing for all $t \in \mathbb{R}_+$, we observe that

$$z'(t) \le [c_1g(t) + c_2f(t)]z(t) + R(t)z^2(t), \tag{1.2.212}$$

which, implies

$$z(t) \le \frac{c_1c_2Q(t)}{1 - c_1c_2\int_0^t R(s)Q(s)ds}. \tag{1.2.213}$$

The desired inequality in (1.2.208) now follows by using $y(t) \leq z(t)$.

If c_1, c_2 are non-negative, we carry out the above arguments with $c_1+\varepsilon$ and $c_2+\varepsilon$ instead of c_1 and c_2, where $\varepsilon > 0$ is an arbitrary small constant, and subsequently pass to the limit as $\varepsilon \to 0$ to obtain (1.2.208). The proof is thus complete. □

Remark 1.2.14 Note that in the special case when $g(t) = 0$ and $c_2 = 1$ or $f(t) = 0$ and $c_1 = 1$, the inequality given in Theorem 1.2.25 reduces to the well-known Gronwall inequality [79].

The next result is a corollary of Theorem 1.1.60.

Corollary 1.2.11 (The Agarwal-Ryoo-Kim Inequality [17]) *Let u, f_i, p_i, ϕ and $g(u)$ be as in Theorem 1.1.60 and let $p > 1$ be a constant. If, for all $t \in J$,*

$$u^p(t) \leq a(t) + \int_{\phi(\alpha)}^{\phi(t)} p_1(t_1) f_1(t_1) u(t_1) g(u(t_1)) dt_1$$
$$+ \sum_{i=2}^{n} \int_{\phi(\alpha)}^{\phi(t)} p_1(t_1) \Big(\int_{\phi(\alpha)}^{\phi(t_1)} p_2(t_2) \Big(\cdots \Big(\int_{\phi(\alpha)}^{\phi(t_{i-2})} p_{i-1}(t_{i-1})$$
$$\times \Big(\int_{\phi(\alpha)}^{\phi(t_{i-1})} p_i(t_i) f_i(t_i) u(t_i) g(u(t_i)) dt_i \Big) dt_{i-1} \Big) \cdots \Big) dt_2 \Big) dt_1, \tag{1.2.214}$$

then for $t \in [\alpha, T_3]$,

$$u(t) \leq \left\{ G_3^{-1} \left[G_3 \left(a^{(p-1)/p}(t) \right) + \frac{p-1}{p} F_1(t) \right] \right\}^{1/(p-1)} \tag{1.2.215}$$

where $T_3 \in I$ is chosen so that $G_3(a^{(p-1)/p}(t)) + \frac{p-1}{p} F_1(t) \in Dom\,(G_3^{-1})$,

$$G_3(r) = \int_{r_0}^{r} \frac{ds}{g(v^{1/(p-1)})(s)}, \quad r \geq r_0 > 0, \tag{1.2.216}$$

and G_3^{-1} denotes the inverse function of G_3, and $F_1(t)$ is defined in (1.1.363) for any $t \in I$.

In the sequel, we introduce some new nonlinear delay integral inequalities, due to Ma and Yang [365], of Ou-Yang type, which generalize some results of Pachpatte [498] and Yang [438].

Pachpatte [498] discussed the following delay integral inequalities, which generalize the Ou-Yang inequality, by means of the same argument as that used by Tsamatos and Ntouyas [652], for all $t \in \mathbb{R}_+$,

$$x^2(t) \leq c^2 + 2 \int_0^t x(\sigma(s)) \{ f(s) W[x(\sigma(s))] + h(s) \} ds, \tag{1.2.217}$$

$$x^2(t) \leq c^2 + 2\int_0^t x(\sigma(s))\left\{f(s)\left(\int_0^s g(\tau)W[x(\sigma(\tau))]d\tau\right) + h(s)\right\} ds, \quad (1.2.218)$$

$$x^2(t) \leq c^2 + 2\int_0^t x^2(\sigma(s))\left\{f(s)\left(\int_0^s g(\tau)W[\log x(\sigma(\tau))]d\tau\right) + h(s)\right\} ds, \quad (1.2.219)$$

with the initial condition

$$\begin{cases} x(t) = \psi(t), & \text{for all } t \in [a, 0], \\ \psi(\sigma(t)) \leq c & \text{for all } \ t \in \mathbb{R}_+ \ \text{ with } \ \sigma(t) \leq 0. \end{cases} \quad (1.2.220)$$

Pachpatte [498] proved the following result.

Theorem 1.2.26 (The Pachpatte Inequality [498]) *Let* f, g, h *and* $W \in C(\mathbb{R}_+, \mathbb{R}_+)$, $\sigma \in C(\mathbb{R}_+, \mathbb{R})$ *with* $\sigma(t) \leq t$ *and* $-\infty < a := \inf\{\sigma(t) : t \in \mathbb{R}_+\} \leq 0$, $\psi \in C([a, 0], \mathbb{R}_+)$ *and* $x \in C([a, +\infty], \mathbb{R}_+)$. *Furthermore, let* W *be non-decreasing and let* $W(u) > 0$ *hold for all* $u > 0$. *Then*

(i) from (1.2.217)–(1.2.220), we have for all $0 \leq t \leq v_1$,

$$x(t) \leq G^{-1}\left[G\left(c + \int_0^t h(s)ds\right) + \int_0^t f(s)ds\right],$$

(ii) from (1.2.218)–(1.2.220), we have for all $0 \leq t \leq v_2$,

$$x(t) \leq G^{-1}\left[G\left(c + \int_0^t h(s)ds\right) + \int_0^t f(s)\left(\int_0^s g(\tau)d\tau\right)ds\right],$$

(iii) from (1.2.219)–(1.2.220), we have for all $0 \leq t \leq v_3$,

$$x(t) \leq \exp\left(G^{-1}\left[G\left(\log c + \int_0^t h(s)ds\right) + \int_0^t f(s)\left(\int_0^s g(\tau)d\tau\right)ds\right]\right),$$

where

$$G(u) := \int_{u_0}^u \frac{ds}{W(s)}, \quad u \geq u_0 > 0,$$

and G^{-1} *denotes the inverse function of* G, *and the positive numbers* v_1, v_2 *and* v_3 *are chosen so that the quantity in the square brackets of (i), (ii) and (iii) is in the range of* G.

Next, we generalize the conclusions (i)–(iii) of Theorem 1.2.26. A delay integral inequality similar to inequality (1.2.219) is also discussed.

We define $\mathbb{R}_1 = [1, +\infty)$, and denote by $C(M, S)$ the class of all continuous functions defined on set M with range in the set S. The basic assumption in the following Theorems 1.2.27–1.2.29 is as follows:

Assumption (H):

(i) f, g, h, and $n \in C(\mathbb{R}_+, \mathbb{R}_+)$ with $n(t)$ non-decreasing.
(ii) $W \in C(\mathbb{R}_+, \mathbb{R}_+)$ is non-decreasing with $W(u) > 0$ for all $u > 0$,
(iii) $\sigma \in (\mathbb{R}_+, \mathbb{R})$, $\sigma(t) \le t$ for all $t \in \mathbb{R}_+$, with $-\infty < a := \inf\{\sigma(t) : t \in \mathbb{R}_+\} \le 0$,
(iv) $\psi \in C([a, 0], \mathbb{R}_+)$ and $x \in C([a, +\infty), \mathbb{R}_+)$,
(v) $\phi \in C^1(\mathbb{R}, \mathbb{R}_+)$ with ϕ' non-decreasing and $\phi'(u) > 0$ for all $u > 0$.

Consider first the next generalization of inequality (1.2.217), for all $t \in \mathbb{R}_+$,

$$\phi(x(t)) \le n(t) + \int_0^t \phi'[x(\sigma(s))]\{f(s)W[\sigma(s)] + g(s)x(\sigma(s)) + h(s)\}ds, \tag{1.2.221}$$

with the initial condition

$$\begin{cases} x(t) = \psi(t), & \text{for all } t \in [a, 0], \\ \psi(\sigma(t)) \le \phi^{-1}(n(t)), & \text{for every } t \ge 0 \text{ with } \sigma(t) \le 0. \end{cases} \tag{1.2.222}$$

Theorem 1.2.27 (The Ma-Yang Inequality [405]) *Let Assumption (H) hold. Then inequality (1.2.221) with condition (1.2.222) implies, for all* $0 \le t \le \alpha$,

$$x(t) \le G^{-1}\left\{G\left[G\left(\exp\int_0^t g(s)ds\right)\left(\phi^{-1}(n(s)) + \int_0^t h(s)ds\right)\right] + \exp\left(\exp\int_0^t g(s)ds\right)\int_0^t f(s)ds\right\}, \tag{1.2.223}$$

where G *and* G^{-1} *are as defined in Theorem 1.2.26 and the positive number* α *is chosen so that the quantity in the curly brackets of (1.2.223) is in the range of* G.

Proof Let $\varepsilon > 0$ be an arbitrary small constant. Fixing any positive number $T(\le \alpha)$, we define a positive non-decreasing function $u(t)$ by, for all $t \in J = [0, T]$,

$$\phi(u(t)) = n(T) + \varepsilon \int_0^t \phi'[x(\sigma(s))]\{f(s)W[x(\sigma(s))] + g(s)x(\sigma(s)) + h(s)\}ds. \tag{1.2.224}$$

Then $u(t) \geq \phi^{-1}(n(T)+\varepsilon) > 0$ for all $t \in J$ and for all $t \in J$,

$$x(t) \leq u(t). \tag{1.2.225}$$

Thus for every $t \geq 0$ with $\sigma(t) \geq 0$, we have

$$x(\sigma(t)) < u(\sigma(t)) \leq u(t)$$

since $u(t)$ is non-decreasing and $\sigma(t) \leq t$. By condition (1.2.223), for every $t \geq 0$ with $\sigma(t) \leq 0$, we have for all $t \in J$,

$$x(\sigma(t)) = \psi(\sigma(t)) \leq \phi^{-1}(n(t)) \leq \phi^{-1}(n(T)) \leq \phi^{-1}(n(\tau)+\varepsilon) \leq u(t),$$

since ϕ^{-1} is non-decreasing. Hence we always have the relation, for all $t \in J$,

$$x(\sigma(t)) \leq u(t). \tag{1.2.226}$$

By differentiation, we derive from (1.2.224) that

$$\begin{aligned}\phi'(u(t))\frac{du}{dt} &= \phi'(x(\sigma(t)))\{f(t)W[x(\sigma(t))] + g(t)x(\sigma(t)) + h(t)\}\\ &\leq \phi'(u(t))\{f(t)W[u(t)] + g(t)u(t) + h(t)\},\end{aligned}$$

i.e.,

$$\frac{du}{dt} \leq f(t)W[u(t)] + g(t)u(t) + h(t)$$

since $u(t) > 0$ for all $t \in J$, ϕ' is non-decreasing with $\phi'(u) > 0$ for all $u > 0$, and (1.2.226) holds.

Integrating the both sides of the last inequality from 0 to t, then we can obtain for all $t \in J$,

$$u(t) \leq n_1(t) + \int_0^t g(s)u(s)ds,$$

where

$$n_1(t) = \phi^{-1}(n(T)+\varepsilon) + \int_0^t h(s)ds + \int_0^t f(s)W[u(s)]ds.$$

From the last inequality and the well-known Gronwall inequality, it follows that for all $t \in J$,

$$u(t) \leq \left[\phi^{-1}(n(T)+\varepsilon)\int_0^t h(s)ds + \int_0^t f(s)W[u(s)]ds\right]\exp\left(\int_0^t g(s)ds\right)$$

$$\le \left[\phi^{-1}(n(T)+\varepsilon)\int_0^T h(s)ds + \int_0^t f(s)W[u(s)]ds\right]\exp\left(\int_0^T g(s)ds\right). \tag{1.2.227}$$

Setting, for all $t \in J$,

$$v(t) := \left[\phi^{-1}(n(T)+\varepsilon)\int_0^T h(s)ds + \int_0^t f(s)W[u(s)]ds\right]\exp\left(\int_0^T g(s)ds\right), \tag{1.2.228}$$

then by (1.2.227), we have for all $t \in J$,

$$u(t) \le v(t). \tag{1.2.229}$$

Differentiating (1.2.228) and using (1.2.229), we derive, for all $t \in J$,

$$\frac{dv(t)}{dt} \le \left(\exp\left(\int_0^T g(s)ds\right)\right)f(t)W[v(t)],$$

or

$$dG[v(t)] \equiv \frac{dv(t)}{W[v(t)]} \le \left(\exp\left(\int_0^T g(s)ds\right)\right)f(t)dt,$$

since by (1.2.228), $v(t) > 0$ for all $t \in J$ and the condition (ii) in assumption (H). Integrating the both sides of the last relation from 0 to t, and in view of $v(0) = \left(\phi^{-1}(n(T)+\varepsilon)\int_0^T h(s)ds\right)\exp\left(\int_0^t g(s)ds\right)$ from (1.2.228), we have for all $t \in J$,

$$G[v(t)] \le G\left\{\left(\exp\left(\int_0^t g(s)ds\right)\right)\left(\phi^{-1}(n(T)+\varepsilon) + \int_0^t h(s)ds\right)\right\}$$
$$+ \int_0^t \left(\exp\left(\int_0^T g(s)ds\right)\right)f(s)ds.$$

Taking $t = T$ in the last inequality and then letting $\varepsilon \to 0$, we can obtain

$$G[v(T)] \le G\left[\left(\exp\int_0^t g(s)ds\right)\left(\phi^{-1}(n(T)) + \int_0^t h(s)ds\right)\right]$$
$$+ \left(\exp\int_0^T g(s)ds\right)\int_0^T f(s)ds.$$

Since $T \in (0, \alpha]$ is arbitrary, from the last relation we have for all $t \in \mathbb{R}_+$,

$$G[v(t)] \leq G\left[\left(\exp \int_0^t g(s)ds\right)\left(\phi^{-1}(n(t)) + \int_0^t h(s)ds\right)\right] + \left(\exp \int_0^t g(s)ds\right)\int_0^t f(s)ds, \tag{1.2.230}$$

or for all $t \in \mathbb{R}_+$,

$$v(t) \leq G^{-1}\left\{G\left[\left(\exp \int_0^t g(s)ds\right)\left(\phi^{-1}(n(t)) + \int_0^t h(s)ds\right)\right] + \left(\exp \int_0^t g(s)ds\right)\int_0^t f(s)ds\right\}.$$

Hence, by (1.2.225), (1.2.229) and (1.2.230), we conclude for all $0 < t \leq \alpha$,

$$x(t) \leq G^{-1}\left\{G\left[\left(\exp \int_0^t g(s)ds\right)\left(\phi^{-1}(n(t)) + \int_0^t h(s)ds\right)\right] + \left(\exp \int_0^t g(s)ds\right)\int_0^t f(s)ds\right\}. \tag{1.2.231}$$

By (1.2.221), (1.2.231) holds also when $t = 0$. □

Now we consider the next generalization of inequality (1.2.218), for all $t \in \mathbb{R}_+$,

$$\phi(x(t)) \leq n(t) + \int_0^t \phi'(x(\sigma(s)))\left\{f(s)\left(\int_0^s g(\tau)W[x(\sigma(\tau))]d\tau\right) + h(s)x(\sigma(s)) + k(s)\right\}ds. \tag{1.2.232}$$

Theorem 1.2.28 (The Ma-Yang Inequality [405]) *Let $k(t) \in C(\mathbb{R}_+, \mathbb{R}_+)$ and Assumption (H) holds. Then inequality (1.2.232) with condition (1.2.222) implies, for all $0 < t \leq \beta$,*

$$x(t) \leq G^{-1}\left\{G\left[\left(\exp \int_0^t h(s)ds\right)\left(\phi^{-1}(n(t) + \int_0^t k(s)ds\right)\right] + \left(\exp \int_0^t h(s)ds\right)\int_0^t f(s)\left(\int_0^s g(\tau)d\tau\right)ds\right\}, \tag{1.2.233}$$

where G and G^{-1} are as defined in Theorem 1.2.26, and the positive number β is chosen so that the quantity in the curly brackets of (1.2.233) is in the range of G.

Proof Fixing any positive number $T(\leq \beta)$ and taking an arbitrary positive small constant ε, we define on interval $j = [0, T]$, a function $v(t)$ by, for all $t \in J$,

$$\phi(v(t)) : = n(T) + \varepsilon + \int_0^t \phi'(x(\sigma(s))) \left\{ f(s) \left(\int_0^s g(\tau) W[x(\sigma(\tau))] d\tau \right) + h(s) x(\sigma(s)) + k(s) \right\}. \tag{1.2.234}$$

Thus from (1.2.232) and (1.2.234), we can derive for all $t \in J$,

$$x(t) \leq v(t). \tag{1.2.235}$$

Using the same argument as used in the proof of Theorem 1.2.27, we can obtain for all $t \in J$,

$$x(\sigma(t)) \leq v(t).$$

Differentiating (1.2.234) and using the last relation, we can derive for all $t \in J$,

$$\frac{dv(t)}{dt} \leq k(t) + h(t)v(t) + f(t) \int_0^t g(\tau) W[v(t)] d\tau.$$

Integrating the both sides of the last inequality from 0 to t, and using $v(0) = \phi^{-1}(n(T) + \varepsilon)$, then we derive

$$v(t) \leq \left[\phi^{-1}(n(T) + \varepsilon) + \int_0^t k(s) ds + \int_0^t f(s) \left(\int_0^t g(\tau) W[v(\tau)] d\tau \right) ds \right] + \int_0^t h(s) v(s) ds.$$

Using the Gronwall inequality to the last inequality, we can get

$$\begin{aligned} u(t) &\leq \left[\phi^{-1}(n(T)+\varepsilon) + \int_0^t k(s) ds + \int_0^t f(s) \left(\int_0^s g(\tau) W[v(\tau)] d\tau \right) ds \right] \exp \int_0^t h(s) ds \\ &\leq H(T)\theta_\varepsilon(T) + H(T) \int_0^t f(s) \left(\int_0^s g(\tau) W[v(\tau)] d\tau \right) ds, \quad t \in J, \end{aligned} \tag{1.2.236}$$

where $H(t) = \exp\left(\int_0^t h(s)ds\right)$, $\theta_\varepsilon = \phi^{-1}(n(T)+\varepsilon) + \int_0^t k(s)ds$. Setting

$$X(t) := H(T)\theta_\varepsilon(T) + H(T)\int_0^t f(s)\left(\int_0^s g(\tau)W[v(\tau)]d\tau\right)ds,$$

by (1.2.236), we can conclude for all $t \in J$,

$$v(t) \le X(t). \tag{1.2.237}$$

Differentiating $X(t)$ and using (1.2.237), we may obtain for all $t \in J$,

$$\begin{aligned}\frac{dX(t)}{dt} &= H(T)f(t)\int_0^t g(\tau)W[v(\tau)]d\tau \\ &\le H(T)f(t)\int_0^t g(\tau)W[X(\tau)]d\tau \\ &\le H(T)f(t)\int_0^t g(\tau)fd\tau W[X(t)].\end{aligned}$$

Because $X(t)$ is positive and $W(u) > 0$ for all $u > 0$, the last relation can be rewritten in the form: for all $t \in J$,

$$dG[X(t)] = \frac{dX(t)}{W[X(t)]} \le H(T)f(t)\left(\int_0^t g(\tau)d\tau\right)dt.$$

Integrating the both sides of the last inequality from 0 to t, then we obtain, for all $t \in J$,

$$G[X(t)] \le G[H(t)\theta_\varepsilon(T)] + H(T)\int_0^t f(s)\left(\int_0^s g(\tau)d\tau\right)ds.$$

Since $T \in (0, \beta)$, taking $t = T$ and letting $\varepsilon \to 0^+$, then we derive from the last inequality that

$$X(T) \le G^{-1}\left\{G[H(T)\theta_0(T)] + H(T)\int_0^T f(s)\left(\int_0^s g(\tau)d\tau\right)ds\right\}.$$

Because T is any number from $(0, \beta]$, by (1.2.235), (1.2.237), and the last inequality, we can obtain the validity of (1.2.233) on $(0, \beta]$. By (1.2.232), inequality (1.2.233) holds also when $t = 0$. □

Now, we consider the following nonlinear delay inequality which is a variant of the inequality (1.2.219), for all $t \in \mathbb{R}_+$,

$$x^r(t) \le c^r + \int_0^t x^r(\sigma(s))\left(f(s)x^q(\sigma(s)) + g(s)\right)ds. \tag{1.2.238}$$

Theorem 1.2.29 (The Ma-Yang Inequality [405]) *Let $c > 0,\ q > 0,\ r > 0$ be constants, and $f,\ h,\ x$ and σ are defined as in Theorem 1.2.26. Then inequality (1.2.238) with condition (1.2.220) implies for all $0 \le t \le v$,*

$$x(t) \le \left(\exp \int_0^t \frac{g(s)}{r} ds\right) \left\{ \frac{1}{c^q} - \int_0^t \frac{qf(s)}{r} \left(\exp \int_0^s \frac{qg(\tau)}{r} d\tau\right) ds \right\}^{-1/q} \tag{1.2.239}$$

where v is a positive number satisfying

$$\frac{1}{c^q} > \int_0^v \frac{qf(s)}{r} \left(\exp\left(\int_0^s \frac{qg(\tau)}{r} d\tau\right)\right) ds.$$

Proof Define, for all $t \in \mathbb{R}_+$,

$$W^r(t) := c^r + \int_0^t x^r(\sigma(s))\{f(s)x^q(\sigma(s)) + g(s)\}ds. \tag{1.2.240}$$

By (1.2.238), we have, for all $t \in \mathbb{R}_+$,

$$x(t) \le w(t). \tag{1.2.241}$$

Applying the same argument as used in the proof of Theorem 1.2.27, we can obtain, for all $t \in \mathbb{R}_+$,

$$x(\sigma(t)) \le w(t).$$

Differentiating (1.2.240) and using the last relation, we derive, for all $t \in \mathbb{R}_+$,

$$\frac{dw(t)}{dt} \le \frac{g(t)}{r} w(t) + \frac{f(t)}{r} w^{1+q}(t). \tag{1.2.242}$$

In view of $w(0) = c$, by a well-known comparison theorem for ODEs, from (1.2.242) we infer that for all $t \in I$,

$$w(t) \le y(t), \tag{1.2.243}$$

where $I = (0, \rho)$ is the maximal existence interval of the solution $y(t)$ to the following initial value problem of the Bernoulli equation:

$$\frac{dy(t)}{dt} = \frac{g(t)}{r} y + \frac{f(t)}{r} y^{1+q}, \quad \text{for all } t \in \mathbb{R}_+,\ y(0) = c.$$

The unique solution of the last equation is, for all $0 \le t \le v$,

$$y(t) = \left(\exp \int_0^t \frac{g(s)}{r} ds\right) \left\{ c^{-q} - \int_0^t \frac{qf(s)}{r} \left(\exp \int_0^s \frac{qg(\tau)}{r} d\tau\right) ds \right\}^{-1/q}.$$

Hence the desired inequality (1.2.239) follows from (1.2.241), (1.2.243), and the last relation immediately. □

Letting $\phi(u) = u^p$ in Theorem 1.2.27, then we obtain the following corollary.

Corollary 1.2.12 (The Ma-Yang Inequality [405]) *Let $p \geq 1$ be a constant and assumption (H) holds. Then the nonlinear delay inequality, for all $t \in \mathbb{R}_+$,*

$$x^p(t) \leq n(t) + \int_0^t x^{p-1}(\sigma(s)) \Big\{ f(s)W[x(\sigma(s))] + g(s)x(\sigma(s)) + h(s) \Big\} ds, \tag{1.2.244}$$

which condition (1.2.224) implies: for all $0 \leq t \leq \gamma$,

$$x(t) \leq G^{-1}\left(G(\xi_p(t)) + \left(\exp \int_0^t \frac{g(s)}{p} ds\right) \int_0^t \frac{f(s)}{p} ds \right), \tag{1.2.245}$$

where $\xi_p(t) = \left(\exp \int_0^t \frac{g(s)}{p} ds\right)\left(n^{1/p}(t) + \int_0^t \frac{f(s)}{p} ds\right)$, and the positive number γ is chosen so that the quantity in the curly brackets of (1.2.245) is in the range of G.

Remark 1.2.15

(i) In Corollary 1.2.12, letting $p = 2$, $n(t) = c^2$, $f(t) = 2a(t)$, $g(t) \equiv 0$, and $h(t) = 2b(t)$, then it follows conclusion (i) of Theorem 1.2.26.
(ii) The special case of inequality (1.2.244) when $W(u) = u$, $g(t) \equiv 0$, and $\sigma(t) \equiv t$ was studied by Yang in [694].

Corollary 1.2.13 (The Ma-Yang Inequality [405]) *Let $x(t) \in C([a, +\infty), \mathbb{R}_1)$, $n(t) \in C(\mathbb{R}_+, \mathbb{R}_+)$, $p > 0$ be a constant and assumption (H) holds. Then the nonlinear delay integral inequality, for all $t \in \mathbb{R}_+$,*

$$x^p(t) \leq n(t) + \int_0^t x^p(\sigma(s)) \Big\{ f(s)W[\log x(\sigma(s))] + g(s)\log x(\sigma(s)) + h(s) \Big\} ds, \tag{1.2.246}$$

with condition (1.2.224) implies, for all $0 \leq t \leq \delta$,

$$x(t) \leq \exp\left[G^{-1}\left\{ G[\bar{\xi}_p(t)] + \left(\exp \int_0^t \frac{g(s)}{p} ds\right) \int_0^t \frac{f(s)}{p} ds \right\} \right], \tag{1.2.247}$$

where $\bar{\xi}_p(t) = \left(\exp \int_0^t \frac{g(s)}{p} ds\right)\left(\frac{1}{p}\log n(t) + \int_0^t \frac{f(s)}{p} ds\right)$, G and G^{-1} are defined as in Theorem 1.2.26, and the positive number δ is chosen so that the quantity in the curly brackets of (1.2.247) is in the range of G.

Proof Taking $u(t) = \log x(t)$, then inequality (1.2.246) reduces to for all $t \in \mathbb{R}_+$,

$$e^{Pu(t)} \le n(t) + \int_0^t e^{Pu(\sigma(s))}\Big\{f(s)W[u(\sigma(s))] + g(s)u(\sigma(s)) + h(s)\Big\}ds,$$

which is a special case of inequality (1.2.221) when $\phi(u) = \exp(Pu)$. By Theorem 1.2.26, we get the desired inequality (1.2.247) directly. □

Remark 1.2.16 The inequality (1.2.246) with $p = 2$ is different from the inequality (1.2.219) of Theorem 1.2.26.

Letting $\phi(u) = u^p$ in Theorem 1.2.28, then we obtain the next corollary.

Corollary 1.2.14 (The Ma-Yang Inequality [405]) *Let $p \ge 1$ be a constant and assumption (H) holds, then the inequality, for all $t \in \mathbb{R}_+$,*

$$x^p(t) \le n(t) + \int_0^t x^{p-1}(\sigma(s))\left\{ f(s)\left(\int_0^s g(\tau)W[x(\sigma(\tau))]d\tau\right) + h(s)x(\sigma(s)) + k(s)\right\} ds, \tag{1.2.248}$$

with condition (1.2.222) implies: for all $0 \le t \le \eta$,

$$x(t) \le G^{-1}\left(G[H_p(t)\theta_p(t)] + H_p(t)\int_0^t \frac{f(s)}{p}\left(\int_0^s g(\tau)d\tau\right)ds\right), \tag{1.2.249}$$

where $H_p(t) = \exp\left(\int_0^t \frac{h(s)}{p}ds\right)$, $\theta_p(t) = n^{1/p}(t) + \int_0^t \frac{k(s)}{p}ds$, and the positive number η is chosen so that the quantity of the curly brackets of (1.2.249) is in the range of G.

Remark 1.2.17 In Corollary 1.2.14, letting $p = 2$, $n(t) = c^2$, $f(t) = 2a(t)$, $h(t) \equiv 0$, and $k(t) = 2b(t)$, then we derive the conclusion (ii) of Theorem 1.2.26.

Corollary 1.2.15 (The Ma-Yang Inequality [405]) *Let $x(t) \in C([a, +\infty), \mathbb{R}_1)$, $n(t) \in C(\mathbb{R}_+, \mathbb{R}_+)$, $p > 0$, be a constant, and assumption (H) holds. Then the nonlinear delay inequality, for all $t \in \mathbb{R}_+$,*

$$x^p(t) \le n(t) + \int_0^t x^p(\sigma(s))\left\{ f(s)\left(\int_0^s g(\tau)W[\log x(\sigma(\tau))]d\tau\right) + h(s)\log x(\sigma(s)) + k(s)\right\} ds, \tag{1.2.250}$$

with condition (1.2.222) implies

$$x(t) \le \exp\left[G^{-1}\left\{G[H_p(t)\bar{\theta}_p(t)] + H_p(t)\int_0^t \frac{f(s)}{p}\left(\int_0^s g(\tau)d\tau\right)ds\right\}\right] \tag{1.2.251}$$

where $\bar{\theta}_p(t) = \frac{1}{p}\log n(t) + \int_0^t \frac{k(s)}{p}ds$, $H_p(t)$ *is defined as in Corollary 1.2.14, and the positive number* μ *is chosen so that the quantity of the curly brackets of (1.2.251) is in the range of G.*

Proof Taking $u(t) = \log x(t)$, then inequality (1.2.251) reduces to

$$e^{Pu(t)} \le n(t) + \int_0^t e^{Pu(\sigma(s))}\left\{\int_0^s f(s)\,(g(\tau)W[u(\sigma(\tau))]d\tau)\right.$$
$$\left. + h(s)u(\sigma(s)) + k(s)\right\}ds, \quad t \in \mathbb{R}_+.$$

This is a special case of inequality (1.2.233) when $\phi(u) = \exp(Pu)$. An application of Theorem 1.2.28 to the inequality yields the desired inequality (1.2.251). □

Remark 1.2.18 In Corollary 1.2.15, letting $p = 2,\ n(t) = c^2,\ f(t) = 2a(t),\ h(t) \equiv 0$, and $k(t) = 2b(t)$, then it follows conclusion (iii) of Theorem 1.2.28.

Remark 1.2.19 In Theorem 1.2.29, if r obeys the more restrictive condition $r \ge 1$, then (1.2.238) can be considered as a particular case of inequality (1.2.244) when $p = r,\ h(t) \equiv 0,\ n(t) = c^r$, and $W(\xi) = \xi^{q+1}$; i.e., for all $t \in \mathbb{R}_+$,

$$x^r(t) \le c^r + \int_0^t x^{r-1}(\sigma(s))\left\{f(s)x^{q+1}(\sigma(s)) + g(s)\right\}ds.$$

By definition, we have $G(u) = \frac{1}{-q}(u^{-q} - u_0^{-q})$ and hence

$$G^{-1}(v) = [u_0^{-q} - qv]^{-1/q}.$$

An application of Corollary 1.2.12 to the last inequality yields, for all $0 \le t \le \bar{v}$,

$$x(t) \le \left(\exp\int_0^t \frac{g(s)}{r}ds\right)$$
$$\times\left\{\frac{1}{c^q} - \left(\exp\int_0^t \frac{g(s)(q+1)}{r}ds\right)\int_0^t \frac{qf(s)}{r}ds\right\}^{-1/q}, \tag{1.2.252}$$

where $\bar{v}$ is positive number satisfying

$$\frac{1}{c^q} > \left(\int_0^{\bar{v}} \frac{qf(s)}{r} ds \right) \left(\exp \left(\int_0^{\bar{v}} \frac{g(s)(q+1)}{r} ds \right) \right).$$

Obviously, in many situations, the bound in (1.2.239) is not only better than that given in (1.2.252), but also the validity of (1.2.252) when $0 < r < 1$ cannot be established by using Corollary 1.2.12.

1.3 The One-Dimensional Dafermos Inequality

Dafermos established the following generalization of Ou-Yang's inequality in the process of establishing a connection between stability and the second law of thermodynamics.

Theorem 1.3.1 (The Dafermos Inequality [180]) *Assume that the non-negative functions* $y(t) \in L^\infty[0, T]$ *and* $g(t) \in L^1[0, T]$ *satisfy the inequality, for all* $x \in [0, T]$,

$$y^2(x) \leq M^2 y^2(0) + \int_0^x [2\alpha y^2(t) + 2Ng(t)y(t)]dt, \tag{1.3.1}$$

where α, M, N *are non-negative constants. Then for all* $x \in [0, T]$,

$$y(x) \leq Me^{\alpha x} y(0) + Ne^{\alpha x} \int_0^x g(t)dt. \tag{1.3.2}$$

Proof Let $z(x) = M^2 y^2(0) + 2\int_0^x [\alpha y^2(t) + Ng(t)y(t)]dt$. Then, $z'(x) = 2\alpha y^2(x) + 2Ng(x)y(x)$. Thus, using inequality (1.3.1), we have $z'(x) \leq 2\alpha z(x) + 2Ng(x)\sqrt{z(x)}$. Hence,

$$\frac{d}{dx}\left(e^{-\alpha x}\sqrt{z(x)}\right) \leq Ne^{-\alpha x} g(x) \leq Ng(x).$$

Thus now integrating the above inequality, inequality (1.3.2) follows readily. □

Remark 1.3.1 In Theorem 1.3.1, if $\alpha = 0$, then the result reduces to an inequality of Ou-Yang [438], i.e., Theorem 1.2.1..

Remark 1.3.2 It is clear from the proof that inequality (1.3.2) is not best possible. Its right-hand side can be improved by replacing the integral $\int_0^t g(\tau)d\tau$ *by* $\int_0^t e^{-\alpha\tau} g(\tau)d\tau$.

Theorem 1.3.2 (The Snow Inequality [620]) *Let $x(t)$ be a real-valued, positive, continuous function and $f(t)$ be a real-valued non-negative continuous function defined on $I = [0, +\infty)$, suppose that $p \geq 2$. If it holds that for all $t \in I$,*

$$x^p(t) \leq x_0 + \int_0^t f(s)x(s)ds, \tag{1.3.3}$$

where x_0 is a positive constant. Then for all $t \in I$,

$$x(t) \leq \left(x_0^{q/p} + \frac{q}{p}\int_0^t f(s)ds\right)^{1/q}, \tag{1.3.4}$$

where $p - q = 1$.

Proof Differentiating $x^p(t)$ with respect to t, we may have

$$px^{p-1}(t)x'(t) \leq f(t)x(t),$$

then

$$px^{p-2}(t)x'(t) \leq f(t).$$

Now, integrating both sides the last inequality from 0 to t, we may obtain

$$\frac{p}{q}x^q(t) - \frac{p}{q}x^q(0) \leq \int_0^t f(s)ds.$$

But from (1.3.4) $x^p(0) \leq x(0)$, thus $x^q(0) \leq x_0^{q/p}$. Hence, for all $t \in I$,

$$\frac{p}{q}x^q(t) \leq \frac{p}{q}x_0^{q/p} + \int_0^t f(s)ds,$$

which gives us the desired bound in (1.3.3). □

We now apply Theorem 1.3.2 to establish the following interesting and useful integral inequality.

Theorem 1.3.3 (The Snow Inequality [620]) *Let $x(t)$ and $f(t)$ be real-valued non-negative continuous functions defined on $I = [0, +\infty)$, and $n(t)$ be a positive, monotonic, non-decreasing continuous function defined on I and if $p \geq 2$. If it holds that for all $t \in I$,*

$$x^p(t) \leq n^p(t) + \int_0^t f(s)x(s)ds, \tag{1.3.5}$$

then for all $t \in I$

$$x(t) \leq n(t)\left[1 + \frac{p}{q}\int_0^t f(s)n^{-q}(s)ds\right]^{1/q}, \tag{1.3.6}$$

where $p - q = 1$.

Proof Since $n(t)$ is positive, monotonic, non-decreasing, we derive from (1.3.5) that

$$[x(t)/n(t)]^p \leq 1 + \int_0^t f(s)[x(s)/n^p(t)]ds,$$

i.e.,

$$[x(t)/n(t)]^p \leq 1 + \int_0^t f(s)n^{-q}(s)[x(s)/n(s)]ds.$$

Let

$$m(t) = x(t)/n(t),\ m(0) \leq 1. \tag{1.3.7}$$

Hence, *for all* $t \in I$,

$$m^p(t) \leq 1 + \int_0^t f(s)n^{-q}(s)m(s)ds.$$

From Theorem 1.3.2, we have, for all $t \in I$,

$$m(t) \leq [1 + \frac{p}{q}\int_0^t f(s)n^{-q}(s)ds]^{1/q}. \tag{1.3.8}$$

Thus the desired bound in (1.3.6) follows from (1.3.7) and (1.3.8). This thus completes the proof. □

Some interesting and useful integral inequalities are embodied in the following several theorems.

Theorem 1.3.4 (The Snow Inequality [620]) *Let* $x(t), f(t)$ *and* $g(t)$ *be real-valued non-negative continuous functions defined on* $I = [0, +\infty)$, *for which the inequality holds, for all* $t \in I$,

$$x(t) \leq x_0 + \int_0^t f(s)[x^p(s) + \int_0^s g(\tau)x(\tau)d\tau]ds, \tag{1.3.9}$$

where x_0 *is a non-negative constant and* $0 \leq p < 1$. *Then for all* $t \in I$,

$$x(t) \leq x_0 + \int_0^t f(s)k_1(s)\exp(\int_0^s g(\tau)x(\tau)d\tau)ds, \tag{1.3.10}$$

where

$$k_1(t) = \left[x_0^{pq} + pq \int_0^t f(s)\exp(-q\int_0^s g(\tau)d\tau)ds\right]^{1/q}, \tag{1.3.11}$$

where $p + q = 1$.

Proof Differentiating $x(t)$ with respect to t, we have for all $t \in I$,

$$x'(t) \le f(t)[x^p(t) + \int_0^t g(s)x(s)ds].$$

Define the function $y(t)$ by, for all $t \in I$,

$$y(t) = x^p(t) + \int_0^t g(s)x(s)ds, \quad y(0) = x_0^p,$$

Hence for all $t \in I$,

$$x'(t) \le f(t)y(t). \tag{1.3.12}$$

Differentiating $y(t)$ with respect to t, and using (1.3.9) and the fact that $x(t) \le y(t)$, we have

$$y'(t) = px^{p-1}(t)x'(t) + g(t)x(t) \le pf(t)y^p(t) + g(t)y(t),$$

which implies, for all $t \in I$,

$$y(t) \le k_1(t)\exp\left(\int_0^t g(s)ds\right),$$

where $k_1(t)$ is as given in (1.3.11). From (1.3.12), we have, for all $t \in I$,

$$x'(t) \le f(t)k_1(t)\exp(\int_0^t g(s)ds). \tag{1.3.13}$$

Now, integrating both sides of (1.3.13) from 0 to t, we obtain the desired bound in (1.3.10). This completes the proof. □

In the special case when $p = 0$, Theorem 1.3.4 takes the following form which is found to be convenient in some applications.

Corollary 1.3.1 (The Snow Inequality [620]) *Let $x(t)$, $f(t)$ and $g(t)$ be real-valued non-negative continuous functions defined on $I = [0, +\infty)$, for which the* *inequality holds, for all* $t \in I$,

$$x(t) \leq x_0 + \int_0^t f(s)(1 + \int_0^s g(\tau)x(\tau)d\tau)ds,$$

where x_0 *is a non-negative constant. Then for all* $t \in I$,

$$x(t) \leq x_0 + \int_0^t f(s) \exp\left(\int_0^s g(\tau)d\tau\right) ds.$$

Theorem 1.3.5 (The Snow Inequality [620]) *Let* $x(t)$, $f(t)$ *and* $g(t)$ *be real-valued non-negative continuous functions defined on* $I = [0, +\infty)$, *for which the inequality holds, for all* $t \in I$,

$$x(t) \leq x_0 + \int_0^t f(s)[x^p(s) + \int_0^s g(\tau)x^p(\tau)d\tau]ds, \tag{1.3.14}$$

where x_0 *is a non-negative constant and* $0 \leq p < 1$. *Then for all* $t \in I$,

$$x(t) \leq x_0 + \int_0^t f(s)k_2(s)ds, \tag{1.3.15}$$

where for all $t \in I$,

$$k_2(t) = \left(x_0^{pq} + q\int_0^t (pf(s) + g(s))ds\right)^{1/q}, \tag{1.3.16}$$

and $p + q = 1$.

Proof Differentiating $x(t)$ with respect to t, we have, for all $t \in I$,

$$x'(t) \leq f(t)[x^p(t) + \int_0^t g(s)x^p(s)ds].$$

Define a function $y(t)$ by

$$y(t) = x^p(t) + \int_0^t g(s)x^p(s)ds, \ y(0) = x^p(0) \leq x_0^p.$$

Hence, for all $t \in I$,

$$x'(t) \leq f(t)y(t). \tag{1.3.17}$$

Differentiating $y(t)$ with respect t and using (1.3.14) and the fact that

$$x^p(t) \leq y^p(t),$$

we have

$$y'(t) = px^{p-1}(t)x'(t) + g(t)x^p(t) \leq [pf(t) + g(t)]y^p(t)$$

which implies, for all $t \in I$,

$$y(t) \leq k_2(t),$$

where $k_2(t)$ is as given in (1.3.16). From (1.3.17), we have, for all $t \in I$,

$$x'(t) \leq f(t)k_2(t). \tag{1.3.18}$$

Now, integrating both sides of (1.3.18) from 0 to t, we obtain the desired bound in (1.3.15). This completes the proof. □

Theorem 1.3.6 (The Snow Inequality [620]) *Let $x(t)$, $f(t)$ and $g(t)$ be real-valued non-negative continuous functions defined on $I = [0, +\infty)$, and $n(t)$ be a positive, monotonic, non-decreasing continuous function defined on I, for which the inequality holds for all $t \in I$,*

$$x(t) \leq n(t) + \int_0^t f(s)[x(s) + \int_0^s g(\tau)x^p(\tau)d\tau]ds, \tag{1.3.19}$$

and $0 \leq p < 1$. Then for all $t \in I$,

$$x(t) \leq n(t)\left[1 + \int_0^t f(s)k_3(s)\exp(\int_0^s f(\tau)d\tau)ds\right], \tag{1.3.20}$$

where for all $t \in I$,

$$k_3(t) = \left[1 + q\int_0^t g(s)n^{-q}(s)\exp\left(-q\int_0^s f(\tau)d\tau\right)ds\right]^{1/q}, \tag{1.3.21}$$

and $p + q = 1$.

Proof Since $n(s)$ is a positive, monotonic, non-decreasing continuous function, we derive from (1.3.19) that

$$[x(t)/n(t)] \leq 1 + \int_0^t f(s)\left[x(s)/n(t) + \int_0^s g(\tau)[x^p(\tau)/n(t)]d\tau\right]ds$$
$$= 1 + \int_0^t f(s)\left[x(s)/n(s) + \int_0^s g(\tau)n^{-q}(\tau)[x(\tau)/n(\tau)]^p d\tau\right]ds.$$

Let

$$m(t) = x(t)/n(t),\ m(0) \leq 1, \tag{1.3.22}$$

then we obtain

$$m(t) \leq 1 + \int_0^t f(s) \left(m(s) + \int_0^s g(\tau) n^{-q}(\tau) m^p(\tau) d\tau \right) ds.$$

Define a function $R(t)$ by

$$R(t) = m(t) + \int_0^t g(s) n^{-q}(s) m^p(s) ds, \quad R(0) \leq 1,$$

thus for all $t \in I$,

$$m(t) \leq 1 + \int_0^t f(s) R(s) ds,$$

$$m'(t) \leq f(t) R(t). \tag{1.3.23}$$

Differentiating $R(t)$ with respect to t, and using (1.3.19) and the fact that $m^p(t) \leq R^p(t)$, we have for all $t \in I$,

$$\begin{aligned} R'(t) &= m'(t) + g(t) n^{-q}(t) m^p(t) \\ &\leq f(t) R(t) + g(t) n^{-q}(t) R^p(t), \end{aligned} \tag{1.3.24}$$

which implies for all $t \in I$,

$$R(t) \leq k_3(t) \exp \left(\int_0^s f(s) ds \right),$$

where $k_3(t)$ is as given in (1.3.21). From (1.3.23), we derive for all $t \in I$,

$$m'(t) \leq f(t) k_3(t) \exp \left(\int_0^t f(s) ds \right).$$

By integrating from 0 to t, we obtain for all $t \in I$,

$$m(t) \leq 1 + \int_0^t f(s) k_3(s) \exp \left(\int_0^s f(\tau) d\tau \right) ds. \tag{1.3.25}$$

The desired bound in (1.3.20) follows from (1.3.22) and (1.3.25), this completes the proof of the theorem. □

Remark 1.3.3 If $n(t) = n_0$ (which is a positive constant), then the integral inequality in [441] follows.

Theorem 1.3.7 (The El-Owaidy-Ragab-Abdeldaim Inequality [437]) *Let $x(t)$, $f(t)$ and $g(t)$ be real-valued non-negative continuous functions defined on*

$J = [\alpha, \beta]$, suppose $p \geq 0, p \neq 1$, and $n(t)$ be a positive, monotonic, non-decreasing continuous function defined on J. If the inequality holds, for all $t \in J$,

$$x(t) \leq n(t) + \int_{\alpha}^{t} f(s)x(s)ds + \int_{\alpha}^{t} g(s)x^{p}(s)ds, \tag{1.3.26}$$

then for all $\alpha \leq t < \beta_1$,

$$x(t) \leq n(t)k_4(t), \tag{1.3.27}$$

where,

$$k_4(t) = \exp\left(\int_{\alpha}^{t} f(s)ds\right) + \left[1 + q\int_{\alpha}^{t} g(s)n^{-q}(s)\exp(-q\int_{\alpha}^{s} f(\tau)d\tau)ds\right], \tag{1.3.28}$$

where $p + q = 1$, and β_1 is chosen so that $q > 0$ on $[\alpha, \beta_1) \subset J$, ($\beta_1 = \beta$ if $q > 0$).

Proof Since $n(t)$ is positive, monotonic, non-decreasing continuous function, we observe from (1.3.26) that

$$[x(t)/n(t)] \leq 1 + \int_{\alpha}^{t} f(s)[x(s)/n(t)]ds + \int_{\alpha}^{t} g(s)[x^{p}(s)/n(t)]ds,$$

i.e.,

$$[x(t)/n(t)] \leq 1 + \int_{\alpha}^{t} f(s)[x(s)/n(t)]ds + \int_{\alpha}^{t} g(s)n^{-q}(s)[x(s)/n(s)]^{p}ds.$$

Let

$$m(t) = x(t)/n(t), \; m(\alpha) \leq 1. \tag{1.3.29}$$

Thus

$$m(t) \leq 1 + \int_{\alpha}^{t} f(s)m(s)ds + \int_{\alpha}^{t} g(s)n^{-q}(s)m^{p}(s)ds.$$

Now we derive from Theorem 1.1.6 that for all $t \in J$,

$$m(t) \leq k_4(t), \tag{1.3.30}$$

where k_4 is as given in (1.3.28). The desired bound in (1.3.27) follow from (1.3.29) and (1.3.30). This thus completes the proof. □

Now, we now apply Theorem 1 and Theorem 4 in [452] to establish the following integral inequalities.

Theorem 1.3.8 (The El-Owaidy-Ragab-Abdeldaim Inequality [437]) *Let $x(t), g(t), h(t)$ and $h(t)$ and be real-valued positive continuous functions defined on $I = [0, +\infty)$, $W(t, u)$ be a positive, continuous, monotonic, non-decreasing, sub-additive and sub-multiplicative function in $u > 0$, for each fixed $t \in I$, the functions $m(t) > 0, E(t) \geq 0$ be non-decreasing in t, and continuous on I, $E(0) = 0$, and suppose further that the inequality holds for all $t \in I$,*

$$x(t) \leq m(t) + h(t)E\left(\int_0^t q(s)W(s, x(s))ds\right) + \int_0^t f(s)\Bigg[x(s)$$
$$+ \int_0^s g(\tau)\left[m(\tau) + h(\tau)E\left(\int_0^\tau q(\theta)W(\theta, x(\theta))d\theta\right)\right]^q x^p(\tau)d\tau\Bigg]ds, \tag{1.3.31}$$

where $0 \leq p < 1, p + q = 1$. Then, for all $t \in I$,

$$x(t) \leq k_5(t)\Bigg[m(t) + h(t)E\Bigg(G^{-1}\Bigg[G\left(\int_0^t q(s)W(s, k_5(s)m(s))ds\right)$$
$$+ \int_0^t q(s)W(s, h(s)k_5(s))ds\Bigg]\Bigg)\Bigg], \tag{1.3.32}$$

where

$$k_5(t) = \left[1+\int_0^t f(s)\left[1+q\int_0^s g(\tau)\exp\left(-q\int_0^t f(\theta)d\theta\right)\right]^{1/q} \exp\left(\int_0^s f(\tau)d\tau\right)ds\right], \tag{1.3.33}$$

$$G(u) = \int_{u_0}^u [ds/W(s, E(s))], \quad u \geq u_0 > 0 \tag{1.3.34}$$

and G^{-1} is the inverse of G and $t \in [0, b] \subset I$ so that

$$G\left(\int_0^t q(s)W(s, k_5(s))m(s)ds\right) + \int_0^t q(s)W(s, h(s)k_5(s))ds \in Dom\ (G^{-1}).$$

Proof Define the function $n(t)$ by

$$n(t) = m(t) + h(t)E\left(\int_0^t q(s)W(s, x(s))ds\right). \tag{1.3.35}$$

Then (1.3.30) can be restated as

$$x(t) \leq n(t) + \int_0^t f(s)\left[x(s) + \int_0^s g(\tau)n^q(\tau)x^p(\tau)d\tau\right]ds.$$

Since $n(t)$ is positive, continuous, monotonic, non-decreasing on I, we have from Theorem 1 in [452] that

$$x(t) \leq n(t)k_5(t), \tag{1.3.36}$$

where $k_5(t)$ is as given in (1.3.33). Now from (1.3.35) and (1.3.36), it follows

$$x(t) \leq k_5(t)\Big[m(t) + h(t)E\Big(\int_0^t q(s)W(s, x(s))ds\Big)\Big].$$

Let

$$x(t) \leq k_5(t)[m(t) + h(t)E(\nu(t))] \tag{1.3.37}$$

where, for all $t \in I$,

$$\nu(t) = \int_0^t q(s)W(s, x(s))ds.$$

Differentiating with respect to t, we have

$$\nu'(t) = q(t)W(t, x(t)) \leq q(t)W(t, k_5(t)[m(t) + h(t)E(\nu(t))]),$$

since W is a sub-additive and sub-multiplicative function for all $u > 0$,

$$\nu'(t) \leq q(t)W(t, k_5(t)m(t)) + q(t)W(t, h(t)k_5(t))W(t, E(\nu(t))),$$

i.e., using (1.3.34) and the fact that $n(t) \geq m(t)$,

$$[\nu'(t)/W(t, E(\nu(t)))] \leq [[q(t)W(t, k_5(t)m(t))]/W(t, E(\nu(t)))] + q(t)W(t, h(t)k_5(t)),$$

which further reduces to

$$G(\nu(t)) \leq G\left(\int_0^t q(s)W(s, k_5(s)m(s))ds\right) + \int_0^t q(s)W(s, h(s)k_5(s))ds.$$

Hence

$$\nu(t) \leq G^{-1}\Big[G\Big(\int_0^t q(s)W(s, k_5(s)m(s))\Big)ds + int_0^t q(s)W(s, h(s)k_5(s))ds\Big]. \tag{1.3.38}$$

The desired bound in (1.3.32) follows from (1.3.37) and (1.3.38), and thus this completes the proof. □

Theorem 1.3.9 (The El-Owaidy-Ragab-Abdeldaim Inequality [437]) *Let $x(t), g(t), f(t), h(t)$ and $q(t)$ be real-valued positive continuous functions defined on $I = [0, +\infty)$, $W(t, u), m(t), E(t)$ are as defined in Theorem 1.3.8, and suppose further that the following inequality holds for all $t \in I$,*

$$x(t) \le m(t) + h(t)E\left(\int_0^t q(s)W(s, x(s))ds\right) + \int_0^t f(s)x(s)ds$$
$$+ \int_0^t g(s)\left[m(s) + h(s)E\left(\int_0^s q(\tau)W(\tau, x(\tau))d\tau\right)\right]^q x^p(s)ds, \quad (1.3.39)$$

and $0 \le p < 1, p + q = 1$. Then for all $t \in I$,

$$x(t) \le k_6(t)\left[m(t) + h(t)E\left(G^{-1}\left[G\left(\int_0^t q(s)W(s, k_6(s)m(s))ds\right)\right.\right.\right.$$
$$\left.\left.\left. + \int_0^t q(s)W(s, h(s)k_6(s))ds\right)\right]\right], \quad (1.3.40)$$

where for all $t \in I$,

$$k_6(t) = \exp\left(\int_0^t f(s)ds\right)\left[1 + q\int_0^t g(s)\exp\left(-q\int_0^s f(\tau)d\tau\right)ds\right]^{1/q}, \quad (1.3.41)$$

and $G(u)$ is as given in (1.3.34) such that

$$G\left(\int_0^t q(s)W(s, k_6(s)m(s))ds + \int_0^t q(s)W(s, h(s)k_6(s))ds\right) \in Dom\ (G^{-1}).$$

Proof Define the function $n(t)$ by

$$n(t) = m(t) + h(t)E\left(\int_0^t q(s)W(s, x(s))ds\right). \quad (1.3.42)$$

Then (1.3.39) can be restated as

$$x(t) \le n(t) + \int_0^t [f(s)x(s) + g(s)n^q(s)x^p(s)]ds.$$

Since $n(t)$ is positive, continuous, monotonic, non-decreasing on I, we conclude from Theorem 4 in [452] that

$$x(t) \le n(t)k_6(t), \quad (1.3.43)$$

where $k_6(t)$ is as in (1.3.41). Now from (1.3.42) and (1.3.43), we have

$$x(t) \leq k_6(t)[m(t) + h(t)E(\int_0^t q(s)W(s, x(s))ds)]. \tag{1.3.44}$$

Let

$$x(t) = k_6(t)[m(t) + h(t)E(v(t))], \tag{1.3.45}$$

where for all $t \in I$,

$$v(t) = \int_0^t q(s)W(s, x(s))ds.$$

Differentiating the above equality with respect to t, we derive

$$\begin{aligned} v'(t) &= q(t)W(t, x(t)) \\ &\leq q(t)W(t, k_6(t)[m(t) + h(t)E(v(t))]), \end{aligned}$$

since W is sub-additive and sub-multiplicative function for $u > 0$, using (1.3.34) and the fact that $n(t) \geq m(t)$, we have

$$[v'(t)/W(t, E(v(t)))] \leq [q(t)W(t, k_6(t)m(t))]/W(t, E(v(t))) + q(t)W(t, h(t)k_6(t)),$$

which further reduces to

$$G(v(t)) \leq G\left(\int_0^t q(s)W(s, k_6(s)m(s))ds + \int_0^t q(s)W(s, h(s)k_6(s))ds\right).$$

Hence

$$v(t) \leq G^{-1}\left[G\left(\int_0^t q(s)W(s, k_6(s)m(s))ds + \int_0^t q(s)W(s, k_6(s)h(s)\right) ds\right]. \tag{1.3.46}$$

The desired bound in (1.3.40) follows from (1.3.44) and (1.3.46), this completes the proof of the theorem. □

Theorem 1.3.10 (The El-Owaidy-Ragab-Abdeldaim Inequality [437]) *Let $x(t), f(t), g(t), h(t)$ and $q(t)$ be real-valued positive continuous functions defined on $I = [0, +\infty)$, $W(t, u), m(t)$ and $E(t)$ are as defined in Theorem 1.3.8, and suppose further that the following inequality holds for all $t \in I$,*

$$x(t) \leq m(t) + h(t)E\left(\int_0^t q(s)W(s, x(s))ds\right) + \int_0^t f(s)\left[x(s) + \int_0^s g(\tau)x(\tau)d\tau\right]ds, \tag{1.3.47}$$

then for all $t \in I$,

$$x(t) \le k_7(t)\Big[m(t) + h(t)E\Big(G^{-1}\Big[G\Big(\int_0^t q(s)W(s, k_7(s)m(s))ds\Big) \tag{1.3.48}$$

$$+ \int_0^t q(s)W(s, h(s)k_7(s))ds\Big]\Big)\Big], \tag{1.3.49}$$

where

$$k_7(t) = 1 + \int_0^t f(s)\exp\Big(\int_0^s (f(\tau) + g(\tau))d\tau\Big)ds, \tag{1.3.50}$$

and G is as defined in Theorem 1.3.9, where

$$G\Big(\int_0^t q(s)W(s, k_7(s)m(s))ds\Big) + \int_0^t q(s)W(s, h(s)k_7(s))ds \in Dom\ (G^{-1}).$$

Proof Define the function $n(t)$ by

$$n(t) = m(t) + h(t)E\Big(\int_0^t q(s)W(s, x(s))ds\Big).$$

Then (1.3.47) can be restated as

$$x(t) \le n(t) + \int_0^t f(s)[x(s) + \int_0^s g(\tau)x(\tau)d\tau]ds.$$

Since $n(t)$ is positive, continuous, monotonic, non-decreasing on I, we have from Corollary 1.2.4 in Qin [557] that for all $t \in I$,

$$x(t) \le n(t)k_7(t),$$

where $k_7(t)$ is as given in (1.3.50). Now, we can complete the proof by following the argument as in the proof of Theorem 1.3.9. □

Remark 1.3.4 If $W(t, x(t)) = W(x(t))$ in the Theorem 1.3.10 and $h(t) = 1$, Theorem 1.3.10 is the same as Theorem 3 in [456]. For $W(t, x(t)) = \omega(x(t)), m(t) = x_0$, x_0 is a positive constant), $h(t) = 1$ and $E(u(t)) = u(t)$, Theorem 1.3.10 is the same as Theorem 2 of [456]. In [21], the authors studied the general form of the integral inequalities with linear and nonlinear terms on the right-hand side. However, the integral inequalities considered in Theorems 1.3.7–1.3.10 are different from those considered in [21]. We note that the integral inequalities obtained in Theorems 1.3.7–1.3.10 allow us to study the stability, boundedness and asymptotic behavior of the solutions of a class of more general differential and integral equations similar to those obtained in [21, 446, 674].

1.4 The One-Dimensional Gollwitzer Inequality and Its Generalization

The following results are due to Gollwitzer [250].

Theorem 1.4.1 (The Gollwitzer Inequality [250]) *Let* $u(t), f(t), g(t)$ *and* $h(t)$ *be non-negative, continuous functions on the interval* $[a, b]$; $G(u)$ *be a continuous, strictly increasing, convex and sub-multiplicative function for all* $u \geq 0$; $G(0) = 0$, $\lim_{u\to 0} G(u) = +\infty$; $\alpha(t)$, $\beta(t)$ *be continuous on* $[a, b]$; $\alpha(t), \beta(t) > 0$, $\alpha(t) + \beta(t) = 1$; *and for all* $a \leq t \leq b$,

$$u(t) \leq f(t) + g(t)G^{-1}\left(\int_a^t G(u(s))h(s)ds\right). \tag{1.4.1}$$

Then for all $t \in [a, b]$,

$$\int_a^t G(u(s))h(s)ds \leq \int_a^t \alpha(s)G(f(s)\alpha(s)^{-1})h(s)\exp\left(\int_s^t \beta(s)G(g(s)\beta(s)^{-1})h(x)dx\right)ds. \tag{1.4.2}$$

Furthermore, if, for all $a \leq x \leq t \leq b$,

$$u(t) \geq u(x) - g(x)G^{-1}\left(\int_a^t G(u(s))h(s)ds\right), \tag{1.4.3}$$

then, for all $a \leq x \leq t \leq b$,

$$u(t) \geq \alpha(t)G^{-1}\left(\alpha(t)^{-1}G(u(x))\exp\left(-\beta(t)G(g(t)\beta(t)^{-1})\int_a^t h(s)ds\right)\right). \tag{1.4.4}$$

Proof Let $\alpha(t)$, $\beta(t) > 0$, $\alpha(t) + \beta(t) = 1$. Then

$$u(t) \leq \alpha(t)(f(t)\alpha(t)^{-1}) + \beta(t)(g(t)\beta(t)^{-1})G^{-1}\left(\int_a^t G(u(s))h(s)ds\right).$$

Since G is convex, sub-multiplicative and monotonic, for all $a \leq t \leq b$,

$$G(u(t)) \leq \alpha(t)G(f(t)\alpha(t)^{-1}) + \beta(t)G(g(t)\beta(t)^{-1})\int_a^t G(u(s)h(s)ds.$$

Thus (1.4.2) follows from Theorem 1.2.7 in Qin [557]. From (1.4.3), we derive for all $a \leq x \leq t \leq b$,

$$u(x) \leq \alpha(t)(u(t)\alpha(t)^{-1}) + \beta(t)g(t)\beta(t)^{-1}G^{-1}(\int_x^t G(u(s))h(s)ds).$$

Since G is convex, sub-multiplicative and monotonic, we conclude for all $a \le t \le b$,

$$\alpha(t)G(f(t)\alpha(t)^{-1}) \ge G(u(x)) - \beta(t)G(g(t)\beta(t)^{-1}) \int_x^t G(u(s)h(s)ds,$$

and (1.4.4) follows immediately from Theorem 1.2.10 in Qin [557]. □

Corollary 1.4.1 (The Gollwitzer Inequality [250]) *If* $G(u) = u^p$, $1 \le p < +\infty$, *then for all* $a \le t \le b$,

$$\int_a^t u^p(s)h(s)ds \le \int_a^t \alpha(s)^{1-p} f^p(s)h(s) \exp\left(\int_s^t \beta(x)^{1-p} g^p(x)h(x)dx\right) ds, \tag{1.4.5}$$

and for $a \le x \le t \le b$

$$u(t) \ge (\alpha(t))^{1-1/p} u(x) \exp\left(\frac{-(\beta(t))^{1-p}}{p} g^p(t) \int_x^t h(s)ds\right). \tag{1.4.6}$$

In order to compare the estimate in Corollary 1.4.1 with that given by Willett [671], we take $g(t) \equiv 1, f(t) \equiv C$, $\alpha = \beta = \frac{1}{2}$ and obtain, from Corollary 1.4.1, for all $t \ge 0$,

$$u(t) \le 1 + \left(\exp(2^{p-1} \int_a^t h(s)ds) - 1\right)^{1/p}.$$

Taking $p = 2$ and $h(t) \equiv 1$, we have

$$u(t) \le C\left[\exp(-t) + (1 - \exp(-2t))^{1/2}\right] \exp(t), \tag{1.4.7}$$

and from Willett's estimate [671], we may obtain

$$u(t) \le C[1 + (1 - \exp(-t))^{1/2}] \exp(t). \tag{1.4.8}$$

The estimate given in (1.4.7) is definitely sharper for large t, while for small values of t, the estimate given in (1.4.8) is sharper. Thus the two estimates are, in general, not comparable.

The left-hand side of (1.4.5) is independent of $\alpha(t)$. It is unknown whether or not there is an optimum function $\alpha(t)$ which minimizes the right-hand side of (1.4.5) for any reasonable class of functional parameters f, g, h and G. We also note that if $0 < g(t) < 1$, then the sub-multiplicative hypothesis on G can be omitted in Theorem 1.4.1.

If $G(u) = u$ in Theorem 1.4.1, we have the well-known Gronwall inequality and a case similar to the Langenhop inequality [328]. If $G(u) = u^p$ in Theorem 1.4.1, $p \ge 1$, then Willett [671] has studied (1.4.1) in connection with a singular perturbation problem. The purpose here is to introduce new estimates for $u(t)$ if G is a convex or concave function.

If G is concave, the previous techniques are clearly not applicable. The following theorem gives us partial results in this case.

Theorem 1.4.2 (The Gollwitzer Inequality [250]) *Let $u(t), h(t)$ be non-negative, continuous functions on the interval $[a, b]$; $G(u)$ be a continuous, concave function for all $u \geq 0$, and continuously differentiable for all $u > 0$; $G'(u) > 0$ for all $u > 0$, $G(0) = 0$, $\lim_{u\to 0} G(u) = +\infty$; $C \geq 0$ a constant, and for all $a \leq t \leq b$,*

$$u(t) \leq C + G^{-1}\left(\int_a^t G(u(s))h(s)ds\right). \tag{1.4.9}$$

Then $a \leq t \leq b$,

$$u(t) \leq G^{-1}\left(G(C)\exp\left(\int_a^t h(s)ds\right)\right). \tag{1.4.10}$$

Furthermore, if, for all $a \leq x \leq t \leq b$,

$$u(t) \geq u(x) - G^{-1}\left(\int_x^t G(u(s))h(s)ds\right), \tag{1.4.11}$$

then $a \leq x \leq t \leq b$,

$$u(t) \geq G^{-1}\left(G(u(x))\exp(-\int_x^t h(s)ds)\right). \tag{1.4.12}$$

Proof It is sufficient to assume that C is positive, since a standard limiting argument can be used to treat the remaining case. Consider (1.4.9) and define $\psi(t)$ as, for all $a \leq t \leq b$,

$$\psi(t) = C + G^{-1}(\alpha + \int_a^t G(u(s))h(s)ds), \quad \alpha > 0. \tag{1.4.13}$$

We note that $\psi(t)$ majorizes the right-hand side of (1.4.9), and hence $\psi(t) \geq u(t)$. Since $G(u)$ is concave, the derivative $G'(u)$ is non-increasing for all $u > 0$. Since $\psi(t) - C > 0$ (note that we use the constant α), for all $a \leq t \leq b$,

$$G'(\psi(t)) \leq G'(\psi(t) - C). \tag{1.4.14}$$

Furthermore, by the fundamental theorem of integral calculus,

$$\ln G(\psi(t)) - \ln G(C + G^{-1}(\alpha)) = \int_a^t G'(\psi(t))G(\psi(t))^{-1}\psi'(s)ds, \tag{1.4.15}$$

where, for all $a \leq t \leq b$,

$$\psi'(s) = G(u(s))h(s)\left(G'(\psi(s) - C)\right)^{-1}.$$

Since $\psi(t) \geq u(t)$ and (1.4.14) holds, the integrand in (1.4.15) is majorized by $h(s)$. Hence, for all $a \leq t \leq b$,

$$\ln\left(G(\psi(t))/G(C + G^{-1}(\alpha))\right) \leq \int_a^t h(s)ds, \tag{1.4.16}$$

or for all $a \leq t \leq b$,

$$G(u(t)) \leq G(C)\exp\left(\int_a^t h(s)ds\right),$$

since the continuity of G permits us to let α approach zero. Thus the first assertion (1.4.10) has been proved.

Consider (1.4.11) and define $\phi(x)$ as, for all $a \leq x \leq t$,

$$\phi(s) = u(t) + +G^{-1}(\int_x^t G(u(\tau))h(\tau)d\tau). \tag{1.4.17}$$

We note that $\phi(x) \geq u(x)$. Assume that $u(t)$ is positive on $[a, b]$. Since G is concave and continuously differentiable, we can use the techniques given in the first part of the proof to show that for all $a \leq x \leq t \leq b$,

$$\ln G(\phi(t)) - \ln G(\phi(x)) \geq -\int_x^t h(s)ds, \tag{1.4.18}$$

whence, for all $a \leq x \leq t \leq b$,

$$G(u(t)) \geq G(u(x))\exp(-\int_x^t h(s)ds). \tag{1.4.19}$$

The estimate given in (1.4.11) is now clear. If $u(t)$ is not positive on $[a, b]$, we can replace $u(t)$ by $\{u(t) + \epsilon\}$ in (1.4.10), $\epsilon > 0$ in (1.4.19) to complete the theorem. □

Corollary 1.4.2 (The Gollwitzer Inequality [250]) *If $G(u) = u^p$, $0 \leq p \leq 1$, then for all $a \leq t \leq b$,*

$$u(t) \leq C\exp\left(p^{-1}\int_a^t h(s)ds\right), \tag{1.4.20}$$

and for all $a \leq x \leq t \leq b$,

$$u(t) \geq u(x)\exp\left(-p^{-1}\int_x^t h(s)ds\right). \tag{1.4.21}$$

In Theorem 1.4.3 below, we shall introduce a general version of the inequality obtained by Gollwitzer in 1969.

Theorem 1.4.3 (The Gollwitzer Inequality [250]) *Let $f(t), u(t), g(t), h(t)$ and $k(t)$ are real-valued non-negative continuous functions defined on I; $G(u)$ be a continuous, strictly increasing, convex and sub-multiplicative function for all $u \geq 0; G(0) = 0, \lim_{u\to+\infty} G(u) = +\infty; \alpha(t), \beta(t)$ be continuous functions on $I; \alpha(t), \beta(t) > 0, \alpha(t) + \beta(t) = 1$; and there holds for all $t \in I$,*

$$u(t) \leq f(t) + g(t)G^{-1}\left(\int_0^t h(s)G(u(s))ds + \int_0^t h(s)\beta(s)G(g(s)\beta^{-1}(s))\right.$$
$$\left.\times\left(\int_0^s k(\tau)G(u(\tau))d\tau\right)ds\right). \tag{1.4.22}$$

Then for all $t \in I$,

$$u(t) \leq f(t) + G^{-1}\left(g(t)\left(\int_0^t h(s)\left\{\alpha(s)G(f(s)\alpha^{-1}(s)) + \beta(s)G(g(s)\beta^{-1}(s)\right)\right.\right.$$
$$\times \exp\left(\int_0^s \beta(\tau)G(g(\tau)\beta^{-1}(\tau))(h(\tau) + k(\tau)\right)\int_0^s \alpha(\tau)G(f(\tau)\alpha^{-1}(\tau))$$
$$\left.\left.+(h(\tau)k(\tau))\exp\left(-\int_0^\tau \beta(\eta)G(g(\eta)\beta^{-1}(\eta))(h(\eta) + k(\eta))d\eta\right)d\tau\right\} ds\right)\right). \tag{1.4.23}$$

Proof We may rewrite (1.4.22) as

$$u(t) \leq \alpha(t)f((t)\alpha^{-1}(t)) + \beta(t)(g(t)\beta^{-1}(t))G^{-1}\Big(\int_0^t h(s)G(u(s))ds$$
$$+ \int_0^t h(s)\beta(s)G(g(s)\beta^{-1}(s))\Big(\int_0^s k(\tau)G(u(\tau))d\tau\Big)ds\Big).$$

Since G is convex, sub-multiplicative and monotonic, we have

$$G(u(t)) \leq \alpha(t)Gf((t)\alpha^{-1}(t)) + \beta(t)G(g(t)\beta^{-1}(t))G^{-1}\Big(\int_0^t h(s)G(u(s))ds$$
$$+ \int_0^t h(s)\beta(s)G(g(s)\beta^{-1}(s))\Big(\int_0^s k(\tau)G(u(\tau))d\tau\Big)ds\Big).$$

Now applying Theorem 1.4.2 yields the desired bound in (1.4.23). The proof is thus complete. □

Setting $G(u) = u^p, 1 \leq p < +\infty$, in Theorem 1.4.3, we arrive at the following corollary.

Corollary 1.4.3 (The Gollwitzer Inequality [250]) *Let $f(t), u(t), g(t), h(t)$ and $k(t)$ are real valued non-negative continuous functions defined on I, let $\alpha(t), \beta(t)$ be positive continuous functions on I such that $\alpha(t), \beta(t) > 0, \alpha(t) + \beta(t) = 1$, let $1 \leq p < +\infty$ and suppose that the following inequality holds for all $t \in I$,*

$$u(t) \leq f(t) + g(t)\Big(\int_0^t h(s)u^p(s)ds + \int_0^t h(s)\beta(s)(g(s)\beta^{-1}(s))^p \times\Big(\int_0^s k(\tau)u^p(\tau)d\tau\Big)ds\Big),$$

then for all $t \in I$,

$$u(t) \leq f(t) + g(t)\Big(\int_0^t h(s)\Big\{\alpha(s)G(f(s)\alpha^{-1}(s))^p + \beta(s)(g(s)\beta^{-1}(s))^p \times \exp\Big(\int_0^s \beta(\tau)(g(\tau)\beta^{-1}(\tau))^p(h(\tau) + k(\tau)\Big) \int_0^s \alpha(\tau)(f(\tau)\alpha^{-1}(\tau))^p + (h(\tau)k(\tau)) \exp\Big(-\int_0^\tau \beta(\eta)(g(\eta)\beta^{-1}(\eta))^p(h(\eta) + k(\eta))d\eta\Big)d\tau\}ds\Big)^{1/p}.$$

The next result is due to Dhongade and Deo [198].

Theorem 1.4.4 (The Dhongade-Deo Inequality [198]) *Suppose that*

(i) the functions $f(x), g(x), h(x)$ and $\theta(x)$ are defined as

(a) $\theta(x), h(x) : (0, +\infty) \to (0, \infty)$,
(b) $f(x) : (0, +\infty) \to (0, +\infty)$ and monotonic non-decreasing in x,
(c) $g(x) : (0, +\infty) \to [1, +\infty)$,

and θ, h, f, and g are continuous functions on $(0, +\infty)$.

(ii) $\Omega, \psi \in \mathcal{F}$ and Ω is sub-multiplicative.
If for all $x \in I$,

$$\theta(x) \leq f(x) + g(x)\psi\Big(\int_0^x h(s)\Omega(\theta(s))ds\Big), \tag{1.4.24}$$

then for all $x \in I'$,

$$\theta(x) \leq f(x)g(x)\Big[1 + \psi\Big\{F^{-1}\Big(\int_0^x h(s)\Omega(g(s))ds\Big)\Big\}\Big], \tag{1.4.25}$$

where F^{-1} is the inverse of F defined by

$$F(u) = \int_{u_0}^{u} \frac{dt}{\Omega(1 + \psi(t))}, \quad 0 < u_0 \leq u \tag{1.4.26}$$

and $x \in I' \subset I$ so that

$$\int_0^x h(s)\Omega(g(s))ds \in Dom\,(F^{-1}).$$

Proof Since $f(x)$ is monotonic, non-decreasing, $g(x) \geq 1$ and $\Omega,\ \psi \in \mathcal{F}$, we can derive from (1.4.34) that for all $x \in I$,

$$\begin{aligned}\frac{\theta(x)}{f(x)} &\leq 1 + g(x)\psi(x)\Big[\int_0^x h(s)\Omega\Big(\frac{\theta(s)}{f(s)}\Big)ds\Big]\\ &\leq g(x)\Big[1 + \psi\Big(\int_0^x h(s)\Omega\Big(\frac{\theta(s)}{f(s)}\Big)ds\Big)\Big].\end{aligned} \tag{1.4.27}$$

Define, for all $x \in I$,

$$R(x) = \int_0^x h(s)\Omega\Big(\frac{\theta(s)}{f(s)}\Big)ds.$$

Then it follows from (1.4.27) that

$$\Omega\Big(\frac{\theta(x)}{f(x)}\Big) \leq \Omega(g(x))\Omega(1 + \psi(R(x))),$$

since Ω is non-decreasing and sub-multiplicative. Now, multiplying both sides by $h(x)$ and using the definition of $R(x)$, we may obtain

$$\frac{R'(x)}{\Omega[1 + \psi(R(x))]} \leq h(x)\Omega(g(x))$$

which, combined with (1.4.26), reduces to

$$\frac{dF(R)}{dx} \leq h(x)\Omega(g(x)).$$

Now integrating the above inequality from 0 to x, we can get

$$R(x) \leq F^{-1}\Big(\int_0^x h(s)\Omega(g(s))ds\Big).$$

Thus (1.4.25) now follows from (1.4.27) and the above estimate of $R(x)$. □

The following theorem, due to Dhongade and Deo [198], provides pointwise estimate of the solution of the integral equation (1.4.30) under suitable conditions on the kernel $k(x, s)$.

Theorem 1.4.5 (The Dhongade-Deo Inequality [198]) *Suppose that*

(i) the functions $\Omega, \psi \in \mathcal{F}$ and are sub-multiplicative,
(ii) $k(x, s)$ $(x \geq s)$ is defined and continuous on $I \times I$, such that

$$k(x, x) = 0, \tag{1.4.28}$$

$$\frac{\partial k(x, s)}{\partial x} \leq g(x)h(s), \tag{1.4.29}$$

where $g(x), h(s)$ are continuous functions on $(0, +\infty)$. If $y(x)$ is a solution of

$$y(x) = f(x) + \psi\Big(\int_0^x k(x, s)\Omega(y(s))ds\Big) \tag{1.4.30}$$

existing on I, then for all $x \in I'$,

$$|y(x)| \leq f(x)\bar{m}(x)\Big[1 + \psi\Big\{F^{-1}\Big(\int_0^x h(s)\Omega(\bar{m}(s))ds\Big)\Big\}\Big], \tag{1.4.31}$$

where F^{-1} is the inverse of F defined in Theorem 1.4.4 and $x \in I' \subset I$ so that

$$\int_0^x h(s)\Omega(\bar{m}(s))ds \in Dom\ (F^{-1})$$

and

$$\bar{m}(x) = 1 + \psi\Big(\int_0^x g(s)ds\Big). \tag{1.4.32}$$

Proof Since $y(x)$ is a solution of (1.4.30), we have, for all $x \in I$,

$$|y(x)| \leq f(x) + \psi\Big(\int_0^x k(x, s)\Omega(|y(s)|)ds\Big). \tag{1.4.33}$$

Define, for all $x \in I$,

$$R(x) = \int_0^x k(x, s)\Omega(|y(s)|)ds,$$

$$R'(x) = k(x, x)\Omega(|y(x)|) + \int_0^x \frac{\partial k(x, s)}{\partial x}\Omega(|y(s)|)ds.$$

Integrating the above equation from 0 to x, we may obtain

$$R(x) \leq \int_0^x g(s)\Big(\int_0^x h(t)\Omega(|y(t)|)dt\Big)ds.$$

Now replacing in the limits s by x, the inequality still holds and becomes the product of two integrals. In view of the definition of $R(x)$, and the fact that $\psi \in \mathcal{F}$, we can write (1.4.33) as follows:

$$|y(x)| \leq f(x) + \bar{m}(x)\psi\Big(\int_0^x h(s)\Omega(|y(s)|)ds\Big).$$

This is just of the form (1.4.24), which, by the application of Theorem 1.4.4, gives us the desired result (1.4.31). □

In the following theorem, due to Dhongade and Deo [198], we introduce a more general inequality which contains n-linear terms and one nonlinear term.

Theorem 1.4.6 (The Dhongade-Deo Inequality [198]) *Suppose that*

(i) the functions $f(x), g_i(x), h_i(x), (i = 1, 2, \ldots, n+1)$ be defined as in Theorem 1.2.44 in Qin [557],
(ii) the functions Ω, ψ be defined as Theorem 1.2.44 in Qin [557]. If, for all $x \in I$,

$$\theta(x) \leq f(x) + \sum_{i=1}^{n} g_i(x)\int_0^x h_i(s)\theta(s)ds + g_{n+1}(x)\psi\Big(\int_0^x h_{n+1}(s)\Omega(\theta(s))ds\Big), \tag{1.4.34}$$

then for all $x \in I'$,

$$\theta(x) \leq E^n(f)E^n(g_{n+1}(x))\Big[1 + \psi\Big\{F^{-1}\Big(\int_0^x h_{n+1}(s)\Omega(E^n g_{n+1}(s))\Big)ds\Big\}\Big], \tag{1.4.35}$$

where E^n is defined inductively as Theorem 1.2.44 in Qin [557] and F^{-1} has the same meaning as in Theorem 1.4.4 and $x \in I' \subset I$ such that

$$\int_0^x h_{n+1}(s)\Omega(E^n g_{n+1}(s))ds \in Dom\,(F^{-1}).$$

Proof Define

$$T(x) = f(x) + g_{n+1}(x)\psi\Big(\int_0^x h_{n+1}(s)\Omega(\theta(s))ds\Big).$$

Then (1.4.34) can be written as, for all $x \in I$,

$$\theta(x) \leq T(x) + \sum_{i=1}^{n} g_i(x) \int_0^x h_i(s)\theta(s)ds.$$

In view of Theorem 1.2.44 in Qin [557], this inequality gives us the estimate

$$\begin{aligned}\theta(x) &\leq E^n\Big(T(x)\Big)\\ &= E^n\Big[f(x) + g_{n+1}(x)\psi\Big(\int_0^x h_{n+1}(s)\Omega(\theta(s))ds\Big)\Big]\\ &= E^n(f) + E^n\Big[g_{n+1}(x)\psi\Big(\int_0^x h_{n+1}(s)\Omega(\theta(s))ds\Big)\Big]\\ &= E^n(f) + g_{n+1}(x)\psi\Big(\int_0^x h_{n+1}(s)\Omega(\theta(s))ds\Big)\\ &\quad\times\Big[E^n g_{n-1}(x)\exp\Big(\int_0^x h_n^{(s)}E^{n-1}(g_n(s))ds\Big)\Big]\end{aligned}$$

whence

$$\theta(x) \leq E^n(f) + E^n(g_{n+1}(x))\psi\Big(\int_0^x h_{n+1}(s)\Omega(\theta(s))ds\Big).$$

This inequality is of the form (1.4.33). Thus the bound (1.4.35) now follows from Theorem 1.4.4. □

The pointwise estimate of the solution of the integral equation, for all $0 \leq x < +\infty$,

$$y(x) = f(x) + \int_0^x k(x,s)y(s)ds + \psi\Big(\int_0^x k^*(x,s)\Omega(y(s))ds\Big), \quad (1.4.36)$$

which contains two different kernels is obtained in the following theorem.

Theorem 1.4.7 (The Dhongade-Deo Inequality [198]) *Suppose that*

(i) *functions, $\Omega, \psi \in \mathcal{F}$ and are sub-multiplicative,*
(ii) *$k(x,s), (x \geq s), k(x,x), \partial k(x,s)/\partial x$ are defined as in Theorem 1.2.45 of Qin [557],*
(iii) *$k^*(x,s), (x \geq s)$ is continuous on $I \times I$ and further*

$$k^*(x,x) = 0, \quad (1.4.37)$$

$$\frac{\partial k^*(x,s)}{\partial x} \leq g_{n+1}(x)h_{n+1}(s), \quad (1.4.38)$$

where $g_{n+1}(x)$ and $h_{n+1}(x)$ are defined and continuous on $(0,+\infty)$ and $[1,+\infty)$, respectively, and if $y(x)$ is a solution of (1.4.36) existing on I, then for all $x \in I'$,

$$|y(x)| \leq z(x)W(x)\Big[1 + \psi\Big\{F^{-1}\Big(\int_0^x h_{n+1}(s)\Omega(W(s))ds\Big)\Big\}\Big], \quad (1.4.39)$$

where F^{-1} is the inverse of F defined as in Theorem 1.4.4 and $x \in I' \subset I$ such that

$$\int_0^x h_{n+1}(s)\Omega(W(s))ds \in Dom\ (F^{-1})$$

and

$$\begin{cases} W(x) = 1 + \bar{z}(x)\psi\Big(\int_0^x g_{n+1}(s)ds\Big), & (1.4.40)\\ z(x) = E^n\Big\{f(x) + \int_0^x f(x)m(s)\exp\Big(\int_s^x m(t)dt\Big)ds\Big\}, & (1.4.41)\\ \bar{z}(x) = \delta(E^{n-1}Q_n)\exp\Big(h_n E^{n-1}(Q_n(s))ds\Big), & (1.4.42) \end{cases}$$

where

$$\delta = 1 + \int_0^x m(s)\exp\Big(\int_s^x m(t)dt\Big)ds, \quad Q_i(x) = \int_0^x g_i(x)\exp\Big(\int_s^x m(t)dt\Big)ds.$$

Proof Since $y(x)$ is a solution of (1.4.36), we have, for all $x \in I$,

$$|y(x)| \leq f(x) + \int_0^x k(x,s)|y(s)|ds + \psi\Big(\int_0^x k^*(x,s)\Omega(|y(s)|)ds\Big).$$

Let, for all $x \in I$,

$$T(x) = f(x) + \psi\Big(\int_0^x k^*(x,s)\Omega(|y(s)|)ds\Big).$$

Then the above inequality takes the form

$$|y(x)| \leq T(x) + \int_0^x k(x,s)|y(s)|ds.$$

This inequality is of the form (1.2.333) of Theorem 1.2.45 in Qin [557]. Applying Theorem 1.2.45 in Qin [557], we have

$$|y(x)| \leq E^n(p(x)),$$

where

$$p(x) = T(x) + \int_0^x m(s)T(s)\exp\Big(\int_s^x m(t)dt\Big)ds.$$

Substituting for $p(x)$, we can get

$$\begin{aligned}|y(x)| &\le E^n\Big[T(x) + \int_0^x m(s)T(s)\exp\Big(\int_s^x m(t)dt\Big)ds\Big] \\ &= E^n\Big[f(x) + \psi\Big(\int_0^x k^*(x,s)\Omega(|y(s)|)ds\Big)\Big] \\ &\quad + \int_0^x m(s)\Big\{f(s) + \psi\Big(\int_0^x k^*(x,s)\Omega(|y(s)|)\exp(\int_t^x m(\tau)d\tau)ds\Big)\Big\}.\end{aligned}$$

Rearranging the terms and applying (1.4.41) and (1.4.42), we may obtain

$$|y(x)| \le z(x) + \bar{z}(x)\psi\Big(\int_0^x k^*(x,s)\Omega(|y(s)|)ds\Big). \tag{1.4.43}$$

Define

$$R(x) = \int_0^x k^*(x,s)\Omega(|y(s)|)ds$$

so that

$$R'(x) \le k^*(x,x)\Omega(|y(x)|) + \int_0^x \frac{\partial k^*(x,s)\Omega(|y(s)|)}{\partial x}ds.$$

Using (1.4.37) and (1.4.38), we have

$$R'(x) \le g_{n+1}(x)\int_0^x h_{n+1}(s)\Omega(|y(s)|)ds.$$

Integrating from 0 to x, we may get

$$R(x) \le \int_0^x g_{n+1}(s)\Big(\int_0^s h_{n+1}(t)\Omega(|y(t)|)dt\Big)ds,$$

which, in view of (1.4.43), yields

$$|y(x)| \le z(x) + z(x)\psi\Big[\int_0^x g_{n+1}(s)\Big(\int_0^s h_{n+1}(t)\Omega(|y(t)|)dt\Big)ds\Big].$$

This, by using (1.4.40) and the fact that ψ is sub-multiplicative, further reduces to

$$|y(x)| \leq z(x) + W(x)\Big(\int_0^x h_{n+1}(s)\Omega(|y(s)|)ds\Big)$$

which is of the form (1.4.24). Now applying Theorem 1.4.4 gives us the desired result (1.4.39). □

Let us consider the following example of the type (1.4.36), for all $x \in I$,

$$y(x) = e^x + \int_0^x e^{x-s}y(s)ds + \int_0^x \sin(x-s)^2 y^{1/3}ds.$$

We note that

$$k(x,x) = 1 = m(x), \quad k^*(x,x) = 0.$$

Since

$$\frac{\partial k}{\partial x} = e^{x-s},$$

$$\frac{\partial k^*}{\partial x} = \cos(x-s)^2 2(x-s) \leq 2xs.$$

We assume that

$$g_1(x) = e^x, \quad h_1(s) = e^{-s}, \quad g_2(x) = x, \quad h_2(s) = 2s.$$

Thus from (1.4.40)–(1.4.42), we can obtain $\delta = e^x, Q_1(x) = xe^x, Q_2(x) = e^x - 1 - x$. Hence

$$\begin{cases} z(x) = (1+x)xe^{2x+x^2/2}(e^x - 1 - x)\exp\Big(\int_0^x 2se^{s+s^2/2}(e^s - 1 - s)ds\Big), \\ w(x) = 1 + \dfrac{x^3}{2}e^{2x+x^2/2}(e^x - 1 - x)\exp\Big(\int_0^x 2se^{s+s^2/2}(e^s - 1 - s)ds\Big). \end{cases}$$

Furthermore,

$$F^{-1}(u) = \frac{2}{3}(u-1)^{3/2}.$$

Now, substituting the above values in (1.4.39), we can get the estimate of the solution $y(x)$.

Theorem 1.4.8 (The Pachpatte Inequality [455]) *Let $x(t)$, $f(t)$, $g(t)$, and $h(t)$ be real-valued positive continuous functions defined on I, let $W(u)$ be a positive* *continuous, monotonic, non-decreasing and sub-multiplicative function for all* $u > 0$, $W(0) = 0$, *and suppose further that the inequality holds for all* $t \in I$,

$$x(t) \le x_0 + g(t)(\int_0^t f(s)x(s)ds) + \int_0^t h(s)W(x(s))ds, \tag{1.4.44}$$

where x_0 *is a positive constant. Then for all* $0 \le t \le b$,

$$\begin{aligned} x(t) \le G^{-1}\Big[G(x_0) + \int_0^t h(s) \\ \times W(1 + g(s)(\int_0^s f(\tau)\exp(\int_\tau^s g(k)f(k)dk)d\tau))ds\Big] \\ \times\Big[1 + g(t)(\int_0^t f(s)\exp(\int_s^t g(\tau)f(\tau)d\tau)ds)\Big], \end{aligned} \tag{1.4.45}$$

where

$$G(r) = \int_{r_0}^r \frac{ds}{W(s)}, \quad r \ge r_0 > 0, \tag{1.4.46}$$

and G^{-1} *is the inverse function of* G, *and* t *is in the sub-interval* $[0, b]$ *of* I *so that*

$$G(x_0) + \int_0^t h(s)W(1 + g(s)(\int_0^s f(\tau)\exp(\int_\tau^s g(k)f(k)dk)d\tau))ds \in Dom\ (G^{-1}).$$

Proof Define

$$n(t) = x_0 + \int_0^t h(s)W(x(s))ds, \quad n(0) = x_0. \tag{1.4.47}$$

Then (1.4.44) can be restated as

$$x(t) \le n(t) + g(t)(\int_0^t f(s)x(s)ds).$$

Since $n(t)$ is positive, monotonic, non-decreasing on I, we have from Theorem 1.2.7 in Qin [557]

$$x(t) \le n(t)(1 + g(t)(\int_0^t f(s)\exp(\int_s^t g(\tau)f(\tau)d\tau)ds)). \tag{1.4.48}$$

Further,

$$W(x(t)) \le W(n(t))W(1 + g(t)(\int_0^t f(s)\exp(\int_s^t g(\tau)f(\tau)d\tau)ds))$$

since W is sub-multiplicative. Hence,

$$\frac{h(t)W(x(t))}{W(n(t))} \leq h(t)W(1 + g(t)(\int_0^t f(s)\exp(\int_s^t g(\tau)f(\tau)d\tau)ds)).$$

Because of (1.4.46) and (1.4.47), this reduces to

$$\frac{d}{dt}G(n(t)) \leq h(t)W(1 + g(t)(\int_0^t f(s)\exp(\int_s^t g(\tau)f(\tau)d\tau)ds)).$$

Now, integrating from 0 to t, we obtain

$$G(n(t)) - G(n(0)) \leq \int_0^t h(s)W(1 + g(s)(\int_0^s f(\tau)\exp(\int_{\tau)}^s g(k)f(k)dk)d\tau))ds. \tag{1.4.49}$$

Thus the desired bound in (1.4.45) follows from (1.4.48) and (1.4.49). The sub-interval $[0, b]$ is obvious. □

Theorem 1.4.9 (The Pachpatte Inequality [455]) *Let $x(t)$, $f(t)$, $g(t)$, and $h(t)$ be real-valued positive continuous functions defined on I; let $W(u)$ be a positive, continuous, monotonic, non-decreasing, sub-additive and sub-multiplicative function for all $u > 0$, $W(0) = 0$; let the functions $p(t) > 0$, $\psi(t) \geq 0$ be non-decreasing in t and continuous on I, $\psi(0) = 0$; and suppose further that the inequality holds for all $t \in I$,*

$$x(t) \leq p(t) + g(t)\left(\int_0^t f(s)x(s)ds\right) + \psi\left(\int_0^t h(s)W(x(s))ds\right).$$

Then for all $0 \leq t \leq b$,

$$\begin{aligned} x(t) \leq \Bigg[& p(t) + \psi(G^{-1}\Bigg[G\Bigg(\int_0^t h(s) \\ & \times W(p(s)\left\{1 + g(s)\left(\int_0^s f(\tau)\exp\left(\int_\tau^s g(k)f(k)dk\right)d\tau\right)\right\})ds\Bigg) \\ & + \int_0^t h(s)W\left(1 + g(s)\left(\int_0^s f(\tau)\exp\left(\int_\tau^s g(k)f(k)dk\right)d\tau\right)\right)ds\Bigg]\Bigg)\Bigg] \\ & \times\left[1 + g(t)(\int_0^t f(s)\exp(\int_s^t g(\tau)f(\tau)d\tau)ds)\right], \end{aligned}$$

where

$$G(r) = \int_{r_0}^r \frac{ds}{W(\psi(s))}, \quad r \geq r_0 > 0,$$

and G^{-1} *is the inverse of* G, *and* $t \in [0, b]$ *of* I *so that*

$$G\left(\int_0^t h(s)W(p(s)\{1+g(s)(\int_0^s f(\tau)\exp(\int_\tau^s g(k)f(k)dk)d\tau)\})ds\right)$$
$$+\int_0^t h(s)W\left(1+g(s)\left(\int_0^s f(\tau)\exp\left(\int_\tau^s g(k)f(k)dk\right)d\tau\right)\right)ds \in Dom\,(G^{-1}).$$

Proof The proof of this theorem follows by an argument similar to that in the proof of Theorem 1.4.8, together with Theorem 1.2.7 in Qin [557]. □

The following theorems are the corollaries of Theorem 1.4.9.

Theorem 1.4.10 (The Pachpatte Inequality [512]) *Let* u, a, b, g, h *be real-valued non-negative continuous functions defined on* $\mathbb{R}_+$ *and* $p > 1$ *be a real constant.*

(a_1) *If for all* $t \in \mathbb{R}_+$,

$$u^p(t) \le a(t) + b(t)\int_0^t (g(s)u^p(s) + h(s)u(s))\,ds, \tag{1.4.50}$$

then for all $t \in \mathbb{R}_+$,

$$u(t) \le \left\{a(t) + b(t)\int_0^t \left(g(s)a(s) + h(s)(\frac{p-1}{p} + \frac{a(s)}{p})\right)\right.$$
$$\left.\times\exp(\int_s^t b(\sigma)(g(\sigma) + \frac{h(\sigma)}{p})\,d\sigma)ds\right\}^{1/p}. \tag{1.4.51}$$

(a_2) *Let* $c(t)$ *be a real-valued positive continuous and non-decreasing function defined on* $\mathbb{R}_+$. *If for all* $t \in \mathbb{R}_+$,

$$u^p(t) \le c^p(t) + b(t)\int_0^t (g(s)u^p(s) + h(s)u(s))\,ds, \tag{1.4.52}$$

then for all $t \in \mathbb{R}_+$,

$$u(t) \le c(t)\left\{1 + b(t)\int_0^t \left(g(s) + h(s)c^{1-p}(s)\right)\right.$$
$$\left.\times\exp\left(\int_s^t b(\sigma)(g(\sigma) + \frac{h(\sigma)}{p}c^{1-p}(\sigma))\,d\sigma\right)ds\right\}^{1/p}. \tag{1.4.53}$$

(a_3) *Let* $k(t, s)$ *and its partial derivative* $\frac{\partial}{\partial t}k(t, s)$ *be real-valued non-negative continuous function for* $0 \le s \le t < +\infty$. *If for all* $t \in \mathbb{R}_+$,

$$u^p(t) \le a(t) + b(t)\int_0^t k(t, s)(g(s)u^p(s) + h(s)u(s))\,ds, \tag{1.4.54}$$

then for all $t \in \mathbb{R}_+$,

$$u(t) \le \left\{ a(t) + b(t) \int_0^t B(\sigma) \exp\left(\int_\sigma^t A(\tau)\, d\tau \right) d\sigma \right\}^{1/p}, \tag{1.4.55}$$

where for all $t \in \mathbb{R}_+$,

$$\begin{cases} A(t) = k(t,t)b(t)\left(g(t) + \dfrac{h(t)}{p} \right) + \displaystyle\int_0^t \frac{\partial}{\partial t} k(t,s) b(s) \left(g(s) + \frac{h(s)}{p} \right) ds, & (1.4.56) \\ B(t) = k(t,t)\left(g(t)a(t) + h(t)\left(\dfrac{p-1}{p} + \dfrac{a(t)}{p} \right) \right) & \\ \qquad + \displaystyle\int_0^t \frac{\partial}{\partial t} k(t,s) \left(g(s)a(s) + h(s) \left(\frac{p-1}{p} + \frac{a(s)}{p} \right) \right) ds. & (1.4.57) \end{cases}$$

Proof (a_1) Define a function $z(t)$ by

$$z(t) = \int_0^t (g(s)u^p(s) + h(s)u(s))\, ds. \tag{1.4.58}$$

Then $z(0) = 0$ and (1.4.50) can be rewritten as

$$u^p(t) \le a(t) + b(t)z(t) \tag{1.4.59}$$

From (1.4.59) and using the Young inequality (see, e.g., [395]),

$$x^{1/p} y^{1/q} \le \frac{x}{p} + \frac{y}{q}, \tag{1.4.60}$$

where $x \ge 0$, $y \ge 0$, and $\frac{1}{p} + \frac{1}{q} = 1$ with $p > 1$, we derive that

$$\begin{aligned} u(t) &\le (a(t) + b(t)z(t))^{1/p} (1)^{1/p/(p-1)} \\ &\le \frac{p-1}{p} + \frac{a(t)}{p} + \frac{b(t)}{p} z(t). \end{aligned} \tag{1.4.61}$$

Differentiating (1.4.58) and using (1.4.59) and (1.4.61), we can get

$$\begin{aligned} z'(t) &\le b(t)\left(g(t) + \frac{h(t)}{p} \right) z(t) \\ &\quad + \left[g(t)a(t) + h(t)\left(\frac{p-1}{p} + \frac{a(t)}{p} \right) \right]. \end{aligned} \tag{1.4.62}$$

Thus the inequality (1.4.62) implies

$$z(t) \le \int_0^t \left[g(s)a(s) + h(s)\left(\frac{p-1}{p} + \frac{a(s)}{p}\right)\right] \times \exp\left\{ \int_s^t b(\sigma)\left\{ g(\sigma) + \frac{h(\sigma)}{p}\right) d\sigma \right) ds. \tag{1.4.63}$$

The required inequality (1.4.51) follows from (1.4.63) and (1.4.59).

(a_2) Since $c(t)$ is a positive, continuous, and non-decreasing function for all $t \in \mathbb{R}_+$, from (1.4.52) we can derive

$$\left(\frac{u(t)}{c(t)}\right)^p \le 1 + b(t)\int_0^t \left[g(s)\left(\frac{u(s)}{c(s)}\right)^p + h(s)c^{1-p}(s)\left(\frac{u(s)}{c(s)}\right)\right] ds.$$

Now applying the inequality given (a_1) yields the desired result in (1.4.53).

(a_3) Define a function $z(t)$ by

$$z(t) = \int_0^t k(t,s)[g(s)u^p(s) + h(s)u(s)]\, ds. \tag{1.4.64}$$

Then as in the proof of part (a_1), from (1.4.54) we see that the inequalities (1.4.59) and (1.4.61) hold. Differentiating (1.4.64) and using (1.4.59), (1.4.61), and the fact that $z(t)$ is monotonic non-decreasing in t, we can get

$$\begin{aligned} z'(t) &= k(t,t)[g(t)u^p(t)+h(t)u(t)] + \int_0^t \frac{\partial}{\partial t}k(t,s)[g(s)u^p(s)+h(s)u(s)]\, ds \\ &\le k(t,t)\left[g(t)(a(t)+b(t)z(t)) + h(t)\left(\frac{p-1}{p} + \frac{a(t)}{p} + \frac{b(t)}{p}z(t)\right)\right] \\ &\quad + \int_0^t \frac{\partial}{\partial t}k(t,s)\left[g(s)(a(s)+b(s)z(s)) + h(s)\left(\frac{p-1}{p} + \frac{a(s)}{p} + \frac{b(s)}{p}z(s)\right)\right] ds \\ &\le \left[k(t,t)b(t)\left(g(t) + \frac{h(t)}{p}\right) + \int_0^t \frac{\partial}{\partial t}k(t,s)b(s)\left(g(s) + \frac{h(s)}{p}\right) ds\right] z(t) \\ &\quad + k(t,t)\left(g(t)(a(t) + h(t)\left(\frac{p-1}{p} + \frac{a(t)}{p}\right)\right) \\ &\quad + \int_0^t \frac{\partial}{\partial t}k(t,s)\left(g(s)a(s) + h(s)\left(\frac{p-1}{p} + \frac{a(s)}{p}\right)\right) ds \\ &= A(t)z(t) + B(t), \end{aligned} \tag{1.4.65}$$

which implies

$$z(t) \le \int_0^t B(\sigma)\exp\left(\int_\sigma^t A(\tau)\, d\tau\right) d\sigma. \tag{1.4.66}$$

Inserting (1.4.66) into $u^p(t) \le a(t) + b(t)z(t)$, we can get the required inequality in (1.4.55). □

Theorem 1.4.11 (The Pachpatte Inequality [512]) *Let u, a, b , g be real-valued non-negative continuous functions defined on $\mathbb{R}_+$ and $p > 1$ be a real constant.*

(b_1) Let $f : \mathbb{R}_+^2 \to \mathbb{R}_+$ be a continuous function such that for all $t \in \mathbb{R}_+$ and all $x \ge y \ge 0$,

$$0 \le f(t,x) - f(t,y) \le m(t,y)(x-y), \tag{1.4.67}$$

where $m : \mathbb{R}_+^2 \to \mathbb{R}_+$ is a continuous function. If for all $t \in \mathbb{R}_+$,

$$u^p(t) \le a(t) + b(t)\int_0^t f(s,u(s))\,ds, \tag{1.4.68}$$

then for all $t \in \mathbb{R}_+$,

$$u(t) \le \left\{a(t) + b(t)\int_0^t f\left(s, \frac{p-1}{p} + \frac{a(s)}{p}\right)\right.$$
$$\left.\times \exp(\int_s^t m\left(\sigma, \frac{p-1}{p} + \frac{a(\sigma)}{p}\right)\frac{b(\sigma)}{p}\,d\sigma)\,ds\right\}^{1/p}. \tag{1.4.69}$$

(b_2) Let $f : \mathbb{R}_+^2 \to \mathbb{R}_+$ be a continuous function such that for all $t \in \mathbb{R}_+$ and all $x \ge y \ge 0$,

$$0 \le f(t,x) - f(t,y) \le m(t,y)\phi^{-1}(x-y), \tag{1.4.70}$$

where $m : \mathbb{R}_+^2 \to \mathbb{R}_+$ is a continuous function and $\phi^{-1}$is the inverse function of ϕ and for $x, y \in \mathbb{R}_+$,

$$\phi^{-1}(xy) \le \phi^{-1}(x)\phi^{-1}(y). \tag{1.4.71}$$

If for all $t \in \mathbb{R}_+$,

$$u^p(t) \le a(t) + b(t)\phi\left(\int_0^t f(s,u(s))\,ds\right), \tag{1.4.72}$$

then for all $t \in \mathbb{R}_+$,

$$u(t) \le \left\{a(t) + b(t)\phi\left(\int_0^t f\left(s, \frac{p-1}{p} + \frac{a(s)}{p}\right)\right)\right.$$
$$\left.\times \exp\left(\int_s^t m\left(\sigma, \frac{p-1}{p} + \frac{a(\sigma)}{p}\right)\phi_{-1}\left(\frac{b(\sigma)}{p}\right)d\sigma\right)ds\right\}^{1/p}. \tag{1.4.73}$$

(b_3) Let $W(r)$ be a real-valued continuous non-decreasing sub-additive and sub-multiplicative function defined on $\mathbb{R}_+$ and $W(r) > 0$ on $(0, +\infty)$. If for all $t \in \mathbb{R}_+$,

$$u^p(t) \le a(t) + b(t) \int_0^t g(s) W(u(s))\, ds, \tag{1.4.74}$$

then for all $0 \le t \le t_1$,

$$u(t) \le \left\{ a(t) + b(t) G^{-1} \left[G(D(t)) + \int_0^t g(s) W\left(\frac{b(s)}{p}\right) ds \right] \right\}^{1/p}, \tag{1.4.75}$$

where for all $t \in \mathbb{R}_+$,

$$\begin{cases} D(t) = \displaystyle\int_0^t g(s) W\left(\frac{p-1}{p} + \frac{a(s)}{p}\right) ds, & (1.4.76) \\ G(r) = \displaystyle\int_{r_0}^r \frac{ds}{W(s)}, \quad r \ge r_0 > 0, & (1.4.77) \end{cases}$$

and G^{-1} is the inverse function of G, and $t_1 \in \mathbb{R}_+$ is chosen so that for all $t \in [0, t_1]$,

$$G(D(t)) + \int_0^t g(s) W\left(\frac{b(s)}{p}\right) ds \in Dom\,(G^{-1}).$$

Proof (b_1) Define a function $z(t)$ by

$$z(t) = \int_0^t f(s, u(s))\, ds. \tag{1.4.78}$$

Then as in the proof of Theorem 1.4.10 (part (a_1)), from (1.4.68) we see that the inequalities (1.4.59) and (1.4.61) hold. From (1.4.78), (1.4.61), and the condition (1.4.67) it follows that

$$\begin{aligned} z'(t) &= f(t, u(t)) \\ &\le f\left(t, \frac{p-1}{p} + \frac{a(t)}{p} + \frac{b(t)}{p} z(t)\right) - f\left(t, \frac{p-1}{p} + \frac{a(t)}{p}\right) \\ &\quad + f\left(t, \frac{p-1}{p} + \frac{a(t)}{p}\right) \\ &\le m\left(t, \frac{p-1}{p} + \frac{a(t)}{p}\right) \frac{b(t)}{p} z(t) + f\left(t, \frac{p-1}{p} + \frac{a(t)}{p}\right), \end{aligned} \tag{1.4.79}$$

which implies

$$z(t) \le \int_0^t f\left(s, \frac{p-1}{p} + \frac{a(s)}{p}\right) \times \exp\left(\int_s^t m\left(\sigma, \frac{p-1}{p} + \frac{a(s)}{p}\right)\frac{b(\sigma)}{p}\,d\sigma\right) ds. \quad (1.4.80)$$

From (1.4.80) and (1.4.59), the desired inequality (1.4.69) follows.

(b_2) Define a function $z(t)$ by (1.4.78) and following the arguments as in the proof of Theorem 1.4.10 (part (a_1)), we see that corresponding to the inequalities (1.4.59) and (1.4.61) hold, and

$$u^p(t) \le a(t) + b(t)\phi(z(t)) \quad (1.4.81)$$

and

$$u(t) \le \frac{p-1}{p} + \frac{a(t)}{p} + \frac{b(t)}{p}\phi(z(t)). \quad (1.4.82)$$

Thus from (1.4.78), (1.4.82), and the condition (1.4.70), (1.4.71) it follows that

$$\begin{aligned} z'(t) &= f(t, u(t)) \\ &\le f\left(t, \frac{p-1}{p} + \frac{a(t)}{p} + \frac{b(t)}{p}\phi(z(t))\right) - f\left(t, \frac{p-1}{p} + \frac{a(t)}{p}\right) \\ &\quad + f\left(t, \frac{p-1}{p} + \frac{a(t)}{p}\right) \\ &\le m\left(t, \frac{p-1}{p} + \frac{a(t)}{p}\right)\phi^{-1}\left(\frac{b(t)}{p}\phi(z(t))\right) \\ &\quad + f\left(t, \frac{p-1}{p} + \frac{a(t)}{p}\right) \\ &\le m\left(t, \frac{p-1}{p} + \frac{a(t)}{p}\right)\phi^{-1}\left(\frac{b(t)}{p}\right)z(t) + f\left(t, \frac{p-1}{p} + \frac{a(t)}{p}\right), \end{aligned} \quad (1.4.83)$$

which implies

$$z(t) \le \int_0^t f\left(s, \frac{p-1}{p} + \frac{a(s)}{p}\right) \times \exp\left(\int_s^t m\left(\sigma, \frac{p-1}{p} + \frac{a(\sigma)}{p}\right)\phi^{-1}\left(\frac{b(\sigma)}{p}\right) d\sigma\right) ds. \quad (1.4.84)$$

The required inequality (1.4.73) follows from (1.4.81) and (1.4.84).

(b_3) Define a function $z(t)$ by

$$z(t) = \int_0^t g(s)W(u(s))ds. \tag{1.4.85}$$

Then as in the proof of Theorem 1.4.10 (part (a_1)), from (1.4.74) we derive that the inequalities (1.4.59) and (1.4.61) hold. From (1.4.85), (1.4.61), and the conditions on W it follows that

$$z(t) \leq D(t) + \int_0^t g(s)W\left(\frac{b(s)}{p}\right) W(z(s))\, ds, \tag{1.4.86}$$

where $D(t)$ is defined by (1.4.76). The rest of the proof can be completed by following the proof of Theorem 1.4.9. We omit here the details. □

Remark 1.4.1 We note that in the special cases when (1) $g = 0$, (2) $g = 0$, $p = 2$ in Theorem 1.4.10, and (3) $p = 2$ in Theorem 1.4.11, we can get new inequalities which may be convenient in certain applications.

As mentioned previously, the integral inequalities established in Theorems 1.4.8 and 1.4.9 are the further generalizations of the corresponding inequalities obtained in [197]. However, the integral inequalities established in [456] are different from those obtained in Theorems 1.4.8 and 1.4.9.

Theorem 1.4.12 (The Pachpatte Inequality [455]) *Let $x(t)$, $f(t)$, $g(t)$, $h(t)$, and $q(t)$ be real-valued positive continuous functions defined on I; let $W(t,u)$ be a positive, continuous, monotonic, non-decreasing in u, $u > 0$, for each fixed $t \in I$; let the functions $p(t) > 0$, $\psi(t) \geq 0$ be non-decreasing in t and continuous on I, $\psi(0) = 0$; and suppose further that the inequality holds for all $t \in I$,*

$$x(t) \leq p(t) + g(t)\left(\int_0^t f(s)x(s)ds\right) + h(t)\psi\left(\int_0^t q(s)W(s,x(s))ds\right). \tag{1.4.87}$$

Then for all $t \in I$,

$$x(t) \leq k(t)[p(t) + h(t)\psi(r(t))], \tag{1.4.88}$$

where

$$k(t) = 1 + g(t)\left(\int_0^t f(s)\exp(\int_s^t g(\tau)f(\tau)d\tau)ds\right), \tag{1.4.89}$$

and $r(t)$ is the maximal solution of

$$r'(t) = q(t)W(t, k(t)[p(t) + h(t)\psi(r(t))]), \quad r(0) = 0, \tag{1.4.90}$$

existing on I.

Proof Define

$$n(t) = p(t) + h(t)\psi(\int_0^t q(s)W(s, x(s))ds). \tag{1.4.91}$$

Then (1.4.87) can be rewritten as

$$x(t) \leq n(t) + g(t)\left(\int_0^t f(s)x(s)ds\right).$$

Since $n(t)$ is positive, monotonic, non-decreasing on I, we derive from Theorem 1.2.7 in Qin [557],

$$x(t) \leq k(t)n(t), \tag{1.4.92}$$

where $k(t)$ is as given in (1.4.89). Now from (1.4.91) and (1.4.92), it follows

$$x(t) \leq k(t)[p(t) + h(t)\psi(v(t))], \tag{1.4.93}$$

where

$$v(t) = \int_0^t q(s)W(s, x(s))ds, \quad v(0) = 0.$$

Therefore, it follows that

$$v'(t) \leq q(t)W(t, k(t)[p(t) + h(t)\psi(v(t))]). \tag{1.4.94}$$

Now applying Lemma 1.1.23 to (1.4.94) and (1.4.90) yields

$$v(t) \leq r(t), \tag{1.4.95}$$

where $r(t)$ is the maximal solution of (1.4.90) such that $r(0) = v(0) = 0$. Thus from (1.4.93) and (1.4.95), the desired bound in (1.4.88) follows. □

Theorem 1.4.12, in the special case when the first integral term on the right-hand side in (1.4.87) is absent, $h(t) = 1$, $q(t) = 1$, $\psi(u) = u$, and $p(t)$ is constant, was first established in [657]. This theorem may easily be modified to include the case in which W depends on three arguments t, s, and x (see, e.g., [323]). Moreover, we also obtain as a special case a useful generalization of the Gronwall-Bellman inequality due to Bihari [82].

Theorem 1.4.13 (The Pachpatte Inequality [455]) *Let $x(t)$, $f(t)$ be real-valued positive continuous functions defined on I; let $n(t)$ be a positive, monotonic, non-decreasing continuous function defined on I; let $\Phi \in \mathcal{F}$; and let $H(u)$ be a positive,* *continuous, monotonic, non-decreasing sub-additive and sub-multiplicative function for all* $u > 0$, $H(0) = 0$, *and* H^{-1} *denotes the inverse function of* H, *for which the inequality holds, for all* $t \in I$,

$$x(t) \le n(t) + H^{-1}[\Phi(\int_0^t f(s)H(x(s))ds)]. \tag{1.4.96}$$

Then for all $0 \le t \le b$,

$$x(t) \le n(t)H^{-1}[1 + \Phi(G^{-1}[G(0) + \int_0^t f(s)ds])], \tag{1.4.97}$$

where

$$G(r) = \int_{r_0}^{r} \frac{ds}{1 + \Phi(s)}, \quad r \ge r_0 > 0, \tag{1.4.98}$$

and G^{-1} *is the inverse of* G, *and* $t \in [0, b]$ *of* I *so that*

$$G(0) + \int_0^t f(s)ds \in Dom\,(G^{-1}).$$

Proof Since H is sub-additive and sub-multiplicative, we derive from (1.4.96)

$$H(x(t)) \le H(n(t)) + \Phi(\int_0^t f(s)H(x(s))ds). \tag{1.4.99}$$

Since $H(n(t))$ is positive, monotonic, non-decreasing and $\Phi \in \mathcal{F}$, we conclude

$$\begin{aligned} \frac{H(x(t))}{H(n(t))} &\le 1 + \Phi(\int_0^t f(s)\frac{H(x(s))}{H(n(t))}ds) \\ &\le 1 + \Phi(\int_0^t f(s)\frac{H(x(s))}{H(n(s))}ds). \end{aligned} \tag{1.4.100}$$

Define a function $v(t)$ such that

$$v(t) = \int_0^t f(s)\frac{H(x(s))}{H(n(s))}ds, \quad v(0) = 0.$$

Then

$$v'(t) = f(t)\frac{H(x(t))}{H(n(t))},$$

which, in view of (1.4.100), implies

$$v'(t) \leq f(t)(1 + \Phi(v(t))). \tag{1.4.101}$$

Dividing both sides of (1.4.101) by $1+\Phi(v(t))$, using (1.4.98), and integrating from 0 to t, we may obtain

$$G(v(t)) - G(v(0)) \leq \int_0^t f(s)ds. \tag{1.4.102}$$

Then from (1.4.100) and (1.4.102), it follows

$$H(x(t)) \leq H(n(t))[1 + \Phi(G^{-1}[G(0) + \int_0^t f(s)ds])]. \tag{1.4.103}$$

Now applying H^{-1} to both sides of (1.4.103), we can obtain the desired bound in (1.4.97). □

We now apply Theorem 1.4.13 to establish the following more general inequalities.

Theorem 1.4.14 (The Pachpatte Inequality [455]) *Let $x(t)$, $f(t)$, and $g(t)$ be real-valued positive continuous functions defined on I; let $\Phi \in \mathcal{F}$; let H, H^{-1} be defined as in Theorem 1.4.13; let W be the same function as defined in Theorem 1.4.8, and suppose further that the inequality holds for all $t \in I$,*

$$x(t) \leq x_0 + H^{-1}[\Phi(\int_0^t f(s)H(x(s))ds)] + \int_0^t g(s)W(x(s))ds$$

where x_0 is a positive constant. Then for all $0 \leq t \leq b$,

$$x(t) \leq \Omega^{-1}\left[\Omega(x_0) + \int_0^t g(s)W\left(H^{-1}[1 + \Phi\left(G^{-1}\left[G(0) + \int_0^s f(\tau)d\tau\right]\right)]\right)ds\right]$$
$$\times H^{-1}\left[1 + \Phi(G^{-1}[G(0) + \int_0^t f(s)ds])\right]$$

where G, G^{-1} are as defined in Theorem 1.4.13, Ω is defined by

$$\Omega(r) = \int_{r_0}^r \frac{ds}{W(s)}, \quad r \geq r_0 > 0,$$

and Ω^{-1} is the inverse function of Ω, and $t \in [0, b]$ of I such that

$$G(0) + \int_0^t f(s)ds \in Dom\ (G^{-1})$$

and

$$\Omega(x_0) + \int_0^t g(s)W(H^{-1}[1 + \Phi(G^{-1}[G(0) + \int_0^s f(\tau)d\tau])])ds \in Dom\ (\Omega^{-1}).$$

Proof The proof is similar to that of Theorem 1.4.13. □

We next establish a more general form of Theorem 1.4.14 which may be used in certain situations.

Theorem 1.4.15 (The Pachpatte Inequality [455]) *Let $x(t)$, $f(t)$, and $g(t)$ be real-valued positive continuous functions defined on I; let $\Phi \in \mathcal{F}$; let H, H^{-1} be as defined in Theorem 1.4.13; let W be the same functions as defined in Theorem 1.4.8; let the functions $p(t) > 0$, $\psi(t) \geq 0$ be non-decreasing in t and continuous on I, $\psi(0) = 0$; and suppose further that the inequality holds for all $t \in I$,*

$$x(t) \leq p(t) + H^{-1}[\Phi(\int_0^t f(s)H(x(s))ds)] + \psi(\int_0^t g(s)W(x(s))ds).$$

Then for all $0 \leq t \leq b$,

$$\begin{aligned}x(t) \leq \Bigg[p(t) &= \psi(\Omega^{-1}[\Omega\left(\int_0^t g(s)W(p(s)\{H^{-1}[1 + \Phi(G^{-1}[G(0) + \int_0^s f(\tau)d\tau])]\})ds\right) \\ &+ \int_0^t g(s)W(H^{-1}[1 + \Phi(G^{-1}[G(0) + \int_0^s f(\tau)d\tau])])ds]) \Bigg] \\ &\times H^{-1}[1 + \Phi(G^{-1}[G(0) + \int_0^t f(s)ds])]\end{aligned}$$

where G, G^{-1} are as defined in Theorem 1.4.13, Ω is defined by

$$\Omega(r) = \int_{r_0}^r \frac{ds}{W(\psi(s))}, \quad r \geq r_0 > 0,$$

and Ω^{-1} is the inverse function of Ω, and $t \in [0, b]$ of I so that

$$G(0) + \int_0^t f(s)ds \in Dom\ (G^{-1})$$

and

$$\begin{aligned}&\Omega\left(\int_0^t g(s)W(p(s)\{H^{-1}\left[1 + \Phi\left(G^{-1}\left[G(0) + \int_0^s f(\tau)d\tau\right]\right)\right]\})ds\right) \\ &+ \int_0^t g(s)W(H^{-1}[1 + \Phi(G^{-1}[G(0) + \int_0^s f(\tau)d\tau])])ds \in Dom\ (\Omega^{-1}).\end{aligned}$$

Proof The detail of the proof of Theorem 1.4.15 follows by arguments similar to those in the proof of Theorem 1.4.9, by making use of Theorem 1.4.13. We omit the details. □

Theorem 1.4.16 (The Pachpatte Inequality [455]) *Let $x(t)$, $f(t)$, $h(t)$, and $q(t)$ be real-valued positive continuous functions defined on I; let $\Phi \in \mathcal{F}$; let H, H^{-1} be as defined in Theorem 1.4.13; let $W(t,u)$, $p(t)$, $\psi(t)$ be as defined in Theorem 1.4.12; and suppose further that the inequality holds for all $t \in I$,*

$$x(t) \le p(t) + H^{-1}\left[\Phi(\int_0^t f(s)H(x(s))ds)\right] + h(t)\psi\left(\int_0^t q(s)W(s,x(s))ds\right)$$

Then for all $t \in I_0$,

$$x(t) \le k_1(t)[p(t) + h(t)\psi(r(t))], \tag{1.4.104}$$

where

$$k_1(t) = H^{-1}\left[1 + \Phi\left(G^{-1}\left[G(0) + \int_0^t f(s)ds\right]\right)\right],$$

and G and G^{-1} are as in Theorem 1.4.13, I_0 is the largest sub-interval of I on which the right-hand side of (1.4.104) exists, and $r(t)$ is the maximal solution of

$$r'(t) = q(t)W(t, k_1(t)[p(t) + h(t)\psi(r(t))]), \quad r(0) = 0,$$

existing on I.

Proof The proof is the same as that of Theorem 1.4.12, and we leave the details to the reader. □

We note that there is no essential difficulty in obtaining the bounds for inequalities of the form, for all $t \in I$,

$$x(t) \le n(t) + \Phi\left(\int_0^t f(s)H(x(s))ds\right),$$

where x, n, f, and Φ are as given in Theorem 1.4.13, $H \in \mathcal{F}$, and G is defined by

$$G(r) = \int_{r_0}^r \frac{ds}{H(1 + \Phi(s))}, \quad r \ge r_0 > 0,$$

by following partial the arguments as in the proofs of Theorem 1.2.7 in Qin [557] and Theorem 1.4.13. In view of this remark, we can use this inequality to obtain inequalities similar to those obtained in Theorems 1.4.14–1.4.16. Since this translation is quite straightforward in view of the above results, we leave it to the reader as an exercise.

Chapter 2
Nonlinear One-Dimensional Discrete (Difference) Inequalities

In this chapter, we shall introduce some nonlinear discrete (difference) integral inequalities.

2.1 Nonlinear One-Dimensional Discrete Bellman-Gronwall Inequalities

Ladyzhenskaya, Solonnikov and Ural'ceva [319] established the following two discrete forms of the Bellman-Granwall inequalities.

Theorem 2.1.1 (The Ladyzhenskaya-Solonnikov Inequality [319]) *Suppose a sequence y_i $(i = 0, 1, \cdots,)$ of non-negative numbers satisfies the recursion relation*

$$y_{i+1} \le Cb^i y_i^{1+\varepsilon}, \quad i = 0, 1, \cdots, \tag{2.1.1}$$

with some positive constants C, ε and $b \ge 1$. Then, for $i = 0, 1, \cdots$,

$$y_i \le C^{[(1+\varepsilon)^i - 1]/\varepsilon} b^{[(1+\varepsilon)^i - 1]/\varepsilon^2 - i/\varepsilon} y_0^{(1+\varepsilon)^i}. \tag{2.1.2}$$

In particular, if $y_0 \le \theta = C^{-1/\varepsilon} b^{-1/\varepsilon^2}$ and $b > 1$, then

$$y_i \le \theta b^{-i/\varepsilon} \tag{2.1.3}$$

and consequently $y_i \to 0$ as $i \to +\infty$.

Proof This conclusion can be proved directly by induction. We leave the details to the reader. □

Y. Qin, *Integral and Discrete Inequalities and Their Applications*,
DOI 10.1007/978-3-319-33304-5_2

Theorem 2.1.2 (The Ladyzhenskaya-Solonnikov Inequality [319]) *Suppose that non-negative numbers y_i and z_i $(i = 0, 1, \cdots,)$ are connected by the system of recursion inequalities*

$$\begin{cases} y_{i+1} \le Cb^i\Big(y_i^{1+\delta} + z_i^{1+\varepsilon}y_i^{\delta}\Big), & (2.1.4) \\ z_{i+1} \le Cb^i\Big(y_i + z_i^{1+\varepsilon}\Big), & (2.1.5) \end{cases}$$

where C, b, ε and δ are some fixed positive numbers with $b \ge 1$. Then

$$y_i \le \lambda b^{-i/d}, \quad z_i \le \left(\lambda b^{-i/d}\right)^{1/(1+\varepsilon)} \tag{2.1.6}$$

where

$$d = \min\left(\delta, \varepsilon/(1+\varepsilon)\right), \quad \lambda = \min\left\{(2C)^{-1/\delta}b^{-1/(\delta d)}, (2C)^{(1+\varepsilon)/\varepsilon}b^{-1/(\varepsilon b)}\right\}$$

as long as $y_0 \le \lambda$ and $z_0 \le \lambda^{1/(1+\varepsilon)}$.

Proof Indeed inequalities (2.1.6) are by condition valid for $i = 0$. Suppose they hold for y_i and z_i, then in view of (2.1.4)–(2.1.5),

$$\begin{cases} y_{i+1} \le Cb^i 2(\lambda b^{-i/d})^{1+\delta} = 2C\lambda^{1+\delta}b^{i(1-(1+\delta)/d)}, \\ z_{i+1} \le 2C\lambda b^{i(1-1/d)}. \end{cases}$$

But, as is easily calculated, the right-hand sides of these inequalities do not exceed $\lambda b^{-(i+1)/d}$ and $(\lambda b^{-(i+1)/d})^{1/(1+\varepsilon)}$ respectively, and hence inequalities (2.1.6) also hold for y_{i+1} and z_{i+1}. The proof is thus complete. □

The next result concerns a nonlinear discrete inequality

$$u(k) \le p(k)\Big(q + \sum_{i=1}^{r} H_i(k,u)\Big), \quad u(k) \le p(k) + q(k)\sum_{i=1}^{r} E_i(k,u), \tag{2.1.7}$$

where

$$\begin{cases} H_i(k,u) = \sum_{l_1=a}^{k-1} f_{i1}(l_1)u^{\alpha_{i1}}(l_1)\cdots\sum_{l_i=a}^{l_{i-1}-1} f_{ii}(l_i)u^{\alpha_{ii}}(l_i), \\ E_i(k,u) = \sum_{l_1=a}^{k-1} f_{i1}(l_1)\sum_{l_2=a}^{l_1-1} f_{i2}(l_2)\cdots\sum_{l_i=a}^{l_{i-1}-1} f_{ii}(l_i)u(l_i) \end{cases} \tag{2.1.8}$$

and $\alpha_{ij}, 1 \le j \le i, 1 \le i \le r$, are non-negative constants and the constant $q > 0$.

In the following result, we shall denote $\alpha_i = \sum_{j=1}^{i}\alpha_{ij}$ and $\alpha = \max_{1\leq i\leq r}\alpha_i$, $\mathbb{N}_a = \{a, a+1, \ldots\}$ where $a \in \mathbb{N}_0 = \{0\} \bigcup \mathbb{N}$. Let $f(k)$ be a function defined on $\mathbb{N}_0$, then we define the difference operator Δ as $\Delta f(k) = f(k+1) - f(k)$, and the higher-order differences for a positive integer m as $\Delta^m f(k) = \Delta(\Delta^{m-1}f(k))$.

Theorem 2.1.3 (The Agarwal Inequality [10]) *Let for all $k \in \mathbb{N}_a$ the inequality (2.1.7) hold. Then, for all $k \in \mathbb{N}_a$,*

$$u(k) \leq qp(k)\prod_{l=a}^{k-1}(1 + \Delta Q(l)), \quad \text{if} \quad \alpha = 1, \tag{2.1.9}$$

$$u(k) \leq p(k)(q^{1-\alpha} + (1-\alpha)Q(k))^{1/1-\alpha}, \quad \text{if} \quad \alpha \neq 1 \tag{2.1.10}$$

where

$$Q(k) = \sum_{i=1}^{r} H_i(k, p)q^{\alpha_i - \alpha}$$

and when $\alpha > 1$, we assume that $q^{1-\alpha} + (1-\alpha)Q(k) > 0$ for all $k \in \mathbb{N}_a$.

Proof The inequality (2.1.7) can be rewritten as $u(k) \leq p(k)v(k)$, where

$$v(k) = q + \sum_{i=1}^{r} H_i(k, u).$$

Thus using the non-decreasing nature of $v(k)$, we have

$$\Delta v(k) \leq \sum_{i=1}^{r} \Delta H_i(k, p)v^{\alpha_i}(k).$$

Since $v(k) \geq q$, we can get

$$\Delta v(k) \leq \sum_{i=1}^{r} \Delta H_i(k, p)q^{\alpha_i - \alpha}v^{\alpha}(k) = \Delta Q(k)v^{\alpha}(k). \tag{2.1.11}$$

If $\alpha = 1$, (2.1.9) immediately follows by using $v(a) = q$, and the fact that $u(k) \leq p(k)v(k)$.

If $\alpha \neq 1$, we have

$$\frac{\Delta v^{1-\alpha}(k)}{1-\alpha} = \int_k^{k+1} \frac{dv(t)}{v^{\alpha}(t)} \leq \frac{\Delta v(k)}{v^{\alpha}(k)}$$

and from (2.1.11), it follows

$$\frac{\Delta v^{1-\alpha}(k)}{1-\alpha} \leq \Delta Q(k). \tag{2.1.12}$$

Summing the inequality (2.1.12), we derive

$$v(k) \leq \left(q^{1-\alpha} + (1-\alpha)Q(k)\right)^{1/1-\alpha}$$

and thus the result (2.1.10) follows from $u(k) \leq p(k)v(k)$. □

Theorem 2.1.4 (The Agarwal Inequality [10]) *Let for all $k \in \mathbb{N}_a$ the following inequality hold*

$$u(k) \leq p(k) + q(k)\left(\sum_{l=a}^{k-1} f(l)u^r(l)\right)^{1/r}, \tag{2.1.13}$$

where $1 \leq r < +\infty$. Then for all $k \in \mathbb{N}_a$,

$$u(k) \leq p(k) + q(k)\frac{\left(\sum_{l=a}^{k-1} f(l)p^r(l)e(l+1)\right)^{1/r}}{1-(1-e(k))^{1/r}}, \tag{2.1.14}$$

where

$$e(k) = \prod_{l=a}^{k-1}(1+f(l)q^r(l))^{-1}. \tag{2.1.15}$$

Proof Note that the function $e(k)$ is the solution of the initial value problem

$$\Delta e(k) = -f(k)q^r(k)e(k+1), \quad e(a) = 1. \tag{2.1.16}$$

Define the function $v(k)$ by

$$v(k) = e(k)\sum_{l=a}^{k-1} f(l)u^r(l). \tag{2.1.17}$$

Thus from (2.1.13) and (2.1.16), it follows

$$v(k+1) - v(k) \leq \left(p(k)f^{1/r}(k)e^{1/r}(k+1) + \frac{q(k)f^{1/r}(k)v^{1/r}(k)}{(1+f(k)q^r(k))^{1/r}}\right)^r - \frac{f(k)q^r(k)v(k)}{1+f(k)q^r(k)}. \tag{2.1.18}$$

Now summing (2.1.18) from a to $k-1$, transposing the second sum from the right-hand side to left-hand side, forming the rth root on both sides, and applying Minkowski's inequality for sums to the right-hand side, we can obtain

$$\left\{v(k)+\sum_{l=a}^{k-1}\frac{f(l)q^r(l)v(l)}{1+f(l)q^r(l)}\right\}^{1/r}\le\left\{\sum_{l=a}^{k-1}f(l)p^r(l)e(l+1)\right\}^{1/r}+\left\{\sum_{l=a}^{k-1}\frac{f(l)q^r(l)v(l)}{1+f(l)q^r(l)}\right\}^{1/r}. \tag{2.1.19}$$

Transposing the second term of the right-hand side of (2.1.19) to left-hand side, we can obtain the left-hand side of the form $w(t)=(c+t)^{1/r}-t^{1/r}(c\ge 0, r\ge 1)$. Since $w'(t)\le 0$ for all $t\ge 0$, we may replace t by a larger quantity without destroying inequality (2.1.19). In this regard, we note that

$$\begin{aligned}\sum_{l=a}^{k-1}\frac{f(l)q^r(l)v(l)}{1+f(l)q^r(l)}&=\sum_{l=a}^{k-1}\frac{f(l)q^r(l)e(l)}{1+f(l)q^r(l)}\Big(\sum_{\tau=a}^{l-1}f(\tau)u^r(\tau)\Big)\\&=\sum_{l=a}^{k-1}f(l)q^r(l)e(l+1)\Big(\sum_{\tau=a}^{l-1}f(\tau)u^r(\tau)\Big)\\&\le\sum_{l=a}^{k-1}q^r(l)e(l+1)\Big(\sum_{l=a}^{l-1}f(l)u^r(l)\Big)\\&=(1-e(k))\sum_{l=a}^{k-1}f(l)u^r(l)\\&=\frac{v(k)}{e(k)}-v(k).\end{aligned}$$

Hence (2.1.19) implies that

$$\left(\frac{v(k)}{e(k)}^{1/r}\right)^{1/r}-\left(\frac{v(k)}{e(k)}^{1/r}-v(k)-v(k)\right)^{1/r}\le\Big(\sum_{l=a}^{k-1}f(l)p^r(l)e(l+1)\Big)^{1/r},$$

i.e.,

$$\left(\frac{v(k)}{e(k)}^{1/r}\right)^{1/r}=\Big(\sum_{l=a}f(l)u^r(l)\Big)^{1/r}\le\frac{\left(\sum_{l=a}^{k-1}f(l)p^r(l)e(l+1)\right)^{1/r}}{1-(1-e(k))^{1/r}}. \tag{2.1.20}$$

Therefore using (2.1.20) in (2.1.13), (2.1.14) follows. □

Theorem 2.1.5 (The Agarwal Inequality [10]) *Let for all $k \in \mathbb{N}_a$ the following inequality hold*

$$u(k+1) \leq qu^p(k). \tag{2.1.21}$$

Then for all $k \in \mathbb{N}_a$,

$$u(k) \leq \begin{cases} q^{\frac{1-p^{k-a}}{1-p}} u^{p^{k-a}}(a), & \text{if} \quad p \neq 1 \\ q^{k-a} u(a), & \text{if} \quad p = 1. \end{cases} \tag{2.1.22}$$

Proof The proof is easily verified, which is left to the reader as an exercise. □

Theorem 2.1.6 (The Agarwal Inequality [10]) *Let for all $k \in \mathbb{N}_a$ the following inequality hold*

$$u(k) \leq p(k)\Big(q + \sum_{i=1}^{r_1} E_i(k,u) + \sum_{i=1}^{r_2} H_i(k,u)\Big), \tag{2.1.23}$$

where $E_i(k,u), H_i(k,u)$ are defined in (2.1.8) respectively, and the constant $q > 0$. Then, if $\alpha_i = \sum_{j=1}^{i} \alpha_{ij}$ and $\max_{1 \leq i \leq r_2} \alpha_i = \alpha \neq 1$, for all $k \in N_a$,

$$u(k) \leq p(k)v(k)\Big(q^{1-\alpha} + (1-\alpha)Q(k)\Big)^{1/1-\alpha}, \tag{2.1.24}$$

where

$$v(k) = \prod_{l=a}^{k-1}\Big(1 + \sum_{l=a}^{r_1} \Delta E_i(l,p)\Big), \quad Q_k = \sum_{i=1}^{r_2} H_i(k, pv) q^{\alpha_i - \alpha} \tag{2.1.25}$$

and when $\alpha > 1$, we assume that $q^{1-\alpha} + (1-\alpha)Q(k) > 0$ for all $k \in N_a$.

Proof We leave the proof to the reader as an exercise. □

In what follows, for the sequence $v : \mathbb{N} \to (0, +\infty)$, the geometrical mean G_n, and the harmonical mean H_n operators are defined as follows:

$$G_n v = \left(\prod_{i=0}^{N} v_i\right)^{\frac{c_n}{n}}, \quad H_n v = \frac{n}{\sum_{i=1}^{n} \frac{c_i}{v_i}}, \quad n = 1, 2, \ldots,$$

where $c = \{c_n\}_{n=1}^{\infty}$ is a known sequence with positive elements.

Theorem 2.1.7 (The Blandzi-Popenda-Agarwal Inequality [368]) *Let b, c, u: $\mathbb{N} \to (0, +\infty)$, and let the following inequality hold for all $n \in \mathbb{N}$,*

$$u_{n+1} \leq b_n G_n u. \tag{2.1.26}$$

Then for all $n \in \mathbb{N}$,

$$u_{n+1} \leq b_n \left(\prod_{i=0}^{N-1} b_i^{\left(\frac{c_n}{n} \prod_{j=i+1}^{n-1}(1+c_j/j)\right)} \right) u_1^{\left(\frac{c_n}{n} \prod_{j=1}^{n-1}(1+c_j/j)\right)}. \tag{2.1.27}$$

Proof We shall find the solution of the equation, for all $n \in \mathbb{N}$,

$$v_{n+1} = b_n G_n v. \tag{2.1.28}$$

Since

$$G_{n+1} w = (G_n w)^{\frac{c_n+1}{n+1} / \frac{c_n}{n}} w_{n+1}^{\frac{c_n+1}{n+1}},$$

from (2.1.28) it follows that

$$G_{n+1} v = (G_n v)^{\frac{c_n+1}{n+1}(1+n/c_n)} b_n^{\frac{c_n+1}{n+1}}. \tag{2.1.29}$$

Let

$$z_n = G_n v, \quad \alpha_n = b_n^{\frac{c_n+1}{n+1}}, \quad \beta_n = \frac{c_n + 1}{n+1}(1 + \frac{n}{c_n}),$$

then by (2.1.29) we get, for all $n \in \mathbb{N}$,

$$z_{n+1} = \alpha_n z_n^{\beta_n}. \tag{2.1.30}$$

The solution of (2.1.30) can be written as

$$z_{n+1} = \left(\prod_{i=1}^{n} \alpha_i^{\prod_{j=i+1}^{n} \beta_j} \right) z_1^{\prod_{j=1}^{n} \beta_j}. \tag{2.1.31}$$

However, since

$$\prod_{j=i+1}^{n} \beta_j = \left(1 + \frac{i+1}{c_{i+1}}\right) \frac{c_{n+1}}{n+1} \prod_{j=i+2}^{n} \left(1 + \frac{c_j}{j}\right), \tag{2.1.32}$$

it follows that

$$\alpha_i^{\prod_{j=i+1}^{n} \beta_j} = b_i^{\frac{c_{n+1}}{n+1} \prod_{j=i+1}^{n}(1+c_j/j)}. \tag{2.1.33}$$

Next, since (2.1.32) is also true for $i = 0$, in view of $z_1 = v_1^{c_1}$, we have

$$z_1^{\prod_{j=1}^{n} \beta_j} = v_1^{\frac{c_{n+1}}{n+1} \prod_{j=1}^{n}(1+c_j/j)}. \tag{2.1.34}$$

Now using (2.1.33) and (2.1.34) in (2.1.31), we get

$$z_n = \left(\prod_{i=1}^{n} b_i^{\left(c_n/n \prod_{j=i+1}^{n-1}(1+c_j/j)\right)} \right) v_1^{\left(c_n/n \prod_{j=1}^{n-1}(1+c_j/j)\right)}.$$

But, $z_n = G_n v$, so the solution of (2.1.28) is stated as

$$v_{n+1} = b_n \left(\prod_{i=1}^{n-1} b_i^{\left(c_n/n \prod_{j=i+1}^{n-1}(1+c_j/j)\right)} \right) v_1^{\left(c_n/n \prod_{j=1}^{n-1}(1+c_j/j)\right)}.$$

The required inequality now follows from the observation that the operator T defined as $T\{w\} = b_n G_n w$ is monotonic on the set of sequences of positive real numbers. □

Theorem 2.1.8 (The Blandzi-Popenda-Agarwal Inequality [368]) *Let b, c, u: $\mathbb{N} \to (0, +\infty)$ and the following inequality hold for all $n \in \mathbb{N}$,*

$$u_{n+1} \leq b_n H_n u. \tag{2.1.35}$$

Then for all $n \in \mathbb{N}$,

$$u_{n+1} \leq b_n \frac{n!}{c_1} \prod_{i=1}^{n-1} \frac{b_i}{i b_i + c_{i+1}} u_1. \tag{2.1.36}$$

Proof The method of the proof is similar to that of Theorem 2.1.7. Consider the equation, for all $n \in \mathbb{N}$,

$$u_{n+1} = b_n H_n w. \tag{2.1.37}$$

Using (2.1.37) in the relation

$$H_{n+1} w = \frac{n+1}{(n/H_n w) + (c_{n+1}/w_{n+1})},$$

we may obtain, for all $n \in \mathbb{N}$,

$$H_{n+1}w = \frac{n+1}{nb_n + c_{n+1}} b_n H_n w, \tag{2.1.38}$$

which must be satisfied by a solution of (2.1.37). Solving (2.1.38) with the unknown $H_n w$ and using (2.1.37) gives the explicit form of w_{n+1}, as, for all $n \in \mathbb{N}$,

$$w_{n+1} = \left(b_n \frac{n!}{c_1} \prod_{i=1}^{n-1} \frac{b_i}{ib_i + c_{i+1}} \right) w_1.$$

We shall now show that the estimate (2.1.36) is the best possible. For this, let X be the set of positive sequences. We define the sequence of operators T_n on X as follows: for all $n \in \mathbb{N}$,

$$T_n w = b_n H_n w.$$

Let $x = \{x_n\}_1^\infty \prec \{y_n\}_1^\infty = y$, i.e., $x_n < y_n$ for all $n \in \mathbb{N}$, then we have

$$T_n x = \frac{nb_n}{\sum_{i=1}^n c_i/x_i} \le \frac{nb_n}{\sum_{i=1}^n c_i/y_i} = T_n y$$

and hence, each T_n is monotonically increasing. Thus, if $x = \{x(n, 1, x_1)\}_{n=1}^\infty$ is any solution of the inequality (2.1.35) (with the initial value x_1), and $y = \{y(n, 1, x_1)\}_{n=1}^\infty$ is the solution of the equation (2.1.37), which is also a solution of (2.1.35), then $x_i < y_i$, $i = 1, 2, \dots, k$, for each fixed k leads to

$$x_{k+1} \le T_k x \le T_k y = y_{k+1}.$$

However, since $x_1 = y_1$, an inductive argument proves that $x_n \le y_n$, for all $n \in \mathbb{N}$. Hence, if $U(x_1)$ is the set of solutions of the inequality (2.1.35) with the initial value x_1, then this set has a greatest element, which is the solution of the corresponding equation (2.1.37). □

In next result, we shall use the displacement operator E^{-k}, which is defined for $v : \mathbb{N} \to \mathbb{R}$ as follows:

$$E^{-k} v_n = \begin{cases} v_{n-k} & n > k, \\ 0, & n \le k. \end{cases}$$

This operator can also be defined for sequences with values in an arbitrary set X, with 0 as some special element of the set X satisfying the above properties.

Theorem 2.1.9 (The Blandzi-Popenda-Agarwal Inequality [368]) *Let b, u: $\mathbb{N} \to (0, +\infty)$, $c : \mathbb{N} \to \mathbb{R}$, and let the following inequality hold for all $n \in \mathbb{N}$,*

$$u_n \le E^{-k}(b_n u_n). \tag{2.1.39}$$

Then

$$u_{kj+s} \le \sum_{i=1}^{j} \left(\prod_{r=i}^{j-1} b_{rk+s} \right) c_{ik+s}, \quad s = 1, 2, \ldots, k, \quad j \in \mathbb{N} \cup \{0\} = \mathbb{N}_0. \tag{2.1.40}$$

Proof For each fixed s, the proof follows inductively. □

Remark 2.1.1 An inequality similar to (2.1.39) has appeared in an earlier work of Popenda [449].

Remark 2.1.2 Theorem 2.1.9 can be generalized for other type of sequences, e.g., matrix sequences. Indeed, for any two matrices $X = (x^{pq})$, $Y = (y^{pq})$, $1 \le p, q \le m$, we write $X \le Y$ if $x^{pq} \le y^{pq}$ for all $1 \le p, q \le m$. Let $\left(u_n^{pq}\right)_{n=1}^{\infty}$, $\left(b_n^{pq}\right)_{n=1}^{\infty}$, $\left(c_n^{pq}\right)_{n=1}^{\infty}$ be sequences of matrices of order $m \times m$, with $u_n^{pq} \ge 0$, $u_n^{pq} \ge 0$ for all $n \in \mathbb{N}$, $1 \le p, q \le m$. Furthermore, let the following inequality hold for all $n \in \mathbb{N}$,

$$\left(u_n^{pq}\right) \le \left(c_n^{pq}\right) + E^{-k}\left(\left(b_n^{pq}\right)\left(u_n^{pq}\right)\right).$$

Then

$$u_{kj+s}^{pq} \le \sum_{i=1}^{j} \left(\prod_{r=i}^{j-1} L(b_{rk+s}^{pq}) \right) (c_{ik+s}^{pq}), \quad s = 1, 2, \ldots, k, \quad j \in \mathbb{N},$$

where

$$\prod_{r=i}^{j} L(x_r^{pq}) = (x_j^{pq})(x_{j-1}^{pq}) \ldots (x_i^{pq}).$$

Ammari and Tucsnak [32] recently established the following discrete inequality with a uniform bound (see, e.g., Ruach, Zhang and Zuazua [575]).

Theorem 2.1.10 (The Ammari-Tucsnak Inequality [32]) *Let $\{a_k\}_{k=1}^{\infty}$ be a sequence of positive numbers satisfying for all $k \ge 0$,*

$$a_{k+1} \le a_k - C a_{k+1}^{2+\alpha}, \tag{2.1.41}$$

for some constants $C > 0$ and $\alpha > -1$. Then there is a constant $M = M(C, \alpha) > 0$ such that for all $k \geq 0$,

$$a_k \leq \frac{M}{(k+1)^{1/(1+\alpha)}}. \tag{2.1.42}$$

Proof Let $F_k = \frac{M}{(k+1)^{1/(1+\alpha)}}$ where $M > 0$ is to be determined later on. After a simple calculation, we obtain

$$\frac{1}{M} lim_{k\to+\infty}[(F_k - F_{k+1})k(k+2)^{1/(1+\alpha)}] = \frac{1}{1+\alpha}, \tag{2.1.43}$$

so there is a constant $k_0 > 0$ such that, for all $k \geq k_0$,

$$F_k - F_{k+1} \leq \frac{2M}{(1+\alpha)k(k+2)^{1/(1+\alpha)}}, \tag{2.1.44}$$

which implies that, for all $k \geq k_1 = \max\{k_0, 2\}$,

$$F_k - F_{k+1} \leq \frac{4}{(1+\alpha)M^{1+\alpha}} F_{k+1}^{2+\alpha}. \tag{2.1.45}$$

If we now suppose that

$$\frac{4}{(1+\alpha)M^{1+\alpha}} < C, \quad \frac{M}{(k_1+1)^{1/(1+\alpha)}} \geq a_k, \tag{2.1.46}$$

then we infer from (2.1.45), for all $k \geq k_1$,

$$F_k - F_{k+1} \leq C F_{k+1}^{2+\alpha}. \tag{2.1.47}$$

Thus it obviously suffices to show that for all $k \geq k_1$,

$$a_k \leq F_k. \tag{2.1.48}$$

We shall prove this by induction over k. In fact, if $k = k_1$, (2.1.48) follows directly from (2.1.46). If we suppose that (2.1.48) holds true for $k \leq m$, by combining (2.1.41) and (2.1.47), we obtain

$$a_{m+1} + C a_{m+1}^{2+\alpha} \leq F_{m+1} + C F_{m+1}^{2+\alpha}$$

which readily implies that $a_{m+1} \leq F_{m+1}$. □

Theorem 2.1.11 (The Yang Inequality [687]) *Let $x(n), p(n)$, and $f(n,s), g(n,s), h(n,s)$ be the same as defined in Theorem 2.1.31 in Qin [557]; and let $H(u)$ be a real-valued non-negative, non-decreasing function defined on $\mathbb{R}_+$, and there exists* *a known real-valued function* $\beta(y) \geq 0$ *on* $\mathbb{R}_+$ *such that there holds for all* $u \geq 0$ *and* $v > 0$,

$$\frac{1}{v}H(u) \leq \beta(v)H(u/v).$$

Suppose that the following inequality holds for all $n \in \mathbb{N}$,

$$x(n) \leq p(n) + \sum_{s=n_0}^{n-1}\Big(f(n,s)x(s) + g(n,s)\sum_{k=n_0}^{n-1}h(s,k)H(x(k))\Big). \tag{2.1.49}$$

Then for all $n \in \mathbb{N},\ n_0 \leq n \leq b$,

$$x(n) \leq p(n)\Big[1 + \sum_{s=n_0}^{n-1}M(n,s)G^{-1}\Big\{G(1) + \sum_{k=n_0}^{s-1}[M(n,k) + \beta(p(n))h(n,k)]\Big\}\Big], \tag{2.1.50}$$

where $M(n,s)$ *is defined by (2.1.142) of Theorem 2.1.30 in Qin [557],* G^{-1} *denotes the inverse function of* G *defined by, for all* $n \in \mathbb{N}$,

$$\Delta G(r(n)) \equiv G(r(n+1)) - G(r(n)) = \frac{\Delta r(n)}{r(n) + H(r(n))}, \tag{2.1.51}$$

and $r(n)$ *is defined for all* $n \in \mathbb{N}$; $b \in \mathbb{N}$ *satisfies, for all* $n_0 \leq n \leq b$,

$$G(1) + \sum_{k=n_0}^{s-1}[M(n,k) + \beta(p(n))h(n,k)] \in Dom\ (G^{-1}),$$

Proof Obviously, the inequality (2.1.50) holds when $n = n_0$. Fixing an arbitrary integer $r : n_0 < r \leq b$. Since $p(n)$ is non-decreasing and $p(r) > 0$, in view of the hypothesis on $H(u)$, we can obtain from (2.1.49), for all $n_0 \leq n \leq r$,

$$\begin{cases} z(n) \leq 1 + \displaystyle\sum_{s=n_0}^{n-1}f(r,s)z(s) + \sum_{s=n_0}^{n-1}g(r,s)\sum_{k=n_0}^{s-1}h(r,k)\beta(p(r))H(z(k)), \\ z(n) = x(n)/p(r). \end{cases} \tag{2.1.52}$$

Define a function $L(n)$ by the right-hand side of (2.1.52), then $L(n_0) = 1$, and since $z(n) \le L(n)$, we have, for all $n_0 \le n \le r$,

$$\begin{aligned}\Delta L(n) &\le M(r,n)\Big[z(n) + \sum_{k=n_0}^{n-1} h(r,k)\beta(p(r))H(z(k))\Big]\\ &\le M(r,n)v(n),\end{aligned} \tag{2.1.53}$$

where

$$v(n) = L(n) + \sum_{k=n_0}^{n-1} h(r,k)\beta(p(r))H(L(k)).$$

Using (2.1.53) and the inequality $v(n) \ge L(n)$, we may obtain

$$\begin{aligned}\Delta v(n) &= \Delta L(n) + h(r,n)\beta(p(r))H(L(n))\\ &\le M(r,n)v(n) + h(r,n)\beta(p(r))H(v(n))\\ &\le [M(r,n) + h(r,n)\beta(p(r))][v(n) + H(v(n))].\end{aligned} \tag{2.1.54}$$

In view of $H(u) \ge 0, v(n_0) = 1$, and $v(n)$ is non-decreasing, it follows from that, for all $n_0 \le n \le r$,

$$\Delta G(v(n)) = \frac{\Delta v(n)}{v(n) + H(v(n))} \le M(r,n) + \beta(p(r))h(r,n)$$

i.e.,

$$G(v(n+1)) \le G(v(n)) + M(r,n) + \beta(p(r))h(r,n). \tag{2.1.55}$$

Substituting successively $n = n_0, n_0 + 1, \ldots, r - 1$, in (2.1.54) and in view of $v(n_0) = 1$, we derive for all $n_0 \le n \le r$,

$$G(v(n)) \le G(1) + \sum_{s=n_0}^{n-1} [M(r,s) + \beta(p(r))h(r,s)]. \tag{2.1.56}$$

Substituting this bound for $v(n)$ in (2.1.53) and then substituting $n = n_0, n_0 + 1, \ldots, r-1$, successively in the obtained inequality and using $L(n_0) = 1$, we obtain

$$L(n) \le 1 + \sum_{s=n_0}^{n-1} M(r,s)G^{-1}\Big[G(1) + \sum_{s=n_0}^{n-1}[M(r,s) + \beta(p(r))h(r,s)]\Big]. \tag{2.1.57}$$

Thus we derive from (2.1.52) and (2.1.57),

$$x(n) \leq p(r)\Big\{1 + \sum_{s=n_0}^{n-1} M(r,s)G^{-1}\Big[G(1) + \sum_{s=n_0}^{n-1}[M(r,s) + \beta(p(r))h(r,s)]\Big]\Big\}. \tag{2.1.58}$$

Finally, letting $n = r$ in (2.1.58) and since r is an arbitrary integer from the set $\{n | n \in \mathbb{N}, n \leq b\}$, then the proof is now complete. □

Remark 2.1.3 In Theorem 2 of [453], a particular case of inequality (2.1.49) in which $f(n,s) = g(n,s) = h(n,s) = g(s)$ and $\beta(y) \equiv 1$, was discussed. Further, we note here that since in [453] it was assumed that the function $H(u)$ belongs to the so-called class S, hence an additional restriction on $p(n)$, that is, $p(n) \geq 1$, must be added there in order to ensure the desired result. We note also that, the function $H(u) = u^{1+q}$, where $q \geq 0$ being a constant, may be chosen as an example which satisfies all of the conditions in Theorem 2.1.11 with $\beta(y) = y^q$. In addition, we note that the inequality (2.1.49) is more general than those inequalities discussed in Theorems 5 and 6 in [610].

The following result is an extension of Theorem 2.1.31 in Qin [557].

Theorem 2.1.12 (The Yang Inequality [687]) *Let $x(n), p(n)$ and $f(n,s), g(n,s), h(n,s)$ be the same as defined in Theorem 2.1.31 in Qin [557]; and let $k(n,s)$ be a real-valued non-negative function defined on $\mathbb{N} \times \mathbb{N}$, and which is non-decreasing in n for $s \in \mathbb{N}$ fixed. Let $W(u)$ be a positive, non-decreasing and sub-multiplicative function for all $u > 0$, $W(0) = 0$. Suppose that the following inequality holds for all $n \in \mathbb{N}$,*

$$\begin{aligned} x(n) \leq p(n) &+ \sum_{s=n_0}^{n-1} f(n,s)x(s) + \sum_{s=n_0}^{n-1} k(n,s)W(x(s)) \\ &+ \sum_{s=n_0}^{n-1} g(n,s)\Big(\sum_{k=n_0}^{s-1} h(r,k)x(k)\Big). \end{aligned} \tag{2.1.59}$$

Then we have for all $n_0 \leq n \leq q$,

$$\begin{aligned} x(n) \leq G^{-1}\Big\{G(p(n)) &+ \sum_{s=n_0}^{n-1} k(n,s)W\Big(1 + \sum_{q=n_0}^{s-1} M(n,s)\prod_{p=n_0}^{q-1} R(n,p)\Big)\Big\} \\ &\times\Big(1 + \sum_{s=n_0}^{n-1} M(n,s)\prod_{p=n_0}^{s-1} R(n,p)\Big), \end{aligned} \tag{2.1.60}$$

where $M(n,s)$ is defined by (2.1.142) of Theorem 2.1.30 in Qin [557],

$$R(n,p) = 1 + M(n,p) + h(n,p),$$

and G^{-1} is the inverse function of G is defined by, for all $n \in \mathbb{N}$,

$$\triangle G(r(n)) = \frac{\triangle r(n)}{W(r(n))}, \tag{2.1.61}$$

here $r(n)$ is defined for all $n \in \mathbb{N}$ and the integer $q \in \mathbb{N}$ is chosen so that the expression in the bracket {} in (2.1.60) belongs to Dom (G^{-1}) when $n_0 \le n \le q$.

Proof Letting $n = n_0$ in (2.1.60), we get $x(n_0) \le p(n_0)$ which follows from the given inequality (2.1.59). Fixing an arbitrary integer $r : r \in \mathbb{N}, n_0 < r \le q$, and setting

$$J(n) = p(r) + \sum_{s=n_0}^{n-1} k(r,s)W(x(s)),$$

then we derive from (2.1.59), for all $n_0 \le n \le r$,

$$x(n) \le J(n) + \sum_{s=n_0}^{n-1} f(r,s)x(s) + \sum_{s=n_0}^{n-1} g(r,s)\Big(\sum_{p=n_0}^{s-1} h(r,p)x(p)\Big). \tag{2.1.62}$$

Since $J(n)$ is positive and non-decreasing, applying Theorem 2.1.30 in Qin [557] to (2.1.62) yields, for all $n_0 \le n \le r$,

$$x(n) \le J(n)\Big[1 + \sum_{s=n_0}^{n-1} M(r,s) \prod_{p=n_0}^{s-1} R(r,p)\Big]. \tag{2.1.63}$$

Furthermore,

$$W(x(n)) \le W(J(n))W\Big(1 + \sum_{s=n_0}^{n-1} M(r,s) \prod_{p=n_0}^{s-1} R(r,p)\Big),$$

since the function W is sub-multiplicative. Hence we have

$$\begin{aligned}\triangle G(J(n)) &= \frac{k(r,n)W(x(n))}{W(J(n))} \\ &\le k(r,n)W\Big(1 + \sum_{s=n_0}^{n-1} M(r,s) \prod_{p=n_0}^{s-1} R(r,p)\Big),\end{aligned}$$

i.e., for all $n_0 \le n \le r$,

$$G(J(n+1)) \le G(J(n)) + k(r,n)W\Big(1 + \sum_{s=n_0}^{n-1} M(r,s) \prod_{p=n_0}^{s-1} R(r,p)\Big). \quad (2.1.64)$$

Substituting successively $n = n_0, n_0 + 1, \ldots, r - 1$, in (2.1.64) and using $J(n_0) = p(r)$, then we obtain for all $n_0 \le n \le r$,

$$G(J(n)) \le G(p(r)) + \sum_{s=n_0}^{n-1} k(r,s)W\Big(1 + \sum_{s=n_0}^{n-1} M(r,s) \prod_{p=n_0}^{s-1} R(r,p)\Big).$$

Therefore, for all $n_0 \le n \le r$,

$$J(n) \le G^{-1}\Big\{G(p(r)) + \sum_{s=n_0}^{n-1} k(r,s)W\Big(1 + \sum_{s=n_0}^{n-1} M(r,s) \prod_{p=n_0}^{s-1} R(r,p)\Big)\Big\}.$$

Substituting this value for $J(n)$ in (2.1.63) and then letting $n = r$ in the obtained inequality, since the choice of r is arbitrary, then the inequality (2.1.60) is proved. □

Remark 2.1.4 If in Theorem 2.1.12, $p(n) = x_0$, here x_0 is a positive constant and $f(n,s) = g(n,s) = f(s), h(n,s) = g(s)$ and $k(n,s) = q(s)$, then we may derive Theorem 3.1 of Pachpatte [453]. We note also that the discrete inequality in Theorem 7 in Singare and Pachpatte [610] is a special case of (2.1.59).

The inequalities established in the following theorems can be used in certain situations.

Theorem 2.1.13 (The Pachpatte Inequality [515]) *Let $u(n)$, $a(n)$, $b(n)$ be real-valued non-negative functions defined for all $n \in \mathbb{N}_0$ and $L : \mathbb{N}_0 \times \mathbb{R}_+ \to \mathbb{R}_+$ be a function which satisfies the condition for all $u \ge v \ge 0$,*

$$0 \le L(n,u) - L(n,v) \le M(n,v)(u-v),$$

where $M(u,v)$ is a real-valued non-negative function defined for all $n \in \mathbb{N}_0$, $v \in \mathbb{R}_+$. If for all $n \in \mathbb{N}_0$,

$$u(n) \le a(n) + b(n) \sum_{s=n+1}^{+\infty} L(s, u(s)), \quad (2.1.65)$$

then for all $n \in \mathbb{N}_0$,

$$u(n) \le a(n) + b(n)e(n) \prod_{s=n+1}^{+\infty} [1 + M(s, a(s)b(s))], \tag{2.1.66}$$

where for all $n \in \mathbb{N}_0$,

$$e(n) = \sum_{s=n+1}^{+\infty} L(s, a(s)). \tag{2.1.67}$$

Proof Define a function $z(n)$ by

$$z(n) = \sum_{s=n+1}^{+\infty} L(s, u(s)), \tag{2.1.68}$$

then from (2.1.65) it follows

$$u(n) \le a(n) + b(n)z(s). \tag{2.1.69}$$

From (2.1.68) and (2.1.69) and the hypotheses on L, we observe that

$$\begin{aligned} z(n) &\le \sum_{s=n+1}^{+\infty} [L(s, a(s) + b(s)z(s)) - L(s, a(s)) + L(s, a(s))] \\ &\le e(n) + \sum_{s=n+1}^{+\infty} M(s, a(s))b(s)z(s). \end{aligned} \tag{2.1.70}$$

when $e(s)$ is defined by (2.1.67). Clearly, $e(n)$ is real-valued non-negative and non-increasing in $n \in \mathbb{N}_0$. Now applying Theorem 2.1.30 in Qin [557] to (2.1.70) yields

$$z(n) \le e(n) \prod_{s=n+1}^{+\infty} [1 + M(s, a(s))b(s)] \le h(n+1)W(g(n+1))W(H(v(n+1))). \tag{2.1.71}$$

Thus the desired inequality in (2.1.66) follows from (2.1.69) and (2.1.71). □

Theorem 2.1.14 (The Pachpatte Inequality [515]) *Let* $u(n)$, $a(n)$, $b(n)$ *be real-valued non-negative functions defined for all* $n \in \mathbb{N}_0$ *and* $L : \mathbb{N}_0 \times \mathbb{R}_+ \to \mathbb{R}_+$ *be a function which satisfies the condition for all* $u \ge v \ge 0$,

$$0 \le L(n, u) - L(n, v) \le M(n, v)\phi^{-1}(u - v),$$

where $M(u, v)$ is defined as in Theorem 2.1.13, $\phi : \mathbb{R}_+ \to \mathbb{R}_+$ be continuous and strictly increasing function with $\phi(0) = 0$, ϕ^{-1} is the inverse function of ϕ and for all $u, v \in \mathbb{R}_+$,

$$\phi^{-1}(uv) \leq \phi^{-1}(u)\phi^{-1}(v).$$

If for all $n \in \mathbb{N}_0$,

$$u(n) \leq a(n) + b(n)\phi\left(\sum_{s=n+1}^{+\infty} L(s, u(s))\right), \tag{2.1.72}$$

then for all $n \in \mathbb{N}_0$,

$$u(n) \leq a(n) + b(n)\phi\left(e(n)\prod_{s=n+1}^{+\infty}[1 + M(s, a(s)b(s))]\right), \tag{2.1.73}$$

where $e(n)$ is defined by (2.1.67).

Proof Define a function $z(n)$ by (2.1.68), then from (2.1.72) we derive

$$u(n) \leq a(n) + b(n)z(s). \tag{2.1.74}$$

From (2.1.68) and (2.1.74) and the hypotheses on L and ϕ, it follows that

$$\begin{aligned} z(n) &\leq \sum_{s=n+1}^{+\infty}[L(s, a(s) + b(s)\phi(z(s))) - L(s, a(s)) + L(s, a(s))] \\ &\leq e(n) + \sum_{s=n+1}^{+\infty} M(s, a(s))\phi^{-1}(b(s)\phi(z(s))) \\ &\leq e(n) + \sum_{s=n+1}^{+\infty} M(s, a(s))\phi^{-1}(b(s))z(s), \end{aligned}$$

when $e(s)$ is defined by (2.1.67). Now by following the last argument as in the proof of Theorem 2.1.13, we may get the required estimate (2.1.73). □

Theorem 2.1.15 (The Agarwal-Thandapani Inequality [18]) *Let the following inequality be satisfied:*

$$\Delta^k u(t) \leq p(t) + \sum_{j=0}^{k}\sum_{s=0}^{t-1} h_j(s)\Delta^k u(s)\Delta^j u(s), \tag{2.1.75}$$

where $p(t)$ is positive and non-decreasing. Then

$$\Delta^k u(t) \leq \frac{p(t)e^{-1}(t)}{1-\sum_{s=0}^{t-1}\phi_{11}(s)e^{-1}(s+1)},$$

where

$$e(t)=\prod_{s=0}^{t-1}[1+\phi_{12}(s)]^{-1}, \quad \phi_{11}(t)=p(t)\sum_{j=0}^{k}h_{k-j}(t)\frac{(t)^{(j)}}{j!},$$

$$\phi_{12}(t)=\sum_{j=0}^{k-1}\sum_{i=0}^{j}\Delta^j u(0)h_j(t)\frac{(t)^{(j-i)}}{(j-i)!},$$

with

$$1-\sum_{s=0}^{t-1}\phi_{11}(s)e^{-1}(s+1)>0.$$

Proof Since $p(t)$ is positive and non-decreasing, we have

$$\frac{\Delta^k u(t)}{p(t)} \leq 1+\sum_{j=0}^{k}\sum_{s=0}^{t-1}h_j(s)\frac{\Delta^k u(s)}{p(s)}\Delta^j u(s). \tag{2.1.76}$$

We define $R(t)$ as the right-hand side of (2.1.76), then

$$\begin{aligned}\Delta R(t) &\leq \sum_{j=0}^{k}h_j(t)R(t)\Delta^j u(t)=h+k(t)p(t)R^2(t)\\ &+\sum_{j=0}^{k-1}h_j(t)R(t)[\sum_{i=j}^{k-1}\frac{(t)^{i-j}}{(i-j)!}\Delta^i u(0)\\ &+\frac{1}{(k-j-1)!}\sum_{s=0}^{t-k+j}(t-s-1)^{(k-j-1)}p(s)R(s)].\end{aligned}$$

Now using the non-decreasing nature of $R(t)$ and $p(t)$, we can find

$$\Delta R(t) \leq \phi_{11}(t)R^2(t)+\phi_{12}R(t),$$

or

$$R(t+1)-[1+\phi_{12}(t)]R(t) \leq \phi_{11}(t)R^2(t).$$

Multiplying the above inequality by $e(t+1)$, we obtain

$$\Delta[R(t)e(t)] \leq \phi_{11}(t)e^{-1}(t+1)[R(t)e(t+1)]^2. \tag{2.1.77}$$

Since $R(t)$ is non-decreasing and $e(t)$ is non-increasing, we conclude when $\Delta[R(t)e(t)] \geq 0$,

$$-\Delta[R(t)e(t)]^{-1} = \int_t^{t+1} \frac{d[R(s)e(s)]}{[R(s)e(s)]^2} \leq \frac{\Delta[R(t)e(t)]}{[R(t)e(t+1)]^2},$$

and hence from (2.1.77), it follows

$$-\Delta[R(t)e(t)]^{-1} \leq \phi_{11}(t)e^{-1}(t+1). \tag{2.1.78}$$

Similarly, for all $t \in \mathbb{N}_0$, when $\Delta[R(t)e(t)] \leq 0$, we have $-\Delta[R(t)e(t)]^{-1} \leq 0$, hence obviously (2.1.78) follows.

Now summing the inequality (2.1.78) from 0 to $t_1 - 1 \in \mathbb{N}_0$, we conclude

$$R(t_1) \leq \frac{e^{-1}(t)}{1 - \sum_{s=0}^{t_1-1} \phi_{11}(s)e^{-1}(s+1)},$$

and now substituting this in (2.1.78), we can get the desired result. □

Remark 2.1.5 As in Theorem 2.1.15, it is easy to find estimates in terms of known functions for the following inequalities:

(a) Let $g_i(t) \geq 1$ $(i = 0, 1, 2, \ldots, k)$; then

$$\Delta^k u(t) \leq p(t) + \sum_{j=0}^{k} g_j(t) \sum_{s=0}^{t-1} h_j(s)\Delta^k u(s)\Delta^j u(s).$$

(b) Let $0 \leq l \leq k-1$; then

$$\Delta^k u(t) \leq p(t) + \sum_{j=0}^{k} \sum_{s=0}^{t-1} h_j(s)\Delta^l u(s)\Delta^j u(s).$$

Theorem 2.1.16 (The Agarwal-Thandapani Inequality [18]) *Let the following inequality be satisfied:*

$$\Delta^k u(t) \leq C + \sum_{j=0}^{k} \sum_{s=0}^{t-1} h_j(s)[\Delta^k u(s)]^{\alpha}[\Delta^j u(s)]^{\alpha_j}, \tag{2.1.79}$$

where α, α_j $(j = 0, 1, \ldots, k)$ are non-negative numbers and C is a positive constant.

Then

$$\Delta^k u(t) \le \frac{C}{[1 + (\alpha+\beta-1)\sum_{s=0}^{t-1} \phi_{13}(t)]^{1/\alpha+\beta-1}}, \tag{2.1.80}$$

where $\beta = \max(\alpha_0, \alpha_1, \ldots, \alpha_k)$ *such that* $\alpha+\beta > 1$, *and*

$$\phi_{13}(t) = C^{\alpha-1}\left[C^{\alpha_k} h_k(t) + \sum_{j=0}^{k-1} h_j(t)(\sum_{i=j}^{k-1} \Delta^i u(0)\frac{(t)^{(i-j)}}{(i-j)!} + C\frac{(t)^{(k-j)}}{(k-j)!})^{\alpha_1}\right],$$

if

$$1 - (\alpha+\beta-1)\sum_{s=0}^{t-1} \phi_{13}(s) > 0.$$

Proof In fact, the inequality (2.1.79) can be rewritten as

$$\frac{\Delta^k u(t)}{C} \le R(t), \tag{2.1.81}$$

where

$$R(t) = 1 + \sum_{j=0}^{k}\sum_{s=0}^{t-1} \frac{h_j(s)}{C}[\Delta^k u(s)]^{\alpha}[\Delta^j u(s)]^{\alpha_j},$$

whence by using the fact that $R(t) \ge 1$ and non-decreasing on $\mathbb{N}_0$,

$$\begin{aligned}\Delta R(t) &\le C^{\alpha-1}R^{\alpha}(t)\left[h_k(t)C^{\alpha_k}R^{\alpha_k}(t) + \sum_{j=0}^{k-1} h_j(t)(\sum_{i=j}^{k-1}\Delta^i u(0)\frac{(t)^{i-j}}{(i-j)!}\right.\\ &\quad\left. + \frac{1}{(k-j-1)!}\sum_{s=0}^{t-k+j}(t-s-1)^{(k-j-1)}CR(s))^{\alpha_1}\right]\\ &\le \phi_{13}R^{\alpha+\beta}(t),\end{aligned}$$

or

$$\frac{\Delta R(t)}{R^{\alpha+\beta}(t)} \le \phi_{13}(t).$$

Now noting that

$$\frac{\Delta[R(t)]^{1-(\alpha+\beta)}}{1-(\alpha+\beta)} = \int_t^{t+1} \frac{d[R(s)]}{[R(s)]^{\alpha+\beta}} \le \frac{\Delta R(t)}{[R(t)]^{\alpha+\beta}},$$

we can conclude

$$\Delta[R(t)]^{1-(\alpha+\beta)} \ge [1-(\alpha+\beta)]\phi_{13}(t).$$

Summing the above inequality and substituting the obtained estimate in (2.1.81), we can conclude the desired estimate (2.1.80). □

Theorem 2.1.17 (The Agarwal-Thandapani Inequality [18]) *Assume the following inequality holds,*

$$\Delta^k u(t) \le p(t) + \sum_{r=1}^{n-1} E_r(t, \sum_{j=0}^{k} \Delta^j u) + E_n(t, (\Delta^k u)^\alpha),$$

where $p(t)$ is positive and non-decreasing, $E_r(t,*)$ is defined in Theorem 2.1.55 of Qin [557], and also the number $1 \neq \alpha \ge 0$. Then

$$\Delta^k u(t) \le p(t)e^{-1}(t)\left\{1 + (1-\alpha)\sum_{s=0}^{t-1} \Delta E_n(s, p^{\alpha-1})e^{1-\alpha}(s+1)\right\}^{1/(1-\alpha)},$$

where

$$e(t) = \prod_{s=0}^{t-1}[1 + \sum_{r=1}^{n-1} \Delta E_r(t, \phi_{14})]^{-1}$$

with

$$\phi_{14}(t) = \frac{1}{p(t)}\sum_{j=0}^{k-1}\sum_{i=0}^{j} \Delta^j u(0)\frac{(i)^{(j-i)}}{(j-i)!} + \sum_{j=0}^{k} \frac{(t)^{(j)}}{j!}.$$

Proof The proof is similar to that of Theorem 2.1.16. □

Theorem 2.1.18 (The Agarwal-Thandapani Inequality [18]) *Assume the following inequality holds,*

$$\Delta^k u(t) \le p(t) + \sum_{i=1}^{n} g_i(t) \sum_{s=0}^{t-1} h_i(s) W\left(\sum_{j=0}^{k} \Delta^j u(s)\right), \tag{2.1.82}$$

where

(i) $p(t)$ *is positive and non-decreasing,*
(ii) $g_i(t) \geq 1,\ i = 1, 2, \ldots, n.$
(iii) W *is positive, continuous, non-decreasing, and sub-multiplicative.*

Then

$$\Delta^k u(t) \leq p(t) \prod_{j=1}^{n} g_j(t) G^{-1}\left(G(1) + \sum_{s=0}^{t-1} \sum_{i=1}^{n} \frac{h_i(s)}{p(s)} W(\phi_{15}(t))\right),$$

where

$$\phi_{15}(t) = p(t) \prod_{i=1}^{n} g_i(t) + \sum_{j=0}^{k-1} \sum_{i=0}^{j} \Delta^i u(0) \frac{(t)^{(j-i)}}{(j-i)!}$$
$$+ \sum_{j=0}^{k-1} \sum_{s=0}^{t-j-1} \frac{(t-s-1)^{(j)}}{j!} p(s) \prod_{i=1}^{n} g_i(s),$$
$$G(u) = \int_{u_0}^{u} \frac{ds}{W(s)}, \qquad 0 < u_0 \leq u,$$

as long as

$$G(1) + \sum_{s=0}^{t-1} \sum_{i=1}^{n} \frac{h_i(s)}{p(s)} W(\phi_{15}(s)) \in Dom\,(G^{-1}).$$

Proof The inequality (2.1.82) can be rewritten as

$$\Delta^k u(t) \leq p(t) \prod_{j=1}^{n} g_j(t) R(t), \tag{2.1.83}$$

where

$$R(t) = 1 + \sum_{i=1}^{n} \sum_{s=0}^{t-1} \frac{h_i(s)}{p(s)} W\left(\sum_{j=0}^{k} \Delta^j u(s)\right), \quad R(0) = 1.$$

Thus we obtain, since $R(t)$ is non-decreasing and $R(t) \geq 1$,

$$\Delta R(t) = \sum_{i=1}^{n} \frac{h_i(t)}{p(t)} W\left(\sum_{j=0}^{k} \Delta^j u(t)\right)$$
$$\leq \sum_{i=1}^{n} \frac{h_i(t)}{p(t)} W(\phi_{15}(t))W(R(t)).$$

From the definition of G, we may derive

$$G(R(t+1)) - G(R(t)) = \int_{R(t)}^{R(t+1)} \frac{ds}{W(s)} \leq \frac{\Delta R(t)}{W(R(t))},$$

whence

$$G(R(t+1)) - G(R(t)) \leq \sum_{i=1}^{n} \frac{h_i(t)}{p(t)} W(\phi_{15}(t)).$$

Now, summing the above inequality from 0 to $t_1 - 1 \in \mathbb{N}_0$, we conclude

$$G(R(t_1)) - G(1) \leq \sum_{s=0}^{t_1-1} \sum_{i=1}^{n} \frac{h_i(t)}{p(t)} W(\phi_{15}(s)),$$

or

$$R(t_1) \leq G^{-1}\left[G(1) + \sum_{s=0}^{t_1-1} \sum_{i=1}^{n} \frac{h_i(s)}{p(s)} W(\phi_{15}(s))\right].$$

Thus substituting this in (2.1.83), we can obtain the desired result. □

Remark 2.1.6 For several particular cases of Theorems 2.1.17 and 2.1.18, see, e.g., [290, 460, 467].

Theorem 2.1.19 (The Pachpatte Inequality [499]) *Let $y(n), f(n), g(n)$ be real-valued non-negative functions defined on $\mathbb{N}_0$ and c_1, c_2 be non-negative real constants. If for all $n \in \mathbb{N}_0$,*

$$y(n) \leq \left(c_1 + \sum_{s=0}^{n-1} f(s)y(s)\right)\left(c_2 + \sum_{s=0}^{n-1} g(s)y(s)\right), \tag{2.1.84}$$

and $c_1c_2\sum_{s=0}^{n-1}A(s)B(s) < 1$ *for all* $n \in \mathbb{N}_0$, *then for all* $n \in \mathbb{N}_0$,

$$y(n) \le \frac{c_1c_2B(n)}{1 - c_1c_2\sum_{s=0}^{n-1}A(s)B(s)}, \tag{2.1.85}$$

where for all $n \in \mathbb{N}_0$,

$$\begin{cases} A(n) = g(n)\sum_{s=0}^{n-1}f(s) + f(n)\sum_{s=0}^{n-1}g(s), \\ B(n) = \prod_{s=0}^{n-1}[1 + c_1g(s) + c_2f(s)]. \end{cases}$$

Proof First assume that c_1, c_2 are positive and define a function $z(n)$ by

$$z(n) = \left(c_1 + \sum_{s=0}^{n-1}f(s)y(s)\right)\left(c_2 + \sum_{s=0}^{n-1}g(s)y(s)\right). \tag{2.1.86}$$

From (2.1.86) and using the formula

$$\Delta[a(n)b(n)] = a(n)\Delta b(n) + b(n+1)\Delta a(n),$$

and the facts that $y(n) \le z(n)$ and $z(n)$ is monotone non-decreasing in $n \in \mathbb{N}_0$, we can obtain

$$\Delta z(n) \le [c_1g(n) + c_2f(n)]z(n) + A(n)z^2(n),$$

i.e.,

$$z(n+1) - [1 + c_1g(n) + c_2f(n)]z(n) \le A(n)z^2(n).$$

Now following the arguments used in the proof of Theorem 2.1.36 in Qin [557], we can conclude

$$z(n) \le \frac{c_1c_2B(n)}{1 - c_1c_2\sum_{s=0}^{n-1}A(s)B(s)}. \tag{2.1.87}$$

Thus the required inequality (2.1.85) now follows by using $y(n) \le z(n)$.

If c_1, c_2 are non-negative, we carry out the above arguments with $c_1+\varepsilon$ and $c_2+\varepsilon$ instead of c_1 and c_2, where $\varepsilon > 0$ is an arbitrary small constant, and subsequently pass to the limit as $\varepsilon \to 0^+$ to obtain (2.1.85). The proof is thus complete. □

Remark 2.1.7 We note that the inequality in Theorem 2.1.19 reduces to the discrete analogue of Gronwall's inequality, i.e., Theorem 1.2.25 (see also, [461, 463]) when $g(n) = 0$ and $c_2 = 1$ or $f(n) = 0$ and $c_1 = 1$.

2.2 The One-Dimensional Bihari Inequalities and Their Generalizations

Theorem 2.2.1 (The Agarwal-Thandapani Inequality [19]) *Let the following inequality be satisfied*

$$u(t) \leq p(t) + q(t) \sum_{\Omega=1}^{n} E_{\Omega}(t,u) \tag{2.2.1}$$

where

$$E_{\Omega}(t,u) = \sum_{t_1=0}^{t-1} f_{\Omega_1}(t_1) \sum_{t_2=0}^{t_1-1} f_{\Omega_2}(t_2) \cdots \sum_{t=0}^{t_{\Omega-1}-1} f_{\Omega\Omega}(t_\Omega) u(t_\Omega) \tag{2.2.2}$$

for all $t \in \mathbb{N}$. *Then for all* $t \in \mathbb{N}$

$$u(t) \leq p(t) + q(t) \sum_{s=0}^{t-1} \Big(\sum_{\Omega=1}^{n} \Delta E_{\Omega}(s,p)\Big) \times \prod_{\tau=s+1}^{t-1} \Big[1 + \sum_{\Omega=1}^{n} \Delta E_{\Omega}(\tau,q)\Big]. \tag{2.2.3}$$

Proof Define $m(t)$ as follows

$$m(t) = \sum_{\Omega=1}^{n} E_{\Omega}(t,u), \quad m(0) = 0$$

and hence

$$\Delta m(t) = \sum_{\Omega=1}^{n} \Delta E_{\Omega}(t,u)$$

where

$$\Delta E_{\Omega}(t,u) = f_{\Omega_1}(t_1) \sum_{t_2=0}^{t-1} f_{\Omega_2}(t_2) \cdots \sum_{t=0}^{t_{\Omega-1}-1} f_{\Omega\Omega}(t_\Omega) u(t_\Omega)$$

From the assumptions on the functions $\Delta m(t)$, hence $m(t)$ is non-decreasing on $\mathbb{N}$. Hence we find

$$\begin{aligned}\Delta m(t) &\le \sum_{\Omega=1}^{n} \Delta E_\Omega(t, p+qm)\\ &\le \sum_{\Omega=1}^{n} \Delta E_\Omega(t,p) + \sum_{\Omega=1}^{n} \Delta E_\Omega(t,qm)\\ &\le \sum_{\Omega=1}^{n} \Delta E_\Omega(t,p) + m(t)\sum_{\Omega=1}^{n} \Delta E_\Omega(t,q)\end{aligned}$$

and thus

$$m(t+1) - [1 + \sum_{\Omega=1}^{n} \Delta E_\Omega(t,q)]m(t) \le \sum_{\Omega=1}^{n} \Delta E_\Omega(t,p).$$

Multiplying the above inequality by $\prod_{s=0}^{t}[1+\sum_{\Omega}^{n} \Delta E_\Omega(s,q)]^{-1}$ and summing over from 0 to $t-1$, we get

$$\begin{aligned}m(t)\prod_{s=0}^{t-1}[1+\sum_{\Omega=1}^{n} \Delta E_\Omega(s,q)]^{-1} &\le \sum_{s=0}^{t-1}(\sum_{\Omega=1}^{n} \Delta E_\Omega(s,p))\\ &\times \prod_{\tau=s+1}^{t-1}[1+\sum_{\Omega=1}^{n} \Delta E_\Omega(\tau,q)]^{-1}\end{aligned}$$

whence

$$m(t) \le \sum_{s=0}^{t-1}(\sum_{\Omega=1}^{n} \Delta E_\Omega(s,p)) \times \prod_{\tau=s+1}^{t-1}[1+\sum_{\Omega=1}^{n} \Delta E_\Omega(\tau,q)].$$

Substituting this estimate in (2.2.1), we obtain the desired inequality (2.2.3). □

Theorem 2.2.2 (The Agarwal-Thandapani Inequality [19]) *Let the following inequality be satisfied for all $t \in \mathbb{N}$,*

$$u(t) \le p(t)[u_0 + \sum_{\Omega=1}^{n-1} E_\Omega(t,u) + E_n(t,u^\alpha)] \tag{2.2.4}$$

where $u_0 \geq 0$ *and* $0 \leq \alpha < 1$. *Then for all* $t \in \mathbb{N}$

$$u(t) \leq p(t)e^{-1}(t)\left\{u_0^{1-\alpha} + (1-\alpha)\sum_{s=0}^{t-1} \Delta E_n(s, p^\alpha)[e(s+1)]^{1-\alpha}\right\}^{1/1-\alpha} \tag{2.2.5}$$

where

$$e(t) = \prod_{s=0}^{t-1}[1 + \sum_{\Omega=1}^{n-1} \Delta E_\Omega(s, p)]^{-1}.$$

Proof Let $R(t)$ be the term inside the bracket of right-hand side of (2.2.4). Then

$$\begin{aligned} u(t) &\leq p(t)R(t), \\ \Delta R(t) &= \sum_{\Omega=1}^{n-1} \Delta E_\Omega(t, u) + \Delta E_n(t, u^\alpha) \\ &\leq \sum_{\Omega=1}^{n-1} \Delta E_\Omega(t, pR) + \Delta E_n(t, p^\alpha R^\alpha) \\ &\leq \sum_{\Omega=1}^{n-1} \Delta E_\Omega(t, p)R(t) + \Delta E_n(t, p^\alpha)R^\alpha(t) \end{aligned}$$

or

$$R(t-1) - [1 + \sum_{\Omega=1}^{n-1} \Delta E_\Omega(t, p)]R(t) \leq \Delta E_n(t, p^\alpha)R^\alpha(t).$$

Multiplying the above inequality by $e(t+1)$, we obtain

$$\Delta[R(t)e(t)] = R(t+1)e(t+1) - R(t)e(t) \tag{2.2.6}$$

$$\leq \Delta E_n(t, p^\alpha) \times e^{1-\alpha}(t+1)[R(t)e(t+1)]^\alpha. \tag{2.2.7}$$

For all $t \in \mathbb{N}$ when $\Delta[R(t)e(t)] \geq 0$, we have

$$\begin{aligned} \frac{\Delta[R(t)e(t)]^{1-\alpha}}{1-\alpha} &= \int_t^{t+1} \frac{d[R(s)e(s)]}{[R(s)e(s)]^\alpha} \\ &\leq \frac{\Delta[R(t)e(t)]}{[R(t)e(t)]^\alpha} \end{aligned}$$

and from (2.2.6), we obtain

$$\frac{\Delta[R(t)e(t)]^{1-\alpha}}{1-\alpha} \le \Delta E_n(t, p^{\alpha}) e^{1-\alpha}(t+1). \tag{2.2.8}$$

Similarly, for all $t \in \mathbb{N}$ when $\Delta[R(t)e(t)] \le 0$, we have $\frac{\Delta[R(t)e(t)]^{1-\alpha}}{1-\alpha} \le 0$. Hence (2.2.7) follows. Summing up both the sides of (2.2.7) from 0 to $t-1$, we get the desired result. □

Theorem 2.2.3 (The Agarwal Inequality [10]) *Let for all $k \in \mathbb{N}_a$ the following inequality be satisfied*

$$u(k) \le p(k) + \sum_{i=1}^{r_1} E_i(k, u) + \sum_{i=1}^{r_2} p_i(k) \sum_{l=a}^{k-1} q_i W_i(u(l)), \tag{2.2.9}$$

where (i) $p(k) \ge 1$ and non-decreasing; (ii) $p_i(k) \ge 1, 1 \le i \le r_2$; (iii) $W_i \in \mathcal{F}_1, 1 \le i \le r_2$. Then for all $k \in \mathbb{N}_a$,

$$u(k) \le p(k) v(k) e(k) \prod_{i=1}^{r_2} J_i(k), \tag{2.2.10}$$

where

$$e(k) = \prod_{i=1}^{r_2} P_i(k),\ v(k) = \prod_{l=a}^{k-1} \Big(1 + \sum_{i=1}^{r_1} \Delta E_i(l, e)\Big),$$

$$J_0(k) = 1,\ J_i(k) = G_j^{-1}\Big(G_j(1) + \sum_{l=a}^{k-1} q_j(l) v(l) e(l) \prod_{i=1}^{j-1} J_i(l)\Big),\ 1 \le j \le r_2$$

and

$$G_j(w) = \int_{w_0}^{w} \frac{dt}{W_j(t)}, \quad w \ge w_0 \ge 1$$

with

$$G_j(1) + \sum_{l=a}^{k-1} q_j(l) v(l) e(l) \prod_{i=1}^{j-1} J_i(l) \in Dom\ (G_j^{-1}), \quad 1 \le j \le r_2.$$

Proof From the hypotheses, inequality (2.2.9) implies that

$$\frac{u(k)}{e(k)} \leq p^*(k) + \sum_{i=1}^{r_1} E_i\left(k, e\frac{u}{e}\right),$$

where

$$p^*(k) = p(k) + \sum_{i=1}^{r_2} \sum_{l=a}^{k-1} q_i(l) W_i(u(l)).$$

Since p^* is non-decreasing, as $u(k) \leq p(k)q(k)\Pi_{l=a}^{k-1}(1 + q(l)f(l))$, we can get

$$\frac{u(k)}{e(k)} \leq p^*(k)v(k). \tag{2.2.11}$$

Now by using the definition of class $\mathcal{F}_1$, (2.2.11) implies that

$$w(k) \leq 1 + \sum_{i=1}^{r_2} \sum_{l=a}^{k-1} q_i(l)e(l)v(l)W_i(w(l)),$$

where $w(k) = \frac{u(k)}{p(k)v(k)e(k)}$. Thus it is sufficient to show that $w(k) \leq \prod_{i=1}^{r_2} J_i(k)$, which will be proved this by induction. For $r_2 = 1$, we have

$$w(k) \leq 1 + \sum_{l=a}^{k-1} q_1(l)e(l)v(l)W_1(w(l)). \tag{2.2.12}$$

Let $z(k)$ be the right-hand side of (2.2.12), then using non-decreasing nature of W_1, we obtain

$$\Delta z(k) \leq q_1(k)e(k)v(k)W_1(z(k)), \quad z(a) = 1. \tag{2.2.13}$$

Next, from the definition on G_1, it follows that

$$\Delta G_1(z(k)) = \int_{z(k)}^{z(k+1)} \frac{dt}{W_1(t)} \leq \frac{\Delta z(k)}{W_1(z(k))}. \tag{2.2.14}$$

Using (2.2.14) in (2.2.13) and summing, we may obtain

$$z(k) \leq G_1^{-1}\left(G_1(1) + \sum_{l=a}^{k-1} q_1(l)e(l)v(l)\right) = J_1(k)$$

which shows that the result is true for $r_2 = 1$. Now assuming that the result is true for some j such that $1 \leq j \leq r_2 - 1$, then to prove for $j + 1$, we have

$$w(k) \leq \Big(1 + \sum_{l=a}^{k-1} q_{j+1}(l)e(l)v(l)W_{j+1}(w(l))\Big) + \sum_{i=1}^{j}\sum_{l=a}^{k-1} q_i(l)e(l)v(l)W_i(w(l)).$$

Since the part inside the bracket is greater than 1 and non-decreasing, we may obtain

$$w(k) \leq \Big(1 + \sum_{l=a}^{k-1} q_{j+1}(l)e(l)v(l)W_{j+1}(w(l))\Big)\prod_{i=1}^{j} J_i(k),$$

which also gives us

$$\frac{w(k)}{\prod_{i=1}^{j} J_i(k)} \leq 1 + \sum_{l=a}^{k-1} q_{j+1}(l)e(l)v(l)\prod_{i=1}^{j} J_i(l)W_{j+1}\Big(\frac{w(l)}{\prod_{i=1}^{j} J_i(k)}\Big).$$

From this $w(k) \leq \prod_{i=1}^{j+1} J_i(k)$ follows by using the same arguments as for the case $r_2 = 1$. This hence completes the proof. □

Theorem 2.2.4 (The Agarwal Inequality [10]) *In addition to the hypotheses of Theorem 2.2.3, let $p_i(k)$, $1 \leq i \leq r_2$ be non-decreasing for all $k \in \mathbb{N}_a$. Then, for all $k \in \mathbb{N}_a$,*

$$u(k) \leq p(k)v^*(k)\prod_{i=1}^{r_2} J_i^*(k),$$

where $v^*(k)$ is the same as $v(k)$ in Theorem 2.2.3 with $e(k) = 1$;

$$J_0^*(k) = 1, J_j^*(k) = p_j(k)G_j^{-1}\Big(G_j(1) + \sum_{l=a}^{k-1} q_j(l)v^*(l)p_j(l)\prod_{i=1}^{j-1} J_i^*(l)\Big),\ 1 \leq j \leq r_2,$$

as long as

$$G_j(1) + \sum_{l=a}^{k-1} q_j(l)v^*(l)p_j(l)\prod_{i=1}^{j-1} J_i^*(l) \in Dom\left(G_j^{-1}\right),\ 1 \leq j \leq r_2,$$

and G_j, $1 \leq j \leq r_2$, are the same as in Theorem 2.2.3.

Proof The proof is similar to that of Theorem 2.1.3. □

Theorem 2.2.5 (The Agarwal Inequality [10]) *Assume the following inequality holds for all* $k \in \mathbb{N}_a$,

$$u(k) \leq p(k) + \sum_{i=1}^{r_1} E_i(k, u) + \sum_{i=1}^{r_2} E_i(k, W_1(u)), \tag{2.2.15}$$

where (i) $p(k) \geq 1$ *and is non-decreasing, (ii)* $W_1 \in \mathcal{F}_1$. *Then for all* $k \in \mathbb{N}_a$,

$$u(k) \leq p(k)v^*(k)G_1^{-1}\Big(G_1(1) + \sum_{i=1}^{r_2} E_i(k, v^*)\Big)$$

as long as

$$G_1(1) + \sum_{i=1}^{r_2} E_i(k, v^*) \in Dom\left(G_1^{-1}\right),$$

where G_1 *is the same as in Theorem 2.2.3.*

Proof The proof is similar to that of Theorem 2.2.3. □

Theorem 2.2.6 (The Agarwal Inequality [10]) *Let in Theorem 2.2.5 hypotheses (i) and (ii) be replaced by (i)* $p(k)$ *is positive and non-decreasing, (ii)* W_1 *is positive, continuous, non-decreasing and sub-multiplicative on* $[0, +\infty)$. *Then for all* $k \in \mathbb{N}_a$,

$$u(k) \leq p(k)v^*(k)G_1^{-1}\Big(G_1(1) + \sum_{i=1}^{r_2} E_i\Big(k, \frac{W_1(pv^*)}{p}\Big)\Big) \tag{2.2.16}$$

as long as

$$G_1(1) + \sum_{i=1}^{r_2} E_i\Big(k, \frac{W_1(pv^*)}{p}\Big) \in Dom\left(G_1^{-1}\right),$$

where G_1 *is the same as in Theorem 2.2.3.*

Proof Noting that $u(k) \leq p(k)q(k)\Pi_{l=a}^{k-1}(1 + q(l)f(l))$, we get

$$u(k) \leq \Big(p(k) + \sum_{i=1}^{r_2} E_i(k, W_1(u))\Big)v^*(k),$$

which implies

$$\frac{u(k)}{p(k)v^*(k)} \leq 1 + \sum_{i=1}^{r_2} E_i\Big(k, W_1\Big(\frac{u}{pv^*}pv^*\Big)/p\Big). \tag{2.2.17}$$

Let $w(k)$ be the right-hand side of (2.2.17), then

$$\Delta w(k) = \sum_{i=1}^{r_2} \Delta E_i\Big(k, W_1\Big(\frac{u}{pv^*}pv^*\Big)/p\Big)$$
$$\leq \sum_{i=1}^{r_2} \Delta E_i(k, W_1(pv^*)/p) W_1(w(k))$$

where we have used the same arguments as in Theorem 2.2.3. Thus we obtain

$$w(k) \leq G_1^{-1}\Big(G_1(1) + \sum_{i=1}^{r_2} \Delta E_i(k, W_1(pv^*)/p)\Big)$$

which yields the inequality (2.2.16). □

Theorem 2.2.7 (The Agarwal Inequality [10]) *Assume the following inequality holds for all* $k \in \mathbb{N}_a$,

$$u(k) \leq p(k) + q(k)h\Big(\sum_{l=a}^{k-1} f(l)W(u(l))\Big), \tag{2.2.18}$$

where the function h, W *are continuous, positive and non-decreasing in* $[0, +\infty)$*. Furthermore, in addition,* W *is sub-additive and sub-multiplicative. Then for all* $k \in \mathbb{N}_a$,

$$u(k) \leq p(k) + q(k)h\Big\{G^{-1}\Big(G\Big(\sum_{l=a}^{k-1} f(l)W(p(l))\Big) + \sum_{l=a}^{k-1} f(l)W(q(l))\Big)\Big\}, \tag{2.2.19}$$

where

$$G(w) = \int_{w_0}^{w} \frac{dt}{W(h(t))}, \quad w \geq w_0 \geq 0, \tag{2.2.20}$$

with

$$G\Big(\sum_{l=a}^{k-1} f(l)W(p(l))\Big) + \sum_{l=a}^{k-1} f(l)W(q(l)) \in Dom\,(G^{-1}). \tag{2.2.21}$$

Proof We leave the proof to the reader as an exercise. □

Theorem 2.2.8 (The Agarwal Inequality [10]) *Assume the following inequality holds for all* $k \in \mathbb{N}_a$,

$$u(k) \leq p(k) + q(k)W^{-1}\Big(\sum_{l=a}^{k-1} f(l)W(u(l))\Big), \tag{2.2.22}$$

where the function W *is increasing, convex and sub-multiplicative on* $[0, +\infty)$ *and* $W(0) = 0, \lim_{u\to+\infty} W(u) = +\infty$. *Then for all* $k \in \mathbb{N}_a$,

$$u(k) \leq p(k) + q(k)W^{-1}\Big(\sum_{l=a}^{k-1} \alpha(l)W(up(l)\alpha^{-1}(l)) \prod_{\tau=l+1}^{k-1} (1 + \beta(\tau)W(q(\tau)\beta^{-1}(\tau))f(\tau))\Big), \tag{2.2.23}$$

where the functions $\alpha(k), \beta(k)$ *are positive and* $\alpha(k) + \beta(k) = 1$ *for all* $k \in \mathbb{N}_a$.

Proof We leave the proof to the reader as an exercise. □

Theorem 2.2.9 (The Agarwal Inequality [10]) *Assume the following inequality holds for all* $k \in \mathbb{N}_a$,

$$u(k) \leq p(k) + \sum_{l=a}^{k-1} q(k, l)W(u(l)), \tag{2.2.24}$$

where the function W *is continuous, positive and non-decreasing on* $[0, +\infty)$. *Then for all* $k \in \mathbb{N}_a$,

$$u(k) \leq G^{-1}\Big(G(P(k)) + \sum_{l=a}^{k-1} Q(k, l)\Big), \tag{2.2.25}$$

where $P(k) = max\{p(\tau) : \tau \in \mathbb{N}(a, k)\}$ *and* $Q(k, l) = max\{q(\tau, l) : \tau \in \mathbb{N}(a, k)\}$, *and*

$$G(P(k)) + \sum_{l=a}^{k-1} Q(k, l) \in Dom\ (G^{-1}). \tag{2.2.26}$$

Proof We leave the proof to the reader as an exercise. □

Theorem 2.2.10 (The Agarwal Inequality [10]) *Assume such that* $k \leq r$, *and the following inequality holds for all* $k, r \in \mathbb{N}_a$,

$$u(r) \geq u(k) - q(r) \sum_{l=k+1}^{r} f(l)W(u(l)), \tag{2.2.27}$$

where the function W is continuous, positive and non-decreasing on $[0, +\infty)$. Then for all $k, r \in \mathbb{N}_a, k \le r$,

$$u(r) \ge G^{-1}\Big(G(u(k)) - q(r)\sum_{l=k+1}^{r} f(l)\Big), \tag{2.2.28}$$

where $G(w) = \int_{w_0}^{w} \frac{dt}{W(t)}, w \ge w_0 \ge 0$, for arbitrary $w_0 \ge 0$, and

$$G(u(k)) - q(r)\sum_{l=k+1}^{r} f(l) \in \mathrm{Dom}\,(G^{-1}). \tag{2.2.29}$$

Proof We leave the proof to the reader as an exercise. □

Theorem 2.2.11 (The Agarwal Inequality [10]) *Assume such that $k \le r$, the following inequality holds for all $k, r \in \mathbb{N}_a$,*

$$u(r) \ge u(k) - q(r)W^{-1}\Big(\sum_{l=k+1}^{r} f(l)W(u(l))\Big), \tag{2.2.30}$$

where the function W is positive, increasing, convex and sub-multiplicative on $(0, +\infty)$ and $\lim_{u\to+\infty} W(u) = +\infty$. Then for all $k, r \in \mathbb{N}_a, k \le r$,

$$u(r) \ge \alpha(r)W^{-1}\Big(\alpha^{-1}(r)W(u(k))\prod_{l=k+1}^{r}(1 + \beta(r)W(q(r)\beta^{-1}(r))f(l))^{-1}\Big), \tag{2.2.31}$$

where the function $\alpha(k), \beta(k)$ are positive and $\alpha(k) + \beta(k) = 1$ for all $k \in \mathbb{N}_a$.

Proof We leave the proof to the reader as an exercise. □

We now establish two nonlinear extensions of Theorem 2.1.34 in Qin [557] which are useful for some situations.

Theorem 2.2.12 (The Yang Inequality [692]) *Assume all hypotheses of Theorem 2.1.34 in Qin [557] hold and $H : [0, +\infty) \to [0, +\infty)$ is strictly increasing and sub-additive, with $H(0) \equiv 0$. Suppose that the following nonlinear discrete inequality holds for all $n \in \mathbb{N}$,*

$$x(n) \le p(n) + H^{-1}\left\{\sum_{j=1}^{q}\sum_{i=1}^{r_j} J_i^{(j)}(n; H(x))\right\}, \tag{2.2.32}$$

where H^{-1} denotes the inverse of H. Then for all $n \in \mathbb{N}_1 \subset \mathbb{N}$, we have

$$x(n) \leq H^{-1}\left\{H(p(n))\prod_{j=1}^{q}\left(\prod_{i=1}^{r_j} B_j^{(i)}(n)\right)\right\}, \tag{2.2.33}$$

where $B_j^{(i)}(n)$ are the same as defined in Theorem 2.1.34 in Qin [557], and $\mathbb{N}_1$ is chosen so that when $n \in \mathbb{N}_1$,

$$H(p(n))\prod_{i=1}^{q}\left(\prod_{i=1}^{r_j} B_j^{(i)}(n)\right) \in Dom\,(H^{-1}).$$

Proof We may easily derive from (2.2.32) that, for all $n \in \mathbb{N}$,

$$\begin{aligned} y(n) &\leq H\left(p(n) + H^{-1}\left\{\sum_{j=1}^{q}\sum_{i=1}^{r_j} J_i^{(j)}(n; y)\right\}\right) \\ &\leq H(p(n)) + \sum_{j=1}^{q}\sum_{i=1}^{r_j} J_i^{(j)}(n; y), \end{aligned} \tag{2.2.34}$$

where $y(n) \equiv H(x(n))$, since H is non-decreasing and sub-additive. Applying Theorem 2.1.34 in Qin [557] to (2.2.34) yields, for all $n \in \mathbb{N}$,

$$H(x(n)) \leq H(p(n))\prod_{j=1}^{q}\left(\prod_{i=1}^{r_j} B_j^{(i)}(n)\right), \tag{2.2.35}$$

where $B_j^{(i)}(n)$ are the same as in Theorem 2.1.34 in Qin [557]. Now, the desired inequality (2.2.33) follows from (2.2.35) immediately, since H^{-1} is non-decreasing. The choice of $\mathbb{N}_1$ is obvious. □

Theorem 2.2.13 (The Yang Inequality [692]) *Assume all hypotheses of Theorem 2.2.12 are satisfied. Suppose further that H is also sub-multiplicative. If the following inequality holds for all $n \in \mathbb{N}$,*

$$x(n) \leq p(n) + g(n)H^{-1}\left\{\sum_{j=1}^{q}\sum_{i=1}^{r_j} J_i^{(j)}(n; H(x))\right\}, \tag{2.2.36}$$

where $g(n)$ is a real-valued non-negative function on $\mathbb{N}$, then for all $n \in \mathbb{N}_2 \subset \mathbb{N}$, we have for all $n \in \mathbb{N}_2$,

$$x(n) \le H^{-1}\left\{H(p(n))\prod_{j=1}^{q}\left(\prod_{i=1}^{r_j}\tilde{B}_j^{(i)}(n)\right)\right\}, \tag{2.2.37}$$

where $\tilde{B}_j^{(i)}(n)$ are obtained from $B_j^{(i)}(n)$ by replacing the functions $f_{i1}^{(j)}(n,s)$ by $H(g(n))f_{i1}^{(j)}(n,s)$, respectively, and $\mathbb{N}_2$ is chosen so that for all $n \in \mathbb{N}_2$,

$$H(p(n))\prod_{i=1}^{q}\left(\prod_{i=1}^{r_j}\tilde{B}_j^{(i)}(n)\right) \in Dom\ (H^{-1}).$$

Proof We observe from (2.2.36) that, for all $n \in \mathbb{N}$,

$$H(x(n)) \le H(p(n)) + H(g(n))\sum_{j=1}^{q}\sum_{i=1}^{r_j} J_i^{(j)}(n; H(x)),$$

since H is non-decreasing, sub-additive, and sub-multiplicative. The last inequality can be rewritten as, for all $n \in \mathbb{N}$,

$$y(n) \le H(p(n)) = \sum_{j=1}^{q}\sum_{i=1}^{r_j} \tilde{J}^{(j)}(n; y(n)), \tag{2.2.38}$$

where $y(n) \equiv H(x(n))$, and $\tilde{J}^{(j)}(n;y)$ are obtained from $\tilde{J}^{(j)}(n;y)$ by changing the functions $\tilde{J}_{i_1}^{(j)}(n;s)$ to $H(g(n))\tilde{J}_{i_1}^{(j)}(n;s)$, respectively. Now applying Theorem 2.1.34 in Qin [557] to (2.2.38) yields, for all $n \in \mathbb{N}$,

$$H(x(n)) \le H(p(n))\prod_{j=1}^{q}\left(\prod_{i=1}^{r_j}\tilde{B}_j^{(i)}(n)\right), \tag{2.2.39}$$

where $\tilde{B}_j^{(i)}(n)$ are as defined in (2.2.37). Thus, the bound on $x(n)$ in (2.2.37) follows from (2.2.39) immediately, since H^{-1} is non-decreasing. The choice of $\mathbb{N}_2$ is obvious. □

We note that if we set $H(z) \equiv z$ in Theorems 2.2.12 and 2.2.13, we obtain Theorem 2.1.34 in Qin [557]. We notice that we can apply Theorem 2.1.34 in Qin [557] to extend some results of Yang [687] to contain finite difference equations that involve multiple summations.

Theorem 2.2.14 (The Pachpatte Inequality [515]) *Let $u(n)$, $a(n)$, $b(n)$ be real-valued non-negative functions defined for all $n \in \mathbb{N}_0$. Let $W(r)$ be a real-valued* *continuous, positive, non-decreasing, sub-additive and sub-multiplicative function on* $\mathbb{R}_+$ *and* $H(r)$ *be a real-valued, continuous, positive and non-decreasing function on* $\mathbb{R}_+$. *If for all* $n \in \mathbb{N}_0$,

$$u(n) \leq f(n) + g(n)H\left(\sum_{s=n+1}^{+\infty} h(s)W(u(s))\right), \tag{2.2.40}$$

then for all $n,\ n_1 \in \mathbb{N}_0, 0 \leq n \leq n_1$,

$$u(n) \leq f(n) + g(n)H\left(G^{-1}\left[G(\sum_{s=n+1}^{+\infty} W(f(s)) + \sum_{s=n+1}^{+\infty} h(s)W(g(s)))\right]\right), \tag{2.2.41}$$

where

$$G(r) = \int_{r_0}^{r} \frac{ds}{W(H(s))}, \quad r > 0, \tag{2.2.42}$$

$r_0 > 0$ *is arbitrary,* G^{-1} *is the inverse function of* G *and for all* $n,\ n_1 \in \mathbb{N}_0$, $0 \leq n \leq n_1$,

$$\sum_{s=n+1}^{\infty} h(s)W(f(s)) + \sum_{s=n+1}^{\infty} h(s)W(g(s)) \in Dom\ (G^{-1}).$$

Proof Define a function $z(n)$ by

$$z(n) = \sum_{s=n+1}^{+\infty} h(s)W(u(s)), \tag{2.2.43}$$

then from (2.2.40) it follows

$$u(n) \leq f(n) + g(n)H(z(s)). \tag{2.2.44}$$

Hence by (2.2.43) and (2.2.44), we get

$$\begin{aligned} z(n) &\leq \sum_{s=n+1}^{+\infty} h(s)W(f(s) + g(s)H(z(s))) \\ &\leq \sum_{s=n+1}^{+\infty} h(s)[W(f(s)) + W(g(s))W(H(z(s)))] \\ &\leq \sum_{s=n+1}^{+\infty} h(s)W(f(s)) + \sum_{s=n+1}^{+\infty} W(g(s))W(H(z(s))). \end{aligned} \tag{2.2.45}$$

Define a function $v(n) = \epsilon + y(n)$, where $y(n)$ is defined by the right-hand side of (2.2.45) and $\epsilon > 0$ is an arbitrarily small constant. Then $z(n) \leq v(n)$ and

$$\begin{aligned} v(n) - v(n+1) &= h(n+1)W(g(n+1))W(H(z(n+1))) \\ &\leq h(n+1)W(g(n+1))W(H(v(n+1))). \end{aligned} \tag{2.2.46}$$

Thus from (2.2.42) and (2.2.44) it follows

$$\begin{aligned} G(v(n)) - G(v(n+1)) &= \int_{v(n+1)}^{v(n)} \frac{ds}{W(H(s))} \\ &\leq \frac{[v(n) - v(n+1)]}{W(H(v(n+1)))} \\ &\leq h(n+1)W(g(n+1)). \end{aligned} \tag{2.2.47}$$

Substituting $n = s$ and taking the sum over s from n to $p-1$ ($p \geq n+1$ is arbitrary in $\mathbb{N}_0$), we may obtain

$$G(v(n)) - G(v(n+1)) \leq \sum_{s=n+1}^{p} h(s)W(g(s)). \tag{2.2.48}$$

Noting that $\lim_{p\to+\infty} v(p) = \sum_{s=1}^{+\infty} h(s)W(f(s)) + \epsilon$ and by taking $p \to +\infty$ in (2.2.48), we may get

$$v(n) \leq G^{-1}[G(\sum_{s=1}^{+\infty} h(s)W(f(s)) + \epsilon) + \sum_{s=n+1}^{+\infty} h(s)W(g(s))]. \tag{2.2.49}$$

The desired inequality in (2.2.41) follows from the fact that $z(n) \leq v(n)$, $\epsilon \to 0$ in (2.2.49) and (2.2.44). The sub-domain $0 \leq n \leq n_1$ is obvious. □

Let us denote by $S(\mathbb{N}_0, \mathbb{R}_+)$ the class of all sequences $\{x_n\}$ with $x_n \geq 0$ for all $n \in \mathbb{N}_0$, the class $C(\mathbb{N}_0, \mathbb{R}_+)$ consists of all real-valued, non-negative and continuous functions defined on $\mathbb{R}_+$.

Theorem 2.2.15 (The Yang Inequality [695]) *Let $\varphi \in C(\mathbb{R}_+, \mathbb{R}_+)$ be strictly increasing function with $\varphi(+\infty) = +\infty$ and $\psi \in C(\mathbb{R}_+, \mathbb{R}_+)$ be non-decreasing. Let $c \geq 0$ be a real constant. If $v, f \in S(\mathbb{N}_0, \mathbb{R}_+)$, and the discrete inequality holds for all $n \in \mathbb{N}_0$,*

$$\varphi(v_n) \leq c + \sum_{s=0}^{n-1} f_s\psi(v_s), \tag{2.2.50}$$

then for all $n \in \mathbb{N}_0, n \le M$,

$$v_n \le \phi^{-1}\{G^{-1}[G(c) + \sum_{s=0}^{n-1} f_s]\}, \tag{2.2.51}$$

where G^{-1} *and* φ^{-1} *are the inverse functions of* G, φ, *respectively, and*

$$G(z) := \int_{z_0}^{z} \frac{ds}{\psi[\varphi^{-1}(s)]}, \quad z \ge z_0, \ z > 0,$$

and M *is chosen so that if* $n \le M$, $n \in \mathbb{N}_0$,

$$G(c) + \sum_{s=0}^{n-1} f_s \in Dom\ (G^{-1}). \tag{2.2.52}$$

Proof We define a sequence $y = \{y_n\} \in S(N_0, \mathbb{R}_+)$ by

$$y_n := c + \sum_{s=0}^{n-1} f_s \psi(v_s), \ n \in \mathbb{N}_0, \tag{2.2.53}$$

Then, for any positive number ε, $\{y_n + \varepsilon\}$ is a non-decreasing sequence consists of positive numbers. From inequality (2.2.50) we have

$$v_n \le \varphi^{-1}[y_n] \le \varphi^{-1}[y_n + \varepsilon], \quad n \in \mathbb{N}_0, \tag{2.2.54}$$

since φ^{-1} is non-negative and increasing on $\mathbb{R}_+$) with $\varphi^{-1}(+\infty) = +\infty$.

Using (2.2.54) we derive from the sequence $\{y_n + \varepsilon\}$ that

$$\Delta(y_n + \varepsilon) = \Delta(y_n) = f_n \psi(v_n) \le f_n \psi\{\varphi^{-1}(y_n + \varepsilon)\}, \quad n \in \mathbb{N}_0, \tag{2.2.55}$$

because ψ is non-negative and non-decreasing on $\mathbb{R}_+$). In view of the fact that for all $n \in \mathbb{N}_0$,

$$\psi\{\varphi^{-1}(y_n + \varepsilon)\} \ge \{\varphi^{-1}(y_0 + \varepsilon)\} = \psi\{\varphi^{-1}(c + \varepsilon)\} > 0,$$

from (2.2.55) we obtain

$$\frac{\Delta(y_n + \varepsilon)}{\psi\{\varphi^{-1}(y_n + \varepsilon)\}} \le f_n, \quad n \in \mathbb{N}_0.$$

Hence, by the definition of definite integrals, from the last relation, we have for all $n \in \mathbb{N}_0$,

$$\int_c^{y_n+\varepsilon} \frac{dz}{\psi\{\varphi^{-1}(z)\}} \leq \sum_{s=0}^{n-1} \frac{\Delta(y_s+\varepsilon)}{\psi\{\varphi^{-1}(y_s+\varepsilon)\}} \leq \sum_{s=0}^{n-1} f_s,$$

since $\psi\{\varphi^{-1}(z)\}$ is positive and non-decreasing. By $G(z) := \int_{z_0}^{z} \frac{ds}{\psi[\varphi^{-1}(s)]}$, the last expression yields for all $n \in \mathbb{N}_0$,

$$G(y_n+\varepsilon) \leq G(c) + \sum_{s=0}^{n-1} f_s.$$

Letting $\varepsilon \to 0$ in the last inequality, we get for all $n \in \mathbb{N}_0$,

$$G(y_n) \leq G(c) + \sum_{s=0}^{n-1} f_s. \tag{2.2.56}$$

By the choice of $M \in \mathbb{N}_0$ in (2.2.52), we infer from the last relation, for all $n \in \mathbb{N}_0$, $n \leq M$,

$$y_n \leq G^{-1}\{G(c) + \sum_{s=0}^{n-1} f_s\}. \tag{2.2.57}$$

Thus the desired inequality (2.2.51) follows from (2.2.55) and (2.2.57). □

In the following, we discuss some systems of two discrete inequalities of Gronwall-Bellman type.

The expression $\sum_{s=0}^{t-1} b(s)$ represents the solution of the linear difference equation $\Delta x(t) = b(t)$ for all $t \in \mathbb{R}$ under the initial condition $x(0) = 0$. Also the expression $\prod_{s=0}^{t-1} C(s)$ represents the solution of the linear difference equation $x(t+1) = C(t)x(t)$ for all $t \in \mathbb{R}$ under the initial condition $x(0) = 1$. It is assumed that $\sum_{s=0}^{t-1} b(s) = 0$ and $\prod_{s=0}^{t-1} C(s) = 1$.

Theorem 2.2.16 (The Salem Inequality [579]) *Assume the following system of two nonlinear inequalities holds,*

$$\begin{cases} u_1(t) \leq a_1(t) + p_1(t)\sum_{s=0}^{t-1} H(u_1(s)) + q_1(t)\sum_{s=0}^{t-1} H(u_2(s)), & (2.2.58) \\ u_2(t) \leq a_2(t) + p_2(t)\sum_{s=0}^{t-1} H(u_1(s)) + q_2(t)\sum_{s=0}^{t-1} H(u_2(s)) & (2.2.59) \end{cases}$$

where $a_1(t), a_2(t), p_1(t), p_2(t), q_1(t)$, *and* $q_2(t)$ *are positive and nonlinear functions, H is positive, non-decreasing, continuous, sub-additive, and sub-multiplicative. Then*

$$\begin{cases} u_1(t) \le a_1(t) + p_2(t)\phi_4(t) + q_1(t)\psi_4(t), \\ u_2(t) \le a_2(t) + p_2(t)\phi_4(t) + q_2(t)\psi_4(t) \end{cases}$$

where

$$\begin{cases} \phi_4(t) = \sum_{s=0}^{t-1} \left\{ H(a_1(s) - p_1(s) - q_1(s)) + (H(p_1(s)) + H(p_1(s)))H(\psi(s)) \right\}, \\ \psi_4(t) = \sum_{s=0}^{t-1} \left\{ H(a_2(s) - p_2(s) - q_2(s)) + (H(p_2(s)) + H(p_2(s)))H(\psi(s)) \right\}, \\ \psi(t) = G^{-1}\left\{ G(2) + \sum_{s=0}^{t-1}(A(s) + B(s)) \right\}, \\ A(t) = H(a_1(t) - p_1(t) - q_1(t)) + H(a_2(t) - p_2(t) - q_2(t)), \\ B(t) = H(p_1(t)) + H(q_1(t)) + H(p_2(s)) + H(q_2(s)), \end{cases}$$

and

$$G(r) = \int_{r_0}^{r} \frac{ds}{s + H(s)}, \quad 0 < r_0 \le r,$$

with

$$G(2) + \sum_{s=0}^{t-1} \left(A(s) + B(s) \right) \in Dom\,(G^{-1}).$$

Proof Inequalities (2.2.58) and (2.2.59) can be rewritten as

$$\begin{cases} u_1(t) \le (a_1(t) - p_1(t) - q_1(t)) + p_1(t)R_1(t) + q_1(t)R_2(t), & (2.2.60) \\ u_2(t) \le (a_2(t) - p_2(t) - q_2(t)) + p_2(t)R_1(t) + q_2(t)R_2(t) & (2.2.61) \end{cases}$$

where

$$R_1(t) = \sum_{s=0}^{t-1} H(u_1(s)) + 1, \quad R_2(t) = \sum_{s=0}^{t-1} H(u_2(s)) + 1.$$

Then we obtain

$$\begin{cases} \Delta R_1(t) \le H(a_1(t) - p_1(t) - q_1(t)) + H(p_1(t))H(R_1(t)) + H(q_1(t))H(R_2(t)), & (2.2.62) \\ \Delta R_2(t) \le H(a_2(t) - p_2(t) - q_2(t)) + H(p_2(t))H(R_1(t)) + H(q_2(t))H(R_2(t)). & (2.2.63) \end{cases}$$

Since $R_1(t)$ and $R_2(t)$ are non-decreasing and $R_1(0) = 1, R_2(0) = 1$, then from the definition of G, we conclude

$$\begin{aligned} & G(R_1(t-1) + R_2(t+1)) - G(R_1(t) + R_2(t)) \\ & = \int_{R_1(t)+R_2(t)}^{R_1(t-1)+R_2(t+1)} \frac{ds}{s + H(s)} \\ & \le \frac{\Delta(R_1(t) + R_2(t))}{R_1(t) + R_2(t) + H(R_1(t) + R_2(t))} \\ & \le A(t) + B(t). \end{aligned}$$

Summing up from 0 to $t-1$, we get

$$R_1(t) + R_2(t) \le \psi(t). \tag{2.2.64}$$

From (2.2.62) and (2.2.64), we derive

$$\Delta R_1(t) \le H(a_1(t) - p_1(t) - q_1(t)) + H(p_1(t)) + H(q_1(t))H(\psi(t)),$$

and summing up from 0 to $t-1$, we obtain

$$R_1(t) \le \phi_4(t). \tag{2.2.65}$$

Also from (2.2.63) and (2.2.64), we may derive

$$R_2(t) \le \psi_4(t). \tag{2.2.66}$$

Hence we can derive (2.2.58)–(2.2.59) from (2.2.60), (2.2.61), (2.2.65), and (2.2.66). □

Theorem 2.2.17 (The Salem Inequality [579]) *Let the following system of two nonlinear inequalities hold*

$$\begin{cases} u_1(t) \leq a_1(t) + p_1(t) \sum_{r=1}^{n} E_1^r(t, H(u_1(s))) + q_1(t) \sum_{r=1}^{n} E_2^r(t, H(u_2(s))), & (2.2.67) \\ u_2(t) \leq a_2(t) + p_2(t) \sum_{r=1}^{n} E_1^r(t, H(u_1(s))) + q_2(t) \sum_{r=1}^{n} E_2^r(t, H(u_2(s))), & (2.2.68) \end{cases}$$

where

(i) $a_1(t), a_2(t), p_1(t), p_2(t), q_1(t)$, *and* $q_2(t)$ *are real-valued, positive and non-decreasing,* $t \in \mathbb{R}$,

(ii) $E_1^r(t, u) = \sum_{t_1=0}^{t-1} f_{r1}(t_1) \sum_{t_2=0}^{t_1-1} f_{r2}(t_2) \cdots \sum_{t_2=0}^{t_{r-1}-1} f_{rr}(t_r) u(t_r)$,

(iii) $f_{rj}(t_j)$ *are real-valued and non-negative functions,* $t_j \in \mathbb{R}$, *and* $H(u(t))$ *is defined in Theorem 2.2.16.*

Then

$$\begin{cases} u_1(t) \leq a_1(t) + p_1(t)\phi_5(t) + q_1(t)\psi_5(t), & (2.2.69) \\ u_2(t) \leq a_2(t) + p_2(t)\phi_5(t) + q_2(t)\psi_5(t), & (2.2.70) \end{cases}$$

where

$$\begin{cases} \phi_5(t) = \sum_{s=0}^{t-1} \left\{ B_1(a_1(s) - p_1(s) - q_1(s)) + (B_1(p_1(s)) + B_1(p_1(s)))H(\psi(s)) \right\}, \\ \psi_5(t) = \sum_{s=0}^{t-1} \left\{ B_2(a_2(s) - p_2(s) - q_2(s)) + (B_2(p_2(s)) + B_2(p_2(s)))H(\psi(s)) \right\}, \\ B_1(b(t)) = \sum_{r=1}^{n} \Delta E_1^r(t, H(b(t))), \quad B_2(b(t)) = \sum_{r=1}^{n} \Delta E_2^r(t, H(b(t))), \\ \psi(t) = G^{-1} \left\{ G(2) + \sum_{s=0}^{t-1} A_1(s) \right\}, \\ A_1(t) = A(a_1 - p_1 - q_1, a_2 - p_2 - q_2) + A(p_1, p_2) + A(q_1, q_2), \quad A(a, b) = B_1(a) + B_2(b), \end{cases}$$

and

$$G(r) = \int_{r_0}^{r} \frac{ds}{s + H(s)}, \quad 0 < r_0 \leq r,$$

with

$$G(2) + \sum_{s=0}^{t-1} A_1(s) \in Dom\,(G^{-1}).$$

Theorem 2.2.18 (The Salem Inequality [579]) *Let the following system of two inequalities hold*

$$u_1(t) \le a_1(t) + p_1(t)\sum_{s=0}^{t-1} e_1(s)u_1(s) + p_2(t)\sum_{s=0}^{t-1} e_2(s)u_2(s)$$
$$+p_3(t)\sum_{s=0}^{t-1} e_3(s)H(u_1(s)) + p_4(t)\sum_{s=0}^{t-1} e_4(s)H(u_2(s)) \qquad (2.2.71)$$

and

$$u_2(t) \le a_2(t) + q_1(t)\sum_{s=0}^{t-1} h_1(s)u_1(s) + q_2(t)\sum_{s=0}^{t-1} h_2(s)u_2(s)$$
$$+q_3(t)\sum_{s=0}^{t-1} h_3(s)H(u_1(s)) + q_4(t)\sum_{s=0}^{t-1} h_4(s)H(u_2(s)) \qquad (2.2.72)$$

where all given functions are real-valued, positive, non-decreasing, and continuous functions, H is defined in Theorem 2.2.16, and for all $t \in \mathbb{R}$,

$$\begin{cases} a_1 \ge p_1 + p_2 + p_3 + p_4, \\ a_2 \ge q_1 + q_2 + q_3 + q_4. \end{cases} \qquad (2.2.73)$$

Then

$$\begin{cases} u_1(t) \le a_1(t) + p_1(t)\sum_{s=0}^{t-1} e_1(s)a_1(s)\psi_6(s) + p_2(t)\sum_{s=0}^{t-1} e_2(s)a_2(s)\psi_7(s) \\ \qquad + p_3(t)\sum_{s=0}^{t-1} e_3(s)\phi_6(s) + p_4(t)\sum_{s=0}^{t-1} e_4(s)\phi_7(s) \\ u_2(t) \le a_2(t) + q_1(t)\sum_{s=0}^{t-1} h_1(s)a_1(s)\psi_6(s) + q_2(t)\sum_{s=0}^{t-1} h_2(s)a_2(s)\psi_7(s) \\ \qquad + q_3(t)\sum_{s=0}^{t-1} h_3(s)\phi_6(s) + q_4(t)\sum_{s=0}^{t-1} h_4(s)\phi_7(s) \end{cases} \qquad (2.2.74)$$

where

$$\begin{cases} \phi_6(t) = H(a_1 - p_1 - p_2 - p_3 - p_4) + (H(p_1) + H(p_2) + H(p_3) + H(p_4))H(\psi_6), \\ \phi_7(t) = H(a_2 - q_1 - q_2 - q_3 - q_4) + (H(q_1) + H(q_2) + H(q_3) + H(q_4))H(\psi_7), \\ \psi_6(t) = 4 + \sum_{s=0}^{t-1}[A_1(s) + B_1(s)\psi(s) + c_1(s)H(\psi(s))], \\ \psi_7(t) = 4 + \sum_{s=0}^{t-1}[A_2(s) + B_2(s) + c_2(s)H(\psi(s))], \\ \psi(t) = G^{-1}\left(G(8) + \sum_{s=0}^{t-1}(A(s) + B(s) + C(s))\right), \\ A_1(t) = e_1(a_1 - P) + e_2(a_2 - Q) + e_3H(a_1 - P) + e_4H(a_2 - Q), \\ B_1(t) = e_1P + e_2Q, \\ C_1(t) = e_3(H(p_1) + H(p_2) + H(p_3) + H(p_4)) + e_4(H(q_1) + H(q_2) + H(q_3) + H(q_4)), \\ A_2(t) = h_1(a_1 - P) + h_2(a_2 - Q) + h_3H(a_1 - P) + h_4H(a_2 - Q), \\ B_2(t) = h_1P + h_2Q, \\ C_2(t) = h_3(H(p_1) + H(p_2) + H(p_3) + H(p_4)) + h_4(H(q_1) + H(q_2) + H(q_3) + H(q_4)), \\ A(t) = A_1(t) + A_2(t), \quad B(t) = B_1(t) + B_2(t), \quad C(t) = C_1(t) + C_2(t), \\ P = p_1 + p_2 + p_3 + p_4, \quad Q = q_1 + q_2 + q_3 + q_4, \end{cases}$$

and

$$G(r) = \int_{r_0}^{r} \frac{ds}{s + H(s)}, \quad 0 < r_0 \leq r,$$

as long as

$$G(8) + \sum_{s=0}^{t-1}(A(s) + B(s) + C(s)) \in Dom\,(G^{-1}). \tag{2.2.75}$$

Proof The proof is similar to the proofs of Theorem 2.2.16. □

2.3 The One-Dimensional Ou-Yang Inequality and Its Generalization

In what follows, we use the notations m, n, q, q to denote the elements of $\mathbb{Z}$. Let $\mathbb{R}_1 = [1, +\infty)$ is a subset of $\mathbb{R}$. For $t_1 > t_2$, $t_1, t_2 \in \mathbb{Z}$ and any function $h : \mathbb{Z} \to \mathbb{R}$, we use the usual convention $\sum_{s=t_1}^{t_2} h(s) = 0$ and $\prod_{s=t_1}^{t_2} h(s) = 1$.

We assume that all the sums and products converge on the respective domains of their definitions.

The next three results are discrete forms of the Ou-Yang inequality, which are due to Pachpatte [493].

Theorem 2.3.1 (The Pachpatte Inequality [493]) *Let $u(n)$ and $f(n)$ be functions defined on $\mathbb{Z}$ into $\mathbb{R}_+$ and $c \geq 0$ be a constant. If the following inequality holds, for all $n \in \mathbb{Z}$,*

$$u^2(n) \leq c + \sum_{s=n+1}^{+\infty} f(s)u(s), \tag{2.3.1}$$

then, for all $n \in \mathbb{Z}$,

$$u(n) \leq \sqrt{c} + \frac{1}{2} \sum_{s=n+1}^{+\infty} f(s). \tag{2.3.2}$$

Proof We first assume that $c > 0$ and define a function $z(n)$ by the right-hand side of (2.3.1), then

$$z(n) - z(n-1) = f(n+1)u(n+1). \tag{2.3.3}$$

Using the inequality $u(n+1) \leq \sqrt{z(n+1)}$ in (2.3.3), we may get

$$z(n) - z(n-1) \leq f(n+1)\sqrt{z(n+1)} \tag{2.3.4}$$

Using the facts that $\sqrt{z(n+1)} > 0$, $\sqrt{z(n+1)} \leq \sqrt{z(n)}$ for all $n \in \mathbb{Z}$ and (2.3.4), we get

$$\begin{aligned} \sqrt{z(n)} - \sqrt{z(n+1)} &= \frac{z(n) - z(n-1)}{\sqrt{z(n)} + \sqrt{z(n+1)}} \\ &\leq \frac{z(n) - z(n-1)}{\sqrt{z(n+1)} + \sqrt{z(n+1)}} \\ &\leq \frac{1}{2} f(n+1). \end{aligned} \tag{2.3.5}$$

Now setting $n = s$ in (2.3.5) and summing over $s = n, n+1, \ldots, m-1$, we may obtain

$$\sqrt{z(n)} - \sqrt{z(m)} \leq \frac{1}{2} \sum_{s=n+1}^{m} f(s). \tag{2.3.6}$$

Noting that $\sqrt{z(m)} = \sqrt{c}$, the desired inequality in (2.3.2) follows as $n \to +\infty$ in (2.3.6) and using the fact that $u(n) \leq \sqrt{z(n)}$ for all $n \in \mathbb{Z}$.

Now suppose $c = 0$. Then from (2.3.1), it follows that for any ϵ,

$$u^2(n) \leq \epsilon + \sum_{s=n+1}^{+\infty} f(s)u(s) \tag{2.3.7}$$

which yields

$$u(n) \leq \sqrt{\epsilon} + \frac{1}{2} \sum_{s=n+1}^{+\infty} f(s). \tag{2.3.8}$$

Since $u(n) \geq 0$ and $\epsilon > 0$ is arbitrary, it follows from (2.3.8) that

$$u(n) \leq \frac{1}{2} \sum_{s=n+1}^{+\infty} f(s). \tag{2.3.9}$$

This thus completes the proof. □

Theorem 2.3.2 (The Pachpatte Inequality [493]) *Let $u(n)$ be a function defined on $\mathbb{Z}$ into $\mathbb{R}_1$ and $f(n)$ be a function defined in Theorem 2.3.1 and $c > 1$ be a constant, if for all $n \in \mathbb{Z}$,*

$$u(n) \leq c + \sum_{s=n+1}^{+\infty} f(s)u(s) \log u(s), \tag{2.3.10}$$

then for all $n \in \mathbb{Z}$,

$$u(n) \leq c^{\sum_{s=n+1}^{+\infty}(1+f(s))}. \tag{2.3.11}$$

Proof Define a function $v(n)$ by the right-hand side of (2.3.10), then

$$v(n) - v(n-1) = f(n+1)u(n+1) \log u(n+1). \tag{2.3.12}$$

Using the fact that $u(n+1) \leq v(n+1)$ in (2.3.12), we may get

$$v(n) \leq v(n+1)[1 + f(n+1) \log v(n+1)] \tag{2.3.13}$$

By setting $n = s$ in (2.3.13) and then substituting $s = n,\ n+1$ successively, we can get

$$v(n) \leq v(m) \prod_{s=n+1}^{m} [1 + f(s) \log v(s)]. \tag{2.3.14}$$

Nothing that $\lim_{m\to+\infty} v(m) = c$ and by letting $m \to +\infty$ in (2.3.14), we get

$$v(n) \le c \prod_{s=n+1}^{m} [1 + f(s) \log v(s)]$$
$$\le c \exp\left(\sum_{s=n+1}^{+\infty} f(s) \log v(s)\right). \tag{2.3.15}$$

From (2.3.15) it follows that

$$\log v(n) \le \log c + \sum_{s=n+1}^{+\infty} f(s) \log v(s). \tag{2.3.16}$$

Define a function $z(n)$ by the right-hand side of (2.3.16), then

$$z(n) - z(n+1) = f(n+1) \log v(n+1). \tag{2.3.17}$$

Using the inequality $\log v(n+1) \le z(n+1)$ in (2.3.17), we can get

$$z(n) \le [1 + f(n+1)]z(n+1) \tag{2.3.18}$$

which, by following the above argument, implies

$$z(n) \le \log c\{\prod_{s=n+1}^{+\infty} [1 + f(s)]\}$$
$$= \log c^{\prod_{s=n+1}^{+\infty}[1+f(s)]}. \tag{2.3.19}$$

Using (2.3.19) in (2.3.16), we can derive

$$\log v(n) \le \log c^{\prod_{s=n+1}^{+\infty}[1+f(s)]}$$

which implies

$$v(n) \le c^{\prod_{s=n+1}^{+\infty}[1+f(s)]}. \tag{2.3.20}$$

Now using (2.3.20) in (2.3.10), we can get the required inequality in (2.3.11). The proof is thus complete. □

The following inequality applies in some situations where Theorems 2.1.3 and 2.1.4 do not apply, which can be also regarded as discrete variables.

Theorem 2.3.3 (The Pachpatte Inequality [493]) *Let $u(n)$, $f(n)$ and $g(n)$ be functions defined on $\mathbb{Z}$, $c > 0$ be a constant. If for all $n \in \mathbb{Z}$,*

$$u^2(n) \le c + \sum_{s=n+1}^{+\infty} f(s)u(s)[u(s) + \sum_{t=s+1}^{+\infty} g(t)u(t)], \tag{2.3.21}$$

then for all $n \in \mathbb{Z}$,

$$u(n) \le \sqrt{c}\Big[1 + \frac{1}{2}\sum_{s=n+1}^{+\infty} f(s) \prod_{t=s+1}^{+\infty} [1 + \frac{1}{2}f(t) + g(t))\Big]. \tag{2.3.22}$$

Proof We first assume that $c > 0$ and define a function $z(n)$ by the right-hand side of (2.3.21), then

$$z(n) - z(n-1) = f(n+1)u(n+1)[u(n+1) + \sum_{t=n+2}^{+\infty} g(t)u(t)]. \tag{2.3.23}$$

Using the fact that $u(n) \le \sqrt{z(n)}$ for all $n \in \mathbb{Z}$ in (2.3.23), we get

$$z(n) - z(n+1) \le f(n+1)\sqrt{z(n+1)}[\sqrt{z(n+1)} + \sum_{t=n+2}^{+\infty} g(t)\sqrt{z(t)}]. \tag{2.3.24}$$

Now following the arguments as in the proof of the inequality of Theorem 2.3.1, we may get

$$\sqrt{z(n)} - \sqrt{z(n+1)} \le \frac{1}{2}f(n+1)[\sqrt{z(n+1)} + \sum_{t=n+2}^{+\infty} g(t)\sqrt{z(t)}]. \tag{2.3.25}$$

Now setting $n = s$ in (2.3.25) and summing over $s = n, n+1, \ldots, m$, we may obtain

$$\sqrt{z(n)} \le \sqrt{z(m)} + \frac{1}{2}\sum_{s=n+1}^{m} f(n+1)[\sqrt{z(s)} + \sum_{t=s+1}^{+\infty} g(t)\sqrt{z(t)}]. \tag{2.3.26}$$

Noting that $\lim_{m\to+\infty} \sqrt{z(m)} = \sqrt{c}$ and by letting $m \to +\infty$ in (2.3.26), we can get

$$\sqrt{z(n)} \le \sqrt{c} + \frac{1}{2}\sum_{s=n+1}^{m} f(n+1)[\sqrt{z(s)} + \sum_{t=s+1}^{+\infty} g(t)\sqrt{z(t)}]. \tag{2.3.27}$$

Define a function $v(n)$ by the right-hand side of (2.3.27), then

$$v(n) - v(n+1) = \frac{1}{2}f(n+1)[\sqrt{z(n+1)} + \sum_{t=n+2}^{+\infty} g(t)\sqrt{z(t)}]. \tag{2.3.28}$$

Using the fact that $\sqrt{z(n)} \leq v(n)$ for all $n \in \mathbb{Z}$ in (2.3.28), we get

$$v(n) - v(n+1) = \frac{1}{2}f(n+1)[v(n+1) + \sum_{t=n+2}^{+\infty} g(t)v(t)]. \tag{2.3.29}$$

Define a function $w(n)$ by

$$w(n) = v(n+1) + \sum_{t=n+2}^{+\infty} g(t)v(t), \tag{2.3.30}$$

then

$$w(n) - w(n+1) = v(n+1) - v(n+2) + g(n+2)v(n+2). \tag{2.3.31}$$

Now using of (2.3.29) and the fact that $v(n+2) \leq w(n+1)$ from (2.3.30) in (2.3.31), we get

$$w(n) \leq [1 + \frac{1}{2}f(n+2) + g(n+2)]w(n+1) \tag{2.3.32}$$

which, by following the arguments in the proof of Theorem 2.3.2 given above, implies

$$w(n) \leq \sqrt{c} \prod_{t=n+2}^{+\infty} [1 + \frac{1}{2}f(t) + g(t)]. \tag{2.3.33}$$

Now using (2.3.33) in (2.3.29), we have

$$v(n) - v(n+1) \leq \frac{1}{2}\sqrt{c}f(n+1) \prod_{t=n+2}^{+\infty} [1 + \frac{1}{2}f(t) + g(t)]. \tag{2.3.34}$$

Now setting $n = s$ in (2.3.34) and summing over $s = n, n+1, \ldots, m-1$, we may get

$$v(n) \leq v(m) + \frac{1}{2}\sqrt{c} \sum_{s=n+1}^{+\infty} f(s) \prod_{t=n+2}^{+\infty} [1 + \frac{1}{2}f(t) + g(t)]. \tag{2.3.35}$$

Noting that $\lim_{m\to+\infty} v(m) = \sqrt{c}$, the desired inequality in (2.3.22) follows by letting $m \to +\infty$ in (2.3.35) and using the facts that $u(n) \leq \sqrt{z(n)} \leq v(n)$ for all $n \in \mathbb{Z}$. The proof of the case when $c = 0$ can be completed by following the arguments as in the proof of Theorem 2.3.1 given above with suitable changes and hence the proof of inequality (2.3.22) is complete. □

Theorem 2.3.4 (The Pachpatte Inequality [500]) *Let u, f, g, h be real-valued non-negative functions defined on $\mathbb{N}_0$ and c be a non-negative real constant.*

(b_1) If for all $n \in \mathbb{N}_0$,

$$u^2(n) \leq c^2 + 2\sum_{s=0}^{n-1}[f(s)u^2(s) + h(s)u(s)], \tag{2.3.36}$$

then for all $n \in \mathbb{N}_0$,

$$u(n) \leq q(n)\prod_{s=0}^{n-1}[1 + f(s)], \tag{2.3.37}$$

with for all $n \in \mathbb{N}_0$,

$$q(n) = c + \sum_{s=0}^{n-1} h(s). \tag{2.3.38}$$

(b_2) If for all $n \in \mathbb{N}_0$,

$$u^2(n) \leq c^2 + 2\sum_{s=0}^{n-1}\left[f(s)u(s)\left(u(s) + \sum_{\tau=0}^{s-1} g(\tau)u(\tau)\right) + h(s)u(s)\right], \tag{2.3.39}$$

then for all $n \in \mathbb{N}_0$,

$$u(n) \leq q(n)\left[1 + \sum_{s=0}^{n-1} f(s)\left(\prod_{\tau=0}^{s-1}[1 + f(\tau) + g(\tau)]\right)\right], \tag{2.3.40}$$

where $q(n)$ is defined by (2.3.38).

(b_3) If for all $n \in \mathbb{N}_0$,

$$u^2(n) \leq c^2 + 2\sum_{s=0}^{n-1}\left[f(s)u(s)\left(\sum_{\tau=0}^{s-1} g(\tau)u(\tau)\right) + h(s)u(s)\right], \tag{2.3.41}$$

then for all $n \in \mathbb{N}_0$,

$$u(n) \le q(n) \prod_{s=0}^{n-1} \left[1 + f(s) \left(\sum_{\tau=0}^{s-1} g(\tau) \right) \right], \tag{2.3.42}$$

where $q(n)$ is defined by (2.3.38).

Proof We only give the assertion in (b_2), while the proofs of (b_1) and (b_3) can be done in the same way.

(b_2) Let $\varepsilon > 0$ be an arbitrary small constant and define a function $z(n)$ by

$$z(n) = (c+\varepsilon)^2 + 2\sum_{s=0}^{n-1} \left[f(s)u(s) \left(u(s) + \sum_{\tau=0}^{s-1} g(\tau)u(\tau) \right) + h(s)u(s) \right]. \tag{2.3.43}$$

From (2.3.43) and using the fact that $u(n) \le \sqrt{z(n)}$, we get

$$\Delta z(n) \le 2\sqrt{z(n)} \left[f(n) \left(\sqrt{z(n)} + \sum_{\tau=0}^{n-1} g(\tau)\sqrt{z(\tau)} \right) + h(n) \right]. \tag{2.3.44}$$

It is easy to verify

$$\Delta(\sqrt{z(n)}) = \frac{z(n+1) - z(n)}{\sqrt{z(n+1)} + \sqrt{z(n)}} \le \frac{\Delta z(n)}{2\sqrt{z(n)}} \tag{2.3.45}$$

where we have used the fact that $\sqrt{z(n)} \le \sqrt{z(n+1)}$. Using (2.3.44) in (2.3.45), we get

$$\Delta(\sqrt{z(n)}) \le \left[f(n) \left(\sqrt{z(n)} + \sum_{\tau=0}^{n-1} g(\tau)\sqrt{z(\tau)} \right) + h(n) \right]. \tag{2.3.46}$$

By taking $n = s$ in (2.3.46) and summing both sides of (2.3.46) from $s = 0$ to $n-1$, we have

$$\sqrt{z(n)} \le q_\varepsilon(n) + \sum_{s=0}^{n-1} f(s) \left(\sqrt{z(s)} + \sum_{\tau=0}^{s-1} g(\tau)\sqrt{z(\tau)} \right), \tag{2.3.47}$$

where $q_\varepsilon(n)$ is defined by (2.3.38) by replacing c by $c+\varepsilon$. Since $q_\varepsilon(n)$ is positive and monotone non-decreasing for all $n \in \mathbb{N}_0$, from (2.3.47) it follows

$$\frac{\sqrt{z(n)}}{q_\varepsilon(n)} \le 1 + \sum_{s=0}^{n-1} f(s) \left(\frac{\sqrt{z(s)}}{q_\varepsilon(s)} + \sum_{\tau=0}^{s-1} g(\tau)\frac{\sqrt{z(\tau)}}{q_\varepsilon(\tau)} \right). \tag{2.3.48}$$

The inequality (2.3.48) implies the estimate (see [442], p. 349)

$$\sqrt{z(n)} \le q_\varepsilon(n)\left[1+\sum_{s=0}^{n-1} f(s)\left(\prod_{\tau=0}^{s-1}[1+f(\tau)+g(\tau)]\right)\right]. \tag{2.3.49}$$

Now using the inequality $u(n) \le \sqrt{z(n)}$ in (2.3.49), and then letting $\varepsilon \to 0^+$, we get the desired inequality in (2.3.40). □

Theorem 2.3.5 (The Pachpatte Inequality [500]) *Let u, f, h, v, g_i, h_i $(i = 1, 2, 3, 4)$ be real-valued non-negative functions defined on $\mathbb{N}_0$ and c_1, c_2, μ be non-negative real constants.*

(1) If for all $n \in \mathbb{N}_0$,

$$u^2(n) \le \left(c_1^2 + 2\sum_{s=0}^{n-1} f(s)u(s)\right)\left(c_2^2 + 2\sum_{s=0}^{n-1} h(s)u(s)\right), \tag{2.3.50}$$

then for all $n \in \mathbb{N}_0$,

$$u(n) \le q_0(n)\prod_{m=0}^{n-1}\left[1+2\left[h(m)\sum_{s=0}^{m-1} f(s) + f(m)\sum_{s=0}^{m} h(s)\right]\right], \tag{2.3.51}$$

where for all $n \in \mathbb{N}_0$,

$$q_0(n) = c_1c_2 + \sum_{s=0}^{n-1}[c_1^2 h(s) + c_2^2 f(s)]. \tag{2.3.52}$$

(2) If for all $n \in \mathbb{N}_0$,

$$u^2(n) \le c_1 + \sum_{s=0}^{n-1}[g_1(s)u^2(s) + h_1(s)u(s)] + \sum_{s=0}^{n-1}[g_2(s)\overline{v}^2(s) + h_2(s)\overline{v}(s)], \tag{2.3.53}$$

$$v^2(n) \le c_2 + \sum_{s=0}^{n-1}[g_3(s)\overline{u}^2(s) + h_3(s)\overline{u}(s)] + \sum_{s=0}^{n-1}[g_4(s)v^2(s) + h_4(s)v(s)], \tag{2.3.54}$$

where $\overline{u}(n) = e^{-\mu n}u(n), \overline{v}(t) = e^{\mu n}v(n)$ *for all* $n \in \mathbb{N}_0$*, then for all* $n \in \mathbb{N}_0$*,*

$$u(n) \leq e^{\mu n}q^*(n)\prod_{s=0}^{n-1}[1 + G(s)], \tag{2.3.55}$$

$$v(n) \leq q^*(t)\prod_{s=0}^{n-1}[1 + G(s)], \tag{2.3.56}$$

where

$$q^*(n) = \sqrt{2(c_1 + c_2)} + \sum_{s=0}^{n-1} H(s), \tag{2.3.57}$$

in which, for all $n \in \mathbb{N}_0$*,*

$$\begin{cases} G(n) = \max\{[g_1(n) + g_3(n)], [g_2(n) + g_4(n)]\}, \\ H(n) = \max\{[h_1(n) + h_3(n)], [h_2(n) + h_4(n)]\}. \end{cases}$$

Proof we only prove the assertion in (1), the proof of the result in (2) can be done in the same way.

(1) Let $\varepsilon > 0$ be an arbitrary small constant and define a function $z(n)$ by

$$z(n) = \left((c_1 + \varepsilon)^2 + 2\sum_{s=0}^{n-1} f(s)u(s)\right)\left((c_2 + \varepsilon)^2 + 2\sum_{s=0}^{n-1} h(s)u(s)\right). \tag{2.3.58}$$

From (2.3.58) and using

$$\Delta[a(n)b(n)] = a(n)\Delta b(n) + b(n+1)\Delta a(n),$$

and the inequality $u(n) \leq \sqrt{z(n)}$, we may derive

$$\begin{aligned} \Delta z(n) \leq\ & 2\sqrt{z(n)}[(c_1 + \varepsilon)^2 h(n) + (c_2 + \varepsilon)^2 f(n)] \\ & +2\left[h(n)\left(\sum_{s=0}^{n-1} f(s)\sqrt{z(s)}\right) + f(n)\left(\sum_{s=0}^{n} h(s)\sqrt{z(s)}\right)\right]. \end{aligned} \tag{2.3.59}$$

It is easy to verify that the inequality (2.3.45) holds. Now from (2.3.45) and (2.3.59), it follows

$$\Delta(z(n)) \leq [(c_1+\varepsilon)^2 h(n) + (c_2+\varepsilon)^2 f(n)] \\ +2\left[h(n)\left(\sum_{s=0}^{n-1} f(s)\sqrt{z(s)}\right) + f(n)\left(\sum_{s=0}^{n} h(s)\sqrt{z(s)}\right)\right]. \tag{2.3.60}$$

By keeping $n = m$ in (2.3.60) and then taking the sum over $m = 0, 1, 2, \cdots, n-1$, we may have

$$\sqrt{z(n)} \leq q_{0\varepsilon}(n) + 2\sum_{m=0}^{n-1}\left[h(m)\left(\sum_{s=0}^{m-1} f(s)\sqrt{z(s)}\right)\right. \\ \left. +f(m)\left(\sum_{s=0}^{m} h(s)\sqrt{z(s)}\right)\right], \tag{2.3.61}$$

where $q_{0\varepsilon}(n)$ is defined by (2.3.52) by replacing c_1 and c_2 by $c_1+\varepsilon$ and $c_2+\varepsilon$, respectively. Since $q_{0\varepsilon}(n)$ is positive and monotone non-decreasing in $n \in \mathbb{N}_0$, from (2.3.61) it follows

$$\frac{\sqrt{z(n)}}{q_{0\varepsilon}(n)} \leq 1 + 2\sum_{m=0}^{n-1}\left[h(m)\left(\sum_{s=0}^{m-1} f(s)\frac{\sqrt{z(s)}}{q_{0\varepsilon}(s)}\right)\right. \\ \left. +f(m)\left(\sum_{s=0}^{m} h(s)\frac{\sqrt{z(s)}}{q_{0\varepsilon}(s)}\right)\right]. \tag{2.3.62}$$

Define

$$v(n) = 1 + 2\sum_{m=0}^{n-1}\left[h(m)\left(\sum_{s=0}^{m-1} f(s)\frac{\sqrt{z(s)}}{q_{0\varepsilon}(s)}\right) + f(m)\left(\sum_{s=0}^{m} h(s)\frac{\sqrt{z(s)}}{q_{0\varepsilon}(s)}\right)\right]. \tag{2.3.63}$$

From (2.3.63) and using $\sqrt{z(n)}/q_{0\varepsilon}(n) \leq v(n)$ and the fact that $v(n)$ is monotone non-decreasing for all $n \in \mathbb{N}_0$, we derive

$$v(n+1) \leq \left[1 + 2\left[h(n)\left(\sum_{s=0}^{n-1} f(s)\right) + f(n)\left(\sum_{s=0}^{n-1} h(s)\right)\right]\right] v(n). \tag{2.3.64}$$

Keeping $n = m$ in (2.3.64) and substituting $m = 0, 1, 2, \cdots, n-1$ successively, we have

$$v(n) \leq \prod_{m=0}^{n-1}\left[1 + 2\left(h(m)\sum_{s=0}^{m-1} f(s) + f(m)\sum_{s=0}^{m} h(s)\right)\right]. \tag{2.3.65}$$

Using (2.3.65) in (2.3.62) and the fact that $u(n) \leq \sqrt{z(n)}$ and then letting $\varepsilon \to 0^+$, we get the desired inequality in (2.3.51). □

Theorem 2.3.6 (The Pachpatte Inequality [500]) *Let $u(n) \geq u_0 \geq 0$ be a real-valued continuous function defined for all $n \in \mathbb{N}_0$, u_0 is a real constant. Let f, g, h be real-valued non-negative continuous functions defined on $n \in \mathbb{N}_0$ and c be a non-negative real constant. Let $W(u)$ be a continuous non-decreasing function defined on an interval $I = [u_0, +\infty)$ and $W(u) > 0$ on $(u_0, +\infty)$, $W(u_0) = 0$.*

(1) If for all $n \in \mathbb{N}_0$,

$$u^2(n) \leq c^2 + 2\sum_{s=0}^{n-1}[f(s)u(s)W(u(s)) + h(s)u(s)]ds, \tag{2.3.66}$$

then for all $n, n_1 \in \mathbb{N}_0, 0 \leq n \leq n_1$,

$$u(n) \leq \Omega^{-1}\left[\Omega(q(n)) + \sum_{s=0}^{n-1} f(s)\right], \tag{2.3.67}$$

where $q(n)$ is defined by (2.3.38), Ω, Ω^{-1} are defined as in Theorem 1.2.13, and $n \in \mathbb{N}_0$ is chosen so that for all $n \in \mathbb{N}_0$ with $0 \leq n \leq n_1$,

$$\Omega(q(n)) + \sum_{s=0}^{n-1} f(s) \in Dom\,(\Omega^{-1}).$$

(2) If for all $n \in \mathbb{N}_0$,

$$u^2(n) \leq c^2 + 2\sum_{s=0}^{n-1}\left[f(s)u(s)\left(u(s) + \sum_{\tau=0}^{s-1} g(\tau)W(u(\tau))d\tau\right) + h(s)u(s)\right], \tag{2.3.68}$$

then for all $n, n_2 \in \mathbb{N}_0, 0 \leq n \leq n_2$,

$$u(n) \leq q(n) + \sum_{s=0}^{n-1} f(s)E^{-1}\left[E(q(s)) + \sum_{\tau=0}^{s-1}[1 + f(\tau) + g(\tau)d\tau]\right], \tag{2.3.69}$$

where $q(n)$ is defined by (2.3.38), E, E^{-1} are as defined in Theorem 1.2.13, and $n \in \mathbb{N}_0$ is chosen so that for all $n \in \mathbb{N}_0$ with $0 \le n \le n_2$,

$$E(q(n)) + \sum_{\tau=0}^{n-1}[1 + f(\tau) + g(\tau)] \in Dom\,(E^{-1}).$$

(3) If for all $n \in \mathbb{N}_0$,

$$u^2(t) \le c^2 + 2\sum_{s=0}^{n-1}\left[f(s)u(s)\left(\sum_{\tau=0}^{s-1} g(\tau)W(u(\tau))\right) + h(s)u(s)\right], \tag{2.3.70}$$

then for all $n, n_3 \in \mathbb{N}_0, 0 \le n \le n_3$,

$$u(n) \le \Omega^{-1}\left[\Omega(q(n)) + \sum_{s=0}^{n-1} f(s)\left(\sum_{\tau=0}^{s-1} g(\tau)\right)\right], \tag{2.3.71}$$

where $q(n)$ is defined by (2.3.38), Ω, Ω^{-1} are as defined in Theorem 1.2.13 and $n \in \mathbb{N}_0$ is chosen so that for all $n \in \mathbb{N}_0$ with $0 \le n \le n_3$,

$$\Omega(q(n)) + \sum_{s=0}^{n-1}[f(s)\left(\sum_{\tau=0}^{s-1} g(\tau)\right) \in Dom\,(\Omega^{-1}).$$

Theorem 2.3.7 (The Yang Inequality [695]) *Suppose that $v, f \in S(\mathbb{N}_0, \mathbb{R}_+)$ and the following discrete inequality holds for all $n \in \mathbb{N}_0$,*

$$v_n^p \le c + \sum_{s=0}^{n-1} f_s v_s^q. \tag{2.3.72}$$

If $p = q$, then for all $n \in \mathbb{N}_0$,

$$v_n \le \left\{c \exp(\sum_{s=0}^{n-1} f_s)\right\}^{1/p}. \tag{2.3.73}$$

If $p > q$, then for all $n \in \mathbb{N}_0$,

$$v_n \le \left[c^{1-q/p} + (1 - q/p)\sum_{s=0}^{n-1} f_s\right]^{1-q/p}. \tag{2.3.74}$$

Theorem 2.3.8 (The Yang Inequality [695]) *Let v_n, g_n and h_n be sequences of non-negative real numbers, $n \in \mathbb{N}_0$, and $k \ge 0$ be a real constant. If the nonlinear* *discrete inequality holds for all* $n \in \mathbb{N}_0$,

$$v_n^2 \leq k_n^2 + 2\sum_{s=0}^{n-1}[h_s v_s + g_s v_s^2], \tag{2.3.75}$$

then for all $n \in \mathbb{N}_0$,

$$v_n \leq \left(k + \sum_{s=0}^{n-1} h_s\right)\sum_{s=0}^{n-1}(1 + g_s).$$

Theorem 2.3.9 (The Yang Inequality [695]) *Let* $a > 0, b > 0$ *and* $c > 0$ *be real constants. If* $v, f \in S(\mathbb{N}_0, \mathbb{R}_+)$, *the following discrete inequality holds for all* $n \in \mathbb{N}_0$,

$$v_n^2 \leq c + \sum_{s=0}^{n-1} f_s[av_s + bv_s^2], \tag{2.3.76}$$

then for all $n \in \mathbb{N}_0$,

$$v_n \leq \frac{a}{b}\left\{(1 + \frac{b}{a}\sqrt{c})\exp\left(\frac{b}{2}\sum_{s=0}^{n-1} f_s\right) - 1\right\}. \tag{2.3.77}$$

Proof The result is a corollary of Theorem 2.2.15. □

Remark 2.3.1 (i) Letting $p = q = 1$ in Theorem 2.3.7, from (2.3.72) we obtain the Bellman inequality and its discrete analogue, respectively; (ii) Letting $p = 2, q = 1, c = k^2$ and $F(t) = 2g(t)$ in Theorem 2.3.7, then from (2.3.72), we obtain Ou-Yang's integral inequality and its discrete analogue, respectively.

Remark 2.3.2 We compare the bound obtained by applying Theorem 2.3.9 to inequality $v_n^2 \leq k_n^2 + \sum_{s=0}^{n-1} max[h_s v_s + g_s v_s^2], \quad n \in \mathbb{N}_0$, from which we have for all $n \in \mathbb{N}_0$,

$$v_n^2 \leq k^2 + \sum_{s=0}^{n-1} max(g_s, h_s)[v_s + v_s^2]. \tag{2.3.78}$$

A suitable application of Theorem 2.3.9, with $a = b = 1, c = k^2$ to (2.3.78) yields for all $n \in \mathbb{N}_0$,

$$v_n \leq \left\{(1 + k)\exp\left[\frac{1}{2}\sum_{s=0}^{n-1} max(g_s, h_s) - 1\right]\right\}. \tag{2.3.79}$$

The bound (2.3.79) is different from as well as incomparable with the bound $v_n \leq [k + \sum_{s=0}^{n-1} h_s] \prod_{s=0}^{n-1}(1 + g_s)$, for all $n \in \mathbb{N}_0$.

Theorem 2.3.10 (The Willett-Wong Inequality [673]) *Suppose that $u_0(n)$, $w(n)$, $v(n-1)$, and $u(n)$ $(n = 1, 2, \cdots)$ are non-negative sequences of numbers with $v(0) = 0$. If the following inequality holds for $1 \leq p < +\infty$, $n = 0, 1, \ldots,$*

$$u(n+1) \leq u_0(n+1) + w(n+1)\left(\sum_{j=0}^{n} v(j)u^p(j)\right)^{1/p}, \quad n = 0, 1, \ldots \tag{2.3.80}$$

then for all $n = 0, 1, \ldots,$

$$\left(\sum_{j=0}^{n} v(j)u^p(j)\right)^{1/p} \leq \frac{\left(\sum_{j=0}^{n} v(j)u_0^p(j)e(j)\right)^{1/p}}{1 - (1 - e(n))^{1/p}} \tag{2.3.81}$$

where

$$e(n) = \prod_{i=0}^{n}\left(1 + v(i)w^p(i)\right)^{-1}. \tag{2.3.82}$$

Proof Define a sequence of numbers

$$\psi(k) = e(k)\sum_{j=0}^{k} v(j)u^p(j), \quad k = 0, 1, \ldots, n, \tag{2.3.83}$$

where

$$\begin{cases} e(k) - e(k-1) = -v(k)w^p(k)e(k), & k = 0, 1, \ldots, n, \\ e(0) = 1. \end{cases} \tag{2.3.84}$$

The solution $e(k)$ of (2.3.84) is given by (2.3.82). It follow from (2.3.80), (2.3.83) and (2.3.84) that

$$\psi(k) - \psi(k-1) \leq \left(v^{1/p}(k)u_0(k)e^{1/p}(k) + \frac{v^{1/p}(k)w(k)\psi^{1/p}(k-1)}{(1 + v(k)w^p(k))^{1/p}}\right)^p - \frac{v(k)w^p(k)\psi(k-1)}{1 + v(k)w^p(k)}, \quad k = 1, 2, \ldots, n. \tag{2.3.85}$$

Next, summing (2.3.85) from $k = 1$ to $k = n$, transposing the second sum in the right member, from the p-th root of both sides, and applying Minkowski's inequality for sums to the right-hand side, we obtain

$$\left(\psi(n) + \sum_{k=1}^{n} \frac{v(k)w^p(k)\psi(k-1)}{1 + v(k)w^p(k)}\right)^{1/p}$$

$$\leq \left(\sum_{k=1}^{n} v(k)u_0^p(k)e(k)\right)^{1/p} + \left(\sum_{k=1}^{n} \frac{v(k)w^p(k)\psi(k-1)}{1 + v(k)w^p(k)}\right)^{1/p}. \quad (2.3.86)$$

Transpose the second term of the right-hand side of (2.3.86) to obtain a left member of the form $f(x) = (c + x)^{1/p} - x^{1/p}(c \geq 0,\ p \geq 1)$. Since $f'(x) \leq 0$ for all $x \geq 0$, we may replace x by a large quantity without destroying inequality (2.3.86). In this regard, we note that

$$\sum_{k=1}^{n} \frac{v(k)w^p(k)\psi(k-1)}{1 + v(k)w^p(k)} = \sum_{k=1}^{n} \frac{v(k)w^p(k)\psi(k-1)}{1 + v(k)w^p(k)} \sum_{j=0}^{k-1} v(j)u^p(j) \leq$$

$$\sum_{k=1}^{n} v(k)w^p(k)e(k) \sum_{j=0}^{n} v(j)u^p(j) = (1 - e(n)) \sum_{j=0}^{n} v(j)u^p(j). \quad (2.3.87)$$

Estimate (2.3.81) follows by substituting (2.3.87) and (2.3.83) into (2.3.86). □

Theorem 2.3.11 (The Willett-Wong Inequality [673]) *Suppose that $v(n)$, $w(n)$ and $u(n+1)$ $(n = 0, 1, \ldots)$ are non-negative sequences of numbers with $v(0) = w(0) = 0$, and that u_0 and p are constants with $u_0 > 0$ and $p \geq 0,\ p \neq 1$. If the inequality holds for all $n = 0, 1, \ldots$,*

$$u(n+1) \leq u_0 + \sum_{j=0}^{n} v(j)u(j) + \sum_{j=0}^{n} w(j)u^p(j), \quad (2.3.88)$$

then for all $n = 0, 1, \ldots$,

$$e(n)u(n+1) \leq \left(u_0^q + q\sum_{k=0}^{n} w(k)e^q(k)\right)^{1/q}, \quad q = 1 - p, \quad (2.3.89)$$

where, for all $n = 0, 1, \ldots$,

$$e(n) = \prod_{j=0}^{n} \Big(1 + v(j)\Big)^{-1}. \quad (2.3.90)$$

Proof Define a sequence of numbers ψ by the right-hand side of (2.3.88), i. e.,

$$\psi(k) = u_0 + \sum_{j=0}^{k} v(j)u(j) + \sum_{j=0}^{k} w(j)u^p(j), \quad k = 0, 1, \ldots, n. \tag{2.3.91}$$

Then, noting that $u^p(k+1) \le \psi^p(k)$ for $p \ge 0$,

$$\psi(k+1) - \psi(k) \le v(k+1)\psi(k) + w(k+1)\psi^p(k). \tag{2.3.92}$$

Transposing $v(k+1)\psi(k)$ in (2.3.92) and multiplying by $e(k+1)$, where

$$\begin{cases} e(k+1) - e(k) = -v(k+1)e(k+1), \\ e(0) = 1, \end{cases} \tag{2.3.93}$$

We derive

$$\psi(k+1)e(k+1) - \psi(k)e(k) \le w(k+1)e^q(k+1)[\psi(k)e(k+1)]^p. \tag{2.3.94}$$

Since $\psi(k)$ is monotone increasing, $e(k)$ is monotone decreasing, and $q-1 \le 0$, we may get for all values x between $\psi(k)e(k)$ and $\psi(k+1)e(k+1)$,

$$[\psi(k)e(k+1)]^{q-1} \ge x^{q-1}.$$

Therefore if we apply the Mean Value Theorem to the function $f(x) = x^q/q$, we conclude that

$$\frac{[\psi(k+1)e(k+1)]^q - [\psi(k)e(k)]^q}{q} \le [\psi(k)e(k+1)]^{q-1}[\psi(k+1)e(k+1) - \psi(k)e(k)]. \tag{2.3.95}$$

From (2.3.94)–(2.3.95), and $q = 1-p$, it follows

$$[\psi(k+1)e(k+1)]^q - [\psi(k)e(k)]^q \le w(k+1)e^q(k+1),$$

from which (2.3.89) follows for all values $q \le 1,\ q \ne 0$ by summing. □

Note that most of the results above have been used in their weak forms in several aspects of differential equations, as we have already indicated to some extent. The corresponding strengthening of Gronwall's inequality. To be more specific, Theorem 2.3.11 may be used to extend the discussions given in [335].

For all $t_1 > t_2, t_1, t_2 \in \mathbb{N}_0$ and any function $u(t)$ defined on $\mathbb{N}_0$, we use the usual convention, $\sum_{s=t_1}^{t_2} u(s) = 0$ and $\prod_{s=t_1}^{t_2} u(s) = 1$. The operators L_j are recursively defined by

$$L_0 u(t) = u(t), \quad L_j u(t) = \frac{1}{r_j(t)} \Delta L_{j-1} u(t), \quad j = 1, 2, \ldots, n,$$

with $r_n(t) = 1$, where $u(t)$ and $r_j(t) > 0$ are some functions defined on $\mathbb{N}_0$. For all $t \in \mathbb{N}_0$ and some functions $r_j(t) > 0, j = 1, 2, \ldots, n-1$ and $q(t)$, we set

$$A(t, r, q(s_n)) = A(t, r_1, \ldots, r_{n-1}, q(s_n)) = \sum_{s_1=0}^{t-1} r_1(s_1) \cdots \sum_{s_{n-1}=0}^{s_{n-2}-1} r_{n-1}(s_{n-1}) \sum_{s_n=0}^{s_{n-1}-1} q(s_n),$$

where $s_0 = t$ and

$$\begin{aligned}\bar{A}(s_1, r, q(s_n)) &= \bar{A}(s_1, r_1, \ldots, r_{n-1}, q(s_n)) \\ &= r_1(s_1) \sum_{s_1=0}^{s_1-1} r_2(s_2) \cdots \sum_{s_{n-1}=0}^{s_{n-2}-1} r_{n-1}(s_{n-1}) \sum_{s_n=0}^{s_{n-1}-1} q(s_n).\end{aligned}$$

Let the product $\mathbb{N}_0 \times \cdots \times \mathbb{N}_0$ (n times) be denoted by $\mathbb{N}_0^n$. A point $(x_1, \ldots, x_n)$ in $\mathbb{N}_0^n$ is denoted by x. For any function $w(x)$ defined on $\mathbb{N}_0^n$, we define the operators $\Delta_1 w(x) = w(x_1 + 1, x_2, \ldots, x_n) - w(x), \ldots, \Delta_n w(x) = w(x_1, \ldots, x_{n-1}, x_n + 1) - w(x)$. For all $x, s \in \mathbb{N}_0^n$ and some function $q(x)$, we set

$$M(x, q(s)) = M(x_1, \ldots, x_n, q(s_1, \ldots, s_n)) = \sum_{s_1=0}^{x_1-1} \cdots \sum_{s_{n-1}=0}^{x_{n-1}-1} \sum_{s_n=0}^{x_n-1} q(s),$$

and

$$\bar{M}(s_1, x_2, \ldots, x_n, q(s)) = \sum_{s_1=0}^{x_1-1} \cdots \sum_{s_{n-1}=0}^{x_{n-1}-1} \sum_{s_n=0}^{x_n-1} q(s).$$

The next result is due to Pachpatte [495].

Theorem 2.3.12 (The Pachpatte Inequality [495]) *Let $f(t) \geq 0, g(t) \geq 0, r_i(t) > 0$, for $i = 1, 2, \ldots, n-1$, be real-valued functions defined on $\mathbb{N}_0$ and c be a non-negative real constant.*

(A_1) *Let $u(t) \geq 0$ be a non-negative real-valued function defined on $\mathbb{N}_0$. If for all $t \in \mathbb{N}_0$,*

$$u^2(t) \leq c^2 + 2A(t, r, f(s_n)u^2(s_n) + g(s_n)u(s_n)), \qquad (2.3.96)$$

then for all $t \in \mathbb{N}_0$,

$$u(t) \leq p(t) \prod_{s_1=0}^{t-1} \left(1 + \bar{A}(t, r, f(s_n))\right), \tag{2.3.97}$$

where for all $t \in \mathbb{N}_0$,

$$p(t) = c + A(t, r, g(s_n)). \tag{2.3.98}$$

(A_2) *Let* $u(t) \geq u_0 \geq 0$ *be a real-valued function defined on* $\mathbb{N}_0$; u_0 *is a real constant. Let* $W(u)$ *be a continuous non-decreasing real-valued function defined on an interval* $I = [u_0, +\infty)$ *and* $W(u) > 0$ *on* $(u_0, +\infty)$, $W(u_0) = 0$. *If for all* $t \in \mathbb{N}_0$,

$$u^2(t) \leq c^2 + 2A(t, r, f(s_n)u(s_n)W(u(s_n)) + g(s_n)u(s_n)), \tag{2.3.99}$$

then for all $0 \leq t \leq t_1$,

$$u(t) \leq \Omega^{-1}(\Omega(p(t)) + A(t, r, f(s_n))), \tag{2.3.100}$$

where $p(t)$ *is as defined in (2.3.98), and*

$$\Omega(h) = \int_{h_0}^{h} \frac{ds}{W(s)}, \quad h \geq h_0 > u_0, \tag{2.3.101}$$

Ω^{-1} *is the inverse of* Ω *and* $t_1 \in \mathbb{N}_0$ *can be chosen so that for all* $t \in \mathbb{N}_0$,

$$\Omega(p(t)) + A(t, r, f(s_n)) \in Dom\ (\Omega^{-1})$$

lying in $0 \leq t \leq t_1$.

(A_3) *Let* $u(t) \geq 0$ *be a real-valued function defined on* $\mathbb{N}_0$ *and the function* $L : \mathbb{N}_0 \times \mathbb{R}_+ \to \mathbb{R}_+$ *satisfies the condition: for all* $t \in \mathbb{N}_0$ *and* $v \geq w \geq 0$,

$$0 \leq L(t, v) - L(t, w) \leq k(t, w)(v - w), \tag{2.3.102}$$

where k *is a real-valued non-negative function defined for all* $t \in \mathbb{N}_0, w \geq 0$. *If for all* $t \in \mathbb{N}_0$,

$$u^2(t) \leq c^2 + 2A(t, r, f(s_n)u(s_n)L(s_n, u(s_n)) + g(s_n)u(s_n)), \tag{2.3.103}$$

then for all $t \in \mathbb{N}_0$,

$$u(t) \leq p(t) + q(t)\prod_{s_1=0}^{t-1}(1 + \bar{A}(t, r, f(s_n)k(s_n, p(s_n)))), \qquad (2.3.104)$$

where $p(t)$ *is as defined in (2.3.98) and for all* $t \in \mathbb{N}_0$,

$$q(t) = A(t, r, f(s_n)L(s_n, p(s_n))). \qquad (2.3.105)$$

Proof (A_1) We first assume that $c > 0$ and define a function $z(t)$ by

$$z(t) = c^2 + 2A(t, r, f(s_n)u^2(s_n) + g(s_n)u(s_n)). \qquad (2.3.106)$$

From (2.3.106) it follows

$$L_n z(t) = 2(f(t)u^2(t) + g(t)u(g)). \qquad (2.3.107)$$

Using the fact that $u(t) \leq \sqrt{z(t)}$ in (2.3.107), we have

$$L_n z(t) \leq 2\sqrt{z(t)}(f(t)\sqrt{z(t)} + g(t)). \qquad (2.3.108)$$

From (2.3.108) and using the fact that $z(t) \leq z(t+1)$, we derive that

$$\triangle\Big(\frac{L_{n-1}z(t)}{\sqrt{z(t)}}\Big) \leq 2(f(t)\sqrt{z(t)} + g(t)). \qquad (2.3.109)$$

Now setting $t = s_n$ in (2.3.109) and summing over $s_n = 0, 1, 2, \ldots, t-1$, we obtain

$$\frac{L_{n-1}z(t)}{\sqrt{z(t)}} \leq 2\sum_{s_n=0}^{t-1}(f(s_n)\sqrt{z(s_n)} + g(s_n)). \qquad (2.3.110)$$

Here we have used the fact that $L_{n-1}z(0) = 0$. Again as above, from (2.3.110), we derive

$$\triangle\Big(\frac{L_{n-2}z(t)}{\sqrt{z(t)}}\Big) \leq 2r_{n-1}(t)\sum_{s_n=0}^{t-1}(f(s_n)\sqrt{z(s_n)} + g(s_n)),$$

which readily yields

$$\frac{L_{n-2}z(t)}{\sqrt{z(t)}} \leq 2\sum_{s_{n-1}=0}^{t-1} r_{n-1}(s_{n-1})\sum_{s_n=0}^{s_{n-1}-1}(f(s_n)\sqrt{z(s_n)} + g(s_n)). \quad (2.3.111)$$

Here we have used the fact that $L_{n-2}z(0) = 0$. Continuing in this way, we may obtain

$$\frac{\Delta z(t)}{\sqrt{z(t)}} \leq 2r_1(t)\sum_{s_2=0}^{t-1} r_2(s_2)\cdots\sum_{s_{n-1}=0}^{s_{n-2}-1} r_{n-1}(s_{n-1})\sum_{s_n=0}^{s_{n-1}-1}(f(s_n)\sqrt{z(s_n)}+g(s_n)). \tag{2.3.112}$$

By using the facts that $\sqrt{z(t)} > 0$, $\Delta z(t) \geq 0$, $\sqrt{z(t)} \leq \sqrt{z(t+1)}$ for all $t \in \mathbb{N}_0$ and (2.3.112), we observe

$$\begin{aligned}\Delta\left(\sqrt{z(t)}\right) &= \frac{\Delta z(t)}{\sqrt{z(t+1)}+\sqrt{z(t)}} \leq \frac{\Delta z(t)}{2\sqrt{z(t)}}\\ &\leq 2r_1(t)\sum_{s_2=0}^{t-1} r_2(s_2)\cdots\sum_{s_{n-1}=0}^{s_{n-2}-1} r_{n-1}(s_{n-1})\sum_{s_n=0}^{s_{n-1}-1}(f(s_n)\sqrt{z(s_n)}+g(s_n)).\end{aligned} \tag{2.3.113}$$

Now, setting $t = s_1$ in (2.3.113) and summing over $s_1 = 0, 1, 2, \ldots, t-1$, we obtain

$$\sqrt{z(t)} \leq p(t) + A(t, r, f(s_n)\sqrt{z(s_n)}). \tag{2.3.114}$$

Since $p(t)$ is positive and monotone non-decreasing in t, from (2.3.114) we derive

$$\frac{\sqrt{z(t)}}{p(t)} \leq 1 + A\left(t, r, f(s_n)\frac{\sqrt{z(s_n)}}{p(s_n)}\right). \tag{2.3.115}$$

Define a function $v(t)$ by

$$v(t) \leq 1 + A\left(t, r, f(s_n)\frac{\sqrt{z(s_n)}}{p(s_n)}\right). \tag{2.3.116}$$

From (2.3.116) we derive

$$L_n v(t) = f(t)\frac{\sqrt{z(t)}}{p(t)}. \tag{2.3.117}$$

Using the fact that $\sqrt{z(t)}/p(t) \leq v(t)$ in (2.3.117), we have

$$L_n v(t) \leq f(t)v(t). \tag{2.3.118}$$

From (2.3.118), and using the facts that $v(t) \leq v(t+1)$, we obtain

$$\Delta\left(\frac{L_{n-1}v(t)}{v(t)}\right) \leq f(t). \tag{2.3.119}$$

Now, setting $t = s_n$ in (2.3.119), and summing over $s_n = 0, 1, 2, \ldots, t-1$, we obtain

$$\frac{L_{n-1}v(t)}{v(t)} \leq \sum_{s_n=0}^{t-1} f(s_n). \tag{2.3.120}$$

Here we have used the fact that $L_{n-1}v(0) = 0$. Again, as above, from (2.3.120), we deduce

$$\Delta\left(\frac{L_{n-2}v(t)}{v(t)}\right) \leq r_{n-1}(t)\sum_{s_n=0}^{t-1} f(s_n),$$

from which it follows

$$\frac{\Delta v(t)}{v(t)} \leq r_1(t)\sum_{s_2=0}^{t-1} r_2(s_2)\cdots\sum_{s_{n-1}=0}^{s_{n-2}-1} r_{n-1}(s_{n-1})\sum_{s_n=0}^{s_{n-1}-1} f(s_n),$$

i.e.,

$$v(t+1) \leq v(t)\Big(1 + \bar{A}(s_1, r, f(s_n))\Big). \tag{2.3.121}$$

Now, setting $t = s_1$ in (2.3.121) and substituting $s_1 = 0, 1, 2, \ldots, t-1$ successively, we obtain

$$v(t) \leq \prod_{s_1=0}^{t-1}(1 + \bar{A}(s_1, r, f(s_n))). \tag{2.3.122}$$

Using (2.3.122) in (2.3.115), and the fact that $u(t) \leq \sqrt{z(t)}$, we may get the required inequality in (2.3.97).

If c is non-negative, we carry out the above procedure with $c + \epsilon$ instead of c, where $\epsilon > 0$ is an arbitrary small constant, and subsequently pass to the limit as $\epsilon \to 0$ to obtain (2.3.97). The proof of (A_1) is complete.

(A_2) Assume that $c > 0$ and define a function $z(t)$ by

$$z(t) = c^2 + 2A(t, r, f(s_n)u(s_n W(u(s_n))) + g(s_n)u(s_n)). \tag{2.3.123}$$

From (2.3.123), and using the fact that $u(t) \le \sqrt{z(t)}$, we get

$$L_n z(t) \le 2\sqrt{z(t)}(f(t)W(\sqrt{z(t)}) + g(t)). \tag{2.3.124}$$

Now following the same steps as in the proof of Part (A_1), below (2.3.108) up to (2.3.114), we have

$$\sqrt{z(t)} \le p(t) + A\Big(t, r, f(s_n)W\Big(\sqrt{z(s_n)}\Big)\Big). \tag{2.3.125}$$

For an arbitrary fixed $T \in \mathbb{N}_0$, it follows form (2.3.125) that for all $0 \le t \le T,\ t,\ T \in \mathbb{N}_0$,

$$\sqrt{z(t)} \le p(T) + A\Big(t, r, f(s_n)W\Big(\sqrt{z(s_n)}\Big)\Big). \tag{2.3.126}$$

Define, for all $0 \le t \le T,\ t,\ T \in \mathbb{N}_0$,

$$v(t) = p(T) + A\Big(t, r, f(s_n)W\Big(\sqrt{z(s_n)}\Big)\Big). \tag{2.3.127}$$

From (2.3.127) and using the fact that $\sqrt{z(t)} \le v(t)$, we get

$$L_n v(t) \le f(t)W\big(v(t)\big). \tag{2.3.128}$$

From the definition of $v(t)$, we derive that $v(t) \le v(t+1)$ for all $t \in \mathbb{N}_0$. Using this fact in (2.3.128), we observe that

$$\Delta\left(\frac{L_{n-1}v(t)}{W\big(v(t)\big)}\right) \le f(t). \tag{2.3.129}$$

Now following the asme steps as given in the proof of Part (A_1), below (2.3.119) up to (2.3.121), we have

$$\frac{\Delta v(t)}{W\big(v(t)\big)} \le r_1(t)\sum_{s_2=0}^{t-1} r_2(s_2)\cdots \sum_{s_{n-1}=0}^{s_{n-2}-1} r_{n-1}(s_{n-1}) \sum_{s_n=0}^{s_{n-1}-1} f(s_n). \tag{2.3.130}$$

From (2.3.101) and (2.3.130), we conclude

$$\Omega(v(t+1)) - \Omega(v(t)) = \int_{v(t)}^{v(t+1)} \frac{ds}{W(s)} \le \frac{\Delta v(t)}{W\big(v(t)\big)}$$
$$\le r_1(t) \sum_{s_2=0}^{t-1} r_2(s_2) \cdots \sum_{s_{n-1}=0}^{s_{n-2}-1} r_{n-1}(s_{n-1}) \sum_{s_n=0}^{s_{n-1}-1} f(s_n). \tag{2.3.131}$$

Now, setting $t = s_1$ in (2.3.131) and summing over $s_1 = 0, 1, 2, \ldots, T-1$, we obtain

$$\Omega(v(T)) - \Omega(p(T)) = A(T, r, f(s_n)). \tag{2.3.132}$$

Since T is arbitrary, the inequality (2.3.132) holds for $t = T$, for all $t \in \mathbb{N}_0$ and hence,

$$v(t) \le \Omega^{-1}(\Omega(p(t)) + A(t, r, f(s_n))). \tag{2.3.133}$$

Using (2.3.133) in (2.3.127), and the fact that $u(t) \le \sqrt{z(t)}$, we get the required inequality in (2.3.100). The sub-domain of $\mathbb{N}_0$ for t is obvious.

The proof of the case when c is non-negative can be completed as mentioned in the proof of Part (A_1). This completes the proof of Part (A_2).

(A_3) Assume that c is positive and defined a function $z(t)$ by

$$z(t) = c^2 + 2A(t, r, f(s_n)u(s_n L(s_n, u(s_n))) + g(s_n)u(s_n)). \tag{2.3.134}$$

From (2.3.134), and using the fact that $u(t) \le \sqrt{z(t)}$, we get

$$L_n z(t) \le 2\sqrt{z(t)}(f(t)L(t, \sqrt{z(t)}) + g(t)). \tag{2.3.135}$$

Now, following the same steps as in the proof of Part (A_1), below (2.3.108) up to (2.3.114), we have

$$\sqrt{z(t)} \le p(t) + A\Big(t, r, f(s_n)L\Big(s_n, \sqrt{z(s_n)}\Big)\Big). \tag{2.3.136}$$

Define

$$v(t) = A\Big(t, r, f(s_n)L\Big(s_n, \sqrt{z(s_n)}\Big)\Big). \tag{2.3.137}$$

From (2.3.137), and using the fact that $\sqrt{z(t)} \leq p(t) + v(t)$ and (2.3.102), we obtain

$$\begin{aligned} L_n v(t) &= f(t)L\Big(t, \sqrt{z(t)}\Big) \\ &\leq f(t)L(t, p(t) + v(t)) \\ &= f(t)(L(t, p(t) + v(t)) - L(t, p(t))) + f(t)L(t, p(t)) \\ &\leq f(t)k(t, p(t))v(t) + f(t)L(t, p(t)). \end{aligned} \tag{2.3.138}$$

Thus from (2.3.138), it follows that

$$v(t) \leq q_\epsilon(t) + A(t, r, f(s_n)k(s_n, p(s_n))v(s_n)), \tag{2.3.139}$$

where $q_\epsilon(t) = \epsilon + q(t)$ in which $q(t)$ is as defined by (2.3.105) and $\epsilon > 0$ is an arbitrary small constant. Since $q_\epsilon(t)$ is positive and monotone non-decreasing for all $t \in \mathbb{N}_0$, from (2.3.139), we can derive

$$\frac{v(t)}{q_\epsilon(t)} \leq 1 + A(t, r, f(s_n)k(s_n, p(s_n))\frac{v(s_n)}{q_\epsilon(s_n)}). \tag{2.3.140}$$

Therefore the inequality (2.3.140) implies

$$v(t) \leq q_\epsilon(t)\prod_{s_1=0}^{t-1}(1 + \bar{A}(s_1, r, f(s_n)k(s_n, p(s_n)))). \tag{2.3.141}$$

The desired inequality (2.3.104) now follows by using (2.3.141) in (2.3.136) and then letting $\epsilon \to 0$ in the resulting inequality and using the fact that $u(t) \leq \sqrt{z(t)}$.

The proof of the case when c is non-negative can be completed as mentioned in the proof of Part (A_1). This completes the proof. □

Let $t = (t_1, t_2, \cdots, t_n)$ be a non-negative vector, and denote $[0, t_1] \times [0, t_2] \times \cdots \times [0, t_n]$ by $[0, t]$. The set of all bounded m-vector-valued functions defined on $[0, t]$ which are non-negative will be denoted by $B^{\geq 0}([0, t], m)$. For such a function u and positive numbers q and p, we shall denote by u^q the function obtained by taking the qth power of each component of u, and by $|u|^p$ the pth root of the sum of the components of u^p.For two such functions u_1 and u_2, we shall write $u_1 \geq u_2$ if the inequality $u_1(s) \geq u_2(s)$ holds (componentwise) for all $s \in [0, t]$. Further, a linear operator K on $B^{\geq 0}([0, t], m)$ will be called monotone if $u_1 \geq u_2$ implies that $Ku_1 \geq Ku_2$. The components of the linear operator K will be denoted by K_{ij}, $1 \leq i, j \leq m$. Let $T = \{t_0, t_1, \cdots, t_s\}$ denote a set of increasing time instances and, for a given function u defined T, we denote the value $u(t_i)$ by the shorthand notation $u(i)$. Furthermore, we shall use the shorthand notation I_1 and I_1' for $(j_1, i_2, \cdots, i_n)$ and $(j_1', i_2, \cdots, i_n)$, respectively, and I_1^{+1} for $(j_1 + 1, i_2, \cdots, i_n)$.

Theorem 2.3.13 (The Pang-Agarwal Inequality [528]) *Let $q > 1, p > 0, y$ be a non-negative m-vector-valued function on T^n, and let K, L be monotone linear operators on functions on $\{T|\{t_s\}\} \times T^{n-1}$. Furthermore, let c be a constant vector such that for all $i \in T^n$,*

$$y^q(i) \le c + q\sum_{j_1=0}^{i_1-1}[Ly^q(I_1) + Ky(I_1)].$$

Then the following inequality holds

$$|y|_p(i) \le \kappa(q/p,m)^{1/q}\left\{\prod_{j_1=0}^{i_1-1}(1+\overline{L}(I_1))|c|_1^{1/q} + \kappa(q,m)^{1/q(q-1)}\right.$$
$$\left.\times\left\{\kappa(q-1,i_1)\sum_{j_1=0}^{i_1-1}\prod_{j_1'=j_1+1}^{i_1-1}[1+\overline{L}(I_1')]^{q-1}\overline{K}(I_1)\right\}^{1/(q-1)}\right\}. \tag{2.3.142}$$

Proof We put

$$z(i) = c + q\sum_{j_1=0}^{i_1-1}[Ly^q(I_1) + Ky(I_1)].$$

Then, as in Theorem 7.4.1 (see below), we have

$$\Delta_1|z|_1(I_1) \le q\overline{L}(I_1)|z|_1(I_1) + q\overline{K}(I_1)\kappa(q,m)^{1/q}|z|_1^{1/q}(I_1).$$

Thus it follows that

$$|z|_1^{1/q}(I_1^{+1}) \le \left(1+q\overline{L}(I_1)\right)^{1/q}\left\{|z|_1^{1/q}(I_1) + \left[\frac{\overline{K}I_1\kappa(q,m)^{1/q}}{1+q\overline{L}(I_1)}\right]^{1/(q-1)}\right\}.$$

Solving this recursively, we may obtain

$$|z|_1^{1/q}(i) \le \prod_{j_1=0}^{i_1-1}(1+q\overline{L}(I_1))^{1/q}|c|_1^{1/q} + \kappa(q,m)^{1/q(q-1)}$$
$$\times\sum_{j_1=0}^{i_1-1}\left\{\prod_{j_1=j_1+1}^{i_1-1}(1+q\overline{L}(I_1'))^{(q-1)/q}\frac{\overline{K}(I_1)}{(1+q\overline{L}(I_1))^{1/q}}\right\}^{1/(q-1)}$$

$$\le \prod_{j_1=0}^{i_1-1}(1+\overline{L}(I_1))|c|_1^{1/q}+\kappa(q,m)^{1/q(q-1)}$$
$$\times\left\{\kappa(q-1,i_1)\sum_{j_1=0}^{i_1-1}\prod_{j_1=j_1+1}^{i_1-1}(1+\overline{L}(I_1'))^{q-1}\overline{K}(I_1)\right\}^{1/(q-1)}.$$

Thus inequality (2.3.142) the follows immediately by $|y|_p \le \kappa(q/p,m)^{1/q}|y|_q \le \kappa(q/p,m)^{1/q}|z|_1^{1/q}$. □

Remark 2.3.3 It is clear that the right-hand side of (2.3.142) is bounded above by

$$\prod_{j_1=0}^{i_1-1}(1+\overline{L}(I_1))\kappa(q/p,m)^{1/q}\left\{|c|_1^{1/q}+\left[\kappa(q,m)^{1/q}\kappa(q-1,i_1)\sum_{j_1=0}^{i_1-1}\overline{K}(I_1)\right]^{1/(q-1)}\right\}. \tag{2.3.143}$$

For the case $p=1, q=2$, (2.3.143) reduces to

$$\prod_{j_1=0}^{i_1-1}(1+\overline{L}(I_1))\left\{m^{1/2}|c|_1^{1/2}+m\sum_{j_1=0}^{i_1-1}\overline{K}(I_1)\right\}.$$

Thus, inequality (2.3.142) includes and in fact improves (2.3.131). We also note that, inequality (2.3.142) covers inequalities obtained by Pachpatte in [496].

Theorem 2.3.14 (The Pachpatte Inequality [506]) *Let $u(t), a(t), b(t), h(t)$ be real-valued non-negative functions defined for all $t \in \mathbb{N}_0$ and let c be a non-negative constant.*

(a_1) If for all $t \in \mathbb{N}_0$,

$$u^2(t) \le c^2 + \sum_{s=0}^{t-1}(u(s+1)+u(s))[a(s)u(s)+h(s)], \tag{2.3.144}$$

then for all $t \in \mathbb{N}_0$,

$$u(t) \le p(t)\prod_{s=t_1}^{t_2}[1+a(s)], \tag{2.3.145}$$

where for all $t \in \mathbb{N}_0$,

$$p(t) = c + \sum_{s=0}^{t-1} h(s). \tag{2.3.146}$$

(a_2) *If for all* $t \in \mathbb{N}_0$,

$$u^2(t) \le c^2 + \sum_{s=0}^{t-1}(u(s+1) + u(s)) \times \left[a(s)\left(u(s) + \sum_{\sigma=0}^{s-1} b(\sigma)u(\sigma)\right) + h(s)\right], \tag{2.3.147}$$

then for all $t \in \mathbb{N}_0$,

$$u(t) \le p(t)\left[1 + \sum_{s=0}^{t-1} a(s) \sum_{\sigma=0}^{s-1}[1 + a(\sigma) + b(\sigma)]\right] \tag{2.3.148}$$

where $p(t)$ *is defined by (2.3.144).*

(a_3) *If for all* $t \in \mathbb{N}_0$,

$$u^2(t) \le c^2 + \sum_{s=0}^{t-1}(u(s+1) + u(s))\left[a(s)\left(\sum_{\sigma=0}^{s-1} b(\sigma)u(\sigma)\right) + h(s)\right], \tag{2.3.149}$$

then for all $t \in \mathbb{N}_0$,

$$u(t) \le p(t)\prod_{s=0}^{t-1}\left[1 + a(s)\left(\sum_{\sigma=0}^{s-1} b(\sigma)\right)\right] \tag{2.3.150}$$

where $p(t)$ *is defined by (2.3.146).*

Proof (a_1) We first assume that $c > 0$ and define a function $z(t)$ by the right-hand side of (2.3.144). Then $z(0) = c^2$, $u(t) \le \sqrt{z(t)}$, and

$$\begin{aligned} \Delta z(t) &= (u(t+1) + u(t))[a(t)u(t) + h(t)] \\ &\le (\sqrt{z(t+1)} + \sqrt{z(t)})[a(t)\sqrt{z(t)} + h(t)]. \end{aligned} \tag{2.3.151}$$

Using the facts that $\sqrt{z(t)} > 0$, $\Delta z(t) \ge 0, t \in \mathbb{N}_0$, and (2.3.151), we observe that

$$\Delta\left(\sqrt{z(t)}\right) = \frac{\Delta z(t)}{\sqrt{z(t+1)} + \sqrt{z(t)}} \le a(t)\sqrt{z(t)} + h(t). \tag{2.3.152}$$

From (2.3.152) it follows

$$\sqrt{z(t)} \le p(t) + \sum_{s=0}^{t-1} a(s)\sqrt{z(s)}. \qquad (2.3.153)$$

Since $p(t)$ is a positive and non-decreasing function for all $t \in \mathbb{N}_0$, from (2.3.153) we infer

$$\frac{\sqrt{z(t)}}{p(t)} \le 1 + \sum_{s=0}^{t-1} a(s)\frac{\sqrt{z(s)}}{p(s)}. \qquad (2.3.154)$$

Now applying Corollary 2.1.4 in Qin [557] to (2.3.154) yields

$$\sqrt{z(t)} \le p(t) + \prod_{s=0}^{t-1}[1 + a(s)]. \qquad (2.3.155)$$

Using (2.3.155) in $u(t) \le \sqrt{z(t)}$, we can give the desired inequality in (2.3.145).

If c is non-negative, we carry out the above product with $c+\epsilon$ instead of c, where $\epsilon > 0$ is an arbitrary small constant, and subsequently pass to the limit as $\epsilon \to 0$ to obtain (2.3.145).

The proofs of assertions in (a_2) and (a_3) can be done in the same way. □

Theorem 2.3.15 (The Pachpatte Inequality [506]) *Let $u(t), a(t), b(t), h(t)$ and c be as in Theorem 2.3.14. Let $g(u)$ be a continuous non-decreasing function defined on $\mathbb{R}_+$ and $g(u) > 0$ for all $u > 0$.*

(b_1) If for all $t \in \mathbb{N}_0$,

$$u^2(t) \le c^2 + \sum_{s=0}^{t-1} a(s)(u(s+1) + u(s))g(u(s)) \qquad (2.3.156)$$

then, for all $t, t_1 \in \mathbb{N}_0, 0 \le t \le t_1$,

$$u(t) \le G^{-1}\left[G(c) + \sum_{s=0}^{t-1} a(s)\right], \qquad (2.3.157)$$

where

$$G(r) = \int_{r_0}^{r} \frac{ds}{g(s)}, \quad r \ge r_0 > 0, \qquad (2.3.158)$$

$r_0 > 0$ is arbitrary, G^{-1} is the inverse function of G, and $t_1 \in \mathbb{N}_0$ is chosen so that for all $t \in \mathbb{N}_0$ such that $0 \le t \le t_1$,

$$G(c) + \sum_{s=0}^{t-1} a(s) \in Dom\,(G^{-1}).$$

(b_2) If for all $t \in \mathbb{N}_0$,

$$u^2(t) \le c^2 + \sum_{s=0}^{t-1} a(s)(u(s+1) + u(s)) \left(u(s) + \sum_{\sigma=0}^{s-1} a(\sigma) g(u(\sigma)) \right) \tag{2.3.159}$$

then, for all $t, t_2 \in \mathbb{N}_0, 0 \le t \le t_2$,

$$u(t) \le c + \sum_{s=0}^{t-1} a(s) E^{-1} \left[E(c) + \sum_{\sigma=0}^{s-1} a(\sigma) \right], \tag{2.3.160}$$

where

$$E(r) = \int_{r_0}^{r} \frac{ds}{s + g(s)}, \qquad r \ge r_0 > 0, \tag{2.3.161}$$

$r_0 > 0$ is arbitrary, E^{-1} is the inverse function of E, and $t_2 \in \mathbb{N}_0$ is chosen so that for all $t \in \mathbb{N}_0$ such that $0 \le t \le t_2$,

$$E(c) + \sum_{\sigma=0}^{t-1} a(\sigma) \in Dom\,(E^{-1}).$$

(b_3) If for all $t \in \mathbb{N}_0$,

$$u^2(t) \le c^2 + \sum_{s=0}^{t-1} a(s)(u(s+1) + u(s)) \left(\sum_{\sigma=0}^{s-1} b(\sigma) g(u(\sigma)) \right) \tag{2.3.162}$$

then, for all $t, t_3 \in \mathbb{N}_0, 0 \le t \le t_3$,

$$u(t) \le G^{-1} \left[G(c) + \sum_{s=0}^{t-1} a(s) \left(\sum_{\sigma=0}^{s-1} b(\sigma) \right) \right], \tag{2.3.163}$$

where G, G^{-1} *are as in part* (b_1) *and* $t_3 \in \mathbb{N}_0$ *is chosen so that for all* $t \in \mathbb{N}_0$ *such that* $0 \le t \le t_3$,

$$G(c) + \sum_{s=0}^{t-1} a(s) \left(\sum_{\sigma=0}^{s-1} b(\sigma) \right) \in Dom\ (G^{-1}).$$

Proof We only prove (b_2), while the proofs of (b_1) and (b_3) can be done similarly.

(b_2) Assume that $c > 0$ and defined a function $z(t)$ by the right-hand side of (2.3.159). Then $z(0) = c^2, u(t) \le \sqrt{z(t)}$, and

$$\Delta z(t) = a(t)(u(t+1) + u(t)) \left(u(t) + \sum_{\sigma=0}^{t-1} a(\sigma) g(u(\sigma)) \right)$$

$$\times a(t)(\sqrt{z(t+1)} + \sqrt{z(t)}) \left(\sqrt{z(t)} + \sum_{\sigma=0}^{t-1} a(\sigma) g\left(\sqrt{z(\sigma)}\right) \right). \tag{2.3.164}$$

Using the facts that $\sqrt{z(t)} > 0,\ \Delta z(t) \ge 0, t \in \mathbb{N}_0$, and (2.3.164), we derive

$$\Delta\left(\sqrt{z(t)}\right) = \frac{\Delta z(t)}{\sqrt{z(t+1)} + \sqrt{z(t)}}$$
$$\le a(t) \left(\sqrt{z(t)} + \sum_{\sigma=0}^{t-1} a(\sigma) g\left(\sqrt{z(\sigma)}\right) \right). \tag{2.3.165}$$

Define a function $v(t)$ by

$$v(t) = \sqrt{z(t)} + \sum_{\sigma=0}^{t-1} a(\sigma) g\left(\sqrt{z(\sigma)}\right). \tag{2.3.166}$$

Then $v(0) = \sqrt{z(0)} = c$, $\sqrt{z(t)} \le v(t)$ from (2.3.166), $\Delta\left(\sqrt{z(t)}\right) \le a(t)v(t)$, and

$$\Delta v(t) = \Delta\left(\sqrt{z(t)}\right) + a(t) g\left(\sqrt{z(t)}\right)$$
$$\le a(t)(v(t) + g(v(t))). \tag{2.3.167}$$

From (2.3.161) and (2.3.166) it follows that

$$E(v(t+1)) - E(v(t)) = \int_{v(t)}^{v(t+1)} \frac{ds}{s+g(s)} \leq \frac{\Delta v(t)}{v(t)+g(v(t))} \leq a(t). \tag{2.3.168}$$

Taking $t = \sigma$ in (2.3.167) and summing up over σ from 0 to $t-1$, we get

$$E(v(t)) \leq E(c) + \sum_{\sigma=0}^{t-1} a(\sigma). \tag{2.3.169}$$

Substituting the bound on $v(t)$ from (2.3.168) into (2.3.164), we get

$$\Delta\left(\sqrt{z(t)}\right) \leq a(t)E^{-1}\left[\sum_{\sigma=0}^{t-1} a(\sigma)\right]. \tag{2.3.170}$$

From (2.3.169) it follows that

$$\sqrt{z(t)} \leq c + \sum_{s=0}^{s-1} a(s)E^{-1}\left[\sum_{\sigma=0}^{t-1} a(\sigma)\right]. \tag{2.3.171}$$

Using (2.3.170) in $u(t) \leq \sqrt{z(t)}$, we can give the required inequality in (2.3.170). The proof of the case when c is non-negative can be completed as mentioned in the proof of (a_1) given in Theorem 2.3.14. □

Theorem 2.3.16 (The Pachpatte Inequality [506]) *Let $u(t), v(t), a(t), b(t)$ be real-valued non-negative functions defined for all $t \in \mathbb{N}_0$.*

(c_1) If for all $s, t \in \mathbb{N}_0, 0 \leq s \leq t$,

$$u^2(t) \geq v^2(s) - \sum_{\sigma=s+1}^{t} a(\sigma)\Big(v(\sigma) + v(\sigma-1)\Big)v(\sigma) \tag{2.3.172}$$

then for all $s, t \in \mathbb{N}_0, 0 \leq s \leq t$,

$$u(t) \geq v(s)\left[\prod_{\sigma=s+1}^{t} [1 + a(\sigma)]\right]^{-1}. \tag{2.3.173}$$

(c_2) *If for all* $s, t \in \mathbb{N}_0, 0 \le s \le t$,

$$u^2(t) \le v^2(s) - \sum_{\sigma=s+1}^{t} a(\sigma)\Big(v(\sigma) + v(\sigma-1)\Big)\left(v(\sigma) + \sum_{\tau=\sigma+1}^{t} b(\tau)v(\tau)\right) \tag{2.3.174}$$

then for all $s, t \in \mathbb{N}_0, 0 \le s \le t$,

$$u(t) \ge v(s)\left[1 + \sum_{\sigma=s+1}^{t} a(\sigma) \prod_{\tau=\sigma+1}^{t} [1 + a(\tau) + b(\tau)]\right]^{-1}. \tag{2.3.175}$$

(c_3) *If for all* $s, t \in \mathbb{N}_0, 0 \le s \le t$,

$$u^2(t) \ge v^2(s) - \sum_{\sigma=s+1}^{t} a(\sigma)\Big(v(\sigma) + v(\sigma-1)\Big)\left(\sum_{\tau=\sigma+1}^{t} b(\tau)v(\tau)\right) \tag{2.3.176}$$

then for all $s, t \in \mathbb{N}_0, 0 \le s \le t$,

$$u(t) \ge v(s)\left[\prod_{\sigma=s+1}^{t}\left[1 + a(\sigma)\left(\sum_{\tau=\sigma+1}^{t} b(t)\right)\right]\right]^{-1}. \tag{2.3.177}$$

Proof We only prove (c_3), the proofs of (c_1) and (c_2) are similar.

(c_3) We may rewrite (2.3.176) as for all $s, t \in \mathbb{N}_0, 0 \le s \le t$,

$$v^2(s) \le u^2(t) + \sum_{\sigma=s+1}^{t} a(\sigma)(v(\sigma) + v(\sigma-1))\left(\sum_{\tau=\sigma+1}^{t} b(\tau)v(\tau)\right). \tag{2.3.178}$$

We first assume that $u(t)$ is positive for fixed $t \in \mathbb{N}_0$ and define a function $z(s)$ by the right-hand side of (2.3.178). Then $z(t) = u^2(t)$, $v(s) \le \sqrt{z(s)}$, and

$$\begin{aligned} z(s) - z(s+1) &= a(s+1)(v(s+1) + v(s))\left(\sum_{\tau=\sigma+2}^{t} b(\tau)v(\tau)\right) \\ &\le a(s+1)\left(\sqrt{z(s+1)} + \sqrt{z(s)}\right)\left(\sum_{\tau=\sigma+2}^{t} b(\tau)\sqrt{z(\tau)}\right). \end{aligned} \tag{2.3.179}$$

Using the facts that $\sqrt{z(s)} > 0, z(s) - z(s+1) \geq 0$ for all $s, t \in \mathbb{N}_0, 0 \leq s \leq t$, and (2.3.179), we observe that

$$\sqrt{z(s)} - \sqrt{z(s+1)} = \frac{z(s) - z(s+1)}{\sqrt{z(s)} + \sqrt{z(s+1)}} \leq a(s+1)\left(\sum_{\tau=\sigma+2}^{t} b(\tau)\sqrt{z(\tau)}\right). \quad (2.3.180)$$

Taking $s = \sigma$ in (2.3.180) and summing up over σ from 0 to $t-1$, we get

$$\sqrt{z(s)} \leq u(t) + \sum_{\sigma=s+1}^{t} a(\sigma)\left(\sum_{\tau=\sigma+1}^{t} b(\tau)\sqrt{z(\tau)}\right). \quad (2.3.181)$$

Define a function $m(s)$ by the right-hand side of (2.3.181). Then $m(t) = u(t)$, $\sqrt{z(s)} \leq m(s)$, $m(s)$ is decreasing with respect to $s \in \mathbb{N}_0$ for all $0 \leq s \leq t$, and

$$\begin{aligned} m(s) - m(s+1) &= a(s+1)\left(\sum_{\tau=s+2}^{t} b(\tau)\sqrt{z(t)}\right) \\ &\leq a(s+1)\left(\sum_{\tau=s+2}^{t} b(\tau)m(\tau)\right) \\ &\leq a(s+1)\left(\sum_{\tau=s+2}^{t} b(\tau)\right) m(s+1), \end{aligned}$$

i.e.,

$$m(s) \leq \left[1 + a(s+1)\left(\sum_{\tau=s+2}^{t} b(\tau)\right)\right] m(s+1). \quad (2.3.182)$$

Taking $s = \sigma$ and substituting $\sigma = s, s+1, s = 2, \cdots, t-1$ successively into (2.3.182), we obtain

$$m(s) \leq u(t) \prod_{\sigma=s+1}^{t} \left[1 + a(\sigma)\left(\sum_{\tau=\sigma+1}^{t} b(\tau)\right)\right]. \quad (2.3.183)$$

Using (2.3.183) in $v(s) \le \sqrt{z(s)} \le m(s)$, we have

$$v(s) \le u(t) \prod_{\sigma=s+1}^{t} \left[1 + a(\sigma) \left(\sum_{\tau=\sigma+1}^{t} b(\tau) \right) \right]. \tag{2.3.184}$$

Since $t \in \mathbb{N}_0$ is arbitrary, the required inequality (2.3.177) follows from (2.3.184). The proof of the case when $p(t)$ is non-negative for fixed $t \in \mathbb{N}_0$ can be completed as mentioned in the proof of part (a_1) in Theorem 2.3.14 given above. □

Theorem 2.3.17 (The Pachpatte Inequality [506]) *Let $u(t), a(t), b(t), h(t)$ be as in Theorem 2.3.14.*

(d_1) If for all $t \in \mathbb{N}_0$,

$$u^2(t) \le c^2 + \sum_{s=0}^{t-1} a(s)(u(s+1) + u(s)) \left(u(s) + \sum_{\sigma=0}^{s-1} [b(\sigma)u(\sigma) + h(\sigma)] \right), \tag{2.3.185}$$

then for all $t \in \mathbb{N}_0$,

$$u(t) \le q(t) \left[1 + \sum_{s=0}^{t-1} a(s) \prod_{\sigma=0}^{s-1} [1 + a(\sigma) + b(\sigma)] \right], \tag{2.3.186}$$

where for all $t \in \mathbb{N}_0$,

$$q(t) = c + \sum_{s=0}^{t-1} a(s) \left(\sum_{\sigma=0}^{s-1} h(\sigma) \right). \tag{2.3.187}$$

(d_2) If for all $t \in \mathbb{N}_0$,

$$u^2(t) \le c^2 + \sum_{s=0}^{t-1} a(s)(u(s+1) + u(s)) \left(\sum_{\sigma=0}^{s-1} [b(\sigma)u(\sigma) + h(\sigma)] \right), \tag{2.3.188}$$

then for all $t \in \mathbb{N}_0$,

$$u(t) \le q(t) \prod_{s=0}^{t-1} \left[1 + a(s) \left(\sum_{\sigma=0}^{s-1} b(\sigma) \right) \right], \tag{2.3.189}$$

where $q(t)$ is defined by (2.3.187).

(d_3) *Let $g(u)$ be as in Theorem 2.3.15. If for all $t \in \mathbb{N}_0$,*

$$u^2(t) \le c^2 + \sum_{s=0}^{t-1} a(s)(u(s+1)+u(s))g\left(u(s) + \sum_{\sigma=0}^{s-1} b(\sigma)g(u(\sigma))\right), \tag{2.3.190}$$

then, for all $t, t_1 \in \mathbb{N}_0, 0 \le t \le t_1$,

$$u(t) \le c + \sum_{s=0}^{t-1} a(s)g\left(G^{-1}\left[G(c) + \sum_{\sigma=0}^{s-1}[a(\sigma)+b(\sigma)]\right]\right), \tag{2.3.191}$$

where G, G^{-1} are as in (b_1) of Theorem 2.3.15 and $t_1 \in \mathbb{N}_0$ is chosen so that for all $t \in \mathbb{N}_0$ such that $0 \le t \le t_1$,

$$G(c) + \sum_{\sigma=0}^{t-1}[a(\sigma)+b(\sigma)] \in Dom\,(G^{-1}).$$

Proof The proofs of the inequalities in Theorem 2.3.17 can be completed by the following the proofs of Theorems 2.3.15–2.3.17. □

Theorem 2.3.18 (The Pachpatte Inequality [506]) *Let the function $u(t) \ge 1$ be defined for all $t \in \mathbb{N}_0$. Let $a(t), b(t)$ be non-negative functions defined for all $t \in \mathbb{N}_0$ and let $c \ge 1$ be a constant.*

(e_1) *If for all $t \in \mathbb{N}_0$,*

$$u^2(t) \le c^2 + \sum_{s=0}^{t-1} a(s)\Big(u(s+1)+u(s)\Big)u(s)\log u(s), \tag{2.3.192}$$

then for all $t \in \mathbb{N}_0$,

$$u(t) \le c^{A(t)} \tag{2.3.193}$$

where for all $t \in \mathbb{N}_0$,

$$A(t) = \prod_{s=0}^{t-1}(1+a(s)). \tag{2.3.194}$$

(e_2) *If for all $t \in \mathbb{N}_0$,*

$$u^2(t) \le c^2 + \textstyle\sum_{s=0}^{t-1} a(s)(u(s+1) + u(s))u(s) \\ \times \left(\log u(s) + \textstyle\sum_{\sigma=0}^{s-1} b(\sigma) \log u(\sigma)\right), \tag{2.3.195}$$

then for all $t \in \mathbb{N}_0$,

$$u(t) \le c^{B(t)} \tag{2.3.196}$$

where for all $t \in \mathbb{N}_0$,

$$B(t) = 1 + \sum_{s=0}^{t-1} a(s) \prod_{\sigma=0}^{s-1} [1 + a(\sigma) + b(\sigma)]. \tag{2.3.197}$$

(e_3) *If for all $t \in \mathbb{N}_0$,*

$$u^2(t) \le c^2 + \sum_{s=0}^{t-1} a(s)(u(s+1) + u(s))u(s) \left(\sum_{\sigma=0}^{s-1} b(\sigma) \log u(\sigma)\right), \tag{2.3.198}$$

then for all $t \in \mathbb{N}_0$,

$$u(t) \le c^{Q(t)} \tag{2.3.199}$$

where for all $t \in \mathbb{N}_0$,

$$Q(t) = \prod_{s=0}^{t-1} \left[1 + a(s) \sum_{\sigma=0}^{s-1} b(\sigma)\right]. \tag{2.3.200}$$

Proof (e_1) Define a function $z(t)$ by the right-hand side of (2.3.192). Then $z(0) = c^2$, $u(t) \le \sqrt{z(t)}$, and

$$\Delta z(t) = a(t)(u(t+1) + u(t))u(t) \log u(t) \\ \le a(t)\left(\sqrt{z(t+1)} + \sqrt{z(t)}\right)\sqrt{z(t)} \log(\sqrt{z}(t)). \tag{2.3.201}$$

By using the facts that $\sqrt{z(t)} > 0$, $\Delta z(t) \ge 0$, and (2.3.201), we conclude

$$\Delta(\sqrt{z(t)}) = \frac{\Delta z(t)}{\sqrt{z(t+1)} + \sqrt{z(t)}} \\ \le a(t)\sqrt{z(t)} \log \sqrt{z(t)}. \tag{2.3.202}$$

From (2.3.202) it follows

$$\sqrt{z(t+1)} \le [1 + a(t) \log \sqrt{z(t)}] \sqrt{z(t)}. \tag{2.3.203}$$

Setting $t = s$ in (2.3.203) and substituting $s = 0, 1, 2, \cdots, t-1$, successively, we get

$$\begin{aligned} \sqrt{z(t)} &\le c \prod_{s=0}^{t-1} \left[1 + a(s) \log \sqrt{z(s)} \right] \\ &\le c \exp \left(\sum_{s=0}^{t-1} a(s) \log \sqrt{z(s)} \right). \end{aligned} \tag{2.3.204}$$

From (2.3.204) we infer

$$\log \sqrt{z(t)} \le \log c + \sum_{s=0}^{t-1} \left(a(s) \log \sqrt{z(s)} \right). \tag{2.3.205}$$

Now applying Corollary 2.1.4 in Qin [557] yields

$$\begin{aligned} \log \sqrt{z(t)} &\le (\log c) A(t) \\ &= \log c^{A(t)} \end{aligned} \tag{2.3.206}$$

where $A(t)$ is defined by (2.3.194). From (2.3.206) we deduce that

$$\sqrt{z(t)} \le c^{A(t)}. \tag{2.3.207}$$

Using (2.3.207) in $u(t) \le \sqrt{z(t)}$, we get the required inequality (2.3.193). □

Theorem 2.3.19 (The Pachpatte Inequality [506]) *Let $u(t), a(t), b(t)$ and c be as in Theorem 2.3.18. Let $g(u)$ be as in Theorem 2.3.15.*

(p_1) If for all $t \in \mathbb{N}_0$,

$$u^2(t) \le c^2 + \sum_{s=0}^{t-1} a(s) \Big(u(s+1) + u(s) \Big) u(s) g(\log u(s)), \tag{2.3.208}$$

then for all $t, t_1 \in \mathbb{N}_0, 0 \le t \le t_1$,

$$u(t) \le \exp \left[G^{-1} \left[G(\log c) + \sum_{s=0}^{t-1} a(s) \right] \right], \tag{2.3.209}$$

where G, G^{-1} are as defined in (b_1) of Theorem 2.3.15 and $t_1 \in \mathbb{N}_0$ is chosen so that for all $t \in \mathbb{N}_0$ such that $0 \le t \le t_1$,

$$G(\log c) + \sum_{s=0}^{t-1} a(s) \in Dom\,(G^{-1}).$$

(p_2) *If for all $t \in \mathbb{N}_0$,*

$$u^2(t) \le c^2 + \sum_{s=0}^{t-1} a(s)(u(s+1) + u(s))u(s) \times \left(\log u(s) + \sum_{\sigma=0}^{s-1} b(\sigma) \log u(\sigma)\right), \tag{2.3.210}$$

then for all $t, t_2 \in \mathbb{N}_0, 0 \le t \le t_2$,

$$u(t) \le c \exp\left[\sum_{s=0}^{t-1} a(s) E^{-1}\left[E(\log c) + \sum_{\sigma=0}^{s-1} a(\sigma)\right]\right], \tag{2.3.211}$$

where E, E^{-1} are as defined in (b_2) of Theorem 2.3.15 and $t_2 \in \mathbb{N}_0$ is chosen so that for all $t \in \mathbb{N}_0$ such that for all $0 \le t \le t_2$,

$$E(\log c) + \sum_{s=0}^{t-1} a(\sigma) \in Dom\,(E^{-1}).$$

(p_3) *If for all $t \in \mathbb{N}_0$,*

$$u^2(t) \le c^2 + \sum_{s=0}^{t-1} a(s)(u(s+1) + u(s))u(s) \left(\sum_{\sigma=0}^{s-1} b(\sigma) g(\log u(\sigma))\right), \tag{2.3.212}$$

then for all $t, t_3 \in \mathbb{N}_0, 0 \le t \le t_3$,

$$u(t) \le \exp\left[G^{-1}\left[G(\log c) + \sum_{s=0}^{s-1} a(s) \left(\sum_{\sigma=0}^{s-1} b(\sigma)\right)\right]\right], \tag{2.3.213}$$

where G, G^{-1} are as defined in (b_1) of Theorem 2.3.15 and $t_3 \in \mathbb{N}_0$ is chosen so that for all $t \in \mathbb{N}_0$ such that for all $0 \le t \le t_3$,

$$G(\log c) + \sum_{s=0}^{t-1} a(s) \left(\sum_{\sigma=0}^{s-1} b(\sigma)\right) \in Dom\,(G^{-1}).$$

Proof We only prove (p_2), the proof of (p_1) can be done similarly.

(p_2) Defined a function $z(t)$ by the right-hand side of (2.3.210). Then $z(0) = c^2, u(t) \le \sqrt{z(t)}$, and

$$\begin{aligned}\Delta z(t) &= a(t)(u(t+1)+u(t))u(t) \times \left(\log u(t) + \sum_{\sigma=0}^{t-1} a(\sigma)g(\log u(\sigma))\right) \\ &\le a(t)(\sqrt{z(t+1)} + \sqrt{z(t)}\sqrt{z(t)} \times \left(\log\sqrt{z(t)} + \sum_{\sigma=0}^{t-1} a(\sigma)g(\log\sqrt{z(\sigma)})\right).\end{aligned} \tag{2.3.214}$$

Using the facts that $\sqrt{z(t)} > 0$, $\Delta z(t) \ge 0$, for all $t \in \mathbb{N}_0$, and (2.3.214), we get

$$\begin{aligned}\Delta(\sqrt{z(t)}) &= \frac{\Delta z(t)}{\sqrt{z(t+1)} + \sqrt{z(t)}} \\ &\le a(t)\sqrt{z(t)}\left(\log\sqrt{z(t)} + \sum_{\sigma=0}^{t-1} a(\sigma)g(\log\sqrt{z(\sigma)})\right).\end{aligned} \tag{2.3.215}$$

Thus from (2.3.215) it follows

$$\sqrt{z(t+1)} \le \left[1 + a(t)\left(\log\sqrt{z(t)} + \sum_{\sigma=0}^{t-1} a(\sigma)g(\log\sqrt{z(\sigma)})\right)\right]. \tag{2.3.216}$$

Letting $t = s$ in (2.3.216) and substituting $s = 0, 1, 2, \cdots, t-1$ successively, we obtain

$$\begin{aligned}\sqrt{z(t)} &\le c\prod_{s=0}^{t-1}\left[1 + a(s)\left(\log\sqrt{z(s)} + \sum_{\sigma=0}^{s-1} a(\sigma)g(\log\sqrt{z(\sigma)})\right)\right] \\ &\le c\exp\left(\sum_{s=0}^{t-1} a(s)\left(\log\sqrt{z(s)} + \sum_{\sigma=0}^{s-1} a(\sigma)g(\log\sqrt{z(\sigma)})\right)\right).\end{aligned} \tag{2.3.217}$$

From (2.3.217) we derive that

$$\log\sqrt{z(t)} \le \log c + \sum_{s=0}^{t-1} a(s)\left(\log\sqrt{z(s)} + \sum_{\sigma=0}^{s-1} a(\sigma)g(\log\sqrt{z(\sigma)})\right). \tag{2.3.218}$$

Define a function $v(t)$ by the right-hand side of (2.3.218). Then $v(0) = \log c$, $\log\sqrt{z(t)} \le v(t)$, and

$$\Delta v(t) = a(t)\left(\log\sqrt{z(t)} + \sum_{\sigma=0}^{t-1} a(\sigma)g(\log\sqrt{z(\sigma)})\right)$$

$$\le a(t)\left(v(t) + \sum_{\sigma=0}^{t-1} a(\sigma)g(v(\sigma))\right). \tag{2.3.219}$$

Define a function $m(t)$ by

$$m(t) = v(t) + \sum_{\sigma=0}^{t-1} a(\sigma)g(v(\sigma)). \tag{2.3.220}$$

Then $m(0) = v(0) = \log c$, $\Delta v(t) \le a(t)m(t)$ from (2.3.219), $v(t) \le m(t)$ from (2.3.220), and

$$\Delta m(t) \le a(t)(m(t) + g(m(t))). \tag{2.3.221}$$

The remaining proof can be completed by following the proof of inequality (b_2) in Theorem 2.3.15 given above, and hence we omit the details. □

Lemma 2.3.1 (The Pachpatte Inequality [502]) *Let $w(t,r)$ be a real-valued continuous function defined for all $t \in \mathbb{R}_+$, $0 \le r < +\infty$. Let $u(t)$ be a real-valued differentiable function defined for all $t \in \mathbb{R}_+$ such that for all $t \in \mathbb{R}_+$,*

$$u'(t) \le w(t, u(t)).$$

Let $r(t)$ be a maximal solution of

$$r'(t) = w(t, r(t)),\ r(0) = r_0,$$

such that $u(0) \le r_0$. Then for all $t \in \mathbb{R}_+$,

$$u(t) \le r(t).$$

Theorem 2.3.20 (The Pachpatte Inequality [502]) *Let y, f, g be real-valued non-negative functions defined on $\mathbb{N}_0$ and c be a non-negative real constant. Let $w(t,r)$ be a real-valued non-negative function defined for all $n \in \mathbb{N}_0$, $0 \le r < +\infty$, and monotone non-decreasing with respect to r for any fixed $n \in \mathbb{N}_0$.*

(1) *If for all* $n \in \mathbb{N}_0$,

$$y^2(n) \le c^2 + 2\sum_{s=0}^{n-1} y(s)w(s, y(s)), \tag{2.3.222}$$

then for all $n \in \mathbb{N}_0$,

$$y(n) \le r(n), \tag{2.3.223}$$

where $r(n)$ *is a solution of*

$$\Delta r(n) = w(n, r(n)), \quad r(0) = c. \tag{2.3.224}$$

(2) *If for all* $n \in \mathbb{N}_0$,

$$y^2(n) \le c^2 + 2\sum_{s=0}^{n-1} y(s)[f(s)y(s) + w(s, y(s))], \tag{2.3.225}$$

then for all $n \in \mathbb{N}_0$,

$$y(n) \le L(n)r(n), \tag{2.3.226}$$

where

$$L(n) = \prod_{s=0}^{n-1}[1 + f(s)], \tag{2.3.227}$$

and $r(n)$ *is a solution of*

$$\Delta r(n) = w(n, L(n)r(n)), \quad r(0) = c. \tag{2.3.228}$$

(3) *If for all* $n \in \mathbb{N}_0$,

$$y^2(n) \le c^2 + 2\sum_{s=0}^{n-1} y(s)[f(s)(y(s) + \sum_{t=0}^{s-1} g(t)y(t)) + w(s, y(s))], \tag{2.3.229}$$

then for all $n \in \mathbb{N}_0$,

$$y(n) \le P(n)r(n), \tag{2.3.230}$$

where

$$P(n) = 1 + \sum_{s=0}^{n-1} f(s) \prod_{t=0}^{s-1} [1 + f(t) + g(t)], \tag{2.3.231}$$

and $r(n)$ *is a solution of*

$$\Delta r(n) = w(n, P(n)r(n)), \quad r(0) = c. \tag{2.3.232}$$

(4) *If for all* $n \in \mathbb{N}_0$,

$$y^2(n) \le c^2 + 2\sum_{s=0}^{n-1} y(s)[f(s)(\sum_{t=0}^{s-1} g(t)y(t)) + w(s, y(s))], \tag{2.3.233}$$

then for all $n \in \mathbb{N}_0$,

$$y(n) \le Q(n)r(n), \tag{2.3.234}$$

where

$$Q(n) = \prod_{s=0}^{n-1} [1 + f(s) \sum_{t=0}^{s-1} g(t)], \tag{2.3.235}$$

and $r(n)$ *is a solution of*

$$\Delta r(n) = w(n, Q(n)r(n)), \quad r(0) = c. \tag{2.3.236}$$

Proof We only prove (3) and (4), the proofs of (1) and (2) can be done similarly.

(3) We assume that $c > 0$ and define a function $z(n)$ by

$$z(n) = c^2 + 2\sum_{s=0}^{n-1} y(s)\left[f(s)\left(y(s) + \sum_{t=0}^{s-1} g(t)y(t) \right) + w(s, y(s)) \right]. \tag{2.3.237}$$

From (2.3.237) and using the fact that $y(n) \le \sqrt{z(n)}$ we observe that

$$\Delta z(n) \le 2\sqrt{z(n)}\left[f(n)\left(\sqrt{z(n)} + \sum_{t=0}^{n-1} g(t)\sqrt{z(t)} \right) + w(n, \sqrt{z(n)}) \right]. \tag{2.3.238}$$

It is easy to observe that

$$\Delta(\sqrt{z(n)}) = \frac{z(n+1) - z(n)}{\sqrt{z(n+1)} + \sqrt{z(n)}} \leq \frac{\Delta z(n)}{2\sqrt{z(n)}}. \tag{2.3.239}$$

Here in the last step of (2.3.239) we have used the fact that $\sqrt{z(n)} \leq \sqrt{z(n+1)}$. By using (2.3.238) in (2.3.239) we get

$$\Delta(\sqrt{z(n)}) \leq f(n)\left(\sqrt{z(n)} + \sum_{t=0}^{n-1} g(t)\sqrt{z(t)}\right) + w(s, \sqrt{z(s)}. \tag{2.3.240}$$

From (2.3.240) it is easy to observe that

$$\sqrt{z(n)} \leq c + \sum_{s=0}^{n-1} f(s)\left(\sqrt{z(s)} + \sum_{t=0}^{s-1} g(t)\sqrt{z(t)}\right) + \sum_{s=0}^{n-1} w(s, \sqrt{z(s)}). \tag{2.3.241}$$

Define a function $m(n)$ by

$$m(n) = c + \sum_{s=0}^{n-1} w(s, \sqrt{z(s)}). \tag{2.3.242}$$

By using (2.3.242), the inequality (2.3.241) can be written as

$$\sqrt{z(n)} \leq m(n) + \sum_{s=0}^{n-1} f(s)\left(\sqrt{z(s)} + \sum_{t=0}^{s-1} g(t)\sqrt{z(t)}\right). \tag{2.3.243}$$

Since $m(n)$ is positive and monotone non-decreasing for $n \in N_0$, the inequality (2.3.243) implies the estimate

$$\sqrt{z(n)} \leq P(n)m(n), \; n \in N_0, \tag{2.3.244}$$

where $P(n)$ is defined by (2.3.231). From (2.3.242) and using (2.3.244) we observe that for all $n \in N_0$

$$\Delta m(n) \leq w(n, P(n)m(n)). \tag{2.3.245}$$

Now applying Lemma 2.3.1 to (2.3.245) and (2.3.232) yields for all $n \in N_0$

$$m(n) \leq r(n), \tag{2.3.246}$$

where $r(n)$ is a solution of (2.3.232). Using (2.3.246) in (2.3.244) we have for all $n \in N_0$

$$\sqrt{z(n)} \leq P(n)r(n). \tag{2.3.247}$$

Now by using the fact that $y(n) \leq \sqrt{z(n)}$ in (2.3.247) we get the required inequality in (2.3.230). If c is non-negative, we can carry out the above produce with $c + \varepsilon$ instead of c, where $\varepsilon > 0$ is an arbitrary small constant, and subsequently pass to the limit $\varepsilon \to 0$ to obtain (2.3.230). This completes the proof of (3).

(4) By assuming that $c > 0$ and defining a function $z(n)$ by the right side of (2.3.233) and following the same steps as in the proof of (3) upto (2.3.240) we (2.3.240) we have

$$\Delta(\sqrt{z(n)}) \leq f(n) \sum_{t=0}^{n-1} g(t)\sqrt{z(t)} + w(s, \sqrt{z(s)}). \tag{2.3.248}$$

From (2.3.248) and using the fact that $\sqrt{z(n)}$ is monotone non-decreasing for $n \in N_0$, it is easy observe that

$$\begin{aligned} \sqrt{z(n)} &\leq c + \sum_{s=0}^{n-1} f(s)\Big(\sum_{t=0}^{s-1} g(t)\sqrt{z(t)}\Big) + \sum_{s=0}^{n-1} w(s, \sqrt{z(s)}) \\ &\leq c + \sum_{s=0}^{n-1} f(s)\sqrt{z(s)}\Big(\sum_{t=0}^{s-1} g(t)\Big) + \sum_{s=0}^{n-1} w(s, \sqrt{z(s)}). \end{aligned} \tag{2.3.249}$$

Now by following the same steps as in the proof of (3) below (2.3.241) with suitable modifications we get the required inequality in (2.3.234). This completes the proof of (4). □

Theorem 2.3.21 (The Pachpatte Inequality [506]) *Let $u(t), a(t), b(t)$, and c be as in Theorem 2.3.10. Let $w(t, r)$ be a real-valued non-negative function defined for all $t \in \mathbb{N}_0, 0 \leq r < +\infty$, and non-decreasing with respect to r for any fixed $t \in \mathbb{N}_0$.*

(q_1) If for all $t \in \mathbb{N}_0$,

$$u^2(t) \leq c^2 + \sum_{s=0}^{t-1} a(s)\Big(u(s+1) + u(s)\Big)\Big(u(s) + w(s, u(s))\Big), \tag{2.3.250}$$

then for all $t \in \mathbb{N}_0$,

$$u(t) \leq A(t)r(t) \tag{2.3.251}$$

where $A(t)$ is defined by (2.3.194) in Theorem 2.3.18 and $r(t)$ is a solution of the difference equation, for all $t \in \mathbb{N}_0$,

$$\Delta r(t) = a(t)w(t, A(t)r(t)), \quad r(0) = c. \tag{2.3.252}$$

(q_2) If for all $t \in \mathbb{N}_0$,

$$u^2(t) \leq c^2 + \sum_{s=0}^{t-1} a(s)(u(s+1) + u(s)) \times \left[\left((u(s) + \sum_{\sigma=0}^{s-1} b(\sigma)u(\sigma) \right) + w(s, u(s)) \right], \tag{2.3.253}$$

then for all $t \in \mathbb{N}_0$,

$$u(t) \leq B(t)r(t) \tag{2.3.254}$$

where $B(t)$ is defined by (2.3.197) in Theorem 2.3.18 and $r(t)$ is a solution of the difference equation, for all $t \in \mathbb{N}_0$,

$$\Delta r(t) = a(t)w(t, B(t)r(t)), \quad r(0) = c. \tag{2.3.255}$$

(q_3) If for all $t \in \mathbb{N}_0$,

$$u^2(t) \leq c^2 + \sum_{s=0}^{t-1} a(s)(u(s+1) + u(s)) \times \left[\left(\sum_{\sigma=0}^{s-1} b(\sigma)u(\sigma) \right) + w(s, u(s)) \right], \tag{2.3.256}$$

then for all $t \in \mathbb{N}_0$,

$$u(t) \leq Q(t)r(t) \tag{2.3.257}$$

where $Q(t)$ is defined by (2.3.200) in Theorem 2.3.18 and $r(t)$ is a solution of the difference equation, for all $t \in \mathbb{N}_0$,

$$\Delta r(t) = a(t)w(t, Q(t)r(t)), \quad r(0) = c. \tag{2.3.258}$$

Proof We only prove (q_3) here, the proofs of (q_1) and (q_2) can be done similarly.

(q_3) Let $c > 0$ and define a function $z(t)$ by the right-hand side of (2.3.256). Then $z(0) = c^2, u(t) \leq \sqrt{z(t)}$, and

$$\begin{aligned}\Delta z(t) &= a(t)(u(t+1)+u(t))\left[\left(\sum_{\sigma=0}^{t-1} b(\sigma)u(\sigma)\right)+w(t,u(t))\right]\\ &\leq a(t)(\sqrt{z(t+1)}+\sqrt{z(t)})\times\left[\left(\sum_{\sigma=0}^{t-1} b(\sigma)\sqrt{z(\sigma)}+w(t,\sqrt{z(t)})\right)\right].\end{aligned} \tag{2.3.259}$$

Using the facts that $\sqrt{z(t)} > 0$, $\Delta z(t) \geq 0$ and (2.3.259) and the following similar arguments as in the proofs of the inequalities given above and closely looking at the proof of Theorem 2.3.20 with suitable modifications, one can very easily complete the rest of the proof, and we leave the details to the reader. □

Let $\mathbb{N}_{n_0} = \{n_0, n_0+1, n_0+2, \ldots\}$, where $n_0 \geq 0$ is an integer. For any function $k(n,s) : \mathbb{N}_{n_0} \to \mathbb{R}_+$, we define the operator Δ_1 by $\Delta_1 k(n,s) = k(n+1,s) - k(n,s)$ for all $n_0 \leq s \leq n$. We use the usual convention of writing $\Sigma_{s\in\Phi}u(s) = 0$ and $\Pi_{s\in\Phi}u(s) = 1$, Φ it is the empty set.

Theorem 2.3.22 (The Pachpatte Inequality [512]) *Let u, a, b, g, h be real-valued non-negative continuous functions defined on $\mathbb{N}_{n_0}$ and $p > 1$ be a real constant.*

(c_1) *If for all $n \in \mathbb{N}_{n_0}$,*

$$u^p(n) \leq a(n) + b(n)\sum_{s=n_0}^{n-1}[g(s)u^p(s) + h(s)u(s)], \tag{2.3.260}$$

then for all $n \in \mathbb{N}_{n_0}$,

$$\begin{aligned}u(n) \leq \Bigg\{&a(n) + b(n)\sum_{s=n_0}^{n-1}(g(s)a(s) + h(s)(\frac{p-1}{p} + \frac{a(s)}{p}))\\ &\times\prod_{\sigma=s+1}^{n-1}[1 + b(\sigma)(g(\sigma) + \frac{h(\sigma)}{p})]\Bigg\}^{1/p}.\end{aligned} \tag{2.3.261}$$

(c_2) *Let $c(n)$ a real-valued positive and non-decreasing function defined on $\mathbb{N}_{n_0}$. If for all $n \in \mathbb{N}_{n_0}$,*

$$u^p(n) \leq c^p(n) + b(n)\sum_{s=n_0}^{n-1}[g(s)u^p(s) + h(s)u(s)], \tag{2.3.262}$$

then for all $n \in \mathbb{N}_{n_0}$,

$$u(n) \le c(n)\left\{1 + b(n)\sum_{s=n_0}^{n-1}[g(s) + h(s)c^{1-p}(s)]\right.$$
$$\left.\times \prod_{\sigma=s+1}^{n-1}[1 + b(\sigma)(g(\sigma) + \frac{h(\sigma)}{p}c^{1-p}(\sigma))]\right\}^{1/p}. \quad (2.3.263)$$

(c_3) *Let* $k(n,s)$, $\Delta_1 k(n,s)$ *be a real-valued non-negative functions for all* $s, n \in \mathbb{N}_{n_0}$, $n_0 \le s \le n$. *If for all* $n \in \mathbb{N}_{n_0}$,

$$u^p(n) \le a(n) + b(n)\sum_{s=n_0}^{n-1} k(n,s)[g(s)u^p(s) + h(s)u(s)], \quad (2.3.264)$$

then for all $n \in \mathbb{N}_{n_0}$,

$$u(n) \le \left\{a(n) + b(n)\sum_{s=n_0}^{n-1}\bar{B}(\sigma)\prod_{\tau=\sigma+1}^{n-1}[1 + \bar{A}(\tau)]\right\}^{1/p}, \quad (2.3.265)$$

where for all $n \in \mathbb{N}_{n_0}$,

$$\begin{cases} \bar{A}(n) = k(n+1,n)b(n)\left(g(n) + \dfrac{h(n)}{p}\right) + \displaystyle\sum_{s=n_0}^{n-1}\Delta_1 k(n,s)b(s)\left(g(s) + \dfrac{h(s)}{p}\right), & (2.3.266) \\ \bar{B}(n) = k(n+1,n)\left(g(n)a(n) + h(n)\left(\dfrac{p-1}{p} + \dfrac{a(n)}{p}\right)\right) \\ \quad + \displaystyle\sum_{s=n_0}^{n-1}\Delta_1 k(n,s)\left(g(s)a(s) + h(s)\left(\dfrac{p-1}{p} + \dfrac{a(s)}{p}\right)\right). & (2.3.267) \end{cases}$$

Proof (c_1) Define a function $z(t)$ by

$$z(t) = \sum_{s=n_0}^{n-1}[g(s)u^p(s) + h(s)u(s)]. \quad (2.3.268)$$

Then $z(n_0) = 0$ and (2.3.260) can be written as

$$u^p(n) \le a(n) + b(n)z(n). \quad (2.3.269)$$

From (2.3.269), we obtain,

$$u(n) \leq \left(\frac{p-1}{p} + \frac{a(n)}{p}\right) + \frac{b(n)}{p} z(n). \tag{2.3.270}$$

From (2.3.268) and using (2.3.269), (2.3.270), we get (see, e.g., [444]),

$$z(n+1) - \left[1 + b(n)\left(g(n) + \frac{h(n)}{p}\right)\right] z(n)$$
$$\leq \left[g(n)a(n) + h(n)\left(\frac{p-1}{p} + \frac{a(n)}{p}\right)\right]. \tag{2.3.271}$$

Multiplying both sides of (2.3.271) by $\prod_{\sigma=n_0}^{n-1}[1 + b(\sigma)(g(\sigma) + \frac{h(\sigma)}{p})]^{-1}$, taking $n = s$ and summing up both sides of the resulting inequality from n_0 to $n-1$, we get (see, e.g., [444]),

$$z(n) \prod_{\sigma=n_0}^{n-1} \left[1 + b(\sigma)\left(g(\sigma) + \frac{h(\sigma)}{p}\right)\right]^{-1}$$
$$\leq \sum_{s=n_0}^{n-1} \left[g(s)a(s) + h(s)\left(\frac{p-1}{p} + \frac{a(s)}{p}\right)\right]$$
$$\times \prod_{\sigma=n_0}^{n-1} \left[1 + b(\sigma)\left(g(\sigma) + \frac{h(\sigma)}{p}\right)\right]^{-1} \tag{2.3.272}$$

which implies

$$z(n) \leq \sum_{s=n_0}^{n-1} \left[g(s)a(s) + h(s)\left(\frac{p-1}{p} + \frac{a(s)}{p}\right)\right]$$
$$\times \prod_{\sigma=s+1}^{n-1} \left[1 + b(\sigma)\left(g(\sigma) + \frac{h(\sigma)}{p}\right)\right]. \tag{2.3.273}$$

Using (2.3.272) in (2.3.269), we can get the required inequality (2.3.261).

(c_2) Since $c(n)$ is a positive and non-decreasing function on $\mathbb{N}_{n_0}$, from (2.3.262) it follows

$$\left(\frac{u(n)}{c(n)}\right)^p \leq 1 + b(n) \sum_{s=n_0}^{n-1} \left[g(s)\left(\frac{u(s)}{c(s)}\right)^p + h(s)c^{1-p}(s)\left(\frac{u(s)}{c(s)}\right)\right]. \tag{2.3.274}$$

Now applying the inequality given in (c_1) yields the desired inequality (2.3.263).

(c_3) Define a function $z(n)$ by

$$z(n) = \sum_{s=n_0}^{n-1} k(n,s)[g(s)u^p(s) + h(s)u(s). \tag{2.3.275}$$

Then as in the proof of part (c_1), from (2.3.264) we see that the inequalities (2.3.269) and (2.3.270) hold. From (2.3.273) and using (2.3.269), (2.3.270), and the fact that the function $z(n)$ is monotone non-decreasing in n, we derive

$$\begin{aligned} z(n+1) - z(n) &= k(n+1,n)[g(n)u^p(n) + h(n)u(n)] \\ &\quad + \sum_{s=n_0}^{n-1} \Delta_1 k(n,s)[g(s)u^p(s) + h(s)u(s)] \\ &\leq k(n+1,n)\left[g(n)(a(n) + b(n)z(n)) + h(n)\right. \\ &\quad \left.\times \left(\frac{(p-1)}{p} + \frac{a(n)}{p} + \frac{b(n)}{p}z(n)\right)\right] \\ &\quad + \sum_{s=n_0}^{n-1} \Delta_1 k(n,s)\left[g(s)(a(s) + b(s)z(s)) + h(s)\right. \\ &\quad \left.\times \left(\frac{(p-1)}{p} + \frac{a(s)}{p} + \frac{b(s)}{p}z(s)\right)\right] \\ &\leq \bar{A}(n)z(n) + \bar{B}(n). \end{aligned} \tag{2.3.276}$$

Note that the inequality (2.3.276) can be written as

$$z(n+1) - \left[1 + \bar{A}(n)\right] z(n) \leq \bar{B}(n). \tag{2.3.277}$$

The inequality (2.3.277) implies the estimate

$$z(n) \leq \sum_{s=n_0}^{n-1} \bar{B}(\sigma) \prod_{\tau=\sigma+1}^{n-1} \left[1 + \bar{A}(\tau)\right]. \tag{2.3.278}$$

Thus from (2.3.278) and (2.3.269), the desired inequality (2.3.265) follows. □

Theorem 2.3.23 (The Pachpatte Inequality [512]) *Let u, a, b, g be real-valued non-negative continuous functions defined on $\mathbb{N}_{n_0}$ and $p > 1$ be a real constant.*

(d_1) *Let $L : \mathbb{N}_{n_0} \times \mathbb{R}_+ \to \mathbb{R}_+$ be a function such that for all $n \in \mathbb{N}_{n_0}$, and all $y \in \mathbb{R}_+$,*

$$0 \leq L(n,x) - L(n,y) \leq M(n,y)(x-y), \tag{2.3.279}$$

where $M(n, y)$ is a real-valued non-negative function defined for all $n \in \mathbb{N}_0, y \in \mathbb{R}_+$. If for all $n \in \mathbb{N}_{n_0}$,

$$u^p(n) \leq a(n) + b(n) \sum_{s=n_0}^{n-1} L(s, u(s)), \tag{2.3.280}$$

then for all $n \in \mathbb{N}_{n_0}$,

$$u(n) \leq \left\{ a(n) + b(n) \sum_{s=n_0}^{n-1} L(s, \frac{p-1}{p} + \frac{a(s)}{p}) \times \prod_{\sigma=s+1}^{n-1} [1 + M(\sigma, \frac{p-1}{p} + \frac{a(\sigma)}{p}) \frac{b(\sigma)}{p}] \right\}^{\frac{1}{p}}. \tag{2.3.281}$$

(d_2) *Let $L : \mathbb{N}_{n_0} \times \mathbb{R}_+ \to \mathbb{R}_+$ be a function satisfying the condition: for all $n \in \mathbb{N}_{n_0}$ and all $x \geq y \geq 0$,*

$$0 \leq L(n, x) - L(n, y) \leq M(t, y)\phi^{-1}(x - y), \tag{2.3.282}$$

where $M(n, y)$ is defined as in (d_1), $\psi : \mathbb{R}_+ \to \mathbb{R}_+$ is a continuous and strictly increasing function with $\psi(0) = 0$, ψ^{-1} is the inverse function of ψ and for all $x, y \in \mathbb{R}_+$,

$$\psi^{-1}(xy) \leq \psi^{-1}(x)\psi^{-1}(y). \tag{2.3.283}$$

If for all $n \in \mathbb{N}_{n_0}$,

$$u^p(n) \leq a(n) + b(n)\psi\left(\sum_{s=n_0}^{n-1} L(s, u(s))\right), \tag{2.3.284}$$

then for all $n \in \mathbb{N}_{n_0}$,

$$u(n) \leq \left\{ a(n) + b(n)\psi(\sum_{s=n_0}^{n-1} L(s, \frac{p-1}{p} + \frac{a(s)}{p}) \times \prod_{\sigma=s+1}^{n-1} [1 + M(\sigma, \frac{p-1}{p} + \frac{a(\sigma)}{p})\psi^{-1}(\frac{b(\sigma)}{p})]) \right\}^{1/p}. \tag{2.3.285}$$

(d_3) *Let $W(r)$ be a real-valued continuous non-decreasing sub-additive and sub-multiplicative function defined on $\mathbb{R}_+$ and $W(r) > 0$ on $(0, +\infty)$. If for all $n \in \mathbb{N}_{n_0}$,*

$$u^p(n) \leq a(n) + b(n) \sum_{s=n_0}^{n-1} g(s) W(u(s)), \tag{2.3.286}$$

then for all $n, n_1 \in \mathbb{N}_{n_0}, n_0 \geq n \geq n_1$,

$$u(n) \leq \left\{ a(n) + b(n) G^{-1} \left[G(\bar{D}(n)) + \sum_{s=n_0}^{n-1} g(s) W\left(\frac{b(s)}{p}\right)\right]\right\}^{1/p}, \tag{2.3.287}$$

where for all $n \in \mathbb{N}_{n_0}$,

$$\bar{D}(n) = \sum_{s=n_0}^{n-1} g(s) W\left(\frac{p-1}{p} + \frac{a(s)}{p}\right), \tag{2.3.288}$$

and $G(r) = \int_{r_0}^{r} \frac{ds}{W(s)}$, $r > 0$, $r_0 > 0$, G^{-1} is the inverse function of G, and $n_1 \in \mathbb{N}_{n_0}$ is chosen so that for all $n_1 \in \mathbb{N}_{n_0}$ lying in $n_0 \leq n \leq n_1$,

$$G(\bar{D}(n)) + \sum_{s=n_0}^{n-1} g(s) W\left(\frac{b(s)}{p}\right) \in Dom\,(G^{-1}).$$

Proof The proof follows by the similar proofs of Theorems 2.3.21–2.3.22 given above. Here we omit the details. □

Remark 2.3.4 In the special cases of Theorems 2.3.22 and 2.3.23, we may have new discrete inequalities which can be used in certain applications in the theory of finite difference equations and numerical analysis.

Next, we shall consider the discrete analogy of the inequality

$$u^p(n) \leq f(n) + p\sum_{s=0}^{n-1} u^{p-1}(n)\Big(h(n,s) + k(n,s)u(s) + j(n,s)\sum_{t=0}^{s-1} g(s,t)u(t)dt\Big), \tag{2.3.289}$$

where $n \in \mathbb{N}_0, \mathbb{N}_0 = \mathbb{N} \cap \{0\}, f, u : \mathbb{N}_0 \to \mathbb{R}_+, g(n,s), h(n,s), j(n,s)$ and $k(n,s)$: $\mathbb{N}_0 \times \mathbb{N}_0 \to \mathbb{R}_+$.

Theorem 2.3.24 (The Yang Inequality [696]) *Let $p \geq 1$ be a constant, $f, u : \mathbb{N}_0 \to \mathbb{R}_+$, with f non-decreasing. Let further $g(n,s)$, $h(n,s)$, $j(n,s)$ and $k(n,s) : \mathbb{N}_0 \times \mathbb{N}_0 \to \mathbb{R}_+$ be non-decreasing in n for every s fixed. If the discrete inequality (2.3.289) holds, then for all $n \in \mathbb{N}_0$,*

$$u(n) \leq \gamma(n)\Big(f^{1/p}(n) + \sum_{s=0}^{n-1} h(n,s)\Big)\prod_{s=0}^{n-1}\Big(1 + \gamma(n)j(n,s) + g(n,s)\Big), \tag{2.3.290}$$

where

$$\gamma(n) := \prod_{s=0}^{n-1}(1 + k(n,s)). \tag{2.3.291}$$

Proof Fix an arbitrary natural number M, and define, for all $n \in \tilde{N}$,

$$v(n) := f(M) + p\sum_{s=0}^{n-1} u^{p-1}(s)\Big(h(M,s) + k(M,s)u(s) + j(M,s)\sum_{t=0}^{s-1} g(s,t)u(t)dt\Big), \tag{2.3.292}$$

where $\tilde{N} := \{0, 1, \dots, M\}$. Then by (2.3.289), we have for all $n \in \tilde{N}$,

$$u(n) \leq v^{1/p}(n), \tag{2.3.293}$$

since $f(n)$ is non-decreasing and $h(n,s), k(n,s)$ and $j(n,s)$ are non-decreasing in n for s fixed. From (2.3.292) it follows that for all $n \in \tilde{N}$,

$$\begin{aligned}\Delta v(t) &= pu^{p-1}(n)\Big(h(M,n) + k(M,n)u(n) + j(M,n)\sum_{t=0}^{n-1} g(n,t)u(t)\Big)\\ &\leq pv^{(p-1)/p}(n)\Big(h(M,n) + k(M,n)v^{1/p}(n) + j(M,n)\sum_{t=0}^{n-1} g(n,t)v^{1/p}(t)\Big).\end{aligned} \tag{2.3.294}$$

Because the function $\alpha(x) := x^{(1-p)/p}$ $(x \in \mathbb{R}_+)$ is decreasing when $p \geq 1$, we easily derive by the mean value theorem that for all $n \in \tilde{N}$,

$$\Delta v^{1/p}(n) \leq \frac{1}{p} v^{\frac{1-p}{p}}(n)\Delta v(n)h(M,n) + k(M,n)v^{1/p}(n) + j(M,n)\sum_{s=0}^{n-1} g(n,s)v^{1/p}(n)dt. \tag{2.3.295}$$

Letting $n = s$ in the last relation and summing over $s = 0, \dots, n-1$, we obtain for all $n \in \tilde{N}$,

$$\eta(n) \leq \Big(f^{1/p}(M) + \sum_{s=0}^{n-1} h(M,s)\Big) + \sum_{s=0}^{n-1} k(M,s)\eta(s) + \sum_{s=0}^{n-1} j(M,s)\sum_{t=0}^{s-1} g(s,t)\eta(t), \tag{2.3.296}$$

where $\eta(n) := v^{1/p}(n)$ and $f^{1/p}(M) = v^{1/p}(0)$. Applying Corollary 2.1.15 in Qin [557] to (2.3.296) yields

$$v^{1/p}(n) \leq \wedge(M,n)\Big(f^{1/p}(M) + \sum_{s=0}^{n-1} h(M,s)\Big)\prod_{s=0}^{n-1}\Big(1 + \wedge(M,n)j(M,s) + g(n,s)\Big), \tag{2.3.297}$$

where $n \in \tilde{N}, \wedge(M,n) := \prod_{s=0}^{n-1}(1 + k(M,s))$. Letting $n = M$ in (2.3.297), from (2.3.293) we derive

$$u(M) \leq \gamma(M)\Big(f^{1/p}(M) + \sum_{s=0}^{M-1} h(M,s)\Big)\prod_{s=0}^{M-1}\Big(1 + \gamma(M)j(M,s) + g(M,s)\Big). \tag{2.3.298}$$

This proves (2.3.290) for $n = M$. Since the inequality (2.3.290) holds trivially for $n = 0$ and M is an arbitrary natural number, thus the proof is complete. □

Theorem 2.3.25 (The Pachpatte Inequality [519]) *Let $u(n), f(n), h(n,s)$, $n, s \in \mathbb{N}_0$, $0 \leq s \leq n < +\infty$, be real-valued non-negative functions. $c \geq 0$ $p > 1$ are real constants. Let $g \in C(\mathbb{R}_+, \mathbb{R}_+)$ be non-decreasing function, $g(u) > 0$, $G(r) = \int_{r_0}^{r} \frac{ds}{g(s^{1/p})}$, G^{-1} is the inverse function of G. If for all $n \in \mathbb{N}_0$,*

$$u^p(n) \leq c + \sum_{s=0}^{n-1}[f(s)g(u(s)) + \sum_{\sigma=0}^{s-1} h(s,\sigma)g(u(\sigma))], \tag{2.3.299}$$

then for all $n, n_1 \in \mathbb{N}_0, 0 \le n \le n_1$,

$$u(n) \le \left(G^{-1}[G(c) + B(n)]\right)^{1/p}, \tag{2.3.300}$$

where

$$B(n) = \sum_{s=0}^{n-1}[f(s) + \sum_{\sigma=0}^{s-1} h(s,\sigma)], \tag{2.3.301}$$

and $n_1 \in \mathbb{N}_0$ *is chosen so that for all* $n \in \mathbb{N}_0$ *lying in* $0 \le n \le n_1$,

$$G(c) + B(n) \in Dom\ (G^{-1}).$$

Proof First we assume that $c > 0$ and define a function $z(n)$ by the right-hand side of (2.3.299). Then $z(n) > 0,\ z(0) = c,\ u(n) \le (z(n))^{1/p}$ and

$$\begin{aligned} z(n+1) - z(n) &= f(n)g(u(n)) + \sum_{\sigma=0}^{n-1} h(n,\sigma)g(u(\sigma)) \\ &\le g((z(n))^{1/p})\left(f(n) + \sum_{\sigma=0}^{n-1} h(n,\sigma)\right). \end{aligned} \tag{2.3.302}$$

From $G(r) = \int_{r_0}^{r} \frac{ds}{g(s^{1/p})}$, and (2.3.302), we conclude

$$\begin{aligned} G(z(n+1)) - G(z(n)) &= \int_{z(n)}^{z(n+1)} \frac{ds}{g(s^{1/p})} \le \frac{z(n+1) - z(n)}{g((z(n))^{1/p})} \\ &\le f(n) + \sum_{\sigma=0}^{n-1} h(n,\sigma). \end{aligned} \tag{2.3.303}$$

By taking in $n = s$ in (2.3.303) and summing up over s from 0 to $n-1$, we can get

$$G(z(n)) \le G(c) + B(n). \tag{2.3.304}$$

From (2.3.304), we have

$$z(n) \le G^{-1}[G(c) + B(n)]. \tag{2.3.305}$$

Using (2.3.305) in $u(n) \le (z(n))^{1/p}$, we have the required inequality (2.3.300). The proof of the case when $c \ge 0$ can be completed similarly, $0 \le n \le n_1$ is obvious.

□

Corollary 2.3.1 (The Pachpatte Inequality [519]) *Let* $u(x,y), f(x,y) \in C(\mathbb{R}_+^2, \mathbb{R}_+), h(x,y,s,t) \in C(\mathbb{R}_+^2, \mathbb{R}_+, \mathbb{R}_+, \mathbb{R}_+)$, $0 \le s \le x < +\infty$, $0 \le t \le y < +\infty$,. c, p *be as in Theorem 2.3.25. If for all* $n \in \mathbb{N}_0$,

$$u^p(t) \le c + \sum_{s=0}^{n-1}[f(s)u(s) + \sum_{\sigma=0}^{s-1} h(s,\sigma)u(\sigma)], \tag{2.3.306}$$

then for all $n \in \mathbb{N}_0$,

$$u(n) \le \left[c^{(p-1)/p} + \frac{p-1}{p} B(n)\right]^{1/(p-1)}, \tag{2.3.307}$$

where $B(n) = \sum_{s=0}^{n-1}[f(s) + \sum_{\sigma=0}^{s-1} h(s,\sigma)]$.

Proof We shall omit the proof here. □

Remark 2.3.5 We note that, the discrete version of Bihari's inequality was established by Hull and Luxemberg in [290]. For other useful nonlinear discrete inequalities, see [512, 513, 673].

2.4 The One-Dimensional Dafermos Inequality

In this section, we introduce the Dafermos inequality which is a generalization of the Ou-Yang inequality.

Let $T = \{t_0, t_1, \cdots, t_s\}$ denote a set of increasing time instances. The following theorem is the discrete version of Theorem 1.3.1, due to Pang and Agarwal [528].

Theorem 2.4.1 (The Pang-Agarwal Inequality [528]) *Let* α, M, N *be non-negative constants and let* y *and* g *be non-negative functions defined on* T *and* $T|\{t_s\}$, *respectively, such that for all* $a \le i \le s$,

$$y^2(i) \le M^2 y^2(0) + 2\sum_{j=0}^{i-1}[\alpha y^2(j) + Ng(j)y(j)]. \tag{2.4.1}$$

Then there holds for all $0 \le i \le n$,

$$y(i) \le (1+\alpha)^i \left[My(0) + \sum_{j=0}^{i-1} Ng(j) \right]. \tag{2.4.2}$$

Proof Let $z(i) = M^2y^2(0)+2\sum_{j=0}^{i-1}[\alpha y^2(j)+Ng(j)y(j)]$. Then $\Delta z(i) = 2[\alpha y^2(j)+Ng(j)y(j)] \le 2[\alpha z(i) + Ng(i)\sqrt{z(i)}]$. Thus

$$\begin{aligned}\sqrt{z(i+1)} &\le \sqrt{(2\alpha+1)z(i) + 2Ng(i)\sqrt{z(i)}}\\ &\le \sqrt{2\alpha+1}\left[(\sqrt{z(i)})^2 + \frac{2Ng(i)}{2\alpha+1}\sqrt{z(i)} + \left(\frac{Ng(i)}{2\alpha+1}\right)^2\right]^{1/2}\\ &\le \sqrt{2\alpha+1}\left[\sqrt{z(i)} + \frac{2Ng(i)}{2\alpha+1}\right].\end{aligned} \tag{2.4.3}$$

Solving the above inequality (2.4.3) recursively, we obtain

$$\begin{aligned}y(i) \le \sqrt{z(i)} &\le (2\alpha+1)^{i/2}My(0) + N\sum_{j=0}^{i-1}(2\alpha+1)^{(i-j-2)/2}g(j)\\ &\le (1+\alpha)^i\left[My(0) + \sum_{j=0}^{i-1}Ng(j)\right].\end{aligned}$$

Remark 2.4.1 We note that the above inequality can be improved by replacing the sum $\sum_{j=0}^{i-1}Ng(j)$ on the right-hand side by $\sum_{j=0}^{i-1}N(2\alpha+1)^{-(j+2)/2}g(j)$.

2.5 The One-Dimensional Nakao Inequalities

In this section, we shall introduce a series of Nakao inequalities (see, e.g., Nakao [411, 412, 414]). These inequalities are connected with difference inequalities which are not only very important for the study of asymptotic behavior of global solutions, but also seem to be interesting in themselves. One of advantages of the Nakao inequalities is that any form of the Nakao inequalities can furnish a decay rate.

Theorem 2.5.1 (The Nakao Inequality [414]) *Let $\phi(t)$ be a bounded positive function on $\mathbb{R}_+$ satisfying, for some constants k and $\alpha > 0$, for all $t \ge 0$,*

$$k\phi(t)^{\alpha+1} \le \phi(t) - \phi(t+1). \tag{2.5.1}$$

Then we have for all $t \ge 1$

$$\phi(t) \le \left\{\alpha k(t-1) + M^{-\alpha}\right\}^{-1/\alpha}, \tag{2.5.2}$$

where $M = \sup_{t\in[0,1]}\phi(t)$.

Proof Put $\phi(t)^{-\alpha} = y(t)$. Then by (2.5.1),

$$\begin{aligned} y(t+1) - y(t) &= \int_0^1 \frac{d}{d\theta}[\theta\phi(t+1) + (1-\theta)\phi(t)]^{-\alpha} d\theta \\ &\geq -\alpha \int_0^1 [\theta\phi(t+1) + (1-\theta)\phi(t)]^{-\alpha-1} d\theta [\phi(t+1) - \phi(t)] \\ &\geq \alpha k \phi(t)^{\alpha+1} \int_0^1 [\phi(t)]^{-\alpha-1} d\theta \\ &= \alpha k. \end{aligned} \tag{2.5.3}$$

For any $t \geq 1$, choose the integer n as $n \leq t < n+1$, and we have from above (2.5.3),

$$y(t) \geq y(t-n) + n\alpha k \geq y(t-n) + (t-1)\alpha k \tag{2.5.4}$$

and hence

$$\phi(t)^{-\alpha} \geq (t-1)\alpha k + \phi(t-n)^{-\alpha} \tag{2.5.5}$$

or

$$\begin{aligned} \phi(t) &\leq [\alpha k(t-1) + \phi(t-n)^{-\alpha}]^{-1/\alpha} \\ &\leq [\alpha k(t-1) + M^{-\alpha}]^{-1/\alpha}. \end{aligned} \tag{2.5.6}$$

The proof thus is complete. □

Theorem 2.5.2 (The Nakao Inequality [414]) *Let $\phi(t)$ be as in Theorem 2.5.1, which satisfies (2.5.1) with $\alpha = 0$. Then we have for all $t \geq 1$,*

$$\phi(t) \leq Me^{-k't} \tag{2.5.7}$$

where $k' = -\log(1-k) > 0$.

Proof By (2.5.1) with $\alpha = 0$,

$$\phi(t+1) \leq (1-k)\phi(t) \quad (which \;\; implies \;\; k < 1). \tag{2.5.8}$$

Therefore, for all $t \geq 1$, we have for the integer n with $n \leq t < n+1$,

$$\begin{aligned} \phi(t) &\leq \frac{1}{1-k}\phi(t-1) \leq \left[\frac{1}{1-k}\right]^n \phi(t-n) \\ &\leq M(1-k)^{-t} = Me^{t\log(1-k)} \end{aligned}$$

which proves the theorem. □

Theorem 2.5.3 (The Nakao Inequality [414]) *Suppose that $\phi(t)$ is a bounded non-negative function on $\mathbb{R}_+$ satisfying*

$$\max_{s\in[t,t+1]} \phi(s)^{1+\alpha} \leq K_0\{\phi(t) - \phi(t+1)\} + g(t) \tag{2.5.9}$$

where $K_0 > 0$ is a constant, $g(t)$ a non-negative function, α a non-negative constant. Then we have

(i) if $\lim_{t\to+\infty} g(t) = 0$, then

$$\lim_{t\to+\infty} \phi(t) = 0. \tag{2.5.10}$$

Moreover,

(ii) if we assume that $\alpha > 0$ and $g(t) \leq K_1|t|^{-\theta-1}$ with constants $\theta > 1/\alpha, K_1 \geq 0$, then for all $t > 0$,

$$\phi(t) \leq C_3 t^{-1/\alpha}, \tag{2.5.11}$$

and

(iii) if $\alpha = 0$ and $g(t) \leq K_2 e^{-\theta t}$ with constants $\theta > 0, K_2 \geq 0$, then

$$\phi(t) \leq C_4 e^{-\theta_1 t} \tag{2.5.12}$$

where $\theta_1 = \min\left(\theta, \log \frac{K_0}{K_0-1}\right)$, and C_3, C_4 are positive constants depending on other known constants.

Proof First, we prove (2.5.10). Suppose that our assertion was false. Then there would exist a real sequence $\{t_n\}_{n=1}^{\infty}$ and $\epsilon_0 > 0$ such that

$$t_n > 2n, \quad \phi(t_n) \geq \epsilon_0 > 0. \tag{2.5.13}$$

Also by our assumption for $g(t)$, we can choose an integer $N > 0$ so large that for all $t \geq N$,

$$g(t) \leq \frac{1}{2}\epsilon_0^{1+\alpha}. \tag{2.5.14}$$

By (2.5.9), (2.5.13) and (2.5.14), we have

$$\epsilon_0^{1+\alpha} \leq K_0(\phi(t_N - 1) - \phi(t_N)) + g(t_N - 1) \tag{2.5.15}$$

and

$$0 < \frac{1}{2}\epsilon_0^{1+\alpha} \leq K_0\left\{\phi(t_N - 1) - \phi(t_N)\right\}. \tag{2.5.16}$$

Therefore we can use again (2.5.9), (2.5.13) and (2.5.14) to obtain

$$\frac{1}{2}\epsilon_0^{1+\alpha} \leq K_0\Big\{\phi(t_N-2)-\phi(t_N-1)\Big\}. \tag{2.5.17}$$

Repeating this procedure, we have for $j = 1, \dots, N$,

$$\frac{1}{2}\epsilon_0^{1+\alpha} \leq K_0\Big\{\phi(t_N-j)-\phi(t_N-j+1)\Big\}. \tag{2.5.18}$$

Summing up the above inequalities over j yields

$$\frac{1}{2}N\epsilon_0^{1+\alpha} \leq K_0\Big\{\phi(t_N-j)-\phi(t_N)\Big\}. \tag{2.5.19}$$

The inequality (2.5.19) is absurd, because the left-hand side tends to $+\infty$ as N goes to $+\infty$ while the right-hand side remains bounded by the boundedness of $\phi(t)$.

Next, we prove (2.5.11). Put $\phi_0(t) = vt^{-\theta}$, $v > 0$, and $w(t) = \phi(t) + \phi_0(t)$. Then we have, for any $t > 0$,

$$\begin{aligned}\max_{s\in[t,t+1]} |w(s)|^{1+\alpha} &= \max_{s\in[t,t+1]} |\phi(s)-\phi_0(s)|^{1+\alpha}\\ &\leq 2^{1+\alpha}\max_{s\in[t,t+1]}[\phi(s)^{1+\alpha}+\phi_0(s)^{1+\alpha}]\\ &\leq 2^{1+\alpha}K_0[w(t)-w(t+1)]+I(t)\end{aligned} \tag{2.5.20}$$

where

$$I(t) = 2^{1+\alpha}[-K_0vt^{-\theta}+K_0v(t+1)^{-\theta}+v^{1+\alpha}t^{-\theta(1+\alpha)}+g(t)]. \tag{2.5.21}$$

We shall show $I(t) < 0$ for sufficiently large t. Indeed, we write

$$\begin{aligned}I(t) = vK_02^{1+\alpha}(t+1)^{-\theta}\Big\{1-(\frac{t+1}{t})^{\theta}\\ +\frac{v^\alpha}{K_0}(t+1)^\theta t^{-\theta(1+\alpha)}+\frac{1}{vK_0}(t+1)^\theta g(t)\Big\}.\end{aligned} \tag{2.5.22}$$

Here it is easily seen that there exists a positive integer $T_1 > 0$ such that for all $t > T_1$,

$$\left(\frac{t+1}{t}\right)^\theta - 1 \geq \frac{1}{2}\theta t^{-1}. \tag{2.5.23}$$

By (2.5.23) and the assumption for $g(t)$, we have, for any $t \geq T_1$,

$$I(t) \leq C(t+1)^{-\theta}t^{-1}\Big[-\frac{\theta}{2}+\frac{1}{vK_0K_1}+\frac{v^\alpha}{K_0}(t+1)^{\theta}t^{-\theta(1+\alpha)+1}\Big]. \tag{2.5.24}$$

Furthermore, since $\theta > \frac{1}{\alpha}$, we have

$$\lim_{t\to+\infty}(t+1)^{\theta_t-\theta(1+\alpha)+1}=0. \tag{2.5.25}$$

Therefore if we choose v so large that $(vK_0K_1)^{-1} < \frac{\theta}{2}$, and choose $T(\geq T_1)$ sufficiently large, then we have, for all $t > T$,

$$I(t) < 0. \tag{2.5.26}$$

Consequently, for any $t > T$, we have from (2.5.20)

$$\max_{s\in[t,t+1]} w(s)^{1+\alpha} \leq 2^{1+\alpha}K_0[w(t)-w(t+1)]. \tag{2.5.27}$$

Now, putting $w(t)^{-\alpha} = y(t)$, it follows from (2.5.27) for all $t > T$,

$$\begin{aligned}
y(t+1)-y(t) &= \int_0^1 \frac{d}{d\theta}\{\theta w(t+1)+(1-\theta)w(t)\}^{-\alpha}d\theta \\
&= \alpha\int_0^1 \{\theta w(t+1)+(1-\theta)w(t)\}^{-1-\alpha}d\theta(w(t)-w(t+1)) \\
&\geq \alpha 2^{-1-\alpha}K_0.
\end{aligned} \tag{2.5.28}$$

Hence, for all $t > T$, choose the integer n such that $n+T \leq t < n+T+1$, we obtain

$$\begin{aligned}
y(t) &\geq y(t-n)+n\alpha 2^{-1-\alpha}K_0 \\
&\geq \min_{s\in[T,T+1]} y(s)+n\alpha 2^{-1-\alpha}K_0
\end{aligned} \tag{2.5.29}$$

or

$$w(t)^{-\alpha} \geq [\max_{s\in[T,T+1]} w(s)]^{-\alpha}+n\alpha 2^{-1-\alpha}K_0 \tag{2.5.30}$$

and hence

$$w(t) \leq \Big[\frac{2^{1+\alpha}K_0}{2^{1+\alpha}K_0[\max_{s\in[T,T+1]} w(s)]^{-\alpha}+\alpha(t-T-1)}\Big]^{\frac{1}{\alpha}}. \tag{2.5.31}$$

From the definition of $w(t)$ and the estimate (2.5.31), we obtain (2.5.11).

Finally, we consider the case $\alpha = 0$. If $K_0 \leq 1$, then we have from (2.5.9)

$$\phi(t+1) \leq g(t) \leq K_2 e^{-\theta t} \tag{2.5.32}$$

and there is nothing to prove. Hence, suppose $K_0 > 1$. Then by (2.5.9), we have

$$\phi(t+1) \leq \frac{K_0 - 1}{K_0}\phi(t) + \frac{K_2}{K_0}e^{-\theta t} \tag{2.5.33}$$

and

$$\begin{aligned}\phi(t) &\leq (\frac{K_0 - 1}{K_0})^n \phi(t-n) + \Sigma_{i=1}^n (\frac{K_0 - 1}{K_0})^{l-1} \frac{K_1}{K_0} e^{-\theta(t-1)} \\ &\leq (\frac{K_0 - 1}{K_0})^n \phi(t-n) + \frac{K_1}{K_0} e^{-\theta(t-1)} \{1 - (\frac{K_0 - 1}{K_0} e^{\theta})^{n-1}\} \\ &\quad \times (1 - \frac{K_0 - 1}{K_0} e^{\theta})^{-1}\end{aligned} \tag{2.5.34}$$

where n is the integer such that $t \leq n < t + 1$. Furthermore, by (2.5.34), we have

$$\phi(t) \leq C\Big\{e^{-(\log \frac{K_0}{K_0 - 1})t} + e^{-\theta t}\Big\} \leq Ce^{-\theta_1 t} \tag{2.5.35}$$

which proves (2.5.12). □

Theorem 2.5.4 (The Nakao Inequality [414]) *Suppose that $\phi(t)$ is a bounded non-negative function on $\mathbb{R}_+$ satisfying*

$$\max_{s \in [t, t+1]} \phi(s)^{1+\alpha} \leq C_0(1+t)^r \Big\{\phi(t) - \phi(t+1)\Big\} + g(t) \tag{2.5.36}$$

where $C_0 > 0$ is a constant, $g(t)$ a non-negative function, and α a non-negative constant. Then we have

(i) if $\alpha > 0, r = 1$ and $\lim_{t \to +\infty} [\log t]^{1+1/\alpha} g(t) = 0$, then for all $t > 0$,

$$\phi(t) \leq C_1 \Big[\log(1+t)\Big]^{-1/\alpha}, \tag{2.5.37}$$

(ii) if $\alpha > 0, 0 \leq r < 1$ and $\lim_{t \to +\infty} t^{(1-r)(1+1/\alpha)} g(t) = 0$, then for all $t > 0$,

$$\phi(t) \leq C_2 t^{-(1-r)/\alpha}, \tag{2.5.38}$$

(iii) if $\alpha = 0, r = 1$ *and* $g(t) \le K_1 t^{-\theta-1}$ *with constants* $\theta > 0, K_1 \ge 0$*, then for all* $t > 0$,

$$\phi(t) \le C_3(1+t)^{-\theta'} \tag{2.5.39}$$

where $\theta' = \min(\theta, C_0^{-1})$,

(iv) if $\alpha = 0, 0 \le r < 1$ *and* $g(t) \le K_2 t^{-\theta} \exp\Big[-\frac{1}{(C_0+1)(1-r)}(t+1)^{1-r}\Big]$ *with* $\theta > 1$*, then for all* $t > 0$,

$$\phi(t) \le C_4 \exp\Big[-\frac{1}{(C_0+1)(1-r)}t^{1-r}\Big]. \tag{2.5.40}$$

In the above, $C_i \quad (i = 1, 2, \cdots)$ *are constants depending on* $\phi(0)$ *and other known constants.*

The above Nakao inequality (see, e.g., Nakao [414]) has several generalizations which we shall state as follows.

Theorem 2.5.5 (The Nakao Inequality [414]) *Suppose that* $\phi(t)$ *is a non-negative continuous non-increasing function on* $\mathbb{R}_+$ *satisfying the inequality, for all* $t \ge 0$,

$$\phi(t+T) \le C\sum_{i=1}^{2}(1+t)^{\theta_i}[\phi(t) - \phi(t+T)]^{\epsilon_i}, \tag{2.5.41}$$

with some $T > 0, C > 0, 0 < \epsilon_i \le 1$ *and* $\theta_i \le \epsilon_i \quad (i = 1, 2)$*. Then* $\phi(t)$ *has the following decay properties:*

(i) If $0 < \epsilon_i < 1$ *with* $\epsilon_1 + \epsilon_2 < 1$ *and* $\theta_i < \epsilon_i, \ i = 1, 2$*, then for all* $t \ge 0$,

$$\phi(t) \le C_0(1+t)^{-\gamma} \tag{2.5.42}$$

with $\gamma = \min_{i=1,2}\{(\epsilon_i - \theta_i)/(1-\epsilon_i)\}$*, where we consider as* $(\epsilon_i - \theta_i)/(1-\epsilon_i) = \infty$ *if* $\epsilon_i = 1$.

(ii) If $\theta_1 = \epsilon_1 < 1$ *and* $\theta_2 < \epsilon_2 \le 1$*, then for all* $t \ge 0$,

$$\phi(t) \le C_0\Big\{\log(2+t)\Big\}^{-\epsilon_1/(1-\epsilon_1)}. \tag{2.5.43}$$

(iii) If $\theta_1 = \epsilon_1 < 1$ *and* $\epsilon_2 = \theta_2 \le 1$*, then for all* $t \ge 0$,

$$\phi(t) \le C_0\Big\{\log(2+t)\Big\}^{-\tilde{\gamma}} \tag{2.5.44}$$

with $\tilde{\gamma} = \min_{i=1,2}\{\epsilon_i/(1-\epsilon_i)\}$.

(iv) If $\epsilon_1 = \epsilon_2 = 1$, *then for all* $t \geq 0$,

$$\begin{cases} \phi(t) \leq C_0 \exp\{-\lambda t^{1-\theta}\} \ \ if \ \ \theta < 1, & (2.5.45) \\ \phi(t) \leq C_0(1+t)^{-\lambda} \ \ if \ \ \theta = 1 & (2.5.46) \end{cases}$$

for some $\lambda > 0, \alpha > 0$, *where we set* $\theta = \min\{\theta_1, \theta_2\}$. *In the above,* C_0 *denotes constants depending on* $\phi(0)$ *and other known constants.*

Proof The case: $\epsilon_1 = \epsilon_2$ and $\eta_1 = \eta_2$ was proved in [411] in more detailed form. The proof, however, is not applicable to our situation and we employ a different technique.

Let us prove (2.5.42). We take $M \geq max_{0 \leq s \leq 1}\phi(s) \equiv \phi(0)$ and assume that

$$\sup_{0 \leq t \leq T} \phi(t)(1+t)^{\alpha} = \phi(T)(1+T)^{\alpha} = M \tag{2.5.47}$$

for some $T \geq 1$. Then, by the inequality (2.5.41), we have

$$\begin{aligned} M(1+T)^{-\alpha} &= \phi(T) \leq C\Sigma_{i=1}^{2} T^{\theta_i} M^{\epsilon_i} (T^{-\alpha} - (1+T)^{-\alpha})^{\epsilon_i} \\ &\leq C\Sigma_{i=1}^{2} T^{\theta_i} M^{\epsilon_i} \alpha T^{-(\alpha+1)\epsilon_i} \end{aligned} \tag{2.5.48}$$

and

$$M \leq C\Sigma_{i=1}^{2} T^{\theta_i - (\alpha+1)\epsilon_i - \alpha} M^{\epsilon_i} \alpha^{\epsilon_i} = C\Sigma_{i=1}^{2} M^{\epsilon_i} \alpha^{\epsilon_i} \tag{2.5.49}$$

which is a contradiction if we take large M. This means that there exists a constant $C_0 = C(\phi(0)) > 0$ such that for all $t \geq 0$,

$$\phi(t)(1+t)^{\alpha} \leq C_0, \tag{2.5.50}$$

which proves (2.5.42).

Next, we consider the case (2.5.43). In this case, we assume, for $M \geq \phi(0)$ and $T \geq 1$, that

$$sup_{0 \leq t \leq T} \phi(t)(\log(2+t))^{\alpha} = \phi(T)[\log(2+T)]^{\alpha} = M \tag{2.5.51}$$

with $\alpha = \frac{\epsilon_1}{1-\epsilon_1}$. Then, we have again by (2.5.41),

$$\begin{aligned} M[\log(2+T)]^{-\alpha} &= \phi(T) \leq C\Sigma_{i=1}^{2} T^{\theta_i} M^{\epsilon_i} [(\log(1+T))^{-\alpha} - (\log(2+T))^{-\alpha}]^{\epsilon_i} \\ &\leq C\Sigma_{i=1}^{2} T^{\theta_i} M^{\epsilon_i} \alpha (1+T)^{-\epsilon_i} [\log(1+T)]^{-(\alpha+1)\epsilon_i} \end{aligned} \tag{2.5.52}$$

whence

$$M \le C\Big\{[\log(2+T)]\alpha - (\alpha+1)\epsilon_1 M^{\epsilon_1} + M^{\epsilon_2}\Big\}$$
$$= C\Sigma_{i=1}^{2} M^{\epsilon_i} \tag{2.5.53}$$

which is again a contradiction if we choose large $M > 0$. This implies (2.5.43). The proof of the case (2.5.44) is essentially included in the above case. Finally, in the case (iv), we have

$$\phi(t) \le C(1+t)^{\eta}[\phi(t) - \phi(t+1)] \tag{2.5.54}$$

which implies (2.5.45)–(2.5.46). □

Remark 2.5.1 It is clear from the proof that Theorem 2.5.5 is generalized to the difference inequality of the form, for all $t \ge 0$,

$$\phi(t+1) \le C\Sigma_{i=1}^{m}(1+t)^{\eta_i}(\phi(t) - \phi(t+1))^{\epsilon_i}. \tag{2.5.55}$$

For example, if $0 < \epsilon_i < 1$ and $\eta_i < \epsilon_i$, we conclude from this inequality that

$$\phi(t) \le C_0(1+t)^{-\alpha} \tag{2.5.56}$$

with $\alpha = min\{\frac{\epsilon_i - \eta_i}{1-\epsilon_i}\}$.

Remark 2.5.2 When $\epsilon_1 = \epsilon_2$ and $\theta_1 = \theta_2$, more detailed results are proved in Nakao [411, 412].

Remark 2.5.3 The above theorem can be easily generalized to the following difference inequality of the form

$$\phi(t+1) \le C\sum_{i=1}^{m}(1+t)^{\theta_i}[\phi(t) - \phi(t+1)]^{\epsilon_i}. \tag{2.5.57}$$

For example, if $0 < \epsilon_i < 1$ and $\theta_i < \epsilon_i$, we obtain from (2.5.57) that

$$\phi(t) \le C_0(1+t)^{-\eta} \tag{2.5.58}$$

with $\eta = \min_{1\le i\le m}\{(\epsilon_i - \theta_i)/(1-\epsilon_i)\}$.

Corollary 2.5.1 (The Nakao Inequality [414]) *Let $\phi(t)$ be a non-negative function on $\mathbb{R}_+$ satisfying*

$$\sup_{t\le s\le t+1} \phi(s)^{1+\gamma} \le K_0(1+t)^{\gamma}\Big\{\phi(t) - \phi(t+1)\Big\} \tag{2.5.59}$$

for some constants $K_0 > 0, \gamma > 0, \beta < 1$. Then $\phi(t)$ has the decay properties: for all $t > 0$,

$$\phi(t) \le C_0(1+t)^{-\frac{(1-\beta)}{\gamma}}; \tag{2.5.60}$$

and if $\gamma = 0$, then for all $t > 0$,

$$\phi(t) \le C_0 \exp\{-\lambda t^{1-\beta}\} \tag{2.5.61}$$

where $C_0 > 0, \lambda > 0$ are constants.

Theorem 2.5.6 (The Nakao Inequality [414]) *Let $\phi(t)$ be a non-negative function on $\mathbb{R}_+$ satisfying*

$$\sup_{t \le s \le t+T} \phi(s)^{1+\gamma} \le g(t)[\phi(t) - \phi(t+T)] \tag{2.5.62}$$

with constants $T > 0, \gamma > 0$ and $g(t)$ is a non-decreasing function. Then $\phi(t)$ has the decay property: for all $t \ge T$,

$$\phi(t) \le \{\phi(0)^{-\gamma} + \gamma \int_T^t g(s)^{-1} ds\}^{-1/\gamma}. \tag{2.5.63}$$

In particular, if $\gamma = 0$ and $g(t) = $ constant in the above, then we have, for all $t > 0$,

$$\phi(t) \le C\phi(0)\exp(-\lambda t) \tag{2.5.64}$$

for some constant $\lambda > 0$.

Chapter 3
Nonlinear One-Dimensional Discontinuous Integral Inequalities

3.1 Nonlinear One-Dimensional Discontinuous Generalizations of the Gronwall-Bellman Inequalities

In this section, we shall introduce some nonlinear discontinuous Gronwall-Bellman inequalities and their generalizations.

Let $[a, b]$ denote an interval, possibly infinite but which we shall assume closed at any finite endpoint and let $\alpha(x)$ and $\beta(x)$ denote functions whose range is in $[a, b]$ whenever the domain is $[a, b]$. The functional equation considered here is

$$\varphi(x) = F[x, \varphi(s), \alpha(x) \leq s \leq \beta(x)] \tag{3.1.1}$$

which, for convenience of notation, we may write as

$$\varphi(x) = F(x, \varphi(\cdot)).$$

Theorem 3.1.1 (The Hanson-Waltman Inequality [269]) *For x in $[a, b]$, let F have the following properties:*

(i) $F(x, \varphi(\cdot))$ *is measurable.*
(ii) $F(x, \varphi_1(\cdot)) \leq F(x, \varphi_2(\cdot))$ *whenever* $\varphi_1(x) \leq \varphi_2(x)$.
(iii) *There exists a measurable function* $\Omega(x)$ *such that* $F(x, \Omega(\cdot)) \leq \Omega(x)$.
(iv) *There exists a measurable function* $\psi(x) \leq \Omega(x)$ *such that* $F(x, \psi(x)) \leq \psi(x)$.

Then there exists a measurable function $\varphi(x)$ which is a solution of equation (3.1.1) almost everywhere and satisfies

$$\psi(x) \leq \varphi(x) \leq \Omega(x).$$

Y. Qin, *Integral and Discrete Inequalities and Their Applications*,
DOI 10.1007/978-3-319-33304-5_3

Proof Let $\mathcal{H}$ be the collection of equivalence classes of Lebesgue measurable functions L on $[a, b]$ such that $\psi(x) \leq L(x) \leq \Omega(x)$ almost everywhere. The functions L_1 and L_2 will be identified if $\{x : L_1(x) \neq L_2(x)\}$ has Lebesgue measure zero. Define $T : \mathcal{H} \to \mathcal{H}$ by

$$T(L(x)) = F(x, L(\cdot)).$$

We shall show that $\mathcal{H}$ is a complete lattice under the relation $L \geq L'$ if $L(x) \leq L'(x)$ a.e. and that T is an isotone mapping of $\mathcal{H}$ into itself. The proof of this will essentially follow from a lattice fixed point theorem. □

For a Lebesgue subset A of $[a, b]$, define

$$\mu(A) = \int_A e^{-x^2} \min\Big[1, \frac{1}{|\psi(x)|}, \frac{1}{|\Omega(x)|}\Big] dx.$$

It is easy to verify that μ is a measure on the Lebesgue subsets of $[a, b]$ which has the same sets of measure zero as does Lebesgue measure. In addition, $\mathcal{H}$ is contained in the collection of real-valued Lebesgue measurable functions which are absolutely integrable with respect to μ and $\mathcal{H}$ has a maximal and a minimal element (Ω and ψ respectively). It follows from Theorem 22 of [212] that $\mathcal{H}$ is complete lattice in the sense of [84] (this can also be done using Lemma 2 of [416]). Note that completeness in the sense of [212] is conditional completeness in the sense of [84], but since $\mathcal{H}$ is bounded, the two concepts are the same for $\mathcal{H}$.

We see that if $\psi(x) \leq L_1(x) \leq L_2(x) \leq \Omega(x)$ a.e., then

(a) $(TL_1)(x) = F(x, L_1(\cdot)) \leq F(x, L_2(\cdot)) = (TL_2)(x)$ so $TL_1 \leq TL_2$ and T is isotone.
(b) $(TL_1)(x) \leq (T\Omega)(x) \leq \Omega$ a.e.
(c) $(TL_1)(x) \geq (T\psi)(x) \geq \psi$ a.e.

Thus $T : \mathcal{H} \to \mathcal{H}$ and is isotone. By a fixed point theorem for complete lattices (Theorem 8 of [84]), we conclude that there is an element L of $\mathcal{H}$ such that

$$L(x) = F(x, L(\cdot)) \;\; a.e.$$

As an application of Theorem 3.1.1, we obtain the following result, due to Viswanstham [657].

Corollary 3.1.1 (The Viswanatham Inequality [657]) *If*

$$\varphi(x) \leq \eta + \int_{x_0}^{x} f(s, \varphi(s))ds$$

where $f(x, y)$ is bounded, measurable, and monotonic increasing in y in the region defined by $|x - x_0| \leq a;\ |y - \eta| \leq b$, where a and b are positive real numbers; $\varphi(x)$

is bounded in the region $|x - x_0| \leq a$*, then there exists a continuous solution of*

$$z(x) = \eta + \int_{x_0}^{x} f(x, z(s))ds$$

valid in the interval $x_0 \leq x \leq x_0 + \alpha, \alpha = \min[a, b/M]$ *where* M *is the bound on* $|f(x,y)|$*, such that* $\varphi(x) \leq z(x)$*.*

Proof Taking

$$F(x, \varphi(s), x_0 \leq s \leq x) = \eta + \int_{x_0}^{x} f(s, \varphi(s))ds,$$

all the conditions of Theorem 3.1.1 are satisfied except (iii). Here, take $\Omega = \eta + M\alpha$,

$$F(x, \eta + M\alpha) = \eta + \int_{x_0}^{x} f(s, \eta + M\alpha)ds \leq \eta + M\alpha.$$

Theorem 3.1.1 yields a measurable function $L(x)$

$$L(x) = \eta + \int_{x_0}^{x} f(s, L(s))ds \quad \text{almost everywhere.}$$

Letting

$$\varphi(x) = \eta + \int_{x_0}^{x} f(s, L(s))ds,$$

then $\varphi(x)$ is continuous, equal to $L(x)$ almost everywhere, and

$$\varphi(x) = \eta + \int_{x_0}^{x} f(s, L(s))ds = \eta + \int_{x_0}^{x} f(s, \varphi(s))ds, \text{ almost everywhere.}$$

□

As a second application, we now consider

$$\varphi(x) = \int_{\alpha(x)}^{\alpha(x)+1} k(t)\varphi(t)dt \tag{3.1.2}$$

with the boundary condition $\lim_{x\to+\infty}\varphi(x) = 1$ where $\lim_{x\to+\infty}\alpha(x) = +\infty$.

Corollary 3.1.2 (The Viswanatham Inequality [657]) *Suppose*

(1) k *is a Lebesgue measurable function on* $[a, +\infty)$ *such that* $0 \leq k(x) \leq 1$ *except for a set* N *of Lebesgue measure zero and* $\lim_{x\to+\infty, x\bar{\in}N} k(x)$ *exists.*

(2) y is a Lebesgue measurable function on $[a, +\infty)$ *such that* $0 \le y(x) \le 1$ *except for a set N of Lebesgue measure zero and* $\lim_{x \to +\infty, x \bar{\in} N} y(x) = 0$,

$$1 - y(t) \le \int_{\alpha(x)}^{\alpha(x)+1} k(t)(1 - y(t))dt \text{ a. e.}$$

Then there exists a continuous solution φ *of equation (3.1.2) satisfying*

$$1 - y(x) \le \varphi(x) \le 1, \quad \text{a. e.}$$

Proof Taking $\Omega = 1$, the theorem immediately provides a solution, almost everywhere satisfying the inequalities. Proceeding as in Corollary 3.1.1, we can obtain a continuous solution. This result is implicit in [615]. □

Remark 3.1.1 We should note that although Corollaries 3.1.1 and 3.1.2 can be obtained by successive iterations (beginning with the lower bound), simple examples show that Theorem 3.1.1 cannot be proved in this way.

Theorem 3.1.2 (The Ma-Debnath Inequality [361]) *Let* $c > 0$ *and* $a \in L^1(I, \mathbb{R}_+)$, *and assume that the function* $w : I = [0, T] \to [1, +\infty)$ *satisfies for all* $0 \le t \le T$,

$$w(t) \le c\Big(1 + \int_0^t a(s)w(s) \log w(s)ds\Big), \tag{3.1.3}$$

then for all $0 \le t \le T$,

$$w(t) \le c \exp\Big(\int_0^t a(s)ds\Big). \tag{3.1.4}$$

Proof Set $v(t) = c\Big(1 + \int_0^t a(s)w(s) \log w(s)ds\Big)$, then $v(0) = c, \;\; w(t) \le v(t)$, and thus

$$v'(t) \le ca(t)v(t) \log v(t).$$

This differential inequality can be integrated directly, giving the right-hand side of (3.1.4), and the lemma follows. □

Next, we study the integral inequality

$$u(x) \le \int_0^x k(x - s)[u(s)]^\beta ds, \quad 0 < x, 0 < \beta, \tag{3.1.5}$$

where $k > 0$ is a given locally integrable function. It is clear that $u(x) \equiv 0$ is a trivial solution of inequality (3.1.5). Therefore, we are interested further in non-trivial continuous, non-negative solutions u of (3.1.5).

Recall that this inequality arises in the study of uniqueness problem for a more general integral equation, for all $t \geq 0$,

$$y(t) = \int_0^t h(t, s, y(s))ds + f(t),$$

in some Banach space. For example, if we consider two solutions y_1 and y_2, take $x(t) = \|y_1(t) - y_2(t)\|$ and assume that

$$\|h(t, s, y_1(s)) - h(t, s, y_2(s))\| \leq k(t - s)\|y_1(s) - y_2(s)\|^{\beta},$$

then we obtain inequality (3.1.5) for $x(t)$.

First, we note that if $1 \leq \beta$, then (3.1.5) has no non-trivial solutions. This is due to the fact that the integral operator

$$Tu(x) = \int_0^x k(x - s)[u(s)]^{\beta} ds, \quad \beta \geq 1$$

is Lipschitz continuous in the class of non-negative, continuous functions. Therefore, we assume that $0 < \beta < 1$. We also note that the existence of a non-trivial solution to (3.1.5) is equivalent to the existence of such a non-trivial solution to the corresponding equation

$$u(x) = \int_0^x k(x - s)[u(s)]^{\beta} ds, \quad 0 < x,\ 0 < \beta < 1. \tag{3.1.6}$$

To see this, we consider any non-trivial solution $v(x)$ of (3.1.5). To deal with non-decreasing functions, we define

$$\bar{v}(x) = \sup\{v(s) : \quad 0 \leq s \leq x\}.$$

Since, the integral operator T has the following monotonicity properties: for any $0 \leq w_1(x) \leq w_2(x)$,

$$Tw_1(x) \leq Tw_2(x),$$

and $Tw(x)$ is non-decreasing for any non-decreasing function $0 < w(x)$, we easily see that $\bar{v}(x)$ is also a non-trivial solution to (3.1.5). Furthermore, it follows from the inequality

$$\bar{v}(x) \leq \int_0^x k(x - s)[\bar{v}(s)]^{\beta} ds \leq K(x)[\bar{v}(x)]^{\beta},$$

where $K(x) = \int_0^x k(s)ds$ that

$$\bar{v}(x) \le K(x)^{1/(1-\beta)}.$$

Now we construct a function sequence

$$v_0(x) = K(x)^{1/(1-\beta)}, \quad v_{n+1}(x) = Tv_n(x), \quad n = 1, 2, \ldots.$$

We may verify directly that $Tv_0(x) < v_0(x)$ and hence we obtain

$$v_{n+1}(x) = Tv_n(x) \le v_n(x) \quad \text{for} \quad n = 1, 2, \ldots$$

Thus $v_n(x)$ is a non-increasing sequence of continuous functions. Since

$$\bar{v}(x) \le v_0(x) \quad \text{and} \quad \bar{v}(x) \le T\bar{v}(x) \le Tv_0(x) \le v_0(x),$$

we obtain $\bar{v}(x) < v_n(x)$ for $n = 1, 2, \ldots$. Now we consider the limit function

$$u(x) = \lim_{n\to+\infty} v_n(x) = \lim_{n\to+\infty} Tv_n(x) \ge \bar{v}(x).$$

Such a $u(x)$ is a non-trivial solution of (3.1.6).

Obviously, equation (3.1.6) is a very special case of the equation

$$u(x) = \int_0^x k(x-s)g(u(s))ds, \tag{3.1.7}$$

where g is a continuous and non-decreasing function.

Recall that the problem of the existence of non-trivial solutions for equation (3.1.7) was studied and some necessary and sufficient conditions, given in [127, 405, 431] were formulated in the form of so-called generalized Osgood conditions. One of the most strength results was obtained for the logarithmically concave kernels k. For example, for such kernels, the following condition,

$$\int_0^\delta (K^{-1})'\left(\frac{s}{g(s)}\right)\frac{ds}{g(s)} < +\infty,$$

where K^{-1} is inverse to K and $\delta > 0$ is sufficiently small, is necessary for the existence of non-trivial solutions to equation (3.1.7). Moreover, in the case $k(x) = x^{\alpha-1}$ or $\exp(-x^{-\alpha}), \alpha > 0$, this condition is also sufficient, see [256, 403, 404]. Unfortunately, if $g(u) = u^\beta, 0 < \beta < 1$, then this condition is satisfied for any k. On the other hand, it is known that if $k(x) = \exp(-\exp(x^{-\alpha}))$, then (3.1.7) has a non-trivial solution if and only if $0 < \alpha < 1$, see [430, 632]. We shall characterize those kernels k, for which the inequality (3.1.5) or equivalently (3.1.6) has non-trivial solutions. The next result is due to Mydlarczyk [406].

Theorem 3.1.3 ([406]) *The inequality* (3.1.5) *has a non-trivial solution if and only if* $0 < \beta < 1$ *and*

$$\int_0^\delta (K^{-1})' \frac{ds}{s(-\log s)} < +\infty,$$

where $\delta > 0$ *is a sufficiently small number.*

Proof (The necessity) Consider the non-trivial solution u of (3.1.6) constructed above. We note that (3.1.6) has also other non-trivial solutions. For example, the functions $u_c(x) = 0$ for all $0 \le x < c$ and $u(x) = u(x-c)$ for all $x \ge c$ $(c > 0)$ are such solutions. Manipulating with c, if necessary we can choose u such that $u(0) = 0$ and $u(x) > 0$ for all $x > 0$. It follows from the construction described above that u is non-decreasing. Furthermore, the integration by parts gives us

$$u(x) = \int_0^x K(x-s)d[u(s)]^\beta, \tag{3.1.8}$$

which implies that u is absolutely continuous and increasing. Finally, substituting $s = u(r)$ into (3.1.8), we have

$$x = \int_0^x K(u^{-1}(x) - u^{-1}(s))d(s^\beta),$$

where u^{-1} is inverse to u. Let $\phi(x) = x^{1/\beta} < x < 1$. Splitting the integral above into two parts, we can obtain

$$x \le K(u^{-1}(x))\phi(x)^\beta + K(u^{-1}(x) - u^{-1}(\phi(x)))x^\beta. \tag{3.1.9}$$

Since $K(u^{-1}(x)) \to 0$ as $x \to 0$, it follows from (3.1.9) that

$$\frac{1}{2}x^{1-\beta} \le K(u^{-1}(x) - u^{-1}(\phi(x))),$$

or

$$K^{-1}\left(\frac{1}{2}x^{1-\beta}\right) \le u^{-1}(x) - u^{-1}(\phi(x)) \tag{3.1.10}$$

for all $0 < x < \delta$, where $\delta > 0$ is sufficiently small.

Now, we note that for any $0 < x < \delta$, the sequence

$$x_0 = x, \quad x_{n+1} = \phi(x_n), \quad n = 1, 2, \ldots$$

is decreasing and convergent to zero.

Since

$$\int_{x_{n+1}}^{x_n} K^{-1}\left(\frac{1}{2}s^{1-\beta}\right)\frac{ds}{s(-\log s)} \leq (-\log \beta)K^{-1}\left(\frac{1}{2}{x_n}^{1-\beta}\right),$$

it follows from (3.1.10) that

$$\int_0^x K^{-1}\left(\frac{1}{2}s^{1-\beta}\right)\frac{ds}{s(-\log s)} < +\infty,$$

for all $0 < x < \delta$, which proves easily our assertion.

The Sufficiency We shall construct one of the solutions to (3.1.5). Let $\psi(x) = x^{2/(1+\beta)} < x < 1$. We claim that the function F given by its inverse

$$F^{-1}(x) = \gamma \int_0^x K^{-1}\left(s^{(1-\beta)/2}\right)\frac{ds}{s(-\log s)}, \quad \gamma = 1/\log(2/(1+\beta))$$

is such a solution.

Indeed, first, we note that

$$\begin{aligned}\int_0^x K(F^{-1}(x) - F^{-1}(s))d(s^\beta) &\geq \int_0^{\psi(x)} K(F^{-1}(x) - F^{-1}(s))d(s^\beta)\\ &\geq K(F^{-1}(x) - F^{-1}(\psi(x)))\psi(x)^\beta.\end{aligned}$$

We observe also that

$$\begin{aligned}F^{-1}(x) - F^{-1}(\psi(x)) &= \gamma \int_{\psi(x)}^x K^{-1}\left(s^{(1-\beta)/2}\right)\frac{ds}{s(-\log s)}\\ &\geq \gamma K^{-1}\left(\psi(x)^{(1-\beta)/2}\right)\int_{\psi(x)}^x \frac{ds}{s(-\log s)}\\ &= K^{-1}\left(\psi(x)^{(1-\beta)/2}\right).\end{aligned}$$

It follows from two inequalities above that for all $0 < x < 1$,

$$\int_0^x K(F^{-1}(x) - F^{-1}(s))d(s^\beta) \geq \psi(x)^{(1-\beta)/2} = x. \tag{3.1.11}$$

Now the substitution $\tau = F(s)$ into the integral above shows that

$$\int_0^x K(x - s)d(F(\tau)^\beta) \geq F(x).$$

Finally, the integration by parts shows that $F(x)$ satisfies (3.1.5), which finishes the proof. □

Remark 3.1.2 We directly verify that for the kernels $k(x) = \exp(-\exp(x^{-\alpha}))$ mentioned above, the following inequalities $k(0.5x) < K(x) < k(2x)$ hold at the vicinity of zero. Now, we easily see that the condition in Theorem 3.1.3 is satisfied in this case, if and only if $0 < \alpha < 1$.

Remark 3.1.3 A substitution $s = \tau^{\alpha}$ $(0 < \alpha < 1)$ into the integral above changes the condition in Theorem 3.1.3 to the following

$$\int_0^{\delta} (K^{-1})(\tau^{\alpha}) \frac{d\tau}{\tau(-\log \tau)} < +\infty.$$

3.2 Nonlinear One-Dimensional Discontinuous Bihari Inequalities and Their Generalizations

Theorem 3.2.1 (The Butler-Rogers Inequality [126]) *Let $x(t), a(t), b(t)$ be positive functions of t, bounded in $c \leq t \leq d$, let $k(t,s)$ be non-negative, bounded on the triangular regions $s \leq t \leq d, c \leq s \leq d$; assume further that $x(t)$ is measurable and $k(t,s)$ is a measurable function of s for each t with $c \leq s \leq d$. Let $f(u), g(u)$ be positive functions for all $u \geq 0$, with f strictly increasing and g non-decreasing. Then defining*

$$A(t) = \sup_{c \leq s \leq t} a(s), \quad B(t) = \sup_{c \leq s \leq t} b(s), \quad K(t,s) = \sup_{s \leq \sigma \leq t} k(\sigma, s), \tag{3.2.1}$$

if the inequality holds for all $c \leq t \leq d$,

$$f(x(t)) \leq a(t) + b(t) \int_c^t k(t,s) g(x(s)) ds, \tag{3.2.2}$$

then for all $c \leq t \leq d' \leq d$,

$$x(t) \leq f^{-1}\left[\Omega^{-1}(\Omega(A(t)) + B(t) \int_c^t K(t,s) ds)\right], \tag{3.2.3}$$

where

$$\Omega(u) = \int_{\varepsilon}^{u} \frac{dw}{g(f^{-1}(w))}, \quad u \geq \varepsilon > 0, \tag{3.2.4}$$

and

$$d' = \max\left\{c \leq \tau \leq d : \Omega(A(\tau)) + B(\tau) \int_c^{\tau} K(\tau, s) ds \leq \Omega(f(+\infty))\right\}. \tag{3.2.5}$$

Proof Let $c \leq t \leq T \leq d'$. It will be obvious from the proof that d' is defined in such a way that the right-hand side of (3.2.3) is meaningful. Then (3.2.2) implies that

$$f(x(t)) \leq A(T) + B(T) \int_c^t K(T,s)g(x(s))ds. \tag{3.2.6}$$

Denoting the right-hand side of (3.2.6) by $V(T,t)$, this is equivalent to

$$f(x(t)) \leq V(T,t).$$

By the monotonicity of g and f, we have,

$$g(x(t)) \leq g(f^{-1}(V(T,t))). \tag{3.2.7}$$

Differentiating with respect to t and using (3.2.7), we conclude

$$\frac{\frac{\partial V}{\partial t}(T,t)}{g(f^{-1}(V(T,t)))} \leq B(T)K(T,t), \quad a.e.$$

Integrating from $t = c$ to $t = T$, we have

$$\Omega(V(T,T)) \leq \Omega(V(T,c)) + B(T) \int_c^T K(T,s)ds,$$

$$V(T,T) \leq \Omega^{-1}\left((V(T,c)) + B(T) \int_c^T K(T,s)ds\right),$$

$$f(x(T)) \leq V(T,T) \leq \Omega^{-1}\left((A(T)) + B(T) \int_c^T K(T,s)ds\right),$$

which is (3.2.3) with a change of dummy variable. □

Note that if $a(t), b(t)$ are non-decreasing functions of t, and $k(t,s)$ is a non-decreasing function of t for each s with $c \leq s \leq d$, (3.2.3) can be replaced by

$$x(t) \leq f^{-1}[\Omega^{-1}\{\Omega(a(t)) + b(t) \int_c^t k(t,s)ds\}], \quad c \leq t \leq d' \leq d,$$

with appropriate modifications if only some of these functions are non-decreasing.

Corollary 3.2.1 (The Butler-Rogers Inequality [126]) *If in Theorem 3.2.1, $f(u) = u^p, g(u) = u^q$, where $p, q > 0$, then the inequality (3.2.2) implies the inequalities*

(i) for all t with $c \le t \le d'$,

$$x(t) \le \left[A(t)^{1-(q/p)} + (1 - \frac{q}{p})B(t) \int_c^t K(t,s)ds\right]^{1/(p-q)},$$

if $p \ne q$ (in the case $q < p; p \ge 1, d'$ is equal to d).

(ii) for all t with $c \le t \le d$,

$$x(t) \le \left[A(t)\exp\Big(B(t) \int_c^t K(t,s)ds\Big)\right]^{1/p}$$

if $p = q$.

We omit the proof and remark that the inequalities in the corollary were obtained in [402] for $p = 1, 0 \le q \le 1$ and $K(t,s) = t - s$.

Corollary 3.2.2 (The Butler-Rogers Inequality [126]) *Let $x(t), a(t), k(t,s)$ satisfy the hypotheses of Theorem 3.2.1. Let $p \ge z$ be a positive integer. If the inequality holds for all $c \le t \le d$,*

$$x(t) \le a(t) + b(t)\Big(\int_c^t k(t,s)x^p(s)ds\Big)^{1/p}, \tag{3.2.8}$$

then for all $c \le t \le d$,

$$x(t) \le A(t)e(p-1)\exp\left(\frac{1}{p}B^p(t)\int_c^t K(t,s)ds\right). \tag{3.2.9}$$

Proof The inequality (3.2.8) implies for all $c \le t \le T \le d$,

$$x(t) - A(T) \le b(t)\Big(\int_c^t k(t,s)x^p(s)ds\Big)^{1/p}.$$

Thus, for all $c \le t \le T \le d$,

$$y^p(t) \le b^p(t)\Big(\int_c^t k(t,s)x^p(s)ds\Big). \tag{3.2.10}$$

where

$$y(t) = \begin{cases} x(t) - A(T), & \text{if } \quad x(t) - A(T) \ge 0, \\ 0, & \text{otherwise.} \end{cases}$$

Applying Theorem 3.2.1 with $f(u) = u^p$, $g(u) = (u + A(T))^p$, and noting that $d' = T$ in this case, we conclude that inequality (3.2.10) implies for all $c \le t \le T$,

$$\Omega(y^p(t)) \le \Omega(0) + B^p(t)\int_c^t K(t,s)ds, \tag{3.2.11}$$

where

$$\Omega(u) = \int_0^u \frac{dw}{(w^{1/p} + A(T))^p}.$$

Thus

$$\Omega(y^p(t)) = \int_0^{y^p(t)} \frac{dw}{(w^{1/p} + A(T))^p} = p\int_0^y \frac{z^{p-1}dz}{(z + A(T))^p}.$$

Letting

$$I_p = \int_0^y \frac{z^{p-1}dz}{(z + A(T))^p},$$

it follows, by integrating by parts,

$$I_p = I_{p-1} - \frac{y^{p-1}}{(p-1)(y + A(T))^{p-1}}.$$

Therefore, we have by induction

$$\Omega(y^p(t)) = p\left[\log\Big(\frac{y + A(T)}{A(T)}\Big) - \sum_{m=1}^{p-1}\frac{y^m}{m(y + A(T))^m}\right]. \tag{3.2.12}$$

From (3.2.11) and (3.2.12), we derive for all $c \le t \le T$,

$$\log\frac{y(t) + A(T)}{A(T)} - \sum_{m=1}^{p-1}\frac{1}{m} \le \frac{1}{p}B^p(t)\int_c^t K(t,s)ds.$$

Thus, for all $c \le t \le T$,

$$y(t) + A(T) \le A(T)e(p-1)\exp\left[\frac{1}{p}B^p(t)\int_s^t K(t,s)ds\right].$$

Hence, taking $t = T$, and using the fact that $x(T) \le y(T) + A(T)$, we obtain the desired result. □

Corollary 3.2.3 (The Tatar Inequality [639]) *Let a, b, K, ψ be non-negative continuous functions on the interval $I = (0, T)$ $(0 < T \le +\infty)$, let $\omega : (0, +\infty) \to \mathbb{R}$ be a continuous, non-negative, and non-decreasing function with $\omega(0) = 0$ and*

$\omega(u) > 0$ for all $u > 0$, and let $A(t) := \max_{0\le s\le t} a(s)$ and $B(t) := \max_{0\le s\le t} b(s)$. Assume that for all $t \in I$,

$$\psi(t) \le a(t) + b(t)\int_0^t K(s)\omega(\psi(s))ds. \tag{3.2.13}$$

Then for all $t \in (0, T_1)$,

$$\psi(t) \le H^{-1}[H(A(t)) + B(t)\int_0^t K(s)ds], \tag{3.2.14}$$

where $H(v) := \int_{v_0}^{v} d\tau/\omega(\tau)$ $(v \ge v_0 > 0)$, H^{-1} is the inverse of H, and $T_1 > 0$ is such that $H(A(t)) + B(t)\int_0^t K(s)ds \in Dom\ (H^{-1})$ for all $t \in (0, T_1)$.

It is of interest to compare Corollary 3.2.2 with those of Willett [671] and Gollwitzer [249]. If we take $a(t) \equiv \alpha, b(t) \equiv 1, k(t,s) \equiv 1, c = 0$, we can obtain from Corollary 3.2.2 that inequality (3.2.9) reduces to

$$x(t) \le \alpha e(p-1)\exp\left(\frac{t}{p}\right), \tag{3.2.15}$$

whereas the estimate of Willett yields

$$x(t) \le \alpha\left[1 + (e^t - 1)\sum_{k=0}^{p-1}(1 - e^{-t})^{k/p}\right], \tag{3.2.16}$$

and that of Gollwitzer yields

$$x(t) \le \alpha[1 + \exp(2^{p-1}t - 1)^{1/p}]. \tag{3.2.17}$$

Although (3.2.16) and (3.2.17) are better estimates than (3.2.15) for small values of t, for large values of t, (3.2.15) is considerably better than either of (3.2.16) or (3.2.17).

Next, we shall introduce the results in [639] and consider the following impulsive integral inequality:

$$\begin{cases} u(t) \le a(t) + b(t)\displaystyle\int_0^t k_1(t,s)u^m(s)ds \\ \qquad + c(t)\displaystyle\int_0^t k_2(t,s)u^n(s-\tau)ds + d(t)\sum_{0<t_k<t}\eta_k u(t_k), \quad \text{for all } t \ge 0, \\ u(t) \le \phi(t), \quad t \in [-\tau, 0],\ \tau > 0, \end{cases} \tag{3.2.18}$$

where $a(t)$, $b(t)$, $c(t)$, and $d(t)$ are non-negative continuous functions, $m, n > 1$, $\eta_k \ge 0$, the points t_k (called "instants of impulse effect") are in the increasing order, and $\lim_{k\to+\infty} t_k = +\infty$. The kernels $k_i(t,s),\ i = 1,2$, are of the form

$$k_i(t,s) = (t-s)^{\beta_i-1} s^{\gamma_i} F_i(s), \quad i = 1,2 \tag{3.2.19}$$

where $\beta_i > 0,\ \gamma_i > -1,\ F_i(t),\ i = 1,2$, and $\phi(t)$ are non-negative continuous functions. For this reason, we say that we are in the presence of an impulsive nonlinear singular version of the Gronwall inequality with delay.

Let the points $t_k \in (0,+\infty),\ k = 1,2,\ldots$ are fixed such that $t_{k+1} > t_k$ and $\lim_{k\downarrow+\infty} t_k = +\infty$.

We consider the set $PC(X,Y)$ of all functions $u:\ X \to Y,\ (X \subset \mathbb{R}, Y \subset \mathbb{R}^n)$ which are piecewise continuous in X with points of discontinuity of the first kind at the points $t_k \in X$, i.e., there exist the limits $\lim_{t\downarrow t_k} u(t) = u(t_k+) < +\infty$ and $\lim_{t\uparrow t_k} u(t) = u(t_k-) = u(t_k) < +\infty$.

Impulsive integral equations, impulsive integro-differential equations, and impulsive differential equations arise naturally in various fields such as population dynamics and optimal control (see the monographs [41, 327, 586]). It seems that the first treatment of impulsive systems should date back to the monograph by Krylov and Bogolyubov [314].

Recall that Samoilenko and Perestyuk [584] first used the following impulsive integral inequality: for all $t \geq 0$,

$$u(t) \leq a + \int_c^t b(s)u(s)ds + \sum_{c<t_k<t} \eta_k u(t_k), \tag{3.2.20}$$

to investigate problems of the form

$$\begin{cases} x' = f(t,x), & t \neq t_k, \\ \Delta x = I_k(x), & t = t_k. \end{cases} \tag{3.2.21}$$

Then, a similar inequality with constant delay was considered by Bainov and Hristova in [40]. Hristova [284] treated a more general inequality with nonlinear functions in u. However, in all these works, the functions (kernels) involved in the integrals are regular, even in the case of integrals of convolution or nonconvolution types (see, e. g., [42, 507]).

We shall consider the case of singular kernels of the form (3.2.19), which arises for instance when we study impulsive evolution problems of the form

$$\begin{cases} \dfrac{du}{dt} + Au = f(t,u,u_t), & t > 0,\ t \neq t_k, \\ u(0) = u_0 \in X, & \\ \Delta u(t_k) = u(t_k^+) - u(t_k^{-1}), & k = 1,2,\ldots, \end{cases} \tag{3.2.22}$$

where A is a sectorial operator (see, e.g., [721] where the case without delay and with globally Lipschitzian right-hand side was treated).

We recall here that nonlinear singular versions of the Gronwall-Bihari inequality were already considered in [308, 309, 379, 637] and Medved' in [384, 385] to investigate problems of the form (3.2.22) and perturbed problems of (3.2.22), but without impulse effects.

Next we first prepare some lemmas and notation to be used latter.

Lemma 3.2.1 (The Tatar Inequality [639]) *For all $\beta > 0$ and $\gamma > -1$, we have, for all $t \geq 0$,*

$$\int_0^t (t-s)^{\beta-1} s^{\gamma} ds = Ct^{\beta+\gamma}, \tag{3.2.23}$$

where $C = C(\beta, \gamma) = \Gamma(\beta)/\Gamma(\beta+\gamma+1)$.

Proof The equality can obtained from the definition of Γ-function, here we omit it. □

Lemma 3.2.2 (The Tatar Inequality [639]) *If $\beta,\ \gamma,\ \delta > 0$, then we have, for any $t > 0$,*

$$t^{1-\beta} \int_0^t (t-s)^{\beta-1} s^{\gamma-1} e^{-\delta s} ds \leq C, \tag{3.2.24}$$

where $C = C(\beta,\ \gamma,\ \delta)$ is a positive constant independent of t. In fact,

$$C = max\{1,\ 2^{1-\beta}\}\Gamma(\gamma)\left(1 + \frac{\gamma}{\beta}\delta^{-\gamma}\right). \tag{3.2.25}$$

Proof See [308] for the detailed proof. □

Let $V(\tau) := 1 + \int_0^{\tau} F_2^2(s)\varphi^{2n}(s-\tau)ds,\ r := \max\{m, n\} > 1,\ t_0 := 0$.

For p and q such that $1/p + 1/q = 1$, we define

$$\begin{aligned} f_p(t) := \sup\Big\{ & a^q(t),\ C^{q/p}(p\beta_1 - p + 1,\ p\gamma_1)b^q(t)t^{q(\beta_1+\gamma_1)-1}, \\ & C^{q/p}(p\beta_2 - p + 1,\ p\gamma_2)c^q(t)t^{q(\beta_2+\gamma_2)-1},\ d^q(t)\Big\}, \end{aligned} \tag{3.2.26}$$

with $C(p\beta_1 - p + 1,\ p\gamma_1)$ and $C(p\beta_2 - p + 1,\ p\gamma_2)$ the constants from Lemma 3.2.1, and T_p be the supremum of all values of t such that

$$\begin{aligned} &\sum_{i=1}^{k} \int_{t_{i-1}}^{t_i} (i+2)^{(q-1)r} \amalg_{j=1}^{i-1} (1 + (j+2)^{q-1}\eta_j^q f(t_j))^r \\ &\times\{F_1^q(s)f^m(s) + F_2^q(s)f^n(s-\tau)\}ds + (k+3)^{(q-1)r} \\ &\times \amalg_{j=1}^{k} \left(1 + (j+2)^{q-1}\eta_j^q f(t_j)\right)^r \int_{t_k}^{t} \{F_1^q(s)f^m(s)f^n(s-\tau)\}\, ds < \frac{V(\tau)^{1-r}}{(r-1)}, \end{aligned} \tag{3.2.27}$$

if $p = q = 2$, put $f(t) = f_2(t)$ and $T := T_2$.

Without loss of generality, we shall suppose that the t_k are such that $\tau < t_{k-1} - t_k \leq 2\tau,\ k = 0, 1, 2, \ldots$ For the general case, see Remark 3.2.1 below.

Theorem 3.2.2 (The Tatar Inequality [639]) *Let the above assumption on the different parameters and functions hold. Suppose that u is in $PC([-\tau, +\infty], [0, +\infty])$ and satisfies (3.2.18), then*

(a) if $\beta_i > 1/2$ and $\gamma_i > -1/2,\ i = 1, 2$, it holds that for all $t \in (t_k,\ t_{k+1}]$,

$$
\begin{aligned}
u(t) \leq & \left[(k+3)f(t)\prod_{l=1}^{k}(1+(k+2)\eta_l^2 f(t_l))\right]^{1/q} \\
& \times\Bigg[V(\tau)^{1-r} - (r-1)\sum_{i=1}^{k}\int_{t_{i-1}}^{t_i}(i+2)^r \amalg_{j=1}^{i-1}\left(1+(j+2)\eta_j^2 f(t_j)\right)^r \\
& \times\left(F_1^2(s)f^m(s) + F_2^2(s)f^n(s-\tau)\right)ds - (r-1)(k+3)^r \\
& \times\prod_{j=1}^{k}(1+(j+2)\eta_j^2 f(t_j))^r\int_{t_k}^{t}\{F_1^2(s)f^m(s) + F_2^2(s)f^n(s-\tau)\}ds\Bigg]^{1/2(1-r)}
\end{aligned}
\tag{3.2.28}
$$

as long as the expression between the second brackets is positive on $(0, T)$;

(b) if $0 < \beta_i \leq 1/2$ and $-1 < \gamma_i \leq -1/2$, then it holds that for all $t \in (t_k, t_{k+1}]$,

$$
\begin{aligned}
u(t) \leq & \left[(k+3)^{q-1}f_p(t)\prod_{l=1}^{k}(1+(k+2)^{q-1}\eta_l^q f(t_l))\right]^{1/q} \\
& \times\Bigg[V(\tau)^{1-r} - (r-1)\sum_{i=1}^{k}\int_{t_{i-1}}^{t_i}(i+2)^{(q-1)r}\prod_{j=1}^{i-1}(1+(j+2)^{q-1}\eta_j^q f(t_j))^r \\
& \times\{F_1^q(s)f_p^m(s) + F_2^q(s)f_p^n(s-\tau)\}ds - (r-1)(k+3)^{(q-1)r} \\
& \times\prod_{j=1}^{k}(1+(j+2)^{q-1}\eta_j^q f(t_j))^r\int_{t_k}^{t}\{F_1^q(s)f_p^m(s) + F_2^2 f_p^n(s-\tau)\}ds\Bigg]^{1/q(1-r)}
\end{aligned}
\tag{3.2.29}
$$

as long as the expression between the second brackets is positive, that is, on $(0, T_p)$.

Proof We shall use a mathematical induction.

(a) **Step 1.** We start by proving the validity of (3.2.28) in the interval $[0, t_1]$ (in fact, the argument we present is valid within the interval $(0, T)$, this fact will be mentioned in every occasion by indicating the right interval over which the estimate is valid) for all $t \in [0, \tau] \subset [0, t_1]$ (see assumptions on t_k), we have

$$u(t) \leq a(t) + b(t) \int_0^t (t-s)^{\beta_1 - 1} s^{\gamma_1} F_1(s) u^m(s) ds$$
$$+c(t) \int_0^t (t-s)^{\beta_2 - 1} s^{\gamma_2} F_2(s) u^n(s-\tau) ds. \qquad (3.2.30)$$

If $\beta_i > 1/2$ and $\gamma_i > -1/2,\ i = 1, 2$, then by the Cauchy-Schwartz inequality and Lemma 3.2.1, we obtain

$$u(t) \leq a(t) + C^{1/2}(2\beta_1 - 1,\ 2\gamma_1) b(t) t^{\beta_1 + \gamma_1 - 1/2} \left(\int_0^t F_1^2(s) u^{2m}(s) ds\right)^{1/2}$$
$$+C^{1/2}(2\beta_2 - 1,\ 2\gamma_2) c(t) t^{\beta_2 + \gamma_2 - 1/2} \left(\int_0^t F_2^2(s) u^{2m}(s-\tau) ds\right)^{1/2}, \qquad (3.2.31)$$

where $C(2\beta_1 - 1,\ 2\gamma_1)$ and $C(2\beta_2 - 1,\ 2\gamma_2)$ are the constants from Lemma 3.2.1. Squaring both sides of (3.2.31), we conclude

$$u^2(t) \leq 3a^2(t) + 3C(2\beta_1 - 1, 2\gamma_1) b^2(t) t^{2(\beta_1 + \gamma_1) - 1} \int_0^t F_1^2(s) u^{2m}(s) ds$$
$$+3C(2\beta_2 - 1,\ 2\gamma_2) c^2(t) t^{2(\beta_2 + \gamma_2) - 1} \int_0^t F_2^2(s) u^{2n}(s-\tau) ds. \qquad (3.2.32)$$

Therefore

$$u^2(t) \leq 3f(t) \left(1 + \int_0^t F_1^2(s) u^{2m}(s) ds + \int_0^t F_2^2(s) u^{2m}(s-\tau) ds\right)$$
$$\leq 3f(t) \left(1 + \int_0^t F_1^2(s) u^{2m}(s) ds + \int_0^\tau F_2^2 \varphi^{2n}(s-\tau) ds\right). \qquad (3.2.33)$$

Putting

$$v_1(t) := 1 + \int_0^\tau F_2^2(s) \varphi^{2n}(s-\tau) ds + \int_0^t F_1^2(s) u^{2m}(s) ds, \qquad (3.2.34)$$

we see that $v_1(t)$ is a non-decreasing positive differentiable function on $[0, \tau],\ v_1(0) = 1 + \int_0^\tau F_2^2 \varphi^{2n}(s-\tau) ds =: V(\tau)$,

$$u^2(t) \leq 3f(t) v_1(t), \qquad (3.2.35)$$

$$v_1'(t) = F_1^2(t) u^{2m}(t) \leq 3^m F_1^2(t) f^m(t) v_1^m(t) \leq 3^r F_1^2(t) f^m(t) v_1^r(t). \qquad (3.2.36)$$

Now applying (3.2.36) (or using Corollary 3.2.3 directly) leads to

$$v_1(t) \leq \left[V(\tau)^{1-\tau} - 3^r(r-1) \int_0^t F_1^2(s) f^m(s) ds \right]^{1/(1-r)} \qquad (3.2.37)$$

provided that $\int_0^t F_1^2(s) f^m(s) ds < V(\tau)^{1-r}/3^r(r-1)$. Therefore, for all $t \in [0, \tau]$,

$$u(t) \leq \sqrt{3f(t)} \left[V(\tau)^{1-r} - 3^r(r-1) \int_0^t F_1^2(s) f^m(s) ds \right]^{1/2(1-r)} \qquad (3.2.38)$$

if $\int_0^t F_1^2(s) f^m(s) ds < V(\tau)^{1-\tau}/3^r(r-1)$.

Let $t \in (\tau, t_1]$. Then, from (3.2.33) and (3.2.34), we deduce

$$u^2(t) \leq 3f(t) \left(v_1(\tau) + \int_\tau^t F_1^2(s) u^{2m}(s) ds + \int_\tau^t F_2^2(s) u^{2n}(s-\tau) ds \right). \qquad (3.2.39)$$

Let

$$w_1(t) := v_1(\tau) + \int_\tau^t F_1^2(s) u^{2m}(s) ds + \int_\tau^t F_2^2(s) u^{2m}(s-\tau) ds. \qquad (3.2.40)$$

Then $w_1(t)$ is a non-decreasing positive differentiable function on $(\tau, t_1]$,

$$w_1(\tau) = v_1(\tau) \leq w_1(t), \quad u^2(t) \leq 3f(t) w_1(t), \qquad (3.2.41)$$

$$w_1'(t) = F_1^2(t) u^{2m}(t) + F_2^2(t) u^{2n}(t-\tau). \qquad (3.2.42)$$

Since $0 < t - \tau \leq \tau$ (see below Remark 3.2.1) and from (3.2.34), (3.2.35), (3.2.41) and (3.2.42),

$$u^2(t-\tau) \leq 3f(t-\tau) v_1(t-\tau) \leq 3f(t-\tau) v_1(\tau) \leq 3f(t-\tau) w_1(t), \qquad (3.2.43)$$

we can write that

$$\begin{aligned} w_1'(t) &\leq F_1^2(t)(3f(t) w_1(t))^m + F_2^2(t)(3f(t-\tau) w_1(t))^n \\ &\leq 3^r [F_1^2(t) f^m(t) + F_2^2(t) f^n(t-\tau)] w_1^r(t). \end{aligned} \qquad (3.2.44)$$

Integrating (3.2.44) from τ to t and using (3.2.37), we can conclude

$$\begin{aligned} w_1(t) &\leq \left[w_1(\tau)^{1-r} - 3^r(r-1) \int_\tau^t [F_1^2(s) f^m(s) + F_2^2(s) f^n(s-\tau)] ds \right]^{1/(1-r)} \\ &\leq \left[V(\tau)^{1-r} - 3^r(r-1) \int_0^\tau F_1^2(s) f^m(s) ds \right. \end{aligned}$$

$$-3^r(r-1)\int_\tau^t [F_1^2(s)f^m(s) + F_2^2(s)f^n(s-\tau)]ds\Bigg]^{1/(1-r)} \tag{3.2.45}$$

$$\le \left[V(\tau)^{1-r} - 3^r(r-1)\int_0^t [F_1^2(s)f^m(s) + F_2^2(s)f^n(s-\tau)]ds\right]^{1/(1-r)},$$

whence, for all $t \in (\tau, t_1]$,

$$u(t) \le \sqrt{3f(t)}\left[V^{1-r} - 3^r(r-1)\int_0^t [F_1^2(s)f^m(s) + F_2^2(s)f^n(s-\tau)]ds\right]^{1/2(1-r)} \tag{3.2.46}$$

if

$$\int_0^t \left[F_1^2(s)f^m(s) + F_2^2(s)f^n(s-\tau)\right] ds < \frac{V^{1-r}}{3^r(r-1)}. \tag{3.2.47}$$

We define the function $\psi_1 : [0, t_1] \to \mathbb{R}$ by

$$\psi_1(t) := \begin{cases} v_1(t), & t \in [0, \tau], \\ w_1(t), & t \in [\tau, t_1]. \end{cases} \tag{3.2.48}$$

It follows easily that (3.2.28) holds over $[0, t_1]$ (recall that $t_0 := 0$).

Step 2. Let $t \in (t_1, t_2]$. If all $t \in (t_1, t_1 + \tau]$, then

$$u(t) \le a(t) + b(t)\int_0^t (t-s)^{\beta_1 - 1}s^{\gamma_1}F_1(s)u^m(s)ds$$
$$+c(t)\int_0^t (t-s)^{\beta_2-1}s^{\gamma_2}F_2(s)u^n(s-\tau)ds + \eta_1 d(t)u(t_1). \tag{3.2.49}$$

Squaring both sides of (3.2.49) after applying the Cauchy-Schwartz inequality and Lemma 3.2.1, as in previous Step 1 from (3.2.31) to (3.2.33), we conclude

$$u^2(t) \le 4f(t)\left(1 + \int_0^t F_1^2(s)u^{2m}(s)ds + \int_0^t F_2^2(s)u^{2n}(s-\tau)ds + \eta_1^2 u^2(t_1)\right)$$
$$\le 4f(t)\Bigg(v_1(\tau) + \int_\tau^{t_1} F_1^2(s)u^{2m}(s)ds + \int_\tau^{t_1} F_2^2(s)u^{2n}(s-\tau)ds$$
$$+ \int_{t_1}^t F_1^2(s)u^{2m}(s)ds + \int_{t_1}^t F_2^2(s)u^{2n}(s-\tau)ds + \eta_1^2 u^{2n}(t_1)\Bigg). \tag{3.2.50}$$

Here we have used definition (3.2.34) of $v_1(t)$. Thanks to (3.2.40) and (3.2.41), we obtain that

$$\begin{aligned} u^2(t) &\le 4f(t)\left(w_1(t_1) + \int_{t_1}^t F_1^2 u^{2m}(s)ds + \int_{t_1}^t F_2^2(s)u^{2n}(s-\tau)ds + 3\eta_1^2 f(t_1)w_1(t_1)\right) \\ &\le 4f(t)\left[1 + 3\eta_1^2 f(t_1)\right]\left(w_1(t_1) + \int_{t_1}^t F_1^2(s)u^{2m}(s)ds + \int_{t_1}^t F_2^2(s)u^{2n}(s-\tau)ds\right). \end{aligned} \tag{3.2.51}$$

Defining

$$v_2(t) := w_1(t_1) + \int_{t_1}^t F_1^2(s)u^{2m}(s)ds + \int_{t_1}^t F_2^2(s)u^{2n}(s-\tau)ds, \tag{3.2.52}$$

we derive that $v_2(t)$ is a non-decreasing positive differentiable function on $(t_1, t_1+\tau]$,

$$v_2(t_1) = w_1(t_1) \le v_2(t), \quad u^2(t) \le 4f(t)[1 + 3\eta_1^2 f(t_1)]v_2(t). \tag{3.2.53}$$

Since $t - \tau \le t_1$, by (3.2.33), (3.2.39), (3.2.40) and (3.2.52), we see that

$$u^2(t-\tau) \le 3f(t-\tau)\psi_1(t-\tau) \le 3f(t-\tau)w_1(t_1) \le 3f(t-\tau)v_2(t), \tag{3.2.54}$$

which, with (3.2.52) and (3.2.53), implies

$$\begin{aligned} v_2'(t) &= F_1^2(t)u^{2m}(t) + F_2^2(t)u^{2n}(t-\tau) \\ &\le \left\{4^m[1 + 3\eta_1^2 f(t_1)]^m f^m(t)F_1^2(t) + 3^n F_2^2(t)f^n(t-\tau)\right\} v_2^r(t). \end{aligned} \tag{3.2.55}$$

Integrating (3.2.55) from t_1 to t, and using (3.2.45), we can get

$$\begin{aligned} v_2(t) &\le \Bigg[v_2(t_1)^{1-r} - (1-r) \\ &\quad \times \int_{t_1}^t \{4^m[1 + 3\eta_1^2 f(t_1)]^m f^m(s)F_1^2(s) + 3^n F_2^2(s)f^n(s-\tau)\}ds\Bigg]^{1/(1-r)} \\ &\le \Bigg[V(\tau)^{1-r} - 3^r(r-1)\int_0^{t_1} [F_1^2(s)f^m(s) + F_2^2(s)f^n(s-\tau)]ds - (r-1) \\ &\quad \times \int_{t_1}^t \left(4^m[1 + 3\eta_1^2 f(t_1)]^m f^m(s)F_1^2(s) + 3^n F_2^2(s)f^n(s-\tau)\right) ds\Bigg]^{1/(1-r)}, \end{aligned} \tag{3.2.56}$$

which yields for all $t \in (t_1, t_1 + \tau]$,

$$u(t) \le 2\sqrt{[1 + 3\eta_1^2 f(t_1)]f(t)} \times \Big[V(\tau)^{1-r} - 3^r(r-1)\int_0^{t_1} [F_1^2(s)f^m(s) + F_2^2(s)f^n(s-\tau)]ds - (r-1)\int_{t_1}^t \int_{t_1}^t \{4^m[1 + 3\eta_1^2 f(t_1)]^m f_1^m(s) + 3^n F_2^2(s)f^n(s-\tau)\}ds \Big]^{1/2(1-r)}, \tag{3.2.57}$$

if

$$3^r \int_0^{t_1} [F_1^2(s)f^m(s) + F_2^2 f^n(s-\tau)]ds + \int_{t_1}^t \left\{4^m[1 + 3\eta_1^2 f(t_1)]^m f^m(s)F_1^2(s) + 3^n F_2^2(s)f^n(s-\tau)\right\} ds \le \frac{V^{1-r}}{r-1}. \tag{3.2.58}$$

Now for all $t \in (t_1 + \tau, t_2]$, from (3.2.34), (3.2.40), (3.2.41), (3.2.52), and

$$u^2(t) \le 4f(t)\left(1 + \int_0^t F_1^2(s)u^{2m}(s)ds + \int_0^t F_2^2(s)u^{2n}(s-\tau)ds + \eta_1^2 u^2(t_1)\right), \tag{3.2.59}$$

we derive

$$\begin{aligned} u^2(t) &\le 4f(t)\Big(v_2(t_1+\tau) + \int_{t_1+\tau}^t + \int_{t_1+\tau}^t F_1^2(s)u^{2m}(s)ds \\ &\quad + \int_{t_1+\tau}^t F_2^2(s)u^{2n}(s-\tau)ds + 3\eta_1^2 f(t_1)v_2(t_1+\tau)\Big) \\ &\le 4f(t)[1 + 3\eta_1^2 f(t_1)]\Big(v_2(t_1+\tau) + \int_{t_1+\tau}^t F_1^2(s)u^{2m}(s)ds \\ &\quad + \int_{t_1+\tau}^t F_2^2(s)u^{2n}(s-\tau)ds\Big), \end{aligned} \tag{3.2.60}$$

because $w_1(t_1) \le v_2(t_1) \le v_2(t_1+\tau)$. At this stage, we denote

$$w_2(t) := v_2(t_1+\tau) + \int_{t_1+\tau}^t F_1^2(s)u^{2m}(s)ds + \int_{t_1+\tau}^t F_2^2(s)u^{2n}(s-\tau)ds. \tag{3.2.61}$$

Then, clearly $w_2(t)$ is a non-decreasing positive differentiable function on $(t_1 + \tau, t_2]$, $w_2(t_1 + \tau) = v_2(t_1 + \tau) \le w_2(t)$, and

$$w_2'(t) = F_1^2(t)u^{2m}(t) + F_2^2(t)u^{2n}(t - \tau). \tag{3.2.62}$$

Note that by (3.2.60) and (3.2.61), we have

$$u^2(t) \le 4f(t)[1 + 3\eta_1^2 f(t_1)]w_2(t), \tag{3.2.63}$$

and since $t_1 < t - \tau < t_1 + \tau$, it follows from (3.2.51) that

$$\begin{aligned} u^2(t-\tau) &\le 4f(t-\tau)[1 + 3\eta_1^2 f(t_1)]v_2(t-\tau) \\ &\le 4f(t-\tau)[1 + 3\eta_1^2 f(t_1)]v_2(t_1+\tau) \\ &\le 4f(t-\tau)[1 + 3\eta_1^2 f(t_1)]w_2(t). \end{aligned} \tag{3.2.64}$$

Thus

$$w_2'(t) \le 4^r[1 + 3\eta_1^2 f(t_1)]^r \left(f^m(t)F_1^2(t) + f^n(t-\tau)F_2^2(t)\right) w_2^r(t). \tag{3.2.65}$$

Again by integrating (3.2.65), we conclude

$$\begin{aligned} w_2(t) &\le \left[w_2^{1-r}(t_1+\tau) - 4^r(r-1)[1 + 3\eta_1^2 f(t_1)]^r \int_{t_1+\tau}^t \left(f^m(s)F_1^2(s) + f^n(s-\tau)F_2^2(s)\right) ds\right]^{1/(1-r)} \\ &\le \Bigg[V(\tau)^{1-r} - 3^r(r-1)\int_0^{t_1} \left[F_1^2(s)f^m(s) + F_2^2(s)f^n(s-\tau)\right] ds \\ &\quad -(r-1)\int_{t_1}^{t_1+\tau} \left(4^m[1 + 3\eta_1^2 f(t_1)]^m f^m(s)F_1^2(s) + 3^n F_2^2(s)f^n(s-\tau)\right) ds \\ &\quad -4^r(r-1)[1 + 3\eta_1^2 f(t_1)]^r \int_{t_1+\tau}^t \{f^m(s)F_1^2(s) + f^n(s-\tau)F_2^2(s)\} ds\Bigg]^{1/(1-r)} \end{aligned}$$

or

$$\begin{aligned} w_2(t) \le \Bigg[&V(\tau)^{1-r} - 3^r(r-1)\int_0^{t_1} \left[F_1^2(s)f^m(s) + F_2^2(s)f^n(s-\tau)\right] ds \\ &-4^r(r-1)\left(1 + 3\eta_1^2 f(t_1)\right)^r \int_{t_1}^t \{f^m(s)F_1^2(s) + f^n(s-\tau)F_2^2(s)\} ds\Bigg]^{1/(1-r)}. \end{aligned} \tag{3.2.66}$$

Hence,

$$
\begin{aligned}
u(t) \le\ & 2\sqrt{f(t)[1+3\eta_1^2 f(t_1)]} \\
& \times\Bigg[V(\tau)^{1-r} - 3^r(r-1)\int_0^{t_1}\left[F_1^2(s)f^m(s)+F_2^2(s)f^n(s-\tau)\right]ds \\
& -4^r(r-1)[1+3\eta_1^2 f(t_1)]^r\int_{t_1}^{t}\left(f^m(s)F_1^2(s)+f^n(s-\tau)F_2^2(s)\right)ds\Bigg]^{1/2(1-r)},
\end{aligned}
\tag{3.2.67}
$$

provided that the expression between brackets is positive. We define $\psi_2 : (t_1, t_2] \to \mathbb{R}$ by

$$
\psi_2(t) := \begin{cases} v_2(t), & t \in [t_1, t_1+\tau], \\ w_2(t), & t \in [t_1+\tau, t_2]. \end{cases} \tag{3.2.68}
$$

It is clear that (3.2.28) holds on $(t_1, t_2]$.

Step 3. Finally, suppose that (3.1.24) is valid over $(t_k, t_{k+1}]$, then if $t \in (t_{k+1}, t_{k+2}]$, we define

$$
\psi_{k+2}(t) := \begin{cases} v_{k+2}(t), & t \in [t_{k+1}, t_{k+1}+\tau], \\ w_{k+2}(t), & t \in [t_{k+1}+\tau, t_{k+2}] \end{cases} \tag{3.2.69}
$$

with

$$
\begin{cases}
v_{k+2} := w_{k+1}(t_{k+1}) + \displaystyle\int_{t_{k+1}}^{t} F_1^2(s)u^{2m}(s)ds + \int_{t_{k+1}}^{t} F_2^2(s)u^{2n}(s-\tau)ds, \\
w_{k+2}(t) := v_{k+2}(t_{k+1}+\tau) + \displaystyle\int_{t_{k+1}+\tau}^{t} F_1^2(s)u^{2m}(s)ds + \int_{t_{k+1}+\tau}^{t} F_2^2(s)u^{2n}(s-\tau)ds.
\end{cases}
\tag{3.2.70}
$$

In a similar manner as in Steps 1 and 2, we can see that (3.2.28) is valid over $(t_{k+1}, t_{k+2}]$.

(b) If $0 < \beta_i \le 1/2$ and $-1 < \gamma_i \le -1/2$, then instead of the Cauchy-Schwartz inequality, using the Hölder inequality with

$$
1 < p < \min\{\frac{1}{1-\beta_i}, -\frac{1}{\gamma_i}, i = 1, 2\}, \tag{3.2.71}
$$

and $q > 1$ such that $1/p + 1/q = 1$, we conclude

$$u(t) \le a(t) + C^{1/p}(p\beta_1 - p + 1,\ p\gamma_1)b(t)t^{\beta_1+\gamma_1-1/q}\left(\int_0^t F_1^q(s)u^{qm}(s)ds\right)^{1/q}$$
$$+C^{1/p}(p\beta_2 - p + 1,\ p\gamma_2)c(t)t^{\beta_2+\gamma_2-1/q}\left(\int_0^t F_2^q(s)u^{qm}(s-\tau)ds\right)^{1/q}. \tag{3.2.72}$$

Then we raise both sides to the power q and use the inequality,

$$\left(\sum_{i=1}^n a_i\right)^r \le n^{r-1}\left(\sum_{i=1}^n a_i^r\right), \quad n \in \mathbb{N},\ r,\ a_i \in \mathbb{R}^+,\ i = 1,\ldots,n. \tag{3.2.73}$$

The rest of the proof remains the same. The proof is now complete. □

Remark 3.2.1 Besides the case treated in the proof, that is, when $\tau < t_{k+1} - t_k \le 2\tau,\ k = 0, 1, 2, ...$, there are several other cases, but each and every one of them can fit in the one considered above or one of the following cases.

Case 1. There exists an $n_k > 1$ such that $t_{k+1} - t_k \ge n_k\tau$, that is, $t_k < t_k + n_k\tau \le t_{k+1}$. In this case, we argue in a similar manner over $(t_k, t_k + \tau],\ (t_k + \tau, t_k + 2\tau], \ldots, (t_k + (n_k - 1)\tau, t_k + n_k\tau]$ and then over $(t_k + n_k\tau, t]$ with $t > t_k + n_k\tau$. Therefore the function $\psi_{k+1}(t)$ will have $n_k + 1$ components.
Case 2. There exists $k_0 \ge 1$ such that $(t_{k_0}, t_{k_0+1}]$ does not contain any $t_{k_0} + n\tau,\ n = 1, 2, \ldots$, that is, $t_{k_0+1} - t_{k_0} < \tau$. Here we deal with this interval in a step using only a function of the form $v_{k_0+1}(t)$, that is, $\psi_{k_0+1}(t) := v_{k_0+1}(t)$.
Case 3. $\tau \in (t_{k_1}, t_{k_1+1}]$ with $k_1 > 0$, that is, $\tau \notin (0, t_1]$ as in **Case 1.** Again, in this situation, we consider only functions of the form $v_k(t)$ until we reach the interval $(t_{k_1}, t_{k_1+1}]$ where we consider both $v_{k_1+1}(t)$ and $w_{k_1+1}(t)$.

Remark 3.2.2 Obviously, if $k_i(t, s) = (t - s)^{\beta_i-1}s^{\gamma_i}e^{-\delta_i s}F_i(s),\ \delta_i > 0,\ i = 1, 2, \ldots$, the proof still works. However, using Lemma 3.2.2 instead of Lemma 3.2.1 throughout the proof, we can have much larger intervals over which the estimates are valid.

Remark 3.2.3 It is clear that Theorem 3.2.2 can be easily extended to other nonlinearities than the polynomial ones, iterated integrals and the case of several variables. One may use the Gronwall-Bihari lemma (Corollary 3.2.3) in case of a non-decreasing nonlinearity. See also [42, 507] for other classes of nonlinearities.

Next, we shall introduce the result due to Hristova [284].

Definition 3.2.1 We shall say that the function $G(u)$ belongs to the class W_1 if

(1) $G \in C(\mathbb{R}_+, \mathbb{R}_+)$.
(2) $G(u)$ is a non-decreasing function.

Definition 3.2.2 We shall say that the function $G(u)$ belongs to the class $W_2(\varphi)$ if

(1) $G \in W_1$.
(2) There exists a function $\varphi \in C(\mathbb{R}_+, \mathbb{R}_+)$ such that $G(uv) \leq \varphi(u)G(v)$ for all $u, v \geq 0$.

We note that if the function $G \in W_1$ and satisfies the inequality $G(uv) \leq G(u)G(v)$ for all $u, v \geq 0$, then $G \in W_2(G)$.

Furthermore, we shall use the following notations $\sum_{i=1}^{k} \alpha_k = 0$ and $\Pi_{i=1}^{k}\alpha_k = 1$.

We shall first consider integral inequalities with delay for piecewise continuous functions.

Theorem 3.2.3 (The Hristova Inequality [284]) *Let the following conditions hold:*

(1) The functions $f_1, f_2, f_3, p, g \in C(\mathbb{R}_+, \mathbb{R}_+)$.
(2) The function $\psi \in C([-h, 0], \mathbb{R}_+)$.
(3) The function $Q \in W_2(\varphi)$ and $Q(u) > 0$ for all $u > 0$.
(4) The function $G \in W_1$.
(5) The function $u \in PC([-h, +\infty), \mathbb{R}_+)$ and satisfies the following inequalities, for all $t \geq 0$,

$$u(t) \leq f_1(t) + f_2(t)G(c + \int_0^t p(s)Q(u(s))ds + \int_0^t g(s)Q(u(s-h))ds)$$
$$+f_3(t) \sum_{0<t_k<t} \beta_k u(t_k), \tag{3.2.74}$$

and for all $t \in [-h, 0]$,

$$u(t) \leq \psi(t), \tag{3.2.75}$$

where $c \geq 0$, $\beta_k \geq 0$, $(k = 1, 2, \ldots)$. Then for all $t \in (t_k, t_{k+1}] \cap [0, \gamma)$, $k = 1, 2, \ldots$, we have the inequality

$$u(t) \leq \rho(t)\Pi_{i=1}^{k}(1 + \beta_i\rho(t_i))$$
$$\times(1 + G(H^{-1}\{H(A) + \sum_{i=1}^{k} \int_{t_{i-1}}^{t_i} p(s)\varphi[\rho(s)\Pi_{j=1}^{i-1}(1 + \beta_i\rho(t_j))]ds$$
$$+ \int_{t_k}^{t} p(s)\varphi[\rho(s)\Pi_{j=1}^{k}(1 + \beta_j\rho(t_j))]ds$$
$$+\Lambda(t)\sum_{i=1}^{k} \int_{t_{i-1}}^{t_i} g(s)\ \varphi[\rho(s-h)\Pi_{j:0<t_j<s-h}(1 + \beta_j\rho(t_k))]ds$$
$$+\Lambda \int_{t_k}^{t} g(s)\varphi[\rho(s-h)\Pi_{j:0<t_j<s-h}(1 + \beta_j\rho(t_k))]ds\})), \tag{3.2.76}$$

where

$$\begin{cases} \Lambda(t) & = \begin{cases} 0 & \textit{for all } t \in [0,h], \\ 1 & \textit{for all } t > h, \end{cases} \quad (3.2.77)\\ \rho & = \max\{f_i(t) : i = 1,2,3\}, \quad A = c + hB_1 Q(B_2), \\ B_1 & = \max\{g(t) : t \in [0,h]\}, \quad B_2 = \max\{\psi(t) : t \in [-h,0]\}, \\ H(u) & = \int_{u_0}^{u} \frac{ds}{Q(1+G(s))}, \quad u \geq u_0 \geq 0, \quad (3.2.78) \end{cases}$$

$$\gamma = \sup\Bigg\{ t \geq 0 : H(A) + \sum_{i=1}^{k} \int_{t_{i-1}}^{t_i} p(s)\varphi[\rho(s)\Pi_{j=1}^{i-1}(1+\beta_j\rho(t_j))]ds$$

$$+ \int_{t_k}^{t} p(s)\varphi[\rho(s)\Pi_{j=1}^{k}(1+\beta_j\rho(t_j))]ds$$

$$+\Lambda(t)\sum_{i=1}^{k}\int_{t_{i-1}}^{t_i} g(s)\varphi[\rho(s-h)\Pi_{0<t_j<s-h}(1+\beta_j\rho(t_k))]ds$$

$$+\Lambda(t)\int_{t_k}^{t} g(s)\varphi[\rho(s-h)\Pi_{0<t_j<s-h}(1+\beta_j\rho(t_k))]ds \in Dom\,(H^{-1}),$$

$$\text{for } \tau \in (t_k, t_{k+1}] \cap [0,t], \; k = 0,1,\ldots \Bigg\},$$

and H^{-1} *is the inverse function of* $H(u)$.

Proof

Case 1. Let $t_1 \geq h$. Let $t \in (0,h] \cap [0,\gamma) \neq \varnothing$.

It follows from the inequalities (3.2.74) and (3.2.75) that for all $t \in (0,h] \cap [0,\gamma)$, it holds

$$u(t) \leq \rho(t)(1 + G(A + \int_0^t p(s)Q(u(s))ds)). \quad (3.2.79)$$

Define the function $v_0^{(0)} : [0,h] \cap [0,\gamma) \to \mathbb{R}_+$ by the equality

$$v_0^{(0)}(t) = A + \int_0^t p(s)Q(u(s))ds. \quad (3.2.80)$$

The function $v_0^{(0)}(t)$ is a non-decreasing differentiable function on $[0,h] \cap [0,\gamma)$ and satisfies

$$u(t) \leq \rho(t)(1 + G(v_0^{(0)}(t))). \quad (3.2.81)$$

Thus the inequality (3.2.81) and the definition (3.2.78) of the function $H(u)$ yield

$$\frac{d}{dt}H(v_0^{(0)}(t)) = \frac{(v_0^{(0)}(t))'}{Q(1+G(v_0^{(0)}(t)))} \leq p(t)\varphi(\rho(t)). \tag{3.2.82}$$

Now integrating the inequality (3.2.82) from 0 to t for all $t \in [0,h] \cap [0,\gamma)$, and using $v_0^{(0)}(0) = A$ in order, we obtain

$$H(v_0^{(0)}(t)) \leq H(A) + \int_0^t p(s)\varphi(\rho(s))ds. \tag{3.2.83}$$

The inequalities (3.2.81) and (3.2.83) imply the inequality (3.2.76) holds for all $t \in [0,h] \cap [0,\gamma)$.

Let $t \in (h,t^1] \cap [0,\gamma) \neq \varnothing$. Then

$$\begin{aligned} u(t) &\leq \rho(t)\left(1 + G(v_0^{(0)}(h) + \int_h^t p(s)Q(u(s))ds + \int_h^t g(s)Q(u(s-h))ds)\right) \\ &= \rho(t)(1 + G(v_0^{(1)}(t)), \end{aligned} \tag{3.2.84}$$

where $v_0^{(1)} : [h,t^1] \cap [0,\gamma) \to \mathbb{R}_+$ is defined by the equality

$$v_0^{(1)}(t) = v_0^{(0)}(h) + \int_h^t p(s)Q(u(s))ds + \int_h^t g(s)Q(u(s-h))ds. \tag{3.2.85}$$

Using the fact that the function $v_0^{(1)}(t)$ is non-decreasing continuous and $v_0^{(0)}(t-h) \leq v_0^{(0)}(h) \leq v_0^{(1)}(t)$ for all $h < t \leq \min\{2h, t^1\}$, we can prove as above that

$$\begin{aligned} H(v_0^{(1)}(t)) &\leq H(v_0^{(0)}(h)) + \int_h^t p(s)\varphi(\rho(s))ds + \int_h^t g(s)\varphi(\rho(s-h))ds \\ &\leq H(A) + \int_0^t p(s)\varphi(\rho(s))ds + \int_h^t g(s)\varphi(\rho(s-h))ds. \end{aligned} \tag{3.2.86}$$

The inequalities (3.2.84), (3.2.86) prove (3.2.76) holds on $t \in (h,t^1] \cap [0,\gamma)$.

Define the function

$$v_0(t) = \begin{cases} v_0^{(0)}(t) & \text{for all } t \in [0,h], \\ v_0^{(1)}(t) & \text{for all } t \in (h,t^1]. \end{cases}$$

Now let $t \in (t^1,t^2] \cap [0,\gamma) \neq \varnothing$. Define the function $v_1 : [t^1,t^2] \cap [0,\gamma) \to \mathbb{R}_+$ by

$$v_1(t) = v_0(t_1) + \int_{t_1}^t p(s)Q(u(s))ds + \int_{t_1}^t g(s)Q(u(s-h))ds. \tag{3.2.87}$$

Therefore, the function $v_1(t)$ is non-decreasing differentiable on $t \in (t_1, t_2] \cap [0, \gamma)$, $v_1(t) \geq v_0(t_1)$ and

$$\begin{aligned} u(t) &\leq \rho(t)(1 + G(v_1(t)) + \beta_1 u(t_1)) \\ &\leq \rho(t)(1 + G(v_1(t)) + \beta_1 \rho(t_1)(1 + G(v_0(t_1)))) \\ &\leq \rho(t)(1 + G(v_1(t)))(1 + \beta_1 \rho(t_1)). \end{aligned} \tag{3.2.88}$$

Consider the following two cases:

Case 1.1. Let $h \leq t_2 - t_1$ and $t \in (t_1 + h, t_2] \cap [0, \gamma)$. Then from (3.2.88), we derive

$$\begin{aligned} u(t-h) &\leq \rho(t-h)(1 + G(v_1(t-h)))(1 + \beta_1 \rho(t_1)) \\ &\leq \rho(t-h)(1 + G(v_1(t)))(1 + \beta_1 \rho(t_1)) \\ &= \rho(t-h)(1 + G(v_1(t)))\Pi_{0<t_k<t-h}(1 + \beta_k \rho(t_k)). \end{aligned} \tag{3.2.89}$$

Case 1.2. Let $h > t_2 - t_1$ or $t \in (t_1, t_1 + h] \cap [0, \gamma)$. Then

$$u(t-h) \leq \rho(t-h)(1 + G(v_0(t-h))). \tag{3.2.90}$$

Using (3.2.90) and $v_0(t-h) \leq v_1(t)$, we obtain

$$\begin{aligned} u(t-h) &\leq \rho(t-h)(1 + G(v_1(t))) \\ &= \rho(t-h)(1 + G(v_1(t)))\Pi_{0<t_k<t-h}(1 + \beta_k \rho(t_k)). \end{aligned} \tag{3.2.91}$$

Thus from (3.2.88), (3.2.90), (3.2.91) and the properties of the function $Q(u)$, we derive

$$\begin{aligned} v_1'(t) &= p(t)Q(u(t)) + g(t)Q(u(t-h)) \\ &\leq \{p(t)\varphi(\rho(t)(1 + \beta_1 \rho(t_1))) \\ &\quad + g(t)\varphi(\rho(t-h)\Pi_{0<t_k<t-h}(1 + \beta_k \rho(t_k)))\} \\ &\quad \times Q(1 + G(v_1(t))). \end{aligned} \tag{3.2.92}$$

We obtain from Definition (3.2.2) and the inequality (3.2.92) that

$$\begin{aligned} \frac{d}{dt} H(v_1(t)) &= \frac{(v_1(t))'}{Q(1 + G(v_1(t)))} \\ &\leq p(t)\varphi(\rho(t)(1 + \beta_1 \rho(t_1))) \\ &\quad + g(t)\varphi(\rho(t-h)\Pi_{0<t_k<t-h}(1 + \beta_k \rho(t_k))). \end{aligned} \tag{3.2.93}$$

Integrating (3.2.93) from t_1 to t, using (3.2.86), we obtain

$$\begin{aligned}H(v_1(t)) &\le H(v_0(t_1)) + \int_{t_1}^{t} p(s)\varphi(\rho(s)(1+\beta_1\rho(t_1)))ds\\&\quad+\int_{t_1}^{t} g(s)\varphi(s-h)(\Pi_{0<t_k<t-h}(1+\beta_k\rho(t_k)))ds\\&\le H(A)+\int_0^{t_1} p(s)\varphi(\rho(s))ds+\int_{t_1}^{t} p(s)\varphi(\rho(s)(1+\beta_1\rho(t_1)))ds\\&\quad+\int_{t_1}^{t} g(s)\varphi\rho(s-h)(\Pi_{0<t_k<t-h}(1+\beta_k\rho(t_k)))ds.\end{aligned}\tag{3.2.94}$$

The inequalities (3.2.88) and (3.2.94) imply the inequality (3.2.76) holds for all $t \in (t_1, t_2] \cap [0, \gamma)$.

We define functions $v_k : [t_k, t_{k+1}] \cap [0, \gamma) \to \mathbb{R}_+$ by the equalities

$$v_k(t) = v_{k-1}(t_k) + \int_{t_k}^{t} p(s)Q(u(s))ds + \int_{t_k}^{t} g(s)Q(u(s-h))ds. \tag{3.2.95}$$

Noting the functions $v_k(t)$ are non-decreasing functions, $v_k(t) \ge v_{k-1}(t_k)$ and for all $t \in (t_k, t_{k+1}] \cap [0, \gamma)$, the following inequalities hold

$$\begin{aligned}u(t) &\le \rho(t)(1+G(v_k(t))+\sum_{i=1}^{k}\beta_i u(t_i))\\&\le \rho(t)\{1+G(v_k(t))+\sum_{i=1}^{k-1}\beta_i u(t_i)\\&\qquad+\beta_k\rho(t_k)(1+G(v_{k-1}(t_k))+\sum_{i=1}^{k-1}\beta_i u(t_i))\}\\&\le \rho(t)(1+G(v_k(t))+\sum_{i=1}^{k-1}\beta_i u(t_i))(1+\beta_k\rho(t_k))\\&\le \ldots \le \rho(t)\{\Pi_{i=1}^{k}(1+\beta_i\rho(t_i))\}(1+G(v_k(t))).\end{aligned}\tag{3.2.96}$$

Using the mathematical induction, we can prove that (3.2.76) is true for all $t \in (t_k, t_{k+1}] \cap [0, \gamma), k = 1, 2, \ldots$.

Case 2. Let there exist a natural number m such that $t_m \le h < t_{m+1}$. As in **Case 1**, we prove inequality (3.2.76) holds using the functions $v_k \in C([t_k, t_{k+1}] \cap$

$[0, \gamma), \mathbb{R}_+)$ defined by the equalities,

$$\begin{cases} v_k(t) = v_{k-1}(t_k) + \int_{t_k}^{t} p(s)Q(u(s))ds, \ \textit{for} \ k = 0, 1, \ldots, m, \\ v_k(t) = v_{k-1}(t_k) + \int_{t_k}^{t} p(s)Q(u(s))ds \\ \qquad + \int_{t_k}^{t} g(s)Q(u(s-h))ds, \quad k > m. \end{cases} \tag{3.2.97}$$

Therefore, the proof is now complete. □

In the partial case when the function $\varphi(s)$ in Definition 3.2.2 is multiplicative, the following result is true.

Corollary 3.2.4 (The Hristova Inequality [284]) *Assume the conditions of Theorem 3.2.3 hold and the function φ satisfy the inequality $\varphi(ts) \leq \varphi(t)\varphi(s)$ for all $t, s \geq 0$.*

Then for all $t \in (t_k, t_{k+1}] \cap [0, \gamma_3)$, the following inequality holds

$$\begin{aligned} u(t) \leq \rho(t)\Pi_{i=1}^{k}(1 + \beta_i\rho(t_i))\{1 + G(H^{-1}(H(A) + \varphi(\Pi_{i=1}^{k}(1 + \beta_i\rho(t_i))) \\ \times \int_0^t (p(s)\varphi(\rho(s)) + \Lambda(s)g(s)\varphi(\rho(s-h)))ds))\}, \end{aligned} \tag{3.2.98}$$

where the functions $\Lambda(t)$ and $H(u)$ are defined by the equalities (3.2.77) and (3.2.78), respectively, and

$$\begin{aligned} \gamma_3 = \sup\Big\{ t \geq 0 : H(A) + \varphi(\Pi_{i=1}^{k}(1 + \beta_i\rho(t_i))) \\ \times \int_0^t (p(s)\varphi(\rho(s)) + \Lambda(s)g(s)\varphi(\rho(s-h)))ds \in Dom\ (H^{-1}) \\ \textit{for all} \ \tau \in [0, t]\Big\}. \end{aligned} \tag{3.2.99}$$

In the case when the function $f_1(t) = 0$ in inequality (3.2.76), we can obtain another bound in which the function $H(u)$ is different and in some cases easier to be used.

Theorem 3.2.4 (The Hristova Inequality [284]) *Assume the following conditions are fulfilled:*

(1) The functions $f_1, f_2, p, g \in C(\mathbb{R}_+, \mathbb{R}_+)$.
(2) The function $\psi \in C([-h, 0], \mathbb{R}_+)$.
(3) The function $Q \in W_2(\varphi)$ and $Q(u) > 0$ for all $u > 0$.

(4) The function $G \in W_1$.
(5) The function $u \in PC([-h, +\infty), \mathbb{R}_+)$ *and satisfies the inequalities*

$$\begin{cases} u(t) \leq f_1(t)G(c + \int_0^t p(s)Q(u(s))ds + \int_0^t g(t)Q(u(s-h))ds) \\ \qquad +f_2(t) \sum_{0<t_k<t} \beta_k u(t_k), \ \ for \ \ all \ \ t \geq 0, & (3.2.100) \\ u(t) \leq \varphi(t), \ \ for \ \ all \ \ t \in [-h, 0], & (3.2.101) \end{cases}$$

where $c \geq 0, \beta_k \geq 0, k = 1, 2, \ldots$.

Then for all $t \in (t_k, t_{k+1}] \cap [0, \gamma)$, $(k = 1, 2, \ldots)$, *we have*

$$\begin{aligned} u(t) \leq\ & \rho(t)\Pi_{i=1}^{k}(1 + \beta_i\rho(t_i)) \\ & \times G(H^{-1}\{H(A) + \sum_{i=1}^{k} \int_{t_{i-1}}^{t_i} p(s)\varphi[\rho(s)\Pi_{j=1}^{i-1}(1 + \beta_j\rho(t_j))]da \\ & + \int_{t_k}^{t} p(s)\varphi[\rho(s)\Pi_{i=1}^{k}(1 + \beta_j\rho(t_j))]ds \\ & +\Lambda(t) \sum_{i=1}^{k} \int_{t_{i-1}}^{t_i} g(s)\varphi[\rho(s-h)\Pi_{j:0<t_j<s-h}(1 + \beta_j\rho(t_k))]ds \\ & +\Lambda(t) \int_{t_k}^{t} g(s)\varphi[\rho(s-h)\Pi_{j:0<t_j<s-h}(1 + \beta_j\rho(t_k))]ds\}), \end{aligned} \quad (3.2.102)$$

where $\Lambda(t)$ *is defined by equality (3.2.77), the constants* A, B_1, B_2, γ *are the same as in Theorem 3.2.1,* $\rho(t) = \max\{f_i(t) : i = 1, 2\}$,

$$H(u) = \int_{u_0}^{u} \frac{ds}{Q(G(s))}, \quad u \geq u_0 > 0. \quad (3.2.103)$$

Proof The proof is similar to that of Theorem 3.2.1. □

As a partial case of Theorem 3.2.1, we can obtain the following result about integral inequalities for pointwise continuous functions without delay.

Theorem 3.2.5 (The Hristova Inequality [284]) *Assume the following conditions hold:*

(1) The functions $f_1, f_2, f_3, p \in C(\mathbb{R}_+, \mathbb{R}_+)$.
(2) The function $Q \in W_2(\varphi)$ *and* $Q(u) > 0$ *for all* $u > 0$.
(3) The function $G \in W_1$.

(4) *The function* $u \in PC(\mathbb{R}_+, \mathbb{R}_+)$ *and satisfies the inequality for all* $t \geq 0$,

$$u(t) \leq f_1(t) + f_2(t)G\{c + \int_0^t p(s)Q(u(s))ds\}$$
$$+f_3(t) \sum_{0<t_k<t} \beta_k u(t_k), \tag{3.2.104}$$

where $c \geq 0, \beta_k \geq 0, k = 1, 2, \ldots$.

Then for all $t \in (t_k, t_{k+1}] \cap [0, \gamma_1), k = 0, 1, 2, \ldots$, *we have*

$$u(t) \leq \rho(t)\Pi_{i=1}^k(1 + \beta_i(\rho(t_i))$$
$$\times(1 + G(H^{-1}\{H(c) + \sum_{i=1}^k \int_{t_{i-1}}^{t_i} p(s)\varphi[\rho(s)\Pi_{j=1}^{i-1}(1 + \beta_j\rho(t_j))]ds$$
$$+ \int_{t_k}^t p(s)\varphi[\rho(s)\Pi_{j=1}^k(1 + \beta_j\rho(t_j))]ds\})), \tag{3.2.105}$$

where the function $H(u)$ *is defined by (3.2.78),* $\rho(t) = \max\{f_i(t) : i = 1, 2\}$,

$$\gamma_1 = \sup\left\{t \geq 0 : H(c) + \sum_{i=1}^k \int_{t_{i-1}}^{t_i} p(s)\varphi[\rho(s)\Pi_{j=1}^{i-1}(1 + \beta_j\rho(t_j))]ds\right.$$
$$\left. + \int_{t_k}^t p(s)\varphi[\rho(s)\Pi_{j=1}^k(1 + \beta_j\rho(t_j))]ds \in Dom\ (H^{-1}),\ for\ \ all\ \ \tau \in [0, t]\right\}.$$

Remark 3.2.4 We note that the obtained inequalities are generalizations of many known results. For example, in the case when $f_1(t) = 0, f_2(t) = 0, \beta_k = 0, G(u) = u, Q(u) = u, h = 0, g(t) = 0$, the result in Theorem 3.2.3 reduces to the classical Gronwall-Bellman inequality.

Now we shall consider different types of nonlinear integral inequalities in which the unknown function is powered, which can be regarded as a variant of the Ou-Yang inequality.

Theorem 3.2.6 (The Hristova Inequality [284]) *Assume the following conditions hold:*

(1) *The functions* $f, g, h, r \in C(\mathbb{R}_+, \mathbb{R}_+)$.
(2) *The function* $\psi \in C([-h, 0], \mathbb{R}_+)$ *and* $\psi(t) \leq c$ *for all* $t \in [-h, 0]$ *where* $c \geq 0$. *The constants* $p > 1, 0 \leq q \leq p$.

(3) The function $u \in PC(\mathbb{R}_+, \mathbb{R}_+)$ and satisfies the inequalities

$$\begin{cases} u^p(t) \le c + \int_0^t [f(s)u^p(t) + g(s)u^q(s)u^{p-q}(s-h) \\ \qquad + h(s)u(s) + r(s)u(s-h)]ds + \sum_{0<t_k<t} \beta_k u^p(t_k), \quad \text{for all } t \ge 0, & (3.2.106) \\ u(t) \le \psi(t), \quad \text{for all } t \in [-h, 0]. & (3.2.107) \end{cases}$$

Then for all $t \in (t_k, t_{k+1}], k = 0, 1, 2, \ldots,$ there holds

$$u(t) \le \sqrt[p]{\Pi_{i=1}^k (1+\beta_i)} \times \sqrt[p]{c + \frac{p-1}{p}\int_0^t (h(s) + r(s))ds} \times \sqrt[p]{\exp\left(\int_0^t \left(f(s) + g(s) + \frac{h(s)+r(s)}{p}\right)ds\right)}. \quad (3.2.108)$$

Proof

Case 1. Let $t_1 \ge h$.

Let $t \in (0, h]$. We define the function $v_0^{(0)} : [-h, h] \to \mathbb{R}_+$ by the equalities

$$v_0(t) = \begin{cases} c + \int_0^t [f(s)u^p(t) + g(s)u^{p-q}(s-h) + h(s)u(s) \\ \quad + r(s)u(s-h)]ds, \quad \text{for all } t \in [0, h], \\ \varphi^p(t), \quad \text{for all } t \in [-h, 0). \end{cases}$$

The function $v_0^{(0)}(t)$ is a non-decreasing differentiable function on $[0, h], u^p(t) \le v_0^{(0)}(t)$ and using the inequality $x^m y^n \le \frac{x}{m} + \frac{y}{n}, n+m = 1$, we obtain for all $t \in [0, h]$,

$$u(t) \le \sqrt[p]{v_0^{(0)}(t)} \le \frac{v_0^{(0)}(t)}{p} + \frac{p-1}{p}, \quad (3.2.109)$$

and for all $t \in [0, h]$,

$$\begin{aligned} u(t-h) &\le \frac{\psi(t-h)}{p} + \frac{p-1}{p} \\ &\le \frac{c}{p} + \frac{p-1}{p} \le \frac{v_0^{(0)}(t)}{p} + \frac{p-1}{p}. \end{aligned} \quad (3.2.110)$$

Therefore there holds

$$\begin{aligned} (v_0^{(0)}(t))' &= f(t)u^p(t) + g(t)u^q(t)u^{p-q}(t-h) + h(t)u(t) + r(t)u(t-h) \\ &\le f(t)v_0^{(0)}(t) + g(t)v_0^{(0)}(t)^{q/p} v_0^{(0)}(t-h)^{(p-q)/p} \end{aligned}$$

$$+h(t)(\frac{v_0^{(0)}(t)}{p}+\frac{p-1}{p})+r(t)(\frac{v_0^{(0)}(t)}{p}+\frac{p-1}{p})$$
$$\le (f(t)+g(t)+\frac{h(t)+r(t)}{p})v_0^{(0)}(t)+(h(t)+r(t))\frac{p-1}{p}. \quad (3.2.111)$$

According to Theorem 1.1.4 in Qin [557] from (3.2.111), we obtain

$$v_0^{(0)}(t) \le \left(c+\frac{p-1}{p}\int_0^t (h(s)+r(s))ds\right)$$
$$\times \exp\left(\int_0^t (f(t)+g(t)+\frac{h(s)+r(s)}{p})ds\right). \quad (3.2.112)$$

From inequality (3.2.112), we derive (3.2.108) holds for all $t \in [0,h]$.

Let $t \in (h,t_1]$. Define the function $v_0^{(0)} : [h,t_1] \to [0,+\infty)$ by the equation

$$v_0^{(1)}(t) = v_0^{(0)}(h) + \int_h^t [f(s)u^p(t)+g(s)u^q(s)u^{p-q}(s-h)+h(s)u(s)+r(s)u(s-h)]ds.$$

From the definition of the function $v_0^{(1)}(t)$ and (3.2.106), it follows that for all $t \in (h,t_1]$,

$$u^p(t) \le v_0^{(0)}(t). \quad (3.2.113)$$

Case 1.1. Let $h < t \le \min\{t_1, 2h\}$. Then $t-h \in (0,h]$ and

$$u(t-h) \le (v_0^{(0)}(t-h))^{1/p} \le (v_0^{(0)}(t))^{1/p} \le \frac{v_0^{(1)}(t)}{p}+\frac{p-1}{p}.$$

Case 1.2. Let $t_1 > 2h$ or $t \in (2h,t_1]$. Then

$$u(t-h) \le (v_0^{(0)}(t-h))^{1/p} \le (v_0^{(0)}(t-h))^{1/p} \le \frac{v_0^{(1)}(t)}{p}+\frac{p-1}{p}$$

and

$$(v_0^{(1)}(t))' \le \left(f(t)+g(t)+\frac{h(t)+r(t)}{p}\right)v_0^{(1)}(t)+(h(t)+r(t))\frac{p-1}{p}. \quad (3.2.114)$$

From the inequalities (3.2.113), (3.2.114) and applying Theorem 1.1.4 in Qin [557], we obtain

$$
\begin{aligned}
(v_0^{(1)}(t))' &\le \left(v_0^{(0)}(h) + \frac{p-1}{p}\int_h^t (h(s)+r(s))ds\right) \\
&\quad \times \left(\int_h^t (f(t)+g(t)+\frac{h(s)+r(s)}{p})ds\right) \\
&\le \left(c + \frac{p-1}{p}\int_0^t (h(s)+r(s))ds\right) \\
&\quad \times \left(\int_0^t (f(t)+g(t)+\frac{h(s)+r(s)}{p})ds\right). \qquad (3.2.115)
\end{aligned}
$$

Thus from the inequalities (3.2.113) and (3.2.114), we prove (3.2.108) holds on $(h, t_1]$.

Define a function

$$
v_0(t) = \begin{cases} v_0^{(0)}(t), & for\ all\ t \in [0, h], \\ v_0^{(1)}(t), & for\ all\ t \in [h, t_1]. \end{cases}
$$

Now let $t \in (t_1, t_2]$.

Define the function $v_1 : [t_1, t_2] \to \mathbb{R}_+$ by the equation

$$
\begin{aligned}
v_1(t) = v_0(t_1) &+ \int_h^t [f(s)u^p(t) + g(s)u^q(s)u^{p-q}(s-h) + h(s)u(s) \\
&+ r(s)u(s-h)]ds + \beta_1 u^p(t_1). \qquad (3.2.116)
\end{aligned}
$$

We note that $v_0(t) \le v_1(t), u^p(t) \le v_1(t), u^p(t_1) \le v_0(t_1), u(t-h) \le \sqrt[p]{v_1(t)}, u(t-h) \le \sqrt[p]{v_1(t)}$ and $\sqrt[p]{v_1(t)} \le \frac{v_1(t)}{p} + \frac{p-1}{p}$ for all $t \in (t_1, t_2]$.

The function $v_1(t)$ satisfies

$$
\begin{aligned}
v_1(t) &\le \left((1+\beta_1)v_0(t_1) + \frac{p-1}{p}\int_{t_1}^t (h(s)+r(s))ds\right) \\
&\quad \times \exp\left(\int_{t_1}^t (f(t)+g(t)+\frac{h(s)+r(s)}{p})ds\right) \\
&\le \left(1+\beta_1)(c + \frac{p-1}{P}\int_0^t (h(s)+r(s))ds\right) \\
&\quad \times \exp\left(\int_0^t (f(t)+g(t)+\frac{h(s)+r(s)}{p})ds\right). \qquad (3.2.117)
\end{aligned}
$$

From the inequalities $u(t) \le \sqrt[p]{v_1(t)}$ and (3.2.117), (3.2.108) follows for all $t \in (t_1, t_2]$. Using the mathematical induction, we can prove (3.2.108) for all $t \ge 0$.

Case 2. Let there exist a natural number m such that $t_m \le h < t_{m+1}$. As in **Case 1**, we can prove (3.2.108) holds, using functions $v_k(t), k = 1, 2, \ldots$, defined by

$$v_k(t) = v_{k-1}(t_k) + \int_{t_k}^{t} [f(s)u^p(t) + g(s)u^q(s)u^{p-q}(s-h) + h(s)u(s) \\ + r(s)u(s-h)]ds + \beta_k u^p(t_k). \qquad (3.2.118)$$

□

Remark 3.2.5 Some of the inequalities proved by Pachpatte in [496, 498, 500] are partial cases of Theorems 3.2.1 and Theorem 3.2.4.

Next, we shall study the inequality

$$u(t) \le \phi(t) + \int_{t_0}^{t} K(t, s, u(s))ds, \qquad (3.2.119)$$

for establishing the estimate $u(t) \le \sigma_\phi(t)$, where $\sigma_\phi(t)$ is the solution of Volterra's integral equation

$$\sigma(t) = \phi(t) + \int_{t_0}^{t} K(t, s, \sigma(s))ds$$

as mentioned in [375].

Recall that the works [102–109, 114, 116, 117, 154, 292, 582, 583, 589] were dedicated precisely to this question (i.e., the solvability of Chaplygin's problem). In detail, there were investigated inequalities of a type

$$u(t) \le \phi(t) + \int_{t_0}^{t} K(t, s, u(s))ds + \sum_{t_0 < t_k < t} \psi(t, t_k)\beta_k(u(t_k - 0)), \quad (3.2.120)$$

where $u(t)$ is a non-negative piecewise continuous function with first kind discontinuities at the points $\{t_i\}, t_0 < t_1 < \cdots, \lim_{i \to +\infty} t_i = +\infty$.

We shall present the conditions of solvability for Chaplygin's problem for the integro-sum inequality of Wendroff's type:

$$u(x) \le \phi(x) + \int\int_{G_n} H(y, u(y))dy + \sum_{j=1}^{n-1} \int_{\Gamma_j \cap G_n} \beta_j(x)u(x)d\mu_{\phi_j}, \quad (3.2.121)$$

where $x = (x_1, x_2)$, $G_n \subset \mathbb{R}^2$, μ is some measure concentrated on the curves $\{\Gamma_j\}$, $j = \{1, 2, \ldots, +\infty\}$.

We first introduce the results on Gronwall-Bellman-Bihari type integral inequalities for discontinuous functions.

Theorem 3.2.7 (The Borysenko Inequality [102]) *Let $V(t)$ be a non-negative piecewise continuous function at all $t \geq t_0$, with first kind discontinuities at the points t_i, and satisfying the integral inequality*

$$V(t) \leq C + \int_{t_0}^{t} P(\tau)V^m(\tau)d\tau + \sum_{t_0<t_i<t} \beta_i V(t_i - 0),\ m > 0,\ m \neq 1, \tag{3.2.122}$$

where $t_1 < t_2 < \cdots, \lim_{i\to+\infty} t_i = +\infty,\ C \geq 0, \beta_i \geq 0, P(t) \geq 0,$ for all $t \geq t_0$. Then the following estimates hold:

(1) if $0 < m < 1$, then for all $t \geq t_0$,

$$V(t) \leq \prod_{t_0<t_i<t} (1 + \beta_i) \left(C^{1-m} + (1 - m) \int_{t_0}^{t} P(\tau)d\tau \right)^{1/(1-m)}, \tag{3.2.123}$$

(2) if $m > 1$, then for all $t \in [t_0, +\infty]$,

$$\begin{aligned} V(t) \leq & \ C \prod_{t_0<t_i<t}(1 + \beta_i) \\ & \times \left[1 - (m - 1)C^{m-1} \prod_{t_0<t_i<t}^{m-1}(1 + \beta_i) \int_{t_0}^{t} P(\tau)d\tau\right]^{-1/(m-1)}, \end{aligned} \tag{3.2.124}$$

with

$$\int_{t_0}^{t} P(\tau)d\tau < \frac{C^{1-m}}{(m - 1)\prod_{t_0<t_i<t}^{m-1}(1 + \beta_i)}.$$

Remark 3.2.6 For applications of the results of Theorem 3.2.7, we refer to [102, 103, 105, 589].

Theorem 3.2.8 (The Perestyuk-Chernikova Inequality [588]) *Let $u(t)$ be a non-negative piecewise continuous function at all $t \geq t_0$, with first kind discontinuities at $t = \tau_i$ and satisfying the inequality, for all $t \geq t_0$,*

$$u(t) \leq C + \int_{t_0}^{t} V(\tau)\Phi(u(\tau))d(\tau) + \sum_{t_0<\tau_i<t} \beta_i u(\tau_i), \tag{3.2.125}$$

where $C \geq 0, \beta_i \geq 0, V(t)$ is a positive continuous function, $\Phi(u)$ is a positive continuous non-decreasing function for all $0 < u < \bar{u}$ $(\bar{u} < +\infty)$.

Then the function $u(t)$ satisfies for all $\tau_i < t \leq \tau_i + 1$,

$$u(t) \leq \Psi_i^{-1} \left(\int_{\tau_i}^{t} V(\tau)d\tau \right), \tag{3.2.126}$$

if

$$\int_{\tau_i}^{t} V(\tau)d\tau < \psi_i^{-1}(\overline{u} - 0),$$

where

$$\Psi_i^{-1}(u) = \int_{c_i}^{u} \frac{du}{\Phi(u)}, \; c_i = (1 + \beta_i)\Psi_i^{-1}\left(\int_{\tau_i-1}^{\tau} V(\tau)d\tau\right)$$

$$\Psi_0(u) = \int_{c}^{u} \frac{du}{\Phi(u)}, \; i = 1, 2, \cdots, \tau_0 = t_0.$$

Remark 3.2.7 For the applications of Theorem 3.2.8 with $\Phi(u) = u^m, \; m = 1$, we refer to [585, 589].

Theorem 3.2.9 (The Borysenko Inequality [104]) *Let us consider the integro-sum equation of the following form*

$$\sigma(t) = \phi(t) + \int_{t_0}^{t} K(t, s, \sigma(s))ds + \sum_{t_0<\tau_k<t} \Psi(t, t_k)\mu_k(\sigma(t_k - 0)), \tag{3.2.127}$$

where $\sigma(t), \phi(t), \Psi(t, t_k)$ are continuous non-negative functions $(k = 1, 2, \cdots)$ for all $t \geq t_0$, except for $\sigma(t)$, which has first kind discontinuities at the points t_k and

$$t_0 < t_1 < \cdots, \; \lim_{i\to+\infty} t_i = +\infty.$$

The function $K(t, s, u)$, which is non-negative at $t \geq s \geq t_0$, is determined in the domain $t \geq s \geq t_0$, $|u| \leq k$ and at fixed t and s, it is non-decreasing with respect to u; the functions $\mu_k(\sigma)$ are continuous non-negative and non-decreasing with respect to σ.

Then for all $t \in [t_0, +\infty]$, there holds

$$u(t) \leq \sigma_\phi(t) \tag{3.2.128}$$

where $\sigma_\phi(t)$ is some solution of (3.2.127), continuous in each interval $[t_k, t_{k+1}]$, $k = 0, 1, \cdots$; $u(t)$ is a piecewise continuous function with first kind discontinuities at t_i points; this function satisfies the integro-sum inequality:

$$u(t) \leq \phi(t) + \int_{t_0}^{t} K(t, s, \sigma(s))ds + \sum_{t_0<\tau_k<t} \Psi(t, t_k)\mu_k(\sigma(t_k - 0)), \tag{3.2.129}$$

where $\sigma(t_k - 0) = \lim_{t\to t_k-} \sigma(t)$.

Corollary 3.2.5 (The Borysenko Inequality [104]) *Assume that $V(t)$ is a non-negative piecewise continuous function at all $t \geq t_0$ with first kind discontinuities at the points t_i, and satisfies the integrosum inequality*

$$V(t) \leq \Psi(t) + \int_0^t P(\tau)V^m(\tau)d\tau + \sum_{t_0<\tau_k<t} \beta_i V(t_i - 0),\ m > 1,$$

where $t_1 < t_2 < \cdots, \lim_{i\to+\infty} t_i = +\infty, \Psi(t)$ is a positive monotonously non-decreasing function at all $t \geq t_0, \beta_i \geq 0, P(t) \geq 0$.

Then the following estimates hold:

(1) if $0 < m < 1$, then for all $t \geq t_0$,

$$V(t) \leq \Psi(t) \prod_{t_0<t_i<t} (1 + \beta_i) \left(1 + (1-m) \int_{t_0}^t \Psi^{m-1}(\tau)P(\tau)d\tau\right)^{1/(1-m)};$$

(2) if $m = 1$, then for all $t \geq t_0$,

$$V(t) \leq \Psi(t) \prod_{t_0<t_i<t} (1 + \beta_i) \exp\left(\int_{t_0}^t P(\tau)d\tau\right);$$

(3) if $m > 1$, then for all $t \geq t_0$,

$$V(t) \leq \Psi(t) \prod\nolimits_{t_0<t_i<t} \times \left[1 - (m-1) \prod\nolimits_{t_0>t_i<t}^{m-1} (1 + \beta_i) \int_0^t \Psi^{m-1}(\tau)P(\tau)d\tau\right]^{1/(1-m)},$$

with

$$\int_0^t \Psi^{m-1}(\tau)P(\tau)d\tau < \left[(m-1) \prod_{t_0>t_i<t}^{m-1} (1 + \beta_i)\right]^{-1}.$$

Definition 3.2.3 $W(x) \in \overline{\mathcal{F}}$ if and only if

(a) $W(\alpha\beta) \leq W(\alpha)W(\beta)$;
(b) $W : [0, +\infty] \to [0, +\infty], W(0) = 0$;
(c) W is non-decreasing.

Theorem 3.2.10 (The Borysenko Inequality [106]) *Consider an integro-sum functional inequality:*

$$v(x) \leq \varphi(x) + q(x) \int_{x_0}^x f(\tau)W(v(p(\tau)))d\tau + \sum_{x_0<x_i<x} \beta_i v(x_i - 0), \quad (3.2.130)$$

where $x \geq x_0, q(x) \geq 1, \varphi(x)$ is positive non-decreasing, $\beta_i = const \geq 0$; $v(x)$ is a piecewise continuous function with first kind discontinuities at the points $x_i : x_0 <$

$x_1 < \cdots$, $\lim_{n\to+\infty} x_n = +\infty, f \geq 0$, *where* $W(x) \in \overline{\mathcal{F}}$ *(see Definition 3.1.1 in Qin [557]).*

Suppose that $p(s) \in \mathcal{F}$. *Then for arbitrary* $x \in [x_0, T],\ T \leq +\infty$, *the following inequality holds for all* $x \in [x_i, x_{i+1}]$,

$$v(x) \leq \varphi(x)q(x)\Phi_i^{-1}\left(\int_{x_i}^{x} \frac{f(t)}{\varphi(t)} W^*(t)dt\right),$$

with

$$\int_{x_i}^{x} f(t)\varphi^{-1}(t)W^*(t)dt \in Dom\,(\Phi_i^{-1}),\ \Phi_0(\xi) = \int_1^{\xi} \frac{d\eta}{W(\eta)},$$

$$\Phi_i(\xi) = \int_{l_i}^{\xi} \frac{d\eta}{W(\eta)},\quad i = 1, 2, \cdots$$

$$l_i = (1 + \beta_i q(x_i))\Phi_{i-1}^{-1}\left(\int_{x_i-1}^{x_i} f(y)\varphi^{-1}(y)W^*(y)dy\right),\ i = 1, 2, \cdots.$$

Here $W^*(v) = W(g(p(v)\varphi(p(v))))$.

Consider a class of functions $\mathcal{F}_2 : f \in \mathcal{F}_2 \Longleftrightarrow$:

(a) $f(x)$ is positive, continuous, and non-decreasing for $x > 0$; (b) $\forall t > 1, u \leq 0 \Longrightarrow t^{-1}f(u) \leq f(t^{-1}u)$; (c) $f(0) = 0$.

Theorem 3.2.11 (The Borysenko Inequality [106]) *Assume that a piecewise continuous non-negative function* $\varphi(x)$ *with first kind discontinuities at the points* $\{x_i\}$ *satisfies the inequality* (3.2.130)*, where* φ, q, p *satisfy the conditions of Theorem 3.2.10, function* W *belongs to the class of functions* $\mathcal{F}_2$. *Then for all* $x \in [x_0, x^*]$, *and for all* $x \in [x_i, x_{i+1}]$,

$$v(x) \leq \varphi(x)q(x)\overline{\Phi}_i^{-1}\left(\int_{x_i}^{x} G(\tau)d\tau\right),\quad i = 0, 1, 2, \cdots,$$

where

$$\begin{cases} \overline{\Phi}_0(\sigma) = \displaystyle\int_1^{\xi} W^{-1}(\sigma)d\sigma,\ \overline{\Phi}_i(\sigma) = \int_{l_i}^{\xi} W^{-1}(\sigma)d\sigma,\ i = 1, 2, \cdots; \\ l_i = (1 + \beta_i q(x_i))\Phi_{i-1}^{-1}\left(\displaystyle\int_{x_i-1}^{x_i} G(\tau)d\tau,\ G(t)f(t)q(p(t))\right), \\ x^* = \sup\left\{x : \displaystyle\int_{x_i-1}^{x} G(\tau)d\tau \in Dom\,(\overline{\Phi}_{i-1}^{-1}), i = 1, 2, \cdots\right\}. \end{cases}$$

Theorem 3.2.12 (The Samoilenko-Borysenko Inequality [583]) *Consider the integro-sum inequality in the following form:*

$$u(x) \leq u_0 + q_1(x)\int_{x_0}^{x} f(s)W_1(u(p(s)))ds + q_2(x)\int_{x_0}^{x} g(s)W_2(u(\sigma(s)))ds + \sum_{x_0<x_i<x}\beta_i u(x_i - 0), \tag{3.2.131}$$

where $f(x), g(x), p(x), \sigma(x)$ *are non-negative continuous functions:*

for all $x > x_0$, $p(s) \leq x, \sigma(x) \leq x, q_1(x) \geq 1, q_2(x) \geq 1$; $W_1(x) \in \overline{\mathcal{F}}, W_2(x) \in \mathcal{F}_2$; $u_0 = const. \geq 1$, $\beta_i = const. \geq 0$, $u(x)$ *is a piecewise continuous non-negative function with first kind discontinuities at the points* $\{x_i\}$, *and satisfies conditions of Theorem 3.2.10. Then for all* $x \geq x_0$,

$$u(x) \leq q_1(x)q_2(x)S_i(x)F^{-1}[\int_{x_0}^{x}\Psi(\overline{x})d\overline{x}], \tag{3.2.132}$$

where

$$\begin{cases} S_i = G_i^{*-1}[\int_{x_i}^{x} f(s)q_1(p(s))ds], \\ G_i^*(\xi) = \int_{l_i}^{\xi} W_1^{-1}(\sigma)d\sigma, \ i = 1, 2, \cdots; \ G_0^*(\xi) = \int_{u_0}^{\xi} W_1^{-1}(v)dv; \\ l_i = (1+\beta_i)S_{i-1}(x_i), \ F(\eta) = \int_{u_0}^{\eta} W_2^{-1}(s)ds, \end{cases}$$

and F^{-1}, G_i^{*-1} *are the inverse of the functions* F *and* G_i^*, *respectively, and*

$$\begin{cases} \int_{x_i}^{x} f(\tau)q_1(p(\tau))d\tau \in Dom\,(G_i^{*-1}), \quad i = 1, 2, \cdots, \\ \int_{x_0}^{x} \Psi(\tau)d\tau \in Dom\,(F^{-1}), \quad i = 1, 2, \cdots. \end{cases}$$

Here $\Psi(x) = g(x)W_2[q_1(\sigma(x))q_2(\sigma(x))]S_i(\sigma(x))$.

Theorem 3.2.13 (The Borysenko Inequality [109]) *Assume that* $V(t)$ *is a non-negative piecewise continuous on* $J = [t_0, +\infty]$ *function with first kind discontinuities at points* $\{t_i\} : t_1 < t_2 < \cdots, \lim_{i\to+\infty} t_i = +\infty$, *satisfying the integro-sum inequality*

$$V(t) \leq \psi(t) + \int_{t_0}^{t} q(\tau)V(\tau)d\tau + \sum_{t_0<t_i<t} a_i V^m(t_i - 0),$$

where $\psi(t)$ is a positive monotonously non-decreasing on J function, $q(t) \geq 0, a_i \geq 0, m > 0$. Then for all $t \geq t_0$,

$$\begin{cases} V(t) \leq \psi(t) \prod_{t_0<t_i<t} (1 + a_i\psi^{m-1}(t_i)) \exp\left(\int_{t_0}^{t} q(s)ds\right), \text{ if } 0 < m < 1, & (3.2.133) \\ V(t) \leq \psi(t) \prod_{t_0<t_i<t} (1 + a_i\psi^{m-1}(t_i)) \exp\left(m\int_{t_0}^{t} q(s)ds\right), \text{ if } m \geq 1. & (3.2.134) \end{cases}$$

Remark 3.2.8 Theorem 3.2.13 generalizes the fundamental results for discontinuous functions obtained by Bellman.

Theorem 3.2.14 (The Borysenko Inequality [109]) *Assume that $V(t)$ is a non-negative piecewise continuous on $J = [t_0, +\infty]$ function with first kind discontinuities at the points $\{t_i\}$ satisfying the inequality*

$$\overline{V}(t) \leq \overline{\psi}(t) + \int_{t_0}^{t} \overline{q}(\tau)\overline{V}^m(\tau)d\tau + \sum_{t_0<t_i<t} \overline{V}^m(t_i - 0), \tag{3.2.135}$$

where $\overline{\psi}(t) > 0, \overline{q}(t) \geq 0, \overline{a}_i \geq 0, m > 0, m \neq 1, \overline{\psi}(t)$ is non-decreasing for all $t \in J$.

Then

(1) if $0 < m < 1$, then for all $t \geq t_0$,

$$\overline{V}(t) \leq \overline{\psi}(t) \prod_{t_0<t_i<t} (1 + \overline{a}_i m\overline{\psi}^{m-1}(t_i))\left[1 - (1-m)\int_{t_0}^{t} \overline{\psi}^{m-1}(\tau)\overline{q}(\tau)d\tau\right]^{1/(1-m)}, \tag{3.2.136}$$

(2) if $m > 1$, then for all $t \geq t_0$,

$$\overline{V}(t) \leq \overline{\psi}(t) \prod_{t_0<t_i<t} (1 + \overline{a}_i m\overline{\psi}^{m-1}(t_i))\left[1 - (1-m)\left[\prod_{t_0<t_i<t} (1 + \overline{a}_i m\overline{\psi}^{m-1}(t_i))\right]^{m-1} \times \int_{t_0}^{t} \overline{q}(\tau)\overline{\psi}^{m-1}(\tau)d\tau\right]^{-1/(m-1)},$$

with

$$\begin{cases} \int_{t_0}^{t} \overline{q}(\tau)\overline{\psi}(\tau)d\tau \leq \frac{1}{m}, & (3.2.137) \\ \prod_{t_0<t_i<t} \left(1 + \overline{a}_i m\overline{\psi}^{m-1}(t_i)\right) < (\frac{m}{m-1})^{1/(m-1)}. & (3.2.138) \end{cases}$$

Remark 3.2.9 Theorem 3.2.14 generalizes the result for discontinuous functions obtained by Bihari.

Theorem 3.2.15 (The Borysenko-Ciarletta-Iovane Inequality [113]) *Assume that $\phi(t)$ is a non-negative function with first kind discontinuities at points $\{t_i\}$: $t_1 < t_2 < \cdots, \lim_{i\to+\infty} t_i = +\infty$, satisfying the following integro-sum inequality*

$$\begin{aligned}\varphi(t) \leq C + \int_{t_0}^{t} q(s)\varphi(s)ds + \int_{t_0}^{t} q(s)\int_{t_0}^{t} q(\sigma)(\varphi^m(\sigma)d\sigma)ds \\ +\Sigma_{t_0<t_i<t}\beta_i\varphi(t_i-0), \quad if \quad m>0,\end{aligned} \tag{3.2.139}$$

where $C \geq 0,\ q(t) \geq 0, g(t) \geq 0,\ \beta = const. \geq 0$. Then

(1) if $0 < m < 1$, then for all $t \geq t_0$,

$$\begin{aligned}\varphi(t) \leq \exp\left(\textstyle\int_{t_0}^{t} q(\tau)d\tau\right)\Bigg[\left(C\textstyle\prod_{t_0<t_i<t}(1+\beta_i)\right)^{1-m} \\ +(1-m)\textstyle\int_{t_0}^{t} g(s)\exp\left((m-1)\textstyle\int_{t_0}^{s} g(\sigma)d\sigma\right)ds\Bigg]^{1/(1-m)};\end{aligned} \tag{3.2.140}$$

(2) if $m = 1$, then for all $t \geq t_0$,

$$\varphi(t) \leq C\prod_{t_0<t_i<t}(1+\beta_i)\exp\left(\int_{t_0}^{t}(q(\tau)+g(\tau))d\tau\right); \tag{3.2.141}$$

(3) if $m > 1$, then for all $t \geq t_0$,

$$\begin{aligned}\varphi(t) \leq C\prod_{t_0<t_i<t}(1+\beta_i)\exp\left(\int_{t_0}^{t} q(\tau)d\tau\right) \\ \times\Bigg[1-(m-1)\prod_{t_0<t_i<t}(1+\beta_i)^{m-1}C^{m-1} \\ \times\int_{t_0}^{t} g(s)\exp[(m-1)\int_{t_0}^{s} g(\sigma)d\sigma]ds\Bigg]^{-1/(1-m)},\end{aligned} \tag{3.2.142}$$

with

$$\int_{t_0}^{t} g(s)\exp\left((m-1)\int_{t_0}^{s} g(\sigma)d\sigma\right)ds < \left[(m-1)\prod_{t_0<t_i<t}(1+\beta_i)^{m-1}C^{m-1}\right]^{-1}. \tag{3.2.143}$$

Theorems 3.2.7–3.2.14 can be proved, using the inductive method and the methodology of the integral inequalities theory. We shall illustrate this fact by proving Theorem 3.2.15.

Proof of Theorem 3.2.15 Suppose that $t \in [t_0, t_1]$. Then

$$\varphi(t) \leq C + \int_{t_0}^{t} q(s)\varphi(s)ds + \int_{t_0}^{t} q(s)(\int_{t_0}^{t} g(\sigma)\varphi^m(\sigma)d\sigma)ds.$$

Define $V(t) := C + \int_{t_0}^{t} q(s)\varphi(s)ds + \int_{t_0}^{t} q(s)\left(\int_{t_0}^{s} g(\sigma)\varphi^m(\sigma)d\sigma\right)ds$.

Obviously, $\varphi(t_0) = V(t_0) = C, \varphi(t) \leq V(t)$, for all $t \geq t_0$. Then

$$\begin{aligned}\frac{dV}{dt} &= q(t)\varphi(t) + q(t)\int_{t_0}^{t} g(\sigma)\varphi^m(\sigma)d\sigma \\ &\leq q(t)[V(t) + \int_{t_0}^{t} g(\sigma)V^m(\sigma)d\sigma]. \end{aligned} \tag{3.2.144}$$

Let $W(t) = V(t) + \int_{t_0}^{t} g(\sigma)V^m(\sigma)d\sigma$. Then $W(t_0) = V(t_0) = C$, $V(t) \leq W(t)$, for all $t \geq t_0$. We obtain

$$\frac{dW}{dt} \leq q(t)W(t) + g(t)W^m(t),$$

which implies for $0 < m < 1$, for all $t \geq t_0$,

$$\begin{aligned}\varphi(t) \leq \exp\left(\int_{t_0}^{t} q(\tau)d\tau\right)\Big\{C^{1-m} + (1-m)\int_{t_0}^{t} g(s) \\ \times \exp\left((m-1)\int_{t_0}^{s} g(\sigma)d\sigma\right)ds\Big\}^{1/(1-m)},\end{aligned} \tag{3.2.145}$$

or for $m = 1$, for all $t \geq t_0$;

$$\varphi(t) \leq C\exp\left(\int_{t_0}^{t} (q(\tau) + g(\tau))d\tau\right), \tag{3.2.146}$$

or for $m > 1$, for all $t \geq t_0$,

$$\begin{aligned}\varphi(t) \leq C\exp\left(\int_{t_0}^{t} q(\tau)d\tau\right)\Big\{C^{1-m} + (1-m)\int_{t_0}^{t} g(s) \\ \times \exp\left((m-1)\int_{t_0}^{s} g(\sigma)d\sigma\right)ds\Big\}^{-1/(m-1)},\end{aligned} \tag{3.2.147}$$

with

$$\int_{t_0}^{t} g(\tau)\exp\left((m-1)\int_{t_0}^{t} g(\sigma)d\sigma\right)d\tau < \left((m-1)C^{m-1}\right)^{-1}.$$

Thus, from (3.2.147) for all $t \in [t_0, t_1], \varphi(t)$ satisfies the inequalities (3.2.139). (3.2.141). In fact, applying the scheme described in [102, 585, 589] for interval $[t_k, t_{k+1}], k = 1, 2, \cdots$, and the estimates for the function $\varphi(t)$ on the interval $[t_{k-1}, t_k]$, we can obtain the estimates (by using inductive method) (3.2.139)–(3.2.141) on all of interval J. □

Let T be a time scale. For $a, b \in T$ with $a < b$, we define the time scales interval by

$$[a,b]_T = \{t \in T : a \le t \le b\}.$$

The next lemma is a useful tool for the proofs of the next theorems.

Lemma 3.2.3 (The Ferreira-Torres Inequality [229]) *Let $a, b \in T$, consider the time scales interval $[a,b]_T$, and a delta differential function $r : [a,b]_T \to \mathbb{R}_0 = (0, +\infty)$ with $r^{\Delta}(t) \ge 0$ on $[a,b]_T^k$. Define*

$$G(x) = \int_{x_0}^{x} \frac{ds}{g(s)}, \quad x \ge x_0 > 0,$$

where $g \in C(\mathbb{R}_0, \mathbb{R}_0)$, is positive and non-decreasing on $(0, +\infty)$. Then for each $t \in [a,b]_T$, we have

$$G(r(t)) \le G(r(a)) + \int_a^t \frac{r^{\Delta}(\tau)}{g(r(\tau))}\Delta\tau.$$

Proof Since g is positive and non-decreasing on $(0, +\infty)$, we have, successively, that for all $t \in [a,b]_T^k$ and $h \in [0,1]$,

$$\begin{aligned} r(t) &\le r(t) + h\mu(t)r^{\Delta}(t), \\ g(r(t)) &\le g(r(t) + h\mu(t)r^{\Delta}(t)), \\ \frac{1}{g(r(t)+h\mu(t)r^{\Delta}(t))} &\le \frac{1}{g(r(t))}, \\ \int_0^1 \frac{1}{g(r(t)+h\mu(t)r^{\Delta}(t))}dh &\le \int_0^1 \frac{1}{g(r(t))}dh = \frac{1}{g(r(t))}, \\ \left\{\int_0^1 \frac{1}{g(r(t)+h\mu(t)r^{\Delta}(t))}dh\right\} r^{\Delta}(t) &\le \frac{r^{\Delta}(t)}{g(r(t))}. \end{aligned} \tag{3.2.148}$$

By Δ-integrating the last inequality in (3.2.148) from a to t and having in mind that the chain rule guarantees that

$$(G \circ r)^{\Delta}(t) = \left\{ \int_0^1 G'(r(t) + h\mu(t)r^{\Delta}(t))dh \right\} r^{\Delta}(t)$$
$$= \left\{ \int_0^1 \frac{1}{g(r(t) + h\mu(t)r^{\Delta}(t))} dh \right\} r^{\Delta}(t),$$

we obtain the desired result, except at $t = b$ in the case that $\rho(b) < b$. To handle this case, we just need to integrate the last inequality in (3.2.148) from a to b. □

Theorem 3.2.16 (The Ferreira-Torres Inequality [229]) *Let $u(t)$ and $f(t)$ be non-negative rd-continuous functions in the time scales interval $T_* := [a, b]_T$ and T_*^k, respectively. Let $k(t, s) : T_* \times T_*^k \to \mathbb{R}$ be continuous at (t, t), where $t \in T_*^k$ with $t > a$, in such a way that $k(t, s)$ and $k^{\Delta_1}(t, s)$ are non-negative for every $t, s \in T_*$ with $s \leq t$, for which they are defined (it is assumed that k is not identically zero on $T_*^k \times T_*^{k^2}$). Let $\Phi \in C(\mathbb{R}_0, \mathbb{R}_0)$ be a non-decreasing, sub-additive and sub-multiplicative function, such that $\Phi(u) > 0$ for all $u > 0$ and let $W \in C(\mathbb{R}_0, \mathbb{R}_0)$ be a non-decreasing function such that for all $u > 0$, we have $W(u) > 0$. Assume that $a(t)$ is a positive rd-continuous function and non-decreasing for all $t \in T_*$. If*

$$u(t) \leq a(t) + \int_a^t f(s)u(s)\Delta s + \int_a^t f(s)W\left(\int_a^s k(s, \tau)\Phi(u(\tau))\Delta\tau\right)\Delta s, \quad (3.2.149)$$

for all $\tau, s, t \in T_*, a \leq \tau \leq s \leq t \leq b$, then for all $t \in T_*$ satisfying

$$\Psi(\zeta) + \int_a^{\rho(t)} k(\rho(t), s)\Phi(p(s))\Phi\left(\int_a^s f(\tau)\Delta\tau\right)\Delta s \in Dom\,(\Psi^{-1}),$$

we have

$$u(t) \leq p(t)a(t)$$
$$+p(t)\int_a^t f(s)W\left[\Psi^{-1}\left(\Psi(\zeta) + \int_a^s k(s, \tau)\Phi(p(\tau))\Phi\left(\int_a^\tau f(\theta)\Delta\theta\right)\Delta\tau\right)\right]\Delta s,$$
$$(3.2.150)$$

where

$$p(t) = 1 + \int_a^t f(s)e_f(t, \sigma(s))\Delta s, \quad (3.2.151)$$

$$\zeta = \int_a^{\rho(b)} k(\rho(b), s)\Phi(p(s)a(s))\Delta s,$$

$$\Psi(x) = \int_{x_0}^x \frac{1}{\Phi(W(s))} ds, \quad x \geq x_0 > 0, \quad (3.2.152)$$

and Ψ^{-1} is the inverse of Ψ.

Proof Define the function $z(t)$ in T_* by

$$z(t) = a(t) + \int_a^t f(s)W\left(\int_a^s k(s,\tau)\Phi(u(\tau))\Delta\tau\right)\Delta s. \tag{3.2.153}$$

Then (3.2.149) can be restated as

$$u(t) \le z(t) + \int_a^t f(s)u(s)\Delta s.$$

Clearly, $z(t)$ is rd-continuous in $t \in T_*$. Using Gronwall's inequality, we get

$$u(t) \le z(t) + \int_a^t f(s)z(s)e_f(t,\sigma(s))\Delta s.$$

Moreover, it is easy to see that $z(t)$ is non-decreasing in all $t \in T_*$. We get

$$u(t) \le z(t)p(t), \tag{3.2.154}$$

where $p(t)$ is defined by (3.2.151). Define, for all $t \in T_*^k$,

$$v(t) = \int_a^t k(t,s)\Phi(u(s))\Delta s.$$

From (3.2.154), and taking into account the properties of Φ, we observe that

$$\begin{aligned}
v(t) &\le \int_a^t k(t,s)\Phi\left[p(s)\left(a(s) + \int_a^s f(\tau)W(v(\tau))\Delta\tau\right)\right]\Delta s \\
&\le \int_a^t k(t,s)\Phi(p(s)a(s))\Delta s + \int_a^t k(t,s)\Phi\left(p(s)\int_a^s f(\tau)W(v(\tau))\Delta\tau\right)\Delta s \\
&\le \int_a^{\rho(b)} k(\rho(b),s)\Phi(p(s)a(s))\Delta s + \int_a^t k(t,s)\Phi\left(p(s)\int_a^s f(\tau)\Delta\tau\right)\Phi(W(v(s)))\Delta s \\
&= \zeta + \int_a^t k(t,s)\Phi\left(p(s)\int_a^s f(\tau)\Delta\tau\right)\Phi(W(v(s)))\Delta s.
\end{aligned}$$

Since p and a are positive functions, we have that $\Phi(a(s)p(s)) > 0$ for all $s \in T_*$. Since $k^{\Delta_1} \ge 0$, we must have $\zeta > 0$, hence $r(t)$ is a positive function on T_*^k. In addition, $r(t)$ is delta differentiable on $T_*^{k^2}$ with

$$\begin{aligned}
r^{\Delta}(t) &= k(\sigma(t),t)\Phi\left(p(t)\int_a^t f(\tau)\Delta\tau\right)\Phi(W(v(t))) \\
&\quad + \int_a^t k^{\Delta_1}(t,s)\Phi\left(p(s)\int_a^s f(\tau)\Delta\tau\right)\Phi(W(v(s)))\Delta s
\end{aligned}$$

$$\leq \Phi(W(r(t)))\left[k(\sigma(t),t)\Phi\left(p(t)\int_a^t f(\tau)\Delta\tau\right)\right.$$
$$\left.+\int_a^t k^{\Delta_1}(t,s)\Phi\left(p(s)\int_a^s f(\tau)\Delta\tau\right)\Delta s\right]. \qquad (3.2.155)$$

Dividing both sides of inequality (3.2.155) by $\Phi(W(r(t)))$, we obtain

$$\frac{r^{\Delta}(t)}{\Phi(W(r(t)))} \leq \left[\int_a^t k(t,s)\Phi\left(p(s)\int_a^s f(\tau)\Delta\tau\right)\Delta s\right]^{\Delta}.$$

Let us consider the function Ψ defined by (3.2.152). Delta-integrating this last inequality from a to t and using Lemma 3.2.3, we obtain

$$(r(t)) \leq \Psi(r(a)) + \int_a^t k(t,s)\Phi\left(p(s)\int_a^s f(\tau)\Delta\tau\right)\Delta s,$$

which further yields, for all $t \in T_*^k$,

$$r(t) \leq \Psi^{-1}\left(\Psi(\zeta) + \int_a^t k(t,s)\Phi(p(s))\Phi\left(\int_a^s f(\tau)\Delta\tau\right)\Delta s\right). \qquad (3.2.156)$$

Combining (3.2.156), (3.2.154) and (3.2.153), we obtain the desired inequality (3.2.150). □

Remark 3.2.10 One is interest to study the situation when k is not identically zero on $T_*^k \times T_*^{k^2}$, which comprises the new cases, not considered previously in the literature. The case $k(t,s) \equiv 0$, was studied in Theorem 3.1 of [26] and is not discussed here.

If we let $T = \mathbb{R}$ in Theorem 3.2.16, we get ([193], Theorem 2.1). If, in turn, we consider $T = \mathbb{Z}$, then we obtain the following result.

Corollary 3.2.6 (The Ferreira-Torres Inequality [229]) *Let $u(t)$ and $f(t)$ be non-negative functions in the time scales interval $T_* := [a,b]_{\mathbb{Z}}$ and $[a,b-1]_{\mathbb{Z}}$, respectively. Let $k(t,s)$ be defined as in Theorem 3.2.16 in such a way that $k(t,s)$ and $k^{\Delta_1}(t,s) = k(\sigma(t),s) - k(t,s)$ are non-negative for every $t,s \in T_*$, with $s \leq t$ for which they are defined (it is assumed that k is not identically zero on $[a,b-1]_{T_*} \times [a,b-2]_{T_*}$). Let $\Phi \in C(\mathbb{R}_0,\mathbb{R}_0)$ be a non-decreasing, sub-additive and sub-multiplicative function such that $\Phi(u) > 0$, for all $u > 0$ and let $W \in C(\mathbb{R}_0,\mathbb{R}_0)$ be a non-decreasing function such that for all $u > 0$, we have $W(u) > 0$. Assume that $a(t)$ is a positive and non-decreasing function for all $t \in T_*$. If*

$$u(t) \leq a(t) + \sum_{s=a}^{t-1} f(s)u(s) + \sum_{s=a}^{t-1} f(s)W\left(\sum_{\tau=a}^{s-1} k(s,\tau)\Phi(u(\tau))\right),$$

for $\tau, s, t \in T_*, a \le \tau \le s \le t \le b$, *then for all* $t \in T_*$ *satisfying*

$$\Psi(\zeta) + \sum_{s=a}^{t-2} k(t-1,s)\Phi(p(s))\Phi\left(\sum_{\tau=a}^{s-1} f(\tau)\right) \in Dom\ (\Psi^{-1}),$$

we have

$$u(t) \le p(t)\left\{a(t) + \sum_{s=a}^{t-1} f(s)W\left[\Psi^{-1}\left(\Psi(\zeta) + \sum_{\tau=a}^{s-1} k(s,\tau)\Phi(p(\tau))\Phi\left(\sum_{\theta=a}^{\tau-1} f(\theta)\right)\right)\right]\right\},$$

where

$$p(t) = 1 + \sum_{s=a}^{t-1} f(s)e_f(t, s+1),$$

$$\zeta = \sum_{s=a}^{b-1} k(b-1,s)\Phi(p(s)a(s)),$$

$$\Psi(x) = \int_{x_0}^{x} \frac{1}{\Phi(W(s))} ds, \quad x >\ge x_0 > 0,$$

and Ψ^{-1} *is the inverse of* Ψ.

For the particular case $T + \mathbb{R}$, Theorem 3.2.17 below generalizes the result obtained by Oguntuase in [428] (Theorems 2.3 and 2.9).

Theorem 3.2.17 (The Ferreira-Torres Inequality [229]) *Suppose that* $u(t)$ *is a non-negative rd-continuous function in the time scales interval* $T_* = [a,b]_T$ *and that* $h(t), f(t)$ *are non-negative rd-continuous functions in the time scales interval* T_*^k. *Assume that* $b(t)$ *is a non-negative rd-continuous function and not identically zero on* $T_*^{k^2}$. *Let* $\Phi(u)$, $W(u)$ *and* $a(t)$ *be as defined in Theorem 3.2.16. If*

$$u(t) \le a(t) + \int_a^t f(s)u(s)\Delta s + \int_a^t f(s)h(s)W\left(\int_a^s b(\tau)\Phi(u(\tau))\Delta\tau\right)\Delta s,$$

for all $\tau, s, t \in T_*, a \le \tau \le s \le t \le b$, *then for all* $t \in T_*$ *satisfying*

$$\Psi(\xi) + \int_a^{p(t)} b(\tau)\Phi(p(\tau))\Phi\left(\int_a^\tau f(\theta)h(\theta)\Delta\theta\right)\Delta\tau \in Dom\ (\Psi^{-1}),$$

we have

$$u(t) \le p(t)a(t)$$
$$+p(t)\int_a^t f(s)h(s)W\left[\Psi^{-1}\left(\Psi(\xi) + \int_a^s b(\tau)\Phi(p(\tau))\Phi\left(\int_a^\tau f(\theta)h(\theta)\Delta\theta\right)\Delta\tau\right)\right]\Delta s,$$

where $p(t)$ is defined by (3.2.151), Ψ is defined by (3.2.152), and

$$\xi = \int_a^{p(b)} b(s)\Phi(p(s)a(s))\Delta s.$$

Proof Similarly to the proof of Theorem 3.2.16, we can prove the theorem. □

Some new integro-functional inequalities of Bellman-Bihari type will be given as follows.

Suppose that $\tau(s) \in E$ is a class of continuous functions $\tau : \mathbb{R} \to \mathbb{R}$, such that $\tau(s) \le s$, $\lim_{|s|\to+\infty} \tau(s) = +\infty$.

Theorem 3.2.18 (The Iovane-Borysenko Inequality [292]) *Assume that $\varphi(t)$ is a non-negative, at $t \ge t_0$, piecewise continuous function with first kind discontinuities at the points $\{t_i\}$ ($t_0 < t_1 < t_2 < \cdots$, $\lim_{i\to+\infty} t_i(s) = +\infty$), satisfying the integro-sum inequality, for all $t \ge t_0$,*

$$\varphi(t) \le n(t) + \int_{t_0}^{t} g(s)\varphi(\tau(s))ds + \sum_{t_0<t_i<t} \beta_i \varphi^m(t_i - 0); \qquad (3.2.157)$$

where $n(t)$ is a positive non-decreasing function for all $t \ge t_0$, $g(s) \ge 0$, parameter $m > 0$, $\beta_i = const. \ge 0$, then

(1) if $0 < m < 1$, then for all $t \ge t_0$,

$$\varphi(t) \le n(t) \prod_{t_0<t_i<t} (1 + \beta_i n^{m-1}(t_i)) \exp\left(\int_{t_0}^{t} g(s)\frac{n(\tau(s))}{n(s)}ds\right); \quad (3.2.158)$$

(2) if $m \ge 1$, then for all $t \ge t_0$,

$$\varphi(t) \le n(t) \prod_{t_0<t_i<t} (1 + \beta_i n^{m-1}(t_i)) \exp\left(m\int_{t_0}^{t} g(s)\frac{n(\tau(s))}{n(s)}ds\right), \quad (3.2.159)$$

with $\varphi(t_i - 0) = \lim_{t\to t_i-} \varphi(t)$.

Remark 3.2.11 From the estimates (3.2.158)–(3.2.159) in some particular cases, well known results in the theory of integral inequalities for continuous and discontinuous functions can be obtained. If $n(t) = c = const. > 0$, $\tau(s) = s$, $\beta_i = 0$, a classical result of Gronwall and Bellman follows from (3.2.158), (3.2.159) [70]. If $\beta_i = 0$, the results in [27] are obtained; if $m = 1$, $n(t) = c = const.$, $\tau(s) = s$, the results in [585] can be reached, if $m = 1$, the result in [589] is obtained; if $\tau(s) = s$, the results in [109] are obtained. For the discrete case, when $n(t) = c$, $g(s) = 0$, $m = 1$, the results in [10] are obtained.

Theorem 3.2.19 (The Iovane-Borysenko Inequality [292]) *Let $\tau(s) \in E$ and the non-negative function $\varphi(t)$ satisfy the inequality for all $t \geq t_0$,*

$$\varphi(t) \leq \psi(t) + q(t) \int_{t_0}^{t} g(s)\varphi^m(\tau(s))ds + \sum_{t_0<t_i<t} \beta_i \varphi^m(t_i - 0), \tag{3.2.160}$$

where $\{t_i\}$, satisfying the conditions of Theorem 3.2.18, are first kind discontinuity points of the function $\varphi(t)$; $\psi(t)$ is a positive non-decreasing function at all $t \geq t_0, q(t) \geq 1, g(t) \geq 0$, for all $t \geq t_0$, the parameter $m > 0, m \neq 1$, and $\beta_i =$ const. ≥ 0. Then

(1) if $0 < m < 1$, then for all $t \geq t_0$,

$$\varphi(t) \leq \psi(t) \prod_{t_0<t_i<t} \left(1 + \beta_i \psi^{m-1}(t_i) q^m(t_i)\right)$$
$$\times \left[1 + (1-m) \int_{t_0}^{t} g(s)\psi^{m-1}(s) q^m(\tau(s)) [\frac{\psi(\tau(s))}{\tau(s)}]^m ds\right]^{1/(1-m)}, \tag{3.2.161}$$

(2) if $m > 1$, then for all $t \geq t_0$,

$$\varphi(t) \leq \psi(t) \prod_{t_0<t_i<t} \left(1 + \beta_i m \psi^{m-1}(t_i) q^m(t_i)\right)$$
$$\times \left\{1 + (1-m) \left[\prod_{t_0<t_i<t} (1 + \beta_i \psi^{m-1}(t_i) q^m(t_i))\right]^{m-1}\right.$$
$$\left.\times \int_{t_0}^{t} g(s)\psi^{m-1}(s) q^m(\tau(s)) [\frac{\psi(\tau(s))}{\tau(s)}]^m ds\right\}^{1/(1-m)}, \tag{3.2.162}$$

with

$$\begin{cases} \displaystyle\int_{t_0}^{t} g(s)\psi^{m-1}(s) q^m(\tau(s)) \left(\frac{\psi(\tau(s))}{\tau(s)}\right)^m ds \leq \frac{1}{m}, \\ \displaystyle\prod_{t_0<t_i<t} \left(1 + \beta_i \psi^{m-1}(t_i) q^m(t_i)\right) < \left(1 + \frac{1}{m-1}\right)^{1/(m-1)}. \end{cases} \tag{3.2.163}$$

Remark 3.2.12 If $\beta_i = 0, \psi(t) = c = const > 0, q(t) = 1, \tau(s) = s$, from Theorem 3.2.19 the result presented by Bihari in [82] follows; if $\beta_i = 0$, the result of Theorem 3.2.19 coincides with the result given by Akinyele in [27]; if $q(t) = 1, \tau(s) = s$ from the result of Theorem 3.2.19, the result given by Borysenko in [109] follows.

Theorem 3.2.20 (The Iovane-Borysenko Inequality [292]) *Let $\varphi(t)$ be a non-negative piecewise continuous function, with first kind discontinuities at the points $\{t_i\} : t_1 < t_2 < \cdots, \lim_{i\to+\infty} t_i = +\infty$, if satisfy the following integro-sum inequality holds for $m > 0$,*

$$\varphi(t) \le n(t) + q(t)\Big[\int_{t_0}^{t} f(s)\varphi(\sigma(s))ds + \int_{t_0}^{t} f(s)(\int_{t_0}^{s} g(t)\varphi(\tau(t))dt)ds + \sum_{t_0<t_i<t} \beta_i\varphi^m(t_i-0)\Big], \tag{3.2.164}$$

where $n(t)$ is a non-decreasing function, $n(t) > 0$, $q(t) \ge 1$, $f(s) \ge 0$, $\sigma(t) \in E$, $g(t) \ge 0$, $\beta_i \ge 0$. Then the following estimates hold:

(1) if $0 < m \le 1$, then

$$\varphi(t) \le n(t)q(t) \prod_{t_0<t_i<t} \big(1 + \beta_i q^m(t_i)n^{m-1}(t_i)\big) \times \exp\Big[\int_{t_0}^{t} \frac{f(\xi)g(\sigma(\xi))n(\sigma(\xi)) + g(\xi)q(\tau(\xi))n(\tau(\xi))}{n(\sigma(\xi))}d\xi\Big]; \tag{3.2.165}$$

(2) if $m \ge 1$, then

$$\varphi(t) \le n(t)q(t) \prod_{t_0<t_i<t} \big(1 + q^m(t_i)n^{m-1}(t_i)\big) \times \Big[m\int_{t_0}^{t} \frac{f(\xi)g(\sigma(\xi))n(\sigma(\xi)) + g(\xi)q(\tau(\xi))n(\tau(\xi))}{n(\sigma(\xi))}d\xi\Big]. \tag{3.2.166}$$

Moreover, we introduce a new integro-sum inequality of Bellman-Bihari type with retardation.

Theorem 3.2.21 (The Gallo-Piccirillo Inequality [243]) *Suppose that a non-negative piecewise continuous function $V(t)$ at $t \ge t_0 \ge 0$, with discontinuities of the first kind in the points t_k $(t_0 < t_1, \cdots, \lim_{i\to+\infty} t_i = +\infty)$ satisfies inequality*

$$V(t) \le \psi(t) + g(t)\int_{t_0}^{t} q(s)V(\tau(s))ds + p(t)\sum_{t_0<t_i<t} a_i V^m(t_i-0) \tag{3.2.167}$$

where $\psi(t) > 0$ is a non-decreasing function at $t \ge t_0, g(t) \ge 1, p(t) \ge 1, q(s) \in C(\mathbb{R}_+, \mathbb{R}_+), \tau(s) \in \Im$-class of functions: $\tau : \mathbb{R} \to \mathbb{R}, \tau(s) \le s, \lim_{|s|\to+\infty}\tau(s) = +\infty, (\tau = \tau(s)$ is the "delaying" argument), $a_i \ge 0, m > 0$. Then the following *estimates hold for all* $t \geq t_0$,

(1) if $m \in [0, 1]$, *then*

$$V(t) \leq \psi(t)g(t)p(t) \prod_{t_0<t_i<t} \left(1 + a_i\psi^{m-1}(t_i - 0)g^{m-1}(t_i - 0)p^{m-1}(t_i - 0)\right)$$
$$\times \exp\left(\int_{t_0}^{t} q(s)g(\tau(s))p(\tau(s))\frac{\psi(\tau(s))}{\psi(s)}ds\right), \qquad (3.2.168)$$

(2) if $m \geq 1$, *then*

$$V(t) \leq \psi(t)g(t)p(t) \prod_{t_0<t_i<t} \left(1 + a_i\psi^{m-1}(t_i - 0)g^{m}(t_i - 0)p^{m}(t_i - 0)\right)$$
$$\times \exp\left(m\int_{t_0}^{t} q(s)g(\tau(s))p(\tau(s))\frac{\psi(\tau(s))}{\psi(s)}ds\right). \qquad (3.2.169)$$

Proof Noting that $\psi(t)$ is a non-decreasing and positive function, we have

$$\frac{V(t)}{\psi(t)} \leq 1 + g(t)\int_{t_0}^{t} q(s)\frac{V(\tau(s))}{\psi(s)}ds + p(t)\sum_{t_0<t_i<t} a_i\psi^{m-1}(t_i - 0)\left[\frac{V(t_i - 0)}{\psi(t_i - 0)}\right]^m.$$

Setting $u(t) = V(t)/\psi(t), u(t_0) = 1$, we can get

$$u(t) \leq g(t)p(t)\left[1 + \int_{t_0}^{t} q(s)\frac{V(\tau(s))}{\psi(s)}ds + \sum_{t_0<t_i<t} a_i\psi^{m-1}(t_i - 0)\left[\frac{V(t_i - 0)}{\psi(t_i - 0)}\right]^m\right].$$

Denoting

$$W(t) = 1 + g(t)\int_{t_0}^{t} q(s)\frac{V(\tau(s))}{\psi(s)}ds + \sum_{t_0<t_i<t} a_i\psi^{m-1}(t_i - 0)\left[\frac{V(t_i - 0)}{\psi(t_i - 0)}\right]^m,$$

we have

$$\begin{cases} V(t_i - 0) \leq \psi(t_i - 0)g(t_i - 0)p(t_i - 0)W(t_i - 0), \quad \textit{for } i = 1, 2, \ldots, \\ V(t) \leq \psi(t)g(t)p(t)W(t), \quad W(t_0) = 1, \\ V(\tau(s)) \leq \psi(\tau(s))g(\tau(s))p(\tau(s))W(\tau(s)) \leq \psi(\tau(s))g(\tau(s))p(\tau(s))W(s). \end{cases}$$

Then

$$W(t) \leq 1 + \int_{t_0}^{t} q(s)g(\tau(s))p(\tau(s))\frac{\psi(\tau(s))}{\psi(s)}W(s)ds$$
$$+ \sum_{t_0<t_i<t} a_i\psi^{m-1}(t_i - 0)g^m(t_i - 0)p^m(t_i - 0)[W(t_i - 0)]^m.$$

Setting in the last inequality $Q(s) = q(s)g(\tau(s))p(\tau(s))\frac{\psi(\tau(s))}{\psi(s)}$ and $A_i = a_i\psi^{m-1}(t_i - 0)g^m(t_i - 0)p^m(t_i - 0)$, we can obtain

$$W(t) \leq 1 + \int_{t_0}^{t} Q(s)W(s)ds + \sum_{t_0<t_i<t} A_i[W(t_i - 0)]^m.$$

Let us consider intervals $I_i = [t_{i-1}, t_i], i = 1, 2, \cdots$.

For all $t \in I_1$, then we have $W(t) \leq 1 + \int_{t_0}^{t} Q(s)W(s)ds$. Hence by the classical Gronwall-Bellman inequality (see, Theorem 3.1.2 in Qin [557]), we conclude $W(t) \leq \exp\left(\int_{t_0}^{t} Q(s)ds\right)$. It is obvious that for all $t \in I_1$,

$$V(t) \leq \psi(t)g(t)p(t)\exp\left(\int_{t_0}^{t} Q(s)ds\right).$$

Now let us consider the next interval I_2. Then

$$W(t) \leq 1 + A_1\exp\left(m\int_{t_0}^{t_1} Q(s)ds\right) + \exp\left(\int_{t_0}^{t_1} Q(s)ds\right) + \int_{t_1}^{t} Q(s)W(s)ds.$$

Thus we have for all $t \in I_2$,

$$W(t) \leq (1 + A_1)\exp\left(\int_{t_0}^{t_1} Q(s)ds\right) + \int_{t_1}^{t} Q(s)W(s)ds, \quad \textit{if } m \in [0, 1], \tag{3.2.170}$$

$$W(t) \leq (1 + A_i)\exp\left(m\int_{t_0}^{t_1} Q(s)ds\right) + \int_{t_1}^{t} Q(s)W(s)ds, \quad \textit{if } m \geq 1. \tag{3.2.171}$$

Noting that $W(t)$ satisfies that $V(t) \leq \psi(t)g(t)p(t)W(t)$, for all $t \geq t_0$, from inequalities (3.2.170) and (3.2.171), it follows that estimates (3.2.168) and (3.2.169) hold for $V(t)$. Consequently, by using a similar procedure for all $t \in I_k$, we can complete the proof of theorem. □

Remark 3.2.13 Theorem 3.2.21 is a new analogue of the classical Gronwall-Bellman result for discontinuous functions with Hölder type discontinuities in the points $\{t_i\}$. For particular cases, from the estimates (3.2.168) and (3.2.169), we may obtain well-known results from the integral inequalities theory for continuous and piecewise continuous functions.

For $a_i = 0, g(x) = 1$, the estimates (3.2.168) and (3.2.169) of Theorem 3.2.21 give a functional integral inequality with delay generalization which unifies and embodies Gronwall-Bellman inequality [79], and Dhongate-Deo inequality [198].

Estimates (3.2.168) and (3.2.169) are also similar to the results by Pachpatte [456] and by Akinyele [28].

For $\tau(s) = s, \psi(t) = c, g(t) = p(t) = 1$, the estimates (3.2.168) and (3.2.169) reduce to a result by Samoilenko and Perestyuk [585].

For $p(t) = g(t) = 1, m = 1$, the results of Theorem 3.2.21 are similar with the result by Borysenko [109]. For $p(t) = g(t) = 1, \tau(s) = s, m = 1$, (3.2.168) and (3.2.169) coincide with results by Lakshmikantham, Bainov, Simeonov [327].

For $p(t) = g(t) = 1, \tau(s) = s$, the result of Theorem 3.2.21 reduces to the results by Borysenko [109], Borysenko, Gallo and Toscano [115].

If $p(t) = g(t) = 1$, from the result of Theorem 3.2.21, the result by Iovane [111] follows.

Corollary 3.2.7 (The Gallo-Piccirillo Inequality [243]) *If $V(t)$ satisfies conditions of Theorem 3.2.21, then the following estimates hold for all $t \geq t_0$:*

(1) if $m \in (0, 1]$, then

$$V(t) \leq \psi(t)g(t)p(t) \exp\left(\int_{t_0}^{t} q(s)g(\tau(s))p(\tau(s))\frac{\psi(\tau(s))}{\psi(s)}ds\right) + \sum_{t_0<t_i<t} a_i\psi^{m-1}(t_i-0)g^m(t_i-0)p^m(t_i-0), \tag{3.2.172}$$

(2) if $m \in (1, +\infty)$, then

$$V(t) \leq \psi(t)g(t)p(t) \exp\left(m\int_{t_0}^{t} q(s)g(\tau(s))p(\tau(s))\frac{\psi(\tau(s))}{\psi(s)}ds\right) + \sum_{t_0<t_i<t} a_i\psi^{m-1}(t_i-0)g^m(t_i-0)p^m(t_i-0). \tag{3.2.173}$$

Remark 3.2.14 From estimates (3.2.172) and (3.2.173), in the particular case (*i.e.*, $\psi(t) = const.$, $p(t) = g(t) = 1, m = 1$), the result by Ahmed [25] follows (see also Gao et al. [244]).

Theorem 3.2.22 (The Gallo-Piccirillo Inequality [243]) *Assume the nonnegative piecewise continuous function $V(t)$ satisfies the integro-sum inequality*

$$V(t) \leq \psi(t) + g(t)\int_{t_0}^{t} q(s)V^n(\tau(s))ds + p(t)\sum_{t_0<t_i<t} a_iV^m(t_i-0), \tag{3.2.174}$$

where all functions on the right-hand side of (3.2.174) satisfy conditions of Theorem 3.2.21 and n is a positive constant. Then the following estimates hold for *all* $t \geq t_0$*:*

(1) if $m = n \in (0, 1]$*, then*

$$V(t) \leq \psi(t)g(t)p(t) \prod_{t_0<t_i<t} (1 + a_i\psi^{m-1}(t_i - 0)g^m(t_i - 0)p^m(t_i - 0))$$
$$\times \left[1 + (1-m)\int_{t_0}^t q(s)\psi^{m-1}(s)g^m(\tau(s))p^m(\tau(s))\left[\frac{\psi(\tau(s))}{\psi(s)}\right]^m ds\right]^{\frac{1}{1-m}}; \tag{3.2.175}$$

(2) if $n = m > 1$*, then*

$$V(t) \leq \psi(t)g(t)p(t) \prod_{t_0<t_i<t} \left(1 + ma_i\psi^{m-1}(t_i - 0)g^m(t_i - 0)p^m(t_i - 0)\right)$$
$$\times\left[1 - (m-1)\left[\prod_{t_0<t_i<t} (1 + ma_i\psi^{m-1}(t_i - 0)g^{m-1}(t_i - 0)p^{m-1}(t_i - 0))\right]^{m-1}\right.$$
$$\left.\times \int_{t_0}^t q(s)\psi^{m-1}(s)g^m(\tau(s))p^m(\tau(s))\left[\frac{\psi(\tau(s))}{\psi(s)}\right]^m ds\right]^{\frac{1}{1-m}}, \tag{3.2.176}$$

with

$$\begin{cases} \int_{t_0}^t q(s)\psi^{m-1}(s)g^m(\tau(s))p^m(\tau(s))\left[\frac{\psi(\tau(s))}{\psi(s)}\right]^m ds \leq \frac{1}{m}, & (3.2.177) \\ \prod_{t_0<t_i<t} (1 + ma_i\psi^{m-1}(t_i - 0)g^{m-1}(t_i - 0)p^{m-1}(t_i - 0)) < \left(\frac{m}{m-1}\right)^{\frac{1}{m-1}}. & (3.2.178) \end{cases}$$

Proof Obviously from (3.2.174) it follows

$$\frac{V(t)}{\psi(t)} \leq 1 + g(t)\int_{t_0}^t q(s)\frac{V^n(\tau(s))}{\psi(s)}ds + \sum_{t_0<t_i<t} a_i\frac{V^m(t_i - 0)}{\psi(t_i - 0)}$$
$$\leq g(t)p(t)\left[1 + \int_{t_0}^t q(s)\psi^{n-1}(s)\left[\frac{V(\tau(s))}{\psi(s)}\right]^n ds + \sum_{t_0<t_i<t} a_i\psi^{m-1}(t_i - 0)[\frac{V(t_i - 0)}{\psi(t_i - 0)}]^m\right].$$

Denoting by

$$\tilde{W}(t) = 1 + \int_{t_0}^t q(s)\psi^{n-1}(s)\left[\frac{V(\tau(s))}{\psi(s)}\right]^n ds + \sum_{t_0<t_i<t} a_i\psi^{m-1}(t_i - 0)\left[\frac{V(t_i - 0)}{\psi(t_i - 0)}\right]^m,$$

so that $\tilde{W}(t_0) = 1$, we obtain

$$\begin{cases} V(t_i - 0) \leq \psi(t_i - 0)g(t_i - 0)p(t_i - 0)\tilde{W}(t_i - 0), \quad for\ i = 1, 2, \cdots, \\ V(t) \leq \psi(t)g(t)p(t)W(t), \quad \tilde{W}(t_0) = 1, \\ V(\tau(s)) \leq \psi(\tau(s))g(\tau(s))p(\tau(s))\tilde{W}(\tau(s)) \leq \psi(\tau(s))g(\tau(s))p(\tau(s))\tilde{W}(s). \end{cases}$$

Then

$$\begin{aligned} \tilde{W}(t) \leq 1 + &\int_{t_0}^{t} q(s)g(\tau(s))p(\tau(s))[\frac{\psi(\tau(s))}{\psi(s)}]^n\tilde{W}(s)ds \\ &+ \sum_{t_0<t_i<t} a_i\psi^{m-1}(t_i - 0)g^m(t_i - 0)p^m(t_i - 0)[\tilde{W}(t_i - 0)]^m. \end{aligned}$$

Now let us set $\tilde{Q} = q(s)g(\tau(s))p(\tau(s))[\frac{\psi(\tau(s))}{\psi(s)}]^n$, $\tilde{A}_i = A_i$; so we get

$$\tilde{W}(t) \leq 1 + \int_{t_0}^{t} \tilde{Q}(s)\tilde{W}(s)ds + \sum_{t_0<t_i<t} \tilde{A}_i[\tilde{W}(t_i - 0)]^m.$$

Noting that the Bihari inequality on the interval I_1, for the function $\widetilde{W}(t)$, we can obtain for all $t \in I_1$, if $n \in [0, 1]$,

$$\widetilde{W}(t) \leq \left[1 + (1 - n)\int_{t_0}^{t} \widetilde{Q}(\tau)d\tau\right]^{\frac{1}{1-n}},$$

and if $n > 1$,

$$\widetilde{W}(t) \leq [1 - (n - 1)\int_{t_0}^{t} \widetilde{Q}(\tau)d\tau]^{-\frac{1}{1-n}}, \tag{3.2.179}$$

with for all $t \in I_1$,

$$\int_{t_0}^{t} \widetilde{Q}(\tau)d\tau < \frac{1}{n - 1}.$$

According to conditions (3.2.177) and (3.2.178), the inequality automatically holds.

Let $t \in I_2$. Then we have $(m = n)$

$$\begin{aligned} \widetilde{W}(t) \leq \widetilde{A}_1 &\left[1 + (1 - m)\int_{t_0}^{t_1} \widetilde{Q}(\tau)d\tau\right]^{\frac{m}{1-m}} \\ &+ \left[1 + (1 - m)\int_{t_0}^{t_1} \widetilde{Q}(\tau)d\tau\right]^{\frac{1}{1-m}} + \int_{t_1}^{t} \widetilde{Q}(\tau)\widetilde{W}(\tau)d\tau \end{aligned}$$

which yields for all $t \in I_2$: if $m \in [0,1]$,

$$\widetilde{W}(t) \leq (1+\widetilde{A}_1)\left[1+(1-m)\int_{t_0}^{t}\widetilde{Q}(\tau)d\tau\right]^{\frac{1}{1-m}},$$

and if $m > 1$,

$$\widetilde{W}(t) \leq (1+\widetilde{A}_1 m)\left\{1-(m-1)[1+A_1 m]^{m-1}\int_{t_0}^{t}\widetilde{Q}(\tau)d\tau\right\}^{\frac{1}{1-m}},$$

with for all $t \in I_2$,

$$\int_{t_0}^{t}\widetilde{Q}(\tau)d\tau < \frac{1}{(m-1)[1+A_1 m]^{m-1}}. \tag{3.2.180}$$

According to (3.2.177) and (3.2.178), as in (3.2.179), the inequality (3.2.180) always holds. Thus (3.2.175) and (3.2.176) hold on I_2. We complete the proof reasoning like in Theorem 3.2.21. □

Corollary 3.2.8 (The Gallo-Piccirillo Inequality [243]) *If $\psi(t) = c = const. > 0$, $p(t) = q(t) = 1$, $\tau(s)$, $m = n$, then the function $V(t)$ satisfies for all $t \geq t_0$,*
if $m \in (0,1)$, then

$$V(t) \leq \prod_{t_0<t_i<t}(1+a_i c^{m-1})\left[c^{1-m}+(1-m)\int_{t_0}^{t}q(\tau)d\tau\right]^{\frac{1}{1-m}}; \tag{3.2.181}$$

and if $m > 1$, then

$$V(t) \leq c\prod_{t_0<t_i<t}(1+a_i c^{m-1})\left[1-(m-1)\left[c\prod_{t_0<t_i<t}(1+a_i c^{m-1})\right]^{m-1}\int_{t_0}^{t}q(\tau)d\tau\right]^{-\frac{1}{m-1}}, \tag{3.2.182}$$

with

$$\int_{t_0}^{t}q(\tau)d\tau \leq \frac{c^{1-m}}{m}, \qquad \prod_{t_0<t_i<t}\left(1+a_i m c^{m-1}\right) < \left(\frac{m}{m-1}\right)^{\frac{1}{m-1}}. \tag{3.2.183}$$

Remark 3.2.15 If $p(t) = 1$, from Theorem 3.2.22 it follows the result obtained by Iovane [111]. If $p(t) = q(t) = 1$, $\tau(s) = s$, the present estimates are similar to the results obtained by Borysenko [109], Borysenko, Gallo and Toscano [115]. Moreover, if $a_i = 0$, $\psi(t) = c$ =const. > 0, $p(t) = q(t) = 1$, $\tau(s) = s$ from Theorem 3.2.22, the classical result by Bihari [82] follows; if $\beta_i = 0$, estimates (3.2.175) and (3.2.176) coincide with the result by Akinyele [28].

Pachpatte [491] obtained one generalization of Theorem 3.1.2 which can be stated as follows.

Theorem 3.2.23 (The Pachpatte Inequality [491]) *Let* $p_i \in L^1(I, \mathbb{R}_0), \mathbb{R}_0 = (0, +\infty)$. *Let* g *be a continuously differentiable function defined on* $\mathbb{R}_+$, *and* $g > 0$ *and* $g' \geq 0$ *on* $\mathbb{R}_0$. *If* $u : I \to \mathbb{R}_1 = [1, +\infty)$ *satisfies for all* $t \in [0, T]$,

$$u(t) \leq u_0 + H[t, p_1, p_2, \ldots, p_{n-1}, pug(\log u)], \tag{3.2.184}$$

where $u_0 \geq 1$ *is a constant,*

$$H[t, p_1, p_2, \ldots, p_{n-1}, pug(\log u)] = \int_0^t p_1(t_1) \int_0^{t_1} p_2(t_2) \cdots \int_0^{t_{n-2}} p_{n-1}(t_{n-1}) \int_0^{t_{n-1}} pug(\log u) dt_n \cdots dt_2 dt_1, \tag{3.2.185}$$

then for all $t,\ t_1 \in [0, T], 0 \leq t \leq t_1$,

$$u(t) \leq \exp\left(G^{-1}\Big[G(\log u_0) + H[t, p_1, p_2, \ldots, p_{n-1}, p]\Big]\right), \tag{3.2.186}$$

where

$$G(r) = \int_{r_0}^{r} \frac{ds}{g(s)}, \quad r \geq r_0 > 0, \tag{3.2.187}$$

and G^{-1} *is the inverse of* G *and* t_1 *is chosen so that for all* $t \in [0, t_1]$, *with* $t_1 \in [0, T]$.

$$G(\log u_0) + H[t, p_1, p_2, \ldots, p_{n-1}, p] \in Dom\,(G^{-1}). \tag{3.2.188}$$

Proof Since the proof of the theorem is similar to that of the next theorem, we omit the proof. □

We define the differential operators L_i, $0 \leq i \leq n$, by

$$L_0 x(t) = x(t), \quad L_i x(t) = \frac{1}{p_i(t)} \frac{d}{dt}(L_{i-1} x(t)), \quad 1 \leq i \leq n, \tag{3.2.189}$$

with $p_n(t) = 1$, where $x(t)$ and $p_i(t) > 0$ are some functions defined on I. For all $t \in I$ and some functions $q_j(t) > 0$, $j = 1, \ldots, n-1$ and $q(t) \geq 0$ defined on I, we define

$$H[t, q_1, q_2, \ldots, q_{n-1}, q] = \int_0^t q_1(t_1) \int_0^{t_1} q_2(t_2) \cdots \int_0^{t_{n-1}} q(t_n) dt_n dt_{n-1} \cdots dt_1, \tag{3.2.190}$$

where $t_0 = t$.

A more generalized version (3.2.190) of the inequality than that given in Theorem 3.2.23 is given in the following theorem.

Theorem 3.2.24 (The Ma-Debnath Inequality [361]) *Let $p_i \in L^1(I, \mathbb{R}_0)$ and $p \in L^1(I, \mathbb{R}_+)$. Let further $g(t) \in C^1(\mathbb{R}_+, \mathbb{R}_+)$ with satisfying $g > 0$ and $g' \geq 0$ on $\mathbb{R}_0$ and $\phi \in C^2(\mathbb{R}_+, \mathbb{R}_+)$ with $\phi'(u) > 0$ for all $u > 0$ and non-decreasing. If $u : I \to \mathbb{R}_+$ satisfies for all $t \in I$,*

$$\phi(u(t)) \leq u_0 + H[t, p_1, p_2, \ldots, p_{n-1}, p\phi'(u)g(\log u)], \tag{3.2.191}$$

where $u_0 \geq 1$ is a constant, then for all $t \in I_1 = [0, t_1] \subset I$,

$$u(t) \leq \exp\left(G^{-1}\left[G(\phi^{-1}(\log u_0)) + H[t, p_1, p_2, \ldots, p_{n-1}, p]\right]\right), \tag{3.2.192}$$

where

$$G(r) = \int_{r_0}^{r} \frac{ds}{g(s)}, \quad r \geq r_0 > 0, \tag{3.2.193}$$

and G^{-1} is the inverse of G and t_1 is chosen so that for all $t \in I_1$,

$$G(\phi^{-1}(\log u_0)) + H[t, p_1, p_2, \ldots, p_{n-1}, p] \in Dom\,(G^{-1}). \tag{3.2.194}$$

Proof Let $\epsilon > 0$ be an arbitrary small constant and define on I a non-decreasing function

$$v_\epsilon(t) = u_0 + \epsilon + H[t, p_1, p_2, \ldots, p_{n-1}, p\phi'(u)g(u)]. \tag{3.2.195}$$

Thus from (3.2.191) and (3.2.195), we derive

$$u(t) \leq \phi^{-1}(v_\epsilon(t)), \quad \text{forall} t \in I,$$

$$v_\epsilon(0) = u_0 + \epsilon > 0, \tag{3.2.196}$$

$$L_n v_\epsilon(t) = p(t)\phi'(u(t))g(u(t)). \tag{3.2.197}$$

Since ϕ' and g are non-decreasing, by (3.2.196) and (3.2.197), we obtain

$$L_n v_\epsilon(t) \leq p(t)\phi'[\phi^{-1}(v_\epsilon(t))]g[\phi^{-1}(v_\epsilon(t))]. \tag{3.2.198}$$

Using the fact that $v_\epsilon(t)$ is positive,

$$\frac{d}{dt}(\phi'[\phi^{-1}(v_\epsilon(t))]) = \phi''[\phi^{-1}(v_\epsilon(t))]\frac{1}{\phi'(v_\epsilon(t))}\frac{dv_\epsilon(t)}{dt} \geq 0, \tag{3.2.199}$$

and $L_n v_\epsilon(t) \geq 0$ for all $t \in I$, it follows from (3.2.199) that

$$\frac{L_n v_\epsilon(t)}{\phi'[\phi^{-1}(v_\epsilon(t))]} \leq p(t)g[\phi^{-1}(v_\epsilon(t))] + \frac{L_{n-1}v_\epsilon(t) \times (d/dt)(\phi'[\phi^{-1}(v_\epsilon(t))])}{\phi'^2[\phi^{-1}(v_\epsilon(t))]}, \tag{3.2.200}$$

i.e.,

$$\frac{d}{dt}\frac{L_{n-1}v_\epsilon(t)}{\phi'[\phi^{-1}(v_\epsilon(t))]} \leq p(t)g[\phi^{-1}(v_\epsilon(t))]. \tag{3.2.201}$$

Integrating (3.2.201) from 0 to t and using the fact that $L_{n-1}v_\epsilon(0) = 0$, we get

$$\frac{L_{n-1}v_\epsilon(t)}{\phi'[\phi^{-1}(v_\epsilon(t))]} \leq \int_0^t p(t_n)g[\phi^{-1}(v_\epsilon(t_n))]dt_n. \tag{3.2.202}$$

It also allows from (3.2.202) that

$$\frac{d}{dt}\frac{L_{n-2}v_\epsilon(t)}{\phi'[\phi^{-1}(v_\epsilon(t))]} \leq p_{n-1}(t)\int_0^t p(t_n)g[\phi^{-1}(v_\epsilon(t_n))]dt_n, \tag{3.2.203}$$

which, by integrating from 0 to t and using the fact that $L_{n-2}v_\epsilon(0) = 0$, leads to

$$\frac{L_{n-2}v_\epsilon(t)}{\phi'[\phi^{-1}(v_\epsilon(t))]} \leq \int_0^t p_{n-1}(t_{n-1})\int_0^{t_{n-1}} p(t_n)g[\phi^{-1}(v_\epsilon(t_n))]dt_n dt_{n-1}. \tag{3.2.204}$$

Repeating the above argument successively, we obtain

$$\begin{aligned}\frac{(d/dt)v_\epsilon(t)}{\phi'[\phi^{-1}(v_\epsilon(t))]} &\leq p_1(t)\int_0^t p_2(t_2)\int_0^{t_2} p_3(t_3)\cdots\int_0^{t_{n-2}} p_{n-1}(t_{n-1}) \\ &\quad \times \int_0^{t_{n-1}} p_n(t_n)g[\phi^{-1}(v_\epsilon(t_n))]dt_n dt_{n-1}\cdots dt_3 dt_2. \end{aligned} \tag{3.2.205}$$

For any invertible and continuously differentiable function $a(t)$, by changing the variable $\eta = a^{-1}(\xi)$, we get

$$\int \frac{d\xi}{a'[a^{-1}(\xi)]} = \int \frac{a'(\eta)}{a'(\eta)}d\eta = \eta + c = a^{-1}(\xi) + c. \tag{3.2.206}$$

Using the above fact and integrating (3.2.205) from 0 to t, we obtain

$$\phi^{-1}(v_\epsilon(t)) \leq \phi^{-1}(u_0 + \epsilon) + H[t, p_1, p_2, \ldots, p_{n-1}, pg[\phi^{-1}(v_\epsilon(t))]]. \tag{3.2.207}$$

Define a function $w(t)$ by

$$w(t) = \phi^{-1}(u_0 + \epsilon) + H[t, p_1, p_2, \dots, p_{n-1}, pg[\phi^{-1}(v_\epsilon(t))]], \quad (3.2.208)$$

then we have for all $t \in I$,

$$\phi^{-1}(v_\epsilon) \le w(t), \quad (3.2.209)$$

and $w(0) = \phi^{-1}(u_0 + \epsilon) > 0$.

It follows from (3.2.207) and (3.2.208) that

$$L_n w(t) = p(t)g[\phi^{-1}(v_\epsilon(t))] \le p(t)g(w(t)). \quad (3.2.210)$$

From (3.2.210) and using the fact that $g(w(t))$ is positive, $g'(w(t)) \ge 0$, $L_{n-1}w(t) \ge 0$ for all $t \in I$, we see

$$\frac{L_n w(t)}{g(w(t))} \le p(t) + \frac{[(d/dt)g(w(t))] \cdot L_{n-1}w(t)}{g^2(w(t))}, \quad (3.2.211)$$

i.e.,

$$\frac{d}{dt}\left[\frac{L_{n-1}w(t)}{g(w(t))}\right] \le p(t). \quad (3.2.212)$$

Starting with (3.2.212) and using the same steps as used from (3.2.201) to (3.2.205), we can derive

$$\frac{(d/dt)w(t)}{g(w(t))} \le p_1(t)\int_0^t p_2(t_2)\int_0^{t_2} p_3(t_3)\cdots\int_0^{t_{n-2}} p_{n-1}(t_{n-1}) \times \int_0^{t_{n-1}} p(t_n)dt_n dt_{n-1}\cdots dt_3 dt_2. \quad (3.2.213)$$

Now, by the definition of G, keeping $t = t_1$ and integrating both sides of (3.2.213) from 0 to t, we conclude

$$G(w(t)) \le G(w(0)) + H[t, p-1, p_2, \dots, p_{n-1}, p], \quad (3.2.214)$$

i.e.,

$$w(t) \le G^{-1}[G(\phi^{-1}(u_0 + \epsilon)) + H[t, p_1, p_2, \dots, p]], \quad t \in I_1. \quad (3.2.215)$$

Thus the desired bound in (3.2.192) now follows from (3.2.196), (3.2.209), and (3.2.215), together with the limit $\epsilon \to 0^+$. □

Corollary 3.2.9 (The Ma-Debnath Inequality [361]) *Suppose that the functions* $p, p_1, \ldots, p_{n-1}$ *and* g *are defined as in Theorem 3.2.24, and* $u_0 \geq 1$ *and* $k > 0$ *are constants. If* $u : I \to \mathbb{R}_1$ *satisfies for all* $t \in I$,

$$u^k(t) \leq u_0 + H[t, p_1, p_2, \ldots, p_{n-1}, pu^k g(\log u)], \tag{3.2.216}$$

then for all $t \in I_2$, *we have*

$$u(t) \leq \exp G^{-1}\Big[G(\frac{1}{k}\log u_0) + H[t, p_1, p_2, \ldots, p_{n-1}, \frac{1}{k}p]\Big], \tag{3.2.217}$$

where G *and* G^{-1} *are defined as in Theorem 3.2.24 and* $I_2 = [0, t_2) \subset I$, t_2 *is chosen so that*

$$G(\frac{1}{k}\log u_0) + H[t, p_1, p_2, \ldots, p_{n-1}, \frac{1}{k}p] \in Dom\,(G^{-1}). \tag{3.2.218}$$

Proof Changing the variable $u = \exp(v)$ in (3.2.216) leads to

$$\exp(kv) \leq u_0 + H[t, p_1, p_2, \ldots, p_{n-1}, p\exp(kv)g(v)] \tag{3.2.219}$$

$$= u_0 + H\Big[t, p_1, p_2, \ldots, p_{n-1}, \frac{p}{k}(\exp(kv))'g(v)\Big] \tag{3.2.220}$$

which is a special case of Theorem 3.2.24 when $\phi = \exp(kv)$. By Theorem 3.2.24, we derive from (3.2.219) that for all $t \in I_2$,

$$v \leq G^{-1}\left[G\Big(\frac{1}{k}\log u_0\Big) + H[t, p_1, p_2, \ldots, p_{n-1}, \frac{1}{k}p]\right]. \tag{3.2.221}$$

Thus inequality (3.2.217) follows. □

Remark 3.2.16 When $k = 1$ in Corollary 3.2.9, we can derive the assertion of Theorem 3.2.23.

Let $\phi = u^k$ ($k > 1$ is a constant) in Theorem 3.2.24, then this leads to the following corollary.

Corollary 3.2.10 (The Ma-Debnath Inequality [361]) *Suppose that the functions* u, p, p_1, $\ldots, p_{n-1}$ *and* g *are defined as in Theorem 3.2.24 and* $k > 1$ *is a constant, if the inequality holds for all* $t \in I$,

$$u^k(t) \leq u_0 + H[t, p_1, p_2, \ldots, p_{n-1}, kpu^{k-1}g(u)], \tag{3.2.222}$$

then for all $t \in I_3$,

$$u(t) \leq G^{-1}[G(u_0^{1/k}) + H[t, p_1, p_2, \ldots, p_{n-1}, kp]], \tag{3.2.223}$$

where G and G^{-1} are defined as in Theorem 3.2.24 and $I_3 = [0, t_3] \subset I$, t_3 is chosen so that

$$G(u_0^{1/k}) + H[t, p_1, p_2, \ldots, p_{n-1}, kp] \in Dom\ (G^{-1}). \tag{3.2.224}$$

Remark 3.2.17 Setting by $n = 1$, $k = 2$, $g \equiv 1$ in Corollary 3.2.10, we may arrive at Ou-Yang's integral inequality given in [438].

Remark 3.2.18 By choosing other suitable special functions ϕ, we may get other interesting inequalities which cannot be derived from Theorem 3.2.23.

3.3 Nonlinear One-Dimensional Discontinuous Ou-Yang Inequality and Its Generalizations

In this section, we introduce one-dimensional nonlinear discontinuous Ou-Yang inequality and its generalizations.

We first briefly introduce the time scales calculus, which can be found in [90, 91].

A time scale T is an arbitrary non-empty closed subset of $\mathbb{R}$. The forward jump operator σ on T is defined by for all $t \in T$,

$$\sigma(t) := \inf\{s \in T : s > t\} \in T. \tag{3.3.1}$$

In this definition, we put $\inf \varnothing - \sup T$, where $\varnothing$ is the empty set. If $\sigma(t) > t$, then we say that t is right-scattered. If $\sigma(t) = t$ and $t < \sup T$, then we say that t is right-dense. The backward jump operator, left-scattered and left-dense points are defined in a similar way. The graininess $\mu : T \to [0, +\infty)$ is defined by $\mu(t) := \sigma(t) - t$. The set T^k is derived from T as follows: If T has a left-scattered maximum m, then $T^k = T - \{m\}$; otherwise, $T = T^k$.

Remark 3.3.1 Clearly, we see that $\sigma(t) = t$ if $T = \mathbb{R}$ and $\sigma(t) = t + 1$ if $T = \mathbb{Z}$.

For $f : T \to \mathbb{R}$ and all $t \geq t_0, t \in T^k$, we define $f^\alpha(t)$ to be the number (provided it exists) such that given any $\varepsilon > 0$, there is a neighborhood U of t with, for all $s \in U$,

$$|[f(\sigma(t)) - f(s)] - f^\alpha[\sigma(t) - s]| \leq \varepsilon|\sigma(t) - s|. \tag{3.3.2}$$

We call $f^\alpha(t)$ the delta derivative of f at t.

Remark 3.3.2 $f^\alpha(t)$ is the usual derivative f' if $T = \mathbb{R}$ and the useful forward difference Δf (defined by $\Delta f = f(t+1) - f(t)$) if $T = \mathbb{Z}$.

We say that $f : T \to \mathbb{R}$ is rd-continuous provided f is continuous at each right-dense point of T and has a finite left-sided limit at each left-dense point of T. As usual, the set of rd-continuous functions is denoted by C_{rd}. A function $F : T \to \mathbb{R}$ is called an antiderivative of $f : T \to \mathbb{R}$ provided $F^{\alpha}(t) = f(t)$ holds for all $t \in T^k$. In this case, we define the Cauchy integral of f by, for all $a, b \in T$,

$$\int_a^b f(t)\Delta t = F(b) - F(a). \tag{3.3.3}$$

We say that $p : T \to \mathbb{R}$ is regressive provided $1 + \mu(t)p(t) \neq 0$ for all $t \in T$. We denote by $\mathcal{R}$ the set of all regressive and rd-continuous functions. We define the set of all positively regressive functions by $\mathcal{R}^+ = \{p \in \mathcal{R} : 1 + \mu(t)p(t) > 0, \text{ for all } t \in T\}$.

For all $h > 0$, we define the cylinder transformation $\xi_h : C_h \to Z_h$ by

$$\xi_h(z) = \frac{1}{h} Log(1 + zh), \tag{3.3.4}$$

where Log is the principal logarithm function, and

$$C_h = \left\{ z \in \mathbb{C} : z \neq -\frac{1}{h} \right\}, \quad Z_h = \left\{ z \in \mathbb{C} : -\frac{\pi}{h} < Im(z) \leq \frac{\pi}{h} \right\}. \tag{3.3.5}$$

For $h = 0$, we define $\xi_0(z) = z$ for all $z \in \mathbb{C}$.

If $p \in \mathcal{R}$, then we define the exponential function by, for all $s, t \in T$,

$$e_p(t, s) = \exp\left(\int_s^t \xi_{\mu(\tau)}(p(\tau))\Delta\tau \right). \tag{3.3.6}$$

Theorem 3.3.1 ([346]) *If $p \in \mathcal{R}$ and fix $t_0 \in T$, then the exponential function $e_p(\cdot, t_0)$ is for the unique solution of the initial value problem*

$$x^{\alpha} = p(t)x, \quad x(t_0) = 1 \quad \text{on} \quad T. \tag{3.3.7}$$

Theorem 3.3.2 ([346]) *If $p \in \mathcal{R}$, then*

(i) $e_p(t, t) \equiv 1$ *and* $e_0(t, s) \equiv 1$;
(ii) $e_p(\sigma(t), s) = (1 + \mu(t)p(t))e_p(t, s)$;
(iii) *If $p \in \mathcal{R}^+$, then for all $t \in T$, $e_p(t, t_0) > 0$.*

Remark 3.3.3 Clearly, the exponential function is given by, for all $s, t \in \mathbb{R}$,

$$e_p(t, s) = e^{\int_s^t p(\tau)d\tau}, \quad e_{\alpha}(t, s) = e^{\alpha(t-s)}, \quad e_{\alpha}(t, 0) = e^{\alpha t} \tag{3.3.8}$$

where $\alpha \in \mathbb{R}$ is a constant and $p : T \to \mathbb{R}$ is a continuous function if $T = \mathbb{R}$, and the exponential function is given by, for all $s, t \in \mathbb{Z}$ with $s < t$,

$$e_p(t,s) = \prod_{\tau=s}^{t-1}[1+p(\tau)], \quad e_\alpha(t,s) = (1+\alpha)^{(t-s)}, \quad e_\alpha(t,0) = (1+\alpha)^t, \tag{3.3.9}$$

where $\alpha \neq -1$ is a constant and $p : \mathbb{Z} \to \mathbb{R}$ is a sequence satisfying $p(t) \neq -1$ for all $t \in \mathbb{Z}$ if $T = Z$.

Theorem 3.3.3 ([346]) *If $p \in \mathcal{R}$ and fix $a, b, c \in T$, then*

$$\int_a^b p(t)e_p(c, \sigma(t))\Delta t = e_p(c,a) - e_p(c,b). \tag{3.3.10}$$

Theorem 3.3.4 ([346]) *Let $t_0 \in T^k$ and $w : T \times T^k \to \mathbb{R}$ be continuous at all $(t,t), t \in T^k$ with all $t > t_0$. Assume that $w_1^\alpha(t,\cdot)t$ is rd-continuous on $[t_0, \sigma(t)]$. If for any $\varepsilon > 0$, there exists a neighborhood U of t, independent of $\tau \in [t_0, \sigma(t)]$, such that for all $s \in U$,*

$$|w(\sigma(t),\tau) - w(s,\tau) - w_1^\alpha(\sigma(t) - s)| \leq \varepsilon\, |\sigma(t) - s|\,, \tag{3.3.11}$$

where w_1^α denotes the derivative of w with respect to the first variable, then

$$v(t) := \int_{t_0}^t w(t,\tau)\Delta\tau \tag{3.3.12}$$

implies

$$v^\alpha(t) = \int_{t_0}^t w_1^\alpha t, \tau)\Delta\tau + w(\sigma(t), t). \tag{3.3.13}$$

The following theorem found in [90] is a foundational result in dynamic inequalities.

Theorem 3.3.5 (Comparison Theorem [90]) *Suppose $u, b \in C_{rd}$, $a \in \mathcal{R}^+$. If for all $t \geq t_0,\ t \in T^k$,*

$$u^\Delta(t) \leq a(t)u(t) + b(t), \tag{3.3.14}$$

then for all $t \geq t_0,\ t \in T^k$,

$$u(t) \leq u(t_0)e_\alpha(t,t_0) + \int_{t_0}^t e_\alpha(t, \sigma(t))b(\tau)\Delta\tau. \tag{3.3.15}$$

In the sequel, we shall introduce integral inequalities on time scales. We always assume that $p \geq q > 0$, p and q are real constants, and $t \geq t_0,\ t_0 \in T^k$.

The following lemma is useful in the following.

Lemma 3.3.1 ([346]) *Let for all $K > 0$, $a \geq 0$. Then*

$$a^{\frac{q}{p}} \leq \left(\frac{q}{p}K^{\frac{p-q}{p}}a + \frac{q-p}{p}K^{\frac{q}{p}}\right). \tag{3.3.16}$$

Proof If $a = 0$, then we easily see that the inequality (3.3.16) holds. Thus we only prove that the inequality (3.3.16) holds in the case of $a > 0$.

Letting

$$f(K) = \frac{q}{p}K^{\frac{p-q}{p}}a + \frac{q-p}{p}K^{\frac{q}{p}}, \quad K > 0, \tag{3.3.17}$$

we have

$$f'(K) = \frac{q(p-q)}{p^2}K^{\frac{q-2p}{p}}(K-a). \tag{3.3.18}$$

On the other hand,

$$\begin{cases} f'(K) \geq 0,\ K > a, \\ f'(K) = 0,\ K = a, \\ f'(K) \leq 0,\ 0 < K < a. \end{cases} \tag{3.3.19}$$

Therefore,

$$f(K) \geq f(a) = a^{\frac{q}{p}}. \tag{3.3.20}$$

The proof is thus complete. □

Theorem 3.3.6 (The Li-Sheng Inequality [346]) *Assume that $u, a, b, g, h \in C_{rd}$, $u(t), a(t), b(t), g(t)$, and $h(t)$ are non-negative. If for all $t \in T^k$,*

$$u^p(t) \leq a(t) + b(t)\int_{t_0}^{t}[g(\tau)u^p(\tau) + h(\tau)u^q(\tau)]\Delta\tau, \tag{3.3.21}$$

then for all $t \in T^k$, for any $K > 0$,

$$u(t) \leq \left\{a(t) + b(t)\int_{t_0}^{t}\left[g(\tau)a(\tau) + h(\tau)\left(\frac{K(p-q) + qa(\tau)}{pK^{(p-q)/p}}\right)\right]e_F(t, \sigma(\tau))\Delta\tau\right\}^{\frac{1}{p}}, \tag{3.3.22}$$

where

$$F(t) = b(t)\left(g(t) + \frac{qh(t)}{pK^{(p-q)/p}}\right). \tag{3.3.23}$$

Proof Define a function $z(t)$ by, for all $t \in T^k$,

$$z(t) = \int_{t_0}^{t} [g(\tau)u^p(\tau) + h(\tau)u^q(\tau)]\Delta\tau. \tag{3.3.24}$$

Then $z(t_0) = 0$ and (3.3.21) can be restated as, for all $t \in T^k$,

$$u^p(t) \leq a(t) + b(t)z(t). \tag{3.3.25}$$

Using Lemma 3.3.1, from (3.3.25), it follows, for any $K > 0$,

$$\begin{aligned} u^q(t) &\leq (a(t) + b(t)z(t))^{q/p} \\ &\leq \frac{K(p-q) + qa(t)}{pK^{(p-q)/p}} + \frac{qb(t)z(t)}{pK^{(p-q)/p}}. \end{aligned} \tag{3.3.26}$$

Thus combining (3.3.24)–(3.3.26), we can get for all $t \in T^k$,

$$\begin{aligned} z^{\alpha}(t) &= g(t)u^p(t) + h(t)u^q(t) \\ &\leq g(t)[a(t) + b(t)z(t)] + h(t)\left(\frac{K(p-q) + qa(t)}{pK^{(p-q)/p}} + \frac{qb(t)z(t)}{pK^{(p-q)/p}}\right) \\ &= \left[a(t)g(t) + h(t)\frac{K(p-q) + qa(t)}{pK^{(p-q)/p}}\right] + F(t)z(t), \end{aligned} \tag{3.3.27}$$

where $F(t)$ is defined as in (3.3.23).

It is easy to see that $F(t) \in \mathcal{R}^+$. Therefore, using Theorem 3.3.5 and noting $z(t_0) = 0$, from (3.3.27) we derive, for all $t \in T^k$,

$$z(t) \leq \int_{t_0}^{t} \left[a(t)g(t) + h(t)\frac{K(p-q) + qa(t)}{pK^{(q-p)/p}}\right] e_F(t, \sigma(\tau))\Delta\tau. \tag{3.3.28}$$

Clearly, the desired inequality (3.3.22) follows from (3.3.25) and (3.3.28). This completes the proof. □

Corollary 3.3.1 (The Li-Sheng Inequality [346]) *Let $T = \mathbb{R}$ and assume that $u(t), a(t), b(t), g(t), h(t) \in C(\mathbb{R}_+, \mathbb{R}_+)$. If the inequality holds for all $t \in \mathbb{R}_+$,*

$$u^p(t) \leq a(t) + b(t)\int_0^t [g(s)u^p(s) + h(s)u^q(s)]ds, \tag{3.3.29}$$

then for all $t \in \mathbb{R}_+$, *for any* $K > 0$,

$$u(t) \leq \left\{ a(t) + b(t) \int_0^t \left[a(\tau)g(\tau) + h(\tau)\frac{K(p-q) + qa(t)}{pK^{(q-p)/p}} \right] \times \exp\left(\int_\tau^t F(s)ds \right) d\tau \right\}^{1/p}, \tag{3.3.30}$$

where $F(t)$ *is defined as in* (3.3.23) *of Theorem 3.3.6.*

The next result can be regarded as a generalization of the Ou-Yang inequality.

Corollary 3.3.2 (The Li-Sheng Inequality [346]) *Let* $T = \mathbb{Z}$ *and assume that* $u(t), a(t), b(t), g(t)$, *and* $h(t)$ *are non-negative functions defined for* $t \in \mathbb{N}_0 = \mathbb{N} \cup \{0\}$. *If the inequality holds for all* $t \in \mathbb{N}_0$,

$$u^p(t) \leq a(t) + b(t) \sum_{s=0}^{t-1} [g(s)u^p(s) + h(s)u^q(s)]ds, \tag{3.3.31}$$

then for all $t \in \mathbb{N}_0$, *for any* $K > 0$,

$$u(t) \leq \left\{ a(t) + b(t) \sum_{\tau=0}^{t-1} \left[a(\tau)g(\tau) + h(\tau)\frac{K(p-q) + qa(t)}{pK^{(q-p)/p}} \right] \times \prod_{s=\tau+1}^{t-1} (1 + F(s))d\tau \right\}^{1/p}, \tag{3.3.32}$$

where $F(t)$ *is defined as in* (3.3.23) *of Theorem 3.3.6.*

Remark 3.3.4 Letting $p > 1, K = q = 1$ in Corollaries 3.3.1 and 3.3.2, we easily obtain Theorem 1.4.10 (a_1) and Theorem 2.3.22 (c_1) established by Pachpatte [512], respectively.

Corollary 3.3.3 (The Li-Sheng Inequality [346]) *Assume that* $u, h \in C_{rd}, u(t)$ *and* $h(t)$ *are non-negative,* $\beta \geq 0$ *is a real constant. If for all* $t \in T^k$,

$$u^p(t) \leq \beta + \int_{t_0}^t h(\tau)u^q(\tau)\Delta\tau, \tag{3.3.33}$$

then for all $t \in T^k$, *for any* $K > 0$,

$$u(t) \leq \left\{ \frac{1}{q}[(K(p-q) + q\beta)e_F(t, t_0) - K(p-q)] \right\}^{\frac{1}{p}}, \tag{3.3.34}$$

where

$$\bar{F}(t) = \frac{qh(t)}{pK^{(p-q)/p}}. \tag{3.3.35}$$

Proof Using Theorem 3.3.6, it follows from (3.3.33) that, for all $t \in T^k$, for any $K > 0$,

$$\begin{aligned} u(t) &\le \left\{\beta + \int_{t_0}^{t} h(\tau)\frac{K(p-q)+q\beta}{pK^{(p-q)/p}} e_{\bar{F}}(t,\sigma(\tau))\Delta\tau\right\}^{1/p} \\ &= \left\{\beta + \left(\frac{K(p-q)}{q}+\beta\right)\int_{t_0}^{t} h(\tau)\bar{F}(\tau)e_{\bar{F}}(t,\sigma(\tau))\Delta\tau\right\}^{1/p} \\ &= \left\{\beta + \left(\frac{K(p-q)}{q}+\beta\right)[e_{\bar{F}}(t,t_0) - e_{\bar{F}}(t,t)]\right\}^{1/p} \\ &= \left\{\beta + \left(\frac{K(p-q)}{q}+\beta\right)e_{\bar{F}}(t,t_0) - \frac{K(p-q)}{q} - \beta\right\}^{1/p} \\ &= \left\{\frac{1}{q}[(q\beta + K(p-q))e_{\bar{F}}(t,t_0) - K(p-q)]\right\}^{1/p}, \end{aligned} \tag{3.3.36}$$

where the second equation holds because of Theorem 3.3.3, and the third equation holds because of Theorem 3.3.1 (i). This thus completes the proof. □

Theorem 3.3.7 (The Li-Sheng Inequality [346]) *Assume that $u, a, b, g, h_i \in C_{rd}$, $u(t), a(t), b(t), g(t)$, and $h_i(t)$ are non-negative, and $i = 1, 2, \ldots, n$, and assume further there exists a sequence of positive real numbers $q_1, q_2, \ldots, q_n$ such that $p \ge q_i > 0, i = 1, 2, \ldots, n$. If for all $t \in T^k$,*

$$u^p(t) \le a(t) + b(t)\int_{t_0}^{t}\left[g(\tau)u^p(\tau)\sum_{i=1}^{n} h_i(\tau)u^{q_i}(\tau)\right]\Delta\tau, \tag{3.3.37}$$

then for all $t \in T^k$, for any $K > 0$,

$$u(t) \le \left\{a(t) + b(t)\int_{t_0}^{t}\left[g(\tau)a(\tau) + \sum_{i=1}^{n} h_i(\tau)\left(\frac{K(p-q_i)+q_i a(\tau)}{pK^{(p-q_i)/p}}\right)\right]e_{F^*}(t,\sigma(\tau))\Delta\tau\right\}^{\frac{1}{p}}, \tag{3.3.38}$$

where

$$F^*(t) = b(t)\left(g(t) + \sum_{i=1}^{n}\frac{q_i h_i(t)}{pK^{(p-q_i)/p}}\right). \tag{3.3.39}$$

Proof Define a function $z(t)$ by, for all $t \in T^k$,

$$z(t) = \int_{t_0}^{t} [g(\tau)u^p(\tau) + \sum_{i=1}^{n} h_i(\tau)u^{q_i}(\tau)]\Delta\tau. \tag{3.3.40}$$

Then $z(t_0) = 0$ and as in the proof of Theorem 3.3.6, we have (3.3.25) and for any $K > 0$,

$$u^{q_i}(t) \leq \frac{K(p-q) + q_i a(t)}{pK^{(p-q_i)/p}} + \frac{q_i b(t)z(t)}{pK^{(p-q_i)/p}}, \quad i = 1, 2, \ldots, n. \tag{3.3.41}$$

Therefore, for all $t \in T^k$,

$$\begin{aligned} z^{\alpha}(t) &= g(t)u^p(t) + \sum_{i=1}^{n} h_i(t)u^{q_i}(t) \\ &\leq g(t)[a(t) + b(t)z(t)] + \sum_{i=1}^{n} h_i(t)\left(\frac{K(p-q_i) + q_i a_i(t)}{pK^{(p-q_i)/p}} + \frac{q_i b(t)z(t)}{pK^{(p-q_i)/p}}\right) \\ &= \left[a(t)g(t) + \sum_{i=1}^{n} h_i(t)\frac{K(p-q_i) + q_i a_i(t)}{pK^{(p-q_i)/p}}\right] + F^*(t)z(t), \end{aligned} \tag{3.3.42}$$

where $F^*(t)$ is defined as in (3.3.39).

The remainder of the proof is similar to that of Theorem 3.3.6 and we omit it here. □

Theorem 3.3.8 (The Li-Sheng Inequality [346]) *Assume that* $u, a, b, g, h \in C_{rd}$, $u(t), a(t), b(t), g(t)$, *and* $h(t)$ *are non-negative, and* $w(t, s)$ *is defined as in Theorem 3.3.4 such that* $w(t, s) \geq 0$ *and* $w_1^{\alpha}(t, s) \geq 0$ *for all* $t, s \in T$ *with* $s \leq t$. *Assume further for any* $\varepsilon > 0$, *there exists a neighborhood* U *of* t, *independent of* $\tau \in [t_0, \sigma(t)]$, *such that for all* $s \in U$,

$$|[w(\sigma(t), \tau) - w(s, \tau) - w_1^{\alpha}(\sigma(t) - s)][g(\tau)u^p(\tau) + h(\tau)u^q(\tau)]| \leq \varepsilon\, |\sigma(t) - s|\,. \tag{3.3.43}$$

If for all $\tau \in T^k$,

$$u^p(t) \leq a(t) + b(t)\int_{t_0}^{t} w(t, \tau)[g(\tau)u^p(\tau) + h(\tau)u^q(\tau)]\Delta\tau, \tag{3.3.44}$$

then for all $t \in T^k$, *for any* $K > 0$,

$$u(t) \leq \left\{a(t) + b(t)\int_{t_0}^{t} e_A(t, \sigma(\tau))B(\tau)\Delta\tau\right\}^{1/p}, \tag{3.3.45}$$

where, for all $t \in T^k$,

$$\begin{cases} A(t) = w(t, \sigma(\tau))b(t)\left(g(t) + \dfrac{qh(t)}{pK^{(p-q)/p}}\right) + \displaystyle\int_{t_0}^{t} w_1^{\alpha}(t,\tau)b(\tau)\left(g(\tau) + \dfrac{qh(\tau)}{pK^{(p-q)/p}}\right)\Delta\tau, \\ B(t) = w(\sigma(t), t)\left[a(t)g(t) + h(t)\left(\dfrac{K(p-q) + qa(t)}{pK^{(p-q)/p}}\right)\right] \\ \qquad + \displaystyle\int_{t_0}^{t} w_1^{\alpha}(t,\tau)\left[a(\tau)g(\tau) + \left(\dfrac{K(p-q) + qa(t)}{pK^{(p-q)/p}}\right)\right]\Delta\tau. \end{cases} \tag{3.3.46}$$

Proof Define a function $z(t)$ by, $t \in T^k$,

$$z(t) = \int_{t_0}^{t} k(t,\tau)\Delta\tau, \tag{3.3.47}$$

where, for all $t \in T^k$,

$$k(t,\tau) = w(t,\tau)[g(\tau)u^p(\tau) + h(\tau)u^q(\tau)]. \tag{3.3.48}$$

Then $z(t_0) = 0$. As in the proof of Theorem 3.3.6, we easily obtain (3.3.25) and (3.3.26).

It follows from (3.3.48) that

$$k(\sigma(t), t) = w(\sigma(t), t)[g(t)u^p(t) + h(t)u^q(t)], \tag{3.3.49}$$

$$k_1^{\alpha}(t,\tau) = w_1^{\alpha}(t,\tau)[g(\tau)u^p(\tau) + h(\tau)u^q(\tau)]. \tag{3.3.50}$$

Therefore, noting the condition (3.3.43), using Theorem 3.3.4, and combining (3.3.47)–(3.3.50), (3.3.25), and (3.3.26), we conclude for all $t \in T^K$,

$$\begin{aligned} z^{\alpha}(t) &= k(\sigma(t), t) + \int_{t_0}^{t} k_1^{\alpha}(t,\tau)\Delta\tau \\ &= w(\sigma(t), t)[g(t)u^p(t) + h(t)u^q(t)] + \int_{t_0}^{t} w_1^{\alpha}(t,\tau)[g(\tau)u^p(\tau) + h(\tau)u^q(\tau)]\Delta\tau \\ &\le w(\sigma(t), t)\left[a(t)g(t) + h(t)\left(\frac{K(p-q) + qa(t)}{pK^{(p-q)/p}}\right) + b(t)\left(g(t) + \frac{qh(t)}{pK^{(p-q)/p}}\right)z(t)\right] \\ &\quad + \int_{t_0}^{t} w_1^{\alpha}(t,\tau)\left[a(\tau)g(\tau) + h(\tau)\left(\frac{K(p-q) + qa(\tau)}{pK^{(p-q)/p}}\right)\right. \\ &\quad \left. + b(\tau)\left(g(\tau) + \frac{qh(\tau)}{pK^{(p-q)/p}}\right)z(\tau)\right]\Delta\tau \\ &\le \left[w(\sigma(t), t)b(t)\left(g(t) + \frac{qh(t)}{pK^{(p-q)/p}}\right)\right. \end{aligned}$$

$$
\begin{aligned}
&+ + \int_{t_0}^{t} w_1^{\alpha}(t,\tau)b(\tau)\left(g(\tau)+\frac{qh(\tau)}{pK^{(p-q)/p}}\right)\Delta\tau\Bigg] z(t)\\
&+w(\sigma(t),t)\left[a(t)g(t)+h(t)\left(\frac{K(p-q)+qa(t)}{pK^{(p-q)/p}}\right)\right]\\
&+\int_{t_0}^{t} w_1^{\alpha}(t,\tau)\left[a(\tau)g(\tau)+h(\tau)\left(\frac{K(p-q)+qa(\tau)}{pK^{(p-q)/p}}\right)\right]\Delta\tau\\
&= A(t)z(t)+B(t). \qquad (3.3.51)
\end{aligned}
$$

Therefore, using Theorem 3.3.5 and noting $z(t_0) = 0$, we can get for all $t \in T^k$,

$$
z(t) \leq \int_{t_0}^{t} e_A(t,\sigma(\tau))B(\tau)\Delta\tau. \qquad (3.3.52)
$$

It is easy to see that the desired inequality (3.3.45) follows from (3.3.10) and (3.3.52). The proof is thus complete. □

The next result can be also regarded as a generalisation of the Ou-Yang inequality.

Corollary 3.3.4 (The Li-Sheng Inequality [346]) *Let $T = \mathbb{R}$ and assume that $u(t), a(t), b(t), g(t), h(t) \in C(\mathbb{R}_+, \mathbb{R}_+)$. If $w(t,s)$ and its partial derivative $(\partial w(t,s)/\partial t$ are real-valued non-negative continuous functions for all $t, s \in \mathbb{R}_+$. with $s \leq t$. If the inequality holds for all $t \in \mathbb{R}_+$,*

$$
u^p(t) \leq a(t) + b(t)\int_0^t w(t,\tau)[g(\tau)u^p(\tau)+h(\tau)u^q(\tau)]d\tau, \qquad (3.3.53)
$$

then for all $t \in \mathbb{R}_+$, for any $K > 0$,

$$
u(t) \leq \left\{a(t) + b(t)\int_0^t \exp\left(\int_{\tau}^{t} \bar{A}(s)ds\right)\bar{B}(\tau)d\tau\right\}^{1/p}, \qquad (3.3.54)
$$

where for all $t \in \mathbb{R}_+$,

$$
\begin{cases}
\bar{A}(t) = w(t,t)b(t)\left(g(t)+\dfrac{qh(t)}{pK^{(p-q)/p}}\right) + \displaystyle\int_0^t \frac{\partial w(t,\tau)}{\partial t} b(\tau)\left(g(\tau)+\frac{qh(\tau)}{pK^{(p-q)/p}}\right)d\tau,\\
\bar{B}(t) = w(t,t)\left[a(t)g(t)+h(t)\left(\dfrac{K(p-q)+qa(t)}{pK^{(p-q)/p}}\right)\right]\\
\qquad + \displaystyle\int_0^t \frac{\partial w(t,\tau)}{\partial t}\left[a(\tau)g(\tau)+h(\tau)\left(\frac{K(p-q)+qa(\tau)}{pK^{(p-q)/p}}\right)\right]d\tau.
\end{cases} \qquad (3.3.55)
$$

Remark 3.3.5 Let $p > 1, K = q = 1$. Then the inequality established in Corollary 3.3.4 reduces to Theorem $3(c_1)$ in [512].

Corollary 3.3.5 (The Li-Sheng Inequality [346]) *Let $T = \mathbb{Z}$ and assume that $u(t), a(t), b(t), g(t)$ and $h(t)$ are non-negative functions defined for all $t \in \mathbb{N}_0$. If $w(t,s)$ and $\triangle_1 w(t,s)$ are real-valued non-negative functions for all $t, s \in \mathbb{N}_0$ with $s \le t$. If the inequality holds for all $t \in \mathbb{N}_0$,*

$$u^p(t) \le a(t) + b(t)\sum_{\tau=0}^{t-1} w(t,\tau)[g(\tau)u^p(\tau) + h(\tau)u^q(\tau)], \tag{3.3.56}$$

then for all $t \in \mathbb{N}_0$, for any $K > 0$,

$$u(t) \le \left\{a(t) + b(t)\sum_{\tau=0}^{t-1}\widetilde{B}(\tau)\prod_{s=\tau+1}^{t-1}(1+\widetilde{A}(s))\right\}^{1/p}, \tag{3.3.57}$$

where $\triangle_1 w(t,s) = w(t+1,s) - w(t,s)$ for all $t, s \in \mathbb{N}_0$ with $s \le t$, for all $t \in \mathbb{N}_0$,

$$\begin{cases} \widetilde{A}(t) = w(t+1,t)b(t)\left(g(t) + \dfrac{qh(t)}{pK^{(p-q)/p}}\right) + \displaystyle\sum_{\tau=0}^{t-1}\triangle_1 w(t,\tau)b(\tau)\left(g(\tau) + \dfrac{qh(\tau)}{pK^{(p-q)/p}}\right), \\ \widetilde{B}(t) = w(t+1,t)\left[a(t)g(t) + h(t)\left(\dfrac{K(p-q)+qa(t)}{pK^{(p-q)/p}}\right)\right] \\ \qquad + \displaystyle\sum_{\tau=0}^{t-1}\triangle_1 lw(t,\tau)\left[a(\tau)g(\tau) + h(\tau)\left(\dfrac{K(p-q)+qa(\tau)}{pK^{(p-q)/p}}\right)\right]. \end{cases} \tag{3.3.58}$$

Remark 3.3.6 Let $p > 1, K = q = 1$. Then the inequality established in Corollary 3.3.5 reduces to Theorem 3(c_3) in [512].

Corollary 3.3.6 (The Li-Sheng Inequality [346]) *Suppose that $0 \le \alpha$ is a constant, $u(t)$ and $w(t,s)$ are defined as in Theorem 3.3.8. If for any $\varepsilon > 0$, there exists a neighborhood U of t, independent of $\tau \in [t_0, \sigma(t)]$, such that for all $s \in U$,*

$$|u^q(t)[w(\sigma(t),\tau) - w(s,\tau) - w_1^\alpha(t,\tau)(\sigma(t)-s)]| \le \varepsilon\,|\sigma(t)-s|\,. \tag{3.3.59}$$

If for all $t \in T^k$,

$$u^p(t) \le \alpha + \int_{t_0}^t w(t,\tau)u^q(\tau)\Delta\tau, \tag{3.3.60}$$

then for all $t \in T^k$, for any $K > 0$,

$$u(t) \le \left\{\frac{1}{q}[(K(p-q)+q\alpha)e_{\widetilde{A}}(t,t_0) - K(p-q)]\right\}^{1/p}, \tag{3.3.61}$$

where, for all $t \in T^k$.

$$\widehat{A}(t) = \frac{q}{pK^{(p-q)/p}} \left(w(\sigma(t), t) + \int_{t_0}^{t} w_1^{\alpha}(t, \tau) \Delta \tau \right). \tag{3.3.62}$$

Proof Letting $b(t) = 1, g(t) = 0$ and $h(t) = 1$ in Theorem 3.3.8, we obtain for all $t \in T^k$,

$$\begin{aligned} A(t) &= \frac{q}{pK^{(p-q)/p}} \left(w(\sigma(t), t) + \int_{t_0}^{t} w_1^{\alpha}(t, \tau) \Delta \tau \right) := \widehat{A}(t). \\ B(t) &= \frac{K(p-q) + q\alpha}{pK^{(p-q)/p}} \left(w(\sigma(t), t) + \int_{t_0}^{t} w_1^{\alpha}(t, \tau) \Delta \tau \right) \\ &= \frac{K(p-q) + q\alpha}{pK^{(p-q)/p}} \widehat{A}(t). \end{aligned} \tag{3.3.63}$$

Therefore, by Theorem 3.3.8, noting (3.3.63), we easily obtain for all $t \in T^k$, for any $K > 0$,

$$\begin{aligned} u(t) &\le \left(\alpha + \int_{t_0}^{t} e_A(t, \sigma(t)) B(\tau) \Delta \tau \right)^{1/p} \\ &= \left(\alpha + \int_{t_0}^{t} e_{\widehat{A}}(t, \sigma(t)) \frac{K(p-q) + q\alpha}{q} \widehat{A}(\tau) \Delta \tau \right)^{1/p} \\ &= \left(\alpha + \frac{K(p-q) + q\alpha}{q} \int_{t_0}^{t} e_{\widehat{A}}(t, \sigma(t)) \widehat{A}(\tau) \Delta \tau \right)^{1/p} \\ &= \left(\alpha + \frac{K(p-q) + q\alpha}{q} [e_{\widehat{A}}(t, t_0) - e_{\widehat{A}}(t, t)] \right)^{1/p} \\ &= \left(\frac{K(p-q) + q\alpha}{q} e_{\widehat{A}}(t, t_0) - \frac{K(p-q)}{q} \right)^{1/p}. \end{aligned} \tag{3.3.64}$$

The proof is thus complete. □

Theorem 3.3.9 (The Li-Sheng Inequality [346]) *Assume that* $u, a, b, g, h_i \in C_{rd}$, $u(t), a(t), b(t), g(t)$, *and* $h_i(t)$ *are non-negative,* $i = 1, 2, \ldots, n$, *and there exists a sequence of positive real numbers* $q_1, q_2, \ldots, q_n$ *such that* $p \ge q_i > 0, i = 1, 2, \ldots, n$. *Let* $w(t, s)$ *be defined as in Theorem 3.3.4 such that* $w(t, s) \ge 0$ *and* $w_1^{\alpha}(t, s) \ge 0$ *for all* $t, s \in T$ *with* $s \le t$. *Assume further for any* $\varepsilon > 0$, *there exists a neighborhood* U *of* t, *independent of* $\tau \in [t_0, \sigma(t)]$, *such that for all* $s \in U$,

$$\left| [w(\sigma(t), \tau) - w(s, \tau) - w_1^{\alpha}(\sigma(t) - s)][g(\tau) u^p(\tau) + \sum_{i=1}^{n} h_i(\tau) u^{q_i}(\tau)] \right| \le \varepsilon \left| \sigma(t) - s \right|. \tag{3.3.65}$$

If for all $t \in T^k$,

$$u^p(t) \le a(t) + b(t)\int_{t_0}^{t} w(t,\tau)[g(\tau)u^p(\tau) + \sum_{i=1}^{n} h_i(\tau)u^{q_i}(\tau)]\Delta\tau, \tag{3.3.66}$$

then for all $t \in T^k$, *for any* $K > 0$,

$$u(t) \le \left\{a(t) + b(t)\int_{t_0}^{t} e_{A^*}(t,\sigma(\tau))B^*(\tau)\Delta\tau\right\}^{1/p}, \tag{3.3.67}$$

where for all $t \in T^k$,

$$\begin{cases} A^*(t) = w(t,\sigma(t))b(t)\left(g(t) + \sum_{i=1}^{n} \frac{q_i h_i(t)}{pK^{(p-q_i)/p}}\right) \\ \qquad + \int_{t_0}^{t} w_1^{\alpha}(t,\tau)b(\tau)\left(g(\tau) + \sum_{i=1}^{n} \frac{q_i h_i(\tau)}{pK^{(p-q_i)/p}}\right)\Delta\tau, \\ B^*(t) = w(\sigma(t),t)\left[a(t)g(t) + \sum_{i=1}^{n} h_i(t)\left(\frac{K(p-q_i) + q_i a(t)}{pK^{(p-q_i)/p}}\right)\right] \\ \qquad + \int_{t_0}^{t} w_1^{\alpha}(t,\tau)\left[a(\tau)g(\tau) + \sum_{i=1}^{n} h_i(\tau)\left(\frac{K(p-q) + qa(t)}{pK^{(p-q)/p}}\right)\right]\Delta\tau. \end{cases} \tag{3.3.68}$$

Proof Similarly to the proof of Theorem 3.3.8, we easily obtain the following result. □

Theorem 3.3.10 (The Li-Sheng Inequality [346]) *Assume that* $u, a, b \in C_{rd}$, $u(t)$, $a(t)$ *and* $b(t)$ *are non-negative. Let* $f : T^k \times \mathbb{R} \to \mathbb{R}_+$ *be a continuous function such that for all* $t \in T^k$ *and all* $x \ge y \ge 0$,

$$0 \le f(t,x) - f(t,y) \le \phi(t,y)(x-y), \tag{3.3.69}$$

where $\phi : T^k \times \mathbb{R} \to \mathbb{R}_+$ *is a continuous function. If for all* $t \in T^k$,

$$u^p(t) \le a(t) + b(t)\int_{t_0}^{t} f(\tau, u^q(\tau))\Delta\tau, \tag{3.3.70}$$

Then for all $t \in T^k$, *for any* $K > 0$,

$$u(t) \le \left(a(t) + b(t)\int_{t_0}^{t} e_M(t,\sigma(\tau))f\left(\tau, \frac{K(p-q) + qa(\tau)}{pK^{(p-q)/p}}\right)\Delta\tau\right)^{1/p}, \tag{3.3.71}$$

where, for all $t \in T^k$.

$$M(t) = \phi\left(t, \frac{K(p-q)+qa(\tau)}{pK^{(p-q)/p}}\right)\frac{qb(t)}{pK^{(p-q)/p}}. \tag{3.3.72}$$

Proof Define a function $z(t)$ by, for all $t \in T^k$.

$$z(t) = \int_{t_0}^{t} f(\tau, u^q(\tau))\Delta\tau. \tag{3.3.73}$$

Then $z(t_0) = 0$ and (3.3.70) can be written as (3.3.25). As in the proof of Theorem 3.3.6, from (3.3.25), we easily obtain (3.3.26). Obviously, it follows from (3.3.73), (3.3.26), and (3.3.69) that for all $t \in T^k$,

$$\begin{aligned}
z^{\alpha}(t) &= f(t, u^q(t)) \\
&\le f\left(t, \frac{K(p-q)+qa(t)}{pK^{(p-q)/p}} + \frac{qb(t)}{pK^{(p-q)/p}}z(t)\right) - f\left(t, \frac{K(p-q)+qa(t)}{pK^{(p-q)/p}}\right) \\
&\quad + f\left(t, \frac{K(p-q)+qa(t)}{pK^{(p-q)/p}}\right) \\
&\le \phi\left(t, \frac{K(p-q)+qa(\tau)}{pK^{(p-q)/p}}\right)\frac{qb(t)}{pK^{(p-q)/p}}z(t) + f\left(t, \frac{K(p-q)+qa(t)}{pK^{(p-q)/p}}\right) \\
&= M(t)z(t) + f\left(t, \frac{K(p-q)+qa(t)}{pK^{(p-q)/p}}\right),
\end{aligned} \tag{3.3.74}$$

where $M(t)$ is defined as in (3.3.72). Using Theorem 3.3.5 and noting $z(t_0) = 0$, from (3.3.74), we conclude, for all $t \in T^k$.

$$z(t) \le \int_{t_0}^{t} e_M(t, \sigma(\tau)) f\left(t, \frac{K(p-q)+qa(t)}{pK^{(p-q)/p}}\right)\Delta\tau. \tag{3.3.75}$$

It is easy to see that the desired inequality (3.3.71) follows from(3.3.25) and (3.3.75). The proof thus is complete. □

Remark 3.3.7 Let $p > 1, K = q = 1$. We easily see that Theorem 3.3.10 reduces to Theorem 1.4.11 (b_1) if $T = \mathbb{R}$, and Theorem 2.3.23 (d_1) if $T = \mathbb{Z}$.

By Theorem 3.3.10, we can establish the following more general result.

Theorem 3.3.11 (The Li-Sheng Inequality [346]) *Assume that* $u, a, b \in C_{rd}$, $u(t), a(t)$ *and* $b(t)$ *are non-negative, and* $f_i : T^k \times \mathbb{R} \to \mathbb{R}_+$ *be a continuous function such that*

$$0 \le f_i(t, x) - f_i(t, y) \le \phi_i(t, y)(x - y), \tag{3.3.76}$$

for all $t \in T^k$ *and all* $x \geq y \geq 0$, *where* $\phi_i : T^k \times \mathbb{R} \to \mathbb{R}_+$ *is a continuous function,* $i = 1, 2, \ldots, n$. *Assume further there exists a sequence of positive real numbers* $q_1, q_2, \ldots, q_n$ *such that* $p \geq q_i > 0, i = 1, 2, \ldots, n$. *If for all* $t \in T^k$,

$$u^p(t) \leq a(t) + b(t) \sum_{i=1}^{n} \int_{t_0}^{t} f_i(\tau, u^{q_i}(\tau))\Delta\tau, \tag{3.3.77}$$

then $t \in T^k$, *for any* $K > 0$,

$$u(t) \leq \left(a(t) + b(t) \sum_{i=1}^{n} \int_{t_0}^{t} e_{M^*}(t, \sigma(\tau)) f_i \left(\tau, \frac{K(p - q_i) + q_i a(\tau)}{pK^{(p-q_i)/p}} \right) \Delta\tau \right)^{1/p}, \tag{3.3.78}$$

where, for all $t \in T^k$,

$$M^*(t) = \sum_{i=1}^{n} \phi_i \left(t, \frac{K(p - q_i) + q_i a(\tau)}{pK^{(p-q_i)/p}} \right) \frac{q_i b(t)}{pK^{(p-q_i)/p}}. \tag{3.3.79}$$

Chapter 4
Applications of Nonlinear One-Dimensional Continuous, Discontinuous Integral Inequalities and Discrete Inequalities

In this chapter, we shall choose some models to apply the results in Chaps. 1–3.

4.1 Applications of Theorem 1.1.1

4.1.1 An Application of Theorem 1.1.1 to A Nonlinear Differential Equation

Consider the nonlinear differential equation

$$u'' + f(t, u) = 0. \tag{4.1.1}$$

Cohen [157] proved the following theorems.

Theorem 4.1.1 (The Cohen Inequality [157]) *Suppose $f(t, u)$ satisfies the following conditions:*

(H1) $f(t, u)$ *is continuous on* $D := \{(t, u) :\ t \geq 0,\ -\infty < u < +\infty\}$,
(H2) *the derivative* $f_u(t, u)$ *exists on* D *and* $f_u(t, u) > 0$ *on* D,
(H3) $|f(t, u)| < f_u(t, u)|u|$ *on* D.

In addition, suppose that

$$\int_1^{+\infty} tf_u(t, 0)dt < +\infty. \tag{4.1.2}$$

Then equation (4.1.1) has solutions which are asymptotic to $a + bt$ *as* $t \to +\infty$, *where* $a,\ b$ *are constants and* $b \neq 0$.

Y. Qin, *Integral and Discrete Inequalities and Their Applications*,
DOI 10.1007/978-3-319-33304-5_4

We should point out that in the proof of Theorem 4.1.1, Cohen used Bellman's method [69] and Gronwall's inequality.

In the sequel, we shall use the same method and Bihari's inequality in Theorem 1.1.1, to generalize Theorem 4.1.1.

Theorem 4.1.2 (The Cohen Inequality [157]) *Let $f(t,u)$ be continuous on D. If there are two non-negative continuous functions $v(t)$, $\varphi(t)$ for all $t \geq 0$, and a continuous function $g(u)$ for all $u \geq 0$, such that:*

(i) $\int_1^{+\infty} v(t)\varphi(t)dt < +\infty$,
(ii) for all $u > 0$, $g(u)$ is positive and non-decreasing,
(iii) $|f(t,u)| \leq v(t)\varphi(t)g(|u|/t)$ *for all* $t \geq 1$, $-\infty < u < +\infty$,

then the equation (4.1.1) has solutions which are asymptotic to $a + bt$, where a, b are constants and $b \neq 0$.

Proof Integrating (4.1.1) twice on $[1,t]$, we have

$$u(t) = c_1 + c_2 t - \int_1^t (t-s)f(s,u(s))ds. \tag{4.1.3}$$

Choose $c_1 > 1$ and let $c_3 = c_1 + |c_2|$. Then for all $t > 1$, we have

$$\begin{aligned}\frac{|u(t)|}{t} &\leq c_3 + \int_1^t f(s,u(s))ds \\ &\leq c_3 + \int_1^t v(s)\varphi(s)g(|u(s)|/s)ds.\end{aligned}$$

By Bihari's inequality in Theorem 1.1.1, we have

$$\frac{|u(t)|}{t} \leq G^{-1}(G(c_3) + \int_1^t v(s)\varphi(s)ds). \tag{4.1.4}$$

Here $G(x) = \int_1^x dt/g(t)$, $G^{-1}(x)$ is the inverse function of $G(x)$. From $g(t) > 0$, we know that $G(x)$ is increasing; hence $G^{-1}(x)$ exists, and is also increasing.

Now let $c_4 = G(c_3) + \int_1^{+\infty} v(s)\varphi(s)ds$. Since $G^{-1}(x)$ is increasing, we have

$$\frac{|u(t)|}{t} \leq G^{-1}(c_4). \tag{4.1.5}$$

Differentiating (4.1.3), we have

$$u'(t) = c_2 - \int_1^t f(s,u(s))ds. \tag{4.1.6}$$

By conditions (i), (ii), (iii) and (4.1.5), we have

$$\int_1^t |f(s, u(s))|ds \le \int_1^t v(s)\varphi(s)g(|u(s)|/s)ds$$
$$\le g(G^{-1}(c_4)) \int v(s)\varphi(s)ds < +\infty.$$

Therefore $u'(t) \to c_2 - \int_1^{+\infty} f(s, u(s))ds$ as $t \to +\infty$. If we choose c_2 sufficiently large, then $u'(t) > 1$. Hence $\lim_{t\to+\infty} u'(t) \ne 0$. This proves the theorem. □

Remark 4.1.1 If we assume $v(t) = f_u(t, 0)$, $\varphi(t) = t$, $g(u) = u$ in Theorem 4.1.2, we obtain Theorem 4.1.1.

We give an example to which Theorem 4.1.2 applies, but Theorem 4.1.1 does not.

Example 4.1.1

$$u'' + t^{-4}u^2 \cos u = 0. \tag{4.1.7}$$

Since $f_u(t, 0) = 0$, (H3) does not hold and Theorem 4.1.1 dose not apply. Let $v(t) = t^{-4}$, $\varphi(t) = t^2$, $g(u) = u^2$. Then conditions (i), (ii) and (iii) are satisfied and equation (4.1.7) has solutions asymptotic to $a + bt$ as $t \to +\infty$.

4.1.2 Applications of Theorem 1.1.1 to Some Differential Equations

The purpose of this subsection is to establish–by means of the solutions of the equations $y' = f(x, y) + \varepsilon_1$ and $y' = f(x, y) + \varepsilon_2$ (ε_1 and ε_2 are constants) with the same or different initial conditions, provided that $f(x, y)$ satisfies the "Osgood condition"

$$|f(x, y_2) - f(x, y_1)| \le \omega(|y_2 - y_1|) \tag{4.1.8}$$

in a domain G where $\omega(u)$ is subjected to certain conditions. Further we shall give simple proofs for the uniqueness theorems of Nagumo, Osgood etc.

The next result is the uniqueness theorems of Nagumo.

Theorem 4.1.3 (The Nagumo Inequality [407]) *If the function $f(x, y)$ is continuous in a domain $G(x, y)$ and all points ξ, η of G have a neighborhood where*

$$|x - \xi||f(x, y_2) - f(x, y_1)| \le |y_2 - y_1| \tag{4.1.9}$$

holds whenever the points (x, y_1) *and* (x, y_2) *belong to this neighborhood, then the equation*

$$y' = f(x, y) \tag{4.1.10}$$

has at most solution $\varphi(x)$ *satisfying the initial condition* $\varphi(x) = \eta$.

Proof Assuming there exists another solution $\psi(x)$ too with the same initial condition $\varphi(x) = \eta$, we have

$$\varphi(x) = \eta + \int_\xi^x f(t, \varphi(t))dt, \qquad \psi(x) = \eta + \int_\xi^x f(t, \psi(t))dt \tag{4.1.11}$$

whence for all $x > \xi$,

$$|\varphi(x) - \psi(x)| \leq \int_\xi^x |f(t, \varphi(t)) - f(t, \psi(t))|dt. \tag{4.1.12}$$

Taking (4.1.9) into account, we get

$$|\varphi(x) - \psi(x)| \leq \int_\xi^x |\tfrac{\varphi(t)-\psi(t)}{t-\xi}|dt \tag{4.1.13}$$

for sufficiently small $x - \xi > 0$. Here the integrand has a limit for $t = \xi$, since

$$\lim_{t\to\xi+} \tfrac{\varphi(t)-\psi(t)}{t-\xi} = \lim_{t\to\xi+} \tfrac{\varphi'(t)-\psi'(t)}{1} = f(\xi - \eta) - f(\xi - \eta) = 0 \tag{4.1.14}$$

i.e., the function $\frac{\varphi(t)-\psi(t)}{t-\xi}$ may be completed to a continuous function by taking 0 for its value at $t = \xi$. Therefore we can determine that for all $\varepsilon > 0$, there exists a number $\delta > 0$ such that $|\frac{\varphi(t)-\psi(t)}{t-\xi}| < \varepsilon$ if $|t - \xi| < \delta$ and $\xi + \delta < x$. Then we derive from (4.1.13)

$$|\varphi(x) - \psi(x)| \leq \int_\xi^{\xi+\delta} |\tfrac{\varphi(t)-\psi(t)}{t-\xi}|dt + \int_{\xi+\delta}^x |\tfrac{\varphi(t)-\psi(t)}{t-\xi}|dt \leq \varepsilon\delta + \int_{\xi+\delta}^x \tfrac{|\varphi(t)-\psi(t)|}{t-\xi}dt. \tag{4.1.15}$$

Applying the Bellman-Gronwall's inequality, i.e., Theorem 1.1.2 in Qin [557], we obtain

$$|\varphi(x) - \psi(x)| \leq \varepsilon\delta e^{\int_{\xi+\delta}^x \frac{1}{t-\xi}dt} = \varepsilon\delta \tfrac{x-\xi}{\delta} = \varepsilon(x - \xi) \tag{4.1.16}$$

for a certain right-hand neighborhood of ξ and any positive number ε. For $x \leq \xi$, we conclude in the same way by replacing $|\int_\xi^x u(t)dt|$ by $\int_\xi^x u(t)dt$ $(u(t) \geq 0)$. In this neighborhood of ξ, we have $\varphi \equiv \psi$ and in a known way, we can conclude that this holds for their whole existence intervals.

Let us remark that if instead of condition (4.1.9), we suppose

$$|x - \xi||f(x, y_2) - f(x, y_1)| \leq M(|y_2 - y_1|) \quad with \quad M > 1, \tag{4.1.17}$$

then by a similar calculation, we obtain

$$|\varphi(x) - \psi(x)| \le \varepsilon\delta|\tfrac{x-\xi}{\delta}|^M = \varepsilon\tfrac{|x-\xi|^M}{\delta^{M-1}} \tag{4.1.18}$$

and since this depends on δ, the further conclusion is impossible. Therefore the constant $M = 1$ cannot be increased in this way.

Next we shall show the uniqueness of Osgood Theorem.

Theorem 4.1.4 (The Osgood Inequality [436]) *If, in a domain $G(x, y)$ the function $f(x, y)$ satisfies the condition*

$$|f(x, y_2) - f(x, y_1)| \le \omega(|y_2 - y_1|) \tag{4.1.19}$$

where $\omega(u)$ is continuous for all $u \ge 0$, $\omega(u) > 0$ for all $u > 0$ and $\omega(0) = 0$, further, if $\int_0^u \frac{dt}{\omega(t)}$ is divergent for all $u > 0$, then the equation $y' = f(x, y)$ has almost one solution $\varphi(\xi) = \eta$ where (ξ, η) is a point of G.

Proof In the proof we shall restrict ourselves to the case where $\omega(u)$ is non-decreasing. Suppose there exist two different solutions $\varphi(x), \psi(x)$ with $\varphi(\xi) = \psi(\xi) = \eta$. That is to say, there exist points $\bar{\xi}, \bar{\xi} \ge \xi$ where $\varphi(\bar{\xi}) \ne \psi(\bar{\xi})$. Let the lower bound of these $\bar{\xi}$ be ξ_0. We have then $\xi_0 \ge \xi$ and $\varphi(\xi_0) = \psi(\xi_0)$, but $\varphi(x) \ne \psi(x)$ for $\xi_0 < x \le \xi_0 + \gamma$ with a certain number $\gamma > 0$.

By virtue of the hypotheses,

$$\varphi'(x) - \psi'(x) = f(x, \varphi(x)) - f(x, \psi(x)) \le \omega(|\varphi(x) - \psi(x)|) \tag{4.1.20}$$

whence

$$\frac{\omega(\Delta)}{\omega(V)} \le 1 \qquad or \qquad \frac{V'}{\omega(V)} \le 1. \tag{4.1.21}$$

Introducing the notation

$$\int_{u_0}^{u} \frac{dt}{\omega(t)} = \Omega(u), \tag{4.1.22}$$

we have

$$\frac{d\Omega(V)}{dx} \le 1 \tag{4.1.23}$$

whence by integration, for $\delta > 0, \quad \xi_0 + \delta < x$,

$$\Omega(V(x)) \le \Omega(V(\xi_0 + \delta)) + x - (\xi_0 + \delta). \tag{4.1.24}$$

If $\delta \to 0^+$, then

$$V(\xi_0 + \delta) \to 0, \quad \Omega(V(\xi_0 + \delta)) \to \int_{u_0}^{0} \frac{dt}{\omega(t)} = -\infty, \qquad u_0 > 0, \tag{4.1.25}$$

but $\Omega(V(x))$ is a finite number for all $x > \xi_0$, i.e., (4.1.24) leads to contradiction. We obtain the same contradiction for all $x < \xi$. Therefore $\varphi(x) \equiv \psi(x)$ in some neighborhood of ξ, consequently, also in their whole existence intervals too. □

If $\int_0^u \frac{dt}{\omega(t)}$ is convergent for all $u > 0$, then we get from (4.1.24) for $\delta \to 0^+$,

$$\Omega(V) \le \Omega(0) + x - \xi_0. \tag{4.1.26}$$

Assuming, e.g., $u_0 = 0$, it is

$$\Omega(0) = 0, \quad \Omega(V(x)) \le x - \xi_0 \tag{4.1.27}$$

whence, for all $x \ge \xi_0$,

$$\Delta = |\varphi(x) - \psi(x)| \le V \le \Omega^{-1}(x - \xi_0). \tag{4.1.28}$$

Therefore, if uniqueness does not hold, the difference of two solutions is subdue to this estimate.

Theorem 4.1.5 (The Bihari Inequality [82]) *If $\varphi(x)$ and $\psi(x)$ are the solutions of the equations*

$$y' = f(x, y) + \varepsilon_1, \qquad y' = f(x, y) + \varepsilon_2, \qquad \varepsilon_1 \ge 0, \varepsilon_2 \ge 0 \tag{4.1.29}$$

satisfying the initial conditions $\varphi(\xi_1) = \eta_1, \psi(\xi_2) = \eta_2$ and existing in $a \le x \le b$, if further

$$|f(x, y_2) - f(x, y_1)| \le \omega(|y_2 - y_1|), \qquad |f(x - y)| \le A \quad in \quad G, \tag{4.1.30}$$

then

$$|\varphi(x) - \psi(x)| \le \Omega^{-1}(\Omega(k) + |x - \xi_2|), \tag{4.1.31}$$

with

$$k = |\xi_1 - \xi_2|A + |\eta_1 - \eta_2| + (b - a)(\varepsilon_1 + \varepsilon_2). \tag{4.1.32}$$

Proof We have, in fact,

$$\varphi(x) = \eta_1 + \int_{\xi_1}^{x} f(t, \varphi(t)) + \varepsilon_1(x - \xi_1), \;\; \psi(x) = \eta_2 + \int_{\xi_2}^{x} f(t, \psi(t)) + \varepsilon_2(x - \xi_2), \tag{4.1.33}$$

whence for all $x \geq \xi_2$,

$$|\varphi(x) - \psi(x)| \leq |\eta_1 - \eta_2| + (\varepsilon_1 + \varepsilon_2)(b - a) \\ + \left| \int_{\xi_1}^{\xi_2} |f(t, \varphi(t))| dt + \int_{\xi_2}^{x} |f(t, \varphi(t)) - f(t, \psi(t))| dt \right|. \quad (4.1.34)$$

Further for all $x \geq \xi_2$,

$$|\varphi(x) - \psi(x)| \leq |\eta_1 - \eta_2| + (\varepsilon_1 + \varepsilon_2)(b - a) + |\xi_1 - \xi_2| A + \int_{\xi_2}^{x} \omega(|\varphi(t) - \psi(t)|) dt, \quad (4.1.35)$$

or

$$\Delta \leq k + \int_{\xi_2}^{x} \omega(\Delta) dt \quad (4.1.36)$$

with $\Delta = |\varphi(x) - \psi(x)|$. Hence, applying the Bihari inequality (Theorem 1.1.1) with $F(t) \equiv 1, M = 1$, we conclude for all $x \geq \xi_2$,

$$\Delta = \Omega^{-1}(\Omega(k) + x - \xi_2). \quad (4.1.37)$$

This formula holds also if $\int_0^u \frac{dt}{\omega(t)}$ is divergent for all $u > 0$. In the case, if $(\xi_2, \eta_2) \to (\xi_1, \eta_1)$ and $\varepsilon_1, \varepsilon_2 \to 0$, then $k \to 0$, $\Omega(k) \to -\infty$ and

$$\Omega^{-1}(\Omega(k) + x - \xi_2) \to \Omega^{-1}(-\infty) = 0, \quad (4.1.38)$$

i.e., $\Delta \to 0$. Therefore, if the function $f(x, y)$ satisfies the "Osgood condition" in all points of G, the solution of the equation $y' = f(x, y)$ is a continuous function of the initial values ξ, η and of the parameter ε. □

From this fact, we can deduce the following general dependence theorem:

Theorem 4.1.6 (The Bihari Inequality [82]) *Let $f(x, y)$ satisfy (4.1.19) in $G(x, y)$ and let $\varphi(x, 0, \xi_0, \eta_0)$ be the (unique) solution of $y' = f(x, y)$ with $\varphi(\xi_0, 0, \xi_0, \eta_0) = \eta_0$, existing in $a \leq x \leq b$, and $\varphi(x, \varepsilon, \xi, \eta)$ the unique solution of $y' = f(x, y) + \varepsilon$ with $\varphi(\xi, \varepsilon, \xi, \eta) = \eta$ and $a < \alpha < \beta < b$, then $\varphi(x, \varepsilon, \xi, \eta)$ exists in $\alpha < x < \beta$ for sufficiently small $\varepsilon > 0$ and $|\xi - \xi_0| + |\eta - \eta_0|$, further, $\varphi(x, \varepsilon, \xi, \eta)$ tends to $\varphi(x, 0, \xi_0, \eta_0)$ uniformly in $\alpha < x < \beta$ if $(\xi, \eta) \to (\xi_0, \eta_0)$ and $\varepsilon \to 0$.*

The same theorem may be proved if the right-hand side of the differential equation depends on an arbitrary parameter μ, i.e., we have the equation $y' = f(x, y, \mu)$ and we have $f(x, y, \mu) \to f(x, y, \mu_0)$ uniformly in G if $\mu \to \mu_0$. If, especially, we take $\omega(u) = Mu^\alpha$ where $M > 0, 0 < \alpha < 1$, then $\int_0^u \frac{dt}{\omega(t)}$ is convergent for all $u > 0$, therefore we cannot conclude that the Lipschitz condition with the exponent α

$$|f(x,y_2)-f(x,y_1)| \le M|y_2-y_1|^\alpha, \quad M>0, 0<\alpha<1 \tag{4.1.39}$$

implies uniqueness (moreover, we can find easily examples where this condition is satisfied and there are more solutions).

By the above results, we obtain as application the following bound estimate:

$$|\varphi(x)-\psi(y)| \le \left(k^{1-\alpha}+M(1-\alpha)|x-\xi_2|\right)^{\frac{1}{1-\alpha}}, \quad a \le x \le b, \tag{4.1.40}$$

where

$$k = |\xi_2-\xi_1|A + |\eta_2-\eta_1| + (\varepsilon_1+\varepsilon_2)(b-a). \tag{4.1.41}$$

Assuming identical initial conditions and $\varepsilon_1 = \varepsilon_2 = 0$, we have

$$|\varphi(x)-\psi(y)| \le (M(1-\alpha)|x-\xi_2|)^{\frac{1}{1-\alpha}}, \quad 0 \le \alpha \le 1, a \le x \le b, \tag{4.1.42}$$

e.g., the so-called maximal and minimal integrals have a difference subdue to this bound estimation. It is also easy to see that the limit of the right-hand side is 0 for $\alpha \to 1^-$.

Consider now the condition

$$|x-\xi|^\alpha |f(x,y_2)-f(x,y_1)| \le \omega(|y_2-y_1|), \quad 0 \le \alpha \le 1 \tag{4.1.43}$$

in place of

$$|f(x,y_2)-f(x,y_1)| \le \omega(|y_2-y_1|). \tag{4.1.44}$$

We assume that this condition is valid for a certain neighborhood of all points (ξ,η) of G. The function $\omega(u)$ is the same as in the Osgood theorem but we assume further that $\omega'(0)$ exists. In this case (one equation and identical initial conditions)

$$\lim_{t\to\xi+}\frac{\omega(\Delta)}{(t-\xi)^\alpha} = \lim_{t\to\xi+}\frac{D_+\omega(|\varphi(t)-\psi(t)|)}{\alpha(t-\xi)^{\alpha-1}} = \frac{1}{\alpha}\omega'(0)[\pm|\varphi'(\xi)-\psi'(\xi)|]\cdot 0 = 0 \tag{4.1.45}$$

(D_+ means the right-hand derivative), that is, the function $\frac{\omega(\Delta)}{(t-\xi)^\alpha}$ may be completed to a continuous function for $t = \xi$ too. By a similar way as above, we get (two equations and arbitrary initial conditions)

$$\begin{cases} \Delta \le k + \int_{\xi_2}^{x} \frac{1}{(t-\xi_2)^\alpha}dt, & \text{for sufficiently small } x-\xi_2 \ge 0, \\ k = |\xi_1-\xi_2|A + |\eta_1-\eta_2| + (\varepsilon_1+\varepsilon_2)(b-a). \end{cases} \tag{4.1.46}$$

The hypotheses of the generalized Bellman lemma are not satisfied here (e.g., $\frac{1}{(t-\xi)^{\alpha}}$ is not continuous in $\xi_2 \le t \le x$). However, we can easily obtain—by a little modification of the procedure (as in the case of the Nagumo Theorem)—the following inequality

$$\Delta \le \Omega^{-1}(\Omega(k) + \frac{(x-\xi_2)^{1-\alpha}}{1-\alpha}), \tag{4.1.47}$$

i.e., also condition (4.1.43) assures the uniqueness of the solution under identical initial conditions and $\varepsilon_1 = \varepsilon_2 = 0$. If these are not identical or $\varepsilon_1 \neq \varepsilon_2$, or else $\int_0^u \frac{dt}{\omega(t)}$ is convergent for all $u > 0$, then we obtain a bound for Δ in the neighborhood of ξ_2 in which (4.1.43) is valid. The uniqueness theorem discussed here is not a special case of Perron's theorem.

The next result is the uniqueness theorem of Perron [532].

Theorem 4.1.7 (The Perron Inequality [532]) *Let the function $f(x,y)$, defined in the domain $|x-\xi| < a$, $|y-\eta| < b$ (domain $G(x,y)$), satisfy the condition*

$$|f(x,y_2) - f(x,y_1)| \le \omega(|x-\xi|, |y_2-y_1|) \tag{4.1.48}$$

for all points (x,y_1) and (x,y_2) of G, and let $\omega(x,u)$ be a continuous function for all $0 \le x < a$, $u \ge 0$. If $\varphi(x)$ and $\psi(x)$ are integrals of the equation

$$y' = f(x,y) \tag{4.1.49}$$

with $\varphi(\xi) = \psi(\xi) = \eta$ and belonging to $G(x,y)$, then

$$|\varphi(x) - \psi(x)| \le Z(x), \tag{4.1.50}$$

where $Z(x)$ means the maximal integral of the equation

$$u' = \omega(x-\xi, u) \tag{4.1.51}$$

with $Z(x) = 0$. Therefore, if $\omega(x,0) = 0$ and $u \equiv 0$ is the unique solution of (4.1.51) with $u(\xi) = 0$, then $\varphi(x) \equiv \psi(x)$ in $G(x,y)$.

Bompiani [92] has found this theorem before Perron but has made use of the restriction that $\omega(x,u)$ is non-decreasing function of u. We shall make use of the same restriction, but the proof will be very simple.

Proof Since, for all $x \ge \xi$,

$$\varphi'(x) - \psi'(x) = f(x,\varphi(x)) - f(x,\psi(x)) \le \omega(|x-\xi|, |\varphi(x)-\psi(x)|), \tag{4.1.52}$$

we have, for all $x \geq \xi$,

$$|\varphi(x) - \psi(x)| \leq \int_{\xi}^{x} \omega(t - \xi, |\varphi(t) - \psi(t)|)dt, \tag{4.1.53}$$

or

$$\Delta \leq \int_{\xi}^{x} \omega(t - \xi, \Delta)dt \quad where \quad \Delta = \Delta(x) = |\varphi(x) - \psi(x)|. \tag{4.1.54}$$

Denoting the right-hand side of (4.1.54) by $V(x) = V$, we have $\Delta \leq V$. But

$$V'(x) = \omega(x - \xi, \Delta(x)) \tag{4.1.55}$$

and, on account of the monotonity of $\omega(x, u)$ in u, for all $x \geq \xi$,

$$V'(x) \leq \omega(x - \xi, V(x)), \tag{4.1.56}$$

i.e., $V(x)$ is a lower function of the equation

$$u' = \omega(x - \xi, u) + \varepsilon, \quad \varepsilon > 0 \tag{4.1.57}$$

for all $x \geq \xi$ and $V(\xi) = 0$, and therefore, for all $x \geq \xi$,

$$V(x) \leq z_{\varepsilon}(x), \tag{4.1.58}$$

where $z_{\varepsilon}(x)$ means the minimal integral of (4.1.57) for all $x \geq \xi$ with $z_{\varepsilon}(\xi) = 0$ and—as known—$z_{\varepsilon}(x)$ tends uniformly to $z(x)$ if $\varepsilon \to 0^+$. Consequently,

$$0 \leq \Delta(x) \leq V(x) \leq z(x). \tag{4.1.59}$$

Similar proof holds for all $x \leq \xi$. It is easy to see that the theorem of Osgood is a special case of this theorem. Hence, similarly, we have, for all $x \geq \xi_2$,

$$\Delta \leq k + \int_{\xi_2}^{x} \omega(t - \xi_2, \Delta)dt, \tag{4.1.60}$$

provided that (4.1.48) holds with $\xi = \xi_2$ and $\varphi(x)$ and $\psi(x)$ are in G. Then

$$0 \leq \Delta(x) \leq \overline{Z}(x) \quad \text{in} \quad \text{G}, \tag{4.1.61}$$

where $\overline{Z}(x)$ means the maximal integral of (4.1.51) with $\overline{Z}(\xi_2) = k$. These results may be extended to systems of differential equations too. □

We now finally consider a theorem due to Tamarkine [635] and corrected by Lavrentiev [331].

Theorem 4.1.8 (The Lavrentiev-Tamarkine Inequality [331, 635]) *Let $f(x,y)$ be continuous for $|x| < a, |y| < b$ (domain D) with $a > 0, b > 0$ satisfying the condition*

$$|f(x,y) - f(x,\varphi(x))| \geq \omega(|y - \varphi(x)|) \tag{4.1.62}$$

where $\varphi(x)$ is an integral curve (passing through the origin and defined for $|x| < a$) of the differential equation

$$y' = f(x,y) \tag{4.1.63}$$

and $\omega(u)$ is an increasing continuous function for all $u \geq 0$, furthermore, $\omega(0) = 0$ and the integral $\int_0^u \frac{dt}{\omega(t)}$ is convergent for all $u > 0$. Then the equation (4.1.63) has at least two (and thus an infinite number of) integral curves passing through the origin. Moreover, as Lavrentiev notes, the same is valid concerning all the points of the curve $\varphi(x)$ (in D).

Proof We give a simple proof of the above theorem, further we give a lower minimal integral $g(x)$ of (4.1.63) with the initial conditions $G(0) = g(0) = 0$.

On account of (4.1.62), the function $h(x,y) = f(x,y) - f(x,\varphi(x))$ can not vanish in the connected domain $|x| < a, y > \varphi(x)$ and therefore has a constant sign. We distinguish case (a) $h(x,y) > 0$ and case (b) $h(x,y) < 0$.

Let a fixed integral of the equation

$$y' = f(x,y) + \varepsilon, \quad \varepsilon > 0 \tag{4.1.64}$$

passing through the origin be denoted by $\psi_\varepsilon(x)$. As known, we can determine that for all $a_1 < a$, there exists a number $\varepsilon > 0$ such that $\psi_\varepsilon(x)$ exists for all $|x| < a$, and satisfies $|\psi_\varepsilon(x)| < b$, and we have

$$\psi_\varepsilon(x) > G(x) \geq \varphi(x), \quad 0 < x < a_1, \tag{4.1.65}$$

$$\psi_\varepsilon(x) < g(x) \leq \varphi(x), \quad -a_1 < x < 0. \tag{4.1.66}$$

We obtain immediately

$$\psi_\varepsilon(x) - \varphi(x) = \int_0^x [f(t,\psi_\varepsilon(t)) - f(t,\varphi(t))]dt + \varepsilon x. \tag{4.1.67}$$

Take now the case (a) and $0 < x < a_1$, then

$$\psi_\varepsilon(x) - \varphi(x) > \int_0^x [f(t,\psi_\varepsilon(t)) - f(t,\varphi(t)]dt \geq \int_0^x \omega(\psi_\varepsilon(t) - \varphi(t))dt, \; 0 < x < a_1 \tag{4.1.68}$$

whence, applying the Bihari Inequality (Theorem 1.1.1),

$$\psi_\varepsilon(x) - \varphi(x) > \Omega^{-1}(x), \ \ 0 < x < a_1, \quad where \ \ \Omega(u) = \int_0^u \frac{dt}{\omega(t)}, \tag{4.1.69}$$

consequently (since $\psi_\varepsilon(x) \Rightarrow G(x)$ for $\varepsilon \to 0^+$ and $0 \leq x < a_1$)

$$G(x) - \varphi(x) \geq \Omega^{-1}(x), \quad 0 \leq x < a_1 \tag{4.1.70}$$

for all $0 < a_1 < a$ and thus for $0 \leq x < a$. We have a fortiori

$$G(x) - g(x) \geq \Omega^{-1}(x), \quad 0 \leq x < a. \tag{4.1.71}$$

Considering the case (b) and again $0 < x < a_1$, equation (4.1.69) gives us for all $0 < x < a_1$,

$$0 < \psi_\varepsilon(x) - \varphi(x) < \varepsilon x < \varepsilon a_1 < \varepsilon a, \tag{4.1.72}$$

whence, for all $0 < x < a$,

$$G(x) - \varphi(x) \equiv 0. \tag{4.1.73}$$

Regard now both cases (a) and (b) for $-a < x < 0$. By the linear transformation $x = -\xi$, equation (4.1.63) reduces to

$$\frac{dy}{d\xi} = -f(-\xi, y(\xi)) = F(\xi, y), \tag{4.1.74}$$

and here the function $F(\xi, y)$ satisfies the condition (4.1.62) and $h(\xi, y) = F(\xi, y) - F(\xi, \varphi(-\xi))$ has an opposite sign as $h(x, y) = f(x, y) - f(x, \varphi(x))$. Therefore we have in case (a), for all $-a < x \leq 0$,

$$G(x) \equiv \varphi(x), \tag{4.1.75}$$

and in case (b), for all $-a < x \leq 0$,

$$G(x) - \varphi(x) \geq \Omega^{-1}(-x). \tag{4.1.76}$$

Instead of the domain $|x| < a, y > \varphi(x)$ (domain D_1), we could have considered the domain $|x| < a, y < \varphi(x)$ (domain D_2) without any change in the above reasoning.

We have four cases corresponding to the signs of $h(x, y)$ in D_1 and D_2, respectively:

(1) $G - \varphi \geq \Omega^{-1}, \quad 0 \leq x < a, \varphi - g \geq \Omega^{-1}(-x), \ -a < x \leq 0,$ $(++)$,
(2) $G - \varphi \geq \Omega^{-1}, \quad \varphi - g \equiv 0,$ $(+-)$,

(3) $G-\varphi \equiv 0, \qquad \varphi-g \geq \Omega^{-1}(-x), \qquad (-+),$
(4) $G-\varphi \equiv 0, \qquad \varphi-g \equiv 0, \qquad (--).$

e.g., in the fourth case and for $\varphi(x) \equiv k$, uniqueness $G(x) \equiv g(x)$ follows. □

4.1.3 Applications of Theorem 1.1.1 to General Differential Equations

In this section, we shall use Theorem 1.1.1 to establish upper and lower bounds on the norm of a solution of the equation

$$\frac{dz}{dx} = F(x, z). \tag{4.1.77}$$

Upper bounds obtained by application of Bellman's lemma (Theorem 1.1.2 in Qin [557]) and its generalization by Bihari [82] (i.e., Theorem 1.1.1) have been much used in the study of solutions of equations such as (4.1.77). Similar methods permit the determination of analogous lower bounds which seem to be unknown until now. Next we shall present a result which might indicate why this is the case.

Concerning (4.1.77), we make the following assumptions:

(1) x is a real variable, z and F are finite dimensional complex vectors with n components z_i and F_i, respectively,
(2) F is continuous in (x, z) for all z and all $x \in [a, b]$, i.e., $a \leq x \leq b$ with $a < b$;
(3) for some norm, say $|z| = \sum_{i=1}^{n} |z_i|$, F satisfies

$$|F(x, z)| \leq v(x)g(|z|) \tag{4.1.78}$$

where
(4) $v(x)$ is continuous and $v(x) \geq 0$ for all $x \in [a, b]$;
(5) $g(u)$ is continuous and non-decreasing for all $u \geq 0$ and $g(u) > 0$ for all $u > 0$.

Theorem 4.1.9 (The Langenhop Inequality [328]) *Let $z(x)$ be continuous, satisfy $|z(x)| > 0$ and be a solution of (4.1.77) for $x \in [a, b]$, where F satisfies conditions (1)–(5) above. Then for all $x \in [a, b]$,*

$$|z(x)| \leq G^{-1}\Big(G(|z(a)|) + \int_a^x v(s)ds\Big) \tag{4.1.79}$$

where

$$G(u) = \int_{u_0}^{u} (g(t))^{-1}dt, \quad u \geq u_0 \geq 0,$$

and, for all $x \in [a, b]$,

$$|z(x)| \geq G^{-1}\left(G(|z(a)|) - \int_a^x v(s)ds\right) \tag{4.1.80}$$

where $G(|z(a)|) - \int_a^x v(s)ds$ *is in the domain of* G^{-1}.

Proof We have $\frac{dz}{dx} = F(x, z)$ so that, for all $x, y \in [a, b]$,

$$z(x) = z(y) + \int_y^x F(s, z(s))ds. \tag{4.1.81}$$

Let $u(x) = |z(x)|$. Then from (4.1.81) and (4.1.78), we have for $y \geq x$,

$$u(x) \leq u(y) + \int_y^x v(s)g(u(s))ds \tag{4.1.82}$$

and

$$u(x) \geq u(y) - \int_y^x v(s)g(u(s))ds. \tag{4.1.83}$$

We shall prove that (4.1.80) follows from (4.1.83) and merely indicate how (4.1.79) follows from (4.1.82) since this latter case is essentially Bihari's result (i.e., Theorem 1.1.1).

For fixed x in the interval $a < x \leq b$, we define for $a \leq y \leq x$,

$$\phi(y) = u(x) + \int_y^x v(s)g(u(s))ds. \tag{4.1.84}$$

Then $\phi'(y) = -v(y)g(u(y))$, so from (4.1.83) and conditions (4) and (5), it follows that

$$\phi'(y) + v(y)g(u(y)) \geq 0. \tag{4.1.85}$$

Since $\phi(y) > 0$, condition (5) and the definition of G along with (4.1.85) imply that

$$\frac{d}{dy}G(\phi(y)) + v(y) \geq 0, \tag{4.1.86}$$

which leads to

$$G(\phi(x)) - G(\phi(y)) + \int_y^x v(s)ds \geq 0, \quad a \leq y \leq x \leq b. \tag{4.1.87}$$

By (4.1.84), we have $\phi(x) = u(x)$ so (4.1.87), (4.1.83) and the monotonicity of G imply that

$$G(u(x)) = G(u(y)) - \int_y^x v(s)ds. \tag{4.1.88}$$

Setting $y = a$ and using the monotonicity of G^{-1}, we have (4.1.80) for those $x \in [a, b]$ for which $G(u(a)) - \int_a^x v(s)ds$ is in the domain of G^{-1}.

The proof of (4.1.79) may be carried out similarly except we may immediately set $y = a$ in (4.1.82) and define $\psi(x) = u(a) + \int_a^x v(s)g(u(s))ds$. Then $\psi'(x) = v(x)g(u(x))$ and the remaining details are analogous to the above. □

Inequality (4.1.79) and particularly its specialization in the case $g(u) \equiv u$ (Bellman's lemma, Theorem 1.1.2 in Qin [557].) have been used extensively in the theory of differential equations. We shall show this for the special case $g(u) \equiv u$. This may be stated as follows. Let $u(x)$ and $v(x)$ be continuous on the interval $[a, b]$ and let $u(x) \geq 0$ and $v(x) \geq 0$ on this interval, but $v(x) \not\equiv 0$. Then the condition

$$u(x) \geq u(a) - \int_a^x v(s)u(s)ds \tag{4.1.89}$$

does not imply

$$u(x) \geq u(a) \exp\left(-\int_a^x v(s)ds\right). \tag{4.1.90}$$

We shall prove an equivalent result.

Theorem 4.1.10 (The Langenhop Inequality [328]) *Let $v(x)$ be continuous and such that $v(x) \geq 0$, but $v(x) \not\equiv 0$ on $[a, b]$ where $a < b$. Then there exist continuous functions $u(x)$, $w(x)$ which are non-negative on $[a, b]$ and an $x_0 \in [a, b]$ such that*

$$u(x) = u(a) - \int_a^x v(s)u(s)ds + w(x) \tag{4.1.91}$$

on $[a, b]$, but

$$u(x_0) < u(a)E(x_0, a) \tag{4.1.92}$$

where

$$E(x, y) = \exp\left(\int_x^y v(s)ds\right).$$

Remark 4.1.2 If $u(x)$ is not non-negative, the result is much more easily established, but the inequality (4.1.89) arises naturally in differential equation theory and in this case $u(x) \geq 0$.

Proof From the hypotheses regarding v, it follows that there is an interval $[a_1, b_1]$ such that $a_1 < b_1$ and for all $a_1 \le x \le b_1$,

$$v(x) > 0. \tag{4.1.93}$$

Consider the function h defined by

$$h(x) = \int_{a_1}^{x} E(s, b_1)ds - \int_{x}^{b_1} E(s, b_1)ds. \tag{4.1.94}$$

Clearly $h(x)$ is continuous and $h(a_1) < 0$. On the other hand, it follows from (4.1.93) that $h((a_1 + b_1)/2) > 0$. Hence there exists an x_1 such that

$$h(x_1) = 0, \;\; a_1 < x_1 < (a_1 + b_1)/2. \tag{4.1.95}$$

Having chosen x_1 as in (4.1.95), we select $u(a)$ such that

$$u(a) > (a_1 + b_1 - 2x_1)E(a, b_1). \tag{4.1.96}$$

Obviously, $u(a) > 0$. Now we define a function r by the equations

$$r(x) = \begin{cases} 0, & a \le x \le a_1 \\ x - a_1, & a_1 \le x \le x_1 \\ 2x_1 - a_1 - x, & x_1 \le x \le b_1 \\ (2x_1 - a_1 - b_1)E(x, b_1), & b_1 \le x \le b. \end{cases} \tag{4.1.97}$$

The function $w(x)$ is now defined to be the solution of

$$w(x) = r(x) + \int_{a}^{x} w(s)v(s)E(x, s)ds \tag{4.1.98}$$

and in turn $u(x)$ is defined to be the corresponding solution of (4.1.91) with $u(a)$ chosen to satisfy (4.1.96).

It may be verified directly that $u(x)$ given by

$$u(x) = u(a)E(x, a) + w(x) - \int_{a}^{x} w(s)v(s)E(x, s)ds \tag{4.1.99}$$

satisfies (4.1.91), or conversely this solution may be obtained by solving the equation $(d/dx)[u(x) - w(x)] = -v(x)u(x)$ which arises from (4.1.91). From (4.1.98) and (4.1.99), it then follows

$$u(x) = u(a)E(x, a) + r(x) \tag{4.1.100}$$

and since $r(b_1) < 0$, the inequality (4.1.92) follows for $x_0 = b_1$.

It remains to show then that $w(x) \geq 0$ and $u(x) \geq 0$. The explicit form of the solution $w(x)$ of (4.1.98) is

$$w(x) = r(x) + \int_a^x r(s)v(s)E(s,x)ds \tag{4.1.101}$$

which may be obtained in the same way as was indicated for (4.1.99). Since $r(x)$ is continuous, we may rewrite (4.1.101), after integrating by parts, in the form of a Stieltjes integral

$$w(x) = \int_a^x E(s,x)dr(s). \tag{4.1.102}$$

Consider now the function $w(x)E(x,b_1) = \int_a^x E(s,b_1)dr(s)$. It is clear from the form of $r(x)$ as given in (4.1.97) that this function is non-decreasing on the interval $[a,x_1]$, non-increasing on the interval $[x_1,b_1]$ and again non-decreasing on the interval $[b_1,b]$. Hence the minimum value of this function is attained either $x = a$ or $x = b_1$. But its value at $x = a$ is clearly zero while its value at $x = b_1$ is also zero by choice of x_1(see, e.g., (4.1.94) and (4.1.95)). Hence this function is non-negative on $[a,b]$ and so also then is $w(x)$.

Turning now to $u(x)$, we observe from (4.1.100) and (4.1.97) that certainly $u(x) \geq 0$ on the interval $[a,x_1]$. On the interval $[x_1,b_1]$, $r(x)$ is decreasing as is the term $u(a)E(x,a)$ so the minimum value of $u(x)$ on this interval is attained at $x = b_1$. Thus we need to verify $u(x) \geq 0$ only on the interval $[b_1,b]$. Here we have

$$u(x) = [u(a) - (a_1 + b_1 - 2x_1)E(a,b_1)]E(x,a) \tag{4.1.103}$$

which is clearly positive by (4.1.99). Hence $u(x) \geq 0$ for all $x \in [a,b]$ and the theorem is proved. □

4.2 Applications of Theorems 1.1.1 and 1.1.5 to Nonlinear Integro-Differential Equations

In this section, we shall apply the results in Theorems 1.1.1 and 1.1.5 to study asymptotic behavior of solutions of the nonlinear integro-differential equations

$$x'(t) - x'(0) + \int_0^t a(s)g(x(s))ds = h(t), \quad t \in [0,+\infty). \tag{4.2.1}$$

In particular, we shall establish bounds for solutions of (4.2.1) in terms of the forcing term $h(t)$. In case $a \in C\big([0,+\infty)\big)$, $h \in C\big([0,+\infty)\big)$, and $g \in C\big((-\infty,+\infty)\big)$, equation (4.2.1) is equivalent to the integral equation

$$x(t) = x(0) + tx'(0) + H(t) - \int_0^t (t-s)a(s)g(x(s))ds \tag{4.2.2}$$

and the differential equation

$$u''(t) + a(t)g(u(t)) + H(t) = 0, \tag{4.2.3}$$

where for all $t \in [0, +\infty)$,

$$u = x - H, \quad \textit{and} \quad H(t) = \int_0^t h(s)ds.$$

If h is in addition absolutely continuous on bounded intervals, then (4.2.1) is equivalent to the differential equation

$$x''(t) = a(t)g(x(t)) = e(t), \tag{4.2.4}$$

where $e(t) = h'(t)$.

Equation (4.2.1) and its equivalent forms (4.2.2) and (4.2.3) are treated by Hastings [275], who discussed the asymptotic behavior of $x(t)$ as $t \to +\infty$. More general equations than (4.2.1) have been widely discussed, see for example Nohel [424], in particular delay integro-differential equations, see for example Levin and Nohel [337]. Equation (4.2.4) has been discussed in a slightly more general form by Liang [349]. In case $a(t) = const.$, bounds for solutions of (4.2.4) were obtained by Putzer [554] who also showed that these bounds are in a certain sense best possible. When $e(t) \equiv 0$, the literature on equation. (4.2.4) is voluminous, for boundedness results, see, e.g., Waltman [659], Wong [717], and the references cited there. Equation (4.2.1) arises from the study of nonlinear oscillator in acoustics; see Potter [552] for an account of probabilistic and practical aspects of (4.2.1). Bounds and asymptotic bounds for more general second-order systems than (4.2.1) were given by Hastings [273].

We assume throughout that a, h and g are continuous with respect to their domains of definition. Whenever applicable, integrals $\int_a^b f(s)ds$ will be abbreviated as $\int_a^b f$.

Theorem 4.2.1 (The Muldowney-Wong Inequality [402]) *If f is a positive non-decreasing continuous function on $(0, +\infty)$ such that $|g(x)| \leq f(|x|)$, $x \in \mathbb{R}$, $x \neq 0$, then any solution $x(t)$ of equation (4.2.1) may be continued throughout $[0, \tau_2)$ and satisfies for all $t \in [0, \tau_2)$,*

$$|x(t)| \leq \Phi^{-1}\left(\Phi(|x(0)| + \int_0^t |dx_0|) + \int_0^t (t-s)|a(s)|ds\right), \tag{4.2.5}$$

where Φ and Φ^{-1} are as defined in Theorem 1.1.1,

$$x_0(t) = x(0) + tx'(0) + H(t), \quad \int_0^t |dx_0| = \int_0^t |x'(0) + h(s)|ds$$

is the total variation on $[0, t]$ *of* x_0*, and* $[0, \tau_2)$ *is the largest interval on which the right-hand side of (4.2.5) is defined.*

Proof Equation (4.2.2) implies for all $t \in [0, T]$,

$$\begin{aligned} |x(t)| &\leq |x(0)| + \int_0^t |dx_0| + \int_0^t (t-s)|a(s)|f(|x(s)|)ds \\ &\leq |x(0)| + \int_0^T |dx_0| + \int_0^t (T-s)|a(s)|f(|x|)ds \end{aligned}$$

where $[0, T]$ is any interval on which $x(t)$ exists. Hence by Theorem 1.1.1, for all $t \in [0, T)$ provided that $0 \leq T \leq \tau_2$,

$$|x(t)| \leq \Phi^{-1}(\Phi(|x(0)| + \int_0^T |dx_0|) + \int_0^t (T-s)|a(s)|f(|x(s)|)ds). \tag{4.2.6}$$

If $T \in [0, \tau_2)$, then (4.2.2) and (4.2.6) imply that $\lim_{t \to T} = x(T)$ exists, (4.2.5) holds at $t = T$ and x may be continued to the right of T. Since T is arbitrary in $[0, \tau_2)$, we conclude that $x(t)$ may be continued throughout $[0, \tau_2)$ and it also satisfies (4.2.5) for all $t \geq 0$. □

Corollary 4.2.1 *If, in Theorem 4.2.1,* $f(|x|) = |x|^{1-\alpha}$, $0 < \alpha \leq 1$, *then any solution* $x(t)$ *of (4.2.1) satisfies*

$$|x(t)| \leq \left[(|x(0)| + \int_0^t |dx_0|)^\alpha + \alpha \int_0^t (t-s)|a(s)|ds \right]^{1/\alpha} \tag{4.2.7}$$

for all $t \geq 0$.

Corollary 4.2.2 *If in Theorem 4.2.1,* $f(|x|) = |x|^p, p \geq 1$, *then any solution of (4.2.1) satisfies (4.2.7) for* $p > 1$ *provided the right-hand side of (4.2.7) is defined and for* $p = 1$, *we have for all* $t \geq 0$,

$$|x(t)| \leq (|x(0)| + \int_0^t |dx_0| \exp(\int_0^t (t-s)|a(s)|ds). \tag{4.2.8}$$

Combining Corollaries 4.2.1 and 4.2.2, we can sharpen an asymptotic bound given in [275] (Theorem 3) as follows: We remark that the original statement in [275] is in error and is corrected in [274]. A similar but erroneous result is given in Zhang [717], corrected and extended in Theorem 3 by [678].

Theorem 4.2.2 (The Muldowney-Wong Inequality [402]) *If* $|g(x)| \leq |x|^{1-\alpha}$ *for* $0 \leq \alpha < 1$, *then every solution of (4.2.1) satisfies, as* $t \to +\infty$,

$$x(t) = o(t + t \int_0^t |h| + t(\int_0^t s^{1-\alpha}|a(s)|ds)^{1/\alpha}). \tag{4.2.9}$$

Proof Let $X(t) = [1/(t+1)]|x(t)|$. We may easily estimate (4.2.2) as follows

$$X(t) \le |x(0)| + |\dot{x}(0)| + \int_0^t |h(s)|ds + \int_0^t (s+a)^{1+\alpha}|a(s)|(X(s))^{1-\alpha}ds.$$

Applying Corollary 4.2.1, we easily obtains (4.2.9). □

In case $\alpha = 1$, we need only to remark that the bound given in (4.2.8) of Corollary 4.2.2 implies the asymptotic bound of [274] (Theorem III). When $g(x) = x$, results of this type can in fact be traced back to Hille [281] (Theorem 3) where essentially similar arguments were used. The result here also includes a result of Waltman [660] as a special case (see, [676]).

We note that the bounds given in (4.2.5), (4.2.7), (4.2.8), (4.2.9) involves the non-decreasing function

$$|x(0)| + \int_0^t |dx_0|$$

as an upper bound for $|x_0(t)|$. This is necessary in order to use Bihari's inequality in Theorem 1.1.1, where η is a constant. In case we wish to establish sharper bounds involving $|H(t)|$ instead of $\int_0^t |h(s)|ds$, this difficulty may be overcome by using the following extension of Bihari's inequality.

Theorem 4.2.3 (The Muldowney-Wong Inequality [402]) *If* $|g(x)| \le f(|x|)$, *and* f *is given and in Theorem 1.1.5; then any solution of (4.2.1) may be continued throughout* $[0, \tau_4)$ *and satisfies*

$$|x(t)| \le |x_0(t)| + \Phi^{-1}\left[\Phi\left(\int_0^t (t-s)|a(s)|f(|x_0(s)|)ds\right) + \int_0^t (t-s)|a(s)|ds\right], \tag{4.2.10}$$

where Φ *and* Φ^{-1} *are as in Theorem 1.1.1 and* $[0, \tau_4)$ *is the largest interval on which the right-hand side of (4.2.10) exists.*

Proof Using Theorem 1.1.5, we can prove, similar to Theorems 4.2.1 and 4.2.2, the following sharper bounds involving $|H(t)|$ instead of $\int_0^t |h(s)|ds$. □

Corollary 4.2.3 *If* $|g(x)| \le |x|^{1-\alpha}$, $0 < \alpha \le 1$, *then any solution of (4.2.2) satisfies*

$$|x(t)| \le |x_0(t)| + \left\{\left(\int_0^t (t-s)|a(s)||x_0(s)|^{1-\alpha}ds\right)^{\alpha} + \alpha\int_0^t (t-s)|a(s)|ds\right\}^{1/\alpha} \tag{4.2.11}$$

for all $a \ge 0$.

We note that a similar result as that of (4.2.8) in Corollary 4.2.2 in the linear case when $\alpha = 1$ can be formulated. We omit the details.

Theorem 4.2.4 (The Muldowney-Wong Inequality [402]) *If* $|g(x)| \le |x|^{1-\alpha}$, $0 < \alpha \le 1$, *then every solution of (4.2.1) satisfies as* $t \to +\infty$,

$$x(t) = o\left(t + |H(t)| + t\int_0^t (|H|^{1-\alpha}) + t(\int_0^t s^{1-\alpha}|a(s)|ds)^{1/\alpha}\right). \quad (4.2.12)$$

We note that (4.2.12) is in some instances stronger than (4.2.6), in particular, when

$$\int_0^{+\infty} |a(s)|ds < +\infty.$$

The bound established in the above theorem seems to be what Theorem III of [275] is originally intended for (cf. [274]).

Next, we shall establish bounds for solutions of (4.2.1) under the effect of both $|a|$ and $|g|$ where no sign restrictions are imposed. Hence we shall establish bounds of solutions of (4.2.1) by making sign and other restrictions on a and g. In this case, we may introduce suitable energy functions involving solution $x(t)$ and use it in establishing bounds and asymptotic bounds for $x(t)$. No proofs will be given for assertions about the continuation of solutions of (4.2.1) since they would essentially be the same as the proof of a similar assertion in Theorem 4.2.1.

It is shown in the main theorem of [275] that if $g(x)$ is a non-decreasing odd function which is positive for all $x > 0$ and if $a(t)$ is absolutely continuous on bounded intervals with

$$\int_0^{+\infty} \frac{|a'|}{a} < +\infty,$$

then any solution $x(t)$ of equation (4.2.1) can be continued throughout $[0, +\infty)$ and

$$x(t) = o\left(1 + \int_0^t |h(s)|ds\right)$$

as $t \to +\infty$.

In the following theorem, we shall relax the condition on $a(t)$.

Theorem 4.2.5 (The Muldowney-Wong Inequality [402]) *Let the following conditions hold:*

(i) $a(t)$ *is positive and absolutely continuous on bounded intervals, and*
(ii) $g(x)$ *is an odd non-decreasing continuous function on* $\mathbb{R}$ *such that*

$$xg(x) > 0 \quad whenever \quad x \neq 0. \quad (4.2.13)$$

Then any solution $x(t)$ of equation (4.2.1) may be continued throughout $[0, +\infty)$ and there exists a non-negative constant γ, depending on $x'(0)$, such that

$$|x(t)| \leq \left(\gamma + |x(0)| + 3\int_0^t |h(s)|ds\right) \exp\left(\int_0^t \frac{a'_-}{a} ds\right). \qquad (4.2.14)$$

Proof Let $[\alpha, \beta)$ be any sub-interval of $[0, +\infty)$ which contains no zeros of $x'(t) - h(t)$. On this interval, (4.2.1) is equivalent to the differential equation

$$u''(t) + a(t)g(u(t) + \int_\alpha^t h) = 0, \qquad (4.2.15)$$

where $u(t) = x(t) - \int_\alpha^t h(\tau)d\tau$ for any fixed $T \in [\alpha, \beta)$, and all $t \in [\alpha, T]$, we also have

$$u'(t)\left\{u''(t) + a(t)g(u(t) - (sgn\ u'(t))\int_\alpha^T |h(\tau)|d\tau)\right\} \leq 0. \qquad (4.2.16)$$

On the interval $[\alpha, T]$, define the following Lyapunov-like function for all $t \in [\alpha, T]$,

$$V(t, T) = \frac{(u''(t))^2}{2a(t)} + G(u(t) - (sgn\ u'(t))\int_0^T |h(s)|ds) \qquad (4.2.17)$$

where $G(x) = \int_0^{|x|} g(s)ds$, then it easily follows from (4.2.16) and (4.2.17) that

$$\begin{aligned} V'(t, T) &= u'(t)\left(\frac{u''(t)}{a(t)} + g\left(u(t) - (sgn\ u'(t))\int_0^T |h(\tau)|d\tau\right)\right) - \frac{a'(t)(u'(t))^2}{a(t)\,2a(t)} \\ &\leq \frac{a'_-(t)}{a(t)} V(t, T), \end{aligned}$$

hence

$$V(t, T) \leq V(\alpha, T) \exp\left(\int_\alpha^t \frac{a'_-}{a} ds\right). \qquad (4.2.18)$$

In particular, when $t = T$ in (4.2.18), we have

$$V(t, t) \leq V(\alpha, t) \exp\left(\int_\alpha^t \frac{a'_-}{a} ds\right). \qquad (4.2.19)$$

Consequently,

$$G\left(|u(t)| + \int_0^t |h(\tau)|d\tau\right) \le \left[\frac{(u'(\alpha))^2}{2a(\alpha)} + G\left(|u(\alpha)| + \int_\alpha^t |h(\tau)|d\tau\right)\right] \exp\left(\int_\alpha^t \frac{a'_-}{a}ds\right)$$
$$\le G(\gamma + |u(\alpha)| + \int_\alpha^t |h(\tau)|d\tau) \exp\left(\int_\alpha^t \frac{a'_-}{a}ds\right),$$

where $G(\gamma) = [(u'(\alpha))^2/2a(\alpha)]$, (notice that $\gamma = 0$ if $x'(\alpha) = h(\alpha)$). Note that $G(x_1) \le \lambda G(x_2)$, $\lambda \ge 1$, implies that $|x_1| \le \lambda |x_2|$. Hence,

$$\left||u(t)| - \int_\alpha^t |h(\tau)|d\tau\right| \le (\gamma + |u(\alpha) + \int_\alpha^t |h(\tau)|d\tau) \exp\left(\int_\alpha^t \frac{a'_-(\tau)}{a(\tau)}d\tau\right),$$

which implies

$$|x(t)| \le \left[\gamma(x'(0)) + |x(\alpha)| + 3\int_0^t |h(\tau)|d\tau\right] \exp\left(\int_\alpha^t \frac{a'_-(\tau)}{a(\tau)}d\tau\right). \quad (4.2.20)$$

Suppose that $x' - h$ has finite number of zeros in any compact interval. Let $t_0 = 0$, and $\{t_n\}$ be the successive zeros on $[0, +\infty)$ of $x' - h$. Then by (4.2.20), there holds for all $t \in [t_0, t_1]$,

$$|x(t)| \le [\gamma(x'(0)) + |x(0)| + 3\int_0^t |h(\tau)|d\tau] \exp\left(\int_\alpha^t \frac{a'_-(\tau)}{a(\tau)}d\tau\right). \quad (4.2.21)$$

We can show that (4.2.14) holds by induction. Suppose that (4.2.21) holds if $t \in [t_{n-1}, t_n]$, then for all $t \in [t_n, t_{n+1}]$,

$$|x(t)| \le [|x(t_n)| + 3\int_{t_n}^t |h(\tau)|d\tau] \exp\left(\int_\alpha^t \frac{a'_-(\tau)}{a(\tau)}d\tau\right)$$
$$\le [\gamma(x'(0)) + |x(0)| + 3\int_0^{t_n} |h(\tau)|d\tau] \exp\left(\int_\alpha^t \frac{a'_-(\tau)}{a(\tau)}d\tau\right)$$
$$+[3\int_{t_n}^t |h(\tau)|d\tau] \exp\left(\int_\alpha^t \frac{a'_-(\tau)}{a(\tau)}d\tau\right)$$
$$\le [\gamma(x'(0)) + |x(0)| + 3\int_0^t |h(\tau)|d\tau] \exp\left(\int_\alpha^t \frac{a'_-(\tau)}{a(\tau)}d\tau\right).$$

Hence (4.2.21) holds for all $t \ge 0$. On the other hand, suppose that $\tau \in [0, +\infty)$ is a finite limit of zeros of $x' - h$. Note that when $\dot{u}(t_n) = 0$,

$$\int_0^{t_n} ag(x) = x'(0);$$

and also that between any two zeros of $u'(t)$, there is a zero of $a(t)g(x(t))$ and hence a zero of $x(t)$ on account of (4.2.13). In particular $x(\tau) = 0$, (cf. [659]), and so the induction argument may be continued to the right of τ as before. This completes the proof. □

Corollary 4.2.4 *Under the assumptions of Theorem 4.2.5 , every solution of (4.2.1) satisfies:*

$$x(t) = o(1 + \int_0^t |h(\tau)|d\tau) \exp\left(\int_\alpha^t \frac{a'_-(\tau)}{a(\tau)} d\tau\right), \quad as \quad t \to +\infty. \tag{4.2.22}$$

The bound given by (4.2.22) in Corollary 4.2.4 extends Hastings' main result ([275], Theorem 1; cf. also [274]).

Theorem 4.2.6 (The Muldowney-Wong Inequality [402]) *Let the following conditions hold:*

(i) $a(t)$ *is positive and absolutely continuous on bounded interval, and*
(ii) $|g(x)| \leq f(|x|)$, *if* $x \neq 0$, *where* f *is a positive non-decreasing continuous function on* $(0, +\infty)$,
(iii) $-\infty < y_0 = \inf\{G(x) : x \in (-\infty, +\infty)\}$, *where* $G(x) = \int_0^x g(s)ds$, *and*
(iv) $f(K(y)) \leq ky^{1-\alpha}$ *for all* $y \geq y_1 \geq 0$ *where* $K(y) = \sup\{|x| : y \geq G(x)\}$ *and* k, α *are constants with* $k \geq 0$, $\alpha \leq 1$.

Then every solution $x(t)$ *of equation (4.2.1) may be continued throughout* $[0, \tau_6)$ *and satisfies, for all* $t \in [0, \tau)$,

$$|x(t)| \leq K[E(t)W_\alpha(t)], \tag{4.2.23}$$

and for all $t \in [0, \tau_6)$,

$$|x'(t)| \leq |h(t)| + (2a(t)E(t)W_\alpha(t))^{1/2}, \tag{4.2.24}$$

where

$$\begin{cases} E(t) = \exp\left(\int_\alpha^t \frac{a'_-(\tau)}{a(\tau)} d\tau\right), \quad W_0(t) = \eta \exp(\int_0^t |h(\tau)|d\tau), \\ W_\alpha(t) = (\eta^\alpha + \alpha k \int_0^t (|h|E^{-\alpha}))^{1/\alpha}, \quad \alpha \neq 0, \\ \eta = \frac{1}{2a(0)}(x'(0) - h(0))^2 + G(x(0)) + y_1 - y_0, \end{cases}$$

and $[0, \tau_6)$ *is the largest interval on which the right-hand side of (4.2.23) and (4.2.24) are defined (e.g., if* $0 \leq \alpha \leq 1$, *then* $\tau_6 = +\infty$*).*

Proof Consider $u(t) = x(t) - H(t)$ and define for all $t \geq 0$,

$$V(t) = \frac{(u'(t))^2}{2a(t)} + G(u(t) + H(t) + y_1 - y_0. \qquad (4.2.25)$$

Note that $V(t) \geq y_1 \geq 0$. Differentiating (4.2.25), we have

$$V' = u'(\frac{u''}{a} + g(u + H)) - \frac{1}{2}a'(\frac{u'}{a})^2 + hg(u + H).$$

Since $\ddot{u} + ag(u + H) = 0$ from (4.2.3), and $-\frac{1}{2}a'(u'/a)^2 \leq (a'_-/a)V$ from (4.2.25); also $|u + H| \leq K(V)$ from (4.2.25), (iii) and (iv), hence

$$|hg(u + H)| \leq |h|f(|u + H|)|h|f(K(V)) \leq |h|kV^{1-\alpha}$$

so that

$$V' \leq \dot{a}_-/aV + k|h|V^{1-\alpha}.$$

Therefore by Corollary 1.1.1, $V(t) \leq E(t)W_\alpha(t)$, for all $t \in [0, \tau_6)$, and (4.2.23), (4.2.24) follow since $|x| = |u + H| \leq K(EW_\alpha)$ (K non-decreasing), and

$$|x' - h| = |u'| \leq (2aV)^{1/2} \leq (2aEW_\alpha)^{1/2}.$$

□

Corollary 4.2.5 *Under the assumptions of Theorem 4.2.6, every solution of (4.2.1) satisfies, for all $t \in [0, \tau_6)$,*

$$|x(t)| \leq |x(0)| + \int_0^t [|h(\tau)| + (2aE(\tau)W_\alpha(\tau))^{1/2}]d\tau. \qquad (4.2.26)$$

This follows by a simple integration of (4.2.24).

Corollary 4.2.6 *If $a(t)$ satisfies condition (i) of Theorem 4.2.6 and there exist constants λ, b and c such that, for all x,*

$$0 < b \leq |x|^{-\lambda}xg(x) \leq c,$$

where $\lambda > 1$; then for all $t \geq 0$,

$$\begin{cases} |x(t)| \leq \left[(\frac{\lambda}{b}\eta)^{1/\lambda} + \frac{c}{b}\int_0^t(|h(\tau)|E^{-1/\lambda}(\tau))d\tau\right](E(t))^{1/\lambda}, \\ |x'(t)| \leq |h(t)| + \left[\eta^{1/\lambda} + \frac{c}{b}(\frac{b}{\lambda})^{1/\lambda}\int_0^t(|h|E^{-1/\lambda})\right]^{\lambda/2}(2a(t)E(t))^{1/2}, \end{cases} \qquad (4.2.27)$$

where

$$\eta = \frac{1}{2a(0)}(x'(0) - h(0))^2 + \frac{c}{\lambda}|x(0)|^\lambda, \quad E(t) = \exp\left(\int_\alpha^t \frac{a'_-(\tau)}{a(\tau)}d\tau\right).$$

Bounds given in (4.2.27) follows from Theorem 4.2.6 by taking $f(|x|) = c|x|^{\lambda-1}$ and $G(x) \geq (b/\lambda)|x|^\lambda$ in (4.2.23) and (4.2.24). In this case, $K(y) \leq [(\lambda/b)y]^{1/\lambda}$ for all $y \geq 0$ and $f(K(y)) \leq c[(\lambda/b)y]^{1-(1/\lambda)}$; hence $k = c(\lambda/b)^{1-(1/\lambda)}$, $\alpha = (1/\lambda)$ and $\tau_6 = +\infty$ (since $0 < \alpha < 1$).

Corollary 4.2.6 is in practice considerably stronger than Corollary 4.2.4. In fact, the condition of Corollary 4.2.6 need only to hold for $|x| \geq x_0 > 0$, whereas in Theorem 4.2.6 and its corollaries it is crucial that $g(x) = 0$ if and only if $x = 0$. Nevertheless, Theorem 4.2.6 is not itself a stronger result than Theorem 4.2.5 in general since there exist functions g satisfying (4.2.2) which do not satisfy (iv) of Theorem 4.2.6 for any α or are such that $\alpha < 0$ so that Theorem 4.2.6 fails to hold throughout $[0, +\infty)$. One such function $g(x)$ may be defined as follows: for $k = 0, 1, 2, \ldots$ denote

$$d_k = 10^{-10^{10^{k+1}}}$$

and define $g(0) = 0$ and by induction

$$g(k) = 10^{10^{10^k}}, \quad g(k + d_k)10^{10^{10^{k+1}}},$$

$g(x) = g(k)$, $k - 1 + d_{k-1} \leq x \leq k$, and linear between k and $k + d_k$. It is not difficult to see that

$$G(k + d_k) = O(k10^{10^{10^k}})$$

and $g(x) \neq O(G(x))^\alpha$ for any $\alpha \geq 0$. However if $g(x)$ is defined by $g(-x) = -g(x)$ for $x < 0$, it satisfies condition (4.2.13).

Corollary 4.2.7 *If $g(x)$ is an odd non-decreasing function which is positive when x is positive and for all $x \geq x_0 > 0$,*

$$g(x) \leq k[G(x)]^{1-\alpha}, \tag{4.2.28}$$

where $\alpha \leq 1$, and $a(t)$ satisfies condition (i) of Theorem 4.2.6, then conclusions (4.2.23), (4.2.24) hold for every solution of equation (4.2.1) with $K = G^{-1}$, where G^{-1} is the inverse function of $G = G(x)$, $x \geq 0$.

In this case, (4.2.28) and (iv) of Theorem 4.2.6 are equivalent. A sufficient condition for an odd non-decreasing function $g(x)$ which is positive when x is positive to satisfy (4.2.28) is that $g(x) \leq x[g(vx)]^{1-\alpha}$, $0 < v < 1$ for all $x \geq x_0 > 0$. In this case, $k = c(x_0(1 - v))^{1-\alpha}$. For example, if $g(x)$ is sub-additive or concave, this latter condition holds with $c = 2, v = \frac{1}{2}, \alpha = 0$. Examples of functions satisfying (4.2.28) and fail to satisfy (4.2.13) are:

$$g(x) = \begin{cases} e^{x-1} - 1, & x \geq 1, \\ 0, & 0 \leq x < 1, \end{cases}$$

and

$$g(x) = \begin{cases} \log x, & x \geq 1, \\ 0, & 0 \leq x < 1, \end{cases}$$

with $g(-x) = -g(x)$ for all $x \geq 0$. Finally we close our discussion with a proof of Theorem 4.2.6.

4.3 Application of Corollaries 1.1.2 and 1.1.3 to Asymptotic Behavior Solutions to the Second Order Differential Equation

In this section, we shall study asymptotic behavior of solutions of the equation

$$u'' + f(t, u, u') = 0 \tag{4.3.1}$$

when u' is absent has been discussed by Cohen [157], Tong [645] and Trench [648].

In this section, we shall prove the following theorem.

Theorem 4.3.1 ([181]) *Assume the following hypotheses hold,*

(i) The function $f(t, u, v)$ is continuous on $D = \{(t, u, v) : t \geq 1, u, v \in \mathbb{R}\}$.
(ii) $|f(t, u, u')| \leq \phi(t)g(|u|/t) + \psi(t)|u'|$ for all $(t, u, u') \in D$, where $\phi(t)$ and $\psi(t)$ are non-negative continuous functions on $[1, +\infty)$.
(iii) $g(u)$ is a non-negative, continuous, non-decreasing function on $[0, +\infty)$, and satisfies for all $\alpha \geq 1, u \geq 0$,

$$g(\alpha u) \leq \phi_1(\alpha)g(u)$$

where $\phi_1(\alpha) > 0$ is continuous for all $\alpha \geq 1$.
(iv) $\int_1^{+\infty} \psi(t)dt = k_1 < +\infty, \int_1^{+\infty} \phi(t)dt = k_2 < +\infty$. We also assume that there exists a constant $K \geq 1$ such that

$$E(t) \int_1^{+\infty} \phi(s) \frac{\phi_1(KE(s))}{E^2(s)} ds \leq K \int_1^{+\infty} \frac{ds}{g(s)}, \tag{4.3.2}$$

where $E(s) \equiv \exp(\int_1^s \psi(r)dr)$.

Then for any solution $u(t)$ of (4.3.1) with initial conditions $u(1) = c_1, u'(1) = c_2$ such that $|c_1| + |c_2| \leq K$,

$$\lim_{t\to+\infty} \int_1^t f(s, u(s), u'(s))ds = \alpha(c_1, c_2) < +\infty$$

always exists, and if we set $a = c_2 - \alpha(c_1, c_2)$, then $u(t) = b + at + o(t)$ as $t \to +\infty$, for any constant b.

Proof Because of (i) and standard existence theorem [178], (4.3.1) does have solutions $u \in C^1(I)$, where $I = [1, +\infty)$ corresponding to arbitrary given initial values $u(1) = c_1, u'(1) = c_2$.

Integrating (4.3.1) twice from 1 to t, we get

$$u'(t) = c_2 - \int_1^t f(s, u(s), u'(s))ds, \tag{4.3.3}$$

and for all $t \geq 1$,

$$u(t) = c_1 + c_2(t-1) - \int_1^t (t-s)f(s, u(s), u'(s))ds. \tag{4.3.4}$$

If we put

$$A(t) = \int_1^t \phi(s)g(\frac{|u(s)|}{s})ds, \quad B(t) = \int_1^t \psi(s)|u'(s)|ds, \tag{4.3.5}$$

it follows from (ii), (4.3.3), (4.3.4) that

$$\begin{cases} |u'(t)| \leq |c_2| + A(t) + B(t), & (4.3.6) \\ \dfrac{|u(t)|}{t} \leq K + A(t) + B(t) & (4.3.7) \end{cases}$$

Then

$$\begin{cases} A'(t) \leq \phi(t)g(K + A(t) + B(t)), \\ B'(t) \leq \psi(t)g(K + A(t) + B(t)), \end{cases}$$

so adding, and setting $C(t) = K + A(t) + B(t)$, we get for all $t \geq 1$,,

$$C'(t) \leq \phi(t)g(C(t)) + \psi(t)C(t),$$

or multiplying by the $\exp\left(-\int_1^t \psi(s)ds\right)$,

$$\frac{d}{dt}\left[C(t)\exp\left(-\int_1^t \psi(s)ds\right)\right] \leq g(C(t))\phi(t)\exp\left(-\int_1^t \psi(s)ds\right).$$

Integrating the above inequality, we obtain, for all $t \geq 1$,

$$C(t) \leq KE(t) + E(t) \int_1^t \frac{\phi(s)g(C(s))}{E(s)} ds, \tag{4.3.8}$$

where $E(t) = \exp\left(\int_1^t \psi(s)ds\right)$. Applying Corollary 1.1.2 to (4.3.8), we get

$$C(t) \leq KE(t)G^{-1}[K^{-1}E(t) \int_1^t D(s)ds], \tag{4.3.9}$$

where

$$G(r) = \int_1^r \frac{ds}{g(s)}, \quad D(s) = \phi(s)\frac{\phi_1(KE(s)}{E^2(s)},$$

which holds for all $t \geq 1$ by Corollary 1.1.2 because, for all $t \geq 1$, by (4.3.2),

$$K^{-1}E(t) \int_1^t D(s)ds \in Dom\ (G^{-1})$$

But

$$K^{-1}D(s) \leq k_0(K)\phi(s), \tag{4.3.10}$$

where

$$k_0(K) = \max\left\{\frac{\phi_1(u)}{u} : K \leq u \leq Ke^{k_1}\right\}. \tag{4.3.11}$$

Hence (4.3.10) will hold provided that

$$k_0(K)k_2e^{k_1} \leq \int_1^{+\infty} \frac{ds}{g(s)}.$$

From (4.3.6), (4.3.7) and (4.3.9) it follows that for all $t \geq 1$,

$$\begin{cases} |u'(t)| \leq C(t) \leq KE(t)G^{-1}[K^{-1}E(t) \int_1^t D(s)ds], & (4.3.12) \\ \dfrac{|u(t)|}{t} \leq C(t) \leq KE(t)G^{-1}[K^{-1}E(t) \int_1^t D(s)ds]. & (4.3.13) \end{cases}$$

From (iv) and (4.3.10) it follows that

$$KE(t)G^{-1}[K^{-1}E(t) \int_1^t D(s)ds] \leq k_3(K), \tag{4.3.14}$$

where

$$k_3(K) = Ke^{k_1}G^{-1}[k_2e^{k_1}k_0(K)], \tag{4.3.15}$$

and $k_0(K)$ is defined by (4.3.11).

Thus from (4.3.12) and (4.3.13), we derive

$$|u'(t)| \le k_3(K), \quad \frac{|u(t)|}{t} \le k_3(K). \tag{4.3.16}$$

Therefore we have from (ii) that for all $t \ge 1$,

$$\int_1^t |f(s, u(s), u'(s))|ds \le \int_1^t \phi(s)g(\frac{|u(s)|}{s})ds + \int_0^t \psi(s)|u'(s)|ds$$
$$\le k_2 g(k_3(K)) + k_1 k_2(K).$$

This proves that the integral $\int_1^t f(s, u(s), u'(s))ds$ is absolutely convergent and consequently that

$$\lim_{t\to+\infty} \int_1^t f(s, u(s), u'(s))ds = \alpha(c_1, c_2) < +\infty$$

always exists.

Also by (4.3.3),

$$\lim_{t\to+\infty} u'(t) = c_2 - \alpha(c_1, c_2) = \alpha$$

exists. Hence by Hospital's rule, we also have

$$\lim_{t\to+\infty} \frac{u(t)}{t} = \lim_{t\to+\infty} u'(t) = \alpha.$$

But then, for any constant b,

$$\lim_{t\to+\infty} \frac{u(t) - (b + at)}{t} = a - 0 - a = 0.$$

This completes the proof. □

Example 4.3.1 Consider the equation

$$u'' + (2t)^{-4}u^2 \cos u + t^{-2}u' \sin^3 u = 0, \quad t \ge 1.$$

Here we have $g(u) = u^2, \phi(t) = (4t)^{-2}, \psi(t) = t^{-2}$ and $\phi_1(\alpha) = \alpha^2$. From (4.3.2) it follows that all solutions $u(t)$ corresponding to initial conditions

$u(1) = c_1, u'(1) = c_2$ having $|c_1| + |c_2| \le 16e^{-1}$ are asymptotic to $b + at$ as $t \to +\infty$.

Example 4.3.2 Consider the equation

$$u'' + \frac{u^2 L(u)}{(t+u)t^n} + \frac{u' M(u)}{t^m} = 0, \quad t \ge 1, \tag{4.3.17}$$

where $L(u)$ and $M(u)$ are continuous functions such that $|L(u)| \le N, |M(u)| \le N$ for all $u \ge 0, N > 0$ is a constant, $n \ge 3$ and $m \ge 2$ are positive integers. Here we have $g(u) = u^2/(1+u), \phi(t) = t^{1-n}, \psi(t) = t^{-m}$ and $\phi_1(\alpha) = \alpha^2$.

Since $\int_1^{+\infty} \frac{1+u}{u^2} du$ diverges, it follows that all solution $u(t)$ of (4.3.17) are asymptotic to $b + at$ as $t \to +\infty$.

Theorem 4.3.2 ([181]) *Let (i), (ii) and (iv) be the same as in Theorem 4.3.1, while (iii) is replaced by the following:*

(iii)' $g(u)$ is a non-negative, continuous, monotonic, non-decreasing function and satisfies a Lipschitz condition

$$|g(u+v) - g(u)| \le \lambda v \tag{4.3.18}$$

for all $u, v \ge 0$, where λ is a positive constant. Furthermore, we assume that $g(0) = 0$. Then the conclusion of Theorem 4.3.1 remains true.

Proof In a similar argument as in the proof of Theorem 4.3.1, we obtain (4.3.8), which in view of (4.3.18) and Corollary 1.1.3, implies

$$C(t) \le KE(t)\left[1 + \int_1^t \phi(s)\frac{g(KE(s))}{KE(s)} \exp\left(\int_s^t \lambda\phi d\tau\right) ds\right]. \tag{4.3.19}$$

Using (iv) and the fact that $0 \le g(u)/u \le \lambda$ holds for all $u > 0$, it follows from (4.3.19) that for all $t \ge 1$,

$$\begin{aligned} C(t) &\le Ke^{k_1}[1 + \int_0^t \lambda\phi(s) \exp(\lambda \int_s^t \phi d\tau) ds] \\ &= Ke^{k_1} \cdot e^{\lambda k_2}. \end{aligned} \tag{4.3.20}$$

Hence from (4.3.6) and (4.3.7) it follows that for all $t \ge 1$,

$$\begin{aligned} |u'(t)| &\le |c_2| + A(t) + B(t) \le C(t) \le k_3, \\ \frac{|u(t)|}{t} &\le K + A(t) + B(t) \le C(t) \le k_3, \end{aligned}$$

where $k_3 = Ke^{k_1 + \lambda k_2}$.

Therefore we infer from (ii) that for all $t \geq 1$,

$$\int_1^t |f(s, u(s), u'(s)|ds \leq k_2 g(k_3) + k_1 k_3.$$

The proof can be completed in the same way as in Theorem 4.3.1. □

4.4 An Application of Theorem 1.1.21 to Nonlinear Volterra Integral Equations

Theorem 4.4.1 (The Kong-Zhang Inequality [311]) *Consider equation*

$$y(x) = f(x) + \int_0^x k(x,s)y(s)ds + \psi\left(\int_0^x k^*(x,s)w(y(s))ds\right), \text{ for all } x \in \mathbb{R}_+.$$

Suppose

1) $f(x) \geq 0$ *is continuous on* $\mathbb{R}_+, w(u) \in \mathcal{F}, \psi(u) \geq 0$ *is non-decreasing, sub-multiplicative and continuous on* $\mathbb{R}_+$*;*
2) $k(x,s)$ *is defined as in Theorem 4.4.1 in Qin [557];*
3) $k^*(x,s)(x \geq s)$ *is non-negative and continuous on* $\mathbb{R}_+ \times \mathbb{R}_+$*, and*

$$k^*(x,x) = 0, \frac{\partial k^*(x,s)}{\partial x} \leq q_{n+1}(x)h_{n+1}(s), \tag{4.4.1}$$

where $q_{n+1}(x)$ *and* $h_{n+1}(x)$ *are continuous on* $\mathbb{R}_+$*.*

Then for all $x \in [0, b)$*,*

$$|y(x)| \leq A_n(p) + A_n(g_{n+1})\psi\left\{F^{-1}\left[F\left(\int_0^x h_{n+1}\bar{A}_n(g_{n+1})w\left(\frac{A_n(p)}{A_n(g_{n+1})}\right)ds\right)\right.\right.$$
$$\left.\left.+\int_0^x h_{n+1}\bar{A}(g_{n+1})ds\right]\right\},$$

where $p(x), g_i(x)$ $(i = 1, 2, \ldots, n)$*,* $A_n(u)$ *and* b *are the same as in Theorem 4.4.1 in Qin [557], and for all* $x \in \mathbb{R}_+$*,*

$$\begin{cases} g_{n+1}(x) &= r(X) + \int_0^x r(s)m(s)\exp\left(\int_s^x m(t)dt\right)ds, \\ r(x) &= \psi\left(\int_0^x q_{n+1}(s)ds\right), \\ \bar{A}_{n+1}(g_{n+1}) &= \max\{A_{n+1}(g_{n+1}), 1\}. \end{cases}$$

Proof Denote

$$R(x) = \int_0^x k^*(x,s)w(y(s))ds,$$

so that

$$R'(x) \le k^*(x,s)w(y(s)) + \int_0^x \frac{\partial k^*(x,s)w(y(s))ds}{\partial x}.$$

By using (4.4.1), we get

$$R'(x) \le q_{n+1}(x)\int_0^x h_{n+1}w(|y(s)|)ds. \tag{4.4.2}$$

Integrating (4.4.2) from 0 to x, we get

$$R(x) \le \int_0^x q_{n+1}(s)ds \int_0^x h_{n+1}w(|y(s)|)ds,$$

hence

$$\psi(R(s)) \le r(x)\psi\left(\int_0^x h_{n+1}(s)w(|y(s)|)ds\right).$$

Let $T(x) = f(x) + \psi(R(x))$, then

$$T(x) \le f(x) + r(x)\psi\left(\int_0^x h_{n+1}(s)w(|y(s)|)ds\right),$$

and

$$|y(x)| \le T(x) + \int_0^x k(x,s)|y(s)|ds.$$

According to Theorem 4.4.1 in Qin [557],

$$|y(x)| \le A_n(\bar{p}),$$

where

$$\begin{aligned}
\bar{p}(x) &= T(x) + \int_0^x T(s)m(s)\exp\left(\int_s^x m(t)dt\right)ds \\
&\le \left[f + \int_0^x fm\exp\left(\int_s^x mdt\right)ds\right] \\
&\quad + \left[r + \int_0^x rm\exp\left(\int_s^x mdt\right)ds\right]\psi\left(\int_0^x h_{n+1}w(|y|)ds\right) \\
&= p(x) + g_{n+1}\psi\left(h_{n+1}(s)w(|y(s)|)ds\right).
\end{aligned}$$

Hence from (2) of Lemma 1.2.1 in Qin [557]

$$|y(x)| \le A_n(p) + A_n(g_{n+1})\psi\left(\int_0^x h_{n+1}w(|y|)ds\right).$$

Using Theorem 1.1.21, we come to the conclusion. □

4.5 An Application of Theorem 1.1.22 to a Kind of Differential Equations

Consider equation

$$y'(x) + \alpha(x)y(x) = F(x, y(x)), \tag{4.5.1}$$

where y and F are n-dimensional vectors, $\alpha(x)$ is non-negative and continuous on $\mathbb{R}_+$, F is continuous on $\mathbb{R}_+ \times \mathbb{R}_+$.

Theorem 4.5.1 (The Kong-Zhang Inequality [311]) *Suppose F in (4.5.1) satisfies*

$$\|F(x, y(x))\| \le \beta(x)\|y(x)\| + \gamma(x)w(\|y(x)\|), \tag{4.5.2}$$

where $\|\cdot\|$ denotes a vector norm, $\beta(x)$ and $\gamma(x)$ are non-negative and continuous on $\mathbb{R}_+$. Then all solutions $y(x)$ to equation (4.5.1), there holds that for all $x \in [0, b)$,

$$\|y(x)\| \le \exp\int_0^x (\beta(s) - \alpha(s))\,ds \cdot \left\{k + G^{-1}[G\left(w(k)\int_0^x \gamma(s)\exp\left(\int_0^s \tau(t)dt\right)ds\right) + \int_0^x \gamma(s)\exp\left(\int_0^s \tau(t)ds\right)ds]\right\}, \tag{4.5.3}$$

where

$$G(u) = \int_{u_0}^u \frac{ds}{w(s)},\quad u \ge u_0 > 0,\quad k = \|y(0)\|,\quad \tau(x) = \max\{\alpha(x), \beta(x)\}.$$

Proof From (4.5.1), we derive

$$\left[\exp\left(\int_0^x \alpha(s)ds\right)y(x)\right]' = \exp\left(\int_0^x \alpha(s)ds\right)F(x, y(x)),$$

$$y(x) = \exp\left(-\int_0^x \alpha(s)ds\right)\left[y(0) + \int_0^x \exp\left(\int_0^s \alpha(t)dt\right)F(s, y(s))ds\right],$$

which yields

$$\|y(x)\| \le \exp\left(-\int_0^x \alpha(s)ds\right)\left[k + \int_0^x \exp\left(\int_0^s \alpha(t)dt\right)\right.$$
$$\left.\times(\beta(s)\|y(s)\| + \gamma(s)w(\|y(s)\|))ds\right].$$

Let

$$\begin{cases} f(x) = k\exp\left(-\int_0^x \alpha(s)ds\right), \quad g_1(x) = g_2(x) = \exp\left(-\int_0^x \alpha(s)ds\right), \\ h_1(x) = \beta(x)\exp\left(\int_0^x \alpha(s)ds\right), \quad h_2(x) = \gamma(x)\exp\left(\int_0^x \alpha(s)ds\right). \end{cases}$$

Then

$$\begin{cases} A_1(f) = k\exp\left(-\int_0^x \alpha(s)ds\right) + \exp\left(-\int_0^x \alpha(s)ds\right)\int_0^x k\beta(s)\exp\left(\int_s^x \beta(t)dt\right)ds \\ \quad = k\exp\left(\int_0^x (\beta(s) - \alpha(s))\right)ds, \\ A_1(g_2) = \exp\left(\int_0^x (\beta(s) - \alpha(s))ds\right), \\ h_2\bar{A}_1(g_2) = \gamma(x)\exp\left(\int_0^x \tau(s)ds\right), \end{cases}$$

where $\tau(x) = \max\{\alpha(x), \beta(x)\}$. By Theorem 1.1.22, we conclude

$$\|y(x)\| \le k\exp\left(\int_0^x (\beta(s) - \alpha(s)\right)ds + \exp\left(\int_0^x (\beta(s) - \alpha(s))ds\right)$$
$$\times G^{-1}\left[G\left(\int_0^x \gamma(s)\exp\left(\int_0^s \tau(t)ds\right)w(k)ds\right) + \int_0^x \gamma(s)\exp\left(\int_0^s \tau(t)dt\right)ds\right].$$

□

Thus, from Theorem 4.5.1, we can easily obtain the following corollary.

Corollary 4.5.1 *In addition to the conditions in Theorem 4.5.1, let*

1) $\int_0^{+\infty} \tau(x)dx < +\infty$,
2) $\int_0^{+\infty} \gamma(x)dx < +\infty$. *Then every solution of equation (4.5.1) is bounded.*

Corollary 4.5.2 *In addition to the conditions in Theorem 4.5.1 and Corollary 4.7.1, let* $w(0) = 0$ *and*

$$\int_0^\delta \frac{ds}{w(s)} = +\infty, \quad \delta > 0. \tag{4.5.4}$$

Then the zero solution of equation (4.5.1) is stable.

In fact, from (4.5.1)–(4.5.2), we see that equation (4.5.1) has zero solution. Condition (1.1.133) in Lemma 1.1.9 implies $G(u) \to -\infty$, as $u \to 0$, i.e., $G^{-1}(u) \to -\infty$. Let k be sufficiently small in (4.5.3). Considering (4.5.4), we can conclude $\|y(x)\|$ must be sufficiently small.

4.6 An Application of Theorem 1.1.26 to Nonlinear Vector Integral Equations

Let $y(t), y_0(t)$ be functions of a real variable t with values in a Banach space X. When we consider integral operators of the form

$$Sy(t) = y_0(t) + \int_a^t h(t, s, y(s))ds, \tag{4.6.1}$$

$$Ty(t) = y_0(t) + \int_a^b h(t, s, y(s))ds, \tag{4.6.2}$$

it is found that these operators behave quite differently with respect to the existence of fixed points. In particular, the application of the method of successive approximations gives rise to different problems. All of this is classical when $h(t, s, y)$ is a linear function of y (see, e.g., [649, 713]).

We first shall assume that $h(t, s, y)$ satisfies a Lipschitz condition of the form

$$|h(t, s, y) - h(t, s, x)| \leq \xi(t, s)|y - x|, \tag{4.6.3}$$

where $|\cdot|$ denotes the norm in X. Assuming such a condition is not unusual when successive approximation techniques are applied. We shall further suppose that the function ξ determines a real number $|\|\xi\||$, which is called the "double norm " of ξ and is defined by

$$|\|\xi\|| = \left(\int_a^b \left(\int_a^b \xi^q(t, s)ds\right)^{p/q} dt\right)^{1/p}, \tag{4.6.4}$$

where p and q are conjugate indices, i.e., $p^{-1} + q^{-1} = 1$. For further information on double norms, we may refer to [713].

Without restriction on the magnitude of $|\|\xi\||$ and under reasonable conditions, the sequence $\{S^n y_0\}$, obtained by iteration, is a convergent sequence whose limit is a solution of $Sy = y$. Detailed discussions of this problem can be found in Erdelyi [221] or Tricomi [649], when $h(t, s, y)$ satisfies (1.1.56) and $p = 1,\ q = +\infty$ and $p = 2 = q$, respectively. The present result due to Willett [671], extends their results to a general Banach function space, the only restriction on p being $p \neq +\infty$.

For the equation $Ty = y$, the method of successive approximations depends on the requirement that T be a contraction mapping between approximate spaces. Such a requirement places additional restrictions on $|||\xi|||$. Tricomi [649] and Trjitzinsky [650], for example, assume that $|||\xi||| < 1$.

The equations $Sy = y$ and $Ty = y$ are special cases of a nonlinear integral equation of the form

$$y(t) = u(t) + \int_a^b g(t, s, y(s))ds + \int_a^t f(t, s, y(s))ds. \tag{4.6.5}$$

We can also consider equation (4.6.5) to be a special case of the equation $Ty = y$. However, application of the method of successive approximations directly to (4.6.5) is in general a successful technique only under unnecessarily restrictive condition on f.

This difficulty may be overcome by utilizing the properties of S. Under reasonable assumptions on f (involving no limitation on the norm of the Lipschitz function associated with f), u and g, the integral equation

$$y(t) = u(t) + \int_a^t f(t, s, y(s))ds + \int_a^b g(t, s, x(s))ds. \tag{4.6.6}$$

defines implicitly an operator $U : x(t) \to y(t)$. Under additional conditions, primarily on g, we may show that U is a contraction operator, and the unique fixed point of U will be a solution of equation (4.6.5).

It is easy to construct simple examples which are within the scope of present technique, and which the general theorems produced by a direct application of the method of successive approximations fail to include. For example, consider the equation $Ty = y$ when $h(t, s, y)$ satisfies the Lipschitz condition given by equation (4.6.3) with

$$\xi(t, s) = \begin{cases} \exp[(t - s)/\varepsilon], & \text{if } s > t, \\ 1, & \text{if } s < t. \end{cases} \tag{4.6.7}$$

It thus follows that $|||\xi||| > (b - a)/2$ for all p, q and $\varepsilon > 0$. Thus, $|||\xi|||$ can be made arbitrary large uniformly in p, q and ε by making $(b - a)$ sufficiently large. Neither Trjitzinsky's result nor the result found in Tricomi includes such problems; nor does it seem likely that the direct generalization of these results to L^p spaces for other values of p includes such cases. On the other hand, the technique here applies for any p, q, a and b , if ε is sufficiently small and positive.

In what follows, t and s denote real variables confined to an interval I, which has a as left endpoint and b as right endpoint. I may be closed, open, or half-open, and bounded or unbounded. R is the set $\{(t, s) : t \in T, s \in I\}$. Lower case Latin letters, other than t, s, a and b, usually denote vector functions with values in the

Banach space X, whereas Greek letters usually denote numerical functions. The vector function $x(t)$ is "measurable on I" if there exists a sequence of countably-valued functions converging pointwise almost everywhere on I to $x(t)$. "Measurable on R" is defined similarly. "Almost everywhere " means with respect to linear or plane Lebesgue measure. In the sequel, it goes without saying that most equations and statements hold almost everywhere in the approximate set.

For all $t \in I$ and $x(t)$ measurable on I, $\int_a^t x = \int_a^t x(s)ds$, if it exists, will be the Bochner integral over $a \leq s \leq t$. When no confusion can arise, we shall refrain from writing differentials after integrals.

Let $|x(t)|$ be the norm of $x(t)$ in X; for $x(t)$ measurable on I and p a real number, $1 \leq p < +\infty$, define

$$\|x\|_p = \left(\int_a^b |x(t)|^p dt \right)^{1/p}.$$

The norm $\|x\|_p$ is well-defined because the numerical function $|x(t)|$ is measurable on I if $x(t)$ is measurable on I. In order to include the boundary case $p = +\infty$, define $\|x\|_\infty$ to be the essential supremum of $|x(t)|$ on I. If X is the set of real numbers, and $x(t)$ is a numerical function, $\|x\|_p$ can be identified as the well-known L^p norm of $x(t)$. Hence, for numerical functions no confusion should arise if we denote their L^p norm by $\|\cdot\|_p$ and their absolute value by $|\cdot|$.

Besides L^p, we need two other function spaces: $B_p = \{x(t) : x(t)$ measurable on I and $\|x\|_p < +\infty\}$, and $W = \{x(t) \in B_p : |x(t) - u(t)| \leq \delta(t)\}$, where $\delta(t)$ and $u(t)$ are given functions from L^p and B_p, respectively. For a discussion of the B_p spaces, we can consult Zaanen [713], or other works. We note the following two facts here: $x(t) \in B_p$ if, and only if, $x(t)$ is measurable and $|x(t)| \in L^p$ and B_p is a Banach space with respect to $\|\cdot\|_p$.

W is also a complete metric space with respect to the B_p norm, for let $\{x(t)\}$ be a Cauchy sequence in W and $x(t)$ be its unique limit in B_p. Then, as $n \to +\infty$,

$$\| |x - x_n| \|_p = \|x - x_n\|_p \to 0.$$

Hence, by a well known property of L^p convergence, there exists a subsequence $\{|x(t) - x_{nk}(t)|\}$ which converges pointwise to zero almost everywhere in I. We conclude that $x(t) \in W$, because

$$|x(t) - u(t)| \leq |x(t) - x_{nk}(t)| + |x_{nk}(t) - u(t)| \leq \varepsilon + \delta(t),$$

where $\varepsilon \to 0^+$ as $k \to +\infty$.

For any measurable numerical function $\xi(t,s)$ on $\mathbb{R}$, by $\|\xi(t,\cdot)\|_p$, or just $\|\xi\|_p$, we shall mean the numerical function whose value at $t \in I$ is

$$\|\xi(t,\cdot)\|_p = \left(\int_a^b |\xi(t,s)|^p ds \right)^{1/p}.$$

The double norm of $\xi(t,s)$, which is defined as $\|\|\xi(t,\cdot)\|_q\|_p$, will be denoted by $|\|\xi\||$. When $p \neq 1$, $|\|\cdot\||$ is given be equation (4.6.4).

We shall need next two lemmas. The first lemma is the well-known principle of contraction mapping applied to the intersection of W with a closed ball in B_p. The second lemma is Theorem 1.1.26 [156].

Lemma 4.6.1 (Contraction Mapping Principle) *Let V be a closed ball in B_p with center $u(t)$ and radius ρ. Let U be a contraction mapping of $V\bigcap W$ into W, i.e., assume that $\|Uy - Ux\|_p \leq \nu\|y - x\|_p$ for any pair of points x, y of $V\bigcap W$, where ν is a constant such that $0 \leq \nu < 1$. Then, if $\|Uu - u\|_p \leq \rho(1-\nu)$, there is one and only one point $x \in V\bigcap W$ such that $z = Uz$.*

In order to determine conditions under which the mapping U (see, equation (4.6.6)) is well-defined on a set $V\bigcap W$ in B_p, we shall consider first the operator S defined by, for all $y(s) \in W$,

$$Sy(t) = y_0(t) + \int_a^t f(t,s,y(s))ds, \tag{4.6.8}$$

We assume the following assumptions hold.

(H1) $u(t) \in B_p$, and $f(t,s,y(s))$ is measurable on R for each $y(s)$ in W. (Define $f(t,s,y) = 0$ if $s > t$.)

(H2) There exist measurable numerical functions $\lambda(t,s)$ and $\alpha(t)$ on R and I, respectively, such that $|\|\lambda\|| < +\infty$, $|\|\alpha\|| < +\infty$, and

$$|f(t,s,u(s))| \leq \lambda(t,s)\alpha(s), \text{ for all } (t,s) \in R, \tag{4.6.9}$$

$$|f(t,s,y(s)) - f(t,s,x(s))| \leq \lambda(t,s)|y(s) - x(s)|, \text{ for all } (t,s) \in R, \text{ and } x(s), y(s) \in W. \tag{4.6.10}$$

Let $\sigma(t)$ be a numerical function satisfying

$$\sigma(t) \geq \sum_{h=0}^{+\infty}\left[\left(\int_a^t \|\lambda(s,\cdot)\|_q^p ds\right)^k \Big/ k!\right]^{1/p}. \tag{4.6.11}$$

Theorem 4.6.1 (The Willett Inequality [671]) *If $y_0(t) \in W$ and* **(H1)**, **(H2)** *hold for sufficiently large $\delta(t)$, e.g., for any $\delta(t)$ satisfying, for almost every $t \in I$,*

$$\delta(t) \geq |y_0(t) - u(t)| + \|\lambda(t,\cdot)\|_q\sigma(t)\left(\int_a^t \alpha^p(\tau)d\tau\right)^{1/p} \tag{4.6.12}$$

then the operator S, defined by equation (4.6.8), has a unique fixed point in W.

Proof Define $S^0u(t) = y_0(t)$, and assume equation (4.6.12) holds. We assert that $S^{n+1}u(t) \in W$, and

$$|S^{n+1}u(t) - S^n u(t)|^p \le \|\lambda\|_q^p \left(\int_a^t \alpha^p(\tau)d\tau\right)^{1/p} \left(\int_a^t \|\lambda\|_q^p ds\right)^n \Big/ n!, \qquad (4.6.13)$$

for $n = 0, 1, 2, \ldots$

The proof of this assertion is by induction. Suppose $m > 0$ and the assertion holds for $n = 0, 1, \ldots, m-1$. Since $S^m u(t) \in W$, $S^{m+1}u(t) = S(S^m u(t))$ is well-defined. Since $m > 0$ by the induction assumption, we obtain from the definitions of $S^{m+1}u(t)$ and $S^m u(t)$ and the assumptions that

$$|S^{m+1}u(t) - S^m u(t)|^p \le \int_a^t \lambda(t)|S^m u(s) - S^{m-1}u(s)|^p ds.$$

Next, by using Hölder's inequality and the induction hypothesis, we can derive equation (4.6.13) for $n = m$ in the following way,

$$\begin{aligned} |S^{m+1}u(t) - S^m u(t)|^p &\le \|\lambda\|_q^p \int_a^t |S^m u(s) - S^{m-1}u(s)|^p \\ &\le \|\lambda\|_q^p \left[\int_a^t \|\lambda(s,\cdot)\|_q^p \left(\int_a^s \|\lambda\|_q^p(\tau)d\tau\right)^{m-1} \left(\int_a^s \alpha^p(\tau)d\tau\right) ds\right] \Big/ (m-1)! \\ &\le \|\lambda\|_q^p \left(\int_a^t \alpha^p(\tau)d\tau\right) \left(\int_a^t \|\lambda\|_q^p d\tau\right)^m \Big/ m. \end{aligned}$$

From

$$S^{m+1}u(t) - u(t) = \sum_{n=0}^{m} [S^{n+1}u(t) - S^n u(t)] + y_0(t) - u(t)$$

and (4.6.13) for $n \le m$, it follows that

$$|S^{m+1}u(t) - u(t)| \le |y_0(t) - u(t)| + \|\lambda\|_q \left(\int_a^t \alpha^p(\tau)d\tau\right)^{1/p} \sum_{n=0}^{m} \left[\left(\int_a^t \|\lambda\|_q^p d\tau\right)^n \Big/ n!\right]^{1/p}.$$

If $\delta(t)$ satisfies equation (4.6.12), where $\sigma(t)$ is defined by (4.6.11), then $|S^{m+1}u(t) - u(t)| \le \delta(t)$; and we conclude that $S^{m+1}u(t) \in W$.

By assumption, $S^0u(t) = y_0(t) \in W$. We get directly from the definition of $Su(t)$ that

$$|Su(t) - y_0(t)| = \left|\int_a^t f(t, s, u(s))ds\right|.$$

Inequality (4.6.13) for $n = 0$ and $Su(t) \in W$ follow from here in the same way as the case $n = m$. This completes the induction proof.

Equation (4.6.13) implies that

$$\|S^{n+1}u(t) - S^n u(t)\|_p \leq \|\alpha\|_p \|\lambda\|^{n+1} \Big/ [(n+1)!]^{1/p},$$

and so $\{S^n u(t)\}$ is a Cauchy sequence in W. Let $y(t)$ be the unique limit point in W of this sequence.

Let $y(t)$ and $x(t)$ be any two fixed points in W for S. Then,

$$|y(t) - x(t)| \leq \int_a^t \lambda(t,s)|y(s) - x(s)|ds \leq \|\lambda\|_q \left(\int_a^t |y(s) - x(s)|^p ds \right)^{1/p}.$$

By Theorem 1.1.26,

$$\int_a^t |y(s) - x(s)|^p ds = 0$$

or, in the other words, $y(s) = x(s)$ almost everywhere in I.

Equation (4.6.13) also implies that $\{S^n u(t)\}$ is a Cauchy sequence in X for almost every $t \in I$. By uniqueness, the limit function of $\{S^n u(t)\}$ in X must be $y(t)$. Thus it follows that $y(t)$ is a fixed point for S,

$$\begin{aligned} |y(t) - Sy(t)| &\leq |y(t) - S^n u(t)| + |S^n u(t) - Sy(t)| \\ &\leq |y(t) - S^n u(t)| + \|\lambda\|_q + \|S^{n-1}u(t) - y(t)\|_p \\ &\to 0 \quad \text{as } n \to +\infty, \text{ for almost every } t \in I. \end{aligned}$$

Thus the proof is complete. □

We are now prepared to discuss equation (4.6.5) in the manner described above. In addition to **(H1)–(H2)**, we assume further the following assumptions hold.

(H3) $g(t, s, y(s))$ is measurable on R for each $y(s)$ in W. There exist measurable numerical functions $\mu(t, s)$ and $\beta(t)$ on R and I, respectively, such that $|||\mu||| < +\infty$, $\|\beta\|_p < +\infty$, and

$$\begin{cases} |g(t,s,u(s))| \leq \mu(t,s)\beta(s), \quad \textit{for all } (t,s) \in R, & (4.6.14) \\ |g(t,s,y(s)) - g(t,s,x(s))| \leq \mu(t,s)|y(s) - x(s)|, & \\ \qquad \textit{for all } (t,s) \in R, \ x(s), y(s) \in W. & (4.6.15) \end{cases}$$

(H4) There exists a number $\nu > 0$ such that

$$\frac{\|e^{1/p}\|\mu\|_q\|_p}{1 - [1 - e(b)]^{1/p}} \leq \nu < 1,$$

where for all $t \in I$,

$$e(t) = \exp\left(-\int_a^t \|\lambda(s,\cdot)\|_q^p ds\right). \tag{4.6.16}$$

Let γ be a non-negative number such that

$$\gamma \geq \frac{\left(\int_a^b \left[\lambda(t,s)\alpha(s)ds + \int_a^b \mu(t,s)\beta(s)ds\right]^p e(t)dt\right)^{1/p}}{1-[1-e(b)]^{1/p}}. \tag{4.6.17}$$

Theorem 4.6.2 (The Willett Inequality [671]) *If* **(H1)–(H4)** *hold for sufficiently large $\delta(t)$, e.g., for any $\delta(t)$ satisfying, for all $t \in I$,*

$$\delta(t) \geq \sigma(t)\|\lambda(t,\cdot)\|_q \left(\int_a^t \alpha^p(\tau)d\tau\right)^{1/p} + (1-\nu)^{-1}\gamma\|\mu(t,\cdot)\|_q + \int_a^b \mu(t,s)\beta(s)ds, \tag{4.6.18}$$

then equation (4.6.5) has a unique solution $y(t)$ in W, and

$$\|y-u\|_p \leq \gamma(1-\nu)^{-1}. \tag{4.6.19}$$

Proof Let V be a closed ball in B_p with center $u(t)$ and radius $\rho = (1-\nu)^{-1}\gamma$. Theorem 4.6.1 will imply that the mapping U, which is defined in (4.6.6), is well-defined for each $x \in V \bigcap W$ and that $Ux \in W$, if equation (4.6.12) holds for

$$y_0(t) = u(t) + \int_a^b g(t,s,x(s))ds, \quad x(s) \in V \cap W.$$

By (H4) and the triangle and Hölder inequalities, we obtain

$$\begin{aligned} |y_0(t) - u(t)| &\leq \int_a^b \mu(t,s)|x(s)-y(s)|ds + \int_a^b \mu(t,s)\beta(t,s)ds \\ &\leq \|\mu(t,\cdot)\|_q\|x-u\|_p + \int_a^b \mu(t,s)\beta(t,s)ds. \end{aligned}$$

Since $\|x-u\|_p \leq \rho = (1-\nu)^{-1}\gamma$, it follows that

$$|y_0(t) - u(t)| \leq \delta(t) - \|\lambda(t,\cdot)\|\sigma(t)\left(\int_a^t \alpha^p(\tau)d\tau\right)^{1/p},$$

if $\delta(t)$ satisfies inequality (4.6.18). Thus, equation (4.6.12) holds for each $x(t) \in V \bigcap W$.

Next, we shall prove $\|Ux_1 - Ux_2\|_p \leq \nu\|x_1 - x_2\|_p$ for any pair of points $x_1, x_2 \in V \bigcap W$, where ν is defined in (H4). Let $y_1 = Ux_2$, $y_2 = Ux_2$; y_1 and y_2 are now well defined. By utilizing (H2) and (H3) and the usual inequalities, we obtain

$$|y_2(t) - y_1(t)| \leq \|x_2 - x_1\|_p \|\mu(t,\cdot)\|_q + \|\lambda(t,\cdot)\|_q \left(\int_a^t |y_2(s) - y_1(s)|^p ds\right)^{1/p}.$$

By Theorem 1.1.26,

$$\left(\int_a^t |y_2(s) - y_1(s)|^p ds\right)^{1/p} \leq \|x_2 - x_1\|_p \frac{\left(\int_a^t \|\mu(s,\cdot)\|_q^p e(s)ds\right)^{1/p}}{1 - [1 - e(t)]^{1/p}},$$

where $e(t)$ is defined by equation (4.6.16). Thus, $\|y_2 - y_1\|_p \leq \|x_2 - x_1\|_p \nu$, if ν is taken as in (H4).

In order to apply the contraction mapping principle to U, we have left only to show that $\|Uu - u\|_p \leq \nu(1 - \rho)$. Let $Uu = v$. Then,

$$|v(t) - u(t)| \leq \int_a^t \lambda(t,s)|v(s) - u(s)|ds + \int_a^t \lambda(t,s)\alpha(s)ds$$
$$+ \int_a^b \mu(t,s)\beta(s)ds \leq \|\lambda(t,\cdot)\|_q \left(\int_a^t |v(s) - u(s)|^p ds\right)^{1/p} + \eta(t),$$

where $\eta(t)$ denotes the sum of the last two integrals. It is clear that $\eta(t) \in L^p$. We obtain next by using Theorem 1.1.26 that

$$\left(\int_a^t |v - u|^p ds\right)^{1/p} \leq \frac{\left(\int_a^t \eta^p e(s)ds\right)^{1/p}}{1 - [1 - e(t)]^{1/p}}$$

hence, $\|v - u\|_p \leq \gamma$, where γ satisfies equation (4.6.17). Since $\gamma = \rho(1 - \nu)$, we conclude that $\|v - u\|_p \leq \rho(1 - \nu)$.

All the assumptions of the contraction mapping principle in Lemma 4.6.1 have been shown to hold for U in the present situation. Hence, we conclude that there exists a unique $y(t)$ in $V \bigcap W$ such that $y(t) = Uy(t)$, or, in other words, $y(t)$ is a solution of integral equation (4.6.5). This thus completes the proof of the theorem. □

4.7 An Application of Theorem 1.1.35 and Corollary 1.1.7 to Integro-differential Equations

In this section, we shall employ Theorem 1.1.35 to study the integro-differential equation

$$(a(t)x')' + b(t)x' + c(t)x = r(t) \int_0^t g(s)x(s)ds + \theta(t,x,x') \qquad (4.7.1)$$

as $t \to +\infty$. Grace and Lalli [251] studied the asymptotic behavior of equation (4.7.1) and proved that under certain conditions on the functions $r(t), g(t)$ and $\theta(t, x, x')$, there is a solution of equation (4.7.1) satisfying any given initial conditions which tends to a solution of the linear differential equation

$$(a(t)x')' + b(t)x' + c(t)x = 0, \tag{4.7.2}$$

for which the general solution is known.

As in [251], we shall denote

$$\xi(t) \geq \max(|Z_1(t)|, |Z_2(t)|), \quad \eta(t) \geq \max(|Z_1'(t)|, |Z_2'(t)|),$$

where $Z_1(t)$ and $Z_2(t)$ are any two linearly independent solutions of equation (4.7.2).

Assumption 1 Let us assume that if $|x(t)| \leq \xi(t)u$ and $|x'(t)| \leq \eta(t)u$, then there exist continuous non-negative functions $r_1(t), f(t)$ and a continuous non-decreasing $w(u)$ for all $u \in [0, +\infty)$ with the property that $G(u) = \int_{u_0}^{u}(ds/w(s)) \to +\infty$ as $u \to +\infty$ and such that either

(a) $|\theta(t, x, x')| \leq r_1(t)u$ or
(b) $|\theta(t, x, x')| \leq r_1(t)u + f(t)w(u)$ holds.

In the sequel we shall let

$$\xi(t) \geq max(|Z_1(t)|, \ |Z_2(t)|), \ \eta(t) \geq max(|Z_1'(t)|, \ |Z_2'(t)|),$$

where $Z_1(t)$, $Z_2(t)$ are any two linearly independent solutions of (4.7.2). We shall now prove the following results:

Theorem 4.7.1 (The Grace-Lalli Inequality [251]) *In addition to Assumption 1(a), assume the following conditions hold:*

(i) $a(t), b(t)$ and $c(t)$ are continuous, and $a(t) > 0$ for all $t \in I$,
(ii) $r(t)$ and $g(t)$ are continuous and non-negative for all $t \in I$,
(iii) $[\xi(t)r(t)]/a(t)W(t)$ and $\xi(t)g(t) \in \mathcal{L}(0, \infty)$, where

$$W(t) = Z_1'Z_2 - Z_1Z_2' > 0.$$

Then for every pair (x_0, x_0') of numbers there is a solution of (4.7.1) which can be written in the form

$$x(t) = A(t)Z_1(t) + B(t)Z_2(t), \tag{4.7.3}$$

satisfying the initial conditions and $x'(0) = x_0'$ with $lim_{t\to\infty}A(t) = l$ and $lim_{t\to\infty}B(t) = m$.

Proof Let us assume that $x(t)$ is a solution of (4.7.1) and is written in the form (4.7.3). We shall require that

$$A'(t)Z_1 + B'(t)Z_2 = 0. \tag{4.7.4}$$

Differentiating (4.7.3) with respect to t, we get

$$\begin{cases} x'(t) = A(t)Z_1'(t) + B(t)Z_2'(t), \\ x''(t) = A'(t)f_1(t) + B'(t)Z_2'(t) + A(t)Z_1''(t) + B(t)f_2'(t). \end{cases}$$

Using the fact that Z_1 and Z_2 are solutions of (4.7.2) and that x is a solution of (4.7.1) we can reduce the last equation to

$$A'(t)Z_1'(t) + B'(t)Z_2'(t) = h(t), \tag{4.7.5}$$

where

$$h(t) = \Bigg[r(t) \int_0^t g(s)\Big\{A(s)Z_1(s) + B(s)Z_2(s)\Big\}ds$$
$$+\theta\Big(t, A(t)Z_1(t) + B(t)Z_2(t), A(t)Z_1'(t) + B(t)Z_2'(t)\Big)\Bigg]/a(t).$$

Solving (4.7.4) and (4.7.5) for $A'(t)$ and $B'(t)$ we get

$$A'(t) = [Z_2(t)h(t)]/W(t), \quad B'(t) = [-Z_1(t)h(t)]/W(t). \tag{4.7.6}$$

Integrating (4.7.6) from 0 to $t > 0$ we get

$$\begin{cases} A(t) = A(0) + \displaystyle\int_0^t \frac{Z_2(s)h(s)}{W(s)}ds, \\ B(t) = B(0) - \displaystyle\int_0^t \frac{Z_1(s)h(s)}{W(s)}ds. \end{cases} \tag{4.7.7}$$

Using Assumption 1(a), (4.7.7) yields

$$|A(t)| + |B(t)| \le |A(0)| + |B(0)| + 2\int_0^t \frac{\xi(s)}{a(s)W(s)}\Big\{r(s)\int_0^s g(\tau)\xi(\tau)\Big\{|A(\tau)| + |B(\tau)|\Big\}d\tau$$
$$+r(s)\Big(|A(s)| + |B(s)|\Big)\Big\}\,ds. \tag{4.7.8}$$

Letting $|A(t) + B(t)| = K(t)$, we get

$$K(t) \leq K(0) + \int_0^t 2\phi(s)K(s)ds + 2\int_0^t \phi(s)\left\{\int_0^s g(\tau)\xi(\tau)K(\tau)d\tau\right\}ds,$$

where

$$\phi(s) = \frac{\xi(s)r(s)}{a(s)W(s)}.$$

Using Corollary 1.1.7 with $h(t) = 0$ or $W(u) = 0$, we obtain the following estimate:

$$K(t) \leq K(0)\left[1 + \int_0^t 2\phi(s)\exp\int_0^\tau \Big(2\phi(\tau) + g(\tau)\xi(\tau)d\tau\Big)ds\right],$$

from which the boundedness of $K(t)$ follows. Since $A(0)$ and $B(0)$ are arbitrary constants and hence can be selected as solutions of the system

$$\begin{cases} A(0)Z_1(0) + B(0)Z_2(0) = x_0, \\ A(0)Z_1'(0) + B(0)Z_2'(0) = x_0'. \end{cases}$$

From the fact that $A(t)$ and $B(t)$ are bounded it follows that the limits of $A(t)$ and $B(t)$ exist as $t \to \infty$. This completes the proof. □

Theorem 4.7.2 (The Grace-Lalli Inequality [251]) *In addition to Assumption 1 (b), if we assume that*

$$\frac{\xi(t)f(t)}{a(t)W(t)} \in \mathcal{L}(0,\infty),$$

then the conclusion of Theorem 4.7.1 holds, provided w is sub-multiplicative and $w(0) = 0$.

Proof As in the proof of Theorem 4.7.1 we obtain, using Assumption 1(b),

$$|A(t)| + |B(t)| \leq |A(0)| + |B(0)| + 2\int_0^t \frac{\xi(s)}{a(s)W(s)}\left\{r(s)\left\{\int_0^s g(\tau)\xi(\tau)\Big(|A(\tau)| + |B(\tau)|\Big)d\tau\right\}\right.$$

$$\left. + r(s)\Big(|A(s)| + |B(s)|\Big) + f(s)w\Big(|A(s)| + |B(s)|\Big)\right]ds. \tag{4.7.9}$$

With $K(t)$ defined as before, we obtain from (4.7.9) the following inequality:

$$K(t) \leq K(0) + 2\int_0^t \frac{r(s)\xi(s)}{a(s)W(s)} K(s)ds + 2\left\{ \int_0^s g(\tau\xi(\tau)K(\tau)d\tau \right\}$$
$$+2\int_0^t \frac{\xi(s)}{a(s)W(s)} f(s)w(K(s))ds.$$

Using Corollary 1.1.7, we get

$$K(t) \leq G^{-1}\left[G(K(0)) + \int_0^t \frac{2\xi(s)}{a(s)W(s)} f(s)w\left\{ 1 + \int_0^t \frac{2r(\tau)\xi(\tau)}{a(\tau)W(\tau)} \exp\left(\int_0^\tau \psi(v)dv \right) ds \right]$$
$$\times\left(1 + \int_0^t \frac{2r(s)\xi(s)}{a(s)W(s)} \exp\left(\int_0^s \psi(v)dv \right) ds \right),$$

where

$$\psi(v) = \frac{2r(v)\xi(v)}{a(v)W(v)} + g(v)\xi(v),$$

from which it follows that $K(t)$ is bounded. The rest of the proof is similar to that of Theorem 4.7.1. □

Theorem 4.7.3 (The Agerwal Inequality [5]) *In addition to Assumption 1 (a), let the following conditions hold:*

(i) $a(t)$, $b(t)$ *and* $c(t)$ *are continuous, and* $a(t) > 0$ *for all* $t \in I$,
(ii) $r(t)$ *and* $g(t)$ *are continuous and non-negative for all* $t \in I$,
(iii) for all $t \in I$,

$$\int_0^t \frac{\xi}{W(s)a(s)} \left[r_1(s) + r(s)\int_0^s \xi(\tau)g(\tau)d\tau \right] ds < +\infty, \qquad (4.7.10)$$

where

$$W(t) = Z_1'Z_2 - Z_1Z_2' > 0.$$

Then for every pair (x_0, x_0') *of numbers, there is a solution of equation (4.5.1) which can be written in the form*

$$x(t) = A(t)Z_1(t) + B(t)Z_2(t),$$

satisfying the initial conditions $x(0) = x_0$ *and* $x'(0) = x_0$ *with* $\lim_{t\to+\infty} A(t) = 1$ *and* $\lim_{t\to+\infty} B(t) = m$.

Proof Applying Corollary 1.2.16 in Qin [557] and using a similar argument to Theorem 4.7.1, we can prove the theorem. □

Theorem 4.7.4 (The Agerwal Inequality [5]) *In addition to Assumption 1(b) and Theorem 4.7.3, if we further assume that for all $t \in I$,*

$$\int_0^t \frac{\xi(s)f(s)}{W(s)a(s)}ds < +\infty,$$

then the conclusion of Theorem 4.7.3 holds, provided that w is sub-multiplicative and $w(0) = 0$.

Proof The proof is an application of Theorem 1.1.35 and similar to that given in Theorem 4.7.2. In case $r_1(t) = r(t)$ as considered by Grance and Lalli [251] condition (4.7.10) takes the form

$$\int_0^t \frac{\xi(s)r(s)}{W(s)a(s)}[1 + \int_0^s \xi(\tau)g(\tau)d\tau]ds < +\infty, \quad t \in I, \tag{4.7.11}$$

which is automatically satisfied if

$$\int_0^t \frac{\xi(s)r(s)}{W(s)a(s)}ds < +\infty, \quad \int_0^t \xi(s)g(s)ds < +\infty, \quad t \in I, \tag{4.7.12}$$

as required in their proofs. In several situations, condition (4.7.11) is satisfied, whereas condition (4.7.12) is not, for example, we may consider the equation

$$x'' + 2x' + x = e^{-2t}\int_0^t \frac{e^s}{1+s}x(s)ds + e^{-zt}x,$$

then $W(t) = e^{-2t}$, $\xi(t) = \eta(t) = (1+t)e^{-t}$, $a(t) = 1$, $r(t) = e^{-2t}$, $g(t) = e^t/(1+t)$ and it is easy to verify that (4.7.11) is satisfied, whereas $\int_0^t \xi(s)g(s)ds \not< +\infty$, for all $t \in I$. □

4.8 Applications of Theorems 1.1.41 and 1.1.43 to Qualitative Analysis of Nonlinear Differential Equations

In this section, we shall apply Theorems 1.1.41 and 1.1.43 to investigate the qualitative analysis of two applications.

First, we consider the system of nonlinear differential equations

$$\frac{du}{dt} = F_1(t, u(t), \int_{t_0}^t K_1(t, u(s))ds), \tag{4.8.1}$$

for all $t \in I = [t_0, t_1] \subset \mathbb{R}_+$, where $u \in C(I, \mathbb{R}^n)$, $F_1 \in C(I \times \mathbb{R}^n \times \mathbb{R}^n, \mathbb{R}^n)$ and $K_1 \in C(I \times \mathbb{R}^n, \mathbb{R}^n)$. We shall assume that the Cauchy problem

$$\begin{cases} \dfrac{du}{dt} = F_1(t, u(t), \displaystyle\int_{x_0}^{t} K_1(t, u(s))ds), \\ u(t_0) = u_0 \in \mathbb{R}^n, \end{cases} \tag{4.8.2}$$

has a unique solution, for every $t_0 \in I$ and $u_0 \in \mathbb{R}^n$. We shall denote this solution by $u(\cdot, t_0, u_0)$.

The following theorem deals with the estimate on the solution of the nonlinear Cauchy problem (4.8.2).

Theorem 4.8.1 (The Denche-Khellaf Inequality [193]) *Assume that the functions F_1 and K_1 in (4.8.2) satisfy the conditions*

$$||K_1(t, u)|| \leq h(t)\phi(||u||), \text{ for all } t \in I, \tag{4.8.3}$$

$$||F_1(t, u, v)|| \leq ||u|| + ||v||, \text{ for all } u, v \in \mathbb{R}^n, \tag{4.8.4}$$

where h and ϕ are as defined in Theorem 1.1.41. Then we have the estimate, for all $t_0 \leq t \leq t_2$,

$$||u(t, t_0, u_0)|| \leq e^{t-t_0}(||u_0|| + \int_{t_0}^{t} h(s)E_1(s, ||u_0||)ds), \tag{4.8.5}$$

where

$$\begin{cases} E_1(t, ||u_0||) = \psi^{-1}(\psi(v) + \displaystyle\int_{t_0}^{t} \phi(e^{\tau - x_0} \int_{t_0}^{\tau} h(\sigma)d\sigma)d\tau, & (4.8.6) \\ \psi(t) = \displaystyle\int_{a}^{t} ds\phi(s), \; t \geq a > 0, & (4.8.7) \\ v = \displaystyle\int_{t_0}^{t_1} ||u_0||\phi(e^{s-x_0})ds, & (4.8.8) \end{cases}$$

and t_2 is chosen so that $\psi(v) + \int_{t_0}^{t} \phi(e^{\tau - x_0} \int_{t_0}^{\tau} h(\sigma)d\sigma)d\tau$ is in Dom (ψ^{-1}), for all $t_0 \leq t \leq t_2$.

Proof Let $t_0 \in I, u_0 \in \mathbb{R}^n$ and $u(t, t_0, u_0)$ be the solution of the Cauchy problem (4.8.2). Then we get

$$u(t, t_0, u_0) = u_0 + \int_{t_0}^{t} F_1(s, u(s, t_0, u_0), \int_{t_0}^{s} K_1(s, u(\tau, t_0, u_0))d\tau)ds. \tag{4.8.9}$$

Using (4.8.3) and (4.8.4) in (4.8.9), we conclude

$$\begin{aligned}||u(t,t_0,u_0)|| &\le ||u_0|| + \int_{t_0}^{t} f(s)[||u(s,t_0,u_0)|| + \int_{t_0}^{s} ||K_1(s,u(\tau,t_0,u_0))||d\tau]ds \\ &\le ||u_0|| + \int_{t_0}^{t} f(s)(||u(s,t_0,u_0)|| + h(s)\int_{t_0}^{s} \phi(||u(\tau,t_0,u_0)||)d\tau)ds.\end{aligned} \tag{4.8.10}$$

Now, applying Theorem 1.1.41 with $a(t) = ||u_0||, f(t) = b(t) = 1$ and $W(u) = u$ to (4.8.10) yields (4.8.5).

If, in addition, we assume that the function F_1 satisfies the general condition

$$||F_1(t,u,v)|| \le f(t)(g(||u||) + W(||v||)), \tag{4.8.11}$$

where f, g and W are as defined in Theorem 1.1.43, we obtain an estimate for $u(t,t_0,u_0)$, and from any particular conditions of (4.8.11) and (4.8.3), we can get some useful results similar to Theorem 4.8.1. □

4.9 Applications of Theorems 1.1.47 and 1.1.48 to Integral Equations and Functional Differential Equations

In this section, we shall show that the global existence of solutions to certain integral equations and functional differential equations by using Theorems 1.1.47 and 1.1.48.

We first consider the integral equation

$$u(t) = k(t) + \int_0^{\alpha(t)} f(s)\omega(u(s))ds, \quad t \ge 0, \tag{4.9.1}$$

where $k, f, w \in C(\mathbb{R}_+, \mathbb{R}_+)$ with $\omega(0) = 0$ and $\alpha \in C^1(\mathbb{R}_+, \mathbb{R}_+)$ is non-decreasing with $\alpha(t) \le t$ on $\mathbb{R}_+$. Under these assumptions, (4.9.1) has a solution $u \in C([0,T), \mathbb{R}_+)$ on some maximal interval of existence $[0,T)$. Moreover, if $T < +\infty$, then

$$\limsup_{t\to T} |u(t)| = +\infty. \tag{4.9.2}$$

Now we prove the following theorem.

Theorem 4.9.1 ([355]) *Assume that for all* $x, y \in \mathbb{R}_+$,

$$|\omega(x) - \omega(y)| \le z(|x-y|),$$

with $z \in C(\mathbb{R}_+, \mathbb{R}_+)$ *non-decreasing,* $z(x) > 0$ *for all* $x > 0$.

If

$$\int_0^1 \frac{ds}{z(s)} = \int_1^{+\infty} \frac{ds}{z(s)} = +\infty,$$

then (4.9.1) has a unique solution $u(t)$ *defined on* $\mathbb{R}_+$. *Moreover, if* k *is bounded on* $\mathbb{R}_+$ *and if either* α *is bounded on* $\mathbb{R}_+$ *or* $\int_0^{+\infty} f(s)ds < +\infty$, *then this solution is bounded on* $\mathbb{R}_+$.

Proof As the existence of a solution on some maximal interval $[0, T)$ is guaranteed [175], let us first prove the uniqueness statement. Suppose that on some interval $[0, t_0]$ with $t_0 > 0$, (4.9.1) has two solutions $u_1, u_2 \in C([0, t_0], \mathbb{R}_+)$. From the corresponding two equations, we obtain for all $0 \le t \le t_0$,

$$u_1(t) - u_2(t) = \int_0^{\alpha(t)} f(s)[\omega(u_1(s)) - \omega(u_2(s))]ds.$$

Denote $v(t) = |u_1(t) - u_2(t)|$ for all $t \in [0, t_0]$. Using the hypotheses, we deduce, $0 \le t \le t_0$,

$$v(t) \le \int_0^{\alpha(t)} f(s)z(v(s))ds. \tag{4.9.3}$$

Let

$$G(r) = \int_1^r \frac{ds}{z(s)}, \quad r > 0,$$

and note that $G(0) = -\infty$ and $G(+\infty) = +\infty$. From (4.9.3) it is clear that for all $\epsilon > 0$, and for all $0 \le t \le t_0$,

$$v(t) \le \epsilon + \int_0^{\alpha(t)} f(s)z(v(s))ds,$$

which, by Theorem 1.1.47, implies, for all $0 \le t \le t_0$,

$$v(t) \le G^{-1}\left(G(\epsilon) + \int_0^{\alpha(t)} f(s)ds\right).$$

For every fixed $t \in [0, t_0]$, let $\epsilon \to 0$ in the above inequality to infer $v(t) = 0$. Therefore, the uniqueness of the solution is proved. □

Let us now show that the solution is global, i.e., $T = +\infty$, where T is the maximal time of existence. If $T < +\infty$, then relation (4.9.2) holds. With $k_0 = \max_{0 \le t \le T} k(t)$, we obtain from (4.9.1) that for all $0 \le t < T$,

$$u(t) \le k_0 + \int_0^{\alpha(t)} f(s)z(u(s))ds, \tag{4.9.4}$$

as $\omega(u(s)) = \omega(u(s)) - \omega(0) \le z(u(s))$ for $0 \le s < T,\ u$ being non-negative.

Applying Theorem 1.1.47 to (4.9.4), we deduce that for all $0 \le t < T$,

$$u(t) \le G^{-1}(G(k_0) + \int_0^{\alpha(t)} f(s)ds), \tag{4.9.5}$$

which contradicts (4.9.2) and therefore the global existence is proved.

If k is bounded, with $k_0 = \sup_{t \in \mathbb{R}_+} k(t)$, we obtain that (4.9.4) holds on $\mathbb{R}_+$. From (4.9.4), by means of Theorem 1.1.47, relation (4.9.5) is obtained for all $t \in \mathbb{R}_+$. The boundedness assertion is now clear. □

Remark 4.9.1 The uniqueness and global existence statements in Theorem 4.9.1 follow also from Bihari's inequality [82]. However, if α and k are bounded on $\mathbb{R}_+$ and $\int_0^{+\infty} f(s)ds = +\infty$, the result here yields the boundedness of the solution on $\mathbb{R}_+$. This conclusion cannot be obtained by using Bihari's result.

Consider now the functional differential equation

$$x'(t) = F(t, x(t), x(\alpha(t)), x'(q(t))), \quad x(0) = x_0, \tag{4.9.6}$$

with $F \in C(\mathbb{R}_+ \times \mathbb{C}^{3n}, \mathbb{C}^n)$, and $\alpha, q \in C^1(\mathbb{R}_+, \mathbb{R}_+)$, and $\alpha(t) \le t,\ q(t) < t$ for all $t > 0$. By a result in [210], we have that there exists a maximal existence interval of $[0, T)$ of a solution to equation (4.9.6). Moreover, if $T < +\infty$, then

$$\limsup_{t \to T} |x(t)| = +\infty. \tag{4.9.7}$$

Using the integral inequality in Theorem 1.1.48, we shall give sufficient conditions for the global existence of the solutions of equation (4.9.6).

Theorem 4.9.2 ([355]) *Assume that $\alpha : \mathbb{R}_+ \to \mathbb{R}_+$ is an increasing diffeomorphism of $\mathbb{R}_+$ and for all $t \ge 0,\ (x, y, z) \in \mathbb{C}^{3n}$,*

$$|F(t, x, y, z)| \le a(t)\omega(|x|) + b(t)\omega(|y|) + c(t)d(|z|),$$

where $a,\ b,\ c,\ d,\ \omega \in C(\mathbb{R}_+, \mathbb{R}_+)$ and $\omega(u) > 0$ for all $u > 0$, ω is non-decreasing with $\int_1^{+\infty} \frac{ds}{\omega(s)} = +\infty$. Then all solution of equation (4.9.6) are global in time.

Proof If the assertion is not true, there is some $x_0 \in \mathbb{C}^n$ such that the problem (4.9.6) has a solution $x(t)$ which blows up in a finite time T. If $M = sup_{u\in[0,T]}q(u)$, let

$$k_0 = |x_0| + \sup_{t\in[0,M]} d(|x'(t)|)\left(\int_0^T c(s)ds\right).$$

setting $u(t) = |x(t)|$, $t \in [0, T)$, we deduce from (4.9.6) after integration, and hypothesis on F that for all $0 \le t < T$,

$$u(t) \le k_0 + \int_0^t a(s)\omega(u(s))ds + \int_0^t b(s)\omega(u(\alpha(s)))ds.$$

Changing variables in the second integral of the right-hand side in the above inequality, we obtain for all $t_0 \le t < T$,

$$u(t) \le k_0 + \int_0^t a(s)\omega(u(s))ds + \int_0^{\alpha(t)} b(\alpha^{-1}(s))\omega(u(s))(\alpha^{-1})'(s)ds.$$

From Theorem 1.1.48, we can now infer that $u(t)$ is bounded on $[0, T)$, which contradicts (4.9.7) and thus completes the proof. □

Remark 4.9.2

(i) In [597], a continuation theorem for (4.9.6) is obtained under the hypothesis that F has at most linear growth in all the variables other than t. We allow a much larger class of nonlinearities.
(ii) In the particular case when F in problem (4.9.6) does not depend upon the last variable, Theorem 4.9.2 yields the global existence of solutions for retarded differential equations and we recover some results from [165, 270].

Example 4.9.1 Consider the generalized pantograph equation [293]

$$x'(t) = Ax(t) + Bx(pt) + Cx'(pt), \quad t > 0,\ x(0) = x_0,$$

where A, B, C are complex matrices and $p \in (0, 1)$. Theorem 4.9.2 shows that all solutions are global in time.

Example 4.9.2 Consider the generalized Liénard system with time delay

$$x' = y - F(x), \quad y' = g(t, x(t - \tau(t))), \tag{4.9.8}$$

where $F \in C(\mathbb{R}, \mathbb{R})$, $g \in C(\mathbb{R}_+ \times \mathbb{R}, \mathbb{R})$, and $\tau(t) \le t$ on $\mathbb{R}_+$. If $\alpha(t) \equiv t - \tau(t)$ is an increasing diffeomorphism of $\mathbb{R}_+$ and for all $(t, x) \in \mathbb{R}_+ \times \mathbb{R}$,

$$|F(x)| \le \omega(|x|), \quad |g(t, x)| \le a(t)\omega(|x|),$$

with $a, \omega \in C(\mathbb{R}_+, \mathbb{R}_+)$, $\omega(u) > 0$ for all $u > 0$, ω non-decreasing and such that $\int_1^{+\infty} \frac{ds}{\omega(s)}$, then all solutions of equation (4.9.8) are global.

Indeed, we have that

$$\begin{cases} |(y - F(x), 0)| \leq |(x, y)| + \omega(|(x, y)|), & \text{for all } (x, y) \in \mathbb{R}^2, \\ |(0, g(t, x))| \leq a(t)\omega(|(x, y)|), & \text{for all } (t, x, y) \in \mathbb{R}_+ \times \mathbb{R}^2, \end{cases}$$

and hypotheses on ω guarantee that $\int_1^{+\infty} \frac{ds}{s+\omega(s)} = +\infty$ (see, e.g., [165]). Thus the above statement follows from Theorem 4.9.2.

Remark 4.9.3 The problem of the global existence of the solutions to problem (4.9.8) was investigated in [631], where the global existence criteria were given provided

$$\{t > 0 : \tau(t) = 0\} = \{0\}. \tag{4.9.9}$$

For $g(x) = F(x) = xln(1 + |x|), \quad x \in \mathbb{R}$, and $\tau[t] = \frac{1}{2}ln(t^2 - 2t + 2), \quad t \geq 1$, we prove that all solutions of problem (4.9.8) are global. Note that the results from [631] are not applicable in this case, as condition (4.9.9) is not satisfied.

4.10 Applications of Theorem 1.1.48 to Retarded and Impulsive Differential Equations

In this section, we shall apply Theorem 1.1.58 to the qualitative analysis of solutions to retarded differential equations. Delay differential equations arise in the theory of control, mathematical biology, mathematical economics, and the theory of systems which communicate through lossless channels (see, e.g., [266]). The oscillatory behavior of the solutions of delay differential equations was studied (see [261, 534] and the citations therein).

We shall give sufficient conditions under which the retarded differential equation

$$x'(t) = F(t, x(a(t))), \quad t \geq 0, \tag{4.10.1}$$

has a positive solution on $\mathbb{R}_+$. Here $a \in C^1(\mathbb{R}_+, \mathbb{R}_+)$ is a diffeomorphism of $\mathbb{R}_+$ with $a(t) \leq t$ on $\mathbb{R}_+$.

Theorem 4.10.1 (The Lipovan Inequality [355]) *Assume that $F : \mathbb{R}_+ \times \mathbb{R} \to \mathbb{R}$ is a continuous function for which there exists a constant $\delta > 0$ and a continuous function $a(t) > 0$ for all $t \in \mathbb{R}_+$ such that for all $t \in \mathbb{R}_+$ and for all $0 < x < \delta$,*

$$|F(t, x)| \leq a(t)x.$$

If for all $t \geq 0$,

$$\int_{a(t)}^{t} a(s)ds < \frac{1}{e}, \quad \int_0^{+\infty} a(s)ds < +\infty,$$

then for every initial data x_0 *such that* $0 < x_0 < \delta \exp(-e \int_0^{+\infty} a(s)ds)$, *(4.10.1) has a positive solution* $x(t)$ *on* $\mathbb{R}_+$ *which satisfies, for all* $t \geq 0$,

$$x_0 \exp\left(-e \int_0^t a(s)ds\right) < x(t) \leq x_0 \exp\left[\int_0^t a(s) \exp(e \int_{a(s)}^s a(\sigma)d\sigma)ds\right].$$

Proof Under our conditions, it is known (see [266], Chaps. 3, 4) that for every $x_0 \in \mathbb{R}$, there exists a solution $x(t)$ of equation (4.10.1) with initial data $x(0) = x_0$, defined on some maximal interval $[0, T)$ (note that $a(0) = 0$). Moreover, if $T < +\infty$, then

$$\limsup_{t \to T} |x(t)| = +\infty. \tag{4.10.2}$$

Following an idea introduced in [261], let us perform the transformation for all $0 \leq t < T$,

$$y(t) = x(t) \exp\left(e \int_0^t a(s)ds\right).$$

Observe that

$$y(0) = x(0) = x_0$$

and for all $t \in [0, T)$,

$$y'(t) = x'(t) \exp\left(e \int_0^t a(s)ds\right) + ex(t)a(t) \exp\left(e \int_0^t a(s)ds\right). \tag{4.10.3}$$

In particular,

$$y'(t) = F(0, x_0) + ex_0 a(0) \geq a(0)x_0(e - 1) > 0.$$

Therefore, for any $t > 0$ near zero, we have

$$x_0 < y(t) < \delta \exp\left(e \int_0^t a(s)ds\right). \tag{4.10.4}$$

We shall show that (4.10.4) holds on $(0, T)$. If not, there are two possible cases:

(A) There exists a $t_1 \in (0, T)$ such that (4.10.4) holds for all times $t \in (0, t_1)$ and $y(t_1) = \delta \exp\left(e \int_0^{t_1} a(s)ds\right)$.

(B) There exists a $t_2 \in (0, T)$ such that (4.10.4) holds for all times $t \in (0, t_2)$ and $y(t_2) = x_0$.

Let us first show that case (A) does not occur.

Assume that (A) holds. From (4.10.4) it follows that for all $0 \leq t < t_1$,

$$0 < x(t) < \delta.$$

Using the hypothesis on F, we therefore obtain from (4.10.3) that for all $0 \le t < t_1$,

$$y'(t) \le a(t)x(a(t)) \exp\left(e \int_0^t a(s)ds\right) + ex(t)a(t) \exp\left(e \int_0^t a(s)ds\right).$$

Considering the definition by y, we infer for all $0 \le t < t_1$,

$$y'(t) \le a(t)y(a(t)) \exp\left(e \int_{a(t)}^t a(s)ds\right) + ea(t)y(t).$$

Integrating on $[0, t]$, we deduce that for all $0 \le t < t_1$,

$$y(t) \le x_0 + e \int_0^t a(s)y(s)ds + \int_0^t a(s)y(a(s)) \exp\left(e \int_{a(s)}^s a(\sigma)d\sigma\right) ds.$$

A change of variables transforms the above inequality into for all $0 \le t < t_1$,

$$y(t) \le x_0+e \int_0^t a(s)y(s)ds+ \int_0^{a(t)} a(a^{-1}(s))y(s) \exp(e \int_s^{a^{-1}(s)} a(\sigma)d\sigma)(a^{-1})'(s)ds.$$

Applying Theorem 1.1.48, we obtain that for all $0 \le t < t_1$,

$$y(t) \le x_0 \exp\left[e \int_0^t a(s)ds + \int_0^t a(s) \exp(e \int_{a(s)}^s a(\sigma)d\sigma)ds\right], \tag{4.10.5}$$

after performing a change of variables. Now letting $t \to t_1$ in the above relation, we get

$$\begin{aligned} y(t_1) &\le x_0 \exp\left[e \int_0^{t_1} a(s)ds + \int_0^{t_1} a(s) \exp(e \int_{a(s)}^s a(\sigma)d\sigma)ds\right] \\ &\le x_0 \exp\left[e \int_0^{+\infty} a(s)ds + \int_0^{t_1} a(s) \exp(e \int_{a(s)}^s a(\sigma)d\sigma)ds\right] \\ &\le \delta \exp\left[\int_0^{t_1} a(s) \exp(e \int_{a(s)}^s a(\sigma)d\sigma)ds\right] \\ &\le \delta \exp\left(e \int_0^{t_1} a(s)ds\right). \end{aligned}$$

The obtained contradiction proves that case (A) never holds. Therefore we have, for all $0 \le t < T$,

$$x(t) < \delta, \tag{4.10.6}$$

and then (4.10.5) also holds on $[0, T)$. It follows that, for all $0 \leq t < T$,

$$x(t) \leq x_0 \exp\left[\int_0^t a(s)\exp(e\int_{a(s)}^s a(\sigma)d\sigma)ds\right].$$

Consider now case (B). We shall prove that, for all $0 \leq t < t_2$,

$$y'(t) > 0. \tag{4.10.7}$$

As $y'(t) > 0$ it is clear that (4.10.7) holds for any $t \geq 0$ near zero. If (4.10.7) does not hold, there exists a $t_3 \in (0, t_2)$ such that $y'(t_3) = 0$ and $y'(t) > 0$ for all $t \in (0, t_3)$. Taking into account (4.10.3) and (4.10.6), we may obtain for all $0 < t < t_3$ that

$$\begin{aligned}
y'(t) &= F(t, x(a(t)))\exp\left(e\int_0^t a(s)ds\right) + ex(t)a(t)\exp\left(e\int_0^t a(s)ds\right)\\
&\geq -a(t)x(a(t))\exp\left(e\int_0^t a(s)ds\right) + ea(t)y(t)\\
&= -a(t)y(a(t))\exp\left(e\int_{a(t)}^t a(s)ds\right) + ea(t)y(t).
\end{aligned}$$

Letting $t \to t_3$ in the above inequality and taking into account the monotonicity of y on $[0, t_3)$, we can deduce that

$$y'(t) \geq a(t_3)y(t_3)[e - \exp(e\int_{a(t)}^t a(s)ds)] > 0.$$

This contradicts our assumption $y'(t_3) = 0$ and completes the proof that case (B) does not hold.

In conclusion, (4.10.4) holds on $[0, T)$. This shows in particular that $x(t)$ does not blow up in a finite time. By (4.10.2) we infer that $T = +\infty$ and the proof is thus complete. □

For the existence of positive solutions to equation (4.10.1), we have the following result.

Theorem 4.10.2 (The Lipovan Inequality [355]) *Assume that $F : \mathbb{R}_+ \times \mathbb{R} \to \mathbb{R}$ is a continuous function for which there exists a constant $\delta > 0$ and a continuous function $a(t) > 0$ for all $t \in \mathbb{R}_+$ such that for all $t \in \mathbb{R}_+$ and $0 < x < \delta$,*

$$0 > F(t, x) \geq -a(t)x.$$

If $a \in C(\mathbb{R}_+, \mathbb{R}_+)$ with $a(t) \le t$ for all $t \ge 0$, and $\lim_{t \to +\infty} a(t) = +\infty$, while for all $t \ge 0$,

$$\int_{a(t)}^{t} a(s)ds \le \frac{1}{e},$$

then equation (4.10.1) has a positive solution $x(t)$ on $\mathbb{R}_+$ satisfying, for all $t \ge 0$,

$$x_0 \exp\left(-e \int_0^t a(s)ds\right) < x(t).$$

There are cases when the above result yields the existence of a positive solution to equation (4.10.1), but Theorem 4.10.2 is not applicable.

Example 4.10.1 Take $a(t) \equiv ln(t+1)$ and $F(t,x) \equiv (t+1)^{-4}x^2$. By Theorem 4.10.1, (4.10.1) has in this case a positive solution on $\mathbb{R}_+$.

With respect to the conditions we imposed in Theorem 4.10.2 to guarantee the existence of a positive solution, the following example is relevant.

Example 4.10.2 Consider the delay differential equation

$$x'(t) + p(t)x(a(t)) = 0, \tag{4.10.8}$$

where $p \in C(\mathbb{R}_+, \mathbb{R}_+)$ and $a : \mathbb{R}_+ \to \mathbb{R}_+$ is an increasing diffeomorphism of $\mathbb{R}_+$ such that $a(t) \le t$ on $\mathbb{R}_+$.

By a result in [534], if

$$\int_0^{+\infty} p(s)ds < +\infty, \qquad \lim_{t \to +\infty} inf \int_{a(t)}^{t} p(s)ds > \frac{1}{e},$$

then all solutions of equation (4.10.8) are oscillatory.

Let us now show that Theorem 4.10.1 can be applied to prove the existence of non-oscillatory solutions to certain impulsive differential equations.

Let $a : \mathbb{R}_+ \to \mathbb{R}_+$ be an increasing diffeomorphism of $\mathbb{R}_+$ such that $a(t) \le t$ on $\mathbb{R}_+$ and such that $a(t) = t - m\tau$ far out where $\tau > 0$ and $m \in \mathbb{N}$.

Consider the delay impulsive differential equation

$$y'(t) + p(t)y((a(t))) = 0, \ y(t_k^+) - y(t_k) = by(t_k), \ k \ge 1, \tag{4.10.9}$$

where $b > -1$, and $t_1 > 0$, $t_{k+1} - t_k = \tau$, and $p \in C(\mathbb{R}_+, \mathbb{R}_+)$.

By a result in [684], all solutions of equation (4.10.9) are oscillatory if and only if all solutions of the delay differential equation

$$x'(t) + p(t)(1+b)^{-m}x((a(t))) = 0$$

are oscillatory. In view of Theorem 4.10.1, we can obtain the following theorem.

Theorem 4.10.3 ([355]) *If*

$$\int_0^{+\infty} |p(s)|ds < \frac{(1+b)^m}{e},$$

then the impulsive differential equation (4.10.9) has a non-oscillatory solution.

4.11 Applications of Corollary 1.1.15 and Theorem 1.1.60 to Integro-Differential Equations

In this section, we shall use Corollary 1.1.15 and Theorem 1.1.65 to study the following integro-differential equation

$$px^{p-1}(t)x'(t) = F\left(t, x(t-\tau(t)), \int_\alpha^{t-\tau(t)} G(t_1, x(t_1-\tau(t_1)))dt_1\right) \tag{4.11.1}$$

for all $t \in I$, where $p > 1$ is constant, let $F \in C(I \times \mathbb{R}^2, \mathbb{R})$, $G \in C^1(I \times \mathbb{R}, \mathbb{R})$ and $\tau \in C^1(I, I)$ be non-increasing with $t - \tau(t) \geq 0$, $t - \tau(t) \in C^1(I, I)$ and $\tau(\alpha) = 0$.

The following result provides a bound on the solutions of equation (4.11.1).

Theorem 4.11.1 ([17]) *Assume that $F : I \times \mathbb{R}^2 \to \mathbb{R}$ is a continuous function, and there exist continuous non-negative functions $b_i(t)$, $i = 1, 2$, such that*

$$\begin{cases} |F(t,u,v)| \leq b_1(t)g(|u|) + b_1(t)|v|, & (4.11.2) \\ |G(s,w)| \leq b_2(s)g(|w|), & (4.11.3) \end{cases}$$

where the function g is the same as in Theorem 1.1.59. Let $M = \max_{x \in I}(\frac{1}{1-\tau'(x)})$. If $x(\eta)$ is any solution of the problem (4.11.1), then

$$|x(\eta)| \leq \left[G_1^{-1}(G_1(|x(\alpha)|^p) + B[Mb_1(\overline{\eta}_1), M^2 b_2(\overline{\eta}_1)b_2(\overline{\eta}_2)])\right]^{\frac{1}{p}}, \tag{4.11.4}$$

where the functions G_1, G_1^{-1} are as in Corollary 1.1.15, $\overline{\eta}_1 = \eta_1 + \tau(t_1)$, $\overline{\eta}_2 = \eta_2 + \tau(t_2)$, for η_1, $\eta_2 \in I$, and

$$B[Mb_1(\overline{\eta}_1), M^2 b_2(\overline{\eta}_1)b_2(\overline{\eta}_2)] = \int_{\phi(\alpha)}^{\phi(\eta)} Mb_1(\overline{\eta}_1)d\eta_1 + \int_{\phi(\alpha)}^{\phi(\eta)}\int_{\phi(\alpha)}^{\phi(\eta_1)} M^2 b_2(\overline{\eta}_1)b_3(\overline{\eta}_2)d\eta_2 d\eta_1, \tag{4.11.5}$$

where $\phi(\gamma) = \gamma - \tau(\gamma)$ for $\gamma \in I$.

Proof It is obvious that the solution $x(\eta)$ of the problem (4.11.1) satisfies the equivalent integral equation

$$x^p(\eta) = x^p(\alpha) + \int_\alpha^\eta F(t, x(t-\tau(t)), \int_\alpha^{t_1-\tau(t_1)} G(t_2, x(t_2-\tau(t_2)))dt_2)dt_1. \tag{4.11.6}$$

From (4.11.2)–(4.11.3), and making the change of variables, we have

$$\begin{aligned} |x(\eta)|^p &\le |x(\alpha)|^p + \int_\alpha^\eta b_1(t_1)g(|x(t_1-\tau(t_1))|) \\ &\quad + \int_\alpha^\eta \int_\alpha^{t_1-\tau(t_1)} b_1(t_1)b_2(t_2)g(|x(t_2-\tau(t_2))|)dt_2dt_1 \\ &\le |x(\alpha)|^p + \int_\alpha^{\eta-\tau(\eta)} Mb_1(\overline{\eta}_1)g(|x(\eta_1)|)d\eta_1 \\ &\quad + \int_\alpha^{\eta-\tau(eta)} \int_\alpha^{\eta_1-\tau(\eta_1)} M^2b_1(\overline{\eta}_1)b_2(\overline{\eta}_2)g(|x(\eta_2)|)d\eta_2d\eta_1 \end{aligned} \tag{4.11.7}$$

where $\overline{\eta}_1 = \eta_1 + \tau(t_1)$, $\overline{\eta}_2 = \eta_2 + \tau(t_2)$, for η_1, $\eta_2 \in I$. Now applying the inequality in Corollary 1.1.15 to (4.11.7) yields the desired result. □

Next we consider the following integro-differential equation:

$$(h(t)x'(t))' = F(t, x(t) - \tau(t), \int_\alpha^t G(t_1, x(t_1 - \tau(t_1)))dt_1), \tag{4.11.8}$$

for all $t \in I$, $F \in C(I \times \mathbb{R}^2, \mathbb{R})$, $G \in C^1(I \times \mathbb{R})$, and h is positive and continuous in I. The following theorem provides an upper bound on the solutions of equation (4.12.8).

Theorem 4.11.2 ([17]) *Assume that $F : I \times \mathbb{R}^2 \to \mathbb{R}$ is a continuous function, and there exist continuous non-negative functions $f_i(t)$, $i = 2, 3$, such that*

$$\begin{cases} |F(t, u, v)| \le f_2(t)[|u|g(|u|) + |v|], \\ |G(s, w)| \le f_3(s)|w|g(|w|), \end{cases} \tag{4.11.9}$$

where the function g is the same as in Theorem 1.1.60. If $x(t)$ is any solution of the problem (4.11.8), then

$$|x(t)| \le \exp[G_e^{-1}(G_e(\ln(a)) + C[f_2(\overline{s}_2), f_2(\overline{s}_2)f_3(\overline{s}_3)])] - 1, \tag{4.11.10}$$

where $G_e(r) = \int_{r_0}^{r} \frac{ds}{g(e^s)}$ *for all* $r \geq r_0 > 0$, $a = 1+|x(\alpha)|+h(\alpha)|x'(\alpha)| \int_{\phi(\alpha)}^{\phi(t)} \frac{1}{h(\bar{s}_1)} ds_1$ *and*

$$\begin{aligned} C[f_1,f_2,f_3] &= \int_{\phi(\alpha)}^{\phi(t)} \frac{1}{h(\bar{s}_1)} \int_{\phi(\alpha)}^{\phi(t)_1} f_2(\bar{s}_2) M^2 ds_2 ds_1 \\ &+ \int_{\phi(\alpha)}^{\phi(t)} \frac{1}{h(\bar{s}_1)} \int_{\phi(\alpha)}^{\phi(t_1)} f_2(\bar{s}_2) \int_{\phi(\alpha)}^{\phi(t_2)} f_3(\bar{s}_3) M^3 ds_3 ds_2 ds_1 \end{aligned} \tag{4.11.11}$$

and $\bar{s}_1 = s_1 + \tau(t_1)$, $\bar{s}_2 = s_2 + \tau(t_2)$, *and* $\bar{s}_3 = s_3 + \tau(t_2)$ *for* $s_1, s_2, s_3 \in I$.

Proof We easily verify that the solution of $x(t)$ of the problem (4.11.8) satisfies the equivalent integral equation

$$\begin{aligned} x(t) &= x(\alpha) + h(\alpha)x'(\alpha) \int_{\alpha}^{t} \frac{1}{h_1(t_1)} dt_1 \\ &+ \int_{\alpha}^{t} \frac{1}{h(t_1)} \int_{\alpha}^{t_1} F(t_2, x(t_2) - \tau(t_2)), \int_{\alpha}^{t} G(t_3, x(t_3 - \tau(t_3))) dt_3) dt_2 dt_1. \end{aligned} \tag{4.11.12}$$

From (4.11.9), and making the change of variables, we have

$$\begin{aligned} |x(t)| + 1 &\leq 1 + x(\alpha) + h(\alpha)|x'(\alpha)| \int_{\phi(\alpha)}^{\phi(t)} \frac{1}{h(\bar{s}_1)} ds_1 \\ &+ \int_{\phi(\alpha)}^{\phi(t)} \frac{1}{h(\bar{s}_1)} \int_{\phi(\alpha)}^{\phi(t_1)} f_2(\bar{s}_2) M^2 |x(s_2)| g(|s(s_2)|) ds_2 ds_1 \\ &+ \int_{\phi(\alpha)}^{\phi(t)} \frac{1}{h(\bar{s}_1)} \int_{\phi(\alpha)}^{\phi(t_1)} f_2(\bar{s}_2) \int_{\phi(\alpha)}^{\phi(t_2)} f_3(\bar{s}_3) M^3 |x(s_3)| g(|x(s_3)|) ds_3 ds_2 ds_1, \end{aligned} \tag{4.11.13}$$

where $s_i = t_i - \tau(t_i)$, $\bar{s}_i = s_i + \tau(t_i)$, $s_i \in I$ for $i = 1, 2, 3$. Now when $\phi(u) = u$ and $f_1 = f_4 = \cdots = f_n = 0$, applying the inequality in Theorem 1.1.60 to (4.11.13) yields the desired result. □

4.12 Applications of Theorem 1.1.73 to Retarded Differential Equations

Consider the retarded differential equation

$$\begin{cases} x'(t) = F(t, x(t), x(\alpha(t)), x'(\beta(t))), \\ x(0) = x_0, \end{cases} \tag{4.12.1}$$

where $F \in C(\mathbb{R}^+ \times \mathbb{R}^n \times \mathbb{R}^n \times \mathbb{R}^n, \mathbb{R}^n)$, $\alpha, \beta \in C^1(\mathbb{R}_+, \mathbb{R}_+)$ and $\alpha(t) \le t$, $\beta(t) < t$ for all $t > 0$.

Theorem 4.12.1 ([13]) *Suppose that* $\alpha : \mathbb{R}_+ \to \mathbb{R}_+$ *is an increasing diffeomorphism and for all* $t \ge 0$, $x, y, z \in \mathbb{R}^n$,

$$|F(t,x,y,z)| \le f_1(t)w_1(|x|) + f_2(t)w_2(|y|) + f_3(t)h(|z|), \tag{4.12.2}$$

where $f_i, h, w_j \in C(\mathbb{R}_+, \mathbb{R}_+)$, $w_j(u) > 0$ *for all* $u > 0$, w_j *are non-decreasing,* $\int_{u_j}^{+\infty} \frac{dz}{w_j(z)} = +\infty$ *for all* $u_j > 0$, $i = 1,2,3$, $j = 1,2$, *and* $w_1 \infty w_2$. *Then all solutions of problem (4.12.1) exist on* $\mathbb{R}^+$.

Proof As in [210], let $[0, T)$ be the maximal interval of existence of a solution $x(t)$ for problem (4.12.1), which satisfies the initial condition $x(0) = x_0$. If $T < +\infty$, then

$$\limsup_{t \to T} |x(t)| = +\infty. \tag{4.12.3}$$

Let $M =: \sup_{t\in[0,t]}\{\beta(t)\}$. Then $M < T$ since $\beta(t) < t$. Let $u(t) = |x(t)|$. Integrating (4.12.1), we obtain, for all $0 \le t < T$,

$$u(t) \le c + \int_0^t f_1(s)w_1(u(s))ds + \int_0^t f_2(s)w_2(u(\alpha(s)))ds,$$

where

$$c := |x_0| + \sup_{t\in[0,M]} \{h(|\dot{x}|)\}(\int_0^T f_3(s)ds).$$

It is equivalent to

$$u(t) \le c + \int_0^t f_1(s)w_1(u(s))ds + \int_{\alpha(0)}^{\alpha(t)} f_2(\alpha^{-1}(s))(\alpha^{-1})'(s)w_2(u(s)), \quad 0 \le t < T,$$

because $\alpha : \mathbb{R}_+ \to \mathbb{R}_+$ is an increasing diffeomorphism. Applying Theorem 1.1.73, where $a(t)$, $f_1(t,s)$, $f_2(t,s)$, $b_1(t)$, $b_2(t)$ are replaced by $c, f_1(s), f_2(\alpha^{-1}(s))(\alpha^{-1})'(s), t, \alpha(t)$, respectively, we get, for all $t_0 \le t \le T_1$,

$$u(t) \le W_2^{-1}[W_2(r_2(t)) + \int_{\alpha(t_0)}^{\alpha(t)} f_2(\alpha^{-1}(s))(\alpha^{-1})'(s)ds], \tag{4.12.4}$$

where T_1 is given as in Theorem 1.1.73 and $r_2(t) = W_1^{-1}[W_1(c) + \int_{t_0}^{t} f_1(s)ds]$. Note that $\int_{u_j}^{+\infty} \frac{dz}{w_j(z)} = +\infty, j = 1, 2$. As in Remark 1.1.23, (4.12.4) holds on $[t_0, T_1]$ for any $T_1 \in [t_0, T)$. So (4.12.4) also holds on $[t_0, T)$. Observe that the right hand side of (4.12.4), as a function in t, is bounded on the compact interval $[t_0, T]$ because of continuity. So is $u(t)$ on $[t_0, T)$, which contradicts (4.12.3). This completes the proof. □

The case that $w_1 = w_2$ in Theorem 4.1.1 is just Proposition 3 in [355]. For example, the problem

$$\begin{cases} x'(t) = a\sqrt[3]{x(t)} + bx(\gamma t) + cx'(\gamma(t)), & t > 0, \\ x(0) = x_0, \end{cases} \tag{4.12.5}$$

where a, b, c are constants and $\gamma \in (0, 1)$, satisfies conditions in Theorem 4.1.1. Hence all solutions of this problem exist globally on $\mathbb{R}_+$.

Second, we consider the functional differential equation

$$x' = \frac{1}{t} + \exp(-t)\sqrt{|x(t)| + 1} + t\exp(-t)Fx(t), \tag{4.12.6}$$

where $x : [0, +\infty) \to \mathbb{R}$ is a differentiable function and F is a continuous operator on $\mathbb{R}$ such that $|Fx| \le c_0|x|$ for a constant $c_0 > 0$. In particular, (4.12.6) becomes an integro-differential equation and a retarded functional differential equation as we take $Fx(t) = \int_{t_0}^{t} H(t, s, x(s))ds$ and $Fx(t) = x(t - \tau)$ respectively. General theory on such equations can be found in [266, 326].

Integrating (4.12.6), we obtain, for all $t \ge t_0 > 0$,

$$|x(t)| \le a(t) + \int_{t_0}^{t} f_1(t, s)w_1(|x(s)|)ds + \int_{t_0}^{t} f_2(t, s)w_2(|x(s)|)ds, \tag{4.12.7}$$

where $a(t) = |x(t_0)| + lnt - lnt_0$, $w_1(u) = \sqrt{u + 1}, w_2(u)$, $f_1(t, s) = \exp(-s)$ and $f_2(t, s) = c_0 s\exp(-s)$. Note that $w_1 \propto w_2$ because $\frac{w_2(u)}{w_1(u)} = \frac{u}{\sqrt{u+1}}$ is non-decreasing for all $u > 0$. Moreover, for all $u_1, u_2 > 0$, we have

$$\begin{cases} W_1(u) \int_{u_1}^{u} \frac{dz}{\sqrt{z + 1}} = 2(\sqrt{u + 1} - \sqrt{u_1 + 1}), \quad W_1^{-1}(u) = (\frac{u}{2} + \sqrt{u_1 + 1})^2 - 1, \\ W_2(u) \int_{u_2}^{u} \frac{dz}{z} = ln\frac{u}{u_2}, \quad W_2^{-1}(u) = u_2\exp(u), \\ r_2(t) = \left[\sqrt{|x(t_0)| + lnt - lnt_0 + 1} + \frac{1}{2}(\exp(-t_0) - \exp(-t))\right]^2 - 1. \end{cases}$$

As in Remark 1.1.23, $T_1 = +\infty$ because $\int_{u_1}^{+\infty} \frac{dz}{w_1(z)} = \int_{u_1}^{+\infty} \frac{dz}{\sqrt{z+1}} = +\infty$ and $\int_{u_2}^{+\infty} \frac{dz}{w_2(z)} = \int_{u_2}^{+\infty} \frac{dz}{c_0 z} = +\infty$. Applying Theorem 1.1.73 to (4.12.7), we get for all $t \geq t_0$,

$$|x(t)| \leq \left\{ \left[\sqrt{|x(t_0)| + lnt - lnt_0 + 1} + \frac{1}{2}(\exp(-t_0) - \exp(-t)) \right]^2 - 1 \right\}$$
$$\times \exp\left[c_0(t_0+1)\exp(-t_0) - c_0(t+1)\exp(-t)\right]. \tag{4.12.8}$$

Observe that $\exp(-t)$ and $(t+1)\exp(-t)$ tend to 0 as $t \to +\infty$ in (4.12.8). It follows that there exist positive constants K_1 and K_2 such that for all $t \geq t_0$,

$$|x(t)| \leq K_1 + K_2 \ln t, \tag{4.12.9}$$

which gives an estimate for $x(t)$.

Now, we shall use Theorem 1.1.73 to study almost periodicity of a weak hyperbolic invariant manifold. We note that the function $u(t) = \exp(-\rho t)|\phi_{t_0} + \delta(\xi)_{t+t_0+\delta} - \phi_{t_0}(\xi)_{t+t_0}|$ is estimated in Lemma 5.4 of [718], where $\phi_{t_0}(\xi)_t$ is the general form of solutions on the weak hyperbolic invariant manifold. $u(t)$ is proved to satisfy the following inequality (4.12.10) and the following lemma is proved in [718] and plays an important role in the estimation.

Lemma 4.12.1 ([13]) *Suppose $t_0 \in \mathbb{R}$ is given and suppose the non-negative bounded continuous function $u(t) : [t_0, +\infty) \to \mathbb{R}$ satisfies the inequality for all $t \geq t_0$,*

$$u(t) \leq k + b\exp(-\gamma t) - b_0 t \exp(-\gamma t) + c\int_{t_0}^{t} \exp(-\gamma(t-s))u(s)ds$$
$$+ d\int_{t}^{+\infty} \exp(-\omega(t-s))u(s)ds, \tag{4.12.10}$$

where $\gamma > 0, \omega < 0, k \geq 0, c > 0, d \geq 0$ and $\gamma, \omega, k, b, b_0, c, d$ are constants. If $\lambda := \frac{c}{\gamma} + \frac{d}{-\omega} < 1$, then we have for all $t \geq t_0$,

$$u(t) \leq \frac{k}{1-\lambda} + \frac{b_0}{c}\exp(-\gamma t) + \frac{|\bar{b}|}{1-\lambda}\exp\left(-\gamma t + \frac{c}{1-\lambda}(t-t_0)\right), \tag{4.12.11}$$

where $\bar{b} = b - \frac{b_0}{c} - b_0 t_0 + \frac{b_0 d}{c(\gamma-\omega)} - \frac{kc}{(1-\lambda)\gamma}\exp(\gamma t_0)$.

We consider a special form of (4.12.10) where $d = 0$ and $k > 0$, i.e., for all $t \geq t_0$,

$$u(t) \leq k + b\exp(-\gamma t) - b_0 t \exp(-\gamma t) + c\int_{t_0}^{t} \exp(-\gamma(t-s))\,u(s)ds. \tag{4.12.12}$$

Clearly, it involves a function term outside integrals that the result in [355] cannot be applied. However, by Theorem 1.1.73, we obtain

Theorem 4.12.2 ([13]) *Suppose that (4.12.12) holds and conditions in Lemma 4.12.1 are satisfied with $d = 0$ and $k > 0$. Then we have for all $t \geq t_*$,*

$$u(t) \leq (k(\exp \gamma t) + b - b_0 t) \exp\Big((c-\gamma)t - ct_0\Big), \tag{4.12.13}$$

for some $t_* \geq t_0$.

Proof Multiplying (4.12.12) by $\exp(\gamma t)$, we get, for all $t \geq t_0$,

$$v(t) \leq a(t) + c\int_{t_0}^{t} v(s)ds, \tag{4.12.14}$$

where $v(t) := \exp(\gamma t)u(t)$ and $a(t) := k\exp(\gamma t) + b - b_0 t$. Since $\gamma > 0$ and $k > 0$, we can see that $\lim_{t\to+\infty} a(t) = +\infty$ and $\lim_{t\to+\infty} a'(t) = \lim_{t\to\infty}(k\gamma\exp(\gamma t) - b_0) = +\infty$. Therefore, there exists a $t_* \in [t_0, +\infty)$ such that, for all $t \in [t_*, +\infty)$,

$$a(t) \geq 0, \quad a'(t) \geq 0. \tag{4.12.15}$$

In what follows, it suffices to discuss (4.12.14) for all $t \in [t_*, +\infty)$.

Clearly, (4.12.14) is in the form of (1.1.486), where $t_0 = t_*,\ t_1 = +\infty, b_1(t) = t, f_1(t,s) = c$, and $w_1(u) = u$. As in Remark 1.1.23, $T_1 = +\infty$ because $\int_{u_1}^{+\infty} \frac{dz}{z} = +\infty$ for all $u_1 > 0$. By Theorem 1.1.73 and the fact (4.12.15), we get, for all $t \geq t_*$,

$$v(t) \leq (k\exp(\gamma t) + b - b_0 t)\exp\left(c(t - t_0)\right), \tag{4.12.16}$$

that is, (4.12.13) is proved. □

Clearly, $\lim_{t\to+\infty} \exp(-\gamma t) = 0$ and $\exp(-\gamma t)$ is bounded on $[t_*, +\infty)$, since $\gamma > 0$. It follows from (4.12.11) and the definition of λ that for all $t \geq t_*$,

$$u(t) \leq M_1 + M_2 \exp(ct), \tag{4.12.17}$$

where both M_1 and M_2 are positive constants. Note that $c < \frac{-\gamma^2 + 2\gamma c}{\gamma - c}$. This implies that (4.12.17) is sharper than (4.12.16) for large t and thus the estimate in Theorem 4.12.2 is sharper than the estimate in Lemma 4.12.1 for large t in this aspect.

4.13 Applications of Theorem 1.2.1 to Solutions of Linear Differential Equations

In this section, we shall use Theorem 1.2.1 to show the boundedness of solutions of the following linear differential equation:

$$y'' + A(t)y = 0, \tag{4.13.1}$$

where the coefficient $A(t)$ could be discussed apparently in three cases as $t \to +\infty$:

(Case I) $\lim_{t\to+\infty} A(t) = a^2$,
(Case II) $\lim_{t\to+\infty} A(t) = -a^2$,
(Case III) $\lim_{t\to+\infty} A(t) = +\infty$.

In the sequel, these cases will be discussed and the corresponding conclusions will respectively improve the results concluded by Bellman.

Utilizing Theorem 1.2.1, we can prove the following theorem:

Theorem 4.13.1 ([438]) *If the coefficient $A(t)$ of the equation (4.13.1) satisfies that*

(i) $A(t) > 0$, $A(t)$, $A'(t)$ are absolutely integrable;
(ii) $\sqrt{A(t)} - \frac{1}{2}\int_0^t \frac{|A'(s)|}{\sqrt{A(s)}}ds > R > 0$,

with a constant $R > 0$. Then all solutions of equation (4.13.1) are bounded as $t \to +\infty$.

Proof Multiplying y' to the both sides of (4.13.1) and integrating it then from 0 to t by parts, we have

$$\frac{y'^2}{2} + A(t)\frac{y^2}{2} - \int_0^t \frac{y^2}{2}A'(s)ds = c,\ c > 0$$

or

$$\frac{y'^2}{2} + A(t)\frac{y^2}{2} = c + \int_0^t \frac{y^2}{2}A'(s)ds.$$

For $A(t) > 0$, then

$$A(t)\frac{y^2}{2} \le c + \int_0^t \frac{y^2}{2}|A'(s)|ds, \tag{4.13.2}$$

which can be rewritten as

$$\left(\sqrt{A(t)}y\right)^2 \le 2c + \int_0^t |\sqrt{A(s)}y||y|\frac{|A'(s)|}{\sqrt{A(s)}}ds.$$

Employing Theorem 1.2.1, we then get

$$|\sqrt{A(t)}y| \leq \sqrt{2c} + \frac{1}{2}\int_0^t \frac{|A'(s)|}{\sqrt{A(s)}}|y|ds. \tag{4.13.3}$$

Then we put $M = \max_{0\leq t<+\infty} |y|$ into (4.13.3), and find that

$$|\sqrt{A(t)}M| \leq \sqrt{2c} + \frac{1}{2}\int_0^t \frac{|A'(s)|}{\sqrt{A(s)}}Mds, \tag{4.13.4}$$

or

$$\left(\sqrt{A}(t) - \frac{1}{2}\int_0^t \frac{|A'(s)|}{\sqrt{A(s)}}ds\right)M \leq \sqrt{2c}.$$

Utilizing (1.2.1) in Theorem 1.2.1, then we have

$$|y| \leq M \leq \frac{\sqrt{2c}}{\sqrt{A(t)} - \frac{1}{2}\int_0^t \frac{|A'(s)|}{\sqrt{A(s)}}ds} < \frac{\sqrt{2c}}{R} < +\infty$$

this completes the proof. □

Corollary 4.13.1 ([66]) *For $y''(t)+A(t)y(t) = 0$, if $A > 0$, $A'(t) \geq 0$, for all $t \geq t_0$, then as $t \to +\infty$, all the solutions are bounded.*

Proof Suppose that $A(0) > 0$, then the condition (*i*) of Theorem 4.13.1 is satisfied. Meanwhile, $A'(t) \geq 0$

$$\begin{aligned}\sqrt{A(t)} - \frac{1}{2}\int_0^t \frac{|A'(s)|}{\sqrt{A(s)}}ds &= \sqrt{A(t)} - \frac{1}{2}\int_0^t \frac{A'(t)}{\sqrt{A(s)}}ds \\ = \sqrt{A(t)} - \frac{1}{2}(2\sqrt{A(s)}|_0^t) &= \sqrt{A(0)} > \frac{1}{2}\sqrt{A(0)} = R > 0.\end{aligned}$$

So the condition (ii) holds, and the proof is thus complete. □

Now we consider the equation

$$y''(t) + (a^2 + \phi(t))y(t) = 0, \tag{4.13.5}$$

Using (4.13.4), we can conclude the following theorem.

Theorem 4.13.2 ([438]) *If (4.13.5) satisfies*

(i) $|\phi(t)| \leq b^2 < a^2$, $\phi(t) > 0$, *for all* $t \geq t_0$,
(ii) $\sqrt{\phi(t)} - \frac{1}{2}\int_0^t \frac{|\phi'(s)|}{\sqrt{a^2-b^2}}ds > R > 0$, $t \to +\infty$,

then as $t \to +\infty$, all the solutions to equation (4.13.5) are bounded.

Proof From the condition (i),

$$|A(t)| = |a^2 + \phi(t)| \geq |a^2| - |\phi(t)| \geq a^2 - b^2,$$

then from (4.13.4), it follows

$$|\sqrt{a^2 + \phi(t)}M| \leq \sqrt{2c} + \frac{1}{2}\int_0^t \frac{|\phi'(s)|}{\sqrt{a^2 - b^2}} M ds,$$

or

$$\left(\sqrt{\phi(t)} - \frac{1}{2}\int_0^t \frac{|\phi'(t_1)|}{\sqrt{a^2 - b^2}} dt_1\right) M \leq \sqrt{2c}.$$

By the condition (ii), $|y| \leq M \leq \frac{\sqrt{2c}}{R} < +\infty$. □

Now we employ the following lemma.

Lemma 4.13.1 ([70]) *If all the solutions with their first-order derivatives of $u''(t) + a(t)u(t) = 0$ are bounded, then all the solutions to the equation*

$$u'' + (a(t) + b(t))u = 0 \tag{4.13.6}$$

are bounded under $\int_0^{+\infty} |b(\tau)| d\tau < +\infty$.

Then it is not difficult to prove the following theorem.

Theorem 4.13.3 ([438]) *If the equation*

$$y''(t) + (a^2(t) + \phi(t) + \psi(t))y(t) = 0, \tag{4.13.7}$$

satisfies that

(i) $|\phi(t)| \leq b^2 < a^2$, $\phi(t) > 0$, *for all* $t > t_0$,
(ii) $\int_0^{+\infty} |\phi(t)| dt < +\infty$, $\int_0^{+\infty} |\psi(t)| dt < +\infty$,
(iii) $[\sqrt{\phi(t)} - \frac{1}{2}\int_0^t \frac{|\phi'(s)|}{\sqrt{a^2 - b^2}} ds] > R > 0, \quad t \to +\infty$,

then all the solutions to equation (4.13.7) are bounded as $t \to +\infty$.

Proof Obviously, we need only to verify the first-order derivatives of solutions to (4.13.5) are bounded as $t \to +\infty$.

Multiplying y' to the both sides of (4.13.5) and integrating it from 0 to t by part, we have

$$\frac{y'^2}{2} + \frac{a^2}{2}y^2 + \int_0^t \phi(s)y(s)y'(s)ds = c_1,$$

then

$$y'^2(t) \le 2c_1 + 2\int_0^t |\phi(s)||y(s)||y'(s)|ds.$$

By Theorem 1.2.1 and Theorem 4.13.2, we can get by condition (ii)

$$|y'(t)| \le \sqrt{2c} + \int_0^t |\phi(s)y(s)|ds \le \sqrt{2c} + \int_0^t |\phi(s)|ds < +\infty.$$

Employing Lemma 4.13.1, the proof is complete. □

We can clearly see that Theorems 4.13.2 and 4.13.3 have been improved as $\phi(t) > 0$.

Now we extend Theorem 4.13.1 further. Suppose that $A(t) = f(t) + g(t)$ and $f(t) > 0$, so (4.13.1) can be rewritten as

$$y''(t) + [f(t) + g(t)]y(t) = 0. \tag{4.13.8}$$

Multiplying y' to the both sides of (4.13.8) and integrating it from 0 to t by parts, then

$$\frac{y'^2(t)}{2} - \frac{y'^2(0)}{2} + f(t)\frac{y^2}{2} - \int_0^t \frac{y^2(s)}{2} f'(s)ds + \int_0^t g(s)y(s)y'(s)ds = 0,$$

i.e., as

$$\frac{y'^2(t)}{2} + f(t)\frac{y^2(t)}{2} = \frac{y'^2(0)}{2} + \int_0^t \frac{y^2(s)}{2} f'(s)ds - \int_0^t g(s)y(s)y'(s)ds.$$

For $f(t) > 0$, then

$$f(t)y^2(t) \le y'^2(0) + \int_0^t y^2(s)|f'(s)|ds + \int_0^t 2|g(s)||y(s)||y'(s)|ds$$

which can be rewritten as

$$[\sqrt{f(t)}y]^2 \le y'^2(0) + \int_0^t |\sqrt{f(s)}||y(s)|\frac{|y(s)||f'(s)| + 2|g(s)||y'(s)|}{\sqrt{f(s)}}ds. \tag{4.13.9}$$

Employing Theorem 1.2.1 to (4.13.9), we then get

$$\sqrt{f'(t)}|y(t)| \le |y'(0)| + \frac{1}{2}\int_0^t \frac{|y(s)||f'(s)| + 2|g(s)||y'(s)|}{\sqrt{f(s)}}ds.$$

Integrating (4.13.8) from 0 to t, we can get

$$|y'(t)| \leq |y'(0)| + \int_0^t (|f(s)| + |g(s)|)|y(s)|ds,$$

which, gives us

$$\sqrt{f(t)}|y(t)| \leq |y'(0)| + \frac{1}{2}\int_0^t \frac{|f'(s)|}{\sqrt{f(s)}}|y(s)|ds \tag{4.13.10}$$
$$+ \int_0^t \frac{|g(s)||y'(0)|}{\sqrt{f(s)}}dt_1 + \int_0^t \frac{|g(s)| \int_0^t [f(\tau) + |g(\tau)|]|y(\tau)|d\tau}{\sqrt{f(s)}}ds.$$

Then we put $M = \max_{0\leq t<\infty} |y|$ into (4.13.10), and suppose that

(*i*) $\int_0^t \frac{|g(t_1)|}{\sqrt{f(t_1)}}|y|dt_1 < +\infty,$

(*ii*) $\sqrt{f(t_1)}dt_1 - \frac{1}{2}\int_0^t \frac{|f'(t_1)|}{\sqrt{f(t_1)}}dt_1 - \int_0^t \frac{|g(t_1)|}{\sqrt{f(t_1)}}[\int_0^{t_1} [f(t_2) + |g(t_2)|]dt_2]dt_1 = \tau(t) > R > 0,$

as $t \to +\infty$.

Then from (4.13.10) it follows that

$$\tau(t)M \leq |y'(0)| + \int_0^t \frac{|y'(0)||g(t_1)|}{\sqrt{f(t_1)}}dt_1,$$

$$|y| \leq M < \frac{1}{\tau(t)}[y'(0) + \int_0^t \frac{|y'(0)||g(t_1)|}{\sqrt{f(t_1)}}dt_1] < +\infty.$$

□

As a result, we conclude Theorem 4.13.4 as follows:

Theorem 4.13.4 ([438]) *If the equation* $y''(t) + (f(t) + g(t))y(t) = 0$ *under the conditions:*

(*i*) $\int_0^t \frac{|g(s)|}{\sqrt{f(s)}}|y(s)|ds < +\infty,$

(*ii*) $\sqrt{f(s)}ds - \frac{1}{2}\int_0^t \frac{|f'(s)|}{\sqrt{f(t)}} - \int_0^t \frac{|g(s)|}{\sqrt{f(s)}}[\int_0^s [f(\tau) + |g(\tau)|]d\tau]ds > R > 0,$

then as $t \to +\infty$, *all the solutions are bounded.*

If setting $g(t) \equiv 0$, $A(t) = f(t)$, we then get Theorem 4.13.1 again.

4.14 Applications of Theorems 1.2.7 and 1.2.8 to Integral and Differential Equations

In this section, we shall use Theorems 1.2.7 and 1.2.8 to obtain the estimates of the solutions of certain integral and differential equations.

Example 4.14.1 As a first application, we obtain the estimate on the solution of a nonlinear one-dimensional integral equation of the form

$$u^p(t) = f(t) + \int_a^t k(t,s)g(s,u(s))ds, \tag{4.14.1}$$

where $f : \mathbb{R}_+ \to \mathbb{R}$, $k : \mathbb{R}_+ \times \mathbb{R}_+ \to \mathbb{R}$, $g : \mathbb{R}_+ \times \mathbb{R} \to \mathbb{R}$ are continuous functions and $p > 1$ is a constant. Okrasinski [429] studied the problems of existence and uniqueness of the solutions of the variant of (4.14.1) written in the form

$$u^p = K * u + L, \quad p > 1, \tag{4.14.2}$$

where K, L are known smooth functions depending on physical parameters and the convolution on the right-hand side is well defined. For an interesting discussion concerning the occurrence of (4.14.2) in the theory of water percolation phenomena and its physical meaning, see [429] and references cited therein.

Here we assume that every solution $u(t)$ of equation (4.14.1) under discussion exists on an interval $\mathbb{R}_+$. We suppose that the functions f, k, g in (4.14.1) satisfy the following conditions

$$|f(t)| \le c_1, \quad |k(t,s)| \le c_2, \quad |g(t,u)| \le r(t)|u|, \tag{4.14.3}$$

where c_1, c_2 are non-negative constants and $r : \mathbb{R}_+ \to \mathbb{R}_+$ is a continuous function. From (4.14.1) and using (4.14.2), it easily follows

$$|u(t)|^p \le C_1 + \int_0^t c_2 r(s)|u(s)|ds. \tag{4.14.4}$$

Now applying Theorem 1.2.7 with $n = 1$ yields

$$|u(t)| \le \left[c_1^{(p-1)/p} + \frac{p-1}{p} \int_0^t c_2 r(s)ds \right]^{1/(p-1)}. \tag{4.14.5}$$

The right-hand side of (4.14.5) gives us the bound on the solution $u(t)$ of equation (4.14.1) in terms of the known quantities.

We now consider (4.14.1) under the following conditions on the functions f, k and g in equation (4.14.1):

$$|f(t)| \le c_1 e^{-pt}, \quad |k(t,s)| \le e^{-pt}, \quad |g(t,u)| \le r(t)|u|, \tag{4.14.6}$$

where $c_1, r(t)$ are as defined above, $h : \mathbb{R}_+ \to \mathbb{R}_+$ is a continuous function, and

$$\int_0^{+\infty} h(s)r(s)e^{-s}ds < +\infty. \tag{4.14.7}$$

From (4.14.1) and using (4.14.6), we derive

$$\left(e^t|u(t)|\right)^p \leq c_1 + \int_0^t h(s)r(s)e^{-s}\left(e^s|u(s)|\right)ds. \tag{4.14.8}$$

Now applying Theorem 1.2.7 with $n = 1$ yields

$$e^t|u(t)| \leq \left[c_1^{(p-1)/p} + \frac{p-1}{p}\int_0^t h(s)r(s)e^{-s}ds\right]^{1/(p-1)}. \tag{4.14.9}$$

From (4.14.9) and (4.14.7), we conclude

$$|u(t)| \leq c^* e^{-t}, \tag{4.14.10}$$

where $c^* \geq 0$ is a constant. From (4.14.10), we see that the solution $u(t)$ of equation (4.14.1) approaches zero as $t \to +\infty$.

Example 4.14.2 We obtain the estimate on the solution of the following second-order differential equation of the form

$$(a(t)\psi(u(t))u'(t))' + r(t)u(t) = 0, \tag{4.14.11}$$

with the given initial conditions

$$u(0) = c_1, \quad u'(0) = c_2, \tag{4.14.12}$$

where $a(t), r(t)$ are real-valued continuous functions defined on $\mathbb{R}_+$, and $a(t) > 0$ for all $t \in \mathbb{R}_+$, $\psi(u) > 0$ for all $u \neq 0$ and is defined by $\psi(u) = u^{p-1}$ for $p > 1$ a fixed real number, and $c_1 > 0, c_2 \geq 0$ are real constants. The problem of existence of a solution for the more general version of (4.14.11) with given boundary conditions was studied in [89, 138]. For a detailed account of many physical situations to which such equations are relevant can be found in [9]. Here it is assumed that every solution $u(t)$ of problem (4.14.11)–(4.14.12) exists on an interval $\mathbb{R}_+$ and is non-trivial.

Integrating (4.14.11) twice from 0 to t and using the initial conditions in (4.14.12), we know that the problem (4.14.11)–(4.14.12) is equivalent to the following integral equation

$$u^p(t) = c_1^p + pa(0)c_1^{p-1}c_2\int_0^t \frac{1}{a(s)}ds - p\int_0^t \frac{1}{a(s)}\left\{\int_0^s r(\tau)u(\tau)\right\}ds. \tag{4.14.13}$$

We suppose that the function $a(t)$ in (4.14.11) satisfies the condition

$$\int_0^{+\infty} \frac{1}{a(s)} ds < +\infty. \tag{4.14.14}$$

From (4.14.13) and using (4.14.14), we get

$$|u(t)|^p \leq c^* + \int_0^t \frac{p}{a(s)} \left(\int_0^s |r(\tau)||u(\tau)|d\tau \right) ds, \tag{4.14.15}$$

where $c^* > 0$ is a constant. Now applying Theorem 1.2.7 with $n = 2$ yields

$$|u(t)| \leq \left[c^{*(p-1)/p} + \frac{p-1}{p} \int_0^t \frac{p}{a(s)} \left(\int_0^s |r(\tau)|d\tau \right) ds \right]^{1/(p-1)}. \tag{4.14.16}$$

The right-hand side of (4.14.16) gives us the estimate on the solution $u(t)$ of problem (4.14.11)–(4.14.12) in terms of the known quantities.

Example 4.14.3 We note that the inequality in Theorem 1.2.8 can be used to obtain the estimate on the solution of the following system of integral equations

$$\begin{cases} u^p(t) = f_1(t) + \int_0^t k_1(t,s) g_1(s, u(s), v(s)) ds, \\ v^p(t) = f_2(t) + \int_0^t k_2(t,s) g_2(s, u(s), v(s)) ds, \end{cases} \tag{4.14.17}$$

under some suitable conditions on the functions involved in (4.14.17) where $p > 1$ is a constant. Furthermore, we also note that Theorem 1.2.8 can be used to obtain the estimate on the solution of the system of differential equations,

$$\begin{cases} (a_1(t)\psi(u(t))u'(t))' + q_1(t)u(t) + q_2(t)v(t) = 0, \\ (a_2(t)\psi(v(t))v'(t))' + q_3(t)v(t) + q_4(t)v(t) = 0, \end{cases} \tag{4.14.18}$$

with initial data

$$u(0) = c_1, \quad u'(0) = c_2, \quad v(0) = c_3, \quad v'(0) = c_4, \tag{4.14.19}$$

by using some suitable conditions on the functions involved in (4.14.18), where $c_1 > 0, c_2 \geq 0, c_3 > 0, c_4 \geq 0$ are constants and ψ is as defined in **Example 4.14.2**.

4.15 Applications of Theorems 1.2.11 and 2.3.4 to Integro-Differential Equations and Finite Difference Equations

In this section, we present some applications of Theorems 1.2.10 and 2.3.1 to obtain bounds on the solutions of certain integro-differential equations and finite difference equations. These applications are given as examples.

First, we shall establish a bound on the solution of the integro-differential equation

$$x'(t) - F\left(t, x(t), \int_0^t K[t, s, x(s)]ds\right) = h(t), \quad x(0) = x_0, \tag{4.15.1}$$

where $h : \mathbb{R}_+ \to \mathbb{R}$, $K : {\mathbb{R}_+}^2 \times \mathbb{R} \to \mathbb{R}$, $F : \mathbb{R}_+ \times \mathbb{R}^2 \to \mathbb{R}$ are continuous functions. Here we assume that the solution $x(t)$ of problem (4.15.1) exists on $\mathbb{R}_+$.

Multiplying both sides of (4.15.1) by $x(t)$, substituting $t = s$, and then integrating it from 0 to t, we arrive at

$$x^2(t) = x_0^2 + 2\int_0^t \left[x(s)F\left(s, x(s), \int_0^s K[s, \tau, x(\tau)]d\tau\right) + h(s)x(s)\right]ds. \tag{4.15.2}$$

We assume that

$$\begin{cases} |K[t, s, x(s)]| \le f(t)g(s)|x(s)|, & (4.15.3) \\ |F(t, x(t), v)| \le f(t)|x(t)| + |v|, & (4.15.4) \end{cases}$$

where f, g are real-valued non-negative continuous functions defined on $\mathbb{R}_+$. From (4.15.2)–(4.15.4), we derive

$$|x(t)|^2 \le |x_0|^2 + 2\int_0^t \left[f(s)|x(s)|\left(|x(s)| + \int_0^s g(\tau)|x(\tau)|d\tau\right) + |h(s)||x(s)|\right]ds. \tag{4.15.5}$$

Now applying the inequality (a_2) in Theorem 1.2.11 yields

$$|x(t)| \le p_1(t)\left[1 + \int_0^t f(s)\exp\left(\int_0^s [f(\tau) + g(\tau)]d\tau\right)ds\right], \tag{4.15.6}$$

where for all $t \in \mathbb{R}_+$,

$$p_1(t) = |x_0| + \int_0^t |h(s)|ds.$$

The inequality (4.15.6) gives us the bound on the solution $x(t)$ of problem (4.15.1) in terms of the known functions. □

Now we shall obtain a bound on the solution of the finite difference equation

$$\Delta y^2(n) = 2y(n)[\phi(n, y(n)) + h(n)], \quad y(0) = y_0, \tag{4.15.7}$$

where h and ϕ are real-valued functions defined on $\mathbb{N}_0$ and $\mathbb{N}_0 \times \mathbb{R}$, respectively.

We assume that

$$|\phi(n, y(n))| \leq f(n)|y(n)|, \tag{4.15.8}$$

where $f(n)$ is a real-valued non-negative function defined on $\mathbb{N}_0$. It is easy to observe that if $y(n)$ is a solution of problem (4.15.7), then it is also a solution of the equivalent sum-difference equation

$$y^2(n) = y_0^2 + 2\sum_{s=0}^{n-1} y(s)[\phi(s, y(s)) + h(s)]. \tag{4.15.9}$$

Using (4.15.8) in (4.15.9), we derive

$$|y(n)|^2 \leq |y_0|^2 + 2\sum_{s=0}^{n-1}[f(s)|y(s)|^2 + |h(s)||y(s)|]. \tag{4.15.10}$$

Now applying the inequality (b_1) in Theorem 2.3.4 yields

$$|y(n)| \leq q_1(n)\prod_{s=0}^{n-1}[1 + f(s)], \tag{4.15.11}$$

where for all $n \in \mathbb{N}_0$,

$$q_1(n) = |y_0| + \sum_{s=0}^{n-1}|h(s)|.$$

The inequality (4.15.11) gives us the bound on the solution $y(n)$ of problem (4.15.7) in terms of the known functions.

4.16 Applications of Theorem 1.2.15 to Integro-Differential Equations

In this section, we shall use Theorem 1.2.15 to establish some estimates of integro-differential equation.

Example 4.16.1 Consider first the integral inequality

$$w^4(t) \le t + \int_0^t (t+s)^2 |w(s)|^3 ds, \quad \text{for all } t \in \mathbb{R}_+, \tag{4.16.1}$$

where $w : \mathbb{R}_+ \to \mathbb{R}_+$ is continuous. Clearly this inequality has infinite many continuous solutions on $\mathbb{R}_+$ such as $w(t) = 0, w(t) = \frac{1}{C}t^{1/4}$ and $w(t) = \frac{t}{M}$, here $C \ge 1$ and $M \gg 1$ being constants. If $w(t)$ is a solution of (4.16.1) on $\mathbb{R}_+$, then so is the function $-w(t)$. An application of Theorem 1.2.15 to (4.16.1) yields for all $t \in \mathbb{R}_+$,

$$|w(t)| \le t^{1/4} + \int_0^t \frac{1}{4}(t+s)^2 ds = t^{1/4} + \frac{7t^3}{12}.$$

Example 4.16.2 Consider the integro-differential equation

$$x'(t) = \Phi\Big(t, x(t), \int_0^t K(t,s,x(s))ds\Big), \quad t \in \mathbb{R}_+, \tag{4.16.2}$$

with the initial condition $x(0) = x_0$, where $\Phi : \mathbb{R}_+ \times \mathbb{R}^2 \to \mathbb{R}, K : \mathbb{R}_+^2 \times \mathbb{R} \to \mathbb{R}$ are continuous functions. Assume that the solution $x(t)$ of the last initial value problem exists on $\mathbb{R}_+$ and the following conditions hold

$$\begin{cases} |K(t,s,u)| \le e(t,s)|u| + w(t,s), \\ |\Phi(t,u,v)| \le f_1(t)|u| + f_2(t)|v| + f_3(t), \end{cases}$$

where $t, s \in \mathbb{R}_+$ and f_i $(i = 1,2,3)$ are real-valued, non-negative and continuous functions on $\mathbb{R}_+$, and $e(t,s)$ and $w(t,s)$ are real-valued, non-negative and continuous functions defined on $\mathbb{R}_+^2$, with $e(t,s)$ non-decreasing in t for every s fixed. Multiplying both sides of (4.16.2) by $x(t)$ and integrating from 0 to t, then we obtain for all $t \in \mathbb{R}_+$,

$$x^2(t) = x_0^2 + 2\int_0^t x(s)\Phi\Big(x, x(s), \int_0^s K(s,v,x(v))dv\Big)ds.$$

By using the assumed conditions, we derive from the last relation that

$$|x(t)|^2 \le |x_0|^2 + 2\int_0^t |x(s)|\Big(h(s) + f_1(s)|x(s)| + f_2(s)\int_0^s e(s,m)|x(m)|dm\Big)ds, \tag{4.16.3}$$

where $h(s) := f_2(s)\int_0^s w(s,m)dm + f_3(s)$. The last inequality is a special case of inequality (1.2.103) in Theorem 1.2.15, applying Theorem 1.2.15 to (4.16.3), then the desired bound on $x(t)$ follows.

4.17 Applications of Theorem 1.2.19 to Differential Equations with Time Delay

Consider the functional differential equation

$$\begin{cases} X'(t) = H(t, X(t), X(\alpha(t))), \\ X(0) = X_0, \end{cases} \tag{4.17.1}$$

with $X_0 \in \mathbb{R}^n, H \in C(\mathbb{R}_+ \times \mathbb{R}^{2n}, \mathbb{R}^n)$, and $\alpha \in C^1(\mathbb{R}_+, \mathbb{R}_+)$ satisfying $\alpha(t) \le t$ for all $t > 0$. A result in [210] guarantees that for every $X_0 \in \mathbb{R}^n$, problem (4.17.1) has a solution. Without additional hypotheses on H, the uniqueness of solutions is not granted. However, every solution of problem (4.17.1) has a maximal existence time $T > 0$ and if $T < +\infty$, then

$$\limsup_{t\to T} \|X(t)\|_{\mathbb{R}^n} = +\infty.$$

We can now applying Theorem 1.2.19 to the following problems.

Example 4.17.1 Consider first the generalized Liénard equation with time delay

$$\begin{cases} x' = y - F(x), \\ y' = G(t, x(t-\tau(t))), \end{cases} \tag{4.17.2}$$

where $F \in C^1(\mathbb{R}, \mathbb{R}), G \in C(\mathbb{R}_+ \times \mathbb{R}, \mathbb{R}), \tau \in C^1(\mathbb{R}_+, \mathbb{R}_+)$, and $\tau(t) \le t$ on $\mathbb{R}_+$. If $\alpha(t) \equiv t - \tau(t)$ is an increasing diffeomorphism of $\mathbb{R}_+$ and

$$\begin{cases} -xF(x) \le |x|v(|x|), \quad \text{for all}\, x \in \mathbb{R}, \\ G^2(t,x) \le h(t)|x|v(|x|), \quad \text{for all}\, (t,x) \in \mathbb{R}_+ \times \mathbb{R}, \end{cases}$$

for some non-decreasing function $v \in C(\mathbb{R}_+, \mathbb{R}_+)$ with the properties $v(u) > 0$ for all $u > 0$ and $\int_1^{+\infty}(1/v(s))ds = +\infty$, then all solutions of equation (4.17.2) are global.

Indeed, if $(x(t), y(t))$ is a solution to (4.17.2) defined on the maximal existence interval $[0, T)$, let $u(t) = \sqrt{x^2(t) + y^2(t)}$ for $t \in [0, T)$. Then, using (4.17.2) and hypotheses on the functions F and G, we obtain, for all $0 < t < T$,

$$\begin{aligned}\frac{d}{dt}u^2(t) &= 2xx' + 2yy' = 2xy - 2xF(x) + 2yG(t, x(\alpha))\\ &\leq x^2 + y^2 - 2xF(x) + y^2 + G^2(t, x(\alpha))\\ &\leq 2u^2 + 2uv(u) + h(t)|x(\alpha)|v(|x(\alpha)|).\end{aligned}$$

With $w(u) := u + v(u)$, an integration on $[0, t]$ with $t < T$ yields

$$\begin{aligned}u^2(t) &\leq u^2(0) + 2\int_0^t u(s)w(u(s))ds + \int_0^t h(s)|x(\alpha(s))|v(|x(\alpha(s))ds\\ &\leq u^2(0) + 2\int_0^t u(s)w(u(s))ds + \int_0^t h(s)|x(\alpha(s))|w(|x(\alpha(s))|)ds\\ &= u^2(0) + 2\int_0^t u(s)w(u(s))ds + \int_0^{\alpha(t)} \frac{h(\alpha^{-1}(r))}{\alpha'(\alpha_{-1}(r))}|x(r)|w(|x(r)|)dr\\ &\leq u^2(0) + 2\int_0^t u(s)w(u(s))ds + 2\int_0^{\alpha(t)} \frac{h(\alpha^{-1}(r))}{\alpha'(\alpha^{-1}(r))}u(r)w(u(r))dr,\end{aligned}$$

after performing the change of variables $r = \alpha(s)$ at some intermediate step. Above α^{-1} is the inverse of the diffeomorphism α. Our hypotheses on v guarantee that $\int_1^{+\infty}(1/w(r))dr = +\infty$ (see, e.g., [165]). Therefore, if

$$W(r) = \int_1^r \frac{ds}{w(s)}, \qquad r > 0,$$

then by Theorem 1.2.19, we deduce for all $0 \leq t < T$,

$$\begin{aligned}u(t) &\leq W^{-1}\left[W(u(0)) + t + \int_0^{\alpha(t)} \frac{h(\alpha^{-1}(r))}{\alpha'(\alpha_{-1}(r))}dr\right]\\ &= W^{-1}\left[W(u(0)) + t + \int_0^t h(s)ds\right],\end{aligned}$$

which proves that $u(t)$ does not blow-up in a finite time. Therefore, $T = +\infty$ and all solutions of problem (4.17.2) exist globally.

Remark 4.17.1 The problem of the global existence of the solutions to problem (4.17.2) was also considered in [165, 355, 631]. Observe that if

$$G(t,x) \equiv x, \quad F(x) \equiv x^3 \ on\ \mathbb{R},$$

and

$$\tau(t) = \begin{cases} 0, & \text{for all } t \in [0,1], \\ \frac{1}{2}\ln(t^2 - 2t + 2), & \text{for all } t \geq 1, \end{cases}$$

then the present results ensure the global existence of all solutions to equation (4.17.2). To see this, it suffices to take $v(u) \equiv u, h(t) \equiv 1$, and to check that $\alpha(t) \equiv t - \tau(t)$ is an increasing diffeomorphism of $\mathbb{R}_+$. However, none of the global existence results presented in [165, 355, 631] are applicable in this case. Indeed, we cannot find a non-decreasing function $w \in C(\mathbb{R}_+, \mathbb{R}_+)$ with $w(u) > 0$ for all $u > 0$ and $\int_1^{+\infty}(1/w(s))ds = +\infty$, such that for all $x \in \mathbb{R}$,

$$|F(x)| = |x|^3 \leq w(|x|).$$

Therefore in the presented case, we cannot conclude global existence from the results in [165, 355]. Also, since in [631], it was assumed that

$$\{t > 0 : \tau(t) = 0\} = \emptyset$$

and we have $\tau \equiv 0$ on $[0, 1]$, we see that the case is not covered by the result in [631].

Example 4.17.2 Consider the Rayleigh equation with time delay

$$\begin{cases} x' = y, \\ y' = -F(y) - G(x(\alpha(t))), \end{cases} \tag{4.17.3}$$

where $F, G \in C(\mathbb{R}, \mathbb{R}), \alpha \in C^1(\mathbb{R}_+, \mathbb{R}_+)$, and $\alpha(t) \leq t$ for all $t \geq 0$. If $\alpha(t)$ is an increasing diffeomorphism of $\mathbb{R}_+$ and

$$-xF(x) \leq |x|v(|x|), \quad G^2(x) \leq |x|v(|x|), \quad x \in \mathbb{R},$$

for some non-decreasing function $v \in C(\mathbb{R}_+, \mathbb{R}_+)$ with the properties $v(u) > 0$ for all $u > 0$ and $\int_1^{+\infty}(1/v(s))ds = +\infty$, then all solutions of problem (4.17.3) are global.

Indeed, if $(x(t), y(t))$ is a solution to problem (4.17.3) defined on the maximal existence interval $[0, T)$, let $u(t) = \sqrt{x^2(t) + y^2(t)}$ for all $t \in [0, T)$. Then for all $0 < t < T$,

$$\begin{aligned}\frac{d}{dt}u^2(t) &= 2xx' + 2yy' = 2xy - 2yF(y) - 2yG(x(\alpha))\\ &\leq x^2 + y^2 - 2yF(y) + y^2 + G^2(x(\alpha))\\ &\leq 2u^2 + 2uv(u) + (u(\alpha))v(u(\alpha)),\\ &\leq 2u[u + v(u)] + 2(u(\alpha))[(u(\alpha)) + v(u(\alpha))],\end{aligned}$$

if we take into account our hypotheses on F and G. Denoting $w(u) := u + v(u)$ and integrating the above inequality on $[0, t]$ with $t < T$, we obtain

$$\begin{aligned}u^2(t) &\leq u^2(0) + 2\int_0^t u(s)w(u(s))ds + 2\int_0^t u(\alpha(s))w(u(\alpha(s)))ds\\ &\leq u^2(0) + 2\int_0^t u(s)w(u(s))ds + 2\int_0^{\alpha(t)} \frac{1}{\alpha'(\alpha^{-1}(r))}u(r)w(u(r))dr,\end{aligned}$$

after performing the change of variables $r = \alpha(s)$. Our hypotheses on v ensure

$$\int_1^{+\infty} \frac{ds}{s + v(s)} = +\infty$$

(see, e.g., [165]). If

$$W(r) = \int_1^r \frac{ds}{w(s)}, \quad r > 0,$$

then Theorem 1.2.19 implies for all $0 \leq t < T$,

$$\begin{aligned}u(t) &\leq W^{-1}\left[W(u(0)) + t + \int_0^{\alpha(t)} \frac{1}{\alpha'(\alpha^{-1}(r))}dr\right]\\ &= W^{-1}[W(u(0)) + 2t].\end{aligned}$$

Hence $u(t)$ is bounded on $[0, T)$ if $T < +\infty$. We conclude that all solutions of problem (4.17.3) are global.

Remark 4.17.2 Letting $\alpha(t) \equiv t$ in Example 4.17.2, we obtain a criterion for the global existence of solutions to the Rayleigh equation $x'' + F(x') + G(x) = 0$. This problem was also considered in [164, 622]. Note that in the case

$$F(x) \equiv x \quad \textit{and} \quad G(x) \equiv x\sin(x) \quad \textit{on } \mathbb{R},$$

the present results (with $v(u) \equiv u$) guarantee that all solutions of the Rayleigh equation are defined globally in time. However, since the function $x \mapsto xG(x)$ changes sign in every neighborhood of infinity, this conclusion cannot be obtained by applying the results in [164] or [622].

4.18 An Application of Corollary 1.2.6 to an Integral Equation

We consider an integral equation

$$x^p(t) = a(t) + \int_t^{+\infty} F[s, x(s), x(\phi(s))]ds. \tag{4.18.1}$$

Assume that

$$|F(x, y, u)| \leq f(x)|u|^p + g(x)|u| \tag{4.18.2}$$

and

$$|a(t)| \leq c, c > 0, p \geq q > 0, p \neq 1, \tag{4.18.3}$$

where f, g are non-negative continuous real-valued functions, and $\phi \in C^1(\mathbb{R}_+, \mathbb{R}_+)$ is non-decreasing with $\phi(t) \geq t$ on $\mathbb{R}_+$. From (4.18.1)–(4.18.3), we derive

$$|x(t)|^p \leq c + \int_t^{+\infty} (f(s)|x(\phi(s))|^q + g(s)|x(\phi(s))|)ds. \tag{4.18.4}$$

Making the change of variables from the above inequality and taking $M = \sup_{t\in\mathbb{R}_+} \frac{1}{\phi'(t)}$, we get

$$|x(t)|^p \leq c + M \int_t^{+\infty} \left(\bar{f}(s)|x(s)|^q + \bar{g}(s)|x(s)|\right) ds \tag{4.18.5}$$

where $\bar{f}(s) = f(\phi^{-1}(s)), \bar{g}(s) = g(\phi^{-1}(s))$. From Corollary 1.2.6, we obtain

$$\begin{cases} |x(t)| \leq \left(c^{1-\frac{1}{p}} + \dfrac{M(p-1)}{p} \displaystyle\int_{\phi(t)}^{+\infty} \bar{g}(s)ds\right)^{\frac{p}{p-1}} \exp\left(\dfrac{M}{p} \displaystyle\int_{\phi(t)}^{+\infty} \bar{f}(s)ds\right), & when\ p = q, \\ |x(t)| \leq \left[\left(c^{1-\frac{1}{p}} + \dfrac{M(p-1)}{p} \displaystyle\int_{\phi(t)}^{+\infty} \bar{g}(s)ds\right)^{\frac{p-q}{p-1}} + \dfrac{M(p-q)}{p} \displaystyle\int_{\phi(t)}^{+\infty} \bar{f}(s)ds\right]^{\frac{1}{p-q}}, & when\ p > q. \end{cases}$$

If the integrals of $f(s), g(s)$ are bounded, then we have the bound of the solution of equation (4.18.1).

4.19 Applications of Theorem 1.2.20 to Differential Equations

In this section, we use Theorem 1.2.20 to give some estimates on solutions to the following differential equation involving several retarded arguments

$$x^{p-1}(t)x'(t) = f(t, x(t-h_1(t)), \cdots, x(t-h_n(t))), \tag{4.19.1}$$

for all $t \in I := [t_0, T)$, with the given initial condition

$$x(t_0) = x_0 \tag{4.19.2}$$

where $p > 1$ and x_0 are constants, $f \in C(I \times \mathbb{R}^n, \mathbb{R})$ and for $i = 1, \cdots, n$, let $h_i \in C(I, \mathbb{R}_+)$ be non-increasing and such that $t - h_i(t) \geq 0, t - h_i(t) \in C^1(I, I), h_i'(t) < 1, h_i(t_0) = 0$.

Theorem 4.19.1 (The Pachpatte Inequality [523]) *Suppose that*

$$|f(t, u_1, \cdots, u_n)| \leq \sum_{i=1}^{n} b_i(t)|u_i|, \tag{4.19.3}$$

where $b_i(t)$ are as in Theorem 1.2.20. Let

$$Q_i = \max_{t \in I} \frac{1}{1 - h_i'(t)}, \quad i = 1, \cdots, n. \tag{4.19.4}$$

If $x(t)$ is any solution of problem (4.19.1)–(4.19.2), then for all $t \in I$,

$$|x(t)| \leq \left\{ |x_0|^{p-1} + (p-1) \sum_{i=1}^{n} \int_{t_0}^{t-h_i(t)} \bar{b}_i(\sigma) d\sigma \right\}^{\frac{1}{p-1}}, \tag{4.19.5}$$

where $\bar{b}_i(\sigma) = Q_i b_i(\sigma + h_i(s))$, $\sigma, s \in I$.

Proof The solution $x(t)$ of problem (4.19.1)–(4.19.2) can be written as

$$x^P(t) = x_0^p + p \int_{t_0}^{t} f(s, x(s - h_1(s), \cdots, x(s - h_n(s))ds. \tag{4.19.6}$$

Using (4.19.6), (4.19.3), (4.19.4) and making the change of variables, we have, for all $t \in I$,

$$|x(t)|^p \leq |x_0^p| + p\int_{t_0}^{t}\sum_{i=1}^{n} b_i(s)|x(s-h_i(s))|ds$$

$$\leq |x_0^p| + p\sum_{i=1}^{n}\int_{t_0}^{t-h_i(t)} \bar{b}_i(s)|x(\sigma)|ds. \tag{4.19.7}$$

Now applying the inequality part (2) (when $a_i = 0$) in Theorem 1.2.20, to (4.19.7) yields the required estimate in (4.19.5). □

Remark 4.19.1 From Theorem 4.19.1, it follows that the inequalities given in [356] cannot be used to obtain an estimate on the solution of the problem (4.19.1)–(4.19.2).

4.20 Applications of Theorem 1.2.22 and Corollary 1.2.7 to Differential Equations with Time Delay

We shall use Theorem 1.2.22 to prove the global existence of solutions to certain differential equations with time delay.

Consider the functional differential equation involving several retarded arguments with the initial condition

$$\begin{cases} \phi'(x(t))x'(t) = F(t, x(t-h_1(t)), \cdots, x(t-h_n(t))), \ t \in I, \\ x(t_0) = x_0, \end{cases} \tag{4.20.1}$$

where x_0 is a constant, $F \in C(I \times \mathbb{R}^n, \mathbb{R})$, $h_i(t) \in C(I, \mathbb{R}_+)$, $i = 1, \cdots, n$, be non-increasing such that $t - h_i(t) \geq 0$, $t - h_i(t) \in C^1(I, I)$, $h_i'(t) < 1$, and $\phi \in C^1(\mathbb{R}, \mathbb{R})$ is an increasing function with $\phi(|x|) \leq |\phi(x)|$. The following theorem deals with a bound on the solution of problem (4.20.1).

Theorem 4.20.1 ([14]) *Assume that $F : I \times \mathbb{R}^n \to 0$ is a continuous function for which there exist continuous non-negative function $f_i(t)$, $g_i(t)$, $i = 1, \cdots, n$, for all $t \in I$ such that*

$$|F(t, u_1, \cdots, u_n)| \leq \sum_{i=1}^{n} |u_i|^q \left[f_i(t)\psi(|u_i|) + g_i(t)\right], \tag{4.20.2}$$

where $q > 0$ is a constant and ψ is as in Theorem 1.2.22. Let

$$Q_i = \max_{t \in I} \frac{1}{1 - h_i'(t)}, \quad i = 1, \cdots, n. \tag{4.20.3}$$

If $x(t)$ is any solution of problem (4.20.1), then we have for all $t \in I$,

$$|x(t)| \leq \phi^{-1}\left\{G^{-1}\left[\Psi^{-1}(\Psi(\overline{k}(t_0)) + \sum_{i=1}^{n}\int_{t_0-h_i(t_0)}^{t-h_i(t)} \overline{f}_i(\sigma)d\sigma)\right]\right\}, \quad (4.20.4)$$

where G, Ψ are as in Theorem 1.2.22 and

$$\overline{k}(t_0) = G(|\phi(x_0)|) + \sum_{i=1}^{n}\int_{t_0-h_i(t_0)}^{t-h_i(t)} \overline{g}_i(\sigma)d\sigma, \quad (4.20.5)$$

$\overline{f}_i(\sigma) = Q_i f_i(\sigma + h_i(s))$, $\overline{g}_i(\sigma) = Q_i g_i(\sigma + h_i(s))$ *for all s, $\sigma \in I$.*

Proof It is easy to see that the solution $x(t)$ of the problem (4.20.1) satisfies the equivalent integral equation

$$\phi(x(t)) = \phi(x(t_0)) + \int_{t_0}^{t} F(s, x(s-h_1(s)), \cdots, x(s-h_n(s)))\, ds. \quad (4.20.6)$$

From (4.20.2) and (4.20.6), it follows for all $t, s \in I$,

$$\begin{aligned}|\phi(x(t))| &= |\phi(x(t_0))| + \int_{t_0}^{t} |F(s, x(s-h_1(s)), \cdots, x(s-h_n(s)))|\,ds \\ &\leq |\phi(x_0)| + \int_{t_0}^{t}\sum_{i=1}^{n} |x(s-h_i(s))|^q \left[f_i(t)\psi(|x(s-h_i(s))|) + g_i(t)\right] ds.\end{aligned} \quad (4.20.7)$$

By making the change of variables on the right-hand side of the inequality (4.20.7) and rewriting, we have

$$|\phi(x(t))| \leq |\phi(x_0)| + \int_{t_0}^{t}\sum_{i=1}^{n} |x(\sigma)|^q \left[f_i(\sigma)\psi(|x(\sigma)|) + g_i(\sigma)\right] d\sigma, \quad (4.20.8)$$

where $\overline{f}_i(\sigma) = Q_i f_i(\sigma + h_i(s))$, $\overline{g}_i(\sigma) = Q_i g_i(\sigma + h_i(s))$ for all s, $\sigma \in I$.

Now applying the inequality in Theorem 1.2.22 to the inequality (4.20.8) yields the desired result. □

Remark 4.20.1 Consider the functional differential equation involving several retarded arguments with the initial condition:

$$\begin{cases} px^{p-1}(t)x'(t) = F(t, x(t-h_1(t)), \cdots, x(t-h_n(t))), \ t \in I, \\ x(t_0) = x_0, \end{cases} \quad (4.20.9)$$

where $p > 0$ and x_0 are constants, $F \in C(I\times\mathbb{R}^2, \mathbb{R})$, $h_i(t) \in C(I, \mathbb{R}_+)$, $i = 1, \cdots, n$, be non-increasing such that $t - h_i(t) \geq 0$, $t - h_i(t) \in C^1(I, I)$, $h_i'(t) < 1$.

Assume that $F : I \times \mathbb{R}^n \to \mathbb{R}$ is a continuous function for which there exist continuous non-negative function $f_i(t)$, $g_i(t)$, $i = 1, \cdots, n$, for all $t \in I$ such that

$$|F(t, u_1, \cdots, u_n)| \leq \sum_{i=1}^{n} |u_i|^q [f_i(t)\psi(|u_i|) + g_i(t)], \tag{4.20.10}$$

where $q > 0$ $(p > q)$ is a constant and ψ is as in Theorem 1.2.22. If $x(t)$ is any solution of the problem (4.20.9), then it satisfies the equivalent integral equation

$$x^p(t) = x^p(t_0) + \int_{t_0}^{t} F(s, x(s - h_1(s)), \cdots, x(s - h_n(s)))ds. \tag{4.20.11}$$

From (4.20.10) and (4.20.11), it follows for all $t, s \in I$,

$$\begin{aligned} |x^p(t)| &\leq |x(t_0)|^p + \int_{t_0}^{t} |F(s, x(s - h_1(s)), \cdots, x(s - h_n(s)))|ds \\ &\leq |x_0|^p + \int_{t_0}^{t} \sum_{i=1}^{n} |x(s - h_i(s))|^q [f_i(t)\psi(|x(s - h_i(s))|) + g_i(t)]ds. \end{aligned} \tag{4.20.12}$$

By making the change of variables on the right-hand side of the inequality (4.20.12) and rewriting, we get

$$|x(t)|^p \leq |x_0|^p + \int_{t_0}^{t} \sum_{i=1}^{n} |x(\sigma)|^q [f_i(\sigma)\psi(|x(\sigma)|) + g_i(\sigma)]d\sigma, \tag{4.20.13}$$

where $\overline{f}_i(\sigma) = Q_i f_i(\sigma + h_i(s))$, $\overline{g}_i(\sigma) = Q_i g_i(\sigma + h_i(s))$ for all $s, \sigma \in I$. Now applying the inequality in Corollary 1.2.7 to the inequality (4.20.13) yields, for all $t \in I$,

$$x(t) \leq \left[\Psi_0^{-1}(\Psi_0(\overline{k}_1(t_0)) + \frac{p-q}{p} \sum_{i=1}^{n} \int_{\alpha_i(t_0)}^{\alpha_i(t)} \overline{f}_i(\sigma)d\sigma) \right]^{\frac{1}{p-q}} \tag{4.20.14}$$

where Ψ_0 is as in Corollary 1.2.7,

$$\overline{k}_1(t_0) = x_0^{\frac{(p-q)}{p}} + \frac{p-q}{p} \sum_{i=1}^{n} \int_{\alpha_i(t_0)}^{\alpha_i(t)} \overline{g}_i(\sigma)d\sigma, \tag{4.20.15}$$

and $\overline{f}_i(\sigma) = Q_i f_i(\sigma + h_i(s))$, $\overline{g}_i(\sigma) = Q_i g_i(\sigma + h_i(s))$ for all $s, \sigma \in I$.

The following theorem provides a uniqueness on the solution of the problem (4.20.9).

Theorem 4.20.2 *Assume that $F : I \times \mathbb{R}^3 \to \mathbb{R}$ is a continuous function for which there exist continuous non-negative function $f_i(t)$, $i = 1, \cdots, n$, for all $t \in I$ such that*

$$|F(t, u_1, \cdots, u_n) - F(t, v_1, \cdots, v_n)| \leq \sum_{i=1}^{n} f_i(t)|u_i - v_i|^q, \tag{4.20.16}$$

where $p > 1$ is a constant, then the problem (4.20.9) has at most one solution on I.

Proof Let $x(t)$ and $\overline{x}(t)$ be two distinct solutions of the problem (4.20.9), we have

$$\begin{aligned} x^p(t) - \overline{x}^p(t) = \int_{t_0}^{t} [F(s, x(s - h_1(s)), \cdots, x(s - h_n(s))) \\ -F(s, \overline{x}(s - h_1(s)), \cdots, \overline{x}(s - h_n(s)))]ds. \end{aligned} \tag{4.20.17}$$

From (4.20.16) and (4.20.17), it follows for all $t, s \in I$,

$$|x^p(t) - \overline{x}^p(t)| \leq \int_{t_0}^{t} \left(\sum_{i=1}^{n} f_i(s)|x^p(s - h_i(s)) - \overline{x}^p(s - h_i(s))| \right) ds. \tag{4.20.18}$$

By making the change of variables on the right-hand side of the inequality (4.20.18) and rewriting, we have

$$(|x^p(t) - \overline{x}^p(t)|^{\frac{1}{p}})^p \leq \sum_{i=1}^{n} \int_{\beta_i(t_0)}^{\beta_i(t)} \sum_{i=1}^{n} \left[|x^p(\sigma) - \overline{x}^p(\sigma)|^{\frac{1}{p}} \right]^{p-1} \overline{f}_i |x^p(\sigma) - \overline{x}^p(\sigma)|^{\frac{1}{p}} d\sigma, \tag{4.20.19}$$

where $\beta_i(t) = t - h_i(t), \overline{f}_i(\sigma) = Q_i f_i(\sigma + h_i(s))$ for all $s, \sigma \in I$. Now when $\psi(u) = u$, $q = p - 1$, applying the inequality in Corollary 1.2.7 to the function $|x^p(t) - \overline{x}^p(t)|^{\frac{1}{p}}$ and the inequality (4.20.19), we conclude that for all $t \in I$,

$$|x^p(t) - \overline{x}^p(t)|^{\frac{1}{p}} \leq 0. \tag{4.20.20}$$

Hence $x(t) = \overline{x}(t)$. □

4.21 An Application of Theorem 1.2.25 to the Epidemic Model

This section deals with a new integral inequality and its applications to the study of qualitative behavior of the solutions of certain epidemic model.

In 1981, Gripenberg [255] studied the qualitative behavior of the equation

$$x(t) = k\left(p(t) - \int_0^t A(t-s)x(s)ds\right)\left(f(t) + \int_0^t a(t-s)x(s)ds\right). \tag{4.21.1}$$

This equation arises in the study of the spread of an infectious disease that does not induce permanent immunity. For detailed meanings of the various functions arising in (4.27.1), see [255] and also [37, 211, 254, 662]. In [255], the author studied the existence of a unique bounded continuous and non-negative solution of problem (4.21.1) for all $t \in \mathbb{R}_+$ under appropriate assumptions on A and a and also obtained sufficient conditions for the convergence of the solution to a limit when $t \to +\infty$. Aside from the various physical meanings of the functions arising in (4.21.1), we believe that equations like (4.21.1) are of great interest and that further investigation of the qualitative behavior of their solutions even under the usual hypotheses on the functions in problem (4.21.1) is much more interesting.

Over the years integral inequalities have become a major tool in the analysis of various integral equations that occur in nature or are built by man. Although a great many papers have been written on various types of integral inequalities, it seems that the bounds provided by the existing results on integral inequalities do not apply directly to the study of the qualitative behavior of the solutions of equation (4.21.1). This amounts to finding a suitable inequality in order achieve a diversity of desired goals.

In what follows, we assume that the functions x, p, f, A, a in (4.21.1) are real-valued, continuous, and defined on $\mathbb{R}_+$ and k is a positive real constant and restrict our consideration to the solutions of (4.21.1) which exist on $\mathbb{R}_+$.

Theorem 4.21.1 ([499]) *Assume that*

$$|p(t)| \le c_1, \quad |f(t)| \le c_2, \tag{4.21.2}$$

$$|A(t-s)| \le M_1 q(s), \quad |a(t-s)| \le M_2 h(s), \tag{4.21.3}$$

for all $s, t \in \mathbb{R}_+$, $0 \le s \le t$, where c_1, c_2, M_1, M_2 are non-negative real constants, and q and h are real-valued non-negative continuous functions defined on $\mathbb{R}_+$. If $c_1 c_2 k \int_0^t R_1(s)Q_1(s)ds < 1$ for all $t \in \mathbb{R}_+$ and for all $t \in \mathbb{R}_+$,

$$E(t) = \frac{c_1 c_2 k Q_1(t)}{1 - c_1 c_2 k \int_0^t R_1(s)Q_1(s)ds} < +\infty, \tag{4.21.4}$$

where $R_1(t)$ and $Q_1(t)$ are as defined in (1.2.209) and (1.2.210) by replacing c_1 by c_1k, $f(t)$ by $M_1kq(t)$, $g(t)$ by $M_2h(t)$, then every solution $x(t)$ of problem (4.21.1) existing on $\mathbb{R}_+$ is bounded.

Proof From (4.21.1) and using the hypotheses (4.21.2)–(4.21.3), it is easy to observe

$$|x(t)| \leq \left(c_1k + \int_0^t M_1kq(s)|x(s)|ds\right)\left(c_2 + \int_0^t M_2h(s)|x(s)|ds\right). \tag{4.21.5}$$

Now an application of Theorem 1.2.25 to (4.21.5) yields for all $t \in \mathbb{R}_+$,

$$|x(t)| \leq E(t). \tag{4.21.6}$$

In view of the hypothesis (4.21.4), the estimate in (4.21.6) implies the boundedness of the solution $x(t)$ of equation (4.21.1) on $\mathbb{R}_+$. The proof is hence complete. □

Theorem 4.21.2 ([499]) *Assume that*

$$|p(t)| \leq c_1e^{-\lambda t}, \quad |f(t)| \leq c_2e^{-\lambda t}, \tag{4.21.7}$$

$$|A(t-s)| \leq M_1q(s)e^{-\lambda(t-2s)}, \quad |a(t-s)| \leq M_2h(s)e^{-\lambda(t-2s)}, \tag{4.21.8}$$

for all $s, t \in \mathbb{R}_+$, $0 \leq s \leq t$, where c_1, c_2, M_1, M_2, q, h are as defined in Theorem 4.21.1 and $\lambda \geq 0$ is a real constant. If (4.21.4) holds, then all solutions of (4.21.1) approach zero as $t \to +\infty$.

Proof From (4.21.1) and using the hypotheses (4.21.7), (4.21.8), it follows

$$|x(t)| \leq e^{-2\lambda t}\left(c_1k + \int_0^t M_1kq(s)|x(s)|e^{2\lambda s}ds\right)\left(c_2 + \int_0^t M_2h(s)|x(s)|e^{2\lambda s}ds\right). \tag{4.21.9}$$

Multiplying both sides of (4.21.9) by $e^{2\lambda t}$, applying Theorem 1.2.25 with $y(t) = |x(t)|e^{2\lambda t}$, then multiplying the resulting inequality by $e^{-2\lambda t}$, we obtain, for all $t \in \mathbb{R}_+$,

$$|x(t)| \leq E(t)e^{-2\lambda t}. \tag{4.21.10}$$

In view of the hypothesis (4.21.4), the inequality in (4.21.10) yields the desired result, and the proof is thus complete. □

Theorem 4.21.3 ([499]) *Assume that*

$$|p(t)| \leq c_1e^{\lambda t}, \quad |f(t)| \leq c_2e^{\lambda t}, \tag{4.21.11}$$

$$|A(t-s)| \leq M_1q(s)e^{\lambda(t-2s)}, \quad |a(t-s)| \leq M_2h(s)e^{\lambda(t-2s)}, \tag{4.21.12}$$

for all $s,t \in \mathbb{R}_+$, $0 \le s \le t$, where $c_1, c_2, M_1, M_2, \lambda, q, h$ are as defined in Theorem 4.21.2. If (4.21.4) holds for all $t \in \mathbb{R}_+$, then all solutions of (4.21.1) are slowly growing.

Proof From (4.21.1) and using the hypotheses (4.21.11)–(4.21.12), we derive

$$|x(t)| \le e^{2\lambda t}\left(c_1 k + \int_0^t M_1 k q(s)|x(s)|e^{-2\lambda s}ds\right)\left(c_2 + \int_0^t M_2 h(s)|x(s)|e^{-2\lambda s}ds\right). \tag{4.21.13}$$

Multiplying both sides of (4.21.13) by $e^{-2\lambda t}$, applying Theorem 1.2.25 with $y(t) = |x(t)|e^{-2\lambda t}$, then multiplying the resulting inequality by $e^{2\lambda t}$, we obtain for all $t \in \mathbb{R}_+$,

$$|x(t)| \le E(t)e^{2\lambda t}. \tag{4.21.14}$$

In view of the hypothesis (4.21.4), the inequality in (4.21.14) demonstrates that the solution of equation (4.21.1) grows more slowly than any positive exponential. The proof is now complete. □

4.22 An Application of Corollary 1.2.11 to Delay Differential Equations

Consider the delay integral equation

$$x^p(t) = F\left(t, x(\sigma(t)), \int_0^t K(t,s,x(\sigma(s)))ds\right), \quad t \in \mathbb{R}_+ \tag{4.22.1}$$

with the initial condition

$$x(t) = \psi(t), \quad t \in [a,0], \tag{4.22.2}$$

with

$$\psi(\sigma(t)) \le n^{1/p}(t), \quad \text{for every } t \ge 0 \text{ with } \sigma(t) \le 0, \tag{4.22.3}$$

where $F \in C(\mathbb{R}_+ \times \mathbb{R}^2, \mathbb{R})$, $K \in (\mathbb{R}_+^2 \times \mathbb{R}, \mathbb{R})$, $p \ge 1$ is a constant, and x, σ, and a are as defined in Theorem 1.2.27.

Assume that

$$\begin{cases} |F(t,u,v)| \le n(t) + m|v|, & \text{for all } t \in \mathbb{R},\ u,v \in \mathbb{R}, \quad (4.22.4) \\ |K(t,u,v)| \le |v|^{p-1} f(s) W(|v|) + g(s)|v| + h(s), & \text{for all } t,s \in \mathbb{R}_+,\ v \in \mathbb{R}_+, \end{cases} \tag{4.22.5}$$

where $m > 0$ is a constant, and the functions n, f, g, and W are defined as in Theorem 1.2.27.

For every continuous solution $x(t)$ of (4.22.1) satisfying the condition (4.22.2)–(4.22.3), from (4.22.1), (4.22.4), and (4.22.5), we obtain, for all $t \in J(x)$,

$$\begin{aligned}|x(t)|^p &= \left|F\left(t, x(\sigma(t)), \int_0^t K[t, s, x(\sigma(s))]ds\right)\right| \\ &\leq n(t) + m\int_0^t |x(\sigma(s)|^{p-1}\Big\{f(s)W(|x(\sigma(s))|) \\ &\quad + g(s)|x(\sigma(s))| + h(s)\Big\}ds,\end{aligned}$$

where $J(x)$ denotes the maximal existent interval of $x(t)$.

An application of Corollary 1.2.11 to the above inequality yields for all $t \in J(x) \cap [0, \gamma]$,

$$\begin{aligned}|x(t)| \leq G^{-1}\Bigg(G\left[\left(\exp\int_0^t \frac{mg(s)}{p}ds\right)\left(n^{1/p}(t) + \int_0^t \frac{mh(s)}{p}ds\right)\right] \\ + \exp\left(\int_0^t \frac{mg(s)}{p}ds\right)\int_0^t \frac{mf(s)}{p}ds\Bigg),\end{aligned}$$

where G and G^{-1} are as defined in Theorem 1.2.26, and $\gamma > 0$ is chosen so that the quantity in the curly brackets is in the range of G.

4.23 An Application of Theorem 1.4.10 to Differential Equations

In this section, we shall apply Theorem 1.4.10 (part(a1)) to obtain the explicit bound on the solution of a certain differential equation

$$u^{p-1}(t)u'(t) + F(t, u(t)) = r(t), u(0) = u_0, \tag{4.23.1}$$

where $p > 1$ is a fixed real number, u_0 is a real constant, and $u, r : \mathbb{R}_+ \to \mathbb{R}$, $F : \mathbb{R}_+ \times \mathbb{R} \to \mathbb{R}$ are continuous functions. It is easy to verify that the problem (4.23.1) is equivalent to the integral equation

$$\frac{u^p(t)}{p} - \frac{u_0^p}{p} + \int_0^t F(s, u(s))\, ds = \int_0^k r(s)\, ds. \tag{4.23.2}$$

We assume that the function F satisfies the condition

$$|F(t,u)| \le h(t)|u|, \tag{4.23.3}$$

where $h : \mathbb{R}_+ \to \mathbb{R}_+$ is a continuous function. From (4.23.2) and (4.23.3) it follows

$$|u(t)|^p \le \bar{a}(t) + p\int_0^t h(s)|u(s)|\,ds, \tag{4.23.4}$$

where $\bar{a}(t) = |u_0|^p + p\int_0^t |r(s)|\,ds$. Now applying Theorem 1.4.10 (*part* (a_1)) with $g(t) = 0$ yields for all $t \in \mathbb{R}_+$,

$$|u(t)| \le \left\{\bar{a}(t) + p\int_0^t h(s)\left(\frac{p-1}{p} + \frac{\bar{a}(s)}{p}\right)\exp\left(\int_s^t h(\sigma)\,d\sigma\right)ds\right\}^{\frac{1}{p}}. \tag{4.23.5}$$

The right-hand side of (4.23.5) gives us the bound on the solution of (4.23.1) in terms of the known quantities.

4.24 Applications of Theorem 2.1.18 to Higher Order Difference Equations

First we shall consider the $(k+1)$th order difference equation

$$\Delta^{k+1}y(t) = f(t, y(t), \Delta y(t), \ldots, \Delta^k y(t)) \tag{4.24.1}$$

and show that Theorem 2.1.80 in Qin [557] is directly applicable to find the upper estimates for the solutions of (4.24.1) provided

$$|f(t, u_0, u_1, \ldots, u_k)| \le \sum_{j=0}^{k} h_j(t)|u_j|. \tag{4.24.2}$$

In fact any solution of (4.24.1) also satisfies

$$\Delta^k y(t) = \Delta^k y(0) + \sum_{s=0}^{t-1} f(s, y(s), \Delta y(s), \ldots, \Delta^k y(s)),$$

or

$$|\Delta^k y(t)| \le |\Delta^k y(0)| + \sum_{j=0}^{k}\sum_{s=0}^{t-1} h_j(s)|\Delta^j y(s)|.$$

Hence from Theorem 2.1.80 in Qin [557], we obtain

$$|\Delta^k y(t)| \le |\Delta^k y(0)| + \sum_{s=0}^{t-1} \phi_1^*(s) \prod_{\tau=s+1}^{t-1} [1 + \phi_2^*(\tau)],$$

where ϕ_1^* and ϕ_2^* are the same as $\phi_1(t)$ and $\phi_2(t)$ with $p(t) = |\Delta^k y(0)|$ and $q(t) = 1$.
Now from the identity (2.1.274) of Theorem 2.1.80 in Qin [557], we find

$$|y(t)| \le \sum_{i=0}^{k-1} \frac{(t)^{(i)}}{i!} |\Delta^i y(0)| + \frac{1}{(k-1)!} \sum_{s=0}^{t-k} (t-s-1)^{(k-1)}$$
$$\times \left(|\Delta^k y(0)| + \sum_{\tau_1=0}^{s-1} \phi_1^*(\tau_1) \prod_{\tau_2=\tau_1+1}^{s-1} [1 + \phi_2^*(\tau_2)] \right).$$

Similarly, if the function f satisfies

$$|f(t, u_0, u_1, \ldots, u_k)| \le \sum_{j=0}^{k} h_j(s) W(\sum_{i=0}^{k} |u_i|),$$

where W is the same as in Theorem 2.1.18 and $|\Delta^k u(0)| \ge 0$. Similar estimate for the solution of (4.24.1) can be obtained directly on using Theorem 2.1.18.

Several other properties, such as boundedness, uniqueness, and asymptotic behavior, of the solutions of (4.24.1) can be discussed now, with the help of these inequalities, as in ordinary differential equations.

The result obtained in Theorem 2.1.83 in Qin [557] can be used directly for several particular kth order systems of difference inequalities, for example in the system

$$\begin{cases} \Delta^k u_1(t) \le p_1(t) + q_1(t) \sum_{t_1=0}^{t-1} f_{11}(t_1) u_1(t_1) + \sum_{t_1=0}^{t-1} f_{12}(t_1) \Delta^k u_2(t_1), & (4.24.3) \\ \Delta^k u_2(t) \le p_2(t) + q_2(t) \sum_{t_1=0}^{t-1} f_{21}(t_1) u_1(t_1). & (4.24.4) \end{cases}$$

If we substitute (4.24.4) in (4.24.3), then we obtain an inequality of the form (2.1.280) in Qin [557].

4.25 Applications of Theorems 2.2.1 and 2.2.2 to Linear Stochastic Discrete Systems

In this sequel, we use the results of Theorems 2.2.1–2.2.2 to study for several particular systems of discrete inequalities, explicit upper estimates. For example, for two dimensional discrete inequalities

$$|x_i(t)| \leq |k_i| + \sum_{s=0}^{t-1} |f_i(s, x_1(s), x_2(s))|, \quad i = 1, 2$$

which appear in the study of two-dimensional differential systems using Euler's method, if

$$|f_i(s, x_1(s), x_2(s))| \leq b_i(t) + a_{i1}(t)|x_1(t)| + a_{i2}(t)|x_2(t)|,$$

then it follows from Corollary 3.2 in [4] that $|x_i(t)| \leq u_i(t)$ where $u_1(t)$ and $u_2(t)$ are the solutions of the following discrete system

$$\begin{cases} \Delta u_i(t) = b_i(t) + a_{i1}(t)u_1(t) + a_{i2}u_2(t), \\ u_i(0) = |k_i|. \end{cases} \tag{4.25.1}$$

Now Theorem 2.2.1 can be used to obtain upper estimates at least. In fact, from (4.25.1) it follows

$$\begin{aligned} u_2(t) &= \prod_{s=0}^{t-1}(1 + a_{22}(s))\Bigg[|k_2| + \sum_{s=0}^{t-1}(b_2(s) + a_{11}(s)u_1(s)) \\ &\quad \times \prod_{\tau=0}^{s}[1 + a_{22}(\tau)]^{-1}\Bigg]. \end{aligned}$$

Now substituting this in the first equation of (4.25.1), we find for $u_1(t)$ the exact form as in Theorem 2.2.1.

Next we shall make a comparative study of some known results. Following the same notations as in [447], we consider the linear stochastic discrete system

$$y_{n+1}(\omega) = A(\omega)y_n(\omega), \quad y_0(\omega) = x_0 \tag{4.25.2}$$

and the perturbed system including an operator T as

$$\begin{cases} x_{n+1}(\omega) = A(\omega)x_n(\omega) + f_n(\omega, x_n(\omega), (Tx_n)(\omega)), \\ x_0(\omega) = x_0. \end{cases} \tag{4.25.3}$$

Let $Y_n(\omega)$ denotes the stochastic fundamental matrix solution of the homogeneous system (4.25.2) such that $Y_0(\omega)$ is the unit matrix.

The following modified versions of [447] which require weaker conditions can be proved using the results obtained and the parallel arguments.

Theorem 4.25.1 ([19]) *Suppose that*

$$|Y_n(\omega)Y_{s+1}^{-1}(\omega)f_s(\omega, x_s(\omega), (Tx_s)(\omega))| \leq a_s(\omega)|x_s(\omega)| + b_s(\omega)|(Tx_s)(\omega)|$$

where $a_n(\omega)$, $b_n(\omega)$ are non-negative random functions defined for all $s \in \mathbb{N}$, $\omega \in \Omega$. Furthermore, suppose that the operator T satisfies the inequality

$$|(Tx_n)(\omega)| \leq \sum_{s=0}^{n-1} c_s(\omega)|x_s(\omega)|$$

where $c_n(\omega)$ is a non-negative random function defined for all $n \in \mathbb{N}$, $\omega \in \Omega$. Then for every bounded random solution $x_n(\omega)$ of equation (4.25.2) on $\mathbb{N}$, the corresponding random solution $x_n(\omega)$ of problem (4.25.3) is bounded on $\mathbb{N}$ provided that

$$\prod_{s=0}^{+\infty}[1 + a_s(\omega) + b_s(\omega)\sum_{\tau=0}^{n-1} c_\tau(\omega)] < +\infty.$$

Theorem 4.25.2 ([19]) *Let us assume*

$$|Y_n(\omega)Y_{s+1}^{-1}(\omega)| \leq Me^{-\alpha(n-s)}, \quad |Y_n(\omega)| \leq Me^{-\alpha n},$$

$$|f_n(\omega, x_n(\omega), (Tx_n)(\omega))| \leq a_n(\omega)|x_n(\omega)| + b_n(\omega)|(Tx_n)(\omega)|$$

$$|(Tx_n)(\omega)| \leq e^{-\alpha n}\sum_{s=0}^{n-1} c_s(\omega)|x_s(\omega)|$$

where $M > 0$, $\alpha > 0$ are constants and $a_n(\omega)$, $b_n(\omega)$, $c_n(\omega)$ are defined in Theorem 4.25.1. Then all random solutions of problem (4.25.3) approach zero as $n \to +\infty$

$$K = \prod_{s=0}^{+\infty}\left[1 + a_s(\omega) + b_s(\omega)\sum_{\tau=0}^{n-1} c_\tau(\omega)e^{-\alpha\tau}\right] < +\infty.$$

Remark 4.25.1 In Theorem 4.25.2, let $-\alpha = \epsilon$ and $K \leq c$ where $c > 0$ is a constant, then the conclusion of Theorem 2.2.2 follows.

4.26 Applications of Theorems 2.2.16 and 2.2.18 to Difference Equations

In this section, we shall use Theorem 2.2.16 and 2.2.18 to difference equations.

Example 4.26.1 We first consider the following sum-difference system of Volterra type

$$\begin{cases} u_1(t) = C_1 + \sum_{s=0}^{t-1}[F_1(t,s,u_1(s),u_2(s)) + K_1(u_1(s)+u_2(s))], \\ u_2(t) = C_2 + \sum_{s=0}^{t-1}[F_2(t,s,u_1(s),u_2(s)) + K_2(u_1(s)+u_2(s))] \end{cases}$$

where $C_1 \geq 4, C_2 \geq 4$, and the functions $F_1, F_2, K_1,$ *and* K_2 satisfy, for all $t \in \mathbb{N}$,

$$\begin{cases} |F_1(t,s,u_1(s),u_2(s))| & \leq e_1(s)|u_1(s)| + e_2(s)|u_2(s)|, \\ |F_2(t,s,u_1(s),u_2(s))| & \leq h_1(s)|u_1(s)| + h_2(s)|u_2(s)|, \\ |K_1(u_1(s),u_2(s))| & \leq e_3(s)H(|u_1(s)|) + e_4(s)H(|u_2(s)|), \\ |K_2(u_1(s),u_2(s))| & \leq h_3(s)H(|u_1(s)|) + h_4(s)H(|u_2(s)|). \end{cases}$$

Hence we get

$$\begin{cases} |u_1(t)| \leq C_1 + \sum_{s=0}^{t-1} e_1(s)|u_1(s)| + \sum_{s=0}^{t-1} e_2(s)|u_2(s)| \\ \qquad + \sum_{s=0}^{t-1} e_3(s)H(|u_1(s)|) + \sum_{s=0}^{t-1} e_4(s)H(|u_2(s)|), \\ |u_2(t)| \leq C_2 + \sum_{s=0}^{t-1} h_1(s)|u_1(s)| + \sum_{s=0}^{t-1} h_2(s)|u_2(s)| \\ \qquad + \sum_{s=0}^{t-1} h_3(s)H(|u_1(s)|) + \sum_{s=0}^{t-1} h_4(s)H(|u_2(s)|). \end{cases}$$

The above two inequalities are exactly of the same form as (2.2.71) and (2.2.72) considered in Theorem 2.2.18, where $p_1 = p_2 = p_3 = p_4 = q_1 = q_2 = q_3 = q_4 = 1$. Thus it is possible to find the estimates for $|u_2(t)|$ in terms of known function.

Example 4.26.2 Consider the following system:

$$\begin{cases} u_1(t) = C_3(t) + \sum_{s=0}^{t-1} k_1(u_1(s), u_2(s)) \\ u_2(t) = C_4(t) + \sum_{s=0}^{t-1} k_1(u_1(s), u_2(s)) \end{cases}$$

where

$$K_1(u_1, u_2) \leq H(|u_1|) + H(|u_2|).$$

Hence we get

$$\begin{cases} |u_1(t)| \leq C_3 + \sum_{s=0}^{t-1} H(|u_1(s)|) + \sum_{s=0}^{t-1} H(|u_2(s)|), \\ |u_2(t)| \leq C_4 + \sum_{s=0}^{t-1} H(|u_1(s)|) + \sum_{s=0}^{t-1} H(|u_2(s)|) \end{cases}$$

The above two inequalities are exactly of the same form as (2.2.58) and (2.2.59) in Theorem 2.2.16 where $a_1 = c_3, a_2 = c_4, andp_1 = p_2 = q_1 = q_2 = 1$. From Theorem 2.2.16, we derive

$$A(t) = H(C_3(t) - 2) + H(C_4(t) - 2), \quad B(t) = 4H(1) = C,$$

$$\begin{cases} \psi(t) &= G^{-1}\left\{G(2) + \sum_{s=0}^{t-1}(4H(1) + H(C_3(s) - 2) + H(C_4(s) - 2))\right\} \\ |u_1(t)| &\leq C_3(t) + \sum_{s=0}^{t-1}\{H(C_3(s) - 2) + 4H(1)\psi(s) + H(C_4(s) - 2)\} \\ |u_2(t)| &\leq C_4(t) + \sum_{s=0}^{t-1}\{H(C_3(s) - 2) + 4H(1)\psi(s) + H(C_4(s) - 2)\} \end{cases}$$

and

$$G(t) = \int_{r_0}^{r} \frac{ds}{s + H(s)}, \quad 0 < r_0 \leq r.$$

4.27 An Application of Theorem 2.3.12 to Finite Difference Equations.

In this section, we shall use Theorem 2.3.12 to obtain bounds on the solutions of certain finite difference equations. Consider the following higher order finite difference equation of the form

$$\triangle\Big(\frac{1}{r_{n-1}(t)}\triangle\Big(\frac{1}{r_{n-2}(t)}\cdots\triangle\Big(\frac{1}{r_1(t)}\triangle z^2(t)\Big)\Big)\Big) = z(t)F(t, z(t)) + G(t, z(t)), \tag{4.27.1}$$

with the given initial conditions

$$z(0) = z_0, \quad \frac{1}{r_{i-1}(0)}\triangle\Big(\frac{1}{r_{i-2}(0)}\cdots\triangle\Big(\frac{1}{r_1(0)}\triangle z^2(t0)\Big)\Big) = 0, \tag{4.27.2}$$

for $i = 2, 3, \ldots, n$. Here, $r_1(t), \ldots, r_{n-1}(t)$ are real-valued positive functions defined on $\mathbb{N}_0$, z_0 is a real constant and F, G are real-valued functions defined on $\mathbb{N}_0 \times \mathbb{R}$. It is easy to observe that the problem (4.27.1)–(4.27.2) is equivalent to the following sum-difference equation

$$z^2(t) = z_0^2 + A\Big(t, r, z(s_n)F(s_n, z(s_n)) + G(s_n, z(s_n))\Big). \tag{4.27.3}$$

If $z(t)$ is a solution of the problem (4.27.1)–(4.27.2), then it satisfies the equation (4.27.3). We assume that

$$|F(t, z)| \leq 2f(t)|z|, \quad |G(t, z)| \leq 2g(t)|z|, \tag{4.27.4}$$

where $f(t), g(t)$ are real-valued non-negative functions defined on $\mathbb{N}_0$. Using (4.27.4) in (4.27.3), we have

$$|z(t)|^2 \leq |z_0|^2 + 2A\Big(t, r, f(s_n)|z(s_n)|^2 + g(s_n)|z(s_n)|\Big). \tag{4.27.5}$$

Now an application of Theorem 2.3.12 yields, for all $t \in \mathbb{N}_0$,

$$|z(t)| \leq p_0(t) \prod_{s_1=0}^{t-1} (1 + \bar{A}(s_1, r, f(s_n))), \tag{4.27.6}$$

where for all $t \in \mathbb{N}_0$,

$$p_0(t) = |z_0| + A(t, r, g(s_n)). \tag{4.27.7}$$

The inequality (4.27.6) obtains the bound on the solution $z(t)$ of the problem (4.27.1)–(4.27.2) in terms of the known functions.

4.28 Applications of Theorem 2.3.14 to Some Finite Difference Equations

We first consider the following second-order difference equation:

$$\Delta^2 u(t) = f(t, u(t)), \tag{4.28.1}$$

with given initial conditions

$$u(0) = c, \quad \Delta u(0) = 0, \tag{4.28.2}$$

where c is a constant and f is a real-valued function defined on $\mathbb{N}_0 \times \mathbb{R}$. In the past few years, many authors have studied the qualitative behavior of the solutions of equation (4.28.1) and its further generalizations with different viewpoints. We shall apply the inequality established in Theorem 2.3.14 to study the boundedness and other properties of the solutions of equation (4.28.1) with initial conditions (4.28.2).

The first result deals with the boundedness of the solutions of equation (4.28.1) with the given initial conditions in (4.28.2).

Theorem 4.28.1 (The Pachpatte Inequality [506]) *Suppose that the function f satisfies the condition, for all $t \in \mathbb{N}_0$ and $|u| < +\infty$,*

$$|f(t, u)| \leq b(t)|u| \tag{4.28.3}$$

where $b(t)$ is a real-valued non-negative function defined for all $t \in \mathbb{N}_0$ and

$$\prod_{s=0}^{t-1}\left[1 + \sum_{\sigma=0}^{s-1} b(\sigma)\right] < +\infty \tag{4.28.4}$$

Then the solution $u(t)$ of problem (4.28.1)–(4.28.2) is bounded and for all $t \in \mathbb{N}_0$,

$$|u(t)| \leq |c| \prod_{s=0}^{t-1}\left[1 + \sum_{\sigma=0}^{s-1} b(\sigma)\right]. \tag{4.28.5}$$

Proof Let $u(t)$ be a solution of problem (4.28.1)–(4.28.2) for all $t \in \mathbb{N}$. From (4.28.1)–(4.28.2), it follows

$$\Delta u(t) = 1 + \sum_{\sigma=0}^{t-1} f(\sigma, u(\sigma)). \tag{4.28.6}$$

Now multiplying both sides of (4.28.6) by $(u(t + 1) + u(t))$, we observe that

$$u^2(t + 1) - u^2(t) = (u(t + 1) + u(t)) \sum_{\sigma=0}^{t-1} f(\sigma, u(\sigma)). \tag{4.28.7}$$

Taking $t = s$ in (4.28.7) and summing up over s form 0 to $t - 1$, we obtain

$$u^2(t) = c^2 + \Sigma_{s=0}^{t-1}(u(s + 1) + u(s)) \sum_{\sigma=0}^{t-1} f(\sigma, u(\sigma)). \tag{4.28.8}$$

From (4.28.8) and (4.28.3), we derive that

$$u^2(t) \leq |c|^2 + \Sigma_{s=0}^{t-1}(|u(s + 1)| + |u(s)|) \sum_{\sigma=0}^{t-1} b(\sigma)|u(\sigma)|. \tag{4.28.9}$$

Now a suitable application of Theorem 2.3.14 yields

$$|u(t)| \leq |c| \prod_{s=0}^{t-1} \left[1 + \sum_{\sigma=0}^{t-1} b(\sigma) \right].$$

and the proof is now complete.

Our next result deals with the dependence of solutions of equation on initial data.

Theorem 4.28.2 (The Pachpatte Inequality [506]) *Let $u_1(t)$ and $u_2(t)$ be the solutions of equation (4.28.1) with the given initial conditions*

$$u_1(0) = c_1, \quad \Delta u_1(0) = 0 \tag{4.28.10}$$

and

$$u_2(0) = c_2, \quad \Delta u_2(0) = 0 \tag{4.28.11}$$

respectively, where c_1, c_2 are real constants. Suppose that the function f satisfies the condition

$$|f((t,u) - f((t,\overline{u})| \leq b(t)|u - \overline{u}|, \tag{4.28.12}$$

where $b(t)$ is a real-valued non-negative function defined on $\mathbb{N}_0$. Then for all $t \in \mathbb{N}_0$,

$$|u_1(t) - u_2(t)| \leq |c_1 - c_2| \prod_{s=0}^{t-1} \left[1 + \sum_{\sigma=0}^{t-1} b(\sigma) \right]. \tag{4.28.13}$$

Proof From the hypotheses, it follows

$$\begin{cases} \Delta u_1(t) = \sum_{\sigma=0}^{t-1} f(\sigma, u_1(\sigma)), & (4.28.14) \\ \Delta u_2(t) = \sum_{\sigma=0}^{t-1} f(\sigma, u_2(\sigma)), & (4.28.15) \end{cases}$$

Let $z(t) = u_1(t) - u_2(t)$ for $t \in N_0$. From (4.28.14)–(4.28.15), we derive

$$\Delta z(t) = \sum_{\sigma=0}^{t-1} \{f(\sigma, u_1(\sigma)) - f(\sigma, u_2(\sigma))\}. \tag{4.28.16}$$

Now multiplying both sides of (4.28.16) by $(z(t+1)+z(t))$, we obtain

$$z^2(t+1) - z^2(t) = (z(t+1)+z(t)) \sum_{\sigma=0}^{t-1} \{f(\sigma, u_1(\sigma)) - f(\sigma, u_2(\sigma))\}. \quad (4.28.17)$$

Taking $t = s$ in (4.28.17) and summing up over s form 0 to $t-1$, we get

$$z^2(t) = z^2(0) + \sum_{s=0}^{t-1}(z(s+1)+z(s)) \sum_{\sigma=0}^{s-1} \{f(\sigma, u_1(\sigma)) - f(\sigma, u_2(\sigma))\}. \quad (4.28.18)$$

From (4.28.18) and (4.28.12), it follows that

$$|z(t)|^2 \le |c_1 - c_2|^2 + \sum_{s=0}^{t-1}(|z(s+1)| + |z(s)|) \sum_{\sigma=0}^{s-1} b(\sigma)|z(\sigma)|. \quad (4.28.19)$$

Applying Theorem 2.3.14 to (4.28.19), we get

$$|u_1(t) - u_2(t)| \le |c_1 - c_2| \prod_{s=0}^{t-1} \left[1 + \sum_{\sigma=0}^{s-1} b(\sigma) \right],$$

which proves our result.

We now consider the following difference equations

$$\begin{cases} \Delta^2 u(t) = f(t, u(t), \mu), & (4.28.20) \\ \Delta^2 u(t) = f(t, u(t), \mu_0), & (4.28.21) \end{cases}$$

with the given initial conditions

$$u(0) = c, \qquad \Delta u(0) = 0, \quad (4.28.22)$$

where c is a constant,f is a real-valued function defined on $\mathbb{N}_0 \times \mathbb{R} \times \mathbb{R}$, and μ, μ_0 are real parameters.

The following theorem shows the dependence of solutions of equations (4.28.20) and (4.28.21) on pure parameters.

Theorem 4.28.3 ([506]) *Suppose that*

$$|f(t, u, \mu) - f(t, \bar{u}, \mu)| \le b(t)|u - \bar{u}| \quad (4.28.23)$$

$$|f(t, u, \mu) - f(t, u, \mu_0)| \le q(t)|u - \mu_0| \quad (4.28.24)$$

where $b(t)$ and $q(t)$ are real-valued non-negative functions defined for all $t \in \mathbb{N}_0$. If $u_1(t)$ and $u_2(t)$ are the solutions of equations (4.28.20)–(4.28.21) with the given *initial conditions (4.28.22), then for all* $t \in \mathbb{N}_0$,

$$|u_1(t) - u_2(t)| \le \left[|\mu - \mu_0| \sum_{s=0}^{t-1} \left(\sum_{\sigma=0}^{t-1} q(\sigma) \right) \right] \prod_{s=0}^{t-1} \left[1 + \sum_{\sigma=0}^{s-1} b(\sigma) \right]. \tag{4.28.25}$$

Proof Let $z(t) = u_1(t) - u_2(t)$ for all $t \in \mathbb{N}_0$. As in the proof of Theorem 4.28.2, from the hypotheses, we obtain

$$\Delta z(t) = \sum_{\sigma=0}^{t-1} \Big\{ f(\sigma, u_1(\sigma), \mu) - f(\sigma, u_2(\sigma), \mu) + f(\sigma, u_2(\sigma), \mu) - f(\sigma, u_2(\sigma), \mu_0) \Big\}. \tag{4.28.26}$$

Multiplying both sides of (4.28.26) by $(z(t+1) + z(t)$, we get

$$z^2(t+1) - z^2(t) = (z(t+1) + z(t)) \sum_{\sigma=0}^{t-1} \Big\{ f(\sigma, u_1(\sigma), \mu) - f(\sigma, u_2(\sigma), \mu) + f(\sigma, u_2(\sigma), \mu) - f(\sigma, u_2(\sigma), \mu_0) \Big\}. \tag{4.28.27}$$

Taking $t = s$ in (4.28.27) and summing up over s from 0 to $t-1$, we get

$$z^2(t) = z^2(0) + \sum_{s=0}^{t-1} (z(s+1) + z(s)) \times \sum_{\sigma=0}^{s-1} \Big\{ f(\sigma, u_1(\sigma), \mu) - f(\sigma, u_2(\sigma), \mu) + f(\sigma, u_2(\sigma), \mu) - f(\sigma, u_2(\sigma), \mu_0) \Big\}. \tag{4.28.28}$$

Form (4.28.28) and (4.28.23)–(4.28.24), we infer

$$|z(t)|^2 \le \sum_{s=0}^{t-1} (|z(s+1)| + |z(s)|) \sum_{\sigma=0}^{s-1} \Big\{ b(\sigma)|z(\sigma)| + q(\sigma)|\mu - \mu_0| \Big\}. \tag{4.28.29}$$

Now applying Theorem 2.3.14 to (4.28.29) yields

$$|u_1(t) - u_2(t)| \le \left[\sum_{s=0}^{t-1} \left(\sum_{\sigma=0}^{t-1} q(\sigma)|\mu - \mu_0| \right) \right] \prod_{s=0}^{t-1} \left[1 + \sum_{\sigma=0}^{s-1} b(\sigma) \right],$$

which proves our result. □

4.29 Applications of Theorems 3.2.5 and 3.2.6 and Corollary 3.2.4 to Nonlinear Impulsive Integro-Differential and Differential Equations

In this section, we shall use Theorems 3.2.5–3.2.6 and Corollary 3.2.4 to obtain bounds for the solutions of different type equations.

Example 4.29.1 Consider the nonlinear impulsive equation with delay

$$\begin{cases} u(t) = f(t) + \left[\int_0^t p(s)\sqrt{u(s)}ds + \int_0^t g(s)\sqrt{u(s-h)}ds\right]^2 \\ \qquad + \sum_{0<t_k<t} \beta_k u(t_k), \quad \text{for all } t \geq 0, & (4.29.1) \\ u(t) = 0, \quad t \in [-h, 0], & (4.29.2) \end{cases}$$

where $\beta_k \geq 0, k = 1, 2, \ldots, p, g \in C(\mathbb{R}_+, \mathbb{R}_+), f \in C(\mathbb{R}_+, [0, 1]),\ h > 0$.
We note the solutions of the problem (4.29.1)–(4.29.2) are non-negative.

Define the functions $G(u) = u^2, Q(u) = \sqrt{u}$. Then $Q \in W_2(\phi)$, where $\phi(u) = \sqrt{u}$.
Consider the function

$$H(u) = \int_0^u \frac{ds}{Q(1 + G(s))} = \int_0^u \frac{ds}{\sqrt{1+s^2}} = \ln(u + \sqrt{1+u^2}). \qquad (4.29.3)$$

Then the inverse function of $H(u)$ will be defined by

$$H^{-1}(u) = \sinh(u) = \frac{1}{2}(e^u - e^{-u}). \qquad (4.29.4)$$

Applying Corollary 3.2.4, we obtain

$$u(t) \leq \Pi_{0<t_k<t}(1+\beta_k)\left\{1 + \left[\sinh\left(\sqrt{\Pi_{0<t_k<t}(1+\beta_k)\int_0^t (p(s) + \Lambda g(s))ds}\right)\right]^2\right\}.$$

Example 4.29.2 Consider the initial value problem for the nonlinear impulsive integro-differential equation

$$\begin{cases} u'(t) = 2f(t)\sqrt{u(t)}\int_0^t f(s)\sqrt{u(s)}ds, \quad t > 0, t \neq t_k, & (4.29.5) \\ u(t_k + 0) = \beta_k u(t_k), & (4.29.6) \\ u(0) = c, & (4.29.7) \end{cases}$$

where $c \geq 0, \beta_k \geq 0, k = 1, 2, \ldots, f \in C(\mathbb{R}_+, \mathbb{R}_+)$.

The solutions of the given problem satisfy the inequality

$$u(t) \leq c + \left[\int_0^t f(s)\sqrt{u(s)}ds\right]^2 + \sum_{0<t_k<t} \beta_k u(t_k), \quad t_k \geq 0. \tag{4.29.8}$$

Define the functions $G(u) = u^2, Q(u) = \sqrt{u}$. Then the functions $H(u)$ and $H^{-1}(u)$ are defined by the equations (4.29.3)–(4.29.4).

We note that the solutions of the problem (4.29.5)–(4.29.7) are non-negative. According to Theorem 3.2.5, the solutions $u(t)$ satisfy

$$\begin{aligned} u(t) &\leq A\Pi_{0<t_k<t}(1 + A\beta_k) \\ &\times \left\{1 + \left[sh\left(\sqrt{A\Pi_{0<t_k<t}(1 + A\beta_k)} \int_0^t f(s)ds\right)\right]^2\right\}, \end{aligned} \tag{4.29.9}$$

where $A = \max\{1, c\}$.

Example 4.29.3 Consider the initial value problem for the nonlinear impulsive differential-difference equation

$$\begin{cases} u'(t) = F(t, u(t), u(t-h)) + r(t), \quad t > 0,\ t \neq t_k, & (4.29.10) \\ u(t_k + 0) = \beta_k u(t_k), & (4.29.11) \\ u(t) = \psi(t), \quad t \in [-h, 0], & (4.29.12) \end{cases}$$

where $\beta_k = const., k = 1, 2, \ldots, r \in C(\mathbb{R}_+, \mathbb{R}), \psi \in C([-h, 0], \mathbb{R}), F \in C(\mathbb{R}_+ \times \mathbb{R}^2, \mathbb{R})$ and there exist functions $p, g \in C(\mathbb{R}_+, \mathbb{R}_+)$ such that $|F(t, u, v)| \leq p(t)\sqrt{|u|} + g(t)\sqrt{|v|}$. Let $|\psi(0) + \int_0^t r(s)ds| \leq 1$, for all $t \geq 0$.

We assume that the solutions $u(t)$ of the initial value problem (4.29.10)–(4.29.12) exist on $\mathbb{R}_+$.

The solution $u(t)$ satisfies the inequalities

$$\begin{cases} |u(t)| \leq \left|\psi(0) + \int_0^t r(s)ds\right| + \int_0^t \left(p(s)\sqrt{|u(s)|} + g(s)\sqrt{|u(s-h)|}\right) ds \\ \qquad + \sum_{0<t_k<t} |\beta_k||u(t_k)|, \ \text{for all } t > 0, & (4.29.13) \\ |u(t)| = |\psi(t)|, \quad \text{for all } t \in [-h, 0]. & (4.29.14) \end{cases}$$

Consider functions $f_1(t) = |\psi(0) + \int_0^t r(s)ds|, G(u) = u, Q(u) = \sqrt{u}, f_2(t) = 1, f_3(t) = 1$. Then $\psi(u) = \sqrt{u}$ and according to the equality (3.2.78) in Corollary 3.2.4, the function $H(u) = 2(\sqrt{1+u} - 1)$ and its inverse is $H^{-1}(u) = (\frac{u}{2} + 1)^2 - 1$.

Applying Corollary 3.2.4 to (4.29.13)–(4.29.14), we get

$$|u(t)| \leq \Pi_{0<t_k<t}(1+|\beta_k|)\left(\sqrt{1+hB_1\sqrt{B_2}}\right.$$
$$\left.+\frac{1}{2}\sqrt{\Pi_{0<t_k<t}(1+|\beta_k|)}\int_0^t (p(s)+\Lambda g(s))ds\right)^2, \quad (4.29.15)$$

where $B_1 = \max\{|g(s)| : s \in [0,h]\}, B_2 = \max\{\varphi(s) : s \in [-h,0]\}$.

Example 4.29.4 Consider the initial value problem for the nonlinear impulsive differential-difference equation

$$\begin{cases} u(t)u'(t) = F(t,u(t),u(t-h)) \\ \qquad +q(t)u(t)+r(t)u(t-h), \quad t>0, \ t \neq t_k, & (4.29.16) \\ u(t_k+0) = \beta_k u(t_k), & (4.29.17) \\ u(t) = \psi(t), \quad t \in [-h,0], & (4.29.18) \end{cases}$$

where $\beta_k = const., k = 1,2,\ldots,r, q \in C(\mathbb{R}_+,\mathbb{R}), \psi \in C([-h,0],\mathbb{R}), F \in C(\mathbb{R}_+ \times \mathbb{R}^2,\mathbb{R})$ and there exist functions $p,g \in C(\mathbb{R}_+,\mathbb{R}_+)$ such that $|F(t,u,v)| \leq p(t)u^2 + g(t)v^2$. Let $|\psi(0)+\int_0^t r(s)ds| \leq 1$, for all $t \geq 0$. We assume that the solutions of the initial value problem (4.29.6)–(4.29.8) exist on $\mathbb{R}_+$.

The solution $u(t)$ satisfies the inequalities

$$\begin{cases} u^2(t) \leq \psi^2(0) + \int_0^t \Big(p(s)u^2(s)+g(s)u^2(s-h)+q(s)u(s)+r(s)u(s-h)\Big)ds \\ \qquad + \sum_{0<t_k<t} \beta_k^2 u^2(t_k), \quad \text{for all } t>0, & (4.29.19) \\ u^2(t) = \psi^2(t), \quad \text{for all } t \in [-h,0]. & (4.29.20) \end{cases}$$

Applying Theorem 3.2.6 to the inequalities (4.29.19)–(4.29.20) we may obtain

$$\begin{cases} u^2(t) \leq \psi^2(0) + \int_0^t \big(p(s)u^2(s)+g(s)u^2(s-h)+q(s)u(s)+r(s)u(s-h)\big)\,ds \\ \qquad + \sum_{0<t_k<t} \beta_k^2 u^2(t_k), \quad \text{for all } t>0, & (4.29.21) \\ u^2(t) = \psi^2(t), \quad \text{for all } t \in [h,0]. & (4.29.22) \end{cases}$$

4.30 An Application of Theorem 3.2.17 to Nonlinear Dynamic Equation

We use Theorem 3.2.17 to the qualitative analysis of a nonlinear dynamic equation. Let $a, b \in T$ and consider the initial value problem

$$u^{\Delta}(t) = F\left(t, u(t), \int_a^t K(t, u(s))\Delta s\right), \quad t \in T_*^k, \quad u(a) = u_a, \qquad (4.30.1)$$

where $T_* = [a, b]_T, u \in C_{rd}^1[T_*], F \in C_{rd}[T_* \times \mathbb{R} \times \mathbb{R}, \mathbb{R}]$ and $K \in C_{rd}[T_* \times \mathbb{R}, \mathbb{R}]$.

In what follows, we shall assume that the problem (4.30.1) has a unique solution, which we denote by $u_*(t)$.

Theorem 4.30.1 ([229]) *Assume that the functions F and K in (4.30.1) satisfy the conditions*

$$|K(t, u)| \leq h(t)\Phi(|u|), \qquad (4.30.2)$$

$$|F(t, u, v)| \leq |u| + |v|, \qquad (4.30.3)$$

where h and Φ are as defined in Theorem 3.2.17. Then, for all $t \in T_*$ such that

$$\Psi(\xi) + \int_0^{\rho(t)} \Phi(p(\tau))\Phi\left(\int_a^{\tau} h(\theta)\Delta\theta\right)\Delta\tau \in Dom\,(\Psi^{-1}),$$

we have

$$|u_*(t)| \leq p(t)\left\{|u_a| + \int_a^t h(s)\Psi^{-1}\left(\Psi(\xi) + \int_a^s \Phi(p(\tau))\Phi\left(\int_a^{\tau} h(\theta)\Delta\theta\right)\Delta\tau\right)\Delta s\right\}, \qquad (4.30.4)$$

where

$$p(t) = 1 + \int_a^t e_1(t, \sigma(s))\Delta s,$$

$$\xi = \int_a^{\rho(b)} \Phi(p(s)|u_0|)\Delta s,$$

$$\Psi(x) = \int_{x_0}^x \frac{1}{\Phi(s)}ds, \quad x \geq x_0 > 0.$$

Proof Let $u_*(t)$ be the solution of (4.29.1). Then, we have

$$u_*(t) = u_a + \int_a^t F\left(s, u_*(s), \int_a^s K(s, u_*(\tau))\Delta\tau\right)\Delta s. \qquad (4.30.5)$$

Using (4.29.2)–(4.29.3) in (4.29.5), we have

$$|u_*| \leq |u_a| + \int_a^t \left(|u_*(s)| + \int_a^t |K(s, u_*(\tau))| \Delta\tau \right) \Delta s$$
$$\leq |u_a| + \int_a^t \left(|u_*(s)| + h(s) \int_a^t \Phi(|u_*(\tau)|) \Delta\tau \right) \Delta s. \quad (4.30.6)$$

A suitable application of Theorem 3.2.17 to (4.29.6), with $a(t) = |u_a|, f(t) = b(t) = a$ and $W(u) = u$, yields (4.9.4). □

4.31 Applications of Theorem 3.2.21 and Corollary 3.2.7 to Impulsive Differential Systems

Let us first consider an impulsive system of the type

$$\begin{cases} \frac{dx}{dt} = f(t, x) \quad t \neq t_i \\ \Delta x|_{t=t_i} = I_i(x) \\ x(t_0^+) = x_0 \end{cases} \quad (4.31.1)$$

where $x \in \mathbb{R}^n, f \in \mathbb{R}^n, I_i(x) \in \mathbb{R}^n$ $(i = 1, 2, \ldots), t \geq t_0 \geq 0, \lim_{i \to +\infty} t_i = +\infty, , t_{i-1} < t_i,$ *for all* $i = 1, 2, \ldots$.

Let us assume that $f(t, x)$ and $I_i(x)$ are defined in some domain $\Omega = \{(t, x) : t \in J = [t_0, T], T \leq +\infty, \|x\| \leq h\}$ and that they satisfy the following nonlinearity conditions

$$\|f(t, x)\| \leq q(t)\|x\|, \quad q : \mathbb{R}_+ \to \mathbb{R}_+; \quad (4.31.2)$$
$$\|I_i(x)\| \leq a_i\|x\|^m, \quad a_i = const. > 0, \ m > 0. \quad (4.31.3)$$

Denoting by $x(t) = x(t, t_0, x_0)$ the solution of Cauchy problem for system (4.31.1), so that $x(t) = x(t, t_0, x_0)$ $(x(t_0) = x_0)$, it is obvious (see, e.g., [118, 327]) that its integro-sum representation is the following

$$x(t, t_0, x_0) = x_0 + \int_{t_0}^t f(\tau, x(\tau, t_0, x_0)) d\tau + \sum_{t_0 < t_i < t} I_i(x(t_i - 0, t_0, x_0)), \quad (4.31.4)$$

Thus system (4.31.1) is equivalent to (4.31.4). By using conditions (4.31.2) and (4.31.3), it is easy to get

$$\|x(t, t_0, x_0) \leq \|x_0\| + \int_{t_0}^{t} q(\tau)\|x(\tau, t_0, x_0)\|d\tau + \sum_{t_0<t_i<t} a_i\|x(t_i - 0, t_0, x_0)\|^m. \tag{4.31.5}$$

By setting$V(t) = \|x(t, t_0, x_0)\|$, we obtain an integro-sum inequality,

$$V(t) \leq \|x_0\| + \int_{t_0}^{t} q(\tau)V(\tau)d\tau + \sum_{t_0<t_i<t} a_i V^m(t_i - 0) \tag{4.31.6}$$

if in (3.2.167) of Theorem 3.2.21, $\psi(t) = \|x_0\|, g(t) = p(t) = 1, \tau(s) = s$. From (4.31.6), (3.2.102) and (3.2.103), we can obtain the following theorem.

Theorem 4.31.1 (The Gallo-Piccirillo Inequality [243]) *Under assumptions (4.31.2), (4.31.3) for system (4.31.1), we have, for all $t \geq t_0$, if $m \in [0, 1]$,*

$$\|x(t, t_0, x_0)\| \leq \|x_0\| \prod_{t_0<t_i<t} \left(1 + a_i\|x_0\|^{m-1}\right) \exp\left[\int_{t_0}^{t} q(\tau)d\tau\right], \tag{4.31.7}$$

or for all $t \geq t_0$, if $m > 1$,

$$\|x(t, t_0, x_0)\| \leq \|x_0\| \prod_{t_0<t_i<t} \left(1 + a_i\|x_0\|^{m-1}\right) \exp\left[m\int_{t_0}^{t} q(\tau)d\tau\right]. \tag{4.31.8}$$

By using Corollary 3.2.7, it is possible to obtain the following statement.

Theorem 4.31.2 ([243]) *Under assumptions of the above theorem, the solutions of system (4.31.1) satisfy the following estimates for all $t \in J$, if $m \in [0, 1]$,*

$$\|x(t, t_0, x_0)\| \leq \|x_0\| \exp\left[\int_{t_0}^{t} q(\tau)d\tau + \sum_{t_0<t_i<t} a_i\|x_0\|^{m-1}\right], \tag{4.31.9}$$

or for all $t \in J$, if $m \geq 1$,

$$\|x(t, t_0, x_0)\| \leq \|x_0\| \exp\left[m\int_{t_0}^{t} q(\tau)d\tau + \sum_{t_0<t_i<t} a_i\|x_0\|^{m-1}\right]. \tag{4.31.10}$$

Now we consider the case of strong nonlinearity in zero, in which we assume:

$$\|f(t,x)\| \le \tilde{q}(t)\|x\|^m, \quad \tilde{q}:\mathbb{R}_+ \to \mathbb{R}_+, m>0, \tag{4.31.11}$$

$$\|I_i(x)\| \le a_i\|x\|^m. \tag{4.31.12}$$

By using conditions (4.31.11)–(4.31.12), from (4.31.4), we obtain the integro-sum inequality

$$\|x(t,t_0,x_0)\| \le \|x_0\| + \int_{t_0}^t \tilde{q}(\tau)\|x(\tau,t_0,x_0)\|^m d\tau + \sum_{t_0<t_i<t} a_i\|x(t_i,t_0,x_0)\|^m. \tag{4.31.13}$$

The following result is valid.

Theorem 4.31.3 ([243]) *Under assumptions (4.31.11)–(4.31.12), the following estimates are fulfilled: for all $t \in J$, $m \in [0,1]$,*

$$\|x(t,t_0,x_0)\| \le \prod_{t_0<t_i<t}\left(1+a_i\|x_0\|^{m-1}\right)\left[\|x_0\|^{1-m} + (1-m)\int_{t_0}^t \tilde{q}(\tau)d\tau\right]^{\frac{1}{1-m}}, \tag{4.31.14}$$

or for every $t \ge t_0$ and $m > 1$,

$$\|x(t,t_0,x_0)\| \le \|x_0\| \prod_{t_0<t_i<t}\left(1+a_i\|x_0\|^{m-1}\right) \times \left[1-(m-1)\left(\|x_0\| \prod_{t_0<t_i<t}\left(1+a_i\|x_0\|^{m-1}\right)\right)^{m-1}\int_{t_0}^t \tilde{q}(\tau)d\tau\right]^{-\frac{1}{m-1}} \tag{4.31.15}$$

with

$$\begin{cases} m\|x_0\|^{m-1}\int_{t_0}^t \tilde{q}(\tau)d\tau \le 1, & (4.31.16)\\ \prod_{t_0<t_i<t}\left(1+a_i m\|x_0\|^{m-1}\right) < \left(1+\frac{1}{m-1}\right)^{\frac{1}{m-1}}. & (4.31.17)\end{cases}$$

The notion of stability by Lyapunov for solutions of system (4.31.1) has been introduced in [118, 327, 585, 589]. The notion of practical stability (stability by Chetaev) for solutions of system (4.31.1) has been introduced in [118].

Let us assume in (4.31.1), $f(t,0) = I_i(0)$, for all $t \in J$, for all $i \in \mathbb{N}$. We say that the trivial solution $x \equiv 0$ of system (4.31.1) is practically stable with relative fixed values $(\lambda, , \Lambda, J)$ if, for an arbitrary solution of system (4.31.1), $x(t, t_0, x_0)$: $x(t_0, t_0, x_0) = x_0 \neq 0$, the estimate $\|x(t, t_0, x_0)\| < \Lambda$, for all $t \in J$ is justified only if $0 \neq \|x_0\| < \lambda$.

From the result of Theorem 4.31.1 the following theorem follows.

Theorem 4.31.4 ([243])

(A) Let us suppose that the right-hand side of system (4.31.1) satisfies the following conditions:

(i) $\|f(t,x)\| \leq q(t)\|x\|$, $q : \mathbb{R}_+ \to \mathbb{R}_+$;

(ii) $\|I_i\| \leq a_i\|x\|^m$;

(iii) there exists a constant ξ *such that for all* $t \in J$,

$$\prod_{t_0 < t_i < t} (1 + a_i\|x_0\|^{m-1}) < \xi < +\infty;$$

(iv) there exists a constant η *such that for all* $t \in J$,

$$\int_{t_0}^{t} q(\tau)d\tau < \eta < +\infty.$$

Then all solutions of system (4.31.1) are bounded.

(B) Suppose that, for system (4.31.1), (i), (ii) and (iii) for $m > 1$, *(iv) are valid. Then trivial solutions is Lyapunov stable.*

(C) Let condition (B) hold and, moreover,

$$\prod_{t_0 < t_i < t} (1 + a_i\lambda^{m-1}) \exp(m\eta) < \frac{\Lambda}{\lambda}.$$

Then the trivial solution of system (4.32.1) is practically stable.

From Theorem 4.31.2, we derive the following theorem.

Theorem 4.31.5 ([243])

(A) Let us assume that (i), (ii) of Theorem 4.31.4 hold and

$$\int_{t_0}^{t} q(\tau)d\tau + \sum_{t_0 < t_i < t} a_i\|x_0\|^{m-1} < +\infty.$$

Then all solution of system (4.31.1) are bounded.

(B) If condition (A) is valid for $m > 1$, *then the trivial solution of system (4.31.1) is Lyapunov stable.*

(C) If condition (B) holds and $\exp[m\int_{t_0}^{t} q(\tau)d\tau + \sum_{t_0<t_i<t} a_i\lambda^{m-1}] < \frac{\Lambda}{\lambda}$, *then the solution* $x \equiv 0$ *of (4.31.1)* (λ, Λ, J) *is stable.*

From Theorem 4.31.3 the following follows.

Theorem 4.31.6 ([243])

(A) Let us suppose that the right-hand side of system (4.31.1) satisfies the following conditions:

(i) $\|f(t,x)\| \le \tilde{q}(t)\|x\|^m$, $\tilde{q}: \mathbb{R}_+ \to \mathbb{R}_+$, $m \in [0,1]$;

(ii) $\|I_i\| \le a_i\|x\|^m$;

(iii) there exists a constant ξ *such that*

$$\prod_{t_0<t_i<t} (1 + a_i\|x_0\|^{m-1}) < \tilde{\xi} < +\infty, \quad \prod_{t_0<t_i<t} (1 + ma_i\|x_0\|^{m-1}) < \tilde{\xi} < +\infty,$$

(iv) there exists a constant $\tilde{\eta}$ *such that* $\int_{t_0}^{t} \tilde{q}(\tau)d\tau < \tilde{\eta} < +\infty$.

Then all solutions of system (4.31.1) are bounded (for the case $m > 1$, *inequalities (4.31.16) and (4.31.17) guarantee that conditions (iii) and (iv) are valid).*

(B) If (i) and (ii) of (A) hold and (4.31.16) takes place, then the trivial solution of system (4.31.1) is Lyapunov stable.

(C) If (B) takes place and

$$\int_{t_0}^{t} \tilde{q}(\tau)d\tau \le \frac{1}{m\lambda^{m-1}}, \quad \prod_{t_0<t_i<t} (1 + ma_i\lambda^{m-1}) < \left(1 + \frac{1}{m-1}\right)^{\frac{1}{m-1}},$$

$$\lambda\left(1 + \frac{1}{m-1}\right)^{\frac{1}{m-1}}\left[1 - \lambda^{m-1}(1 + \frac{1}{m-1})\right]^{-\frac{1}{m-1}} < \Lambda,$$

the solution $x = 0$ *of system (4.31.1) is practically stable.*

4.32 An Application of Corollary 3.2.10 to a Class of Differential Equations

Consider the differential equation

$$\begin{cases} L_n x^k(t) = p(t)x^{k-1}(t)f(x(t)), & t \in I, \\ L_i x^k(0) = C_{i-1}, & i = 1, 2, \ldots, n, \end{cases} \tag{4.32.1}$$

where $k > 1$ and C_{i-1}, $1 \le i \le n$ are constants, p and $f \in C(I, \mathbb{R})$, and L_n is defined as in Theorem 3.2.24. It is easy to observe that (4.32.1) is equivalent to the integral equation

$$x^k(t) = b(t) + H[t, p_1, p_2, \ldots, p_{n-1}, px^{k-1}f(x)], \quad \text{for all } t \in I, \tag{4.32.2}$$

where

$$b(t) = C_0 + \sum_{i=1}^{n-1} C_i H[t, p_1, \ldots, p_i]. \tag{4.32.3}$$

If $|b(t)| \le u_0$ and $|f(u)| \le g(|u|)$, where u_0 and g are defined as in Theorem 3.2.24, then we derive from (4.32.2) that

$$|x(t)|^k \le u_0 + H[t, p_1, p_2, \ldots, p_{n-1}, |p||x|^{k-1}g(|x|)], \quad \text{for all } t \in I. \tag{4.32.4}$$

Applying Corollary 3.2.10 to inequality (4.32.4), we can conclude

$$|x(t)| \le G^{-1}\Big[G(u_0^{1/k}) + H[t, p_1, p_2, \ldots, p_{n-1}, \frac{1}{k}|p|]\Big], \quad \text{for all } t \in I_3 \cap J(x), \tag{4.32.5}$$

where $J(x)$ denotes the maximal existence interval of $x(t)$.

4.33 An Application of Corollary 3.3.2 to a Dynamics Equation on Time Scales

In this section, we present an application of Corollary 3.3.2 to obtain the explicit estimates on the solutions of a dynamic equation on time scales.

Consider the following initial value problem on time scales

$$(u^p(t))^{\Delta} = H(t, u^q(t)), \quad u(t_0) = C, \quad t \in T^k, \tag{4.33.1}$$

where C, p, and q are constants, $p \ge q > 0$, and $H : T^k \times \mathbb{R} \to \mathbb{R}$ is a continuous function.

Assume that for all $t \in T^k$,

$$|H(t, u^q(t))| \le h(t)|u^q(t)|. \tag{4.33.2}$$

If $u(t)$ is a solution of problem (4.33.1), then

$$|u(t)| \leq \left\{ \frac{1}{q} \left[(K(p-q) + q|C|^p) e_{\bar{F}}(t, t_0) - K(p-q) \right] \right\}^{1/p}, \quad \text{for any} \quad K > 0,\ t \in T^k, \tag{4.33.3}$$

where h(t) is a non-negative function, and $\bar{F}(t)$ is defined by (3.3.35) in Corollary 3.3.3.

In fact, the solution $u(t)$ of (4.33.1) satisfies the following equivalent equation

$$u^p(t) = C^p + \int_{t_0}^{t} H(\tau, u^q(\tau))\Delta\tau, \quad \text{for all } t \in T^k. \tag{4.33.4}$$

Noting the assumption (4.33.2), we easily obtain

$$|u(t)|^p \leq |C|^p + \int_{t_0}^{t} h(\tau)|u(\tau)|^q \Delta\tau, \quad \text{for all } t \in T^k. \tag{4.33.5}$$

Now applying Corollary 3.3.2 to (4.33.5) yields (4.33.3).

Chapter 5
Nonlinear Multi-Dimensional Continuous Integral Inequalities

5.1 Nonlinear Two-Dimensional Bellman-Gronwall Inequalities and Their Generalizations

5.1.1 Nonlinear Two-Dimensional Bellman-Gronwall-Wendroff Inequalities, Snow Inequalities and Their Generalizations

Theorem 5.1.1 (The Pachpatte Inequality [438]) *Assume that $u(x,y), a(x,y)$, $b(x,y)$ are non-negative continuous functions defined for all $x, y \in \mathbb{R}_+$ and $F : \mathbb{R}_+^3 \to \mathbb{R}_+$ be a continuous function which satisfies the condition: for all $u \geq v \geq 0$,*

$$0 \leq F(x,y,u) - F(x,y,v) \leq K(x,y,v)(u-v),$$

where $K(x,y,v)$ is a non-negative continuous function defined for all $x, y, v \in \mathbb{R}_+$.

(c_1) Assume that $a(x,y)$ is non-increasing in $x \in \mathbb{R}_+$. If for all $x, y \in \mathbb{R}_+$,

$$u(x,y) \leq a(x,y) + \int_0^x b(s,y)u(s,y)ds + \int_0^x \int_y^{+\infty} F(s,t,u(s,t))dtds, \tag{5.1.1}$$

then for all $x, y \in \mathbb{R}_+$,

$$u(x,y) \leq p(x,y)\Bigg[a(x,y) + B(x,y) \times \exp\left(\int_0^x \int_y^{+\infty} K(s,t,a(s,t))p(s,t)dtds\right)\Bigg], \tag{5.1.2}$$

Y. Qin, *Integral and Discrete Inequalities and Their Applications*,
DOI 10.1007/978-3-319-33304-5_5

where, for all $x, y \in \mathbb{R}_+$

$$B(x, y) = \int_0^x \int_y^{+\infty} F(s, t, p(s, t))a(s, t)dtds, \tag{5.1.3}$$

and $p(x, y)$ *is defined by (5.1.47) in Qin [557].*

(c_2) *Assume that* $a(x, y)$ *is non-increasing in* $x \in \mathbb{R}_+$. *If for all* $x, y \in \mathbb{R}_+$,

$$u(x, y) \le a(x, y) + \int_x^{+\infty} b(s, y)u(s, y)ds + \int_x^{+\infty} \int_y^{+\infty} F(s, t, u(s, t))dtds, \tag{5.1.4}$$

then for all $x, y \in \mathbb{R}_+$,

$$u(x, y) \le \bar{p}(x, y)\left[a(x, y) + \bar{B}(x, y)\exp\left(\int_0^x \int_y^{+\infty} K(s, t, \bar{p}(s, t)a(s, t))\bar{p}(s, t)dtds\right)\right], \tag{5.1.5}$$

where for all $x, y \in \mathbb{R}_+$

$$\bar{B}(x, y) = \int_x^{+\infty} \int_y^{+\infty} F(s, t, \bar{p}(s, t)a(s, t))dtds, \tag{5.1.6}$$

and $\bar{p}(x, y)$ *is defined by (5.1.51) in Qin [557].*

Proof (c_1) Define a function $z(x, y)$ by

$$z(x, y) = a(x, y) + \int_0^x \int_y^{+\infty} F(s, t, u(s, t))dtds. \tag{5.1.7}$$

Then (5.1.1) can be rewritten as

$$u(x, y) \le z(x, y) + \int_0^x b(s, y)u(s, y)ds. \tag{5.1.8}$$

Clearly $z(x, y)$ is a non-negative, continuous and non-decreasing function in $x, x \in \mathbb{R}_+$. Treating $y, y \in \mathbb{R}_+$ fixed in (5.1.8) and using Theorem 1.1.4 in [557] to (5.1.8), we obtain

$$u(x, y) \le z(x, y)p(x, y), \tag{5.1.9}$$

where $p(x, y)$ is defined by (5.1.1). From (5.1.9) and (5.1.8), it follows

$$u(x, y) \le p(x, y)[a(x, y) + v(x, y)], \tag{5.1.10}$$

where

$$v(x,y) = \int_0^x \int_y^{+\infty} F(s,t,u(s,t))dtds. \tag{5.1.11}$$

From (5.1.10)–(5.1.11) and the hypotheses on F, we derive that

$$\begin{aligned} v(x,y) \le & \int_0^x \int_y^{+\infty} \Big[F(s,t,p(s,t)(a(s,t)+v(s,t))) \\ & -F(s,t,p(s,t)a(s,t)) + F(s,t,p(s,t)a(s,t))\Big]dtds \\ \le & B(x,y) + \int_0^x \int_y^{+\infty} K(s,t,p(s,t)a(s,t))p(s,t)v(s,t)dtds. \end{aligned} \tag{5.1.12}$$

Clearly, $B(x,y)$ is non-negative, continuous, non-decreasing in x and non-increasing in y for all $x, y \in \mathbb{R}_+$. Following the proof of Theorem 1.1.4 in Qin [557], we can obtain

$$v(x,y) \le B(x,y) \exp\left(\int_0^x \int_y^{+\infty} K(s,t,p(s,t)a(s,t))p(s,t)dtds\right). \tag{5.1.13}$$

Thus the required inequality (5.1.2) follows from (5.1.10) and (5.1.13). □

The following result is devoted to the Gronwall's inequality for functions of two independent variables. The ideas used here are two-dimensional analogues of ideas applied to functions of one independent variable first by Faedo [225], and later by Sato and Iwasaki [616], Opial [435] and others.

Theorem 5.1.2 (The David-Rasmussen Inequality [569]) *Let $P_0(x_0, y_0)$ and $P_1(x_1, y_1)$ be points in the domain D such that $(x_1 - x_0)(y_1 - y_0) \ge 0$ and such that the closed rectangle R_1 with opposite vertices P_0 and P_1 is contained in D. Let $g(x,y)$, $\phi(x,y)$ and $K(x,y)$, with $K(x,y)$ non-negative, be continuous functions on D. If for all (x,y) in R_1,*

$$\phi(x,y) \le g(x,y) + \int_{x_0}^x \int_{y_0}^y K(t,s)\phi(t,s)dsdt. \tag{5.1.14}$$

then $\phi(x,y) \le \Phi(x,y)$ on R_1, where $\Phi(x,y)$ satisfies the case of equality (5.1.36) in Qin [557].

Proof It is obvious that a continuous function $\Phi(x,y)$ satisfying the case of equality (5.1.36) in Qin [557] exists, i.e., for all $(x,y) \in R_1$,

$$\Phi(x,y) = g(x,y) + \int_{x_0}^x \int_{y_0}^y K(t,s)\Phi(t,s)dsdt.$$

If $Q = (x_1 - x_0)(y_1 - y_0) = 0$, the theorem is trivially true on the degenerate rectangle. If $Q > 0$, define $\psi(x, y) = \phi(x, y) - \Phi(x, y)$ so that on R_1

$$\psi(x, y) \leq \int_{x_0}^{x} \int_{y_0}^{y} K(t, s)\psi(t, s)dsdt.$$

Then by Lemma 5.1.3 in Qin [557], $\psi(x, y) \leq 0$, or $\phi(x, y) \leq \Phi(x, y)$ on R_1. □

Lemma 5.1.1 ([569]) *Let $P_0(x_0, y_0)$ and $P_1(x_1, y_1)$ be points in the domain D such that $(x_1 - x_0)(y_1 - y_0) \geq 0$ and such that the closed rectangle R_1 with opposite vertices P_0 and P_1 is contained in D. Let $g(x, y)$ and $K(x, y, u)$ be functions, with the former continuous on D and with the latter continuous on $D \times R$, non-decreasing in u, and satisfying the Lipschitz condition*

$$|K(x, y, u) - K(x, y, v)| \leq L|u - v|. \tag{5.1.15}$$

If $\{\epsilon_n\}$ is a strictly decreasing sequence of constants with limit zero, define on R_1 the sequence of continuous functions $\{\phi_n(x, y)\}$ such that each satisfies

$$\phi_n(x, y) = g(x, y) + \epsilon_n + \int_{x_0}^{x} \int_{y_0}^{y} K(t, s, \phi_n(t, s))dsdt.$$

Then the limit of $\{\phi_n(x, y)\}$, as $n \to +\infty$, is the maximal solution on R_1 of

$$\phi(x, y) = g(x, y) + \int_{x_0}^{x} \int_{y_0}^{y} K(t, s, \phi(t, s))dsdt. \tag{5.1.16}$$

Proof The Lipschitz condition guarantees the existence of the functions ϕ and ϕ_n used in this proof (see linear case). If $Q = (x_1 - x_0)(y_1 - y_0) = 0$, the proof is trivial. For the case in which $Q > 0$, it will be shown that the sequence $\{\phi_n\}$ has a limit, that it has a subsequence which converges uniformly to Φ, and hence that the limit of $\{\phi_n\}$ as $n \to +\infty$, is Φ. The existence of the limit of $\{\phi_n)$ will be established by showing that the sequence is both strictly decreasing and bounded below. A contradiction argument will show that since $\{\epsilon_n\}$ is strictly decreasing, so is $\{\phi_n\}$; that is, $m > n$ implies $\phi_m < \phi_n$ for all points in R_1. If this were not the case, then since $\phi_m(x_0, y_0) < \phi_n(x_0, y_0)$, the continuity of ϕ_m and ϕ_n implies that there must be a point $P_2(x_2, y_2)$ in R_1 such that:

(1) $P_2(x_2, y_2) \neq P_0(x_0, y_0)$,
(2) $m > n$ implies $\phi_m < \phi_n$ on the closed rectangle R_2 with opposite vertices P_0 and P_2, except
(3) at the point P_2 where $\phi_m(x_2, y_2) = \phi_n(x_2, y_2)$. Note that the point P_2 is by no means unique. Note also that as in Theorem 5.1.2, the term "rectangle" includes

the degenerate case in which $x_2 = x_0$ or $y_2 = y_0$. Now since $K(x, y, u)$ is non-decreasing in u and $\epsilon_m < \epsilon_n$,

$$\begin{aligned}\phi_n(x, y) &= g(x_2, y_2) + \epsilon_n + \int_{x_0}^{x_2}\int_{y_0}^{y_2} K(t, s, \phi_n(t, s))dsdt\\ &> g(x_2, y_2) + \epsilon_m + \int_{x_0}^{x_2}\int_{y_0}^{y_2} K(t, s, \phi_m(t, s))dsdt\\ &= \phi_m(x_2, y_2).\end{aligned}$$

But then $\phi_n(x_2, y_2) > \phi_m(x_2, y_2)$, which contradicts the above (3). Thus $\{\phi_n\}$ must be strictly decreasing on R_1.

Since $\{\phi_n\}$ is strictly decreasing, to show that it has a limit, it is sufficient to establish that it has a lower bound. It will be seen that any solution of

$$\phi(x, y) = g(x, y) + \int_{x_0}^{x}\int_{y_0}^{y} K(t, s, \phi(t, s))dsdt$$

will provide such a bound. In fact, a contradiction argument as above will be used to show that $\phi(x, y) < \phi_n(x, y)$ for all n and all (x, y) in R_1. If there is a function ϕ_n in the sequence for which it is not true, then since $\phi(x_0, y_0) < \phi_n(x_0, y_0)$, continuity implies that there must be a point $P_3(x_3, y_3)$ in R_1 such that:

(1) $P_3(x_3, y_3) \neq P_0(x_0, y_0)$,
(2) $\phi < \phi_n$ on the closed rectangle R_3 with opposite vertices P_0 and P_3, except
(3) at the point P_3 where $\phi(x_3, y_3) = \phi_n(x_3, y_3)$.

Then as before

$$\begin{aligned}\phi(x_3, y_3) &= g(x_3, y_3) + \int_{x_0}^{x_3}\int_{y_0}^{y_3} K(t, s, \phi(t, s))dsdt\\ &< g(x_3, y_3) + \epsilon_n + \int_{x_0}^{x_3}\int_{y_0}^{y_3} K(t, s, \phi_n(t, s))dsdt\\ &= \phi_n(x_3, y_3),\end{aligned}$$

which contradicts the above (3). Now since ϕ is continuous on the compact set R_1, it must be bounded below by some constant M and thus $M \leq \phi(x, y) \leq \phi_n(x, y)$ for all n and all (x, y) in R_1. Therefore $\{\phi_n\}$ is both strictly decreasing and bounded below on R_1, and hence has a limit there.

Since the limit of $\{\phi_n\}$ exists, it must also be the limit of any subsequence of $\{\phi_n\}$. Therefore, to prove that the limit of $\{\phi_n\}$ is Φ, we need only establish the existence of a subsequence which converges to Φ. The existence of such a subsequence follows from an application of a vector version of the Arzela-Ascoli Theorem [45]: If the sequence $\{\phi_n\}$ is both bounded and equi-continuous on R_1, then it has a uniformly convergent subsequence, and that uniform convergence will establish that the subsequence has a limit Φ.

The decreasing sequence $\{\phi_n\}$ is clearly bounded below by M and above by ϕ_1. Since ϕ_1 is continuous and R_1 is compact, there must exist a constant N such that $\phi_1 \leq N$ on R_1. Therefore, $M \leq \phi_n \leq N$ for all n, and so the sequence is bounded on R_1.

To show that the sequence is also equi-continuous, we need to show that for any $\epsilon > 0$, there exists a $\delta > 0$ such that for all (x', y') and (x'', y'') in R_1, $||(x', y') - (x'', y'')|| < \delta$ implies that $|\phi_n(x', y') - \phi_n(x'', y'')| < \epsilon$ for all n. Clearly,

$$\phi_n(x', y') - \phi_n(x'', y'') = g(x', y') - g(x'', y'') + \int_{x_0}^{x'} \int_{y_0}^{y''} K(t, s, \phi_n(t, s))dsdt$$
$$- \int_{x_0}^{x''} \int_{y_0}^{y''} K(t, s, \phi_n(t, s))dsdt. \tag{5.1.17}$$

Simplification of this expression falls into eight cases, taking into account the possible relative positions of P_0, P_1, (x', y') and (x'', y''). Since all eight are extremely similar and lead to the same result (5.1.17), the only one treated here is that in which $x_0 \leq x'' \leq x' \leq x_1$ and $y_0 \leq y'' \leq y' \leq y_1$. It is easily verified that in this case the right-hand side of (5.1.17) can be rewritten as

$$g(x', y') - g(x'', y'') + \int_{x_0}^{x''} \int_{y''}^{y'} K(t, s, \phi_n(t, s))dsdt$$
$$+ \int_{x''}^{x''} \int_{y0}^{y'} K(t, s, \phi(t, s))dsdt.$$

Thus

$$\begin{aligned}
&|\phi_n(x' - y') - \phi_n(x'', y'')| \\
&\leq |g(x', y') - g(x'' - y'')| + P[|x'' - x_0|(y' - y'') + (x' - x'')|y' - y_0|] \\
&\leq |g(x', y') - g(x'' - y'')| + PQ[|y' - y''| + |x' - x''|],
\end{aligned} \tag{5.1.18}$$

where

$$|K(x, y, u)| \leq P \ \ on \ \ R_1 \times [M, N], \quad Q = \max(|x_1 - x_0|, |y_1 - y_0|).$$

Using the "Taxicab" norm, if $\delta_1 < \epsilon/(2PQ)$, then

$$||(x', y') - (x'', y'')|| = |x' - x''| + |y' - y''| < \delta_1$$

implies $PQ[|y'-y''|+|x'-x''|] < PQ\delta_1 < \epsilon/2$; and since $g(x,y)$ is continuous, there is some $\delta_2 > 0$ such that $||(x',y')-(x'',y'')|| < \delta_2$ implies $|g(x',y')-g(x'',y'')| < \epsilon/2$. Now choose $\delta = \min(\delta_1,\delta_2)$ so that $||(x',y')-(x'',y'')|| < \delta$ implies $|\phi_n(x',y')-\phi_n(x'',y'')| < \epsilon/2+\epsilon/2 = \epsilon$. Thus $\{\phi_n\}$ is both bounded and equicontinuous on R_1, and hence has a subsequence $\{\psi_n\}$ which converges uniformly there.

Since $\{\phi_n\}$ and $\{\psi_n\}$ have the same limit, it remains to be shown only that the limit of $\{\psi_n\}$ is Φ, the maximal solution of (5.1.16). Indeed such case is easily seen since the continuity of $K(x,y,u)$ and the uniform convergence of $\{\psi_n\}$ imply

$$\lim_{n\to+\infty}\psi_n(x,y) = g(x,y) + \int_{x_0}^{x}\int_{y_0}^{y} K(t,s,\lim_{n\to+\infty}\psi_n(t,s))dsdt.$$

Thus $\lim_{n\to+\infty}\psi_n$ is a solution of (5.1.16). It is also the maximal solution since every solution ϕ satisfies $\phi < \phi_n$ for all n and all (x,y) in R_1. Thus $\phi(x,y) \le \lim_{n\to+\infty}\phi_n(x,y) = \lim_{n\to+\infty}\psi_n(x,y) = \Phi(x,y)$ as desired. □

Theorem 5.1.3 (The Rasmussen Inequality [569]) *Let $P_0(x_0,y_0)$ and $P_1(x_1,y_1)$ be points in the domain D such that $(x_1-x_0)(y_1-y_0) \ge 0$ and such that the closed rectangle R_1 with opposite vertices P_0 and P_1 is contained in D. Let $g(x,y)$, $\phi(x,y)$ and $K(x,y,u)$ be functions, with the former two continuous on $D\times R$, non-decreasing in u, and satisfying the Lipschitz condition*

$$|K(x,y,u)-K(x,y,v)| \le L|u-v|. \tag{5.1.19}$$

If for all $(x,y)\in R_1$,

$$\phi(x,y) \le g(x,y) + \int_{x_0}^{x}\int_{y_0}^{y} K(t,s,\phi(t,s))dsdt, \tag{5.1.20}$$

then for all $(x,y)\in R_1$,

$$\phi(x,y) \le \Phi(x,y) \tag{5.1.21}$$

where $\Phi(x,y)$ is the maximal solution of equality in (5.1.20).

Proof The Lipschitz condition and Lemma 5.1.1 guarantee the existence of the functions ϕ_n and Φ used in this proof. In the degenerate case where $Q = (x_1-x_0)(y_1-y_0) = 0$, this result is trivially true. To establish the non-trivial case where $Q > 0$, let $\{\epsilon_n\}$ be a strictly decreasing sequence of constants with limit zero. Let $\phi_n(x,y)$ be a solution to

$$\phi_n(x,y) = g(x,y) + \epsilon_n + \int_{x_0}^{x}\int_{y_0}^{y} K(t,s,\phi_n(t,s))dsdt.$$

It will be shown that $\phi(x,y) < \phi_n(x,y)$ for all n and all $(x,y) \in R_1$, and hence that $\phi(x,y) \le \lim_{n\to+\infty} \phi_n(x,y) = \Phi(x,y)$ by Lemma 5.1.1, this will be accomplished using the same contradiction argument employed twice in the preceding lemma: assume that $\phi < \phi_n$ is not true for some point in R_1. Clearly, $\phi(x_0,y_0) < \phi_n(x_0,y_0)$, so by continuity there must be a point $P_2(x_2,y_2)$ in R_1 such that:

(1) $P_2(x_2,y_2) \neq P_0(x_0,y_0)$,
(2) $\phi < \phi_n$ on the closed rectangle R_2 with opposite vertices P_0 and P_2, except
(3) at P_2 where $\phi(x_2,y_2) = \phi_n(x_2,y_2)$.

Since $K(x,y,u)$ is non-decreasing in u and $\epsilon_n > 0$,

$$\begin{aligned}\phi(x_2,y_2) &= g(x_2,y_2) + \int_{x_0}^{x_2}\int_{y_0}^{y_2} K(t,s,\phi(t,s)dsdt \\ &< g(x_2,y_2) + \epsilon_n + \int_{x_0}^{x_2}\int_{y_0}^{y_2} K(t,s,\phi_n(t,s))dsdt \\ &= \phi_n(x_2,y_2).\end{aligned}$$

But this contradicts (3) above and hence $\phi(x,y) < \phi_n(x,y)$ for all n and all $(x,y) \in R_1$. Thus $\phi(x,y) \le \lim_{n\to+\infty} \phi_n(x,y) = \Phi(x,y)$ by Lemma 5.1.1. □

In the sequel,we shall assume the following:

(H_1) $u(x,y), a(x,y), b(x,y), c(x,y), p(x,y)$ and $q(x,y)$ are real-valued non-negative continuous functions defined on a domain D.
(H_2) $P_0(x_0,y_0)$ and $P(x,y)$ be two points in D such that $(x-x_0)(y-y_0) > 0$ and R the rectangular region whose opposite corners are the points P_0 and P.

Theorem 5.1.4 (The Pachpatte Inequality [472]) *Suppose* (H_1) *and* (H_2) *are true. Let* $G(r)$ *be continuous, strictly increasing, convex and sub-multiplicative function for all* $r \ge 0$, $G(0) = 0$, $\lim_{r\to+\infty} G(r) = +\infty$ *for all* (x,y) *in D,* $\alpha(x,y)$, $\beta(x,y)$ *be positive continuous functions defined on a domain in D, and* $\alpha(x,y) + \beta(x,y) = 1$. *Let* $v(s,t;x,y)$ *and* $w(s,t;x,y)$ *be the solutions of the characteristic initial value problem*

$$\begin{cases} L[v] = v_{st} - \left[p(s,t) + \beta(s,t)G(b(s,t)\beta^{-1}(s,t))(c(s,t) + q(s,t))\right]v = 0, \\ v(s,y) = v(x,t) = 1, \end{cases} \tag{5.1.22}$$

and

$$\begin{cases} M[w] = w_{st} - \left[\beta(s,t)G(b(s,t)\beta^{-1}(s,t))(c(s,t) - p(s,t))\right]w = 0, \\ w(s,y) = w(x,t) = 1, \end{cases} \tag{5.1.23}$$

respectively and let D^+ be a connected sub-domain of D which contains P and on which $v > 0$ and $w > 0$. Then, if $R \subset D^+$ and for all $(x, y) \in R$,

$$u(x,y) \leq a(x,y) + b(x,y)G^{-1}\Big[\int_{x_0}^{x}\int_{y_0}^{y} c(s,t)G(u(s,t))dsdt + \int_{x_0}^{x}\int_{y_0}^{y} p(s,t)\left(\int_{x_0}^{s}\int_{y_0}^{t} q(\xi,\eta)G(u(\xi,\eta))d\xi d\eta\right) dsdt\Big], \tag{5.1.24}$$

then for all $(x, y) \in R$,

$$u(x,y) \leq a(x,y) + b(x,y)G^{-1}\Bigg[\int_{x_0}^{x}\int_{y_0}^{y} w(s,t;x,y)\Big\{a(s,t)G(a(s,t)\alpha^{-1}(s,t))c(s,t) + p(s,t)\int_{x_0}^{s}\int_{y_0}^{t} a(\xi,\eta)G(a(\xi,\eta)\alpha^{-1}(\xi,\eta))[c(\xi,\eta)+q(\xi,\eta)] \times v(\xi,\eta;s,t)d\xi d\eta\Big\} dsdt\Bigg]. \tag{5.1.25}$$

Proof We may rewrite (5.1.24) as

$$u(x,y) \leq \alpha(x,y)a(x,y)\alpha^{-1}(x,y) + \beta(x,y)b(x,y)\beta^{-1}(x,y) \times G^{-1}\Big[\int_{x_0}^{x}\int_{y_0}^{y} c(s,t)G(u(s,t))dsdt + \int_{x_0}^{x}\int_{y_0}^{y} p(s,t)\left(\int_{x_0}^{s}\int_{y_0}^{t} q(\xi,\eta)G(u(\xi,\eta))d\xi d\eta\right) dsdt\Big].$$

Since G is convex, sub-multiplicative and monotonic, we infer

$$G(u(x,y)) \leq \alpha(x,y)G(a(x,y)\alpha^{-1}(x,y)) + \beta(x,y)G((b(x,y)\beta^{-1}(x,y)) \times\Big[\int_{x_0}^{x}\int_{y_0}^{y} c(s,t)G(u(s,t))dsdt + \int_{x_0}^{x}\int_{y_0}^{y} p(s,t)\left(\int_{x_0}^{s}\int_{y_0}^{t} q(\xi,\eta)G(u(\xi,\eta))d\xi d\eta\right) dsdt\Big]. \tag{5.1.26}$$

The estimate (5.1.25) follows by first applying Theorem 5.1.32 in Qin [557] with $a(x,y) = \alpha(x,y)\ G(a(x,y)\alpha^{-1}(x,y)), b(x,y) = \beta(x,y)G(b(x,y)\beta^{-1}(x,y))$ and $u(x,y) = G(u(x,y))$ and then applying G^{-1} to both sides of the resulting inequality. □

Theorem 5.1.5 (The Pachpatte Inequality [472]) *Suppose (H_1) and (H_2) are true. Let $G(r)$ be continuous, strictly increasing, sub-additive and sub-multiplicative* *function for all $r > 0$, $G(0) = 0$, for all (x, y) in D, and G^{-1} is the inverse function of G. Let $v(s, t; x, y)$ and $w(s, t; x, y)$ be the solutions of the characteristic initial value problem*

$$\begin{cases} L[v] = v_{st} - [p(s,t) + G(b(s,t))(c(s,t) + q(s,t))]v = 0, \\ v(s,y) = v(x,t) = 1, \end{cases} \tag{5.1.27}$$

and

$$\begin{cases} M[w] = w_{st} - [G(b(s,t))c(s,t) - p(s,t)]w = 0, \\ w(s,y) = w(x,t) = 1, \end{cases} \tag{5.1.28}$$

respectively and let D^+ be a connected sub-domain of D which contains P and on which $v > 0$ and $w > 0$. Then, if $R \subset D^+$ and for all $(x, y) \in R$,

$$u(x,y) \le a(x,y) + b(x,y)G^{-1}\Big[\int_{x_0}^{x}\int_{y_0}^{y} c(s,t)G(u(s,t))dsdt + \int_{x_0}^{x}\int_{y_0}^{y} p(s,t)\left(\int_{x_0}^{s}\int_{y_0}^{t} q(\xi,\eta)G(u(\xi,\eta))d\xi d\eta\right)dsdt\Big], \tag{5.1.29}$$

then for all $(x, y) \in R$,

$$u(x,y) \le G^{-1}\Big[G(a(x,y)) + G(b(x,y))\Big[\int_{x_0}^{x}\int_{y_0}^{y} w(s,t;x,y)\Big\{G(a(s,t))c(s,t) + p(s,t)\int_{x_0}^{s}\int_{y_0}^{t} G(a(\xi,\eta))[c(\xi,\eta) + q(\xi,\eta)]v(\xi,\eta;x,y)d\xi d\eta\Big\}dsdt\Big]\Big]. \tag{5.1.30}$$

Proof Since G is sub-additive, sub-multiplicative and monotonic, we derive from (5.1.29)

$$G(u(x,y)) \le G(a(x,y)) + G(b(x,y))\Big[\int_{x_0}^{x}\int_{y_0}^{y} c(s,t)G(u(s,t))dsdt + \int_{x_0}^{x}\int_{y_0}^{y} p(s,t)\left(\int_{x_0}^{s}\int_{y_0}^{t} q(\xi,\eta)G(u(\xi,\eta))d\xi d\eta\right)dsdt\Big]. \tag{5.1.31}$$

The desired bound in (5.1.30) follows by first applying Theorem 5.1.32 in [557] to (5.1.31) with $a(x, y) = G(a(x, y))$, $b(x, y) = G(b(x, y))$ and $u(x, y) = G(u(x, y))$ and then applying G^{-1} to both sides of the resulting inequality. □

Next we apply Theorem 5.1.33 in Qin [557] to establish the following integral inequalities similar to that proved in Theorems 5.1.4 and 5.1.5 which can be used in some applications.

Theorem 5.1.6 (The Pachpatte Inequality [472]) *Suppose* (H_1) *and* (H_2) *are true. Let* $G(r)$, $\alpha(x,y)$, $\beta(x,y)$ *be the same functions as defined in Theorem 5.1.4. Let* $v(s,t;x,y)$ *and* $w(s,t;x,y)$ *be the solutions of the characteristic initial value problem*

$$\begin{cases} L[v] = v_{st} - \beta(s,t)G(b(s,t)\beta^{-1}(s,t))[c(s,t) + p(s,t) + q(s,t)]v = 0, \\ v(s,y) = v(x,t) = 1, \end{cases} \tag{5.1.32}$$

and

$$\begin{cases} M[w] = w_{st} - \beta(s,t)G(b(s,t)\beta^{-1}(s,t))c(s,t)w = 0, \\ w(s,y) = w(x,t) = 1, \end{cases} \tag{5.1.33}$$

respectively and let D^+ *be a connected sub-domain of* D *which contains* P *and on which* $v > 0$ *and* $w > 0$. *Then, if* $R \subset D^+$ *and for all* $(x,y) \in R$,

$$\begin{aligned} u(x,y) \le a(x,y) + b(x,y)G^{-1}\Big[&\int_{x_0}^{x}\int_{y_0}^{y} c(s,t)G(u(s,t))dsdt \\ &+ \int_{x_0}^{x}\int_{y_0}^{y} p(s,t)\Big(G(u(s,t)) + \beta(s,t)G(b(s,t)\beta^{-1}(s,t)) \\ &\times \int_{x_0}^{s}\int_{y_0}^{t} q(\xi,\eta)G(u(\xi,\eta))d\xi d\eta\Big)dsdt\Big], \end{aligned} \tag{5.1.34}$$

then for all $(x,y) \in R$,

$$\begin{aligned} u(x,y) \le a(x,y) + b(x,y)G^{-1}\Bigg[&\int_{x_0}^{x}\int_{y_0}^{y} w(s,t;x,y)\Bigg\{ \alpha(s,t)G(a(s,t)\alpha^{-1}(s,t)) \\ &\times [c(s,t) + p(s,t)] + \beta(s,t)G(b(s,t)\beta^{-1}(s,t))p(s,t) \\ &\times \int_{x_0}^{x}\int_{y_0}^{y} a(\xi,\eta)G(a(\xi,\eta)\alpha^{-1}(\xi,\eta)) \\ &\times [c(\xi,\eta) + p(\xi,\eta) + q(\xi,\eta)]v(\xi,\eta;s,t)d\xi d\eta\Bigg\} dsdt\Bigg]. \end{aligned} \tag{5.1.35}$$

Theorem 5.1.7 (The Pachpatte Inequality [472]) *Suppose* (H_1) *and* (H_2) *are true. Let* $G(r)$, G^{-1} *be the same functions as defined in Theorem 5.1.5. Let* $v(s,t;x,y)$ *and* $w(s,t;x,y)$ *be the solutions of the characteristic initial value problem*

$$\begin{cases} L[v] = v_{st} - G(b(s,t)[c(s,t) + p(s,t) + q(s,t)]v = 0, \\ v(s,y) = v(x,t) = 1, \end{cases} \tag{5.1.36}$$

and

$$\begin{cases} M[w] = w_{st} - G(b(s,t))c(s,t)w = 0, \\ w(s,y) = w(x,t) = 1, \end{cases} \tag{5.1.37}$$

respectively and let D^+ be a connected sub-domain of D which contains P and on which $v > 0$ and $w > 0$. Then, if $R \subset D^+$ and for all $(x,y) \in R$,

$$\begin{aligned} u(x,y) \le a(x,y) + b(x,y)G^{-1}\Big[\int_{x_0}^{x}\int_{y_0}^{y} c(s,t)G(u(s,t))dsdt \\ + \int_{x_0}^{x}\int_{y_0}^{y} p(s,t)\Big(G(u(s,t)) + G(b(s,t)) \\ \times \int_{x_0}^{s}\int_{y_0}^{t} q(\xi,\eta)G(u(\xi,\eta))d\xi d\eta\Big)dsdt\Big], \end{aligned} \tag{5.1.38}$$

then for all $(x,y) \in R$,

$$\begin{aligned} u(x,y) \le G^{-1}\Bigg[G(a(x,y)) + G(b(x,y))\Bigg[\int_{x_0}^{x}\int_{y_0}^{y} w(s,t;x,y)\Bigg(G(a(s,t))[c(s,t) + p(s,t)] \\ + G(b(s,t))p(s,t)\int_{x_0}^{s}\int_{y_0}^{t} G(a(\xi,\eta))[c(\xi,\eta) + p(\xi,\eta) + q(\xi,\eta)] \\ \times v(\xi,\eta;x,y)d\xi d\eta\Bigg)dsdt\Bigg]\Bigg]. \end{aligned} \tag{5.1.39}$$

We note that in the special case when $p(x,y) = q(x,y) = 0$, Theorems 5.1.4–5.1.7 and Theorems 5.1.32, 5.1.33 in Qin [557] reduces to the further generalizations of the integral inequality recently established by Snow [619]. In the special case when $c(x,y) = 0$, the results in these theorems are new to the literature.

We now apply Theorem 5.1.34 in Qin [557] to establish the following interesting and useful integral inequalities in two independent variables.

Theorem 5.1.8 (The Pachpatte Inequality [473]) *Suppose*

(H_1) $u(x,y), a(x,y), b(x,y), c(x,y), p(x,y), q(x,y), r(x,y), h(x,y)$, and $g(x,y)$ are real-valued non-negative continuous functions defined on a domain D.

(H_2) $P_0(x_0,y_0)$ and $P(x,y)$ are two points in D such that $(x-x_0)/(y-y_0) \ge 0$ and R is the rectangular region whose opposite corners are points P_0 and P.

(H_3) The function $V(s,t;x,y)$ and $W(s,t;x,y)$ are the Riemann functions for the partial differential operators L and T, respectively, and satisfy all the properties of Riemann functions for operators with continuous coefficients.

Let $G(u)$ be continuous, strictly increasing, convex, and sub-multiplicative function for all $u \geq 0, G(0) = 0, \lim_{u\to+\infty} G(u) = +\infty$; for all (x, y) in D, let $\alpha(x, y), \beta(x, y)$ be positive continuous functions defined on a domain D; and let $\alpha(x, y) + \beta(x, y) = 1$. Let $V(s, t; x, y)$ be the solution of the characteristic initial value problem

$$M[V] = 0, \tag{5.1.40}$$

in which M is the adjoint operator of the operator L defined by

$$L[\Psi] = \Psi_{st} + a_1\Psi_s + a_2\Psi_t + a_3\Psi \tag{5.1.41}$$

with $a_1 = -\beta G(b\beta^{-1})cq$, $a_2 = -\beta G(b\beta^{-1})cp$, and $a_3 = -[g + \beta G(b\beta^{-1})c(r + h)]$. Let $W(s, t; x, y)$ be the solution of the characteristic initial value problem

$$N[W] = 0, \tag{5.1.42}$$

in which N is the adjoint operator of the operator T defined by

$$T[\phi] = \phi_{st} + b_1\phi_s + b_2\phi_t + b_3\phi \tag{5.1.43}$$

with $b_1 = -\beta G(b\beta^{-1})cq$, $b_2 = -\beta G(b\beta^{-1})cp$, and $b_3 = -\beta G(b\beta^{-1})c(r - h)$. Let D^+ be a connected sub-domain of D which contains P and on which $V \geq 0$ and $W \geq 0$. If $R \subset D^+$ and for all $(x, y) \in R$,

$$\begin{aligned} u(x, y) \leq a(x, y) + b(x, y)G^{-1}\Big[&p(x, y)\int_{x_0}^{x} c(s, y)G(u(s, y))ds \\ &+q(x, y)\int_{y_0}^{y} c(x, t)G(u(x, y))dt + r(x, y)\int_{x_0}^{x}\int_{y_0}^{y} c(s, t)G(u(s, t))ds\,dt \\ &+h(x, y)\int_{x_0}^{x}\int_{y_0}^{y} g(s, t)(\int_{x_0}^{s}\int_{y_0}^{t} c(\xi, \eta)G(u(\xi, \eta))d\xi\,d\eta)ds\,dt\Big], \end{aligned} \tag{5.1.44}$$

then for all $(x, y) \in R$,

$$\begin{aligned} u(x, y) \leq a(x, y) + b(x, y)G^{-1}\Big[&p(x, y)\int_{x_0}^{x}\int_{x_0}^{x} c(s, y)G(u(s, y))ds \\ &+q(x, y)\int_{y_0}^{y} c(x, t)G(u(x, t))dt + r(x, y)Q_0(x, y) \\ &+h(x, y)\int_{x_0}^{x}\int_{y_0}^{y} g(s, t)Q_0(s, t)ds\,dt\Big], \end{aligned} \tag{5.1.45}$$

where $Q_0(x, y)$ is defined by the right-hand side of (5.1.218) in [557] by replacing $a(x, y)$ by $\alpha(x, y)G(a(x, y)\alpha^{-1}(x, y))$ and $b(x, y)$ by $\beta(x, y)G(b(x, y)\beta^{-1}(x, y))$. Further, if $q(x, y) = 0$, then for all $(x, y) \in R$,

$$u(x,y) \le a(x,y) + b(x,y)G^{-1}\Big[r(x,y)Q_0(x,y) + h(x,y)\int_{x_0}^{x}\int_{y_0}^{y} g(s,t)Q_0(s,t)dsdt$$
$$+p(x,y)\int_{x_0}^{x} c(s,y)f_0(s,y)$$
$$\times \exp(\int_{s}^{x} c(\xi,y)\beta(\xi,y)G(b(\xi,y)\beta^{-1}(\xi,y))p(\xi,y)d\xi)ds\Big], \quad (5.1.46)$$

where $f_0(x, y)$ is defined by the right-hand side of (5.1.220) in [557] replacing $a(x, y)$ by $\alpha(x, y)G(a(x, y)\alpha^{-1}(x, y))$, $b(x, y)$ by $\beta(x, y)G(b(x, y)\beta^{-1}(x, y))$, and $Q(x, y)$ by $Q_0(x, y)$. Again, if $p(x, y) = 0$, then for all $(x, y) \in R$,

$$u(x,y) \le a(x,y) + b(x,y)G^{-1}\Big[r(x,y)Q_0(x,y) + h(x,y)\int_{x_0}^{x}\int_{y_0}^{y} g(s,t)Q_0(s,t)dsdt$$
$$+q(x,y)\int_{y_0}^{y} c(x,t)f_0(x,t)$$
$$\times \exp(\int_{t}^{y} c(x,\eta)\beta(x,\eta)G(b(x,\eta)\beta^{-1}(x,\eta))q(x,\eta)d\eta)dt\Big], \quad (5.1.47)$$

where $Q_0(x, y)$ and $f_0(x, y)$ are as defined above.

Proof We may rewrite (5.1.44) as

$$u(x,y) \le \alpha(x,y)a(x,y)\alpha^{-1}(x,y)$$
$$+\beta(x,y)b(x,y)\beta^{-1}G^{-1}\Big[p(x,y)\int_{x_0}^{x} c(s,y)G(u(s,y))ds$$
$$+q(x,y)\int_{y_0}^{y} c(s,t)G(u(x,t))dt + r(x,y)\int_{x_0}^{x}\int_{y_0}^{y} c(s,t)G(u(s,t))dsdt$$
$$+h(x,y)\int_{x_0}^{x}\int_{y_0}^{y} g(s,t)(\int_{x_0}^{s}\int_{y_0}^{t} c(\xi,\eta)G(u(\xi,\eta))d\xi\, d\eta)dsdt\Big].$$

Since G is convex, sub-multiplicative, and monotonic, we have

$$G(u(x,y)) \le \alpha(x,y)G(a(x,y)\alpha^{-1}(x,y))$$
$$+\beta(x,y)G(b(x,y)\beta^{-1}(x,y))\Big[p(x,y)\int_{x_0}^{x} c(s,y)G(u(s,y))ds$$

$$+q(x,y)\int_{y_0}^{y} c(x,t)G(u(x,t))dt + r(x,y)\int_{x_0}^{x}\int_{y_0}^{y} c(s,t)G(u(s,t))dsdt$$

$$+h(x,y)\int_{x_0}^{x}\int_{y_0}^{y} g(s,t)(\int_{x_0}^{s}\int_{y_0}^{t} c(\xi,\eta)G(u(\xi,\eta))d\xi\, d\eta)dsdt\Big].$$

The estimate (5.1.45) follows by first applying Theorem 5.1.34 in Qin [557] with $a(x,y) = \alpha(x,y)G\Big(a(x,y)\alpha^{-1}(x,y)\Big), b(x,y) = \beta(x,y)G\Big(b(x,y)\beta^{-1}(x,y)\Big)$, and $u(x,y) = G(u(x,y))$ and then applying G^{-1} to both sides of the resulting inequality. The rest of the proof when $q(x,y) = 0$ and $p(x,y) = 0$ follows by the similar argument as in the last part of the proof of Theorem 5.1.34 in Qin [557] in view of the proof of the first part of this theorem with suitable modifications. We omit the details. □

Theorem 5.1.9 (The Pachpatte Inequality [473]) *Suppose (H_1)-(H_3) are true. Let $G(u)$ be a positive, continuous, strictly increasing, sub-additive, and sub-multiplicative function for all $u \geq 0$, $G(0) = 0$ for all $(x,y) \in D$, and G^{-1} is the inverse function of G. Let $V(s,t;x,y)$ be the solution of the characteristic initial value problem (5.1.40) in which M is the adjoint operator of the operator L defined by (5.1.41) with $a_1 = -G(b)cq$, $a_2 = -G(b)cp$, and $a_3 = -[g + G(b)c(r+h)]$. Let $W(s,t;x,y)$ be the solution of the characteristic initial value problem (5.1.42) in which N is the adjoint operator of the operator T defined by (5.1.43) with $b_1 = -G(b)cq$, $b_2 = -G(b)cp$, and $b_3 - G(b)c(r-h)$. Let D^+ be a connected sub-domain of D which contains P and on which $V \geq 0$ and $W \geq 0$. If $R \subset D^+$ and for all $(x,y) \in R$,*

$$u(x,y) \leq a(x,y) + b(x,y)G^{-1}\Big[p(x,y)\int_{x_0}^{x} c(s,y)G(u(s,y))ds$$

$$+q(x,y)\int_{y_0}^{y} c(x,t)G(u(x,t))dt + r(x,y)\int_{x_0}^{x}\int_{y_0}^{y} c(s,t)G(u(s,t))dsdt$$

$$+h(x,y)\int_{x_0}^{x}\int_{y_0}^{y} g(s,t)(\int_{x_0}^{s}\int_{y_0}^{t} c(\xi,\eta)G(u(\xi,\eta))d\xi\, d\eta)dsdt\Big], \quad (5.1.48)$$

then for all $(x,y) \in R$,

$$u(x,y) \leq G^{-1}\Big[G(a(x,y)) + G(b(x,y))(p(x,y)\int_{x_0}^{x} c(s,y)G(u(s,y))ds$$

$$+q(x,y)\int_{y_0}^{y} c(x,t)G(u(x,t))dt + r(x,y)Q_1(x,y)$$

$$+h(x,y)\int_{x_0}^{x}\int_{y_0}^{y} g(s,t)Q_1(s,t)dsdt)\Big], \quad (5.1.49)$$

where $Q_1(x,y)$ is defined by the right-hand side of (5.1.218) in Qin [557] by replacing $a(x,y)$ by $G(a(x,y))$ and $b(x,y)$ by $G(b(x,y))$. Further, if $q(x,y) = 0$, then we have for all $(x,y) \in R$,

$$u(x,y) \le G^{-1}\Big(f_1(x,y) + G(b(x,y))p(x,y)\Big[\int_{x_0}^{x} c(s,y)f_1(s,y) \times \Big(\int_{s}^{x} c(\xi,y)G(b(\xi,y))p(\xi,y)d\xi\Big)ds\Big]\Big), \tag{5.1.50}$$

where $f_1(x,y)$ is defined by the right-hand side of (5.1.220) in Qin [557] by replacing $a(x,y)$ by $G(a(x,y))$, $b(x,y)$ by $G(b(x,y))$, and $Q(x,y)$ by $Q_1(x,y)$. Again, if $p(x,y) = 0$, then

$$u(x,y) \le G^{-1}\Big[f_1(x,y) + G(b(x,y))q(x,y)\Big\{\int_{y_0}^{y} c(x,t)f_1(x,t) \times \exp\Big(\int_{t}^{y} c(x,\eta)G(b(x,\eta))q(x,\eta)d\eta\Big)dt\Big\}\Big], \tag{5.1.51}$$

where $Q_1(x,y)$ and $f_1(x,y)$ are as defined above.

Proof Since G is sub-additive, sub-multiplicative, and monotonic, we infer from (5.1.48)

$$\begin{aligned} G(u(x,y)) \le G(a(x,y)) + G(b(x,y))\Big[p(x,y)\int_{x_0}^{x} c(s,y)G(u(s,y))ds \\ +q(x,y)\int_{y_0}^{y} c(x,t)G(u(x,t))dt + r(x,y)\int_{x_0}^{x}\int_{y_0}^{y} c(s,t)G(u(s,t))dsdt \\ +h(x,y)\int_{x_0}^{x}\int_{y_0}^{y} g(s,t)\Big(\int_{x_0}^{s}\int_{y_0}^{t} c(\xi,\eta)G(u(\xi,\eta))d\xi\, d\eta\Big)dsdt\Big]. \end{aligned} \tag{5.1.52}$$

The desired bound in (5.1.49) follows by first applying Theorem 5.1.34 in Qin [557] to (5.1.52) with $a(x,y) = G(a(x,y))$, $b(x,y) = G(b(x,y))$, and $u(x,y) = G(u(x,y))$ and then applying G^{-1} to both sides of the resulting inequality. Further, by setting $q(x,y) = 0$ and $p(x,y) = 0$ in (5.1.52) and applying Theorem 5.1.34 in Qin [557], we obtain the desired bounds in (5.1.50) and (5.1.51). □

We note that the integral inequalities established in Theorems 5.1.8 and 5.1.9 are the two independent variable generalizations of the integral inequalities established in Theorem 1 by Gollwitzer [250], and Theorem 2 by Pachpatte [451]. Furthermore, we note that the functions $V(s,t;x,y)$ and $W(s,t;x,y)$ involved in Theorems 2.1–2.3 are the well-known Riemann functions relative to the point $P(x,y)$ (see, e.g., [618]). The existence and continuity of the Riemann function is well-known and may be demonstrated by the method of successive approximation (see, e.g., [177]).

The next result, due to Bondage-Pachpatte [98], deals with the two independent variable generalization of the integral inequality established by Gollwitzer [250].

Theorem 5.1.10 (The Pachpatte Inequality [98]) *Suppose $\phi(x,y)$, $b(x,y)$, and $c(x,y)$ are real-valued non-negative continuous functions defined on a domain D. Let $G(r)$ be continuous, strictly increasing, convex and sub-multiplicative function for all $r \geq 0$, $G(0) = 0$, $\lim_{r\to+\infty} G(r) = +\infty$ for all (x,y) in D, $\alpha(x,y)$, $\beta(x,y)$ be positive continuous functions defined on D, and $\alpha(x,y) + \beta(x,y) = 1$. Let $P_0(x_0,y_0)$ and $P(x,y)$ be two points in D such that $(x-x_0)(y-y_0) > 0$ and R be the rectangular region whose opposite corners are the points P_0 and P. Let $v(s,t;x,y)$ be the solution of the characteristic initial value problem*

$$L[v] = v_{st} - c(s,t)\beta(s,t)G(b(s,t)\beta^{-1}(s,t))v = 0, \quad v(x,t) = v(s,y) = 1, \tag{5.1.53}$$

and let D^+ be a connected sub-domain of D which contains P and on which $v > 0$. If $R \subset D^+$ and for all $(x,y) \in R$,

$$\phi(x,y) \leq a(x,y) + b(x,y)G^{-1}\left(\int_{x_0}^{x}\int_{y_0}^{y} c(s,t)G(\phi(s,t))dsdt\right), \tag{5.1.54}$$

then for all $(x,y) \in R$

$$\begin{aligned}\phi(x,y) \leq a(x,y) + b(x,y)G^{-1}\Big(&\int_{x_0}^{x}\int_{y_0}^{y} c(s,t)\alpha(s,t)G(a(s,t)\alpha^{-1}(s,t))\\ &\times v(s,t;x,y)dsdt\Big).\end{aligned} \tag{5.1.55}$$

Proof We may rewrite (5.1.54) as

$$\begin{aligned}\phi(x,y) \leq\ & \alpha(x,y)a(x,y)\alpha^{-1}(x,y)\\ &+\beta(x,y)b(x,y)\beta^{-1}(x,y)G^{-1}\left(\int_{x_0}^{x}\int_{y_0}^{y} c(s,t)G(\phi(s,t))dsdt\right).\end{aligned}$$

Since G is convex, sub-multiplicative and monotonic, we can get

$$\begin{aligned}G(\phi(x,y)) \leq\ & \alpha(x,y)G(a(x,y))\alpha^{-1}(x,y)\\ &+\beta(x,y)G\Big(b(x,y)\beta^{-1}(x,y)\Big)G^{-1}\left(\int_{x_0}^{x}\int_{y_0}^{y} c(s,t)G(\phi(s,t))dsdt\right).\end{aligned} \tag{5.1.56}$$

Define

$$u(x,y) = \int_{x_0}^{x}\int_{y_0}^{y} c(s,t)G(\phi(s,t)), \quad u(x,y_0) = u(x_0,y) = 0, \tag{5.1.57}$$

then

$$u_{xy}(x,y) = c(x,y)G(\phi(x,y)),$$

which, in view of (5.1.56), implies

$$\begin{aligned} L[u] &= u_{xy}(x,y) - c(x,y)\beta(x,y)G(b(x,y)\beta^{-1}(x,y))u(x,y) \\ &\leq c(x,y)[\alpha(x,y)G(a(x,y))\alpha^{-1}(x,y)]. \end{aligned} \tag{5.1.58}$$

The operator L is self-adjoint and hyperbolic, and for any twice continuously differentiable u and v the operator L satisfies the identity

$$vL[u] - uL[v] = -(uv_y)_x + (vu_x)_y. \tag{5.1.59}$$

Let P_0 and P be any points as in the theorem and label the directed sides and corners of the rectangle R.

Using s and t as the independent variables, we integrate the identity (5.1.59) over R and use Green's formula to obtain

$$\int\int_R (vL[u] - uL[v])dsdt = -\int_{C_1+C_2+C_3+C_4} (vu_s ds + uv_t dt).$$

This holds for any functions in C^2 (Fig. 5.1).

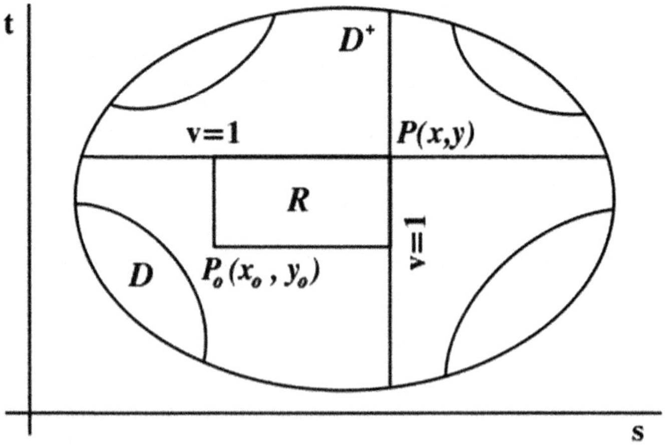

Fig. 5.1 Region and directed path around R

For the particular function u defined earlier, we have $u = 0$ on c_3 and $u = u_s = 0$ on c_4, so the right-hand side of the above identity reduces to

$$-\int_{C_1} vu_s ds - \int_{C_2} uv_t dt. \tag{5.1.60}$$

Now suppose v satisfies

$$\begin{cases} L[v] = v_{st} - c(s,t)\beta(s,t)G(b(s,t)\beta^{-1}(s,t))v = 0, & (5.1.61)\\ v = 1, \quad \text{on } C_1, & (5.1.62)\\ v_t = 0, \quad \text{on } C_2. & (5.1.63)\end{cases}$$

Then (5.1.62) and (5.1.63) imply that

$$v = 1, \quad \text{on } C_2. \tag{5.1.64}$$

Since $v \geq 0$ on R and $u(P_1) = 0$, by using (5.1.58), (5.1.60) becomes

$$u(x,y) \leq \int_{x_0}^{x}\int_{y_0}^{y} c(s,t)\alpha(s,t)G(a(s,t)\alpha^{-1}(s,t))v(s,t;x,y)dsdt.$$

The conclusion (5.1.55) follows from (5.1.56) and (5.1.64). □

The proof of this theorem is obtained by reducing the integral inequality to a differential inequality and then integrating it by Riemann's method for hyperbolic partial differential equations [177]. The function $v(s,t;x,y)$ involved in Theorem 5.1.10 is a Riemann function relative to the point $P(x,y)$ for the self adjoint operator L. There is such a function and a domain D^+ on which $v > 0$ since $v = 1$ on the vertical and horizontal lines through P and since V is continuous (Fig. 5.2).

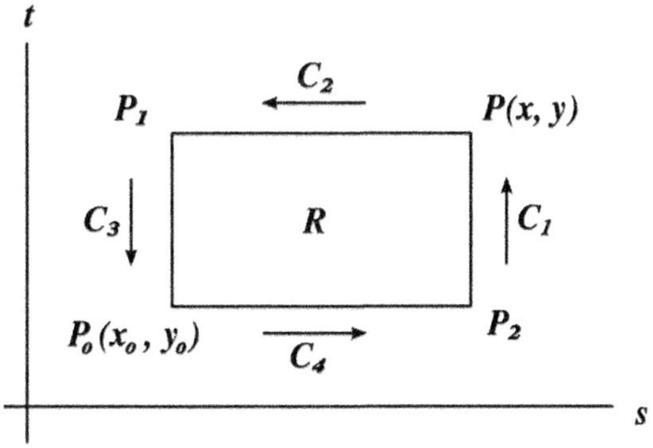

Fig. 5.2 Directed path around R

In Theorem 5.1.11 below, we introduce the following two independent variable generalization of the integral inequality established by Pachpatte [454].

Theorem 5.1.11 (The Pachpatte Inequality [454]) *Suppose $\phi(x,y)$, $a(x,y)$, $b(x,y)$, $c(x,y)$, and $k(x,y)$ are real-valued non-negative continuous functions defined on a domain D. Let $G(r)$, $\alpha(x,y)$, $\beta(x,y)$ be the same functions as defined in Theorem 5.1.10. Let $P_0(x_0,y_0)$ and $P(x,y)$ be two points in D such that $(x-x_0)(y-y_0) > 0$ and R be the rectangular region whose opposite corners are the points P_0 and P. Let $E(s,t;x,y)$ be the solution of the characteristic initial value problem*

$$\begin{cases} L[E] = E_{st} - c(s,t)\beta(s,t)G(b(s,t)\beta^{-1}(s,t))[c(s,t)+k(s,t)]E(s,t) = 0, \\ E(x,t) = E(s,y) = 1, \end{cases} \tag{5.1.65}$$

and let D^+ be a connected sub-domain of D which contains P and on which $v > 0$. If $R \subset D^+$ and for all $(x,y) \in R$,

$$\begin{aligned} \phi(x,y) \le a(x,y) + b(x,y)G^{-1}\Big[&\int_{x_0}^{x}\int_{y_0}^{y} c(s,t)G(\phi(s,t))dsdt \\ &+ \int_{x_0}^{x}\int_{y_0}^{y} c(s,t)\beta(s,t)G(b(s,t)\beta^{-1}(s,t)) \\ &\times \left(\int_{x_0}^{s}\int_{y_0}^{t} k(m,n)G(\phi(m,n))dmdn\right)dsdt\Big], \end{aligned} \tag{5.1.66}$$

then for all $(x,y) \in R$,

$$\begin{aligned} \phi(x,y) \le a(x,y) + b(x,y)G^{-1}\Big[&\int_{x_0}^{x}\int_{y_0}^{y} Q(s,t)dsdt \\ &+ \int_{x_0}^{x}\int_{y_0}^{y} c(s,t)\beta(s,t)G(b(s,t)\beta^{-1}(s,t)) \\ &\times \left(\int_{x_0}^{s}\int_{y_0}^{t} Q(m,n)E(m,n,s,t)dmdn\right)dsdt\Big], \end{aligned} \tag{5.1.67}$$

where

$$\begin{aligned} Q(x,y) = {}& c(x,y)\alpha(x,y)G(a(x,y)\alpha^{-1}(x,y)) + c(x,y)\beta(x,y)G(b(x,y)\beta^{-1}(x,y)) \\ &\times \int_{x_0}^{x}\int_{y_0}^{y} k(m,n)\alpha(m,n)G(a(m,n)\alpha^{-1}(m,n))dmdn. \end{aligned} \tag{5.1.68}$$

Proof We may rewrite (5.1.66) as

$$\phi(x,y) \le \alpha(x,y)a(x,y)\alpha^{-1}(x,y)$$
$$+\beta(x,y)b(x,y)\beta^{-1}(x,y)G^{-1}\left[\int_{x_0}^{x}\int_{y_0}^{y} c(s,t)G(\phi(s,t))dsdt\right.$$
$$\left.+\int_{x_0}^{x}\int_{y_0}^{y} c(s,t)\beta(s,t)G(b(s,t)\beta^{-1}(s,t))\left(\int_{x_0}^{s}\int_{y_0}^{t} k(m,n)G(\phi(m,n))dmdn\right)dsdt\right].$$

Since G is convex, sub-multiplicative and monotonic, we obtain

$$G(\phi(x,y)) \le \alpha(x,y)G(a(x,y)\alpha^{-1}(x,y))$$
$$+\beta(x,y)G(b(x,y)\beta^{-1}(x,y))G^{-1}\Big[\int_{x_0}^{x}\int_{y_0}^{y} c(s,t)G(\phi(s,t))dsdt$$
$$+\int_{x_0}^{x}\int_{y_0}^{y} c(s,t)\beta(s,t)G(b(s,t)\beta^{-1}(s,t))$$
$$\times\left(\int_{x_0}^{s}\int_{y_0}^{t} k(m,n)G(\phi(m,n))dmdn\right)dsdt\Big]. \tag{5.1.69}$$

Define

$$u(x,y) = \int_{x_0}^{x}\int_{y_0}^{y} c(s,t)G(\phi(s,t))dsdt + \int_{x_0}^{x}\int_{y_0}^{y} c(s,t)\beta(s,t)G(b(s,t)\beta^{-1}(s,t))$$
$$\times\left(\int_{x_0}^{s}\int_{y_0}^{t} k(m,n)G(\phi(m,n))dmdn\right)dsdt, \quad u(x,y_0) = ux_0,y) = 0,$$

then

$$u_{xy}(x,y) = c(x,y)\Big[G(\phi(x,y)) + \beta(x,y)G(b(x,y)\beta^{-1}(x,y))$$
$$\times\int_{x_0}^{x}\int_{y_0}^{y} k(m,n)G(\phi(m,n))dmdn\Big],$$

which, in view of (5.1.67), implies

$$u_{xy}(x,y) \le Q(x,y) + c(x,y)\beta(x,y)G(b(x,y)\beta^{-1}(x,y))$$
$$\times\left\{u(x,y) + \int_{x_0}^{x}\int_{y_0}^{y} k(m,n)G(b(m,n)\beta^{-1}(m,n))u(m,n)dmdn\right\}. \tag{5.1.70}$$

Define

$$\begin{cases} r(x,y) = u(x,y) + \int_{x_0}^{x}\int_{y_0}^{y} k(m,n)\beta(m,n)G(b(m,n)\beta^{-1}(m,n))u(m,n)dmdn, \\ r(x,y_0) = r(x_0,y) = 0, \end{cases} \tag{5.1.71}$$

then

$$r_{xy}(x,y) = u_{xy}(x,y) + k(x,y)\beta(x,y)G(b(x,y)\beta^{-1}(x,y))u(x,y). \tag{5.1.72}$$

Using the facts that

$$u_{xy}(x,y) \le Q(x,y) + c(x,y)\beta(x,y)G(b(x,y)\beta^{-1}(x,y))r(x,y),$$

from (5.1.70) and $u(x,y) \le r(x,y)$ from (5.1.71) in (5.1.72), it follows

$$L[r] = r_{xy}(x,y) - \beta(x,y)G(b(x,y)\beta^{-1}(x,y))[c(x,y) + k(x,y)]r(x,y) \le Q(r,y).$$

Now following the last argument as in the proof of Theorem 5.1.10, we can obtain

$$r(x,y) \le \int_{x_0}^{x}\int_{y_0}^{y} Q(s,t)E(s,t;x,y)dsdt.$$

Using this bound on $r(x,y)$ in (5.1.70), we can arrive at

$$u_{xy}(x,y) \le Q(x,y)+c(x,y)\beta(x,y)G(b(x,y)\beta^{-1}(x,y)) \int_{x_0}^{x}\int_{y_0}^{y} Q(s,t)E(s,t;x,y)dsdt.$$

Integrating both sides of the above inequality first with respect to y from y to y_0, and then with respect to x from x_0 to x, we conclude

$$\begin{aligned} u(x,y) \le & \int_{x_0}^{x}\int_{y_0}^{y} Q(s,t)dsdt + \int_{x_0}^{x}\int_{y_0}^{y} c(s,t)\beta(s,t)G(b(s,t)\beta^{-1}(s,t)) \\ & \times \left(\int_{x_0}^{s}\int_{y_0}^{t} k(m,n)E(m,n,s,t)dmdn\right) dsdt. \end{aligned} \tag{5.1.73}$$

Thus the desired bound in (5.1.66) follows from (5.1.67) and (5.1.73). □

We next introduce the two independent variable generalization of the integral inequality established by Pachpatte [471].

Theorem 5.1.12 (The Pachpatte Inequality [471]) *Suppose $\phi(x,y)$, $a(x,y)$, $b(x,y)$, $c(x,y)$, and $k(x,y)$ are real-valued non-negative continuous functions* *defined on a domain D. Let $N(r)$ be a positive, continuous, strictly increasing, sub-additive and sub-multiplicative function for all $r \geq 0$ and N^{-1} is the inverse function of N. Let $P_0(x_0, y_0)$ and $P(x, y)$ be two points in D such that $(x - x_0)(y - y_0) > 0$ and R be the rectangular region whose opposite corners are the points P_0 and P. Let $e(s, t; x, y)$ be the solution of the characteristic initial value problem*

$$\begin{cases} L[e] = e_{st}N(b(s,t))[c(s,t) = K(s,t)]e(s,t) = 0, \\ e(x,t) = e(s,y) = 1, \end{cases} \tag{5.1.74}$$

and let D^+ be a connected sub-domain of D which contains P and on which $v > 0$. If $R \subset D^+$ and for all $(x, y) \in R$,

$$\phi(x,y) \leq a(x,y) + b(x,y)N^{-1}\Big[\int_{x_0}^{x}\int_{y_0}^{y} c(s,t)N(\phi(s,t))dsdt \\ + \int_{x_0}^{x}\int_{y_0}^{y} c(s,t)N(b(s,t))\Big(\int_{x_0}^{s}\int_{y_0}^{t} k(m,n)N(\phi(m,n))dmdn\Big)dsdt\Big], \tag{5.1.75}$$

then for all $(x, y) \in R$,

$$\phi(x,y) \leq N^{-1}\Big[N(a(x,y)) + N(b(x,y))\int_{x_0}^{x}\int_{y_0}^{y} c(s,t)dsdt\Big\{N(a(s,t)) \\ +N(b(s,t))\int_{x_0}^{s}\int_{y_0}^{t}[c(m,n) + k(m,n)]N(b(m,n))e(m,n,s,t)dmdn\Big\}dsdt\Big]. \tag{5.1.76}$$

Proof The proof follows by the similar argument as in the proofs of Theorems 5.1.10 and 5.1.11 with suitable modifications (see also [471]). We omit the details. □

We note that the inequalities here can be extended very easily to the corresponding vector problems as in [620]. We also note that there is no essential difficulty in obtaining n independent variable generalizations of the inequalities established in Theorems 5.1.9–5.1.11 by using the technique used by Young in [658].

We next introduce the following inequality which can be used in more general situation.

Theorem 5.1.13 (The Pachpatte-Pachpatte Inequality [525]) *Let $u(t), a(t), b(t)$, be as in Theorem 1.1.5 in Qin [557] and $L : \mathbb{R}_+^2 \to \mathbb{R}_+$ be a continuous function which satisfies the condition: for all $u \geq v \geq 0$,*

$$0 \leq L(t,u) - L(t,v) \leq M(t,v)(u - v),$$

where $M(t, v)$ is a real-valued non-negative continuous functions defined for all $t, v \in \mathbb{R}_+$. If for all $t \in \mathbb{R}_+$,

$$u(t) \le a(t) + \int_t^{+\infty} b(s)u(s)ds + \int_t^{+\infty} L(s, u(s))ds, \tag{5.1.77}$$

then for all $t \in \mathbb{R}_+$,

$$u(t) \le E(t)\Big[a(t) + A(t)\exp\Big(\int_t^{\infty} M(s, E(s)a(s))E(s)ds\Big)\Big], \tag{5.1.78}$$

where for all $t \in \mathbb{R}_+$,

$$E(t) = \exp\Big(\int_t^{+\infty} b(s)ds\Big), \tag{5.1.79}$$

$$A(t) = \int_t^{\infty} L(s, E(s)a(s))ds. \tag{5.1.80}$$

Proof Define a function $z(t)$ by

$$z(t) = \int_t^{+\infty} L(s, u(s))ds, \tag{5.1.81}$$

then (5.1.77) can be restated as

$$u(t) \le a(t) + z(t) + \int_t^{+\infty} b(s)u(s)ds. \tag{5.1.82}$$

since $a(t) + z(t)$ is non-negative, continuous and non-increasing for all $t \in \mathbb{R}_+$, by applying Theorem 1.1.5 in Qin [557] to (5.1.82), we have

$$u(t) \le (a(t) + z(t))E(t). \tag{5.1.83}$$

From (5.1.81) and (5.1.83) and the hypotheses on L, we observe that

$$\begin{aligned} z(t) &\le \int_t^{+\infty} [L(s, E(s)a(s) + E(s)z(s)) - L(s, E(s)(s)) + L(s, E(s)a(s))]ds \\ &\le A(t) + \int_t^{+\infty} M(s, E(s)a(s))E(s)z(s)ds. \end{aligned} \tag{5.1.84}$$

Clearly, $A(t)$ is non-negative, continuous and non-increasing for all $t \in \mathbb{R}_+$. Now applying Theorem 1.1.5 in Qin [557] to (5.1.84) yields

$$z(t) \le A(t)\exp\Big(\int_t^{\infty} M(s, E(s)a(s))E(s)ds\Big). \tag{5.1.85}$$

Using (5.1.85) in (5.1.83), we get the required inequality in (5.1.78).

The inequalities established in the following Theorems 5.1.14–5.1.16 can be used in certain applications.

Theorem 5.1.14 (The Pachpatte-Pachpatte Inequality [525]) *Let $u(t), a(t), b(t)$ be real-values non-negative continuous functions defined for all $t \in \mathbb{R}_+$ and L, M be as in Theorem 5.1.13. If for all $t \in \mathbb{R}_+$,*

$$u(t) \le a(t) + b(t)\int_t^{+\infty} L(s, u(s))ds, \tag{5.1.86}$$

then for all $t \in \mathbb{R}_+$,

$$u(t) \le a(t) + b(t)e(t)\exp\Big(\int_t^{+\infty} M(s, a(s))b(s)ds\Big), \tag{5.1.87}$$

where for all $t \in \mathbb{R}_+$,

$$e(t) = \int_t^{+\infty} L(s, a(s))ds. \tag{5.1.88}$$

Proof Define a function $z(t)$ by (5.1.81). Then from (5.1.86), we have

$$u(t) \le a(t) + b(t)z(t). \tag{5.1.89}$$

From (5.1.81), (5.1.89) and the hypotheses on L, we observe that

$$\begin{aligned} z(t) &\le \int_t^{+\infty} [L(s, a(s) + b(s)z(s)) - L(s, a(s)(s)) + L(s, a(s))]ds \\ &\le e(t) + \int_t^{+\infty} M(s, a(s))b(s)z(s)ds, \end{aligned} \tag{5.1.90}$$

where $e(t)$ is defined by (5.1.88). Clearly $e(t)$ is real-valued non-negative, continuous and non-increasing in all $t \in \mathbb{R}_+$. An application of Theorem 1.1.5 in Qin [557] to (5.1.90) yields

$$z(t) \le e(t)\exp\Big(\int_t^{+\infty} M(s, a(s))b(s)ds\Big). \tag{5.1.91}$$

Thus the desired inequality in (5.1.87) follows from (5.1.89) and (5.1.91). □

Theorem 5.1.15 (The Pachpatte-Pachpatte Inequality [525]) *Let $u(x, y), a(x, y)$, $b(x, y)$, be as in Theorem 5.1.9 in Qin [557] and $L : \mathbb{R}_+^3 \to \mathbb{R}_+$ be a continuous function which satisfies the condition: for all $u \geq v \geq 0$,*

$$0 \leq L(x, y, u) - L(x, y, v) \leq M(x, y, v)(u - v),$$

where $M(x, y, v)$ is a real-valued non-negative continuous functions defined for all $x, y, v \in \mathbb{R}_+$. If for all $x, y \in \mathbb{R}_+$,

$$u(x, y) \leq a(x, y) + \int_x^{+\infty}\int_y^{+\infty} b(s, t)u(s, t)dtds + \int_x^{+\infty}\int_y^{+\infty} L(s, t, u(s, t))dtds, \tag{5.1.92}$$

then for all $x, y \in \mathbb{R}_+$,

$$u(x, y) \leq F(x, y)\left[a(x, y) + B(x, y)\exp\left(\int_x^{+\infty}\int_y^{+\infty} M(s, t, F(s, t)a(s, t))\right)F(s, t)dtds\right], \tag{5.1.93}$$

where for all $x, y \in \mathbb{R}_+$,

$$F(x, y) = \exp\left(\int_x^{+\infty}\int_y^{+\infty} b(s, t)dtds\right), \quad B(x, y) = \int_x^{+\infty}\int_y^{+\infty} L(s, t, F(s, t)a(s, t))dtds. \tag{5.1.94}$$

Proof The proof follows by the proofs of Theorems 5.1.13–5.1.14 and Theorem 2.1.59 in Qin [557], and using Theorem 5.1.9 in Qin [557]. □

Theorem 5.1.16 (The Pachpatte-Pachpatte Inequality [525]) *Let $u(x, y), a(x, y)$, $b(x, y)$ be real-value non-negative continuous functions defined for all $x, y \in \mathbb{R}_+$ and L, M be as in Theorem 5.1.15. If for all $x, y \in \mathbb{R}_+$,*

$$u(x, y) \leq a(x, y) + b(x, y)\int_x^{+\infty}\int_y^{+\infty} L(s, t, u(s, t))dtds, \tag{5.1.95}$$

then for all $x, y \in \mathbb{R}_+$,

$$u(x, y) \leq a(x, y) + b(x, y)f(x, y)\exp\left(\int_x^{+\infty}\int_y^{+\infty} M(s, t, a(s, t))b(s, t)dtds\right), \tag{5.1.96}$$

where for all $x, y \in \mathbb{R}_+$,

$$f(x,y) = \int_x^{+\infty}\int_y^{+\infty} L(s,t,a(s,t))dtds. \tag{5.1.97}$$

Proof The proof is similar to that of Theorem 5.1.15. □

Theorem 5.1.17 (The Pachpatte-Pachpatte Inequality [525]) *Let $u(x,y), a(x,y)$, $b(x,y)$ be as in Theorem 5.1.16 and $L : \mathbb{R}_+^3 \to \mathbb{R}_+$ be a continuous function which satisfies the condition: for all $u \geq v \geq 0$,*

$$0 \leq L(x,y,u) - L(x,y,v) \leq M(x,y,v)\psi^{-1}(u-v),$$

where $M(x,y,v)$ is defined as in Theorem 5.1.15, $\psi : \mathbb{R}_+ \to \mathbb{R}_+$ be a continuous and strictly increasing function with $\psi(0) = 0$, ψ^{-1} is the inverse function of ψ and for all $u, v \in \mathbb{R}_+$,

$$\psi^{-1}(uv) \leq \psi^{-1}(u)\psi^{-1}(v).$$

If for all $x, y \in \mathbb{R}_+$,

$$u(x,y) \leq a(x,y) + b(x,y)\psi\left(\int_x^{+\infty}\int_y^{+\infty} L(s,t,u(x,y))dtds\right), \tag{5.1.98}$$

then for all $x, y \in \mathbb{R}_+$,

$$u(x,y) \leq a(x,y) + b(x,y)\psi\left(f(x,y)\exp\left(\int_x^{+\infty}\int_y^{+\infty} M(s,t,a(s,t))\psi^{-1}(b(s,t))\right)dtds\right), \tag{5.1.99}$$

where $f(x,y)$ is defined by (5.1.97).

Proof The proof is similar to that of Theorem 5.1.15. □

5.1.2 Nonlinear Two-Dimensional Bihari Inequality and Their Generalizations

In this subsection, we shall introduce some new nonlinear integral inequalities of the Gronwall-Bellman-Ou-Yang-type in two variables, which, on the one hand, generalize and on the other hand furnish a tool for the study of qualitative as well as quantitative properties of solutions of differential equations.

Bainov, Simeonov and Lipovan [42] observed the following Gronwall-Bellman-type inequality.

Theorem 5.1.18 (The Bainov-Simeonov Inequality [42]) *Let $I = [0, a]$, $J = [0, b]$, where $0 < a, b \leq +\infty$. Let $c \geq 0$ be a constant, $\varphi \in C(\mathbb{R}_+, \mathbb{R}_+)$ be non-decreasing with $\varphi(r) > 0$ for all $r > 0$, and $b \in C(I \times J, \mathbb{R}_+)$. If $u \in C(I \times J, \mathbb{R}_+)$ satisfies for all $(x, y) \in I \times J$,*

$$u(x, y) \leq c + \int_0^x \int_0^y b(s, t)\varphi(u(s, t))dtds,$$

then for all $(x, y) \in [0, x_1] \times [0, y_1]$,

$$u(x, y) \leq \Phi^{-1}\left[\Phi(c) + \int_0^x \int_0^y b(s, t)dtds\right]$$

where

$$\Phi(r) := \int_1^r \frac{ds}{\varphi(s)}, \quad r \geq 1,$$

and Φ^{-1} is the inverse of Φ, and $(x_1, y_1) \in I \times J$ is chosen such that $\Phi(c) + \int_0^x \int_0^y b(s, t)dsdt \in Dom\ (\Phi^{-1})$ for all $(x, y) \in [0, x_1] \times [0, y_1]$.

Proof The proof is left to the reader as an exercise. □

Snow [619, 620] and Ghoshal and Masood [246] have obtained some useful generalizations of Theorem 5.1.18 by using Riemann's function. Bondge and Pachpatte [94] have also obtained some useful generalizations of Theorem 5.1.18.

Theorem 5.1.19 (The Bondge-Pachpatte Inequality [94]) *Let $\phi(x, y)$ and $p(x, y)$ be real-valued non-negative continuous functions defined for all $x \geq 0,\ y \geq 0$. Let $H(u)$ be a positive, continuous, monotonic non-decreasing function for all $u > 0,\ H(0) = 0,\ H_y(u) \geq 0$, and suppose further that the inequality holds for all $x \geq 0,\ y \geq 0$,*

$$\phi(x, y) \leq a(x) + b(y) + \int_0^x \int_0^y p(s, t)H(\phi(s, t))dsdt, \tag{5.1.100}$$

where $a(x),\ b(y) > 0,\ a'(x),\ b'(y) \geq 0$ are real-valued continuous functions defined for all $x \geq 0,\ y \geq 0$. Then for all $0 \leq x \leq x_1,\ 0 \leq y \leq y_1$,

$$\phi(x, y) \leq \Omega^{-1}\left[\Omega(a(0) + b(y)) + \int_0^x \frac{a'(s)}{H(a(s) + b(0))}ds + \int_0^x \int_0^y p(s, t)dsdt\right], \tag{5.1.101}$$

where

$$\Omega(r) = \int_{r_0}^r \frac{ds}{H(s)}, \quad r \geq r_0 \geq 0, \tag{5.1.102}$$

and Ω^{-1} *is the inverse function of* Ω, *and* x_1, y_1 *are chosen so that*

$$\Omega(a(0)+b(y))+\int_0^x \frac{a'(s)ds}{H(a(s)+b(0))}ds+\int_0^x\int_0^y p(s,t)dsdt \in Dom\,(\Omega^{-1})$$

for all x, y *lying in the sub-intervals* $0 \le x \le x_1$, $0 \le y \le y_1$ *of real numbers.*

Proof Define a function $u(x,y)$ by the right-hand side of (5.1.100), then

$$\begin{cases} u_{xy}(x,y) = p(x,y)H(\phi(x,y)), \\ u(x,0) = a(x)+b(0), \quad u(0,y) = a(0)+b(y), \end{cases}$$

which, in view of (5.1.100), implies

$$u_{xy}(x,y) \le p(x,y)H(u(x,y)),$$

i.e.,

$$\frac{u_{xy}(x,y)}{H(u(x,y))} \le p(x,y). \tag{5.1.103}$$

From (5.1.103) it follows

$$\begin{cases} \dfrac{H(u(x,y))u_{xy}(x,y)}{H^2(u(x,y))} \le p(x,y) + \dfrac{H_y(u(x,y))u_x(x,y)}{H^2(u(x,y))}, \\ \dfrac{\partial}{\partial y}\left(\dfrac{u_x(x,y)}{H(u(x,y))}\right) \le p(x,y). \end{cases}$$

Integrating both sides of the above inequality with respect to y from 0 to y, we have

$$\frac{u_x(x,y)}{H(u(x,y))} \le \frac{a'(x)}{H(a(x)+b(0))} + \int_0^y p(x,t)dt. \tag{5.1.104}$$

From (5.1.102) and (5.1.104), we derive

$$\Omega_x(u(x,y)) \le \frac{a'(x)}{H(a(x)+b(0))} + \int_0^y p(x,t)dt.$$

Integrating both sides of the above inequality with respect to x from 0 to x, we can get

$$\Omega(u(x,y)) \le \Omega(a(0)+b(y)) + \int_0^x \frac{a'(s)}{H(a(s)+b(0))}ds + \int_0^x\int_0^y p(s,t)dtds. \tag{5.1.105}$$

Now substituting the bound on $u(x,y)$ from (5.1.105) into (5.1.100), we can obtain the desired bound in (5.1.101). The intervals of real numbers for x and y are obvious. □

We next establish the following two independent variable generalization of the integral inequality established by Bondge-Pachpatte [94].

Theorem 5.1.20 (The Bondge-Pachpatte Inequality [94]) *Let $\phi(x,y)$ and $p(x,y)$ be real-valued non-negative continuous functions defined for all $x \ge 0,\ y \ge 0$. Let $H(u),\ H_y(u)$ be the same functions as defined in Theorem 5.1.19, and suppose further that the inequality holds for all $x \ge 0,\ y \ge 0$,*

$$\phi(x,y) \le a(x) + b(y) + \int_0^x \int_0^y p(s,t)\Big(\phi(s,t) + \int_0^s \int_0^t p(m,n)H(\phi(m,n))dmdn\Big)dsdt, \tag{5.1.106}$$

where $a(x),\ b(y) > 0,\ a'(x),\ b'(y) \ge 0$ are real-valued continuous functions defined for all $x \ge 0,\ y \ge 0$. Then for all $0 \le x \le x_2,\ 0 \le y \le y_2$,

$$\phi(x,y) \le a(x) + b(y) + \int_0^x \int_0^y p(s,t)G^{-1}\Big[G(a(0)+b(t)) + \int_0^t \frac{a'(m)dm}{a(m)+b(0)+H(a(m)+b(0))} + \int_0^s \int_0^t p(m,n)dmdn\Big]dsdt, \tag{5.1.107}$$

where

$$G(r) = \int_{r_0}^r \frac{ds}{s+H(s)}, \quad r \ge r_0 \ge 0, \tag{5.1.108}$$

and G^{-1} is the inverse function of G, and $x_2,\ y_2$ are chosen so that

$$G(a(0)+b(y)) + \int_0^t \frac{a'(m)}{a(m)+b(0)+H(a(m)+b(0))}dm + \int_0^s \int_0^t p(m,n)dmdn]dsdt \in Dom\,(G^{-1})$$

for all $x,\ y$ lying in the sub-intervals $0 \le x \le x_2,\ 0 \le y \le y_2$ of real numbers.

Proof Define a function $u(x,y)$ by the right-hand side of (5.1.106), then

$$\begin{cases} u_{xy}(x,y) = p(x,y)(\phi(x,y) + \int_0^x \int_0^y p(m,n)H(\phi(m,n))dmdn), \\ u(x,0) = a(x)+b(0),\ \ u(0,y) = a(0)+b(y) \end{cases}$$

which, in view of (5.1.106), implies

$$u_{xy}(x,y) \le p(x,y)\Big(u(x,y) + \int_0^x \int_0^y p(m,n)H(\phi(m,n))dmdn\Big). \tag{5.1.109}$$

If we put

$$\begin{cases} v(x,y) = u(x,y) + \int_0^x \int_0^y p(m,n)H(\phi(m,n))dmdn, \\ v(x,0) = a(x) + b(0), \quad v(0,y) = a(0) + b(y), \end{cases}$$

then

$$v_{xy}(x,y) = u_{xy}(x,y) + p(x,y)H(u(x,y)). \tag{5.1.110}$$

Using the facts that $u_{xy}(x,y) \le p(x,y)v(x,y)$ from (5.1.108) and $u(x,y) \le v(x,y)$ from (5.1.109) in (5.1.110), we have

$$v_{xy}(x,y) \le p(x,y)[v(x,y) + H(v(x,y))].$$

Now by following the similar argument as in the proof of Theorem 5.1.19 in view of the definition of G, we obtain

$$v(x,y) \le G^{-1}\Big[G(a(0)+b(y)) + \int_0^x \frac{a'(s)ds}{a(s)+b(0)+H(a(s)+b(0))} + \int_0^x \int_0^y p(s,t)dsdt\Big].$$

Substituting the above bound on $v(x,y)$ in (5.1.109) and then integrating both sides first with respect to y from 0 to y and then with respect to x from 0 to x, we conclude

$$\begin{aligned} u(x,y) \le\ & a(x) + b(y) + \int_0^x \int_0^y p(s,t)G^{-1}[G(a(0)+b(t)) \\ & + \int_0^s \frac{a'(m)dm}{a(m)+b(0)+H(a(m)+b(0))} + \int_0^s \int_0^t p(m,n)dmdn]dsdt. \end{aligned}$$

Now substituting this bound on $u(x,y)$ in (5.1.106), we can obtain the desired bound in (5.1.107). The intervals of real numbers for x and y are obvious. □

We next introduce the following integro-differential inequality in two independent variables.

Theorem 5.1.21 (The Bondge-Pachpatte Inequality [94]) *Let $\phi(x,y)$ and $\phi_{xy}(x,y)$ be real-valued non-negative continuous functions defined for all $x \ge 0,\ y \ge 0$; $\phi(x,0) = \phi(0,y) = 0$, and $p(x,y)$ be real-valued continuous function defined for all $x \ge 0,\ y \ge 0$. Let $H(u),\ H_y(u)$ be the same functions as defined in Theorem 5.1.19, and suppose further that the inequality holds for all $x \ge 0,\ y \ge 0$,*

$$\phi_{xy}(x,y) \le a(x) + b(y) + M\Big[\phi(x,y) + \int_0^x \int_0^y p(s,t)H(\phi_{st}(s,t))dsdt\Big], \tag{5.1.111}$$

where $a(x),\ b(y) > 0,\ a'(x),\ b'(y) \geq 0$ *are real-valued continuous functions defined for all* $x \geq 0,\ y \geq 0$, *and* $M \geq 0$ *is a constant. Then for all* $0 \leq x \leq x_3,\ 0 \leq y \leq y_3$,

$$\phi_{xy}(x,y) \leq G^{-1}\Big[G(a(0)+b(y)) + \int_0^x \frac{a'(s)ds}{a(s)+b(0)+H(a(s)+b(0))} + M\int_0^x\int_0^y p(s,t)dsdt\Big] \tag{5.1.112}$$

where G *and* G^{-1} *are as defined in Theorem 5.1.20.* G^{-1} *is the inverse function of* G, *and* $x_3,\ y_3$ *are chosen so that*

$$G(a(0)+b(y)) + \int_0^x \frac{a'(s)}{a(s)+b(0)+H(a(s)+b(0))}ds + M\int_0^x\int_0^y p(s,t)dsdt \in Dom\,(G^{-1}),$$

for all $x,\ y$ *lying in the sub-intervals* $0 \leq x \leq x_3,\ 0 \leq y \leq y_3$ *of real numbers.*

Proof Define a function $u(x,y)$ by the right-hand side of (5.1.111), then

$$\begin{cases} u_{xy}(x,y) = M[\phi_{xy}(x,y) + p(x,y)H(\phi_{xy}(x,y))], \\ u(x,0) = a(x) + b(0),\quad u(0,y) = a(0) + b(y) \end{cases}$$

which, by using (5.1.111) and the fact that $p(x,y) \geq 1$, implies

$$u_{xy}(x,y) \leq Mp(x,y)[u(x,y) + H(u(x,y))].$$

Now, following the similar argument as in the proof of Theorem 5.1.19, we may obtain

$$u(x,y) \leq G^{-1}\Big[G(a(0)+b(y)) + \int_0^x \frac{a'(s)ds}{a(s)+b(0)+H(a(s)+b(0))} + M\int_0^x\int_0^y p(s,t)dsdt\Big].$$

Substituting this bound on $u(x,y)$ in (5.1.111), we can obtain the desired bound in (5.1.112). □

Next we shall introduce the following two independent variable generalization of the integro-differential inequality established by Pachpatte [462] which may be convenient in situations.

Theorem 5.1.22 (The Bondge-Pachpatte Inequality [94]) *Let* $\phi(x,y),\ \phi_{xy}(x,y),\ p(x,y)$ *and* $H(u),\ H_y(u)$ *be the same functions as in Theorem 5.1.21, and suppose further that the inequality holds for all* $x \geq 0,\ y \geq 0$,

$$\phi_{xy}(x,y) \leq a(x) + b(y) + \int_0^x\int_0^y p(s,t)H\Big(\phi(s,t) + \phi_{st}(s,t)\Big)dsdt, \tag{5.1.113}$$

where $a(x),\ b(y) > 0,\ a'(x),\ b'(y) \geq 0$ *are real-valued continuous functions defined for all* $x \geq 0,\ y \geq 0$, *and* $M \geq 0$ *is a constant. Then for all* $0 \leq x \leq x_4,\ 0 \leq y \leq y_4$,

$$\phi_{xy}(x,y) \leq a(x) + b(y) + \int_0^x \int_0^y p(s,t)H(G^{-1}[G(a(0)+b(t)) + \int_0^s \frac{a'(m)}{a(m)+b(0)+H(a(m)+b(0))}dm + \int_0^s \int_0^t p(m,n)dmdn])dsdt, \quad (5.1.114)$$

where G *and* G^{-1} *are as defined in Theorem 5.1.20.* G^{-1} *is the inverse function of* G, *and* $x_4,\ y_4$ *are chosen so that*

$$G(a(0)+b(y)) + \int_0^x \frac{a'(s)}{a(s)+b(0)+H(a(s)+b(0))}ds + \int_0^x \int_0^y p(s,t)dsdt \in Dom\,(G^{-1}),$$

for all $x,\ y$ *lying in the sub-intervals* $0 \leq x \leq x_4,\ 0 \leq y \leq y_4$ *of real numbers.*

Proof Define a function $u(x,y)$ by the right-hand side of (5.1.113), then

$$\begin{cases} u_{xy}(x,y) = p(x,y)H(\phi(x,y) + \phi_{st}(x,y)), \\ u(x,0) = a(x) + b(0),\quad u(0,y) = a(0) + b(y). \end{cases} \quad (5.1.115)$$

From the definition of $u(x,y)$ and (5.1.112), it follows

$$\phi_{xy}(x,y) \leq u(x,y) \quad (5.1.116)$$

and hence which together with (5.1.116), yields

$$\phi(x,y) \leq \int_0^x \int_0^y u(s,t)dsdt. \quad (5.1.117)$$

Using (5.1.116) and (5.1.117) in (5.1.115), we may derive

$$u_{xy}(x,y) \leq p(x,y)H\Big(u(x,y) + \int_0^x \int_0^y u(s,t)dsdt\Big).$$

If we put

$$\begin{cases} v(x,y) = u(x,y) + \int_0^x \int_0^y u(m,n)dmdn, \\ v(x,0) = a(x) + b(0),\quad v(0,y) = a(0) + b(y), \end{cases}$$

then as in Theorem 5.1.20, we can obtain

$$v_{xy}(x,y) \leq p(x,y)[v(x,y) + H(v(x,y))].$$

Since the remainder of the proof is similar to an argument as in Theorem 5.1.20, we may leave the details to the reader. □

The following results, due to Dragomir and Kim [207], concerns integral inequalities involving functions of two independent variables.

All the functions which appear in the following Theorems 5.1.23–5.1.26 are assumed to be real-valued and all the integrals are involved in existence on the domains of their definitions.

In the following results, we shall use the class $\mathcal{F}$ of functions introduced in Chap. 1.

Theorem 5.1.23 (The Dragomir-Kim Inequality [207]) *Let $u(x,y)$, $a(x,y)$, $b(x,y)$, $c(x,y)$, $d(x,y)$ be non-negative continuous functions defined for all $x, y \in \mathbb{R}_+$, let $g \in \mathcal{F}$. Define a function $z(x,y)$ by*

$$z(x,y) = a(x,y) + c(x,y)\int_0^x \int_y^{+\infty} b(s,t)u(s,t)dtds,$$

with $z(x,y)$ is non-decreasing in x and $z(x,y) \geq 1$ for all $x, y \in \mathbb{R}_+$. If for all $\alpha, x, y \in \mathbb{R}_+$, and $\alpha \leq x$,

$$u(x,y) \leq z(x,y) + \int_\alpha^x b(s,y)g(u(s,y))ds, \tag{5.1.118}$$

then for all $x, y \in \mathbb{R}_+$,

$$u(x,y) \leq p(x,y)\left[a(x,y) + c(x,y)e(x,y)\exp\left(\int_0^x \int_y^{+\infty} d(s,t)p(s,t)c(s,t)dtds\right)\right], \tag{5.1.119}$$

where

$$\begin{cases} p(x,y) = G^{-1}\Big(G(1) + \int_\alpha^x b(s,y)ds\Big), & (5.1.120)\\ e(x,y) = \int_0^x \int_y^{+\infty} d(s,t)p(s,t)c(s,t)dtds, & (5.1.121)\\ G(u) = \int_{u_0}^u \frac{ds}{g(s)}, \quad u \geq u_0 > 0 & (5.1.122) \end{cases}$$

and G^{-1} is the function of G such that $G(1) + \int_\alpha^x b(s,y)ds \in Dom\ (G^{-1})$.

Proof Let $z(x, y)$ is a non-negative, continuous, non-decreasing and let $g \in \mathcal{F}$. Then (5.1.118) can be restated as

$$\frac{u(x,y)}{z(x,y)} \leq 1 + \int_{\alpha}^{x} b(s,y)\frac{1}{z(x,y)}g(u(s,y))ds. \tag{5.1.123}$$

Define a function $w(x, y)$ by the right-hand side of (5.1.123), then $\frac{u(x,y)}{z(x,y)} \leq w(x,y)$ and

$$w(x,y) \leq 1 + \int_{\alpha}^{x} b(s,y)g(w(s,y))ds. \tag{5.1.124}$$

Treating $x, y \in \mathbb{R}_+$ fixed in (5.1.124) and using (i) of Theorem 1.1.2 to (5.1.124), we can get

$$w(x,y) \leq G^{-1}(G(1) + \int_{\alpha}^{x} b(s,y)ds). \tag{5.1.125}$$

Using (5.1.125) in $[u(x, y)/z(x, y)] \leq w(x, y)$, we may obtain

$$u(x,y) \leq z(x,y)p(x,y),$$

where $p(x, y)$ is defined by (5.1.120). From the definition of $z(x, y)$, we derive

$$u(x,y) \leq p(x,y)(a(x,y) + c(x,y)v(x,y)), \tag{5.1.126}$$

where

$$v(x,y) = \int_{0}^{x}\int_{y}^{+\infty} d(s,t)u(s,t)dtds.$$

From (5.1.126), we derive

$$\begin{aligned} v(x,y) &\leq \int_{0}^{x}\int_{y}^{+\infty} d(s,t)p(s,t)\left(a(s,t) + c(s,t)v(s,t)\right)dtds \\ &= e(x,y) + \int_{0}^{x}\int_{y}^{+\infty} d(s,t)p(s,t)c(s,t)v(s,t)dtds, \end{aligned}$$

where $e(x, y)$ is defined by (5.1.121). Clearly, $e(x, y)$ is non-negative, continuous, non-decreasing in x, $x \in \mathbb{R}_+$ and non-increasing in y, $y \in \mathbb{R}_+$. Now, by (i) of Lemma 5.1.4 in Qin [557], we obtain

$$u(x,y) \leq e(x,y)\exp\left(\int_{0}^{x}\int_{y}^{+\infty} d(s,t)p(s,t)c(s,t)u(s,t)dtds\right). \tag{5.1.127}$$

Using (5.1.127) in (5.1.126), we can get the required inequality (5.1.119). □

Theorem 5.1.24 (The Dragomir-Kim Inequality [207]) *Let $u(x, y), a(x, y), b(x, y), c(x, y), d(x, y)$ be non-negative continuous functions defined for all $x, y \in \mathbb{R}_+$, let $g \in \mathcal{F}$. Define a function $z(x, y)$ by*

$$z(x,y) = a(x,y) + \int_x^{+\infty}\int_y^{+\infty} d(s,t)u(s,t)dtds,$$

with $z(x, y)$ is non-decreasing in x and $z(x, y) \geq 1$ for all $x, y \in \mathbb{R}_+$. If for all $\beta, x, y \in \mathbb{R}_+$ and $\beta \leq x$,

$$u(x,y) \leq z(x,y) + \int_x^{\beta} b(s,y)g(u(s,y))ds,$$

then for all $x, y \in \mathbb{R}_+$,

$$u(x,y) \leq \bar{p}(x,y)\Bigg[a(x,y) + c(x,y)\bar{e}(x,y) \times \exp\left(\int_x^{+\infty}\int_y^{+\infty} d(s,t)\bar{p}(s,t)c(s,t)dtds\right)\Bigg],$$

where

$$\begin{cases} \bar{p}(x,y) = G^{-1}\Big(G(1) + \int_x^{\beta} b(s,y)ds\Big), & (5.1.128) \\ \bar{e}(x,y) = \int_x^{+\infty}\int_y^{+\infty} d(s,t)\bar{p}(s,t)a(s,t)dtds, & (5.1.129) \end{cases}$$

and G is defined in (5.1.122), G^{-1} is the inverse function of G, such that

$$G(1) + \int_x^{\beta} b(s,y)ds \in Dom\,(G^{-1}).$$

Proof The proof follows by a similar argument to that in the proof of Theorem 5.1.23 with suitable changes. We omit the details. □

Theorem 5.1.25 (The Dragomir-Kim Inequality [207]) *Let $u(x, y), a(x, y), b(x, y), c(x, y)$ be non-negative continuous functions defined for all $x, y \in \mathbb{R}_+$ and $F : \mathbb{R}_+^3 \to \mathbb{R}_+$ be a continuous function which satisfies the condition: for all $u \geq v \geq 0$,*

$$0 \leq F(x,y,u) - F(x,y,v) \leq K(x,y,v)(u-v) \qquad (5.1.130)$$

where $K(x, y, v)$ is a non-negative continuous function defined for all $x, y, v \in \mathbb{R}_+$. And let $g \in \mathcal{F}$. Define a function $z(x, y)$ by

$$z(x, y) = a(x, y) + c(x, y) \int_0^x \int_y^{+\infty} F(s, t, u(s, t))dtds,$$

with non-decreasing in x and $z(x, y) \geq 1$ for all $x, y \in \mathbb{R}_+$. If for all $\alpha, x, y \in \mathbb{R}_+$ and $\alpha \leq x$,

$$u(x, y) \leq z(x, y) + \int_\alpha^x b(s, y)g(u(s, y))ds, \tag{5.1.131}$$

then for all $x, y \in \mathbb{R}_+$,

$$u(x, y) \leq p(x, y)\left[a(x, y) + c(x, y)A(x, y) \exp(\int_0^x \int_y^{+\infty} K(s, t, p(s, t)a(s, t))p(s, t)c(s, t)dtds)\right], \tag{5.1.132}$$

where $p(x, y)$ is defined by (5.1.120) and

$$A(x, y) = \int_0^x \int_y^{+\infty} F(s, t, p(s, t)a(s, t))dtds, \tag{5.1.133}$$

and $G(u) = \int_{u_0}^u \frac{ds}{g(s)}$, $u \geq u_0 > 0$, G^{-1} is the inverse function of G such that

$$G(1) + \int_\alpha^x b(s, y)ds \in Dom\ (G^{-1}).$$

Proof The proof follows by a similar argument to that in the proof of Theorem 5.1.23. Since $z(x, y)$ is a non-negative, continuous, non-decreasing and let $g \in \mathcal{F}$, then

$$u(x, y) \leq z(x, y)p(x, y),$$

where $p(x, y)$ is defined by (5.1.120). From the definition of $z(x, y)$, we can get

$$u(x, y) \leq p(x, y)(a(x, y) + c(x, y)w(x, y)), \tag{5.1.134}$$

where $w(x, y)$ is defined by

$$w(x, y) = \int_0^x \int_y^{+\infty} F(s, t, u(s, t))dtds.$$

From (5.1.131) and (5.1.134), we infer

$$w(x,y) \le \int_0^x \int_y^{+\infty} \Big[F(s,t,p(s,t)(a(s,t)+c(s,t)u(s,t))) + F(s,t,p(s,t)a(s,t)) - F(s,t,p(s,t)a(s,t)) \Big] dtds$$

$$\le A(x,y) + \int_0^x \int_y^{+\infty} K(s,t,p(s,t)a(s,t))p(s,t)c(s,t)w(s,t)dtds,$$

where $A(x,y)$ is defined by (5.1.133). Clearly, $A(x,y)$ is non-negative, continuous, non-decreasing in $x, x \in \mathbb{R}_+$ and non-increasing in $y, y \in \mathbb{R}_+$. Now, by (i) of Lemma 5.1.3, we conclude

$$w(x,y) \le A(x,y) \exp\left(\int_0^x \int_y^{+\infty} K(s,t,p(s,t)a(s,t))p(s,t)c(s,t)w(s,t)dtds \right). \tag{5.1.135}$$

Using (5.1.134) in (5.1.135), we can get the required inequality (5.1.132). □

Theorem 5.1.26 (The Dragomir-Kim Inequality [207]) *Let the assumptions of Theorem 5.1.25 hold. Define a function $z(x,y)$ by*

$$z(x,y) = a(x,y) + c(x,y) \int_x^{+\infty} \int_y^{+\infty} F(s,t,u(s,t))dtds,$$

with non-decreasing in x and $z(x,y) \ge 1$ for all $x, y \in \mathbb{R}_+$. If for all $\beta, x, y \in \mathbb{R}_+$ and $\beta \le x$,

$$u(x,y) \le z(x,y) + \int_x^{\beta} b(s,y)g(u(s,y))ds, \tag{5.1.136}$$

then for all $x, y \in \mathbb{R}_+$,

$$u(x,y) \le \bar{p}(x,y) \left[a(x,y) + c(x,y)\bar{A}(x,y) \exp\left(\int_x^{+\infty} \int_y^{+\infty} \bar{K}(s,t,\bar{p}(s,t)a(s,t))\bar{p}(s,t)c(s,t)dtds \right) \right], \tag{5.1.137}$$

where $\bar{p}(x,y)$ is defined by (5.1.128),

$$\bar{A}(x,y) = \int_x^{+\infty} \int_y^{+\infty} F(s,t,\bar{p}(s,t)a(s,t))dtds, \tag{5.1.138}$$

and G is defined in (5.1.122), G^{-1} is the inverse function of G, such that

$$G(1) + \int_x^\beta b(s, y)ds \in Dom\,(G^{-1}).$$

Proof The proof follows by a similar argument to that in the proof of Theorem 5.1.25 with suitable changes. We omit the details. □

5.1.3 Nonlinear Two-Dimensional Nonlinear Ou-Yang Inequality, Gollwitzer Inequality and Their Generalizations

The next result is due to Pachpatte [519].

Theorem 5.1.27 (The Pachpatte Inequality [519]) *Let $u(x, y), f(x, y) \in C(\mathbb{R}_+^2, \mathbb{R}_+), h(x, y, s, t) \in C(\mathbb{R}_+^2 \times \mathbb{R}_+^2, \mathbb{R}_+)$ for all $0 \le s \le x < +\infty$, $0 \le t \le y < +\infty$. Let c, p, g, G, G^{-1} be as in Theorem 1.1.36. If for all $x, y \in \mathbb{R}_+$,*

$$u^p(x, y) \le c + \int_0^x \int_0^y \Big[f(s, t)g(u(s, t)) + \int_0^s \int_0^t h(s, t, \sigma, \eta)g(u(\sigma, \eta))d\eta d\sigma\Big]dsdt, \tag{5.1.139}$$

then for all $0 \le x \le x_1$, $0 \le y \le y_1$, $x, x_1, y, y_1 \in \mathbb{R}_+$,

$$u(x, y) \le \left[G^{-1}[G(c) + A(x, y)]\right]^{1/p}, \tag{5.1.140}$$

where

$$A(x, y) = \int_0^x \int_0^y \Big[f(s, t) + \int_0^s \int_0^t h(s, t, \sigma, \eta)d\eta d\sigma\Big]dsdt, \tag{5.1.141}$$

and $x_1, y_1 \in \mathbb{R}_+$, are chosen so that

$$G(c) + A(x, y) \in Dom\,(G^{-1}),$$

for all x, y lying in the intervals $0 \le x \le x_1$, $0 \le y \le y_1$ of $\mathbb{R}_+$.

Proof Let $c > 0$ and define a function $z(x, y)$ by the right-hand side of (5.1.139). Then $z(0, y) = z(x, 0) = c,\ u(x, y) \le (z(x, y))^{1/p}$ and

$$\begin{aligned} D_1 z(t) &= \int_0^y [f(x, t)g(u(x, t)) + \int_0^x \int_0^t h(x, t, \sigma, \eta)g(u(\sigma, \eta))d\eta d\sigma]dt \\ &\le \int_0^y [f(x, t)g(z(x, t))^{1/p} + \int_0^x \int_0^t h(x, t, \sigma, \eta)g((z(\sigma, \eta))^{1/p})d\sigma d\eta]dt \\ &\le g((z(x, y))^{1/p}) \int_0^y [f(x, t) + \int_0^x \int_0^t h(x, t, \sigma, \eta)d\sigma d\eta]dt. \end{aligned} \tag{5.1.142}$$

From (1.1.236) and (5.1.142), we deduce

$$D_1 G(z(x,y)) = \frac{D_1 z(x,y)}{g((z(x,y))^{1/p})}$$
$$\le \int_0^y \Big[f(x,t) + \int_0^x \int_0^t h(x,t,\sigma,\eta)d\sigma d\eta\Big]dt. \qquad (5.1.143)$$

Keeping y fixed in (5.1.143), setting $x = s$ and integrating with respect to s from 0 to x and using the fact that $z(0,y) = c$, we get

$$G(z(x,y)) \le G(c) + A(x,y). \qquad (5.1.144)$$

Now substituting the bound on $z(x,y)$ from (5.1.144) in $u(x,y) \le (z(x,y))^{1/p}$, we can obtain the desired bound in (5.1.140). The proof of the case when $c \ge 0$ can be completed as mentioned in the proof of Theorem 1.1.25. The domain $0 \le x \le x_1,\ 0 \le y \le y_1$ is obvious. □

Next, we shall introduce the following corollary whose proof is similar to that of Corollary 1.1.5.

Corollary 5.1.1 *Let u, f, h, c, p be as in Theorem 5.1.26. If for all $x, y \in \mathbb{R}_+$,*

$$u^p(x,t) \le c + \int_0^x \int_0^t \Big[f(s,t)u(s,t) + \int_0^s \int_0^t h(s,t,\sigma,\eta)u(\sigma,\eta)d\sigma d\eta\Big]dtds, \qquad (5.1.145)$$

then for $x, y \in \mathbb{R}_+$,

$$u(x,y) \le \Big[c^{(p-1)/p} + \frac{p-1}{p})A(x,y\Big]^{1/(p-1)}, \qquad (5.1.146)$$

where $A(x,y)$ is defined by (5.1.141).

Remark 5.1.1 We note that, the upper bound on the inequality (5.1.145) when $p = 1$ and $h = 0$ was established by Wendroff [65]. For various generalizations of Wendroff's inequality, see [42, 507].

As an application of Theorem 5.1.39 in Qin [557], we next introduce the following two independent variable generalization of the Gollwitzer's inequality in [249].

Theorem 5.1.28 (The Bondge-Pachpatte Inequality [95]) *Let $\phi(s,t)$, $a(s,t)$, $b(s,t)$, and $u(s,t)$ be as defined in Theorem 5.1.39 in Qin [557]; let $H(r)$ be a positive, continuous, strictly increasing, convex, and sub-multiplicative function for all $r > 0$, $H(0) = 0$, $\lim_{r\to+\infty} H(r) = +\infty$; let $\alpha(s,t)$, $\beta(s,t)$ be positive* *continuous functions defined on* $\mathbb{R}_+ \times \mathbb{R}_+$, *and* $\alpha(s,t) + \beta(s,t) = 1$. *Suppose further that the inequality holds for all* $0 \le x \le s < +\infty$, $0 \le y \le t < +\infty$,

$$u(s,t) \ge \phi(x,y) - a(s,t)\, H^{-1}\Big(\int_x^s \int_y^t b(m,n)\, H(\phi(m,n))\, dm\, dn\Big). \tag{5.1.147}$$

Then for all $0 \le x \le s < +\infty$, $0 \le y \le t < +\infty$,

$$\begin{aligned} u(s,t) &\ge \alpha(s,t)\, H^{-1}(\alpha^{-1}(s,t))\, H(\phi(x,y)) \\ &\quad \times \exp\Big(- \beta(s,t)\, H(a(s,t))\beta^{-1}(s,t) \int_x^s \int_y^t b(m,n)\, dm\, dn\Big). \end{aligned} \tag{5.1.148}$$

Proof We may rewrite (5.1.147) as

$$\begin{aligned} \phi(x,y) &\le \alpha(s,t)\, u(s,t)\alpha^{-1}(s,t) + \beta(s,t)\, a(s,t)\beta^{-1}(s,t) \\ &\quad \times H^{-1}\left(\int_x^s \int_y^t b(m,n)\, H(\phi(m,n))\, dm\, dn\right). \end{aligned}$$

Since H is convex, sub-multiplicative and monotonic, we have

$$\begin{aligned} \alpha(s,t)\, H(u(s,t)\alpha^{-1}(s,t)) &\ge H(\phi(x,y)) - \beta(s,t)\, H(a(s,t)\beta^{-1}(s,t)) \\ &\quad \times \left(\int_x^s \int_y^t b(m,n)\, H(\phi(m,n))\, dm\, dn\right). \end{aligned}$$

Now applying Theorem 5.1.39 in Qin [557] to the above inequality, we may get the desired bound in (5.1.148). □

We next introduce the following two independent variable generalization of the integral inequality established by Langenhop [328].

Theorem 5.1.29 (The Langenhop Inequality [328]) *Let* $u(s,t)$, $a(s,t)$, *and* $b(s,t)$ *be as defined in Theorem 5.1.39 in Qin [557]; let* $W(r)$ *be a positive continuous, monotonic, non-decreasing function for all* $r > 0$, $W(0) = 0$, *and* $(\partial/\partial y)W(r(x,y)) = W_y(r(x,y)) \ge 0$; *and suppose further that the inequality holds for all* $0 \le x \le s < +\infty$, $0 \le y \le t < +\infty$,

$$u(s,t) \ge u(x,y) - a(s,t)\left(\int_x^s \int_y^t b(m,n)\, W(u(m,n))\, dm\, dn\right). \tag{5.1.149}$$

Then for all s_1, $t_1 \in \mathbb{R}_+$, $0 \le x \le s \le s_1$, $0 \le y \le t \le t_1$,

$$u(s,t) \ge \Omega^{-1}\left[\Omega(u(x,y)) - a(s,t)\Big(\int_x^s \int_y^t b(m,n)\, dm\, dn\Big)\right], \tag{5.1.150}$$

where

$$\Omega(r) = \int_{r_0}^{r} \frac{ds}{W(s)}, \; r \geq r_0 > 0, \tag{5.1.151}$$

and Ω^{-1} *is the inverse function of* Ω *such that for all* $0 \leq x \leq s < +\infty$, $0 \leq y \leq t < +\infty$,

$$\Omega(u(x,y)) - a(s,t)\Big(\int_y^t b(m,n)\, dm\, dn\Big) \in Dom\, (\Omega^{-1}).$$

Proof We may rewrite (5.1.149) as

$$u(x,y) \leq u(s,t) + a(s,t)\Big(\int_y^t b(m,n)\, W(u(m,n))\, dm\, dn\Big). \tag{5.1.152}$$

For fixed s and t in the interval $\mathbb{R}_+$, we define for all $0 \leq x \leq s,\ 0 \leq y \leq t$,

$$\begin{cases} r(x,y) = u(s,t) + a(s,t)\Big(\int_y^t b(m,n)\, W(u(m,n))\, dm\, dn\Big), \\ r(x,y) = r(s,y) = u(s,t). \end{cases} \tag{5.1.153}$$

Then from (5.1.153) it follows

$$r_{xy}(x,y) = a(s,t)\, b(x,y)\, W(u(x,y)),$$

which, by (5.1.152), implies

$$r_{xy}(x,y) \leq a(s,t)\, b(x,y)\, W(r(x,y)),$$

i.e.,

$$\frac{r_{xy}(x,y)}{W(r(x,y))} \leq a(s,t)\, b(x,y). \tag{5.1.154}$$

From (5.1.154), we derive that

$$\frac{W(r(x,y))\, r_{xy}(x,y)}{W^2(r(x,y))} \leq a(s,t)\, b(x,y) + \frac{W_y(r(x,y))(r_x(x,y)}{W^2(r(x,y))},$$

i.e.,

$$\frac{\partial}{\partial y}\Big(\frac{r_x(x,y)}{W(r(x,y))}\Big) \leq a(s,t)\, b(x,y).$$

Now integrating both sides of the above inequality with respect to y from y to t, we can get

$$\frac{r_x(x,t)}{W(\, r(x,t))} - \frac{r_x(x,y)}{W(\, r(x,y))} \leq a(s,t) \int_y^t b(x,n)\, dn. \tag{5.1.155}$$

Thus from (5.1.151) and (5.1.155) it follows

$$\Omega_x(\, r(x,y)) - \Omega_x(\, r(x,y)) \leq a(s,t) \int_y^t b(x,n)\, dn.$$

Integrating both sides of the above inequality with respect to x from x to s, we conclude

$$\Omega(\, r(x,y)) \leq \Omega(\, u(s,t)) + a(s,t)\Big(\int_x^s \int_y^t b(m,n)\, dm\, dn \Big),$$

which implies

$$\Omega(\, u(s,t)) \geq \Omega(\, u(x,y)) - a(s,t)\Big(\int_x^s \int_y^t b(m,n)\, dm\, dn \Big). \tag{5.1.156}$$

The desired bound in (5.1.150) follows from (5.1.156). The intervals of real numbers s and t are also obvious. □

We now apply Theorem 5.1.40 in Qin [557] to present the following two independent variable generalization of the integral inequality established by Pachpatte [450].

Theorem 5.1.30 (The Pachpatte Inequality [450]) *Let $\phi(s,t)$, $a(s,t)$, $b(s,t)$, $c(s,t)$, and $u(s,t)$ be as defined in Theorem 5.1.40 in Qin [557]; let $H(r), \alpha(s,t)$ and $\beta(s,t)$ be as defined in Theorem 5.1.29; and suppose further that the inequality holds for all $0 \leq x \leq s < +\infty$, $0 \leq y \leq t < +\infty$,*

$$u(s,t) \geq \phi(x,y) - a(s,t)\, H^{-1}\Bigg[\int_x^s \int_y^t b(m,n)\, H(\phi(m,n))\, dm\, dn$$
$$+ \int_x^s \int_y^t b(m,n) \Big(\int_m^S \int_n^t c(\xi,\zeta)\, H(\phi(\xi,\zeta))\, d\xi\, d\zeta \Big)\, dm\, dn \Bigg], \tag{5.1.157}$$

then for all $0 \le x \le s < +\infty,\ 0 \le y \le t < +\infty$,

$$u(s,t) \ge \alpha(s,t)\, H^{-1}\Bigg[\alpha^{-1}(s,t)\, H(\phi(x,y))\, \Big\{1 + \beta(s,t)\, H(\,a(s,t)\,\beta^{-1}(s,t)) \\ \times \int_x^s \int_y^t b(m,n) \exp\Bigg(\int_m^s \int_n^t [\beta(s,t)\, H(\,a(s,t)\beta^{-1}(s,t))\, b(\zeta,\zeta) + \,c(\xi,\zeta)] \\ \times d\xi\, d\zeta\Bigg)\, dm\, dn\Big\}^{-1}\Bigg]. \tag{5.1.158}$$

Proof We may rewrite (5.1.157) as

$$\phi(x,y) \le \alpha(s,t)\, u(s,t)\alpha^{-1}(s,t) \\ +\beta(s,t)\, a(s,t)\beta^{-1}(s,t)\, H^{-1}\Bigg[\int_x^s \int_y^t b(m,n)\, H(\phi(m,n))\, dm\, dn \\ + \int_x^s \int_y^t b(m,n)\Big(\int_m^s \int_n^t c(\xi,\zeta)\, H(\phi(\xi,\zeta))\, d\xi\, d\zeta\Big)\, dm\, dn\Bigg].$$

Since H is convex, sub-multiplicative and monotonic, we have

$$\alpha(s,t)\, H(\,u(s,t)\alpha^{-1}(s,t)) \\ \ge\ H(\phi(x,y)) - \beta(s,t)\, H(\,a(s,t)\beta^{-1}(s,t))\Bigg[\int_x^s \int_y^t b(m,n)\, H(\phi(m,n))\, dm\, dn\Bigg].$$

Now applying Theorem 5.1.40 in Qin [557] to the above inequality, we can obtain the desired bound in (5.1.158). □

Next, we introduce a two independent variable generalization of the integral inequality established by Pachpatte in [451].

Theorem 5.1.31 (The Pachpatte Inequality [451]) *Let* $u(s,t)$, $a(s,t)$, $b(s,t)$, *and* $c(s,t)$ *be as defined in Theorem 5.1.40 in Qin [557]; let* $G(r)$ *be a positive, continuous, strictly increasing, sub-additive and sub-multiplicative function for all* $r > 0$, $H(0) = 0$; *let* G^{-1} *denote the inverse function of* G; *and suppose further that the inequality for all* $0 \le x \le s < +\infty,\ 0 \le y \le t < +\infty$,

$$u(s,t) \ge\ u(x,y) -\ a(s,t)\, G^{-1}\Bigg[\int_x^s \int_y^t b(m,n)\, G(\,u(m,n))\, dm\, dn \\ + \int_x^s \int_y^t b(m,n)\Bigg(\int_m^s \int_n^t c(\xi,\zeta)\, G(\,u(\xi,\zeta))\, d\xi\, d\zeta\Bigg)\, dm\, dn\Bigg], \tag{5.1.159}$$

then for all $0 \le x \le s < +\infty,\ 0 \le y \le t < +\infty,$

$$u(s,t) \ge u(x,y)\, G^{-1}\Bigg(\Bigg[1 + G(a(s,t)) \int_x^s \int_y^t b(m,n) \times \exp\Bigg(\int_m^s \int_n^t [\, b(\xi,\zeta)\, G(a(s,t)) + c(\xi,\zeta)]\, d\xi\, d\zeta\Bigg)\, dm\, dn\Bigg]^{-1}\Bigg). \tag{5.1.160}$$

Proof In fact, we may rewrite (5.1.159) as

$$u(x,y) \le u(s,t) + a(s,t)\, G^{-1}\Bigg[\int_x^s \int_y^t b(m,n)\, G(u(m,n))\, dm\, dn + \int_x^s \int_y^t b(m,n)\Bigg(\int_m^s \int_n^t c(\xi,\zeta)\, G(u(\xi,\zeta))\, d\xi\, d\zeta\Bigg)\, dm\, dn\Bigg]. \tag{5.1.161}$$

Since G is sub-additive and sub-multiplicative, we infer from (5.1.161)

$$G(u(x,y)) \le G(u(s,t)) + G(a(s,t))\Bigg[\int_x^s \int_y^t b(m,n)\, G(u(m,n))\, dm\, dn + \int_x^s \int_y^t b(m,n)\Big(\int_m^s \int_n^t c(\xi,\zeta)\, G(u(\xi,\zeta))\, d\xi\, d\zeta\Big)\, dm\, dn\Bigg]. \tag{5.1.162}$$

Defining $r(x,y)$ by the right-hand side of (5.1.162) and by following a similar argument to that in the proof of Theorem 5.1.40 in Qin [557], with suitable modifications, we can obtain the desired bound in (5.1.160). □

5.1.4 Nonlinear Two-Dimensional Nonlinear Henry Inequalities

The following several results are concerned with nonlinear integral inequalities with weakly singular kernels for functions in two and n independent variables. These results , due to Medved' [387], are related to the well-known Gronwall-Bihari and Henry inequalities for functions in one variable and the Wendroff inequality for functions in two variables. A modification of the Ou-Yang-Pachpatte inequality and inequalities for functions in n independent variables are also introduced.

Henry proposed in his book [279] a method to estimate solutions of linear integral inequality with weakly singular kernel, which plays the same role in the geometric theory of parabolic partial differential equations (see, e.g., [267, 279, 590]) as the well-known Gronwall inequality in the theory of ordinary differential equations. In [384], a new method to estimate solutions for nonlinear integral inequalities with singular kernels of Bihari type is proposed. The resulting estimate formulas are similar to those for classical integral inequalities (see, e.g., [47, 82, 230, 374, 375, 382, 383, 500]). For instance, the estimate of solution of the Henry inequality is of exponential form in contrary to the Henry's estimate (see, e.g., [279, 590]) by an infinite series of a complicated form. The method has been applied in [385] in the proof of global existence of solutions and a stability theorem for a class of parabolic PDEs. In what follows, we shall introduce the method proposed in [384] to obtain an analogue of the Wendroff inequality (see, e.g., [47, 230, 374, 375]) for functions in two variables. Thandapani and Agarwal [643] proved some results concerning inequalities for functions in n independent variables. Applying the method of desingularization of weakly singular inequalities, we introduce a singular version of one of them, due to [387]. We note that the works [94, 100, 474, 643] contain many results on inequalities of Wendroff type and applying the desingularization method, we can prove their singular versions in a similar way. We also present an estimate of solutions of an analogue of the Ou-Yang inequality whose generalization for the nonlinear case has been given by Pachpatte [500].

We shall study an inequality of the type, for all $(x, y) \in [0, T) \times [0, T)$ $(0 < T \leq +\infty)$,

$$u(x, y) \leq a(x, y) + \int_0^x \int_0^y (x-s)^{\alpha-1}(y-t)^{\beta-1} F(s, t)\omega(u(s, t))dsdt, \tag{5.1.163}$$

where $\alpha > 0, \beta > 0$. Results on integral inequalities in two variables with regular kernels (i.e., with $\alpha = 1, \beta = 1$, F continuous) and $a(x, y)$ constant are contained in the books [47, 230, 374, 375].

Lemma 5.1.2 ([387]) *Let $\omega : \mathbb{R}_+ \mapsto \mathbb{R}$ be a non-negative, non-decreasing C^1 function, $a(x, y)$ be a non-negative C^2-function on $[0, T)^2$, $(0 < T \leq +\infty)$ such that*

$$\frac{\partial^2 a(x, y)}{\partial x \partial y} \geq 0, \quad \frac{\partial a(x, y)}{\partial y} \geq 0, \quad (or\ \frac{\partial a(x, y)}{\partial x} \geq 0)$$

on $[0, T)^2$ $(0 < T \leq +\infty)$. Let $k(x, y)$ be a continuous, non-negative C^2 function and $z(x, y)$ be a continuous, non-negative function on $[0, T)^2$ satisfying for all $(x, y) \in [0, T)^2$,

$$z(x, y) \leq a(x, y) + \int_0^x \int_0^y k(s, t)\omega(z(s, t)dsdt. \tag{5.1.164}$$

Then for all $(x, y) \in [0, T_1)^2$,

$$z(x, y) \leq \Omega^{-1}\Big[\Omega(a(x, y)) + \int_0^x \int_0^y k(s, t)dsdt\Big], \tag{5.1.165}$$

where $T_1 > 0$ *is such that the argument of* Ω^{-1} *in the above inequality belongs to Dom* (Ω^{-1}) *for all* $(x, y) \in [0, T_1)^2$.

Proof Let $V(x, y)$ be the right-hand side of (5.1.164). Then

$$\frac{\partial^2 V(x, y)}{\partial x \partial y} = \frac{\partial^2 a(x, y)}{\partial x \partial y} + k(x, y)\omega(z(x, y)), \tag{5.1.166}$$

$$\frac{\partial^2 \Omega(V(x, y))}{\partial x \partial y} = \Omega'(V(x, y))\frac{\partial^2 V(x, y)}{\partial x \partial y} + \Omega''(v(x, y))\frac{\partial V(x, y)}{\partial x}\frac{\partial V(x, y)}{\partial y}. \tag{5.1.167}$$

Since $\Omega'(V) = \frac{1}{\omega(V)}$ and $\Omega''(V) \leq 0$, we obtain from (5.1.166) and (5.1.167)

$$\begin{aligned} \frac{\partial^2 \Omega(V(x, y))}{\partial x \partial y} &\leq \frac{\partial^2 a(x, y)}{\partial x \partial y}\frac{1}{\omega(V)} + k(x, y) \\ &\leq q\frac{\partial^2 a(x, y)}{\partial x \partial y}\frac{1}{\omega(a(x, y))} + k(x, y). \end{aligned} \tag{5.1.168}$$

However,

$$\begin{aligned} \frac{\partial}{\partial x \partial y}\Omega(a(x, y)) &= \frac{\partial}{\partial x \partial y}\int_0^{a(x,y)} \frac{d\sigma}{\omega(\sigma)} = \frac{\partial}{\partial x}\left(\frac{\partial a(x, y)}{\partial y}\frac{1}{\omega(a(x, y))}\right) \\ &= \frac{\partial^2 a(x, y)}{\partial x \partial y}\frac{1}{\omega(a(x, y))} - \omega'(a(x, y))\frac{\partial a(x, y)}{\partial x}\frac{1}{\omega(a(x, y))^2} \\ &\geq \frac{\partial^2 a(x, y)}{\partial x \partial y}\frac{1}{\omega(a(x, y))} \geq \frac{\partial^2 a(x, y)}{\partial x \partial y}\frac{1}{\omega(a(x, y))}, \end{aligned}$$

i.e.,

$$\frac{\partial}{\partial x \partial y}\Omega(a(x, y)) \geq \frac{\partial^2 a(x, y)}{\partial x \partial y}\frac{1}{\omega(a(x, y))}. \tag{5.1.169}$$

If $\frac{\partial a}{\partial y} \geq 0$, then we can obtain (5.1.168) by estimating $\frac{\partial}{\partial x \partial y}\Omega(a(x, y))$. Thus we obtain from (5.1.167) and (5.1.168),

$$\frac{\partial^2 \Omega(V(x, y))}{\partial x \partial y} \leq \frac{\partial^2 \Omega(a(x, y))}{\partial x \partial y} + k(x, y),$$

which yields

$$\Omega(V(x,y)) \leq \Omega(a(x,y)) + \int_0^x \int_0^y k(s,t)dsdt.$$

From this inequality, we can obtain

$$z(x,y) \leq V(x,y) \leq \Omega^{-1}\left[\Omega(a(x,y)) + \int_0^x \int_0^y k(s,t)dsdt\right].$$

□

Remark 5.1.2 If $a(x, y)$ is a constant, then the lemma is a consequence of [375]. In this case, it suffices to assume that ω is only continuous.

Theorem 5.1.32 (The Medved' Inequality [387]) *Let $a(x, y)$ be a non-negative, C^2-function,*

$$\frac{\partial^2 a(x,y)}{\partial x \partial y} \geq 0, \ \frac{\partial a(x,y)}{\partial x} \geq 0 \ (or \ \frac{\partial a(x,y)}{\partial y} \geq 0) \tag{5.1.170}$$

on $[0,T)^2 = [0,T) \times [0,T)$ $(0 < T \leq +\infty)$, $u(x,y)$, $F(x,y)$ be continuous, non-negative functions on $[0,T)^2$ satisfying the inequality (5.1.163), where $\omega : \mathbb{R}_+ \to \mathbb{R}$ is a non-negative C^1-function. Then the following assertions holds:

(i) Suppose $\alpha > \frac{1}{2}, \beta > \frac{1}{2}$ and satisfies the condition (q) with $q = 2$ in Qin [557]. Then for all $(x, y) \in [0, T_1)^2 = [0, T_1) \times [0, T_1)$,

$$u(x,y) \leq e^{x+y}\left\{\Omega^{-1}[(2a^2(x,y)) + 2K\int_0^x \int_0^y F^2(s,t)R(s+t)dsdt]\right\}^{\frac{1}{2}} \tag{5.1.171}$$

where

$$K = \frac{\Gamma(2\beta-1)\Gamma(2\alpha-1)}{4^{\alpha+\beta-1}},$$

and Γ is the Gamma function, $\Omega(v) = \int_{v_0}^{v} \frac{dy}{w(y)}$, $v \geq v_0 > 0$, Ω^{-1} is the inverse of Ω and $T_1 > 0$ is such that the argument of Ω^{-1} in (5.1.171) belongs to Dom (Ω^{-1}) for all $(x, y) \in [0, T_1)^2$.

(ii) Suppose $\alpha = \beta = \frac{l}{z+1}$ for some real number $z \geq 1$ and ω satisfies the condition (q) in Qin [557] with $q = z + 2$. Then for all $(x, y) \in [0, T_2)^2$,

$$u(x,y) \leq e^{x+y}\left\{\Omega^{-1}\left[\Omega(2a^2(x,y)) + M_z\int_0^x \int_0^y F^q(s,t)R(s+t)dsdt\right]\right\}^{\frac{1}{q}}, \tag{5.1.172}$$

where

$$p = \frac{z+2}{z+1},\ M_z = \left(\frac{\Gamma(2-p\delta)}{p^{(1-p\delta)}}\right)^{\frac{2}{p}},\ \delta = 1-\beta = \frac{z}{z+1},$$

and $T_2 > 0$ *is such that the argument of* Ω^{-1} *belongs to Dom* (Ω^{-1}) *for all* $(x, y) \in [0, T_2)^2$.

Proof First let us prove the assertion (i). Using the Cauchy-Schwartz inequality, we derive from (5.1.163)

$$\begin{aligned} u(x,y) &\le a(x,y) + \int_0^x \int_0^y (x-s)^{\alpha-1} e^s (y-t)^{\beta-1} e^t [e^{-(s+t)} F(s,t)\omega(u(s,t))] ds dt \\ &\le a(x,y) + \left[\int_0^x \int_0^y (x-s)^{2\alpha-2} e^{2s} (y-t)^{2\beta-2} e^{2t} ds dt\right]^{\frac{1}{2}} \\ &\quad \times \left[\int_0^x \int_0^y e^{-(s+t)} F^2(s,t)\omega^2(u(s,t)) ds dt\right]^{\frac{1}{2}}. \end{aligned} \tag{5.1.173}$$

For the first integral in (5.1.173), we have

$$\begin{aligned} \int_0^x \int_0^y (x-s)^{2\alpha-2} e^{2s} (y-t)^{2\beta-2} e^{2t} ds dt &= e^{2(x+y)} \int_0^x \sigma^{2\alpha-2} e^{-2\sigma} \int_0^y \eta^{2\beta-2} e^{-2\eta} d\sigma d\eta \\ &= \frac{e^{2(x+y)}}{2^{2(\alpha+\beta)-2}} \int_0^x \sigma^{2\alpha-2} e^{-\sigma} \int_0^y \eta^{2\beta-2} e^{-\xi} d\sigma d\xi \\ &\le \frac{e^{2(x+y)}}{2^{2(\alpha+\beta)-2}} \Gamma(2\beta-1)\Gamma(2\alpha-1). \end{aligned}$$

Therefore we obtain from (5.1.173),

$$u(x,y) \le a(x,y) K^{\frac{1}{2}} \left[\int_0^x \int_0^y F(s,t)^2 e^{-2(s+t)} \omega(u(s,t))^2 ds dt\right]^{\frac{1}{2}}$$

where K is as in Theorem 5.1.32. Using the inequality (5.1.163) with $n = 2,\ r = 2$ and applying the condition (q) in Qin [557] with $q = 2$, we can obtain

$$v(x,y) \le \alpha(x,y) + 2K \int_0^x \int_0^y F^2(s,t) R(s+t) \omega(v(s,t)) ds dt \tag{5.1.174}$$

where

$$v(x,y) = \left(e^{-(x+y)} u(x,y)\right)^2,\ \alpha(x,y) = 2a(x,y)^2. \tag{5.1.175}$$

Applying Lemma 5.1.2 to the inequality (5.1.174), we obtain

$$v(x,y) \le \Omega^{-1}\left[\Omega(a(x,y)) + 2K\int_0^x\int_0^y F^2(s,t)R(s+t)dtds\right].$$

Using (5.1.175), we get

$$u(x,y) \le e^{x+y}\left\{\Omega^{-1}\left[\Omega(2a^2(x,y)) + 2K\int_0^x\int_0^y F^2(s,t)R(t+s)dtds\right]\right\}^{\frac{1}{2}}.$$

which gives us (5.1.171).

(ii) Let$p = \frac{z+2}{z+1}$, $q = z+2$. Then

$$u(x,y) \le a(x,y) + \left[\int_0^x\int_0^y (x-s)^{-p\delta}e^{ps}(y-t)^{-p\delta}e^{pt}dsdt\right]^{\frac{1}{p}}$$
$$\times\left[\int_0^x\int_0^y e^{-q(s+t)}F^q(s,t)\omega^q(u(s,t))dtds\right]^{\frac{1}{q}}.$$

We note that

$$\int_0^x\int_0^y (x-s)^{-p\delta}e^{ps}(y-t)^{-p\delta}e^{pt}dsdt = \int_0^x (x-s)^{-p\delta}e^{ps}\int_0^y (y-t)^{-p\delta}e^{-pt}dtds$$
$$\le \frac{e^y}{p^{1-p\delta}}\Gamma(1-p\delta)\int_0^y (x-s)^{-p\delta}e^p sds \le \frac{e^{x+y}}{p^{2(1-p\delta)}}\Gamma(1-p\delta)^2.$$

Thus we conclude

$$u(x,y) \le a(x,y) + Ke^{x+y}\left[\int_0^x\int_0^y F^q(s,t)R(t+s)\omega(e^{-q(s+t)}u^q(s,t))dsdt\right]^{\frac{1}{q}}$$

which yields

$$v(x,y) \le a(x,y) + 2K^2\int_0^x\int_0^y F^q(s,t)R(t+s)\omega(v(s,t))dsdt,$$

where

$$\alpha(x,y) = 2a^2(x,y),\ v(x,y) = (e^{-(x+y)}u(x,y))^q,\ M_z = \left(\frac{\Gamma(1-p\delta)}{p^{1-p\delta}}\right)^{\frac{2}{p}}$$

and this implies (5.1.172). If $\alpha \ne \beta, \alpha, \beta < \frac{1}{2}$, then there are some technical problems and we omit this case. □

Theorem 5.1.33 (The Medved' Inequality [387]) *Let functions a, F be as in Theorem 5.1.32 and $u(x, y)$ be a continuous, non-negative function on $[0, T)^2$ satisfying the inequality*

$$u(x,y) \le a(x,y) + \int_0^x \int_0^y (x-s)^{\beta-1} \times (y-t)^{\beta-1} s^{\gamma-1} t^{\gamma-1} F(s,t) u(s,t) ds dt, \tag{5.1.176}$$

where $\beta > 0,\ \gamma > 0$. Then the following assertions hold:

(i) *If $\beta > \frac{1}{2},\ \gamma > 1 - \frac{1}{2p}$, then for all $(x, y) \in [0, T)^2$,*

$$u(x,y) \le e^{x+y} \Phi(x,y), \tag{5.1.177}$$

where

$$\Phi(x,y) = 2^{1-\frac{1}{2q}} \exp\Big[\frac{4^{q-1}}{q} K^q L^q \int_0^x \int_0^y F^{2q}(s,t) e^{q(s+t)} ds dt\Big], \tag{5.1.178}$$

and K is as Theorem 5.1.32,

$$L = \Big(\frac{\Gamma((2\gamma-2)p+1)}{p^{(2\gamma-2)p+1}}\Big)^{\frac{2}{q}},\ p \ge 1,\ q \ge 1,\ \frac{1}{p} + \frac{1}{q} = 1.$$

(ii) *Let $\beta = \frac{1}{z+1}$ for some real number $z \ge 1, p = \frac{z+2}{z+1},\ q = z+1,\ \gamma > 1 - \frac{1}{kq}$, where $k > 1$. Then*

$$u(x,y) \le e^{x+y} \Psi(x,y), \tag{5.1.179}$$

where

$$\psi(x,y) = 2^{1-\frac{1}{rq}} a(x,y) \exp\left[\frac{Q^{rq}}{rq} \int_0^x \int_0^y e^{r(s+t)} F(s,t)^{rq} ds dt\right],$$

and $\gamma > 1$ is such that $1/K + 1/r = 1$, $Q = M_z P$, M_z is as in Theorem 5.1.32, $P = [\Gamma(sq(\gamma-1)+1)]^{2/k}$ and $\alpha = -z/(z+1) = \beta - 1$.

Proof

(i) From (i) and (5.1.176) it follows

$$\begin{aligned} u(x,y) &\le a(x,y) + \Big[\int_0^x \int_0^y (x-s)^{2\gamma-2} e^{2s} (y-t)^{2\beta-2} e^{2t} ds dt\Big]^{1/2} \\ &\quad \times \Big[\int_0^x \int_0^y s^{2\gamma-2} t^{2\gamma-2} F^2(s,t) (e^{-(s+t)} u(s,t))^2 ds dt\Big]^{1/2} \\ &\le a(x,y) + e^{x+y} K^{1/2} \times \Big[\int_0^x \int_0^y s^{2\gamma-2} t^{2\gamma-2} F^2(s,t) (e^{-(s+t)} u(s,t))^2 ds dt\Big]^{1/2}, \end{aligned}$$

where K is as in Theorem 5.1.32, which yields

$$\nu(x,y) \le c(x,y) + 2K \int_0^x \int_0^y s^{2\gamma-2} t^{2\gamma-2} F^2(s,t)\nu(s,t)dsdt \tag{5.1.180}$$

where

$$\nu(x,y) = (e^{-(x+y)}u(x,y))^2, \ \ c(x,y) = 2a^2(x,y). \tag{5.1.181}$$

From (5.1.180), we infer

$$\begin{aligned}\nu(x,y) \le c(x,y) + 2K\bigg[\int_0^x \int_0^y s^{(2\gamma-2)p} t^{(2\gamma-2)p} e^{-p(s+t)} dsdt\bigg]^{1/p} \\ \times\bigg[\int_0^x \int_0^y F^{2q}(s,t)e^{q(s+t)}\nu^q(s,t)dsdt\bigg]^{1/q}.\end{aligned} \tag{5.1.182}$$

For the first integral in (5.1.182), we have

$$\begin{aligned}\int_0^x \int_0^y s^{(2\gamma-2)p} t^{(2\gamma-2)p} e^{-p(s+t)} dsdt &= \frac{1}{(p^{(2\gamma-2)p+1})^2}\int_0^{px} \sigma^{(2\gamma-2)p} e^{-\sigma} \int_0^{py} \tau^{(2\gamma-2)p} e^{-\tau} d\tau d\sigma \\ &\le \left(\frac{\Gamma((2\gamma-2)p+1)}{p^{(2\gamma-2)p+1}}\right)^2\end{aligned}$$

which, combined with (5.1.182), implies

$$\nu(x,y) \le c(x,y) + 2KL \int_0^x \int_0^y F^{2q}(s,t)e^{q(s+t)}\nu^q(s,t)dsdt, \tag{5.1.183}$$

which yields

$$\nu^q(x,y) \le 2^{q-1}\Big[c(x,y)^q + 2^q K^q L^q \int_0^x \int_0^y F^{2q}(s,t)e^{q(s+t)}\nu^q(s,t)dsdt\Big]. \tag{5.1.184}$$

Also from the assumptions of theorem it follows that

$$\frac{\partial c(x,y)}{\partial x \partial y} \ge 0, \ \frac{\partial c(x,y)}{\partial x} \ge 0, \ (or \ \frac{\partial c(x,y)}{\partial y} \ge 0).$$

Thus from Lemma 5.1.2 and (5.1.184), we obtain

$$\nu(x,y) \le 2^{q-1}c^q(x,y)\exp\left[\frac{4^q}{2}K^q L^q \int_0^x \int_0^y F^{2q}(s,t)e^{q(s+t)}dsdt\right]$$

and the equalities (5.1.181) yield (5.1.184).

(ii) From the inequality (5.1.176), we infer

$$u(x,y) \le a(x,y) + \left[\int_0^x \int_0^y (x-s)^{-p\delta}(y-t)^{-p\delta} e^p(s+t)dsdt\right]^{1/p}$$

$$\times \left[\int_0^x \int_0^y s^{q(\gamma-1)} t^{q(\gamma-1)} e^{-q(s+t)} F^q(s,t) u^q(s,t)dsdt\right]^{1/q}$$

$$\le a(x,y) + e^{x+y}\left(\frac{\Gamma(1-ap)}{p^{1-ap}}\right)^{2/p} \left[\int_0^x \int_0^y s^{kq(\gamma-1)} t^{kq(\gamma-1)} e^{-(s+t)}\right]^{1/k}$$

$$\times \left[\int_0^x \int_0^y e^{r(s+t)} F^{rq}(s,t) e^{-(s+t)} u^{rq}(s,t)dsdt\right]^{1/rq}$$

$$\le a(x,y) + e^{x+y} Q \left[\int_0^x \int_0^y e^{r(s+t)} F^{rq}(s,t) e^{-(s+t)} u^{rq}(s,t)dsdt\right]^{1/rq},$$

where $Q = M_z P$, M_z is as in Theorem 5.1.32, P is as in theorem and r, k are as in the assertion (ii). The above inequality yields

$$v(x,y) \le 2^{qr-1}\left[a^{rq}(x,y) + Q^{rq}\int_0^x \int_0^y e^{r(s+t)} F^{rq}(s,t) v(s,t)dsdt\right],$$

where

$$v(x,y) = \left(e^{-(x+y)} u(x,y)\right)^{rq}. \tag{5.1.185}$$

Therefore we have

$$v(x,y) \le 2^{qr-1} a^{rq}(x,y) \exp\left[Q^{rq}\int_0^x \int_0^y e^{r(s+t)} F^{rq}(s,t)dsdt\right].$$

and using (5.1.185), we obtain (5.1.179). □

The next theorem is an analog of the Ou-Yang-Pachpatte inequality (see, e.g., [384, 500]).

Theorem 5.1.34 (The Medved' Inequality [387]) *Let* $T > 0$, F *and* ω *be as in Theorem 5.1.32 and* a *be a positive constant. Let* $u(x,y)$ *be a continuous, non-negative function on* $[0,T)^2$ *satisfying the inequality for all* $(x,y) \in [0,T)^2$,

$$u^2(x,y) \le a + \int_0^x \int_0^y (x-s)^{\alpha-l}(y-t)^{\beta-1} F(s,t)\omega(u(s,t))dsdt. \tag{5.1.186}$$

Then the following assertions hold:

(i) *Suppose* $\alpha > 1/2, \beta > 1/2$, *and* ω *satisfies the condition (q) in Qin [557] with* $q = 2$. *Then for all* $(x, y) \in [0, T)^2$,

$$u(x, y) \leq e^{x+y}\Phi(x, y), \tag{5.1.187}$$

where

$$\Phi(x, y) = \left[\Lambda^{-1}(\Lambda(2a^2) + 2K\int_0^x\int_0^y F^2(s, t)R(s + t)dsdt)\right]^{1/4},$$

and K *is the number from Theorem 5.1.32 and* $\Lambda(v) = \int_{v_0}^{v} d\delta/\omega(\sqrt{\sigma})$, $v \geq v_0 > 0$, $T_1 > 0$ *is such that the argument of* Λ^{-1} *belongs to Dom* (Λ^{-1}) *for all* $(x, y) \in [0, T_1)^2$.

(ii) *Suppose* $\alpha = \beta = 1/(z + 1)$ *for some real number* $z \geq 1$ *and let* $p = (z + 2)/(z + 1), q = z + 2$. *Assume that* ω *satisfies the condition (q) in Qin [557] with* $q = z + 2$. *Then for all* $(x, y) \in [0, T_2)^2$,

$$u(x, y) \leq e^{x+y}\Psi(x, y), \tag{5.1.188}$$

where for all $(x, y) \in [0, T_2)$,

$$\Psi(x, y) = \left[\Lambda^{-1}(\lambda(2^{q-1}a^q)) + 2^{q-1}M_z^q\int_0^x\int_0^y F^q(s, t)R(s + t)dsdt\right]^{1/2q},$$

and $T_2 > 0$ *is such that the argument of* Λ^{-1} *in the above inequality belongs to Dom* (Λ^{-1}) *for all* $(x, y) \in [0, T_2)^2$, M_z *is as in Theorem 5.1.32.*

Proof

(i) Using the Cauchy-Schwartz inequality and Jensen inequality, we can obtain

$$\begin{aligned}
u^2(x, y) &\leq a + \int_0^x\int_0^y (x - s)^{\alpha-1}(y - t)^{\beta-1}e^{s+t}\omega(u(x, y))dsdt \\
&\leq a + \left(\int_0^x\int_0^y (x - s)^{2\alpha-2}(y - t)^{2\beta-2}e^{2(s+t)}dsdt\right)^{1/2} \\
&\quad\times\left(\int_0^x\int_0^y F^2(s, t)R(s + t)\omega(e^{-2(s+t)}u^2(s, t))dsdt\right)^{1/2} \\
&\leq a + Ke^{-(x+y)}\left(\int_0^x\int_0^y F^2(s, t)R(s + t)\omega(e^{-2(s+t)}u^2(s, t))dsdt\right)^{1/2},
\end{aligned}$$

where K is as in Theorem 5.1.32. Applying the Jensen inequality similarly as in the proof of Theorem 5.1.32, we obtain

$$e^{-(x+y)}u^2(x,y) \le 2a^2 + 2K\int_0^x\int_0^y F^2(s,t)R(s+t)\omega(e^{-(s+t)}u(s,t))dsdt,$$

where K is as in Theorem 5.1.32, which yields

$$v^2(x,y) \le c + 2K\int_0^x\int_0^y F^2(s,t)R(s+t)\omega(v(s,t))dsdt, \tag{5.1.189}$$

where

$$v(x,y) = \left(e^{-(x+y)}u(x,y)\right)^2, \quad c = 2a^2. \tag{5.1.190}$$

Let $V(x,y)$ be the right-hand side of (5.1.189). Then

$$v(x,y) \le \sqrt{V(x,y)}, \quad \omega(v(x,t)) \le \omega(\sqrt{V(x,y)}). \tag{5.1.191}$$

We note that

$$\frac{\partial^2 V(x,y)}{\partial x\partial y} = 2KF^2(x,y)R(x+y)\omega(v(x,y)), \tag{5.1.192}$$

and

$$\begin{aligned}\frac{\partial^2}{\partial x\partial y}\int_0^{V(x,y)}\frac{dt}{\omega(\sqrt{t})} &= \frac{\partial}{\partial x}\left(\frac{1}{\omega(\sqrt{V(x,y)})}\right)\frac{\partial V}{\partial y}\\ &= \frac{\partial^2 V(x,y)}{\partial x\partial y}\frac{1}{\omega(\sqrt{V(x,y)})} - \frac{\partial V(x,y)}{\partial y}\frac{\partial V(x,y)}{\partial x}\frac{\omega'(\sqrt{V(x,y)})}{2\sqrt{V(x,y)}\omega(\sqrt{V(x,y)})^2}\\ &\le \frac{\partial^2 V(x,y)}{\partial x\partial y}\frac{1}{\omega(\sqrt{V(x,y)})},\end{aligned}$$

i.e.,

$$\frac{\partial^2}{\partial x\partial y}\Lambda(V(x,y)) \le \frac{\partial^2 V(x,y)}{\partial x\partial y}\frac{1}{\omega(\sqrt{V(x,y)})} \tag{5.1.193}$$

Thus from the above inequality and (5.1.192), we derive

$$\frac{\partial}{\partial x\partial y}\Lambda(V(x,y)) \le 2K\int_0^x\int_0^y F(s,t)^2R(s+t)dsdt,$$

and using (5.1.190)–(5.1.191), we can obtain the inequality (5.1.186).

(ii) Following the proof of the assertion (ii) of Theorem 5.1.32, we can show that

$$w^2(x,y) \leq \alpha + 2K^2 \int_0^x \int_0^y F^q(s,t)R(s+t)\omega(w(s,t))dsdt \tag{5.1.194}$$

where

$$\alpha = 2a^2, \quad w(x,y) = (e^{-(x+y)}u(x,y))^q.$$

Applying the same procedure to (5.1.194) as we have used in the proof of the assertion (ii) as well as that one from the proof of (ii) of Theorem 5.1.32, we can prove the inequality (5.1.188). □

Next, we shall introduce the results due to Shastri and Kasture [599]. To this end, we consider a Bihari-type inequality of the form, for all $x \geq 0, y \geq 0$

$$\phi(x,y) \leq a(x,y) + \int_0^x \int_0^y c(s,t)W[\phi(s,t)]dsdt. \tag{5.1.195}$$

Recall that there are many works dealing with (5.1.195) or its one dimensional analogue or even with more general inequalities in one or two dimensions [59, 197, 198, 200, 451, 642, 643, 702, 703, 707]. However, in all earlier work, particularly in two dimensions, one or more of the following stringent conditions are imposed on W:
(1) W is sub-additive;
(2) W is sub-multiplicative;
(3) W is convex;
(4) $\frac{1}{v}W[u] \leq W[u/v]$, for all $u \geq 0, v > 0$; (5) There exists a function ϕ continuous on $[0,+\infty)$ such that for all $\alpha > 0,\ u, v \geq 0$,

$$W[u+\alpha v] \leq W[u] + \phi(\alpha)W[v].$$

The next result is to relax such conditions on W, which limit the class of admissible nonlinear functions W in (5.1.195). Other conditions assumed by earlier authors for the study of (5.1.195) are:
(6) The functions ϕ, a and c are real-valued conditions and non-negative for all $x \geq 0, y \geq 0$;
(7) $W[u]$ is a real-valued positive continuous non-decreasing, function for all $u > 0$.

We retain these assumptions (6) and (7) and add an assumption on the function $a(x,y)$, not taken by earlier authors.
(8) The derivatives $a_x(x,y), a_y(x,y)$ and $a_{xy}(x,y)$ of the function $a(x,y)$ exist, and are continuous for all $x \geq 0, y \geq 0$ and $a_x(x,y) \geq 0, a_y(x,y) \geq 0$ while $a_{xy}(x,y) \leq 0$ there.

We next introduce point-wise estimates for ϕ satisfying (5.1.195), due to [599], subject to conditions (6), (7), (8) above. The importance of such results in the study of the qualitative behaviour of the solutions of differential and integral equations including the existence via monotone methods [477], uniqueness and continuous dependence on initial conditions [703], and stability [655], is well-illustrated by earlier authors.

Theorem 5.1.35 (The Shastri-Kasture Inequality [599]) *Assume (5.1.195) hold subject to conditions (6), (7) and (8). Then for all* $x \geq 0, y \geq 0$,

$$\phi(x,y) \leq \Omega^{-1}[\Omega\{a(0,y)\} + \Omega\{a(x,0)\} - \Omega\{a(0,0)\} + \int_0^x \int_0^y c(s,t)dsdt], \tag{5.1.196}$$

where

$$\Omega(r) = \int_{r_0}^{r} \frac{ds}{W(s)}, \tag{5.1.197}$$

and Ω^{-1} *is the inverse function of* Ω, *and it is assumed that the quantity in square bracket on the right-hand side of (5.1.196) belongs to the domain of* Ω^{-1}.

Proof Define

$$u(x,y) = a(x,y) + \int_0^x \int_0^y c(s,t)W[\phi(s,t)]dsdt$$

so that

$$u(0,y) = a(0,y), u_x(x,0) = a_x(x,0), u_y(0,y) = a_y(0,y).$$

Then

$$\phi(x,y) \leq u(x,y) \tag{5.1.198}$$

and

$$u_x(x,y) = a_x(x,y) + \int_0^y c(x,t)W[\phi(x,t)]dt. \tag{5.1.199}$$

Since $a_{xy}(x,y) \leq 0$, we have

$$\begin{aligned} u_{xy}(x,y) &\leq c(x,y)W[\phi(x,y)] \\ &\leq c(x,y)W[u(x,y)]. \end{aligned}$$

Thus

$$\frac{u_{xy}(x,y)}{W[u(x,y)]} \leq c(x,y).$$

Since u_x, u_y are non-negative, we conclude

$$\frac{W[u(x,y)]u_{xy}(x,y)}{W^2[u(x,y)]} \le c(x,y) + \frac{u_x(x,y)u_y(x,y)W'[u(x,y)]}{W^2[u(x,y)]}.$$

Now keeping x fixed, setting $y = t$ and integrating with respect to t from 0 to y, we get

$$\frac{u_x(x,y)}{W[u(x,y)]} \le \frac{a_x(x,0)}{W[a(x,0)]} + \int_0^y c(x,t)dt.$$

Using the definition of Ω in (5.1.197), we obtain

$$\Omega_x[u(x,y)] \le \frac{a_x(x,0)}{W[a(x,0)]} + \int_0^y c(x,t)dt.$$

Keeping y fixed, setting $x = s$ and integrating with respect to s from 0 to x, we obtain

$$\begin{aligned}\Omega[u(x,y)] \le \Omega[a(0,y)] + \Omega[a(x,0)] - \Omega[a(0,0)]\\ + \int_0^x \int_0^y c(s,t)dsdt.\end{aligned}$$

Since Ω is strictly increasing, so is Ω^{-1}. Therefore

$$\begin{aligned}u(x,y) \le \Omega^{-1}[\Omega[a(x,0)] + \Omega[a(0,y)] - \Omega[a(0,0)]\\ + \int_0^x \int_0^y c(s,t)dsdt].\end{aligned}$$

Substituting this bound on $u(x,y)$ in (5.1.198), we can get the desired estimate (5.1.196). □

Note that there is a large class of functions satisfying the condition (8). For example, if F, f, g are continuously differentiable non-negative functions on $(0,+\infty)$ such that $F'(\xi), f'(\xi), g'(\xi) \ge 0$, while $F''(\xi) \le 0$ for all $\xi \in (0,+\infty)$, then the composite function

$$a(x,y) \le F[f(x) + g(y)] \tag{5.1.200}$$

satisfies the conditions (8).

$$\begin{cases} F(\xi) = K\xi^{\alpha}, 0 \le \alpha \le 1, \xi \in (0,+\infty), \\ F(\xi) = K\log(1+\xi) \end{cases}$$

where $K > 0$ is a constant, are some examples of F, while the set of real-valued functions on $[0, +\infty)$ is rich with non-negative non-decreasing differentiable function $[0, +\infty)$. Thus a function having the form (5.1.200) in general, and a constant function in particular, can be always found to majorise the given free term $a(x, y)$ on a closed and bounded sub-domain of the first quadrant.

The characteristic initial value problem for a hyperbolic differential equation [658]

$$\phi_{xy}(x, y) = c(x, y)W[\phi(x, y)] \tag{5.1.201}$$

when converted to an integral equation, generates a free term of the form $[f(x) + g(y)]$, which is of the form (5.1.200) if f and g are positive and non-decreasing, or else the free term can be majored easily by a function of the form (5.1.200).

Again the functions of the form $c(x, y)W[\phi(x, y)]$ appearing on the right-hand side of (5.1.201) can be used as majoring functions in the study of the qualitative theory of nonlinear differential and integral equations [658]. Thus Theorem 5.1.35 has a wide range of applicability.

We next introduce the following inequalities in two independent variables similar to those given in Theorems 1.2.7 and 1.2.8.

Theorem 5.1.36 (The Pachpatte Inequality [496]) *Let $F(x, y)$ and $g(x, y)$ be real-valued non-negative continuous functions defined for all $x, y \in \mathbb{R}_+$, and let $p > 1$ be a constant. If for all $x, y \in \mathbb{R}_+$,*

$$F^p(x, y) \leq c + B[x, y, gF], \tag{5.1.202}$$

where $c \geq 0$ is a constant, then for all $x, y \in \mathbb{R}_+$,

$$F(x, y) \leq \left[c^{\frac{(p-1)}{p}} + (\frac{p-1}{p})B[x, y, g]\right]^{\frac{1}{p-1}}. \tag{5.1.203}$$

Proof In order to establish the inequality (5.2.203), we first assume that $c > 0$ and define a function $z(x, y)$ by

$$z(x, y) = c + B[x, y, gF]. \tag{5.1.204}$$

From (5.2.204) it is easy to observe that

$$D_2^m D_1^n z(x, y) = g(x, y)F(x, y). \tag{5.1.205}$$

Using the fact that $F(x, y) \leq \sqrt[p]{z(x, y)}$ in (5.2.205), we have

$$D_2^m D_1^n z(x, y) \leq g(x, y)\sqrt[p]{z(x, y)}. \tag{5.1.206}$$

From (5.2.206) and the facts that $z(x,y)$ is positive, and $D_2[\sqrt[p]{z(x,y)}]$ and $D_2^{m-1}D_1^n z(x,y)$ are non-negative for all $x, y \in \mathbb{R}_+$, we observe that

$$\frac{D_2^m D_1^n z(x,y)}{\sqrt[p]{z(x,y)}} \leq g(x,y) + \frac{D_2[\sqrt[p]{z(x,y)}]D_2^{m-1}D_1^n z(x,y)}{[\sqrt[p]{z(x,y)}]^2},$$

i.e.,

$$D_2\left(\frac{D_2^{m-1}D_1^n z(x,y)}{\sqrt[p]{z(x,y)}}\right) \leq g(x,y). \tag{5.1.207}$$

Keeping x fixed in (5.2.207), setting $y = t$ and then, by integrating with respect to t from y and using the fact that $D_2^{m-1}D_1^n z(x,0) = 0$, we have

$$\frac{D_2^{m-1}D_1^n z(x,y)}{\sqrt[p]{z(x,y)}} \leq \int_0^y g(x,t)dt. \tag{5.1.208}$$

Again as above, from (5.2.208) and the facts that $z(x,y)$ is positive and $D_2(\sqrt[p]{z(x,y)})$ and $D_2^{m-1}D_1^n z(x,y)$ are non-negative for $x, y \in \mathbb{R}_+$, we observe that

$$D_2\left(\frac{D_2^{m-1}D_1^n z(x,y)}{\sqrt[p]{z(x,y)}}\right) \leq \int_0^y g(x,t)dt. \tag{5.1.209}$$

By keeping x fixed in (5.2.209), setting $y = t_1$, then integrating with respect to t_1 from 0 to y, and using the fact that $D_2^{m-1}D_1^n z(x,0) = 0$, we have

$$D_2\left(\frac{D_2^{m-1}D_1^n z(x,y)}{\sqrt[p]{z(x,y)}}\right) \leq \int_0^y \int_0^{t_1} g(x,t)dtdt_1.$$

Computing in this way, we obtain

$$\frac{D_1^n z(x,y)}{\sqrt[p]{z(x,y)}} \leq \int_0^y \int_0^{t_{m-1}} \cdots \int_0^{t_1} g(x,t)dtdt_1 \cdots dt_{m-1}. \tag{5.1.210}$$

From (5.2.210) and the facts that $z(x,y)$ is positive and $D_1[\sqrt[p]{z(x,y)}]$ and $D_1^{n-1}z(x,y)$ are non-negative for all $x, y \in \mathbb{R}_+$, we observe that

$$\frac{D_1^n z(x,y)}{\sqrt[p]{z(t)}} \leq \int_0^y \int_0^{t_{m-1}} \cdots \int_0^{t_1} g(x,t)dtdt_1 \cdots dt_{m-1} + \frac{D_1[\sqrt[p]{z(x,y)}]D_1^{n-1}z(x,y)}{[\sqrt[p]{z(x,y)}]^2}. \tag{5.1.211}$$

Now keeping y fixed in (5.2.211), setting $x = s$, then integrating with respect to s from 0 to x, and using the fact that $D_1^{n-1}z(0, y) = 0$, we have

$$\frac{D_1^n z(x, y)}{\sqrt[p]{z(x, y)}} \leq \int_0^x \int_0^y \int_0^{t_{m-1}} \cdots \int_0^{t_1} g(s, t)dtdt_1 \cdots dt_{m-1}ds.$$

Computing in this way, we obtain

$$\frac{D_1^n z(x, y)}{\sqrt[p]{z(x, y)}} \leq \int_0^x \int_0^{s_{n-2}} \cdots \int_0^{s_1} \int_0^y \int_0^{t_{m-1}} \cdots$$
$$\times \int_0^{t_1} g(s, t)dtdt_1 \cdots dt_{m-1}dsds_1 \cdots ds_{n-2}. \qquad (5.1.212)$$

Now by keeping y fixed in (5.2.212), setting $x = s_{n-1}$, then integrating with respect to s_{n-1} from 0 to x, and using the fact that $z(0, y) = c$, we have

$$[\sqrt[p]{z(x, y)}]^{p-1} - [\sqrt[p]{c}]^{p-1} \leq (\frac{p-1}{p})B[x, y, g]. \qquad (5.1.213)$$

From (5.2.213) and using the fact $F(x, y) \leq \sqrt[p]{z(x, y)}$, we observe

$$F(x, y) \leq \left[c^{(p-1)/p} + \frac{p-1}{p}B[x, y, g]\right]^{1/(p-1)}. \qquad (5.1.214)$$

The proof of the case when $c = 0$ can be completed by following the arguments in the proof of Theorem 1.2.7 given above, and hence the proof is complete. □

Theorem 5.1.37 (The Pachpatte Inequality [496]) *Let $u(x, y) \geq 0, v(x, y) \geq 0, h_i(x, y) > 0$ for $i = 1, 2, 3, 4$ be real-valued continuous functions defined for all $x, y \in \mathbb{R}_+$ and let $p > 1$ be a constant. If c_1, c_2 and μ are non-negative constants such that for all $x, y \in \mathbb{R}_+$,*

$$u^p(x, y) \leq c_1 + B[x, y, h_1 u] + B[x, y, h_2\overline{v}], \qquad (5.1.215)$$
$$v^p(x, y) \leq c_2 + B[x, y, h_3\overline{u}] + B[x, y, h_4 v], \qquad (5.1.216)$$

where $\overline{u}(x, y) = \exp(-p\mu(x + y))u(x, y)$ and $\overline{v}(x, y) = \exp(p\mu(x + y))v(x, y)$ for all $x, y \in \mathbb{R}_+$, then for all $x, y \in \mathbb{R}_+$,

$$u(x, y) \leq \exp(\mu(x + y))$$
$$\times \left[\{2^{p-1}(c_1 + c_2)\}^{\frac{(p-1)}{p}} + 2^{p-1}(\frac{p-1}{p})A[x, y, h]\right]^{\frac{1}{p-1}}, \qquad (5.1.217)$$

$$v(x, y) \leq \left[\{2^{p-1}(c_1 + c_2)\}^{\frac{(p-1)}{p}} + 2^{p-1}(\frac{p-1}{p})B[x, y, h]\right]^{\frac{1}{p-1}}, \qquad (5.1.218)$$

where for all $t \in \mathbb{R}_+$,

$$h(x,y) = \max\{[h_1(x,y) + h_3(x,y)], [h_2(x,y) + h_4(x,y)]\}. \tag{5.1.219}$$

Proof The proof follows by the same arguments as those given in the proof of Theorem 1.2.8 above and by applying Theorem 5.1.36 with suitable modification, and hence we omit here. □

Let $I = [t_0, T)$, $J_1 = [x_0, X)$ and $J_2 = [y_0, Y)$ be given subsets of $\mathbb{R}$, $\Delta = J_1 \times J_2$.

Theorem 5.1.38 (The Pachpatte Inequality [520]) *Let* $u(x,y), a(x,y) \in C(\Delta, \mathbb{R}_+)$, $b(x,y,s,t) \in C(\Delta^2, \mathbb{R}_+)$, *for all* $x_0 \le s \le x \le X, y_0 \le t \le y \le Y$, $\alpha(x) \in C^1(J_1, J_1)$, $\beta(y) \in C^1(J_2, J_2)$ *be non-decreasing with* $\alpha(x) \le x$ *on* J_1, $\beta(y) \le y$ *on* J_2 *and* $k \le 0$ *be a constant.*

(b_1) *If for all* $(x,y) \in \Delta$,

$$u(x,y) \le k + \int_{\alpha(x_0)}^{\alpha(x)} \int_{\beta(y_0)}^{\beta(y)} [a(s,t)u(s,t) + \int_{\alpha(x_0)}^{s} \int_{\beta(y_0)}^{t} b(s,t,\sigma,\eta)u(\sigma,\eta)d\eta d\sigma]dtds, \tag{5.1.220}$$

then for all $(x,y) \in \Delta$,

$$u(x,y) \le k\exp(A(x,y)), \tag{5.1.221}$$

where for all $(x,y) \in \Delta$,

$$A(x,y) = \int_{\alpha(x_0)}^{\alpha(x)} \int_{\beta(y_0)}^{\beta(y)} [a(s,t) + \int_{\alpha(x_0)}^{s} \int_{\beta(y_0)}^{t} b(s,t,\sigma,\eta)d\eta d\sigma]dtds. \tag{5.1.222}$$

(b_2) *Let* g *be as part* (a_2) *of Theorem 1.2.36 in Qin [557]. If for all* $(x,y) \in \Delta$,

$$u(x,y) \le k + \int_{\alpha(x_0)}^{\alpha(x)} \int_{\beta(y_0)}^{\beta(y)} [a(s,t)g(u(s,t)) + \int_{\alpha(x_0)}^{s} \int_{\beta(y_0)}^{t} b(s,t,\sigma,\eta)g(u(\sigma,\eta))d\eta d\sigma]dtds, \tag{5.1.223}$$

then for all $x_0 \le x \le x_1, y_0 \le y \le y_1$,

$$u(x,y) \le G^{-1}[G(k) + A(x,y)] \tag{5.1.224}$$

where $A(x, y)$ is defined by (5.1.222), G, G^{-1} are as defined part (a_2) of Theorem 1.2.36 in Qin [557] and $x_1 \in J_1, y_1 \in J_2$ are chosen so that

$$G(k) + A(x, y) \in Dom\,(G^1),$$

for all x and y lying in $[x_0, x_1]$ and $[y_0, y_1]$ respectively.

Proof (b_1) Let $k > 0$ and define a function $z(x, y)$ by the right-hand side of (5.1.220). Then $z(x, y) > 0, z(x_0, y) = z(x, y0) = k, u(x, y) \le z(x, y)$ and

$$\begin{aligned} D_1 z(x, y) &= [\int_{\beta(y_0)}^{\beta(y)} [a(\alpha(x), t)u(\alpha(x), t) + \int_{\alpha(x_0)}^{\alpha(x)} \int_{\beta(y_0)}^{t} b(\alpha(x), t, \sigma, \eta)u(\sigma, \eta)d\eta d\sigma]dt]\alpha'(x) \\ &\le [\int_{\beta(y_0)}^{\beta(y)} [a(\alpha(x), t)z(\alpha(x), t) \\ &\quad + \int_{\alpha(x_0)}^{\alpha(x)} \int_{\beta(y_0)}^{t} b(\alpha(x), t, \sigma, \eta)z(\sigma, \eta)d\eta d\sigma]dt]\alpha'(x). \end{aligned} \tag{5.1.225}$$

From (5.1.225) it is easy to observe that

$$\frac{D_1 z(x, y)}{z(x, y)} \le [\int_{\beta(y_0)}^{\beta(y)} [a(\alpha(x), t) + \int_{\alpha(x_0)}^{\alpha(x)} \int_{\beta(y_0)}^{t} b(\alpha(x), t, \sigma, \eta)d\eta d\sigma]dt]\alpha'(x). \tag{5.1.226}$$

Keeping y fixed in (5.1.226), setting $x = \epsilon$ and integrating it with respect to ϵ from x_0 to x and making the change of variables, we get

$$z(x, y) \le k \exp(A(x, y)). \tag{5.1.227}$$

Using (5.1.227) in $u(x, y) \le z(x, y)$, we get the required inequality in (5.1.221). The case $k \ge 0$ follows as mentioned in the proof of (a_1) of Theorem 1.2.36 in Qin [557].

(b_2) The proof can be completed by following the proof of (a_2) of Theorem 1.2.36 in Qin [557] and closely looking at the proof of (b_1) . Here we omit the details. □

Theorem 5.1.39 (The Pachpatte Inequality [518]) *Let $a, b \in C(\Delta, \mathbb{R}_+), \alpha \in C^1(J_1, J_1)$, $\beta \in C^1(J_2, J_2)$ be non-decreasing with $\alpha(x) \le x$ on J_1, $\beta(y) \le y$ on J_2 and $k \ge 0$, $c \ge 1$, and $p > 1$ are constants. Let $g_i \in C(\mathbb{R}_+, \mathbb{R}_+)$ be non-decreasing functions with $g_i(u) > 0$ for all $u > 0$.*

(d_1) If $u \in C(\Delta, \mathbb{R}_+)$ and for all $(x, y) \in \Delta$,

$$u(x, y) \le k + \int_{x_0}^{x} \int_{y_0}^{y} a(s, t)g_1(u(s, t))dtds + \int_{\alpha(x_0)}^{\alpha(x)} \int_{\beta(y_0)}^{\beta(y)} b(s, t)g_2(u(s, t))dtds, \tag{5.1.228}$$

then for all $x_0 \le x \le x_1, y_0 \le y \le y_1$,

(*i*) *in case* $g_2(u) \le g_1(u)$,

$$u(x,y) \le G_1^{-1}[G_1(k) + A(x,y) + B(x,y)], \tag{5.1.229}$$

(*ii*) *in case* $g_1(u) \le g_2(u)$,

$$u(x,y) \le G_2^{-1}[G_2(k) + A(x,y) + B(x,y)] \tag{5.1.230}$$

where G_i, G_i^{-1} *are as part* (b_1) *of Theorem 1.2.30 in Qin [557], and*

$$A(x,y) = \int_{x_0}^{x} \int_{y_0}^{y} a(s,t)dtds, \tag{5.1.231}$$

$$B(x,y) = \int_{\alpha(x_0)}^{\alpha(x)} \int_{\beta(y_0)}^{\beta(y)} b(s,t)dtds, \tag{5.1.232}$$

and $x_1 \in J_1, y_1 \in J_2$ *are chosen so that for* $i = 1, 2$,

$$G_i(k) + A(x,y) + B(x,y) \in Dom\,(G_i^{-1}),$$

for all x and y lying in $[x_0, x_1]$ *and* $[y_0, y_1]$, *respectively.*

(d_2) *If* $u \in C(\Delta, \mathbb{R}_1)$ *and for all* $(x,y) \in \Delta$,

$$u(x,y) \le c + \int_{x_0}^{x} \int_{y_0}^{y} a(s,t)u(s,t)g_1(\log u(s,t))dtds$$
$$+ \int_{\alpha(x_0)}^{\alpha(x)} \int_{\beta(y_0)}^{\beta(y)} b(s,t)u(s,t)g_2(\log u(s,t))dtds, \tag{5.1.233}$$

then for all $x_0 \le x \le x_2, y_0 \le y \le y_2$,

(*i*) *in case* $g_2(u) \le g_1(u)$,

$$u(x,y) \le \exp\left(G_1^{-1}[G_1(\log c) + A(x,y) + B(x,y)]\right), \tag{5.1.234}$$

(*ii*) *in case* $g_1(u) \le g_2(u)$,

$$u(x,y) \le \exp\left(G_2^{-1}[G_2(\log c) + A(x,y) + B(x,y)]\right); \tag{5.1.235}$$

where $G_i, G_i^{-1}, A(x,y), B(x,y)$ *are as in* (d_1) *and* $x_2 \in J_1, y_2 \in J_2$ *are chosen so that for* $i = 1, 2$,

$$G_i(\log c) + A(x,y) + B(x,y) \in Dom\,(G_i^{-1}),$$

for all x and y lying in $[x_0, x_2]$ *and* $[y_0, y_2]$*, respectively.*

(d_3) *If* $u \in C(\Delta, \mathbb{R}_+)$ *and for all* $(x, y) \in \Delta$,

$$u^p(x,y) \le k + \int_{x_0}^{x}\int_{y_0}^{y} a(s,t)g_1(u(s,t))dtds + \int_{\alpha(x_0)}^{\alpha(x)}\int_{\beta(y_0)}^{\beta(y)} b(s,t)g_2(u(s,t))dtds, \quad (5.1.236)$$

then for all $x_0 \le x \le x_3, y_0 \le y \le y_3$,

(*i*) *in case* $g_2(u) \le g_1(u)$,

$$u(x,y) \le \left[H_1^{-1}[H_1(k) + A(x,y) + B(x,y)]\right]^{1/p}, \quad (5.1.237)$$

(*ii*) *in case* $g_1(u) \le g_2(u)$,

$$u(x,y) \le \left[H_2^{-1}[H_2(k) + A(x,y) + B(x,y)]\right]^{1/p}, \quad (5.1.238)$$

where H_i, H_i^{-1} *are as part* (b_3) *of Theorem 1.2.30 in Qin [557], and* $A(x,y), B(x,y)$ *are defined by (5.1.231) and (5.1.232), and* $x_3 \in J_1, y_3 \in J_2$ *are chosen so that for* $i = 1, 2$,

$$H_i(k) + A(x,y) + B(x,y) \in Dom\ (H_i^{-1}),$$

for all x and y lying in $[x_0, x_3]$ *and* $[y_0, y_3]$*, respectively.*

Proof Since the proofs resemble one another, we only give the details for (d_3), the proofs of the remaining inequalities can be completed in the same manner.

(d_3) Let $k > 0$ and define a function $z(x,y)$ by the right-hand side of (5.1.236). Then $z(x,y) > 0, z(x_0,y) = z(x,y_0) = k$, and $u(x,y) \le (z(x,y))^{1/p}$, and

$$\begin{aligned} D_1 z(x,y) &= \int_{y_0}^{y} a(x,t)g_1(u(x,t))dt + \left(\int_{\beta(y_0)}^{\beta(y)} b(\alpha(x),t)g_2(u(\alpha(x),t))dt\right)\alpha'(x) \\ &\le \int_{y_0}^{y} a(x,t)g_1(\{z(x,t)\}^{1/p})dt + \left(\int_{\beta(y_0)}^{\beta(y)} b(\alpha(x),t)g_2(\{z(\alpha(x),t)\}^{1/p})dt\right)\alpha'(x) \\ &\le g_1(\{z(x,y)\}^{1/p})\int_{y_0}^{y} a(x,t)dt + g_2(\{z(\alpha(x),\beta(y))\}^{1/p})\left(\int_{\beta(y_0)}^{\beta(y)} b(\alpha(x),t)dt\right)\alpha'(x). \end{aligned} \quad (5.1.239)$$

(i) When $g_2(u) \leq g_1(u)$, then from (5.1.239), we derive that

$$\frac{D_1 z(x,y)}{g_1(\{z(x,y)\}^{1/p}} \leq \int_{y_0}^{y} a(x,t)dt + (\int_{\beta(y_0)}^{\beta(y)} b(\alpha(x),t)dt)\alpha'(x). \tag{5.1.240}$$

From (1.1.240) of Theorem 1.2.30 in Qin [557] and (5.1.240), we infer

$$\begin{aligned} D_1 H_1(z(x,y)) &= \frac{D_1 z(x,y)}{g_1(\{z(x,y)\}^{1/p}} \\ &\leq \int_{y_0}^{y} a(x,t)dt + \left(\int_{\beta(y_0)}^{\beta(y)} b(\alpha(x),t)dt\right)\alpha'(x). \end{aligned} \tag{5.1.241}$$

Keeping y fixed in (5.1.241), setting $x = \sigma$, then integrating with respect to σ from x_0 to $x, x \in J_1$, and making the change of variable, we conclude

$$H_1(z(x,y)) \leq H_1(k) + A(x,y) + B(x,y). \tag{5.1.242}$$

Using the bound on $z(x,y)$ from (5.1.242) in $u(x,y) \leq \{z(x,y)\}^{1/p}$, we get (5.1.237). The case $k \geq 0$ follows as mentioned in the proof of part (a_1) of Theorem 1.2.29 in Qin [557]. The sub-domain for x, y is obvious. The proof of the case when $g_1(u) \leq g_2(u)$ can be completed similarly. □

Theorem 5.1.40 (The Pachpatte Inequality [521]) *Let $u, a, b_i \in C(\Delta, \mathbb{R}_+)$ and $\alpha_i \in C^1(J_1, J_1)$, $\beta_i \in C^1(J_2, J_2)$ be non-decreasing with $\alpha_i(x) \leq x$ on J_1, $\beta_i(y) \leq y$ on J_2 for $i = 1, \cdots, n$ and $k \geq 0$ be a constant. Let $g \in C(\mathbb{R}_+, \mathbb{R}_+)$ be non-decreasing and sub-multiplicative function with $g(u) > 0$ for all $u > 0$.*

(B1) If for all $x \in J_1, y \in J_2$,

$$u(x,y) \leq k + \int_{x_0}^{x} a(s,y)u(s,y)ds + \sum_{i=1}^{n} \int_{\alpha_i(x_0)}^{\alpha_i(x)} \int_{\beta_i(y_0)}^{\beta_i(y)} b_i(s,t)g(u(s,t))dtds, \tag{5.1.243}$$

then for all $x_0 \leq x \leq x_1, y_0 \leq y \leq y_1; x, x_1 \in J_1, y, y_1 \in J_2$,

$$u(x,y) \leq q(x,y)G^{-1}\left[G(k) + \sum_{i=1}^{n} \int_{\alpha_i(x_0)}^{\alpha_i(x)} \int_{\beta_i(y_0)}^{\beta_i(y)} b_i(s,t)g(q(s,t))dtds\right] \tag{5.1.244}$$

where for all $x \in J_1, y \in J_2$,

$$q(x,y) = \exp\left(\int_{x_0}^{x} a(\xi,y)d\xi\right), \tag{5.1.245}$$

and G^{-1} *is the inverse function of*

$$G(r) = \int_{r_0}^{r} \frac{ds}{g(s)}, \quad r \geq r_0 > 0, \tag{5.1.246}$$

$r_0 > 0$ *is arbitrary and* $x_1 \in J_1$, $y_1 \in J_2$ *are chosen so that*

$$G(k) + \sum_{i=1}^{n} \int_{\alpha_i(x_0)}^{\alpha_i(x)} \int_{\beta_i(y_0)}^{\beta_i(y)} b_i(s,t)g(q(s,t))dtds \in Dom\ (G^{-1}),$$

for all x and y lying in $[x_0, x_1]$ *and* $[y_0, y_1]$ *respectively.*

(B2) If for all $x \in J_1$, $y \in J_2$,

$$u(x,y) \leq k + \int_{y_0}^{y} a(x,t)u(x,t)dt + \sum_{i=1}^{n} \int_{\alpha_i(x_0)}^{\alpha_i(x)} \int_{\beta_i(y_0)}^{\beta_i(y)} b_i(s,t)g(u(s,t))dtds, \tag{5.1.247}$$

then for all $x_0 \leq x \leq x_2, y_0 \leq y \leq y_2; x, x_2 \in J_1, y, y_2 \in J_2$,

$$u(x,y) \leq r(x,y)G^{-1}\left[G(k) + \sum_{i=1}^{n} \int_{\alpha_i(x_0)}^{\alpha_i(x)} \int_{\beta_i(y_0)}^{\beta_i(y)} b_i(s,t)g(r(s,t))dtds\right] \tag{5.1.248}$$

where G, G^{-1} *are as in part (B1),* $r(x,y) = \exp\left(\int_{y_0}^{y} a(x,\eta)d\eta\right)$, *and* $x_2 \in J_1, y_2 \in J_2$ *are chosen so that*

$$G(k) + \sum_{i=1}^{n} \int_{\alpha_i(x_0)}^{\alpha_i(x)} \int_{\beta_i(y_0)}^{\beta_i(y)} b_i(s,t)g(r(s,t))dtds \in Dom\ (G^{-1}),$$

for all x and y lying in $[x_0, x_2]$ *and* $[y_0, y_2]$ *respectively.*

Proof We give the details of the proof of (B1) only. The proof of the remaining inequalities can be completed by closely looking at the proofs of the above mentioned inequalities with suitable modifications.

(B1) Define a function $z(x,y)$ by

$$z(x,y) = k + \sum_{i=1}^{n} \int_{\alpha_i(x_0)}^{\alpha_i(x)} \int_{\beta_i(y_0)}^{\beta_i(y)} b_i(s,t)g(u(s,t))dtds. \tag{5.1.249}$$

Then (5.1.243) can be stated as

$$u(x,y) \leq z(x,y) + \int_{x_0}^{x} a(s,y)u(s,y)ds. \tag{5.1.250}$$

Using Theorems 1.1.4–1.1.5 in Qin [557] to (5.1.250), we have for all $x \in J_1, y \in J_2$,

$$u(x,y) \le q(x,y)z(x,y), \tag{5.1.251}$$

where $q(x,y)$ and $z(x,y)$ are defined by (5.1.245) and (5.1.249). From (5.1.249) and (5.1.251) and the hypotheses on g, we deduce

$$\begin{aligned} z(x,y) &\le k + \sum_{i=1}^{n} \int_{\alpha_i(x_0)}^{\alpha_i(x)} \int_{\beta_i(y_0)}^{\beta_i(y)} b_i(s,t) g(q(s,t)z(s,t))dtds \\ &\le k + \sum_{i=1}^{n} \int_{\alpha_i(x_0)}^{\alpha_i(x)} \int_{\beta_i(y_0)}^{\beta_i(y)} b_i(s,t) g(q(s,t)) g(z(s,t))dtds. \end{aligned} \tag{5.1.252}$$

Let $k > 0$ and define a function $v(x,y)$ by the right-hand side of (5.1.252). Then it is easy to check that $v(x,y) > 0$, $v(x_0,y) = v(x,y_0) = k$, $z(x,y) \le v(x,y)$ and

$$\begin{aligned} D_1 v(x,y) &= \sum_{i=1}^{n} \left(\int_{\beta_i(y_0)}^{\beta_i(y)} b_i(\alpha_i(x),t) g(q(\alpha_i(x),t)) g(z(\alpha_i(x),t))dt \right) \alpha_i'(x) \\ &\le \sum_{i=1}^{n} \left(\int_{\beta_i(y_0)}^{\beta_i(y)} b_i(\alpha_i(x),t) g(q(\alpha_i(x),t)) g(v(\alpha_i(x),t))dt \right) \alpha_i'(x) \\ &\le g(v(x,y)) \sum_{i=1}^{n} \left(\int_{\beta_i(y_0)}^{\beta_i(y)} b_i(\alpha_i(x),t) g(q(\alpha_i(x),t))dt \right) \alpha_i'(x). \end{aligned} \tag{5.1.253}$$

From (5.1.246) and (5.1.253), we infer

$$D_1 G(v(x,y)) = \frac{D_1 v(x,y)}{g(v(x,y))} \le \sum_{i=1}^{n} \left(\int_{\beta_i(y_0)}^{\beta_i(y)} b_i(\alpha_i(x),t) g(q(\alpha_i(x),t))dt \right) \alpha_i'(x). \tag{5.1.254}$$

Keeping y fixed in (5.1.254), setting $x = \sigma$ and integrating it with respect to σ from x_0 to x, $x \in J_1$ and making the change of variables, we get

$$G(v(x,y)) \le G(k) + \sum_{i=1}^{n} \int_{\alpha_i(x_0)}^{\alpha_i(x)} \int_{\beta_i(y_0)}^{\beta_i(y)} b_i(s,t) g(q(s,t))dtds. \tag{5.1.255}$$

Since $G^{-1}(v)$ is increasing, from (5.1.255), we derive

$$v(x,y) \le G^{-1}\left[G(k) + \sum_{i=1}^{n} \int_{\alpha_i(x_0)}^{\alpha_i(x)} \int_{\beta_i(y_0)}^{\beta_i(y)} b_i(s,t) g(q(s,t))dtds \right]. \tag{5.1.256}$$

Using (5.1.256) in $z(x, y) \le v(x, y)$ and then the bound on $z(x, y)$ in (5.1.251), we can get the required inequality in (5.1.244). The case $k \ge 0$ can be completed by replacing k by $k + \varepsilon$ ($\varepsilon > 0$ arbitrary), then passing to the limit of $\varepsilon \to 0^+$. □

Theorem 5.1.41 (The Pachpatte Inequality [523]) *Let $u, a_i, b_i \in C(\Delta, R_+)$, and $\alpha_i \in C^1(I_1, I_1)$, $\beta_i \in C^1(I_2, I_2)$ be non-decreasing with $\alpha_i(x) \le x$ on I_1, $\beta_i \le y$ for $i = 1, 2, \cdots, n$. Let $p > 1$ and $c \ge 0$ be constants,*

(1) If for all $(x, y) \in \Delta$,

$$u^p(x, y) \le c + p \sum_{i=1}^{n} \int_{\alpha_i(x_0)}^{\alpha_i(x)} \int_{\beta_i(y_0)}^{\beta_i(y)} [a_i(s, t)u^p(s, t) + b_i(s, t)u(s, t)]\, dtds, \tag{5.1.257}$$

then for all $(x, y) \in \Delta$,

$$u(x, y) \le \left\{ B(x, y) \exp\left((p-1) \int_{\alpha_i(x_0)}^{\alpha_i(x)} \int_{\beta_i(y_0)}^{\beta_i(y)} a_i(\sigma, \tau) d\tau d\sigma \right) \right\}^{1/(p-1)}, \tag{5.1.258}$$

where for all $(x, y) \in \Delta$,

$$B(x, y) = \{c\}^{(p-1)/p} + (p-1) \int_{\alpha_i(x_0)}^{\alpha_i(x)} \int_{\beta_i(y_0)}^{\beta_i(y)} b_i(\sigma, \tau) d\tau d\sigma. \tag{5.1.259}$$

(2) Let w be as in Theorem 1.2.20, part (2). If for all $(x, y) \in \Delta$,

$$u^p(x, y) \le c + p \sum_{i=1}^{n} \int_{\alpha_i(x_0)}^{\alpha_i(x)} \int_{\beta_i(y_0)}^{\beta_i(y)} [a_i(s, t)u(s, t)w(u(s, t)) + b_i(s, t)u(s, t)]\, dtds, \tag{5.1.260}$$

then for all $x_0 \le x \le x_1, y_0 \le y \le y_1$,

$$u(x, y) \le \left\{ G^{-1} \left[G(B(x, y)) + (p-1) \sum_{i=1}^{n} \int_{\alpha_i(x_0)}^{\alpha_i(x)} \int_{\beta_i(y_0)}^{\beta_i(y)} a_i(\sigma, \tau) d\tau d\sigma \right] \right\}^{1/(p-1)}, \tag{5.1.261}$$

where $B(x, y)$ is defined by (5.1.259), G, G^{-1} are as in Theorem 1.2.20 part (2) and $x_1 \in I_1, y_1 \in I_2$ are chosen so that

$$G(B(x, y)) + (p-1) \sum_{i=1}^{n} \int_{\alpha_i(x_0)}^{\alpha_i(x)} \int_{\beta_i(y_0)}^{\beta_i(y)} a_i(\sigma, \tau) d\tau d\sigma \in Dom\ (G^{-1}),$$

for all x, y lying in the interval $x_0 \le x \le x_1, y_0 \le y \le y_1$.

Proof We only give the details of the proof for (2); the proof of (1) is similar. Let $c > 0$ and define a function $z(x, y)$ by the right-hand side of (5.1.260). Then $z(x, y) > 0, z(x_0, y) = z(x, y_0) = c, z(x, y)$ is non-decreasing in $(x, y) \in \Delta, u(x, y) \leq \{z(x, y)\}^{1/p}$ and

$$\begin{aligned} D_2D_1z(x,y) &= p\sum_{i=1}^{n}[a_i(\alpha_i(x),\beta_i(y))u(\alpha_i(x),\beta_i(y))w(u(\alpha_i(x),\beta_i(y))) \\ &\quad +b_i(\alpha_i(x),\beta_i(y))u(\alpha_i(x),\beta_i(y))]\beta_i'(y)\alpha_i'(x), \\ &\leq p\sum_{i=1}^{n}[a_i(\alpha_i(x),\beta_i(y))\{z(\alpha_i(x),\beta_i(y))\}^{1/p}w(\{z(\alpha_i(x),\beta_i(y))\}^{1/p}) \\ &\quad +b_i(\alpha_i(x),\beta_i(y))\{z(\alpha_i(x),\beta_i(y))\}^{1/p}]\beta_i'(y)\alpha_i'(x), \\ &\leq p\sum_{i=1}^{n}[a_i(\alpha_i(x),\beta_i(y))w(\{z(\alpha_i(x),\beta_i(y))\}^{1/p}) \\ &\quad +b_i(\alpha_i(x),\beta_i(y))]\{z(x,y)\}^{1/p}\beta_i'(y)\alpha_i'(x), \end{aligned} \tag{5.1.262}$$

From (5.1.262), we observe that

$$\begin{aligned} \frac{D_2D_1z(x,y)}{\{z(x,y)\}^{\frac{1}{p}}} &\leq p\sum_{i=1}^{n}[a_i(\alpha_i(x),\beta_i(y))w(\{z(\alpha_i(x),\beta_i(y))\}^{1/p}) \\ &\quad +b_i(\alpha_i(x),\beta_i(y))]\beta_i'(y)\alpha_i'(x) + \frac{D_1z(x,y)\left[D_2\{z(x,y)\}^{1/p}\right]}{\left[\{z(x,y)\}^{1/p}\right]^2}, \end{aligned} \tag{5.1.263}$$

i.e., for $(x, y) \in \Delta$,

$$\begin{aligned} D_2\left(\frac{D_1z(x,y)}{\{z(x,y)\}^{1/p}}\right) &\leq p\sum_{i=1}^{n}[a_i(\alpha_i(x),\beta_i(y))w(\{z(\alpha_i(x),\beta_i(y))\}^{1/p}) \\ &\quad +b_i(\alpha_i(x),\beta_i(y))]\beta_i'(y)\alpha_i'(x). \end{aligned} \tag{5.1.264}$$

By keeping x fixed in (5.1.264), we set $y = t$ and then, by integrating with respect to t from y_0 to y and using the fact that $D_1z(x, y_0) = 0$, we have

$$\begin{aligned} \frac{D_1z(x,y)}{\{z(x,y)\}^{1/p}} &\leq p\int_{y_0}^{y}\sum_{i=1}^{n}[a_i(\alpha_i(x),\beta_i(t))w(\{z(\alpha_i(x),\beta_i(t))\}^{1/p}) \\ &\quad +b_i(\alpha_i(x),\beta_i(t))]\beta_i'(t)\alpha_i'(x)dt. \end{aligned} \tag{5.1.265}$$

Now by keeping y fixed in (5.1.265) and setting $x = s$ and integrating with respect to s from x_0 to x we have

$$\{z(x,y)\}^{1/p} \le \{c\}^{(p-1)/p} + (p-1) \times \int_{x_0}^{x}\int_{y_0}^{y}\sum_{i=1}^{n}[a_i(\alpha_i(s),\beta_i(t))w(\{z(\alpha_i(s),\beta_i(t))\}^{1/p}) + b_i(\alpha_i(s),\beta_i(t))]\beta_i'(t)\alpha_i'(s)dtds. \tag{5.1.266}$$

By making the change of variables on the right hand side of (5.1.266) and rewriting we have

$$\{z(x,y)\}^{1/p} \le B(x,y) + (p-1)\sum_{i=1}^{n}\int_{\alpha_i(x_0)}^{\alpha_i(x)}\int_{\beta_i(y_0)}^{\beta_i(y)} a_i(\sigma,\tau)w\left(\{z(\sigma,\tau)\}^{1/p}\right)d\sigma d\tau. \tag{5.1.267}$$

Now fix $\lambda \in I_1, \mu \in I_2$ such that $x_0 \le x \le x_1, y_0 \le y \le \mu \le y_1$. Then from (5.1.267) we observe that

$$\{z(x,y)\}^{1/p} \le B(\lambda,\mu) + (p-1)\sum_{i=1}^{n}\int_{\alpha_i(x_0)}^{\alpha_i(x)}\int_{\beta_i(y_0)}^{\beta_i(y)} a_i(\sigma,\tau)w\left(\{z(\sigma,\tau)\}^{1/p}\right)d\sigma d\tau. \tag{5.1.268}$$

for $x_0 \le x \le x_1, y_0 \le y \le \mu \le y_1$. Define a function $v(x,y)$ by the right hand side of (5.1.268). Then $v(x,y) > 0, v(x_0,y) = v(x,y_0) = B(\lambda,\mu), v(x,y)$ is non-decreasing for $x_0 \le x \le \lambda, y_0 \le y \le \mu, \{z(x,y)\}^{1/p} \le v(x,y)$ and

$$v(x,y) \le B(\lambda,\mu) + (p-1)\sum_{i=1}^{n}\int_{\alpha_i(x_0)}^{\alpha_i(x)}\int_{\beta_i(y_0)}^{\beta_i(y)} a_i(\sigma,\tau)w\left(\{v(\sigma,\tau)\}^{1/(p-1)}\right)d\tau d\sigma.$$

for $x_0 \le x \le \lambda, y_0 \le y \le \mu$. Now by following the proof of Theorem 5.1.40 (see also [518], [521]) we get

$$v(x,y) \le G^{-1}\left[G(B(\lambda,\mu)) + (p-1)\sum_{i=1}^{n}\int_{\alpha_i(x_0)}^{\alpha_i(x)}\int_{\beta_i(y_0)}^{\beta_i(y)} a_i(\sigma,\tau)d\tau d\sigma\right]. \tag{5.1.269}$$

for $x_0 \le x \le \lambda \le x_1, y_0 \le y \le \mu \le y_1$. Since (λ,μ) is arbitrary, we get the desired inequality in (5.1.261) from (5.1.269) and the fact that

$$u(x,y) \le \{z(x,y)\}^{1/p} \le \left\{[v(x,y)]^{p/(p-1)}\right\}^{1/p} = \{v(x,y)\}^{1/(p-1)}.$$

The proof of the case when $c \geq 0$ can be completed as mentioned in the proof of Theorem 1.2.20, part (1). The domain $x_0 \leq x \leq x_1, y_0 \leq y \leq y_1$ is obvious. □

Remark 5.1.3 We note that the inequalities established in Theorem 5.1.41 can be extended very easily for functions involving more than two independent variables (see [507]). If we take $p = 2, n = 1, \alpha_1 = \alpha, \beta_1 = \beta, a_1 = f, b_1 = g$ in Theorem 5.1.41, then we get the two independent variable generalizations of the inequalities given in [356] (see Corollary 2 and Theorem 1). For a slight variant of the inequality in Theorem 5.1.41 given in [356] and its two independent variable version, see [518].

In the sequel, we shall establish a more general form of integral inequality

$$u^p(x,y) \leq a(x,y) + \sum_{i=1}^{n} \int_{b_1(x_0)}^{b_1(x)} \int_{c_1(y_0)}^{c_1(y)} f_i(x,y,t,s)\varphi_i(u(t,s))dsdt \tag{5.1.270}$$

for all $(x,y) \in [x_0,x_1) \times [y_0,y_1)$, where $a(x,y)$ is a function and $\varphi_i's$ is may not be monotone. We employ a technique of monotonization to construct a sequence of functions in integral which each possesses stronger monotonicity than previous one.

Throughout, $x_0, x_1, y_0, y_1 \in \mathbb{R}$ are given numbers. Let $I := [x_0, x_1), J := [y_0, y_1)$ and $\Lambda := I \times J \subset \mathbb{R}^2$. Suppose that

(P_1) all φ_i $(i = 1, \ldots, n)$ are continuous functions on $\mathbb{R}_+$ and positive on $(0, +\infty)$,
(P_2) $a(x,y) \geq 0$ on Λ,
(P_3) $b_i : I \to I$ $(i = 1, \ldots, n)$ and $c_i : J \to J$ $(i = 1, \ldots, n)$ are C^1 and non-decreasing such that $b_i(x) \leq x$ on I and $c_i(y) \leq y$ on J,
(P_4) all f_i $(i = 1, \ldots, n)$ are non-negative functions $\Lambda \times \Lambda$. We technically consider a sequence of functions $w_i(s)$, which can be calculated recursively by

$$\begin{cases} w_1(s) := \max_{\tau \in [0,s]}\{\varphi_1(\tau)\}, \\ w_{i+1}(s) := \max_{\tau \in [0,s]}\{\varphi_{i+1}(\tau)/w_i(\tau)\}w_i(s), \quad i = 1, \ldots, n. \end{cases} \tag{5.1.271}$$

Then for given constant $u_i > 0$, the function

$$W_{p,i}(u, u_i) := \int_{u_i}^{u} \frac{ds}{w_i(s^{1/p})}$$

is well-defined for all $u > 0$ and strictly increasing. When there is no confusion, we simply let $W_{p,i}(u)$ denote $W_{p,i}(u, u_i)$ and $W_{p,i}^{-1}$ denote its inverse. As explained in Remark 2 in [13], different choices of u_i in $W_{p,i}$ do not affect the results below.

Definition 5.1.1 Let $A \subset \mathbb{R}$ be a set. For $\omega_1, \omega_2 : A \to \mathbb{R}_+$ two functions, we shall denote $\omega_1 \propto \omega_2$ if $\frac{\omega_2}{\omega_1}$ is non-decreasing on A.

Theorem 5.1.42 (The Wang Inequality [664]) *Suppose that $(P_1 - P_4)$ hold and $u(x, y)$ is a non-negative function on Λ satisfying (5.1.270). Then for all $(x, y) \in [x_0, X_1) \times [y_0, Y_1)$,*

$$u(x, y) \le \{W_{p,n}^{-1}(\Xi_n(x, y))\}^{1/p} \tag{5.1.272}$$

where

$$\Xi_i(x, y) := W_{p,i}(r_i(x, y)) + \int_{b_i(x_0)}^{b_i(x)} \int_{c_i(y_0)}^{c_i(y)} \max_{(\tau,\xi)\in[x_0,x]\times[y_0,y]} f_i(\tau, \xi, t, s)dsdt,$$

$i = 1, 2, \dots, n$, $r_i(x, y)$ is determined recursively by

$$\begin{cases} r_1(x, y) := a(x_0, y_0) + \int_{x_0}^{x} |a_x(t, y_0)|dt + \int_{y_0}^{y} |a_y(x, s)|ds, \\ r_i(x, y) := W_{p,i-1}^{-1}(\Xi_{i-1}(x, y)), \end{cases} \tag{5.1.273}$$

and $(X_1, Y_1) \in \Lambda$ is arbitrarily given on the boundary of the planar region

$$R := \left\{(x, y) \in \Lambda : \Xi_i(x, y) \le \int_{u_i}^{+\infty} \frac{ds}{w_i(s^{1/p})},\ i = 1, \dots, n\right\}. \tag{5.1.274}$$

Proof First of all, we monotonize some given functions f_i, φ_i in the integral. Obviously, the sequence $(w_i(s))$ defined by $\varphi_i s$ in (5.1.271) consists of non-decreasing non-negative functions and satisfy $w_i(s) \ge \varphi_i(s),\ i = 1, \dots, n$. Moreover,

$$w_i \propto w_{i+1}, \quad i = 1, \dots, n-1, \tag{5.1.275}$$

for comparison of monotonicity of functions, because the ratios $w_{i+1}(s)/w_i(s),\ i = 1, \dots, n-1$, are all non-decreasing. Furthermore, let

$$\tilde{f}_i(x, y, t, s) := \max_{(\tau,\xi)\in[x_0,x]\times[y_0,y]} f_i(\tau, \xi, t, s),$$

which is also non-decreasing in x and y for each fixed s and t and satisfies $\tilde{f}_i(x, y, t, s) \ge f_i(x, y, t, s) \ge 0$ for all $i = 1, \dots, n$. With the above defined functions w_i and $\tilde{f}_i$, from (5.1.270) we get for all $(x, y) \in \Lambda$,

$$u^p(x, y) \le a(x, y) + \sum_{i=1}^{n} \int_{b_1(x_0)}^{b_1(x)} \int_{c_1(y_0)}^{c_1(y)} \tilde{f}_i(x, y, t, s)w_i(u(t, s))dsdt. \tag{5.1.276}$$

Firstly, we discuss the case that $a(x, y) > 0$ for all $(x, y) \in \Lambda$. It means that $r_1(x, y) > 0$ for all $(x, y) \in \Lambda$. In such a circumstance $r_1(x, y)$ is positive and non-decreasing on Λ and

$$r_1(x, y) \geq a(x_0, y_0) + \int_{x_0}^{x} a_x(t, y_0)dt + \int_{y_0}^{y} a_y(x, y)ds = a(x, y).$$

Consider the auxiliary inequality to (5.1.276)

$$u^p(x, y) \leq r_1(x, y) + \sum_{i=1}^{n} \int_{b_1(x_0)}^{b_1(x)} \int_{c_1(y_0)}^{c_1(y)} \tilde{f}_i(X, Y, t, s) w_i(u(t, s)) ds dt \tag{5.1.277}$$

for all $(x, y) \in [x_0, X) \times [y_0, Y)$, where $x_0 \leq X \leq X_1$ and $y_0 \leq Y \leq Y_1$ are chosen arbitrarily, and claim that

$$u(x, y) \leq \{W_{p,n}^{-1}(\Upsilon_n(X, Y, x, y))\}^{1/p} \tag{5.1.278}$$

for all $x, X \in [x_0, X_2)$ with $x \leq X$ and y, Y in $[y_0, Y_2)$ with $y \leq Y$, where

$$\Upsilon_i(X, Y, x, y) := W_{p,i}(\tilde{r}_i(X, Y, x, y)) + \int_{b_1(x_0)}^{b_1(x)} \int_{c_1(y_0)}^{c_1(y)} \tilde{f}_i(X, Y, t, s) ds dt$$

$i = 1, \ldots, n$, $\tilde{r}_i(X, Y, x, y)$ is defined recursively by

$$\begin{cases} \tilde{r}_1(X, Y, x, y) := r_1(x, y), \\ \tilde{r}_i(X, Y, x, y) := W_{p,i-1}^{-1}(\Xi_{i-1}(X, Y, x, y)), \end{cases} \tag{5.1.279}$$

and (X_2, Y_2) are both functions of X, Y such that $(X_2(X, Y), Y_2(X, Y)) \in \Lambda$ lies on the boundary of the planar region

$$R_1(x, y) := \left\{(x, y) \in \Lambda : \Upsilon_i(X, Y, x, y) \leq \int_{u_i}^{+\infty} \frac{ds}{w_i(s^{1/p})},\ i = 1, \ldots, n\right\}.$$

We can choose X_2, Y_2 appropriately such that for all $(X, Y) \in [x_0, X_1) \times [y_0, Y_1)$,

$$X_2(X, Y) \geq X_1, \quad Y_2(X, Y) \geq Y_1. \tag{5.1.280}$$

In fact, from the fact of (X_1, Y_1) being on the boundary of R, we see that

$$\Upsilon_i(X_1, Y_1, X_1, Y_1) = \Xi_i(X_1, Y_1) = \int_{u_i}^{+\infty} \frac{ds}{w_i(s^{1/p})}. \tag{5.1.281}$$

Moreover, the monotonicity that $\tilde{r}_i(X, Y, x, y)$ and $\tilde{f}_i(X, Y, x, y)$ are both non-decreasing in each variable implies that $\Upsilon_i(X, Y, x, y)$ is also non-decreasing in each variable. Therefore, it follows from (5.2.351) that the rectangles $[x_0, X_1) \times [y_0, Y_1)$, $[x_0, X_2) \times [y_0, Y_2)$ and Λ are nestled one by one, i.e.,

$$[x_0, X_1) \times [y_0, Y_1) \subset [x_0, X_2) \times [y_0, Y_2) \subset \Lambda. \tag{5.1.282}$$

Obviously,

$$[x_0, X_1) \times [y_0, Y_1) \subset R,\ [x_0, X_2(X, Y)) \times [y_0, Y_2(X, Y)) \subset R_1(X, Y),$$

so that $r_i,\ \tilde{r}_i (i = 1, 2, \ldots, n)$ are well-defined.

Now prove (5.1.278) by induction. Let $\beta_1(x, y)$ denote the function on the right-hand side of (5.1.277), which is a non-negative and non-decreasing function on $[x_0, Y) \times [y_0, Y)$. Then (5.1.277) is equivalent to

$$u^p(x, y) \leq \beta_1(x, y) \quad \textit{for all } (x, y) \in [x_0, Y) \times [y_0, Y). \tag{5.1.283}$$

By (P_3), $b_1' \geq 0$ and $b_1(x)$ for all $x \in [x_0, X)$. Moreover, w_1 is non-decreasing. Then

$$\begin{aligned}
\frac{\frac{\partial}{\partial x}\beta_1(x, y)}{w_1(\beta_1^{1/p}(x, y))} &\leq \frac{\frac{\partial}{\partial x} r_1(x, y)}{w_1(r_1^{1/p}(x, y))} + \frac{b_1'(x)}{w_1(\beta_1^{1/p}(x, y))}\int_{c_1(y_0)}^{c_1(y)} \tilde{f}_1(X, Y, b_1(x), s) w_1(u(b_1(x), s)) ds \\
&\leq \frac{\frac{\partial}{\partial x} r_1(x, y)}{w_1(\beta_1^{1/p}(x, y))} + \frac{b_1'(x)}{w_1(\beta_1^{1/p}(x, y))}\int_{c_1(y_0)}^{c_1(y)} \tilde{f}_1(X, Y, b_1(x), s) w_1(\beta_1^{1/p}(x, s)) ds \\
&\leq \frac{\frac{\partial}{\partial x} r_1(x, y)}{w_1(\beta_1^{1/p}(x, y))} + b_1'(x)\int_{c_1(y_0)}^{c_1(y)} \tilde{f}_1(X, Y, b_1(x), s) ds.
\end{aligned}$$

Integrating both sides of the above inequality from x_0 to x, we obtain for all $(x, y) \in [x_0, Y) \times [y_0, Y)$,

$$\begin{aligned}
W_{p,1}(\beta_1(x, y)) &\leq W_{p,1}(r_1(x, y)) + \int_{x_0}^{x} b_1'(t)\int_{c_1(y_0)}^{c_1(y)} \tilde{f}_1(X, Y, b_1(t), s) ds dt \\
&= W_{p,1}(r_1(x, y)) + \int_{b_1(x_0)}^{b_1(x)}\int_{c_1(y_0)}^{c_1(y)} \tilde{f}_1(X, Y, b_1(t), s) ds dt,
\end{aligned} \tag{5.1.284}$$

the right-hand side of which is contained in the domain of $W_{p,1}^{-1}$ by the definition of X_2, Y_2 and (5.1.282). It follows from (5.1.283), (5.1.284) and the monotonicity of $W_{p,1}^{-1}$ that for all $x_0 \leq x \leq X < X_2, y_0 \leq y \leq Y < Y_2$,

$$u(x, y) \leq \beta_1^{1/p} \leq \left\{ W_{p,1}^{-1}[W_{p,1}(r_1(x, y)) + \int_{b_1(x_0)}^{b_1(x)}\int_{c_1(y_0)}^{c_1(y)} \tilde{f}_1(X, Y, b_1(t), s) ds dt] \right\}^{1/p},$$

implying that (5.1.278) is true for $n = 1$. Next, we make the inductive assumption that (5.1.278) is true for $n = k$. Consider

$$u^p(x,y) \le r_1(x,y) + \sum_{i=1}^{k+1} \int_{x_0}^{x} b_1'(t) \int_{c_1(y_0)}^{c_1(y)} \tilde{f}_1(X,Y,t,s) w_i(u(t,s)) ds dt \tag{5.1.285}$$

for all $x_0 \le x \le X, y_0 \le y \le Y$. Let $\beta_2(x,y)$ denote the non-negative and non-decreasing function on the right-hand side of (5.1.285) and rewrite (5.1.285) as for all $(x,y) \in [x_0, Y) \times [y_0, Y)$,

$$u^p(x,y) \le r_1(x,y).$$

Let $\phi_{i+1}(u) := w_{i+1}/w_1(u), i = 1,\ldots,k$. Similarly to the above statement for $n = 1$, by the fact that $b_i' \ge 0$ and $b_i(x) \le x$ for all $x \in [x_0, X)$, given by (P_3), and the monotonicity of w_i, we have for all $(x,y) \in [x_0, Y) \times [y_0, Y)$,

$$\begin{aligned}
\frac{\frac{\partial}{\partial x}\beta_2(x,y)}{w_1(\beta_2^{1/p}(x,y))} &\le \frac{\frac{\partial}{\partial x} r_1(x,y)}{w_1(r_2^{1/p}(x,y))} + \sum_{i=1}^{k+1} \frac{b_i'(x)}{w_1(\beta_2^{1/p}(x,y))} \int_{c_i(y_0)}^{c_i(y)} \tilde{f}_i(X,Y,b_i(x),s) w_i(u(b_i(x),s)) ds \\
&= \frac{\frac{\partial}{\partial x} r_1(x,y)}{w_1(r_2^{1/p}(x,y))} + b_1'(x) \int_{c_i(y_0)}^{c_i(y)} \tilde{f}_1(X,Y,b_1(x),s) ds \\
&\quad + \sum_{i=1}^{k} b_{i+1}'(x) \int_{c_{i+1}(y_0)}^{c_{i+1}(y)} \tilde{f}_{i+1}(X,Y,b_{i+1}(x),s) \phi_{i+1}(\beta_2^{1/p}(b_{i+1}(x),s)) ds.
\end{aligned}$$

Integrating the above from x_0 to x, we get for all $(x,y) \in [x_0, Y) \times [y_0, Y)$,

$$\begin{aligned}
W_{p,1}(\beta_2(x,y)) \le W_{p,1}(r_1(x,y)) &+ \int_{b_1(x_0)}^{b_1(x)} \int_{c_1(y_0)}^{c_1(y)} \tilde{f}_1(X,Y,t,s) ds dt \\
&+ \sum_{i=1}^{k} \int_{b_{i+1}(x_0)}^{b_{i+1}(x)} \int_{c_{i+1}(y_0)}^{c_{i+1}(y)} \tilde{f}_{i+1}(X,Y,t,s) \phi_{i+1}(\beta_2^{1/p}(t,s)) ds.
\end{aligned}$$

Let

$$\begin{cases} \xi^p(x,y) := W_{p,1}(\beta_2(x,y)), & (5.1.286) \\ \theta_1 := W_{p,1}(r_1(x,y)) + \displaystyle\int_{b_1(x_0)}^{b_1(x)} \int_{c_1(y_0)}^{c_1(y)} \tilde{f}_1(X,Y,t,s) ds dt. & (5.1.287) \end{cases}$$

It follows that

$$\xi^p \le \theta_1(x,y) + \sum_{i=1}^{k} \int_{b_{i+1}(x_0)}^{b_{i+1}(x)} \int_{c_{i+1}(y_0)}^{c_{i+1}(y)} \tilde{f}_{i+1}(X,Y,t,s) \phi_{i+1}[(W_{p,1}^{-1}(\xi^p(x,y)))^{1/p}] ds, \tag{5.1.288}$$

the same form as (5.1.277) for $n = k$, for all $(x, y) \in [x_0, Y) \times [y_0, Y)$ and we are ready to use the inductive assumption for (5.1.278). In order to demonstrate the basic condition of monotonicity, let $h(s) := (W_{p,1}^{-1}(sp))^{1/p}$ which is clearly a continuous and non-decreasing function on $[0, +\infty)$. Thus each $\phi_i(h(s))$ is continuous and non-decreasing on $[0, +\infty)$ and satisfies $\phi_i(h(s)) > 0$ for all $s > 0$. Moreover,

$$\frac{\phi_{i+1}(h(s))}{\phi_i(h(s))} = \frac{w_{i+1}(h(s))}{w_i(h(s))} = \max_{\tau\in[0,h(s)]}\{\frac{\varphi_{i+1}(\tau)}{w_i(\tau)}\},$$

which is also continuous and non-decreasing on $[0, +\infty)$ and positive on $(0, +\infty)$, implying that

$$\phi_i(h(s)) \propto \phi_{i+1}(h(s)), i = 2, ..., k.$$

Therefore, the inductive assumption for (5.1.278) can be used to (5.1.288) and we obtain

$$\xi(x, y) \le \left\{\Phi_{p,k+1}^{-1}(\eta_{k+1}(X, Y, x, y))\right\}^{1/p} \tag{5.1.289}$$

for all $x_0 \le x < min(X, X_3)$ and $y_0 \le y < min(Y, Y_3)$, where

$$\begin{cases} \Phi_{p,i} := \displaystyle\int_{\varpi(u_i)}^{u} \frac{ds}{\phi_i(h(s))}, \quad u > 0, & (5.1.290)\\ \varpi(u) := (W_{p,1}(u))^{1/p}, & (5.1.291)\\ \eta_1(X, Y, x, y) := \Phi_{p,i}(r_{i-1}(x, y)) + \displaystyle\int_{b_i(x_0)}^{b_i(x)}\int_{c_i(y_0)}^{c_i(y)} \tilde{f}_1(X, Y, t, s)dsdt, & (5.1.292)\\ \theta_i := \Phi_{p,i}^{-1}(\eta_i(X, Y, x, y)), & (5.1.293) \end{cases}$$

$i = 2, .., , k + 1$, and X_3, Y_3 are functions of (X, Y) such that $(X_3(X, Y), Y_3(X, Y)) \in \Lambda$ lies on the boundary of the planar region

$$R_2(X, Y) := \left\{(x, y) \in \Lambda : \eta_i(X, Y, x, y) \le \int_{\varpi(u_i)}^{\varpi(+\infty)} \frac{ds}{\phi_i(h(s))}, \ i = 2, \ldots, k + 1\right\}.$$

Here $\varpi(+\infty)$ denotes either the limit $\lim_{u\to+\infty}\varpi(u)$ if it converges or $+\infty$. Note that

$$\Phi_{p,i}(u) = \int_{u_i}^{W_{p,i}^{-1}(u^p)} \frac{ds}{w_i(s^{1/p})} = W_{p,i}(W_{p,1}^{-1}(u^p)), \quad i = 2, .., k + 1. \tag{5.1.294}$$

Thus (5.1.289), where we note those functions defined in (5.1.286), (5.1.287) and (5.1.292), can be equivalently written as, for all $x_0 \le x < \min(X, X_3), y_0 \le y < \min(Y, Y_3)$,

$$u(x,y) \le \beta_2^{1/p}(x,y) = (W_{p,1}^{-1}(\xi^p(x,y)))^{1/p}$$
$$\le \left\{ W_{p,k+1}^{-1}[W_{p,k+1}(W_{p,1}^{-1}(\theta_k(x,y))) + \int_{b_{k+1}(x_0)}^{b_{k+1}(x)} \int_{c_{k+1}(y_0)}^{c_{k+1}(y)} \tilde{f}_{k+1}(X,Y,t,s)dsdt] \right\}^{1/p}. \tag{5.1.295}$$

We further claim that the term $W_{p,1}^{-1}(\theta_i(x,y))$ in the formula (5.1.295) is just the same as $\tilde{r}_{i+1}(X,Y,x,y)$, defined in (5.1.279), for all $i = 1, \ldots, k$. For convenience, let $\tilde{\theta}_i(x,y)$ denote that term. It is trivial to see that $\tilde{\theta}_1(x,y) = \tilde{r}_2(X,Y,x,y)$. Assume that the claimed result is true for some i. Then, using (5.1.289) and noting some definitions of functions in (5.1.292) and (5.1.293), we have

$$\tilde{\theta}_{i+1}(x,y) = W_{p,1}^{-1}\left\{ \Phi_{p,i+1}^{-1}[\Phi_{p,i+1}(\theta_i(x,y)) + \int_{b_{i+1}(x_0)}^{b_{i+1}(x)} \int_{c_{i+1}(y_0)}^{c_{i+1}(y)} \tilde{f}_{k+1}(X,Y,t,s)dsdt] \right\}$$
$$= W_{p,i+1}^{-1}[W_{p,i+1}(W_{p,1}^{-1}(\theta_i(x,y))) + \int_{b_{i+1}(x_0)}^{b_{i+1}(x)} \int_{c_{i+1}(y_0)}^{c_{i+1}(y)} \tilde{f}_{k+1}(X,Y,t,s)dsdt]$$
$$= W_{p,i+1}^{-1}[W_{p,i+1}(\tilde{r}_{i+1}(X,Y,x,y)) + \int_{b_{i+1}(x_0)}^{b_{i+1}(x)} \int_{c_{i+1}(y_0)}^{c_{i+1}(y)} \tilde{f}_{k+1}(X,Y,t,s)dsdt]$$
$$= \tilde{r}_{i+2}(X,Y,x,y).$$

Thus the claimed result is proved. Hence (5.1.295) can be equivalently written as

$$u(x,y) \le \left\{ W_{p,k+1}^{-1}[W_{p,k+1}(\tilde{r}_{k+1}(X,Y,x,y)) + \int_{b_{k+1}(x_0)}^{b_{k+1}(x)} \int_{c_{k+1}(y_0)}^{c_{k+1}(y)} \tilde{f}_{k+1}(X,Y,t,s)dsdt] \right\}^{1/p}. \tag{5.1.296}$$

Similarly, from (5.1.292) and (5.1.294),

$$\eta_i(X,Y,x,y) = W_{p,i}(\tilde{r}_{k+1}(X,Y,x,y)) + \int_{b_i(x_0)}^{b_i(x)} \int_{c_i(y_0)}^{c_i(y)} \tilde{f}_i(X,Y,t,s)dsdt$$
$$= \Upsilon_i(X,Y,x,y). \tag{5.1.297}$$

Note that $\int_{\varpi(u_i)}^{\varpi(+\infty)} \frac{ds}{\phi_i(h(s))} = \int_{u_i}^{+\infty} ds/w_i(s^{1/p})$. Then, comparing the definition of R_2 with that of R_1 and noting (5.1.297), we see that X_3, Y_3 can be chosen appropriately such that for all $(X,Y) \in [x_0, X_1) \times [y_0, Y_1)$,

$$X_3(X,Y) = X_2(X,Y), \quad Y_3(X,Y) = Y_2(X,Y). \tag{5.1.298}$$

It means that (5.1.296) holds for all $x_0 \le x < X \le X_2, y_0 \le y < Y \le Y_2$. It actually proves (5.1.278) by induction. Having (5.1.278), we start from the original inequality (5.1.270) and see that

$$u^p(X,Y) \le r_1(X,Y) + \sum_{i=1}^{n} \int_{b_i(x_0)}^{b_i(X)} \int_{c_i(y_0)}^{c_i(Y)} \tilde{f}_1(X,Y,t,s) w_i(u(t,s)) ds dt,$$

i.e., the auxiliary inequality (5.2.348) holds for $x = X, y = Y$. By (5.2.349), we get

$$\begin{aligned} u(X,Y) &\le \{W_{p,n}^{-1}[W_{p,n}(\tilde{r}_n(X,Y,X,Y)) + \int_{b_i(x_0)}^{b_i(X)} \int_{c_i(y_0)}^{c_i(Y)} \tilde{f}_i(X,Y,t,s) ds dt]\}^{1/p}, \\ &= \{W_{p,n}^{-1}[W_{p,n}(r_n(X,Y)) + \int_{b_i(x_0)}^{b_i(X)} \int_{c_i(y_0)}^{c_i(Y)} \tilde{f}_i(X,Y,t,s) ds dt]\}^{1/p} \end{aligned}$$

for all $x_0 \le X \le X_1, y_0 \le Y \le Y_1$ since $X_2 \ge X_1, Y_2 \ge Y_1$ and $\tilde{r}_n(X,Y,X,Y) = r_n(X,Y)$. This proves (5.2.343).

The remainder case is that $a(x,y) = 0$ for some $(x,y) \in \Lambda$. Let

$$r_{1,\varepsilon}(x,y) := r_1(x,y) + \varepsilon,$$

where $\varepsilon > 0$ is an arbitrary small number. Obviously, $r_{1,\varepsilon}(x,y) > 0$, we get

$$u(x,y) \le \left\{ W_{p,n}^{-1}[W_{p,n}(r_{n,\varepsilon}(x,y)) + \int_{b_n(x_0)}^{b_n(X)} \int_{c_n(y_0)}^{c_n(Y)} \tilde{f}_n(X,Y,t,s) ds dt] \right\}^{1/p},$$

for all $x_0 \le x < X_1, y_0 \le y < Y_1$. Letting $\varepsilon \to 0^+$, we obtain (5.2.343) because of continuity of $r_{i,\varepsilon}$ in ε and continuity of $W_{p,i}$ and $W_{p,i}^{-1}$ for $i = 1, \ldots, n$. This completes the proof. □

Remark 5.1.4 Remark that X_1, Y_1 are defined by (5.2.345). In particular, (5.2.343) is true for all $(x,y) \in \Lambda$ when all $w_i, i = 1, 2, \ldots, n$, satisfy $\int_{u_i}^{+\infty} ds/w_i(s^{1/p}) = \infty$, so we may take $X_1 = x_1, Y_1 = y_1$.

If we choose $n = 2, \varphi_1(s) := s^q, \varphi_2(s) := s^q \psi(s), f_i(x,y,t,s) := (p/(p-q))g_i(t,s)$, where $i = 1, 2$ and $0 < q < p$, and restrict $a(x,y)$ to be a constant a, then we can give a different estimate from [142] for the unknown function u in the inequality

$$\begin{aligned} u^p(x,y) \le a &+ \frac{p}{p-q} \int_{b_1(x_0)}^{b_1(x)} \int_{c_1(y_0)}^{c_1(y)} g_1(t,s) u^q(t,s) ds dt \\ &+ \frac{p}{p-q} \int_{b_2(x_0)}^{b_2(x)} \int_{c_2(y_0)}^{c_2(y)} g_2(t,s) u^q(t,s) \psi(u(t,s)) ds dt, \end{aligned}$$

where $a \geq 0$ and $p > q > 0$ are constants, $b_i(s)$ and $c_i(s)$ are C^1 non-decreasing functions, $g_i(s)$ are continuous positive functions and ψ is a continuous and non-decreasing positive function. If we choose $p = 1$ and $u(x, y) := v(x)$, let $a(x, y) := a(x), f_i(x, y, t, s) := g_i(x, t), i = 1, \ldots, n$, and restrict all $c_i s$ to satisfy that $c_i(y) - c_i(y_0) = 1$ for all $y \in J$, then inequality (5.2.341) reduces to the same form as that in Theorem 2.1 in [13], where we do not require the monotonicity of sequence of functions φ_i. Obviously, Theorem 5.2.59 is applicable to more general form than Theorem 2.1 in [13].

The following theorem concerns some new Gronwall-Ou-Yang integral inequalities in two independent variables. These results are obtained by Cheung and Ma [145].

We define $\mathbb{R}_1 = [1, +\infty)$, and for any $k \in \mathbb{N}$, $\mathbb{R}^k_+ = (\mathbb{R}^+)^k$. Denote by $C^i(M, S)$ the class of all i-times continuously differentiable functions defined on set M with range in set S $(i = 1, 2, \cdots)$ and $C^0(M, S) = C(M, S)$. The first-order partial derivatives of a function $z(x, y)$ for $x, y \in \mathbb{R}$ with respect to x and y are denoted as usual by $D_1 z(x, y)$ and $D_2 z(x, y)$, respectively. We also assume that all improper integrals appeared in the sequel are always convergent.

Lemma 5.1.3 *Let $u(x, y)$, $a(x, y)$, $c(x, y)$ and $d(x, y)$ be non-negative continuous functions defined for all $x, y \in \mathbb{R}_+$ and $w(u)$ be a non-negative, non-decreasing continuous function for all $u \in \mathbb{R}_+$ with $w(u) > 0$ for all $u > 0$.*

(i) Assume that $a(x, y)$ and $c(x, y)$ are non-decreasing in x and non-increasing in y for all $x, y \in \mathbb{R}_+$. If for all $x, y \in \mathbb{R}_+$,

$$u(x, y) \leq a(x, y) + c(x, y) \int_0^x \int_y^{+\infty} d(s, t) w(u(s, t))\, dt\, ds, \tag{5.1.299}$$

then for all $0 \leq x \leq x_1$, $y_1 \leq y \leq +\infty$,

$$u(x, y) \leq G^{-1} \left[G(a(x, y)) + c(x, y) \int_0^x \int_y^{+\infty} d(s, t)\, dt\, ds \right], \tag{5.1.300}$$

where

$$G(r) := \int_{r_0}^r \frac{dr}{w(r)}, \quad r \geq r_0 > 0, \tag{5.1.301}$$

and G^{-1} is the inverse function of G, and $x_1, y_1 \in \mathbb{R}_+$ are chosen so that

$$G(a(x, y)) + c(x, y) \int_0^x \int_y^{+\infty} d(s, t)\, dt\, ds \in Dom\,(G^{-1}). \tag{5.1.302}$$

(ii) Assume that $a(x, y)$ *and* $c(x, y)$ *are non-increasing in each variable* $x,\ y \in \mathbb{R}_+$. *If for all* $x,\ y \in \mathbb{R}_+$,

$$u(x,y) \le a(x,y) + c(x,y) \int_x^{+\infty} \int_y^{+\infty} d(s,t)w(u(s,t))\, dt\, ds, \tag{5.1.303}$$

then for all $0 \le x \le x_2,\ y_2 \le y \le +\infty$,

$$u(x,y) \le G^{-1}\left[G(a(x,y)) + c(x,y) \int_x^{+\infty} \int_y^{+\infty} d(s,t)\, dt\, ds\right], \tag{5.1.304}$$

where G *and* G^{-1} *are defined as in (i), and* $x_2,\ y_2 \in \mathbb{R}_+$ *are chosen so that*

$$G(a(x,y)) + c(x,y) \int_x^{+\infty} \int_y^{+\infty} d(s,t)\, dt\, ds \in Dom\,(G^{-1}). \tag{5.1.305}$$

Proof

(i) Fixing any numbers $\bar{x}_1$ and $\bar{y}_1$ with $0 < \bar{x}_1 \le x_1$ and $y_1 \le \bar{y}_1 < +\infty$, from (5.1.299) we infer

$$u(x,y) \le a(\bar{x}_1, \bar{y}_1) + c(\bar{x}_1, \bar{y}_1) \int_0^{x} \int_y^{+\infty} d(s,t)w(u(s,t))\, dt\, ds \tag{5.1.306}$$

for all $0 \le x \le \bar{x}_1,\ \bar{y}_1 \le y < +\infty$.

Defining $r_1(x, y)$ as the right-hand side of the last inequality, then $r_1(0, y) = r_1(x, +\infty) = a(\bar{x}_1, \bar{y}_1)$,

$$u(x,y) \le r_1(x,y), \tag{5.1.307}$$

where $r_1(x, y)$ is non-increasing in $y \in [\bar{y}_1, +\infty)$, and

$$\begin{aligned} D_1 r_1(x,y) &= c(\bar{x}_1, \bar{y}_1) \int_y^{+\infty} d(x,t)w(u(x,t))\, dt \le c(\bar{x}_1, \bar{y}_1) \int_y^{+\infty} d(x,t)w(r_1(x,t))\, dt \\ &\le c(\bar{x}_1, \bar{y}_1)w(r_1(x,y)) \int_y^{+\infty} d(x,t)\, dt. \end{aligned} \tag{5.1.308}$$

Dividing both sides of (5.1.308) by $w(r(x, y))$, we obtain

$$\frac{D_1 r_1(x,y)}{w(r_1(x,y))} \le c(\bar{x}_1, \bar{y}_1) \int_y^{+\infty} d(x,t)\, dt. \tag{5.1.309}$$

From (5.1.301) and (5.1.309), we infer

$$D_1G(r_1(x,y)) \le c(\bar{x}_1,\bar{y}_1)\int_y^{+\infty} d(x,t)\,dt. \tag{5.1.310}$$

Now setting $x = s$ in (5.1.310) and then integrating with respect to s from 0 to x, we obtain

$$G(r_1(x,y)) \le G(r_1(0,y)) + c(\bar{x}_1,\bar{y}_1)\int_0^x\int_y^{+\infty} d(s,t)\,dt\,ds. \tag{5.1.311}$$

Noting $G(r_1(0,y)) = G(a(\bar{x}_1,\bar{y}_1))$, we get

$$G(r_1(x,y)) \le G(a(\bar{x}_1,\bar{y}_1)) + c(\bar{x}_1,\bar{y}_1)\int_0^x\int_y^{+\infty} d(s,t)\,dt\,ds. \tag{5.1.312}$$

Taking $x = \bar{x}_1,\ y = \bar{y}_1$ in (5.1.307) and the last inequality, we have

$$u(\bar{x}_1,\bar{y}_1) \le r_1(x,y), \tag{5.1.313}$$

$$G(r_1(\bar{x}_1,\bar{y}_1)) \le G(a(\bar{x}_1,\bar{y}_1)) + c(\bar{x}_1,\bar{y}_1)\int_0^{\bar{x}_1}\int_{\bar{y}_1}^{+\infty} d(s,t)\,dt\,ds. \tag{5.1.314}$$

Since $0 < \bar{x}_1 \le x_1,\ y_1 \le \bar{y}_1 < +\infty$ are arbitrary, from (5.1.314) we deduce for all $0 < x \le x_1,\ y_1 \le y < +\infty$,

$$u(x,y) \le r_1(x,y), \tag{5.1.315}$$

$$G(r_1(x,y)) \le G(a(x,y)) + c(x,y)\int_0^x\int_y^{+\infty} d(s,t)\,dt\,ds, \tag{5.1.316}$$

or

$$r_1(x,y) \le G^{-1}\left[G(a(x,y)) + c(x,y)\int_0^x\int_y^{+\infty} d(s,t)\,dt\,ds\right]. \tag{5.1.317}$$

Hence from (5.1.315) and (5.1.317) it follows that for all $0 < x \le x_1,\ y_1 \le y < +\infty$,

$$u(x,y) \le G^{-1}\left[G(a(x,y)) + c(x,y)\int_0^x\int_y^{+\infty} d(s,t)\,dt\,ds\right]. \tag{5.1.318}$$

By (5.1.299), (5.1.318) holds also when $x = 0$.

(ii) The proof of (ii) is similar to the argument in the proof of Lemma 5.1.3 (i) with suitable modification. We omit the details here. □

Theorem 5.1.43 (The Cheung-Ma Inequality [145]) *Let $a(x,y)$, $c(x,y)$, and $w(u)$ be defined as in Lemma 5.1.3 (i), and $e(x,y) \in C(\mathbb{R}_+^2, \mathbb{R}_+)$. Let $\varphi(u) \in C^1(\mathbb{R}_+, \mathbb{R}_+)$ with $\varphi'(u) > 0$ for all $u > 0$, here φ' denotes the derivative of φ. If for all $x,\ y \in \mathbb{R}_+$,*

$$\varphi(u(x,y)) \le a(x,y) + c(x,y) \int_0^x \int_y^{+\infty} \varphi'(u(s,t))[d(s,t)w(u(s,t)) + e(s,t)]\, dt\, ds, \tag{5.1.319}$$

then

$$u(x,y) \le G^{-1}\left(G[\varphi'(a(x,y)) + E(x,y)] + c(x,y) \int_0^x \int_y^{+\infty} d(s,t)\, dt\, ds \right) \tag{5.1.320}$$

for all $0 \le x \le x_3,\ y_3 \le y < +\infty$, where

$$E(x,y) := c(x,y) \int_0^x \int_y^{+\infty} e(s,t)\, dt\, ds, \tag{5.1.321}$$

and G and G^{-1} are defined as in Lemma 5.1.3, φ^{-1} is the inverse function of φ, and $x_3,\ y_3 \in \mathbb{R}_+$ are chosen so that

$$G[\varphi^{-1}(a(x,y)) + E(x,y)] + c(x,y) \int_0^x \int_y^{+\infty} d(s,t)\, dt\, ds \in Dom\ (G^{-1}). \tag{5.1.322}$$

Proof If $a(x,y) > 0$, fixing any numbers $\bar{x}_3$ and $\bar{y}_3$ $(0 < \bar{x}_3 \le x_3,\ y_3 \le \bar{y}_3 < +\infty)$, from (5.1.319) we derive for all $0 < x \le \overline{x_3},\ \overline{y_3} \le y < +\infty$,

$$\varphi(u(x,y)) \le a(\bar{x}_3, \bar{y}_3) + c(\bar{x}_3, \bar{y}_3) \int_0^x \int_y^{+\infty} \varphi'(u(s,t))[d(s,t)w(u(s,t)) + e(s,t)]\, dt\, ds. \tag{5.1.323}$$

Defining $r_2(x,y)$ as the right-hand side of the last inequality, then for all $0 \le x \le \bar{x}_3,\ \bar{y}_3 \le y < +\infty$,

$$r_2(0,y) = r_2(x,+\infty) = a(\bar{x}_3, \bar{y}_3), \tag{5.1.324}$$

$$u(x,y) \le \varphi^{-1}(r_2(x,y)). \tag{5.1.325}$$

Since $r_2(x, y)$ is non-increasing in y, by (5.1.325), we have

$$\begin{aligned} D_1 r_2(x, y) &= c(\bar{x}_3,\ \bar{y}_3) \int_y^{+\infty} \varphi'(u(x,t))[d(x,t)w(u(x,t)) + e(x,t)]\, dt \\ &\le c(\bar{x}_3,\ \bar{y}_3) \int_y^{+\infty} \varphi'(\varphi^{-1}(r_2(x,t)))[d(x,t)w(\varphi^{-1}(r_2(x,t))) + e(x,t)]\, dt \\ &\le c(\bar{x}_3,\ \bar{y}_3)\varphi'(\varphi^{-1}(r_2(x,y))) \int_y^{+\infty} [d(x,t)w(\varphi^{-1}(r_2(x,t))) + e(x,t)]\, dt. \end{aligned} \tag{5.1.326}$$

Dividing both sides of (5.1.326) by $\varphi'(\varphi^{-1}(r_2(x, y)))$, we get

$$\frac{D_1 r_2(x,y)}{\varphi'(\varphi^{-1}(r_2(x,y)))} \le c(\bar{x}_3, \bar{y}_3) \int_y^{+\infty} [d(x,t)w(\varphi^{-1}(r_2(x,t))) + e(x,t)]\, dt. \tag{5.1.327}$$

Observe that for any continuously differentiable and invertible function $\Phi(\xi)$, by the change of variable $\eta = \Phi^{-1}(\xi)$, we obtain

$$\int \frac{d\xi}{\Phi'(\Phi^{-1}(\xi))} = \int \frac{\Phi'(\eta)}{\Phi'(\eta)} d\eta = \eta + c = \Phi^{-1}(\xi) + c. \tag{5.1.328}$$

Keeping y fixed in (5.1.327), setting $x = s$ and integrating with respect to s from 0 to x, and applying (5.1.328) to the left-hand side, we obtain

$$\begin{aligned} \varphi^{-1}(r_2(x,y)) &\le \varphi^{-1}(r_2(0,y)) + c(\bar{x}_3, \bar{y}_3) \int_0^x \int_y^{+\infty} [d(s,t)w(\varphi^{-1}(r_2(s,t))) + e(s,t)]\, dt\, ds \\ &= \varphi^{-1}(a(\bar{x}_3, \bar{y}_3)) + c(\bar{x}_3, \bar{y}_3) \int_0^x \int_y^{+\infty} [d(s,t)w(\varphi^{-1}(r_2(s,t))) + e(s,t)]\, dt\, ds. \end{aligned} \tag{5.1.329}$$

Applying Lemma 5.1.3 (i) to the last inequality, we get for all $0 \le x \le \bar{x}_3,\ \bar{y}_3 \le y < +\infty$,

$$\begin{aligned} \varphi^{-1}(r_2(x,y)) \le G^{-1}\Bigg(G\Big[&\varphi^{-1}(a(\bar{x}_3, \bar{y}_3)) + c(\bar{x}_3, \bar{y}_3) \int_0^x \int_y^{+\infty} e(s,t)\, dt\, ds \Big] \\ &+ c(\bar{x}_3, \bar{y}_3) \int_0^x \int_y^{+\infty} d(s,t)\, dt\, ds \Bigg). \end{aligned} \tag{5.1.330}$$

By (5.1.325), (5.1.330) and using similar procedures as from (5.1.314) to (5.1.315) in the proof of Lemma 5.1.3 (i), we can get the desired bound of $u(x, y)$ in (5.1.320). By continuity, (5.1.320) also holds for the case $a(x, y) \geq 0$. □

Theorem 5.1.44 (The Cheung-Ma Inequality [145]) *Let $a(x, y)$, $c(x, y)$, $w(u)$ be defined as in Lemma 5.1.3 (ii) and $\varphi(u)$, $e(x, y)$ defined as in Theorem 5.1.43. If for all x, $y \in \mathbb{R}_+$,*

$$\varphi(u(x, y)) \leq a(x, y) + c(x, y) \int_x^{+\infty} \int_y^{+\infty} \varphi'(u(s, t))[d(s, t)w(u(s, t)) + e(s, t)] \, dt \, ds, \tag{5.1.331}$$

then for all $x_4 \leq x < +\infty$, $y_4 \leq y < +\infty$,

$$u(x, y) \leq G^{-1}\left(G[\varphi^{-1}(a(x, y)) + \bar{E}(x, y)] + c(x, y) \int_x^{+\infty} \int_y^{+\infty} d(s, t) \, dt \, ds\right), \tag{5.1.332}$$

where

$$\bar{E}(x, y) := c(x, y) \int_x^{+\infty} \int_y^{+\infty} e(s, t) \, dt \, ds, \tag{5.1.333}$$

G and G^{-1} are defined as in Lemma 5.1.3, φ and φ^{-1} are defined as in Theorem 5.1.43, and x_4, $y_4 \in \mathbb{R}_+$ are chosen so that

$$G[\varphi^{-1}(a(x, y)) + \bar{E}(x, y)] + c(x, y) \int_x^{+\infty} \int_y^{+\infty} d(s, t) \, dt \, ds \in Dom\,(G^{-1}). \tag{5.1.334}$$

Proof The proof follows by an argument similar to that in the proof of Theorem 5.1.43 with suitable modification. We omit the details here. □

Theorem 5.1.45 (The Cheung-Ma Inequality [145]) *Let $a(x, y)$, $c(x, y)$, $e(x, y)$, $w(u)$, $\varphi(u)$, and $\varphi'(u)$ be defined as in Theorem 5.1.43. Let $b(x, y)$, $d(x, y)$, and $f(x, y) \in C(\mathbb{R}_+^2, \mathbb{R}_+)$ and $b(x, y)$, $d(x, y)$ be non-decreasing in x and non-increasing in y. If for all x, y, $\alpha \in \mathbb{R}_+$,*

$$\begin{aligned}\varphi(u(x, y)) \leq{}& a(x, y) + b(x, y) \int_\alpha^x c(s, y)\varphi(u(s, y)) \, ds \\ &+ d(x, y) \int_0^x \int_y^{+\infty} \varphi'(u(s, t))[f(s, t)w(u(s, t)) + e(s, t)] \, dtds,\end{aligned} \tag{5.1.335}$$

with $\alpha \le x$, *then for all* $0 \le x \le x_5,\ y_5 \le y < +\infty$,

$$u(x,y) \le G^{-1}\Bigg(G[\varphi^{-1}(p(x,y)a(x,y)) + p(x,y)E_1(x,y)] + p(x,y)d(x,y)\int_0^x\int_y^{+\infty} f(s,t)\,dt\,ds\Bigg), \tag{5.1.336}$$

where

$$p(x,y) := 1 + b(x,y)\int_\alpha^x c(s,y)\exp\left(\left(\int_s^x b(m,y)c(m,y)\,dm\right)\right) ds, \tag{5.1.337}$$

$$E_1(x,y) := d(x,y)\int_0^x\int_y^{+\infty} e(s,t)\,dt\,ds, \tag{5.1.338}$$

G *and* G^{-1} *are defined as in Lemma 5.1.3,* φ *and* φ^{-1} *are defined as in Theorem 5.1.43, and* $x_5,\ y_5 \in \mathbb{R}_+$ *are chosen so that*

$$G[\varphi^{-1}(p(x,y)a(x,y)) + p(x,y)E_1(x,y)] + p(x,y)d(x,y)\int_0^x\int_y^{+\infty} f(s,t)\,dt\,ds \in Dom\,(G^{-1}). \tag{5.1.339}$$

Proof Define a function $z(x,y)$ by

$$z(x,y) = a(x,y) + d(x,y)\int_0^x\int_y^{+\infty} \varphi'(u(s,t))[f(s,t)w(u(s,t)) + e(s,t)]\,dt\,ds. \tag{5.1.340}$$

Then (5.1.335) can be rewritten as

$$\varphi(u(x,y)) \le z(x,y) + b(x,y)\int_\alpha^x c(s,y)\varphi(u(s,y))\,ds. \tag{5.1.341}$$

Obviously, $z(x,y)$ is non-negative and continuous in $x \in \mathbb{R}_+$. Fixing $y \in \mathbb{R}_+$ in (5.1.341) and using Corollary 1.2.3 (i) in Qin [557], we get

$$\varphi(u(x,y)) \le z(x,y) + b(x,y)\int_\alpha^x z(s,y)c(s,y)\exp\left(\left(\int_s^x b(m,y)c(m,y)\,dm\right)\right) ds. \tag{5.1.342}$$

Since $z(x,y)$ is non-decreasing in $x \in \mathbb{R}_+$, we obtain from the last inequality that

$$\varphi(u(x,y)) \le z(x,y)p(x,y), \tag{5.1.343}$$

where $p(x, y)$ is defined by (5.1.337). From (5.1.343), we derive

$$\varphi(u(x,y)) \le p(x,y)\left(a(x,y)+d(x,y)\int_0^x\int_y^{+\infty}\varphi'(u(s,t))[f(s,t)w(u(s,t))+e(s,t)]\,dt\,ds\right). \tag{5.1.344}$$

Observe that $p(x, y)$, $a(x, y)$, and $d(x, y)$ are continuous, non-decreasing in x and non-increasing in y for $x,\ y \in \mathbb{R}_+$, so also are $p(x, y)a(x, y)$ and $p(x, y)d(x, y)$. Now applying Theorem 5.1.43 to (5.1.344), we can directly get the desired bound $u(x, y)$ in (5.1.336). □

Theorem 5.1.46 (The Cheung-Ma Inequality [145]) *Let $u(x, y)$, $f(x, y)$, $e(x, y)$, $\varphi(u)$, and $w(u)$ be defined as in Theorem 5.1.45. Let $a(x, y)$, $b(x, y)$, $c(x, y)$, and $d(x, y)$ be non-negative continuous and non-increasing in each variable $x,\ y \in \mathbb{R}_+$. If for all $x,\ y \in \mathbb{R}^+$ with $x \le \beta$,*

$$\begin{aligned}\varphi(u(x,y)) &\le a(x,y)+b(x,y)\int_x^{\beta}c(s,y)\varphi(u(s,y))\,ds\\ &\quad +d(x,y)\int_x^{+\infty}\int_y^{+\infty}\varphi'(u(s,t))[f(s,t)w(u(s,t))+e(s,t)]\,dt\,ds,\end{aligned} \tag{5.1.345}$$

then for all $x_6 \le x < +\infty,\ y_6 \le y < +\infty$,

$$u(x,y)\le G^{-1}\left(G[\varphi^{-1}(\bar{p}(x,y)a(x,y))+\bar{p}(x,y)\bar{E}_1(x,y)]+\bar{p}(x,y)d(x,y)\int_x^{+\infty}\int_y^{+\infty}f(s,t)\,dt\,ds\right), \tag{5.1.346}$$

where

$$\bar{p}(x,y) := 1+b(x,y)\int_x^{\beta}c(s,y)\exp\left(\int_x^{s}b(m,y)c(m,y)\,dm\right)ds, \tag{5.1.347}$$

$$\bar{E}_1(x,y) := d(x,y)\int_x^{+\infty}\int_y^{+\infty}e(s,t)\,dt\,ds, \tag{5.1.348}$$

$G,\ G^{-1},\ \varphi$ and φ^{-1} are defined as in Theorem 5.1.45, and $x_6,\ y_6 \in \mathbb{R}_+$ are chosen so that

$$G[\varphi^{-1}(\bar{p}(x,y)a(x,y))+\bar{p}(x,y)\bar{E}_1(x,y)]+\bar{p}(x,y)d(x,y)\int_x^{+\infty}\int_y^{+\infty}f(s,t)\,dt\,ds\in Dom\,(G^{-1}). \tag{5.1.349}$$

Proof The proof follows by an argument similar to that in the proof of Theorem 5.1.45 with suitable modification. We omit the details here. □

By choosing suitable functions for φ, some interesting new Gronwall-Ou-Yang type inequalities of two variables can be obtained from Theorems 5.1.45 and 5.1.46. For example, the following interesting inequalities are easily obtained.

Corollary 5.1.2 (The Cheung-Ma Inequality [451]) *Let $b(x,y)$, $c(x,y)$, $d(x,y)$, $e(x,y)$, $f(x,y)$, and $w(u)$ be as defined in Theorem 5.1.45. Let $k \geq 1$ be a real number. If for all x, y, $\alpha \in \mathbb{R}_+$ with $\alpha \leq x$,*

$$u^k(x,y) \leq a(x,y) + b(x,y)\int_\alpha^x c(s,y)u^k(s,y)\,ds$$
$$+d(x,y)\int_0^x\int_y^{+\infty} u^{k-1}(s,t)[f(s,t)w(u(s,t)) + e(s,t)]\,dt\,ds, \quad (5.1.350)$$

then for all $0 \leq x \leq x_7$, $y_7 \leq y_7 < +\infty$,

$$u(x,y) \leq G^{-1}\Bigg(G\left[p^{1/k}(x,y)a^{1/k}(x,y) + \frac{1}{k}p(x,y)E_1(x,y)\right]$$
$$+\frac{1}{k}p(x,y)d(x,y)\int_0^x\int_y^{+\infty} f(s,t)\,dt\,ds\Bigg), \quad (5.1.351)$$

where G, G^{-1}, $p(x,y)$ and $E_1(x,y)$ are as defined in Theorem 5.2.62, and x_7, $y_7 \in \mathbb{R}_+$ are chosen so that

$$G[p^{1/k}(x,y)a^{1/k}(x,y) + p(x,y)E_1(x,y)] + p(x,y)d(x,y)\int_0^x\int_y^{+\infty} f(s,t)\,dt\,ds \in Dom\,(G^{-1}).$$
$$(5.1.352)$$

Proof This follows immediately from Theorem 5.1.45 by setting $\varphi(u) = u^k$. □

Corollary 5.1.3 (The Cheung-Ma Inequality [145]) *Let $b(x,y)$, $c(x,y)$, $d(x,y)$, $e(x,y)$, $f(x,y)$, and $w(u)$ be as defined in Theorem 5.1.45. Let $u(x,y)$, $a(x,y) \in C(\mathbb{R}_+, \mathbb{R}_1)$ and $k > 0$ be a real number. If for all x, y, $\alpha \in \mathbb{R}_+$ with $\alpha \leq x$,*

$$u^k(x,y) \leq a(x,y) + b(x,y)\int_\alpha^x c(s,y)u^k(s,y)\,ds$$
$$+d(x,y)\int_0^x\int_y^{+\infty} u^k(s,t)[f(s,t)w(\log u(s,t)) + e(s,t)]\,dt\,ds,$$
$$(5.1.353)$$

then for all $0 \leq x \leq x_8,\ y_8 \leq y < +\infty$,

$$u(x,y) \leq \exp\left\{G^{-1}\left[G\left(\frac{1}{k}\log(p(x,y)a(x,y)) + \frac{1}{k}p(x,y)E_1(x,y)\right)\right.\right.$$
$$\left.\left.+\frac{1}{k}p(x,y)d(x,y)\int_0^x\int_y^{+\infty} f(s,t)\,dt\,ds\right]\right\}, \quad (5.1.354)$$

where $G,\ G^{-1},\ p(x,y)$ *and* $E_1(x,y)$ *are as defined in Theorem 5.1.45, and* $x_8,\ y_8 \in \mathbb{R}_+$ *are chosen so that*

$$G\left(\frac{1}{k}\log(p(x,y)a(x,y)) + \frac{1}{k}p(x,y)E_1(x,y)\right)$$
$$+\frac{1}{k}p(x,y)d(x,y)\int_0^x\int_y^{\infty} f(s,t)\,dt\,ds \in Dom\,(G^{-1}). \quad (5.1.355)$$

Proof Using the change of variable $v(x,y) = \log u(x,y)$, inequality (5.1.353) reduces to

$$e^{kv(x,y)} \leq a(x,y) + b(x,y)\int_\alpha^{+\infty} c(s,y)e^{kv(x,y)}\,ds$$
$$+d(x,y)\int_0^x\int_y^{+\infty} e^{kv(s,t)}[f(s,t)w(v(s,t)) + e(s,t)]\,dt\,ds, \quad (5.1.356)$$

which is a special case of inequality (5.1.335) when $\varphi(v) = \exp(kv)$. By Theorem 5.1.45, the desired inequality (5.1.354) follows. □

Theorem 5.1.47 (The Cheung-Ma Inequality [145]) *Let* $u(x,y),\ a(x,y),\ b(x,y),\ c(x,y),\ d(x,y),\ e(x,y), f(x,y),$ *and* $\varphi(u)$ *be as defined in Theorem 5.1.45, and* $L,\ M \in C(\mathbb{R}^3_+, \mathbb{R}^+)$ *satisfy, for all* $x,\ y,\ v,\ w \in \mathbb{R}_+$ *with* $v \leq w$,

$$0 \leq L(x,y,v) - L(x,y,w) \leq M(x,y,w)(v-w). \quad (5.1.357)$$

If for all $x,\ y,\ \alpha \in \mathbb{R}_+$ *with* $\alpha \leq x$,

$$\varphi(u(x,y)) \leq a(x,y) + b(x,y)\int_\alpha^x c(s,y)\varphi(u(s,y))\,ds$$
$$+d(x,y)\int_0^x\int_y^{+\infty} \varphi'(u(s,t))[f(s,t)L(s,t,u(s,t)) + e(s,t)]\,dt\,ds,$$
$$(5.1.358)$$

then for all $x,\ y \in \mathbb{R}_+$,

$$u(x,y) \leq N_1(x,y) + p(x,y)d(x,y)L_1(x,y)\exp(M_1(x,y)) \quad (5.1.359)$$

where

$$\begin{cases} N_1(x,y) := \varphi^{-1}(p(x,y)a(x,y)) + p(x,y)E_1(x,y), \\ L_1(x,y) := \int_0^x \int_y^{+\infty} f(s,t)L[s,t,N_1(s,t)]\,dt\,ds, \\ M_1(x,y) := \int_0^x \int_y^{+\infty} f(s,t)p(s,t)d(s,t)M[s,t,N_1(s,t)]\,dt\,ds, \end{cases} \tag{5.1.360}$$

and $p(x,y)$, $E_1(x,y)$ *are defined in (5.1.337), (5.1.338), respectively.*

Proof By similar arguments as those used in the proof of Theorem 5.1.45, applying Corollary 1.2.3 (i) in Qin [557] to (5.1.358), we conclude for all $x,\ y \in \mathbb{R}_+$,

$$\begin{aligned} \varphi(u(x,y)) \le\ & p(x,y)a(x,y) \\ & + p(x,y)d(x,y)\int_0^x \int_y^{+\infty} \varphi'(u(s,t))[f(s,t)L(s,t,u(s,t)) + e(s,t)]\,dt\,ds. \end{aligned} \tag{5.1.361}$$

Defining a non-negative continuous function $z(x,y)$ as the right-hand side of (5.1.361), then using similar procedures as in the proof of Theorem 5.1.43, we can derive from (5.1.361) that for all $x,\ y \in \mathbb{R}_+$,

$$u(x,y) \le \varphi^{-1}(z(x,y)),$$

$$\varphi^{-1}(z(x,y)) \le N_1(x,y) + p(x,y)d(x,y)\int_0^x \int_y^{+\infty} f(s,t)L[s,t,\varphi^{-1}(z(s,t))]\,dt\,ds, \tag{5.1.362}$$

where $N_1(x,y)$ is defined in (5.1.360).

Setting

$$\xi(x,y) = \int_0^x \int_y^{+\infty} f(s,t)L[s,t,\varphi^{-1}(z(s,t))]\,dt\,ds, \tag{5.1.363}$$

then from (5.1.362), we derive for all $x,\ y \in \mathbb{R}_+$,

$$\varphi^{-1}(z(x,y)) \le N_1(x,y) + p(x,y)d(x,y)\xi(x,y). \tag{5.1.364}$$

Since $L(x, y, v)$ is non-decreasing with respect to v for fixed $x,\ y$, by (5.1.363) and (5.1.364) with condition (5.1.357), we obtain

$$\begin{aligned}\xi(x,y) &\le \int_0^x \int_y^{+\infty} f(s,t)L[s,t,N_1(s,t)+p(s,t)d(s,t)\xi(s,t)]\,dt\,ds\\ &\le \int_0^x \int_y^{+\infty} f(s,t)L[s,t,N_1(s,t)]\,dt\,ds\\ &\quad+\int_0^x \int_y^{+\infty} f(s,t)p(s,t)d(s,t)M[s,t,N_1(s,t)]\xi(s,t)\,dt\,ds.\end{aligned} \tag{5.1.365}$$

Applying Lemma 5.1.3 (i) (the case when $w(u) = u,\ c(x,y) \equiv 1$) to the last inequality, we conclude

$$\begin{aligned}\xi(x,y) &\le \left(\int_0^x \int_y^{+\infty} f(s,t)L[s,t,N_1(s,t)]\,dt\,ds\right)\\ &\quad\times\exp\left(\int_0^x \int_y^{+\infty} f(s,t)p(s,t)d(s,t)M[s,t,N_1(s,t)]\,dt\,ds\right)\\ &= L_1(x,y)\exp(M_1(x,y)),\end{aligned} \tag{5.1.366}$$

where $L_1(x,y)$ and $M_1(x,y)$ are defined in (5.1.360). The required inequality (5.1.359) now follows from (5.1.362), (5.1.364) and the last inequality. □

Theorem 5.1.48 (The Cheung-Ma Inequality [145]) *Let $u(x,y),\ a(x,y),\ b(x,y),\ c(x,y),\ d(x,y),\ f(x,y)$ and $\varphi(u)$ be defined in Theorem 5.1.46, and $L(x,y,v)$ and $M(x,y,v)$ as defined in Theorem 5.1.47. If for all $\beta,\ x,\ y \in \mathbb{R}_+$ with $x \le \beta$,*

$$\begin{aligned}\varphi(u(x,y)) &\le a(x,y) + b(x,y)\int_x^\beta c(s,y)\varphi(u(s,y))\,ds\\ &\quad+d(x,y)\int_x^{+\infty}\int_y^{+\infty}\varphi'(u(s,t))[f(s,t)L(s,t,u(s,t))+e(s,t)]\,dt\,ds,\end{aligned} \tag{5.1.367}$$

then for all $x,\ y \in \mathbb{R}_+$,

$$u(x,y) \le \bar{N}_1(x,y) + \bar{p}(x,y)d(x,y)\bar{L}_1(x,y)\exp(\bar{M}_1(x,y)) \tag{5.1.368}$$

where

$$\begin{cases} \bar{N}_1(x,y) := \varphi^{-1}(\bar{p}(x,y)a(x,y)) + \bar{p}(x,y)\bar{E}_1(x,y), \\ \bar{L}_1(x,y) := \int_x^{+\infty}\int_y^{+\infty} f(s,t)L[s,t,\bar{N}_1(s,t)]\,dt\,ds, \\ \bar{M}_1(x,y) := \int_x^{+\infty}\int_y^{+\infty} f(s,t)\bar{p}(s,t)d(s,t)M[s,t,\bar{N}_1(s,t)]\,dt\,ds, \end{cases} \tag{5.1.369}$$

and $\bar{p}(x,y)$, $\bar{E}_1(x,y)$ *are defined in (5.1.347), (5.1.348), respectively.*

Proof The proof follows by an argument similar to that of Theorem 5.1.47 with suitable modification. We omit the details here. □

Remark 5.1.5 As in Corollaries 5.1.2 and 5.1.3, other new Ou-Yang type integral inequalities of two variables can be obtained from Theorems 5.1.47 and 5.1.48 by choosing suitable functions for φ. Details are omitted here.

The following two results, due to Zheng, Wu and Deng [723] generalizes the results Theorems 5.1.43–5.1.48 of Cheung and Ma to more general inequalities with more than once distinct nonlinear terms.

Cheung [143], and Dragomir and Kim [208] established additional Ou-Yang type integral inequalities involving functions of two independent variables. Meng and Li [388] generalized the results of Pachpatte [516] to certain new integrals. Cheung and Ma [145] (see, Theorems 5.2.22–5.2.26) discussed the following inequalities

$$\begin{cases} u(x,y) \le a(x,y) + c(x,y)\int_0^x\int_y^{+\infty} d(s,t)\,w(u(s,t))\,dt\,ds, \\ u(x,y) \le a(x,y) + c(x,y)\int_x^{+\infty}\int_y^{+\infty} d(s,t)\,w(u(s,t))\,dt\,ds, \end{cases} \tag{5.1.370}$$

where $a(x,y)$, and $c(x,y)$ have certain monotonicity.

Motivated by the work of Cheung and Ma [145], the following two results on more general integral inequalities with n nonlinear terms will be discussed

$$u(x,y) \le a(x,y) + \sum_{i=1}^{n}\int_0^x\int_y^{+\infty} d_i(x,y,s,t)\,w_i(u(s,t))\,dt\,ds, \tag{5.1.371}$$

$$u(x,y) \le a(x,y) + \sum_{i=1}^{n}\int_x^{+\infty}\int_y^{+\infty} d_i(x,y,s,t)\,w_i(u(s,t))\,dt\,ds, \tag{5.1.372}$$

where we do not require the monotonicity of $a(x,y)$ and $d_i(x,y,s,t)$. Furthermore, we note that some results of Cheung and Ma [145] can be deduced from the present results as some special cases.

As in [13, 151, 541], we define $w_1 \propto w_2$ for $w_1, w_2 : A \subset \mathbb{R} \to \mathbb{R}\backslash\{0\}$ if w_2/w_1 is non-decreasing on A. This concept helps us compare monotonicity of different functions. Suppose that

(C_1) $w_i(u)$ $(i = 1, \ldots, n)$ is a non-negative, non-decreasing, and continuous function for all $u \in A$ with $w_i(u) > 0$ for all $u > 0$ such that $w_1 \propto w_2 \propto \ldots \propto w_n$;

(C_2) $a(x, y)$ is a non-negative and continuous function for all $x, y \in \mathbb{R}_+$;

(C_3) $d_i(x, y, s, t)$ $(i = 1, \ldots, n)$ is a continuous and non-negative function for all $x, y, s, t \in \mathbb{R}_+$ take the notation $W_i(u) := \int_{u_i}^{u}(dz/w_i(z))$, for all $u \geq u_i$, where $u_i > 0$ is a given constant. Clearly, W_i is strictly increasing, so its inverse W_i^{-1} is well-defined, continuous, and increasing in its corresponding domain.

Theorem 5.1.49 (The Zheng-Wu-Deng Inequality [723]) *In addition to the assumptions (C_1), (C_2), and (C_3), suppose that $a(x, y)$ and $d_i(x, y, s, t)$ are bounded in $y \in \mathbb{R}_+$ for each fixed $x, s, t \in \mathbb{R}_+$. If $u(x, y)$ is a continuous and non-negative function satisfying (5.1.371) for all $x, y \in \mathbb{R}^+$, then for all $0 \leq x \leq x_1, y_1 \leq y < +\infty$,*

$$u(x, y) \leq W_n^{-1}\left[W_n(b_n(x, y)) + \int_0^x \int_y^{+\infty} \tilde{d}_n(x, y, s, t)\, dt\, ds \right], \qquad (5.1.373)$$

where b_n is determined recursively by

$$\begin{cases} b_1(x, y) = \tilde{a}(x, y), \\ b_{i+1}(x, y) = W_i^{-1}\left[W_i(b_i(x, y)) + \int_0^x \int_y^{+\infty} \tilde{d}_i(x, y, s, t)\, dt\, ds \right], \\ \tilde{a}(x, y) = \sup_{0 \leq \tau \leq x} \sup_{y \leq \mu \leq +\infty} a(\tau, \mu),\ \tilde{d}_i(x, y, s, t) = \sup_{0 \leq \tau \leq x} \sup_{y \leq \mu \leq +\infty} d_i(\tau, \mu, s, t), \end{cases} \qquad (5.1.374)$$

and $W_1 := 0$, and x_1, y_1 are chosen such that

$$W_i(b_i(x_1, y_1)) + \int_0^{x_1} \int_{y_1}^{+\infty} \tilde{d}_i(x, y, s, t)\, dt\, ds \leq \int_{a_i}^{+\infty} \frac{dz}{w_i(z)} \qquad (5.1.375)$$

for $i = 1, \cdots, n$.

Proof From the assumptions, we know that $\tilde{a}(x, y)$ and $\tilde{d}_i(x, y, s, t)$ are well-defined. Moreover, $\tilde{a}(x, y)$ and $\tilde{d}_i(x, y, s, t)$ are non-negative, non-decreasing in x, non-increasing in y; and satisfy $\tilde{a}(x, y) \geq a(x, y)$ and $\tilde{d}_i(x, y, s, t) \geq d_i(x, y, s, t)$ for $i = 1, \cdots, n$.

We first discuss the case that $a(x, y) > 0$ for all $x,\ y \in \mathbb{R}_+$. Thus, $b_1(x, y)$ is positive, non-decreasing in x, non-increasing in y; and satisfies $b_1(x, y) \geq a(x, y)$ for all $x, y \in \mathbb{R}_+$.

From (5.1.371), we derive

$$u(x,y) \le b_1(x,y) + \sum_{i=1}^{n} \int_0^x \int_y^{+\infty} \tilde{d}_i(x,y,s,t)\, w_i(u(s,t))\, dt\, ds. \qquad (5.1.376)$$

Choose arbitrary $\tilde{x}_1,\ \tilde{y}_1$ such that $0 \ge \tilde{x}_1 \ge x_1, y_1 \ge \tilde{y}_1 < +\infty$. From (5.1.376), it follows for all $0 \le x \le \tilde{x}_1 \le x_1,\ y_1 \le \tilde{y}_1 \le y < +\infty$,

$$u(x,y) \le b_1(\tilde{x}_1,\tilde{y}_1) + \sum_{i=1}^{n} \int_0^x \int_y^{+\infty} \tilde{d}_i(\tilde{x}_1,\tilde{y}_1,s,t)\, w_i(\, u(s,t))\, dt\, ds. \qquad (5.1.377)$$

Having (5.1.377), we claim for all $0 \le x \le \min\{\tilde{x}_1,\ x_2\},\ \max\{\tilde{y}_1, y_2\} \le y \le +\infty$,

$$u(x,y) \le\ W_n^{-1}\left[\ W_n(\,\tilde{b}_n(\tilde{x}_1,\tilde{y}_1,x,y)) + \int_0^x \int_y^{+\infty} \tilde{d}_n(\tilde{x}_1,\tilde{y}_1,s,t)\, dt\, ds\right] \qquad (5.1.378)$$

where

$$\begin{aligned} \tilde{b}_1(\tilde{x}_1,\tilde{y}_1,x,y) &= b_1(\tilde{x}_1,\tilde{y}_1), \\ \tilde{b}_{i+1}(\tilde{x}_1,\tilde{y}_1,x,y) &= W_i^{-1}\left[W_i(\tilde{b}_i(\tilde{x}_1,\tilde{y}_1,x,y)) + \int_c^x \int_y^{+\infty} \tilde{d}_i(\tilde{x}_1,\tilde{y}_1,s,t)\, dt\, ds\right] \end{aligned} \qquad (5.1.379)$$

for $i = 1,\ldots,n-1$ and $x_2,\ y_2 \in \mathbb{R}_+$ are chosen such that

$$W_i(\,\tilde{b}_i(\tilde{x}_1,\tilde{y}1,x_2,y_2)) + \int_0^{x_2} \int_{y_2}^{+\infty} \tilde{d}_i(\tilde{x}_1,\tilde{y}_1,s,t)\, dt\, ds \le \int_{u_i}^{+\infty} \frac{dz}{w_i(z)} \qquad (5.1.380)$$

for $i = 1,\ldots,n$. Note that we may take $x_2 = x_1$ and $y_2 = y_1$. In fact, $\tilde{b}_i(\tilde{x}_1,\tilde{y}_1,x,y)$ and $\tilde{d}_i(\tilde{x}_1,\tilde{y}_1,x,y)$ are non-decreasing in $\tilde{x}_1$, non-increasing in $\tilde{y}_1$ for fixed x,y. Furthermore, it is easy to check that $\tilde{b}_i(\tilde{x}_1,\tilde{y}_1,\tilde{x}_1,\tilde{y}_1) =\ b_i(\tilde{x}_1,\tilde{y}_1)$ for $i = 1,\ldots,n$. If x_2, y_2 are replaced by x_1, y_1 on the left-hand side of (5.1.380), we have from (5.1.375)

$$\begin{aligned} &W_i(\,\tilde{b}_i(\tilde{x}_1,\tilde{y}_1,x_1,y_1)) + \int_0^{x_1} \int_{y_1}^{+\infty} \tilde{d}_i(\tilde{x}_1,\tilde{y}_1,s,t)\, dt\, ds \\ &\le\ W_i(\,\tilde{b}_i(x_1,y_1,x_1,y_1)) + \int_0^{x_1} \int_{y_1}^{+\infty} \tilde{d}_i(x_1,y_1,,s,t)\, dt\, ds \\ &= W_i(\,b_i(x_1,y_1)) + \int_0^{x_1} \int_{y_1}^{+\infty} \tilde{d}_i(x_1,y_1,s,t)\, dt\, ds \le \int_{u_i}^{+\infty} \frac{dz}{w_i(z)}. \end{aligned} \qquad (5.1.381)$$

Thus it means that we can take $x_2 = x_1, y_2 = y_1$.

In the following, we shall use mathematical induction to prove (5.1.378). For $n = 1$, let

$$z(x,y) = \int_0^x \int_y^{+\infty} \tilde{d}_1(\tilde{x}_1,\tilde{y}_1,s,t)\, w_1(\, u(s,t))\, dt\, ds. \tag{5.1.382}$$

Then $z(x,y)$ is differentiable, non-negative, non-decreasing for all $x \in [0,\tilde{x}_1]$, and non-increasing for all $y \in [\tilde{y}_1,+\infty)$ and $z(0,y) = z(x,+\infty) = 0$. From (5.1.377), we infer

$$\begin{cases} u(x,y) \le b_1(\tilde{x}_1,\tilde{y}_1) + z(x,y), \\ D_1 z(x,y) = \displaystyle\int_y^{+\infty} \tilde{d}_1(\tilde{x}_1,\tilde{y}_1,x,t)\, w_1(\, u(x,t))\, dt \\ \qquad \le \displaystyle\int_y^{+\infty} \tilde{d}_1(\tilde{x}_1,\tilde{y}_1,x,t) w_1(b_1(\tilde{x}_1,\tilde{y}_1) + z(x,t))\, dt \\ \qquad \le w_1(\, b_1(\tilde{x}_1,\tilde{y}_1) + z(x,y)) \displaystyle\int_y^{+\infty} \tilde{d}_1(\tilde{x}_1,\tilde{y}_1,x,t)\, dt. \end{cases} \tag{5.1.383}$$

Since w_1 is non-decreasing and $b_1(\tilde{x}_1,\tilde{y}_1) + z(x,y) > 0$, we get

$$\begin{aligned} \frac{D_1(b_1(\tilde{x}_1,\tilde{y}_1) + z(x,y))}{w_1(b_1(\tilde{x}_1,\tilde{y}_1) + z(x,y))} &= \frac{D_1 z(x,y)}{w_1(b_1(\tilde{x}_1,\tilde{y}_1) + z(x,y))} \\ &\le \frac{w_1(b_1(\tilde{x}_1,\tilde{y}_1) + z(x,y)) \int_y^{+\infty} \tilde{d}_1(\tilde{x}_1,\tilde{y}_1,x,t)\, dt}{w_1(b_1(\tilde{x}_1,\tilde{y}_1) + z(x,y))} \\ &= \int_y^{+\infty} \tilde{d}_1(\tilde{x}_1,\tilde{y}_1,x,t)\, dt. \end{aligned} \tag{5.1.384}$$

Integrating both sides of the above inequality from 0 to x, we obtain

$$\begin{aligned} W_1(b_1(\tilde{x}_1,\tilde{y}_1)+z(x,y)) &\le W_1(b_1(\tilde{x}_1,\tilde{y}_1)+z(0,y)) + \int_0^x \int_y^{+\infty} \tilde{d}_1(\tilde{x}_1,\tilde{y}_1,s,t)\, dt\, ds \\ &= W_1(b_1(\tilde{x}_1,\tilde{y}_1)) + \int_0^x \int_y^{+\infty} \tilde{d}_1(\tilde{x}_1,\tilde{y}_1,s,t)\, dt\, ds. \end{aligned} \tag{5.1.385}$$

Thus the monotonicity of W_1^{-1} implies

$$u(x,y) \le b_1(\tilde{x}_1,\tilde{y}_1) + z(x,y) \le W_1^{-1}\left[W_1(b_1(\tilde{x}_1,\tilde{y}_1) + \int_0^x \int_y^{+\infty} \tilde{d}_1(\tilde{x}_1,\tilde{y}_1,s,t)\, dt\, ds \right], \tag{5.1.386}$$

that is, (5.1.447) is true for $n = 1$.

Assume that (5.1.377) is true for $n = m$. Consider

$$u(x,y) \le b_1(\tilde{x}_1,\tilde{y}_1) + \sum_{i=1}^{m+1} \int_0^x \int_y^{+\infty} \tilde{d}_i(\tilde{x}_1,\tilde{y}_1,s,t)\, u_i(u(s,t))\, dt\, ds \tag{5.1.387}$$

for all $0 \le x \le \tilde{x}_1, \tilde{y}_1 \le y < +\infty$. Let

$$z(x,y) = \sum_{i=1}^{m+1} \int_0^x \int_y^{+\infty} \tilde{d}_i(\tilde{x}_1,\tilde{y}_1,s,t)\, w_i(u(s,t))\, dt\, ds. \tag{5.1.388}$$

Then $z(x,y)$ is differentiable, non-negative, non-decreasing for all $x \in [0,\tilde{x}_1]$, and non-increasing for $y \in [\tilde{y}_1,+\infty)$. Obviously, $z(0,y) = z(x,+\infty) = 0$ and $u(x,y) \le b_1(\tilde{x}_1,\tilde{y}_1) + z(x,y)$. Since w_1 is non-decreasing and $b_1(\tilde{x}_1,\tilde{y}_1) + z(x,y) > 0$, we have

$$\begin{aligned}
&\frac{D_1(b_1(\tilde{x}_1,\tilde{y}_1) + z(x,y))}{w_1(b_1(\tilde{x}_1,\tilde{y}_1) + z(x,y))} \\
&\le \frac{\sum_{i=1}^{m+1} \int_y^{+\infty} \tilde{d}_i(\tilde{x}_1,\tilde{y}_1,x,t) w_i(u(x,t))\, dt}{w_1(b_1(\tilde{x}_1,\tilde{y}_1) + z(x,y)} \\
&\le \frac{\sum_{i=1}^{m+1} \int_y^{+\infty} \tilde{d}_i(\tilde{x}_1,\tilde{y}_1,x,t) w_i(b_1(\tilde{x}_1,\tilde{y}_1) + z(x,t))\, dt}{w_1(b_1(\tilde{x}_1,\tilde{y}_1) + z(x,y)} \\
&\le \int_y^{+\infty} \tilde{d}_1(\tilde{x}_1,\tilde{y}_1,x,t)\, dt + \sum_{i=2}^{m+1} \int_y^{+\infty} \tilde{d}_i(\tilde{x}_1,\tilde{y}_1,x,t) \phi_i(b_1(\tilde{x}_1,\tilde{y}_1) + z(x,t))\, dt \\
&\le \int_y^{+\infty} \tilde{d}_1(\tilde{x}_1,\tilde{y}_1,x,t)\, dt + \sum_{i=1}^{m} \int_y^{+\infty} \tilde{d}_{i+1}(\tilde{x}_1,\tilde{y}_1,x,t) \phi_{i+1}(b_1(\tilde{x}_1,\tilde{y}_1) + z(x,t))\, dt,
\end{aligned} \tag{5.1.389}$$

where $\phi_{i+1}(u) = w_{i+1}(u)/w_1(u)$, $i = 1,\ldots,m$. Integrating the above inequality from 0 to x, we obtain for all $0 \le x \le \tilde{x}_1$ and $\tilde{y}_1 \le y < +\infty$,

$$\begin{aligned}
W_1(b_1(\tilde{x}_1,\tilde{y}_1) + z(x,y)) &\le W_1(b_1(\tilde{x}_1,\tilde{y}_1)) + \int_0^x \int_y^{+\infty} \tilde{d}_1(\tilde{x}_1,\tilde{y}_1,s,t)\, dt\, ds \\
&\quad + \sum_{i=1}^{m} \int_0^x \int_y^{+\infty} \tilde{d}_{i+1}(\tilde{x}_1,\tilde{y}_1,s,t) \phi_{i+1}(b_1(\tilde{x}_1,\tilde{y}_1) + z(s,t))\, dt\, ds
\end{aligned} \tag{5.1.390}$$

or

$$\xi(x,y) \le c_1(x,y) + \sum_{i=1}^{m} \int_0^x \int_y^{+\infty} \tilde{d}_{i+1}(\tilde{x}_1, \tilde{y}_1, s, t)\phi_{i+1}(W^{-1}(\xi(s,t)))\, dt\, ds \tag{5.1.391}$$

the same as (5.1.377) for $n = m$, where $\xi(x,y) = W_1(b_1(\tilde{x}_1, \tilde{y}_1) + z(x,y))$ and $c_1(x,y) = W_1(b_1(\tilde{x}_1, \tilde{y}_1)) + \int_0^x \int_y^{+\infty} \tilde{d}_1(\tilde{x}_1, \tilde{y}_1, s, t)\, dt\, ds$.

From the assumption (C_1), each $\phi_{i+1}(W_1^{-1}(u)), i = 1, \ldots, m$, is continuous and non-decreasing for all u. Moreover, $\phi_2(W_1^{-1}) \propto \phi_3(W_1^{-1}) \propto \ldots \propto \phi_m(W_1^{-1})$. By the inductive assumption, we have, for all $0 \le x \le \min\{\tilde{x}_1, x_3\}$, $\max\{\tilde{y}_1, y_3\} \le y < +\infty$,

$$\xi(x,y) \le \Phi_{m+1}^{-1}\left[\Phi_{m+1}(c_m(x,y)) + \int_0^x \int_y^{+\infty} d_{m+1}(\tilde{x}_1, \tilde{y}_1, s, t)\, dt\, ds\right] \tag{5.1.392}$$

where $\Phi_{i+1}(u) = \int_{\tilde{u}_{i+1}}^{u} (dz/\phi_{i+1}(W_1^{-1}(z)))$, $u > 0$, $\tilde{u}_{i+1} = W_1(u_{i+1})$, Φ_{i+1}^{-1} is the inverse of Φ_{i+1}, $i = 1, \ldots, m$,

$$c_{i+1}(x,y) = \Phi_{i+1}^{-1}\left[\Phi_{i+1}(c_i(x,y)) + \int_0^x \int_y^{+\infty} \tilde{d}_{i+1}(\tilde{x}_1, \tilde{y}_1, s, t)\, dt\, ds\right], \; i = 1, \ldots, m, \tag{5.1.393}$$

and $x_3, y_3 \in \mathbb{R}_+$ are chosen such that

$$\Phi_{i+1}(c_i(x_3, y_3)) + \int_0^{x_3} \int_{y_3}^{+\infty} \tilde{d}_{i+1}(\tilde{x}_1, \tilde{y}_1, s, t)\, dt\, ds \le \int_{\tilde{u}_{i+1}}^{W_1(+\infty)} \frac{dz}{\phi_{i+1}(W_1^{-1}(z))} \tag{5.1.394}$$

for $i = 1, \ldots, m$.

Note that

$$\begin{aligned} \Phi_i(u) &= \int_{\tilde{u}_i}^{u} \frac{dz}{\phi_i(W_1^{-1}(z))} = \int_{W_1(u_i)}^{u} \frac{w_1(W_1^{-1}(z))dz}{w_i(W_1^{-1}(z))} \\ &= \int_{u_i}^{W_1^{-1}(u)} \frac{dz}{w_i(z)} = W_i \circ W_1^{-1}(u), \; i = 2, \ldots, m+1. \end{aligned} \tag{5.1.395}$$

From (5.1.392), we infer for all $0 \le x \le \min\{\tilde{x}_1, x_3\}$, $\max\{\tilde{y}_1, y_3\} \le y < +\infty$,

$$\begin{aligned} u(x,y) &\le b_1(\tilde{x}_1,\tilde{y}_1) + z(x,y) = W_1^{-1}(\xi(x,y)) \\ &\le W_{m+1}^{-1}\left[W_{m+1}(W_1^{-1}(c_m(x,y))) + \int_0^x \int_y^{+\infty} \tilde{d}_{m+1}(\tilde{x}_1,\tilde{y}_1,s,t)\,dt\,ds\right]. \end{aligned} \tag{5.1.396}$$

Let $\tilde{c}_i(x,y) = W_1^{-1}(c_i(x,y))$. Then,

$$\begin{aligned} \tilde{c}_1(x,y) &= W_1^{-1}(c_1(x,y)) \\ &= W_1^{-1}\left[W_1(b_1(\tilde{x}_1,\tilde{y}_1)) + \int_0^x \int_y^{+\infty} \tilde{d}_1(\tilde{x}_1,\tilde{y}_1,s,t)\,dt\,ds\right] \\ &= \tilde{b}_2(\tilde{x}_1,\tilde{y}_1,x,y). \end{aligned} \tag{5.1.397}$$

Moreover, with the assumption that $\tilde{c}_m(x,y) = \tilde{b}_{m+1}(\tilde{x}_1,\tilde{y}_1,x,y)$, we get

$$\begin{aligned} \tilde{c}_{m+1}(x,y) &= W_1^{-1}\left[\Phi_{m+1}^{-1}(\Phi_{m+1}(c_m(x,y)) + \int_0^x \int_y^{+\infty} \tilde{d}_{m+1}(\tilde{x}_1,\tilde{y}_1,s,t)\,dt\,ds)\right] \\ &= W_{m+1}^{-1}\left[W_{m+1}(W_1^{-1}(c_m(x,y))) + \int_0^x \int_y^{+\infty} \tilde{d}_{m+1}(\tilde{x}_1,\tilde{y}_1,s,t)\,dt\,ds\right] \\ &= W_{m+1}^{-1}\left[W_{m+1}(\tilde{c}_m(x,y)) + \int_0^x \int_y^{+\infty} \tilde{d}_{m+1}(\tilde{x}_1,\tilde{y}_1,s,t)\,dt\,ds\right] \\ &= W_{m+1}^{-1}\left[W_{m+1}(\tilde{b}_{m+1}(\tilde{x}_1,\tilde{y}_1,x,y)) + \int_0^x \int_y^{+\infty} \tilde{d}_{m+1}(\tilde{x}_1,\tilde{y}_1,s,t)\,dt\,ds\right] \\ &= \tilde{b}_{m+2}(\tilde{x}_1,\tilde{y}_1,x,y) \end{aligned} \tag{5.1.398}$$

which proves that

$$\tilde{c}_i(x,y) = \tilde{b}_{i+1}(\tilde{x}_1,\tilde{y}_1,x,y),\ i = 1,\ldots,m. \tag{5.1.399}$$

Therefore, (5.1.394) becomes

$$\begin{aligned} &W_{i+1}(\tilde{b}_{i+1}(\tilde{x}_1,\tilde{y}_1,x_3,y_3)) + \int_0^{x_3} \int_{y_3}^{+\infty} \tilde{d}_{i+1}(\tilde{x}_1,\tilde{y}_1,s,t)\,dt\,ds \\ &\le \int_{\tilde{u}_{i+1}}^{W_i(+\infty)} \frac{dz}{\phi_{i+1}(W_1^{-1}(z))} = \int_{u_{i+1}}^{\infty} \frac{dz}{w_{i+1}(z)},\quad i = 1,\ldots,m. \end{aligned} \tag{5.1.400}$$

The above inequalities and (5.1.380) imply that we may take $x_2 = x_3, y_2 = y_3$. From (5.1.396), we conclude for all $0 \le x \le \tilde{x}_1 \le x_2,\ y_2 \le \tilde{y}_1 \le y < +\infty$,

$$u(x,y) \le W_{m+1}^{-1}\left[W_{m+1}(\tilde{b}_{m+1}(\tilde{x}_1,\tilde{y}_1,x,y)) + \int_0^x \int_y^{+\infty} \tilde{d}_{m+1}(\tilde{x}_1,\tilde{y}_1,s,t)\,dt\,ds\right]. \tag{5.1.401}$$

This proves (5.1.378) by mathematical induction.

Taking $x = \tilde{x}_1,\ y = \tilde{y}_1,\ x_2 = x_1$, and $y_2 = y_1$, we have for all $0 \le \tilde{x}_1 \le x_1,\ y_1 \le \tilde{y}_1 < +\infty$,

$$u(\tilde{x}_1,\tilde{y}_1) \le W_n^{-1}\left[W_n(\tilde{b}_n(\tilde{x}_1,\tilde{y}_1,\tilde{x}_1,\tilde{y}_1)) + \int_0^{\tilde{x}_1} \int_{\tilde{y}_1}^{+\infty} \tilde{d}_n(\tilde{x}_1,\tilde{y}_1,s,t)\,dt\,ds\right]. \tag{5.1.402}$$

It is easy to verify $\tilde{b}_n(\tilde{x}_1,\tilde{y}_1,\tilde{x}_1,\tilde{y}_1) = b_n(\tilde{x}_1,\tilde{y}_1)$. Thus, (5.1.402) can be rewritten as

$$u(\tilde{x}_1,\tilde{y}_1) \le W_n^{-1}\left[W_n(b_n(\tilde{x}_1,\tilde{y}_1)) + \int_0^{\tilde{x}_1} \int_{\tilde{y}_1}^{+\infty} \tilde{d}_n(\tilde{x}_1,\tilde{y}_1,s,t)\,dt\,ds\right]. \tag{5.1.403}$$

Since $\tilde{x}_1, \tilde{y}_1$ are arbitrary, replace $\tilde{x}_1$ and $\tilde{y}_1$ by x and y respectively, and we have for all $0 \le x \le x_1,\ y_1 \le y < +\infty$,

$$u(x,y) \le W_n^{-1}\left[W_n(b_n(x,y)) + \int_0^x \int_y^{+\infty} \tilde{d}_n(x,y,s,t)\,dt\,ds\right]. \tag{5.1.404}$$

In case $a(x,y) = 0$ for some $x, y \in \mathbb{R}_+$. Let $b_{1,\epsilon}(x,y) := b_1(x,y) + \epsilon$ for all $x, y \in \mathbb{R}_+$, where $\varepsilon > 0$ is arbitrary, and then $b_{1,\varepsilon}(x,y) > 0$. Using the same arguments as above, where $b_1(x,y)$ is replaced with $b_{1,\epsilon}(x,y) > 0$, we get

$$u(x,y) \le W_n^{-1}[W_n(b_{n,\epsilon}(x,y)) + \int_0^x \int_y^{+\infty} \tilde{d}_n(x,y,s,t)\,dt\,ds]. \tag{5.1.405}$$

Letting $\varepsilon \to 0^+$, we obtain (5.1.373) by the continuity of W_i and W_i^{-1} under the notation $W_1(0) := 0$. □

Remark 5.1.6 x_1 and y_1 are confined by (5.1.375). In particular, (5.1.373) is true for all x, y where all w_i $(i = 1, \ldots, n)$ satisfy $\int_{u_i}^{+\infty}(dz/w_i(z)) = +\infty$.

Remark 5.1.7 As in [13, 151, 541], different choices of u_i in W_i do not affect the above present results.

Theorem 5.1.50 (The Zheng-Wu-Deng Inequality [723]) *In addition to the assumptions* $(C_1), (C_2)$, *and* (C_3), *suppose that* $a(x,y)$ *and* $d_i(x,y,s,t)$ *are* *bounded in* $x, y \in \mathbb{R}_+$ *for each fixed* $s, t \in \mathbb{R}_+$. *If* $u(x, y)$ *is a continuous and non-negative function satisfying (5.1.372) for all* $x, y \in \mathbb{R}_+$, *then for all* $x_4 \le x < +\infty$, $y_4 \le y < +\infty$,

$$u(x,y) \le W_n^{-1}\left[W_n(b_n(x,y)) + \int_x^{+\infty}\int_y^{+\infty} \hat{d}_n(x,y,s,t)\,dt\,ds\right] \tag{5.1.406}$$

where $b_n(x, y)$ *is determined recursively by*

$$\begin{cases} b_1(x,y) = \hat{a}(x,y), \\ b_{i+1}(x,y) = W_i^{-1}\left[W_i(b_i(x,y)) + \int_x^{+\infty}\int_y^{+\infty} \hat{d}_i(x,y,s,t)\,dt\,ds\right], \\ \hat{a}(x,y) = \sup_{x\le\tau<+\infty}\sup_{y\le\mu<+\infty} a(\tau,\mu), \\ \hat{d}_i(x,y,s,t) = \sup_{x\le\tau<+\infty}\sup_{y\le\mu<+\infty} d_i(\tau,\mu,s,t), \end{cases} \tag{5.1.407}$$

$W_1(0) := 0$, *and* $x_4, y_4 \in \mathbb{R}_+$ *are chosen such that*

$$W_i(b_i(x_4,y_4)) + \int_{x_4}^{+\infty}\int_{y_4}^{+\infty} \hat{d}_i(x,y,s,t)\,dt\,ds] \le \int_{u_i}^{+\infty}\frac{dz}{w_i(z)} \tag{5.1.408}$$

for $i = 1, \dots, n$.

Proof The proof is similar to the argument in the proof of Theorem 5.1.49 with suitable modification. We omit the details here. □

Remark 5.1.8 Take $d_1(x, y, s, t) = c(x, y)d(s, t)$ and $n = 1$ in (5.1.371). Suppose that $a(x, y)$ and $c(x, y)$ are continuous, non-negative, non-decreasing in x and non-increasing in y; and $d(s, t)$ is non-negative and continuous. We note that

$$b_1(x,y) = a(x,y),\ \tilde{d}_1(x,y,s,t) = c(x,y)d(s,t). \tag{5.1.409}$$

From Theorem 5.1.49, it follows

$$u(x,y) \le W_1^{-1}\left[W_1(a(x,y)) + c(x,y)\int_0^{x}\int_y^{+\infty} d(s,t)\,dt\,ds\right], \tag{5.1.410}$$

which is exactly (5.1.300).

Remark 5.1.9 Take $d_1(x, y, s, t) = c(x, y)d(s, t)$ and $n = 1$ in (5.1.373). Suppose that $a(x, y)$ and $c(x, y)$ are continuous, non-negative, non-increasing in x, y; and $d(s, t)$ is non-negative and continuous. It is easy to check that

$$b_1(x,y) = a(x,y),\ \hat{d}_1(x,y,s,t) = c(x,y)d(s,t). \tag{5.1.411}$$

From Theorem 5.1.50, we get

$$u(x,y) \leq W_1^{-1}\left[W_1(a(x,y)) + c(x,y)\int_x^{+\infty}\int_y^{+\infty} d(s,t)\,dt\,ds\right], \tag{5.1.412}$$

which is (5.1.304).

Integral inequalities involving functions and their derivatives have been established by many authors in the literature during the past several years, see [48, 272, 395]. In particular, integral inequalities of considerable interest involving functions and their derivatives are associated with the names of Wirtinger and Opial [395]. A large number of papers have been written dealing with the various extensions and generations of these two inequalities, see [53, 607] and the references given therein. Pachpatte [480] has established some new integral inequalities of the Opial type in two independent variables which in turn contain as a special case the interesting analog of Opial's inequality given by Yang [698]. We shall introduce the result in [480] and some new integral inequalities of Wirtinger and Opial type involving real-valued functions of two independent variables and their partial derivatives. The method used in the proofs is very elementary and based on some simple observations and applications of the fundamental inequalities. The following established results in this yield in the special cases the two independent variable analogous of some of the Wirtinger and Opial type inequalities established in [481] and Traple in [647].

Next we use the following notations: $D_1p(x,y) = \frac{\partial}{\partial x}p(x,y)$, $D_2p(x,y) = \frac{\partial}{\partial y}p(x,y)$, $D_1D_2p(x,y) = \frac{\partial^2}{\partial x\partial y}p(x,y)$ and $\Delta = [a,b]\times[c,d]$, $\Delta_1 = [a,X]\times[c,Y]$, $\Delta_2 = [a,X]\times[Y,d]$, $\Delta_3 = [X,b]\times[c,Y]$, $\Delta_4 = [X,b]\times[Y,d]$ for all $a \leq X \leq b$, $c \leq X \leq d$.

Theorem 5.1.51 (The Pachpatte Inequality [485]) *Let $p(x,y)$ be real-valued non-negative function defined on Δ. If $f_r(x,y)$, $D_1f_r(x,y)$ and $D_1D_2f_r(x,y)$ are real-valued continuous functions defined on Δ for all $r = 1,\cdots,n$ and if $f_r(a,y) = f_r(b,y) = D_1f_r(x,c) = D_1f_r(x,d) = 0$, for $a \leq x \leq b$, $c \leq y \leq d$, then*

$$\begin{aligned}
&\int_a^b\left(\int_c^d p(x,y)\left[\prod_{r=1}^n |f_r(x,y)|^{m_r}\right]^{2/n} dy\right)dx \\
&\leq \frac{1}{n}K(a,b,c,d,n,m_1,\cdots,m_n)\left\{\int_a^b\left(\int_c^d p(x,y)dy\right)dx\right\} \\
&\quad\times\left\{\int_a^b\left(\int_c^b\left[\sum_{r=1}^n |D_1D_2f_r(x,y)|^{2m_r}\right]dy\right)dx\right\},
\end{aligned} \tag{5.1.413}$$

where $m_r \geq 1$ (for $r = 1, \cdots, n$) are constants and

$$K(a,b,c,d,n,m_1,\cdots,m_n) = \left(\frac{1}{4}\right)^{(2/n)\sum_{r=1}^{n} m_r} \{(b-a)(d-c)\}^{1+(2/n)\sum_{r=1}^{n}(m_r-1)} \tag{5.1.414}$$

is a constant depending on $a, b, c, d, n, m_1, \cdots, m_n$.

Proof It is easy to observe that the following identities hold: for $r = 1, \cdots, n$,

$$f_r(x,y) = \int_a^x \Big(\int_c^y D_1D_2f_r(s,t)dt\Big)ds, \quad \text{for all } (x,y) \in \Delta_1, \tag{5.1.415}$$

$$f_r(x,y) = -\int_a^x \Big(\int_y^d D_1D_2f_r(s,t)dt\Big)ds, \quad \text{for all } (x,y) \in \Delta_2, \tag{5.1.416}$$

$$f_r(x,y) = -\int_x^b \Big(\int_c^y D_1D_2f_r(s,t)dt\Big)ds, \quad \text{for all } (x,y) \in \Delta_3, \tag{5.1.417}$$

$$f_r(x,y) = -\int_x^b \Big(\int_y^d D_1D_2f_r(s,t)dt\Big)ds, \quad \text{for all } (x,y) \in \Delta_4. \tag{5.1.418}$$

From (5.1.415)–(5.1.418) we derive for all $(x,y) \in \Delta$,

$$|f_r(x,y)| \leq \frac{1}{4}\int_a^b \Big(\int_c^d |D_1D_2f_r(s,t)|dt\Big)ds. \tag{5.1.419}$$

From (5.1.419) and using Hölder's inequality twice with indices m_r and $1/(m_r-1)$ for $r = 1, \cdots, n$, we obtain

$$|f_r(x,y)|^{m_r} \leq \Big(\frac{1}{4}\Big)^{m_r}(b-a)(d-c)^{m_r-1}\int_a^b \Big(\int_c^d |D_1D_2f(s,t)|dt\Big)ds. \tag{5.1.420}$$

From (5.1.420) and using the elementary inequalities

$$(b_1\cdots b_n)^{\frac{1}{n}} \leq \frac{1}{n}(b_1+\cdots+b_n) \tag{5.1.421}$$

(for all $b_1, \cdots, b_n \geq 0$ reals and $n \geq 1$) and

$$(b_1+\cdots+b_n)^2 \leq n(b_1^2+\cdots+b_n^2) \tag{5.1.422}$$

(for all $b_1, \cdots, b_n \geq 0$ reals) and the Schwartz inequality twice, we obtain

$$\begin{aligned}
\left[\prod_{i=1}^{n} |f_r(x,y)|^{r_m}\right]^{2/n} &\leq \left(\frac{1}{4}\right)^{(2/n)\sum_{r=1}^{n} m_r} \{(b-a)(d-c)\}^{(2/n)\sum_{r=1}^{n}(m_r-1)} \\
&\quad \times \left[\left(\prod_{r=1}^{n}\left(\int_a^b \left(\int_c^d |D_1D_2f_r(s,t)|dt\right)ds\right)\right)\right] \\
&\leq \left(\frac{1}{4}\right)^{(2/n)\sum_{r=1}^{n} m_r} \{(b-a)(d-c)\}^{(2/n)\sum_{r=1}^{n}(m_r-1)} \\
&\quad \times \left[\frac{1}{n}\sum_{r=1}^{n}\left(\int_a^b \left(\int_c^d |D_1D_2f_r(s,t)|dt\right)ds\right)\right]^2 \\
&\leq \frac{1}{n}\left(\frac{1}{4}\right)^{(2/n)\sum_{r=1}^{n} m_r} \{(b-a)(d-c)\}^{(2/n)\sum_{r=1}^{n}(m_r-1)} \\
&\quad \times \left\{\sum_{r=1}^{n}\left\{\int_a^b \left(\int_c^d |D_1D_2f_r(s,t)|dt\right)ds\right\}^2\right\} \\
&\leq \frac{1}{n}K(a,b,c,d,n,m_1,\cdots,m_n) \\
&\quad \times \int_a^b \left(\int_c^d \left[\sum_{i=1}^{n} |D_1D_2f_r(s,t)|^2 m_r\right]dt\right)ds.
\end{aligned} \tag{5.1.423}$$

Multiplying both sides of (5.1.423) by $p(x,y)$ and integrating the resulting the resulting inequality on Δ, we have

$$\begin{aligned}
&\int_a^b \left(\int_c^d p(x,y)\left[\prod_{r=1}^{n} |f_r(x,y)|^{m_r}\right]^{2/n} dy\right) dx \\
&\leq \frac{1}{n}K(a,b,c,d,n,m_1,\cdots,m_n)\left\{\int_a^b \left(\int_c^d p(x,y)dy\right)dx\right\} \\
&\quad \times \left\{\int_a^b \left(\int_c^b \left[\sum_{r=1}^{n} |D_1D_2f_r(x,y)|^{2m_r}\right]dy\right)dx\right\},
\end{aligned}$$

which is the desired inequality in (5.1.413) and the proof is complete. □

In the special cases when (i) $m_r = 1$ for $r = 1, \cdots, n$, (ii) $n = 2$, (iii) $n = 1$, (iv) $n = 2$ and $m_1 = m_2 = 1$, and (v) $n = 1$ and $m_1 = 1$, the inequality established in

(5.1.413) reduces respectively to the following inequalities

$$\int_a^b \Big(\int_c^d p(x,y)\Big[\prod_{r=1}^n |f_r(x,y)|\Big]^{2/n} dy\Big)dx$$
$$\leq \frac{1}{n}K(a,b,c,d,n,1,\cdots,1)\Big\{\int_a^b \Big(\int_c^d p(x,y)dy\Big)dx\Big\}$$
$$\times \left\{\int_a^b \Big(\int_c^b \Big[\sum_{i=1}^n |D_1D_2f_r(x,y)|^2\Big]dy\Big)dx\right\}, \qquad (5.1.424)$$

$$\int_a^b \Big(\int_c^d p(x,y)|f_1(x,y)|^{m_1}|f_2(x,y)|^{m_2} dy\Big)dx$$
$$\leq \frac{1}{2}K(a,b,c,d,2,m_1,m_2)\Big\{\int_a^b \Big(\int_c^d p(x,y)dy\Big)dx\Big\}$$
$$\times \left\{\int_a^b \Big(\int_c^b \big[|D_1D_2f_1(x,y)|^{2m_1} + |D_1D_2f_2(x,y)|^{2m_2}\big]dy\Big)dx\right\}, \qquad (5.1.425)$$

$$\int_a^b \Big(\int_c^d p(x,y)|f_1(x,y)|^{2m_1} dy\Big)dx \leq K(a,b,c,d,1,m_1)\Big\{\int_a^b \Big(\int_c^d p(x,y)dy\Big)dx\Big\}$$
$$\times \left\{\int_a^b \Big(\int_c^b |D_1D_2f_1(x,y)|^{2m_1} dy\Big)dx\right\}, \qquad (5.1.426)$$

$$\int_a^b \Big(\int_c^d p(x,y)|f_1(x,y)||f_2(x,y)|dy\Big)dx \leq \frac{1}{2}K(a,b,c,d,2,1,1)\Big\{\int_a^b \Big(\int_c^d p(x,y)dy\Big)dx\Big\}$$
$$\times \left\{\int_a^b \Big(\int_c^b \big[|D_1D_2f_1(x,y)|^2 + |D_1D_2f_2(x,y)|^2\big]dy\Big)dx\right\}, \qquad (5.1.427)$$

$$\int_a^b \Big(\int_c^d p(x,y)|f_1(x,y)|^2 dy\Big)dx \leq K(a,b,c,d,2,1,1)\Big\{\int_a^b \Big(\int_c^d p(x,y)dy\Big)dx\Big\}$$
$$\times \left\{\int_a^b \Big(\int_c^b |D_1D_2f_1(x,y)|^2 dy\Big)dx\right\}. \qquad (5.1.428)$$

□

It is interesting to note that the inequalities obtained in (5.1.426) and (5.1.425) are the two independent variable analogous of the Wirtingers type inequalities established by Pachpatte in [481], and the inequality obtained in (5.1.427) is a two independent variable analogue of the Wirtinger type inequality established by Traple in [647].

Theorem 5.1.52 (The Pachpatte Inequality [485]) *Let the functions $p(x,y)$, $f_r(x,y)$, $D_1f_r(x,y)$ and $D_1D_2f_r(x,y)$ be as defined in Theorem 5.1.51. Then*

$$\int_a^b(\int_c^d p(x,y)[\prod_{r=1}^n |f_r(x,y)|^{m_r}]^{1/n}[\sum_{r=1}^n |D_1D_2f_r(x,y)|^{m_r}]dy)dx$$

$$\leq \left\{K(a,b,c,d,n,1,\cdots,1)\int_a^b(\int_c^d p^2(x,y)dy)dx\right\}^{\frac{1}{2}}$$

$$\times\left\{\int_a^b(\int_c^b[\sum_{r=1}^n |D_1D_2f_r(x,y)|^{2m_r}]dy)dx\right\}, \tag{5.1.429}$$

where $m_r \geq 1$ (for $r = 1,\cdots,n$) are constants and $K(a,b,c,d,n,m_1,\cdots,m_2)$ is as defined in (5.1.414).

If we take (i) $m_r = 1$ for $r = 1,\cdots,n$, (ii) $n = 1$ and $m_1 = 1$ in (5.1.428), then we get respectively the following inequalities

$$\int_a^b\left(\int_c^d p(x,y)\left[\prod_{r=1}^n |f_r(x,y)|\right]^{1/n}\left[\sum_{r=1}^n |D_1D_2f_r(x,y)|\right]dy\right)dx$$

$$\leq \left\{K(a,b,c,d,n,m_1,\cdots,m_n)\int_a^b\left(\int_c^d p^2(x,y)dy\right)dx\right\}^{\frac{1}{2}}$$

$$\times\left\{\int_a^b(\int_c^b[\sum_{r=1}^n |D_1D_2f_r(x,y)|^2]dy)dx\right\}, \tag{5.1.430}$$

$$\int_a^b\left(\int_c^d p(x,y)|f_1(x,y)|^{m_1}|D_1D_2f_1(x,y)|^{m_1}dy\right)dx$$

$$\leq \left\{K(a,b,c,d,1,m_1)\int_a^b\left(\int_c^d p^2(x,y)dy\right)dx\right\}^{\frac{1}{2}}$$

$$\times\left\{\int_a^b(\int_c^b |D_1D_2f_1(x,y)|^{2m_1}dy)dx\right\}, \tag{5.1.431}$$

$$\int_a^b\left(\int_c^d p(x,y)|f_1(x,y)||D_1D_2f_1(x,y)|dy\right)dx$$

$$\leq \left\{K(a,b,c,d,1,1)\int_a^b\left(\int_c^d p^2(x,y)dy\right)dx\right\}^{\frac{1}{2}}$$

$$\times\left\{\int_a^b\left(\int_c^b |D_1D_2f_1(x,y)|^2dy\right)dx\right\}. \tag{5.1.432}$$

Proof By virtue of the Schwartz inequality twice and the inequalities (5.1.413) and (5.1.422), we observe that

$$\int_a^b \left(\int_c^d p(x,y) \left[\prod_{r=1}^n |f_r(x,y)| \right]^{1/n} \left[\sum_{i=1}^n |D_1 D_2 f_r(x,y)| \right] dy \right) dx$$

$$\leq \left\{ \int_a^b \left(\int_c^d p^2(x,y) \left[\prod_{r=1}^n |f_r(x,y)|^{m_r} \right]^{2/n} dy \right) dx \right\}^{1/2}$$

$$\times \left\{ \int_a^b \left(\int_c^d \left[\sum_{r=1}^n |D_1 D_2 f_r(x,y)|^{m_r} \right] dy \right) dx \right\}^{1/2}$$

$$\leq \left\{ \frac{1}{n} K(a,b,c,d,n,m_1,\cdots,m_n) \left\{ \int_a^b \left(\int_c^d p(x,y) dy \right) dx \right\} \right\}$$

$$\times \left\{ \int_a^b \left(\int_c^b \left[\sum_{r=1}^n |D_1 D_2 f_r(x,y)|^{2m_r} \right] dy \right) dx \right\}^{1/2}$$

$$\times \left\{ \int_a^b \left(\int_c^b n \left[\sum_{r=1}^n |D_1 D_2 f_r(x,y)|^{2m_r} \right] dy \right) dx \right\}^{1/2}$$

$$= \left\{ K(a,b,,c,d,n,m_1,\cdots,m_n) \int_a^b \left(\int_c^d p^2(x,y) dy \right) dx \right\}$$

$$\times \left\{ \int_a^b \left(\int_c^b n \left[\sum_{r=1}^n |D_1 D_2 f_r(x,y)|^{2m_r} \right] dy \right) dx \right\},$$

which is the desired inequality in (5.1.428) and the proof is complete. □

We note that the inequality obtained in (5.1.431) is a two independent variable analogue of the Opial type inequality established by Traple in [647]. In the special case when $p(x,y)$ is a constant, then from (5.1.431), we have the following Opial type inequality

$$\int_a^b \left(\int_c^d p(x,y) |f_1(x,y)| |D_1 D_2 f_1(x,y)| dy \right) dx$$

$$\leq \frac{(b-a)(d-c)}{4} \int_a^b \left(\int_c^d |D_1 D_2 f_1(x,y)|^2 dy \right) dx. \qquad (5.1.433)$$

Here, we note that the possible constant $(b-a)(d-c)/4$ involved in (5.1.432) is not the best possible constant. The inequality (5.1.432) with the best possible constant $(b-a)(d-c)/8$ is recently established by Yang in [698] by using a different method. However, the present proof of inequality (5.1.432) depends on the inequality established in Theorem 5.1.51 which in turn does not yield the sharp constant obtained in [698].

Throughout all the functions which appear in the inequalities are assumed to be real-valued and all the integrals, sums and products involved exist on the respective domains of their definitions. We need the inequalities, which are the slight variants of the inequalities given in [507].

Theorem 5.1.53 (The Pachpatte Inequality [516]) *Let $u(x, y)$, $a(x, y)$, $b(x, y)$ be non-negative continuous functions defined for all $x, y \in \mathbb{R}_+$ and $F : \mathbb{R}_+^3 \to \mathbb{R}_+$ be a continuous function which satisfies the condition: for all $u \geq v \geq 0$,*

$$0 \leq F(x, y, u) - F(x, y, v) \leq K(x, y, v)(u - v),$$

where $K(x, y, v)$ is a non-negative continuous function defined for all $x, y, v \in \mathbb{R}_+$.

(c_1) Assume that $a(x, y)$ is non-decreasing in all $x \in \mathbb{R}_+$. If for all $x, y \in \mathbb{R}_+$,

$$u(x, y) \leq a(x, y) + \int_0^x b(s, y)u(s, y)ds + \int_0^x \int_y^{+\infty} F(s, t, u(s, t))dtds, \tag{5.1.434}$$

then for all $x, y \in \mathbb{R}_+$,

$$u(x, y) \leq p(x, y)\left[a(x, y) + B(x, y)\exp\left(\int_0^x \int_y^{+\infty} K(s, t, p(s, t)a(s, t))p(s, t)dtds\right)\right], \tag{5.1.435}$$

where for all $x, y \in \mathbb{R}_+$,

$$B(x, y) = \int_0^x \int_y^{+\infty} F(s, t, p(s, t)a(s, t))dtds, \tag{5.1.436}$$

and

$$p(x, y) = \exp\left(\int_0^x b(s, y)ds\right). \tag{5.1.437}$$

(c_2) Assume that $a(x, y)$ is non-increasing in all $x \in \mathbb{R}_+$. If for all $x, y \in \mathbb{R}_+$,

$$u(x, y) \leq a(s, y) + \int_x^{+\infty} b(s, y)u(s, y)ds + \int_x^{+\infty} \int_y^{+\infty} F(s, t, u(s, t))dtds, \tag{5.1.438}$$

then for all $x, y \in \mathbb{R}_+$,

$$u(x, y) \leq \bar{p}(x, y)|\left[a(x, y) + \bar{B}(x, y)\exp\left(\int_x^{+\infty} \int_y^{+\infty} K(s, t, \bar{p}(s, t)a(s, t))\bar{p}(s, t)dtds\right)\right], \tag{5.1.439}$$

where for all $x, y \in \mathbb{R}_+$,

$$\bar{B}(x, y) = \int_x^{+\infty} \int_y^{+\infty} F(s, t, \bar{p}(s, t)a(s, t))dtds, \tag{5.1.440}$$

and

$$\bar{p}(x, y) = \exp\left(\int_x^{+\infty} b(s, y)ds\right). \tag{5.1.441}$$

Proof $(c1)$ Define a function $z(x, y)$ by

$$z(x, y) = a(x, y) + \int_0^x \int_y^{+\infty} F(s, t, u(s, t))dtds. \tag{5.1.442}$$

Then (5.1.433) can be restated as

$$u(x, y) \leq z(x, y) + \int_0^x b(s, y)u(s, y)ds. \tag{5.1.443}$$

Clearly, $z(x, y)$ is a non-negative, continuous and non-decreasing function in x, $x \in \mathbb{R}_+$. Treating y, $y \in \mathbb{R}_+$ fixed in (5.1.442) and using Theorem 1.1.4 in Qin [557] to (5.1.442), we obtain

$$u(x, y) \leq z(x, y)p(x, y), \tag{5.1.444}$$

where $p(x, y)$ is defined by (5.1.436). From (5.1.443) and (5.1.442), we have

$$u(x, y) \leq p(x, y)[a(x, y) + v(x, y)], \tag{5.1.445}$$

where

$$v(x, y) = \int_0^x \int_y^{+\infty} F(s, t, u(s, t))dtds. \tag{5.1.446}$$

From (5.1.445), (5.1.444) and the hypotheses on F, it follows that

$$\begin{aligned} v(x, y) \leq & \int_0^x \int_y^{+\infty} [F(s, t, p(s, t)(a(s, t) + v(s, t))) - F(s, t, p(s, t)a(s, t)) \\ & + F(s, t, p(s, t)a(s, t))]dtds \\ \leq & B(x, y) + \int_0^x \int_y^{+\infty} K(s, t, p(s, t)a(s, t))p(s, t)v(s, t)dtds. \end{aligned} \tag{5.1.447}$$

Clearly, $B(x, y)$ is non-negative, continuous and non-decreasing in x and non-increasing in y for $x, y \in \mathbb{R}_+$. Following the proof of $(a1)$ of Theorem 5.1.7 in Qin [557], we get

$$v(x, y) \leq B(x, y) \exp\left(\int_0^x \int_y^{+\infty} K(s, t, p(s, t)a(s, t))p(s, t)dtds\right). \tag{5.1.448}$$

The required inequality (5.1.434) follows from (5.1.444) and (5.1.447). □

In the subsequent discussion, we assume the following hold,

(A_1) $u(x, y), a(x, y), b(x, y), c(x, y), p(x, y)$ and $q(x, y)$ are real-valued non-negative continuous functions defined on a domain D.

(A_2) $P_0(x_0, y_0)$ and $P(x, y)$ be two points in D such that $(x - x_0)(y - y_0) > 0$ and R the rectangular region whose opposite corners are the points P_0 and P.

A useful two independent variable generalization of Theorem 1.2.11 in Qin [557] is embodied in the following theorem.

Theorem 5.1.54 (The Pachpatte Inequality [472]) *Suppose (A_1) and (A_2) are true. Let $v(s, t; x, y)$ and $w(s, t; x, y)$ be the solutions of the characteristic initial value problem*

$$\begin{cases} L[v] = v_{st} - [p(s, t) + b(s, t)(c(s, t) + q(s, t))]v = 0, \\ v(s, y) = v(x, t) = 1, \end{cases} \tag{5.1.449}$$

and

$$\begin{cases} M[\omega] = \omega_{st} - [b(s, t)c(s, t) - p(s, t)]v = 0, \\ \omega(s, y) = \omega(x, t) = 1, \end{cases} \tag{5.1.450}$$

respectively and let D^+ be a connected sub-domain of D which contains P and on which $v > 0$ and $\omega > 0$. Then, if $R \subset D^+$,

$$u(x, y) \leq a(x, y) + b(x, y)\left(\int_{x_0}^x \int_{y_0}^y (c(s, t)u(s, t))\mathrm{d}s\mathrm{d}t + \int_{x_0}^x \int_{y_0}^y p(s, t)\left(\int_{x_0}^s \int_{y_0}^t (q(\xi, \eta)u(\xi, \eta))\mathrm{d}\xi\mathrm{d}\eta\right)\mathrm{d}s\mathrm{d}t\right), \tag{5.1.451}$$

then

$$u(x, y) \leq a(x, y) + b(x, y)\left[\int_{x_0}^x \int_{y_0}^y (\omega(s, t; x, y)(a(s, t)c(s, t) + p(x, t)\int_{x_0}^s \int_{y_0}^t (a(\xi, \eta) \times [c(\xi, \eta) + q(\xi, \eta)]\, v(\xi, \eta; x, y))\mathrm{d}\xi\mathrm{d}\eta)\right)\mathrm{d}s\mathrm{d}t\right] \tag{5.1.452}$$

Proof Define a function $\phi(x, y)$ such that

$$\begin{cases} \phi(x, y) = \int_{x_0}^{x} \int_{y_0}^{y} (c(s, t)u(s, t))\mathrm{d}s\mathrm{d}t + \int_{x_0}^{x} \int_{y_0}^{y} (p(s, t)(\int_{x_0}^{s} \int_{y_0}^{t} (q(\xi, \eta)u(\xi, \eta))\mathrm{d}\xi\mathrm{d}\eta))\mathrm{d}s\mathrm{d}t, \\ \phi(x_0, y) = \phi(x, y_0) = 0, \end{cases}$$

then we have

$$\phi_{xy}(x, y) = c(x, y)u(x, y) + p(x, y) \int_{x_0}^{x} \int_{y_0}^{y} q(\xi, \eta)u(\xi, \eta)\mathrm{d}\xi\mathrm{d}\eta$$

which, in view of (5.1.450), implies

$$\begin{aligned} \phi_{xy}(x, y) \ &\le c(x, y)[a(x, y) + b(x, y)\phi(x, y)] + p(x, y)(\int_{x_0}^{x} \int_{y_0}^{y} q(\xi, \eta)[a(\xi, \eta) \\ &\quad + b(\xi, \eta)\phi(\xi, \eta)\mathrm{d}\xi\mathrm{d}\eta]). \end{aligned}$$

Adding $p(x, y)\phi(x, y)$ to both sides of the above inequality, we have

$$\begin{aligned} \phi_{xy}(x, y) + p(x, y)\phi(x, y) &\le c(x, y)[a(x, y) + b(x, y)\phi(x, y)] \\ &\quad + p(x, y)(\phi(x, y) + \int_{x_0}^{x} \int_{y_0}^{y} q(\xi, \eta)[a(\xi, \eta) + b(\xi, \eta)\phi(\xi, \eta)\mathrm{d}\xi\mathrm{d}\eta]). \end{aligned} \quad (5.1.453)$$

If we put

$$\begin{cases} \psi(x, y) = \phi(x, y) + \int_{x_0}^{x} \int_{y_0}^{y} q(\xi, \eta)[a(\xi, \eta) + b(\xi, \eta)\phi(\xi, \eta)\mathrm{d}\xi\mathrm{d}\eta, \\ \psi(x_0, y) = \psi(x, y_0) = 0, \end{cases} \quad (5.1.454)$$

then we obtain

$$\psi_{xy} = \phi_{xy} + q(x, y)[a(x, y) + b(x, y)\phi(x, y)]. \quad (5.1.455)$$

Using $\phi_{xy} \le c(x, y)[a(x, y) + b(x, y)\phi(x, y)] + p(x, y)\psi(x, y))$ from and $\phi(x, y) \le \psi(x, y))$ from (5.1.453) in (5.1.454), we have

$$\psi_{xy} \le a(x, y)[c(x, y) + q(x, y)] + [p(x, y) + b(x, y)(c(x, y) + q(x, y))]\psi(x, y)$$

i.e.,

$$L[\psi] = \psi_{xy} - [p(x, y) + b(x, y)(c(x, y) + q(x, y))]\psi(x, y) \le a(x, y)[c(x, y) + q(x, y)]. \quad (5.1.456)$$

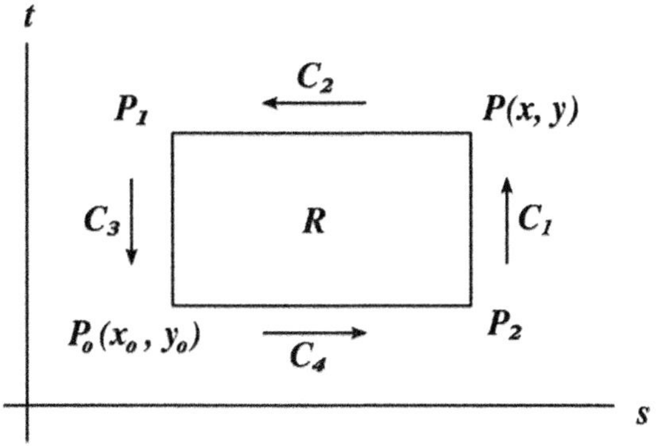

Fig. 5.3 Directed path around R

The operator L is self-adjoint and hyperbolic. For any twice continuously differentiable ψ and υ the operator L satisfies the identity

$$\upsilon L[\psi] - \psi L[\upsilon] = -(\psi \upsilon_y)_x + (\upsilon \psi_x)_y. \tag{5.1.457}$$

Let P_0 and P be any points as in theorem and label the directed sides and corners of the rectangle R as shown in Fig. 5.3. Using s and t as the independent variables, we integrate the identity (5.1.456) over R and use Green's theorem to obtain

$$\int_R (\upsilon L[\psi] - \psi L[\upsilon]) \mathrm{d}s \mathrm{d}s = -\int_{C_1+C_2+C_4+C_3} \upsilon \psi_s \mathrm{d}s + \psi \upsilon_t \mathrm{d}t$$

$$= -\int_{C_1+C_4} (\upsilon \psi_s) \mathrm{d}s - \int_{C_2+C_3} \psi \upsilon_s \mathrm{d}t$$

which holds for any functions in C^2.

For the particular function ψ defined earlier, we have $\psi = 0$ on C_3 and $\psi = \psi_s = 0$ on C_4, so the right-hand side in the above identity reduces to

$$-\int_{C_1} \upsilon \psi_s ds - \int_{C_2} \psi \upsilon_s dt. \tag{5.1.458}$$

Now suppose υ satisfies

$$\begin{cases} L[\upsilon] = \upsilon_(st) - [p(s,t) + b(s,t)(c(s,t) + q(s,t)]\upsilon = 0 & (5.1.459) \\ \upsilon = 1 \quad on \quad C_1 & (5.1.460) \\ \upsilon_t = 0 \quad on \quad C_2. & (5.1.461) \end{cases}$$

Then (5.1.459)and (5.1.460) imply that

$$\upsilon = 1 \quad on \quad C_2. \tag{5.1.462}$$

Since $\upsilon \geq 0$ on R and $\psi(P_1) = 0$. By using (5.1.455), identity (5.1.457) becomes

$$\psi(P) \leq \int_R (\upsilon[a(s,t)[c(s,t) + q(s,t)]])dsdt,$$

i.e.,

$$\psi(x,y) \leq \int_{x_0}^{x} \int_{y_0}^{y} (a(s,t)[c(s,t) + q(s,t)]\upsilon(s,t;x,y)dsdt.$$

Substituting this bound on $\psi(x,y)$ in (5.1.452), we obtain

$$\begin{aligned} M[\phi] &= \phi(xy) - [b(x,y)c(x,y) - p(x,y)]\phi(x,y) \\ &\leq [a(x,y)c(x,y) - p(x,y) \int_{x_0}^{x} \int_{y_0}^{y} (a(s,t)[c(s,t) + q(s,t)]\upsilon(s,t;x,y)dsdt. \end{aligned}$$

Again by following the same argument as above, we obtain

$$\begin{aligned} \phi(x,y) \leq \int_{x_0}^{x} \int_{y_0}^{y} & \omega(s,t;x,y)[a(s,t)c(s,t) \\ & + p(s,t) \int_{x_0}^{s} \int_{y_0}^{t} a(\xi,\eta)[c(\xi,\eta) + q(\xi,\eta)]\upsilon(\xi,\eta;s,t)\mathrm{d}\xi\mathrm{d}\eta]\mathrm{d}s\mathrm{d}t. \end{aligned}$$

Now substituting this bound on $\phi(x,y)$ in (5.1.450), we obtain the desired bound in (5.1.451). □

The proof of this theorem is obtained by reducing the integral inequality (5.1.450) to a partial differential inequality and then integrating it by Riemann's method for hyperbolic partial differential equations [618]. The functions $\upsilon(s,t;x,y)$ and $\omega(s,t;x,y)$ involved in theorem are Riemann functions relative to the point $P(x,y)$ for the self adjoint operators L and M respectively. There are such functions and a domain D^+ on which $\upsilon > 0$ since $\upsilon = 1$ and $\omega > 0$since $\omega = 1$ on the vertical and horizontal lines through P and since υ and ω are continuous. The existence and continuity of the Riemann function is well-known and may be demonstrated by the method of successive approximation (see [177]).

Another interesting and useful generalization of Theorem 1.2.11 in Qin [557] is embodied in the following theorem (Fig. 5.4).

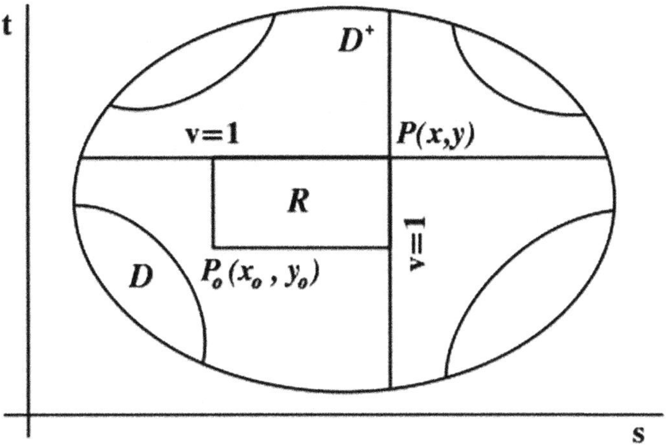

Fig. 5.4 Region and directed path around R

Theorem 5.1.55 (The Pachpatte Inequality [472]) *Suppose* (A_1) *and* (A_2) *are true. Let* $\upsilon(s,t;x,y)$ *and* $\omega(s,t;x,y)$ *be the solutions of the characteristic initial value problem*

$$\begin{cases} L[\upsilon] = \upsilon_{st} - b(s,t)[c(s,t) + p(s,t) + q(s,t)]\upsilon = 0, \\ \upsilon(s,y) = \upsilon(x,t) = 1 \end{cases} \tag{5.1.463}$$

and

$$M[\omega] = \omega_{st} - b(s,t)c(s,t)\omega = 0, \quad \omega(s,y) = \omega(x,t) = 1, \tag{5.1.464}$$

respectively and let D^+ *be a connected sub-domain of* D *which contains* P *and on which* $\upsilon > 0$ *and* $\omega > 0$. *Then, if* $R \subset D^+$,

$$u(x,y) \leq a(x,y) + b(x,y)[\int_{x_0}^{x}\int_{y_0}^{y} c(s,t)u(s,t)\mathrm{d}s\mathrm{d}t$$
$$+ \int_{x_0}^{x}\int_{y_0}^{y} p(s,t)(u(s,t) + b(s,t)\int_{x_0}^{s}\int_{y_0}^{t} q(\xi,\eta)u(\xi,\eta)\mathrm{d}\xi\mathrm{d}\eta)\mathrm{d}s\mathrm{d}t], \tag{5.1.465}$$

then

$$u(x,y) \leq a(x,y) + b(x,y)\Bigg[\int_{x_0}^{x}\int_{y_0}^{y} \omega(s,t;x,y)(a(s,t)(c(s,t) + p(s,t))$$
$$+b(s,t)p(s,t)\int_{x_0}^{s}\int_{y_0}^{t} a(\xi,\eta)[c(\xi,\eta) + p(\xi,\eta) + q(\xi,\eta)]\upsilon(\xi,\eta;s,t)\mathrm{d}\xi\mathrm{d}\eta)\mathrm{d}s\mathrm{d}t\Bigg]. \tag{5.1.466}$$

The proof of this theorem follows by an argument similar to that in the proof of Theorem 5.1.54 with suitable modifications. We omit the details. We now apply Theorem 5.1.54 to establish the following interesting and useful integral inequalities which in turn are the further generalizations of the integral inequalities recently established by Gollwitzer [250] and Pachpatte [451].

Theorem 5.1.56 (The Pachpatte Inequality [472]) *Suppose(A_1) and (A_2) are true. Let G(r) be continuous, strictly increasing, convex and sub-multiplicative function for all $r \geq 0$, G(0)=0, $\lim_{r\to+\infty} G(r) = +\infty$ for all (x,y) in D, $\alpha(x,y)$, $\beta(x,y)$ be positive continuous functions defined on a domain D, and $\alpha(x,y) + \beta(x,y) = 1$. Let $\upsilon(s,t;x,y)$ and $\omega(s,t;x,y)$ be the solutions of the characteristic initial value problem*

$$\begin{cases} L[\upsilon] = \upsilon_{st} - [p(s,t) + \beta(s,t)G(b(s,t)\beta^{-1}(s,t))(c(s,t) + q(s,t))] = 0, \\ \upsilon(s,y) = \upsilon(x,t) = 1, \end{cases} \tag{5.1.467}$$

and

$$\begin{cases} M[\omega] = m_{st} - [\beta(s,t)G(b(s,t)\beta^{-1}(s,t))(c(s,t) - p(s,t))]\omega = 0, \\ \omega(s,y) = \omega(x,t) = 1, \end{cases} \tag{5.1.468}$$

respectively and let D^+ be a connected sub-domain of D which contains P and on which $\upsilon > 0$ and $\omega > 0$. Then, if $R \subset D^+$,

$$u(x,y) \leq a(x,y) + b(x,y)G^{-1}\left[\int_{x_0}^{x}\int_{y_0}^{y} c(s,t)G(u(s,t))\mathrm{d}s\mathrm{d}t + \int_{x_0}^{x}\int_{y_0}^{y} p(s,t)(\int_{x_0}^{s}\int_{y_0}^{t} q(\xi,\eta)G(u(\xi,\eta))\mathrm{d}\xi\mathrm{d}\eta)\mathrm{d}s\mathrm{d}t\right], \tag{5.1.469}$$

then

$$\begin{aligned} u(x,y) &\leq a(x,y) + b(x,y)G^{-1}\left[\int_{x_0}^{x}\int_{y_0}^{y} \omega(s,t;x,y)(G(a(s,t)\alpha(s,t))c(s,t)\right. \\ &+p(s,t)\int_{x_0}^{s}\int_{y_0}^{t} \alpha(\xi,\eta) \\ &\left.\times G(a(\xi,\eta)\alpha^{-1}(\xi,\eta))[c(\xi,\eta) + q(\xi,\eta)]\upsilon(\xi,\eta;s,t)\mathrm{d}\xi\mathrm{d}\eta)\right]\mathrm{d}s\mathrm{d}t. \end{aligned} \tag{5.1.470}$$

Proof We may rewrite (5.1.468) as

$$\begin{aligned} u(x,y) &\leq \alpha(x,y)a(x,y)\alpha^{-1}(x,y) \\ &+\beta(x,y)b(x,y)\beta^{-1}G^{-1}\left[\int_{x_0}^{x}\int_{y_0}^{y} p(s,t)(\int_{x_0}^{s}\int_{y_0}^{t} q(\xi,\eta)G(u(\xi,\eta))\mathrm{d}\xi\mathrm{d}\eta)\mathrm{d}s\mathrm{d}t\right]. \end{aligned}$$

Since G is convex, sub-multiplicative and monotonic, we have

$$\begin{aligned} G(u(x,y)) \ &\leq \alpha(x,y)G(a(x,y)\alpha^{-1}(x,y)) \\ &\quad +\beta(x,y)G((x,y)\beta^{-1}(x,y))\left[\int_{x_0}^{x}\int_{y_0}^{y} c(s,t)G(u(s,t))\mathrm{d}s\mathrm{d}t\right. \\ &\quad \left.\times\int_{x_0}^{x}\int_{y_0}^{y} p(s,t)(\int_{x_0}^{s}\int_{y_0}^{t} q(\xi,\eta)G(u(\xi,\eta))\mathrm{d}\xi\mathrm{d}\eta)\mathrm{d}s\mathrm{d}t\right]. \end{aligned} \tag{5.1.471}$$

The estimate given in (5.1.469) follows by first applying Theorem 5.1.54 with $a(x,y) = \alpha(x,y)G(a(x,y)\alpha^{-1}(x,y))$,

$b(x,y) = \beta(x,y)G(b(x,y)\beta^{-1}(x,y))$ and $u(x,y) = G(u(x,y))$ and then applying $G^{-1}$to both sides of the resulting inequality. □

Theorem 5.1.57 (The Pachpatte Inequality [472]) *Suppose (A_1) and (A_2) are true. Let $G(r)$ be a positive, continuous, strictly increasing, sub-additive and sub-multiplicative function for all $r \geq 0$, $G(0) = 0$, for all $(x,y) \in D$, and G^{-1} is the inverse function of G. Let $\upsilon(s,t;x,y)$ and $\omega(s,t;x,y)$ be the solutions of the characteristic initial value problem.*

$$\begin{cases} L[\upsilon] = \upsilon_{st} - [p(s,t) + G(b(s,t))(c(s,t) + q(s,t))]\upsilon = 0, \\ \upsilon(s,y) = \upsilon(x,t) = 1, \end{cases} \tag{5.1.472}$$

and

$$\begin{cases} M[\omega] = \omega_{st} - [G(b(s,t))c(s,t) - p(s,t)]\omega = 0, \\ \omega(s,y) = \omega(x,t) = 1, \end{cases} \tag{5.1.473}$$

respectively and let D^+ be a connected sub-domain of D which contains P and on which $\upsilon > 0$ and $\omega > 0$. Then, if $R \subset D^+$,

$$\begin{aligned} u(x,y) \leq a(x,y) + b(x,y)G^{-1}[&\int_{x_0}^{x}\int_{y_0}^{y} c(s,t)G(u(s,t))\mathrm{d}s\mathrm{d}t \\ &+\int_{x_0}^{x}\int_{y_0}^{y} p(s,t)\int_{x_0}^{s}\int_{y_0}^{t} q(\xi,\eta)G(u(\xi,\eta))\mathrm{d}\xi\mathrm{d}\eta)\mathrm{d}s\mathrm{d}t], \end{aligned} \tag{5.1.474}$$

then

$$\begin{aligned} u(x,y) \leq G^{-1}[G(a(x,y)) + G(b(x,y))[&\int_{x_0}^{x}\int_{y_0}^{y} \omega(s,t;x,y)(G(a(s,t))c(s,t) \\ &+p(s,t)\int_{x_0}^{s}\int_{y_0}^{t} G(a(\xi,\eta))[c(\xi,\eta) + q(\xi,\eta)]\upsilon(\xi,\eta;s,t)\mathrm{d}\xi\mathrm{d}\eta)\mathrm{d}s\mathrm{d}t]]. \end{aligned} \tag{5.1.475}$$

Proof Since G is sub-additive, sub-multiplicative and monotonic, we have from (5.1.473)

$$G(u(x,y)) \leq G(a(x,y)) + G(b(x,y))[\int_{x_0}^{x}\int_{y_0}^{y} c(s,t)G(u(s,t))\mathrm{d}s\mathrm{d}t + \int_{x_0}^{x}\int_{y_0}^{y} p(s,t)\int_{x_0}^{s}\int_{y_0}^{t} q(\xi,\eta)G(u(\xi,\eta))\mathrm{d}\xi\mathrm{d}\eta)\mathrm{d}s\mathrm{d}t]. \tag{5.1.476}$$

The desired bound in (5.1.474) follows by first applying Theorem 5.1.54 to (5.1.475) with $a(x,y) = G(a(x,y)), b(x,y) = G(b(x,y))$ and $u(x,y) = G(u(x,y))$ and then applying G^{-1} to both sides of the resulting inequality. □

Now we apply Theorem 5.1.55 to establish the following integral inequalities similar to that proved in Theorems 5.1.56 and 5.1.57 which can be used in some applications. The proofs of Theorems 5.1.56 and 5.1.57 can be adapted readily into this context.

Theorem 5.1.58 (The Pachpatte Inequality [472]) *Suppose* (A_1) *and* (A_2) *are true. Let* $G(r), \alpha(x,y), \beta(x,y)$ *be same functions as defined in Theorem 5.1.56. Let* $\upsilon(s,t;x,y)$ *and* $\omega(s,t;x,y)$ *be the solutions of the characteristic initial value problem*

$$\begin{cases} L[\upsilon] = \upsilon_{st} - \beta(s,t)G(b(s,t)\beta^{-1}(s,t))[(c(s,t) + p(s,t) + q(s,t))]\upsilon = 0, \\ \upsilon(s,y) = \upsilon(x,t) = 1, \end{cases} \tag{5.1.477}$$

and

$$\begin{cases} M[\omega] = \omega_{st} - [\beta(s,t)G(b(s,t)\beta^{-1}(s,t))c(s,t)\omega = 0, \\ \omega(s,y) = \omega(x,t) = 1, \end{cases} \tag{5.1.478}$$

respectively and let D^{+} *be a connected sub-domain of* D *which contains* P *and on which* $\upsilon > 0$ *and* $\omega > 0$. *Then, if* $R \subset D^{+}$,

$$u(x,y) \leq a(x,y) + b(x,y)G^{-1}\left[\int_{x_0}^{x}\int_{y_0}^{y} c(s,t)G(u(s,t))\mathrm{d}s\mathrm{d}t + \int_{x_0}^{x}\int_{y_0}^{y} p(s,t)(G(u(s,t)) + \beta(s,t)G(b(s,t)\beta^{-1}(s,t))\int_{x_0}^{s}\int_{y_0}^{t} q(\xi,\eta)G(u(\xi,\eta))\mathrm{d}\xi\mathrm{d}\eta)\mathrm{d}s\mathrm{d}t\right], \tag{5.1.479}$$

then

$$u(x,y) \leq a(x,y) + b(x,y)G^{-1}\left[\int_{x_0}^{x}\int_{y_0}^{y} \omega(s,t;x,y)(\alpha(s,t)G(a(s,t)\alpha^{-1}(s,t))(c(s,t) + p(s,t)) + \beta(s,t)G(b(s,t)\beta^{-1}(s,t)p(s,t))\int_{x_0}^{s}\int_{y_0}^{t} \alpha(s,t)G(b(s,t)\alpha^{-1}(s,t)) \times[c(\xi,\eta) + p(\xi,\eta) + q(\xi,\eta)]\upsilon(\xi,\eta;s,t)\mathrm{d}\xi\mathrm{d}\eta)\mathrm{d}s\mathrm{d}t\right]. \tag{5.1.480}$$

Theorem 5.1.59 (The Pachpatte Inequality [472]) *Suppose (A_1) and (A_2) are true. Let G, G^{-1} be same functions as defined in Theorem 5.1.57. Let $\upsilon(s,t;x,y)$ and $\omega(s,t;x,y)$ be the solutions of the characteristic initial value problem*

$$\begin{cases} L[\upsilon] = \upsilon_{st} - G(b(s,t))[(c(s,t)+p(s,t)+q(s,t))]\upsilon = 0, \\ \upsilon(s,y) = \upsilon(x,t) = 1, \end{cases} \tag{5.1.481}$$

and

$$\begin{cases} M[\omega] = \omega_{st} - G(b(s,t))c(s,t)\omega = 0, \\ \omega(s,y) = \omega(x,t) = 1, \end{cases} \tag{5.1.482}$$

respectively and let D^+ be a connected sub-domain of D which contains P and on which $\upsilon > 0$ and $\omega > 0$. Then, if $R \subset D^+$,

$$u(x,y) \le a(x,y) + b(x,y)G^{-1}\left[\int_{x_0}^{x}\int_{y_0}^{y} c(s,t)G(u(s,t))\mathrm{d}s\mathrm{d}t \right.$$
$$\left. + \int_{x_0}^{x}\int_{y_0}^{y} p(s,t)(G(u(s,t)) + G(b(s,t)))\int_{x_0}^{s}\int_{y_0}^{t} q(\xi,\eta)G(u(\xi,\eta))\mathrm{d}\xi\mathrm{d}\eta)\mathrm{d}s\mathrm{d}t\right], \tag{5.1.483}$$

then

$$u(x,y) \le G^{-1}\left[G(a(x,y)) + G(b(x,y))\left(\int_{x_0}^{x}\int_{y_0}^{y} \omega(s,t;x,y)(G(a(s,t))[c(s,t)+p(s,t)] \right.\right.$$
$$\left.\left. + G(b(s,t))p(s,t)\int_{x_0}^{s}\int_{y_0}^{t} G(a(\xi,\eta))[c(\xi,\eta)+q(\xi,\eta)]\upsilon(\xi,\eta;s,t)\mathrm{d}\xi\mathrm{d}\eta)\mathrm{d}s\mathrm{d}t\right)\right]. \tag{5.1.484}$$

We note that in the special case when $p(x,y) = q(x,y) = 0$, Theorems 5.1.54–5.1.59 reduces to the further generalizations of the integral inequality recently established by Snow [619]. In the special case when $c(x,y) = 0$, the results in Theorems 5.1.54–5.1.59 are new to the literature.

Assume $x_0, y_0 \in \mathbb{R}$ are two fixed numbers. Let $I := [x_0, X) \subset \mathbb{R}$, $J := [y_0, Y) \subset \mathbb{R}$, and $\Delta := I \times J \subset \mathbb{R}^2$. Note that here we allow X or Y to be $+\infty$. As usual, $C^i(U,V)$ will denote the set of all i-times continuously differentiable functions of U into V, and $C^0(U,V) := C(U,V)$. Partial derivatives of a function $z(x,y)$ are denoted by z_x, z_y, z_{xy}, etc. The identity function will be denoted as id and so in particular, id_U is the identity function of U onto itself.

For any $\varphi, \psi \in C(\mathbb{R}_+, \mathbb{R}_+)$ and any constant $\beta > 0$, define

$$\begin{cases} \Phi_\beta(r) := \int_1^r \frac{ds}{\varphi(s^{\frac{1}{\beta}})}, \quad \Psi_\beta(r) := \int_1^r \frac{ds}{\psi(s^{\frac{1}{\beta}})}, \ r > 0, \\ \Phi_\beta(0) := \lim_{r\to 0^+} \Phi_\beta(r), \quad \Psi_\beta(0) := \lim_{r\to 0^+} \Psi_\beta(r). \end{cases}$$

Note that we allow $\Psi_\beta(0)$ and $\Phi_\beta(0)$ to be $-\infty$ here.

Theorem 5.1.60 (The Cheung Inequality [142]) *Let $c \geq 0$ and $p > 0$ be constants. Let $b \in C(\Delta, \mathbb{R}_+)$, $\gamma \in C^1(I, I), \delta \in C^1(J, J)$, and $\varphi \in C(\mathbb{R}_+, \mathbb{R}_+)$ be functions satisfying*

(i) γ, δ are non-decreasing and $\gamma \leq id_I$, $\delta \leq id_J$; and
(ii) φ is non-decreasing with $\varphi(r) > 0$ for all $r > 0$.

If $u \in C(\Delta, \mathbb{R}_+)$ satisfies for all $(x, y) \in \Delta$,

$$u^p(x, y) \leq c + \int_{\gamma(x_0)}^{\gamma(x)} \int_{\delta(y_0)}^{\delta(y)} b(s, t)\varphi(u(s, t))dtds, \tag{5.1.485}$$

then for all $(x, y) \in [x_0, x_1] \times [y_0, y_1]$,

$$u(x, y) \leq \{\Phi_p^{-1}[\Phi_p(c) + B(x, y)]\}^{1/p}, \tag{5.1.486}$$

where

$$B(x, y) := \int_{\gamma(x_0)}^{\gamma(x)} \int_{\delta(y_0)}^{\delta(y)} b(s, t)dtds,$$

and Φ_p^{-1} is the inverse of Φ_p, $(x_1, y_1) \in \Delta$ is chosen in such a way that $\Phi_p(c) + B(x, y) \in Dom\ (\Phi_p^{-1})$ for all $(x, y) \in [x_0, x_1] \times [y_0, y_1]$.

Proof It suffices to consider the case $c > 0$, for the case $c = 0$ can then be arrived at by continuity argument. Denote by $g(x, y)$ the right-hand side of (5.1.484). Then $g > 0, u \leq g^{1/p}$ on Δ, we have and g is non-decreasing in each variable. Hence for any $(x, y) \in \Delta$,

$$\begin{aligned} g_x(x, y) &= \gamma'(x) \int_{\delta(y_0)}^{\delta(y)} b(\gamma(x), t)\varphi(u(\gamma(x), t))dt \\ &\leq \gamma'(x) \int_{\delta(y_0)}^{\delta(y)} b(\gamma(x), t)\varphi(g(\gamma(x), t))dt \\ &\leq \gamma'(x)\varphi(g(\gamma(x), \delta(y))) \int_{\delta(y_0)}^{\delta(y)} b(\gamma(x), t)dt \\ &\leq \gamma'(x)\varphi(g(x, y)) \int_{\delta(y_0)}^{\delta(y)} b(\gamma(x), t)dt. \end{aligned}$$

By the definition of Φ_p, we get

$$\begin{aligned}(\Phi_p \circ g)_x(x,y) &= \left.\frac{d\Phi_p}{dr}\right|_{g(x,y)} \cdot g_x(x,y)\\ &\leq \frac{1}{\varphi(g(x,y))} \cdot \gamma'(x)\varphi(g(x,y)) \int_{\delta(y_0)}^{\delta(y)} b(\gamma(x),t)dt\\ &= \left(\int_{\delta(y_0)}^{\delta(y)} b(\gamma(x),t)dt\right)\gamma'(x).\end{aligned}$$

Integrating with respect to x on $[x_0, x]$ gives us

$$\begin{aligned}\Phi_p(g(x,y)) - \Phi_p(g(x_0,y)) &\leq \int_{x_0}^{x} (\int_{\delta(y_0)}^{\delta(y)} b(\gamma(\xi),t)dt)\gamma'(\xi)d\xi\\ &= \int_{\gamma(x_0)}^{\gamma(x_1)} \int_{\delta(y_0)}^{\delta(y_1)} b(s,t)dtds,\end{aligned}$$

or for all$(x,y) \in \Delta$,

$$\Phi_p(g(x,y)) \leq \Phi_p(c) + B(x,y).$$

As Φ_p^{-1} is increasing on *Dom* (Φ_p^{-1}), this yields for all $(x,y) \in [x_0,x_1] \times [y_0,y_1]$,

$$u(x,y) \leq g^{1/p}(x,y) \leq \left(\Phi_p^{-1}[\Phi_p(c) + B(x,y)]\right)^{1/p}.$$

Thus the proof is complete. □

Remark 5.1.10

(i) In many cases, the non-decreasing function φ satisfies $\int_1^{+\infty}(ds/\varphi(s^{1/p})) = +\infty$. Examples of such functions are $\varphi \equiv 1$, $\varphi(s) = s^p, \varphi(s) = s^{p/2}$, etc. In such cases, $\Phi_p(+\infty) = +\infty$, so we may take $x_1 = X, y_1 = Y$. In particular, inequality (5.1.485) holds for all $(x,y) \in \Delta$.

(ii) Theorem 5.1.60 reduces to Theorem 2.1 of Cheung [143] when $p = 1$, and reduces further to Theorem 5.1.18 if we set $\gamma(x) = x, \delta(y) = y$.

(iii) Theorem 5.1.60 is also a generalization of the main result in Lipovan [355] to the case of two independent variables. In fact, if we set $p = 1$ and $\delta(y) = \delta(y_0)$ for all $y \in J$, Theorem 5.1.60 reduces to Theorem 1.1.47. If we further require $\gamma(x) = x$ for all $x \in I$, Theorem 5.1.60 further reduces to the famous Bihari's inequality [82].

Theorem 5.1.61 (The Cheung Inequality [142]) *Let $k \geq 0$ and $p > 1$ be constants. Let $a, b \in C(\Delta, \mathbb{R}_+), \alpha, \gamma \in C^1(I, I), \beta, \delta \in C^1(J, J)$, and $\varphi \in C(\mathbb{R}_+, \mathbb{R}_+)$ be functions satisfying*

(i) $\alpha, \beta, \gamma, \delta$ are non-decreasing with $\alpha, \gamma \leq id_I$, $\beta, \delta \leq id_J$; and
(ii) φ is non-decreasing with $\varphi(r) > 0$ for all $r > 0$. If $u \in C(\Delta, \mathbb{R}_+)$ satisfies for all $(x, y) \in \Delta$,

$$u^p(x,y) \leq k + \frac{p}{p-1}\int_{\alpha(x_0)}^{\alpha(x)}\int_{\beta(y_0)}^{\beta(y)} a(s,t)u(s,t)dtds$$
$$+\frac{p}{p-1}\int_{\gamma(x_0)}^{\gamma(x)}\int_{\delta(y_0)}^{\delta(y)} b(s,t)u(s,t)\varphi(u(s,t))dtds, \qquad (5.1.487)$$

then for all $(x, y) \in [x_0, x_1] \times [y_0, y_1]$,

$$u(x,y) \leq \left(\Phi_{p-1}^{-1}\left[\Phi_{p-1}(k^{1-1/p} + A(x,y)) + B(x,y)\right]\right)^{1/p-1}, \qquad (5.1.488)$$

where

$$A(x,y) := \int_{\alpha(x_0)}^{\alpha(x)}\int_{\beta(y_0)}^{\beta(y)} b(s,t)dtds, \quad B(x,y) := \int_{\gamma(x_0)}^{\gamma(x)}\int_{\delta(y_0)}^{\delta(y)} b(s,t)dtds,$$

and $(x_1, y_1) \in \Delta$ is chosen in such a way that $\Phi_{p-1}\left(k^{1-1/p} + A(x,y)\right) + B(x,y) \in Dom\,(\Phi_{p-1}^{-1})$ for all $(x,y) \in [x_0, x_1] \times [y_0, y_1]$.

Proof It suffices to consider the case $k > 0$, for the case $k = 0$ can then be arrived at by continuity argument. So assume $k > 0$. Denote by $f(x, y)$ the right-hand side of (5.1.486). Then $f > 0, u \leq f^{1/p}$ on Δ, and f is non-decreasing in each variable. Hence for any $(x, y) \in \Delta$, we have

$$f_x(x,y) = \frac{p}{p-1}\alpha'(x)\int_{\beta(y_0)}^{\beta(y)} a(\alpha(x),t)u(\alpha(x),t)dt$$
$$+\frac{p}{p-1}\gamma'(x)\int_{\delta(y_0)}^{\delta(y)} b(\gamma(x),t)u(\gamma(x),t)\varphi(u(\gamma(x),t))dt$$
$$\leq \frac{p}{p-1}\alpha'(x)\int_{\beta(y_0)}^{\beta(y)} a(\alpha(x),t)f^{1/p}(\alpha(x),t)dt$$
$$+\frac{p}{p-1}\gamma'(x)\int_{\delta(y_0)}^{\delta(y)} b(\gamma(x),t)f^{1/p}(\gamma(x),t)\varphi(f^{1/p}(\gamma(x),t))dt$$
$$\leq \frac{p}{p-1}\alpha'(x)f^{1/p}(\alpha(x),\beta(y))\int_{\beta(y_0)}^{\beta(y)} a(\alpha(x),t)dt$$

$$+\frac{p}{p-1}\gamma'(x)f^{1/p}(\gamma(x),\delta(y))\int_{\delta(y_0)}^{\delta(y)} b(\gamma(x),t)\varphi(f^{1/p}(\gamma(x),t))dt$$

$$\leq \frac{p}{p-1}\alpha'(x)f^{1/p}\int_{\beta(y_0)}^{\beta(y)} a(\alpha(x),t)dt$$

$$+\frac{p}{p-1}\gamma'(x)f^{1/p}\int_{\delta(y_0)}^{\delta(y)} b(\gamma(x),t)\varphi(f^{1/p}(\gamma(x),t))dt$$

Since $f^{1/p} > 0$, we get

$$\frac{p-1}{p}\frac{f_x(x,y)}{f^{1/p}(x,y)} \leq \Big(\int_{\beta(y_0)}^{\beta(y)} a(\alpha(x),t)dt\Big)\alpha'(x)$$

$$+\Big(\int_{\delta(y_0)}^{\delta(y)} b(\gamma(x),t)\varphi(f^{1/p}(\gamma(x),t))dt\Big)\gamma'(x).$$

Thus integrating with respect to x on $[x_0, x]$ yields

$$f^{1/p}(x,y) - f^{1/p}(x_0,y) \leq \int_{x_0}^{x}\Big(\int_{\beta(y_0)}^{\beta(y)} a(\alpha(\xi),t)dt\Big)\alpha'(\xi)d\xi$$

$$+\int_{x_0}^{x}\Big(\int_{\delta(y_0)}^{\delta(y)} b(\gamma(\xi),t)\varphi(f^{1/p}(\gamma(\xi),t))dt\Big)\gamma'(\xi)d\xi$$

$$= \int_{\alpha(x_0)}^{\alpha(x)}\int_{\beta(y_0)}^{\beta(y)} a(s,t)dtds + \int_{\gamma(x_0)}^{\gamma(x)}\int_{\delta(y_0)}^{\delta(y)} a(s,t)dtds,$$

or for all $(x, y) \in \Delta$,

$$f^{1-1/p}(x,y) \leq k^{1-1/p} + A(x,y) + \int_{\gamma(x_0)}^{\gamma(x)}\int_{\delta(y_0)}^{\delta(y)} a(s,t)dtds.$$

Hence for any fixed $\bar{x}, \bar{y} \in [x_0, x_1] \times [y_0, y_1]$, since A is non-decreasing in each variable, we have, for all $(x, y) \in [x_0, \bar{x}] \times [y_0, \bar{y}]$,

$$f^{1-1/p}(x,y) \leq k^{1-1/p} + A(\bar{x},\bar{y}) + \int_{\gamma(x_0)}^{\gamma(x)}\int_{\delta(y_0)}^{\delta(y)} a(s,t)dtds.$$

Applying Theorem 5.1.60 to the function $f^{1-1/p}(x, y)$, we conclude for all $(x, y) \in [x_0, \bar{x}] \times [y_0, \bar{y}]$,

$$f^{1-1/p}(x,y) \leq \Phi_{1-1/p}^{-1}[\Phi_{1-1/p}(k^{1-1/p} + A(\bar{x},\bar{y})) + B(x,y)]$$

In particular, this gives us

$$u(\bar{x},\bar{y}) \le f^{1/p}(\bar{x},\bar{y}) = [f^{1-1/p}(\bar{x},\bar{y})]^{1/(p-1)}$$
$$\le \left\{ \Phi_{1-1/p}^{-1}[\Phi_{1-1/p}(k^{1-1/p} + A(\bar{x},\bar{y})) + B(\bar{x},\bar{y})] \right\}.$$

Since $(\bar{x},\bar{y}) \in [x_0, x_1] \times [y_0, y_1])$ is arbitrary, this completes the proof of the theorem. □

Remark 5.1.11

(i) Similar to (i) of the previous remark, in many cases $\Phi_{p-1}(+\infty) = +\infty$ and so in such cases, inequality (5.1.487) holds for all $(x, y) \in \Delta$.
(ii) Similarly to (ii) of the previous remark, if we set $\beta(y) = \beta(y_0)$ and $\delta(y) = \delta(y_0)$ for all $y \in J$ in Theorem 5.1.61, we easily arrive at the following one-dimensional result.

Theorem 5.1.61 can easily be applied to generate other useful nonlinear integral inequalities in more general situations. For example, we have the following theorem.

Theorem 5.1.62 (The Cheung Inequality [142]) *Let $k \ge 0$ and $p > q > 0$ be constants. Let $a, b \in C(\Delta, \mathbb{R}_+), \alpha, \gamma \in C^1(I, I), \beta, \delta \in C^1(J, J)$, and $\varphi \in C(\mathbb{R}_+, \mathbb{R}_+)$ be functions satisfying*

(i) $\alpha, \beta, \gamma, \delta$ are non-decreasing with $\alpha, \gamma \le id_I$, $\beta, \delta \le id_J$; and
(ii) φ is non-decreasing with $\varphi(r) > 0$ for all $r > 0$.

If $u \in C(\Delta, \mathbb{R}_+)$ satisfies for all $(x, y) \in \Delta$,

$$u^p(x,y) \le k + \frac{p}{p-q}\int_{\alpha(x_0)}^{\alpha(x)}\int_{\beta(y_0)}^{\beta(y)} a(s,t)u^q(s,t)dtds$$
$$+\frac{p}{p-q}\int_{\gamma(x_0)}^{\gamma(x)}\int_{\delta(y_0)}^{\delta(y)} b(s,t)u^q(s,t)\varphi(u(s,t))dtds \qquad (5.1.489)$$

then for all $(x, y) \in [x_0, x_1] \times [y_0, y_1]$,

$$u(x,y) \le \left(\Phi_{p-q}^{-1}[\Phi_{p-q}(k^{1-q/p} + A(x,y)) + B(x,y)]\right)^{1/p-q} \qquad (5.1.490)$$

where

$$A(x,y) := \int_{\alpha(x_0)}^{\alpha(x)}\int_{\beta(y_0)}^{\beta(y)} b(s,t)dtds, \quad B(x,y) := \int_{\gamma(x_0)}^{\gamma(x)}\int_{\delta(y_0)}^{\delta(y)} b(s,t)dtds,$$

and $(x_1, y_1) \in \Delta$ is chosen in such a way that $\Phi_{p-q}(k^{1-q/p} + A(x,y)) + B(x,y) \in Dom\,(\Phi_{p-q}^{-1})$ for all $(x, y) \in [x_0, x_1] \times [y_0, y_1]$.

Proof For any $r > 0$, define

$$\psi(r) := \varphi(r^{1/q}). \qquad (5.1.491)$$

Then clearly ψ satisfies condition (ii) of Theorem 5.1.61. By (5.1.488), we get for all $(x, y) \in \Delta$,

$$u^p(x, y) \leq k + \frac{p}{p-q}\int_{\alpha(x_0)}^{\alpha(x)}\int_{\beta(y_0)}^{\beta(y)} a(s, t)u^q(s, t)dtds,$$
$$+\frac{p}{p-q}\int_{\gamma(x_0)}^{\gamma(x)}\int_{\delta(y_0)}^{\delta(y)} b(s, t)u^q(s, t)\psi(u(s, t))dtds.$$

Writing $v = u^q$, this becomes for all $(x, y) \in \Delta$,

$$v^{p/q}(x, y) \leq k + \frac{p/q}{p/q-1}\int_{\alpha(x_0)}^{\alpha(x)}\int_{\beta(y_0)}^{\beta(y)} a(s, t)v(s, t)dtds$$
$$+\frac{p/q}{p/q-1}\int_{\gamma(x_0)}^{\gamma(x)}\int_{\delta(y_0)}^{\delta(y)} b(s, t)v(s, t)\varphi(v(s, t))dtds.$$

Since $p/q > 1$, it follows from Theorem 5.1.61 that for all $(x, y) \in [x_0, x_1] \times [y_0, y_1]$,

$$v(x, y) \leq \left(\Psi_{p/q-1}^{-1}[\Phi_{p/q-1}(k^{1-q/p} + A(x, y)) + B(x, y)]\right)^{1/(p/q-1)}$$
$$= \left(\Psi_{(p-q)/q}^{-1}[\Phi_{(p-q)/q}(k^{(p-q)/p} + A(x, y)) + B(x, y)]\right)^{q/(p-q)}$$

where $(x_1, y_1) \in \Delta$ is chosen in such a way that for all $(x, y) \in [x_0, x_1] \times [y_0, y_1]$,

$$\Phi_{(p-q)/q}(k^{(p-q)/p} + A(x, y)) + B(x, y) \in Dom\ (\Phi_{(p-q)/q}^{-1}).$$

Now it is easy to check by the definition of ψ in (5.1.490) that

$$\Psi_{(p-q)/q}(r) = \Phi_{p-q}(r),$$

which implies for all $(x, y) \in [x_0, x_1] \times [y_0, y_1]$,

$$v(x, y) \leq \left(\Phi_{p-q}^{-1}[\Phi_{p-q}(k^{(p-q)/p} + A(x, y)) + B(x, y)]\right)^{q/(p-q)}$$

or for all $(x, y) \in [x_0, x_1] \times [y_0, y_1]$,

$$u(x, y) = v^{1/q}(x, y) \leq \left(\Phi_{p-q}^{-1}[\Phi_{p-q}(k^{(p-q)/p} + A(x, y)) + B(x, y)]\right)^{1/(p-q)}$$

where $(x_1, y_1) \in \Delta$ is chosen in such a way that $\Phi_{p-q}(k^{(p-q)/p} + A(x, y)) + B(x, y) \in Dom\ (\Phi_{p-q}^{-1})$ for all $(x, y) \in [x_0, x_1] \times [y_0, y_1]$. □

A special case of Theorem 5.1.62 is the following corollary.

Corollary 5.1.4 (The Cheung Inequality [142]) *Let $k \geq 0$ and $p > 1$ be constants. Let $a, b \in C(\Delta, \mathbb{R}_+), \alpha, \gamma \in C^1(I, I), \beta, \delta \in C^1(J, J)$, and $\varphi \in C(\mathbb{R}_+, \mathbb{R}_+)$ be functions satisfying*

(i) $\alpha, \beta, \gamma, \delta$ are non-decreasing with $\alpha, \gamma \leq id_I,\ \beta, \delta \leq id_J$; and
(ii) φ is non-decreasing with $\varphi(r) > 0$ for all $r > 0$.

If $u \in C(\Delta, \mathbb{R}_+)$ satisfies for all $(x, y) \in \Delta$,

$$u^p(x, y) \leq k + p \int_{\alpha(x_0)}^{\alpha(x)} \int_{\beta(y_0)}^{\beta(y)} a(s, t) u^{p-1}(s, t) dt ds,$$
$$+ p \int_{\gamma(x_0)}^{\gamma(x)} \int_{\delta(y_0)}^{\delta(y)} b(s, t) u^{p-1}(s, t) \varphi(u(s, t)) dt ds,$$

then for all $(x, y) \in [x_0, x_1] \times [y_0, y_1]$,

$$u(x, y) \leq \left\{\Phi_1^{-1}[\Phi_1(k^{1/p} + A(x, y)) + B(x, y)]\right\}$$

where

$$A(x, y) := \int_{\alpha(x_0)}^{\alpha(x)} \int_{\beta(y_0)}^{\beta(y)} b(s, t) dt ds,\quad B(x, y) := \int_{\gamma(x_0)}^{\gamma(x)} \int_{\delta(y_0)}^{\delta(y)} b(s, t) dt ds,$$

and $(x_1, y_1) \in \Delta$ is chosen in such a way that $\Phi_1(k^{1/p} + A(x, y)) + B(x, y) \in Dom\ (\Phi_1^{-1})$ for all $(x, y) \in [x_0, x_1] \times [y_0, y_1]$.

Proof The assertion follows immediately from Theorem 5.1.62 by taking $q = p - 1 > 0$. □

In particular, we have the following useful corollary.

Corollary 5.1.5 (The Cheung Inequality [142]) *Let $k \geq 0$ and $p > 1$ be constants. Let $a, b \in C(\Delta, \mathbb{R}_+), \alpha, \gamma \in C^1(I, I), \beta, \delta \in C^1(J, J)$ be functions such that $\alpha, \beta, \gamma, \delta$ are non-decreasing with $\alpha, \gamma \leq id_I,\ \beta, \delta \leq id_J$. If $u \in C(\Delta, \mathbb{R}_+)$*

satisfies for all $(x, y) \in \Delta$,

$$u^p(x,y) \le k + p\int_{\alpha(x_0)}^{\alpha(x)}\int_{\beta(y_0)}^{\beta(y)} a(s,t)u^{p-1}(s,t)dtds$$
$$+p\int_{\gamma(x_0)}^{\gamma(x)}\int_{\delta(y_0)}^{\delta(y)} b(s,t)u^p(s,t)dtds$$

then we have for all $(x, y) \in \Delta$,

$$u(x,y) \le \left(k^{1/p} + A(x,y)\right)\exp\left(B(x,y)\right) \tag{5.1.492}$$

where $A(x, y)$ *and* $B(x, y)$ *are as defined in Theorem 5.1.61.*

Proof Assume first that $k > 0$. Let $\varphi = id$ on $\mathbb{R}_+$. Then all conditions of Corollary 5.1.4 are satisfied. Note that in this cases $\Phi_1 = ln$ and so $\Phi_1^{-1} = exp$ is defined everywhere on $\mathbb{R}$. By Corollary 5.1.4, we get for all $(x, y) \in \Delta$,

$$u(x,y) \le \exp\left[ln(k^{1/p} + A(x,y)) + B(x,y)\right] = \left(k^{1/p} + A(x,y)\right)\exp\left(B(x,y)\right).$$

Note that the above inequality holds for all $k > 0$, by continuity argument it also holds for $k = 0$. □

Remark 5.1.12 Corollary 5.1.5 generalizes the results of Pachpatte (Theorem 1.2.11 (a_1)), Dafermos (Theorem 1.3.1), and Ou-Yang (Theorem 1.2.1).

Corollary 5.1.6 (The Cheung Inequality [142]) *Let* $k \ge 0$ *and* $p > 1$ *be constants. Let* $b \in C(\Delta, \mathbb{R}_+), \gamma \in C^1(I, I), \delta \in C^1(J, J)$ *be functions such that* γ, δ *are non-decreasing with* $\gamma \le id_I$, $\delta \le id_J$. *If* $u \in C(\Delta, \mathbb{R}_+)$ *satisfies for all* $(x, y) \in \Delta$,

$$u^p(x,y) \le k + p\int_{\gamma(x_0)}^{\gamma(x)}\int_{\delta(y_0)}^{\delta(y)} b(s,t)u^p(s,t)dtds,$$

then we have for all $(x, y) \in \Delta$,

$$u(x,y) \le k^{1/p}\exp\left(B(x,y)\right), \tag{5.1.493}$$

where $B(x, y)$ *are as defined in Theorem 5.1.61.*

Proof The proof follows immediately from Corollary 5.2.8 by taking $a \equiv 0$. □

Remark 5.1.13 Corollary 5.1.6 generalizes the corollary in Lipovan [355] to the case of two independent variables. In fact, if we set $p = 2$ and $\delta(y) = \delta(y_0)$ for all $y \in J$, Corollary 5.1.6 reduces to the above mentioned corollary. In particular, if we further require $\gamma(x) = x$ for all $x \in I$, Corollary 5.1.6 further reduces to the famous Gronwall-Bellman inequality [65, 259].

Remark 5.1.14 It is evident that Theorem 5.1.62 and Corollary 5.1.4–5.1.6 can easily be generalized to obtain explicit bounds for functions satisfying certain integral inequalities involving more retarded arguments. It is also clear that these results can be extended to functions of more than two variables in the obvious way. Details of these are rather algorithmic and so will not be given here.

The following theorem deals with the two independent variable versions of the inequalities established in Theorem 1.2.15 which can be used in certain applications.

Theorem 5.1.63 (The Pachpatte Inequality [523]) *Let $u, a_i, b_i \in C(\Delta, \mathbb{R}_+)$, and $\alpha_i \in C^1(I_1, I_1)$, $\beta_i \in C^1(I_2, I_2)$ be non-decreasing with $\alpha_i(x) \le x$ on I_1, $\beta_i \le y$ for $i = 1, 2, \cdots, n$. Let $p > 1$ and $c \ge 0$ be constants.*

(d_1) If for all $(x, y) \in \Delta$,

$$u^p(t) \le c + p \sum_{i=1}^{n} \int_{\alpha_i(t_0)}^{\alpha_i(t)} \int_{\beta_i(y_0)}^{\beta_i(y)} [a_i(s,t)u(s,t)w(u(s,t)) + b_i(s,t)u(s,t)]\, dtds, \tag{5.1.494}$$

then for all $(x, y) \in \Delta$,

$$u(x,y) \le \left\{ B(x,y) \exp\left((p-1) \sum_{i=1}^{n} \int_{\alpha_i(x_0)}^{\alpha_i(x)} \int_{\beta_i(y_0)}^{\beta_i(y)} a_i(\sigma,\tau) d\sigma d\tau \right) \right\}^{\frac{1}{p-1}}, \tag{5.1.495}$$

where for all $(x, y) \in \Delta$,

$$B(x,y) = c^{\frac{p-1}{p}} + (p-1) \sum_{i=1}^{n} \int_{\alpha_i(x_0)}^{\alpha_i(x)} \int_{\beta_i(y_0)}^{\beta_i(y)} b_i(\sigma,\tau) d\tau d\sigma. \tag{5.1.496}$$

(d_2) Let w be as in part (2) of Theorem 1.1.32. If for all $(x, y) \in \Delta$,

$$u^p(x,y) \le c + p \sum_{i=1}^{n} \int_{\alpha_i(t_0)}^{\alpha_i(t)} \int_{\beta_i(y_0)}^{\beta_i(y)} [a_i(s,t)u(s,t)w(u(s,t)) + b_i(s,t)u(s,t)]\, dtds, \tag{5.1.497}$$

then for all $x_0 \le x \le x_1, y_0 \le y \le y_1$,

$$u(x,y) \le \left\{ G^{-1}\left[G(B(x,y)) + (p-1) \sum_{i=1}^{n} \int_{\alpha_i(x_0)}^{\alpha_i(x)} \int_{\beta_i(y_0)}^{\beta_i(y)} a_i(\sigma,\tau) d\tau d\sigma \right] \right\}^{\frac{1}{p-1}}, \tag{5.1.498}$$

where $B(x,y)$ is defined by (5.1.495), G, G^{-1} are as in part (2) of Theorem 1.2.15, and $x_1 \in I_1, y_1 \in I_2$ are chosen so that

$$G(r) = \int_{r_0}^{r} \frac{ds}{w(s^{\frac{1}{p-1}})}, \quad r \geq r_0 > 0, \tag{5.1.499}$$

and $r_0 > 0$ is arbitrary and $t_1 \in I$ is chosen so that

$$G(B(x,y)) + (p-1)\sum_{i=1}^{n}\int_{\alpha_i(x_0)}^{\alpha_i(x)}\int_{\beta_i(y_0)}^{\beta_i(y)} a_i(\sigma,\tau)d\tau d\sigma \in Dom\,(G^{-1}),$$

for all x, y lying in the interval $x_0 \leq x \leq x_1, y_0 \leq y \leq y_1$.

Proof We only give the proof of (d_2), the proof of (d_1) can be done in the same manner. Let $c > 0$ and define a function $z(x,y)$ by the right hand side of (5.1.497). Then $z(x,y) > 0, z(x_0,y) = z(x,y_0) = c, z(x,y)$ is non-decreasing in $(x,y) \in \Delta, u(x,y) \leq \{z(x,y)\}^{\frac{1}{p}}$ and

$$\begin{aligned}
D_2D_1z(x,y) &= p\sum_{i=1}^{n}[a_i(\alpha_i(x),\beta_i(y))u(\alpha_i(x),\beta_i(y))w(u(\alpha_i(x),\beta_i(y)))\\
&\qquad +b_i(\alpha_i(x),\beta_i(y))u(\alpha_i(x),\beta_i(y))]\beta_i'(y)\alpha_i'(x)\\
&\leq p\sum_{i=1}^{n}[a_i(\alpha_i(x),\beta_i(y))\{z(\alpha_i(x),\beta_i(y))\}^{\frac{1}{p}}w(\{z(\alpha_i(x),\beta_i(y))\}^{\frac{1}{p}})\\
&\qquad +b_i(\alpha_i(x),\beta_i(y))\{z(\alpha_i(x),\beta_i(y))\}^{\frac{1}{p}}]\beta_i'(y)\alpha_i'(x)\\
&\leq p\sum_{i=1}^{n}[a_i(\alpha_i(x),\beta_i(y))w(\{z(\alpha_i(x),\beta_i(y))\}^{\frac{1}{p}})\\
&\qquad +b_i(\alpha_i(x),\beta_i(y))]\{z(x,y)\}^{\frac{1}{p}}\beta_i'(y)\alpha_i'(x).
\end{aligned} \tag{5.1.500}$$

From (5.1.499) we derive that for all $(x,y) \in \Delta$,

$$\begin{aligned}
\frac{D_2D_1z(x,y)}{\{z(x,y)\}^{\frac{1}{p}}} &\leq p\sum_{i=1}^{n}\Big[a_i(\alpha_i(x),\beta_i(y))w(\{z(\alpha_i(x),\beta_i(y))\}^{\frac{1}{p}})\\
&\qquad +b_i(\alpha_i(x),\beta_i(y))\Big]\beta_i'(y)\alpha_i'(x) + \frac{D_1z(x,y)\left[D_2\{z(x,y)\}^{\frac{1}{p}}\right]}{\left[\{z(x,y)\}^{\frac{1}{p}}\right]^2},
\end{aligned} \tag{5.1.501}$$

i.e.,

$$D_2\left(\frac{D_1 z(x,y)}{\{z(x,y)\}^{\frac{1}{p}}}\right) \le p\sum_{i=1}^{n}[a_i(\alpha_i(x),\beta_i(y))w(\{z(\alpha_i(x),\beta_i(y))\}^{\frac{1}{p}}) \\ +b_i(\alpha_i(x),\beta_i(y))]\beta_i'(y)\alpha_i'(x). \tag{5.1.502}$$

Keeping x fixed in (5.1.501), setting $y = t$ and integrating with respect to t from y_0 to y and using the fact that $D_1 z(x, y_0) = 0$, we have

$$\frac{D_1 z(x,y)}{\{z(x,y)\}^{\frac{1}{p}}} \le p\int_{y_0}^{y}\sum_{i=1}^{n}[a_i(\alpha_i(x),\beta_i(t))w(\{z(\alpha_i(x),\beta_i(t))\}^{\frac{1}{p}}) \\ +b_i(\alpha_i(x),\beta_i(t))]\beta_i'(t)\alpha_i'(x)dt. \tag{5.1.503}$$

Now keeping y fixed in (5.1.502) and setting $x = s$ and integrating with respect to s from x_0 to x, we have

$$\{z(x,y)\}^{\frac{1}{p}} \le c^{\frac{p-1}{p}} + (p-1)\int_{x_0}^{x}\int_{y_0}^{y}\sum_{i=1}^{n}[a_i(\alpha_i(s),\beta_i(t))w(\{z(\alpha_i(s),\beta_i(t))\}^{\frac{1}{p}}) \\ +b_i(\alpha_i(s),\beta_i(t))]\beta_i'(t)\alpha_i'(s)dtds. \tag{5.1.504}$$

Making the change of variables on the right-hand side of (5.1.503) and rewriting, we have

$$(z(x,y))^{\frac{1}{p}} \le B(x,y) + (p-1)\sum_{i=1}^{n}\int_{\alpha_i(x_0)}^{\alpha_i(x)}\int_{\beta_i(y_0)}^{\beta_i(y)} a_i(\sigma,\tau)w\left(\{z(\sigma,\tau)\}^{\frac{1}{p}}\right)d\sigma d\tau. \tag{5.1.505}$$

Now fix $\lambda \in I_1, \mu \in I_2$ such that $x_0 \le x \le x_1, y_0 \le y \le \mu \le y_1$. Then from (5.1.504), we observe that for all $x_0 \le x \le x_1, y_0 \le y \le \mu \le y_1$,

$$(z(x,y))^{\frac{p-1}{p}} \le B(\lambda,\mu) + (p-1)\sum_{i=1}^{n}\int_{\alpha_i(x_0)}^{\alpha_i(x)}\int_{\beta_i(y_0)}^{\beta_i(y)} a_i(\sigma,\tau)w\left(\{z(\sigma,\tau)\}^{\frac{1}{p}}\right)d\sigma d\tau. \tag{5.1.506}$$

Define a function $v(x,y)$ by the right-hand side of (5.1.505). Then $v(x,y) > 0, v(x_0,y) = v(x,y_0) = B(\lambda,\mu), v(x,y)$ is non-decreasing for all $x_0 \le x \le \lambda$,

$y_0 \le y \le \mu$, $\{z(x,y)\}^{\frac{1}{p}} \le v(x,y)$ and for all $x_0 \le x \le \lambda, y_0 \le y \le \mu$,

$$v(x,y) \le B(\lambda,\mu) + (p-1)\sum_{i=1}^{n}\int_{\alpha_i(x_0)}^{\alpha_i(x)}\int_{\beta_i(y_0)}^{\beta_i(y)} a_i(\sigma,\tau)w\left(\{v(\sigma,\tau)\}^{\frac{1}{p-1}}\right)d\tau d\sigma.$$

Now by following the proof of part (B_1) in Theorem 5.1.40 (see also [518]), we get for all $x_0 \le x \le \lambda \le x_1, y_0 \le y \le \mu \le y_1$,

$$v(x,y) \le G^{-1}\left[G(B(\lambda,\mu)) + (p-1)\sum_{i=1}^{n}\int_{\alpha_i(x_0)}^{\alpha_i(x)}\int_{\beta_i(y_0)}^{\beta_i(y)} a_i(\sigma,\tau)d\tau d\sigma\right]. \quad (5.1.507)$$

Since (λ,μ) is arbitrary, we get the desired inequality in (5.1.497) from (5.1.506) and the fact that

$$u(x,y) \le \{z(x,y)\}^{\frac{1}{p}} \le \left\{[v(x,y)]^{\frac{p}{p-1}}\right\}^{\frac{1}{p}} = \{v(x,y)\}^{\frac{1}{p-1}}.$$

The proof of the case when $c \ge 0$ can be completed as mentioned in the proof of part (1) in Theorem 1.2.15. The domain $x_0 \le x \le x_1, y_0 \le y \le y_1$ is obvious. □

Remark 5.1.15 We note that the inequalities established in Theorem 5.1.63 can be extended very easily for functions involving more than two independent variables (see, e.g., [507]). If we take $p = 2, n = 1, \alpha_1 = \alpha, \beta_1 = \beta, a_1 = f, b_1 = g$ in Theorem 5.1.63, then we get the two independent variable generalizations of the inequalities given in [356] (see, e.g., Corollary 2 and Theorem 1). For a slight variant of the inequality in Theorem 5.1.63 given in [356] and its two independent variable version, see, e.g., [518].

5.2 Nonlinear Multi-Dimensional Bellman-Gronwall Inequality and Their Generalizations

5.2.1 The Opial Inequalities, LaSalle Inequalities, Gollwitzer, Langenhop, Bondge and Pachpatte Inequalities and Their Generalizations

Rasmussen [569] obtained a nonlinear two-dimensional version of the inequality by using ideas previously applied to functions of one independent variable by Opial [435] and others.

In the sequel, we shall introduce the result from Headley [277] show that these techniques can be further exploited to obtain nonlinear extensions to any number of independent variables.

Let G be an open connected (possibly unbounded) set contained in N-dimensional Euclidean space $\mathbb{R}^N$. For any two points x and y in G, with $x = (x_1, \cdots, x_N)$ and $y = (y_1, \cdots, y_N)$, define the set $G(x, y)$ to be the closed rectangular parallelepiped with one diagonal joining the points x and y; that is,

$$G(x,y) = \left\{ t \in \mathbb{R}^N \middle| t_j = (1-\lambda_j)y_j + \lambda_j x_j,\ 0 \le \lambda_j \le 1,\ j = 1, 2, \cdots, N \right\}.$$

We remark that the identity $G(x, y) = G(y, x)$ is an immediate consequence of the definition of $G(x, y)$. This symmetry will enable us to drop the requirement in [569] that the line joining the points x and y have non-negative (though not necessarily finite) slope.

For fixed ξ in G, define the integral operator K by

$$(Kv)(x) = \int_{G(x,\xi)} k(t, v(t_0))\, dt, \tag{5.2.1}$$

where v and k are real-valued functions (k being continuous on $G \times \mathbb{R}$), x is a point of G, the set $G(x, \xi)$ is contained in G, and dt is Lebesgue measure on $\mathbb{R}^N$.

Theorem 5.2.1 ([277]) *Let ξ and y be points in a (possibly unbounded) domain $G \subset \mathbb{R}^N$ such that $G(\xi, y) \subset G$. Let g and k be real-valued functions, with g continuous on G, and with k continuous on $G \times \mathbb{R}$ and non-decreasing with respect to its last argument. Let ϵ_n $(n = 1, 2, \cdots)$ be a strictly decreasing sequence of real numbers with limit zero. Suppose that there exists a family $\{v_n | n = 1, 2, \cdots\}$ of functions continuous on $G(y, \xi)$ such that, for $n = 1, 2, \cdots$, and all x in $G(y, \xi)$,*

$$v_n(x) = g(x) + \epsilon_n + (Kv_n)(x). \tag{5.2.2}$$

Let U be the maximal solution on $G(y, \xi)$ of the nonlinear Volterra integral equation

$$u(x) = g(x) + (Ku)(x). \tag{5.2.3}$$

Then $\lim_{n\to+\infty} v_n = U$ *on $G(y, \xi)$.*

Proof If $\xi_j = y_j$ for some j, then the parallelepiped $G(y, \xi)$ has volume zero and the result is trivially true. We therefore suppose that $\xi_j \neq y_j$ for $j = 1, 2, \cdots, N$. We shall show that the sequence $\{v_n\}$ is strictly decreasing and satisfies the hypothesis of the Ascoli-Arzela Theorem [236]. Accordingly, we first note that $v_m(\xi) - v_n(\xi) = \epsilon_m - \epsilon_n < 0$ whenever $m > n$. If the sequence $\{v_n\}$ were not strictly decreasing, then it would follow from the continuity of the functions v_m and v_n that, for some z in $G(y, \xi)$, with $z \neq \xi$, we would have $v_m < v_n$ on the set $G(z, \xi) - \{z\}$, whilst $v_m(z) = v_n(z)$. But then it follows the definition of v_n and the monotonicity of $\{\epsilon_n\}$

and k that

$$\begin{aligned} v_n(z) &= g(z) + \epsilon_n + (kv_n)(z) \\ &> g(z) + \epsilon_m + (kv_m)(z) = v_m(z), \end{aligned}$$

whenever $m > n$. This contradicts the definition of z and shows that $v_m(x) < v_n(x)$ whenever $m > n$ and $x \in G(y, \xi)$. It follows that the sequence $\{v_n\}$ $(n = 1, 2, \cdots)$ is bounded above by $M_1 = \max\{v_1(x)|x \in G(y, \xi)\}$. Let u be any solution of (5.2.3) on $G(y, \xi)$. Then $u(\xi) < v_n(\xi)$ for $n = 1, 2, \cdots$.

We now show that $u(x) < v_n(x)$ for $n = 1, 2, \cdots$ and all x in $G(y, \xi)$. If this were not true, then there would exist some function v_m satisfying (5.2.2) and some point η in $G(y, \xi) - \{\eta\}$, whilst $v_m(\eta) = u(\eta)$. But then it follows from (5.2.2)–(5.2.3) and the monotonicity of k that

$$u(\eta) = g(\eta) + Ku(\eta) < g(\eta) + \epsilon_m + (Kv_m)(\eta) = v_m(\eta),$$

which contradicts the definition of η and shows that $u(x) < v_n(x)$ for each positive integer n and all x in $G(y, \xi)$. Consequently, the sequence $\{v_n\}$ is bounded below by

$$M_2 = \min\Big\{u(x)|x \in G(y, \xi)\Big\}.$$

To prove equicontinuity, let ϵ be any positive number, and let y' and y'' be any two points in $G(y, \xi)$, with $y' = (y'_1, \cdots, y'_N)$ and $y'' = (y''_1, \cdots, y''_N)$. For $n = 1, 2, \cdots$, the first two terms on the right-hand side of the identity

$$v_n(y') - v_n(y'') = (Kv_n)(y') - (Kv_n)(y'') + g(y') - g(y'') \tag{5.2.4}$$

may be written as

$$(Kv_n)(y') - (Kv_n)(y'') = \int_E K(t, v_n(t))\, dt - \int_F K(t, v_n(t))\, dt, \tag{5.2.5}$$

where

$$E = G(y', \xi) - G(y'', \xi), \quad F = G(y'', \xi) - G(y', \xi),$$

the negative signs denoting relative complementation.

The set E may be decomposed into N disjoint (possibly degenerate) rectangular parallelepipeds $E_1, \cdots, E_N$ as follows

$$\begin{cases} E_1 = \Big\{x \in E|\min(y'_1, y''_1) \le x_1 \le \max(y'_1, y''_1)\Big\}, \\ E_j = \Big\{x \in E - E_{j-1}|\min(y'_j, y''_j) \le x_j \le \max(y'_j, y''_j)\Big\}, \quad j = 2, 3, \ldots, N. \end{cases}$$

Similarly, the set F may be decomposed into N disjoint (possibly degenerate) rectangular parallelepipeds $F_1, \cdots, F_N$ defined by

$$\begin{cases} F_1 = \Big\{x \in F \,|\, \min(y'_1, y''_1) \le x_1 \le \max(y'_1, y''_1)\Big\}, \\ F_j = \Big\{x \in E - E_{j-1} \,|\, \min(y'_j, y''_j) \le y_j \le \max(y'_j, y''_j)\Big\}, \quad j = 2, 3, \cdots, N. \end{cases}$$

For $j = 1, 2, \cdots, N$, consider the parallelepipeds E_j and F_j. Each edge parallel to the x_j-axis has length $|y'_j - y''_j|$, and each edge parallel to the x_r-axis ($r \neq j$) has length not greater than $\max(|y'_r - \xi_r|, |y''_r - \xi_r|)$, and therefore not greater than $|y_r - \xi_r|$. It thus follows from (5.2.5) that, for $n = 1, 2, \cdots$,

$$|(Kv_n)(y') - (Kv_n)(y'')| \le 2AB \sum_{j=1}^{N} |y'_j - y''_j|,$$

where

$$\begin{cases} A = \sup\Big\{k(x, u) | x \in G(y, \xi), M_2 \le u \le M_1\Big\}, \\ B = \max\Big\{y_r - \xi_r | r = 1, 2, \cdots, N\Big\}. \end{cases}$$

We now introduce the norm

$$\|y' - y''\| = \sum_{j=1}^{N} |y'_j - y''_j|.$$

Since g is continuous and therefore uniformly continuous on the compact set $G(y, \xi)$, there exists a positive number δ_1 such that $|g(y') - g(y'')| < \epsilon/2$ whenever $y', y'' \in G(y, \xi)$ and $\|y' - y''\| < \delta_1$. For $A > 0$, choose δ_2 so that $0 < \delta_2 < \epsilon/(4AB)$, and let $\delta = \min(\delta_1, \delta_2)$. It follows from (5.2.4) that, for $n = 1, 2, \cdots, |v_n(y') - v_n(y'')| < \epsilon$ whenever $y', y'' \in G(y, \xi)$ and $\|y' - y''\| < \delta$.

If $A = 0$, this estimate is trivially true for all $\delta < \delta_1$. We have therefore shown that the sequence $\{v_n\}$ is equicontinuous and uniformly bounded on the compact set $G(y, \xi)$. According to the Ascoli-Arzela theorem, there exists a subsequence $\{v_{n_i}\}$ ($i = 1, 2, \cdots$), being decreasing and bounded below, converges on $G(y, \xi)$. It thus follows that $\lim_{n\to+\infty} v_n = \lim_{n\to+\infty} v_{n_i}$. If we let $i \to +\infty$ in the identity

$$v_{n_i}(x) = g(x) + \epsilon_{n_i} + (Kv_{n_i})(x), \quad x \in G(y, \xi),$$

we can see that, in view of the continuity of k and the uniform convergence of the sequence $\{v_{n_i}\}$, the function $\lim_{i\to+\infty} v_{n_i}$ is a solution of (5.2.3). Moreover, since each solution u of (5.2.3) satisfies the inequality $u(x) < v_n(x)$ for all x in $G(y, \xi)$

and $n = 1, 2, \cdots$, so does the maximal solution U; consequently

$$U(x) \leq \lim_{i \to +\infty} v_{n_i}(x) \leq U(x)$$

for all x in $G(y, \xi)$. This completes the proof of the theorem. □

We note that the existence of maximal solutions for the integral equations is a consequence of general results proved in Walter's monograph [658].

Theorem 5.2.2 (The Walter Inequality [658]) *Let ξ and y be points in a (possibly unbounded) domain $G \subset \mathbb{R}^N$ such that $G(\xi, y) \subset G$, let g, v and k be real-valued functions, with g and v continuous on G, and with k continuous on $G \times \mathbb{R}$ and non-decreasing with respect to its last argument. Let v be a solution on $G(y, \xi)$ of the nonlinear integral inequality*

$$v(x) \leq g(x) + (Kv)(x). \tag{5.2.6}$$

Then $v(x) \leq U(x)$ for all x in $G(y, \xi)$, where U is the maximal solution on $G(y, \xi)$ of the integral equation

$$u(x) = g(x) + (ku)(x). \tag{5.2.7}$$

Proof It $\xi_j = y_j$ for some j, then the result is trivially true. We therefore suppose that $\xi_j \neq y_j$ for $j = 1, 2, \cdots, N$. Let $\{\epsilon_n\}$ $(n = 1, 2, \cdots)$ be a strictly decreasing sequence of real numbers with limit zero. For $n = 1, 2, \cdots$, let v_n be a continuous solution on $G(y, \xi)$ of the integral equation

$$v_n(x) = g(x) + \epsilon_n + (Kv_n)(x).$$

We now show that $v(x) < v_n(x)$ for all positive integers n and all x in $G(y, \xi)$. If this were false for some v_m at some point of $G(y, \xi)$, then, in view of the inequalities

$$v(\xi) \leq g(\xi) < g(\xi) + \epsilon_m = v_m(\xi),$$

it follows from the continuity of v and v_m that there must exist some point z in $G(y, \xi)$, with $z \neq \xi$, such that $v(x) < v_m(x)$ on the set $G(y, \xi)/\{z\}$, whilst $v(z) = v_m(z)$. But then

$$v(z) \leq g(z) + (Kv)(z) < g(z) + \epsilon_m + (Kv_m)(z) = v_m(z).$$

This contradicts the definition of z and shows that $v(x) < v_n(x)$ for $n = 1, 2, \cdots$, and all x in $G(y, \xi)$. It follows from Theorem 5.2.1 that, for all x in $G(y, \xi)$,

$$v(x) \leq \lim_{n \to +\infty} v_n(x) \leq U(x).$$

This completes the proof of the theorem. □

It should be noted that G need not be connected. It is enough if ξ, y and $G(y,\xi)$ lie in the same component (=maximal connected subset) of G.

The Gronwall-Bellman inequality has been extended to several variables by a number of mathematicians [94, 135, 153, 159, 190, 246, 277, 456, 619, 620]. For example, Conlan and Diaz [159] generalized the Gronwall-Bellman inequality in n variables in order to prove uniqueness of solutions of a nonlinear partial differential equation. Walter [658] gave a more natural extension of the Gronwall-Bellman inequality in several variables by using the properties of monotone operators. Snow [619] obtained corresponding inequality in two variable scalar and vector-valued functions by using the notion of a Riemann function. Young [709] established Gronwall's inequality in n variables, which coincides with the result given in Walter [658], where a representation of the Riemann function is used. Chandra and Davis [135] generalized the Gronwall-Bellman inequality to systems of n linear inequalities in m variables by arguments that amount to manipulation of the resolvent kernel equation for a monotone operator. Their results encompass some works of Chu and Metcalf [153], Snow [619, 620], Walter [658], Wendroff [47], and Young [709], as well as providing extensions to kernels having more general form and weaker regularity properties. Bondge and Pachpatte [94] and Yeh [701] established some nonlinear integral inequalities of Wendroff type [47]. Shih and Yeh [601] extended the Gronwall-Bellman inequality to several variables by a different approach.

Throughout S will denote an open bounded set in the n dimensional Euclidean space $\mathbb{R}^n$.

Dhongade and Deo [197] introduce the class $\mathcal{F}$ in Definition 1.1.3.

We establish next the following integral inequality which may be used certain situations.

Theorem 5.2.3 (The Yeh-Shih Inequality [707]) *Let $w(x), g(x)$ be real-valued positive continuous functions on S, and $n(x)$ be a positive continuous non-decreasing function on S and H on $\mathcal{F}$. Suppose that the inequality holds for all x in S with $x \geq x^0$,*

$$w(x) \leq n(x) + \int_{x^0}^{x} g(s)\left(w(s) + \int_{x^0}^{s} g(t)H(w(t))dt\right) ds. \tag{5.2.8}$$

Then for $x^0 \leq x \leq x^*$,

$$w(x) \leq n(x)\left\{1 + \int_{x^0}^{x} g(s)G^{-1}\left(G(1) + \int_{x^0}^{s} g(t)dt\right) ds\right\}, \tag{5.2.9}$$

where

$$G(r) = \int_{r_0}^{r} \frac{ds}{s + H(s)}, \quad r \geq r_0 > 0,$$

and G^{-1} *is the inverse of* G *and* x^* *is chosen so that*

$$G(1) + \int_{x^0}^{x} g(t)dt \in Dom\,(G^{-1})$$

for all x *in* S *lying in the parallelopiped* $x^0 \le x \le x^*$.

Proof Since $n(x)$ is positive, non-decreasing and H in $\mathcal{F}$, we infer from (5.2.8),

$$\begin{aligned}\frac{w(x)}{n(x)} &\le 1 + \int_{x^0}^{x} g(s)\left(\frac{w(s)}{n(s)} + \int_{x^0}^{x} \frac{g(t)H(w(t))}{n(t)}dt\right)ds \\ &\le 1 + \int_{x^0}^{x} g(s)\left(\frac{w(s)}{n(s)} + \int_{x^0}^{x} g(t)H\left(\frac{w(t)}{n(t)}\right)dt\right)ds. \qquad (5.2.10)\end{aligned}$$

Define $u(x)$ by the right-hand side of (5.2.10). Then

$$\begin{cases} D_1 \cdots D_n u(x) = g(x)\left(\dfrac{w(x)}{n(x)} + \displaystyle\int_{x^0}^{x} g(t)H(\frac{w(t)}{n(t)})dt\right), \\ u(x) = 1 \text{ on } x_i = x_i^0,\ i = 1, 2, \cdots, n. \end{cases} \qquad (5.2.11)$$

It follows from (5.2.10) and (5.2.11) that

$$D_1 \cdots D_n u(x) \le g(x)\left(u(x) + \int_{x^0}^{x} g(t)H(u(t))dt\right). \qquad (5.2.12)$$

Let

$$v(x) = u(x) + \int_{x^0}^{x} g(t)H(u(t))dt.$$

Then

$$\begin{cases} v(x) = u(x) \text{ on } x_i = x_i^0,\ i = 1, \cdots, n, \\ u(x) \le v(x), \end{cases}$$

and

$$D_1 \cdots D_n v(x) = D_1 \cdots D_n u(x) + g(x)H(v(x)). \qquad (5.2.13)$$

It follows from (5.2.12) and (5.2.13) that

$$D_1 \cdots D_n v(x) \le g(x)(v(x) + H(v(x))),$$

i.e.,

$$\frac{D_1 \cdots D_n v(x)}{v(x) + H(v(x))} \leq g(x).$$

Thus

$$\frac{(v(x) + H(v(x)))D_1 \cdots D_n v(x)}{(v(x) + H(v(x)))^2} \leq g(x) + \frac{D_n(v(x) + H(v(x)))D_1 \cdots D_n v(x)}{(v(x) + H(v(x)))^2},$$

i.e.,

$$D_n \left(\frac{D_1 \cdots D_{n-1} v(x)}{v(x) + H(v(x))} \right) \leq g(x). \tag{5.2.14}$$

Integrating both sides of (5.2.14) with respect to the component x_n of x from x_n^0 to x_n, we get

$$\frac{D_1 \cdots D_{n-1} v(x)}{v(x) + H(v(x))} \leq \int_{x_n^0}^{x_n} g(x_1, \cdots, x_{n-1}, t_n) dt_n.$$

Therefore

$$\frac{(v(x) + H(v(x)))D_1 \cdots D_{n-1} v(x)}{(v(x) + H(v(x))^2} \leq \int_{x_n^0}^{x_n} g(x_1, \cdots, x_{n-1}, t_n) dt_n + \frac{D_{n-1}(v(x) + H(v(x)))D_1 \cdots D_{n-2} v(x)}{(v(x) + H(v(x))^2},$$

i.e.,

$$D_{n-1} \left(\frac{D_1 \cdots D_{n-2} v(x)}{v(x) + H(v(x))} \right) \leq \int_{x_n^0}^{x_n} g(x_1, \cdots, x_{n-1}, t_n) dt_n.$$

Integrating both sides of the above inequality with respect to the component x_{n-1} of x from x_{n-1}^0 to x_{n-1}, we derive

$$\frac{D_1 \cdots D_{n-2} v(x)}{v(x)} \leq \int_{x_{n-1}^0}^{x_{n-1}} \int_{x_n^0}^{x_n} g(x_1, \cdots, x_{n-2}, t_{n-1}, t_n) dt_n dt_{n-1}.$$

Continuing in this way, we may arrive at

$$\frac{D_1 D_2 v(x)}{v(x) + H(v(x))} \leq \int_{x_3^0}^{x_3} \cdots \int_{x_n^0}^{x_n} g(x_1, x_2, t_3, \cdots, t_n) dt_n \cdots dt_3.$$

Thus

$$D_2\left(\frac{D_1 v(x)}{v(x)+H(v(x))}\right) \le \int_{x_3^0}^{x_3} \cdots \int_{x_n^0}^{x_n} g(x_1, x_2, t_3, \cdots, t_n)dt_n \cdots dt_3. \tag{5.2.15}$$

Integrating both sides of (5.2.15) with respect to the component x_2 of x from x_2^0 to x_2, we can obtain

$$\frac{D_1 v(t)}{v(x)+H(v(x))} \le \int_{x_2^0}^{x_2} \cdots \int_{x_n^0}^{x_n} g(x_1, t_2, \cdots, t_n)dt_n \cdots dt_2.$$

Hence

$$D_1 G(v(x)) \le \int_{x_2^0}^{x_2} g(x_1, t_2, \cdots, t_n)dt_n \cdots dt_2.$$

Integrating both sides of the above inequality with respect to the component x_1 of x from x_1^0 to x_1, we can get

$$G(v(x)) - G(1) \le \int_{x^0}^{x} g(t)dt.$$

Hence

$$v(x) \le G^{-1}\left(G(1) + \int_{x^0}^{x} g(t)dt\right).$$

Substituting the above bound on $v(x)$ in (5.2.12), we can conclude

$$D_1 \cdots D_n u(x) \le G^{-1}\left(G(1) + \int_{x^0}^{x} g(t)dt\right).$$

Integrating both sides of the above inequality from x^0 to x, we deduce

$$u(x) \le 1 + \int_{x^0}^{x} g(s)G^{-1}\left(G(1) + \int_{x^0}^{x} g(t)dt\right)ds$$

from which and $w(x) \le n(x)u(x)$, we can obtain the desired bound in (5.2.13). This proves the theorem. □

Remark 5.2.1 The integral inequality obtained in Theorem 5.2.3 extends Pachpatte's result [456] to several variables.
Similarly, we have the following result.

Theorem 5.2.4 (The Yeh-Shih Inequality [707]) *Let $w(x)$ and $g(x)$ be real-valued positive continuous functions on S, and $n(x)$ be a positive, non-decreasing* *continuous function on S and H on $\mathcal{F}$. Suppose that the inequality holds for all x in S with $x \geq x^0$,*

$$w(x) \leq n(x) + \int_{x^0}^{x} g(s)H(w(s))ds.$$

Then for $x^0 \leq x \leq x^*$,

$$w(x) \leq n(x)G^{-1}\left(G(1) + \int_{x^0}^{s} g(s)ds\right) ds,$$

where

$$G(r) = \int_{r_0}^{r} \frac{ds}{H(s)}, \quad r \geq r_0 > 0,$$

and G^{-1} is the inverse of G and x^* is chosen so that $G(1) + \int_{x^0}^{x} g(s)ds \in Dom\ (G^{-1})$.

Remark 5.2.2 For $n = 1$ in Theorem 5.2.4, we obtain Bihari's inequality [82] and Theorem 3 of [197].

We know that the Gronwall Bellman's inequality is covered by LaSalle's inequality. LaSalle's inequality is important in proving uniqueness, boundedness, and convergence of successive approximation [330].

A two independent variable generalization of the Gronwall-Bellman-LaSalle inequality due to Wendorff given in [47] has evoked of Snow [620], Ghoshal and Masood [246], Chandra and Davis [135], Headley [277], Bondge and Pachpatte [94, 95], Chu and Metcalf [153], Shin and Yeh [601], Walter [658], Yeh [701], and Young [709].

In the sequel, we shall introduce n independent variable generalizations of the integral inequalities from [600] established by LaSalle [330], Gollwitzer [249], Langenhop [328], Pachpatte [450, 451], and Bondge and Pachpatte [95] for $n = 1$ or $n = 2$.

First, Gollwizer's inequality [249] and Bondge and Pachpatte's inequality [95] are unified in the following theorem.

Theorem 5.2.5 (The Shih-Yeh Inequality [600]) *Let $w(x), a(x)$ and $b(x)$ be a real-valued, non-negative and continuous functions defined on $\mathbb{R}^n$; let $u(s)$ be a positive real-valued continuous functions defined on $\mathbb{R}^n$. Suppose that the inequality holds for all $0 \leq x \leq s$,*

$$u(s) \geq w(x) - a(s) \int_{x}^{s} b(t)w(t)dt, \tag{5.2.16}$$

where $s \in \mathbb{R}^n$. *Then for all* $0 \leq x \leq s$,

$$u(x) \geq w(x) \exp\left(-a(s) \int_x^s b(t)dt\right). \tag{5.2.17}$$

Proof We first discuss the case when n is even. We may rewrite (5.2.16) as

$$w(x) \leq u(s) + a(s) \int_x^s b(t)w(t)dt. \tag{5.2.18}$$

For fixed s in $\mathbb{R}^n$, we define, for all $0 \leq x \leq s$,

$$r(x) = u(s) + a(s) \int_x^s b(t)w(t)dt. \tag{5.2.19}$$

Then

$$r(x) = u(s) \quad \text{on} \quad x_i = s_i, \quad i = 1, 2, \ldots, n; \tag{5.2.20}$$

and

$$D_1 D_2 \ldots D_n r(x) = a(s)b(x)w(x). \tag{5.2.21}$$

Then by (5.2.18)

$$D_1 D_2 \ldots D_n r(x) \leq a(s)b(x)r(x),$$

which implies

$$\frac{r(x)D_1 \ldots D_n r(x)}{r^2(x)} \leq a(s)b(x) + \frac{D_n r(x)(D_1 \ldots, D_{n-1} r(x))}{r^2(x)},$$

i. e.,

$$D_n \left(\frac{D_1 \ldots D_{n-1} r(x)}{r(x)}\right) \leq a(s)b(x). \tag{5.2.22}$$

Integrating both sides of (5.2.22) with respect to the component x_n of x from x_n to s_n, we have

$$\frac{D_1 \cdots D_{n-1} r(x_1, \cdots, x_{n-1}, s_n)}{r(x_1, \cdots, x_{n-1}, s_n)} - \frac{D_1 \cdots D_{n-1} r(x)}{r(x)} \leq a(s) \int_{x_n}^{s_n} b(x_1, \cdots, x_{n-1}, t_n)dt_n.$$

Since

$$D_1 \cdots D_{n-1} r(x_1, \cdots, x_{n-1}, s_n) = 0,$$

we derive

$$-\frac{D_1\cdots D_{n-1}r(x)}{r(x)} \le a(s)\int_{x_n}^{s_n} b(x_1,\cdots,x_{n-1},t_n)dt_n,$$

which implies

$$-D_{n-1}\left(\frac{D_1\cdots D_{n-2}r(x)}{r(x)}\right) \le a(s)\int_{x_n}^{s_n} b(x_1,\cdots,x_{n-1},t_n)dt_n. \quad (5.2.23)$$

Integrating both sides of (5.2.23) with respect to the component x_{n-1} of x from x_{n-1} to s_{n-1}, we arrive at

$$\frac{D_1\cdots D_{n-2}r(x)}{r(x)} \le a(s)\int_{x_{n-1}}^{s_{n-1}}\int_{x_n}^{s_n} f(x_1,\cdots,x_{n-2},t_{n-1},t_n)dt_n dt_{n-1}.$$

Computing in this way, we can obtain

$$\frac{D_1D_2r(x)}{r(x)} \le a(s)\int_{x_3}^{s_3}\cdots\int_{x^n}^{s_n} b(x_1,x_2,t_3,\cdots,t_n)dt_n\cdots dt_3. \quad (5.2.24)$$

It follows from (5.2.24) that

$$D_2\left(\frac{D_1r(x)}{r(x)}\right) \le a(s)\int_{x_3}^{s_3}\cdots\int_{x_0}^{s_n} b(x_1,x_2,t_3,\cdots,t_n)dt_n\cdots dt_3. \quad (5.2.25)$$

Integrating both sides of (5.2.25) with respect to the component x_2 of x from x_2 to s_2, we have

$$\frac{D_1r(x_1,s_2,x_3,\cdots,x_n)}{r(x_1,s_2,x_3,\cdots,x_n)} - \frac{D_1r(x)}{r(x)} \le a(s)\int_{x_2}^{s_2}\cdots\int_{x_n}^{s_n} b(x_1,t_2,\cdots,t_n)dt_n\cdots dt_2.$$

Thus

$$-\frac{D_1r(x)}{r(x)} \le a(s)\int_{x_2}^{s_2}\cdots\int_{x_n}^{s_n} b(x_1,t_2,\cdots,t_n)dt_n\cdots dt_2.$$

Integrating both sides of the above inequality with respect to the component x_1 of x from x_1 to s_1, we have

$$\log\frac{r(x)}{u(s)} \le a(s)\int_x^s b(t)dt,$$

which implies

$$w(x) \leq r(x) \leq u(s) \exp\left(a(s) \int_x^s b(t)dt\right)$$

and the theorem follows for even n.

Next, we discuss the case n is odd. As in the proof of the case n is even, we have

$$D_1 D_2 \ldots D_n r(x) = -a(s)b(x)w(x),$$

and

$$\frac{r(x)D_1 \ldots D_n r(x)}{r^2(x)} \geq a(s)b(x) + \frac{D_n r(x)(D_1 \ldots, D_{n-1} r(x))}{r^2(x)},$$

i.e.,

$$D_n\left(\frac{D_1 \ldots D_{n-1} r(x)}{r(x)}\right) \geq -a(s)b(x).$$

Integrating both sides of the above inequality with respect to the component x_n of x from x_n to s_n, we conclude

$$-\frac{D_1 \ldots D_{n-1} r(x)}{r(x)} \geq -a(s) \int_{x_n}^{s_n} b(x_1, \ldots, x_{n-1}, t_n)dt_n,$$

which implies

$$D_{n-1}\left(\frac{D_1 \ldots D_{n-2} r(x)}{r(x)}\right) \leq a(s) \int_{x_n}^{s_n} b(x_1, \ldots, x_{n-1}, t_n)dt_n.$$

As in the proof of the case n is even, we can obtain the desired result. □

As an application of Theorem 5.2.5, we can establish the following n independent variable generalization of the equality of LaSalle [330], Gollwitzer [250], and Bondge and Pachpatte [95].

Theorem 5.2.6 (The Shih-Yeh Inequality [600]) *Let $w(s), a(s)$ and $b(s)$ be as defined in Theorem 5.2.5; let $H(r)$ be a positive, continuous, strictly increasing, convex, and sub-multiplicative for all $r > 0$, $H(0) = 0$ and $\lim_{r\to+\infty} H(r) = +\infty$. Suppose that $g(s)$ and $h(s)$ are positive functions defined on $\mathbb{R}^n$ with $g(s)+h(s) = 1$ and the following inequality holds for all $0 \leq x \leq s$, where $s \in \mathbb{R}^n$,*

$$u(s) \geq w(x) - a(s)H^{-1}\left(\int_x^s b(t)H(w(t))dt\right). \tag{5.2.26}$$

Then for all $0 \le x \le s$,

$$u(s) \ge g(s)H^{-1}[g^{-1}(s)H(w(x))\exp(-h(s)H(a(s)h^{-1}(s))\int_x^s b(t)dt)]. \tag{5.2.27}$$

Proof We may rewrite (5.2.26) as

$$w(x) \le g(s)u(s)g^{-1}(s) + h(s)a(s)h^{-1}(s)H^{-1}(\int_x^s b(t)H(w(t))dt).$$

Since H is convex sub-multiplicative and increasing, we have

$$H(w(x)) \le g(s)H(u(s)g^{-1}(s)) + h(s)H(a(s)h^{-1}(s))\int_x^s b(t)H(w(t))dt,$$

i.e.,

$$g(s)H(u(x)g^{-1}(s)) \le H(w(x)) - h(s)H(a(s)h^{-1}(s))\int_x^s b(t)H(w(t))dt.$$

Applying Theorem 5.2.5 to the above inequality, we can get the desired bound in (5.2.27). □

We next introduce the following n independent variable generalization of the integral inequality established by Langenhop [328] and Bondge and Pachpatte [94].

Theorem 5.2.7 (The Shih-Yeh Inequality [600]) *Let* $u(s), a(s)$ *and* $b(s)$ *be as defined in Theorem 5.2.5; let* $W(r)$ *be a positive, continuous, non-decreasing function for all* $r > 0$, $W(0) = 0$ *and* $W'(r) \in C(\mathbb{R}_+, \mathbb{R}_+)$. *Suppose that the inequality holds for all* $0 < x \le s$, *where* $s \in \mathbb{R}^n_+$,

$$u(s) \ge u(x) - a(s)\int_x^s b(t)W(u(t))dt. \tag{5.2.28}$$

Then for all $s^0 \in \mathbb{R}^n_+$, *and all* $0 \le x \le s \le s^0$,

$$u(s) \ge Q^{-1}[Q(u(x)) - a(s)\int_x^s b(t)dt], \tag{5.2.29}$$

where

$$Q(r) = \int_{r_0}^r \frac{ds}{W(s)}, \qquad r \ge r_0 > 0, \tag{5.2.30}$$

and Q^{-1} is the inverse function of Q, and for all $0 \leq x \leq s$,

$$Q(u(x)) - a(s)\int_x^s b(t)dt \in Dom\ (Q^{-1}).$$

Proof We only prove the case when n is even. We can rewrite (5.2.28) as

$$u(x) \leq u(s) + a(s)\int_x^s b(t)W(u(t))dt. \tag{5.2.31}$$

For fixed s in $\mathbb{R}_+^n$, we define, for all $0 \leq x \leq s$,

$$r(x) = u(s) + a(s)\int_x^s b(t)W(u(t))dt.$$

Then

$$r(x) = u(s) \qquad \text{on} \quad x_i = s_i, \quad i = 1, 2, \ldots, n; \tag{5.2.32}$$

$$D_1 D_2 \ldots D_n r(x) = a(s)b(x)W(u(x)), \tag{5.2.33}$$

and

$$u(x) \leq r(x).$$

Since W is non-decreasing, (5.2.33) implies

$$D_1 D_2 \ldots D_n r(x) \leq a(s)b(x)W(r(x)),$$

i.e.,

$$\frac{D_1 D_2 \ldots D_n r(x)}{W(r(x))} \leq a(s)b(x).$$

Thus

$$\frac{W(r(x))D_1 \ldots D_n r(x)}{W^2(r(x))} \leq a(s)b(x) + \frac{D_n W(r(x))D_1 \ldots, D_{n-1} r(x)}{W^2(r(x))},$$

i.e.,

$$D_n\left(\frac{D_1 \ldots D_{n-1} r(x)}{W(r(x))}\right) \leq a(s)b(x). \tag{5.2.34}$$

Integrating both sides of (5.2.34) with respect to the component x_n of x from x_n to s_n, we get

$$\frac{D_1 \ldots D_{n-1} r(x_1, \ldots, x_{n-1}, s_n)}{W(r(x_1, \ldots, x_{n-1}, s_n))} - \frac{D_1 \ldots D_{n-1} r(x)}{W(r(x))} \leq a(s) \int_{x_n}^{s_n} b(x_1, \ldots, x_{n-1}, t_n) dt_n.$$

Since

$$D_1 \ldots D_{n-1} r(x_1, \ldots, x_{n-1}, s_n) = 0,$$

we know

$$-\frac{D_1 \ldots D_{n-1} r(x)}{r(x)} \leq a(s) \int_{x_n}^{s_n} b(x_1, \ldots, x_{n-1}, t_n) dt_n,$$

which implies

$$-D_{n-1} \left(\frac{D_1 \ldots D_{n-2} r(x)}{W(r(x))} \right) \leq a(s) \int_{x_n}^{s_n} b(x_1, \ldots, x_{n-1}, t_n) dt_n.$$

As in the proof of Theorem 5.2.5, we have

$$\frac{D_1 r(x_1, s_2, x_3, \cdots, x_n)}{r(x_1, s_2, x_3, \cdots, x_n)} - \frac{D_1 r(x)}{r(x)} \leq a(s) \int_{x_2}^{s_2} \cdots \int_{x_n}^{s_n} b(x_1, t_2, \cdots, t_n) dt_n \cdots dt_2. \tag{5.2.35}$$

Thus it follows from (5.2.30) and (5.2.35) that

$$-D_1 Q(r(x)) \leq a(s) \int_{x_2}^{s_2} \cdots \int_{x_n}^{s_n} b(x_1, t_2, \cdots, t_n) dt_n \cdots dt_2. \tag{5.2.36}$$

Integrating both sides of the above inequality with respect to the component x_1 of x from x_1 to s_1, we have

$$-Q(r(s_1, x_2, \cdots, x_n)) + Q(x) \leq a(s) \int_x^s b(t) dt.$$

It follows from (5.2.32) that

$$-Q(r(s)) + Q(r(x)) \leq a(s) \int_x^s b(t) dt,$$

i.e.,

$$Q(r(s)) \geq Q(r(x)) - a(s) \int_x^s b(t)dt. \tag{5.2.37}$$

From (5.2.37), we can obtain the desired bound in (5.2.29). □

Next we introduce n independent variable generalizations of the integral inequalities established by Pachpatte [450, 451] and Bondge and Pachpatte [94].

The next result concerns the n independent variable generalization of the integral inequality established by Pachpatte [453] and Bondge and Pachpatte [94].

Theorem 5.2.8 (The Shih-Yeh Inequality [600]) *Let $w(s)$, $a(s)$, $b(s)$, and $c(s)$ be real-valued non-negative continuous functions defined on $\mathbb{R}^n_+$; let $u(s)$ be a positive real-valued continuous functions defined on $\mathbb{R}^n_+$. Suppose that the inequality holds for all $0 \leq x \leq s$, where $s \in \mathbb{R}^n_+$,*

$$u(s) \geq w(s) - a(s) \left[\int_x^s b(m)w(m)dm + \int_x^s b(m)(\int_m^s c(t)w(t)dt)dm \right]. \tag{5.2.38}$$

Then for all $0 \leq x \leq s$,

$$u(x) \geq w(x) \left[1 + a(s) \int_x^s b(m) \exp\left(\left(\int_m^s (a(s)b(t) + c(t) \right) dt)dm \right]^{-1}. \tag{5.2.39}$$

Proof We only proof the case when n is even. We rewrite (5.2.38) as

$$w(x) \leq u(s) + a(s) \left[\int_x^s b(m)w(m)dm + \int_x^s b(m)(\int_m^s c(t)w(t)dt)dm \right]. \tag{5.2.40}$$

For fixed s in $\mathbb{R}^n_+$, we define, for all $0 \leq x \leq s$,

$$r(x) = u(s) + a(s) \left[\int_x^s b(m)w(m)dm + \int_x^s b(m) \left(\int_m^s c(t)w(t)dt \right) dm \right]. \tag{5.2.41}$$

Then

$$r(x) = u(s) \quad \text{on} \quad x_i = s_i, \quad i = 1, 2, \ldots, n; \tag{5.2.42}$$

and

$$w(x) \leq r(x).$$

Hence

$$D_1D_2\dots D_nr(x) = a(s)b(x)\left[w(x) + \int_x^s c(t)w(t)dt\right] \qquad (5.2.43)$$
$$\le a(s)b(x)\left[r(x) + \int_x^s c(t)r(t)dt\right].$$

Define

$$v(x) = r(x) = r(s), \qquad \text{on } x_i = s_i,\ i = 1, 2, \cdots, n \qquad (5.2.44)$$

and

$$D_1\cdots D_nv(x) = D_1\cdots D_nr(x) + c(x)r(x)$$
$$\le [a(s)b(x) + c(x)]v(x).$$

By an argument similar to that in the proof of Theorem 5.2.5, we obtain

$$v(x) \le u(s)\exp\left(\int_x^s (a(s)b(t) + c(t))dt\right).$$

Substituting this bound on $v(x)$ in (5.2.43), we have

$$D_1\dots D_nr(x) \le a(s)b(x)u(x)\exp\left(\int_x^s (a(s)b(t) + c(t))dt\right).$$

Integrating both sides of the above inequality with respect to the component x_n of x from x_n to s_n, we arrive at

$$D_1\cdots D_{n-1}r(x_1,\dots,x_{n-1},s_n) - D_1\dots D_{n-1}r(x)$$
$$\le a(s)u(s)\int_{x_n}^{s_n} b(x_1,\cdots,x_{n-1},m_n)$$
$$\times\exp\left(\int_{x_1}^{s_1}\cdots\int_{x_{n-1}}^{s_{n-1}}\int_{x_n}^{s_n}(a(s)b(t) + c(t))dt\right)dm_n.$$

Integrating both sides of the above inequality with respect to the component x_{n-1} of x from x_{n-1} to s_{n-1}, we derive

$$-D_1\cdots D_{n-2}r(x_1,\cdots,x_{n-2},s_{n-1},x_n) + D_1\cdots D_{n-2}r(x)$$
$$\le a(s)u(s)\int_{x_{n-1}}^{s_{n-1}}\int_{x_n}^{s_n} b(x_1,\cdots,x_{n-2},m_{n-1},m_n)$$
$$\times\exp\left(\int_{x_1}^{s_1}\cdots\int_{x_{n-2}}^{s_{n-2}}\int_{m_{n-1}}^{s_{n-1}}\int_{m_n}^{s_n}(a(s)b(t) + c(t))dt\right)dm_ndm_{n-1}.$$

Computing in this way, we conclude

$$\begin{aligned}
&D_1 r(x_1, s_2, x_3, \ldots, x_n) - D_1 r(x) \\
&\leq a(s)u(s) \int_{x_2}^{s_2} \ldots \int_{x_n}^{s_n} b(x_1, m_2, \ldots, m_n) \\
&\quad \times \exp\left(\int_{x_1}^{s_1} \int_{m_2}^{s_2} \cdots \int_{m_n}^{s_n} (a(s)b(t) + c(t))dt \right) dm_n \cdots dm_2.
\end{aligned}$$

Integrating both sides of the above inequality with respect to the component x_1 of x from x_1 to s_1, we obtain

$$r(x) \leq u(s) \left[1 + a(s) \int_x^s b(m) \exp\left(\int_m^s (a(s)b(t) + c(t))dt \right) dm \right]. \tag{5.2.45}$$

The desired bound in (5.2.39) follows from (5.2.40) and (5.2.45). □

Next we can apply Theorem 5.2.8 to establish the following n independent variable generalization of the integral inequality established by Pachpatte [450] and Bondge and Pachpatte [97].

Theorem 5.2.9 (The Shih-Yeh Inequality [600]) *Let $w(x)$, $a(s)$, $b(s)$, $c(s)$, and $u(x)$ be as defined in Theorem 5.2.8; let $H(r)$, $g(s)$, and $h(s)$ be as defined in Theorem 5.2.6. Suppose that the inequality holds for all $0 \leq x \leq s$, where $s \in \mathbb{R}^n_+$,*

$$\begin{aligned}
u(x) \geq w(x) - a(s)H^{-1}\bigg[&\int_x^s b(m)H(w(m))dm \\
&+ \int_x^s b(m)\Big(\int_m^s c(t)H(w(t))dt\Big)dm \bigg].
\end{aligned} \tag{5.2.46}$$

Then for all $0 \leq x \leq s$,

$$\begin{aligned}
u(s) \geq g(s)H^{-1}\bigg\{ &g^{-1}H(w(x))[1 + h(s)H(a(s)h^{-1}(s)) \int_x^s b(m) \\
&\times \exp \int_m^s |h(s)H(a(s)h^{-1}(s))b(t) + c(t)|dtdm] \bigg\}^{-1}.
\end{aligned} \tag{5.2.47}$$

Proof We can rewrite (5.2.46) as

$$\begin{aligned}
w(x) \leq\ & g(s)u(s)g^{-1}(s) + h(s)a(s)h^{-1}(s)H^{-1} \\
&\times \Big[\int_x^s b(m)H(w(m))dm + \int_x^s b(m)\Big(\int_m^s c(t)H(w(t))dt\Big)dm \Big].
\end{aligned}$$

Since H is convex, sub-multiplicative, and strictly increasing, we get

$$g(s)H(u(s)g^{-1}(s)) \geq H(w(x)) - h(s)H(a(s)h^{-1}(s)) \int_x^s b(m)H(w(m))dm.$$

By Theorem 5.2.8, we can derive the desired bound in (5.2.47). Thus the proof is complete. □

Next we shall introduce the following n-independent-variable generalization of the integral inequality established by Pachpatte [450] and Bondge and Pachpatte [97].

Theorem 5.2.10 (The Shih-Yeh Inequality [600]) *Let $u(s)$, $a(s)$, $b(s)$, and $c(s)$ be as defined in Theorem 5.2.8; let $G(r)$ be a positive, continuous, strictly increasing, sub-additive, and sub-multiplicative function for all $r > 0$, $r \in \mathbb{R}_+$ and $G(0) = 0$; let G^{-1} denote the inverse function of G. Suppose the following inequality holds for all $0 \leq x \leq s$, where $s \in \mathbb{R}_+^n$,*

$$u(s) \geq u(x) - a(s)G^{-1}\left[\int_x^s b(m)G(u(m))dm + \int_x^s b(m)\left(\int_m^s c(t)G(u(t))dt\right)dm\right] \tag{5.2.48}$$

Then for all $0 \leq x \leq s$,

$$u(s) \geq u(x)G^{-1}\left\{\left[1 + G(a(s))\int_x^s b(m) \times \exp\left(\int_m^s (b(t)G(a(s)) + c(t))dt\right)dm\right]^{-1}\right\}. \tag{5.2.49}$$

Proof We may rewrite (5.2.48) as

$$u(x) \leq u(s) + a(s)G^{-1}\left[\int_x^s b(m)G(u(m))dm + \int_x^s b(m)\left(\int_m^s c(t)G(u(t))dt\right)dm\right]. \tag{5.2.50}$$

Since G is sub-additive, we infer from (5.2.50)

$$G(u(x)) \leq G(u(s)) + G(a(s))\Big[\int_x^s b(m)G(u(m))dm + \int_x^s b(m)\Big(\int_m^s c(t)G(u(t))dt\Big)dm\Big]. \tag{5.2.51}$$

By defining $r(x)$ by the right-hand side of (5.2.51) and following a similar argument to that in the proof of Theorem 5.2.9 with suitable modifications, we can obtain the desire bound in (5.2.49). □

The next result, due to Singare-Pachpatte [613], concerns the following n independent variable generalization of Gollwitzer's inequality given in [250] for lower bound on unknown function.

Theorem 5.2.11 (The Singare-Pachpatte Inequality [613]) *Let $\phi(x), a(x), b(x)$ and $u(x)$ be as defined in Theorem 5.4.41 in Qin [557], $H(r)$ be a positive, continuous, strictly increasing, convex and sub-multiplicative function for all $r > 0, H(0) = 0; \lim_{r\to+\infty} H(r) = +\infty$. Let $\alpha(s), \beta(s)$ be positive continuous functions defined on Ω with $\alpha(s) + \beta(s) \equiv 1$. Suppose further that the inequality holds for all $x \leq s; x, s \in \Omega$,*

$$u(s) \geq \phi(x) - a(s)H^{-1}\left(\int_x^s b(\xi)H(\phi(\xi))d\xi\right), \tag{5.2.52}$$

then for all $x \leq s; x, s \in \Omega$,

$$u(s) \geq \alpha(s)H^{-1}\left[\alpha^{-1}(s)H(\phi(x))\exp(-\beta(s)H(a(s)\beta^{-1}(s)))\int_x^s b(\xi)d\xi\right]. \tag{5.2.53}$$

Proof The proof is identical to that given by Gollwitzer [250]. We rewrite (5.2.52) as

$$\phi(x) \leq \alpha(s)u(s)\alpha^{-1}(s) + \beta(s)a(s)\beta^{-1}(s)H^{-1}\left(\int_x^s b(\xi)H(\phi(\xi))d\xi\right). \tag{5.2.54}$$

Since H is convex, sub-multiplicative and monotonic, we get

$$\alpha(s)H(u(s)\alpha^{-1}(s)) \geq H(\phi(s)) - \beta(s)H(a(s)\beta^{-1}(s))\left(\int_x^s b(\xi)H(\phi(\xi))d\xi\right).$$

Now applying Theorem 5.4.41 in Qin [557] yields the desired bound in (5.2.53). □

Remark 5.2.3 We note that in Theorem 5.2.11, if we take $H(u) = u$, then Theorem 5.2.11 reduces to Theorem 5.4.41 in Qin [557].

In the next theorem, we introduce an n independent variable generalization of the integral inequality established by Langenhop [328].

Theorem 5.2.12 (The Singare-Pachpatte Inequality [613]) *Let $u(x), a(x)$ and $b(x)$ be as defined in Theorem 5.4.41 in Qin [557]; let $W(r)$ be a positive, continuous, monotonic non-decreasing function for all $r > 0$, $W(0) = 0$ and $W'(r)$ exist and is continuous, with $W'(r) \geq 0$ for all $r \geq 0$; and suppose further that the inequality holds for all $x \leq s; x, s \in \Omega$,*

$$u(s) \geq u(x) - a(s) \int_x^s b(\xi) W(u(\xi)) d\xi. \tag{5.2.55}$$

Then, for $\Omega_1 \subset \Omega$,

$$u(s) \geq G^{-1}\Big[G(u(x)) - a(s)(\int_x^s b(\xi) d\xi)\Big] \tag{5.2.56}$$

where

$$G(r) = \int_{r_0}^r \frac{ds}{W(s)}, \quad r \geq r_0 > 0, \tag{5.2.57}$$

and r_0 is any fixed positive number; G^{-1} is the inverse of function of G, and Ω_1 is such that

$$G(u(x)) - a(s)(\int_x^s b(\xi) d\xi) \in Dom\ (G^{-1})$$

for all $x \leq s, x, s \in \Omega_1 \subset \Omega$.

Proof We may rewrite (5.2.55) as

$$u(x) \leq u(s) + a(s) \left(\int_x^s b(\xi) W(u(\xi)) d\xi \right). \tag{5.2.58}$$

For fixed $s \in \Omega$, we define for $x \leq s, x \in \Omega$,

$$r(x) = u(s) + a(s) \left(\int_x^s b(\xi) W(u(\xi)) d\xi \right). \tag{5.2.59}$$

Therefore

$$r(s_1, x_2, \cdots, x_n) = \cdots = r(x_1, \cdots, x_{n-1}, s_n) = u(s_1, \cdots, s_n).$$

Then by the same argument as in the proof of Theorem 5.2.11, we obtain from (5.2.59) that

$$D_1 \cdots D_k r(x) = (-1)^k a(s) \int_{x_{k+1}}^{s_{k+1}} \cdots \int_{x_n}^{s_n} b(x_1, \cdots x_k, \xi_{k+1}, \cdots, \xi_n) \times W(u(x_1, \cdots, x_k \xi_{k+1}, \cdots, \xi_n)) d\xi_n \cdots d\xi_{k+1} \tag{5.2.60}$$

and continuing in this way, we obtain

$$D_1 \cdots D_n r(x) = (-1)^n a(s) b(x) W(u(x)). \tag{5.2.61}$$

We distinguish the following two cases.

Case I. If the order n of the derivatives in (5.2.61) is even, then from (5.2.61) we infer

$$D_1 \cdots D_n r(x) = a(s) b(x) W(u(x)) \tag{5.2.62}$$

which, along with (5.2.58), implies

$$D_1 \cdots D_n r(x) \leq a(s) b(x) W(r(x))$$

i.e.,

$$\frac{D_1 \cdots D_n r(x)}{W(r(x))} \leq a(s) b(x). \tag{5.2.63}$$

From (5.2.63), we get

$$\frac{W(r(x))[D_1 \cdots D_n r(x)]}{W^2(r(x))} \leq a(s) b(x) + \frac{W'(r(x)) \cdot D_n(r(x))[D_1 \cdots D_{n-1} r(x)]}{W^2(r(x))}. \tag{5.2.64}$$

For, by (5.2.64) we see that $D_n r(x)$ and $D_1 \cdots D_{n-1} r(x)$ are both non-positive which implies that $D_n r(x)[D_1 \cdots D_{n-1} r(x)]$ is non-negative and hence (5.2.64) is true. Now (5.2.64) is equivalent to

$$D_n \left(\frac{D_1 \cdots D_{n-1} r(x)}{W(r(x))} \right) \leq a(s) b(x).$$

Now keeping $x_1, \cdots, x_{n-1}$ fixed in the above inequality, setting $x_n = \xi_n$ and then integrating with respect to ξ_n from x_n to s_n, we have

$$\frac{D_1 \cdots D_{n-1} r(x)}{W(r(x))} \geq -a(s) \int_{x_n}^{s_n} b(x_1, \cdots, x_{n-1}, \xi_n) d\xi_n. \tag{5.2.65}$$

Again from (5.2.65), we observe that

$$\frac{W(r(x))[D_1\cdots D_{n-1}r(x)]}{W^2(r(x))} \geq -a(s)\int_{x_n}^{s_n} b(x_1,\cdots,x_{n-1},\xi_n)d\xi_n + \frac{W'(r(x))D_{n-1}(r(x))[D_1\ldots D_{n-2}r(x)]}{W^2(r(x))}. \tag{5.2.66}$$

Nothing that, (5.2.66) shows that $D_{n-1}r(x)$ is non-positive and $D_1\cdots D_{n-2}r(x)$ is non-negative, which implies that $D_{n-1}r(x)[D_1\cdots D_{n-2}r(x)]$ is non-positive, (5.2.66) is true. But (5.2.66) is equivalent to

$$D_{n-1}\Big(\frac{D_1\cdots D_{n-2}r(x)}{W(r(x))}\Big) \geq -a(s)\int_{x_n}^{s_n} b(x_1,\cdots,x_{n-1},\xi_n)d\xi_n.$$

Now keeping $x_1,\cdots,x_{n-2},x_n$ fixed in the above inequality, setting $x_{n-1}=\xi_{n-1}$ and then integrating with respect to ξ_{n-1} from x_{n-1} to s_{n-1}, we get,

$$\frac{D_1\cdots D_{n-2}r(x)}{W(r(x))} \leq a(s)\int_{x_{n-1}}^{s_{n-1}}\int_{x_n}^{s_n} b(x_1,\cdots,x_{n-2},\xi_{n-1},\xi_n)d\xi_n d\xi_{n-1}.$$

Proceeding in this way, we finally obtain

$$\frac{D_1 r(x)}{W(r(x))} \geq -a(s)\int_{x_2}^{s_2}\int_{x_n}^{s_n} b(x_1,\xi_2\cdots,\xi_n)d\xi_n\ldots d\xi_2. \tag{5.2.67}$$

From (5.2.57) and (5.2.67), we conclude,

$$D_1 G(r(x)) \geq -a(s)\int_{x_2}^{s_2}\int_{x_n}^{s_n} b(x_1,\xi_2\cdots,\xi_n)d\xi_n\cdots d\xi_2. \tag{5.2.68}$$

Now keeping $x_2,\cdots,x_n$ fixed in (5.2.68), setting $x_1=\xi_1$ and then integrating with respect to ξ_1 from x_1 to s_1, we obtain

$$G(r(x)) \leq G(u(s)) + a(s)\int_x^s b(\xi)d\xi \tag{5.2.69}$$

which implies

$$G(u(s)) \geq G(u(s)) - a(s)\int_x^s b(\xi)d\xi. \tag{5.2.70}$$

Thus the desired bound in (5.2.56) follows from (5.2.70). The sub-domain Ω_1 of Ω is obvious.

Case II. If the order n is odd, then (5.2.61) becomes

$$D_1 \cdots D_n r(x) = -a(s)b(x)W(u(x)),$$

and the proof proceeds exactly as in **Case I**, again leading to (5.2.70). □

Remark 5.2.4 We note that in Theorem 5.2.12, if we take $W(u) = u$, then (5.2.56) reduces to

$$u(s) \geq u(x) \exp\left(-a(s) \int_x^s b(\xi)d\xi\right),$$

and if we set, $W(u) = u^\alpha, 0 < \alpha < 1$, then (5.2.53) reduces to

$$u(s) \geq \left[u^\beta(x) - \beta a(s) \int_x^s b(\xi)d\xi\right]^{1/\beta}$$

where $\alpha + \beta = 1$.

The next result from [643] for the nonlinear case is concerned with the inequality

$$u(x) \leq a(x)\left(c + \sum_{r=1}^{m} H^r(x,u)\right), \tag{5.2.71}$$

where

$$H(x,u) = \int_y^x f_{r1}(x^1)u^{\alpha_{r1}}(x^1) \cdots \int_y^{x^{r-1}} f_{rr}u^{\alpha_{rr}}(x^r)\, dx^r \cdots dx^1$$

and α_{ri}, $1 \leq i \leq r$, $1 \leq r \leq m$ are non-negative real numbers and the constant $c > 0$.

In the following result, we shall denote $a_r = \sum_{i=1}^r a_{rl}$, $a = \max_{1 \leq r \leq m} a_r$.

Theorem 5.2.13 (The Thandapani-Agarwal Inequality [643]) *Assume that inequality (5.2.71) holds in Ω (an open bounded set in $\mathbb{R}^n$). Then we have*

$$\begin{cases} u(x) \leq ca(x)exp\left(\int_y^x Q(s)\, ds\right), & if \;\; \alpha = 1, \quad (5.2.72) \\ u(x) \leq a(x)\left(c^{1-\alpha} + (1-\alpha)\int_y^x Q(s)\, ds\right)^{1/1-\alpha}, & if \;\; \alpha \neq 1, \quad (5.2.73) \end{cases}$$

where

$$Q(x) = \sum_{r=1}^{m} H_x^r(x, a)c^{\alpha_r - \alpha}$$

and when $\alpha > 1$, *we assume,* $c^{1-\alpha} + (1-\alpha)\int_y^x Q(s)\,ds > 0$.

Proof In fact, inequality (5.2.71) can be rewritten as

$$u(x) \le a(x)\phi(x), \tag{5.2.74}$$

where

$$\phi(x) = c + \sum_{r=1}^{m} H^r(x, u).$$

Thus using the non-decreasing nature of $\phi(x)$ and (5.2.74), we easily find

$$\phi_x(x) \le \sum_{r+1}^{m} H_x^r(x, a)[\phi(x)]^{\alpha_r}.$$

Since $\phi(x) \ge c$, we can get

$$\begin{aligned}\phi_x(x) &\le \sum_{r=1}^{m} H_x^r(x, a)c^{\alpha_r - \alpha}\phi^{\alpha}(x)\\ &= Q(x)\phi^{\alpha}(x).\end{aligned}$$

Now following the proof of Theorem 5.4.54 in Qin [557], we easily show

$$\frac{\phi_{x_1}(x)}{\phi_{x_2}(x)} \le \int_{y_2}^{x_2} \cdots \int_{y_n}^{x_n} Q(x_1, s_2, \cdots, s_n)\,ds_2 \cdots ds_n. \tag{5.2.75}$$

Since $\phi(y_1, x_2, \ldots, x_n) = c$, the results (5.1.423)–(5.1.424) follow by integrating (5.2.75). □

For $n = m = 1$, $a(x) = 1, \alpha_{11} = 2$, Theorem 5.2.13 reduces to first result in this direction by Freedman [235]; also for m up to 2 (see, e.g., [94]).

For the next result, we shall need the following class of functions:

In [706], Yeh and Shih considered a class of functions $\mathcal{F}$, whose definition is Definition 1.1.3 in Chap. 1.

Clearly condition (ii) in Definition 1.1.5 implies that $H(u) \equiv H(1)u$. To avoid such a triviality, see, Beeseck [56] redefined the class $\mathcal{F}$ as the class of $\mathcal{F}_1$, see Definition 1.1.6.

Definition 5.2.1 A function $W: [0,+\infty) \to (0,+\infty)$ is said to belong to the class $\mathcal{F}_2$ if

(i) $W(u)$ is positive, non-decreasing, continuous and $W_{x_k}(u(x_1, x_2, \cdots, x_n)) \geq 0$ for all $2 \leq k \leq n$ and all $u \geq 0$,
(ii) $(1/v)W(u) \leq W(u/v)$ for all $u \geq 0,\ v \geq 1$.

This class $\mathcal{F}_2$ has been modified here as given by class $\mathcal{F}_1$ and used for $n = 1$ in [194, 195] to avoid the triviality $W(u) = uW(1)$; see also [56].

Theorem 5.2.14 (The Thandapani-Agarwal Inequality [643]) *Assume that the inequality holds*

$$u(x) \leq a(x) + \sum_{r=1}^{m} E^r(x, u) + \sum_{i=1}^{l} g_i(x) \int_y^x h_i(s) W_i(u(s))\, ds, \qquad (5.2.76)$$

where

(i) $a(x) \geq 1$ *and is non-decreasing,*
(ii) $g_i(x) \geq 1,\ 1 \leq i \leq l,$
(iii) $W_i \in \mathcal{F}_2,\ 1 \leq i \leq l.$

Then we have

$$u(x) \leq a(x)\psi(x)e(x) \prod_{i=1}^{l} F_i(x),$$

where

$$\begin{cases} \psi(x) = \exp\left(\sum_{r=1}^{m} E^r(x, e)\right), \\ F_k(x) = G_k^{-1}\left[G_k(l) + \int_y^x h_k(s)\psi(s)e(s) \prod_{j=1}^{k-1} F_j(s)\, ds\right], \\ F_0(x) = 1, \quad 1 \leq k \leq l, \\ G_k(\theta) = \int_{\theta_0}^{\theta} \frac{ds}{W_k(s)}, \quad 0 \leq \theta_0 \leq \theta, \end{cases}$$

as long as

$$G_k(1) + \int_y^x h_k(s)\psi(s)e(s) \prod_{j=1}^{k-1} F_j(s)\, ds \in Dom\,(G_k^{-1}), \quad 1 \leq k \leq l.$$

Proof From inequality (5.2.76), we derive

$$\frac{u(x)}{e(x)} \leq a^*(x) + \sum_{r=1}^{m} E^r(x, \frac{eu}{e})$$

where

$$a^* = a(x) + \sum_{i=1}^{l} \int_y^x h_i(s) W_i(u(s))\, ds.$$

Since $a^*(x)$ is non-decreasing, from Theorem 5.4.54 in Qin [557] it follows that

$$\frac{u(x)}{e(x)} \le a^*(x)\psi(x)$$

and hence by using the definition of class $\mathcal{F}_2$, we get

$$y(x) \le 1 + \sum_{i=1}^{l} \int_y^x h_i(s)e(s)\phi(s) W_i(y(s))\, ds$$

where

$$y(x) = \frac{u(x)}{a(x)\psi(x)e(x)}.$$

Thus it is sufficient to show that $y(x) \le \prod_{i=1}^{l} F_i(x)$, which will be proved by finite induction. For $l = 1$, we have

$$y(x) \le 1 + \int_y^x h_1(s)e(s)\psi(s) W_1(y(s))\, ds.$$

Let $\phi_1(x)$ be the right-hand side of the above inequality, then using non-decreasing nature of W_1, we find

$$\phi_{1x}(x) \le h_1(x)e(x)\psi(x) W_1(\phi_1(x))$$

or

$$\left(\frac{\phi_{1x_1 \ldots x_{n_1}}(x)}{W_1(\phi_1(x))}\right)_{x_n} \le h_1(x)e(x)\psi(x)$$

and hence as in Theorem 5.4.54 in Qin [557], we get

$$\frac{\phi_{1x_1 \ldots x_{n_1}}(x)}{W_1(\phi_1(x))} \le \int_{y_n}^{x_n} h_1(x_1, \ldots, x_{n-1}, s_n) e(x_1, \ldots, x_{n-1}, s_n) \psi(x_1, \ldots, x_{n-1}, s_n)\, ds_n.$$

Repeating the above procedure, we may obtain

$$\frac{\phi_{1x_1}(x)}{W_1(\phi_1(x))} \le \int_{y_2}^{x_2} \cdots \int_{y_n}^{x_n} h_1(x_1, \dots, x_{n-1}, s_n) e(x_1, \dots, x_{n-1}, s_n) \times \psi(x_1, \dots, x_{n-1}, s_n)\, ds_2 \cdots ds_n. \tag{5.2.77}$$

From the definition of G_1, it follows

$$\begin{aligned} G(\phi_1(x)) - G_1(\phi_1(y_1, x_2, \dots, x_n)) &= \int_{\phi_1(y_1, x_2, \dots, x_n)}^{\phi_1(x)} \frac{ds}{W_1(s)} \\ &= \int_{y_1}^{x_1} \frac{\phi_{1s_1}(s_1, x_2, \dots, x_n)}{W_1(\phi_1(s_1, x_2, \dots, x_n))}\, ds_1. \end{aligned} \tag{5.2.78}$$

By using (5.2.77) in (5.2.78), we can obtain

$$\phi_1(x) \le G_1^{-1}\left[G_1(1) + \int_y^x h_1(s) e(s) \psi(s)\, ds\right] = F_1(x).$$

Now assuming that result is true for some k such that $1 \le k \le l-1$ for $k+1$, we conclude

$$y(x) \le \left[1 + \int_y^x h_{k+1}(s) e(s) \psi(s) W_{k+1}(y(s))\, ds\right] + \sum_{i=1}^{k} \int_y^x h_i(s) e(s) \psi(s) W_i(y(s))\, ds.$$

Since the part inside the bracket is non-decreasing, we easily find

$$y(x) \le \left[1 + \int_y^x h_{k+1}(s) e(s) \psi(s) W_{k+1}(y(s))\, ds\right] \prod_{i=1}^{k} F_i(x)$$

or

$$\frac{y(x)}{\prod_{i=1}^{k} F_i(s)} \le 1 + \int_y^x h_{k+1}(s) e(s) \psi(s) \prod_{i=1}^{k} W_{k+1}\left(y(s) / \prod_{i=1}^{k} F_i(s)\right) ds$$

from which, $y(x) \le \prod_{i=1}^{k+1} F_i(x)$ follows by using the same arguments as for the case $l = 1$. This thus completes the proof. □

Theorem 5.2.15 (The Thandapani-Agarwal Inequality [643]) *In addition to the hypotheses of Theorem 5.2.14, let $g_i(x)$, $q \le i \le l$ be non-decreasing, Then we have*

$$u(x) \le a(x)\psi(x)\prod_{i=1}^{l} F_i(x),$$

where

$$\begin{cases} \psi_1(x) = \exp\left(\sum_{r=1}^{m} E^r(x,1)\right), \\ F_k(x) = g_k(x)G_k^{-1}\left[g_k(1) + \int_y^x h_k(s)\psi_1 9s)g_k(s)\prod_{i=1}^{k-1} F_i(s)\,ds\right], \\ 1 \le k \le l, \quad F_0(x) = 1 \end{cases}$$

as long as

$$G_k(1) + \int_y^x h_k(s)\psi_1(s)g_k(s)\prod_{i=1}^{k-1} F_i(s)\,ds \in Dom\,(G_k^{-1}), \quad 1 \le k \le l.$$

Theorem 5.2.16 (The Thandapani-Agarwal Inequality [643]) *Assume the inequality holds*

$$u(x) \le a(x) + \sum_{r=1}^{m} E^r(x,u) + \sum_{i=1}^{l} E^i(x, W(u)) \tag{5.2.79}$$

where

(i) $a(x) \ge 1$ and is non-decreasing,
(ii) $W \in \mathcal{F}_2$.

Then we have

$$u(x) \le a(x)\psi_1(x)G^{-1}\left[G(1) + \int_y^x \sum_{i=1}^{l} E_s^i(s,\psi_1)\,ds\right] \tag{5.2.80}$$

where $\psi_1(x)$ is the same as in Theorem 5.2.15 and the term inside the bracket of (5.2.80)$\in$ Dom (G^{-1}).

The proofs of Theorem 5.2.15 and 5.2.16 are similar to the proof of Theorem 5.2.14.

Theorem 5.2.17 (The Thandapani-Agarwal Inequality [643]) *Assume that the inequality (5.2.79) holds, where*

(i) $a(x)$ is positive and non-decreasing,
(ii) W is positive, continuous, non-decreasing, sub-multiplicative and $W_{x_k}(u(x_1,\ldots,x_n)) \ge 0$ for all $2 \le k \le n$.

Then we have

$$u(x) \leq a(x)\psi_1(x)G^{-1}\left[G(1) + \int_y^x \sum_{r=1}^{l} E_s^r(s, \frac{W(a\psi_1)}{a})\, ds\right], \tag{5.2.81}$$

where $\psi_1(x)$ *is the same as in Theorem 5.2.15 and the term inside the bracket of (5.2.81)*$\in$ *Dom* (G^{-1}).

Proof We apply Theorem 5.4.54 in Qin [557] for inequality (5.2.79) to obtain

$$u(x) \leq \left[a(x) + \sum_{i=1}^{l} E_x^i(x, W(u))\right]\psi_1(x)$$

or

$$\frac{u(x)}{a(x)\psi(x)} \leq 1 + \sum_{i=1}^{l} E_x^i\left(x, W\left(\frac{u}{a\psi_1}a\psi_1)/a\right)\right). \tag{5.2.82}$$

Let $\phi(x)$ be the right-hand side of (5.2.82), then

$$\phi_x(x) = \sum_{i=1}^{l} E_x^i\left(x, W\left(\frac{u}{a\psi_1}a\psi_1)/a\right)\right).$$

Now using the fact that W is non-decreasing and sub-multiplicative, we get

$$\frac{\phi_x(x)}{W(\phi(x))} \leq \sum_{i=1}^{l} E_x^i(x, W(a\psi_1)/a).$$

Using the same arguments as those in Theorem 5.2.14, we easily find

$$\phi(x) \leq G^{-1}\left[G(1) + \int_y^x \sum_{i=1}^{l} E_s^i(s, W(a\psi_1)/a)\, ds\right]$$

which gives us the required result. □

Some particular cases $n = 2,\ m$ up to 2 with different assumptions on $a(x)$ have been discussed in [94].

Gollwitzer [250] considered a couple of functional integral inequalities in one independent variables and obtained results which incorporate the well-known Bellman lemma [65], a case similar to the Langenhop inequality [328], and an inequality studied by Willett [671]. Bondge and Pachpatte [95] gave some integral inequalities in two independent variables which extend Lemma 2 and part one of

Theorem 1 of Gollwitzer, as well as an inequality of Langenhop. These results are useful tools in the analysis of differential and integral equations.

Next we show that by making use of a version of Bellman lemma in n variables [709], it is possible to extend completely Theorem 1 of Gollwitzer to integrals involving n independent variables. This method, which gives a sharp bound for the solution, is entirely different from the method used by Bondge and Pachpatte [95]. However, by generalizing the procedure used in [95], we are able to extend a Langenhop inequality to n dimension.

Let D be a bounded domain in E^n and denote by x a point $(x_1, \cdots, x_n)$ in D. If x and y are any two points in D, we say $x < y$ if $x_i < y_i$ for $i = 1, \cdots, n$. Further we adopt the notation

$$\int_x^y f(s)ds = \int_{x_n}^{y_n} \cdots \int_{x_1}^{y_1} f(s_1, \cdots, s_n)ds_1 \cdots ds_n$$

In the subsequent discussions, we shall require the following conditions:

(C_1): $a(x), b(x), c(x)$ and $u(x)$ are conditions, non-negative functions in D;

(C_2): $G(u)$ is a continuous, increasing, convex and sub-multiplicative function for all $u \geq 0$ such that $G(0) = 0$ and $G(u) \to +\infty$ as $u \to +\infty$;

(C_3): $\alpha(x)$ and $\beta(x)$ are positive functions in D such that $\alpha(x) + \beta(x) = 1$.

Lemma 5.2.1 (The Young Inequality [710]) *Suppose condition (C_1) holds. Let $v(t; y)$ be the solution of the characteristic initial value problem*

$$\begin{cases} (-1)^n D_1 \cdots D_n v(t; y) - b(t)c(t)v(t; y) = 0 \quad in \;\; D \\ v(t; y) = 1 \quad when \;\; t_i = y_i \end{cases} \tag{5.2.83}$$

where $D_i = \partial/\partial t_i, i = 1, \cdots, n$. Let R be a sub-domain of D containing y in which $v \geq 0$. If for all x in R, and for all $x < y$,

$$u(y) \leq a(y) + b(y) \int_x^y c(s)u(s)ds, \tag{5.2.84}$$

then

$$u(y) \leq a(y) + b(y) \int_x^y a(t)c(t)v(t; y)dt. \tag{5.2.85}$$

This lemma is an n-dimensional analogue of the well known Bellman lemma which, except for the extra factor $b(y)$, was proved in [709]. The function $v(t; y)$ is known as a Riemann function with pole at the point y.

The next lemma is an extension of a special case of Langenhop inequality [328] cast in the form of Lemma 5.2.1.

Lemma 5.2.2 (The Young Inequality [710]) *Suppose condition* (C_1) *holds. Let* $v(s;x)$ *be the Riemann function with pole at* x *satisfying*

$$\begin{cases} D_n v(s;x) - b(y)c(s)v(s;x) = 0 \quad in \quad D \\ v(s;x) = 1 \quad when \quad s_i = x_i \end{cases} \tag{5.2.86}$$

where $D_i = \partial/\partial s_i, i = 1, \cdots, n$. *Let* R *be a sub-domain of* D *containing* x *in which* $v \geq 0$. *If for all* y *in* R, *and for all* $x < y$,

$$a(y) \geq u(x) - b(y) \int_x^y c(s)u(s)ds, \tag{5.2.87}$$

then

$$a(y) \geq u(x) / \left[1 + b(y) \int_x^y c(s)v(s;x)ds \right]. \tag{5.2.88}$$

Proof The proof of this lemma is similar to that of Lemma 5.2.1 and is therefore omitted. □

We remark that in both Lemmas 5.2.1 and 5.2.2, the bound obtained for u is sharp in the sense that if equality prevails in (5.2.84) or (5.2.87), then equality also holds in (5.2.85) or (5.2.88), respectively.

Lemma 5.2.3 (The Young Inequality [710]) *Let condition* (C_1) *hold. If for all* $x < y$,

$$a(y) \geq u(x) - b(y) \int_x^y c(s)u(s)ds, \tag{5.2.89}$$

then for all $x < y$,

$$a(y) \geq u(x) \exp\left(-b(y) \int_x^y c(s)ds \right). \tag{5.2.90}$$

Proof Set

$$v(x) = a(y) + b(y) \int_x^y c(s)u(s)ds \tag{5.2.91}$$

so that $u(x) \leq v(x)$. Then

$$(-1)^n D_1 \cdots D_n v(x) = b(y)c(x)u(x) \leq b(y)c(x)v(x)$$

or

$$\frac{(-1)^n D_1 \cdots D_n v(x)}{v(x)} \leq b(y)c(x) \tag{5.2.92}$$

where $D_i = \partial/\partial x_i, i = 1, \cdots, n$. Since $(-1)^n(D_1 v)(D_2 \cdots D_n v) \geq 0$, we may rewrite (5.2.92) as

$$(-1)^n D_1 \left[\frac{D_2 \cdots D_n v(x)}{v(x)}\right] \leq b(y)c(x).$$

Integrating this from x_1 to y_1, using s_1 as a variable of integration, and noting that $v(x) = a(y)$ when $x_i = y_i$ for any i, $1 \leq i \leq n$, we find

$$(-1)^n \frac{D_2 \cdots D_n v(x)}{v(x)} \leq \int_{x_1}^{y_1} b(y)c(s)ds_1. \tag{5.2.93}$$

Since $(-1)^n(D_2 v)(D_2 \cdots D_n v) \geq 0$, we may repeat the process. Proceeding in this manner, we obtain after $(n-1)$ times,

$$-\frac{D_n v(x)}{v(x)} \leq \int_{x_{n-1}}^{y_{n-1}} \cdots \int_{x_1}^{y_1} b(y)c(s)ds_1 \cdots ds_{n-1}.$$

Integration of this from x_n to y_n gives us

$$-\ln a(y) + \ln v(x) \leq \int_x^y b(y)c(s)ds,$$

which leads to

$$u(x) \leq v(x) \leq a(y) \exp\left(b(y) \int_x^y c(s)ds\right).$$

Hence this yields the desired result (5.2.90). □

We point out that this lemma is a direct extension of Lemma 5.2.2 of Gollowitzer. It provides an alternate bound for $a(y)$ which, however, is not as sharp as that given by Lemma 5.2.2.

The next theorem is an extension of the first part of Theorem 1 in [250].

Theorem 5.2.18 (The Young Inequality [710]) *Suppose conditions* $(C_1)-(C_2)$ *hold. Let* $v(t;y)$ *be the solution of the characteristic initial value problem*

$$\begin{cases} (-1)^n D_1 \cdots D_n v(t;y) - \beta(t)G(b(t)/\beta(t))v(t;y) = 0, \\ v(t;y) = 1 \quad \text{when} \quad t_i = y_i, \end{cases} \tag{5.2.94}$$

where $D_i = \partial/\partial t_i, i = 1, \cdots, n$. *Let* R *be a sub-domain containing* y *in which* $v \geq 0$. *If for all* x *in* R, *and for all* $x < y$,

$$u(y) \leq a(y) + b(y)G^{-1}\left(G(u(t))c(t)dt\right), \tag{5.2.95}$$

then

$$\int_x^y G(u(t))c(t)dt \leq \int_x^y \alpha(t)G(a(t)/\alpha(t))c(t)v(t;y)dt \tag{5.2.96}$$

where G^{-1} *denotes the inverse of* G.

Proof From (5.2.95), we derive

$$u(y) \leq \alpha(y)(a(y)/\alpha(y)) + \beta(y)(b(y)/\beta(y))G^{-1}\left(\int_x^y G(u)c(s)ds\right).$$

Since G is convex, sub-multiplicative and monotonically increasing, we have

$$G(u(y)) \leq \alpha(y)G(a(y)/\alpha(y)) + \beta(y)G(b(y)/\beta(y))\int_x^y G(u)c(s)ds.$$

Hence the inequality (5.2.96) follows from Lemma 5.2.1. □

An estimate for $u(y)$ is obtained by substituting (5.2.96) back in (5.2.95). This estimate is the best possible in the sense that if equality holds in next theorem.

Theorem 5.2.19 (The Young Inequality [710]) *Suppose conditions* $(C_1) - (C_2)$ *hold. Let* $v(s;x)$ *be the solution of the characteristic initial value problem*

$$\begin{cases} (-1)^n D_1 \cdots D_n v(s;x) - \beta(y)G(b(y)/\beta(y))c(s)v(s;x) = 0 \quad in \quad D, \\ v(s;x) = 1 \quad when \quad s_i = x_i, \end{cases} \tag{5.2.97}$$

where $D_i = \partial/\partial s_i, i = 1, \cdots n$. *Let* R *be a sub-domain of* D *containing* x *in which* $v \geq 0$. *If for all* x *in* $\mathbb{R}$, *and for all* $x < y$,

$$u(y) \geq u(x) - b(y)G^{-1}\left(\int_x^y G(u(s))c(s)ds\right), \tag{5.2.98}$$

then

$$u(y) \geq \alpha(y)G^{-1}\left(G(u(x))/[\alpha(y) + \alpha(y)\beta(y)G(b(y)/\beta(y))\int_x^y v(s;x)c(s)ds]\right). \tag{5.2.99}$$

Proof In fact, from (5.2.98), we derive

$$u(x) \leq \alpha(y)(u(y)/\alpha(y)) + \beta(y)(b(y)/\beta(y))G^{-1}\left(\int_x^y G(u(s))c(s)ds\right).$$

From the definition of G it follows that

$$G(u(x)) \leq \alpha(y)G(u(y)/\alpha(y)) + \beta(y)G(b(y)/\beta(y))\int_x^y G(u(s))c(s)ds.$$

Hence, by Lemma 5.2.2

$$\alpha(y)G(u(y)/\alpha(y)) \geq G(u(x))/[1 + \beta(y)G(b(y)/\beta(y))\int_x^y v(s;x)c(s)ds].$$

from which the inequality (5.2.99) follows. □

Noting that under conditions (C_1), (C_2) and (C_3), if (5.2.98) holds, then

$$\alpha(y)G(u(y)/\alpha(y)) \geq G(u(x))\exp[-\beta(y)G(b(y)/\beta(y))\int_x^y c(s)ds]$$

which provides an alternative bound for $u(y)$. This result follows from Lemma 5.2.3. □

The next result extends an inequality of Langenhop [328] to one involving n variables.

Theorem 5.2.20 (The Young Inequality [710]) *Let condition* (C_1) *hold and let* $g(u)$ *be conditions and non-decreasing for all* $u \geq 0$ *and* $g(u) >$ *for all* $u > 0$. *If for all* $x < y$,

$$u(y) \geq u(x) - b(y)\int_x^y c(s)g(u(s))ds, \tag{5.2.100}$$

then

$$u(y) \geq H^{-1}\left(H(u(x)) - b(y)\int_x^y c(s)ds\right) \tag{5.2.101}$$

where H^{-1} *denotes the inverse of the function*

$$H(v) = \int_{v_0}^{v} \frac{dr}{g(r)}, \quad v \geq v_0 > 0. \tag{5.2.102}$$

Proof This theorem can be proved by the same method used in proving Lemma 5.2.3. □

5.2.2 The Wendroff Inequalities, Bihari Inequalities, Ou-Yang Inequalities and Their Generalizations

The following results are some generalizations of Wendroff's integral inequality in n independent variables, which are due to [701]. The two independent variable generalization of this inequality is given by Wendroff [47]. Wendroff's inequality has interested many mathematics, such as Bondge and Pachpatte [94, 95], Chandra and Davis [135], Ghoshal and Masood [246], Headley [277], Pachpatte [471], Snow [619, 620], Defranco [190], and Young [709].

In what follows, we introduce some nonlinear integral inequalities in n independent variables, which are the further generalized results of some inequalities established by Bondge and Pachpatte [94].

Let S be an open bounded set in $\mathbb{R}^n$ and a point $(x_1, \cdots, x_n)$ in S be denoted by x. Let $x^0 = (x_1^0, \cdots, x_n^0)$ and x $(x^0 < x)$ be any two points in S, $D = D_1 D_2 \cdots D_n$ where $D_i = \partial/\partial x_i$ for $i = 1, 2, \cdots, n$.

We now wish to introduce the following n independent variable generalization of the Bondge and Pachpatte type [94] (see, Theorems 5.1.19–5.1.22), which can be used in investigating the behavior of solutions of a class of nonlinear hyperbolic partial and integro-differential equations.

Theorem 5.2.21 (The Yeh Inequality [701]) *Let $w(x)$ and $p(x)$ be real-valued non-negative continuous functions defined for all $x \geq x^0$ and $H(u)$ be a positive, continuous, monotonic non-decreasing function for all $u > 0$ satisfying $H(0) = 0$ and $D_1 H(u) \geq 0$ for $i = 1, 2, \ldots, n$. Suppose that the following inequality holds for all $x \geq x^0$,*

$$w(x) \leq \sum_{i=1}^{n} a_i(x_i) + \int_{x^0}^{x} p(t)\Big(w(t) + \int_{x^0}^{t} p(s)H(w(s))ds\Big)dt \tag{5.2.103}$$

where $a_i(x_i) > 0$ and $a_i'(x) \geq 0$ $(i = 1, 2, \ldots, n)$ are real-valued continuous functions defined for all $x \geq x^0$. Then for all $x^0 \leq x \leq x^*$,

$$\begin{aligned} w(x) \leq \sum_{i=1}^{n} a_i(x_i) + \int_{x^0}^{x} p(t)G^{-1}\Bigg\{ & G(\sum_{i=2}^{n} a_i(t_i) + a_1(x_1^0)) \\ & + \int_{x_1^0}^{t_1} \frac{a_1'(s_1)}{\sum_{i=3}^{n} a_i(x_i) + a_2(x_2^0) + a_1(s_1) + H\Big(\sum_{i=3}^{n} a_i(x_i) + a_2(x_2^0) + a_1(s_1)\Big)} ds_1 \\ & + \int_{x^0}^{t} p(s)ds\Bigg\} dt \equiv F(x), \end{aligned} \tag{5.2.104}$$

where

$$G(r) = \int_{r^0}^{r} \frac{ds}{s + H(s)}, \quad r \geq r^0 > 0, \tag{5.2.105}$$

and G^{-1} *is the inverse function of* G, *and* x^* *is choose so that*

$$G\left(\sum_{i=2}^{n} a_i(t_i) + a_1(x_1^0)\right) + \int_{x^0}^{x} p(s)ds$$
$$+ \int_{x_1^0}^{x_1} \frac{a_1'(s_1)}{\sum_{i=3}^{n} a_i(x_i) + a_2(x_2^0) + a_1(s_1) + H\Big(\sum_{i=3}^{n} a_i(x_i) + a_2(x_2^0) + a_1(s_1)\Big)} ds_1$$
$$\in Dom\,(G^{-1})$$

for all $x \in S$ *lying in the parallelepiped* $x^0 \leq x \leq x^*$.

Proof Define a function $u(x)$ by the right-hand side of (5.2.103), then

$$\begin{cases} Du(x) = p(x)(w(x) + \int_{x^0}^{x} p(s)H(w(x))ds), \\ u(x) = \sum_{i=1}^{n} a_i(x_i) + a_j(x_j^0) - a_j(x_j) \quad x_j = x_j^0, \ 1 \leq j \leq n. \end{cases}$$

It follows from (5.2.103) that

$$Du(x) \leq p(x)\Big(u(x) + \int_{x^0}^{x} p(s)H(u(s))ds\Big). \tag{5.2.106}$$

Let

$$v(x) = u(x) + \int_{x^0}^{x} p(s)H(u(s))ds.$$

Then

$$v(x) = u(x) \quad on \quad x_j = x_j^0, \ 1 \leq j \leq n;$$

$$Dv(x) = Du(x) + p(x)H(u(x)), \ Du(x) \leq p(x)v(x), \ u(x) \leq v(x).$$

Hence

$$Dv(x) \leq p(x)(v(x) + H(v(x))),$$

i.e.,

$$\frac{D_1 \cdots D_n v(x)}{v(x) + H(v(x))} \leq p(x).$$

Thus

$$\frac{(v(x) + H(v(x)))D_1 \cdots D_n v(x)}{(v(x) + H(v(x)))^2} \leq p(x) + \frac{D_n(v(x) + H(v(x)))D_1 \cdots D_{n-1} v(x)}{(v(x) + H(v(x)))^2},$$

i.e.,

$$D_n \left(\frac{D_1 \cdots D_{n-1} v(x)}{v(x) + H(v(x))} \right) \leq p(x). \tag{5.2.107}$$

Integrating both sides of (5.2.107) with respect to the component x_n of x from x_n^0 to x_n, we may derive

$$\frac{D_1 \cdots D_{n-1} v(x)}{v(x) + H(v(x))} \leq \int_{x_n^0}^{x_n} p(x_1, \cdots, x_{n-1}, t_n) dt_n.$$

Therefore

$$\frac{(v(x) + H(v(x)))D_1 \cdots D_{n-1} v(x)}{(v(x) + H(v(x)))^2}$$
$$\leq \int_{x_n^0}^{x_n} p(x_1, \cdots, x_{n-1}, t_n) dt_n + \frac{D_{n-1}(v(x) + H(v(x)))D_1 \cdots D_{n-2} v(x)}{(v(x) + H(v(x)))^2},$$

i.e.,

$$D_{n-1} \left(\frac{D_1 \cdots D_{n-2} v(x)}{v(x) + H(v(x))} \right) \leq \int_{x_n^0}^{x_n} p(x_1, \cdots, x_{n-1}, t_n) dt_n.$$

Integrating both sides of the above inequality with respect to the component x_{n-1} of x from x_{n-1}^0 to x_{n-1}, we have

$$\frac{D_1 \cdots D_{n-2} v(x)}{v(x) + H(v(x))} \leq \int_{x_{n-1}^0}^{x_{n-1}} \int_{x_n^0}^{x_n} p(x_1, \cdots, x_{n-2}, t_{n-1}, t_n) dt_{n-1} dt_n.$$

Continuing in this way, we obtain

$$\frac{D_1 D_2 v(x)}{v(x) + H(v(x))} \leq \int_{x_3^0}^{x_3} \cdots \int_{x_n^0}^{x_n} p(x_1, x_2, t_3, \cdots, t_n) dt_3 \cdots dt_n. \tag{5.2.108}$$

It follows from (5.2.108) that

$$D_2\left(\frac{D_1 v(x)}{v(x)+H(v(x))}\right) \le \int_{x_3^0}^{x_3}\cdots\int_{x_n^0}^{x_n} p(x_1,x_2,t_3,\cdots,x_{n-2},t_n)dt_3\cdots dt_n. \tag{5.2.109}$$

Integrating both sides of (5.2.109) with respect to the component x_2 of x from x_2^0 to x_2, we can get

$$\frac{D_1 v(x)}{v(x)+H(v(x))} - \frac{D_1 v(x_1,x_2^0,x_3,\cdots,x_n)}{v(x_1,x_2^0,x_3,\cdots,x_n)+H(v(x_1,x_2^0,x_3,\cdots,x_n))}$$
$$\le \int_{x_2^0}^{x_2}\cdots\int_{x_n^0}^{x_n} p(x_1,t_2,\cdots,t_n)dt_2\cdots dt_n.$$

Hence

$$D_1 G(v(x)) \le \frac{a_1'(x_1)}{\sum_{i=1}^{n} a_i(x_i)+a_2(x_2^0)-a_2(x_2)+H(\sum_{i=1}^{n} a_i(x_i)+a_2(x_2^0)-a_2(x_2))}$$
$$+\int_{x_2^0}^{x_2}\cdots\int_{x_n^0}^{x_n} p(x_1,t_2,\cdots,t_n)dt_2\cdots dt_n.$$

Integrating both sides of the above inequality with respect to the component x_1 of x from x_1^0 to x_1, we obtain

$$G(v(x)) - G\left(\sum_{i=1}^{n} a_i(x_i)+a_1(x_1^0)\right) \le \int_{x_n^0}^{x_n} p(t)dt$$
$$+\int_{x_1^0}^{x_1} \frac{a_1'(t_1)}{\sum_{i=3}^{n} a_i(x_i)+a_2(x_2^0)+a_1(t_1)+H(\sum_{i=3}^{n} a_i(x_i)+a_2(x_2^0)+a_1(t_1))}dt_1.$$

Hence

$$v(x) \le G^{-1}\left\{G\left(\sum_{i=2}^{n} a_i(x_i)+a_1(x_1^0)\right)+\int_{x^0}^{x} p(t)dt\right.$$
$$\left.+\int_{x_1^0}^{x_1} \frac{a_1'(t_1)}{\sum_{i=3}^{n} a_i(x_i)+a_2(x_2^0)+a_1(t_1)+H(\sum_{i=3}^{n} a_i(x_i)+a_2(x_2^0)+a_1(t_1))}dt_1\right\}.$$

Substituting this bound on $v(x)$ in (5.2.106) and then integrating both sides from x^0 to x, we conclude

$$u(x) \leq F(x),$$

where $F(x)$ is the function as defined in (5.2.104). From this and $w(x) \leq u(x)$, we can obtain the desired bound in (5.2.104). Thus the proof is complete. □

We next introduce an n independent variable generalization of the inequality (see, Theorem 5.1.21) given by Bondge and Pachpatte [94].

Theorem 5.2.22 (The Yeh Inequality [701]) *Let $w(x)$ and $Dw(x)$ be real-valued non-negative continuous functions defined for all $x \geq x^0$, $w(x) = 0$ on $x_j = x_j^0$, $1 \leq j \leq n$; and $p(x) \geq 1$ be a real-valued continuous functions defined for all $x \geq x^0$; let $H(u)$ and $D_iH(u)$ be the same functions as defined in Theorem 5.2.21. Suppose that the following inequality holds for all $x \geq x^0$,*

$$Dw(x) \leq \sum_{i=1}^{n} a_i(x_i) + M\left(w(x) + \int_{x^0}^{x} p(t)H(Dw(t)dt\right) \tag{5.2.110}$$

where $M \geq 0$ is a constant; $a_i(x_i) > 0$ and $a_i'(x_i) \geq 0$ are the same functions as defined in Theorem 5.2.21. Then for all $x^0 \leq x \leq x^*$,

$$\begin{aligned}
Dw(x) \leq G^{-1}\Bigg(& G\left(\sum_{i=2}^{n} a_i(x_i) + a_1(x_1^0)\right) + M\int_{x^0}^{x} p(t)dt \\
&+ \int_{x_1^0}^{x_1} \frac{a_1'(t_1)}{\sum_{i=3}^{n} a_i(x_i) + a_2(x_2^0) + a_1(t_1) + H(\sum_{i=3}^{n} a_i(x_i) + a_2(x_2^0) + a_1(t_1))} dt_1\Bigg) \\
\equiv K(x),&
\end{aligned} \tag{5.2.111}$$

where G and G^{-1} are the same functions as defined in Theorem 5.2.21 and x^* is chosen so that

$$\begin{aligned}
& G\left(\sum_{i=2}^{n} a_i(x_i) + a_1(x_1^0)\right) + M\int_{x^0}^{x} p(t)dt \\
&+ \int_{x_1^0}^{x_1} \frac{a_1'(t_1)}{\sum_{i=3}^{n} a_i(x_i) + a_2(x_2^0) + a_1(t_1) + H(\sum_{i=3}^{n} a_i(x_i) + a_2(x_2^0) + a_1(t_1))} dt_1 \\
&\in Dom\ (G^{-1})
\end{aligned}$$

for all $x \in S$ lying in the parallelepiped $x^0 \leq x \leq x^*$.

Proof Define a function $u(x)$ by the right-hand side of (5.2.110), then

$$\begin{cases} Du(x) = M(Dw(x) + p(x)H(Dw(x))), \\ u(x) = \sum_{i=1}^{n} a_i(x_i) + a_j(x_j^0) - a_j(x_j) \quad on \quad x_j = x_j^0,\ 1 \le j \le n. \end{cases}$$

Using (5.2.110) and $p(x) \ge 1$, we have

$$Du(x) \le Mp(x)(u(x) + H(u(x))).$$

Using the similar as in the proof of Theorem 5.2.21, we get

$$u(x) \le K(x),$$

where $K(x)$ is the function as defined in (5.2.111). Substituting this bound in (5.2.110), we can derive the desired inequality (5.2.111). □

We next introduce the following n independent variable generalization of the integral inequality (see, Theorem 5.1.22) given by Bondge and Pachpatte [94].

Theorem 5.2.23 (The Yeh Inequality [701]) *Let $w(x), Dw(x), p(x), H(u)$ and $D_iH(u)$ be the same functions as in Theorem 5.2.22. Suppose that the following inequality holds for all $x \ge x^0$,*

$$Dw(x) \le \sum_{i=1}^{n} a_i(x_i) + \int_{x^0}^{x} p(t)H(w(t) + Dw(t))dt \tag{5.2.112}$$

where $a_i(x_i) > 0$ and $a_i'(x_i) \ge 0$ are the same functions as in Theorem 5.2.21. Then for all $x^0 \le x \le x^*$,

$$Dw(x) \le \sum_{i=1}^{n} a_i(x_i) + \int_{x^0}^{x} p(t)H\Bigg(G^{-1}\Bigg[G\Bigg(\sum_{i=2}^{n} a_i(t_i) + a_1(x_1^0)\Bigg) + \int_{x^0}^{t} p(s)ds + \int_{x_1^0}^{t_1} \frac{a_1'(s_1)}{\sum_{i=3}^{n} a_i(x_i) + a_2(x_2^0) + a_1(s_1) + H(\sum_{i=3}^{n} a_i(x_i) + a_2(x_2^0) + a_1(s_1))} ds_1\Bigg]\Bigg)dt, \tag{5.2.113}$$

where G and G^{-1} are the same functions as defined in Theorem 5.2.21 and x^* is chosen so that

$$G\Bigg(\sum_{i=2}^{n} a_i(x_i) + a_1(x_1^0)\Bigg) + \int_{x^0}^{x} p(s)ds$$

$$+\int_{x_1^0}^{x_1} \frac{a_1'(s_1)}{\sum_{i=3}^{n} a_i(x_i) + a_2(x_2^0) + a_1(s_1) + H(\sum_{i=3}^{n} a_i(x_i) + a_2(x_2^0) + a_1(s_1))} ds_1$$

$$\in Dom\ (G^{-1})$$

for all $x \in S$ lying in the parallelepiped $x^0 \le x \le x^*$.

Proof Define a function $u(x)$ by the right-hand side of (5.2.112), then

$$Du(x) = p(x)H(w(x) + Dw(x)) \tag{5.2.114}$$

and

$$u(x) = \sum_{i=1}^{n} a_i(x_i) + a_j(x_j^0) - a_j(x_j) \quad on \quad x_j = x_j^0, 1 \le j \le n.$$

Hence

$$Dw(x) \le u(x),$$

which implies

$$w(x) \le \int_{x^0}^{x} u(t)dt.$$

Thus

$$Du(x) \le p(x)H\left(u(x) + \int_{x^0}^{x} u(t)dt\right).$$

Let

$$v(x) = u(x) + \int_{x^0}^{x} u(t)dt.$$

Then

$$v(x) = u(x) \quad on \quad x_j = x_j^0,\ 1 \le j \le n.$$

As in the proof of Theorem 5.2.21, we obtain

$$Dv(x) \le p(x)(v(x) + H(v(x))).$$

The remainder of the proof follows by an argument similar to that in the proof of Theorem 5.2.21 with suitable modifications. We omit the details. □

In the next result, the comparison method will be used to establish a number of fundamental partial integral inequalities in n independent variables.

Here, we mention in particular the papers by Conlan and Diaz [159], Snow [619], Young [709], Rasmussen [569], Headly [277], Chandra and Davis [135], other sources of partial integral inequalities, see the lecture notes of Beesack [54] and the monograph [658] of Walter.

Next, we introduce some useful partial integral inequalities from [476] in n independent variables which are motivated by a well-known integral inequality due to Ważeski [668] and the integral inequalities established in [441, 455, 456].

We use the following notations.

Let Ω be an open bounded set in $\mathbb{R}^n$ and let $D_i = \partial/\partial x_i$, $1 \leq i \leq n$; and denote by D the parallelepiped defined by $x^0 < \xi < x$ (that is, $x_i^0 < \xi_i < x_i$). For $x, y \in \Omega$, $x \leq y$ if and only if $x_i \leq y_i$ for $1 \leq i \leq n$.

We assume

(H_1) The function $f(x)$ is real-valued, positive continuous and non-decreasing in x and defined on Ω.

(H_2) The functions $\phi(x)$, $a(x)$, $b(x)$, $c(x)$ and $g(x)$ are real-valued, non-negative, continuous and defined on Ω.

(H_3) The function $q(x) \geq 1$ is real-valued, continuous and defined on Ω.

(H_4) The function $K(x, y, \phi)$ and $W(x, \phi)$ are real-valued, non-negative, continuous and defined on $\Omega^2 \times \mathbb{R}$ and $\Omega \times \mathbb{R}$, respectively, and non-decreasing in the last variables; and $K(x, y, \phi)$ is uniformly Lipschitz in the last variable.

(H_5) The functions $H : [0, +\infty) \to [0, +\infty)$ is positive, non-decreasing and continuous and satisfies

(i) $(1/v)H(u) \leq H(u/v)$, for all $u > 0$, $v \geq 1$;
(ii) $H(u)$ is sub-multiplicative for all $u \geq 0$.

A useful general version of Ważeski's inequality [668] in n independent variables is embodied in the following theorem.

Theorem 5.2.24 (The Pachpatte Inequality [476]) *Suppose* (H_4) *is true, and let* $\phi(x)$ *and* $a(x)$ *be as defined in* (H_2). *If for all* $x \in \Omega$,

$$\phi(x) \leq a(x) + W\Big(x, \int_{x^0}^{x} K(x, y, \phi(y))dy\Big), \tag{5.2.115}$$

then for all $x \in \Omega$,

$$\phi(x) \leq a(x) + W(x, r(x)), \tag{5.2.116}$$

where $r(x)$ is the solution of the equation

$$r(x) = \int_{x^0}^{x} K\Big(x, y, a(y) + W(y, r(y))\Big)dy, \tag{5.2.117}$$

existing on Ω.

Proof Define

$$u(x) = \int_{x^0}^{x} K(x, y, \phi(y))dy. \tag{5.2.118}$$

Then (5.2.115) can be restated as

$$\phi(x) \le a(x) + W(x, u(x)). \tag{5.2.119}$$

Using the monotonicity assumption on K and (5.2.119) in (5.2.118), we arrive at

$$u(x) \le \int_{x^0}^{x} K(x, y, a(y) + W(y, u(y)))dy. \tag{5.2.120}$$

Now applying Corollary 1.1.16 to (5.2.117) and (5.2.120) yields

$$u(x) \le r(x), \tag{5.2.121}$$

where $r(x)$ is the solution of (5.2.117). Now using (5.2.121) in (5.2.119), we can obtain the desired bound in (5.2.116). □

We next introduce the following n independent variable generalization of the integral inequality established in [441], which combines the features of two inequalities, namely, the n independent variable generalization of Wendorff's inequality [47] and the integral inequality given by Headly [277], and can be used more effectively in the theory of partial integral equations involving n independent variables.

Theorem 5.2.25 (The Pachpatte Inequality [476]) *Suppose $(H_1), (H_3), (H_4)$ are true, and let $\phi(x), g(x)$ and $c(x)$ be as defined in (H_2). If for all $x \in \Omega$,*

$$\begin{aligned}\phi(x) \le f(x) + q(x)&\left[\int_{x^0}^{x} g(y)\phi(y)dy + \int_{x^0}^{x} g(y)q(y)\Big(\int_{x^0}^{y} c(z)\phi(z)dz\Big)dy\right]\\ &+W\Big(x, \int_{x^0}^{x} K(x, y, \phi(y))dy\Big),\end{aligned} \tag{5.2.122}$$

then for all $x \in \Omega$,

$$\phi(x) \le E_0(x)[f(x) + W(x, r(x))], \tag{5.2.123}$$

where

$$E_0(x) = q(x)\left[1 + \int_{x^0}^{x} g(y)q(y)\exp\Big(\int_{x^0}^{y} q(z)|g(z) + c(z)|dz\Big)dy\right], \tag{5.2.124}$$

and $r(x)$ *is a solution of the equation*

$$r(x) = \int_{x^0}^{x} K\Big(x, y, E_0(y)[f(y) + W(y, r(y))]\Big)dy, \tag{5.2.125}$$

existing on Ω.

Proof Define a function $m(x)$ by

$$m(x) = f(x) + W\Big(x, \int_{x^0}^{x} K(x, y, \phi(y))dy\Big). \tag{5.2.126}$$

Then (5.2.122) can be rewritten as

$$\begin{aligned}\phi(x) \le m(x) + q(x)\bigg[&\int_{x^0}^{x} g(y)\phi(y)dy \\ &+ \int_{x^0}^{x} g(y)q(y)\Big(\int_{x^0}^{y} c(z)\phi(z)dz\Big)dy\bigg].\end{aligned} \tag{5.2.127}$$

Since $m(x)$ is positive, non-decreasing and $q(x) \ge 1$, we derive from (5.2.127) that

$$\begin{aligned}\frac{\phi(x)}{m(x)} = q(x)\bigg[1 &+ \int_{x^0}^{x} g(y)\frac{\phi(y)}{m(y)}dy \\ &+ \int_{x^0}^{x} g(y)q(y)\Big(\int_{x^0}^{y} c(z)\frac{\phi(z)}{m(z)}dz\Big)dy\bigg].\end{aligned} \tag{5.2.128}$$

Define a function $u(x)$ such that

$$\begin{cases} u(x) = 1 + \displaystyle\int_{x^0}^{x} g(y)\frac{g(y)}{m(y)}dy + \int_{x^0}^{x} g(y)q(y)\Big(\int_{x^0}^{y} c(z)\frac{\phi(z)}{m(z)}dz\Big)dy, \\ u(x) = 1, \qquad \text{on } x_j = x_j^0, \qquad 1 \le j \le n, \end{cases}$$

then

$$D_1 \ldots D_n u(x) = g(x)\frac{\phi(x)}{m(x)} + g(x)q(x)\int_{x^0}^{x} c(z)\frac{\phi(z)}{m(z)}dz,$$

which, in view of (5.2.128), implies

$$D_1 \dots D_n u(x) = g(x)q(x)\Big[u(x) + \int_{x^0}^{x} c(z)q(z)u(z)dz\Big]. \tag{5.2.129}$$

If we put

$$\begin{cases} v(x) = u(x) + \int_{x^0}^{x} c(z)q(z)u(z)dz, \\ v(x) = u(x), \qquad on\ x_j = x_j^0, \qquad 1 \le j \le n, \end{cases} \tag{5.2.130}$$

then

$$D_1 \dots D_n v(x) = D_1 \dots D_n u(x) + c(x)q(x)u(x). \tag{5.2.131}$$

Using the facts that $D_1 \dots D_n u(x) \le g(x)q(x)v(x)$ from (5.2.129) and $u(x) \le v(x)$ from (5.2.130)–(5.2.131), we arrive at

$$D_1 \dots D_n v(x) \le q(x)[g(x) + c(x)]v(x). \tag{5.2.132}$$

From (5.2.132), we deduce that

$$\frac{v(x)D_1 \dots D_n v(x)}{v^2(x)} \le q(x)[g(x) + c(x)] + \frac{D_n v(x)[D_1 \dots D_{n-1} v(x)]}{v^2(x)},$$

i.e.,

$$D_n\left(\frac{D_1 \dots D_{n-1} v(x)}{v(x)}\right) \le q(x)[g(x) + c(x)].$$

By keeping $x_1, \dots, x_{n-1}$ fixed in the above inequality, setting $x_n = y_n$ and then integrating with respect to y_n from x_n^0 to x_n, we get

$$\frac{D_1 \dots D_{n-1} v(x)}{v(x)} \le \int_{x_n^0}^{x_n} q(x_1, \dots, x_{n-1}, y_n)[g(x_1, \dots, x_{n-1}, y_n) + c(x_1, \dots, x_{n-1}, y_n)]dy_n. \tag{5.2.133}$$

Again, as above, from (5.2.133), we derive

$$D_{n-1}\left(\frac{D_1 \dots D_{n-2} v(x)}{v(x)}\right) \le \int_{x_n^0}^{x_n} q(x_1, \dots, x_{n-1}, y_n) \times[g(x_1, \dots, x_{n-1}, y_n) + c(x_1, \dots, x_{n-1}, y_n)]dy_n.$$

By keeping $x_1, \ldots, x_{n-2}$, and x_n fixed in the above inequality, setting $x_{n-1} = y_{n-1}$ and then integrating with respect to y_{n-1} from x_{n-1}^0 to x_{n-1}, we arrive at

$$\frac{D_1 \ldots D_{n-1} v(x)}{v(x)} \leq \int_{x_{n-1}^0}^{x_{n-1}} \int_{x_n^0}^{x_n} q(x_1, \ldots, x_{n-2}, y_{n-1}, y_n) \Big[g(x_1, \ldots, x_{n-2}, y_{n-1}, y_n) + c(x_1, \ldots, x_{n-2}, y_{n-1}, y_n) \Big] dy_n dy_{n-1}.$$

Computing in this way, we have

$$\frac{D_1 v(x)}{v(x)} \leq \int_{x_2^0}^{x_2} \cdots \int_{x_n^0}^{x_n} q(x_1, y_2, y_3, \cdots, y_n) \times [g(x_1, y_2, y_3 \cdots, y_n) + c(x_1, y_2, y_3 \cdots, y_n)] dy_2 \cdots dy_n.$$

Now keeping $x_2, \ldots, x_n$ fixed in the above inequality, setting $x_1 = y_1$ and then integrating with respect to y_1 from x_1^0 to x_1, we infer

$$v(x) \leq \exp\left(\int_{x^0}^{x} q(y)[g(y) + c(y)] dy \right).$$

Substituting this bound on $v(x)$ in (5.2.129), setting $x_n = y_n$ and then integrating both sides with respect to y_n from x_n^0 to x_n; then setting $x_{n-1} = y_{n-1}$ and integrating with respect to y_{n-1} from x_{n-1}^0 to x_{n-1}; and continuing in this way, finally setting $x_1 = y_1$ and then integrating with respect to y_1 from x_1^0 to x_1, we finally obtain

$$u(x) \leq 1 + \int_{x^0}^{x} g(y) q(y) \exp\left(\int_{x^0}^{y} q(z)[g(z) + c(z)] dz \right) dy.$$

Substituting this bound on $u(x)$ in (5.2.128), we have

$$\phi(x) \leq E_0(x) m(x), \tag{5.2.134}$$

where $E_0(x)$ is as defined in (5.2.124). From (5.2.126) and (5.2.134), it follows

$$\phi(x) \leq E_0(x)[f(x) + W(x, \int_{x^0}^{x} K(x, y, \phi(y)) dy)]. \tag{5.2.135}$$

Now applying Theorem 5.2.24 yields the desired bound in (5.2.123). □

We note that in the special case when $c(z) = 0$, the inequality established in Theorem 5.2.25 reduces to another interesting inequality which can be used in some application.

Another interesting and useful partial integral inequality in n independent variables involving two nonlinear functions on the right-hand side of the inequality is established in the following theorem.

Theorem 5.2.26 (The Pachpatte Inequality [476]) *Suppose* (H_1) *and* $(H_3)-(H_5)$ *are true, and let* $\phi(x)$ *and* $g(x)$ *be as defined in* (H_2). *If holds for all* $x \in \Omega$,

$$\phi(x) \leq f(x) + q(x)\Big(\int_{x^0}^{x} g(y)H(\phi(y))dy\Big) + W\Big(x, \int_{x^0}^{x} K(x, y, \phi(y))dy\Big), \qquad (5.2.136)$$

then for all $x \in \Omega_1 \subset \Omega$, *we have*

$$\phi(x) \leq E_1(x)[f(x) + W(x, r(x))], \qquad (5.2.137)$$

where

$$E_1(x) = q(x)G^{-1}[G(1) + \int_{x^0}^{x} g(y)H(q(y))dy], \qquad (5.2.138)$$

in which

$$G(v) = \int_{v^0}^{v} \frac{ds}{H(s)}, \qquad v \geq v_0 > 0 \qquad (5.2.139)$$

and G^{-1} *is the inverse of* G *such that for all* $x \in \Omega_1$,

$$G(1) + \int_{x^0}^{x} g(y)H(q(y))dy \in Dom\ (G^{-1})$$

and $r(x)$ *is a solution of the equation*

$$r(x) = \int_{x^0}^{x} K(x, y, E_1(y))[f(y) + W(y, r(y))]dy, \qquad (5.2.140)$$

existing on Ω.

Proof Define a function $m(x)$ as in the proof of Theorem 5.2.25, then (5.2.136) can be rewritten as

$$\phi(x) \leq m(x) + q(x)(\int_{x^0}^{x} g(y)H(\phi(y))dy). \qquad (5.2.141)$$

Since $m(x)$ is positive, non-decreasing and non-decreasing $q(x) \geq 1$, and in view of $(H_5)(i)$, we deduce from (5.2.141) that

$$\frac{\phi(x)}{m(x)} \leq q(x)[1 + \int_{x^0}^{x} g(y)H(\frac{\phi(y)}{m(y)})dy]. \qquad (5.2.142)$$

Define

$$\begin{cases} u(x) = 1 + \int_{x^0}^{x} g(y)H(\frac{\phi(y)}{m(y)})dy, \\ u(x) = 1, \qquad \text{on } x_j = x_j^0, \qquad 1 \le j \le n, \end{cases}$$

then we get

$$D_1 \dots D_n u(x) = g(x)H(\frac{\phi(x)}{m(x)}),$$

which, in view of (5.2.142) and sub-multiplicative character of H, implies

$$D_1 \dots D_n u(x) \le g(x)H(q(x))H(u(x)). \tag{5.2.143}$$

From (5.2.143) it follows

$$\frac{H(u(x))D_1 \dots D_n u(x)}{H^2(u(x))} \le g(x)H(q(x))\frac{D_n H(u(x))[D_1 \dots D_{n-1} v(x)]}{H^2(u(x))},$$

i.e.,

$$D_n\Big(\frac{D_1 \dots D_{n-1} u(x)}{H(u(x))}\Big) \le g(x)H(q(x)).$$

Now following a similar argument to that in the proof of Theorem 5.2.25 with suitable modifications, we can obtain

$$\frac{D_1 u(x)}{H(u(x))} \le \int_{x_2^0}^{x_2} \cdots \int_{x_n^0}^{x_n} g(x_1, y_2, y_3, \cdots, y_n) \\ \times H(q(x_1, y_2, y_3 \cdots, y_n))dy_2 \cdots dy_n. \tag{5.2.144}$$

From (5.2.139) and (5.2.144) and keeping $x_2, \dots, x_n$ fixed, we know

$$D_1 u(x) \le \int_{x_2^0}^{x_2} \cdots \int_{x_n^0}^{x_n} g(x_1, y_2, y_3, \cdots, y_n) \\ \times H(q(x_1, y_2, y_3 \cdots, y_n))dy_2 \cdots dy_n.$$

By keeping $x_2, \dots, x_n$ fixed in the above inequality, setting $x_1 = y_1$ and then integrating with respect to y_1 from x_1^0 to x_1, we conclude

$$u(x) \le G^{-1}[G(1) + \int_{x^0}^{x} g(y)H(q(y))dy].$$

The rest of the proof is immediate by analogy with the last argument in the proof of Theorem 5.2.25. The sub-domain Ω_1 of Ω is obvious. □

Next we shall introduce a further generalization of the integral inequality established by Bondge and Pachpatte [94] which can be used in more general situations.

Theorem 5.2.27 (The Pachpatte Inequality [476]) *Suppose (H_1), (H_4) and (H_5) are true, and let $\phi(x)$ and $g(x)$ be as defined in (H_2). If there holds for all $x \in \Omega$,*

$$\phi(x) \leq f(x) + (\int_{x^0}^{x} g(y)\Big(\phi(y) + \int_{x^0}^{y} g(z)H(\phi(z))dz\Big)dy + W\Big(x, \int_{x^0}^{x} K(x, y, \phi(y))dy\Big), \tag{5.2.145}$$

then for all $x \in \Omega_2 \subset \Omega$, we have

$$\phi(x) \leq E_2(x)[f(x) + W(x, r(x))], \tag{5.2.146}$$

where

$$E_2(x) = 1 + \int_{x^0}^{x} g(y)G^{-1}\Big[G(1) + \int_{x^0}^{y} g(z)dz\Big]dy, \tag{5.2.147}$$

in which

$$G(v) = \int_{v^0}^{v} \frac{ds}{s + H(s)}, \qquad v \geq v_0 > 0; \tag{5.2.148}$$

and G^{-1} is the inverse of G such that for all $x \in \Omega_2$,

$$G(1) + \int_{x^0}^{x} g(y)H(q(y))dy \in Dom(G^{-1}),$$

and $r(x)$ is a solution of he equation

$$r(x) = q\int_{x^0}^{x} K(x, y, E_1(y)[f(y) + W(y, r(y))])dy, \tag{5.2.149}$$

existing on Ω.

Proof The details of the proof follow by a similar argument to that in the proof of Theorem 5.2.26, together with the proof of Theorem 5.1.20, and we omit the details. □

Next, we shall use Young's method [709] to establish a new and more general partial integral inequality in n independent variables. The following inequality is established by solving the characteristic initial value problems by the Riemann method.

Theorem 5.2.28 (The Pachpatte Inequality [476]) *Suppose* $(H_1), (H_2), (H_4)$ *are true, and let* $v(y;x)$ *and* $e(y;x)$ *be the solutions of the characteristic initial value problems*

$$\begin{cases} (-1)^n v_{y_1 \ldots y_n}(y;x) - [a(y)b(y) + a(y)g(y) + c(y)]v(y;x) = 0 & \text{in } \Omega, \\ v(y;x) = 1 \quad on\ y_i = x_i, \quad 1 \le i \le n, \end{cases} \tag{5.2.150}$$

and

$$\begin{cases} (-1)^n e_{y_1 \ldots y_n}(y;x) - [a(y)b(y) - c(y)]e(y;x) = 0 & in\ \Omega, \\ e(y;x) = 1 \quad on\ y_i = x_i, \quad 1 \le i \le n, \end{cases} \tag{5.2.151}$$

respectively, and let D^+ *be a connected sub-domain of* Ω *containing* x *such that* $v \ge 0, e \ge 0$ *for all* $y \in D^+$ *and for all* $x \in \Omega$,

$$\begin{aligned} \phi(x) \le f(x) + a(x)\Big[\int_{x^0}^{x} b(y)\phi(y)dy + \int_{x^0}^{x} c(y)\Big(\int_{x^0}^{y} g(z)\phi(z)dz\Big)dy\Big] \\ + W\Big(x, \int_{x^0}^{x} K(x,y,\phi(y))dy\Big), \end{aligned} \tag{5.2.152}$$

then we have for all $x \in \Omega$,

$$\phi(x) \le E_3(x)[f(x) + W(x, r(x))], \tag{5.2.153}$$

where

$$\begin{aligned} E_3(x) = 1 + a(x)\Bigg[\int_{x^0}^{x} e(y;x)\Big\{b(y) \\ + c(y)\Big(\int_{x^0}^{y} [b(z) + g(z)]v(z;y)dz\Big)\Big\}dy\Bigg], \end{aligned} \tag{5.2.154}$$

and $r(x)$ *is a solution of the equation*

$$r(x) = \int_{x^0}^{x} K\Big(x, y, E_3(y)[f(y) + W(y, r(y))]\Big)dy, \tag{5.2.155}$$

existing on Ω.

Proof Define a function $m(x)$ as in the proof of Theorem 5.2.25, then (5.2.152) can be rewritten as

$$\phi(x) \leq m(x) + a(x)[\int_{x^0}^{x} b(y)\phi(y)dy + \int_{x^0}^{x} c(y)(\int_{x^0}^{x} g(z)\phi(z)dz)dy]. \tag{5.2.156}$$

Since $m(x)$ is positive, non-decreasing, we observe from (5.2.156) that

$$\frac{\phi(x)}{m(x)} \leq 1 + a(x)\Big[\int_{x^0}^{x} b(y)\frac{\phi(y)}{m(y)}dy + \int_{x^0}^{x} c(y)\Big(\int_{x^0}^{y} g(z)\frac{\phi(z)}{m(z)}dz\Big)dy\Big]. \tag{5.2.157}$$

Define a function $u(x)$ such that

$$\begin{cases} u(x) = \int_{x^0}^{x} b(y)\frac{g(y)}{m(y)}dy + \int_{x^0}^{x} c(y)(\int_{x^0}^{y} c(z)\frac{\phi(z)}{m(z)}dz)dy, \\ u(x) = 0, \qquad on\ x_i = x_i^0, \qquad 1 \leq i \leq n; \end{cases}$$

then we obtain

$$D_1 \ldots D_n u(x) = b(x)\frac{\phi(x)}{m(x)} + c(x)\int_{x^0}^{x} g(z)\frac{\phi(z)}{m(z)}dz,$$

which, in view of (5.2.157), implies

$$\begin{aligned} D_1 \ldots D_n u(x) + c(x)u(x) \leq\ & b(x)[1 + a(x)u(x)] \\ & + c(x)\Big[\int_{x^0}^{x} g(x)[1 + a(z)u(z)]dz\Big]. \end{aligned} \tag{5.2.158}$$

If we put

$$\begin{cases} \Psi(x) = u(x) + \int_{x^0}^{x}[1 + a(z)u(z)]dz, \\ \Psi(x) = u(x) = 0, \qquad on\ x_i = x_i^0, \qquad 1 \leq i \leq n, \end{cases} \tag{5.2.159}$$

then we obtain

$$D_1 \ldots D_n \Psi(x) = D_1 \ldots D_n u(x) + g(x)[1 + a(x)u(x)]. \tag{5.2.160}$$

Using $D_1 \ldots D_n u(x) \leq b(x)[1 + a(x)u(x)] + c(x)\Psi(x)$ from (5.2.158) and $u(x) \leq \Psi(x)$ from (5.2.159)–(5.2.160), we can arrive at

$$\begin{aligned} L[\Psi] &= D_1 \ldots D_n \Psi(x) - [a(x)b(x) + a(x)g(x) + c(x)]\Psi(x) \\ &\leq [b(x) + g(x)]. \end{aligned} \tag{5.2.161}$$

Furthermore, all pure mixed derivatives of Ψ with respect to $x_1, \ldots, x_{i-1}, x_{i+1}, \ldots, x_n$ up to order $n-1$ vanish on $x_i = x_i^0$, $1 \leq i \leq n$. If w is a function which is n times continuously differentiable in D, then

$$\begin{aligned} wL\Psi &- \Psi Mw \\ &= \sum_{k=1}^{n}(-1)^{k-1}D_k[(D_0D_1\dots D_{k-1}w)(D_{k+1}\dots D_nD_{n+1}\Psi)], \end{aligned} \tag{5.2.162}$$

where

$$Mw = (-1)^n D_1\dots D_n w(x) - [a(x)b(x) + a(x)g(x) + c(x)]w(x)$$

with $D_0 \equiv D_{n+1} = I$ the identity. By integrating (5.2.162) over D, using y as a variable of integration, and noting that Ψ vanishes together with all its mixed derivatives up to order $n-1$ on $y_k = x_k^0$, $1 \le k \le n$, we can obtain

$$\begin{aligned} &\int_D (wL\Psi - \Psi Mw)dy \\ &= \sum_{k=1}^{n}(-1)^{k-1}\int_{y_k=x_k}(D_1\dots D_{k-1}w)(D_{k+1}\dots D_nD_{n+1}\Psi)dy'. \end{aligned} \tag{5.2.163}$$

Now let w be chosen as the function v satisfying (5.2.150). Since $v = I$ on $y_k = x_k$, $1 \le k \le n$, it follows that $D_1\dots D_{k-1} = v(y;x) = 0$ on $y_k = x_k$ for $2 \le k \le n$. Thus (5.2.163) becomes

$$\int_D vL\Psi(y)dy = \int_{y_1=x_1} v(y;x)D_2\dots D_n\Psi(y)dy' = \Psi(x). \tag{5.2.164}$$

By the continuity of v and by the fact that $v = 1$, there is a domain D^+ containing x on which $v \ge 0$. Now multiplying (5.2.161) throughout by v and using (5.2.159) and (5.2.164), we can obtain

$$\Psi(x) \le [b(y) + g(y)]v(y,x)dy.$$

Now substituting this bound on $\Psi(x)$ in (5.2.158), we can obtain

$$\begin{aligned} L[u] &= D_1\dots D_n u(x) - [a(x)b(x) - c(x)]u(x) \\ &\le b(x) + c(x)\left(\int_{x^0}^{x}[b(y) + g(y)]v(y;x)dy\right). \end{aligned}$$

Again following the same argument as above, we can obtain

$$u(x) \le \int_{x^0}^{x} e(y,x)\Big\{b(y) + c(y)\big(\int_{x^0}^{y}[b(z) + g(z)]v(z;y)dz\big)\Big\}dy.$$

Now substituting this bound on $u(x)$ in (5.2.157), we obtain

$$\phi(x) \leq E_3(x)m(x), \tag{5.2.165}$$

where $E_3(x)$ is as defined in (5.2.154). From the definition of $m(x)$ and (5.2.165), we infer

$$\phi(x) \leq E_3(x)\Big[f(x) + W(x, \int_{x^0}^{x} K(x, y, \phi(y))dy)\Big].$$

Now applying Theorem 5.2.24 to the above inequality yields the desired bound in (5.2.153). □

We now introduce an interesting and useful n-independent variable generalization of Theorem 4.31.1. We observe that while Pachpatte's result contains two nonlinear terms in (4.31.1), we shall present a result, due to Akinyele [27] which extends the non-linear terms to any finite number.

Theorem 5.2.29 (The Akinyele Inequality [27]) *Let* (H_1) *and* $(H_3) - (H_4)$ *hold and suppose* ϕ *and* $g_j, j = 1, 2, \ldots, m$, *are as defined in* (H_2). *Assume that* H_j, $j = 1, 2, \ldots, m$, *satisfying* (H_5); *if there holds for all* $x \in \Omega$,

$$\phi(x) \leq f(x) + q(x)\sum_{l=1}^{m}\left(\int_{x^0}^{x} g_l(y)H_l(\phi(y))dy\right) + W(x, \int_{x^0}^{x} K(x, y, \phi(y))dy) \tag{5.2.166}$$

then, for $x \in \Omega_1 \subset \Omega$,

$$\phi(x) \leq \{f(x) + W(x, R(x))\}\Pi_{l=1}^{m}E_l(x), \tag{5.2.167}$$

where the function G_l *are defined as*

$$G_l(u) = \int_{u^0}^{u} \frac{ds}{H_l(s)}, \quad 0 < u^0 \leq u, \quad l = 1, 2, \ldots, m, \tag{5.2.168}$$

with

$$E_1(x) = q(x)G_1^{-1}\left[G_1(1) + \int_{x^0}^{x} g_1(s)H_1(q(s))ds\right] \tag{5.2.169}$$

and

$$E_l(x) = q(x)G_l^{-1}\left[G_l(1) + \int_{x^0}^{x} g_l(s)\Pi_{i=1}^{l-1}E_i(s)H_l(q(s))ds\right], \; l = 1, 2, \ldots, m; \tag{5.2.170}$$

G_l^{-1} is the inverse of G_l such that $G_l(1)+\int_{x^0}^{x} g_l(s)\prod_{i=1}^{l-1}E_i(s)H(q(s))\in Dom\,(G_l^{-1})$ and $R(x)$ is a solution of the integral equation

$$V(x)=\int_{x^0}^{x}K(x,y,\prod_{l=1}^{m}E_l(y)\{f(y)+W(y,V(y))\})dy. \tag{5.2.171}$$

Proof If $m=1$, then (5.2.166) becomes (5.2.136) and Theorem 5.2.26 implies that inequality (5.2.167) is true if (5.2.166) holds. We now proceed by induction and assume that inequality (5.2.166) implies (5.2.167) is true for k where $1\le k\le m-1$. Then this means

$$\phi(x)\le f(x)+q(x)\sum_{l=1}^{k}\left(\int_{x^0}^{x}g_l(y)H_l(\phi(y))dy\right)$$
$$+W(x,\int_{x^0}^{x}K(x,y,\phi(y))dy) \tag{5.2.172}$$

which, further, implies

$$\phi(x)\le\prod_{l=1}^{k}E_l(x)\{f(x)+W(x,R(x))\} \tag{5.2.173}$$

where $E_1=q(x)G_1^{-1}\left[G_1(1)+\int_{x^0}^{x}g_1(y)H_1(q(y))dy\right]$, and

$$E_l(x)=q(x)G_l^{-1}\left[G_l(1)+\int_{x^0}^{x}g_l(y)\prod_{j=1}^{l-1}E_j(y)H_l(q(y))dy\right]$$

for $l=1,2,3,\ldots,k$. G_l^{-1} is the inverse of G_l such that $G_l(1)+\int_{x^0}^{x}g_l(y)\prod_{j=1}^{l-1}E_j(y)H_l(q(y))dy\in Dom\,(G_l^{-1})$ for $l=1,2,\ldots,k$ and $R(x)$ is a solution of the integral equation

$$R(x)=\int_{x^0}^{x}K(x,y,\prod_{l=1}^{k}\{f(y)+W(y,R(y))\})dy. \tag{5.2.174}$$

Now assume that (5.2.166) holds for $m=k+1$; then

$$\phi(x)\le f(x)+q(x)\sum_{l=1}^{k+1}\left(\int_{x^0}^{x}g_l(y)H_l(\phi(y))dy\right)+W(x,\int_{x^0}^{x}K(x,y,\phi(y))dy)$$
$$\le f(x)+q(x)\sum_{l=1}^{k}\left(\int_{x^0}^{x}g_l(y)H_l(\phi(y))dy\right)+q(x)\int_{x^0}^{x}g_{k+1}(y)H_{k+1}(\phi(y))dy$$
$$+W(x,\int_{x^0}^{x}K(x,y,\phi(y))dy). \tag{5.2.175}$$

Define

$$u(x) = f(x) + q(x) \int_{x^0}^{x} g_{k+1}(y) H_{k+1}(\phi(y)) dy.$$

Then (5.2.175) becomes

$$\phi(x) \leq u(x) + q(x) \sum_{l=1}^{k} \left(\int_{x^0}^{x} g_l(y) H_l(\phi(y)) dy \right) + W(x, \int_{x^0}^{x} K(x, y, \phi(y)) dy) \tag{5.2.176}$$

where $u(x)$ is a positive function, continuous and non-decreasing in x. Hence, by assumption, (5.2.176) implies

$$\phi(x) \leq \prod_{l=1}^{k} E_l(x) \{u(x) + W(x, R(x))\} \tag{5.2.177}$$

where $E_l(x)$ is as defined earlier and $R(x)$ is a solution of the integral equation (5.2.174) with $f(x)$ replaced by $u(x)$. Set $P(x) = \prod_{l=1}^{k} E_l(x)$; then $P(x)$ is a positive function and so (5.2.177) becomes

$$\phi(x) \leq P(x)\{f(x) + W(x, R(x))\} + P(x) q(x) \int_{x^0}^{x} g_{k+1}(y) H_{k+1}(\phi(y)) dy.$$

By assumption on q, p, f, W, and H_{k+1}, we have

$$\begin{aligned} &\frac{\phi(x)}{P(x)[f(x) + W(x, R(x))]} \\ &\leq 1 + q(x) \int_{x^0}^{x} g_{k+1}(y) H_{k+1} \left(\frac{\phi(y)}{P(y)[f(y) + W(y, R(y))]} \right) P(y) dy \\ &\leq q(x) \left[1 + \int_{x^0}^{x} P(y) g_{k+1}(y) H_{k+1} \left(\frac{\phi(y)}{P(y)[f(y) + W(y, R(y))]} \right) dy \right]. \end{aligned} \tag{5.2.178}$$

Define $J : \Omega \to \mathbb{R}$ such that

$$J(x) = 1 + \int_{x^0}^{x} g_{k+1}(s) P(s) H_{k+1} \left(\frac{\phi(s)}{P(s)(f(s) + W(s, R(s)))} \right) ds$$

and

$$J(x) = 1 \qquad on \;\; x_j = x_j^0, 1 \leq j \leq n.$$

Then $D_1 D_2 \cdots D_n J(x) = g_{k+1} P(x) \left(\frac{\phi(x)}{P(x)\{f(x)+W(x,R(x))\}} \right)$ and, using (5.2.178) and the sub-multiplicative property of H_{k+1},

$$D_1 D_2 \ldots D_n J(x) \leq g_{k+1}(x) P(x) H_{k+1}(q(x)) H_{k+1}(J(x)).$$

Hence

$$\frac{H_{k+1}(J(x)) \cdot D_1 D_2 \ldots D_n J(x)}{[H_{k+1}(J(x))]^2} \leq g_{k+1}(x) P(x) H_{k+1}(q(x)) + \frac{D_1 D_2 \ldots D_{n-1} J(x) \cdot D_n H_{k+1}(J(x))}{[H_{k+1}(J(x))]^2},$$

that is,

$$D_n \left(\frac{D_1 D_2 \ldots D_{n-1} J(x)}{H_{k+1}[J(x)]} \right) \leq g_{k+1}(x) P(x) H_{k+1}(q(x)). \tag{5.2.179}$$

Keeping $x_1, \ldots, x_{n-1}$ fixed in (5.2.179), setting $s_n = y_n$, and integrating with respect to y_n from x_n^0 to x_n, we have

$$\frac{D_1 \cdots D_{n-1} J(x)}{H_{k+1}(J(x))} \leq \int_{x_n^0}^{x_n} g_{k+1}(x_1 \ldots x_{n-1}, y_n) H_{k+1}(q(x_1 \ldots x_{n-1}, y_n)) dy_n. \tag{5.2.180}$$

Set $\xi = (x_1, \ldots, x_{n-1}, y_n)$ in (5.2.180), and use the same type of arguments to arrive at

$$D_{n-1} \left(\frac{D_1 \ldots D_{n-1} J(x)}{H_{k+1}(J(x))} \right) \leq \int_{x_n^0}^{x_n} g_{k+1}(\xi) P(\xi) H_{k+1}(q(\xi)) dy_n. \tag{5.2.181}$$

Keeping $x_1 \ldots x_{n-2}$ and y_n fixed, setting $x_{n-1} = y_{n-1}$, integrating (5.2.181) from x_{n-1}^0 to x_{n-1} with respect to y_{n-1}, and setting $\eta = (x_1, x_2, \ldots, x_{n-2})$, we have

$$\frac{D_1 \ldots D_{n-2} J(x)}{H_{k+1}(J(x))} \leq \int_{x_{n-1}^0}^{x_{n-1}} \int_{x_n^0}^{x_n} (\eta, y_{n-1}, y_n) P(\eta, y_{n-1}, y_n) H_{k+1}(q(\eta, y_{n-1}, y_n)) dy_{n-1} dy_n.$$

Proceeding in this manner, we arrive at

$$\frac{D_1 J(x)}{H_{k+1}(J(x))} \leq \int_{x_2^0}^{x_2} \cdots \int_{x_n^0}^{x_n} g_{k+1}(x_1, y_2, \ldots, y_n) P(x_1, y_2, \ldots, y_n) H_{k+1}(q(x_1, y_2, \ldots, y_n)) dy_2 \ldots dy_n. \tag{5.2.182}$$

Using (5.2.168) and (5.2.182),

$$D_1 G_{k+1}(J(x)) \leq \int_{x_2^0}^{x_2} \cdots \int_{x_n^0}^{x_n} g_{k+1}(x_1, y_2, \ldots, y_n) P(x_1, y_2, \ldots, y_n) H_{k+1}(q(x_1, y_2, \ldots, y_n)) dy_2 \ldots dy_n.$$

Finally, keeping $y_2 \ldots y_n$ fixed, setting $x_1 = y_1$, and integrating with respect to y_1 from x_1^0 to x_1, we obtain

$$G_{k+1}(J(x)) \leq G_{k+1}(1) + \int_{x^0}^{x} g_{k+1}(y) P(y) H_{k+1}(q(y)) dy$$

which concludes $J(x) \leq G_{k+1}^{-1}[G_{k+1}(1) + \int_{x^0}^{x} g_{k+1}(y) P(y) H_{k+1}(q(y)) dy]$.

Consequently, by (5.2.178),

$$\begin{aligned} \frac{\phi(x)}{P(x)[f(x) + W(x, R(x))]} &\leq q(x) J(x) \\ &\leq q(x) G_{k+1}^{-1}[G_{k+1}(1) + \int_{x^0}^{x} g_{k+1}(y) \prod_{l=1}^{k} E_l(y) H_{k+1}(q(y)) dy] \\ &= E_{k+1}(x), \end{aligned}$$

that is,

$$\begin{aligned} \phi(x) &\leq \prod_{l=1}^{k} E_l(x) \cdot E_{k+1}(x) \{f(x) + W(x, R(x))\} \\ &= \prod_{l=1}^{k+1} E_l(x) \{f(x) + W(x, R(x))\}. \end{aligned}$$

where $R(x)$ is a solution of

$$R(x) = \int_{x^0}^{x} K(x, y, \prod_{l=1}^{k} E_l(y)\{u(y) + W(y, R(y))\}dy.$$

Now $u(x) = f(x) + q(x) \int_{x^0}^{x} g_{k+1}(y)H_{k+1}(\phi(y))dy$, so that

$$u(x) + W(x, R(x)) = f(x) + W(x, R(x)) + q(x) \int_{x^0}^{x} g_{k+1}(y)H_{k+1}(\phi(y))dy.$$

Using inequality (5.2.177) and a property of H_{k+1},

$$\begin{aligned} &u(x) + W(x, R(x)) \\ &\le f(x) + W(x, R(x)) + q(x) \int_{x^0}^{x} g_{k+1}(y)H_{k+1}\left(\prod_{l=1}^{k} E_l(y)\{u(y) + W(y, R(y))\}\right) dy \\ &\le f(x) + W(x, R(x)) + q(x) \int_{x^0}^{x} g_{k+1}(y)\frac{\prod_{l=1}^{k} E_l(y)}{\prod_{l=1}^{k} E_l(y)}H_{k+1}\left(\prod_{l=1}^{k} E_l(y)\{u(y) + W(y, R(y))\}\right) dy \\ &\le f(x) + W(x, R(x)) + q(x) \int_{x^0}^{x} g_{k+1}(y)\prod_{l=1}^{k} E_l(y)H_{k+1}(u(y) + W(y, R(y)))dy. \end{aligned}$$

Setting $m(x) = u(x) + W(x, R(x)), n(x) = f(x) + W(x, R(x))$, and applying Theorem 5.2.26 to the above inequality, we obtain

$$\begin{aligned} m(x) &\le n(x)q(x)[G_{k+1}^{-1}\{G_{k+1}(1) + \int_{x^0}^{x} g_{k+1}(y)\prod_{l=1}^{k} E_l(y)H_{k+1}(q(y))dy\}] \\ &\le E_{k+1}(x)\{f(x) + W(x, R(x))\}. \end{aligned}$$

Hence,

$$u(x) + W(x, R(x)) \le E_{k+1}(x)\{f(x) + W(x, R(x))\}$$

so that

$$\prod_{l=1}^{k} E_l(y)\{u(y) + W(y, R(y))\} \le \prod_{l=1}^{k+1} E_l(y)\{f(y) + W(y, R(y))\}$$

and, by the assumption on K,

$$R(x) \le \int_{x^0}^{x} K(x, y, \prod_{l=1}^{k+1} E_l(y)\{f(y) + W(y, R(y))\})dy.$$

Define $V_0(x) = R(x)$ and, for $j = 1, 2, \ldots,$

$$V_j(x) = \int_{x^0}^{x} K(x, y, \prod_{l=1}^{k+1} E_l(y)\{f(y) + W(y, V_{j-1}(y))\})dy.$$

Then

$$V_1(x) = \int_{x^0}^{x} K(x, y, \prod_{l=1}^{k+1} E_l(y)\{f(y) + W(y, R(y))\})dy \geq R(x).$$

Hence, by the assumptions on K and W, we have

$$R(x) \leq V_1(x) \leq V_2(x) \leq \ldots \leq V_j(x) \leq \ldots$$

and, by the uniform Lipschitz continuity of K in the last variable and the Arzela's theorem, the sequence $\{V_j(x)\}$ converges to a unique solution $V(x)$ of the integral equation

$$V(x) = \int_{x^0}^{x} K(x, y, \prod_{l=1}^{k+1} E_l(y)\{f(y) + W(y, V(y))\})dy \tag{5.2.183}$$

and $R(x) \leq V(x)$ existing on Ω.

Thus, since W is non-decreasing in the last variables,

$$\phi(x) \leq \{f(x) + W(x, R(x))\} \prod_{l=1}^{k+1} E_l(x) \leq \{f(x) + W(x, V(x))\} \prod_{l=1}^{k+1} E_l(x)$$

where $V(x)$ is a solution of equation (5.2.183). We have shown that, if (5.2.166) implies (5.2.167) for $m = k$, then (5.2.166) implies (5.2.167) for $m = k + 1$, so that the proof is complete by the induction hypothesis. □

We next apply Theorem 5.4.46 or Theorem 5.4.47 in Qin [557] to establish Theorems 5.2.30–5.2.35.

Theorem 5.2.30 (The Yeh Inequality [703]) *Suppose that $u(x)$, $a(x)$, $b(x)$, $c(x)$, $p(x)$ and $q(x)$ are real valued non-negative continuous functions defined on Q (an open bounded set of $\mathbb{R}^n$). Let $G(r)$ be real-valued continuous, strictly increasing, convex and sub-multiplication function, $r \geq 0$, $G(0) = 0$, $\lim_{r\to+\infty} G(r) = +\infty$ and G^{-1} be the inverse function of G. Let $A(x)$ and $B(x)$ be positive continuous functions defined on Q and $A(x) + B(x) = 1$ for all $x \in Q$. Let $v(s, x)$ and $w(s, x)$ be* *the solutions of the characteristic initial value problems*

$$\begin{cases} (-1)^n \frac{\partial^n v(s,x)}{\partial s_1 \ldots \partial s_n} - \Big[p(s) + B(s)G(b(s)B^{-1}(s))(c(s)+q(s))\Big]v(s,x) = 0 \quad in \quad Q, \\ v(s;x) = 1 \quad on\ s_i = x_i,\ i = 1,\ldots,n, \end{cases} \tag{5.2.184}$$

and

$$\begin{cases} (-1)^n \frac{\partial^n w(s,x)}{\partial s_1 \ldots \partial s_n} - \Big[B(s)G(b(s))B^{-1}c(s) - p(s)\Big]w(s,x) = 0 \quad in \quad Q, \\ w(s;x) = 1 \quad on\ s_i = x_i,\ i = 1,\ldots,n, \end{cases} \tag{5.2.185}$$

respectively and let D^+ be a connected sub-domain of Q which contains x such that $u \geq 0$, $v \geq 0$ and $w \geq 0$ for all $s \in D^+$. If $D \subset D^+$ and the following inequality holds for all $x \in D$,

$$u(x) \leq a(x) + b(x)G^{-1}\left(\int_{x^0}^{x} c(s)G(u(s))ds + \int_{x^0}^{s} p(s)(\int_{x^0}^{s} q(t)G(u(t))dt)ds\right), \tag{5.2.186}$$

then

$$\begin{aligned} u(x) \leq G\Bigg(& A(x)G(a(x)A^{-1}(x)) + B(x)G(b(x)B^{-1}(x)) \\ &\times\Bigg[\int_{x^0}^{x} w(s,x)\{A(s)G(b(s)A^{-1}(s))c(s) \\ &+p(t)\int_{x^0}^{x} A(t)G(a(t)A^{-1}(t))[c(t)+q(t)]v(t,s)dt\}ds\Bigg]\Bigg). \end{aligned} \tag{5.2.187}$$

Proof We may rewrite (5.2.186) as

$$\begin{aligned} u(x) \leq\ & A(x)a(x)A^{-1}(x) + B(x)b(x)B^{-1}(x)G^{-1}\Bigg\{\int_{x^0}^{x} c(s)G(u(s))ds \\ &+ \int_{x^0}^{x} p(s)(\int_{x^0}^{s} q(t)G(u(t))dt)ds\Bigg\}. \end{aligned}$$

Since G is convex, sub-multiplicative and strictly increasing, we have

$$G(u(x)) \le A(x)G(a(x)A^{-1}(x)) + B(x)G(b(x)B^{-1}(x))\left\{\int_{x^0}^{x} c(s)G(u(s))ds + \int_{x^0}^{x} p(s)(\int_{x^0}^{s} q(t)G(u(t))dt)ds\right\}.$$

It follows from Theorem 5.4.46 in Qin [557] that

$$G(u(x)) \le A(x)G(a(x)A^{-1}(x)) + B(x)G(b(x)B^{-1}(x))\left[\int_{x^0}^{x} w(s,x)\{A(s)G(a(s)A^{-1}(s))c(s)ds + p(s)\int_{x^0}^{s} A(t)G(a(t)A^{-1}(t))[c(t)+q(t)]v(t,s)dt\}ds\right].$$

Applying G^{-1} to the both sides of the above inequality, we can derive the desired result (5.2.187). □

Theorem 5.2.31 (The Yeh Inequality [703]) *Suppose that $u(x)$, $a(x)$, $b(x)$, $c(x)$, $p(x)$ and $q(x)$ are real-valued non-negative continuous functions defined on Q. Let $G(r)$ be a positive, continuous, strictly increasing, sub-additive and sub-multiplicative function for all $r \ge 0$, $G(0) = 0$, and G^{-1} be the inverse function of G. Let $A(x)$ and $B(x)$ be positive continuous functions defined on Q and $A(x) + B(x) = 1$ for all $x \in Q$. Let $v(s,x)$ and $w(s,x)$ be the solutions of the characteristic initial value problems*

$$\begin{cases} (-1)^n \frac{\partial^n v(s,x)}{\partial s_1 \dots \partial s_n} - [p(s) + G(b(s)(c(s)+q(s))]v(s,x) = 0 & \text{in} \quad Q, \\ v(s,x) = 1 \quad \text{on } s_i = x_i,\ i = 1,\dots,n, \end{cases} \tag{5.2.188}$$

and

$$\begin{cases} (-1)^n \frac{\partial^n w(s,x)}{\partial s_1 \dots \partial s_n} - [G(b(s))c(s) - p(s)]w(s,x) = 0 & \text{in} \quad Q, \\ w(s,x) = 1 \quad \text{on } s_i = x_i,\ i = 1,\dots,n, \end{cases} \tag{5.2.189}$$

respectively and let D^+ be a connected sub-domain of Q which contains x such that $u \ge 0$, $v \ge 0$ and $w \ge 0$ for all $s \in D^+$. If $D \subset D^+$ and the following inequality holds for all $x \in D$,

$$u(x) \le a(x) + b(x)G^{-1}\left(\int_{x^0}^{x} c(s)G(u(s))ds + \int_{x^0}^{s} p(s)\left(\int_{x^0}^{s} q(t)G(u(t))dt\right)ds\right), \tag{5.2.190}$$

then

$$u(x) \leq G^{-1}\left(G(a(x)) + G(b(x))\Big[\int_{x^0}^{s} w(s,x)\{G(a(t))[c(t) + q(t)]v(t,s)dt\}ds\Big]\right). \tag{5.2.191}$$

Proof Since G is sub-additive, sub-multiplicative and strictly increasing, it follows from (5.2.190) that

$$G(u(x)) \leq G(a(x)) + G(b(x))\left[\int_{x^0}^{s} c(s)G(u(s))ds + \int_{x^0}^{x} p(s)\Big(\int_{x^0}^{s} q(t)G(u(t))dt\Big)ds\right]. \tag{5.2.192}$$

As in the proof of Theorem 5.2.30, first applying Theorem 5.4.46 in Qin [557] to (5.2.192) and then applying G^{-1} to both sides of the resulting inequality, we can obtain the desired result (5.2.191). □

In the same manner, we can show the following two theorems.

Theorem 5.2.32 (The Yeh Inequality [703]) *Suppose that $u(x)$, $a(x)$, $b(x)$, $c(x)$, $p(x)$ and $q(x)$ are real-valued non-negative continuous functions defined on Q. $G(r)$ are the functions as defined in Theorem 5.2.30. Let $v(s,x)$ and $w(s,x)$ be the solutions of the characteristic initial value problems*

$$\begin{cases} (-1)^n \frac{\partial^n v(s,x)}{\partial s_1 \dots \partial s_n} - B(s)G(b(s)B^{-1}(s))[c(s) + p(s) + q(s)]v(s,x) = 0 & \text{in } Q, \\ v(s,x) = 1 \quad \text{on } s_i = x_i,\ i = 1, \dots, n, \end{cases} \tag{5.2.193}$$

and

$$\begin{cases} (-1)^n \frac{\partial^n w(s,x)}{\partial s_1 \dots \partial s_n} - B(s)G(b(s)B^{-1}(s))c(s)w(s,x) = 0 & \text{in } Q, \\ w(s,x) = 1 \quad \text{on } s_i = x_i,\ i = 1, \dots, n, \end{cases} \tag{5.2.194}$$

respectively and let D^+ be a connected sub-domain of Q which contains x such that $u \geq 0$, $v \geq 0$, and $w \geq 0$ for all $s \in D^+$. If $D \subset D^+$ and the following inequality holds for all $x \in D$,

$$u(x) \leq a(x) + b(x)G^{-1}\Big(\int_{x^0}^{x} c(s)G(u(s))ds + \int_{x^0}^{s} p(s)(G(u(s)) + B(s)G(b(s)B^{-1}(s))(\int_{x^0}^{s} q(t)G(u(t))dt)ds\Big), \tag{5.2.195}$$

then we have

$$u(x) \leq G^{-1}\Bigg(A(x)G(a(x)A^{-1}(x)) + B(x)G(b(x)B^{-1}(x)) \times\Big\{\int_{x^0}^{x} w(s;x)[A(s)G(a(s)A^{-1}(s))[c(s)+p(s)] + B(s)G(b(s)B^{-1}(s))p(s)\int_{x^0}^{x} A(t)G(a(t)A^{-1}(t)) \times[c(t)+p(t)+q(t)]v(t;s)dt]ds\Big\}\Bigg). \tag{5.2.196}$$

Theorem 5.2.33 (The Yeh Inequality [703]) *Suppose that $u(x)$, $a(x)$, $b(x)$, $c(x)$, $p(x)$, $q(x)$, $G(r)$ and $G^{-1}(t)$ are the functions as defined in Theorem 5.2.31. Let $v(s,x)$ and $w(s,x)$ be the solutions of the characteristic initial value problems*

$$\begin{cases} (-1)^n \frac{\partial^n v(s,x)}{\partial s_1 \ldots \partial s_n} - G(b(s))[c(s)+p(s)+q(s)]v(s,x) = 0 & \text{in } Q, \\ v(s,x) = 1 & \text{on } s_i = x_i,\ i = 1,\ldots,n, \end{cases} \tag{5.2.197}$$

and

$$\begin{cases} (-1)^n \frac{\partial^n w(s,x)}{\partial s_1 \ldots \partial s_n} - G(b(s))c(s)w(s,x) = 0 & \text{in } Q, \\ w(s,x) = 1 & \text{on } s_i = x_i,\ i = 1,\ldots,n, \end{cases} \tag{5.2.198}$$

respectively and let D^+ be a connected sub-domain of Q which contains x such that $u \geq 0$, $v \geq 0$, and $w \geq 0$ for all $s \in D^+$. If $D \subset D^+$ and the following inequality holds for all $x \in D$,

$$u(x) \leq a(x) + b(x)G^{-1}\Bigg(\int_{x^0}^{x} c(s)G(u(s))ds + \int_{x^0}^{s} p(s)\Big(G(u(s)) + G(b(s)B^{-1}(s))\int_{x^0}^{s} q(t)G(u(t))dt\Big)ds\Bigg), \tag{5.2.199}$$

then we have

$$u(x) \leq G^{-1}\Bigg(G(a(x)) + G(b(x))\Big[\int_{x^0}^{x} w(s;x)[G(a(s))(c(s)+p(s)) + G(b(s))p(s)\int_{x^0}^{s} G(a(t))(c(t)+p(t)+q(t))v(t,s)dtds\Big]\Bigg). \tag{5.2.200}$$

Theorem 5.2.34 (The Yeh Inequality [703]) *Suppose that $u(x)$, $D_1 \dots D_n u(x)$, $a(x)$ and $b(x)$ are real-valued non-negative continuous functions defined on Q. Let $v(s,x)$ be a solution of the characteristic initial value problems*

$$\begin{cases} (-1)^n \frac{\partial^n v(s,x)}{\partial s_1 \dots \partial s_n} - [1+b(s)]v(s,x) = 0 & in \quad Q, \\ v(s,x) = 1 & on\ s_i = x_i,\ i = 1,\dots,n \end{cases} \tag{5.2.201}$$

and let D^+ be a connected sub-domain of Q which contains x such that $v \geq 0$ for all $s \in D^+$. If $D \subset D^+$ and the following inequality holds for all $x \in D$,

$$D_1 \dots D_n u(x) \leq a(x) + \int_{x^0}^{x} [b(s)u(s) + D_1 \dots D_n u(s)] ds, \tag{5.2.202}$$

then we have

$$\begin{aligned} u(x) \leq h(x) + \int_{x^0}^{x} \Big\{ a(s) + \int_{x^0}^{s} b(t)[a(t) + h(t) \\ + \int_{x^0}^{t} v(m;s)(b(m)(a(m)+h(m)) + a(m))dm]dt \Big\} ds, \end{aligned} \tag{5.2.203}$$

where

$$\begin{aligned} h(x) &= \sum u(x_1^0, x_2, \dots, x_n) - \sum u(x_1^0, x_2^0, x_3, \dots, x_n) \\ &+ \dots + (-1)^{i-1} \sum u(x_1^0, \dots, x_i^0, x_{i+1}, \dots, x_n) \\ &+ \dots + (-1)^{n-1} u(x_1^0, \dots, x_n^0) \geq 0. \end{aligned} \tag{5.2.204}$$

Here

$$\begin{aligned} \sum u(x_1^0, x_2, \dots, x_n) &= u(x_1^0, x_2, \dots, x_n) + u(x_1, x_2^0, x_3, \dots, x_n) \\ &+ \dots + u(x_1, \dots, x_{n-1}, x_n^0); \\ \sum u(x_1^0, x_2^0, \dots, x_n) &= u(x_1^0, x_2^0, \dots, x_n) + u(x_1^0, x_2, x_3^0, x_4, \dots, x_n) \\ &+ \dots + u(x_1, \dots, x_{n-2}, x_{n-1}^0, x_n^0); \\ &\vdots \\ \sum u(x_1^0, \dots, x_{n-1}^0, x_n) &= u(x_1^0, x_2^0, \dots, x_{n-2}^0, x_{n-1}, x_n^0) \\ &+ \dots + u(x_1, x_2^0, \dots, x_n^0); \end{aligned}$$

Proof Let

$$A(x) = \int_{x^0}^{x} b(s)[u(s) + D_1 \cdots D_n u(s)]ds.$$

Then

$$\begin{cases} A(x) = 0 & on\ x_i = x_i^0,\ i = 1, \ldots, n, \\ D_1 \ldots D_n A(x) = b(x)[u(x) + D_1 \ldots D_n u(x)] \end{cases} \tag{5.2.205}$$

and from (5.2.205) it follows

$$D_1 \ldots D_n u(x) \leq a(x) + A(x). \tag{5.2.206}$$

Integrating both sides of (5.2.206) from x^0 to x, we obtain

$$u(x) \leq h(x) + \int_{x^0}^{x} [a(s) + A(s)]ds, \tag{5.2.207}$$

where $h(s)$ is the function as defined in (5.2.204). It follows from (5.2.205)–(5.2.207) that

$$D_1 \ldots D_n A(x) \leq b(x)\Big[h(x) + a(x) + A(x) + \int_{x^0}^{x} (a(s) + A(s))ds\Big]. \tag{5.2.208}$$

Let

$$B(x) = A(x) + \int_{x^0}^{x} (a(s) + A(s))ds.$$

Then

$$\begin{aligned} B(x) &= A(x) \qquad on\ x_i = x_i^0,\ i = 1, \ldots, n, \\ A(x) &\leq B(x), \\ D_1 \ldots D_n B(x) &= D_1 \ldots D_n A(x) + a(x) + A(x) \end{aligned}$$

and

$$D_1 \ldots D_n A(x) \leq b(x)[h(x) + a(x) + B(x)].$$

Thus

$$D_1 \ldots D_n B(x) - [1 + b(x)]B(x) \leq b(x)[a(x) + h(x)] + a(x).$$

As in the proof of Theorem 5.4.46 in Qin [557], we can derive

$$B(x) \le \int_{x^0}^{x} v(s,x)[b(s)(a(s)+h(s))+a(s)]ds,$$

which, combined with (5.2.208), implies

$$D_1 \dots D_n A(x) \le b(x)\Big[a(x)+h(x)+\int_{x^0}^{x} v(s,x)[b(s)(a(s)+h(s))+a(s)]ds\Big]. \quad (5.2.209)$$

Since $A(x)=0$ on $x_i=x_i^0$ for $i=1,\dots,n$, it follows from (5.2.209) that

$$A(x) \le \int_{x^0}^{x} b(s)\Big(a(s)+h(s)+\int_{x^0}^{s} v(t;s)[b(t)(a(t) +h(t))+a(t)]dt\Big)ds.$$

Substituting the above estimate for $A(x)$ in (5.2.206) and integrating both sides from x^0 to x, we can obtain the desired bound in (5.2.203). □

Theorem 5.2.35 (The Yeh Inequality [703]) *Suppose that $D_1 \cdots D_n u(x)$, $a(x)$, $b(x)$, $c(x)$ and $p(x)$ are real-valued non-negative continuous functions defined on Q. Let $v(s,x)$ and $w(s,x)$ be the solutions of the characteristic initial value problems*

$$\begin{cases} (-1)^n \frac{\partial^n v(s;x)}{\partial s_1 \dots \partial s_n} - [1+b(s)+c(s)+p(s)]v(s;x)=0 & \text{in} \quad Q, \\ v(s,x)=1 & \text{on } s_i=x_i,\ i=1,\dots,n, \end{cases} \quad (5.2.210)$$

and

$$\begin{cases} (-1)^n \frac{\partial^n w(s,x)}{\partial s_1 \dots \partial s_n} - [1+b(s)-c(s)]w(s;x)=0 & \text{in} \quad Q, \\ w(s,x)=1 & \text{on } s_i=x_i,\ i=1,\dots,n, \end{cases} \quad (5.2.211)$$

respectively, and let D^+ be a connected sub-domain of Q containing x such that $u \ge 0$, $v \ge 0$ and $w \ge 0$ for all $s \in D^+$. If $D \subset D^+$ and the following inequality holds for all $x \in D$,

$$D_1 \dots D_n u(x) \le a(x) + \int_{x^0}^{x} b(s)[u(s)+D_1\dots D_n u(s)]ds + \int_{x^0}^{x} c(s)\Big(\int_{x^0}^{s} p(t)[u(t)+D_1\dots D_n u(t)]dt\Big)ds,$$

then we have

$$u(x) \leq h(x) + \int_{x^0}^{x} \Big(a(s) + \int_{x^0}^{s} [b(t)(a(t) + c(t) + E(t))$$
$$+c(t) \int_{x^0}^{t} p(m)(a(m) + h(m) + E(m))dm]dt\Big)ds$$

where $h(x)$ is the function as defined in Theorem 5.2.34 and

$$E(x) = \int_{x^0}^{x} w(s;x)\Big\{a(s) + b(s)[a(s) + h(s)]$$
$$+c(s) \int_{x^0}^{s} v(t;s)([a(t) + h(t)][b(t) + p(t)]) + a(t))dt\Big\} ds.$$

Proof The details of the proof closely follow the proofs of Theorem 5.4.46 in Qin [557] and Theorem 5.2.34 with suitable modifications. We omit the details. □

Next, we shall introduce the result from [288]. Let $a,\ b \in \mathbb{R}^n,\ b > a$. We shall introduce the following notations:

$$B(a,b) = I_1(a,b) \times I_2(a,b) \times \cdots \times I_n(a,b), \quad B(0,b) = B^b,$$

$$B_k(a,b) = I_1(a,b) \times \cdots \times I_{k-1}(a,b) \times I_{k+1}(a,b) \times \cdots \times I_n(a,b),$$

where

$$1 \leq k \leq n, \quad I_k(a,b) = [a_k,\ b_k].$$

Theorem 5.2.36 (The Hristova-Bainov Inequality [288]) *Let the following conditions hold*

(1) The functions $u(x),\ g(x),\ f(x) : \mathbb{R}^n \to \mathbb{R}$ are continuous for all $x \in \mathbb{R}^n,\ x > 0$.
(2) The function $w(t) : \mathbb{R} \to \mathbb{R}$ is continuous, non-negative and non-decreasing and, besides, $w(t \cdot \tau) \leq w(t) \cdot w(\tau)$ for all $t,\ \tau > 0$.
(3) The following inequality holds for all $x \in \mathbb{R}^n,\ 0 < x < +\infty$.

$$u(x) \leq u_0 + \int_{B^x} f(s)u(s)ds + \int_{B^x} g(s)w(u(s))ds \tag{5.2.212}$$

where $u_0 > 0$ is a constant.

Then the following inequality holds for all $x \in \tilde{B}$

$$u(x)\exp\Big(-\int_{B^x} f(s)ds\Big) \leq G^{-1}\Big[G(u_0) + \int_{B^x} g(s)w\Big(\exp\int_{B^s} f(\tau)d\tau\Big)ds\Big] \tag{5.2.213}$$

where

$$G(t) = \int_{t_0}^{t} \frac{ds}{w(s)}, \quad 0 < t_0 \leq t,$$

and the function G^{-1} *is the inverse of* G *and*

$$\tilde{B} = \left\{ x : \; G(u_0) + \int_{B^x} g(s) w\left(\exp \int_{B^x} f(\tau) d\tau \right) ds \in Dom\ (G^{-1}) \right\}.$$

Proof Let us define the function

$$p(x) = u_0 + \int_{B^x} g(s) w(u(s)) ds.$$

Then the inequality (5.4.42) of Theorem 5.4.9 in Qin [557] can be rewritten as

$$u(x) \leq p(x) + \int_{B^x} f(s) u(s) ds.$$

By Theorem 5.4.9 in Qin [557], the inequality holds for all $0 < x < +\infty$,

$$u(x) \leq p(x) \exp\left(\int_{B^x} f(s) ds \right).$$

From the condition (2) in Theorem 5.2.36, it follows that

$$w(u(x)) \leq w(p(x)) w\left(\exp\left(\int_{B^x} f(s) ds \right) \right).$$

Then

$$\begin{aligned} \frac{\int_{B_k^x} g(x_k, s') w(u(x_k, s')) ds'}{w(p(x))} &\leq \int_{B_k^x} \frac{g(x_k, s') w(u(x_k, s'))}{w(p(x_k, s'))} ds' \\ &\leq \int_{B_k^x} g(x_k, s') w\Big(\exp \int_{I_k^x} \int_{B_{kf}^s} (\tau) d\tau \Big) ds' \end{aligned}$$

where $k \in \{1, 2, \ldots, n\}$,

$$\begin{aligned} ds' &= ds_1 \cdots ds_{k-1} ds_{k+1} \ldots ds_n, \\ x' &= (x_1, x_2, \cdots, x_{k-1}, x_{k+1}, \ldots, x_n), \quad (x_k, x') \equiv x. \end{aligned}$$

It follows from the definition of the function G that

$$\frac{\partial}{\partial x_k} G(p(x)) = \frac{\int_{B_k^x} g(x_k, s') w(u(x_k, s')) ds'}{w(p(x))}$$
$$\leq \int_{B_k^x} g(x_k, s') w\Big(\exp \int_{I_k^x} \int_{B_k^s} f(\tau) d\tau \Big) ds'. \quad (5.2.214)$$

Integrating the inequality (5.2.214) from 0 to x_k, we obtain

$$G(p(x)) - G(p(D, x')) \leq \int_{B^x} g(s) w(\exp \int_{B^s} f(\tau) d\tau) ds$$

where $x \equiv (x_k, x')$, or

$$G(p(x)) \leq G(u_0) + \int_{B^x} g(s) w(\exp \int_{B^s} f(\tau) d\tau) ds$$

which implies the inequality (5.2.213). □

Theorem 5.2.37 (The Hristova-Bainov Inequality [288]) *Let the following conditions hold*

(1) The functions $u(x), f(x) : \mathbb{R}^n \to \mathbb{R}$ are continuous and positive for all $x > 0$.
(2) The function $w(t) : \mathbb{R} \to \mathbb{R}$ is continuous and non-decreasing for all $t \geq 0$ and, besides, $(1/\tau) w(t) \leq w(t/\tau)$ where $\tau > 0$.
(3) The function $g(x) : \mathbb{R}^n \to \mathbb{R}$ is continuous and non-decreasing for all $x > 0$.

Then, if the following inequality holds for all $0 < x < +\infty$,

$$u(x) \leq g(x) + \int_{B^x} f(s) w(u(s)) ds, \quad (5.2.215)$$

then we have, for all $x \in \tilde{B}$,

$$u(x) \leq g(x) G^{-1}\Big[G(1) + \int_{B^x} f(s) ds \Big], \quad (5.2.216)$$

where

$$G(t) = \int_{t_0}^{t} \frac{dp}{w(p)}, \quad 0 < t_0 \leq t,$$

and the function G^{-1} is the inverse of G with,

$$\tilde{B} = \Big\{ x : \; G(1) + \int_{B^x} f(s) ds \in Dom\ (G^{-1}) \Big\}.$$

Proof From the condition (3) of Theorem 5.2.37 and from the inequality (5.2.215), it follows that

$$\frac{u(x)}{g(x)} \leq 1 + \int_{B^x} \frac{f(s)w(u(s))}{g(s)} ds \leq 1 + \int_{B^x} f(s)w(\frac{u(s)}{g(s)})ds.$$

Applying Bihari's inequality (see, Theorem 1.1.1) to the function $\frac{u(x)}{g(x)}$, we shall obtain the inequality (5.2.216). □

Theorem 5.2.38 (The Hristova-Bainov Inequality [288]) *Assume*

(1) the conditions (1) and (2) of Theorem 5.2.35 hold;
(2) the function $Q(t) : \mathbb{R} \to \mathbb{R}$ is continuous and non-decreasing for all $t > 0$ and besides, $(1/\tau)Q(t) \leq w(t/\tau)$, $\tau > 0$;
(3) the following inequality holds for all $0 < x < +\infty$,

$$u(x) \leq k + \int_{B^x} f(s)Q(u(s))ds + \int_{B^x} g(s)w(u(s))ds \tag{5.2.217}$$

where $k > 0$ is a constant.

Then the following inequality holds for all $x \in \tilde{B}$,

$$u(x)\Big[G^{-1}(G(1) + \int_{B^x} f(s)ds)^{-1}\Big] \leq \tag{5.2.218}$$

$$F^{-1}\Big\{F(k) + \int_{B^x} g(s)w\Big[G^{-1}(G(1) + \int_{B^s} f(\tau)d\tau)\Big]ds\Big\},$$

where for all $0 < t_0 \leq t$,

$$G(t) = \int_{t_0}^{t} \frac{dp}{Q(p)}, \quad F(t) = \int_{t_0}^{t} \frac{dp}{w(p)},$$

and G^{-1} and F^{-1} are the inverse functions of G and F respectively, and

$$\tilde{B} = \Big\{x : x > 0,\ G(1) + \int_{B^x} f(s)ds \in Dom\ (G^{-1}),$$

$$F(k) + \int_{B^x} g(s)w[G^{-1}(G(1) + \int_{B^s} f(\tau)d\tau)]ds \in Dom\ (F^{-1})\Big\}.$$

Proof Let

$$p(x) = k + \int_{B^x} g(s)w(u(s))ds.$$

Then the inequality (5.2.217) can be rewritten as

$$u(x) \leq p(x) + \int_{B^x} f(s)Q(u(s))ds.$$

Since the function $p(x)$ is continuous, positive and non-decreasing for all $x > 0$, then by Theorem 5.2.37, for all $x \in \tilde{\tilde{B}}$,

$$u(x) \leq p(x)G^{-1}\Big[G(1) + \int_{B^x} f(s)ds\Big]$$

where

$$\tilde{\tilde{B}} = \left\{x > 0, G(1) + \int_{B^x} f(s)ds \in Dom\ (G^{-1})\right\}.$$

Hence, taking into account the properties of the function $w(t)$, we obtain

$$w(u(x)) \leq w(p(x))w\Big[G^{-1}(G(1) + \int_{B^x} f(s)ds)\Big],$$

or

$$\frac{\int_{B_k^x} w(u(x_k, s'))g(x_k, s')ds'}{w(p(x))} \leq \int_{B_k^x} g(x_k, s')w[G^{-1}(G(1) + \int_{I_k^x}\int_{B_k^s} f(\tau)d\tau)]ds',$$

where

$$x' = (x_1, \dots, x_{k-1}, x_{k+1}, \dots, x_n).$$

It follows from the definition of the function $F(t)$ that

$$\begin{aligned}\frac{\partial}{\partial x_k}F(p(x)) &= \frac{\int_{B_k^x} w(u(x_k, s'))g(x_k, s')ds'}{w(p(x))}\\ &\leq \int_{B_k^x} g(x_k, s')w\Big[G^{-1}(G(1) + \int_{I_k^x}\int_{B_k^s} f(\tau)d\tau)\Big]ds'.\end{aligned} \tag{5.2.219}$$

Integrating the inequality (5.2.219) from 0 to x_k, we can obtain

$$F(p(x)) - F(p(0,x')) \le \int_{B^x} g(s)w[G^{-1}(G(1) + \int_{B^s} f(\tau)d\tau)]ds. \qquad (5.2.220)$$

Hence, bearing in mind that $p(0,x') = k$, we can obtain (5.2.220). □

We shall now use Corollary 5.4.15 in Qin [557] to establish the following useful n-independent variable integral inequality with delay which generalizes Theorem 2 of [456] and Theorem 1 of [197].

Theorem 5.2.39 (The Akinyele Inequality [28]) *Let $\phi(x)$, $f(x)$, $g(x)$, $q(x)$, $\sigma(x)$ and $\rho(x)$ be as in Theorem 5.4.36 in Qin [557]. Let $h(x)$ be a real-valued non-negative continuous function defined on $B \subset \mathbb{R}^n$, and $H(u)$ be a positive, continuous, monotonic non-decreasing, and sub-multiplicative function for all $u > 0$ and $H(0) = 0$. If for all $x \in B$ with $x \ge x^0$, and $D_k H(u(x)) \ge 0$ for $k = 2, 3, \cdots, n$, and the following inequality holds for all $x \ge x^0$,*

$$\begin{aligned}\phi(x) \le \phi_0 + q(x)\Big[\int_{x^0}^{x} f(s)\phi(\sigma(s))ds \\ + \int_{x^0}^{x} f(s)\Big(\int_{x^0}^{s} g(t)\phi(\sigma(t))dt\Big)ds\Big] + \int_{x^0}^{x} h(s)H(\phi(\rho(s)))ds, \qquad (5.2.221)\end{aligned}$$

then we have

$$\begin{aligned}\phi(x) \le q(x)\exp(\int_{x^0}^{x}(f(t)+g(t))q(\sigma(t))dt) \\ \times\Big\{G^{-1}(G(\phi_0) + \int_{x^0}^{x} h(s)H\Big[q(\rho(s)) \\ \times\exp\Big(\int_{x^0}^{\rho(s)}(f(t)+g(t))q(\sigma(t))dt\Big)\Big]ds)\Big\} \qquad (5.2.222)\end{aligned}$$

and

$$\phi(x) \le E(x)\Big[1 + \int_{x^0}^{x} f(s)q(\Sigma(s))\exp\Big(\int_{x^0}^{s}\{f(t)+g(t)\}q(\sigma(t))dt\Big)ds\Big] \qquad (5.2.223)$$

where

$$E(x) = q(x)G^{-1}[G(\phi_0) + \int_{x^0}^{x} h(y)H(q(\rho(y)))\{1 + \int_{x^0}^{\rho(y)} f(s)q(\sigma(s))A(s)ds\})dy] \qquad (5.2.224)$$

with

$$\begin{cases} A(s) = \exp\Big(\int_{x^0}^{s}(f(t)+g(t))q(\Sigma(t))dt\Big), \\ G(u) = \int_{u^0}^{u}\frac{ds}{H(s)}, \quad u \geq u_0 > 0, \end{cases} \tag{5.2.225}$$

G^{-1} *is the inverse of* G *and* $x \in B$ *is such that* $x \geq x^0$ *so that*

$$G(\phi_0) + \int_{x^0}^{x} h(s)H\Big(q(\rho(s))\exp\Big(\int_{x^0}^{\rho(s)}(f(t)+g(t))q(\sigma(t))dt\Big)\Big)ds \in Dom\,(G^{-1})$$

and

$$G(\phi_0) + \int_{x^0}^{x} h(s)H\Big\{q(\rho(y))\Big\{1+\int_{x^0}^{\rho(y)} q(\sigma(s))f(s) \times \exp(\int_{x^0}^{s}(f(t)+g(t))q(\sigma(t))dt)ds\Big\}\Big\}dy \in Dom\,(G^{-1}).$$

Proof Define

$$n(x) = \phi_0 + \int_{x^0}^{x} h(s)H(\phi(\rho(s)))ds, \tag{5.2.226}$$

then we have

$$n(x) = \phi_0, \quad if \quad x_i = x_i^0 \quad 1 \leq i \leq n. \tag{5.2.227}$$

Hence (5.2.221) becomes

$$\phi(x) \leq n(x) + q(x)\Big[\int_{x^0}^{x} f(s)\phi(\sigma(s))ds + \int_{x^0}^{x} f(s)\Big(\int_{x^0}^{s} g(t)\phi(\sigma(t))dt\Big)ds\Big]. \tag{5.2.228}$$

By definition, $n(x)$ is positive, monotonic non-decreasing on B, so applying of Corollary 5.4.15 in Qin [557] to (5.2.228) gives us

$$\phi(x) \leq n(x)q(x)\exp\Big(\int_{x^0}^{x}(f(t)+g(t))q(\sigma(t))dt\Big) \tag{5.2.229}$$

and

$$\phi(x) \leq n(x)q(x)\Big[1+\int_{x^0}^{x} f(s)q(\sigma(s)) \times \exp\Big(\int_{x^0}^{x}\{f(t)+g(t)\}q(\sigma(t))dt\Big)ds\Big]. \tag{5.2.230}$$

First consider (5.2.229). Since H is multiplicative and n is non-decreasing, we get

$$H(\phi(\rho(x))) \leq H(n(x))H\Big[q(\rho(x))\exp\Big(\int_{x^0}^{\rho(x)}\{f(t)+g(t)\}q(\sigma(t))dt\Big)\Big]$$

which yields

$$\frac{h(x)H(\phi(\rho(x)))}{H(n(x))} \leq h(x)H\left\{q(\rho(x))\exp\left(\int_{x^0}^{\rho(x)}\{f(t)+g(t)\}q(\sigma(t))dt\right)\right\}. \qquad (5.2.231)$$

Using (5.2.226) and (5.2.231), we get

$$\frac{D_1D_2\ldots D_nn(x)}{H(n(x))} \leq h(x)H[q(\rho(x)) \times \exp(\int_{x^0}^{\rho(x)}(f(t)+g(t))q(\sigma(t))dt)]. \qquad (5.2.232)$$

Proceeding to integrate (5.2.232) from x^0 to x step by step, using the arguments similar to that of Theorem 5.4.60 in Qin [557] and using (5.2.225) and (5.2.227), we arrive at

$$G(n(x)) \leq G(\phi_0) + \int_{x^0}^{x} h(y)H\left(q(\rho(y))\exp\Big(\int_{x^0}^{\rho(x)}(f(t)+g(t))q(\sigma(t))dt\Big)\right)dy.$$

Hence

$$n(x) \leq G^{-1}\left[G(\phi_0) + \int_{x^0}^{x} h(y)H\Big(q(\rho(y))\exp(\int_{x^0}^{\rho(x)}\{f(t)+g(t)\}q(\sigma(t))dt)\Big)dy\right]. \qquad (5.2.233)$$

Inserting (5.2.233) into (5.2.229), we have (5.2.222) as desired.

Now consider (5.2.230) and proceed in the same manner to arrive at

$$\frac{D_1D_2\ldots D_nn(x)}{H(n(x))} \leq h(x)H\left[q(\rho(x))\left\{1+\int_{x^0}^{\rho(x)}f(s)q(\rho(x)) \times \exp\left(\int_{x^0}^{s}(f(t)+g(t))q(\sigma(t))dt\right)ds\right\}\right]. \qquad (5.2.234)$$

Integrating (5.2.234) from x^0 to x, we have

$$G(n(x)) \leq G(\phi_0) + \int_{x^0}^{x} h(y)H\Big[q(\rho(y))\Big(1 + \int_{x^0}^{\rho(x)} f(s)q(\rho(x))\times \exp\Big(\int_{x^0}^{s}(f(t) + g(t))q(\sigma(t))dt\Big)ds\Big)\Big]dy$$

which gives us

$$n(x) \leq G^{-1}\Big[G(\phi_0) + \int_{x^0}^{x} h(y)H\Big[q(\rho(y))\Big(1 + \int_{x^0}^{\rho(x)} f(s)q(\rho(x))\times \exp\Big(\int_{x^0}^{s}\{f(t) + g(t)\}q(\sigma(t))dt\Big)ds\Big)\Big]dy\Big]. \tag{5.2.235}$$

Thus inserting (5.2.235) into (5.2.230) yields (5.2.223). □

The following theorem generalizes Theorem 3 in [706] which, in turn, generalizes Theorem 1 in [714] and includes Bihari's inequality [82] and Theorem 3 of [197].

Theorem 5.2.40 (The Akinyele Inequality [28]) *Let $\phi(x)$ and $g(x)$ be non-negative real-valued continuous functions on (5.2.223). Let $n(x)$ be a non-decreasing continuous function on (5.2.223) such that $n(x) \geq 1$ and $q(x)$ is a real-valued continuous function such that $q \geq 1$. Let $\sigma \in \mathcal{F}_1$ and suppose $\Omega : \mathbb{R}_+ \to \mathbb{R}_+$ belongs to class $\mathcal{F}_1$ with $D_k\Omega(u(x)) \geq 0$ for $k = 2, 3, \ldots, n$. If the following inequality holds for all $x \in B$,*

$$\phi(x) \leq n(x) + q(x)\int_{x^0}^{x} g(s)\Omega(\phi(\sigma(s)))ds,$$

then for all $x^0 \leq x \leq x'$,

$$\phi(x) \leq n(x)q(x)G^{-1}\Big[G(1) + \int_{x^0}^{x} g(s)\Omega(\phi(\sigma(s)))ds\Big]$$

where

$$G(r) = \int_{r_0}^{r} \frac{ds}{\Omega(s)} \quad , r \geq r_0 > 0,$$

and G^{-1} is the inverse of G and x' is chosen so that for all $x^0 \leq x \leq x'$

$$G(1) + \int_{x_0}^{s} \frac{g(y)}{n(y)}dy \in Dom\ (G^{-1}).$$

Proof It is easy to verify

$$\frac{\phi(x)}{n(x)} \le q(x)\Big[1 + \int_{x_0}^{x} \frac{g(y)}{n(y)}\Omega(\phi(\sigma(y)))dy\Big].$$

Define

$$u(x) = 1 + \int_{x_0}^{x} \frac{g(y)}{n(y)}\Omega(\phi(\sigma(y)))dy,$$

then

$$u(x) = 1 \quad if \quad x_i = x_i^0, \quad 1 \le i \le n$$

and

$$\begin{aligned}\phi(\sigma(x)) &\le q(\sigma(x))n(\sigma(x))u(x)\\ &\le q(\sigma(x))n(x)u(x).\end{aligned}$$

Hence

$$\begin{aligned}D_1D_2\ldots D_nu(x) &\le \frac{g(x)}{n(x)}\Omega(q(\sigma(x))n(\sigma(x))u(x))\\ &\le \frac{g(x)q(\sigma(x))}{q(\sigma(x))n(x)}\Omega(q(\sigma(x))n(\sigma(x))u(x))\\ &\le g(x)q(\sigma(x))\Omega(u(x)),\end{aligned}$$

and

$$\frac{\Omega(u(x))D_1D_2\ldots D_nu(x)}{|\Omega(u(x))|^2} \le g(x)q(\sigma(x)) + \frac{D_n\Omega(u(x))D_1D_2\ldots D_{n-1}u(x)}{|\Omega(u(x))|^2},$$

i.e.,

$$D_n\left(\frac{D_1D_2\ldots D_{n-1}u(x)}{\Omega(u(x))}\right) \le g(x)q(\sigma(x)).$$

The rest of the argument can be down as in the last theorem to yield the required result. □

Remark 5.2.5 For $n = 1$, Theorem 5.2.40 generalizes integral inequalities due to Bellman [79], Bihari [82], and Dhongade and Deo [198]. For $\Omega(u) = u,\ \sigma(x) = x$, then $n(x)$ can be taken to be positive and $G'(u) = \exp u$; so Theorem 5.2.40 reduces to Theorem 1 of Yeh [702]. If $\sigma(x) = x$ in Theorem 5.2.40, we have Theorem 3 of [706].

Corollary 5.2.1 *If $q(x) = 1$, $\Omega(v) = v$, (5.4.461) of Theorem 5.4.62 in Qin [557] holds, then,*

$$\phi(x) \le n(x)\exp\Big(\int_{x^0}^{x} g(s)\frac{n(\sigma(s))}{n(s)}\Big) \le n(x)\exp\Big(\int_{x^0}^{x} g(s)ds\Big).$$

Theorem 5.2.41 (The Akinyele Inequality [28]) *Let $\phi(x)$, $f(x)$, $g(x,)$ $\sigma(x)$ and $\rho(x)$ be as defined in Theorem 5.4.60 in Qin [557]. Let $q(x)$ be real-valued continuous functions defined on $\mathbb{R}^n$ such that $q(x) \ge 1$ and Ω be as defined in Theorem 5.4.62 in Qin [557]. Let ω be a continuous function of class $\mathcal{F}_1$. If for all $x \in B$, $D_k\Omega(u(x)) \ge 0$ for $k = 2, 3, \ldots, n$,*

$$\phi(x) \le \phi_0 + q_1(x)\int_{x^0}^{s} f(s)\omega(\phi(\sigma(s)))ds + q_2(x)\int_{x^0}^{s} g(s)\Omega(\phi(\rho(s)))ds, \tag{5.2.236}$$

where ϕ_0 is a constant greater than or equal to one, then for all $x^0 \le x < x'$,

$$\begin{aligned}\phi(x) \le\ & q_1(x)q_2(x)\hat{E}_0(x)F^{-1}\\ &\times[F(\phi_0) + \int_{x^0}^{x} g(s)\Omega(q_1(\rho(s))q_2(\rho(s))\hat{E}_n(\rho(s)))ds],\end{aligned} \tag{5.2.237}$$

where for all $x^0 \le x < x'$,

$$\begin{cases} G(v) = \displaystyle\int_{v_0}^{v}\frac{ds}{\omega(s)}, \quad F(u) = \int_{u_0}^{u}\frac{ds}{\Omega(s)}, \quad v \ge v^0 > 0,\ u \ge u^0 > 0,\\ \hat{E}_0(x) = G^{-1}[G(1) + \displaystyle\int_{x^0}^{x} f(s)q_1(\sigma(s))ds] \end{cases} \tag{5.2.238}$$

and G^{-1} and F^{-1} are the inverses of G and F, respectively, and x' is chosen so that

$$G(1) + \int_{x^0}^{x} f(s)q_1(\sigma(s))ds \in Dom\ (G^{-1})$$

and

$$F(\phi_0) + \int_{x^0}^{x} g(s)\Omega(q_1(\rho(s))q_2(\rho(s))\hat{E}_n(\rho(s)))ds \in Dom\ (F^{-1}).$$

Proof Define

$$m(x) = \phi_0 + q_2(x)\int_{x^0}^{x} g(s)\Omega(\phi(\rho(s)))ds.$$

Then $m(t)$ is continuous monotonic non-decreasing, and $m(t) \geq 1$ so that (5.2.236) becomes

$$\phi(x) \leq m(x) + q_1(x) \int_{x^0}^{x} f(s)\omega(\phi(\sigma(s)))ds.$$

By Theorem 5.2.40,

$$\phi(x) \leq m(x)q_1(x)G^{-1}[G(1) + \int_{x^0}^{x} f(s)q_1(\sigma(s))ds].$$

Now

$$m(x) \leq q_2(x)[\phi_0 + \int_{x^0}^{x} g(s)\Omega(\phi(\rho(s)))ds].$$

Define

$$v(x) = \phi_0 + \int_{x^0}^{x} g(s)\Omega(\phi(\rho(s)))ds,$$

then

$$v(x) = \phi_0 \ for \ x_i = x_i^0, \ 1 \leq i \leq n \ and \ m(x) \leq q_2(x)v(x).$$

Then

$$\phi(x) \leq q_1(x)q_2(x)v(x)G^{-1}[G(1) + \int_{x^0}^{x} f(s)q_1(\sigma(s))ds].$$

Using the sub-multiplicative property of Ω, we get

$$\Omega(\phi(\rho(x))) \leq \Omega\left(q_1(\rho(x))q_2(\rho(x))G^{-1}[G(1) + \int_{x^0}^{x} f(s)q_1(\sigma(s))ds]\right)\Omega(v(x)),$$

i.e.,

$$\frac{g(x)\Omega(\phi(\rho(x)))}{\Omega(v(x))} \leq g(x)\Omega\left(q_1(\rho(x))q_2(\rho(x))G^{-1}[G(1) + \int_{x^0}^{x} f(s)q_1(\sigma(s))ds]\right).$$

Hence if we define $\hat{E}_0(x) = G^{-1}[G(1) + \int_{x^0}^{x} f(s)q_1(\sigma(s))ds]$, then

$$\frac{D_1 \ldots D_n v(x)}{\Omega(v(x))} \leq g(x)\Omega(q_1(\rho(x))q_2(\rho(x))\hat{E}_0(\rho(x)). \tag{5.2.239}$$

Integrating (5.2.239) by the same procedure as in Theorem 5.4.62 in Qin [557] form x^0 to x, we have

$$F(v(x)) \leq F(\phi_0) + \int_{x^0}^{x} g(s)\Omega(q_1(\rho(s))q_2(\rho(s))\hat{E}_0(\rho(s))ds \tag{5.2.240}$$

where F is as defined in (5.2.238). From (5.2.240) it follows

$$v(x) \leq F^{-1}\left[F(\phi_0) + \int_{x^0}^{x} g(s)\Omega(q_1(\rho(s))q_2(\rho(s)))\hat{E}_0(\rho(s))ds\right]$$

whence

$$\phi(x) \leq q_1(x)q_2(x)\hat{E}_0(x)F^{-1}\left[F(\phi_0) + \int_{x^0}^{x} g(s)\Omega(q_1(\rho(s))q_2(\rho(s))\hat{E}_n(\rho(s)))ds\right],$$

which is (5.2.237). □

Corollary 5.2.2 (The Akinyele Inequality [28]) *Let $\phi(x)$, $f(x)$, $g(x)$, $\sigma(x)$, $\rho(x)$, $q_1(x)$, $q_2(x)$, ω and Ω be as in Theorem 5.2.40. Let $n(x)$ be a continuous function such that $n(x) \geq 1$. If for all $x \geq x^0$,*

$$\begin{aligned}\phi(x) \leq\ & n(x) + q_1(x)\int_{x^0}^{x} f(s)\omega(\phi(\sigma(s)))ds \\ & + q_2(x)\int_{x^0}^{x} g(s)\Omega(\phi(\rho(s)))ds,\end{aligned}$$

then

$$\begin{aligned}\phi(x) \leq\ & n(x)q_1(x)q_2(x)\hat{E}_0(x) \\ & \times F^{-1}[F(1) + \int_{x^0}^{x} g(s)\Omega(q_1(\rho(s))q_2(\rho(s))\hat{E}_0(\rho(s)))ds]\end{aligned}$$

where $x^0 \leq x$ and G, G^{-1}, F, F^{-1}, and $\hat{E}_0(x)$ are as defined in Theorem 5.2.41.

Proof Define

$$m(x) = n(x) + q_2(x)\int_{x^0}^{x} g(s)\Omega(\phi(\rho(s)))ds.$$

Then by Theorem 5.2.40, we get

$$\phi(x) \leq m(x)q_1(x)G^{-1}\left[G(1) + \int_{x^0}^{x} f(s)q_1(\sigma(s))ds\right]$$

and

$$m(x) \leq n(x)q_2(x)\Big[1 + \int_{x^0}^{x} g(s)\Omega(\phi(\rho(s)))ds\Big].$$

Define

$$v(x) = 1 + \int_{x^0}^{x} g(s)\Omega(\phi(\rho(s)))ds,$$

then

$$v(x) = 1 \ \ for \ \ x_i = x_i^0, \ 1 \leq i \leq n$$

and

$$m(x) \leq n(x)q_2(x)v(x).$$

Proceeding as in the last theorem, we can obtain

$$v(x) \leq F^{-1}\Big[F(1) + \int_{x^0}^{x} g(s)\Omega(q_1(\rho(s))q_2(\rho(s))\hat{E}_0(\rho(s)))ds\Big].$$

Hence

$$\begin{aligned}\phi(x) \leq\ & n(x)q_1(x)q_2(x)\hat{E}_0(x)\\ & \times F^{-1}\Big[F(1) + \int_{x^0}^{x} g(s)\Omega(q_1(\rho(s))q_2(\rho(s))\hat{E}_0(\rho(s)))ds\Big],\end{aligned}$$

which is the required inequality. □

Remark 5.2.6 For $q_1(x) = q_2(x) = 1$, $\sigma(x) = x$ and $\rho(x) = x$, we have an n-independent variable generalization of Theorem 4 of Dhongade and Deo [197]. If in addition $\omega(z) = z$, we obtain a generalization of Theorem 1 of [197]. For $n = 1$, Theorem 5.2.41 is a functional integral inequality generalization of results, Theorems 1 and 3 of [197]. For $\Omega(u) = u^p$, $p \neq 1 > 0$ and $\omega(z) = z$, Theorem 5.2.41 gives us a generalization of a result of Willett and Wong [673]. Corollary 5.2.2 unifies Theorem 5.4.62 in Qin [557] and Theorem 5.2.40 and generalizes Theorem 5.2.41.

If $x = (x_1, \cdots, x_n), y = (y_1, \cdots, y_n) \in \mathbb{R}^n$, we write $x \leq y$ $(x < y)$ if and only if $x_i \leq y_i$ $(x_i < y_i), i = 1, \cdots, n$. If $x < y$, then $[x, y]$ denotes the $n-$dimensional interval $\{z \in \mathbb{R}^n : x \leq z \leq y\}$. Let $x = (x_1, x^1)$, $x^1 = (x_2, \cdots, x_n)$, $dx^1 = dx_2 \cdots dx_n$. If $D \subset \mathbb{R}^n$ and $f : D \to \mathbb{R}$, we say that $f(x)$ is a non-decreasing function in D if $x, y \in D$ and $x \leq y$ imply $f(x) \leq f(y)$.

The generalization of the Bihari inequality in n dimensional case is stated as follows.

Theorem 5.2.42 (The Bainov-Simeonov Inequality [42]) *Let $\alpha, \beta \in \mathbb{R}^n, \alpha < \beta$. If $u(x), b(x)$ be non-negative continuous functions for all $x \in [\alpha, \beta]$ satisfying the inequality for all $x \in [\alpha, \beta]$,*

$$u(x) \leq a + \int_{\alpha}^{x} b(s)g(u(s))ds, \tag{5.2.241}$$

where $a \geq 0$ is a constant and $g(u)$ is a non-decreasing continuous function for all $u \geq 0$ with $g(u) > 0$ for all $u > 0$. Then for all $x \in D$,

$$u(x) \leq G^{-1}\left[G(a) + \int_{\alpha}^{x} b(s)ds\right], \tag{5.2.242}$$

where $G(u) = \int_{u_0}^{u} dz/g(z), u \geq u_0 > 0$, and

$$D = \left\{x \in [\alpha, \beta] : G(a) + \int_{\alpha}^{x} b(s)ds \in Dom\ (G^{-1})\right\}.$$

Proof In fact, (5.2.241) implies

$$u(x) \leq a + \int_{\alpha_1}^{x_1}\left(\int_{\alpha^1}^{x^1} b(s_1, s^1)g(u(s_1, s^1))\right)ds_1 \equiv v(x_1, x^1). \tag{5.2.243}$$

For fixed $x^1 \in [\alpha^1, \beta^1]$, the function $w(x_1) = v(x_1, x^1)$ satisfies the relations

$$w(\alpha_1) = a, \tag{5.2.244}$$

$$\begin{aligned} w'(x_1) &= \int_{\alpha^1}^{x^1} b(x_1, s^1)g(u(x_1, s^1))ds^1 \\ &\leq \int_{\alpha^1}^{x^1} b(x_1, s^1)ds^1 g(w(x_1)) \end{aligned} \tag{5.2.245}$$

since $v(x_1, x^1)$ and $g(u)$ are non-decreasing. Hence by the Bihari Inequality (i.e., Theorem 1.1.1),

$$w(x_1) \leq G^{-1}\left[G(a) + \int_{\alpha_1}^{x_1}\left(\int_{\alpha^1}^{x^1} b(s_1, s^1)ds^1\right)ds_1\right],$$

which, together with (5.2.243), implies (5.2.242). □

Corollary 5.2.3 *If $a(x)$ is a non-decreasing function in $[\alpha, \beta] \subset \mathbb{R}^n$ and for all $x \in [\alpha, \beta]$,*

$$u(x) \le a(x) + \int_\alpha^x b(s)g(u(s))ds, \tag{5.2.246}$$

then for all $x \in [\alpha, \beta]$,

$$u(x) \le G^{-1}\left[G(a(x)) + \int_\alpha^x b(s)ds\right], \tag{5.2.247}$$

and $x \in [\alpha, \beta]$ such that $G(a(x)) + \int_\alpha^x b(s)ds \in Dom\ (G^{-1})$.

The next result is due to Young [712].

Theorem 5.2.43 (The Young Inequality [712]) *Let $\alpha, \beta \in \mathbb{R}^n, \alpha < \beta$. If $u(x), b(x), f(x), g(x)$ be non-negative continuous functions for all $x \in [\alpha, \beta]$, and let $a(x)$ be a non-decreasing function in $[\alpha, \beta]$. Suppose there holds for all $x \in [\alpha, \beta]$,*

$$u(x) \le a(x) + \int_\alpha^x f(s)[u(s) + \int_\alpha^s g(\tau)u(\tau)d\tau]ds + \int_\alpha^x b(s)h(u(s))ds, \tag{5.2.248}$$

where $h(u)$ is a non-decreasing, sub-multiplicative, continuous function for all $u \ge 0$ with $h(u) > 0$ for all $u > 0$. Then

$$u(x) \le r(x)H^{-1}\left(H(a(x)) + \int_\alpha^x b(s)h(r(s))ds\right), \tag{5.2.249}$$

where $H(u) = \int_{u_0}^u dz/h(z),\ u \ge u_0 > 0$,

$$r(x) = 1 + \int_\alpha^x f(s)\exp\left(\int_\alpha^s [f(\tau) + g(\tau)]d\tau\right)ds,$$

and $x \in [\alpha, \beta]$ is such that $H(a(x)) + \int_\alpha^x b(s)h(r(s))ds \in Dom\ (H^{-1})$.

Proof We set $p(x) = a(x) + \int_\alpha^x b(s)h(u(s))ds$. Then (5.2.248) takes the form

$$u(x) \le p(x) + \int_\alpha^x f(s)[u(s) + \int_\alpha^s g(\tau)u(\tau)d\tau]ds$$

which, along with Corollary 5.4.2 in Qin [557], gives us

$$u(x) \le p(x)\left\{1 + \int_\alpha^x f(s)\exp\left(\int_\alpha^s [f(\tau) + g(\tau)]d\tau\right)ds\right\}.$$

Thus

$$u(x) \leq r(x)\left[a(x) + \int_{\alpha}^{x} b(s)h(u(s))ds\right].$$

Since h is sub-multiplicative,

$$\frac{u(x)}{r(x)} \leq a(x) + \int_{\alpha}^{x} b(s)h(r(s))h(u(s)/r(s))ds. \tag{5.2.250}$$

Applying Corollary 5.2.3, we can obtain

$$\frac{u(x)}{r(x)} \leq H^{-1}\left[H(a(x)) + \int_{\alpha}^{x} b(s)h(r(s))ds\right]$$

which implies (5.2.249). □

Let us consider the following integral inequality

$$u(x) \leq f(x) + \sum_{i=1}^{N}(T_i u)(x) + g(x)G\left[\int_0^x h(x,s)Q(u(s))ds\right], \tag{5.2.251}$$

where $x \in \mathbb{R}_+^n$ and the integral operators T_i are defined by

$$\begin{aligned}(T_i u)(x) = \int_0^x k_{i1}(x,s_1)\int_0^{s_1} k_{i2}(s_1,s_2)\int_0^{s_2} k_{i3}(s_2,s_3)\\ \cdots\int_0^{s_{i-1}} k_{ii}(s_{i-1},s_i)u(s_i)ds_i ds_{i-1}\cdots ds_1,\end{aligned} \tag{5.2.252}$$

where $u,f,g : \mathbb{R}_+^n \to \mathbb{R}_+$ and $h, k_{ij} : \mathbb{R}_+^n \times \mathbb{R}_+^n \to \mathbb{R}_+$; $(i,j = 1,\cdots,N, i \geq j)$ are continuous functions; and $G, Q : \mathbb{R}_+ \to \mathbb{R}_+$ are differentiable functions which verify some other assumptions.

We first consider two particular nonlinear cases of (5.2.251).

Theorem 5.2.44 (The Yang Inequality [693]) *Let $u,f,g : \mathbb{R}_+^n \to \mathbb{R}_+$ and $h : \mathbb{R}_+^n \times \mathbb{R}_+^n \to \mathbb{R}_+$ be continuous, and $G, Q : \mathbb{R}_+ \to \mathbb{R}_+$ be differentiable, non-decreasing, $Q(u) > 0$ for all $u > 0$. Suppose that $Q \in H(\varphi)$ and the integral inequality holds, for all $x \in \mathbb{R}_+^n$,*

$$u(x) \leq f(x) + g(x)G\left[\int_0^x h(x,s)Q(u(s))ds\right]. \tag{5.2.253}$$

Then

$$u(x) \leq \rho(x)\left(1 + G\left[H^{-1}\left(\widehat{h}(x,s)\varphi(\rho(s))ds\right)\right]\right), \tag{5.2.254}$$

where $\rho(x) = \max\{f(x), g(x)\}$, H^{-1} *denotes the inverse function of*

$$H(r) := \int_0^r \frac{dr}{Q(1 + G(r))}, \quad r \geq 0 \tag{5.2.255}$$

and $X \in \mathbb{R}_+$ *is chosen so that*

$$\int_0^x \widehat{h}(x, s)\varphi(x, s)\phi(\rho(s))ds < H(+\infty) \quad \text{as long as } 0 \leq x \leq X.$$

Proof Fixing any $t \in \mathbb{R}^n_+$, $X \geq t > 0$, we derive from (5.2.253) that for all $x \in [0, t]$,

$$u(x) \leq \rho(x)(1 + G(x)), \tag{5.2.256}$$

where $\rho(x) = \max\{f(x), g(x)\}$ and

$$r(x) := \int_0^x \widehat{h}(x, s)Q(u(s))ds. \tag{5.2.257}$$

We derive from (5.2.257) by differentiation that for all $x \in [0, t]$,

$$\begin{aligned} D_n \cdots D_1 r(x) &= \widehat{h}(t, x)Q(u(x)) \leq \widehat{h}(t, x)Q[\rho(x)(1 + G(r(x)))] \\ &\leq \widehat{h}(t, x)\varphi(\rho(x))Q[1 + G(r(x))], \end{aligned} \tag{5.2.258}$$

because of (5.2.256) and $Q \in H(\varphi)$. Because $Q(u) > 0$ for all $u > 0$, and $Q', G', D_n r(x)$ and $D_{n-1} \cdots D_1 r(x)$ are non-negative, we obtain from (5.2.258), for all $x \in [0, t]$,

$$D_n \left[\frac{D_{n-1} \cdots D_1 r(x)}{Q(1 + G(r(x)))} \right] \leq \frac{D_n \cdots D_1 r(x)}{Q(1 + G(r(x)))} \leq \widehat{h}(t, x)\varphi(\rho(x)).$$

Letting $x_n = s_n$ in the last inequality and integrating with respect to s_n over interval $[0, x_n]$, we have, for all $x \in [0, t]$,

$$\frac{D_{n-1} \cdots D_1 r(x)}{Q(1 + G(r(x)))} \leq \int_0^{x_n} \widehat{h}(t, x_{n-1}, s_n)\varphi(\rho(x_{n-1}, s_n))ds_n,$$

since $D_{n-1} \cdots D_1 r(x_{n-1}, 0) = 0$. From the last inequality, we can get for all $x \in [0, t]$,

$$D_{n-1} \left[\frac{D_{n-2} \cdots D_1 r(x)}{Q(1 + G(r(x)))} \right] \leq \int_0^{x_n} \widehat{h}(t, x_{n-1}, s_n)\varphi(\rho(x_{n-1}, s_n))ds_n.$$

Letting $x_{n-1} = s_{n-1}$ in the above inequality and integrating with respect to s_{n-1} over interval $[0, x_{n-1}]$, we have, for all $x \in [0, t]$,

$$\frac{D_{n-2}\cdots D_1 r(x)}{Q(1+G(r(x)))} \leq \int_0^{\bar{x}_{n-1}} \widehat{h}(t, x_{n-2}, \bar{s}_{n-1})\varphi(\rho(x_{n-2}, \bar{s}_{n-1}))d\bar{s}_{n-1},$$

since $D_{n-2}\cdots D_1 r(x_{n-2}, 0, x_n) = 0$.

Continuing in the same way, we then arrive at, for all $x \in [0, t]$,

$$\frac{\partial H(r(x))}{\partial x_1} = \frac{D_1 r(x)}{Q(1+G(r(x)))} \leq \int_0^{\bar{x}_2} \widehat{h}(t, x_1, \bar{s}_2)\varphi(\rho(x_1, \bar{s}_2))d\bar{s}_2,$$

where H is given by (5.2.255). Now, letting $x_1 = s_1$ in the last inequality and integrating with respect to s_1 over interval $[0, x_1]$, then we can obtain, for all $x \in [0, t]$,

$$H(r(x)) \leq \int_0^x \widehat{h}(t, s)\varphi(\rho(s))ds,$$

since $r(0, \bar{x}_2) = 0$ and $H(0) = 0$ hold. Putting $x = t$ in this inequality, then we conclude, for all $0 \leq t \leq X$,

$$r(t) \leq H^{-1}\left[\int_0^t \widehat{h}(t, s)\varphi(\rho(s))ds\right], \tag{5.2.259}$$

where H^{-1} is the inverse function of H, and the choice of $X \in \mathbb{R}^n_+$ is obvious. Finally, since $t > 0$ is arbitrary, then the desired inequality (5.2.254) follows from (5.2.256) and (5.2.259). □

Remark 5.2.7

(i) The special case of Theorem 5.2.44 when $n = 1$ is an extension of Theorem 1 of Dannan [181].
(ii) The two-variable integral inequalities considered in Theorems 1 and 2 by Bainov and Hristova [38], in Theorem 2.1 by Shastri and Kasture [599], and in Theorem 1 by Dzabbarov and Mamyedov [214], are all special cases of inequality (5.2.235).
(iii) Theorem 5.2.44 is a generalization of Theorem 2 of Beesack [59] which in turn extends Theorem 3 of Yeh [702].
(iv) We note also that Theorem 5.2.44 generalizes Proposition 3 of Corduneanu [173] which was proved for the special case when $h(x, s)$ is independent of x. Indeed, the restrictive conditions $g(x) \geq 1$, $0 \leq f(x) < 1$ or $0 < f(x) \leq 1$ of [173] are now dropped in Theorem 5.2.44. In addition, the conditions $G \in \mathcal{F}_1$ and $Q \in \mathcal{F}_1$ in [173] are also dropped and relaxed, respectively, with $Q \in H(\varphi)$ being only required.

Theorem 5.2.45 (The Yang Inequality [693]) *Let u, f, g, h, G and Q be defined as in Theorem 5.2.43. Suppose further that $Q(G(0)) > 0$ holds and the inequality holds, for all $x \in \mathbb{R}_+^n$,*

$$u(x) \le g(x)G\left[\int_0^x h(x,s)Q(u(s))ds\right]. \tag{5.2.260}$$

Then for all $0 \le x \le X$, we have

$$u(x) \le g(x)G\left[W^{-1}\left(\int_0^x \widehat{h}(x,s)\varphi(g(s))ds\right)\right], \tag{5.2.261}$$

where W^{-1} denotes the inverse function of

$$W(r) := \int_0^r \frac{dw}{Q(G(w))}, \quad r \ge 0$$

and $X \in \mathbb{R}_+^n$ is chosen so that

$$\int_0^x \widehat{h}(x,s)\varphi(g(s))ds < W(+\infty) \quad \text{as long as} \quad 0 \le x \le X.$$

Proof The proof is similar to that of Theorem 5.2.44. The changes needed are: (i) The relation (5.2.256) is replaced by $u(x) \le g(x)G(r(x))$, and (ii) Since $Q(G(0)) > 0$, we can replace the function H of Theorem 5.2.44 by the function W defined above. □

We now turn to introduce two theorems on the upper bound for the solutions of inequality (5.2.233).

Theorem 5.2.46 (The Yang Inequality [693]) *Let $u, f, g : \mathbb{R}_+^n \to \mathbb{R}_+$ and $h, k_{ij} : \mathbb{R}_+^n \times \mathbb{R}_+^n \to \mathbb{R}_+$ be continuous, where $i, j = 1, 2, \ldots, N$ with $i \ge j$. Let $G, Q : \mathbb{R}_+ \to \mathbb{R}_+$ be differentiable and non-decreasing, with $Q \in H(\varphi)$, $Q(u) > 0$ for all $u > 0$. Suppose the integral inequality holds, for all $x \in \mathbb{R}_+^n$,*

$$u(x) \le f(x) + \sum_{i=1}^N (T_i u)(x) + g(x)G\left[\int_0^x h(x,s)Q(u(s))ds\right], \tag{5.2.262}$$

where the integral operators T_i are defined by

$$(T_i u)(x) = \int_0^x k_{i1}(x,s_1)\int_0^{s_1} k_{i2}(s_1,s_2)\int_0^{s_2} k_{i3}(s_2,s_3) \cdots \int_0^{s_{i-1}} k_{ii}(s_{i-1},s_i)u(s_i)ds_i ds_{i-1}\cdots ds_1, \tag{5.2.263}$$

where $u : \mathbb{R}^n_+ \to \mathbb{R}_+$ *and* $k_{ij} : \mathbb{R}^n_+ \times \mathbb{R}^n_+ \to \mathbb{R}_+$; $(i, j = 1, \cdots, N, i \geq j)$ *are continuous functions. Then for all* $0 \leq x \leq X$, *we have*

$$u(x) \leq R(x) \left\{ 1 + G\left[H^{-1}\left(\int_0^x \widehat{h}(x,s)\varphi(R(s))ds \right) \right] \right\}, \tag{5.2.264}$$

where H^{-1} *is the same as in Theorem 5.2.44,* $R(x) := \max\{f_1(x), g_1(x)\}$, *and*

$$\begin{cases} f_1(x) = \widehat{f}(x) \exp\left\{ \sum_{i=1}^N \int_0^x H_i(x)\widehat{k}_{ii}(x,s)ds \right\}, \\ g_1(x) = \widehat{g}(x) \exp\left\{ \sum_{i=1}^N \int_0^x H_i(x)\widehat{k}_{ii}(x,s)ds \right\}, \end{cases} \tag{5.2.265}$$

and $X_1 \in \mathbb{R}^n_+$ *is chosen so that*

$$\int_0^x \widehat{h}(x,s)\varphi(R(s))ds < H(+\infty) \quad \text{as long as} \quad 0 \leq x \leq X_1.$$

Proof We easily derive from (5.2.262) that for all $x \in \mathbb{R}^n_+$,

$$\begin{cases} u(x) \leq P(x) + \sum_{i=1}^N (T_i u)(x), \\ P(x) := \widehat{f}(x) + \widehat{g}(x)G\left[\int_0^x \widehat{h}(x,s)Q(u(s)) \right]. \end{cases}$$

Clearly, function P is non-negative, non-decreasing and continuous on $\mathbb{R}^n_+$, so we have $\widehat{P}(x) = P(x)$. Thus applying Theorem 5.4.30 in Qin [557] to the last inequality yields, for all $x \in \mathbb{R}^n_+$,

$$u(x) \leq f_1(x) + g_1(x)G\left[\int_0^x \widehat{h}(x,s)Q(u(s)) \right]. \tag{5.2.266}$$

The inequality (5.2.266) is obvious under the condition of Theorem 5.2.44, hence we arrive at the desired conclusion immediately. □

Theorem 5.2.47 (The Yang Inequality [693]) *Let* $u, f, g : \mathbb{R}^n_+ \to \mathbb{R}_+$, $h, k_{ij} : \mathbb{R}^n_+ \times \mathbb{R}^n_+ \to \mathbb{R}_+$ *be continuous, where* $i, j = 1, 2, \ldots, N$ *with* $i \geq j$. *Let* $G, Q : \mathbb{R}_+ \to \mathbb{R}_+$ *be differentiable,* $Q \in H(\varphi)$, $G \in T(\psi)$, $Q(u) > 0$ *for all* $u > 0$. *Suppose further* $f(x) > 0$ *holds and the inequality (5.2.262) holds for all* $x \in \mathbb{R}^n_+$. *Then we have for all* $0 \leq x \leq X_2$,

$$u(x) \leq f_1(x)L(x) \left\{ 1 + G\left[H^{-1}\left(\int_0^x \widehat{M}(x,s)\varphi(L(s))ds \right) \right] \right\}, \tag{5.2.267}$$

where H^{-1} is the same as in Theorem 5.2.44, and

$$L(x) := \max\{1, g_1(x)\psi(f_1(x))\}, \quad M(x,s) := \frac{\widehat{h}(x,s)\varphi(f_1(s))}{f_1(x)}, \tag{5.2.268}$$

where f_1, g_1 are given by (5.2.265) and $X_2 \in \mathbb{R}^n_+$ is chosen that

$$\int_0^x \widehat{M}(x,s)\varphi(L(s))ds < H(+\infty) \quad \text{as long as} \quad 0 \le x \le X_2.$$

Proof In the above, we have deduced from (5.2.262) that for all $x \in \mathbb{R}^n_+$,

$$u(x) \le f_1(x) + g_1(s)G\left[\int_0^x \widehat{h}(x,s)Q(u(s))\right]. \tag{5.2.269}$$

Since $k_{ij}(x,s)$ are non-negative and $f(x)$ is positive, we derive from (5.2.265) that $\widehat{f}_1(x) = f_1(x)$ holds for all $x \in \mathbb{R}^n_+$. Using $G \in F(\psi)$ and $Q \in H(\varphi)$, we then derive from (5.2.266), for all $x \in \mathbb{R}^n_+$,

$$\begin{aligned}\frac{u(x)}{f_1(x)} &\le 1 + \frac{g_1(x)}{f_1(x)}G\left[\int_0^x \widehat{h}(x,s)Q(u(s))ds\right]\\ &\le 1 + g_1(x)\psi(f_1(x))G\left[\frac{1}{f_1(x)}\int_0^x \widehat{h}(x,s)Q(u(s)ds)\right]\\ &\le 1 + g_1(x)\psi(f_1(x))G\left[\int_0^x M(x,s)Q(\frac{u(s)}{f_1(s)}ds\right],\end{aligned}$$

where $M(x,s)$ is defined by (5.2.268). Now, applying Theorem 5.2.44 to the last inequality completes of the proof. □

Note that the function $M(x,s)$ in (5.2.267) can be replaced by $N(x,s) := \widehat{h}(x,s)\psi(f_1(s))/f_1(s)$.

Remark 5.2.8

(i) When $n = 1$, Theorems 5.2.46–5.2.47 are new extensions of Theorem 1 of Dannan [181].
(ii) The special case of (5.2.262) when $N = 1$, $g(x) = 1$, and k_{11}, h are directly variable-separable had been studied in Theorem 3 by Beesack [59].
(iii) For the same reason as mentioned in Remark 5.2.7.
(iv) Theorems 5.2.46–5.2.47 considerably generalize Proposition 4 of Corduneanu [173].

Theorem 5.2.48 (The Oguntuase Inequality [427]) *Let $k(x,t)$ be a good kernel and $u(x)$ be a real valued non-negative continuous function on S. If $g(x)$ be a positive, non-decreasing continuous function on S and belong to class $\mathcal{F}_1$ for which* *the following inequality holds for all* $x \in S$ *with* $x \geq x^0$,

$$u(x) \leq g(x) + \int_{x^0}^{x} k(x,t)\phi(u(t))dt, \tag{5.2.270}$$

then for all $x^0 \leq x \leq x^*$,

$$u(x) \leq g(x)G^{-1}(G(1) + \int_{x^0}^{x} k(t,t)dt), \tag{5.2.271}$$

where, for all $z \geq z^0 > 0$,

$$G(z) = \int_{z^0}^{z} \frac{ds}{\phi(s)},$$

and G^{-1} *is the inverse of* G *and* x^* *is chosen so that*

$$G(1) + \int_{x^0}^{x} k(t,t)dt \in Dom\ (G^{-1}).$$

Proof Since $g(x)$ is positive and non-decreasing, we can rewrite (5.2.270) as

$$\frac{u(x)}{g(x)} \leq 1 + \int_{x^0}^{x} k(x,t)\frac{\phi(u(t))}{g(t)}dt \leq 1 + \int_{x^0}^{x} k(x,t)\phi(\frac{u(t)}{g(t)})dt.$$

Setting $\frac{u(x)}{g(x)} = v(x)$, then we have

$$v(x) \leq 1 + \int_{x^0}^{x} k(x,t)\phi(v(t))dt.$$

Let

$$r(x) = 1 + \int_{x^0}^{x} k(x,t)\phi(v(t))dt.$$

Then

$$v(x) \leq r(x)$$

and $v(x^0) = 1$ or $x_i = x_i^0, i = 1, 2, ..., n$ and

$$D_1...D_n r(x) = k(x,x)\phi(r(x)),$$

i.e.,

$$\frac{D_1 ... D_n r(x)}{\phi(r(x))} \le k(x, x).$$

Since

$$D_n\left(\frac{D_1 \cdots D_{n-1} r(x)}{\phi(v(x))}\right) = \frac{D_1 \cdots D_{n-1} r(x)}{\phi^2(r(x))} - \frac{D_n(r(x))D_1 \cdots D_{n-1} r(x)}{\phi^2(r(x))}$$

and

$$D_n \phi(r(x)) = \phi'(r(x)) D_n r(x) \ge 0,$$

$$D_1 \cdots D_{n-1} r(x) \ge 0.$$

The above inequality implies

$$D_n\left(\frac{D_1 \cdots D_{n-1} r(x)}{\phi(r(x))}\right) \le k(x, x)$$

provided that $\phi'(r(x)) \ge 0$ for all $r(x) \ge 0$. Integrating with respect to x_n from x_n^0 to x_n and taking into account the fact that $D_1 \cdots D_{n-1} r(x) = 0$ for $x_n = x_n^0$, we have

$$\frac{D_1 \cdots D_{n-1} r(x)}{\phi(v(x))} \ge \int_{x_n^0}^{x_n} k(x_1, x_2, ..., x_{n-1}, t_n, x_1, x_2, \cdots, x_{n-1}, t_n) dt_n.$$

Repeating this, we find (after $n-1$ steps)

$$\frac{D_1 r(x)}{\phi(r(x))} \le \int_{x_1^0}^{x_1} \cdots (\int_{x_n^0}^{x_1} k(x_1, \cdots, x_{n-1}, t_n, x_1, \cdots, x_{n-1}, t_n) dt_n) \cdots dt_2.$$

We note that for all $s \ge s^0 > 0$,

$$G(s) = \int_{s^0}^{s} \frac{dz}{\phi(z)}.$$

It thus follows that

$$D_1 G(r(x)) = \frac{D_1 r(x)}{\phi(r(x))},$$

so that

$$D_1G(r(x)) \leq \int_{x_2^0}^{x_2} k(x_1, t-2, \cdots, t_n, x_1, t_2, \cdots, t_n)dt_n \cdots dt_2.$$

Integrating both sides of the above inequality with respect to the component x_1 from x_1^0 to x_1, we get

$$G(r(x_1, \cdots, x_n)) - G(r(t_1, x_2, \cdots, x_n)) \leq \int_{x^0}^{x} k(t, t)dt.$$

Since $r(t_1, x_2, \cdots, x_n) = 1$, we have

$$r(x) \leq G^{-1}(G(1) + \int_{x^0}^{x} k(t, t)dt),$$

which implies

$$v(x) \leq r(x) \leq G^{-1}(G(1) + \int_{x^0}^{x} k(t, t)dt).$$

Using the fact that $\frac{u(x)}{g(x)} = v(x)$, we have

$$u(x) \leq g(x)G^{-1}(G(1) + \int_{x^0}^{x} k(t, t)dt)$$

which gives us the required result. □

If we set $k(x, t) = h(x)f(t)$, then we shall obtain the following result.

Theorem 5.2.49 (The Oguntuase Inequality [427]) *Let $h(x)$, $f(t)$, $u(x)$ be real-valued non-negative continuous functions on S and $g(x)$ be a positive, non-decreasing continuous function on S, and ϕ belong to class $\mathcal{F}_1$. If $h'(x) = 0$ and the following inequality holds for all $x \in S$ with $x \geq x^0$,*

$$u(x) \leq g(x) + h(x) \int_{x^0}^{x} f(t)\phi(u(t))dt, \tag{5.2.272}$$

then for all $x^0 \geq x \geq x^*$,

$$u(x) \leq g(x)G^{-1}(G(1) + h(x) \int_{x^0}^{x} f(t)dt), \tag{5.2.273}$$

where

$$G(z) = \int_{z^0}^{z} \frac{ds}{\phi(s)}, \quad z \geq z^0 > 0,$$

and G^{-1} is the inverse of G and x^* is chosen so that

$$G(1) + h(x) \int_{x^0}^{x} f(t)dt \in Dom\,(G^{-1}).$$

Proof The proof is similar to that of Theorem 5.2.48 and so the details are omitted. □

Remark 5.2.9 If we set $k(x,t) = f(t)$ in Theorem 5.2.49, then Theorem 5.2.49 reduces to

$$u(x) \leq g(x)G^{-1}\Big(G(1) + \int_{x^0}^{x} f(t)dt\Big).$$

Next, we shall introduce the following n independent variable generalization of Bihari's inequality (see, e.g., [82] and Dhongade and Deo [198]).

Theorem 5.2.50 (The Yeh Inequality [702]) *Suppose that*

(a) $w(x),\ h(x) \in C(\mathbb{R}_+^n, \mathbb{R}_+)$,
(b) $f(x) \in C(\mathbb{R}_+^n, \mathbb{R}_0)$ *and non-decreasing in* x, $\mathbb{R}_0 = (0, +\infty)$,
(c) $g(x) \in C(\mathbb{R}_+^n, J)$.

Let $Q, G \in \mathcal{F}_1$, Q be sub-multiplicative and $Q'(u) \in C(\mathbb{R}_+, \mathbb{R}_+)$. If for all $x \in \mathbb{R}_+^n$,

$$w(x) \leq f(x) + g(x)G\left(\int_0^x h(s)Q(w(s))\,ds\right), \tag{5.2.274}$$

then for all $x^0 \in \mathbb{R}_+^n$, $0 \leq x \leq x^0$,

$$w(x) \leq f(x)g(x)\left\{1 + g\left[H^{-1}\left(\int_0^x h(s)Q(g(s))\,ds\right)\right]\right\}, \tag{5.2.275}$$

where H^{-1} is the inverse of H which is defined by

$$H(r) = \int_{r_0}^{r} \frac{dt}{Q(1 + G(t))}, \quad r \geq r_0 > 0, \tag{5.2.276}$$

and

$$\int_0^x h(s)Q(g(s))\,ds \in Dom\ (H^{-1})$$

for all $0 \le x \le x^0$.

Proof Since $f(x)$ is non-decreasing, $g(x) \ge 1$ and $Q, G \in F$, we have from (5.2.274) that for all $x \in \mathbb{R}^n_+$,

$$\frac{w(x)}{f(x)} \le 1 + g(x)G\left(\int_0^x h(s)Q\left(\frac{w(s)}{f(s)}\right) ds\right)$$
$$\le g(x)\left[1 + G\left(\int_0^x h(s)Q\left(\frac{w(s)}{f(s)}\right) ds\right)\right].$$

Let, for all $x \in \mathbb{R}^n_+$,

$$r(x) = \int_0^x h(s)Q\left(\frac{w(s)}{f(s)}\right) ds.$$

Then

$$r(x) = 0 \quad \text{on} \quad x_i = 0 \quad \text{for} \quad i = 1, 2, \ldots, n, \tag{5.2.277}$$
$$\frac{w(x)}{f(x)} \le g(x)[1 + G(r(x))]. \tag{5.2.278}$$

Hence

$$D_1 \cdots D_n r(x) = h(x)Q\left(\frac{w(x)}{f(x)}\right) \le h(x)Q(g(x))Q(1 + G(r(x)))$$

since Q is non-decreasing and sub-multiplicative. Thus

$$\frac{Q(1 + G(r(x)))D_1 \cdots D_n r(x)}{Q^2(1 + G(r(x)))}$$
$$\le h(x)Q(g(x)) + \frac{D_n Q(1 + G(r(x)))D_1 \cdots D_{n-1} r(x)}{Q^2(1 + G(r(x)))},$$

i.e.,

$$D_n\left(\frac{D_1 \cdots D_{n-1} r(x)}{Q(1 + G(r(x)))}\right) \le h(x)Q(g(x)).$$

Integrating both sides of the above inequality with respect to the component x_n of x from 0 to x_n and using (5.2.277), we get

$$\frac{D_1 \cdots D_{n-1} r(x)}{Q(1 + G(r(x)))} \leq \int_0^{x_n} h(x_1, \ldots, x_{n-1}, t_n) Q(g(x_1, \ldots, x_{n-1}, t_n))\, dt_n$$

which implies

$$D_{n-1}\left(\frac{D_1 \cdots D_{n-2} r(x)}{Q(1 + G(r(x)))}\right) \leq \int_0^{x_n} h(x_1, \ldots, x_{n-1}, t_n) Q(g(x_1, \ldots, x_{n-1}, t_n))\, dt_n.$$

Integrating both sides of the above inequality with respect to the component x_{n-1} of x from 0 to x_{n-1}, we have

$$\begin{aligned}\frac{D_1 \cdots D_{n-2} r(x)}{Q(1 + G(r(x)))} \leq & \int_0^{x_{n-1}} \int_0^{x_n} h(x_1, \ldots, x_{n-1}, t_{n-1}, t_n) \\ & \times Q(g(x_1, \ldots, x_{n-1}, t_{n-1}, t_n))\, dt_n dt_{n-1}.\end{aligned}$$

Continuing in this way and using (5.2.276), we arrive at

$$\begin{aligned} D_1 H(r(x)) &= \frac{D_1 r(x)}{Q(1 + G(r(x)))} \\ &\leq \int_0^{x_2} \cdots \int_0^{x_n} h(x_1, t_2, \ldots, t_n) Q(g(x_1, t_2, \ldots, t_n))\, dt_n \cdots dt_2.\end{aligned}$$

Integrating both sides of the above inequality with respect to the component x_1 of x from 0 to x_1, we conclude

$$H(r(x)) - H(r(0, x_2, \ldots, x_n)) \leq \int_0^x h(t) Q(g(t))\, dt$$

which implies

$$r(x) \leq H^{-1}\left[\int_0^x h(t) Q(g(t))\, dt\right]. \tag{5.2.279}$$

Hence the result (5.2.275) follows readily from (5.2.278) and (5.2.279). □

As an application of Theorem 5.4.22 in Qin [557] and Theorem 5.2.50, we have the following theorem.

Theorem 5.2.51 (The Yeh Inequality [702]) *Assume*

(a) *the functions* $w(x)$, $f(x)$, $g_i(x)$, *and* $h_i(x)$ $(i = 1, 2, \ldots, m+1)$ *are defined as Theorem 5.2.50, and*
(b) *the functions* Q *and* G *are defined as in Theorem 5.2.50. If for all* $x \in \mathbb{R}^n_+$,

$$w(x) \leq f(x) + \sum_{i=1}^{m} g_i(x) \int_0^x h_i(s)w(s)\,ds$$
$$\leq g_{m+1}(x)G\left(\int_0^x h_{m+1}(s)Q(w(s))\,ds\right), \qquad (5.2.280)$$

then for all $x^0 \in \mathbb{R}^n_+$, $0 \leq x \leq x^0$,

$$w(x) \leq E^m f(x)E^m g_{m+1}(x)\left\{1 + G\left[H^{-1}\left(\int_0^x h_{m+1}(s)Q(E^m g_{m+1}(s))\,ds\right)\right]\right\} \qquad (5.2.281)$$

where E^m *is defined inductively as in Theorem 5.4.22 in Qin [557],* H^{-1} *has the same means as in Theorem 5.2.50, and*

$$\int_0^x h_{m+1}(s)Q(E^m g_{m+1}(s))\,ds \in Dom\,(H^{-1}).$$

Proof Let, for all $x \in \mathbb{R}^n_+$,

$$T(x) = f(x) + \sum_{i=1}^{m} g_i(x) \int_0^x h_i(s)w(s)\,ds.$$

Then (5.2.280) can be rewritten as

$$u(x) \leq T(x) + \sum_{i=1}^{m} g_i(x) \int_0^x h_i(s)w(s)ds$$

It follows from Theorem 5.4.22 in Qin [557] that

$$\begin{aligned}
w(x) &\leq E^m(T(x)) \\
&= E^m\left[f(x) + g_{m+1}(x)G\left(\int_0^x h_{m+1}(s)Q(w(s))\,ds\right)\right] \\
&= E^m f(x) + E^m\left[g_{m+1}(x)G\left(\int_0^x h_{m+1}Q(w(s))\,ds\right)\right] \\
&= E^m f(x) + g_{m+1}(x)G\left[\int_0^x h_{m+1}(s_Q(w(s))\,ds\right] \\
&\quad \times \left[E^{m-1}g_m(x)\exp\int_0^x h_m(s)E^{m-1}g_m(s)\,ds\right],
\end{aligned}$$

whence

$$w(x) \leq E^m f(x) + E^m g_{m+1}(x) G\left(\int_0^x h_{m+1}(s) Q(w(s))\, ds\right).$$

This inequality is of the form (5.4.125) of Theorem 5.4.22 in Qin [557], which by applying Theorem 5.2.50, gives us the desired result (5.2.281). □

All the functions which appear in the inequalities in Theorem 5.2.52 are assumed to be real-valued of n variables which are non-negative and continuous. All integrals are assumed to exist on their domains of definitions. We assume that $I = [a; b]$ in any bounded open set in the dimensional Euclidean space $\mathbb{R}^n$ and that our integrals are on $\mathbb{R}^n$ $(n \geq 1)$, where $a = (a_1, a_2, ..., a_n)$, $b = (b_1, b_2, ..., b_n) \in \mathbb{R}^n_+$. Let $C(I, \mathbb{R}_+)$ denote the class of continuous functions from I to $\mathbb{R}_+$.

The following theorem deals with n-independent variables versions of the inequalities established in Theorem 1.1.52. We need the inequalities in the following lemma (see, e.g., [305]).

Lemma 5.2.4 (The Denche-Khellaf Inequality [193]) *Let $u(x)$ and $b(x)$ be non-negative continuous functions, defined for all $x \in I$, and let $g \in \mathcal{F}_1$. Assume that $a(x)$ is positive, continuous function, non-decreasing in each of the variable $x \in I$. Suppose the inequality holds for all $x \in I$ with $x \geq a$,*

$$u(x) \leq c + \int_a^x b(t) g(u(t)) dt, \tag{5.2.282}$$

then

$$u(x) \leq G^{-1}\left[G(c) + \int_a^x b(t) dt\right], \tag{5.2.283}$$

for all $x \in I$ such that $G(c) + \int_a^x b(t)dt \in Dom\ (G^{-1})$, where $G(u) = \int_{u_0}^u b(t)dtdz/g(z)$, $u > 0 \geq u_0 > 0$.

Theorem 5.2.52 (The Denche-Khellaf Inequality [193]) *Let $u(x)$, $f(x)$, $a(x)$ be in $C(I, \mathbb{R}_+)$ and let $k(x,t)$, $D_i k(x,t)$ be in $C(I \times I, \mathbb{R}_+)$ for all $i = 1, 2, \cdots, n$. Let $\phi(u(x))$ be real-valued, positive, continuous, strictly non-decreasing, sub-additive and sub-multiplicative function for all $u(x) \geq 0$ and let $W(u(x))$ be real-valued, positive, continuous and non-decreasing function defined for all $x \in I$. Assume that $a(x)$ is positive continuous function and non-decreasing for all $x \in I$. If for all $a \leq s \leq t \leq x \leq b$,*

$$u(t) \leq a(t) + \int_a^x f(t) g(u(t)) dt + \int_a^x f(t) W\left(\int_a^t k(s,t) \phi(u(s)) ds\right) dt, \tag{5.2.284}$$

then for all $a \leq x \leq x^*$,

$$u(t) \leq \beta(x)\Bigg\{a(x) + \int_a^x f(t)W\Bigg[\psi^{-1}(\psi(\eta)$$
$$+ \int_a^t k(s,b)\phi\Big[\beta(s)\int_a^s f(\tau)d\tau\Big]ds)\Bigg]dt\Bigg\}, \tag{5.2.285}$$

where

$$\beta(x) = G^{-1}\left(G(1) + \int_a^x f(s)ds\right), \tag{5.2.286}$$

$$\eta = \int_a^b k(b,s)\phi(\beta(s)a(s))ds, \tag{5.2.287}$$

$$G(u) = \int_{u_0}^u \frac{dz}{g(z)}, \quad u \geq u_0 > 0, \tag{5.2.288}$$

$$\psi(x) = \int_{x_0}^x \frac{ds}{\phi(w(s))}, \quad x \geq x_0 > 0. \tag{5.2.289}$$

Here G^{-1} *is the inverse of* G *and* ψ^{-1} *is the inverse of* ψ, x^* *is chosen so that*

$$G(1) + \int_a^x f(s)ds \in Dom\ (G^{-1}),$$

$$\psi(\eta) + \int_a^t k(s,b)\phi\left[\beta(s)\int_a^s f(\tau)d\tau\right]ds \in Dom\ (\psi^{-1}).$$

Proof Define the function

$$z(x) = a(x) + \int_a^x f(t)W\left(\int_a^t k(t,s)\phi(u(s))ds\right)dt. \tag{5.2.290}$$

Then (5.2.284) can be restated as

$$u(x) \leq z(x) + \int_a^x f(t)g(u(t))dt.$$

We know that $z(x)$ is a positive, continuous, non-decreasing in $x \in I$ and $g \in \mathcal{F}_1$. Then the above inequality can be rewritten as

$$\frac{u(x)}{z(x)} \leq 1 + \int_a^x f(t)g(\frac{u(t)}{z(t)})dt. \tag{5.2.291}$$

By Lemma 5.2.4, we get

$$u(x) \leq z(x)\beta(x), \tag{5.2.292}$$

where $\beta(x)$ is defined by (5.2.286). By (5.2.290), we obtain

$$z(x) = a(x) + \int_a^x f(t)W(v(t))dt, \tag{5.2.293}$$

where

$$v(x) = \int_a^x k(x,t)\phi(u(t))dt. \tag{5.2.294}$$

By (5.2.294) and (5.2.292) , we observe that

$$\begin{aligned} v(x) &\leq \int_a^x k(b,t)\phi[\beta(t)(a(t) + \int_a^t f(s)W(v(s))ds)]dt \\ &\leq \int_a^x k(b,s)\phi(\beta(t)(a(t)))ds + \int_a^t k(b,s)\phi\left(\beta(s)\int_a^s f(\tau)W(v(\tau))d\tau\right)ds, \\ &\leq \eta + \int_a^x k(b,s)\phi\left[\beta(s)\int_a^s f(\tau)d\tau\right]\phi(W(v(s)))ds \end{aligned} \tag{5.2.295}$$

where η is defined by (5.2.287). Since ϕ is sub-additive and sub-multiplicative function, W and $v(x)$ are non-decreasing for all $x \in I$. Define $r(x)$ as the right-hand side of (5.2.295), then $r(a_1, x_2, ..., x_n) = \eta$ and $v(x) \leq r(x)$, $r(x)$ is positive and non-decreasing in each of the variables $x_1, x_2, x_3, ..., x_n$. Hence

$$\frac{Dr(x)}{\phi(W(r(x)))} \leq k(b,x)\phi\left[\beta(s)\int_a^x f(s)ds\right].$$

Since

$$D_n\left(\frac{D_1\cdots D_{n-1}r(x)}{\phi(W(r(x)))}\right) = \frac{Dr(x)}{\phi(W(r(x)))} - \frac{D_n\phi(W(r(x)))D_1\cdots D_{n-1}r(x)}{\phi^2(W(r(x)))},$$

the above inequality implies

$$D_n\left(\frac{D_1\cdots D_{n-1}r(x)}{\phi(W(r(x)))}\right) \leq \frac{Dr(x)}{\phi(W(r(x)))},$$

and

$$D_n\left(\frac{D_1\cdots D_{n-1}r(x)}{\phi(W(r(x)))}\right) \leq k(b,x)\phi[\theta(x)],$$

where $\theta(x) = \beta(x) \int_a^x f(s)ds$. Integrating with respect to x_n from a_n to x_n, we have

$$\frac{D_1 \cdots D_{n-1} r(x)}{\phi(W(r(x)))} \leq \int_{a_n}^{x_n} k(b, x_1, x_2, \cdots, x_{n-1}, s_n)\phi[\theta(x_1, x_2, \cdots, x_{n-1}, s_n)]ds_n.$$

Repeating the above argument, we find that

$$\frac{D_1 r(x)}{\phi(W(r(x)))} \leq \int_{a_2}^{x_2} \cdots \int_{a_{n-1}}^{x_{n-1}} \int_{a_n}^{x_n} k(b, x_1, s_2, \cdots, s_n)\phi[\theta(x_1, s_2, \cdots, s_n)]ds_n ds_{n-1} \cdots ds_2.$$

Integrating both sides of the above inequality with respect to x_1 from a_1 to x_1, we can get

$$\psi(r(x)) - \psi(\eta) \leq \int_a^x k(b, s)\phi[\theta(s)]ds,$$

and

$$r(x) \leq \psi^{-1}(\psi(\eta) + \int_a^x k(b, s)\phi[\beta(x) \int_a^x f(s)ds]ds)$$

which yields

$$v(x) \leq r(x) \leq \psi^{-1}\left(\psi(\eta) + \int_a^x k(b, s)\phi[\beta(x) \int_a^x f(s)ds]ds\right). \tag{5.2.296}$$

By (5.2.292), (5.2.293) and (5.2.296), we can obtain the desired inequality in (5.2.285). □

In the following theorems, we shall introduce the inequalities similar to those given in Theorems 1.2.7 and 1.2.8 involving functions of several independent variables.

Theorem 5.2.53 (The Pachpatte Inequality [496]) *Let $F(x) \geq 0, b(x) \geq 0$, and $b_i(x) > 0$ for $i = 1, \ldots, n-1$ be real-valued continuous functions defined for all $x \in \mathbb{R}_+^n$, and let $p > 1$ be a constant. If for all $t \in \mathbb{R}_+^n$,*

$$F^p(x) \leq c + M[x_1, \ldots, x_n, b_1, \ldots, b_{n-1}, bF], \tag{5.2.297}$$

where $c \geq 0$ is a constant, then for all $x \in \mathbb{R}_+^n$.

$$F(x) \leq \left[c^{\frac{(p-1)}{p}} + \frac{p-1}{p} M[x_1, \ldots, x_n, b_1, \ldots, b_{n-1}, b]\right]^{\frac{1}{p-1}}. \tag{5.2.298}$$

Theorem 5.2.54 (The Pachpatte Inequality [496]) *Let* $u(x) \geq 0, v(x) \geq 0$, *and* $b_i(x) > 0$ *for* $i = 1, \ldots, n-1$ *and* $h_j(x) \geq 0$ *for* $j = 1, 2, 3, 4$ *be real-valued continuous functions defined for all* $x \in \mathbb{R}_+^n$ *and let* $p > 1$ *be a constant. If* c_1, c_2 *and* μ *are non-negative constants such that for all* $x \in \mathbb{R}_+^n$,

$$\begin{cases} u^p(x) \leq c_1 + M[x_1, \ldots, x_n, b_1, \ldots, b_{n-1}, h_1 u] + M[x_1, \ldots, x_n, b_1, \ldots, b_{n-1}, h_2 \overline{v}], & (5.2.299) \\ v^p(x) \leq c_2 + M[x_1, \ldots, x_n, b_1, \ldots, b_{n-1}, h_3 \overline{u}] + M[x_1, \ldots, x_n, b_1, \ldots, b_{n-1}, h_4 v], & (5.2.300) \end{cases}$$

where $\overline{u}(x) = \exp(-p\mu \sum_{i=1}^{n} x_i)u(x)$ *and* $\overline{v}(x) = \exp(p\mu \sum_{i=1}^{n})v(x)$ *for all* $x \in \mathbb{R}_+^n$, *then*

$$u(x) \leq \exp(\mu \sum_{i=1}^{n})[\{2^{p-1}(c_1 + c_2)\}^{\frac{(p-1)}{p}} + 2^{p-1}(\frac{p-1}{p})M[x_1, \ldots, x_n, b_1, \ldots, b_{n-1}, h]]^{\frac{1}{p-1}}, \quad (5.2.301)$$

$$v(x) \leq [\{2^{p-1}(c_1 + c_2)\}^{\frac{(p-1)}{p}} + 2^{p-1}(\frac{p-1}{p})M[x_1, \ldots, x_n, b_1, \ldots, b_{n-1}, h]]^{\frac{1}{p-1}}, \quad (5.2.302)$$

for all $x \in \mathbb{R}_+^n$, *where for all* $x \in \mathbb{R}_+^n$,

$$h(x) = \max\{[h_1(x) + h_3(x)], [h_2(x) + h_4(x)]\}. \quad (5.2.303)$$

Proofs of Theorems 5.2.54 and 5.2.55 The proofs proceed much as the proofs of the theorems given above and follow by closely looking at the proofs of the main results Theorems 5.2.24–5.2.28 with suitable changes. We omit the details. □

Chapter 6
Nonlinear Multi-Dimensional Discrete (Difference) Inequalities

In this chapter, we collect some nonlinear discrete integral and difference inequalities.

The role played by nonlinear discrete inequalities in one and more than one variable in the theory of difference equation and numerical analysis is well-known. During the last few years, there have been a number of works written on the discrete inequalities of the Gronwall inequality and its nonlinear version to the Bhiari type, see [42, 579, 687, 692].

6.1 Nonlinear Multi-Dimensional Discrete Bellman-Gronwall Inequalities and Their Generalizations

6.1.1 Nonlinear Two-Dimensional Discrete Bellman-Gronwall Inequalities and Bihari Inequalities

Theorem 6.1.1 (The Pachpatte Inequality [516]) *Let $u(m,n), a(m,n), b(m,n)$ be non-negative continuous functions defined for all $m, n \in \mathbb{N}_0$ and $L : \mathbb{N}_0^2 \times \mathbb{R}_+ \to \mathbb{R}_+$ be a continuous function which satisfies the condition: for all $u \geq v \geq 0$,*

$$0 \leq L(m,n,u) - L(m,n,v) \leq M(m,n,v)(u-v),$$

where $M(m,n,v)$ is a non-negative continuous function defined for all $m, n \in \mathbb{N}_0, v \in \mathbb{R}_+$.

Y. Qin, *Integral and Discrete Inequalities and Their Applications*,
DOI 10.1007/978-3-319-33304-5_6

(i) Assume that $a(m,n)$ is non-increasing in $m \in \mathbb{N}_0$. If for all $m, n \in \mathbb{N}_0$,

$$u(m,n) \le a(m,n) + \sum_{s=0}^{m-1} b(s,n)u(s,n) + \sum_{s=0}^{m-1} \sum_{t=n+1}^{+\infty} L(s,t,u(s,t)), \tag{6.1.1}$$

then for all $m, n \in \mathbb{N}_0$,

$$u(m,n) \le q(m,n) \left[a(m,n) + H(m,n) \prod_{s=0}^{m-1} \left(1 + \sum_{t=n+1}^{+\infty} M(s,t,q(s,t)a(s,t))q(s,t) \right) \right], \tag{6.1.2}$$

where for all $m, n \in \mathbb{N}_0$,

$$H(m,n) = \sum_{s=0}^{m-1} \sum_{t=n+1}^{+\infty} L(s,t,q(s,t)a(s,t)), \tag{6.1.3}$$

and $q(m,n)$ is defined by

$$q(m,n) = \prod_{s=0}^{m-1} (1 + b(s,n)).$$

(ii) Assume that $a(m,n)$ is non-increasing in $m \in \mathbb{N}_0$. If for all $m, n \in \mathbb{N}_0$,

$$u(m,n) \le a(m,n) + \sum_{s=m+1}^{+\infty} b(s,n)u(s,n) + \sum_{s=m+1}^{+\infty} \sum_{t=n+1}^{+\infty} L(s,t,u(s,t)), \tag{6.1.4}$$

then for all $m, n \in \mathbb{N}_0$,

$$u(m,n) \le \overline{q}(m,n) \left[a(m,n) + \overline{H}(m,n) \prod_{s=m+1}^{+\infty} \left(1 + \sum_{t=n+1}^{+\infty} M(s,t,\overline{q}(s,t)a(s,t))\overline{q}(s,t) \right) \right], \tag{6.1.5}$$

where for all $m, n \in \mathbb{N}_0$,

$$\overline{H}(m,n) = \sum_{s=m+1}^{+\infty} \sum_{t=n+1}^{+\infty} L(s,t,\overline{q}(s,t)a(s,t)), \tag{6.1.6}$$

and $\overline{q}(m,n)$ *is defined by*

$$\overline{q}(m,n) = \prod_{s=m+1}^{+\infty} (1 + b(s,n)).$$

Proof We only give the proof of (i); the proof of (ii) can be completed by following the proof of (i).

(i) The proof follows by closely looking at the proofs of (p_1), (q_1) and (c_1) given above. Here we leave the details to the reader. □

6.1.2 Nonlinear Two-Dimensional Discrete Ou-Yang Inequalities and Their Generalizations

The next result, due to Salem and Raslan [580], is devoted to nonlinear discrete inequalities in two independent variables.

Theorem 6.1.2 (The Salem-Raslan Inequality [580]) *Let* $u(m,n), a(m,n)$, $b(m,n)$ *be non-negative functions and* $a(m,n)$ *non-decreasing for all* $m,n \in \mathbb{N}$. *If for all* $m,n \in \mathbb{N}$,

$$u^{m_1}(m,n) \le a(m,n) + \sum_{s=0}^{m-1}\sum_{t=0}^{n-1} b(s,t)u^{m_2}(s,t), \tag{6.1.7}$$

then we have

$$u(m,n) \le a^{\frac{1}{m}}(m,n) \prod_{t=0}^{n-1}\left[1 + \sum_{s=0}^{m-1} b(s,t)\right]^{\frac{1}{m_1}}, \quad if \ \ m_1 = m_2, \tag{6.1.8}$$

$$u(m,n) \le a^{\frac{1}{m}}(m,n) \prod_{t=0}^{n-1}\left[1 + \sum_{s=0}^{m-1} b(s,t)a^{\frac{m_2-m_1}{m_1}}\right]^{\frac{m_2(n-t-1)}{m_1^2}}, \quad if \ \ m_1 < m_2, \tag{6.1.9}$$

$$u(m,n) \le a^{\frac{1}{m}}(m,n) \prod_{t=0}^{n-1}\left[1 + \sum_{s=0}^{m-1} b(s,t)a^{\frac{m_2-m_1}{m_1}}\right]^{\frac{1}{m_1}}, \quad if \ \ m_1 > m_2. \tag{6.1.10}$$

Proof Define a function $z(m,n)$ by

$$z^{m_1}(m,n) = a(m,n) + \sum_{s=0}^{m-1}\sum_{t=0}^{n-1} b(s,t)u^{m_2}(s,t). \tag{6.1.11}$$

From (6.1.7) and (6.1.11) it follows

$$u(m,n) \le z(m,n). \tag{6.1.12}$$

Since $a(m,n)$ is non-negative and non-decreasing for all $m,n \in \mathbb{N}$, we infer from (6.1.12)

$$\frac{z^{m_1}(m,n)}{a(m,n)} \le 1 + \sum_{s=0}^{m-1}\sum_{t=0}^{n-1} b(s,t)\frac{z^{m_2}(s,t)}{a(s,t)}. \tag{6.1.13}$$

Define function $v(m,n)$ by

$$v(m,n) = 1 + \sum_{s=0}^{m-1}\sum_{t=0}^{n-1} b(s,t)\frac{z^{m_2}(s,t)}{a(s,t)}, \tag{6.1.14}$$

so we obtain from (6.1.13)–(6.1.14) that

$$z^{m_1}(m,n) \le a(m,n)v(m,n). \tag{6.1.15}$$

From (6.1.14), we get

$$\begin{aligned}&[v(m+1,n+1) - v(m,n+1)]\\ &\quad - [v(m+1,n) - v(m,n)] \le b(m,n)a^{\frac{m_2-m_1}{m_1}}(m,n)v^{\frac{m_2}{m_1}}(m,n).\end{aligned} \tag{6.1.16}$$

Now, we consider the following cases:

Case 1. If $m_1 = m_2$, then from (6.1.16), we infer

$$v(m+1,n+1) - v(m,n+1) - v(m+1,n) \le (-1+b(m,n))v(m,n). \tag{6.1.17}$$

Keeping n fixed in (6.1.17), setting $m = 0,1,2,\ldots,m-1$, then we get

$$v(m,n+1) \le \left[1 + \sum_{s=0}^{m-1} b(s,n)\right] v(m,n). \tag{6.1.18}$$

From (6.1.18), it follows

$$v(m,n) \le \prod_{t=0}^{n-1}\left[1 + \sum_{s=0}^{m-1} b(s,t)\right]. \tag{6.1.19}$$

Thus the required result (6.1.8) follows from (6.1.12), (6.1.15) and (6.1.19).

Case 2. If $m_2 > m_1$, then as in **Case 1** from (6.1.16), we derive

$$v(m+1,n+1) - v(m,n+1) - v(m+1,n) + v(m,n) \leq b(m,n)a^{\frac{m_2-m_1}{m_1}}(m,n)v^{\frac{m_2}{m_1}}(m,n), \tag{6.1.20}$$

when n is fixed and $m = 0, 1, 2, \ldots, m-1$, we obtain from (6.1.151) that

$$v(m,n+1) \leq \left[1 + \sum_{s=0}^{m-1} b(s,n)a^{\frac{m_2-m_1}{m_1}}(s,n)\right] v^{\frac{m_2}{m_1}}(m,n). \tag{6.1.21}$$

To complete the proof of **Case 2**, we need the following lemma which can be proved easily.

Lemma 6.1.1 *If*

$$v(m,n+1) \leq (1 + b(m,n))v^p(m,n), \quad p > 1, \tag{6.1.22}$$

then we have

$$v(m,n) \leq \prod_{t=0}^{n-1} \left(1 + b(m,t))^{(n-t-1)p}\right). \tag{6.1.23}$$

Then from (6.1.21)–(6.1.23) it follows

$$v(m,n) \leq \prod_{t=0}^{n-1} \left[1 + \sum_{s=0}^{m-1} b(s,t)a^{\frac{m_2-m_1}{m_1}}(s,t)\right]^{\frac{m_2(n-t-1)}{m_1}}. \tag{6.1.24}$$

The required result (6.1.9) follows from (6.1.12), (6.1.15) and (6.1.24).

Case 3. If $m_2 < m_1$, then we have $v^{\frac{m_2}{m_1}}(m,n) \leq v(m,n)$. As in the last two cases, we can derive

$$v(m,n+1) \leq \left[1 + \sum_{s=0}^{m-1} b(s,n)a^{\frac{m_2-m_1}{m_1}}(s,nt)\right] v(m,n). \tag{6.1.25}$$

Then from (6.1.25), we deduce

$$v(m,n) \leq \prod_{t=0}^{n-1} \left[1 + \sum_{s=0}^{m-1} b(s,t)a^{\frac{m_2-m_1}{m_1}}(s,t)\right]. \tag{6.1.26}$$

From (6.1.12), (6.1.15) and (6.1.26), we have

$$u(m,n) \le a^{\frac{1}{m_1}}(m,n)\prod_{t=0}^{n-1}\left[1+\sum_{s=0}^{m-1} b(s,t)a^{\frac{m_2-m_1}{m_1}}(s,t)\right]^{\frac{1}{m_1}},$$

which is the required result (6.1.10). □

Remark 6.1.1

(1). If $m_1 = m_2 = 1$, then from (6.1.7) and (6.1.8), we can get the same result as that of Theorem 6.1.20.

(2). If $m_1 = 1, m_2 > 1$, then from (6.1.7) and (6.1.8), we can derive that if

$$u(m,n) \le a(m,n) + \sum_{s=0}^{m-1}\sum_{t=0}^{n-1} b(s,t)u^{m_2}(s,t), \tag{6.1.27}$$

then

$$u(m,n) \le a(m,n)\prod_{t=0}^{n-1}\left[1+\sum_{s=0}^{m-1} b(s,t)a^{m_2-1}(s,t)\right]^{m_2(n-t-1)}. \tag{6.1.28}$$

(3). Let $m_2 = 1, m_1 > 1$, then from (6.1.7)–(6.1.10), we can conclude if

$$u^{m_1}(m,n) \le a(m,n) + \sum_{s=0}^{m-1}\sum_{t=0}^{n-1} b(s,t)u(s,t), \tag{6.1.29}$$

then

$$u(m,n) \le a^{\frac{1}{m_1}}(m,n)\prod_{t=0}^{n-1}\left[1+\sum_{s=0}^{m-1} b(s,t)a^{\frac{1-m_1}{m_1}}(s,t)\right]^{\frac{1}{m_1}}. \tag{6.1.30}$$

Theorem 6.1.3 (The Salem-Raslan Inequality [580]) *Let $u(m,n), a(m,n)$, $b(m,n)$ and $c(m,n)$ be non-negative and $a(m,n)$ be non-decreasing for all $m,n \in \mathbb{N}$, if $m_1, m_2 \in \mathbb{R}_+$, and*

$$u^{m_1}(m,n) \le a(m,n) + \sum_{s=0}^{m-1}\sum_{t=0}^{n-1} b(s,t)u(s,t) + \sum_{s=0}^{m-1}\sum_{t=0}^{n-1} c(s,t)u^{m_2}(s,t), \tag{6.1.31}$$

then we have

$$u(m,n) \le a(m,n)\prod_{t=0}^{n-1}\left[1+\sum_{s=0}^{m-1}(b(s,t)+c(s,t))\right], \text{ if } m_1 = m_2 = 1, \quad (6.1.32)$$

$$u(m,n) \le a^{\frac{1}{m_1}}(m,n)\prod_{t=0}^{n-1}\left[1+\sum_{s=0}^{m-1}(b(s,t) +c(s,t))a^{\frac{1-m_1}{m_1}}(s,t)\right]^{\frac{1}{m_1}}, \text{ if } m_1 = m_2 > 1, \quad (6.1.33)$$

$$u(m,n) \le a^{\frac{1}{m_1}}(m,n)\prod_{t=0}^{n-1}\left[1+\sum_{s=0}^{m-1}(b(s,t) +c(s,t))a^{\frac{1-m_1}{m_1}}(s,t)\right]^{\frac{n-t-1}{m_1^2}}, \text{ if } 0 < m_1 = m_2 > 1, \quad (6.1.34)$$

$$u(m,n) \le a^{\frac{1}{m_1}}(m,n)\prod_{t=0}^{n-1}\left[1+\sum_{s=0}^{m-1}(b(s,t)a^{\frac{1-m_1}{m_1}}(s,t) +c(s,t)a^{\frac{m_2-m_1}{m_1^2}}(s,t))\right]^{\frac{m_2(n-t-1)}{m_1^2}}, \text{ if } m_2 > m_1, \quad (6.1.35)$$

$$u(m,n) \le a^{\frac{1}{m_1}}(m,n)\prod_{t=0}^{n-1}\left[1+\sum_{s=0}^{m-1}(b(s,t)a^{\frac{1-m_1}{m_1}}(s,t) +c(s,t)a^{\frac{m_2-m_1}{m_1^2}}(s,t))\right]^{\frac{1}{m_1}}, \text{ if } 1 \le m_2 < m_1, \quad (6.1.36)$$

and

$$u(m,n) \le a^{\frac{1}{m_1}}(m,n)\prod_{t=0}^{n-1}\left[1+\sum_{s=0}^{m-1}(b(s,t)a^{\frac{1-m_1}{m_1}}(s,t) +c(s,t)a^{\frac{m_2-m_1}{m_1^2}}(s,t))\right]^{\frac{n-t-1}{m_1^2}}, \text{ if } 0 < m_2 < m_1 < 1. \quad (6.1.37)$$

Proof The proof is similar to the proof of Theorem 6.1.2. Here we leave the detail to the reader. □

Remark 6.1.2

(1). If $c(m,n)=0, m_1=m_2$, then we get Theorem 6.1.1. in Qin[557]
(2). If $b(m,n)=0$, then we get Theorem 6.1.2.

The following results, due to Cheung, Ma and Josip [146], is to establish some new and more general nonlinear discrete inequalities involving functions of two independent variables, which generalize some results in [141, 147, 363, 500] and can be readily used as handy and powerful tools in the analysis of certain classes of partial finite difference and sum-difference equations.

For any $\varphi, \psi, h \in C((0,+\infty),(0,+\infty))$, and any constant $\beta>0$, define

$$\Phi_h(r) := \int_1^r \frac{ds}{\varphi \circ h^{-1}(s)}, \quad \Psi_h(r) := \int_1^r \frac{ds}{\psi \circ h^{-1}(s)}, \quad r>0,$$

$$\Phi_h(0) := \lim_{r\to 0+} \Phi_h(r), \quad \Psi_h(0) := \lim_{r\to 0+} \Psi_h(r).$$

Note that we allow $\Phi_h(0)$ and $\Psi_h(0)$ to be $-\infty$ here.

The following results follow the notation in Theorems 6.1.4–6.1.6.

Theorem 6.1.4 (The Cheung-Ma-Josip Inequality [146]) *Suppose $u \in F_0(\Omega)$. If $c \geq 0$ is a constant and $b \in F_0(\Omega), \varphi, h \in C(\mathbb{R}_0,\mathbb{R}_0)$, $\mathbb{R}_0 = (0,+\infty)$, are functions satisfying*

(i) h is strictly increasing with $h(0)=0$ and $h(t)\to+\infty$ as $t\to+\infty$;
(ii) φ is non-decreasing with $\varphi(r)>0$ for all $r>0$;
(iii) for any $(m,n)\in\Omega$,

$$h(u(m,n)) \leq c + \sum_{s=m_0}^{m-1}\sum_{t=n_0}^{n-1} b(s,t)\varphi(u(s,t)), \tag{6.1.38}$$

then for all $(m,n)\in\Omega_{(m_1,n_1)}$,

$$u(m,n) \leq h^{-1}\left(\Phi_h^{-1}[\Phi_h(c)+B(m,n)]\right) \tag{6.1.39}$$

where

$$B(m,n) := \sum_{s=m_0}^{m-1}\sum_{t=n_0}^{n-1} b(s,t),$$

and Φ_h^{-1} is the inverse of Φ_h, and $(m_1,n_1)\in\Omega$ is chosen such that $\Phi_h(c)+B(m,n)\in$ Dom (Φ_h^{-1}) for all $(m,n)\in\Omega_{(m_1,n_1)}$.

Proof It suffices to consider the case $c>0$, for then the case $c=0$ can be arrived at by continuity argument. Denote by $g(m,n)$ the right-hand side of (6.1.38). Then, $g>0, u\leq h^{-1}(g)$ on Ω, and g is non-decreasing in each variable. Hence, for any

$(m, n) \in \Omega$, we have

$$\begin{aligned}\Delta_1 g(m,n) &= g(m+1,n) - g(m,n) \\ &= \sum_{t=n_0}^{n-1} b(m,t)\varphi(u(m,t)) \\ &\le \sum_{t=n_0}^{n-1} b(m,t)\varphi(h^{-1}(g(m,t))) \\ &\le \varphi(h^{-1}(g(m,n-1))) \sum_{t=n_0}^{n-1} b(m,t). \end{aligned} \tag{6.1.40}$$

Therefore, by the Mean-Value Theorem for integrals, for each $(m, n) \in \Omega$, there exists an ξ with $g(m, n) \le \xi \le g(m + 1, n)$ such that

$$\begin{aligned}\Delta_1(\Phi_h \circ g)(m,n) &= \Phi_h(g(m+1,n)) - \Phi_h(g(m,n)) \\ &= \int_{g(m,n)}^{g(m+1,n)} \frac{ds}{\varphi \circ h^{-1}(s)} \\ &= \frac{1}{\varphi(h^{-1}(\xi))} \Delta_1 g(m,n). \end{aligned}$$

Since φ is non-decreasing, $\varphi(h^{-1}(\xi)) \ge \varphi(h^{-1}(g(m, n)))$, and by (6.1.40), we derive for all $(m, n) \in \Omega$,

$$\begin{aligned}\Delta_1(\Phi_h \circ g)(m,n) &\le \frac{1}{\varphi(h^{-1}(g(m,n)))} \Delta_1 g(m,n) \\ &\le \frac{h^{-1}(g(m,n-1))}{\varphi(h^{-1}(g(m,n)))} \sum_{t=n_0}^{n-1} b(m,t) \\ &\le \sum_{t=n_0}^{n-1} b(m,t). \end{aligned}$$

Therefore,

$$\sum_{s=m_0}^{m-1} \Delta_1(\Phi_h \circ g)(s,n) \le \sum_{s=m_0}^{m-1} \sum_{t=n_0}^{n-1} b(s,t) = B(m,n).$$

On the other hand, it is easy to verify that

$$\sum_{s=m_0}^{m-1} \Delta_1(\Phi_h \circ g)(s,n) = \Phi_h \circ g(m,n) - \Phi_h \circ g(m_0,n),$$

whence

$$\Phi_h \circ g(m,n) \le \Phi_h \circ g(m_0,n) + B(m,n) = \Phi_h(c) + B(m,n).$$

Since $\Phi_h^{-1}$is increasing on *Dom* (Φ_h^{-1}), this yields for all $(m,n) \in \Omega_{(m_1,n_1)}$,

$$g(m,n) \le \Phi_h^{-1}[\Phi_h(c) + B(m,n)].$$

Hence, the assertion follows. □

Remark 6.1.3

(i) When h = identity, Theorem 6.1.4 exhibits the discrete analogue of Theorem 2.1 in [143].
(ii) When $h(s) = s^\alpha, \alpha > 0$, Theorem 6.1.4 exhibits the discrete analogue of Theorem 2.1 in [140].
(iii) In many cases, h and φ satisfy $\int_1^{+\infty} \frac{ds}{\varphi \circ h^{-1}(s)} = +\infty$. For example, φ = constant > 0, $\varphi = h$, $\varphi = h^{\frac{1}{2}}$, etc., are such functions. In this case, $\Phi(+\infty) = +\infty$ and so we may take $m_1 = M$, $n_1 = N$. In particular, inequality (6.1.39) holds for all $(m,n) \in \Omega$.

Corollary 6.1.1 (The Cheung-Ma-Josip Inequality [146]) *Suppose $u \in F_+(\Omega)$. If $c \ge 0$ is a constant and $b \in F_+(\Omega)$, $h \in C(\mathbb{R}_+, \mathbb{R}_+)$ are functions satisfying*

(i) h is strictly increasing with $h(0) = 0$ and $h(t) \to +\infty$ as $t \to +\infty$;
(ii) for any $(m,n) \in \Omega$,

$$h(u(m,n)) \le c + \sum_{s=m_0}^{m-1} \sum_{t=n_0}^{n-1} b(s,t) h(u(s,t)),$$

then for all $(m,n) \in \Omega$,

$$u(m,n) \le h^{-1}\left(c \exp B(m,n)\right)$$

where $B(m,n)$ is as defined in Theorem 6.1.4.

Proof Suppose first that $c > 0$. Taking $\varphi = h$, we have

$$\Phi_h(r) = \int_1^r \frac{ds}{\varphi \circ h^{-1}(s)} = \int_1^r \frac{ds}{s} = \ln r, \quad r > 0,$$

and so $\Phi_h^{-1} = \exp$, in particular, it is defined everywhere on $\mathbb{R}$. Hence, by Theorem 6.1.4, for all $(m, n) \in \Omega$,

$$u(m,n) \leq h^{-1}\left(\exp[lnc + B(m,n)]\right) = h^{-1}(c \exp B(m,n)).$$

Finally, as this is true for all $c > 0$, by continuity, this should also hold for the case $c = 0$.

In case when Ω degenerates to a 1-dimensional lattice, Theorem 6.1.4 and Corollary 6.1.1 take the following simpler forms, which are generalizations of some results of Pachpatte in [500].

Corollary 6.1.2 (The Cheung-Ma-Josip Inequality [146]) *Suppose $u \in F_+(I)$. If $c \geq 0$ is a constant and $b \in F_+(I), \varphi, h \in C(\mathbb{R}_+, \mathbb{R}_+)$ are functions satisfying*

(i) h is strictly increasing with $h(0) = 0$ and $h(t) \to +\infty$ as $t \to +\infty$;
(ii) φ is non-decreasing with $\varphi(r) > 0$ for all $r > 0$;
(iii) for any $m \in I$,

$$h(u(m)) \leq c + \sum_{s=m_0}^{m-1} b(s)\varphi(u(s)),$$

then for all $m \in [m_0, m_1] \cap I$,

$$u(m) \leq h^{-1}\left\{\Phi_h^{-1}\left[\Phi_h(c) + \sum_{s=m_0}^{m-1} b(s)\right]\right\}$$

where $m_1 \in I$ is chosen such that $\Phi_h(c) + \sum_{s=m_0}^{m-1} b(s) \in Dom\ (\Phi_h^{-1})$ for all $m \in [m_0, m_1] \cap I$.

Proof It follows immediately from Theorem 6.1.4 by setting $\Omega = I \times \{n_0\}$ for some $n_0 \in \mathbb{Z}$ and extending the functions $b(s), u(s)$ to $b(s, n_0), u(s, n_0)$, respectively, in the obvious way. □

Corollary 6.1.3 (The Cheung-Ma-Josip Inequality [146]) *Suppose $u \in F_+(I)$. If $c \geq 0$ is a constant and $b \in F_+(I)$, $h \in C(\mathbb{R}_+, \mathbb{R}_+)$ are functions satisfying*

(i) h is strictly increasing with $h(0) = 0$ and $h(t) \to +\infty$ as $t \to +\infty$;
(ii) for any $m \in I$,

$$h(u(m)) \leq c + \sum_{s=m_0}^{m-1} b(s)h(u(s)),$$

then for all $m, n \in I$,

$$u(m) \leq h^{-1}\left(c \prod_{s=m_0}^{m-1} \exp b(s)\right).$$

Proof This follows from Corollary 6.1.1 by similar arguments as in the proof of Corollary 6.1.2. □

Theorem 6.1.4 and Corollaries 6.1.1–6.1.3 can easily be applied to generate other useful discrete inequalities in many concrete situations. For example, we have the next result.

Corollary 6.1.4 (The Cheung-Ma-Josip Inequality [146]) *Suppose* $u \in F_+(\Omega)$. *If* $c \geq 0$, $p > 1$ *are constants and* $b \in F_+(I)$ *is a function such that for any* $(m, n) \in \Omega$,

$$u^p(m, n) \leq c + \sum_{s=m_0}^{m-1} \sum_{t=n_0}^{n-1} b(s) u^p(s, t),$$

then for all $(m, n) \in \Omega$,

$$u(m, n) \leq c^{\frac{1}{p}} \exp\left(\frac{1}{p} B(m, n)\right)$$

where $B(m, n)$ *is as defined in Theorem 6.1.4.*

Proof This follows immediately from Corollary 6.1.3 by letting $h(s) = s^p$. □

Corollary 6.1.5 (The Cheung-Ma-Josip Inequality [146]) *Suppose* $u \in F_+(I)$. *If* $c \geq 0$, $p > 1$ *are constants and* $b \in F_+(I)$ *is a function such that for any* $m \in I$,

$$u^p(m) \leq c + \sum_{s=m_0}^{m-1} b(s) u^p(s),$$

then for all $m \in I$,

$$u(m) \leq c^{\frac{1}{p}} \exp\left(\frac{1}{p} \sum_{s=m_0}^{m-1} b(s)\right).$$

Proof This follows from Corollary 6.1.4 by arguments similar to that in the proof of Corollary 6.1.2. □

Theorem 6.1.5 (The Cheung-Ma-Josip Inequality [146]) *Suppose* $u \in F_+(\Omega)$. *If* $k > 0$ *is a constant and* $a, b \in F_+(\Omega)$, $\varphi, h \in C((0, +\infty), (0, +\infty))$ *are functions satisfying*

(i) *$h(t)$ and $H(t) := \frac{h(t)}{t}$, $t > 0$, are strictly increasing with $H(t) \to +\infty$ as $t \to +\infty$;*
(ii) *φ is non-decreasing;*
(iii) *for any $(m, n) \in \Omega$,*

$$h(u(m,n)) \le k + \sum_{s=m_0}^{m-1}\sum_{t=n_0}^{n-1} a(s,t)u(s,t) + \sum_{s=m_0}^{m-1}\sum_{t=n_0}^{n-1} b(s,t)u(s,t)\varphi(u(s,t)), \tag{6.1.41}$$

then for all $(m, n) \in \Omega_{(m_1,n_1)}$,

$$u(m,n) \le H^{-1}\left\{\Phi_H^{-1}\left[\Phi_H\left(\frac{k}{h^{-1}(k)} + A(m,n)\right) + B(m,n)\right]\right\} \tag{6.1.42}$$

where

$$A(m,n) := \sum_{s=m_0}^{m-1}\sum_{t=n_0}^{n-1} a(s,t), \qquad B(m,n) := \sum_{s=m_0}^{m-1}\sum_{t=n_0}^{n-1} b(s,t),$$

and $(m_1, n_1) \in \Omega$ is chosen such that $\Phi_H(\frac{k}{h^{-1}(k)} + A(m,n)) + B(m,n)) \in Dom\ (\Phi_H^{-1})$ for all $(m, n) \in \Omega_{(m_1,n_1)}$.

Proof Denote by $f(m, n)$ the right-hand side of (6.1.41). Then, $f > 0$, $u \le h^{-1}(f)$ on Ω, and f is non-decreasing in each variable. Hence, for any $(m, n) \in \Omega$,

$$\begin{aligned}
\Delta_1 f(m,n) &= f(m+1,n) - f(m,n) \\
&= \sum_{t=n_0}^{n-1} a(m,t)u(m,t) + \sum_{t=n_0}^{n-1} b(m,t)u(m,t)\varphi(u(m,t)) \\
&\le \sum_{t=n_0}^{n-1} a(m,t)h^{-1}(f(m,t)) + \sum_{t=n_0}^{n-1} b(m,t)h^{-1}(f(m,t))\varphi(h^{-1}(f(m,t))) \\
&\le h^{-1}(f(m,n-1))\left[\sum_{t=n_0}^{n-1} a(m,t) + \sum_{t=n_0}^{n-1} b(m,t)\varphi(h^{-1}(f(m,t)))\right],
\end{aligned}$$

or

$$\frac{\Delta_1 f(m,n)}{h^{-1}(f(m,n-1))} \le \sum_{t=n_0}^{n-1} a(m,t) + \sum_{t=n_0}^{n-1} b(m,t)\varphi(h^{-1}(f(m,t))).$$

Therefore, for any $(m, n) \in \Omega$,

$$\begin{aligned}
&\sum_{s=m_0}^{m-1} \frac{\Delta_1 f(s,n)}{h^{-1}(f(s,n-1))} \\
&= \frac{f(m,n)}{h^{-1}(f(m-1,n-1))} - \frac{f(m-1,n)}{h^{-1}(f(m-1,n-1))} \\
&\quad + \frac{f(m-1,n)}{h^{-1}(f(m-2,n-1))} - \frac{f(m-2,n)}{h^{-1}(f(m-2,n-1))} + \cdots \\
&\quad + \frac{f(m_0+1,n)}{h^{-1}(f(m_0,n-1))} - \frac{f(m_0,n)}{h^{-1}(f(m_0,n-1))} \\
&= \frac{f(m,n)}{h^{-1}(f(m-1,n-1))} + \sum_{s=1}^{m-m_0-1} f(m-s,n) \left[\frac{1}{h^{-1}(f(m-s-1,n-1))} \right. \\
&\quad \left. - \frac{1}{h^{-1}(f(m-s,n-1))} \right] - \frac{f(m_0,n)}{h^{-1}(f(m_0,n-1))} \\
&\geq \frac{f(m,n)}{h^{-1}(f(m,n))} - \frac{f(m_0,n)}{h^{-1}(f(m_0,n-1))} \\
&= \frac{f(m,n)}{h^{-1}(f(m,n))} - \frac{k}{h^{-1}(k)} \\
&= H(h^{-1}(f(m,n))) - \frac{k}{h^{-1}(k)}.
\end{aligned}$$

Hence, we have for all $(m, n) \in \Omega$,

$$H(h^{-1}(f(m,n))) \leq \frac{k}{h^{-1}(k)} + A(m,n) + \sum_{s=m_0}^{m-1} \sum_{t=n_0}^{n-1} b(s,t)\varphi(h^{-1}(f(s,t))).$$

In particular, since A is non-decreasing in each variable, for any fixed $(\overline{m}, \overline{n}) \in \Omega_{(m_1,n_1)}$, and for all $(m, n) \in \Omega_{(\overline{m},\overline{n})}$,

$$H(h^{-1}(f(m,n))) \leq \frac{k}{h^{-1}(k)} + A(\overline{m},\overline{n}) + \sum_{s=m_0}^{m-1} \sum_{t=n_0}^{n-1} b(s,t)\varphi(h^{-1}(f(s,t))).$$

Now applying Theorem 6.1.4 to the strictly increasing function H, we get for all $(m,n) \in \Omega_{(\overline{m},\overline{n})}$,

$$h^{-1}(f(m,n)) \leq H^{-1}\{\Phi_H^{-1}[\Phi_H(\frac{k}{h^{-1}(k)} + A(\overline{m},\overline{n})) + B(m,n)]\}.$$

In particular, this yields

$$\begin{aligned} u(\overline{m},\overline{n}) &= h^{-1}(h(u(\overline{m},\overline{n}))) \leq h^{-1}(f(\overline{m},\overline{n})) \\ &\leq H^{-1}(\Phi_H^{-1}[\Phi_H(\tfrac{k}{h^{-1}(k)} + A(\overline{m},\overline{n})) + B(\overline{m},\overline{n})]). \end{aligned}$$

Since $(\overline{m},\overline{n}) \in \Omega_{(m_1,n_1)}$ is arbitrary, this completes the proof. □

Remark 6.1.4

(i) When $h(s) = s^p$, $p > 1$, Theorem 6.1.5 is the discrete analogue of Theorem 2.2 in [140].
(ii) Similar to the previous remark, in many cases $\Phi(+\infty) = +\infty$ and in these cases, inequality (6.1.42) holds for all $(m,n) \in \Omega$.

In case when Ω degenerates to a 1-dimensional lattice, Theorem 6.1.5 takes the following simpler form which is a generalization of Corollary 6.1.3 in [140] and a result of Pachpatte in [500].

Corollary 6.1.6 (The Cheung-Ma-Josip Inequality [146]) *Suppose $u \in F_+(I)$. If $k > 0$ is a constant and $a, b \in F_+(I)$, $\varphi, h \in C(\mathbb{R}_+, \mathbb{R}_+)$ are functions satisfying*

(i) φ is non-decreasing;
(ii) for any $m \in I$,

$$h(u(m)) \leq k + \sum_{s=m_0}^{m-1} a(s)u(s) + \sum_{t=m_0}^{m-1} b(s)u(s)\varphi(u(s)),$$

then for all $m \in [m_0, m_1] \cap I$,

$$u(m) \leq H^{-1}\left\{\Phi_H^{-1}\left[\Phi_H\left(\frac{k}{h^{-1}(k)} + \sum_{s=m_0}^{m-1} a(s)\right) + \sum_{s=m_0}^{m-1} b(s)\right]\right\}$$

where $m_1 \in I$ is chosen such that $\Phi_H\left(\frac{k}{h^{-1}(k)} + \sum_{s=m_0}^{m-1} a(s)\right) + \sum_{s=m_0}^{m-1} b(s) \in$ Dom (Φ_H^{-1}) for all $m \in [m_0, m_1] \cap I$.

Proof It is analogous to that of Corollary 6.1.2 by applying Theorem 6.1.5. □

Theorem 6.1.5 can easily be applied to generate other useful discrete inequalities in many concrete situations. For example, we have the next result.

Theorem 6.1.6 (The Cheung-Ma-Josip Inequality [146]) *Suppose $u \in F_+(\Omega)$. Let $k > 0$ be a constant, $a, b_i \in F_+(\Omega), i = 1, 2, \cdots, l$, where $l \geq 1$ is a positive integer, $h \in C(\mathbb{R}_+, \mathbb{R}_+)$ be defined as in Theorem 6.1.4, and $\varphi_i, Q \in C(\mathbb{R}_+, \mathbb{R}_+)$ be functions satisfying*

(i) φ is non-decreasing;
(ii) h, Q and $\hat{H} := \frac{h \circ Q^{-1}}{t}$ are strictly increasing with both $H(t) := h \circ Q^{-1}$ and $\hat{H}(t)$ tend to $+\infty$ as $t \to +\infty$;
(iii) for any $(m, n) \in \Omega$,

$$h(u(m,n)) \leq k + \sum_{s=m_0}^{m-1}\sum_{t=n_0}^{n-1} a(s,t)Q(u(s,t)) + \sum_{s=m_0}^{m-1}\sum_{t=n_0}^{n-1} Q(u(s,t)) \sum_{i=1}^{l} b_i(s,t)\varphi_i(u(s,t)). \tag{6.1.43}$$

Let $\varphi^* = \max\{\varphi_i : i = 1, 2, \cdots, l\}$, then for all $(m, n) \in \Omega_{(m_1,n_1)}$,

$$u(m,n) \leq \hat{H}^{-1}\left(\hat{\Phi}_{\hat{H}}^{-1}\left[\hat{\Phi}_{\hat{H}}\left(\frac{k}{h^{-1}(k)} + A(m,n)\right) + \hat{B}(m,n)\right]\right) \tag{6.1.44}$$

where

$$\hat{\Phi}_{\hat{H}}(r) = \int_1^r \frac{ds}{\hat{\varphi} \circ \hat{H}^{-1}(s)}, \quad r > 0,$$

$\hat{\varphi} = \varphi^* \circ Q^{-1}$, $A(m, n)$ is defined as in Theorem 6.1.5,

$$\hat{B}(m,n) := \sum_{s=m_0}^{m-1}\sum_{t=n_0}^{n-1}\sum_{i=1}^{l} b_i(s,t),$$

and $(m_1, n_1) \in \Omega$ is chosen such that $\hat{\Phi}_{\hat{H}}\left(\frac{k}{h^{-1}(k)} + A(m,n)\right) + \hat{B}(m,n) \in$ Dom $(\hat{\Phi}_{\hat{H}}^{-1})$ for all $(m, n) \in \Omega_{(m_1,n_1)}$.

Proof Let $v(s,t) = Q(u(s,t))$. Then (6.1.43) becomes

$$h \circ Q^{-1}(v(m,n)) \leq k + \sum_{s=m_0}^{m-1}\sum_{t=n_0}^{n-1} a(s,t)v(s,t) + \sum_{s=m_0}^{m-1}\sum_{t=n_0}^{n-1}\left(\sum_{i=1}^{l} b_i(s,t)\right) v(s,t)\varphi^* \circ Q^{-1}(v(s,t))$$

or

$$H(v(m,n)) \leq k + \sum_{s=m_0}^{m-1}\sum_{t=n_0}^{n-1} a(s,t)v(s,t)$$
$$+ \sum_{s=m_0}^{m-1}\sum_{t=n_0}^{n-1}\left(\sum_{i=1}^{l} b_i(s,t)\right) v(s,t)\hat{\varphi}(v(s,t)). \tag{6.1.45}$$

Thus (6.1.44) follows immediately by applying Theorem 6.1.5 to (6.1.45). □

Corollary 6.1.7 (The Cheung-Ma-Josip Inequality [146]) *Suppose $u \in F_+(\Omega)$. If $k > 0$, $p > q > 0$ are constants and $a, b \in F_+(\Omega)$, $\varphi \in C(\mathbb{R}_+, \mathbb{R}_+)$ are functions satisfying*

(i) φ is non-decreasing;
(ii) for any $(m,n) \in \Omega$,

$$u^p(m,n) \leq k + \sum_{s=m_0}^{m-1}\sum_{t=n_0}^{n-1} a(s,t)u^q(s,t)$$
$$+ \sum_{s=m_0}^{m-1}\sum_{t=n_0}^{n-1} b(s,t)u^q(s,t)\varphi(u(s,t)), \tag{6.1.46}$$

then for all $(m,n) \in \Omega_{(m_1,n_1)}$,

$$u(m,n) \leq \left\{\Phi^{-1}[\Phi(k^{1-\frac{q}{p}} + A(m,n)) + B(m,n)]\right\}^{\frac{1}{p-q}} \tag{6.1.47}$$

where

$$\Phi(r) := \int_1^r \frac{ds}{\varphi(s^{\frac{1}{p-q}})}, \quad r > 0,$$

and $A(m,n), B(m,n)$ are defined as in Theorem 6.1.5, and $(m_1,n_1) \in \Omega$ is chosen such that $\Phi(k^{1-\frac{q}{p}} + A(m,n)) + B(m,n) \in Dom\ (\Phi^{-1})$ for all $(m,n) \in \Omega_{(m_1,n_1)}$.

Proof This follows immediately from Theorem 6.1.6 by taking $h(u) = u^p$, $Q(u) = u^q$, $p > q > 0$, and $l = 1$. □

Remark 6.1.5

(i) In case $q = 1$, $h(x) = x^p$, Corollary 6.1.7 reduces to Theorem 6.1.5. Furthermore, if $p = 2$, this further reduces to Theorem 2.2 in [141].
(ii) In case $a(s,t) \equiv 0$, Corollary 6.1.7 reduces to Theorem 2.3 in [363].

The following corollary is an important special case of Corollary 6.1.7.

Corollary 6.1.8 (The Cheung-Ma-Josip Inequality [146]) *Suppose $u \in F_+(\Omega)$. If $k > 0$, $p > 1$ are constants and $a, b \in F_+(\Omega)$, $\varphi \in C(\mathbb{R}_+, \mathbb{R}_+)$ are functions satisfying*

(i) φ is non-decreasing;
(ii) for any $(m,n) \in \Omega$,

$$u^p(m,n) \leq k + \sum_{s=m_0}^{m-1}\sum_{t=n_0}^{n-1} a(s,t)u^{p-1}(s,t) + \sum_{s=m_0}^{m-1}\sum_{t=n_0}^{n-1} b(s,t)u^{p-1}(s,t)\varphi(u(s,t)),$$

then for all $(m,n) \in \Omega_{(m_1,n_1)}$,

$$u(m,n) \leq \Phi_{id}^{-1}[\Phi_{id}(k^{\frac{1}{p}} + A(m,n)) + B(m,n)]$$

where $A(m,n), B(m,n)$ are defined as in Theorem 6.1.5, and $(m_1,n_1) \in \Omega$ is chosen such that $\Phi_{id}(k^{\frac{1}{p}} + A(m,n)) + B(m,n) \in Dom\ (\Phi_{id}^{-1})$ for all $(m,n) \in \Omega_{(m_1,n_1)}$.

Proof The assertion follows immediately from Corollary 6.1.7 by taking $q = p - 1 > 0$. □

In particular, we also have the following useful consequence.

Corollary 6.1.9 (The Cheung-Ma-Josip Inequality [146]) *Suppose $u \in F_+(\Omega)$. If $k > 0$, $p > 1$ are constants and $a, b \in F_+(\Omega)$ are functions such that for any $(m,n) \in \Omega$,*

$$u^p(m,n) \leq k + \sum_{s=m_0}^{m-1}\sum_{t=n_0}^{n-1} a(s,t)u^{p-1}(s,t) + \sum_{s=m_0}^{m-1}\sum_{t=n_0}^{n-1} b(s,t)u^p(s,t),$$

then for all $(m,n) \in \Omega_{(m_1,n_1)}$,

$$u(m,n) \leq \left(k^{\frac{1}{p}} + A(m,n)\right)\exp\left(B(m,n)\right)$$

where $A(m,n), B(m,n)$ are defined as in Theorem 6.1.5.

Proof Let φ be the identity mapping of $\mathbb{R}_+$ onto itself. Then all conditions of Corollary 6.1.8 are satisfied. Note that in this case $\Phi_{id} = \ln$ and so $\Phi_{id}^{-1} = \exp$. In particular, Φ_{id}^{-1} is defined everywhere on $\mathbb{R}$. By Corollary 6.1.8, we have for all

$(m,n) \in \Omega$,

$$u(m,n) \le \exp\left[\ln(k^{\frac{1}{p}} + A(m,n)) + B(m,n)\right] = \left[k^{\frac{1}{p}} + A(m,n)\right]\exp(B(m,n)).$$

□

In case when Ω degenerates into a 1-dimensional lattice, Corollary 6.1.9 takes the following simpler form which generalizes another result of Pachpatte in [500].

Corollary 6.1.10 (The Cheung-Ma-Josip Inequality [146]) *Suppose $u \in F_+(I)$. If $k > 0,\ p > 1$ are constants and $a, b \in F_0(I)$ are functions such that for any $m \in I$,*

$$u^p(m) \le k + \sum_{s=m_0}^{m-1} a(s)u^{p-1}(s) + \sum_{s=m_0}^{m-1} b(s)u^p(s),$$

then for all $m \in I$,

$$u(m) \le \left(k^{\frac{1}{p}} + \sum_{s=m_0}^{m-1} a(s)\right)\prod_{s=m_0}^{m-1}\exp(b(s)).$$

Proof Analogous to that of Corollary 6.1.2, we may apply Corollary 6.1.9. □

Theorem 6.1.7 (The Pachpatte Inequality [493]) *Let $u(m,n)$ and $f(m,n)$ be functions defined on $\mathbb{Z} \times \mathbb{Z}$ into $\mathbb{R}_+$ and $c \ge 0$ be a constant. If*

$$u^2(m,n) \le c + \sum_{s=m+1}^{+\infty}\sum_{t=n+1}^{+\infty} f(s,t)u(s,t), \tag{6.1.48}$$

then

$$u(m,n) \le \sqrt{c} + \frac{1}{2}\sum_{s=m+1}^{+\infty}\sum_{t=n+1}^{+\infty} f(s,t). \tag{6.1.49}$$

Proof We first assume that $c > 0$ and define a function $z(m,n)$ by the right member of (6.1.48), then

$$\begin{aligned} &[z(m,n) - z(m+1,n)] - [z(m,n+1) - z(m+1,n+1)] \\ &= f(m+1,n+1)u(m+1,n+1) \\ &\le f(m+1,n+1)\sqrt{z(m+1,n+1)}. \end{aligned} \tag{6.1.50}$$

Here we have used the fact that $u(m+1,n+1) \leq \sqrt{z(m+1,n+1)}$ to get (6.1.50). By using $\sqrt{z(m,n)} > 0$, $\sqrt{z(m,n+1)} \leq \sqrt{z(m,n)}$ $\sqrt{z(m+1,n+1)} \leq \sqrt{z(m+1,n)}$
$\sqrt{z(m+1,n+1)} \leq \sqrt{z(m,n+1)}$ for all $m,\ nin\mathbb{Z}$, we observe that

$$\left[\sqrt{z(m,n)} - \sqrt{z(m+1,n)}\right] = \frac{[z(m,n) - z(m+1,n)]}{\left[\sqrt{z(m,n)} + \sqrt{z(m+1,n)}\right]}, \tag{6.1.51}$$

$$\begin{aligned}
&\left[\sqrt{z(m,n)} - \sqrt{z(m+1,n)}\right] - \left[\sqrt{z(m,n+1)} - \sqrt{z(m+1,n+1)}\right] \\
&= \frac{\left[z(m,n) - z(m+1,n)\right]}{\left[\sqrt{z(m,n)} + \sqrt{z(m+1,n)}\right]} - \frac{\left[z(m,n+1) - z(m+1,n+1)\right]}{\left[\sqrt{z(m,n+1)} + \sqrt{z(m+1,n+1)}\right]} \\
&\leq \frac{[z(m,n) - z(m+1,n)]}{\left[\sqrt{z(m,n+1)} + \sqrt{z(m+1,n+1)}\right]} - \frac{\left[z(m,n+1) - z(m+1,n+1)\right]}{\left[\sqrt{z(m,n+1)} + \sqrt{z(m+1,n+1)}\right]} \\
&= \frac{[z(m,n) - z(m+1,n)] - [z(m,n+1) - z(m+1,n+1)]}{\left[\sqrt{z(m,n+1)} + \sqrt{z(m+1,n+1)}\right]} \\
&\leq \frac{[z(m,n) - z(m+1,n)] - [z(m,n+1) - z(m+1,n+1)]}{\left[\sqrt{z(m+1,n+1)} + \sqrt{z(m+1,n+1)}\right]}.
\end{aligned} \tag{6.1.52}$$

Using (6.1.50) in (6.1.52), we have

$$\begin{aligned}
&\left[\sqrt{z(m,n)} - \sqrt{z(m+1,n)}\right] - \left[\sqrt{z(m,n+1)} - \sqrt{z(m+1,n+1)}\right] \\
&\leq \frac{1}{2} f(m+1,n+1).
\end{aligned} \tag{6.1.53}$$

Now keeping m fixed in (6.1.53), setting $n = t$ and summing over $t = n, n+1, \ldots, q-1$, we obtain

$$\begin{aligned}
&\left[\sqrt{z(m,n)} - \sqrt{z(m+1,n)}\right] - \left[\sqrt{z(m,n+1)} - \sqrt{z(m+1,n+1)}\right] \\
&\leq \frac{1}{2} \sum_{t=n+1}^{q} f(m+1,t).
\end{aligned} \tag{6.1.54}$$

Noting that $\lim_{q\to\infty} \sqrt{z(m,q)} = \lim_{q\to+\infty} \sqrt{z(m+1,q)} = \sqrt{c}$, and by letting $q \to +\infty$ in (6.1.54), we get

$$\sqrt{z(m,n)} - \sqrt{z(m+1,n)} \le \frac{1}{2} \sum_{t=n+1}^{q} f(m+1,t). \tag{6.1.55}$$

Keeping n fixed in (6.1.55), setting $m = s$ and summing over $s = m, m+1, \ldots, p-1$, we obtain

$$\sqrt{z(m,n)} - \sqrt{z(p,n)} \le \frac{1}{2} \sum_{s=m+1}^{p} \sum_{t=n+1}^{q} f(s,t). \tag{6.1.56}$$

Noting that $\lim_{p\to+\infty} \sqrt{z(p,n)} = \sqrt{c}$, the required inequality in (6.1.49) follows by letting $p \to +\infty$ in (6.1.56) and using the fact that $u(m,n) \le \sqrt{z(m,n)}$ for $m, n \in$. The proof of the case when $c = 0$ can be completed as in the proof of Theorem 2.3.1. □

Theorem 6.1.8 (The Pachpatte Inequality [493]) *Let $u(m,n)$ be a function defined on $\mathbb{Z} \times \mathbb{Z}$ into $\mathbb{R}_1$ and $f(m,n)$ be a function as defined in Theorem 6.1.7 and $c \ge 1$ be a constant. If*

$$u(m,n) \le c + \sum_{s=m+1}^{+\infty} \sum_{t=n+1}^{+\infty} f(s,t)u(s,t)\log u(s,t), \tag{6.1.57}$$

then

$$u(m,n) \le c^{\prod_{s=m+1}^{\infty}[1+\sum_{t=n+1}^{\infty} f(s,t)]}. \tag{6.1.58}$$

Proof Define a function $z(m,n)$ by the right-hand side of (6.1.57). Then

$$\begin{aligned} &[z(m,n) - z(m+1,n)] - [z(m,n+1) - z(m+1,n+1)] \\ &= f(m+1,n+1)u(m+1,n+1)\log u(m+1,n+1) \\ &\le f(m+1,n+1)u(m+1,n+1)\log z(m+1,n+1). \end{aligned} \tag{6.1.59}$$

Here we have used the fact that $u(m+1,n+1) \le z(m+1,n+1)$ to get the last inequality in (6.1.59). From the definition of $z(m,n)$, we see that $z(m+1,n+1) \le u(m+1,n)$ for all $m, n \in \mathbb{Z}$. Using this fact, we observe from (6.1.59) that

$$\begin{aligned} &\frac{[z(m,n) - z(m+1,n)]}{z(m+1,n)} - \frac{[z(m,n+1) - z(m+1,n+1)]}{z(m+1,n+1)} \\ &\le f(m+1,n+1)\log z(m+1,n+1). \end{aligned} \tag{6.1.60}$$

Now keeping m fixed in (6.1.60), setting $n = t$ and summing $t = n, n+1, \ldots, q-1$, we obtain

$$\frac{[z(m,n) - z(m+1,n)]}{z(m+1,n)} - \frac{[z(m,q) - z(m+1,q)]}{z(m+1,q)}$$
$$\leq \sum_{t=n+1}^{q} f(m+1,t) \log z(m+1,t). \tag{6.1.61}$$

Noting that $\lim_{q\to+\infty} z(m,q) = \lim_{q\to+\infty} z(m+1,q) = c$ and by letting $q \to +\infty$ in (6.1.61), we get

$$z(m,n) \leq [1 + \sum_{t=n+1}^{+\infty} f(m+1,t) \log z(m+1,t)]z(m+1,n). \tag{6.1.62}$$

By keeping n fixed in (6.1.62) and setting $m = s$ and then substituting $s = m, m+1, \ldots, p-1$ successively, we obtain

$$z(m,n) \leq \prod_{s=m+1}^{p} \left[1 + \sum_{t=n+1}^{+\infty} f(s,t) \log z(s,t)\right] z(p,n). \tag{6.1.63}$$

Noting that $\lim_{p\to+\infty} z(p,n) = c$ and by letting $p \to +\infty$ in (6.1.63), we get

$$z(m,n) \leq c \prod_{s=m+1}^{+\infty} [1 + \sum_{t=n+1}^{+\infty} f(s,t) \log z(s,t)]$$
$$\leq c \exp\left\{ \prod_{s=m+1}^{+\infty} \sum_{t=n+1}^{+\infty} f(s,t) \log z(s,t) \right\}. \tag{6.1.64}$$

From (6.1.64) we derive

$$\log z(m,n) \leq \log c + \sum_{s=m+1}^{+\infty} \sum_{t=n+1}^{+\infty} f(s,t) \log z(s,t) \tag{6.1.65}$$

From (6.1.65) and by following exactly the same arguments as above with suitable changes up to the inequality (6.1.64), we obtain

$$\log z(m,n) \leq \log c\{ \prod_{s=m+1}^{+\infty} [1 + \sum_{t=n+1}^{+\infty} f(s,t)]\}$$
$$= \log c^{\prod_{s=m+1}^{+\infty}[1+\sum_{t=n+1}^{+\infty} f(s,t)]} \tag{6.1.66}$$

from (6.1.66) it follows

$$z(m,n) \le c^{\prod_{s=m+1}^{+\infty}[1+\sum_{t=n+1}^{+\infty} f(s,t)]}. \tag{6.1.67}$$

Using (6.1.67) in (6.1.57), we get the required inequality in (6.1.58) and the proof is thus complete. □

Theorem 6.1.9 (The Pachpatte Inequality [493]) *Let $u(m,n)$, $f(m,n)$ and $g(m,n)$ be functions defined on $\mathbb{Z} \times \mathbb{Z}$ into $\mathbb{R}_+$, and $c \ge 0$ be a constant. If*

$$u^2(m,n) \le c + \sum_{s=m+1}^{\infty} \sum_{t=n+1}^{\infty} f(s,t)u(s,t) \times \left[u(s,t) + \sum_{k=s+1}^{\infty} \sum_{r=t+1}^{\infty} g(k,r)u(k,r)\right], \tag{6.1.68}$$

then

$$u(m,n) \le \sqrt{c}\left[1 + \sum_{s=m+1}^{+\infty} \sum_{t=n+1}^{+\infty} \frac{1}{2} f(s,t) \prod_{k=s+1}^{+\infty} \left[1 + \sum_{r=t+1}^{+\infty} \frac{1}{2} f(k,r)g(k,r)\right]\right] \tag{6.1.69}$$

Proof The proof is similar to the proofs of Theorems 2.3.3 and 6.1.7–6.1.8 given above with suitable modifications and hence we omit the details. □

We introduce in the following results some new finite difference inequalities in two independent variables, which are due to Pachpatte [514].

Theorem 6.1.10 (The Pachpatte Inequality [514]) *Let $u(m,n), f(m,n), g(m,n)$, $h(m,n)$ be real-valued non-negative functions defined for all $m,n \in \mathbb{N}_0$ and $u(m,n) \ge u_0 > 0, u_0$ is a real constant. Let $W(r)$ be a real-valued continuous, positive, strictly non-decreasing, sub-additive, and sub-multiplicative function on $I = [u_0, +\infty)$ and let $H(r)$ be a real-valued, continuous, positive, and non-decreasing function on I.*

(b_1) *If for all $m,n \in \mathbb{N}_0$,*

$$u(m,n) \le f(m,n) + g(m,n)H\left(\sum_{s=m+1}^{+\infty} \sum_{t=n+1}^{+\infty} h(s,t)W(u(s,t))\right), \tag{6.1.70}$$

then for all $0 \le m \le m_1, 0 \le n \le n_1, m, m_1, n, n_1 \in \mathbb{N}_0$,

$$u(m,n) \le f(m,n) + g(m,n)H\left(G^{-1}\left[G\left(\sum_{s=1}^{+\infty}\sum_{t=1}^{+\infty} h(s,t)W(f(s,t))\right)\right.\right.$$
$$\left.\left. + \sum_{s=m+1}^{+\infty}\sum_{t=n+1}^{+\infty} h(s,t)W(g(s,t))\right]\right), \tag{6.1.71}$$

where

$$G(r) = \int_{r_0}^{r} \frac{ds}{W(H(s))}, \qquad r \ge r_0 \ge u_0, \tag{6.1.72}$$

G^{-1} *is the inverse function of* G *and for* $0 \le m \le m_1, 0 \le n \le n_1, m, m_1, n, n_1 \in \mathbb{N}_0$,

$$G\left(\sum_{s=1}^{+\infty}\sum_{t=1}^{+\infty} h(s,t)W(f(s,t))\right) + \sum_{s=m+1}^{+\infty}\sum_{t=n+1}^{+\infty} h(s,t)W(g(s,t)) \in Dom\,(G^{-1}).$$

(b_2) *If for all* $m, n \in \mathbb{N}_0$,

$$u(m,n) \le f(m,n) + g(m,n)H\left(\sum_{s=0}^{m-1}\sum_{t=n+1}^{+\infty} h(s,t)W(u(s,t))\right), \tag{6.1.73}$$

then for all $0 \le m \le m_2, 0 \le n \le n_2, m, m_2, n, n_2 \in \mathbb{N}_0$,

$$u(m,n) \le f(m,n) + g(m,n)H\left(G^{-1}\left[G\left(\sum_{s=0}^{+\infty}\sum_{t=1}^{+\infty} h(s,t)W(f(s,t))\right)\right.\right.$$
$$\left.\left. + \sum_{s=0}^{m-1}\sum_{t=n+1}^{+\infty} h(s,t)W(g(s,t))\right]\right), \tag{6.1.74}$$

where G, G^{-1} *are defined in part* (b_1), *and for all* $0 \le m \le m_2, 0 \le n \le n_2, m, m_2, n, n_2 \in \mathbb{N}_0$,

$$G\left(\sum_{s=0}^{+\infty}\sum_{t=1}^{+\infty} h(s,t)W(f(s,t))\right) + \sum_{s=0}^{m-1}\sum_{t=n+1}^{+\infty} h(s,t)W(g(s,t)) \in Dom\,(G^{-1}).$$

Proof We only give the details of the proofs of (b_1). The proof of (b_2) can be completed similarly with suitable modifications.

(b_1) Define a function $z(m,n)$ by

$$z(m,n) = \sum_{s=m+1}^{+\infty} \sum_{t=n+1}^{+\infty} h(s,t)W(u(s,t)), \tag{6.1.75}$$

then from (6.1.70) it follows

$$u(m,n) \leq f(m,n) + g(m,n)H(z(m,n)). \tag{6.1.76}$$

From (6.1.75) and (6.1.76), we derive

$$\begin{aligned} z(m,n) &\leq \sum_{s=m+1}^{+\infty} \sum_{t=n+1}^{+\infty} h(s,t)W(f(s,t) + g(s,t)H(z(s,t))) \\ &\leq \sum_{s=m+1}^{+\infty} \sum_{t=n+1}^{+\infty} h(s,t)[W(f(s,t)) + W(g(s,t))W(H(z(s,t)))] \\ &\leq \sum_{s=1}^{+\infty} \sum_{t=1}^{+\infty} h(s,t)W(f(s,t)) \\ &\quad + \sum_{s=m+1}^{+\infty} \sum_{t=n+1}^{+\infty} h(s,t)W(g(s,t))W(H(z(s,t))). \end{aligned} \tag{6.1.77}$$

Define a function $v(m,n)$ by the right-hand side of (6.1.77). Then, $z(m,n) \leq v(m,n)$ and

$$\begin{aligned} &[v(m,n) - v(m+1,n)] - [v(m,n+1) - v(m+1,n+1)] \\ &= h(m+1,n+1)W(g(m+1,n+1))W(H(z(m+1,n+1))) \\ &\leq h(m+1,n+1)W(g(m+1,n+1))W(H(v(m+1,n+1))). \end{aligned} \tag{6.1.78}$$

From (6.1.78) and the fact that $v(m+1,n+1) \leq v(m+1,n)$, we deduce

$$\begin{aligned} &\frac{[v(m,n) - v(m+1,n)]}{W(H(v(m+1,n)))} - \frac{[v(m,n+1) - v(m+1,n+1)]}{W(H(v(m+1,n+1)))} \\ &\leq h(m+1,n+1)W(g(m+1,n+1)). \end{aligned} \tag{6.1.79}$$

Keeping m fixed in (6.1.79), substituting $n = t$, and taking the sum over $t = n, n+1, \ldots, q-1$ ($q \geq n+1$ is arbitrary in $\mathbb{N}_0$), we can obtain

$$\frac{[v(m,n) - v(m+1,n)]}{W(H(v(m+1,n)))} - \frac{[v(m,q) - v(m+1,q)]}{W(H(v(m+1,q)))}$$
$$\leq \sum_{t=n+1}^{q} h(m+1,t)W(g(m+1,t)). \tag{6.1.80}$$

Noting that $\lim_{q\to+\infty} v(m,q) = \lim_{q\to+\infty} v(m+1,q) = \sum_{s=1}^{+\infty}\sum_{t=1}^{+\infty} h(s,t)$ $W(f(s,t))$ and by letting $q \to +\infty$ in (6.1.80), we have

$$\frac{[v(m,n) - v(m+1,n)]}{W(H(v(m+1,n)))} \leq \sum_{t=n+1}^{+\infty} h(m+1,t)W(g(m+1,t)). \tag{6.1.81}$$

From (6.1.72) and (6.1.80), it follows

$$\begin{aligned} G(v(m,n)) - G(v(m+1,n)) &= \int_{v(m+1,n)}^{v(m,n)} \frac{ds}{W(H(s))} \\ &\leq \frac{[v(m,n) - v(m+1,n)]}{W(H(v(m+1,n)))} \\ &\leq \sum_{t=n+1}^{+\infty} h(m+1,t)W(g(m+1,t)). \end{aligned} \tag{6.1.82}$$

Now, keeping n fixed in (6.1.82), substituting $m = s$, and taking the sum over $s = m, m+1, \ldots, p-1$, ($p \geq m+1$ is arbitrary in $\mathbb{N}_0$), we arrive at

$$G(v(m,n)) - G(v(p,n)) \leq \sum_{s=m+1}^{p}\sum_{t=n+1}^{+\infty} h(s,t)W(g(s,t)). \tag{6.1.83}$$

Noting that $\lim_{q\to+\infty} v(p,n) = \sum_{s=1}^{+\infty}\sum_{t=1}^{+\infty} h(s,t)W(f(s,t))$ and by taking $p \to +\infty$ in (6.1.83), we get

$$\begin{aligned} v(m,n) \ &\leq G^{-1}[G(\textstyle\sum_{s=1}^{+\infty}\sum_{t=1}^{+\infty} h(s,t)W(f(s,t))) \\ &\quad + \textstyle\sum_{s=m+1}^{+\infty}\sum_{t=n+1}^{+\infty} h(s,t)W(g(s,t))]. \end{aligned} \tag{6.1.84}$$

The required inequality in (6.1.71) follows from the fact that $z(m,n) \leq v(m,n)$, (6.1.84) and (6.1.76). The sub-domain $0 < m \leq m_1, 0 \leq n \leq n_1$ is obvious. □

Theorem 6.1.11 (The Pachpatte Inequality [516]) *Let $u(m,n), a(m,n), b(m,n)$ be real-valued non-negative functions defined for all $m, n \in \mathbb{N}_0$ and let $L : \mathbb{N}_0^2 \times \mathbb{R}_+ \to \mathbb{R}_+$ be a function which satisfies the condition: for all $u \geq v \geq 0$,*

$$0 \leq L(m,n,u) - L(m,n,v) \leq M(m,n,v)(u-v)$$

where $M(m,n,v)$ is a real-valued non-negative function defined for all $m, n \in \mathbb{N}_0, v \in \mathbb{R}_+$.

(c_1) *If for all $m, n \in \mathbb{N}_0$,*

$$u(m,n) \leq a(m,n) + b(m,n) \sum_{s=m+1}^{+\infty} \sum_{t=n+1}^{+\infty} L(s,t,u(s,t)), \tag{6.1.85}$$

then for all $m, n \in \mathbb{N}_0$,

$$u(m,n) \leq a(m,n)+b(m,n)e(m,n) \prod_{s=m+1}^{+\infty} \left[1 + \sum_{t=n+1}^{+\infty} M(s,t,a(s,t))b(s,t)\right], \tag{6.1.86}$$

where for all $m, n \in \mathbb{N}_0$,

$$e(m,n) = \sum_{s=m+1}^{+\infty} \sum_{t=n+1}^{+\infty} L(s,t,a(s,t)). \tag{6.1.87}$$

(c_2) *If for all $m, n \in \mathbb{N}_0$,*

$$u(m,n) \leq a(m,n) + b(m,n) \sum_{s=0}^{m-1} \sum_{t=n+1}^{+\infty} L(s,t,u(s,t)), \tag{6.1.88}$$

then for all $m, n \in \mathbb{N}_0$,

$$u(m,n) \leq a(m,n)+b(m,n)\overline{e}(m,n) \prod_{s=0}^{m-1} \left[1 + \sum_{t=n+1}^{+\infty} M(s,t,a(s,t))b(s,t)\right], \tag{6.1.89}$$

where for all $m, n \in \mathbb{N}_0$,

$$\overline{e}(m,n) = \sum_{s=0}^{m-1} \sum_{t=n+1}^{+\infty} L(s,t,a(s,t)). \tag{6.1.90}$$

Proof (c_1) Define a function $z(m,n)$ by

$$z(m,n) = \sum_{s=m+1}^{+\infty} \sum_{t=n+1}^{+\infty} L(s,t,u(s,t)). \tag{6.1.91}$$

Then, from (6.1.85) we infer

$$u(m,n) \leq a(m,n) + b(m,n)z(m,n). \tag{6.1.92}$$

From (6.1.91), (6.1.92) and the hypotheses on L, it follows

$$\begin{aligned} z(m,n) \leq & \sum_{s=m+1}^{+\infty} \sum_{t=n+1}^{+\infty} \Bigg[L(s,t,a(s,t) + b(s,t)z(s,t)) \\ & - L(s,t,a(s,t) + L(s,t,a(s,t)) \Bigg] \\ \leq & e(m,n) + \sum_{s=m+1}^{+\infty} \sum_{t=n+1}^{+\infty} M(s,t,a(s,t))b(s,t)z(s,t), \end{aligned} \tag{6.1.93}$$

where $e(m,n)$ is defined by (6.1.87). Clearly $e(m,n)$ is real-valued non-negative and non-increasing in each variable $m,n \in \mathbb{N}_0$. Now, an application of part (a_1) in Theorem 6.1.10 to (6.1.93) yields

$$z(m,n) \leq e(m,n) \prod_{s=m+1}^{+\infty} [1 + \sum_{t=n+1}^{+\infty} M(s,t,a(s,t))b(s,t)]. \tag{6.1.94}$$

The desired inequality in (6.1.86) follows from (6.1.92) and (6.1.94). □

Theorem 6.1.12 (The Pachpatte Inequality [516]) *Let $u(m,n), a(m,n), b(m,n)$ be real-valued non-negative functions defined for all $m,n \in \mathbb{N}_0$ and let $L : \mathbb{N}_0^2 \times \mathbb{R}_+ \to \mathbb{R}_+$ be a function which satisfies the condition: for all $u \geq v \geq 0$,*

$$0 \leq L(m,n,u) - L(m,n,v) \leq M(m,n,v)\phi^{-1}(u-v)$$

where $M(m,n,v)$ is defined as in Theorem 6.1.8, let $\phi : \mathbb{R}_+ \to \mathbb{R}_+$ be a continuous and strictly increasing function with $\phi(0) = 0, \phi^{-1}$ is the inverse function of ϕ and for all $u, v \in \mathbb{R}_+$,

$$\phi^{-1}(uv) \leq \phi^{-1}(u)\phi^{-1}(v).$$

(d_1) *If for all* $m, n \in \mathbb{N}_0$,

$$u(m,n) \leq a(m,n) + b(m,n)\phi\left(\sum_{s=m+1}^{+\infty}\sum_{t=n+1}^{+\infty} L(s,t,u(s,t))\right), \tag{6.1.95}$$

then for all $m, n \in \mathbb{N}_0$,

$$u(m,n) \leq a(m,n) + b(m,n)\phi \times \left(e(m,n)\prod_{s=m+1}^{+\infty}\left[1 + \sum_{t=n+1}^{+\infty} M(s,t,a(s,t))\phi^{-1}(b(s,t))\right]\right), \tag{6.1.96}$$

where $e(m,n)$ *is defined by (6.1.85).*

(d_2) *If for all* $m, n \in \mathbb{N}_0$,

$$u(m,n) \leq a(m,n) + b(m,n)\phi\left(\sum_{s=0}^{m-1}\sum_{t=n+1}^{+\infty} L(s,t,u(s,t))\right), \tag{6.1.97}$$

then for all $m, n \in \mathbb{N}_0$,

$$u(m,n) \leq a(m,n) + b(m,n)\phi \times \left(\bar{e}(m,n)\prod_{s=0}^{m-1}\left[1 + \sum_{t=n+1}^{+\infty} M(s,t,a(s,t))\phi^{-1}(b(s,t))\right]\right), \tag{6.1.98}$$

where $\bar{e}(m,n)$ *is defined by (6.1.90).*

Proof (d_1) Define a function $z(m,n)$ by (6.1.91), then from (6.1.95) we deduce

$$u(m,n) \leq a(m,n) + b(m,n)\phi(z(m,n)). \tag{6.1.99}$$

From (6.1.91), (6.1.99) and the hypotheses on L and ϕ, it follows

$$z(m,n) \leq \sum_{s=m+1}^{+\infty}\sum_{t=n+1}^{+\infty}\Big[L(s,t,a(s,t) + b(s,t)\phi(z(s,t))) - L(s,t,a(s,t) + L(s,t,a(s,t))\Big]$$

$$\leq e(m,n) + \sum_{s=m+1}^{+\infty} \sum_{t=n+1}^{+\infty} M(s,t,a(s,t))\phi^{-1}(b(s,t)\phi(z(s,t)))$$

$$\leq e(m,n) + \sum_{s=m+1}^{+\infty} \sum_{t=n+1}^{+\infty} M(s,t,a(s,t))\phi^{-1}(b(s,t))z(s,t), \tag{6.1.100}$$

where $e(m,n)$ is defined by (6.1.87). Now following the last arguments as in the proof of (c_1), we can derive the desired inequality in (6.1.96). □

The next result, due to Ma [359], is to give some explicit bounds to some new nonlinear discrete inequalities involving two-variable functions, which, on the one hand, generalizes Ou-Yang's inequality to Volterra-Fredholm, on the other hand, give a handy and effective tool for the study of quantitative properties of solutions of sum-difference equations.

For $w \in C(\mathbb{R}_+, \mathbb{R}_+)$, the function G_1 is defined as

$$G_1(v) = \int_{v_0}^{v} \frac{ds}{w(s)}, \quad v \geq v_0 > 0.$$

Theorem 6.1.13 (The Ma Inequality [359]) *Suppose that u and $a \in F_+(\Omega)$, $k \geq 0$ constant and $w \in C(\mathbb{R}_+, \mathbb{R}_+)$ is non-decreasing with $w(r) > 0$ for all $r > 0$;*

$$G_1(+\infty) = \int_{v_0}^{+\infty} \frac{ds}{w(s)} = +\infty$$

and

$$H_1(t) = G_1(2t - k) - G_1(t) \tag{6.1.101}$$

is strictly increasing for all $t \geq k$.

If $u(m,n)$ satisfies for all $(m,n) \in \Omega$,

$$u(m,n) \leq k + \sum_{s=m_0}^{m-1} \sum_{t=n_0}^{n-1} a(s,t)w(u(s,t)) + \sum_{s=m_0}^{M-1} \sum_{t=n_0}^{N-1} a(s,t)w(u(s,t)), \tag{6.1.102}$$

then for all $(m,n) \in \Omega$,

$$u(m,n) \leq G_1^{-1}\left(G_1\left[H_1^{-1}\left(\sum_{s=m_0}^{M-1} \sum_{t=n_0}^{N-1} a(s,t)\right)\right] + \sum_{s=m_0}^{m-1} \sum_{t=n_0}^{n-1} a(s,t)\right) \tag{6.1.103}$$

where G_1^{-1} and H_1^{-1} are inverse functions of G_1 and H_1, respectively.

Proof Let $k > 0$ and define

$$z(m,n) = k + \sum_{s=m_0}^{m-1}\sum_{t=n_0}^{n-1} a(s,t)w(u(s,t)) + \sum_{s=m_0}^{M-1}\sum_{t=n_0}^{N-1} a(s,t)w(u(s,t)).$$

Then we have

$$\begin{cases} u(m,n) \le z(m,n), \quad (m,n) \in \Omega, \\ z(m_0,n) = k + \sum_{s=m_0}^{M-1}\sum_{t=n_0}^{N-1} a(s,t)w(u(s,t)) \end{cases} \tag{6.1.104}$$

and

$$\begin{aligned} \Delta_1 z(m,n) &= \sum_{t=n_0}^{n-1} a(m,t)w(u(m,t)) \le \sum_{t=n_0}^{n-1} a(m,t)w(z(m,t)) \\ &\le w(z(m,t)) \sum_{t=n_0}^{n-1} a(m,t). \end{aligned}$$

Therefore, by the Mean-Value Theorem for integrals, for each $(m,n) \in \Omega$, there exists $\zeta : z(m,n) \le \zeta \le z(m+1,n)$ such that

$$\begin{aligned} \Delta_1 G_1(z(m,n)) &= G_1(z(m+1,n)) - G_1(z(m,n)) \\ &= \int_{z(m,n)}^{z(m+1,n)} \frac{ds}{w(s)} \\ &= \frac{1}{w(\zeta)} \Delta_1 z(m,n). \end{aligned}$$

Since w is non-decreasing, $w(\zeta) \ge w(z(m,n))$, we get for all $(m,n) \in \Omega$,

$$\begin{aligned} \Delta_1 G_1(z(m,n)) &\le \frac{1}{w(z(m,n))} \Delta_1 z(m,n) \\ &\le \sum_{t=n_0}^{n-1} a(m,t). \end{aligned} \tag{6.1.105}$$

Therefore,

$$\sum_{s=m_0}^{m-1} \Delta_1 G_1(z(s,n)) \le \sum_{s=m_0}^{m-1}\sum_{t=n_0}^{n-1} a(s,t).$$

On the other hand, it is easy to check that

$$\sum_{s=m_0}^{m-1} \Delta_1 G_1(z(s,n)) = G_1(z(m,n)) - G_1(z(m_0,n)),$$

which yields

$$G_1(z(m,n)) \le G_1(z(m_0,n)) + \sum_{s=m_0}^{m-1}\sum_{t=n_0}^{n-1} a(s,t).$$

Since $G_1^{-1}$is increasing, the above inequality yields for all $(m,n) \in \Omega$,

$$z(m,n) \le G_1^{-1}\left[G_1(z(m_0,n)) + \sum_{s=m_0}^{m-1}\sum_{t=n_0}^{n-1} a(s,t)\right]. \tag{6.1.106}$$

Form the last inequality, we observe that

$$\begin{aligned} 2z(m_0,n) - k &= k + 2\sum_{s=m_0}^{M-1}\sum_{t=n_0}^{N-1} a(s,t)w(u(s,t)) \\ &= z(M,N) \le G_1^{-1}\left[G_1(z(m_0,N)) + \sum_{s=m_0}^{M-1}\sum_{t=n_0}^{N-1} a(s,t)\right] \\ &= G_1^{-1}\left[G_1(z(m_0,n)) + \sum_{s=m_0}^{M-1}\sum_{t=n_0}^{N-1} a(s,t)\right], \end{aligned}$$

or

$$G_1(2z(m_0,n) - k) - G_1(z(m_0,n)) \le \sum_{s=m_0}^{M-1}\sum_{t=n_0}^{N-1} a(s,t). \tag{6.1.107}$$

Since $H_1(t) = G_1(2t-k) - G_1(t)$ is increasing for all $t > k$, $H_1(t)$ has the inverse function $H_1^{-1}(t)$ and then from the last inequality, we derive

$$z(m_0,n) \le H_1^{-1}\left(\sum_{s=m_0}^{M-1}\sum_{t=n_0}^{N-1} a(s,t\right). \tag{6.1.108}$$

Substituting (6.1.108) into (6.1.106) and combining with (6.1.105), we can obtain the desired inequality (6.1.103). If $k = 0$, we carry out the above procedure with $\varepsilon > 0$ instead of k and subsequently let $\varepsilon \to 0$. □

Theorem 6.1.14 (The Ma Inequality [359]) *Let $u(m,n)$, $a(m,n)$, $w(u), G_1(u)$ and k be as in Theorem 6.1.13. If $u(m,n)$ satisfies (6.1.102) for all $(m,n) \in \Omega$, and*

$$\hat{H}_1(t) = G_1(2t-k) - G_1(t) - \sum_{s=m_0}^{M-1}\sum_{t=n_0}^{N-1} a(s,t)$$

is increasing and $\hat{H}_1(t) = 0$ has a solution c_1 for all $t \geq k$, then for all $(m,n) \in \Omega$,

$$u(m,n) \leq G_1^{-1}\left[G_1(c_1) + \sum_{s=m_0}^{m-1}\sum_{t=n_0}^{n-1} a(s,t)\right] \tag{6.1.109}$$

where G_1 and G_1^{-1} are defined as in Theorem 6.1.13.

Proof By the same steps from (6.1.104) to (6.1.108) in the proofs of Theorem 6.1.13, we have

$$\begin{cases} u(m,n) \leq z(m,n), & (6.1.110) \\ z(m,n) \leq G_1^{-1}\left[G_1(z(m_0,n)) + \sum_{s=m_0}^{m-1}\sum_{t=n_0}^{n-1} a(s,t)\right]. & (6.1.111) \end{cases}$$

and for all $(m,n) \in \Omega$,

$$G_1(2z(m_0,n)-k) - G_1(z(m_0,n)) \leq \sum_{s=m_0}^{M-1}\sum_{t=n_0}^{N-1} a(s,t)) \tag{6.1.112}$$

From the assumption of Theorem 6.1.14 and (6.1.112), we have

$$\hat{H}_1(z(m_0,n)) \leq 0 = \hat{H}_1(c_1).$$

Since $\hat{H}_1$ is increasing, $\hat{H}_1$ has an inverse function $\hat{H}_1^{-1}$, from the last inequality we derive

$$z(m_0,n) \leq c_1.$$

Substituting the last inequality into (6.1.111) and combining with (6.1.110), we can get the desired inequality (6.1.109). □

Corollary 6.1.11 (The Ma Inequality [359]) *Let $u(m,n), a(m,n)$ and k be as in Theorem 6.1.13. If $u(m,n)$ satisfies for all $(m,n) \in \Omega$,*

$$u(m,n) \leq k + \sum_{s=m_0}^{m-1} \sum_{t=n_0}^{n-1} a(s,t)u(s,t) + \sum_{s=m_0}^{M-1} \sum_{t=n_0}^{N-1} a(s,t)u(s,t) \quad (6.1.113)$$

and

$$\Sigma(M,N) = \exp\left(\sum_{s=m_0}^{M-1} \sum_{t=n_0}^{N-1} a(s,t)\right) < 2, \quad (6.1.114)$$

then for all $(m,n) \in \Omega$,

$$u(m,n) \leq \frac{k}{2 - \Sigma(M,N)} \exp\left(\sum_{s=m_0}^{m-1} \sum_{t=n_0}^{n-1} a(s,t)\right). \quad (6.1.115)$$

Proof In Theorem 6.1.13, by letting $w(u) = u$, we obtain

$$G_1(v) = \int_{v_0}^{v} \frac{ds}{w(s)} = \int_{v_0}^{v} \frac{ds}{s} = \ln \frac{v}{v_0}, \quad v \geq v_0 > 0,$$

$$H_1(t) = G_1(2t - k) - G_1(t) = \ln \frac{2t - k}{t}, \ t \geq k,$$

which give us

$$G_1^{-1}(v) = v_0 \exp v, \quad H_1^{-1}(t) = \frac{k}{2 - \exp t}.$$

From inequality (6.1.103), we can derive inequality (6.1.115). □

Corollary 6.1.12 (The Ma Inequality [359]) *Let $u(m,n), a(m,n)$ and k be as in Theorem 6.1.13, $0 < p < 1$ be a constant. If $u(m,n)$ satisfies for all $(m,n) \in \Omega$,*

$$u(m,n) \leq k + \sum_{s=m_0}^{m-1} \sum_{t=n_0}^{n-1} a(s,t)u^p(s,t) + \sum_{s=m_0}^{M-1} \sum_{t=n_0}^{N-1} a(s,t)u^p(s,t) \quad (6.1.116)$$

then for all $(m,n) \in \Omega$,

$$u(m,n) \leq \left[(c_{11})^{1-p} + (1-p) \sum_{s=m_0}^{m-1} \sum_{t=n_0}^{n-1} a(s,t)\right]^{1/(1-p)} \quad (6.1.117)$$

where c_{11} is the solution of equation, for all $t \geq k$,

$$\hat{H}_1(t) = \frac{1}{1-p}[(2t-k)^{1-p} - t^{1-p}] - \sum_{s=m_0}^{M-1}\sum_{t=n_0}^{N-1} a(s,t) = 0. \tag{6.1.118}$$

Proof By Theorem 6.1.14, we only need to prove that (6.1.118) has a solution c_{11} for all $t \geq k$. In fact, for all $t \geq k$,

$$\hat{H}_1'(t) = \frac{(2^{1/p}t)^p - (2t-k)^p}{[t(2t-k)]^p} > 0$$

$$\hat{H}_1(k) = -\sum_{s=m_0}^{M-1}\sum_{t=n_0}^{N-1} a(s,t) < 0$$

and

$$\lim_{t\to+\infty} \hat{H}_1(t) = \lim_{t\to+\infty} \frac{t^{1-p}}{1-p}\left[\left(2-\frac{k}{t}\right)^{p-1} - 1\right] - \sum_{s=m_0}^{M-1}\sum_{t=n_0}^{N-1} a(s,t) = +\infty,$$

so $\hat{H}_1(t) = 0$ has a unique solution $c_{11} > k$. □

Remark 6.1.6 Though (6.1.117) does not give an exact estimation to the solution of (6.1.116), it is enough to get the upper bound to the solution of (6.1.116) in many cases.

Theorem 6.1.15 (The Ma Inequality [359]) *Suppose that $u(m,n), a(m,n), w(u)$ and k are as in Theorem 6.1.13. Let $\varphi(u) \in C^1(\mathbb{R}_+, \mathbb{R}_+)$ with $\varphi'(u) > 0$ and $\varphi'(u)$ is increasing for all $u > 0$, here $\varphi'(u)$ denotes the derivative of φ. If $u(m,n)$ satisfies for all $(m,n) \in \Omega$,*

$$\begin{aligned}\varphi(u(m,n)) \leq k &+ \sum_{s=m_0}^{m-1}\sum_{t=n_0}^{n-1} a(s,t)\varphi'(u(s,t))w(u(s,t)) \\ &+ \sum_{s=m_0}^{M-1}\sum_{t=n_0}^{N-1} a(s,t)\varphi'(u(s,t))w(u(s,t)),\end{aligned} \tag{6.1.119}$$

and

$$H_2(t) = G_1 \circ \varphi^{-1}(2t-k) - G_1 \circ \varphi^{-1}(t)$$

is increasing for all $t \geq k$, *then for all* $(m,n) \in \Omega$,

$$u(m,n) \leq G_1^{-1}\left(G_1\left[H_2^{-1}\left(\sum_{s=m_0}^{M-1}\sum_{t=n_0}^{N-1} a(s,t)\right)\right] + \sum_{s=m_0}^{m-1}\sum_{t=n_0}^{n-1} a(s,t)\right) \quad (6.1.120)$$

where H_2^{-1} *is the inverse of* H_2, G_1 *and* G_1^{-1} *are defined as in Theorem 6.1.13.*

Proof Similar to the proof of Theorem 6.1.13, it suffices to consider the case $k > 0$. Denote by $z_2(m,n)$ the right-hand side of (6.1.119). Then $z_2 > 0$, $u \leq \varphi^{-1}(z_2)$, and z_2 is non-decreasing in each variable. Hence, for any $(m,n) \in \Omega$, we have

$$\begin{aligned}\Delta_1 z_2(m,n) &= \sum_{t=n_0}^{n-1} a(m,t)\varphi'(u(m,t))w(u(m,t))\\ &\leq \sum_{t=n_0}^{n-1} a(m,t)\varphi'(\varphi^{-1}(z_2(m,t)))w(\varphi^{-1}(z_2(m,t)))\\ &\leq \varphi'(\varphi^{-1}(z_2(m,n)))\sum_{t=n_0}^{n-1} a(m,t)w(\varphi^{-1}(z_2(m,t)))\end{aligned}$$

or

$$\frac{\Delta_1 z_2(m,n)}{\varphi'(\varphi^{-1}(z_2(m,n)))} \leq \sum_{t=n_0}^{n-1} a(m,t)w(\varphi^{-1}(z_2(m,t))).$$

On the other hand, using the differential Mean-Value Theorem and the last inequality, we infer

$$\begin{aligned}\Delta_1[\varphi^{-1}(z_2(m,n))] &= \varphi^{-1}(z_2(m+1,n)) - \varphi^{-1}(z_2(m,n))\\ &= \frac{1}{\varphi'(\varphi^{-1}(\theta))}\Delta_1 z_2(m,n) \leq \frac{\Delta_1 z_2(m,n)}{\varphi'(\varphi^{-1}(z_2(m,n)))}\\ &\leq \sum_{t=n_0}^{n-1} a(m,t)w(\varphi^{-1}(z_2(m,t))). \end{aligned} \quad (6.1.121)$$

Keeping n fixed in (6.1.121) and setting $m = s$ and then summing over $s = m_0, m_0 + 1, \cdots, m-1$, we get for all $(m,n) \in \Omega$,

$$\varphi^{-1}(z_2(m,n)) \leq \varphi^{-1}(z_2(m_0,n)) + \sum_{s=m_0}^{m-1}\sum_{t=n_0}^{n-1} a(s,t)w(\varphi^{-1}(z_2(m,t))). \quad (6.1.122)$$

Now by applying Theorem 6.1.13 to the function $\varphi^{-1}(z_2(m,n))$, we obtain

$$\varphi^{-1}(z_2(m,n)) \leq G_1^{-1}\left(G_1[\varphi^{-1}(z_2(m_0,n))] + \sum_{s=m_0}^{m-1}\sum_{t=n_0}^{n-1} a(s,t)\right). \quad (6.1.123)$$

Observing that

$$2z_2(m_0,n) - k = k + 2\sum_{s=m_0}^{M-1}\sum_{t=n_0}^{N-1} a(s,t)\varphi'(u(s,t))w(u(s,t)) = z_2(M,N),$$

we infer from (6.1.123)

$$\begin{aligned}
G_1 \circ \varphi^{-1}(2z_2(m_0,n) - k) &= G_1 \circ \varphi^{-1}(z_2(M,N)) \\
&\leq G_1 \circ \varphi^{-1}(z_2(m_0,N)) + \sum_{s=m_0}^{M-1}\sum_{t=n_0}^{N-1} a(s,t) \\
&= G_1 \circ \varphi^{-1}(z_2(m_0,n)) + \sum_{s=m_0}^{M-1}\sum_{t=n_0}^{N-1} a(s,t)
\end{aligned}$$

or

$$G_1 \circ \varphi^{-1}(2z_2(m_0,n) - k) - G_1 \circ \varphi^{-1}(z_2(M,N)) \leq \sum_{s=m_0}^{M-1}\sum_{t=n_0}^{N-1} a(s,t). \quad (6.1.124)$$

Since $H_2(t) = G_1 \circ \varphi^{-1}(2t-k) - G_1 \circ \varphi^{-1}(t)$ is increasing for all $t \geq k$, $H_2(t)$ has an inverse function H_2^{-1} and from (6.1.124), we derive

$$z_2(m_0,n) \leq H_2^{-1}\left(\sum_{s=m_0}^{M-1}\sum_{t=n_0}^{N-1} a(s,t)\right).$$

Substituting the above inequality into (6.1.123) and by the definition of $z_2(m,n)$, we can obtain the desired inequality (6.1.120). □

By similar argument as in the proofs of Theorem 6.1.14, we can prove the following result immediately.

Theorem 6.1.16 (The Ma Inequality [359]) *Let $u(m,n), a(m,n)$ and k be as in Theorem 6.1.15. If $u(m,n)$ satisfies (6.1.119) for all $(m,n) \in \Omega$, and*

$$\hat{H}_2(t) = G_1 \circ \varphi^{-1}(2t-k) - G_1 \circ \varphi^{-1}(t) - \sum_{s=m_0}^{M-1}\sum_{t=n_0}^{N-1} a(s,t)$$

is increasing and $\hat{H}_2(t) = 0$ *has a solution* $c_2 \ge k$, *then for all* $(m,n) \in \Omega$,

$$u(m,n) \le G_1^{-1}\left[G_1^{-1}(c_2) + \sum_{s=m_0}^{m-1}\sum_{t=n_0}^{n-1} a(s,t)\right]. \tag{6.1.125}$$

When $\varphi = u^p$ ($p \ge 1$ is a constant) in Theorem 6.1.15, we have the following corollary.

Corollary 6.1.13 (The Ma Inequality [359]) *Let* $u(m,n), a(m,n)$ *and* k *be as in Theorem 6.1.15,* $p \ge 1$ *is a constant. If* $u(m,n)$ *satisfies for all* $(m,n) \in \Omega$,

$$u^p(m,n) \le k + \sum_{s=m_0}^{m-1}\sum_{t=n_0}^{n-1} a(s,t)u^p(s,t)w(u(s,t)) + \sum_{s=m_0}^{M-1}\sum_{t=n_0}^{N-1} a(s,t)u^{p-1}(s,t)w(u(s,t)), \tag{6.1.126}$$

and

$$H_{21}(t) = G_1((2t-k)^{1/p}) - G_1(t^{1/p})$$

is increasing for all $(m,n) \in \Omega$, *then for all* $(m,n) \in \Omega$,

$$u(m,n) \le G_1^{-1}\left(G_1\left[H_{21}^{-1}\left(\sum_{s=m_0}^{M-1}\sum_{t=n_0}^{N-1} a(s,t)\right)\right] + \sum_{s=m_0}^{m-1}\sum_{t=n_0}^{n-1} a(s,t)\right). \tag{6.1.127}$$

Corollary 6.1.14 (The Ma Inequality [359]) *Let* $u(m,n) \in F_1(U), a(m,n)$ *and* k *be as in Theorem 6.1.15. If* $u(m,n)$ *satisfies for all* $(m,n) \in \Omega$,

$$u^p(m,n) \le k + \sum_{s=m_0}^{m-1}\sum_{t=n_0}^{n-1} a(s,t)u^p(s,t)w(\ln u(s,t)) + \sum_{s=m_0}^{M-1}\sum_{t=n_0}^{N-1} a(s,t)u^p(s,t)w(\ln u(s,t)), \tag{6.1.128}$$

and

$$H_{22}(t) = G_1\left(\frac{1}{p}\ln(2t-k)\right) - G_1\left(\frac{1}{p}\ln t\right)$$

is increasing for all $(m,n) \in \Omega$, *then for all* $(m,n) \in \Omega$,

$$u(m,n) \leq G_1^{-1}\left\{G_1\left[H_{22}^{-1}\left(\sum_{s=m_0}^{M-1}\sum_{t=n_0}^{N-1} a(s,t)\right)\right] + \sum_{s=m_0}^{m-1}\sum_{t=n_0}^{n-1} a(s,t)\right\}. \quad (6.1.129)$$

Proof Taking $v(m,n) = \ln u(m,n)$, then (6.1.128) reduces to

$$\exp(pv(m,n)) \leq k + \sum_{s=m_0}^{m-1}\sum_{t=n_0}^{n-1} a(s,t)\exp(pv(s,t))w(v(s,t))$$
$$+ \sum_{s=m_0}^{M-1}\sum_{t=n_0}^{N-1} a(s,t)\exp(pv(s,t))w(v(s,t)), \quad (6.1.130)$$

for all $(m,n) \in \Omega$, which is a special case of inequality (6.1.119) when $\varphi(v) = \exp(pv)$. In this special case,

$$H_2(t) = H_{22}(t) = G_1\left(\frac{1}{p}\ln(2t-k)\right) - G_1\left(\frac{1}{p}\ln t\right).$$

By Theorem 6.1.15, we get the desired inequality (6.1.129) directly. □

Remark 6.1.7 Equations (6.1.126) and (6.1.128) are new discrete Volterra-Fredholm-Ou-Yang-type and Volterra-Fredholm-Engler [219]-Haraux [271] type inequality of two-variable, respectively.

Using Theorems 6.1.13 and 6.1.16, we can get more generalized results as follows.

Theorem 6.1.17 (The Ma Inequality [359]) *Suppose that* $u(m,n), a(m,n), w(u)$, $G_1, G_1^{-1}, H_1, H_1^{-1}$ *and* k *are as in Theorem 6.1.13,* $b(m,n) \in F_+(\Omega)$. *If* $u(m,n)$ *satisfies for all* $(m,n) \in \Omega$,

$$u(m,n) \leq k + \sum_{s=m_0}^{m-1}\sum_{t=n_0}^{n-1} a(s,t)w(u(s,t)) + \sum_{s=m_0}^{M-1}\sum_{t=n_0}^{N-1} a(s,t)w(u(s,t)), \quad (6.1.131)$$

then for all $(m,n) \in \Omega$,

$$u(m,n) \leq G_1^{-1}\left\{G_1\left[H_1^{-1}\left(\sum_{s=m_0}^{M-1}\sum_{t=n_0}^{N-1} a^*(s,t)\right)\right] + \sum_{s=m_0}^{m-1}\sum_{t=n_0}^{n-1} a^*(s,t)\right\} \quad (6.1.132)$$

where $a^*(m,n) \in F_+(\Omega)$ *such that both* $a(m,n)$ *and* $b(m,n)$ *are less than or equal to* $a^*(m,n)$.

Proof From (6.1.131) and assumptions, we infer for all $(m,n) \in \Omega$,

$$u(m,n) \le k + \sum_{s=m_0}^{m-1}\sum_{t=n_0}^{n-1} a^*(s,t)w(u(s,t)) + \sum_{s=m_0}^{M-1}\sum_{t=n_0}^{N-1} a^*(s,t)w(u(s,t)).$$

An application of Theorem 6.1.13 to the above inequality yields (6.1.132) immediately. □

Theorem 6.1.18 (The Ma Inequality [359]) *Let $u(m,n), a(m,n), b(m,n)$, $a^*(m,n)$ and k be as in Theorem 6.1.17; $\varphi(u)$ be as in Theorem 6.1.15. $w_i \in C(\mathbb{R}_+, \mathbb{R}_+)$ be non-decreasing functions with $w_i > 0$ for all $u > 0$, $i = 1, 2$. If (m,n) satisfies for all $(m,n) \in \Omega$,*

$$\varphi(u(m,n)) \le k + \sum_{s=m_0}^{m-1}\sum_{t=n_0}^{n-1} a(s,t)\varphi'(u(s,t))w_1(u(s,t)) + \sum_{s=m_0}^{M-1}\sum_{t=n_0}^{N-1} a(s,t)\varphi'(u(s,t))w_2(u(s,t)), \quad (6.1.133)$$

and there is a function $W(u) \in C(\mathbb{R}_+, \mathbb{R}_+)$ that is non-decreasing such that both w_1 and w_2 are less than or equal to W,

$$G_2(v) = \int_{v_0}^{v} \frac{ds}{W(s)}, \quad v \ge v_0 > 0, \quad G_2(+\infty) = \int_{v_0}^{+\infty} \frac{ds}{W(s)} = +\infty,$$

and

$$H_3(t) = G_1 \circ \varphi^{-1}(2t - k) - G_1 \circ \varphi^{-1}(t)$$

is increasing for all $t \ge k$, then for all $(m,n) \in \Omega$,

$$u(m,n) \le G_2^{-1}\left\{ G_2\left[H_3^{-1}\left(\sum_{s=m_0}^{M-1}\sum_{t=n_0}^{N-1} a^*(s,t) \right)\right] + \sum_{s=m_0}^{m-1}\sum_{t=n_0}^{n-1} a^*(s,t) \right\} \quad (6.1.134)$$

where G_2^{-1} and H_3^{-1} are inverse functions of G_2 and H_3, respectively.

Proof From (6.1.133) and the assumptions, we can deduce

$$\varphi(u(m,n)) \le k + \sum_{s=m_0}^{m-1}\sum_{t=n_0}^{n-1} a^*(s,t)\varphi'(u(s,t))W(u(s,t)) + \sum_{s=m_0}^{M-1}\sum_{t=n_0}^{N-1} a^*(s,t)\varphi'(u(s,t))W(u(s,t)).$$

Now applying Theorem 6.1.15 to the above inequality yields the desired inequality (6.1.134). □

By the same argument as in the proof of Theorem 6.1.14, we can show the following result immediately.

Theorem 6.1.19 (The Ma Inequality [359]) *Let $u(m,n), a(m,n), b(m,n), a^*(m,n), w_i$ $(i = 1,2), W$ and k be as in Theorem 6.1.18. If $u(m,n)$ satisfies (6.1.133), and*

$$\hat{H}_3(t) = G_2 \circ \varphi^{-1}(2t-k) - G_2 \circ \varphi^{-1}(t) - \sum_{s=m_0}^{M-1}\sum_{t=n_0}^{N-1} a^*(s,t)$$

is increasing and $\hat{H}_3(t) = 0$ has a solution $c_3 \geq k$, then for all $(m,n) \in \Omega$,

$$u(m,n) \leq G_2^{-1}\left(G_2(c_3) + \sum_{s=m_0}^{m-1}\sum_{t=n_0}^{n-1} a^*(s,t)\right). \tag{6.1.135}$$

Remark 6.1.8 In Theorems 6.1.17–6.1.19, we can choose function $a^*(m,n) = a(m,n) + b(m,n)$ or $\max\{a(m,n), b(m,n)\}$ as well as in function W.

By Theorem 6.1.19, we can get the following interesting result.

Corollary 6.1.15 (The Ma Inequality [359]) *Let $u(m,n), a(m,n), b(m,n), a^*(m,n)$ and k be as in Theorem 6.1.18, $p \geq 1$ and $0 < q < 1$ be constants. If $u(m,n)$ satisfies for all $(m,n) \in \Omega$,*

$$u^p(m,n) \leq k + \sum_{s=m_0}^{m-1}\sum_{t=n_0}^{n-1} a(s,t)u^p(s,t) + \sum_{s=m_0}^{M-1}\sum_{t=n_0}^{N-1} b(s,t)u^{p+q-1}(s,t) \tag{6.1.136}$$

and

$$\exp\left(\sum_{s=m_0}^{M-1}\sum_{t=n_0}^{N-1} a^*(s,t)\right) < 2^{1/p},$$

then for all $(m,n) \in \Omega$,

$$u(m,n) \leq \left\{(1+\hat{c}_3^{1-q})\exp\left[(1-q)\sum_{s=m_0}^{m-1}\sum_{t=n_0}^{n-1} a^*(s,t)\right] - 1\right\}^{1/(1-q)} \tag{6.1.137}$$

where $\hat{c}_3$ *is the solution of*

$$\hat{H}_3(t) = \frac{1}{q}\ln\frac{1+(2t-k)^{(1-q)/p}}{1+t^{(1-q)/p}} - \sum_{s=m_0}^{M-1}\sum_{t=n_0}^{N-1} a^*(s,t) = 0 \qquad (6.1.138)$$

for all $t \geq k$.

Proof In Theorem 6.1.19, by letting $w_1(u) = u,\ w_2(u) = u^q$ and $W = w_1 + w_2$, we obtain

$$G_2(v) = \int_{v_0}^{v} \frac{ds}{w_1(s) + w_2(s)} = \int_{v_0}^{v} \frac{ds}{s + s^q} = \frac{1}{1-q}\ln\frac{1+v^{1-q}}{1+v_0^{1-q}},\quad v \geq v_0 > 0. \qquad (6.1.139)$$

Hence,

$$G_2^{-1}(v) = \left[(1+v_0^{1-q})\exp((1-q)v) - 1\right]^{1/(1-q)}. \qquad (6.1.140)$$

By computation, we have for all $t \geq k$,

$$\hat{H}_3(t) = \frac{1}{1-q}\ln\frac{1+(2t-k)^{(1-q)/p}}{1+t^{(1-q)/p}} - \sum_{s=m_0}^{M-1}\sum_{t=n_0}^{N-1} a^*(s,t),$$

$$\hat{H}_3'(t) = \frac{k + 2t^{1-(1-q)/p} - (2t-k)^{1-(1-q)/p}}{[2t-k+(2t-k)^{1-(1-q)/p}](t+t^{1-(1-q)/p})} > 0. \qquad (6.1.141)$$

$$\hat{H}_3(t) = -\sum_{s=m_0}^{M-1}\sum_{t=n_0}^{N-1} a^*(s,t) < 0 \qquad (6.1.142)$$

and

$$\lim_{t\to+\infty}\hat{H}_3(t) = \lim_{t\to+\infty}\left[\frac{1}{1-q}\ln\frac{1+(2t-k)^{(1-q)/p}}{1+t^{(1-q)/p}} - \sum_{s=m_0}^{M-1}\sum_{t=n_0}^{N-1} a^*(s,t)\right]$$
$$= \ln 2^{1/p} - \sum_{s=m_0}^{M-1}\sum_{t=n_0}^{N-1} a^*(s,t) > 0. \qquad (6.1.143)$$

By (6.1.141)–(6.1.143), we know that (6.1.138) has a solution $\hat{c}_3 > k$. Now by (6.1.135), (6.1.139) and (6.1.140), we can get the desired (6.1.137). □

In the sequel, we shall introduce some new discrete Gronwall-Bellman-Ou-Yang-type inequalities with explicit bounds on unknown functions.

Let $I := [m_0, M) \cap \mathbb{Z}$ and $J := [n_0, N) \cap \mathbb{Z}$ are two fixed lattices of integral points in $\mathbb{R}$, where $m_0, n_0 \in \mathbb{Z}, M, N \in \mathbb{Z} \cup \{+\infty\}$. Let $\Omega := I \times J \subset \mathbb{Z}^2, \mathbb{R}_+ := [0, +\infty), \mathbb{R}_0 := (0, +\infty), \mathbb{R}_1 := [1, +\infty)$, and for any $(s,t) \in \Omega$, the sub-lattice $[m_0, s] \times [n_0, t] \cap \Omega$ of Ω will be denoted as $\Omega_{(s,t)}$.

If U is a lattice in $\mathbb{Z}$ (respectively $\mathbb{Z}^2$), the collection of all $\mathbb{R}$-valued functions on U is denoted by $F(U)$. For the sake of convenience, we extend the domain of definition of each function in $F(U)$ and $F_+(U)$ trivially to the ambient space $\mathbb{Z}$ (respectively $\mathbb{Z}^2$). So for example, a function in $F(U)$ is regarded as a function defined on $\mathbb{Z}$ (respectively $\mathbb{Z}^2$) with support in U . As usual, the collection of all continuous functions of a topological space X into a topological space Y will be denoted by $C(X, Y)$.

If U is a lattice in $\mathbb{Z}$, the difference operator Δ on $f \in F(\mathbb{Z})$ or $F_+(\mathbb{Z})$ is defined as

$$\Delta f(n) := f(n+1) - f(n), \quad \text{for all } n \in U,$$

and if V is a lattice in $\mathbb{Z}^2$, the partial difference operators Δ_1 and Δ_2 on $u \in F(\mathbb{Z}^2)$ or $F_+(\mathbb{Z}^2)$ are defined as

$$\begin{cases} \Delta_1 u(m,n) := u(m+1,n) - u(m,n), & \text{for all } (m,n) \in V, \\ \Delta_2 u(m,n) := u(m,n+1) - u(m,n), & \text{for all } (m,n) \in V. \end{cases}$$

Theorem 6.1.20 (The Cheung-Ren Inequality [147]) *Suppose $u \in F_+(\Omega)$. If $c \geq 0$ is a constant and $b \in F_+(\Omega)$, $w \in C(\mathbb{R}_+, \mathbb{R}_+)$ are functions satisfying*

(i) w is non-decreasing with $w(r) > 0$ for all $r > 0$;
(ii) for any $(m,n) \in \Omega$,

$$u(m,n) \leq c + \sum_{s=m_0}^{m-1} \sum_{t=n_0}^{n-1} b(s,t) w(u(s,t)), \tag{6.1.144}$$

then for all $(m,n) \in \Omega_{(m_1,n_1)}$,

$$u(m,n) \leq \Phi^{-1}[\Phi(c) + B(m,n)] \tag{6.1.145}$$

where

$$B(m,n) := \sum_{s=m_0}^{m-1} \sum_{t=n_0}^{n-1} b(s,t),$$

$$\Phi(r) := \int_1^r \frac{ds}{w(s)}, \quad r > 0,$$

$$\Phi(0) := lim_{r \to 0^+} \Phi(r),$$

Φ^{-1} is the inverse of Φ, and $(m_1, n_1) \in \Omega$ is chosen such that $\Phi(c) + B(m,n) \in Dom(\Phi^{-1})$ for all $(m,n) \in \Omega_{(m_1,n_1)}$.

Proof It suffices to consider the case $c > 0$, for then the case $c = 0$ can be arrived at by continuity argument. Denote by $p(m,n)$ the right hand side of (6.1.144). Then $p > 0$, $u \le p$ on Ω, and p is non-decreasing in each variable. Hence for any $(m,n) \in \Omega$,

$$\begin{aligned}\Delta_1 p(m,n) &= p(m+1,n) - p(m,n)\\ &= \sum_{t=n_0}^{n-1} b(m,t)w(u(m,t))\\ &\le \sum_{t=n_0}^{n-1} b(m,t)w(p(m,t))\\ &\le w(p(m,n-1)) \sum_{t=n_0}^{n-1} b(m,t).\end{aligned}$$

Therefore, by the Mean-Value Theorem for integrals, for each $(m,n) \in \Omega$, there exists $p(m,n) \le \xi \le p(m+1,n)$ such that

$$\begin{aligned}\Delta_1(\Phi \circ p)(m,n) &= \Phi(p(m+1,n)) - \Phi(p(m,n))\\ &= \int_{p(m,n)}^{p(m+1,n)} \frac{ds}{w(s)}\\ &= \frac{1}{w(\xi)} \Delta_1 p(m,n).\end{aligned}$$

Since w is non-decreasing, $w(\xi) \ge w(p(m,n))$ and so for all $(m,n) \in \Omega$,

$$\begin{aligned}\Delta_1(\Phi \circ p)(m,n) &\le \frac{1}{w(p(m,n))} \Phi\Delta_1 p(m,n)\\ &\le \frac{w(p(m,n-1))}{w(p(m,n))} \sum_{t=n_0}^{n-1} b(m,t)\\ &\le \sum_{t=n_0}^{n-1} b(m,t).\end{aligned}$$

Therefore,

$$\sum_{s=m_0}^{m-1} \Delta_1(\Phi \circ p)(s,n) \le \sum_{s=m_0}^{m-1} \sum_{t=n_0}^{n-1} b(s,t) = B(m,n).$$

On the other hand, it is elementary to check that

$$\sum_{s=m_0}^{m-1} \Delta_1(\Phi \circ p)(s,n) = \Phi \circ p(m,n) - \Phi \circ p(m_0,n),$$

thus

$$\begin{aligned}\Phi \circ p(m,n) &\leq \Phi \circ p(m_0,n) + \Phi \circ p(m,n)\\ &= \Phi(c) + B(m,n).\end{aligned}$$

Since Φ^{-1} is increasing on $Dom\, \Phi^{-1}$, this yields for all $(m,n) \in \Omega_{(m_1,n_1)}$

$$p(m,n) \leq \Phi^{-1}[\Phi(c) + B(m,n)].$$

□

Remark 6.1.9 In many cases the non-decreasing function w satisfies $\int_1^{\infty} \frac{ds}{w(s)} = \infty$. For example, $w = C > 0$, $w(s) = s$, $w(s) = \sqrt{s}$, etc., are such functions. In such cases $\Phi(\infty) = \infty$ and so we may take $m_1 = M$, $n_1 = N$. In particular, inequality (6.1.145) holds for all $(m,n) \in \Omega$.

For any $\varphi, \psi \in C((0,+\infty),(0,+\infty))$ and any constant $\beta > 0$, define

$$\Phi_\beta(r) := \int_1^r \frac{ds}{\varphi(s^{\frac{1}{\beta}})}, \quad \Psi_\beta(r) := \int_1^r \frac{ds}{\psi(s^{\frac{1}{\beta}})}, \quad r > 0,$$

$$\Phi_\beta(0) := \lim_{r\to 0+} \Phi_\beta(r), \quad \Psi_\beta(0) := \lim_{r\to 0+} \Psi_\beta(r).$$

Note that we allow $\Phi_\beta(0)$ and $\Psi_\beta(0)$ to be $-\infty$ here.

Among various generalizations of Ou-Yang's inequality, discretization is also an interesting direction. The point is, similar to the noteworthy contributions of the continuous versions of the inequality to the study of differential equations, one naturally expects that discrete versions of the inequality should also play an important role in the study of difference equations. In this respect, fewer results have been established. Recent results in this direction include the works of Pachpatte [500], Pang and Agarwal [528], and the following recent result of Cheung [141].

Theorem 6.1.21 (The Cheung Inequality [141]) *Suppose $u : \Omega \to \mathbb{R}_+$ is a function on a 2-dimensional lattice Ω, $k \geq 0$ is a constant, $a, b : \Omega \to \mathbb{R}_+$, and $w \in C(\mathbb{R}_+, \mathbb{R}_+)$ are functions satisfying*

(i) w is non-decreasing with $w(r) > 0$ for all $r > 0$;

(ii) for any $(m,n) \in \Omega$,

$$u^2(m,n) \leq k^2 + \sum_{s=m_0}^{m-1}\sum_{t=n_0}^{n-1} a(s,t)u(s,t) + \sum_{s=m_0}^{m-1}\sum_{t=n_0}^{n-1} b(s,t)u(s,t)w(u(s,t)), \tag{6.1.146}$$

then for all $(m,n) \in \Omega_{(m_1,n_1)}$,

$$u(m,n) \leq \Phi^{-1}[\Phi(k + A(m,n)) + B(m,n)] \tag{6.1.147}$$

where

$$A(m,n) := \sum_{s=m_0}^{m-1}\sum_{t=n_0}^{n-1} a(s,t), \quad B(m,n) =: \sum_{s=m_0}^{m-1}\sum_{t=n_0}^{n-1} b(s,t),$$

and $\Phi = \int_1^r \frac{ds}{w(s)}$, $r > 0$, *and* $(m_1, n_1) \in \Omega$ *is chosen such that* $\Phi(k + A(m,n)) + B(m,n) \in Dom\ (\Phi^{-1})$ *for all* $(m,n) \in \Omega_{(m_1,n_1)}$.

Proof It suffices to consider the case $k > 0$. Denote by $q(s,t)$ the right-hand side of (6.1.146). Then $q > 0$, $u \leq \sqrt{u}$ on Ω, and q is non-decreasing in each variable. Hence for any $(m,n) \in \Omega$,

$$\begin{aligned} \Delta_1 q(m,n) &= q(m+1,n) - q(m,n) \\ &= \sum_{t=n_0}^{n-1} a(m,t)u(m,t) + \sum_{t=n_0}^{n-1} b(m,t)u(m,t)w(u(m,t)) \\ &\leq \sum_{t=n_0}^{n-1} a(m,t)\sqrt{q(m,t)} + \sum_{t=n_0}^{n-1} b(m,t)\sqrt{q(m,t)}w(\sqrt{q(m,t)}) \\ &\leq \sqrt{q(m,n-1)}\left[\sum_{t=n_0}^{n-1} a(m,t) + \sum_{t=n_0}^{n-1} b(m,t)\sqrt{q(m,t)}w(\sqrt{q(m,t)})\right], \end{aligned}$$

or

$$\frac{\Delta_1 q(m,n)}{\sqrt{q(m,n-1)}} \leq \sum_{t=n_0}^{n-1} a(m,t) + \sum_{t=n_0}^{n-1} b(m,t)w(\sqrt{q(m,t)}).$$

Therefore, for any $(m, n) \in \Omega$,

$$\sum_{s=m_0}^{m-1} \frac{\Delta_1 q(s,n)}{\sqrt{q(s,n-1)}} \leq \sum_{s=m_0}^{m-1}\sum_{t=n_0}^{n-1} a(s,t) + \sum_{s=m_0}^{m-1}\sum_{t=n_0}^{n-1} b(s,t)w(\sqrt{q(s,t)})$$
$$= A(m,n) + \sum_{s=m_0}^{m-1}\sum_{t=n_0}^{n-1} b(s,t)w(\sqrt{q(s,t)}).$$

On the other hand, by the non-decreasing property of q in each variable, it is easy to check that for all $(m, n) \in \Omega$,

$$\begin{aligned}
&\sum_{s=m_0}^{m-1} \frac{\Delta_1 q(s,n)}{\sqrt{q(s,n-1)}} \\
&= \frac{q(m,n)}{\sqrt{q(m-1,n-1)}} - \frac{q(m-1,n)}{\sqrt{q(m-1,n-1)}} + \frac{q(m-1,n)}{\sqrt{q(m-2,n-1)}} \\
&\quad - \frac{q(m-2,n)}{\sqrt{q(m-2,n-1)}} + \ldots + \frac{q(m_0+1,n)}{\sqrt{q(m_0,n-1)}} - \frac{q(m_0,n)}{\sqrt{q(m_0,n-1)}} \\
&= \frac{q(m,n)}{\sqrt{q(m-1,n-1)}} + \sum_{s=1}^{m-m_0-1} q(m-s,n)[\frac{1}{\sqrt{q(m-s-1,n-1)}} \\
&\quad - \frac{1}{q(m-s,n-1)}] - \frac{q(m_0,n)}{\sqrt{q(m_0,n-1)}} \\
&\geq \frac{q(m,n)}{\sqrt{q(m,n)}} - \frac{q(m_0,n)}{\sqrt{q(m_0,n-1)}} \\
&= \sqrt{q(m,n)} - k.
\end{aligned}$$

Hence we have for all $(m, n) \in \Omega$,

$$\sqrt{q(m,n)} \leq k + A(m,n) + \sum_{s=m_0}^{m-1}\sum_{s=n_0}^{n-1} b(s,t)w(\sqrt{q(s,t)}).$$

In particular, since A is non-decreasing in each variable, for any fixed $(\overline{m}, \overline{n}) \in \Omega_{m_1,n_1}$, for all $(m, n) \in \Omega_{(\overline{m},\overline{n})}$,

$$\sqrt{q(m,n)} \leq (k + A(\overline{m},\overline{n})) + \sum_{s=m_0}^{m-1}\sum_{s=n_0}^{n-1} b(s,t)w(\sqrt{q(s,t)}).$$

Now applying Theorem 6.1.20 to the function $\sqrt{q(m,n)}$, we have for all $(m,n) \in \Omega_{(\overline{m},\overline{n})}$,

$$u(m,n) \leq \sqrt{q(m,n)} \leq \Phi^{-1}[\Phi(k + A(\overline{m},\overline{n})) + B(m,n)].$$

In particular, this gives

$$u(\overline{m},\overline{n}) \leq \Phi^{-1}[\Phi(k + A(\overline{m},\overline{n})) + B(\overline{m},\overline{n})].$$

Since $(\overline{m},\overline{n}) \in \Omega_{(\overline{m},\overline{n})}$ is arbitrary, this concludes the proof of the theorem. □

Theorem 6.1.22 (The Cheung-Ren Inequality [147]) *Suppose* $u \in F_+(\Omega)$. *If* $c \geq 0, \alpha > 0$ *are constants and* $b \in F_+(\Omega), \varphi \in C(\mathbb{R}_+, \mathbb{R}_+)$ *are functions satisfying*

(i) φ *is non-decreasing with* $\varphi(r) > 0$ *for all* $r > 0$;
(ii) for any $(m,n) \in \Omega$,

$$u^\alpha(m,n) \leq c + \sum_{s=m_0}^{m-1}\sum_{t=n_0}^{n-1} b(s,t)\varphi(u(s,t)), \tag{6.1.148}$$

then for all $(m,n) \in \Omega_{(m_1,n_1)}$,

$$u(m,n) \leq \{\Phi_\alpha^{-1}[\Phi_\alpha(c) + B(m,n)]\}^{1/\alpha} \tag{6.1.149}$$

where

$$B(m,n) := \sum_{s=m_0}^{m-1}\sum_{t=n_0}^{n-1} b(s,t),$$

Φ_α^{-1} *is the inverse of* Φ_α, *and* $m_1, n_1 \in \Omega$ *is chosen such that* $\Phi_\alpha(c) + B(m,n) \in$ *Dom* (Φ_α^{-1}) *for all* $(m,n) \in \Omega_{(m_1,n_1)}$.

Proof It suffices to consider the case $c > 0$, while the case $c = 0$ can be arrived at by continuity argument. Denote by $g(m,n)$ the right-hand side of (6.1.148). Then $g > 0, u \leq g^{1/\alpha}$ on Ω, and g is non-decreasing in each variable. Hence for any $(m,n) \in \Omega$, we derive

$$\begin{aligned}\Delta_1 g(m,n) &= g(m+1,n) - g(m,n) = \sum_{t=n_0}^{n-1} b(m,t)\varphi(u(m,t)) \\ &\leq \sum_{t=n_0}^{n-1} b(m,t)\varphi(g^{1/\alpha}(m,t)) \leq \varphi(g^{1/\alpha}(m,n-1))\sum_{t=n_0}^{n-1} b(m,t).\end{aligned} \tag{6.1.150}$$

Therefore, by the Mean-Value Theorem for integrals, for each $(m,n) \in \Omega$, there exists $g(m,n) \le \xi \le g(m+1,n)$ such that

$$\begin{aligned}\Delta_1(\Phi_\alpha \circ g)(m,n) &= \Phi_\alpha(g(m+1,n)) - \Phi_\alpha(g(m,n))\\ &= \int_{g(m,n)}^{g(m+1,n)} \frac{ds}{\varphi(s^{1/\alpha})} = \frac{1}{\varphi(\xi^{1/\alpha})}\Delta_1 g(m,n).\end{aligned}$$

Since φ is non-decreasing, $\varphi(\xi^{1/\alpha}) \ge \varphi(g^{1/\alpha}(m,n))$ and so by (6.1.150), we get for all $(m,n) \in \Omega$,

$$\begin{aligned}\Delta_1(\Phi_\alpha \circ g)(m,n) &\le \frac{1}{\varphi(g^{1/\alpha}(m,n))}\Delta_1 g(m,n)\\ &\le \frac{\varphi(g^{1/\alpha}(m,n-1))}{\varphi(g^{1/\alpha}(m,n))}\sum_{t=n_0}^{n-1} b(m,t) \le \sum_{t=n_0}^{n-1} b(m,t).\end{aligned}$$

Therefore,

$$\sum_{s=m_0}^{n-1} \Delta_1(\Phi_\alpha \circ g)(s,n) \le \sum_{s=m_0}^{m-1}\sum_{t=n_0}^{n-1} b(s,t) = B(m,n).$$

On the other hand, it is easy to check that

$$\sum_{s=m_0}^{n-1} \Delta_1(\Phi_\alpha \circ g)(s,n) = \Phi_\alpha \circ g(m,n) - \Phi_\alpha \circ g(m_0,n),$$

which gives us

$$\Phi_\alpha \circ g(m,n) \le \Phi_\alpha \circ g(m_0,n) + B(m,n) = \Phi_\alpha(c) + B(m,n).$$

Since Φ_α^{-1} is increasing on *Dom* (Φ_α^{-1}), this yields, for all $(m,n) \in \Omega_{(m_1,n_1)}$,

$$g(m,n) \le \Phi_\alpha^{-1}[\Phi_\alpha(c) + B(m,n)].$$

Hence the assertion is proved. □

Remark 6.1.10 (i) When $\alpha = 1$, Theorem 6.1.22 reduces to Theorem 6.1.20.
(ii) In many cases, the non-decreasing function φ satisfies

$$\int_1^{+\infty} \frac{ds}{\varphi(s^{1/\alpha})} = +\infty.$$

For example, φ=constant > 0, $\varphi(s) = s^{\alpha}, \varphi(s) = s^{\alpha/2}$, etc., are such functions. In such cases, $\Phi_{\alpha}(+\infty) = +\infty$ and so we may take $m_1 = M, n_1 = N$. In particular, inequality (6.1.149) holds for all $(m,n) \in \Omega$.

Theorem 6.1.23 (The Cheung-Ren Inequality [147]) *Suppose $u \in F_+(\Omega)$. If $k \geq 0, p > 1$ are constants and $a, b \in F_+(\Omega), \varphi \in C(\mathbb{R}_+, \mathbb{R}_+)$ are functions satisfying*

(i) φ is non-decreasing with $\varphi(r) > 0$ for all $r > 0$;
(ii) for any $(m,n) \in \Omega$,

$$u^p(m,n) \leq k + \sum_{s=m_0}^{m-1}\sum_{t=n_0}^{n-1} a(s,t)u(s,t) + \sum_{s=m_0}^{m-1}\sum_{t=n_0}^{n-1} b(s,t)u(s,t)\varphi(u(s,t)), \tag{6.1.151}$$

then for all $(m,n) \in \Omega_{(m_1,n_1)}$,

$$u(m,n) \leq \left\{\Phi_{p-1}^{-1}[\Phi_{p-1}(k^{1-1/p} + A(m,n)) + B(m,n)]\right\}^{1/(p-1)} \tag{6.1.152}$$

where

$$A(m,n) := \sum_{s=m_0}^{m-1}\sum_{t=n_0}^{n-1} a(s,t), \quad B(m,n) =: \sum_{s=m_0}^{m-1}\sum_{t=n_0}^{n-1} b(s,t),$$

and $(m_1,n_1) \in \Omega$ is chosen such that $\Phi_{p-1}(k^{1-1/p} + A(m,n)) + B(m,n) \in$ Dom (Φ_{p-1}^{-1}) for all $(m,n) \in \Omega_{(m_1,n_1)}$.

Proof Similar to the proof of Theorem 6.1.22, it suffices to consider the case $k > 0$. Denote by $f(s,t)$ the right-hand side of (6.1.151). Then $f > 0, u \leq f^{1/p}$ on Ω, and f is non-decreasing in each variable. Hence for any $(m,n) \in \Omega$, we can obtain

$$\begin{aligned}
\Delta_1 f(m,n) &= f(m+1,n) - f(m,n) \\
&= \sum_{t=n_0}^{n-1} a(m,t)u(m,t) + \sum_{t=n_0}^{n-1} b(m,t)u(m,t)\varphi(u(m,t)) \\
&\leq \sum_{t=n_0}^{n-1} a(m,t)f^{1/p}(m,t) + \sum_{t=n_0}^{n-1} b(m,t)f^{1/p}(m,t)\varphi(f^{1/p}(m,t)) \\
&\leq f^{1/p}(m,n-1)[\sum_{t=n_0}^{n-1} a(m,t) + \sum_{t=n_0}^{n-1} b(m,t)\varphi(f^{1/p}(m,t))],
\end{aligned}$$

or

$$\frac{\Delta_1 f(m,n)}{f^{1/p}(m,n-1)} \leq \sum_{t=n_0}^{n-1} a(m,t) + \sum_{t=n_0}^{n-1} b(m,t)\varphi(f^{1/p}(m,t)).$$

Therefore, for any $(m,n) \in \Omega$,

$$\begin{aligned}\sum_{s=m_0}^{m-1} \frac{\Delta_1 f(s,n)}{f^{1/p}(s,n-1)} &\leq \sum_{s=m_0}^{m-1}\sum_{t=n_0}^{n-1} a(s,t) + \sum_{s=m_0}^{m-1}\sum_{t=n_0}^{n-1} b(s,t)\varphi(f^{1/p}(s,t)) \\ &= A(m,n) + \sum_{s=m_0}^{m-1}\sum_{t=n_0}^{n-1} b(s,t)\varphi(f^{1/p}(s,t)).\end{aligned}$$

On the other hand, by the non-decreasing property of f in each variable, it is easy to check that for all $(m,n) \in \Omega$,

$$\begin{aligned}&\sum_{s=m_0}^{m-1} \frac{\Delta_1 f(s,n)}{f^{1/p}(s,n-1)} \\ &= \frac{f(m,n)}{f^{1/p}(m-1,n-1)} - \frac{f(m-1,n)}{f^{1/p}(m-1,n-1)} + \frac{f(m-1,n)}{f^{1/p}(m-2,n-1)} \\ &\quad - \frac{f(m-2,n)}{f^{1/p}(m-2,n-1)} + \cdots + \frac{f(m_0+1,n)}{f^{1/p}(m_0,n-1)} - \frac{f(m_0,n)}{f^{1/p}(m_0,n-1)} \\ &= \frac{f(m,n)}{f^{1/p}(m-1,n-1)} \\ &\quad + \sum_{s=1}^{m-m_0-1} f(m-s,n)\left[\frac{1}{f^{1/p}(m-s-1,n-1)} - \frac{1}{f^{1/p}(m-s,n-1)}\right] \\ &\quad - \frac{f(m_0,n)}{f^{1/p}(m_0,n-1)} \geq \frac{f(m,n)}{f^{1/p}(m,n)} - \frac{f(m_0,n)}{f^{1/p}(m_0,n-1)} \\ &= f^{1-1/p}(m,n) - k^{1-1/p}.\end{aligned}$$

Hence we conclude for all $(m,n) \in \Omega$,

$$f^{1-1/p}(m,n) \leq k^{1-1/p} + A(m,n) + \sum_{s=m_0}^{m-1}\sum_{t=n_0}^{n-1} b(s,t)\varphi(f^{1/p}(s,t)).$$

In particular, since A is non-decreasing in each variable, for any fixed $(\overline{m},\overline{n}) \in \Omega_{(m_1,n_1)}$ and for all $(m,n) \in \Omega_{(\overline{m_1},\overline{n_1})}$,

$$f^{1-1/p}(m,n) = [f^{1/p}(m,n)]^{p-1} \leq (k^{1-1/p} + A(\overline{m},\overline{n})) + \sum_{s=m_0}^{m-1}\sum_{t=n_0}^{n-1} b(s,t)\varphi(f^{1/p}(s,t)).$$

Now applying Theorem 6.1.22 to the function $f^{1/p}(m,n)$, we can derive for all $m,n) \in \Omega_{(\overline{m_1},\overline{n_1})}$,

$$u(m,n) \leq f^{1/p}(m,n) \leq \left\{\Phi_{p-1}^{-1}[\Phi_{p-1}(k^{1-1/p} + A(\overline{m},\overline{n})) + B(m,n)]\right\}^{1/(p-1)}.$$

In particular, this gives

$$u(\overline{m},\overline{n}) \leq \left\{\Phi_{p-1}^{-1}[\Phi_{p-1}(k^{1-1/p} + A(\overline{m},\overline{n})) + B(\overline{m},\overline{n})]\right\}^{1/(p-1)}.$$

Since $(\overline{m},\overline{n}) \in \Omega_{m_1,n_1}$ is arbitrary, this completes the proof of the theorem. □

In case when Ω degenerates into a 1-dimensional lattice, Theorem 6.1.23 takes the following simpler form which is a generalization of a result of Pachpatte in [500].

Corollary 6.1.16 (The Cheung-Ren Inequality [147]) *Suppose $u \in F_+(I)$. If $k \geq 0, p > 1$ are constants and $a, b \in F_+(I), \varphi \in C(\mathbb{R}_+,\mathbb{R}_+)$ are functions satisfying*

(i) φ is non-decreasing with $\varphi(r) > 0$ for all $r > 0$;
(ii) for any $m \in I$,

$$u^p(m) \leq k + \sum_{s=m_0}^{m-1} a(s)u(s) + \sum_{s=m_0}^{m-1} b(s)u(s)\varphi(u(s)),$$

then for all $m \in [m_0, m_1] \cap I$,

$$u(m) \leq \left\{\Phi_{p-1}^{-1}[\Phi_{p-1}(k^{1-1/p} + \sum_{s=m_0}^{m-1} a(s)) + \sum_{s=m_0}^{m-1} b(s)]\right\}^{1/(p-1)}$$

where $m_1 \in I$ is chosen such that $\Phi_{p-1}(k^{1-1/p} + \sum_{s=m_0}^{m-1} a(s)) + \sum_{s=m_0}^{m-1} b(s) \in$ Dom (Φ_{p-1}^{-1}) for all $m \in [m_0, m_1] \cap I$.

Proof It follows immediately from Theorem 6.1.13 by setting $\Omega = I \times \{n_0\}$ for some $n_0 \in \mathbb{Z}$, and extending the functions $a(s), b(s), u(s)$ to $a(s,n_0), b(s,n_0)$ and $u(s,n_0)$ respectively in the obvious way. □

Theorem 6.1.13 can easily be applied to generate other useful discrete inequalities in more general situations. For example, we have the following result.

Theorem 6.1.24 (The Cheung-Ren Inequality [147]) *Suppose* $u \in F_+(\Omega)$. *If* $k \geq 0, p > q > 0$ *are constants and* $a, b \in F_+(\Omega), \varphi \in C(\mathbb{R}_+, \mathbb{R}_+)$ *are functions satisfying*

(i) φ *is non-decreasing with* $\varphi(r) > 0$ *for all* $r > 0$; *and*
(ii) for any $(m, n) \in \Omega$,

$$u^p(m,n) \leq k + \sum_{s=m_0}^{m-1}\sum_{t=n_0}^{n-1} a(s,t)u^q(s,t) + \sum_{s=m_0}^{m-1}\sum_{t=n_0}^{n-1} b(s,t)u^q(s,t)\varphi(u(s,t)), \tag{6.1.153}$$

then for all $(m, n) \in \Omega_{(m_1,n_1)}$,

$$u(m,n) \leq \left\{\Phi_{p-q}^{-1}[\Phi_{p-q}(k^{1-q/p} + A(m,n)) + B(m,n)]\right\}^{1/(p-q)} \tag{6.1.154}$$

where

$$A(m,n) := \sum_{s=m_0}^{m-1}\sum_{t=n_0}^{n-1} a(s,t), \quad B(m,n) =: \sum_{s=m_0}^{m-1}\sum_{t=n_0}^{n-1} b(s,t),$$

and $(m_1, n_1) \in \Omega$ *is chosen such that* $\Phi_{p-q}(k^{1-q/p} + A(m,n)) + B(m,n) \in$ *Dom* (Φ_{p-q}^{-1}) *for all* $(m,n) \in \Omega_{(m_1,n_1)}$.

Proof For any $r > 0$, define

$$\psi(r) := \varphi(r^{1/q}). \tag{6.1.155}$$

Then clearly ψ satisfies condition (i) of Theorem 6.1.23. By (6.1.153), we have for all $(m, n) \in \Omega$,

$$u^p(m,n) \leq k + \sum_{s=m_0}^{m-1}\sum_{t=n_0}^{n-1} a(s,t)u^q(s,t) + \sum_{s=m_0}^{m-1}\sum_{t=n_0}^{n-1} b(s,t)u^q(s,t)\psi(u^q(s,t)).$$

Writing $v = u^q$, this becomes

$$v^{p/q}(m,n) \leq k + \sum_{s=m_0}^{m-1}\sum_{t=n_0}^{n-1} a(s,t)v(s,t) + \sum_{s=m_0}^{m-1}\sum_{t=n_0}^{n-1} b(s,t)v(s,t)\psi(v(s,t)).$$

Since $p/q > 1$, it follows from Theorem 6.1.23 that for all $(m,n) \in \Omega_{(m_1,n_1)}$,

$$v(m,n) \le \left\{\Psi_{p/q-1}^{-1}[\Psi_{p/q-1}(k^{1-1/(q/p)} + A(m,n)) + B(m,n)]\right\}^{1/(p/q-1)}$$
$$= \left\{\Psi_{(p-q)/q}^{-1}[\Psi_{(p-q)/q}(k^{(p-q)/p} + A(m,n)) + B(m,n)]\right\}^{q/(p-q)}.$$

Now it is elementary to check by the definition of ψ in (6.1.155) that

$$\Psi_{(p-q)/q}(r) = \Phi_{p-q}(r),$$

thus we have for all $(m,n) \in \Omega_{(m_1,n_1)}$,

$$v(m,n) \le \{\Phi_{p-q}^{-1}[\Phi_{p-q}(k^{(p-q)/p} + A(m,n)) + B(m,n)]\}^{q/(p-q)}$$

or for all $(m,n) \in \Omega_{(m_1,n_1)}$,

$$u(m,n) = v^{1/q}(m,n)$$
$$\le \left\{\Phi_{p-q}^{-1}[\Phi_{p-q}(k^{(p-q)/p} + A(m,n)) + B(m,n)]\right\}^{1/(p-q)}$$

where $(m_1,n_1) \in \Omega$ is chosen such that $\Phi_{p-q}(k^{(p-q)/p} + A(m,n)) + B(m,n) \in Dom\ \Phi_{p-q}^{-1}$ for all $(m,n) \in \Omega_{(m_1,n_1)}$. □

The following result is an important special case of Theorem 6.1.24.

Corollary 6.1.17 (The Cheung-Ren Inequality [147]) *Suppose $u \in F_+(\Omega)$. If $k \ge 0, p > 1$ are constants and $a,b \in F_+(\Omega), \varphi \in C(\mathbb{R}_+,\mathbb{R}_+)$ are functions satisfying*

(i) φ is non-decreasing with $\varphi(r) > 0$ for all $r > 0$; (ii) for any $(m,n) \in \Omega$,

$$u^p(m,n) \le k + \sum_{s=m_0}^{m-1}\sum_{t=n_0}^{n-1} a(s,t)u^{p-1}(s,t) + \sum_{s=m_0}^{m-1}\sum_{t=n_0}^{n-1} b(s,t)u^{p-1}(s,t)\varphi(u(s,t)),$$

then for all $(m,n) \in \Omega_{(m_1,n_1)}$,

$$u(m,n) \le \Phi_1^{-1}[\Phi_1(k^{1/p} + A(m,n)) + B(m,n)]$$

where

$$A(m,n) := \sum_{s=m_0}^{m-1}\sum_{t=n_0}^{n-1} a(s,t), \quad B(m,n) =: \sum_{s=m_0}^{m-1}\sum_{t=n_0}^{n-1} b(s,t),$$

and $(m_1, n_1) \in \Omega$ *is chosen such that* $\Phi_1(k^{1/p} + A(m,n)) + B(m,n) \in Dom\ (\Phi_1^{-1})$ *for all* $(m,n) \in \Omega_{(m_1,n_1)}$.

Proof The assertion follows immediately from Theorem 6.1.24 by taking $q = p - 1 > 0$. □

In particular, we have the following useful consequence.

Corollary 6.1.18 (The Cheung-Ren Inequality [147]) *Suppose* $u \in F_+(\Omega)$. *If* $k \geq 0, p > 1$ *are constants and* $a, b \in F_+(\Omega)$ *are functions such that for any* $(m,n) \in \Omega$,

$$u^p(m,n) \leq k + \sum_{s=m_0}^{m-1}\sum_{t=n_0}^{n-1} a(s,t)u^{p-1}(s,t) + \sum_{s=m_0}^{m-1}\sum_{t=n_0}^{n-1} b(s,t)u^p(s,t),$$

then we have for all $(m,n) \in \Omega$,

$$u(m,n) \leq \left(k^{1/p} + A(m,n)\right) \exp B(m,n)$$

where $A(m,n), B(m,n)$ *are defined as in Theorem 6.1.19.*

Proof Assume first that $k > 0$. Let φ be the identity mapping of $\mathbb{R}_+$ onto itself. Then all conditions of Corollary 6.1.17 are satisfied. Note that in this cases $\Phi_1 = ln$ and so $\Phi_1^{-1} = \exp$. In particular, Φ_1^{-1} is defined everywhere on $\mathbb{R}$. By Corollary 6.1.17, we have for all $(m,n) \in \Omega$,

$$u(m,n) \leq \exp\left[ln(k^{1/p} + A(m,n) + B(m,n))\right] = [k^{1/p} + A(m,n)] \exp(B(m,n)).$$

Finally, as this is true for all $k > 0$, by continuity, this should also hold for the case $k = 0$. □

In case when Ω degenerates into a one-dimensional lattice, Corollary 6.1.18 takes the following simpler form which generalizes another result of Pachpatte in [520].

Corollary 6.1.19 (The Cheung-Ren Inequality [147]) *Suppose* $u \in F_+(I)$. *If* $k \geq 0, p > 1$ *are constants and* $a, b \in F_+(I)$, *are functions such that for any* $m \in I$,

$$u^p(m) \leq k + \sum_{s=m_0}^{m-1} a(s)u^{p-1}(s) + \sum_{s=m_0}^{m-1} b(s)u^p(s),$$

then we have for all $m \in I$,

$$u(m) \leq \left[k^{1/p} + \sum_{s=m_0}^{m-1} a(s)\right] \prod_{s=m_0}^{m-1} \exp b(s).$$

Proof Analogous to that of Corollary 6.1.16, we apply Corollary 6.1.18. □

Another special case of Corollary 6.1.18 is the following two-dimensional discrete version of Ou-Yang's inequality.

Corollary 6.1.20 (The Cheung-Ren Inequality [147]) *Suppose $u \in F_+(\Omega)$. If $k \geq 0, p > 1$ are constants and $b \in F_+(\Omega)$ is a function such that for any $(m, n) \in \Omega$,*

$$u^p(m,n) \leq k + \sum_{s=m_0}^{m-1}\sum_{t=n_0}^{n-1} b(s,t)u^p(s,t),$$

then we have for all $(m, n) \in \Omega$,

$$u(m,n) \leq k^{1/p} \exp B(m,n)$$

where $A(m,n), B(m,n)$ are defined as in Theorem 6.1.23.

Proof This follows immediately from Corollary 6.1.18 by setting $a \equiv 0$. □

In case Ω when degenerates into a one-dimensional lattice, Corollary 6.1.20 takes the following simpler form which is a generalized one-dimensional discrete analogue of Ou-Yang's inequality.

Corollary 6.1.21 (The Cheung-Ren Inequality [147]) *Suppose $u \in F_+(I)$. If $k \geq 0, p > 1$ are constants and $b \in F_+(I)$ is a function such that for any $m \in I$,*

$$u^p(m) \leq k + \sum_{s=m_0}^{m-1} b(s)u^p(s),$$

then for all $m \in I$,

$$u(m) \leq k^{1/p} \prod_{s=m_0}^{m-1} \exp b(s).$$

Proof It follows from Corollary 6.1.16 by setting $a \equiv 0$, or by imitating the proof of Corollary 6.1.17 and applying Corollary 6.1.20. □

Remark 6.1.11 It is obvious that the results above can be generalized to obtain explicit bounds for functions satisfying certain discrete sum inequalities involving more retarded arguments. It is also clear that these results can be extended to functions on higher dimensional lattices in the obvious way. As details of these are rather algorithmic, they will not be carried out here.

For any real-valued function $u(x, y), x, y \in \mathbb{N}_0$, we define the operators $\Delta_1 u(x, y) = u(x + 1, y) - u(x, y)$, $\Delta_2 u(x, y) = u(x + 1, y) - u(x, y)$. We write $\Delta_1^n = \Delta_1 \times \cdots \times \Delta_1$ (n times), $\Delta_2^m = \Delta_2 \times \cdots \times \Delta_2$ (m times) and $\Delta_2^m \Delta_1^n u(x, y) = \Delta_2^m(\Delta_1^n u(x, y))$. For all $x, y \in \mathbb{N}_0$ and some function $q(x, y)$ defined for all $x, y \in \mathbb{N}_0$, we set

$$B(x, y, q(s_0, t_0)) = \sum_{s_{n-1}=0}^{x-1} \sum_{s_{n-2}=0}^{s_{n-1}-1} \cdots \sum_{s_0=0}^{s_1-1} \sum_{t_{m-1}=0}^{y-1} \sum_{t_{m-2}=0}^{t_{m-1}-1} \cdots \sum_{t_0=0}^{t_1-1} q(s_0, t_0),$$

where $s_1 = x$ and $t_1 = y$ and also, we set

$$\bar{B}(s_{n-1}, y, q(s_0, t_0)) = \sum_{s_{n-2}=0}^{s_{n-1}-1} \sum_{s_{n-3}=0}^{s_{n-2}-1} \cdots \sum_{s_0=0}^{s_1-1} \sum_{t_{m-1}=0}^{y-1} \sum_{t_{m-2}=0}^{t_{m-1}-1} \cdots \sum_{t_0=0}^{t_1-1} q(s_0, t_0).$$

Let the product $\mathbb{N}_0 \times \cdots \times \mathbb{N}_0$ (n times) be denoted by $\mathbb{N}_0^n$. A point $(x_1, \ldots, x_n)$ in $\mathbb{N}_0^n$ is denoted by x.

Theorem 6.1.25 (The Pachpatte Inequality [495]) *Let $f(x, y) \geq 0$, $g(x, y) \geq 0$ be real-valued functions defined for all $x, y \in \mathbb{N}_0$ and c be a non-negative real constant.*

(B_1) *Let $u(x, y) \geq 0$ be a real-valued function defined for all $x, y \in \mathbb{N}_0$. If for all $x, y \in \mathbb{N}_0$,*

$$u^2(x, y) \leq c^2 + 2B(x, y, f(s_0, t_0)u^2(s_0, t_0) + g(s_n)u(s_0, t_0)), \tag{6.1.156}$$

then for all $x, y \in \mathbb{N}_0$,

$$u(x, y) \leq p(x, y) \prod_{s_{n-1}=0}^{x-1} (1 + \bar{B}(s_{n-1}, y, f(s_0, t_0))), \tag{6.1.157}$$

where for all $x, y \in \mathbb{N}_0$,

$$p(x, y) = c + B(x, y, g(s_0, t_0)). \tag{6.1.158}$$

(B_2) *Let $u(x, y) \geq u_0 \geq 0$ be a real-valued function defined for all $x, y \in \mathbb{N}_0$; u_0 is a real constant. Let $W(u)$ be a continuous non-decreasing real-valued function defined on an interval $I = [u_0, +\infty)$ and $W(u) > 0$ on $(u_0, +\infty)$, $W(u_0) = 0$. If for all $x, y \in \mathbb{N}_0$,*

$$u^2(x, y) \leq c^2 + 2B(x, y, f(s_0, t_0)u(s_0, t_0)W(u(s_0, t_0)) + g(s_0, t_0)u(s_0, t_0)), \tag{6.1.159}$$

then for all $0 \leq x \leq x_1$, $0 \leq y \leq y_1$, $x, x_1, y, y_1 \in \mathbb{N}_0$,

$$u(x, y) \leq \Omega^{-1}(\Omega(p(x, y)) + B(x, y, f(s_0, t_0))), \tag{6.1.160}$$

where $p(x, y)$ *is as defined in (6.1.158), and* Ω, Ω^{-1} *are as defined in* (A_2) *in Theorem 2.3.12 and* $x_1, y_1 \in \mathbb{N}_0$ *be chosen so that for all* $x, y \in \mathbb{N}_0$ *such that for all* $0 \le x \le x_1, 0 \le y \le y_1$,

$$\Omega(p(x, y)) + B(x, y, f(s_0, t_0)) \in Dom\ (\Omega^{-1}).$$

(B_3) *Let* $u(x, y) \ge 0$ *be a real-valued function defined for all* $x, y \in \mathbb{N}_0$ *and the function* $L : \mathbb{N}_0^2 \times \mathbb{R}_+ \to \mathbb{R}_+$ *satisfies the condition: for all* $x, y \in \mathbb{N}_0$ *and* $v \ge w \ge 0$,

$$0 \le L(x, y, v) - L(x, y, w) \le k(x, y, w)(v - w), \qquad (6.1.161)$$

where k is a real-valued non-negative function defined for all $x, y \in \mathbb{N}_0, w \ge 0$. *If for all* $x, y \in \mathbb{N}_0$,

$$u^2(x, y) \le c^2 + 2B(x, y, f(s_0, t_0)u(s_0, t_0)L(s_0, t_0, u(s_{0,0} t)) + g(s_0, t_0)u(s_0, t_0)), \qquad (6.1.162)$$

then for all $x, y \in \mathbb{N}_0$,

$$u(x, y) \le p(x, y) + q(x, y) \prod_{s_{n-1}=0}^{x-1} (1 + \bar{B}(s_{n-1}, y, f(s_0, t_0)k(s_0, t_0, p(s_0, t_0)))), \qquad (6.1.163)$$

where $p(x, y)$ *is as defined in (6.1.158) and for all* $x, y \in \mathbb{N}_0$,

$$q(x, y) = B(x, y, f(s_0, t_0)L(s_0, t_0 p(s_0, t_0))). \qquad (6.1.164)$$

Proof (B_1) Assume that $c > 0$, and define a function $z(x, y)$ by

$$z(t) = c^2 + 2B(x, y, f(s, t)u^2(s, t) + g(s_n)u(s, t)). \qquad (6.1.165)$$

From (6.1.165), and using the fact that $u(x, y) \le \sqrt{z(x, y)}$, we derive

$$\Delta_2^m(\Delta_1^n z(x, y)) \le 2\sqrt{z(x, y)}\Big(f(x, y)\sqrt{z(x, y)} + g(x, y)\Big). \qquad (6.1.166)$$

Using the facts that $\sqrt{z(x, y)} \le \sqrt{z(x, y + 1)}$ and $\Delta_2^m(\Delta_1^n z(x, y)) \ge 0$, from (6.1.166) we derive

$$\frac{\Delta_2^{m-1}(\Delta_1^n z(x, y + 1))}{\sqrt{z(x, y + 1)}} - \frac{\Delta_2^{m-1}(\Delta_1^n z(x, y))}{\sqrt{z(x, y)}} \le 2\Big(f(x, y)\sqrt{z(x, y)} + g(x, y)\Big). \qquad (6.1.167)$$

Now, keeping x fixed in (6.1.167), setting $y = t_0$ and summing over $t_0 = 0, 1, \ldots, y-1$ and using the fact that $\Delta_2^{m-1}(\Delta_1^n z(x,0)) = 0$, we obtain

$$\frac{\Delta_2^{m-1}(\Delta_1^n z(x,y))}{\sqrt{z(x,y)}} \leq 2\sum_{t_0=0}^{y-1}\Big(f(x,t_0)\sqrt{z(x,t_0)} + g(x,t_0)\Big). \qquad (6.1.168)$$

From (6.1.168), and using the facts that $\sqrt{z(x,y)} \leq \sqrt{z(x,y+1)}$ and $\Delta_2^{m-2}(\Delta_1^n z(x,y)) \geq 0$, we observe

$$\frac{\Delta_2^{m-2}(\Delta_1^n z(x,y+1))}{\sqrt{z(x,y+1)}} - \frac{\Delta_2^{m-2}(\Delta_1^n z(x,y))}{\sqrt{z(x,y)}} \leq 2\sum_{t_0=0}^{y-1}\Big(f(x,t_0)\sqrt{z(x,t_0)} + g(x,t_0)\Big). \qquad (6.1.169)$$

Now, keeping x fixed in (6.1.169), setting $y = t_1$ and summing over $t_0 = 0, 1, \ldots, y-1$ and using the fact that $\Delta_2^{m-2}(\Delta_1^n z(x,0)) = 0$, we obtain

$$\frac{\Delta_2^{m-2}(\Delta_1^n z(x,y))}{\sqrt{z(x,y)}} \leq 2\sum_{t_1=0}^{y-1}\sum_{t_0=0}^{t_1-1}\Big(f(x,t_0)\sqrt{z(x,t_0)} + g(x,t_0)\Big).$$

Continuing in this way, we obtain

$$\frac{\Delta_1^n z(x,y)}{\sqrt{z(x,y)}} \leq 2\sum_{t_{m-1}=0}^{y-1}\ \sum_{t_{m-2}=0}^{t_{m-1}-1}\cdots\sum_{t_0=0}^{t_1-1}\Big(f(x,t_0)\sqrt{z(x,t_0)} + g(x,t_0)\Big). \qquad (6.1.170)$$

From (6.1.170), and using the facts that $\sqrt{z(x,y)} \leq \sqrt{z(x+1,y)}$ and $\Delta_1^{n-1} z(x,y) \geq 0$, we observe

$$\frac{\Delta_1^{n-1} z(x+1,y)}{\sqrt{z(x+1,y)}} - \frac{\Delta_1^{n-1} z(x,y)}{\sqrt{z(x,y)}} \leq 2\sum_{t_{m-1}=0}^{y-1}\ \sum_{t_{m-2}=0}^{t_{m-1}-1}\cdots\sum_{t_0=0}^{t_1-1}\Big(f(x,t_0)\sqrt{z(x,t_0)}+g(x,t_0)\Big). \qquad (6.1.171)$$

Now, keeping y fixed in (6.1.171), setting $x = s_0$ and summing over $s_0 = 0, 1, \ldots, x-1$ and using the fact that $\Delta_1^{n-1} z(0,y) = 0$, we obtain

$$\frac{\Delta_1^{n-1} z(x,y)}{\sqrt{z(x,y)}} \leq 2\sum_{s_0=0}^{x-1}\ \sum_{t_{m-1}=0}^{y-1}\ \sum_{t_{m-2}=0}^{t_{m-1}-1}\cdots\sum_{t_0=0}^{t_1-1}\Big(f(s_0,t_0)\sqrt{z(s_0,t_0)} + g(s_0,t_0)\Big).$$

Continuing in this way, we obtain

$$\frac{\Delta_1 z(x,y)}{\sqrt{z(x,y)}} \le 2\sum_{s_{n-2}=0}^{x-1}\sum_{s_{n-3}=0}^{s_{n-2}-1}\cdots\sum_{s_0=0}^{s_1-1}\sum_{t_{m-1}=0}^{y-1}\sum_{t_{m-2}=0}^{t_{m-1}-1}\cdots\sum_{t_0=0}^{t_1-1}\Big(f(s_0,t_0)\sqrt{z(s_0,t_0)}+g(s_0,t_0)\Big). \tag{6.1.172}$$

Using the facts that $\sqrt{z(x,y)} > 0$, $\Delta_1 z(x,y) \ge 0$, $\sqrt{z(x,y)} \le \sqrt{z(x+1,y)}$ for all $x, y \in \mathbb{N}_0$ and using (6.1.172), we observe

$$\begin{aligned}\Delta_1\Big(\sqrt{z(x,y)}\Big) &= \frac{\Delta_1 z(x,y)}{\sqrt{z(x+1,y)}+\sqrt{z(x,y)}} \le \frac{\Delta_1 z(x,y)}{2\sqrt{z(x,y)}}\\ &\le \sum_{s_{n-2}=0}^{x-1}\sum_{s_{n-3}=0}^{s_{n-2}-1}\cdots\sum_{s_0=0}^{s_1-1}\sum_{t_{m-1}=0}^{y-1}\sum_{t_{m-2}=0}^{t_{m-1}-1}\cdots\sum_{t_0=0}^{t_1-1}\\ &\quad\times\Big(f(s_0,t_0)\sqrt{z(s_0,t_0)}+g(s_0,t_0)\Big).\end{aligned} \tag{6.1.173}$$

Now, keeping y fixed in (6.1.173), setting $x = s_{n-1}$ and summing over $s_{n-1} = 0, 1, \ldots, x-1$, we obtain

$$\sqrt{z(x,y)} \le p(x,y) + B(x,y,f(s_0,t_0)\sqrt{z(s_0,t_0)}). \tag{6.1.174}$$

Since $p(x,y)$ is positive and monotone non-decreasing in x and y, from (6.1.174), we conclude

$$\frac{\sqrt{z(x,y)}}{p(x,y)} \le 1 + B\Big(x,y,f(s_0,t_0)\frac{\sqrt{z(s_0,t_0)}}{p(s_0,t_0)}\Big). \tag{6.1.175}$$

Define a function $v(x,y)$ by

$$v(x,y) = 1 + B\Big(x,y,f(s_0,t_0)\frac{\sqrt{z(s_0,t_0)}}{p(s_0,t_0)}\Big). \tag{6.1.176}$$

From (6.1.176), we infer

$$\Delta_2^m(\Delta_1^n v(x,y)) = f(x,y)\frac{\sqrt{z(x,y)}}{p(x,y)}. \tag{6.1.177}$$

Using $\frac{\sqrt{z(x,y)}}{p(x,y)} \le v(x,y)$ in (6.1.177), and then the facts that $v(x,y) \le v(x,y+1)$ and $\Delta_2^{m-1}(\Delta_1^n v(x,y)) \ge 0$, we observe that

$$\frac{\Delta_2^{m-1}(\Delta_1^n v(x,y+1))}{\sqrt{v(x,y+1)}} - \frac{\Delta_2^{m-1}(\Delta_1^n v(x,y))}{\sqrt{v(x,y)}} \le f(x,y). \tag{6.1.178}$$

Now following the same steps, below (6.1.167) up to (6.1.172), we obtain

$$\frac{\Delta_1 v(x,y)}{\sqrt{v(x,y)}} \leq \sum_{s_{n-2}=0}^{x-1} \sum_{s_{n-3}=0}^{s_{n-2}-1} \cdots \sum_{s_0=0}^{s_1-1} \sum_{t_{m-1}=0}^{y-1} \sum_{t_{m-2}=0}^{t_{m-1}-1} \cdots \sum_{t_0=0}^{t_1-1} f(s_0,t_0).$$

i.e.,

$$v(x+1,y) \leq v(x,y)\Big(1 + \sum_{s_{n-2}=0}^{x-1} \sum_{s_{n-3}=0}^{s_{n-2}-1} \cdots \sum_{s_0=0}^{s_1-1} \sum_{t_{m-1}=0}^{y-1} \sum_{t_{m-2}=0}^{t_{m-1}-1} \cdots \sum_{t_0=0}^{t_1-1} f(s_0,t_0)\Big). \tag{6.1.179}$$

Now, keeping y fixed in (6.1.179), setting $x = s_{n-1}$ and summing over $s_{n-1} = 0, 1, \ldots, x-1$ and using the fact that $v(0,y) = 1$, we obtain

$$v(x,y) \leq \prod_{s_{n-1}=0}^{x-1} (1 + \bar{B}(s_{n-1}, y, f(s_0,t_0))). \tag{6.1.180}$$

Using (6.1.180) in (6.1.175), and the fact that $u(x,y) \leq \sqrt{z(x,y)}$, we get the required inequality in (6.1.157).

The proof of the case when c is non-negative can be completed as mentioned in the proof of Part (A_1) of Theorem 2.3.12. This completes the proof of Part (B_1). The proof of the inequalities in (B_2) and (B_3) are respectively similar to the proof of Part (A_2) and Part (A_3) of Theorem 2.3.12 and closely resemble the proof of Part (B_1) given above (see also [488]). Here, we omit the details. □

The following result is the discrete analogue of the inequality given in Theorem 5.1.27.

Theorem 6.1.26 (The Pachpatte Inequality [519]) *Let $u(m,n), f(m,n), h(m,n,s,t)$, $0 \leq s \leq m < +\infty$, $0 \leq t \leq n < +\infty$, $m,n,s,t \in \mathbb{N}_0$ be real-valued non-negative functions. Let p, c, g, G, G^{-1} be as in Theorem 1.1.25 If for all $m, n \in \mathbb{N}_0$,*

$$u^p(m,n) \leq c + \sum_{s=0}^{m-1} \sum_{t=0}^{n-1} \left[f(s,t)g(u(s,t)) + \sum_{\sigma=0}^{s-1} \sum_{\eta=0}^{t-1} h(s,t,\sigma,\eta) g(u(\sigma,\eta)) \right], \tag{6.1.181}$$

then for all $0 \leq m \leq m_1,\ 0 \leq n \leq n_1,\ m, m_1, n, n_1 \in \mathbb{N}_0$,

$$u(m,n) \leq \left(G^{-1}[G(c) + B(m,n)]\right)^{1/p}, \tag{6.1.182}$$

where

$$B(m,n)=\sum_{s=0}^{m-1}\sum_{t=0}^{n-1}\left[f(s,t)+\sum_{\sigma=0}^{s-1}\sum_{\eta=0}^{t-1}h(s,t,\sigma,\eta)\right], \tag{6.1.183}$$

and $m_1, n_1 \in \mathbb{N}_0$ *are chosen so that*

$$G(c)+B(m,n)\in Dom\,(G^{-1}),$$

for all m,n *lying in* $0\le m\le m_1,\ 0\le n\le n_1$.

Proof First we assume that $c>0$ and define a function $z(m,n)$ by the right-hand side of (6.1.181). Then $z(0,n)=z(m,0)=c,\ u(m,n)\le (z(m,n))^{1/p}$ and

$$\begin{aligned}
z(m+1,n)-z(m,n) &=\sum_{t=0}^{n-1}\left[f(m,t)g(u(m,t))+\sum_{\sigma=0}^{m-1}\sum_{\eta=0}^{t-1}h(m,t,\sigma,\eta)g(u(\sigma,\eta))\right]\\
&\le\sum_{t=0}^{n-1}\Bigg[f(m,t)g((z(m,t))^{1/p})\\
&\quad+\sum_{\sigma=0}^{m-1}\sum_{\eta=0}^{t-1}h(m,t,\sigma,\eta)g((u(\sigma,\eta))^{1/p})\Bigg]\\
&\le g((z(m,t))^{1/p})\sum_{t=0}^{n-1}\left[f(m,t)+\sum_{\sigma=0}^{m-1}\sum_{n=0}^{t-1}h(m,t,\sigma,\eta)\right].
\end{aligned}\tag{6.1.184}$$

From 1.1.145 and (6.1.184), we derive

$$\begin{aligned}
G(z(m+1,n))-G(z(m,n)) &=\int_{z(m,n)}^{z(m+1,n)}\frac{ds}{g(s^{1/p})}\le\frac{z(m+1,n)-z(m,n)}{g((z(m,n))^{1/p})}\\
&\le\sum_{t=0}^{n-1}\left[f(m,t)+\sum_{\sigma=0}^{m-1}\sum_{n=0}^{t-1}h(m,t,\sigma,\eta)\right].
\end{aligned}\tag{6.1.185}$$

Keeping n fixed in (6.1.185), setting $m=s$ and summing up over s from 0 to $m-1$, we obtain

$$G(z(m,n))\le G(c)+B(m,n). \tag{6.1.186}$$

Now substituting the bound on $z(m,n)$ from (6.1.186) in $u(m,n)\le (z(m,n))^{1/p}$, we obtain the required inequality in (6.1.182). The proof of the case when $c\ge 0$ can be completed as mentioned in the proof of Theorem 1.1.25. The domain $0 \leq m \leq m_1$, $0 \leq n \leq n_1$ is obvious. □

The following corollary is an immediate consequence of Theorem 6.1.26.

Corollary 6.1.22 (The Pachpatte Inequality [519]) *Let u, f, h, c, p be as in Theorem 6.1.26. If for all $m, n \in \mathbb{N}_0$,*

$$u^p(m,n) \leq c + \sum_{s=0}^{m-1}\sum_{t=0}^{n-1}\left[f(s,t)u(s,t) + \sum_{\sigma=0}^{s-1}\sum_{\eta=0}^{t-1} h(s,t,\sigma)u(\sigma,\eta)\right], \qquad (6.1.187)$$

then for all $m, n \in \mathbb{N}_0$,

$$u(m,n) \leq \left[c^{(p-1)/p} + \frac{p-1}{p}B(m,n)\right]^{1/(p-1)}, \qquad (6.1.188)$$

where $B(m,n)$ is defined by (6.1.183).

Remark 6.1.12 We note that the inequalities established in Theorems 5.1.27 and 6.1.7 can be extended very easily to functions of several independent variables. The precise formulations of these results are very close to that of given above and closely looking at the results given in [507] and [513].

6.1.3 *Three-Dimensional Discrete Bihari Inequalities, Wendroff Inequalities and Pachpatte Inequalities*

In the following theorems, we introduce some nonlinear discrete inequalities in three independent variables of the Bihari [54] and Pachpatte [442, 444, 460, 461, 465] type which can be used in the theory of finite difference equations involving three independent variables.

To this end, we use the following notation. For all $x, y, z \in \mathbb{N}_0$, and functions a, b, c with domain $\mathbb{N}_0$, and p with domain $\mathbb{N}_0^3$ and Ω, V with domain $(0, +\infty)$, set

$$\Psi(x,y,z;a,b,c;\Omega,V(u),p) = \Omega[a(0)+b(y)+c(z)] + \sum_{s=0}^{x-1}\left(\frac{\Delta a(s)}{V[a(s)+b(0)+c(z)]} + \sum_{t=0}^{y-1}\sum_{r=0}^{z-1} p(s,t,r)\right).$$

Theorem 6.1.27 (The Pachpatte-Singare Inequality [526]) *Let $u(x,y,z) \geq u_0 > 0$ and $p(x,y,z) \geq 0$ be real-valued functions defined for all $(x,y,z) \in \mathbb{N}_0^3$ and let W be continuous, positive, strictly increasing function on $I = [u_0, +\infty)$, $u_0 > 0$.*

Suppose further that the following inequality holds for all $(x, y, z) \in \mathbb{N}_0^3$,

$$u(x, y, z) \leq a(x) + b(y) + c(z) + \sum_{s=0}^{x-1}\sum_{t=0}^{y-1}\sum_{r=0}^{z-1} p(s, t, r)W(u(s, t, r)), \quad (6.1.189)$$

where $a(x), b(y), c(z) > 0, \Delta a(x), \Delta b(y), \Delta c(z) \geq 0$, *are real-valued functions defined on* $\mathbb{N}_0$. *Then for all* $0 \leq x \leq x_1, 0 \leq y \leq y_1, 0 \leq z \leq z_1$,

$$u(x, y, z) \leq \Omega^{-1}(\Psi(x, y, z; a, b, c; \Omega, W(u), p)), \quad (6.1.190)$$

where

$$\Omega(r) = \int_{r_0}^{r} \frac{ds}{W(s)}, \ r \geq r_0 \ \text{with}\ r_0 \geq u_0 \quad (6.1.191)$$

and Ω^{-1} *is the inverse of* Ω, *and* x_1, y_1, z_1 *are chosen so that*

$$\Psi(x, y, z; a, b, c; \Omega, W(u), p) \in Dom\,(\Omega^{-1}),$$

for all x, y, z *lying in the sub-intervals* $0 \leq x \leq x_1, 0 \leq y \leq y_1, 0 \leq z \leq z_1$ *of* $\mathbb{N}_0$.

Proof Define a function $m(x, y, z)$ by the right-hand side of (6.1.186) so that $m(0, y, z) = a(0) + b(y) + c(z)$, $m(x, 0, z) = a(x) + b(0) + c(z)$, $m(x, y, 0) = a(x)+b(y)+c(0)$. Then following the same argument as the proof of Theorem 6.2.1 in Qin [557], we obtain

$$\Delta^2 m_{xy}(x, y, z + 1) - \Delta^2 m_{xy}(x, y, z) = p(x, y, z)W(u(x, y, z)),$$

which, in view of the definition of $m(x, y, z)$ and the fact that $m(x, y, z) \leq m(x, y, z+1)$, implies

$$\Delta^2 m_{xy}(x, y, z + 1) - \Delta^2 m_{xy}(x, y, z) \leq p(x, y, z)W(m(x, y, z + 1)),$$

i.e.,

$$\frac{\Delta^2 m_{xy}(x, y, z + 1)}{W(m(x, y, z + 1))} - \frac{\Delta^2 m_{xy}(x, y, z)}{W(m(x, y, z + 1))} \leq p(x, y, z). \quad (6.1.192)$$

From (6.1.192), we infer that

$$\frac{\Delta^2 m_{xy}(x, y, z + 1)}{W(m(x, y, z + 1))} - \frac{\Delta^2 m_{xy}(x, y, z)}{W(m(x, y, z))} \leq p(x, y, z). \quad (6.1.193)$$

Now keeping x, y fixed in (6.1.193), setting $z = r$ and summing over $r = 0, 1, \ldots, z-1$, we obtain

$$\frac{\Delta^2 m_{xy}(x,y,z)}{W(m(x,y,z))} \leq \sum_{r=0}^{z-1} p(x,y,r). \tag{6.1.194}$$

From (6.1.194) and in view of the fact that $m(x,y,z) \leq m(x,y+1,z)$, we see that

$$\frac{\Delta m_x(x,y,z+1)}{W(m(x,y,z+1))} - \frac{\Delta m_x(x,y,z)}{W(m(x,y,z))} \leq \sum_{r=0}^{z-1} p(x,y,r). \tag{6.1.195}$$

Keeping x, z fixed in (6.1.195), setting $y = t$ and summing over $t = 0, 1, \ldots, y-1$, we obtain

$$\frac{\Delta^2 m_x(x,y,z)}{W(m(x,y,z))} \leq \frac{\Delta a(x)}{W(a(x)+b(0)+c(z))} + \sum_{t=0}^{y-1}\sum_{r=0}^{z-1} p(x,t,r). \tag{6.1.196}$$

From (6.1.191) and (6.1.196), we derive

$$\begin{aligned} \Omega(m(x+1,y,z)) - \Omega(m(x,y,z)) &= \int_{m(x,y,z)}^{m(x+1,y,z)} \frac{ds}{W(s)} \\ &\leq \frac{\Delta m_x(x,y,z)}{W(m(x,y,z))} \\ &\leq \frac{\Delta a(x)}{W(a(x)+b(0)+c(z))} + \sum_{t=0}^{y-1}\sum_{r=0}^{z-1} p(x,t,r). \end{aligned} \tag{6.1.197}$$

Now keeping y, z fixed in (6.1.197), setting $x = s$ and summing over $s = 0, 1, \ldots, x-1$, we obtain

$$\begin{aligned} &\Omega(m(x,y,z)) - \Omega(a(0)+b(y)+c(z)) \\ &\leq \frac{\Delta a(x)}{W(a(x)+b(0)+c(z))} + \sum_{t=0}^{y-1}\sum_{r=0}^{z-1} p(x,t,r). \end{aligned} \tag{6.1.198}$$

The desire bound in (6.1.190) now follows by substituting the bound for $m(x,y,z)$ from (6.1.198). The sub-intervals of $\mathbb{N}_0$ for x, y and z are obvious. □

Remark 6.1.13 The estimate in (6.1.190) is independent of the choice of $u_0 \in I$ used in defining Ω. One can use this fact to show that the case $u_0 \leq 0, W(u) > 0$ *on* $(u_0, +\infty)$, and $W(u_0) = 0$ can be obtained as a limiting case from Theorem 6.1.27. This will allow $W(u) = u$ on $(0, +\infty)$. For details, see Beesack [54].

Remark 6.1.14 If we compare Theorem 6.1.27 with $W(u) \equiv u$ for $u \geq 1$, with Theorem 6.2.1 in Qin [557], we see that the hypotheses (1) in Theorem 6.2.1 in Qin [557] and (6.1.189) are then the same, but the bounds are now (2) in Theorem 6.2.1 in Qin [557] and

$$u(x,y,z) \leq [a(0)+b(y)+c(z)]\prod_{s=0}^{x-1}\exp\left[\frac{\Delta a(s)}{a(s)+b(0)+c(z)} + \sum_{t=0}^{y-1}\sum_{r=0}^{z-1}p(s,t,r)\right]. \tag{6.1.199}$$

Using the fact that $Exp\, u \geq 1+u$ for all $u \in \mathbb{R}$, it follows that (2) in Theorem 6.2.1 in Qin [557] gives us the better bound than (6.1.199).

The next result is a three independent variable discrete generalization of the integral inequality recently established by Pachpatte [445].

Theorem 6.1.28 (The Pachpatte-Singare Inequality [526]) *Let $u(x,y,z)$, $p(x,y,z)$ and W satisfy the hypotheses of Theorem 6.1.27, and suppose further that the following inequality holds for all $(x,y,z) \in \mathbb{N}_0^3$,*

$$u(x,y,z) \leq a(x)+b(y)+c(z)+\sum_{s=0}^{x-1}\sum_{t=0}^{y-1}\sum_{r=0}^{z-1}p(s,t,r)\left[u(s,t,r) + \sum_{k=0}^{s-1}\sum_{l=0}^{t-1}\sum_{n=0}^{r-1}p(k,l,n)W(u(k,l,n))\right], \tag{6.1.200}$$

where $a(x), b(y), c(z) > 0$, $\Delta a(x)$, $\Delta b(y)$, $\Delta c(z) \geq 0$, are real-valued functions defined on $\mathbb{N}_0$. Then for all $0 \leq x \leq x_2$, $0 \leq y \leq y_2$, $0 \leq z \leq z_2$,

$$u(x,y,z) \leq [a(0)+b(y)+c(z)] + \sum_{s=0}^{x-1}\left[\Delta a(s) + \sum_{t=0}^{y-1}\sum_{r=0}^{z-1}p(s,t,r)Q(s,t,r)\right], \tag{6.1.201}$$

where

$$Q(x,y,z) = G^{-1}\left(\Psi(x,y,z;a,b,c;G,u+W(u),p)\right), \tag{6.1.202}$$

in which

$$G(r) = \int_{r_0}^{r}\frac{ds}{s+W(s)}, \quad r \geq r_0 \geq u_0 \tag{6.1.203}$$

and G^{-1} *is the inverse of* G *and* x_2, y_2, z_2 *are chosen so that*

$$\Psi(x, y, z; a, b, c; G, u + W(u), p) \in Dom\,(G^{-1}),$$

for all x, y, z *lying in the sub-intervals* $0 \le x \le x_2$, $0 \le y \le y_2$, $0 \le z \le z_2$ *of* $\mathbb{N}_0$.

Proof Define a function $m(x, y, z)$ by the right-hand side of (6.1.200), so that $m(0, y, z) = a(0) + b(y) + c(z)$, $m(x, 0, z) = a(x) + b(0) + c(z)$, $m(x, y, 0) = a(x) + b(y) + c(0)$. Then by the same argument as in the proof of Theorem 6.2.2 in Qin [557], we obtain

$$\Delta^2 m_{xy}(x, y, z + 1) - \Delta^2 m_{xy}(x, y, z)$$
$$= p(x, y, z)\left[m(x, y, z) + \sum_{k=0}^{x-1}\sum_{l=0}^{y-1}\sum_{n=0}^{z-1} p(k, l, n)W(m(k, l, n))\right]. \quad (6.1.204)$$

If we put

$$v(x, y, z) = m(x, y, z) + \sum_{k=0}^{x-1}\sum_{l=0}^{y-1}\sum_{n=0}^{z-1} p(k, l, n)W(m(k, l, n)), \quad (6.1.205)$$

so that

$$v(0, y, z) = a(0) + b(y) + c(z),$$
$$v(x, 0, z) = a(x) + b(0) + c(z),$$
$$v(x, y, 0) = a(x) + b(y) + c(0).$$

Then following the same argument as in the proof of Theorem 6.2.2 in Qin [557], we obtain

$$\Delta^2 v_{xy}(x, y, z + 1) - \Delta^2 v_{xy}(x, y, z) \le p(x, y, z)[v(x, y, z) + W(v(x, y, z))].$$

Now following the same steps as in the proof of Theorem 6.1.27, we obtain

$$v(x, y, z) \le G^{-1}\Bigg[G(a(0) + b(y) + c(z))$$
$$+ \sum_{s=0}^{x-1}\left(\frac{\Delta a(s)}{a(s) + b(0) + c(z) + W(a(s) + b(0) + c(s))}\right.$$
$$\left.+ \sum_{t=0}^{y-1}\sum_{r=0}^{z-1} p(s, t, r)\right)\Bigg] = Q(x, y, z).$$

Substituting this bound for $v(x, y, z)$ in (6.1.204), we have

$$\Delta^2 m_{xy}(x, y, z + 1) - \Delta^2 m_{xy}(x, y, z) \leq p(x, y, z)Q(v(x, y, z)),$$

which implies

$$m(x, y, z) \leq [(a(0) + b(y) + c(z)] + \sum_{s=0}^{x-1} \left[\Delta a(s) + \sum_{t=0}^{y-1} \sum_{r=0}^{z-1} p(s, t, r)Q(s, t, r) \right].$$

Substituting this bound for $m(x, y, z)$ in (6.1.200), we can obtain the desired bound in (6.1.201). The sub-intervals of $\mathbb{N}_0$ for x, y and z are obvious. □

Remark 6.1.15 As pointed in Remark 6.2.3 in Qin [557], there are five other alternative conclusions corresponding to permutations of (x, y, z), (a, b, c), in addition to the conclusion (6.1.190) of Theorem 6.1.27. The same is true in case of the conclusion (6.1.201) of Theorem 6.1.28. Further we note that, if (6.1.200) holds, then from the definition of $m(x, y, z)$ and $v(x, y, z)$, we have

$$u(x, y, z) \leq Q(x, y, z), \tag{6.1.206}$$

on $\mathbb{N}_0^3$, where $Q(x, y, z)$ is defined by (6.1.202). In this case, (6.1.206) gives us the simpler, but not necessarily smaller than (6.1.202). If we compare Theorem 6.1.28 with $W(u) = u$ for all $u \geq 1$ with Theorem 6.2.2 in Qin [557] with $p \equiv q$, we see that (6.2.14) of Theorem 6.2.2 in Qin [557] and (6.1.200) coincide. In this case, a simple analysis shows that $R(x, y, z) \leq Q(x, y, z)$ so that the bound obtained in (6.2.15) of Theorem 6.2.2 in Qin [557] is better than (6.1.201).

The following results establish some discrete inequalities involving three independent variables which can be used in the study of discrete versions of partial differential and integral equations involving three independent variables.

Let $\mathbb{N}_{n_0}$ be the set of points $n_0 + k$ $(k \in \mathbb{N}_0)$, where $n_0 \geq 0$ is a given integer. The expression $u(n_0) + \sum_{s=n_0}^{n-1} b(s)$ represents a solution of the linear difference equation $\Delta u(n) = b(n)$ for all $n \in \mathbb{N}_{n_0}$, where Δ is the operator by $\Delta u(n) = u(n+1) - u(n)$.

It is supposed that $\sum_{s=n_0}^{n_0-1} b(s) = 0$. The expression $u(n_0)\Pi_{s=n_0}^{n-1} c(s)$ represents a solution of the linear difference equation $u(n + 1) = c(n)u(n)$ for all $n \in \mathbb{N}_{n_0}$. It is supposed that $\Pi_{s=n_0}^{n_0-1} c(s) = 1$.

In Theorem 6.1.29, we also use the following notions of the operators:

$$\begin{cases} \Delta_x[u(x, y, z)] = \Delta u_x(x, y, z) = u(x + 1, y, z) - u(x, y, z), \\ \Delta_y[u(x, y, z)] = \Delta u_y(x, y, z) = u(x, y + 1, z) - u(x, y, z), \\ \Delta_z[u(x, y, z)] = \Delta u_z(x, y, z) = u(x, y, z + 1) - u(x, y, z), \\ \Delta_y[\Delta u_x(x, y, z)] = \Delta^2 u_{xy}(x, y, z) = \Delta u_x(x, y + 1, z) - \Delta u_x(x, y, z), \end{cases}$$

and so on.

We often use the letters x, y and z to denote the three independent variables which are the members of $\mathbb{N}_{n_0}$.

We now apply Theorem 6.2.5 in Qin [557] to establish the following more general inequality which can be used in some applications.

Theorem 6.1.29 (The Singare-Pachpatte Inequality [612]) *Let $u(x,y,z)$, $b(x,y,z), c(x,y,z)$ and $p(x,y,z)$ be real-valued non-negative functions defined for all $x \geq 0, y \geq 0, z \geq 0$, and let $W(u)$ be continuous, positive strictly increasing function on $I = [u_0, +\infty)$, $u_0 > 0$, and suppose further that the following inequality holds for all $x \geq 0, y \geq 0, z \geq 0$,*

$$u(x,y,z) \leq M + \sum_{s=0}^{x-1}\sum_{t=0}^{y-1}\sum_{r=0}^{z-1} b(s,t,r)\left[u(s,t,r) + \sum_{k=0}^{s-1}\sum_{l=0}^{t-1}\sum_{n=0}^{r-1} c(k,l,n)u(k,l,n)\right]$$
$$+\sum_{s=0}^{x-1}\sum_{t=0}^{y-1}\sum_{r=0}^{z-1} p(s,t,r)W(u(s,t,r)) \tag{6.1.207}$$

where $M > 0$ is a constant. Then for all $0 \leq x \leq x_2, 0 \leq y \leq y_2, 0 \leq z \leq z_2$,

$$u(x,y,z) \leq \Omega^{-1}\left[\Omega(M) + \sum_{s=0}^{x-1}\sum_{t=0}^{y-1}\sum_{r=0}^{z-1} p(s,t,r)W(R(s,t,r))\right]R(x,y,z) \tag{6.1.208}$$

where

$$\Omega(r) = \int_{r_0}^{r} \frac{ds}{W(s)}, \quad r \geq r_0 > 0,$$

and Ω^{-1} is the inverse function of Ω, and

$$\begin{cases} R(x,y,z) = 1 + \sum_{s=0}^{x-1}\sum_{t=0}^{y-1}\sum_{r=0}^{z-1} b(s,t,r)\prod_{k=1}^{s-1} \\ \qquad \times\left[1 + \sum_{l=0}^{t-1}\sum_{n=0}^{r-1}[b(k,l,n) + c(k,l,n)]\right], \\ \Omega(M) + \sum_{s=0}^{x-1}\sum_{t=0}^{y-1}\sum_{r=0}^{z-1} p(s,t,r)W(R(s,t,r)) \in Dom\,(\Omega^{-1}) \end{cases} \tag{6.1.209}$$

for all x, y, z lying in the sub-intervals $0 \leq x \leq x_2, 0 \leq y \leq y_2, 0 \leq z \leq z_2$ of $\mathbb{N}$.

Proof The proof follows by the similar argument as in the proof of Theorem 6.2.4 in Qin [557], by making use of Theorem 6.2.5 in Qin [557]. We omit the details. □

Next we use the following basic notations and definitions.

$$\mathbb{N}_{k+1} = \{1, 2, \cdots, k, k+1\}, \quad k \in \mathbb{N},$$

$$\mathbb{N}_{m+1} = \{1, 2, \cdots, m, m+1\}, \quad m \in \mathbb{N}$$

and

$$Q = \mathbb{N}_{k+1} \times \mathbb{N}_{m+1}, \quad Q_1 = [1, \alpha] \times [1, \beta], \quad Q_2 = [1, \alpha] \times [\beta, m+1],$$

$Q_3 = [\alpha, k+1] \times [1, \beta], Q_4 = [\alpha, k+1] \times [\beta, m+1]$ for all $1 \leq \alpha \leq k+1, 1 \leq \beta \leq m+1$ and $m, k \in \mathbb{N}$. For $c : Q \to \mathbb{R}$, we define the forward difference operators $\Delta c_y(y,z) = c(y+1,z) - c(y,z)$, $\Delta c_z(y,z) = c(y,z+1) - c(y,z)$, $\Delta^2 c_{yz}(y,z) = \Delta c_y(y,z+1) - \Delta c_y(y,z)$ for all $(y,z) \in Q$. We denote by $F(Q)$ the class of functions $c : Q \to \mathbb{R}$ which $\Delta c_y(y,z)$, $\Delta^2 c_{yz}(y,z)$ exist and such that $c(1,z) = c(k+1,z) = 0$ for all $1 \leq z \leq m+1, m \in \mathbb{N}$, $\Delta c_y(y,1) = 0$, $\Delta c_y(y,m+1) = 0$ for all $1 \leq y \leq k, k \in \mathbb{N}$.

Theorem 6.1.30 (The Pachpatte Inequality [482]) *Let $p, q, r \geq 1$ be constants and suppose that $f, g, h \in F(Q)$. Then*

$$\sum_{y=1}^{k}\sum_{z=1}^{m}[|f(y,z)|^p|g(y,z)|^q|h(y,z)|^r + |h(y,z)|^r|f(y,z)|^p]$$

$$\leq \left(\frac{km}{4}\right)^{2p}\sum_{y=1}^{k}\sum_{z=1}^{m}|\Delta^2 f_{yz}(y,z)|^{2p} + \left(\frac{km}{4}\right)^{2} q\sum_{y=1}^{k}\sum_{z=1}^{m}|\Delta^2 g_{yz}(y,z)|^{2q}$$

$$+\left(\frac{km}{4}\right)^{2r}\sum_{y=1}^{k}\sum_{z=1}^{m}|\Delta^2 h_{yz}(y,z)|^{2r}, \tag{6.1.210}$$

$$\sum_{y=1}^{k}\sum_{z=1}^{m}|f(y,z)|^p|g(y,z)|^q|h(y,z)|^r \cdot (|f(y,z)|^p + |g(y,z)|^q + |h(y,z)^r)$$

$$\leq \left(\frac{km}{4}\right)^{4p}\sum_{y=1}^{k}\sum_{z=1}^{m}|\Delta^2 f_{yz}(y,z)|^{4p} + \left(\frac{km}{4}\right)^{4q}\sum_{y=1}^{k}\sum_{z=1}^{m}|\Delta^2 g_{yz}(y,z)|^{4q}$$

$$+\left(\frac{4m}{4}\right)^{4r}\sum_{y=1}^{k}\sum_{z=1}^{k}|\Delta^2 h_{yz}(y,z)|^{4r}. \tag{6.1.211}$$

Proof From the hypotheses, it is easy to observe that the following identities hold

$$f(y,z) = \sum_{s=1}^{y-1}\sum_{t=1}^{z-1}\Delta^2 f_{st}(s,t), \quad \text{for all } (y,z) \in Q_1, \tag{6.1.212}$$

$$f(y,z) = -\sum_{s=1}^{y-1}\sum_{t=z}^{m}\Delta^2 f_{st}(s,t), \quad \text{for all } (y,z) \in Q_2, \tag{6.1.213}$$

$$f(y,z) = -\sum_{s=y}^{k}\sum_{t=1}^{z-1} \Delta^2 f_{st}(s,t), \quad \text{for all } (y,z) \in Q_3, \tag{6.1.214}$$

$$f(y,z) = \sum_{s=y}^{k}\sum_{t=z}^{m} \Delta^2 f_{st}(s,t), \quad \textit{for all } (y,z) \in Q_4. \tag{6.1.215}$$

Thus from (6.1.212)–(6.1.215), we obtain for all $(y,z) \in Q$,

$$4|f(y,z)| \leq \sum_{s=1}^{k}\sum_{t=1}^{m} |\Delta^2 f_{st} f(s,t)|. \tag{6.1.216}$$

Raising both sides of (6.1.216) to the pth power and using Hölder's inequality twice with indices $p, p/(p-1)$ (see, [228]) to the right-hand side, we have

$$|f(y,z)|^p \leq \left(\frac{1}{4}\right)^p (km)^{p-1} \sum_{s=1}^{k}\sum_{t=1}^{m} |\Delta^2 f_{st}(s,t)|^p. \tag{6.1.217}$$

Similarly, we obtain

$$|g(y,z)|^q \leq \left(\frac{1}{4}\right)^q (km)^{q-1} \sum_{s=1}^{k}\sum_{t=1}^{m} |\Delta^2 g_{st}(s,t)|^q, \tag{6.1.218}$$

and

$$|h(y,z)|^r \leq \left(\frac{1}{4}\right)^r (km)^{r-1} \sum_{s=1}^{k}\sum_{t=1}^{m} |\Delta^2 h_{st}(s,t)|^r. \tag{6.1.219}$$

From (6.1.217)–(6.1.219) and using the elementary inequality $b_1b_2+b_2b_3+b_3b_1 \leq b_1^2+b_2^2+b_3^2$, for all b_1, b_2, b_3 reals and repeated application of Schwartz's inequality, we obtain

$$\begin{aligned}
&|f(y,z)|^p|g(y,z)|^q + |g(y,z)|^q|h(y,z)|^r + |h(y,z)|^r|f(y,z)|^p \\
&\leq \{f(y,z)|^p\}^2 + \{g(y,z)|^q\}^2 + \{|h(y,z)|^r\}^2 \\
&\leq \left\{\left(\frac{1}{4}\right)^p (km)^{p-1} \sum_{s=1}^{k}\sum_{t=1}^{m} |\Delta^2 f_{st}(s,t)|^p\right\}^2 \\
&\quad + \left\{\left(\frac{1}{4}\right)^q (km)^{q-1} \sum_{s=1}^{k}\sum_{t=1}^{m} |\Delta^2 g_{st}(s,t)|^q\right\}^2
\end{aligned}$$

$$+\left\{\left(\frac{1}{4}\right)^{r}(km)^{r-1}\sum_{s=1}^{k}\sum_{t=1}^{m}|\Delta^2 h_{st}(s,t)|^r\right\}^2$$

$$\le\left(\frac{1}{4}\right)^{2p}(km)^{2(p-1)}km\sum_{s=1}^{k}\sum_{t=1}^{m}|\Delta^2 f_{st}(s,t)|^{2p}$$

$$+\left(\frac{1}{4}\right)^{2q}(km)^{2(q-1)}km\sum_{s=1}^{k}\sum_{t=1}^{m}|\Delta^2 g_{st}(s,t)|^{2q}$$

$$+\left(\frac{1}{4}\right)^{2r}(km)^{2(r-1)}km\sum_{s=1}^{k}\sum_{t=1}^{m}|\Delta^2 h_{st}(s,t)|^{2r}. \tag{6.1.220}$$

Thus from (6.1.220) it follows that

$$\sum_{y=1}^{k}\sum_{z=1}^{m}[|f(y,z)|^p|g(y,z)|^q+|g(y,z)|^q|h(y,z)|^r+|h(y,z)|^r|f(y,z)|^p]$$

$$\le\left(\frac{km}{4}\right)^{2p}\sum_{y=1}^{k}\sum_{z=1}^{m}|\Delta^2 f_{yz}(y,z)|^{2p}+\left(\frac{km}{4}\right)^{2q}\sum_{y=1}^{k}\sum_{z=1}^{m}|\Delta^2 g_{yz}(y,z)|^{2q}$$

$$+\left(\frac{km}{4}\right)\sum_{y=1}^{k}\sum_{z=1}^{m}|\Delta^2 h_{yz}(y,z)|^{2r},$$

which gives us the desired inequality in (6.1.210).

From (6.1.217)–(6.1.219) and using the elementary inequalities $b_1b_2b_3(b_1+b_2+b_3)\le\frac{1}{3}(b_1b_2+b_2b_3+b_3b_1)^2$, $b_1b_2+b_2b_3+b_3b_1\le b_1^2+b_2^2+b_3^2$, $(b_1+b_2+b_3)^2\le 3(b_1^2+b_2^2+b_3^2)$, for b_1,b_2,b_3 reals (see, [395]) and repeated application of Schwartz's inequality, we obtain

$$|f(y,z)|^p|g(y,z)|^q|h(y,z)|^r\cdot(|f(y,z)|^p+|g(y,z)|^q+|h(y,z)|^r)$$

$$\le\left[\{|f(y,z)|^p\}^2\right]^2+\left[\{|g(y,z)|^q\}^2\right]^2+\left[\{|h(y,z)|^r\}^2\right]^2$$

$$\le\left[\left\{\left(\frac{1}{4}\right)^{p}(km)^{p-1}\sum_{s=1}^{k}\sum_{t=1}^{m}|\Delta^2 f_{st}(s,t)|^p\right\}^2\right]^2$$

$$+\left[\{\left(\frac{1}{4}\right)^{q}(km)^{q-1}\sum_{s=1}^{k}\sum_{t=1}^{m}|\Delta^2 g_{st}(s,t)|^q\}^2\right]^2$$

$$+\left[\left\{\left(\frac{1}{4}\right)^{r}(km)^{r-1}\sum_{s=1}^{k}\sum_{t=1}^{m}|\Delta^2 h_{st}(s,t)|^{r}\right\}^{2}\right]^{2}$$

$$\leq\left(\frac{1}{4}\right)^{4p}(km)^{4(p-1)}(km)^{2}km\sum_{s=1}^{k}\sum_{t=1}^{m}|\Delta^2 f_{st}(s,t)|^{4p}$$

$$+\left(\frac{1}{4}\right)^{4q}(km)^{4(q-1)}(km)^{2}km\sum_{s=1}^{k}\sum_{t=1}^{m}|\Delta^2 g_{st}(s,t)|^{4q}$$

$$+\left(\frac{1}{4}\right)^{4r}(km)^{4(r-1)}(km)^{2}km\sum_{s=1}^{k}\sum_{t=1}^{m}|\Delta^2 h_{st}(s,t)|^{4r}. \tag{6.1.221}$$

From (6.1.221) it follows that

$$\sum_{y=1}^{k}\sum_{z=1}^{m}|f(y,z)|^{p}|g(y,z)|^{q}|h(y,z)|^{r}\cdot(|f(y,z)|^{p}+|g(y,z)|^{q}+|h(y,z)|^{r})$$

$$\leq\left(\frac{1}{4}\right)^{4p}\sum_{y=1}^{k}\sum_{z=1}^{m}|\Delta^2 f_{yz}(y,z)|^{4p}+\left(\frac{1}{4}\right)^{4q}\sum_{y=1}^{k}\sum_{z=1}^{m}|\Delta^2 g_{yz}(y,z)|^{4q}$$

$$+\left(\frac{1}{4}\right)^{4r}\sum_{y=1}^{k}\sum_{z=1}^{m}|\Delta^2 f_{yz}(y,z)|^{4r},$$

which implies the required inequality in (6.1.211) and hence the proof is complete.
□

Remark 6.1.16 If we take $p = q = r = 2$ and $f(y,z) = g(y,z) = h(y,z) = c(y,z)$ in (6.1.210) and (6.1.211), then we get respectively the following Wirtinger type discrete inequalities

$$\sum_{y=1}^{k}\sum_{z=1}^{m}|c(y,z)|^{2}\leq\left(\frac{km}{4}\right)^{2}\sum_{y=1}^{k}\sum_{z=1}^{m}|\Delta^2 c_{yz}(y,z)|^{2}, \tag{6.1.222}$$

and

$$\sum_{y=1}^{k}\sum_{z=1}^{m}|c(y,z)|^{4}\leq\left(\frac{km}{4}\right)^{4}\sum_{y=1}^{k}\sum_{z=1}^{m}|\Delta^2 c_{yz}(y,z)|^{4}. \tag{6.1.223}$$

For discrete inequalities of the type (6.1.222) and (6.1.223) in one independent variables, see [86, 226, 228, 233, 395, 479, 533].

The next result, due to Yeh [705], is to study a discrete inequality of the Gronwall-Bellman type in n independent variables. As far as we know, before Yeh [705] the existing results for $n > 1$ ([611, 612]) related to the following results are only limited to $n = 3$. In the next theorem, we weaken the conditions of the known results for $n = 3$ as far as possible and generalize them to n independent variables in order to get a more compact and elegant form.

For $x = (x_1, \cdots x_n)$, $y = (y_1, \cdots x_n)$, $\hat{1} = (1, \cdots, 1)$, $\hat{0} = (0, \cdots, 0) \in \mathbb{N}^n$, we define

$$\sum_{y=\hat{0}}^{x-\hat{1}} u(y) := \sum_{y_1=0}^{x_1-1} \cdots \sum_{y_n=0}^{x_n-1} u(y_1, \cdots, y_n)$$

and $x := (x_1, \tilde{x})$, where $\tilde{x} := (x_2, \cdots, x_n)$. The natural partial ordering on $\mathbb{N}_0^n$ is defined by

$$x \leq y \text{ if and only if } x_i \leq y_i \text{ for } i = 1, 2, \cdots, n.$$

The difference operators on $\mathbb{N}_0^n$ are defined as follows:

$$\begin{cases} \Delta u_{x_1}(x_1, x_2, \cdots, x_n) := u(x_1 + 1, x_2, \cdots, x_n) - u(x_1, x_2, \cdots, x_n), \\ \Delta u_{x_2}(x_1, x_2, \cdots, x_n) := u(x_1, x_2 + 1, x_3, \cdots, x_n) - u(x_1, x_2, x_3, \cdots, x_n), \\ \cdots\cdots\cdots \\ \Delta u_{x_n}(x_1, x_2, \cdots, x_n) := u(x_1, \cdots, x_{n-1}, x_n + 1) - u(x_1, \cdots, x_{n-1}, x_n), \end{cases}$$

and

$$\Delta^2 u_{x_1 x_2}(x_1, x_2, \cdots, x_n) := \Delta u_{x_1}(x_1, x_2 + 1, x_3, \cdots, x_n) - \Delta u_{x_1}(x_1, x_2, \cdots, x_n),$$

and so on.

We begin with the following theorem.

Theorem 6.1.31 (The Yeh Inequality [704]) *Let $u(x), f(x)$ and $h(x)$ be real-valued non-negative functions defined on $\mathbb{N}_0^n$ and let $h(r) \in C(\mathbb{R}_+, \mathbb{R}_+)$ be a non-decreasing function such that*

$$Q(r) := \int_{r_0}^{r} \frac{ds}{H(s)}$$

exists for all $r \geq r_0 > 0$ with $r_0 > 0$ fixed, but arbitrary. If the following inequality holds for all $x \in \mathbb{N}_0^n$,

$$u(x) \leq f(x) + \sum_{t=\hat{0}}^{x-\hat{1}} h(t)H(u(t)), \tag{6.1.224}$$

then we have for all $\hat{0} \le x \le b$,

$$u(x) \le Q^{-1}[Q(\overline{f}(x)) + \sum_{t=\hat{0}}^{x-\hat{1}} h(t)], \tag{6.1.225}$$

where

(i) Q^{-1} *is the inverse function of* Q;
(ii) $\overline{f}(x) := \max\{f(y) : \hat{0} \le y \le x\}$,
(iii) $b \in \mathbb{N}_0^n$ *is chosen so that, for all* $\hat{0} \le x \le b$,

$$Q(\overline{f}(x)) + \sum_{t=\hat{0}}^{x-\hat{1}} h(t) \in Range\,(Q) = Dom\,(Q^{-1}).$$

Proof Let

$$v(x) := \sum_{t=\hat{0}}^{x-\hat{1}} h(t)H(u(t)),$$

then

$$u(x) \le f(x) + v(x), \tag{6.1.226}$$

$$\Delta^n v_x(x) = h(x)H(u(x)). \tag{6.1.227}$$

Since H is non-decreasing, it follows from (ii) and (6.1.224)–(6.1.227) that

$$\Delta^n v_x(x) \le h(x)H(f(x)) + v(x) \le h(x)H(\overline{f}(x)) + v(x)$$

for arbitrary $X \ge \hat{0}$ and $\hat{0} \le x \le X$. Set $V(x) := \overline{f}(x) + v(x) + \varepsilon$ ($\varepsilon \ge 0$). So $u(x) \le V(x)$ and for all $\hat{0} \le x \le X$,

$$\Delta^n V_x(x) = \Delta^n v_x(x) \le h(x)H(V(x)) \le h(x)H(V(x_1, \cdots, x_{n-1}, x_n+1)), \tag{6.1.228}$$

which implies

$$\frac{\Delta^{n-1} V_{x_1\cdots x_{n-1}}(x_1, \cdots, x_{n-1}, x_n + 1) - \Delta^{n-1} V_{x_1\cdots x_{n-1}}(x)}{H(V(x_1, \cdots, x_{n-1}, x_n + 1))} \le h(x).$$

Since $\Delta^k V_{x_1\cdots x_k}(x) = \Delta^k v_{x_1\cdots x_k}(x) \ge 0$ always, and $= 0$ if $x_i = 0$ for $i = k+1, \cdots, n$, and since $V(x)$ is non-decreasing in each component, it follows from the above inequality that

$$\frac{\Delta^{n-1}V_{x_1\cdots x_{n-1}}(x_1,\cdots,x_{n-1},x_n+1)}{H(V(x_1,\cdots,x_{n-1},x_n+1))}-\frac{\Delta^{n-1}V_{x_1\cdots x_{n-1}}(x)}{H(V(x))}\leq h(x).$$

Keeping $x_1,\cdots,x_{n-1}$ fixed in the above inequality, setting $x_n=t_n$ and summing over $t_n=0,1,\cdots,x_n-1$, we have

$$\frac{\Delta^{n-1}V_{x_1\cdots x_{n-1}}(x)}{H(V(x))}=\frac{\Delta^{n-2}V_{x_1\cdots x_{n-2}}(x_1,\cdots,x_{n-2},x_{n-1}+1,x_n)-\Delta^{n-2}V_{x_1\cdots x_{n-2}}(x))}{H(V(x))}$$
$$\leq\sum_{t_n=0}^{x_n-1}h(x_1,\cdots,x_{n-1},t_n).$$

Since $V(x)\leq V(x_1,\cdots,x_{n-2},x_{n-1}+1,x_n)$, we have

$$\frac{\Delta^{n-2}V_{x_1\cdots x_{n-2}}(x_1,\cdots,x_{n-2},x_{n-1}+1,x_n)}{H(V(x_1,\cdots,x_{n-2},x_{n-1}+1,x_n))}-\frac{\Delta^{n-2}V_{x_1\cdots x_{n-2}}(x)}{H(V(x))}$$
$$\leq\sum_{t_n=0}^{x_n-1}h(x_1,\cdots,x_{n-1},t_n).$$

Keeping $x_1,\cdots,x_{n-2},x_n$ fixed in the above inequality, setting $x_{n-1}=t_{n-1}$ and summing over $t_{n-1}=0,1,\cdots,x_{n-1}-1$, we infer

$$\frac{\Delta^{n-2}V_{x_1\cdots x_{n-2}}}{H(V(x))}\leq\sum_{t_{n-1}=0}^{x_{n-1}-1}\sum_{t_n=0}^{x_n-1}h(x_1,\cdots,x_{n-2},t_{n-1},t_n).$$

Continuing in this way, we have

$$\frac{\Delta V_{x_1}}{H(V(x))}\leq\sum_{\tilde{t}=\tilde{0}}^{\tilde{x}-\tilde{1}}h(x_1,\tilde{t}) \qquad (6.1.229)$$

which implies, for all $\hat{0}\leq x\leq X$,

$$Q(V(x))\leq Q(\overline{f}(X)+\varepsilon)+\sum_{t=\hat{0}}^{x-\hat{1}}H(t).$$

Thus for all $\hat{0} \leq x \leq X$,

$$u(x) \leq V(x) \leq Q^{-1}[Q(\overline{f}(X) + \varepsilon) + \sum_{t=\hat{0}}^{x-\hat{1}} h(t)].$$

Letting $\varepsilon \downarrow 0$, we have for all $\hat{0} \leq x \leq X$,

$$u(x) \leq Q^{-1}[Q(\overline{f}(X)) + \sum_{t=\hat{0}}^{x-\hat{1}} h(t)]. \tag{6.1.230}$$

In particular, (6.1.230) holds for $x = X \leq b$ provided that b is chosen as defined in (iii). Replacing X by x in (6.1.230) finally gives us for all $\hat{0} \leq x \leq b$,

$$u(x) \leq Q^{-1}[Q\overline{f}(x) + \sum_{t=\hat{0}}^{x-\hat{1}} H(t)].$$

This thus completes the proof. □

Corollary 6.1.23 (The Yeh Inequality [704]) *Under the hypotheses of Theorem 6.1.31, if H = identity mapping, then for any $x \in \mathbb{N}_0^n$,*

$$u(x) \leq \overline{f}(x) \prod_{t_1-=0}^{x_1-1} \left[1 + \sum_{t=\tilde{0}}^{\tilde{x}-\tilde{1}} h(t)\right]. \tag{6.1.231}$$

Proof It follows from (6.1.229) that

$$\frac{V(x_1+1,\tilde{x})}{V(x)} \leq 1 + \sum_{\tilde{t}=\tilde{0}}^{\tilde{x}-\tilde{1}} h(x_1,\tilde{t}).$$

Keeping $\tilde{x} = (x_2, \cdots, x_n)$ fixed in this inequality, setting $x_1 = t_1$ and taking the product over $t_1 = 0, 1, \cdots, x_1 - 1$, we get

$$V(x) \leq \left(\overline{f}(X) + \varepsilon\right) \prod_{t_1=0}^{x_1-1} \left[1 + \sum_{\tilde{t}=\tilde{0}}^{\tilde{x}-\tilde{1}} h(t_1,\tilde{t})\right].$$

Letting $\varepsilon \downarrow 0$ and replacing X by x as in the proof of Theorem 6.1.31, we can obtain the required bound in (6.1.231). □

Remark 6.1.17 In case that f is non-decreasing in each x_i, we have $\overline{f} \equiv f$.

Remark 6.1.18 For $n = 1$ and $f(x) \equiv$ constant, Theorem 6.1.31 reduces to the result of Hull and Luxemburg [290] (see also Beesack's lecture notes [54]). The continuous analogue of Theorem 6.1.31 is due to LaSalle [330].

Remark 6.1.19 For $n = 3$, Theorem 6.1.24 improves Theorem 3 of [611] and Corollary 6.1.23 is an improvement of Theorem 1 in [611] and Theorem 1 in [612]. For $n = 1$, Corollary 6.1.23 improves the results of Lemma 3.2 in Miller [393] and Corollary in Sugiyama [628].

The following theorem is an improvement of Theorem 2 of Pachpatte and Singare [611].

Theorem 6.1.32 (The Yeh Inequality [704]) *Let $u(x), f(x), h(x), H(r), Q(r)$ and $Q^{-1}(r)$ be defined as in Theorem 6.1.31 with $H(r)$ sub-additive and sub-multiplicative and let $g(x), k(x)$ be real-valued non-negative functions defined on $\mathbb{N}_0^n$. If the following inequality holds for all $x \in \mathbb{N}^n$,*

$$u(x) \le f(x) + g(x) \sum_{y=\hat{0}}^{x-\hat{1}} h(y) H\left[u(y) + g(y) \sum_{z=\hat{0}}^{y-\hat{1}} k(z) H(u(z))\right], \tag{6.1.232}$$

then for all $\hat{0} \le x \le b$,

$$u(x) \le f(x) + g(x) Q^{-1}\left\{Q\left[\sum_{y=\hat{0}}^{x-\hat{1}} (h(y)+k(y)) H(f(y))\right] + \sum_{y=\hat{0}}^{x-\hat{1}} (h(y)+k(y)) H(g(y))\right\} \tag{6.1.233}$$

where (i) $b \in \mathbb{N}^n$ is chosen so that for all $\hat{0} \le x \le b$,

$$Q\left[\sum_{y=\hat{0}}^{x-\hat{1}} (h(y) + k(y)) H(f(y))\right] + \sum_{y=\hat{0}}^{x-\hat{1}} \Big(h(y) + k(y)) H(g(y)\Big) \in \mathit{Range}\,(Q) = \mathit{Dom}\,(Q^{-1}).$$

Proof Set

$$w(x) := u(x) + g(x) \sum_{y=\hat{0}}^{x-\hat{1}} k(y) H(u(y)),$$

so $u(x) \leq w(x)$ and $H(u(x)) \leq H(w(x))$. It follows from (6.1.232) that

$$w(x) - g(x)\sum_{y=\hat{0}}^{x-\hat{1}} k(y)H(u(y)) = u(x) \leq f(x) + g(x)\sum_{y=\hat{0}}^{x-\hat{1}} H(w(y)),$$

or

$$w(x) \leq f(x) + g(x)\sum_{y=\hat{0}}^{x-\hat{1}} (k(y) + k(y))H(w(y)).$$

For brevity, set $b := h + k$ and $v(x) := \sum_{y=\hat{0}}^{x-\hat{1}} b(y)H(w(y))$. Then

$$\begin{aligned} w(x) &\leq f(x) + g(x)v(x) \\ \Delta^n v_x(x) &= b(x)H(\omega(x)) \leq b(x)H(f(x) + g(x)v(x)). \end{aligned}$$

Since H is also sub-additive and sub-multiplicative, we have

$$\begin{aligned} \Delta^n v_x(x) &\leq b(x)H(f(x)) + b(x)H(g(x))H(v(x)) \\ &= B(x) + C(x)H(v(x)). \end{aligned}$$

Now by repeated summation and using $\Delta^k v_{x_1,\ldots,x_k}(x) = 0$ if $x_i = 0$ for $i = k+1,\cdots,n$, we get for all $\hat{0} \leq x \leq X$,

$$\begin{aligned} v(x) &\leq \sum_{y=\hat{0}}^{x-\hat{1}} B(y) + \sum_{y=\hat{0}}^{x-\hat{1}} C(y)H(v(y)) = B_1(x) + \sum_{y=\hat{0}}^{x-\hat{1}} C(y)H(v(y)) \\ &\leq B_1(X) + \sum_{y=\hat{0}}^{x-\hat{1}} C(y)H(v(y)). \end{aligned}$$

Set

$$V(x) := B_1(X) + \sum_{y=\hat{0}}^{x-\hat{1}} C(y)H(v(y)),$$

so $V(x) = B_1(X)$ if any $x_i = 0$. Then for all $\hat{0} \leq x \leq X$,

$$\Delta^n V_x(x) = C(x)H(v(x)) \leq C(x)H(V(x)). \tag{6.1.234}$$

If we now proceed as in Theorem 6.1.31, we get for all $\hat{0} \le x \le X$,

$$V(x) \le Q^{-1}\left[Q(B_1(X)) + \sum_{y=\hat{0}}^{x-\hat{1}} C(y)\right].$$

Setting $x = X$ and then replacing X by x in the above inequality, we have, for all $\hat{0} \le x \le b$,

$$\begin{aligned} u(x) &\le w(x) \le f(x) + g(x)v(x) \le f(x) + g(x)V(x) \\ &\le f(x) + g(x)Q^{-1}\left\{Q\left[\sum_{y=\hat{0}}^{x-\hat{1}}(h(y) + k(y))H(f(y))\right]\right\} \\ &\quad + \left.\sum_{y=\hat{0}}^{x-\hat{1}}(h(y) + k(y))H(g(y))\right\}, \end{aligned}$$

where b is chosen as defined in (i). This completes the proof. □

Corollary 6.1.24 *Under the hypotheses of Theorem 6.1.32, if $H(s) \equiv s$, then for all $x \in \mathbb{N}_0^n$,*

$$u(x) \le f(x) + g(x)\sum_{y=\hat{0}}^{x-\hat{1}}(h(y) + k(y))f(y)\prod_{t_1=0}^{x_1-1}\left[1 + \sum_{\hat{t}=\hat{0}}^{\hat{x}-\hat{1}}(h + k)g(t_1, \hat{t})\right]. \tag{6.1.235}$$

Proof It follows from (6.1.234) that for all $\hat{0} \le x \le X$,

$$\Delta^n V_x(x) \le C(x)V(x).$$

Hence if we proceed as in the proof of Corollary 6.1.23, then we can obtain the desired bound in (6.1.235). □

Remark 6.1.20 Corollary 6.1.24 is an improvement of Theorem 1 in [610] for $n = 1$.

Remark 6.1.21 For $k \equiv 0$, the inequalities (6.1.233) and (6.1.234) reduce to the inequalities

$$u(x) \le f(x) + g(x)Q^{-1}\left\{Q\left[\sum_{y=\hat{0}}^{x-\hat{1}}(h(y))H(f(y))\right] + \sum_{y=\hat{0}}^{x-\hat{1}}(h(y)H(g(y))\right\}, \tag{6.1.236}$$

and

$$u(x) \leq f(x) + g(x)\left(\sum_{y=\hat{0}}^{x-\hat{1}} h(y)f(y)\right)\left(\prod_{t_1=0}^{x_1-1}\left[1 + \sum_{\tilde{t}=\tilde{0}}^{\tilde{x}-\tilde{1}}(h(t_1,\tilde{t})g(t_1,\tilde{t})\right]\right), \quad (6.1.237)$$

respectively. Inequality (6.1.236) extends a part of Theorem 1 of Pachpatte [465], which says mainly that if for all $n \in \mathbb{N}_0$,

$$u(n) \leq f(n) + g(n)p\left(\sum_{y=0}^{n-1} h(y)H(u(y))\right),$$

then for all $n \in \mathbb{N}_0$,

$$u(n) \leq f(n) + g(n)p\left\{Q^{-1}\left[Q\left(\sum_{y=0}^{n-1} h(y)H(f(y))\right) + \sum_{y=0}^{n-1} h(y)H(g(y))\right]\right\}.$$

In fact, Theorem 1 of [465] can also be extended to n independent variables. Inequality (6.1.237) extends the results of Jones [297] and Sugiyama [628].

Corollary 6.1.25 (The Yeh Inequality [704]) *Under the hypotheses of Theorem 6.1.32, if $g(x) \equiv 1, k(x) \equiv 0$ and is not required to be sub-multiplicative, then for all $\hat{0} \leq x \leq b$,*

$$u(x) \leq f(x) + Q^{-1}\left\{Q\left[\sum_{y=\hat{0}}^{x-\hat{1}} h(y)H(f(y))\right] + \sum_{y=\hat{0}}^{x-\hat{1}} h(y)\right\},$$

where $b \in \mathbb{N}_0^n$ is chosen so that for all $\hat{0} \leq x \leq b$,

$$Q\left[\sum_{y=\hat{0}}^{x-\hat{1}} h(y)H(f(y))\right] + \sum_{y=\hat{0}}^{x-\hat{1}} h(y) \in Range(Q) = Dom\,(Q^{-1}).$$

Theorem 6.1.33 (The Yeh Inequality [704]) *Let $u(x), f(x)$ and $h(x)$ be defined as in Theorem 6.1.31 with $f(x)$ non-decreasing in each x_i and let $Q(s) \in C(\mathbb{R}_+, \mathbb{R}_+)$ be non-decreasing with for all $r \geq 1$ and $s \geq 0$,*

$$\frac{1}{r}Q(s) \leq Q\left(\frac{s}{r}\right).$$

Let $H(s) \in C(\mathbb{R}_+, [1, +\infty))$ *be a strictly increasing, sub-additive and super-multiplicative function. If the inequality holds for all* $x \in \mathbb{N}_0^n$,

$$u(x) \leq f(x) + H^{-1}\left[Q\left(\sum_{y=\hat{0}}^{x-\hat{1}} h(y)H(f(y))\right)\right], \tag{6.1.238}$$

where H^{-1} *is the inverse function of* H, *then, for all* $\hat{0} \leq x \leq b$,

$$u(x) \leq f(x)H^{-1}\left\{1 + Q\left[G^{-1}\left(\sum_{y=\hat{0}}^{x-\hat{1}} h(y)\right)\right]\right\},$$

where for all $r \geq 0$,

$$G(r) := \int_0^r \frac{ds}{1 + Q(s)},$$

G^{-1} *is the inverse function of* G *and* $b \in \mathbb{N}_0^n$ *is chosen so that*

$$\sum_{y=\hat{0}}^{x-\hat{1}} h(y) \in Range\ (G) = Dom\ (G^{-1})$$

and for all $\hat{0} \leq x \leq b$,

$$1 + Q\left[G^{-1}\left(\sum_{y=\hat{0}}^{x-\hat{1}} h(y)\right)\right] \in Range\ (H) = Dom(H^{-1}).$$

Proof Since H is sub-additive, it follows from (6.1.238) that

$$H(u(x)) \leq H(f(x)) + Q\left[\sum_{y=\hat{0}}^{x-\hat{1}} h(y)H(u(y))\right].$$

Since $H(f(x)) \geq 1$ is non-decreasing, we have

$$\frac{H(u(x))}{H(f(x))} \leq 1 + Q\left[\sum_{y=\hat{0}}^{x-\hat{1}} h(y)\frac{H(u(y))}{H(f(y))}\right]. \tag{6.1.239}$$

Define

$$w(x) := \sum_{y=\hat{0}}^{x-\hat{1}} h(y)\frac{H(u(y))}{H(f(y))}.$$

Thus

$$w(x) = 0 \quad on \quad x_i = 0 \quad for \quad i = 1, 2, \cdots, n,$$

and

$$\Delta^n w_x(x) = h(x)\frac{H(u(x))}{H(f(x))}. \tag{6.1.240}$$

It follows from (6.1.239)–(6.1.240) that

$$\Delta^n w_x(x) \leq h(x)[1 + Q(w(x))].$$

Thus

$$\frac{\Delta^n w_x(x)}{1 + Q(w(x))} \leq h(x).$$

As in the proof of Theorem 6.1.31, we get

$$G(w(x_1 + 1, \tilde{x})) - G(w(x)) = \int_{w(x)}^{w(x_1+1,\tilde{x})} \frac{ds}{1 + Q(s)} \leq \frac{\Delta w_{x_1(x)}}{1 + Q(w(x))} \leq \sum_{\tilde{y}=\tilde{0}}^{\tilde{x}-\tilde{1}} h(x_1, \tilde{t}).$$

Keeping $\tilde{x} = (x_2, \cdots, x_n)$ fixed in this inequality , setting $x_1 = t_1$ and summing over $t_1 = 0, 1, \cdots x_1 - 1$, we obtain

$$G(w(x)) - G(w(0, \tilde{x})) \leq \sum_{t=\tilde{0}}^{x-\tilde{1}} h(t).$$

This and $G(0) = 0$ imply

$$w(x) \leq G^{-1}\left(\sum_{t=\tilde{0}}^{x-\tilde{1}} h(t)\right).$$

This and (6.1.239) imply

$$H(u(x)) \le H(f(x))(1 + Q(w(x))) \le H(f(x))\left(1 + Q\left[G^{-1}\left(\sum_{t=\tilde{0}}^{x-\tilde{1}} h(t)\right)\right]\right).$$

Since H is also super-multiplicative and increasing, $H^{-1}$is sub-multiplicative, we get

$$u(x) \le f(x)H^{-1}\left(1 + Q\left[G^{-1}\left(\sum_{t=\tilde{0}}^{x-\tilde{1}} h(t)\right)\right]\right)$$

which completes the proof. □

Remark 6.1.22 For $n = 2$, Theorem 6.1.33 is very close to Theorem 5 of Singare and Pachpatte [612].

Remark 6.1.23 For $n = 1$, the continuous analogues of Theorems 6.1.32 and 6.1.33 are given in Theorem 1 of [451] and Theorem 5 of [455], respectively.

For $u : \mathbb{N}_0^n \to \mathbb{R}_+$, we define

$$\sum_{y=\hat{0}}^{x-\hat{1}} u(y) := \sum_{y_1=0}^{x_1-1} \cdots \sum_{y_n=0}^{x_n-1} u(y_1, \ldots, y_n), \sum_{y=\hat{0}}^{x-\hat{1}} u(y) := 0 \text{ for some } x_i = 0,$$

$$\prod_{y=\hat{0}}^{x-\hat{1}} u(y) := \prod_{y_1=0}^{x_1-1} \cdots \prod_{y_n=0}^{x_n-1} u(y_1, \ldots, y_n), \prod_{y=\hat{0}}^{x-\hat{1}} u(y) := 0 \text{ for some } x_i = 0.$$

The next theorem is due to Yeh [705].

Theorem 6.1.34 (The Yeh Inequality [705]) *Let $u(x), k(x) : \mathbb{N}_0^n \to \mathbb{R}_+$ and $f(x; s) : \mathbb{N}_0^{2n} \to \mathbb{R}_+$ with $s \le x$. If for all $x \in \mathbb{N}_0^n$,*

$$u(x) \le k(x) + \sum_{s=\hat{0}}^{x-\hat{1}} f(x; s)u(s), \tag{6.1.241}$$

then for all $x \in \mathbb{N}_0^n$,

$$u(x) \le K(x) \prod_{s_1=0}^{x_1-1} \left[1 + \sum_{\tilde{s}=\widetilde{0}}^{\tilde{x}-\tilde{1}} F(x; s_1, \tilde{s})\right], \tag{6.1.242}$$

where

$$K(x) := \sup\left\{k(s) : \hat{0} \le s \le x\right\}, \quad F(x;s) := \sup\left\{f(t;s) : \hat{0} \le t \le x\right\}. \tag{6.1.243}$$

Proof For any fixed point Y on $\mathbb{N}_0^n$, it follows from (6.1.241) and (6.1.243) that for all $\hat{0} \le x \le Y$,

$$u(x) \le K(Y) + \sum_{s=\hat{0}}^{x-\hat{1}} F(Y;s)u(s).$$

Setting

$$V(Y;x) := K(Y) + \sum_{s=\hat{0}}^{x-\hat{1}} F(Y;s)u(s) + \varepsilon, \quad \varepsilon > 0, \tag{6.1.244}$$

we infer from (6.1.244),

$$u(x) \le V(Y;x)$$

and

$$\Delta^n V_x(Y;x) = F(Y;x)u(x) \le F(Y;x)V(Y;x). \tag{6.1.245}$$

It follows from (6.1.245) that

$$\frac{\Delta^{n-1}V_{x_1\ldots x_{n-1}}(Y;x_1,\ldots,x_{n-1},x_n+1)}{V(Y;x_1,\ldots,x_{n-1},x_n+1)} - \frac{\Delta^{n-1}V_{x_1\ldots x_{n-1}}(Y;x)}{V(Y;x)} \le F(Y;x).$$

Keeping $x_1, .., x_{n-1}$ fixed in the above inequality, setting $x_n = s_n$, and summing over $s_n = 0, 1, .., Y_n - 1$, we have

$$\frac{\Delta^{n-1}V_{x_1\ldots x_{n-1}}(Y;x_1,\ldots,x_{n-1},Y_n)}{V(Y;x_1,\ldots,x_{n-1},Y_n)} \le \sum_{s_n=0}^{Y_n-1} F(Y;x_1,\ldots,x_{n-1},s_n).$$

Continuing in this way and using the method described in [705], we have

$$\frac{V_{x_1}(Y;x_1,\tilde{Y})}{V(Y;x_1,\tilde{Y})} = \frac{V(Y;x_1+1,\tilde{Y})}{V(Y;x_1,\tilde{Y})} - 1 \le \sum_{\tilde{s}=\tilde{0}}^{\tilde{Y}-\tilde{1}} F(Y;x_1,\tilde{s}).$$

Keeping $\tilde{Y}$ fixed in the above inequality, setting $x_1 = s_1$, and taking the product over $s_1 = 0, 1, \ldots, Y_1 - 1$, we conclude

$$u(Y) \leq V(Y; Y) \leq [K(Y) + \varepsilon] \prod_{s_1=0}^{Y_1-1} \left[1 + \sum_{\tilde{s}=\tilde{0}}^{\tilde{Y}-\tilde{1}} F(Y; s_1, \tilde{s}) \right].$$

Letting $\varepsilon \downarrow 0$ and replacing Y by x, we obtain the desired result (6.1.242). □

A simpler proof of Theorem 6.1.34 is as follows. For any fixed point Y of $\mathbb{N}_0^n$, it follows from (6.1.241) and (6.1.243) that, for all $\hat{0} \leq x \leq Y$,

$$u(x) \leq K(Y) + \sum_{s=\hat{0}}^{x-\hat{1}} F(Y; s) u(s).$$

By Corollary 6.1.24, we have for all $\hat{0} \leq x \leq Y$,

$$u(x) \leq K(Y) \prod_{s_1=0}^{x_1-1} \left[1 + \sum_{\tilde{s}=\tilde{0}}^{\tilde{x}-\tilde{1}} F(Y; s_1, \tilde{s}) \right]. \tag{6.1.246}$$

In particular, (6.1.246) holds for $x = Y$. Replacing Y by x in (6.1.246) gives us the desired result (6.1.242). □

Remark 6.1.24 For $n = 1$, the continuous analogue of Theorem 6.1.34 is due to Butler and Rogers [126].

As an application of Theorem 6.1.34, we have the following theorem.

Theorem 6.1.35 (The Yeh Inequality [705]) *Let u, k, f, K, F be defined as in Theorem 6.1.34. Let $g(x; s) : \mathbb{N}_0^{2n} \to \mathbb{R}_+$ with $s \leq x$. If for all $x \in \mathbb{N}_0^n$,*

$$u(x) \leq k(x) + \sum_{s=\hat{0}}^{x-\hat{1}} f(x; s) \left[u(s) + \sum_{t=\hat{0}}^{s-\hat{1}} g(s; t) u(t) \right], \tag{6.1.247}$$

then

$$u(x) \leq K(x) \prod_{s_1=0}^{x_1-1} \left\{ 1 + \sum_{\tilde{s}=\tilde{0}}^{\tilde{x}-\tilde{1}} \left[F(x; s_1, \tilde{s}) + G(x; s_1, \tilde{s}) \right] \right\} \tag{6.1.248}$$

or

$$u(x) \leq k(x) + \sum_{s=\hat{0}}^{x-\hat{1}} f(x;s)K(s) \prod_{t_1=0}^{s_1-1} \left\{ 1 + \sum_{\tilde{t}=\tilde{0}}^{\tilde{s}-\tilde{1}} \left[F(s;t_1,\tilde{t}) + G(s;t_1,\tilde{t}) \right] \right\}, \tag{6.1.249}$$

where $G(x;s) := \sup \left\{ g(t;s) : \hat{0} \leq t \leq x \right\}$.

Proof Let

$$w(x) := u(x) + \sum_{t=\hat{0}}^{x-\hat{1}} g(x;t)u(t).$$

Then

$$u(x) \leq w(x) \tag{6.1.250}$$

which, along with (6.1.247), implies

$$u(x) = w(x) - \sum_{s=\hat{0}}^{x-\hat{1}} g(x;s)u(s) \leq k(x) + \sum_{s=\hat{0}}^{x-\hat{1}} f(x;s)w(s). \tag{6.1.251}$$

Thus

$$w(x) \leq k(x) + \sum_{s=\hat{0}}^{x-\hat{1}} \left[f(x;s) + g(x;s) \right] w(s).$$

Applying Theorem 6.1.34, we have

$$w(x) \leq K(x) \prod_{s_1=0}^{x_1-1} \left\{ 1 + \sum_{\tilde{s}=\tilde{0}}^{\tilde{x}-\tilde{1}} \left[F(x;s_1,\tilde{s}) + G(x;s_1,\tilde{s}) \right] \right\}. \tag{6.1.252}$$

From (6.1.250) and (6.1.252), we have the desired result (6.1.248). From (6.1.251) and (6.1.252), the desired result (6.1.249) follows. □

Remark 6.1.25 The discrete inequalities established Theorems 6.1.34–6.1.35 can also be extended either to nonlinear cases as shown in [705], or continuous analogues, or both; we omit the details.

Theorem 6.1.36 (The Pachpatte Inequality [495]) *Let* $u(x) \geq 0$ *be a real-valued function defined for all* $x \in \mathbb{N}_0^n$. *If*

(c_1) *Let* $u(x) \geq 0$ *be a non-negative real-valued function defined for all* $x \in \mathbb{N}_0$. *If for all* $x \in \mathbb{N}_0^n$,

$$u^2(x) \leq c^2 + 2M(x, f(s)u^2(s) + g(s)u(s)), \tag{6.1.253}$$

then for all $x \in \mathbb{N}_0^n$,

$$u(x) \leq p(x) \prod_{s_1=0}^{x_1-1} (1 + \bar{M}(s_1, x_2, \ldots, x_n, f(s))), \tag{6.1.254}$$

where for all $x \in \mathbb{N}_0^n$,

$$p(x) = c + M(x, g(s)). \tag{6.1.255}$$

(c_2) *Let* $u(x) \geq u_0 \geq 0$ *be a real-valued function defined for all* $x \in \mathbb{N}_0$; u_0 *is a real constant. Let* $W(u)$ *be a continuous non-decreasing real-valued function defined on an interval* $I = [u_0, +\infty)$ *and* $W(u) > 0$ *on* $(u_0, +\infty)$, $W(u_0) = 0$. *If for all* $x \in \mathbb{N}_0^n$,

$$u^2(x) \leq c^2 + 2M(x, f(s)u(s)W(u(s)) + g(s)u(s)), \tag{6.1.256}$$

then for $x_i, x_i^* \in \mathbb{N}_0$ *and* $0 \leq x_i \leq x_i^*$, $i = 1, \ldots, n$,

$$u(x) \leq \Omega^{-1}(\Omega(p(x)) + M(x, f(s))), \tag{6.1.257}$$

where $p(x)$ *is as defined in (6.1.255),* Ω, Ω^{-1} *are as defined in Theorem 2.3.12* (A_2) *and* $x_i^* \in \mathbb{N}_0$, $i = 1, \ldots, n$ *be chosen so that*

$$\Omega(p(x)) + M(x, f(s)) \in \mathit{Dom}\,(\Omega^{-1}),$$

for all $x_i \in \mathbb{N}_0$, $i = 1, \ldots, n$, *such that* $0 \leq x_i \leq x_i^*$.

(c_3) *Let* $u(x) \geq 0$ *be a real-valued function defined for all* $x \in \mathbb{N}_0^n$ *and the function* $L : \mathbb{N}_0^n \times \mathbb{R}_+ \to \mathbb{R}_+$ *satisfies the condition: for all* $x \in \mathbb{N}_0^n$ *and* $v \geq w \geq 0$,

$$0 \leq L(x, v) - L(x, w) \leq k(x, w)(v - w), \tag{6.1.258}$$

where k *is a real-valued non-negative function defined for all* $x \in \mathbb{N}_0^n$, $w \geq 0$, *If for all* $x \in \mathbb{N}_0^n$,

$$u^2(x) \leq c^2 + 2M(x, f(s)u(s)L(s, u(s)) + g(s)u(s)), \tag{6.1.259}$$

then for all $x \in \mathbb{N}_0$,

$$u(x) \leq p(x) + q(x) \prod_{s_1=0}^{x_1-1} (1 + \bar{M}(s_1, x_2, \ldots, x_n f(s) k(s, p(s)))), \quad (6.1.260)$$

where $p(x)$ *is as defined in (6.1.255) and for all* $x \in \mathbb{N}_0^n$,

$$q(x) = M(x, f(s) L(s, p(s))). \quad (6.1.261)$$

Proof (c_1) Assume that c is positive and define a function $z(x)$ by

$$z(x) \leq c^2 + 2M(x, f(s)u^2(s) + g(s)u(s)). \quad (6.1.262)$$

From (6.1.262), and using the fact that $u(x) \leq \sqrt{z(x)}$, it is easy to observe that

$$\Delta_n \cdots \Delta_1 z(x) \leq 2\sqrt{z(x)}\Big(f(x)\sqrt{z(x)} + g(x)\Big). \quad (6.1.263)$$

Using the fact that $\sqrt{z(x)} \leq \sqrt{z(x_1, \ldots, x_{n-1}, x_n + 1)}$ in (6.1.263), we observe that

$$\Delta_n \cdots \Delta_1 z(x) \leq 2\sqrt{z(x_1, \ldots, x_{n-1}, x_n + 1)}\Big(f(x)\sqrt{z(x)} + g(x)\Big). \quad (6.1.264)$$

From (6.1.264), it follows that

$$\frac{\Delta_{n-1} \cdots \Delta_1 z(x_1, \ldots, x_{n-1}, x_n + 1)}{\sqrt{z(x_1, \ldots, x_{n-1}, x_n + 1)}} - \frac{\Delta_n \cdots \Delta_1 z(x_1, \ldots, x_{n-1}, x_n)}{\sqrt{z(x_1, \ldots, x_{n-1}, x_n)}} \leq 2\Big(f(x)\sqrt{z(x)} + g(x)\Big). \quad (6.1.265)$$

Keeping $x_1, \ldots, x_{n-1}$ fixed in (6.1.265), setting $x_n = s_n$ and summing over $s_n = 0, 1, \ldots, x_n - 1$, we obtain

$$\frac{\Delta_{n-1} \cdots \Delta_1 z(x)}{\sqrt{z(x)}} \leq 2\sum_{s_n=0}^{x_n-1}\Big(f(x_1, \ldots, x_{n-1}, s_n)\sqrt{z(x_1, \ldots, x_{n-1}, s_n)} + g(x_1, \ldots, x_{n-1}, s_n)\Big). \quad (6.1.266)$$

Here we have used the fact that $\Delta_n \cdots \Delta_1 z(x_1, \ldots, x_{n-1}, 0) = 0$. From (6.1.266), and using the fact that $\sqrt{z(x)} \leq \sqrt{z(x_1, \ldots, x_{n-1}+1, x_n)}$, we have

$$\frac{\Delta_{n-2} \cdots \Delta_1 z(x_1, \ldots, x_{n-1}+1, x_n)}{\sqrt{z(x_1, \ldots, x_{n-2}, x_{n-1}+1, x_n)}} - \frac{\Delta_{n-2} \cdots \Delta_1 z(x_1, \ldots, x_{n-2}, x_{n-1}, x_n)}{\sqrt{z(x_1, \ldots, x_{n-1}, x_n)}}$$
$$\leq 2\sum_{s_n=0}^{x_n-1}\Big(f(x_1, \ldots, x_{n-1}, s_n)\sqrt{z(x_1, \ldots, x_{n-1}, s_n)} + g(x_1, \ldots, x_{n-1}, s_n)\Big). \tag{6.1.267}$$

Keeping $x_1, \ldots, x_{n-2}, x_n$ fixed in (6.1.267), setting $x_{n-1} = s_{n-1}$ and summing over $s_{n-1} = 0, 1, \ldots, x_{n-1} - 1$, we obtain

$$\frac{\Delta_{n-2} \cdots \Delta_1 z(x)}{\sqrt{z(x)}} \leq 2\sum_{s_{n-1}=0}^{x_{n-1}-1}\sum_{s_n=0}^{x_n-1} \times \Big(f(x_1, \ldots, x_{n-1}, s_n)\sqrt{z(x_1, \ldots, x_{n-1}, s_n)}$$
$$+ g(x_1, \ldots, x_{n-1}, s_n)\Big).$$

Here we have used the fact that $\Delta_{n-2} \cdots \Delta_1 z(x_1, \ldots, x_{n-2}, 0, x_n) = 0$. Proceeding in this way, we obtain

$$\frac{\Delta_1 z(x)}{\sqrt{z(x)}} \leq 2\sum_{s_2=0}^{x_2-1} \cdots \sum_{s_n=0}^{x_n-1} \times \Big(f(x_1, s_2, \ldots, s_n)\sqrt{z(x_1, s_2, \ldots, s_n)} + g(x_1, s_2, \ldots, s_n)\Big). \tag{6.1.268}$$

Using the facts that $\sqrt{z(x)} > 0$, $\Delta_1 z(x) \geq 0$, $\sqrt{z(x)} \leq \sqrt{z(x_1+1, x_2, \ldots, x_n)}$, we observe that

$$\Delta_1\sqrt{z(x)} = \frac{z(x_1+1, x_2, \ldots, x_n) - z(x)}{\sqrt{z(x_1+1, x_2, \ldots, x_n)} + \sqrt{z(x)}} \leq \frac{\Delta_1 z(x)}{2\sqrt{z(x)}}. \tag{6.1.269}$$

Using (6.1.268) in (6.1.269), we get

$$\Delta_1\sqrt{z(x)} \leq 2\sum_{s_2=0}^{x_2-1} \cdots \sum_{s_n=0}^{x_n-1}\Big(f(x_1, s_2, \ldots, s_n)\sqrt{z(x_1, s_2, \ldots, s_n)}+g(x_1, s_2, \ldots, s_n)\Big). \tag{6.1.270}$$

Now, keeping $x_1, \ldots, x_n$ fixed in (6.1.270), setting $x_1 = s_1$ and summing over $s_1 = 0, 1, \ldots, x_1 - 1$, we obtain

$$\sqrt{z(x)} \leq p(x) + M\Big(x, f(s)\sqrt{z(s)}\Big). \tag{6.1.271}$$

Since $p(x)$ is positive and monotone non-decreasing in x, from (6.1.271) it follows that

$$\frac{\sqrt{z(x)}}{p(x)} \leq 1 + M\Big(x, f(s)\frac{\sqrt{z(s)}}{p(s)}\Big). \tag{6.1.272}$$

Define a function $v(x)$ by

$$v(x) = 1 + M\Big(x, f(s)\frac{\sqrt{z(s)}}{p(s)}\Big). \tag{6.1.273}$$

From (6.1.273), and using the fact that $\frac{\sqrt{z(x)}}{p(x)} \leq v(x)$, it is easy to observe that

$$\Delta_n \cdots \Delta_1 v(x) \leq f(x)v(x). \tag{6.1.274}$$

Using the fact that $v(x) \leq v(x_1, \ldots, x_{n-1}, x_n + 1)$ in (6.1.274), we have

$$\frac{\Delta_{n-1} \cdots \Delta_1 v(x_1, \ldots, x_{n-1}, x_n + 1)}{v(x_1, \ldots, x_{n-1}, x_n + 1)} - \frac{\Delta_{n-1} \cdots \Delta_1 v(x_1, \ldots, x_{n-1}, x_n)}{v(x_1, \ldots, x_{n-1}, x_n)} \leq f(x). \tag{6.1.275}$$

Now, following the same step, below (6.1.265) up to (6.1.268), we obtain

$$\frac{\Delta_1 v(x)}{v(x)} \leq \sum_{s_2=0}^{x_2-1} \cdots \sum_{s_n=0}^{x_n-1} f(x_1, s_2, \ldots, s_n),$$

i.e.,

$$v(x_1 + 1, x_2, \ldots, x_n) \leq v(x)\Big(1 + \sum_{s_2=0}^{x_2-1} \cdots \sum_{s_n=0}^{x_n-1} f(x_1, s_2, \ldots, s_n)\Big). \tag{6.1.276}$$

Now keeping $x_2, \ldots, x_n$ of x fixed in (6.1.276), setting $x_1 = s_1$ and substituting $s_1 = 0, 1, \ldots, x_1 - 1$ successively and using the fact that $v(0, x_2, \ldots, x_n) = 1$, we obtain

$$v(x) \leq \prod_{s_1=0}^{x^1-1} (1 + \bar{M}(s_1, x_2, \ldots, x_n, f(s))). \tag{6.1.277}$$

Using (6.1.277) in (6.1.272), and observing the fact that $u(x) \leq \sqrt{z(x)}$, we get the desired inequality in (6.1.254).

The proof of the case when c is non-negative can be completed as mentioned in the proof of Part (a_1) of Theorem 1.2.11. This completes the proof of Part (c_1). The proof of the inequalities in (c_2) and (c_3) are respectively similar to the proof of Part (a_2) and Part (a_3) of Theorem 2.3.12 and closely resemble the proof of Part (c_1) given above (see also [489]). Here, we omit the details. □

The following results concern some discrete inequalities of the Wendroff type in several variable which include and generalize some known results of Singare and Pachpatte [611, 612] and Yeh [704, 705]. These results are due to Mao [371].

For $i = (i_1, i_2, \cdots, i_n)$, $j = (j_1, j_2, \cdots, j_n)$, $1 = (1, 1, \cdots, 1)$, $0 = (0, 0, \cdots, 0) \in \mathbb{N}_0^n$, we define

$$\sum_{j=0}^{i-1} u(j) := \sum_{j_1=0}^{i_1-1} \cdots \sum_{j_n=0}^{i_n-1} u(j_1, j_2, \cdots, j_n)$$

We let $\sum_{i=0}^{-1} b(i) = 0$. Let $I = (I_1, I_2, \cdots, I_n) \in \mathbb{N}_0^n$ be fixed with $0 \leq I$. We shall use notations $[0, I] := \{i \in \mathbb{N}_0^n : 0 \leq i \leq I\}$, $[j, j+1) := \{x \in \mathbb{R}^n : j \leq x < j+1\}$, $(j, j+1] := \{x \in \mathbb{R}^n : j < x \leq j+1\}$ and so on.

The next theorem gives us a generalization of Yeh [704], who used many techniques to get this result. However, the present proof is very simple.

Theorem 6.1.37 (The Mao Inequality [371]) *Let $\xi(i), \eta(i) : \{0, I\} \to \mathbb{R}_+$ with $\eta(i)$ positive and non-decreasing. Let $\mu(i_1) : \{0, I_1\} \to \mathbb{R}_+$ be non-decreasing. Let $\phi(i,j)$ be a non-negative function defined for all $i, j \in \mathbb{N}_0^n$ with $0 \leq j \leq i \leq I$. Let H be a continuous function defined on $\mathbb{R}_+$ such that $H(v)$ is positive and non-decreasing for all $v > 0$. Then if the inequality holds for all $i \in \{0, I\}$,*

$$\xi(i) \leq \mu(i_1) + \eta(i) + \sum_{j=0}^{i-1} \phi(i-1, j) H(\xi(j)), \tag{6.1.278}$$

then for all $i \in \{0, i^*\}$,

$$\xi(i) \leq G^{-1}\left(G(\mu(0) + \eta(i)) + \sum_{l=0}^{i_1-1} \frac{\mu(l+1) - \mu(l)}{H(\mu(l) + \eta(i))} + \sum_{j=0}^{i-1} \phi(i-1, j) \right), \tag{6.1.279}$$

where G and G^{-1} are as defined in Theorem 7.2.12 below and $i^* \in \{0, I\}$ is such that for all $i \in \{0, i^*\}$,

$$G(\mu(0) + \eta(i) + \sum_{l=0}^{i_1-1} \frac{\mu(l+1) - \mu(l)}{H(\mu(l) + \eta(i))} + \sum_{j=0}^{i-1} \phi(i-1, j)) \in Dom\,(G^{-1}).$$

Proof Define $A(x), u(x), a(x_1), n(x)$ and $f(x,s)$ for all $0 \le s \le x \le I$ as follows:

$$A(x) := (A_1(x_1), \cdots, A_n(x_n)),\quad A_1(x_1) := \sum_{k=1}^{I_1} \chi_{[k,+\infty)}(x_1),$$

$$u(x) := \sum_{j=0}^{I} \xi(j)\chi_{[j,j+1)}(x),\quad a(x_1) := \sum_{l=0}^{I_1} \mu(l)\chi_{[l,l+1)}(x_1),$$

$$n(x) := \sum_{j=0}^{I} \eta(j)\chi_{[j,j+1)}(x),\quad f(x,s) := \sum_{i=j}^{I}\sum_{j=0}^{I} \varphi(i,j)\chi_{(i,i+1]\times(j,j+1]}(x,s)$$

where χ_D denotes the indicator function of D. Thus it follows from (6.1.278) that for all $0 \le x \le I$,

$$u(x) \le a(x_1) + n(x) + \int_0^x f(x,s)H(u(s_-))dA(s), \tag{6.1.280}$$

which, by Theorem 7.2.13 below, yields desired result (6.1.279) immediately. This completes the proof. □

Remark 6.1.26 If $\mu(i_1) = 0$, and $\varphi(i,j) = \varphi(j)$, then Theorem 6.1.37 reduces to Theorem 1 of Yeh [703].

Similarly, we can use Theorem 7.2.14 below to prove the following discrete inequality.

Theorem 6.1.38 (The Mao Inequality [371]) *Let $\xi(i), \eta(i), : \{0, I\} \to \mathbb{R}_+$ with $\eta(i)$ positive and non-decreasing. Let $\varphi_k(i,j)$ $(k = 1, 2, \cdots, m)$ be a non-negative function defined for all $i, j \in \mathbb{N}_0^n$ with $0 \le j \le i \le I$. Let $r_k \in (0, 1]$, $k = 1, 2, \cdots, m$. If for all $i \in \{0, I\}$,*

$$\xi(i) \le \eta(i) + \sum_{k=1}^{m}\sum_{j=1}^{i-1} \varphi_k(i-1,j)(\xi(j))^{r_k}, \tag{6.1.281}$$

then for all $i \in \{0, I\}$,

$$\xi(i) \le \eta(i)\prod_{k=1}^{m} G_k(i), \tag{6.1.282}$$

where

$$G_k(i) = \begin{cases} \left[1 + (1-r_n)\{\prod_{l=1}^{k-1} G_l(i)\}\sum_{j=0}^{i-1}\varphi_k(i-1,j)\right]^{1/(1-r_k)}, & \text{if } 0 < r_k < 1, \\ \exp(\{\prod_{l=1}^{k-1} G_l(i)\}\sum_{j=0}^{i-1}\varphi_k(i-1,j)), & \text{if } r_k = 1, \end{cases}$$

here we use notation

$$\prod_{i=1}^{0} G_l(i) = 1.$$

In order to get some new discrete inequalities with retardation, we denote by F_2 the family of all function $\alpha : \{0, I\} \to \{0, I\}$ such that $\alpha(i) \le i$ for all $i \in \{0, I\}$.

We now use Theorem 7.2.15 below to prove the following theorem which generalizes Theorem 1 of Yeh [703], hence, Singare and Pachpatte [611] (Theorem 3) and [613] (Theorem 1).

Theorem 6.1.39 (The Mao Inequality [371]) *Let $\xi(i), \mu(i_1), \eta(i), \varphi(i,j)$ be defined as in Theorem 6.1.37. Let $\gamma(i), \zeta(i) : \{0, I\} \to [1, +\infty)$ and $\alpha(i) \in F_2, H \in \mathcal{H}$ with corresponding multiplier function Φ. If the inequality holds for all $i \in \{0, I\}$,*

$$\xi(i) \le \gamma(i)(\mu(i) + \eta(i)) + \zeta(i) \sum_{j=0}^{i-1} \varphi(i-1,j) H(\xi(\alpha(j))), \tag{6.1.283}$$

then for all $i \in \{0, I\}$,

$$\begin{aligned} \xi(i) \le \gamma(i)\zeta(i) G^{-1}(G(\mu(0) + \eta(i)) + \sum_{l=0}^{i_1-1} \frac{\mu(l+1) - \mu(l)}{H(\mu(l) + \eta(i))} \\ + \sum_{j=0}^{i-1} [\varphi(i-1,j)/\gamma(j)] \Phi\{\gamma(\alpha(j))\zeta(\alpha(j))\}), \end{aligned} \tag{6.1.284}$$

where G and G^{-1} are as defined in Theorem 7.2.12 below, and $i^* \in \{0, I\}$ is such that, for all $i \in \{0, i^*\}$,

$$\begin{aligned} G(\mu(0) + \eta(i)) + \sum_{l=0}^{i_1-1} \frac{\mu(l+1) - \mu(l)}{H(\mu(l) + \eta(i))} \\ + \sum_{j=0}^{i-1} [\varphi(i-1,j)/\gamma(j)] \Phi(\gamma(\alpha(j))\zeta(\alpha(j))) \in Dom\,(G^{-1}). \end{aligned}$$

Proof Define $\beta(i) \in F_2$ as follows:

$$\beta(i) := \begin{cases} 0, & i \in \{0, 1\}, \\ \alpha(i-1) + 1, & i \in \{1, I\}. \end{cases}$$

Thus inequality (6.1.284) is equivalent to for all $i \in \{0, I\}$,

$$\xi(i) \leq \gamma(i)(\mu(i_1) + \eta(i)) + \zeta(i) \sum_{j=0}^{i-1} \varphi(i-1,j) H(\xi(\beta(j+1)-1)). \qquad (6.1.285)$$

Let $A(x), u(x), a(x_1), n(x)$ and $f(x,s)$ be as defined in the proof of Theorem 6.1.37.
Furthermore, we define, for all $0 \leq x \leq I$,

$$q(x) := \sum_{j=0}^{I} \zeta(j) chi_{[j,j+1)}(x), \quad \sigma(x) := \sum_{j=0}^{I} \beta(j) \chi_{[j,j+1)}(x).$$

Hence it follows from (7.2.76) below that for all $0 \leq x \leq X$,

$$u(x) \leq h(x)(a(x_1) + n(x)) + q(x) \int_0^x f(x,s) H(u(\sigma(s)_-)) dA(s), \qquad (6.1.286)$$

which, by Theorem 7.2.15 below, implies for all $i \in \{0, i^*\}$,

$$\xi(i) \leq \gamma(i)\zeta(i) G^{-1}\Bigg(G(\mu(0) + \eta(i)) + \sum_{l=0}^{i_1-1} \frac{\mu(l+1) - \mu(l)}{H(\mu(l) + \eta(i))}$$
$$+ \sum_{j=0}^{i-1} [\varphi(i-1,j)/\gamma(j)] \Phi(\gamma(\beta(j+1)-1)\zeta(\beta(j+1)-1)) \Bigg).$$

This is desired result (6.1.285) which completes the proof. □

6.2 Nonlinear Multi-Dimensional Nonlinear Discrete Inequalities

For all $x = (x_1, \cdots, x_n) \in \mathbb{N}^n$ and $c : \mathbb{N}^n \to \mathbb{R}$, we define the forward difference operators as follows

$$\begin{cases} \Delta_1 c(x_1, \cdots, x_n) = c(x_1 + 1, x_2, \cdots, x_n) - c(x_1, x_2, \cdots, x_n), \\ \vdots \\ \Delta_n c(x_1, \cdots, x_n) = c(x_1, \cdots, x_{n-1}, x_{n+1}) - c(x_1, \cdots, x_{n-1}, x_n) \end{cases}$$

and the operator D by

$$Dc(x) = (\Delta_1 c(x), \cdots, \Delta_n c(x)),$$

and

$$|Dc(x)| = [|\Delta_1 c(x)|^2 + \cdots + |\Delta_n c(x)|^2]^{1/2}. \tag{6.2.1}$$

Let B be a bounded domain in $\mathbb{N}^n$ with $n \geq 2$ defined by $B = \{x : \hat{1} \leq x \leq a + \hat{1}\}$, where $\hat{1} = (1, \cdots, 1) \in \mathbb{N}^n, x = (x_1, \cdots, x_n) \in \mathbb{N}^n, a = (a_1, \cdots, a_n) \in \mathbb{N}^n$. We denote by $G(B)$ the class of functions $c : B \to \mathbb{R}$ which satisfy the following conditions

$$\begin{cases} c(1, x_2, \cdots, x_n) = c(x_1, 1, x_3, \cdots, x_n) = \cdots = c(x_1, \cdots, x_{n-1}, 1) = 0, \\ c(a_1 + 1, x_2, \cdots\cdots, x_n) = c(x_1, a_2 + 1, x_3, \cdots, x_n) \\ \qquad\qquad = \cdots = c(x_1, \cdots, x_{n-1}, a_n + 1) = 0. \end{cases}$$

For $c : B \to \mathbb{R}$, we use the following notation

$$\sum_B c(x) = \sum_{x_1=1}^{a_1} \cdots \sum_{x_n=1}^{a_n} c(x_1, \cdots, x_n),$$

and use the usual convention $\sum_{y_1=a_1}^{a_1-1} c(y_1, x_2, \cdots, x_n) = 0,\ \sum_{y_n=a_n}^{a_n-1} c(x_1, \cdots, x_{n-1}, y_n) = 0$.

Theorem 6.2.1 (The Pachpatte Inequality [482]) *Let $p, q, r \geq 1$ be constants and suppose that $u, v, w \in G(B)$. Then*

$$\sum_B [|u(y)|^p |v(y)|^q + |v(y)|^q |w(y)|^r + |w(y)|^r |u(y)|^p]$$
$$\leq \frac{1}{n}\left(\frac{\mu}{2}\right)^{2p} \sum_B |Dv(y)|^{2p} + \frac{1}{n}\left(\frac{\mu}{2}\right)^{2q} \sum_B |Dv(y)|^{2q} + \frac{1}{n}\left(\frac{\mu}{2}\right)^{2r} \sum_B |Dw(y)|^{2r}, \tag{6.2.2}$$

$$\sum_B |u(y)|^p |v(y)|^q |w(y)|^r (|u(y)|^p + |v(y)|^q + |w(y)|^r)$$
$$\leq \frac{1}{n}\left(\frac{\mu}{2}\right)^{4p} \sum_B |Du(y)|^{4p} + \frac{1}{n}\left(\frac{\mu}{2}\right)^{4q} \sum_B |Dv(y)|^{4q} + \frac{1}{2}\left(\frac{\mu}{2}\right)^{4r} \sum_B |Dw(y)|^{4r}, \tag{6.2.3}$$

where $\mu = \max\{a_1, \cdots, a_n\}$ and $|Dc(y)|$ is as defined in (6.2.1).

Proof From the hypotheses, it is easy to observe that the following identities hold

$$\begin{cases} nu(x) = \sum_{y_1=1}^{x_1-1} \triangle_1 u(y_1, x_2, \cdots, x_n) + \cdots + \sum_{y_n=1}^{x_n-1} \triangle_n u(x_1, \cdots, x_{n-1}, y_n), \\ nu(x) = -\sum_{y_1=x_1}^{a_1} \triangle_1 u(y_1, x_2, \cdots, x_n) - \cdots - \sum_{y_n=x_n}^{a_n} \triangle_n u(x_1, \cdots, x_{n-1}, y_n). \end{cases} \tag{6.2.4}$$

From (6.2.4) it follows that

$$2n|u(x)| \le \sum_{y_1=1}^{a_1} |\triangle_1 u(y_1, x_2, \cdots, x_n)| + \cdots + \sum_{y_n=1}^{a_n} |\triangle_n u(x_1, \cdots, x_{n-1}, y_n)|. \tag{6.2.5}$$

From (6.2.5) and using the elementary inequality (see, [191, 395])

$$(b_1 + \cdots + b_n)^k \le L_{k,n}(b_1^k + \cdots + b_n^k), \tag{6.2.6}$$

for all $b_1, \cdots, b_n \ge 0$, where $L_{k,n} = n^{k-1}$ $(k > 1)$ and $L_{k,n} = 1$ $(0 \le k \le 1)$, Hölder's inequality with indices $p, p/(p-1)$ (see, e.g., [179]) and using the definition of μ, we obtain

$$|u(x)|^p \le \frac{1}{n}\left(\frac{1}{2}\right)^p \mu^{p-1}\Bigg[| \sum_{y_1=1}^{a_1} |\triangle_1 u(y_1, x_2, \cdots, x_n)|^p + \cdots + \sum_{y_n=1}^{a_n} |\triangle_n u(x_1, \cdots, x_{n-1}, y_n)|^p\Bigg]. \tag{6.2.7}$$

Similarly, we obtain

$$|v(x)|^q \le \frac{1}{n}\left(\frac{1}{2}\right)^q \mu^{q-1}\Bigg[\sum_{y_1=1}^{a_1} |\triangle_1 v(y_1, x_2, \cdots, x_n)|^q + \cdots + \sum_{y_n=1}^{a_n} |\triangle_n v(x_1, \cdots, x_{n-1}, y_n)|^q\Bigg], \tag{6.2.8}$$

and

$$|w(x)|^r \le \frac{1}{n}\left(\frac{1}{2}\right)^r \mu^{r-1}\Bigg[\sum_{y_1=1}^{a_1} |\triangle_1 w(y_1, x_2, \cdots, x_n)|^r + \cdots + \sum_{y_n=1}^{a_n} |\triangle_n w(x_1, \cdots, x_{n-1}, y_n)|^r\Bigg]. \tag{6.2.9}$$

From (6.2.7)–(6.2.9) and using the elementary inequality $b_1b_2+b_2b_3+b_3b_1 \leq b_1^2+b_2^2+b_3^2$ for b_1, b_2, b_3 reals, repeated application of (6.2.6) when $k = 2$, Schwartz's inequality and the definition of μ, we obtain

$$
\begin{aligned}
&|u(x)|^p|v(x)|^q + |v(x)|^q|w(x)|^r + |w(x)|^r|u(x)|^p \\
&\leq \{|u(x)|^p\}^2 + \{|v(x)|^q\}^2 + \{|w(x)|^r\}^2 \\
&\leq \left\{\frac{1}{n}\left(\frac{1}{2}\right)^p \mu^{p-1}[\sum_{y_1=1}^{a_1} |\Delta_1 u(y_1, x_2, \cdots, x_n)|^p + \sum_{y_n=1}^{a_n} |_n u(x_1, \cdots, x_{n-1}, y_n)|^p]\right\}^2 \\
&\quad + \left\{\frac{1}{n}\left(\frac{1}{2}\right)^q \mu^{q-1}[\sum_{y_1=1}^{a_1} |\Delta_1 v(y_1, x_2, \cdots, x_n)|^q + \cdots + \sum_{y_n=1}^{a_n} |_n v(x_1, \cdots, x_{n-1}, y_n)|^q]\right\}^2 \\
&\quad + \left\{\frac{1}{n}\left(\frac{1}{2}\right)^r \mu^{r-1}[\sum_{y_1=1}^{a_1} |\Delta_1 w(y_1, x_2, \cdots, x_n)|^r + \cdots \right. \\
&\quad \left. + \sum_{y_n=1}^{a_n} |_n w(x_1, \cdots, x_{n-1}, y_n)|^r]\right\}^2 \\
&\leq \left(\frac{1}{n}\right)^2 \left(\frac{1}{2}\right)^{2p} \mu^{2(p-1)} n\mu \left[\sum_{y_1=1}^{a_1} |\Delta_1 u(y_1, x_2, \cdots, x_n)|^{2p} + \cdots \right. \\
&\quad \left. + \sum_{y_n=1}^{a_n} |\Delta_n u(x_1, \cdots, x_{n-1}, y_n)^{2p}\right] \\
&\quad + \left(\frac{1}{n}\right)^2 \left(\frac{1}{2}\right)^{2q} \mu^{2(q-1)} n\mu \left[\sum_{y_1=1}^{a_1} |\Delta_1 v(y_1, x_2, \cdots, x_n)|^{2q} + \cdots \right. \\
&\quad \left. + \sum_{y_n=1}^{a_n} |\Delta_n v(x_1, \cdots, x_{n-1}, y_n)^{2q}\right] \\
&\quad + \left(\frac{1}{n}\right)^2 \left(\frac{1}{2}\right)^{2r} \mu^{2(r-1)} n\mu \left[\sum_{y_1=1}^{a_1} |\Delta_1 w(y_1, x_2, \cdots, x_n)|^{2r} \right. \\
&\quad \left. + \cdots + \sum_{y_n=1}^{a_n} |\Delta_n w(x_1, \cdots, x_{n-1}, y_n)^{2r}\right].
\end{aligned}
\tag{6.2.10}
$$

Setting $x_i = y_i$ $(i = 1, \cdots, n)$ in (6.2.10) and taking the sum over both sides of (6.2.10) with respect to $y_1, \cdots, y_n$ on B, using the definition of μ and the repeated application of (6.2.6) when $0 \leq k \leq 1$, we get

$$
\sum_B [|u(y)|^p|v(y)|^q + |v(y)|^q|w(y)|^r + |w(y)|^r|u(y)|^p]
$$

$$\leq \frac{1}{n}\left(\frac{\mu}{2}\right)^{2p} \sum_{B} \{[|\Delta_1 u(y)|^{2p} + \cdots + |\Delta_n u(y)|^{2p}]^{1/p}\}^p$$

$$+\frac{1}{n}\left(\frac{\mu}{2}\right)^{2q} \sum_{B} \{[|\Delta_1 v(y)|^{2q} + \cdots + |\Delta_n v(y)|^{2q}]^{1/q}\}^q$$

$$+\frac{1}{n}\left(\frac{\mu}{2}\right)^{2r} \sum_{B} \{[|\Delta_1 w(y)|^{2r} + \cdots + |\Delta_n w(y)|^{2r}]^{1/r}\}^r$$

$$\leq \frac{1}{n}\left(\frac{\mu}{2}\right)^{2p} \sum_{B} |Du(y)|^{2p} + \frac{1}{n}\left(\frac{\mu}{2}\right)^{2q} \sum_{B} |Dv(y)|^{2q} + \frac{1}{n}\left(\frac{\mu}{2}\right)^{2r} \sum_{B} |Dw(y)|^{2r}.$$

This completes the proof of inequality (6.2.2). The details of the proof of inequality (6.2.3) follows by the similar arguments as in the proof inequality (6.2.2) in view of the proof of inequality (6.1.233) given above with suitable modifications. We omit the details. □

Remark 6.2.1 We note that in the special case when $p = q = r = 1$ and $u(y)v(y) = w(y) = c(y)$, the inequalities established in (6.2.2) and (6.2.3) reduces respectively to the following inequalities

$$\sum_{B} |c(y)|^2 \leq \frac{1}{n}\left(\frac{\mu}{2}\right)^2 \sum_{B} |Dc(y)|^2, \tag{6.2.11}$$

and

$$\sum_{B} |c(y)|^4 \leq \frac{1}{n}\left(\frac{\mu}{2}\right)^4 \sum_{B} |Dc(y)|^4. \tag{6.2.12}$$

It is easy to observe that the two independent variables versions of (6.2.11) and (6.2.12) are different from the inequalities obtained in (6.1.244) and (6.1.245) respectively. For multidimensional discrete inequalities of the type (6.2.11)–(6.2.12), we refer the reader to [484].

Remark 6.2.2 The discrete inequalities of the type (6.1.244)–(6.1.245) and (6.2.11)–(6.2.12) in one independent variable have been established by various authors by using different techniques, see [86, 226, 228, 233, 395, 479, 533]. Here, we note that the inequalities established in (6.1.232)–(6.1.233), (6.1.244)–(6.1.245), (6.2.11)–(6.2.12) are different from those given in [86, 226, 228, 233, 395, 479, 533] and we believe that the inequalities established here are of independent interest.

We shall next introduce the results from Popenda and Agarwal [551] which is an essentially generalization of linear Gronwall discrete inequalities in several independent variables.

We shall use the following notations: $\mathbb{N}_\xi = \{\xi, \xi+1, \cdots\}$ where ξ is a non-negative integer. Let $\nu = (\nu_1, \cdots, \nu_m)$ then $\mathbb{N}_\nu = \mathbb{N}_{\nu_1} \times \cdots \times \mathbb{N}_{\nu_m}$ (the Cartesian product). Let $\alpha = (\alpha_1, \cdots, \alpha_m)$ where $\alpha_i \in \mathbb{N}_0$, then $|\alpha| = \sum_{i=1}^m \alpha_i$ we shall call α a multi-index or m-index.

We shall need the following operators which can be defined both for sequences of integers m as well as multi-indices. For $n = (n_l, \cdots, n_m) \in \mathbb{N}_\nu$, we define $E^j_{/i} = (n_1, \cdots, n_{i-1}, n_i + j, n_{i+1}, \cdots, n_m) P(n; n_\mu = \nu) = (n_1, \cdots \nu_{\mu_1}, n_{\mu_1+1}, \cdots, \nu\mu_k, \cdots, n_m)$ shift operators (acting here on arguments). In particular, $E_{/i} = (n_1, \cdots, n_{i-1}, n_i + 1, n_{i+1}, \cdots, n_m)$,$P(n; n_\mu = \nu) = (n_1, \cdots \cdot \nu_{\mu_1}, n_{\mu_1+1}, \cdots, \nu\mu_k, \cdots, n_m)$, projection operators, in short $P_{\mu/\nu}n$, where $\mu = \mu_1, \cdots, \mu_k$ is any subsequence of the sequence $1, \ldots, m$. For example, if $n = (n_1, n_2, n_3, n_4), \mu = (2,4), \nu = (1,7,5,3)$, then $P(n; n_{(2,4)} = \nu) = (n_1, 7, n_3, 3)$. In particular, $P(n; n_i = a) = P_{i/a}n = (n_1, \cdots \cdot, n_{i-1}, a, ni + l, \cdots, n_m)$. $R(\alpha, k) = (\alpha_1, \ldots, \alpha_{m-k})$ for all $0 \leq k \leq m$ reduction operator. It is clear that $P_{\mu/\nu}n$ can be presented as the compositions.

$P(n; n_\mu = \nu) = P_{\mu_1/\nu_1} \cdots P_{\mu_k/\nu_k}n$. The difference operators on any function $w : \mathbb{N}_\nu \to \mathbb{R}$ are defined as follows,

$$\Delta^{|\alpha|}_\alpha \omega(n) = \Delta^{\alpha_1}_{/1}(\Delta^{\alpha_2}_{/2}(\cdots(\Delta^{\alpha_m}_{/m}\omega(n)))$$

where for all $k \geq 1$,

$$\Delta^k_{/i}\omega(n) = \sum_{j=0}^{k} C^j_k(-1)^{k-j}\omega(n_1, \cdots, n_i + j, n_{i+1}, \cdots, n_m),$$

and on using shift operators

$$\Delta^k_{/i}\omega(n) = \sum_{j=0}^{k} C^j_k(-1)^{k-j}\omega(E^j_{/i}n).$$

It is supposed that $\Delta^0_{/i}\omega(n) = \omega(n)$ so that if in the multi-index α some of $\alpha_i = 0$, then in the definition of $\Delta^{|\alpha|}_\alpha \omega(n)$ suitable partial differences $\Delta^{\alpha_i}_{/i}$ should be omitted. For a sequence $\sigma = (\sigma_1, \cdots, \sigma_j)$, not necessarily of different elements $\sigma_i \in \{1, \cdots, m\}$, we shall use

$$\Delta^j_{/\sigma}\omega(n) = \Delta_{/\sigma_1(\Delta_{/\sigma_2}(\cdots(\Delta_{/\sigma_j}\omega(n))))}.$$

Let us note the difference between Δ^*_β and $\Delta^*_{/\beta}$. For this, let $\beta = (1,2,1)$, then according to the definitions, $\Delta^{|\beta|}_\beta \omega(n) = \Delta^1_{/1}(\Delta^2_{/2}(\Delta^3_{/3}\omega(n)))$ (here β denotes the order of the difference with respect to the ith variable), while$\Delta^3_{/\beta}\omega(n) = \Delta^1_{/1}(\Delta^2_{/2}(\Delta^3_{/3}\omega(n)))$ (here all the differences are of the first order, and β_i denotes to which variable the difference has to be applied). The multiple summation operators we denote

$$S_\alpha(n,\nu;\omega)=\sum_{j_{1,1}=\nu_1}^{n_1-1}\cdots\sum_{j_{1,\alpha_1}}^{j_{1,\alpha_1-1}-1}=\nu_1\cdots\sum_{j_{m,1}=-\nu_m}^{n_m-1}\cdots\sum_{j_{m,\alpha_m}=\nu_m}^{j_{m,\alpha_{m-1}}-1}\omega(j_{1,\alpha_1},\cdots j_{m,\alpha_m}).$$

It is clear that suitable summations have to be omitted if some of $\alpha_i = 0$. In particular, if $\alpha = (0,\cdots,0,\alpha_i,0,\cdots,0)$, then,

$$S_\alpha(n,\nu;\omega)=S_{\alpha_i}(n,\nu;\omega)=\sum_{j_{i,1}}^{n_i-1}=\cdots\sum_{j_{m,\alpha_m}=\nu_m}^{j_{i,\alpha_{i-1}}-1}\omega(n_1,\cdots,n_{i-1},j_{i,\alpha_i},n_{i+1},\cdots,n_m)$$

while

$$S_\alpha(n,\nu;\omega(P_{i/\nu_i}n))=\sum_{j_{i,1}=\nu_i}^{n_i-1}\cdots\sum_{j_{i,\alpha_i}=\nu_i}^{j_{i,\alpha_i-1}-1}\omega(n_1,\cdots,n_{i-1},\nu_i,n_{i-1},\cdots,n_m)$$

$$=\omega(n_1,\cdots,n_{i-1},\nu_i,n_{i-1},\cdots,n_m)\sum_{j_{i,1}=\nu_i}^{n_i-1}\cdots\sum_{j_{i,\alpha_i}=\nu_i}^{j_{i,\alpha_i-1}-1}1$$

$$\omega(n_1,\cdots,n_{i-1},\nu_i,n_{i-1},\cdots,n_m)C^{\alpha_i}_{n-\nu_i}.$$

We shall follow the standard convention that the empty sums are zero. Therefore, if for some i we have $n_i < \nu_i + \alpha_i$, then $S_\alpha(n,\nu;\omega) = 0$.

If $\alpha = (\alpha_1,\cdots,\alpha_m)$ and $\beta = (\beta_1,\cdots,\beta_m)$ are two multi-indices, then $\alpha-\beta = (\alpha_1-\beta_1,\cdots,\alpha_m-\beta_m)$, and

$$\Delta^{|\beta|}_\beta S_\alpha(n,\nu;\omega)=S_{\alpha-\beta}(n,\nu;\omega),\, \textit{if } \alpha_i\ge\beta_i,\ \textit{for all } i\in\{1,\cdots,m\},$$

moreover,

$$\Delta^{\alpha_i}_{/i}S_{\alpha_i}(n,\nu;\omega)=\omega(n),\ \ \Delta^{|\alpha|}_\alpha S_\alpha(n,\nu;\omega)=\omega(n).$$

If $\omega :\to N_\nu \mapsto \mathbb{R}_+$, then from the above $\Delta^{|\beta|}_\beta S_\alpha(n,\nu;\omega)\ge 0$ and $\Delta^{j}_{/i}S_\alpha(n,\nu;\omega)\ge 0$ for all $j\le\alpha_i$. On the other hand,

$$\Delta^k_{/i}S_\alpha(n,\nu;\omega)=\sum_{j=0}^{k}C^j_k(-1)^{k-j}S_\alpha(E^j_{/i}n,\nu;\omega)=0\ \ \textit{if } n_i+k<\nu_i+\alpha_i.$$

If $\beta_i > \alpha_i$, then

$$\Delta_{/i}^{\beta_i} S_{\alpha_i}(n, \nu; \omega) = \Delta_{/i}^{\beta_i - \alpha_i} \omega(n).$$

It is clear that for some values of n, we have $S_\alpha(n, \nu; \omega) = 0$, while $\delta_\beta^{|\beta|} S_\alpha(n, \nu; \omega) \neq 0$. For example, let $\omega : \mathbb{N}_{(\nu_1, \nu_2)} \to \mathbb{R}$ *and* $\alpha = (3, 2)$, then

$$\begin{aligned}
S_\alpha(n, \nu; \omega) &= \sum_{j_{1,1}=\nu_1}^{n_1-1} \sum_{j_{1,2}=\nu_1}^{j_{1,1}-1} \sum_{j_{1,3}=\nu_1}^{j_{1,2}-1} \sum_{j_{2,1}=\nu_2}^{n_2-1} \sum_{j_{2,2}=\nu_2}^{j_{2,1}-1} \omega(j_{1,3}, j_{2,2}), \\
\Delta_{/1} S_\alpha(n, (\nu_1, \nu_2); \omega) &= \sum_{j_{1,2}=\nu_1}^{n_1-1} \sum_{j_{1,3}=\nu_1}^{j_{1,2}-1} \sum_{j_{2,1}=\nu_2}^{n_2-1} \sum_{j_{2,2}=\nu_2}^{j_{2,1}-1} \omega(j_{1,3}, j_{2,2}), \\
\Delta_{/(1,1)}^2 S_\alpha(n, (\nu_1, \nu_2); \omega) &\equiv \Delta_{/1}^2 S_\alpha(n, (\nu_1, \nu_2); \omega) \equiv \Delta_{(2,0)}^2 S_\alpha(n, (\nu_1, \nu - 2); \omega) \\
&= \sum_{j_{1,3}=\nu_1}^{n_1-1} \sum_{j_{2,1}=\nu_2}^{n_2-1} \sum_{j_{2,2}=\nu_2}^{j_{2,1}-1} \omega(j_{1,3}, j_{2,2}),
\end{aligned}$$

and $S_\alpha((\nu_1+2, \nu_2+2), (\nu_1, \nu_2); \omega) = 0$, in fact, $S_\alpha((\nu_1+i, \nu_2+j), (\nu_1, \nu_2); \omega) = 0$, if $i < 3$ or $j < 2$, while $\Delta_{/1} S_\alpha((\nu_1 + 2, \nu_2 + 2), (\nu_1, \nu_2); \omega) = \omega(\nu_1, \nu_2)$. For a given multi-index $\alpha = (\alpha_1, \cdots, \alpha_m)$, we can construct the set Ξ_α of $|\alpha|! \prod_{i=1}^m 1/(\alpha_i!)$ sequences $\Xi_\alpha = \sigma : (\sigma_1, \cdots, \sigma_{|\alpha|})\ 1, \cdots, |\alpha|$ such that $\sigma_j = i$ for some $i \in \{1, \cdots, m\}$ and all $j \in \{1, \cdots, |\alpha|\}$ and card $\sigma_j : \sigma_j = \alpha = \alpha_i$. Here by card A, we shall denote the cardinal of the set A. For example, let $\alpha = (3, 2)$, then $\Xi_\alpha = (1, 1, 1, 2, 2), (1, 2, 1, 1, 2), (2, 1, 1, 2, 1), \cdots$, and card $\Xi_\alpha = 5!/(2!3!)$. We shall say the function f belongs to the class $M(\beta)$ if $f : \mathbb{N}_\nu \to \mathbb{R}_+$ and

(i) $\Delta_{/(\beta_1, \cdots, \beta_s)}^s f(n) \geq 0$ *for* $s = 1, \cdots, r - 1$ *and all* $n \in \mathbb{N}_\nu$, and , (ii) $\Delta_{/\beta_j}^1 f(n) \geq 0$*forj* $= 1, \cdots, r$ *and all* $n \in \mathbb{N}_\nu$, where $\beta = (\beta_1, \cdots, \beta_r)$ and $\beta_i \in 1, \cdots, m$ for all $i \in 1, \cdots, r$. It is clear that if $f_1, f_2 \in M(\beta)$ and $a > 0$, then $f_1 + f_2 \in M(\beta)$ and a$f_1 \in M(\beta)$. Moreover, if $f \in M(\beta)$ is such that $\Delta_{/(\beta_1, \cdots, \beta_r)}^r f(n) = 0$ for all $n \in \mathbb{N}_\nu$, then $f \in M(\gamma)$ for any $\gamma = (\gamma_1, \cdots, \gamma_r, \cdots, \gamma_k)$ such that $\gamma_i = \beta_i$ for all $i \in \{1, \cdots, r\}$. For example, let $\beta = (3, 1, 2, 1)$, then $f \in M(\beta)$ if from condition (i),

$$\begin{aligned}
\Delta_{/(3,1,2)}^3 f(n) &\equiv \delta_{/3}(\Delta_{/1}(\Delta_{/2} f(n))) \geq 0, \\
\Delta_{/(3,1)}^2 f(n) &\equiv \Delta_{/3}(\Delta_{/1} f(n)) \geq 0, \\
\delta_{/3}^1 f(n) &\geq 0
\end{aligned}$$

and by condition (ii),

$$\Delta_{/1}^1 f(n) \equiv \delta_{/1} f(n) \equiv \Delta_{/(1)}^1 f(n) \geq 0,\ \delta_{/2}^1 f(n) \geq 0, \Delta_{/3}^1 f(n) \geq 0.$$

Theorem 6.2.2 (The Popenda-Agarwal Inequality [551]) *Let $u, b, c : \mathbb{N}_\nu \to \mathbb{R}_+$ and there exists a sequence $\sigma \in \Xi_\alpha$ such that $c \in M(\sigma)$. If there holds for all $n \in \mathbb{N}_\nu$,*

$$u(n) \leq c(n) + S_\alpha(n, \nu; bu), \tag{6.2.13}$$

then for all $n \in \mathbb{N}_\nu$,

$$u(n) \leq min_{\sigma\in\Xi_\sigma:c\in M(\sigma)} c(P(n; n_{\sigma_1} = \nu_{\sigma_1})) \prod_{j_{|\alpha|}=\nu_{\sigma_1}}^{n_{\sigma_1}-1} 1 + \Phi_{|\alpha|}(P(n; n_{\sigma_1} = j_{|\alpha|})), \tag{6.2.14}$$

where

$$\Phi_1(n) = \frac{max\{0, \Delta_\alpha^{|\alpha|} c(n)\}}{c(n)} + b(n)$$

and

$$\begin{aligned}\Phi_{k+1}(n) =& \frac{\Delta_{/R(\sigma,k)}^{|\alpha|-k} c(P(n; n_{\sigma_{|\alpha|}-k+1} = \nu_{\sigma_{|\alpha|}-k+1}))}{c(P(n; n_{\sigma_{|\alpha|}-k+1} = \nu_{\sigma_{|\alpha|}-k+1}))} \\ &+ \sum_{j_k=\nu_{\sigma_{|\alpha|}-k+1}}^{n_{\sigma_{|\alpha|}-k+1}} \Phi_k(P(n; n_{\sigma_{|\alpha|}-k+1} = j_k)), \\ & k = 1, \cdots, |\alpha| - 1.\end{aligned}$$

Proof Let $\sigma \in \Xi_\alpha$ be such that $c \in M(\sigma)$ and let for all $n \in \mathbb{N}_\nu$,

$$z(n) = c(n) + S_\alpha(n, \nu; bu). \tag{6.2.15}$$

Then inequality (6.2.13) yields

$$u(n) \leq z(n). \tag{6.2.16}$$

Therefore, for all $n \in \mathbb{N}_\nu$ from (6.2.15), we derive

$$\begin{aligned}\Delta_{/\sigma}^{|\alpha|} z(n) &= \Delta_{/\sigma}^{|\alpha|} c(n) + b(n)u(n) \leq \Delta_{/\sigma}^{|\alpha|} c(n) + b(n)z(n) \\ &\leq max(0, \Delta_{/\sigma}^{|\alpha|} c(n)) + b(n)z(n).\end{aligned}$$

Since $c \in M(\sigma)$ and $z(n) \geq c(n) > 0$, then

$$\frac{\Delta^{|\alpha|}_{/R(\sigma,0)} z(n)}{z(n)} = \frac{\Delta^{|\alpha|}_{/\sigma} z(n)}{z(n)} \leq \frac{max(0, \Delta^{|\alpha|}_{/\sigma} c(n)}{c(n)} + b(n) = \Phi_1(n).$$

Hence,

$$\frac{\Delta^{|\alpha|-1}_{/R(\sigma,1)} z(E_{/\sigma_{|\alpha|}} n) - \Delta^{|\alpha|-1}_{/R(\sigma,1)} z(n)}{z(n)} \leq \Phi_1(n). \tag{6.2.17}$$

Notice that

$$\Delta_{/\sigma_j} z(n) = \Delta_{/\sigma_j} c(n) + \Delta_{/\sigma_j} S_\alpha(n, \nu; bu), j = 1, \cdots, |\alpha|$$

and by condition (ii) of the definition $M(\sigma)$,

$$\Delta_{/\sigma_j} S_\alpha(n, \nu; bu) \geq 0,\ \Delta_{/\sigma_j} c(n) \geq 0.$$

Thus it follows that for all $j = 1, \cdots, |\alpha|$, and for all $n \in \mathbb{N}_\nu$,

$$z(E_{/\sigma_j} n) \geq z(n).$$

Moreover, by condition (i),

$$\Delta^{|\alpha|-1}_{/R(\sigma,1)} z(n) = \Delta^{|\alpha|-1}_{/R(\sigma,1)} c(n) + \Delta^{|\alpha|-1}_{/R(\sigma,1)} S_\alpha(n, \nu; bu) \geq 0.$$

Hence, from (6.2.17) it follows

$$\frac{\Delta^{|\alpha|-1}_{/R(\sigma,1)} z(E_{/\sigma_{|\alpha|}})}{z(E_{/\sigma_{|\alpha|}})} - \frac{\Delta^{|\alpha|-1}_{/R(\sigma,1)} z(n)}{z(n)} \leq \Phi_1(n). \tag{6.2.18}$$

Now substituting in (6.2.18), $n = P(n; n_{\sigma_{|\alpha|}} = j_1)$ and summing with respect to j_1 from $\nu_{\sigma_{|\alpha|}}$, to $n_{\sigma_{|\alpha|}} - 1$, we get

$$\frac{\Delta^{|\alpha|-1}_{/R(\sigma,1)} z(n)}{z(n)} - \frac{\Delta^{|\alpha|-1}_{/R(\sigma,1)} z(P(n; n_{\sigma_{|\alpha|}} = \nu_{\sigma_{|\alpha|}}))}{z(P(n; n_{\sigma_{|\alpha|}} = \nu_{\sigma_{|\alpha|}}))} \leq \sum_{j_1=\nu_{\sigma_{|\alpha|}}}^{n_{\sigma_{|\alpha|}}-1} \Phi_1(P(n; n_{\sigma_{|\alpha|}} = j_1)). \tag{6.2.19}$$

Let $\sigma_{|\alpha|} = \xi \in \{1, \cdots, m\}$, then card $\sigma_i : \sigma_i = \xi$ *and* $\sigma_i \in R(\sigma, 1) = \alpha_\xi - 1$; furthermore, let $\tau = (\tau_1, \cdots, \tau_{|\alpha|})$ where $\tau_i = \sigma_i$ if $\sigma_i \neq \xi, \tau_i = 0$ if $\sigma_i = \xi$, then for all $n \in \mathbb{N}_{\nu_1} \times \cdots \times \mathbb{N}_{\nu_\xi - 1} \times \nu_\xi \times \mathbb{N}_{\nu_\xi + 1} \times \cdots \times \mathbb{N}_{\nu_m}$,

$$\Delta^{|\alpha|-1}_{/R(\sigma,1)} S_\alpha(P(n; n_\xi = \nu_\xi), \nu; bu) = \Delta^{|\alpha|-\alpha_x i}_{/\tau}(\Delta^{\alpha_\xi - 1}_{/\xi} S_\alpha(P(n; n_\xi = \nu_\xi), \nu; bu)) = 0.$$

Therefore,

$$\begin{aligned}\Delta^{|\alpha|-1}_{/R(\sigma,1)} z(P(n; n_\sigma = \nu_{\sigma_{|\alpha|}}))s =& \Delta^{|\alpha|-1}_{/R(\sigma,1)} c(P(n; n_\xi = \nu_\xi)) \\ &+ \Delta^{|\alpha|-1}_{/R(\sigma,1)} S_\alpha(P(n; n_\xi = \nu_\xi), \nu; bu) \\ =& \Delta^{|\alpha|-1}_{/R(\sigma,1)} c(P(n; n_{\sigma_{|\alpha|}} = \nu_{\sigma_{|\alpha|}}))\end{aligned}$$

and because

$$\begin{aligned} z(P(n; n_{\sigma_{|\alpha|}} = \nu_{\sigma_{|\alpha|}})) &= c(P(n; n_{\sigma_{|\alpha|}} = \nu_{\sigma_{|\alpha|}})) + S_\alpha(P(n; n_{\sigma_{|\alpha|}} = \nu_{\sigma_{|\alpha|}}), \nu; bu) \\ &= c(P(n; n_{\sigma_{|\alpha|}} = \nu_{\sigma_{|\alpha|}})),\end{aligned}$$

from (6.2.19), we derive

$$\frac{\Delta^{|\alpha|-1}_{/R(\sigma,1)} z(n)}{z(n)} \le \frac{\Delta^{|\alpha|-1}_{/R(\sigma,1)} c(P(n; n_{\sigma_{|\alpha|}} = \nu_{\sigma_{|\alpha|}}))}{c(P(n; n_{\sigma_{|\alpha|}} = \nu_{\sigma_{|\alpha|}}))} + \sum_{j_1 = \nu_{\sigma_{|\alpha|}}}^{n_{\sigma_{|\alpha|}} - 1} \Phi_1(P(n; n_{\sigma_{|\alpha|}} = j_1)) = \Phi_2(n).$$

The above inequality yields

$$\frac{\Delta^{|\alpha|-2}_{/R(\sigma,2)} z(E_{/\sigma_{|\alpha|}-1} n) - \Delta^{|\alpha|-2}_{/R(\sigma,2)} z(n)}{z(n)} \le \Phi_2(n),$$

i.e., an inequality similar to that of (6.2.17). Now following the same reasoning and inductive hypotheses , we get

$$\frac{\Delta^1_{/\sigma_1} z(n)}{z(n)} \le \Phi_{|\alpha|}(n)$$

which yields

$$\begin{aligned} z(E_{/\sigma_1} n) &\le 1 + \Phi_{|\alpha|}(n) z(n), \\ z(n) &\le z(P(n; n_{\sigma_{|\alpha|}} = \nu_{\sigma_{|\alpha|}})) \prod_{j_{|\alpha|} = \nu_{\sigma_1}}^{n_{\sigma_1} - 1} 1 + \Phi_{|\alpha|}(P(n; n_{\sigma_1} = j_{|\alpha|})) \end{aligned} \tag{6.2.20}$$

and because

$$z(P(n; n_{\sigma_{|\alpha|}} = \nu_{\sigma_{|\alpha|}})) = c(P(n; n_{\sigma_{|\alpha|}} = \nu_{\sigma_{|\alpha|}}))$$

in view of (6.2.16), we have

$$n(n) \leq z(n) \leq c(P(n; n_{\sigma_{|\alpha|}} = \nu_{\sigma_{|\alpha|}})) \prod_{j_{|\alpha|}=\nu_{\sigma_1}}^{n_{\sigma_1}-1} 1 + \Phi_{|\alpha|}(P(n; n_{\sigma_1} = j_{|\alpha|})).$$

Similar estimate can be obtained for each $\sigma \in \Xi_\alpha$, such that $c \in M(\sigma)$. From this observation, the required inequality (6.2.14) follows.

Remark 6.2.3 The estimate (6.2.14) can be rearranged as follows: for all $n \in \mathbb{N}_\nu$,

$$u(n) \leq min_{\sigma \in \Xi_\alpha : c \in M(\sigma)} c(P(n; n_{\sigma_{|\alpha|}} = \nu_{\sigma_{|\alpha|}})) \prod_{j_{|\alpha|}=\nu_{\sigma_1}}^{n_{\sigma_1}-1} 1 + \Phi_{|\alpha|}(P(n; n_{\sigma_1} = j_{|\alpha|}))$$

$$+ S_{E^{-1}_{/\sigma_1}\alpha}((P(n; n_{\sigma_1} = j_{\sigma_{|\alpha|}}), \nu; b),$$

where

$$\Psi_1(n) = \frac{max(0, \Delta_\alpha^{|\alpha|} c(n))}{c(n)}$$

and

$$\Psi_{k+1}(n) = \frac{\Delta^{|\alpha|-1}_{/R(\sigma,k)} c(P(n; n_{\sigma_{|\alpha|-k+1}} = \nu_{\sigma_{|\alpha|-k+1}}))}{c(P(n; n_{\sigma_{|\alpha|-k+1}} = \nu_{\sigma_{|\alpha|-k+1}}))}, \ k = 1, \cdots, |\alpha| - 1.$$

Remark 6.2.4 The method we have used in Theorem 6.1.32 can be applied (with slight modifications) to general type of inequalities such as, for all $n \in \mathbb{N}_\nu$,

$$u(n) \leq c(n) + \sum_{i=1}^{k} S_{\alpha^i}(n, \nu; b_i u),$$

where $\alpha^i = (\alpha_1^i, \cdots, \alpha_m^i)$, $i = 1, \cdots, k$ and c belongs to a suitable class M. In fact, to obtain a bound it suffices to obtain first some linear inequality of the type

$$\delta_\Psi^k z(n) \leq \Lambda(n) z(n) + Y(n) \tag{6.2.21}$$

and then follow the method of Theorem 6.1.32. To illustrate this, we present the following examples.

Example 6.2.1 Consider the inequality

$$u(n_l,n_2) \le c(n_l,n_2) + \sum_{j_{1,1}=1}^{n_1-1} b_1(j_{1,1},n_2)[u(j_{1,1},n_2) + \sum_{j1,2=1}^{j_{1,1}-1}\sum_{j_{1,3}=1}^{j_{1,2}-1}\sum_{j_{2,1}=4}^{n_2-1} b_2(j_{1,3},j_{2,1})u(j_{1,3},j_{2,1})]. \tag{6.2.22}$$

Denoting the right-hand side of (6.2.22) by $z(n_1,n_2)$, we get

$$\begin{aligned}\Delta_{/1}z(n) &= \Delta_{/1}c(n) + b_1(n)[u(n) + \sum_{j1,2=1}^{j_{1,1}-1}\sum_{j_{1,3}=1}^{j_{1,2}-1}\sum_{j_{2,1}=4}^{n_2-1} b_2(j_{1,3},j_{2,1})u(j_{1,3},j_{2,1})]\\ &\le \Delta_{/1}c(n) + b_1(n)[z(n) + \sum_{j1,2=1}^{j_{1,1}-1}\sum_{j_{1,3}=1}^{j_{1,2}-1}\sum_{j_{2,1}=4}^{n_2-1} b_2(j_{1,3},j_{2,1})u(j_{1,3},j_{2,1})].\end{aligned} \tag{6.2.23}$$

Let $b_1 \ge 0, b_2 \ge 0, u \ge 0$, if $\Delta_{/1}c \ge 0$, then $\Delta_{/1}z \ge 0$, if moreover $\Delta_{/1}c \ge 0$, $\Delta_{/1}b_1 \ge 0$, and we want to estimate u such that $\Delta_{/1}u \ge 0$, then by the definition of $z(n_1,n_2)$, we have

$$\begin{aligned}\Delta_{/1}z &= \Delta_{/1}c(n) + \sum_{j_{1,1}=1}^{n_1-1} b_1(j_{1,1},n_2+1)[\Delta_{/2}u_1(j_{1,1},n_2)\\ &\quad + \sum_{j1,2=1}^{j_{1,1}-1}\sum_{j_{1,3}=1}^{j_{1,2}-1} b_2(j_{1,3},j_{2,1})u(j_{1,3},j_{2,1})]\\ &\quad + \sum_{j_{1,1}=1}^{n_1-1}[u(j_{1,1},n_2) + \sum_{j1,2=1}^{j_{1,1}-1}\sum_{j_{1,3}=1}^{j_{1,2}-1}\sum_{j_{2,1}=4}^{n_2-1}\\ &\quad b_2(j_{1,3},j_{2,1})u(j_{1,3},j_{2,1})]\Delta_{/2}b_1(j_{1,1},n_2) \ge 0.\end{aligned} \tag{6.2.24}$$

Now let

$$\omega(n) = z(n) + \sum_{j1,2=1}^{n_1-1}\sum_{j_{1,3}=1}^{j_{1,2}-1}\sum_{j_{2,1}=4}^{n_2-1} b_2(j_{1,3},j_{2,1})z(j_{1,3},j_{2,1}).$$

from which it follows

$$\Delta_{/1}\omega(n) = \Delta_{/1}z(n)+\sum_{j1,2=1}^{j_{1,1}-1}\sum_{j_{1,3}=1}^{j_{1,2}-1} b_2(j_{1,3},j_{2,1})z(j_{1,3},j_{2,1})$$

$$\leq \Delta_{/1}c(n)+b_1(n)\omega(n)+z(n_1-1,n_2-1)\sum_{j}^{j_{1,1}-1} 1,2 = 1\sum_{j_{1,3}=1}^{j_{1,2}-1} b_2(j_{1,3},j_{2,1})$$

$$\leq \Delta_{/1}c(n)+\Big[b_1(n)+\sum_{j1,2=1}^{j_{1,1}-1}\sum_{j_{1,3}=1}^{j_{1,2}-1} b_2(j_{1,3},j_{2,1})\Big]\omega(n),$$

which is of the form (6.2.20). Now we can apply the method of Theorem 6.1.32. The obtained bound for $\omega(n)$ is then used in the inequality $\Delta_{/1}z(n) \leq \Delta_{/1}c(n) + b_1(n)\omega(n)$, which in turn after suitable summations, leads to the bound for $z(n)$, and consequently, the bound for $u(n)$. It is interesting to note that we can get another inequality for $\Delta_{/1}z(n)\Delta_{/1}z$, which follows directly from (6.2.22), namely,

$$\delta_{/1}z(n) \leq \delta_{/1}c(n) + b_1(n)[1 + \sum_{j1,2=1}^{j_{1,1}-1}\sum_{j_{1,3}=1}^{j_{1,2}-1}\sum_{j_{2,1}=4}^{n_2-1} b_2(j_{1,3},j_{2,1})u(j_{1,3},j_{2,1})]z(n)$$

which is also of the form (6.2.20), consequently, the method of Theorem 6.1.32 is applicable. In the above example, we have three summations with respect to the first variable, and one with respect to the second. In fact, we do not suppose that all α_i in α are the same.

Example 6.2.2 Consider the inequality

$$u(n) \leq c(n) + S_\alpha(n, \nu; b_l(j)(u(j) + S_\beta(j, \nu; b_2(i)u(i)))),$$

where $\beta \leq \alpha$, that is $\beta_k \leq \alpha_k$ for all $k = 1, \cdots, m$. Let $c \in M(\sigma)$ and $\sigma \in \Xi - \beta, \Delta_\beta^{|\beta|}c \leq 0$. Let

$$z(n) = c(n) + S_\alpha(n, \nu; b_l(u + S_\beta(j, \nu; b_2u))),$$

then

$$\Delta_\beta^{|\beta|}z(n) = \Delta_\beta^{|\beta|}c(n)+S_{\alpha-\beta}(n, \nu; b_1(u+S_\beta(j, \nu; b_2u))) \leq \Delta_\beta^{|\beta|}c(n)+S_{\alpha-\beta}(n, \nu; b_1\omega),$$

where

$$(n) = z(n) + S_\beta(n, \nu; b_2z).$$

Hence,

$$\begin{aligned}\Delta_{\beta}^{|\beta|}\omega(n) &= \Delta_{\beta}^{|\beta|}z(n) + b_2(n)z(n)\\ &\le \Delta_{\beta}^{|\beta|}c(n)S_{\alpha-\beta}(n, \nu; b_1\omega) + b_2(n)z(n)\\ &\le \Delta_{\beta}^{|\beta|}c(n) + [S_{\alpha-\beta}(n, \nu; b_1) + b_2(n)]\omega(n).\end{aligned}$$

Thus, the resulting inequality is of the form (6.2.20), and so the method presented in Theorem 6.1.32 allows us to get an estimate on w and consequently, (after suitable summations) on z and then on u. If $\beta = \alpha$ as in [487] and $\Delta_{\beta}^{|\beta|}c(n) = 0$, then $\Delta_{\beta}^{|\beta|}\omega(n) \le [b_1(n_0 + b_2(n)]\omega(n)$.

Lemma 6.2.1 ([341]) *Let u_i, v_i be two real sequences. The following identity holds:*

$$\sum_{i=0}^{n-1} v_i\Delta u_i = u_n v_n - u_0 v_0 - \sum_{i=0}^{n-1} u_{i+1}\Delta v_i. \tag{6.2.25}$$

Proof For every natural number i,

$$\begin{aligned}\Delta(u_i v_i) &= u_{i+1}v_{i+1} - u_i v_i\\ &= u_{i+1}\Delta v_i + v_i\Delta u_i,\end{aligned} \tag{6.2.26}$$

i.e.,,

$$v_i\Lambda u_i = \Lambda(u_i v_i) - u_{i+1}\Lambda v_i.$$

Substituting $i = 0, 1, 2, \cdots, n-1$ and adding in the last equality, we have

$$\begin{aligned}\sum_{i=0}^{n-1} v_i\Delta u_i &= \sum_{i=0}^{n-1}\Delta(u_i v_i) - \sum_{i=0}^{n-1} u_{i+1}\Delta v_i\\ &= u_n v_n - u_0 v_0 - \sum_{i=0}^{n-1} u_{i+1}\Delta v_i.\end{aligned} \tag{6.2.27}$$

The proof is thus complete. □

Lemma 6.2.2 ([341]) *Let u_i, $\Delta u_i, \cdots, \Delta^k u_i$ be non-negative sequences and satisfy $u_0 = \Delta u_0 = \cdots = \Delta^{k-1}u_i = 0$, then*

$$u_n \le \frac{1}{(m-1)!}\sum_{j=0}^{n-1}(n-j-1)^{m-1}\Delta^m u_j, \quad m \le k \tag{6.2.28}$$

$$\Lambda^i \leq \frac{1}{(k-i-1)!}\sum_{j=0}^{n-1}(n-j-1)^{k-i-1}\Delta^k u_j, \quad 0 \leq i \leq k-1. \tag{6.2.29}$$

Proof When $m = 1$, (6.2.28) holds obviously. We now suppose that (6.2.28) holds when $m = t$, where $1 \leq t \leq k-1$. From inductive assumption, Lemma 6.2.1, and $\Delta^t u_0 = 0$, we have

$$\begin{aligned}
u_n &\leq \frac{1}{(t-1)!}\sum_{j=0}^{n-1}(n-j-1)^{n-1}\Delta^t u_j \\
&\leq \frac{1}{(t-1)!}\sum_{j=0}^{n-1}(n-j-1)^{n-1}\Delta^t u_j \\
&\times\frac{(n-j-1)^{t-1} + (n-j-1)^{t-2}(n-j) + \cdots + (n-j-1)(n-j)^{t-2} + (n-j)^{t-1}}{t} \\
&= -\frac{1}{t!}\sum_{j=0}^{n-1}\Delta^t u_j\big[(n-j-1)^{t-1} + (n-j-1)^{t-2}(n-j) + \cdots \\
&\quad +(n-j-1)(n-j)^{t-2} + (n-j)^{t-1}\big]\big[(n-j-1)-(n-j)\big] \\
&= -\frac{1}{t!}\sum_{j=0}^{n-1}\Delta^t u_j\big[(n-j-1)^t-(n-j)^t\big] \\
&= -\frac{1}{t!}\sum_{j=0}^{n-1}\Delta^t u_j\Delta[(n-j)^t] \\
&= -\frac{1}{t!}\sum_{j=0}^{n-1}(n-j-1)^t\Delta^{t+1}u_j.
\end{aligned}$$

This completes the proof of (6.2.28). Substituting $\Delta^i u_n$ for u_n and setting $m = k-i$, we can get (6.2.29) The proof is thus complete. □

Theorem 6.2.3 (The Li Inequality [341]) *Let $p_i \geq 0$, $i = 0, 1, 2, \ldots, k-1, p > p_k > 0$, $p > 1, \{u_x\}, \{\Delta u_x\}, \ldots, \{\Delta^k u_x\}$ be non-negative sequences and satisfy $u_0 = \Delta u_0 = \cdots = \Delta^{k-1}u_0, h_x \geq 0, r_x > 0$, then*

$$\sum_{x=0}^{n-1}h_x(\Delta^k u_x)^{p_k}\prod_{i=0}^{k-1}(\Delta^i u_x)^{p_i} \leq \left(\frac{p_k}{p}\right)^{p_k/p}\frac{w_n(h,r)}{Q}\sum_{x=0}^{n-1}r_x(\Delta^k u_x)^p \tag{6.2.30}$$

where

$$p = \sum^{k} p_i, \quad Q = \prod_{i=0}^{k-1}[(k-i-1)!]^{p_i},$$

$$R_{ix} = \sum_{j=0}^{x-1}(x-j-1)^{(k-i-1)p/(p-1)} r_j^{-1/(p-1)},$$

$$w_n(h,r) = \left[\sum_{x=0}^{n-1} h_x^{p/(p-p_k)} r_x^{-p_k/(p-p_k)} \prod_{i=0}^{k-1} R_{ix}^{p_i(p-1)/(p-p_k)}\right]^{(p-p_k)/p}.$$

Proof For every natural number j, define

$$y_j = \sum_{x=0}^{j-1} r_x(\Delta^k u_x)^p,$$

then we have

$$y_0 = 0,\ \Delta y_j = r_j(\Delta y^k u_j)^p,\ \Delta y^k u_j = r_j^{-1/p}(\Delta y_j)^{1/p}. \tag{6.2.31}$$

From Lemma 6.2.2, (6.2.31), and the Hölder inequality, we can obtain

$$\begin{aligned}
\Delta^i u_n &\le \frac{1}{(k-i-1)!}\sum_{j=0}^{n-1}(n-j-1)^{k-i-1}\Delta^k u_j \\
&= \frac{1}{(k-i-1)!}\sum_{j=0}^{n-1}(n-j-1)^{k-i-1} r_j^{-1/p}(\Delta y_j)^{1/p} \\
&\le \frac{1}{(k-i-1)!}\left[\sum_{j=0}^{n-1}(n-j-1)^{(k-i-1)p/(p-1)} r_j^{-1/(p-1)}\right]^{(p-1)/p} \times \left(\sum_{j=0}^{n-1}\Delta y_t\right)^{1/p} \\
&= \frac{1}{(k-i-1)!} R_{in}^{p-1)/p} y_n^{1/p}.
\end{aligned} \tag{6.2.32}$$

Using (6.2.31), (6.2.32), and the Hölder inequality, we have

$$\begin{aligned}
&\sum_{x=0}^{n-1} h_x(\Delta^k u_x)^{p_k} \prod_{i=0}^{k-1}(\Delta^i u_x)^{p_i} \\
&\quad \le \sum_{x=0}^{n-1} h_x r_x^{-p_k/p}(\Delta y_x)^{-p_k/p} \prod_{i=0}^{k-1}\left[\frac{1}{(k-i-1)!} R_{ix}^{(p-1)/p} y_x^{1/p}\right]^{p_i}
\end{aligned}$$

$$
\begin{aligned}
&= \frac{1}{Q} \sum_{x=0}^{n-1} h_x r_x^{-p_k/p} \prod_{i=0}^{k-1} R_{ix}^{(p-1)p_i/p} \cdot y_x^{(p-p_k)/p} (\Delta y_x)^{p_k/p} \\
&\leq \frac{1}{Q} \left[\sum_{x=0}^{n-1} h_x^{p/(p-p_k)} r_x^{-p_k/(p-p_k)} \prod_{i=0}^{k-1} R_{ix}^{(p-1)p_i/p} \right]^{(p-p_k)/p} \\
&\quad \times \left(\sum_{x=0}^{n-1} y_x^{(p-p_k)/p_k} \Delta y_x \right)^{p_k/p} \\
&\leq \frac{w_n(h,r)}{Q} \left(\sum_{x=0}^{n-1} \int_{y_x}^{y_x+1} t^{(p-p_k)/p_k} dt \right)^{p_k/p} \\
&= \frac{w_n(h,r)}{Q} \left(\int_0^{y_n} t^{(p-p_k)/p_k} dt \right)^{p_k/p} \\
&= \frac{w_n(h,r)}{Q} \left(\frac{p_k}{p} \right)^{p_k/p} y_n \\
&= \left(\frac{p_k}{p} \right)^{p_k/p} \frac{w_n(h,r)}{Q} \sum_{x=0}^{n-1} r_x (\Delta^k u_x)^p. \qquad (6.2.33)
\end{aligned}
$$

The proof is hence complete. □

This result of Li [341] grew out of an investigation of his similar result in the continuous case, see Theorem 1 in [341]. In the following, we point out that the direct adaptation of the results from the continuous case to the discrete case is not always true and the inequality given in the above theorem reduces to the trivial one.

Consider the following assumption in the above theorem

$$u_0 = \Delta u_0 = \cdots = \Delta^k u_0 = 0, \qquad (6.2.34)$$

adapted from continuous case (see, e.g., [341], Theorem 1) to the discrete case. From (6.2.34) we observe that

$$u_0 = 0, \quad \Delta u_0 = u_1 - u_0 = 0 \qquad (6.2.35)$$

follows by using (6.2.35), while

$$u_1 = 0, \quad \Delta^2 u_0 = u_2 - 2u_1 + u_0 = 0 \qquad (6.2.36)$$

follows by using (6.2.35) and (6.2.36),

$$u_2 = 0, \qquad (6.2.37)$$

and continuing in this way

$$u_{k-1} = 0, \quad k = 1, 2, \ldots, n. \tag{6.2.38}$$

This observation shows that the sequences $\{u_x\}, \{\Delta u_x\}, \ldots, \{\Delta^k u_x\}$ for $x = 0, 1, 2, \ldots, n-1$ reduce respectively to $\{0\}, \{0\}, \ldots, \{0\}$ and consequently the equality (6.2.30) in the above theorem reduces to the trivial one.

This clearly shows that the condition (6.2.34) which is adapted from the condition used in the continuous case is not suitable for establishing the inequality in the above theorem. In view of the above remarks, the question of the existence of inequality (6.2.30) for non-trivial sequences $\{u_x\}, \{\Delta u_x\}, \ldots, \{\Delta^k u_x\}$ is open.

Now we are concerned with comparing the solutions $u(x)$ of the nonlinear difference equation

$$\Delta_x^n u(x) = f(x, u(x)) \tag{6.2.39}$$

with solutions $\phi(x)$ and $\varphi(x)$ of the corresponding nonlinear difference inequalities

$$\Delta_x^n \phi(x) \leq f(x, \phi(x)) \tag{6.2.40}$$

and

$$\Delta_x^n \varphi(x) \geq f(x, \varphi(x)), \tag{6.2.41}$$

respectively.

Here $x = (x_1, \ldots, x_n) \in \mathbb{N}_0^n$, $\mathbb{N} = \{0, 1, \ldots\}$, $u(x) = (u_1(x), \ldots, u_m(x))$, $f(x, u) = (f_1(x, u), \ldots, f_m(x, u))^T$, and Δ_x^n is the n-fold forward difference operator $\Delta_{x_1} \cdots \Delta_{x_n} (\Delta a(t) = a(t+1) - a(t), t \in \mathbb{N}_0)$. Thus, we are dealing with systems of m equations or inequalities (component-wise) in n independent variables. Throughout, we shall assume that the function $f(x, u)$ defined on the set $A = \{(x, u) : 0 \leq x \leq X, X \in \mathbb{N}_0^n, u \in \mathbb{R}^m\}$.

In what follows, $(i)x$ denotes a point $(x_1, \ldots, x_n)$ in which i variables at zero. There are $\binom{n}{i}$ total such possibilities. Thus, if at the n hyperplane $x = (1)x$, the function $u(x)$ is known, then a recursive argument can be used to ensure the existence and uniqueness of the solutions of (6.2.39). This is obvious from the summation representation

$$u(x) = \sum_{i=1}^{n} (-1)^{i+1} \sum_i u((i)x) + S_{s=0}^{x-1} f(s, u(s)), \tag{6.2.42}$$

where $\sum_i$ represents the summation over the possibilities $(i)x$, and $S_{s=0}^{x-1}$ stands for the n-fold summation $\sum_{s_1=0}^{x_1-1} \cdots \sum_{s_n=0}^{x_n-1} \sum_{t=t_1}^{t_2} a(t) = 0$ for all $t_1 > t_2; t_1, t_2 \in \mathbb{N}_0$. From these notations, it is also clear that the solutions $\phi(x)$ and $\varphi(x)$ of the inequalities (6.2.40)–(6.2.41) have the summation representation

$$\phi(x) \leq \sum_{i=1}^{n}(-1)^{i+1}\sum_{i=1}^{n}\phi((i)x) + S_{s=0}^{x-1}f(s,\phi(s)) \tag{6.2.43}$$

and

$$\varphi(x) \geq \sum_{i=1}^{n}(-1)^{i+1}\sum_{i=1}^{n}\varphi((i)x) + S_{s=0}^{x-1}f(s,\varphi(s)). \tag{6.2.44}$$

Motivated by Riemann's function approach for the linear continuous Gronwall type of inequalities [6, 135, 629] and references therein, we have obtained a discrete analogue of Riemann's function and employed it to study the discrete Gronwall type of inequalities for the case $m = 1$ in [7], whereas for the general m, n in [9]. This approach accommodates (6.2.39)–(6.2.41), or equivalently, (6.2.42)–(6.2.44) only when f is linear of the form $f(x,u) = A(x)u + h(x)$, where $A(x)$ is an $m \times m$ matrix and $h(x)$ an $m \times 1$ vector. This technique easily provides explicit upper estimates [7, 9] and has the advantage that it requires fewer restrictions on the functions which appear in the inequalities than those needed in direct methods, see, e.g., [547, 704, 705]. To obtain upper estimates on Gronwall type of inequalities several other methods for the continuous general m, n, namely the method of splitting, the method of maxima, the iterative methods, etc., have also been proposed and applied in [60, 160, 161, 714]. Some of these methods will be extended for the discrete case in [9]. We shall also introduce several general comparison results. Obviously, the Gronwall type of inequalities are particular cases of these general theorems in which f has a special form so that for the corresponding inequalities, the unknown functions are readily available or can be estimated. These comparison results are natural generalizations of several principles established for the case $n = 1$ in [4, 571, 629, 630]. However, in this multi-dimensional case m, n, the summation representation plays the key role, whereas results for $n = 1$ are proved following the methods similar to the continuous case [325]. We shall employ these comparison results to study the problems of dependence on initial values and parameters of the solutions $u(x)$ of (6.2.39).

Theorem 6.2.4 (The Agarwal Inequality [8]) *Assume that the following hold:*

(i) the functions $u(x), \phi(x)$, and $\varphi(x)$ are solutions of (6.2.39), (6.2.40) and (6.2.41), respectively, which satisfy

$$\begin{aligned}\sum_{i=1}^{n}(-1)^{i+1}\sum_{i}\phi((i)x) &\leq \sum_{i=1}^{n}(-1)^{i+1}\sum_{i}u((i)x)) \\ &\leq \sum_{i=1}^{n}(-1)^{i+1}\sum_{i}\varphi((i)x); \end{aligned}\tag{6.2.45}$$

(ii) *for all fixed* x, $0 \le x \le X$, *and* $1 \le i \le m$, *the function* $f_i(x, u_1, \ldots, u_m)$ *is non-decreasing with respect to all* $u_1, \ldots, u_m$.

Then for all $x, 0 \le x \le X$,

$$\phi(x) \le u(x) \le \varphi(x). \tag{6.2.46}$$

Proof As we have noted $u(x), \phi(x)$, and $\varphi(x)$ have the representation (6.2.42), (6.2.43), and (6.2.44), respectively. Thus, for all $0 \le x = (j)x \le X$, $1 \le j \le n$, (6.2.46) follows from (6.2.45) and the fact that $S_{s=0}^{(j)x-1} f(s, u(s)) = 0$.

If $u(x) \le \varphi(x)$ is not true for all $0 \le x \le X$, then there is some $1 \le k \le m$ and an $x^*, 0 < x^* \le X$, such that $u_k(x^*) > \varphi_k(x^*)$ and $u(x) \le \varphi(x)$ for all $0 \le x < x^*$. However, since f_k is non-decreasing in $u_1, \ldots, u_m$ from (6.2.44) it follows that

$$\begin{aligned}
\varphi_k(x^*) &\ge \sum_{i=1}^{n}(-1)^{i+1}\sum_{i}\varphi_k((i)x^*) + S_{s=0}^{x^*-1} f_k(s, \varphi(s)) \\
&\ge \sum_{i=1}^{n}(-1)^{i+1}\sum_{i} u_k((i)x^*) + S_{s=0}^{x^*-1} f_k(s, u(s)) \\
&= u_k(x^*).
\end{aligned}$$

This contradiction competes the proof of $u(x) \le \varphi(x)$ for all $0 \le x \le X$. The inequality $\phi(x) \le u(x)$ can be proved in the same manner. □

Remark 6.2.5 If strict inequality holds in (6.2.45), then strict inequality holds in (6.2.46).

Remark 6.2.6 It is easy to verify that

$$\sum_{i=1}^{n}(-1)^{i+1}\sum_{i} u((i)x) = \sum_{k=0}^{n-1}(\sum_{s_1=0}^{x_1-1}\cdots\sum_{s_k=0}^{x_k-1}\Delta^k_{s_1,\ldots,s_k} u(s_1, \ldots, s_k, 0, x_{k+2}, \ldots, x_n))$$

and hence inequality (6.2.45) certainly holds if for all $0 \le k \le n-1$,

$$\begin{aligned}
&\Delta^k_{x_1,\ldots,x_k}\phi(x_1, \ldots, x_k, 0, x_{k+2}, \ldots, x_n) \\
&\quad \le \Delta^k_{x_1,\ldots,x_k} u(x_1, \ldots, x_k, 0, x_{k+2}, \ldots, x_n) \\
&\quad \le \Delta^k_{x_1,\ldots,x_k}\varphi(x_1, \ldots, x_k, 0, x_{k+2}, \ldots, x_n).
\end{aligned}$$

Theorem 6.2.5 ([8]) *Assume that the following conditions hold:*

(i) *$u(x, \mu)$ is the solution of the problem*

$$\begin{cases} \Delta_x^n u(x) = f(x, u(x), \mu) & (6.2.47) \\ u((i)x) = a([\bar{x}_i], \mu), & (6.2.48) \end{cases}$$

where μ is an r-dimensional vector, and $[\bar{x}_i]$ represents the points in $\mathbb{N}_0^{(n-1)}$ of non-zero variables in $(i)x$;

(ii) *for all fixed x, $0 \le x \le X$ and $1 \le j \le m$, the function $f(x, u_1, \dots, u_m, \mu_1, \dots, \mu_r)$ is non-decreasing with respect to $u_1, \dots, u_m$ and $\mu_1, \dots, \mu_r$;*

(iii) *for all fixed $[\bar{x}_i], 0 \le [\bar{x}_i] \le [\bar{X}_i]$, and $1 \le j \le m$, the function $\sum_{i=1}^n (-1)^{i+1} a_j([\bar{x}_i], \mu_1, \dots, \mu_r)$ is strictly increasing in $\mu_1, \dots, \mu_r$.*

Then for all $x, 0 \le x \le X$, the solution $u(x, \mu)$ of (6.2.47)–(6.2.48) is a strictly increasing function of μ, i.e., if $\mu^1 \le \mu^2$, then $u(x, \mu^1) < u(x, \mu^2)$.

Furthermore, if (a) for all fixed $x, 0 \le x \le X$, the function $f(x, u, \mu)$ is continuous with respect to u and μ, and (b) for all fixed $[\bar{x}_i], 0 \le [\bar{x}_i] \le [\bar{X}_i]$, the function $a([\bar{x}_i], \mu)$ is continuous with respect to μ, then for all $0 \le x \le X$,

$$\lim_{\mu \to 0} u(x, \mu) = u(x), \tag{6.2.49}$$

where $u(x)$ is the solution of (6.2.39) satisfying

$$u((i)x) = a([\bar{x}_i]). \tag{6.2.50}$$

Moreover, if $0 \le x \le X < +\infty$, then (6.2.49) is uniform in μ.

Proof Let $\mu^1 < \mu^2$, then since

$$u(x, \mu^l) = \sum_{i=1}^n (-1)^{i+1} \sum_i a([\bar{x}_i], \mu^l) + S_{s=0}^{x-1} f(s, u(s, \mu^l), \mu^l); \quad l = 1, 2,$$

conditions (ii) and (iii) imply that

$$u(x, \mu^2) > \sum_{i=1}^n (-1)^{i+1} \sum_i a([\bar{x}_i], \mu^1) + S_{s=0}^{x-1} f(s, u(s, \mu^2), \mu^1)$$

and now for all $x, 0 \le x \le X$, the inequality $u(x, \mu^1) < u(x, \mu^2)$ follows as in the proof of Theorem 6.1.34.

The rest of the proof is a consequence of the continuity assumptions. □

Theorem 6.2.6 (The Agarwal Inequality [8]) *Assume that the following conditions hold:*

(i) condition (ii) of Theorem 6.1.34;
(ii) there exists a function $v(x,u)$ defined on A which is such that for any function $p(x)$ defined for all $x, 0 \le x \le X$,

$$\Delta_x^n v(x, p(x)) \le f(x, v(x, p(x))); \tag{6.2.51}$$

(iii) the function $u(x)$ is a solution of (6.2.39) which satisfies

$$\sum_{i=1}^{n}(-1)^{i+1}\sum_{i} v((i)x, p((i)x)) \le \sum_{i=1}^{n}(-1)^{i+1}\sum_{i} u((i)x). \tag{6.2.52}$$

Then for all $x, 0 \le x \le X$,

$$v(x, p(x)) \le u(x). \tag{6.2.53}$$

Proof Let $q(x) = v(x, p(x))$, then from (6.2.51) it follows that

$$\begin{aligned}\Delta_x^n q(x) &= \Delta_x^n v(x, p(x)) \le f(x, v(x, p(x))) \\ &= f(x, q(x)).\end{aligned}$$

Also, (6.2.52) is the same as

$$\sum_{i=1}^{n}(-1)^{i+1}\sum_{i} q((i)x) \le \sum_{i=1}^{n}(-1)^{i+1}\sum_{i} u((i)x).$$

Thus, for all $x, 0 \le x \le X$, Theorem 6.1.34 gives us that

$$q(x) = v(x, p) \le u(x).$$

□

Theorem 6.2.7 (The Agarwal Inequality [8]) *Assume that the following conditions hold:*

(i) for all (x,u) and (x,v) in A,

$$|f(x,u) - f(x,v)| \le g(x, |u - v|), \tag{6.2.54}$$

where the function $g(x,u)$ is defined on $A^+ = \{(x,u) : 0 \le x \le X, X \in \mathbb{N}_0^n, u \in \mathbb{R}_+^m\}$, and for all fixed x and $1 \le i \le m$, $g_i(x, u_1, \ldots, u_m)$ is non-decreasing with respect to all $u_1, \ldots, u_m$;

(ii) *there exist functions* $u^1(x), u^2(x), \varepsilon^1(x)$, *and* $\varepsilon^2(x)$ *which are defined for all* $x, 0 \leq x \leq X$, *and satisfy the inequalities*

$$|\Delta_x^n u^1(x) - f(x, u^1(x))| \leq \varepsilon^1(x) \tag{6.2.55}$$

and

$$|\Delta_x^n u^2(x) - f(x, u^2(x))| \leq \varepsilon^2(x); \tag{6.2.56}$$

(iii) $u(x)$ *is a solution of the difference equation*

$$\Delta_x^n u(x) = g(x, u(x)) + \varepsilon^1(x) + \varepsilon^2(x) \tag{6.2.57}$$

which satisfies the inequality

$$|\sum_{i=1}^{n}(-1)^{i+1}\sum_{i}(u^1((i)x) - u^2((i)x))| \leq \sum_{i=1}^{n}(-1)^{i+1}\sum_{i} u((i)x). \tag{6.2.58}$$

Then, for all $x, 0 \leq x \leq X$,

$$|u^1(x) - u^2(x)| \leq u(x). \tag{6.2.59}$$

Proof Inequalities (6.2.55) and (6.2.56) imply that

$$|\Delta_x^n(u^1(x) - u^2(x)) - (f(x, u^1(x)) - f(x, u^2(x)))| \leq \varepsilon^1(x) + \varepsilon^2(x)$$

and hence we have

$$\begin{aligned}&|S_{s=0}^{x-1}\Delta_s^n(u^1(s) - u^2(s)) - S_{s=0}^{x-1}(f(s, u^1(s)) - f(s, u^2(s)))|\\&\leq S_{s=0}^{x-1}(\varepsilon^1(s) + \varepsilon^2(s)),\end{aligned}$$

which implies that

$$\begin{aligned}|u^1(x) - u^2(x)| \leq{}& |v\sum_{i=1}^{n}(-1)^{i+1}\sum_{i}(u^1((i)x) - u^2((i)x)|\\&+S_{s=0}^{x-1}|f(s, u^1(s)) - f(s, u^2(s))| + S_{s=0}^{x-1}(\varepsilon^1(s) + \varepsilon^2(s)).\end{aligned}$$

Using (6.2.54) and (6.2.58 in the above inequality, we obtain

$$y(x) \leq \sum_{i=1}^{n}(-1)^{i+1}\sum_{i} u((i)x) + S_{s=0}^{x-1}(g(s, y(s)) + \varepsilon^1(s) + \varepsilon^2(s)), \tag{6.2.60}$$

where $y(x) = |u^1(x) - u^2(x)|$.

Since $u(x)$, the solution of (6.2.57), has the summation representation

$$u(x) = \sum_{i=1}^{n}(-1)^{i+1}\sum_{i} u((i)x) + S_{s=0}^{x-1}(g(s, u(s)) + \varepsilon^1(s) + \varepsilon^2(s)), \quad (6.2.61)$$

the inequality $y(x) \leq u(x)$ follows on comparing (6.2.60) and (6.2.61) as in Theorem 6.1.34. □

Chapter 7
Nonlinear Multi-Dimensional Discontinuous Inequalities

In this chapter, we introduce some nonlinear discontinuous inequalities for multiple variables.

7.1 Nonlinear Multi-Dimensional Discontinuous Integral Bellman-Gronwall Inequalities in Partially Ordered Banach Spaces

7.1.1 Nonlinear Multi-Dimensional Integral Inequalities for Functions Defined in Partially Ordered Topological Spaces

The following results consider integral inequalities for real functions defined in partially ordered topological spaces with a measure, and the integral operators are nonlinear and are essentially of Volterra type.

Among the contributions devoted to linear integral inequalities of Gronwall-Bellman type, we shall mention the paper [43] where integral inequalities for functions defined in metric spaces are considered.

We shall consider real functions defined in the partially ordered topological space T. The following denotations will be used

$$T_x := \{y : y \in T \text{ and } y < x\};$$

$$[T_x] := \begin{cases} \text{the closure of } T_x & \text{if} \quad T_x \neq \varnothing, \\ \{x\} \quad \text{if} \quad T_x = \varnothing. \end{cases}$$

Y. Qin, *Integral and Discrete Inequalities and Their Applications*,
DOI 10.1007/978-3-319-33304-5_7

Concerning the set T, the following conditions $(C_1) - (C_5)$ will be assumed to be satisfied:

(C_1) T is a partially ordered connected topological space with positive measure μ.
(C_2) For every x from T, the segment T_x is a subset of T which is measurable with respect to the measure μ.
(C_3) If $\{x_\alpha\}$ is a generalized sequence of elements of T convergent to x, then $\mu(T_{x_\alpha} \triangle T_x) = \mu((T_{x_\alpha} \setminus T_x) \cup (T_x \setminus T_{x_\alpha}))$ tends to 0.
(C_4) For every element x of T and for any open set U containing $[T_x]$, there exists an open neighborhood W of x, such that if $y \in W$, then $T_y \subset U$.
(C_5) There exists an element x_0 of T, for which $\mu(T_{x_0}) = 0$.

We shall consider integral operators of the type

$$Vf(x) := \int_T K(x, y, f(y))\chi(x, y)d\mu(y) = \int_{T_x} K(x, y, f(y))d\mu(y) \tag{7.1.1}$$

where $\chi(x, y)$ is the characteristic function of the segment T_x, i.e.,

$$\chi(x, y) := \begin{cases} 1 & \text{if } y < x, \\ 0 & \text{otherwise,} \end{cases}$$

while the kernel $K(x, y, z)$ is defined in $T \times T \times \mathbb{R}$ and assumes values in $\mathbb{R}$.

$D(V)$ will denote the class of all real functions f defined in the space T such that, for every fixed x from T, the function

$$F(y) := K(x, y, f(y))\chi(x, y)$$

is integrable with respect to the measure μ, *i.e.*, $D(V)$ is the set of functions f for which $Vf(x)$ holds for every x from T.

Theorem 7.1.1 (The Ronkov-Bainov Inequality [577]) *Let the conditions* (C_5) *be fulfilled for the set T and let the kernel* $K(x, y, z)$ *of the integral operator V every two fixed elements x and y from T be a monotonely increasing function of z. Then, if for two continuous functions f and h from the class D(V), the following inequality holds for every* $x \in T$,

$$f(x) - Vf(x) < h(x) - Vh(x), \tag{7.1.2}$$

then we have for every $x \in T$, $f(x) < h(x)$.

Proof Let

$$T_0 := \{x : x \in T \text{ and } f(y) < h(y) \text{ for every } y < x\}.$$

We shall show that $T_0 = T$ whence it follows the statement of the theorem. Indeed, if $x \in T_0$, then $K(x, y, f(y)) < K(x, y, h(y))$ for every $y < x$, whence $Vf(x) \le Vh(x)$. On the other hand, inequality (7.1.11) implies that

$$f(x) = (f(x) - Vf(x)) + Vf(x) < h(x) - Vh(x) + Vf(x)$$

whence $f(x) < h(x)$.

Since T is a connected topological space, then in order to show that $T_0 = T$, it is sufficient to verify that T_0 is not empty and that it is open and closed at the same time.

By condition (C_5), there exists an element x_0 of T such that $\mu(T_{x_0}) = 0$ whence it follows that $x_0 \in T_0$. Indeed, if $y < x_0$, then $T_y \subset T_{x_0}$ and hence $\mu(T_y) = 0$, which implies $Vf(y) = Vh(y) = 0$. However, in view of (7.1.2), it is obtained that $f(y) < h(y)$ and therefore $x_0 \in T_0$, i.e., T_0 is not empty.

We shall show that T_0 is a closed subset of T. Indeed, let $\{x_\alpha\}$ be a generalized sequence of elements of T_0 convergent to x. If we assume that x does not belong to T_0, then an element $y \in T_x$ will exist, such that $f(y) \ge h(y)$, whence and by virtue of inequality (7.1.2), it follows that $Vf(y) > Vh(y)$. The last inequality, however, is possible only if there exists a set $D \subset T_y \subset T_x$ for every z, on which the inequality $f(z) > h(z)$ holds and $\mu(D) > 0$. Then

$$\mu(D) \le \mu(T_y \setminus T_{x_\alpha}) \le \mu(T_x \Delta T_{x_\alpha}).$$

On the other hand, by condition $(C_3), \mu(T_x \Delta T_{x_\alpha})$ converges to 0 and hence $\mu(D) = 0$.

Thus, the assumption that x does not belong to T_0 ends up with a contradiction. Therefore, $x \in T_0$, whence it follows that the set T_0 is closed.

Now we shall show that T_0 is also an open subset of T. First, we shall show that if $x \in T_0$, then $[T_x] \subset T_0$. Indeed, the fact that $x \in T_0$ implies that $T_x \subset T_0$. If $T_x \in \varnothing$, then $T_x = \{x\} \subset T_0$. If $T_x \neq \varnothing$, then $[T_x]$ is the closure of T_x and since T_0, as we have already demonstrated, is closed, then $[T_x] \subset T_0$. We saw that if $y \in T_0$, then $f(y) < h(y)$. Then, since the functions f and h are continuous, for any $y \in [T_x]$, there exists an open neighborhood U_y of y, such that if $z \in U_y$, then the inequality $f(z) < h(z)$ holds. By U denote the union of all such neighborhoods of the elements of $[T_0]$, i.e.,

$$U := \bigcup_{y \in [T_x]} U_y.$$

Then U is an open set containing $[T_x]$, on which $f < h$. In view of condition (C_4), there exists an open neighborhood W of x, such that if $z \in W$, then $T_z \subset U$. It is not difficult to verify that the neighborhood W of x is contained in T_0, whence it follows that T_0 is open. Indeed, if $z \in W$ and $y < z$, then $y \in T_z \subset U$ and therefore $f(y) < h(y)$, i.e., $z \in T_0$. This completes the proof. □

Remark 7.1.1 It is easy to verify that the assertion of the theorem still holds if, in (7.1.2), we replace the strict inequality by a non-strict one, provided for any element y of the space T, for which $\mu(T_y) = 0$, the strict inequality

$$f(y) < h(y)$$

holds.

Remark 7.1.2 The condition (C_5) may be omitted if there exists an element x_1 of T, such that for every $y < x_1$, the following inequality, $f(y) < h(y)$, holds.

Corollary 7.1.1 ([577]) *Let the real functions f, g, h and ψ be defined in T, and f and h are continuous. Assume that the integral equation*

$$\psi = g + V\psi \tag{7.1.3}$$

possesses a continuous solution φ. Then, if the following inequality holds for every $x \in T$,

$$f(x) < g(x) + Vf(x) \quad (\textit{respectively},\ h(x) > g(x) + Vh(x))$$

then for every $x \in T$,

$$f(x) - Vf(x) < \varphi(x) - V\varphi(x) \quad (\textit{respectively},\ h(x) - Vh(x) > \varphi(x) - V\varphi(x))$$

and hence $f(x) < \varphi(x)$ (respectively, $h(x) > \varphi(x)$) for every $x \in T$.

The integral inequality from Theorem 7.1.1 will be used in the proof of a sufficient condition for the existence of solution of the integral equation (7.1.3) given by the following theorem.

Theorem 7.1.2 ([577]) *Let T be a compact uniform topological space with finite Borelean measure μ, the condition $(C_1)-(C_5)$ hold. We shall assume that the kernel $K(x, y, z)$ of the integral operator V is continuous and for every two fixed elements x and y from T, it is a monotonely increasing function of z. Then, if g is a continuous function defined in T, for which two continuous functions f and h from $D(V)$ exist, such that for every $x \in T$,*

$$f(x) - Vf(x) < g(x) < h(x) - Vh(x) \tag{7.1.4}$$

then the integral equation (7.1.3) possesses a continuous solution φ satisfying the inequality

$$f(x) < \varphi(x) < h(x) \tag{7.1.5}$$

for any $x \in T$.

Proof The method of consecutive approximations will be used to find a solution of the integral equation (7.1.3), and the function f will be taken as initial approximation. More precisely, we will show that if

$$\begin{cases} f_1 = f \\ f_{n+1} = g + Vf_n \end{cases} \tag{7.1.6}$$

then $\varphi := \lim\{f_n\}$ is a solution of equation (7.1.3). First, we shall show that f_{n+1} is correctly defined, i.e. that the symbol Vf_n holds. For this purpose, we shall need the following lemma.

Lemma 7.1.1 ([577]) *Under the assumptions of Theorem 7.1.2, if ψ is a bounded function from the class $D(V)$, then V_ψ is a continuous function.*

Proof Let $x_1 \in T$ and ε be an arbitrary positive number. By I denote a compact $T \times T \times I$. Therefore, a neighborhood U_1 of x_1 exists, such that if $x \in U_1$, then

$$|K(x_1, y, \psi(y)) - K(x, y, \psi(y))| < \frac{\varepsilon}{3\mu(T)}$$

for every $y \in T$ and a constant B also exists, such that $|K(x, y, \psi(y))| < B$ for every two elements x and y of T. In view of condition (C_3), a neighborhood U_2 of x_1 exists, such that if $x \in U_2$, then $\mu(T_{x_1} \Delta T_x) < (\varepsilon/3B)$. Then, if $x \in U := U_1 \cap U_2$, then

$$\begin{aligned} |V\psi(x_1) - V\psi(x)| &= |\int_{T_{x_1}} K(x_1, y, \psi(y))d\mu(y) - \int_{T_x} K(x, y, \psi(y))d\mu(y)| \\ &\leq \int_{T_{x_1} \cap T_x} |K(x_1, y, \psi(y)) - K(x, y, \psi(y))|d\mu(y) \\ &\quad + \int_{T_{x_1} \setminus T_x} |K(x_1, y, \psi(y))|d\mu(y) + \int_{T_x \setminus T_{x_1}} |K(x, y, \psi(y))|d\mu(y) \\ &\leq \frac{\varepsilon}{3\mu(T)} \mu(T_{x_1} \cap T_x) + B\mu(T_{x_1} \setminus T_x) + B\mu(T_x \setminus T_{x_1}) \leq \varepsilon. \end{aligned}$$

Since f_1 is continuous on the compact T and hence it is bounded, then according to Lemma 7.1.1, Vf_1 and $f_2 = g + Vf_1$ also are continuous. By induction, it is analogously proved that for any natural number n, the function f_n is continuous and therefore Vf_n holds.

Now we shall show that for every $x \in T$, the sequence $\{f_n(x)\}$ is monotonely increasing. Indeed, (7.1.4) implies that

$$f_1(x) - Vf_1(x) = f(x) - Vf(x) < g(x).$$

If we assume that $f_k(x) - Vf_k(x) < g(x)$ for some natural number k, then $f_{k+1}(x) - f_k(x) = g(x) + Vf_k(x) - f_k(x) > 0$. Hence $f_{k+1}(x) > f_k(x)$ whence $Vf_{k+1}(x) > Vf_k(x)$. Then, $f_{k+1}(x) - Vf_{k+1}(x) = g(x) + Vf_k(x) - Vf_{k+1}(x) < g(x)$.

Thus, it has been proved that for every natural number n, the inequality $f_n(x) - Vf_n(x) < g(x)$ holds, or, which is the same, $f_n(x) < g(x) + Vf_n(x) = f_{n+1}(x)$.

By Theorem 7.1.1, the inequality

$$f_n(x) - Vf_n(x) < g(x) < h(x) - Vh(x)$$

implies $f_n(x) < h(x)$. Hence, for any $x \in T, \{f_n(x)\}$ is convergent sequence of real numbers. Denote its limit by $\varphi(x)$. The continuity of the kernel $K(x, y, z)$ implies that $K(x, y, \varphi(y)) = \lim_{n\to+\infty} K(x, y, f_n(y))$. Since for every two fixed elements x and y of T, $K(x, y, z)$ is a monotonely increasing function of z, then

$$K(x, y, f_n(y)) < K(x, y, h(y)),$$

whence, in view of the Levi Theorem, it follows that

$$\lim_{n\to+\infty} \int_{T_x} K(x, y, f_n(y))d\mu(y) = \int_{T_x} K(x, y, \varphi(y))d\mu(y)$$

and therefore we can pass to a limit in (7.1.6) for $n \to +\infty$. Having accomplished the passage to a limit which shows that $\varphi(x) = g(x) + V\varphi(x)$, which establishes that φ is a solution of the integral equation (7.1.3).

Since for every n and any $x \in T$, the inequality $f(x) \leq f_n(x) < h(x)$ holds, then $f(x) \leq \varphi(x) \leq h(x)$, whence it follows that the function φ is bounded since the functions f and h are continuous on the compact T. Then, by Lemma 7.1.1, $V\varphi$, and therefore $\varphi = g + V\varphi$ as well, are continuous functions.

By virtue of Corollary 7.1.1, the function φ satisfies inequality (7.1.5) since it is a continuous solution of the integral equation (7.1.3). This completes the proof. □

Remark 7.1.3 If, in (7.1.4), one of the strict inequalities is replaced by a non-strict one, for example, if $f(x) - Vf(x) \leq g(x) < h(x) - Vh(x)$, then the assertion of Theorem 7.1.2 still holds, and in this case, it can be assumed that the continuous solution φ of equation (7.1.9) satisfies the inequality $f(x) \leq \varphi(x) < h(x)$.

Note also that in view of Remark 7.1.2, both inequalities in (7.1.4) can be replaced by non-strict ones, provided that for every $y \in T$, with $\mu(T_y) = 0$, the inequality $f(y) < g(y) < h(y)$ holds.

So, if the conditions of Theorem 7.1.2 are fulfilled, then the integral equation (7.1.3) has at least one continuous solution. A sufficient condition for uniqueness of this solution is supplied by the following theorem.

Theorem 7.1.3 ([577]) *Let the condition of Theorem 7.1.2 be satisfied. Then, if the kernel $K(x, y, z)$ of the integral operator V satisfies the inequality*

$$\chi(x, y)|K(x, y, z_1) - K(x, y, z_2)| \leq BK_1(y)|z_1 - z_2| \tag{7.1.7}$$

for every two elements x and y of T and every two real numbers z_1, z_2 where B is a constant, while $K_1(y)$ is a function, integrable with respect to the measure μ, then the integral equation (7.1.3) has a unique bounded solution.

Proof Let φ be a bounded function which is a solution of equation (7.1.3). Then, since φ is bounded, Lemma 7.1.1 implies that $V\varphi$, and hence $\varphi = g + V\varphi$, are continuous functions.

Let φ and ψ be two bounded solutions of the integral equation (7.1.3). By T_0 denote the following subset of the space T,

$$T_0 := \{x \in T : \varphi(y) = \psi(y) \text{ for every } y < x\}.$$

We shall show that T_0 is non-empty and that it is closed and open, whence it follows that T_0 coincides with T, and this will complete the proof of Theorem 7.1.3. In fact, according to condition (C_5), an element x_0 of T exists, for which $\mu(T_{x0}) = 0$ and hence $\mu(T_y) = 0$ for every $y < x_0$. But then $V\varphi(y) = V\psi(y) = 0$, and hence, since φ, ψ are solutions of equation (7.1.3), then $\varphi(y) = \psi(y) = g(y)$ whence it follows that $x_0 \in T_0$.

Now we shall show that T_0 is closed. Indeed, let $\{x_\alpha\}$ be a generalized sequence of elements of T_0 convergent to x. If we assume that x does not belong to T_0, then $z < x$ will exist, such that $\varphi(z) \neq \psi(z)$ and hence $V\varphi(z) \neq V\psi(z)$. Then a set $D \subset T_x$ will exist, such that if $y \in D$, then $\varphi(z) \neq \psi(z)$ and $\mu(D) > 0$, therefore, $D \subset T_x \setminus T_{x_\alpha}$, whence $D < \mu(D) \leq \mu(T_x \Delta T_{x_\alpha})$, which contradicts the fact that $\mu(T_x \Delta T_{x_\alpha})$ tends to 0. Hence $x \in T_0$, which proves that the set T_0 is closed.

Next we shall also show that the set T_0 is open as well. Indeed, let $x_1 \in T_0$. Then, $T_{x_1} \subset T_0$ and since T_0 is closed, then $[T_{x_1}] \subset T_0$, the fact that φ, ψ is also continuous. But $|\varphi - \psi| \equiv 0$ on T_0, and hence for any $y \in [T_{x_1}]$, there exists an open neighborhood U_y of y such that $|\varphi - \psi| < 1$ onto U_y. By U denote the union of all such neighborhoods of the elements of $[T_{x_1}]$, i.e.,

$$U := \bigcup_{y \in [T_{x_1}]} U_y.$$

Then, U is an open set containing $[T_{x_1}]$ and in view of the condition (C_4), there exists an open neighborhood W_1 of x_1, such that if $x \in W_1$, then $T_x \subset U$.

On the other hand, since $K_1(y)$ is an integrable function with respect to the measure μ, then $\varepsilon > 0$ exists, such that if $\mu(D) < \varepsilon$, then $\int_D K_1(y)d\mu(y) < 1/2B$.

In view of condition (C_3), an open neighborhood W_2 of x_1 exists, such that if $x \in W_2$, then $\mu(T_{x_1} \Delta T_x) < \varepsilon$. Then, if $W := W_1 \cap W_2$, then $W \subset T_0$. Indeed, let us associate the number, for every $x \in W$,

$$S(x) := \begin{cases} \sup\limits_{z\in T_x} |\varphi(z) - \psi(z)|, & \text{if } T_x \neq \varnothing, \\ 0, \quad \text{if } T_x = \varnothing. \end{cases}$$

Obviously, $0 \le S(x) \le 1$. If we assume that $S(x) > 0$ for some $x \in W$, then an element $z \in T_x$ exists, such that $\frac{3}{4}S(x) < |\varphi(z) - \psi(z)|$. But since φ, ψ are solutions of equation (7.1.3), then

$$|\varphi(z) - \psi(z)| = |V\varphi(z) - V\psi(z)| \le \int_{T_z\setminus T_{x_1}} |K(z, y, \varphi(y)) - K(z, y, \psi(y))| d\mu(y).$$

Inequality (7.1.7) and the fact that $T_z \subset T_x$ imply that

$$\int_{T_z\setminus T_{x_1}} |K(z, y, \varphi(y)) - K(z, y, \psi(y))| d\mu(y) \le BS(x) \int_{T_z\setminus T_{x_1}} K_1(y) d\mu(y).$$

Since $x \in W \subset W_2$, then $\mu(T_{x_1} \Delta T_x) < \varepsilon$ and hence

$$BS(x) \int_{T_z\setminus T_{x_1}} K_1(y) d\mu(y) < \frac{1}{2} S(x).$$

We finally obtain that $\frac{3}{4}S(x) < \frac{1}{2}S(x)$ which contradicts the assumption that $S(x) > 0$. Therefore, $S(x) = 0$, whence it follows that $\varphi \equiv \psi$ on T_x for any $x \in W$, i.e., $W \subset T_0$. Thus, it is shown that T_0 is an open set. Thus the proof is complete. □

Note that Theorem 7.1.1 generalizes some well-known theorems for integral inequalities for functions defined in $\mathbb{R}$ or $\mathbb{R}^n$ where, in most general terms, it is stated that if $f < g + Vf$, where V is a Volterra integral operator, then $f < \varphi$, where φ is a solution of the integral equation $\varphi = g + V\varphi$.

Thus, if by T we denote the interval of real numbers $[a, b]$ with the usual ordering and topology , while μ is the Lebesgue measure, then obviously the conditions $(C_1) - (C_5)$ are fulfilled. The consideration of kernels of the type $K(x, y, z) = K_1(y)z$, which define linear integral operators, leads to the well-known and most used integral inequalities of Gronwall-Bellman type. One of the most widely used nonlinear integral inequalities, the Bihari inequality (Theorem 1.1.1) is for integral operators with kernels of the form: $K(x, y, z) = K_1(y)\varphi(z)$, where φ is a monotonely increasing function. Both linear and nonlinear integral inequalities are generalized for functions defined in sub-domains of $\mathbb{R}^n$, and most often the partial ordering considered there is the following: If $x = (x_1, \cdots, x_n)$ and $y = (y_1, \cdots, y_n)$, then $x < y$ if $x_i < y_i$ for every $i = 1, 2, \cdots, n$. The considered sub-domains of $\mathbb{R}^n$

with the standard topology and Lebesgue measure, with this or some other partial ordering, also satisfy the conditions $(C_1) - (C_5)$.

7.1.2 Matrix Inequalities

We now begin to discuss the matrix inequalities.

Let $\mathcal{A}$ denote the linear space of real $n \times n$ symmetric matrices. In $\mathcal{A}$, we can introduce a partial ordering in more than one way. For instance, using, respectively, cones of non-negative matrices and non-negative definite matrices, two different types of orderings can be introduced in $\mathcal{A}$. Since a non-negative definite matrix is a natural generalization of non-negative number, we adopt the second kind of ordering. Then with this ordering in $\mathcal{A}$, we have

$$X, Y \in \mathcal{A}, \quad X \leq Y \quad \text{if and only if} \quad (X - Y) \in H. \tag{7.1.8}$$

A function $P : \mathcal{A} \to \mathcal{A}$ is called monotonic [47] if $X, Y \in \mathcal{A}$ and $X \leq Y$ imply $P(X) \leq P(Y)$ (that is, $P(Y) - P(X)$ is a non-negative definite matrix).

Theorem 7.1.4 (The Chandra-Fleishman Inequality [136]) *Let H be a real symmetric matrix, Let G be a monotone and Lipschitz continuous function from $\mathcal{A}$ into $\mathcal{A}$:*

$$\|G(X) - G(Y)\| \leq \rho\|X - Y\|. \tag{7.1.9}$$

If there holds

$$X(t) \leq H(t) + \int_0^t G(X(s))ds, \tag{7.1.10}$$

then

$$X(t) \leq Y(t) \tag{7.1.11}$$

on their common interval of existence, where $Y(t)$ is the unique solution of the corresponding equality.

Proof For $n = 1, 2, \ldots,$ set

$$Y_n(t) = H(t) + \int_0^t G(Y_{n-1}(s))ds$$

where $Y_0(t) = X(t) \in \mathcal{A}$. Then $\{Y_n\}, n = 1, 2, \ldots$, are all in $\mathcal{A}$. Next, using the monotonicity of G, it is easily verified that

$$X(t) \leq Y_1(t) \leq \cdots \leq Y_n(t).$$

Since G is Lipschitz continuous, $\{Y_n(t)\}$ converges to the unique solution $Y(t)$ of the corresponding equality. This completes the proof. □

The following results used by Bellman [77] may be regarded as corollaries of the above theorem.

Corollary 7.1.2 (The Chandra-Fleishman Inequality [136]) *The following inequality holds*

$$\frac{dX}{dt} \leq F(t) + G(X), \quad X(0) = C, \tag{7.1.12}$$

then

$$X(t) \leq Y(t) \tag{7.1.13}$$

where F and C are real and symmetric, G has the properties above, and $Y(t)$ is the unique solution of the initial value problem

$$\frac{dY}{dt} \leq F(t) + G(Y), \quad Y(0) = C.$$

Proof In fact, integrating the inequality (7.1.12) yields

$$X(t) \leq C + \int_0^t F(s)ds + \int_0^t G(X(s))ds.$$

If we set

$$H(t) = C + \int_0^t F(s)ds \in \mathcal{A},$$

then the result (7.1.13) follows immediately from Theorem 7.1.4. □

Corollary 7.1.3 (The Chandra-Fleishman Inequality [136]) *If the following inequality holds*

$$\frac{dX}{dt} \leq F(t) + RX + XR^T + \sum_i^m Q_i X Q_i^T, \quad X(0) = C, \tag{7.1.14}$$

then

$$X(t) \leq Y(t), \tag{7.1.15}$$

where F and C are real and symmetric, R and Q_i are real constant $n \times n$ matrices, and $Y(t)$ is the unique solution of the corresponding initial value problem.

Proof Following a familiar procedure (multiplying from the left and right by e^{Rt} and $e^{R^T t}$, respectively, etc.), we get

$$\begin{aligned} X(t) &\leq e^{Rt} C e^{R^T t} + \int_0^t e^{R(t-s)} \Big[F(s) + \sum_i^m Q_i X Q_i^T \Big] e^{R^T (t-s)} ds \\ &= H(t) + \int_0^t G(t, s, X(s)) ds \end{aligned}$$

where we have set

$$H(t) = e^{Rt} C e^{R^T t} + \int_0^t e^{R(t-s)} F(s) e^{R^T (t-s)} ds$$

and

$$G(t, s, x) = e^{R(t-s)} \sum_i^m Q_i X Q_i^T e^{R^T (t-s)}.$$

Clearly, $H(t) \in \mathcal{A}$. Also G is Lipschitz continuous in its last argument and monotonic, because if $X \leq Y$, for each i, $i = 1, \ldots, m$, for any Q_i,

$$Q_i Y Q_i^T - Q_i X Q_i^T = Q_i (Y - X) Q_i^T \in H.$$

Again,

$$e^{R(t-s)} \sum_i^m Q_i (Y - X) Q_i^T e^{R^T (t-s)} \in H.$$

Having established the desired properties for H and G, we may now apply Theorem 7.3.10 to obtain the desired conclusion. □

7.2 Nonlinear Multi-Dimensional Discontinuous Integral Inequalities of Wendroff Type

7.2.1 Nonlinear Two-Dimensional Discontinuous Integral Inequalities of Wendroff Type

In the following results, we introduce some new integral inequalities for discontinuous functions of Wendroff type, due to Borysenko and Iovane [112]. From these results, we also deduce new generalizations of results given by Borysenko [102, 104, 107, 109], by Samoilenko and Borysenko [582, 583] for integro-sum inequalities.

We note that in the earlier articles [102, 103, 582], integro-sum inequalities for the piecewise-continuous functions of a certain type

$$\varphi(t) \le \psi(t) + \int_{t_0}^{t} K(t,s,\varphi(s))ds + \sum_{t_0<t_i<t} \mu(t,t_i)\sigma_k(\varphi(t_i-0)), \qquad (7.2.1)$$

were investigated, here $\varphi(t), \mu(t,t_i), \psi(t), \sigma_i(u)$ are continuous non-negative functions ($i = 1,2,\ldots$) for all $t \ge t_0 \ge 0$, except for $\varphi(t)$, which has the first kind of discontinuities at the points $\{t_i\}, i = 1,2,\ldots,: 0 \le t_0 < t_1 < \ldots, \lim_{i\to+\infty} t_i = +\infty$.

The kernel $K(t,s,u)$, which is non-negative at $t \ge s \ge t_0$, is determined in domain $t \ge s \ge t_0, |u| \le k = const > 0$, and at fixed t and s, it is non-decreasing with respect to u; the functions $\sigma_k(u)$ are continuous, non-negative and non-decreasing with respect to u.

In [104], the first author Borysenko obtained an estimate: for all $t \in [0,+\infty]$,

$$\varphi(t) \le \xi_\psi(t),$$

where $\xi_\psi(t)$ is a solution of the integro-sum equation

$$\xi(t) = \psi(t) + \int_{t_0}^{t} K(t,s,\xi(s))ds + \sum_{t_0<t_i<t} \mu(t,t_i)\sigma_k(\xi(t_i-0))$$

which is continuous in each of the intervals $[t_i, t_{i+1}], i = 0, 1 \ldots$, and has some first kind of discontinuities at the points $\{t_i\}$.

From the result [104], the results in [102, 585] follow for integro-sum inequalities. In all the previously described results for integro-sum inequalities before the results in [109] (see [102, 103, 106, 113, 116, 117, 292, 582, 585, 587, 589]), there were considered Lipschitz type nonlinearities for $\sigma_k(u)$. In the works [104, 107, 589], there were obtained generalization integral inequalities of Bellman-Bihari type for functions of two independent variables with jumps (finite) at some fixed points from an open domain $\Omega \subset \mathbb{R}^2_+$.

In a similar way to that in [109] for the one-dimensional case, we investigate the integral inequalities for the functions of two independent variables (Wendroff type) with non-Lipschitz type functions, which characterize the values of discontinuity (the results [104, 107, 589] follow automatically as particular cases of the results of this section).

In the next theorem, we discuss the inequalities of functions with Lipschtiz type discontinuity.

We consider some set $D^* \subset \mathbb{R}^2$, where $D^* = D \setminus \Gamma, D = \bigcup_j D_j, j = 1, 2, \cdots$;

$$\Gamma = \bigcup_j \Gamma_j, \Gamma_j = \{(x,y) : \varphi_j(x,y) = 0, j = 1, 2, \cdots\}, \Gamma_k \cap \Gamma_{k+1} = \varnothing, k = 1, 2, \cdots,$$

here $\varphi_j(x,y)$ are real-valued continuously differentiable functions such that grad $\varphi_j(x,y) > 0$, for all $j = 1, 2, \cdots$;

$$D_1 = \{(x,y) : x \geq 0, y \geq 0, \varphi_1(x,y) < 0\};$$

$$D_k = \{(x,y) : x \geq 0, y \geq 0, \varphi_{k-1}(x,y) > 0, \varphi_k(x,y) < 0, \text{for all } k > 2, k \in \mathbb{N}\};$$

$$G_p = \{(u,v) : (x,y) \in D_p, 0 \leq u \leq x, 0 \leq v \leq y, p \in \mathbb{N}\};$$

μ_{φ_n} is the Lebesgue Stiltjes measure concentrated on the curves $\{\Gamma_n\}$.

Let us consider a real-valued non-negative, discontinuous, non-decreasing function $u(x,y)$ in D^*, which has finite jumps on the curves $\{\Gamma_j\}$.

Let $g(x,y)$ be a positive non-decreasing continuous function in $\mathbb{R}^2_+$, and let us assume that $u(x,y)$ satisfies the following integro-sum inequality in D^*:

$$u(x,y) \leq g(x,y) + \int\int_{G_n} \Phi(\tau, s, u(\tau, s))d\tau ds$$
$$+ \sum_{j=1}^{n-1} \int_{\Gamma_j \cap G_n} W(x, y, u(x,y))d\mu_{\varphi_j}, \tag{7.2.2}$$

where Φ and W defined in D^*, are non-negative, non-decreasing functions for a $3D$ argument, with fixed first and second arguments.

Theorem 7.2.1 ([399]) *Let the integro-sum equation of the following form hold*

$$\sigma(x,y) = g(x,y) + \int\int_{G_n} \Phi(\tau, s, \sigma(\tau, s))d\tau ds$$
$$+ \sum_{j=1}^{n-1} \int_{\Gamma_j \cap G_n} W(x, y, \sigma(x,y))d\mu_{\varphi_j}; \tag{7.2.3}$$

where $\sigma(x, y)$ is a non-negative discontinuous function, which has finite jumps on the curves $\{\Gamma_j\}$. The functions g, Φ, W are identities as in (7.2.2).

If $u(x, y)$ satisfies inequality (7.2.2), for all $x \geq 0, y \geq 0$, then we have, for all $x \geq 0$, $y \geq 0$,

$$u(x, y) \leq \sigma_g(x, y), \tag{7.2.4}$$

where $\sigma_g(x, y)$ is some solution of integro-sum equation (7.2.3), continuous in domain D^*.

The next results follow from Theorem 7.2.1.

Theorem 7.2.2 (The Samoilenko-Borysenko-Laserra-Matarazzo Inequality [587]) *Under assumptions of Theorem 7.2.1, (A) The following estimate holds for all $(x, y) \in D^*$,*

$$u(x, y) \leq g(x, y) \exp(F_1(x, y)) \prod(\beta_j(x, y)), \tag{7.2.5}$$

if $\Phi = f_1(x, y)u(x, y), f_1 \geq 0 : f_1 \in C(\mathbb{R}_+^2)$, $W = \beta_j(x, y)u(x, y)$, $\beta_j \in C(\mathbb{R}_+^2)$, $j = 1, 2, \cdots$.

(B) The following estimate holds,

$$u(x, y) \leq g(x, y) \exp(F_2(x, y)) \prod(\beta_j(x, y)) \times \left(1 + \int_0^x \int_0^y f_3(\tau, s) g^{-1}(\tau, s) \exp(-F_2(\tau, s)) d\tau ds\right) \tag{7.2.6}$$

if $\Phi = f_2(x, y)u(x, y) + f_3(x, y)$, with W as in (A).

(C) The following assertions hold:

(i) The following estimate is true,

$$u(x, y) \leq g(x, y) \prod(\beta_i(x, y)) \left[1 + (1-\alpha) \int_0^x \int_0^y f_4(\tau, s) g^{\alpha-1}(\tau, s) d\tau ds\right]^{1/(1-\alpha)} \tag{7.2.7}$$

if $0 \leq \alpha < 1$, $\Phi = f_4(x, y)u^\alpha(x, y)$, $\alpha = const. > 0, \alpha \neq 1, W = \beta_j(x, y)u(x, y)$.

(ii) The following estimate holds

$$u(x, y) \leq g(x, y) \prod(\beta_i(x, y)) \left[1 + (1-\alpha) \prod^{\alpha-1}(\beta_i(x, y)) \times \int_0^x \int_0^y f_4(\tau, s) g^{\alpha-1}(\tau, s) d\tau ds\right]^{-1/(1-\alpha)} \tag{7.2.8}$$

for $\alpha > 1$ and for an arbitrary $(x, y) \in D^*$ such that

$$\int_0^x \int_0^y f_4(\tau, s) g^{\alpha-1}(\tau, s) d\tau ds < \left[(\alpha - 1) \prod^{\alpha-1} (\beta_i(x, y)) \right]^{-1}.$$

(D) The following assertions hold:

(i) The following estimate holds for all $0 < \alpha < 1$,

$$u(x, y) \leq g(x, y) \prod (\beta_i(x, y)) \exp(F_5(x, y)) \cdot \left[1 + (1 - \alpha) \right.$$
$$\left. \times \int_0^x \int_0^y f_4(\tau, s) g^{\alpha-1}(\tau, s) \exp((\alpha - 1) F_5(\tau, s)) d\tau ds \right]^{1/(1-\alpha)}, \tag{7.2.9}$$

with $\Phi = f_5(x, y) u(x, y) + f_6(x, y) u^{\alpha}(x, y)$.

(ii) The following estimate is true for all $\alpha > 1$ and for an arbitrary $(x, y) \in D^*$,

$$u(x, y) \leq g(x, y) \prod (\beta_i(x, y)) \exp(F_5(x, y))$$
$$\times \left[1 + (1 - \alpha) \prod^{\alpha-1} (\beta_i(x, y)) \int_0^x \int_0^y f_6(\tau, s) g^{\alpha-1}(\tau, s) \right.$$
$$\left. \exp((\alpha - 1) F_5(\tau, s)) d\tau ds \right]^{1/(1-\alpha)} \tag{7.2.10}$$

such that

$$\int_0^x \int_0^y f_6(\tau, s) g^{\alpha-1}(\tau, s) \exp\left((\alpha - 1) F_5(\tau, s)\right) d\tau ds < \left[(\alpha - 1) \prod^{\alpha - 1} (\beta_i(x, y)) \right]^{-1}.$$

Here

$$\begin{cases} F_i(x, y) := \int_0^x \int_0^y f_i(\tau, s) d\tau ds, \quad i = 1, 3, 5, & (7.2.11) \\ \prod (\beta_j(x, y)) := \prod_{j=1}^{n-1} \left(1 + \int_{\Gamma_j \cap G_n} \beta_j(x, y) d\mu_{\varphi_j} \right). & (7.2.12) \end{cases}$$

In the sequel, we begin to discuss integral inequalities for discontinuous functions with discontinuities of non-Lipschitz type

Theorem 7.2.3 (The Mitropolskiy-Iovane-Borysenko Inequality [399]) *Let a non-negative function $\varphi(t,x)$, determined in the domain*

$$\Omega = \left[\bigcup_{k,j\geq 1} \Omega_{kj} = \{(t,x) : t \in [t_{k-1}, t_k], x \in [x_{k-1}, x_k]\},\ k = 1, 2, \cdots, k = 1, 2, \cdots \right],$$

be continuous in Ω, with the exception of the points $\{t_i, x_i\}$ where there is a finite jump:

$$\varphi(t_i - 0, x_i - 0) \neq \varphi(t_i + 0, x_i + 0),\ i = 1, 2, \cdots$$

which satisfies in Ω a certain integro-sum inequality

$$\varphi(t,x) \leq a(t,x) + \int_{t_0}^{t}\int_{x_0}^{x} b(\xi,\eta)\varphi(\xi,\eta)d\xi d\eta + \sum_{(t_0,x_0)<(t_i,x_i)<(t,x)} \gamma_i \varphi^m(t_i - 0, x_i - 0), \qquad (7.2.13)$$

with $m > 0, t_0 \geq 0, x_0 \geq 0$, where $a(t,x) > 0$, for all $(t,x) \in \Omega$ and non-decreasing with respect to (t,x):

$$\textit{for all } p \leq P,\ q \leq Q \Longrightarrow a(p,q) \leq a(P,Q),$$

$$\text{for all } (p,q) \in \Omega,\ (P,Q) \in \Omega;\ \gamma_i = const. \geq 0, \text{ for all } i \in \mathbb{N}, b \geq 0$$

and also satisfies a certain condition:

$$b(\xi,\eta) = 0,\ \textit{if}\ (\xi,\eta) \in \Omega_{ij},\ i \neq j$$

for arbitrary $i,j = 1, 2, \cdots$. Here $(t_k, x_k) < (t_{k+1}, t_{k+1})$, if $t_k < t_{k+1}$, $x_k < x_{k+1}$, $k = 0, 1, 2, \cdots$ and

$$\lim_{i\to+\infty} t_i = +\infty,\quad \lim_{i\to+\infty} x_i = +\infty.$$

Then the function $\varphi(t,x)$ satisfies the following estimates:

$$\begin{cases} \varphi(t,x) \leq a(t,x) \prod_{(t_0,x_0)<(t_i,x_i)<(t,x)} [1 + \gamma_i a^{m-1}(t_i,x_i)] \\ \qquad \times \exp[\int_{t_0}^{t}\int_{x_0}^{x} b(\xi,\eta)d\xi d\eta],\ \textit{if}\ 0 < m \leq 1; \\ \varphi(t,x) \leq a(t,x) \prod_{(t_0,x_0)<(t_i,x_i)<(t,x)} [1 + \gamma_i a^{m-1}(t_i,x_i)] \\ \qquad \times \exp[m\int_{t_0}^{t}\int_{x_0}^{x} b(\xi,\eta)d\xi d\eta],\ \textit{if}\ m \geq 1. \end{cases} \qquad (7.2.14)$$

Proof Define $W(t,x) = \frac{\varphi(t,x)}{a(t,x)}$, $W(t_0,x_0) = 1$. From inequality (7.2.13), it follows that

$$W(t,x) \le 1 + \int_{t_0}^{t}\int_{x_0}^{x} b(\xi,\eta)W(\xi,\eta)d\xi d\eta + \sum_{(t_0,x_0)<(t_i,x_i)<(t,x)} \gamma_i a^{m-1}(t_i,x_i)[W(t_i-0,x_i-0)]^m. \tag{7.2.15}$$

Consider the domain $\Omega_{11} = \{(t,x) : t \in [t_0,t_1], x \in [x_0,x_1]\}$. The inequality (7.2.15) reduces to

$$W(t,x) \le 1 + \int_{t_0}^{t}\int_{x_0}^{x} b(\xi,\eta)W(\xi,\eta)d\xi d\eta,$$

which gives us, for all $(t,x) \in \Omega_{11}$,

$$W(t,x) \le \exp\Big(\int_{t_0}^{t}\int_{x_0}^{x} b(\xi,\eta)d\xi d\eta\Big). \tag{7.2.16}$$

Now assume that $(t,x) \in \Omega_{22} = \{(t,x) : t \in [t_1,t_2], x \in [x_1,x_2]\}$, then

$$\begin{aligned} W(t,x) &\le \gamma_1 a^{m-1}(t_1,x_1)\exp[m\int_{t_0}^{t_1}\int_{x_0}^{x_1} b(\xi,\eta)d\xi d\eta] \\ &\quad + \int_{t_0}^{t_1}\int_{x_0}^{x_1} b(\xi,\eta)d\xi d\eta + \exp[\int_{t_0}^{t_1}\int_{x_0}^{x_1} b(s,t)dsdt] + \int_{t_1}^{t}\int_{x_1}^{x} b(\xi,\eta)w(\xi,\eta)d\xi d\eta \\ &= \gamma_1 a^{m-1}(t_1,x_1)\exp[m\int_{t_0}^{t_1}\int_{x_0}^{x_1} b(\xi,\eta)d\xi d\eta] + \exp\Big(\int_{t_0}^{t_1}\int_{x_0}^{x_1} b(s,t)dsdt\Big) \\ &\quad + \int_{t_1}^{t}\int_{x_1}^{x} b(\xi,\eta)w(\xi,\eta)d\xi d\eta. \end{aligned} \tag{7.2.17}$$

Hence, we get

$$\begin{aligned} W(t,x) &\le (1 + \gamma_1 a^{m-1}(t_1,x_1))\exp[\int_{t_0}^{t_1}\int_{x_0}^{x_1} b(t,s)dtds] \\ &\quad + \int_{t_1}^{t}\int_{x_1}^{x} b(\xi,\eta)w(\xi,\eta)d\xi d\eta, \quad \textit{if } 0 < m \le 1, \end{aligned}$$

$$W(t,x) \le (1+\gamma_1 a^{m-1}(t_i,x_i))\exp[m\int_{t_0}^{t_1}\int_{x_0}^{x_1} b(t,s)dtds]$$
$$+\int_{t_1}^{t}\int_{x_1}^{x} b(\xi,\eta)w(\xi,\eta)d\xi d\eta, \quad \textit{if } m \ge 1,$$

which imply for any $(t,x) \in \Omega_{22}$,

$$\begin{cases} W(t,x) \le (1+\gamma_1 a^{m-1}(t_1,x_1))\exp\Big(\int_{t_0}^{t}\int_{x_0}^{x} b(\tau,s)d\tau ds\Big), & \textit{if } 0 < m \le 1; \\ W(t,x) \le (1+\gamma_1 a^{m-1}(t_1,x_1))\exp\Big(m\int_{t_0}^{t}\int_{x_0}^{x} b(\tau,s)d\tau ds\Big), & \textit{if } m \ge 1. \end{cases}$$

Suppose that (7.2.14) is justified in the domain Ω_{kk}. Then for all $(t,x) \in \Omega_{k+1,k+1}$, the following inequality holds

$$\begin{aligned} W(t,x) &\le \sum_{i=1}^{k}\gamma_i a^{m-1}(t_i,x_i)\prod_{j=1}^{i-1}(1+\gamma_j a^{m-1}(t_i,x_i)) \\ &\quad \times\exp\Big(\int_{t_0}^{t_i}\int_{x_0}^{x_i} b(\xi,\eta)d\xi d\eta\Big) + \sum_{i=1}^{k}\int_{t_{i-1}}^{t_i}\int_{x_{i-1}}^{x_i}\prod_{j=1}^{i-1}(1+\gamma_j a^{m-1}(t_i,x_i)) \\ &\quad \times\exp\Big(\int_{t_0}^{\xi}\int_{x_0}^{\eta} b(\sigma,v)d\sigma dv\Big)d\xi d\eta + \int_{t_k}^{t}\int_{x_k}^{x} b(\xi,\eta)d\xi d\eta \\ &\le \prod_{j=1}^{k}(1+\gamma_j a^{m-1}(t_i,x_i))\exp\Big(\int_{t_0}^{t_k}\int_{x_0}^{x_k} b(\xi,\eta)d\xi d\eta\Big) + \int_{t_k}^{t}\int_{x_k}^{x} b(\xi,\eta)d\xi d\eta, \end{aligned}$$

if $0 < m \le 1$;

$$W(t,x) \le \prod_{j=1}^{k}(1+\gamma_j a^{m-1}(t_i,x_i))\exp\Big(m\int_{t_0}^{t_k}\int_{x_0}^{x_k} b(\xi,\eta)d\xi d\eta\Big) + \int_{t_k}^{t}\int_{x_k}^{x} b(\xi,\eta)d\xi d\eta,$$

if $m \ge 1$. Then for any $(t,x) \in \Omega_{k+1,k+1}$,

$$\begin{cases} W(t,x) \le \prod_{j=1}^{k}(1+\gamma_j a^{m-1}(t_i,x_i))\exp\Big(\int_{t_0}^{t_k}\int_{x_0}^{x_k} b(\xi,\eta)d\xi d\eta\Big), & \textit{if } 0 < m \le 1; \\ W(t,x) \le \prod_{j=1}^{k}(1+\gamma_j a^{m-1}(t_i,x_i))\exp\Big(m\int_{t_0}^{t_k}\int_{x_0}^{x_k} b(\xi,\eta)d\xi d\eta\Big), & \textit{if } m \ge 1. \end{cases}$$

From the last inequalities by taking into account that $W = \frac{\varphi}{a}$, the estimate (7.2.14) follows for the whole domain Ω. □

Remark 7.2.1 If $\gamma_i = 0, a(t,x) = c = const. > 0$, from the result of Theorem 7.2.3, the classical result of Wendroff follows (see [325]). If $m = 1$, in a particular case we have the result [104]. For the one-dimensional case ($\varphi(t,x) = \varphi(t)$), the result of Theorem 7.2.3 coincides with the result which was presented in [109]; if also $a(t,x) = a(t),\ b(u,v) = 0$, the result of Theorem 7.2.3 coincides with the discrete analogous case (for $m = 1$, Corollary 4.12, [10], similar Theorem 4.21, if $m \neq 1$).

Theorem 7.2.4 (The Mitropolskiy-Iovane-Borysenko Inequality [399]) *Let a non-negative function $\varphi(t,z)$, determined in domain Ω, be continuous in Ω except at the points (t_i,x_i)-points of finite jumps and satisfy the following integro-sum inequality*

$$\varphi(t,x) \leq a(t,x) \int_{t_0}^{t} \int_{x_0}^{x} b(\xi,\eta)\varphi^m(\xi,\eta)d\xi d\eta + \sum_{(t_0,x_0)<(t_i,x_i)<(t,x)} \gamma_i\varphi^m(t_i-0,x_i-0), \tag{7.2.18}$$

$m > 0, m \neq 1$, where a, b, γ_i satisfy the conditions of Theorem 7.2.3. Then for any $(t,x) \in \Omega$, the function $\varphi(t,x)$ satisfies the inequalities:

(1) if $0 < m < 1$,

$$\varphi(t,x) \leq a(t,x) \prod_{(t_0,x_0)<(t_i,x_i)<(t,x)} (1+\gamma_i\varphi^m(t_i,x_i))\Big[1+(1-m) \times \int_{t_0}^{t}\int_{x_0}^{x} a^{m-1}(\xi,\eta)b(\xi,\eta)d\xi d\eta\Big]^{1/(1-m)}; \tag{7.2.19}$$

(2) if $m > 1$,

$$\varphi(t,x) \leq a(t,x) + \prod_{(t_0,x_0)<(t_i,x_i)<(t,x)} (1+m\gamma_i\varphi^m(t_i,x_i)) \times\Big\{1-(m-1)\Big[\prod_{(t_0,x_0)<(t_i,x_i)<(t,x)} (1+m\gamma_i\varphi^m(t_i,x_i))\Big]^{m-1} \times \int_{t_0}^{t}\int_{x_0}^{x} a^{m-1}(\xi,\eta)b(\xi,\eta)d\xi d\eta\Big\}^{-\frac{1}{m-1}} \tag{7.2.20}$$

and for an arbitrary $(t, x) \in \Omega$*:*

$$\begin{cases} \int_{t_0}^{t} \int_{x_0}^{x} a^{m-1}(\xi, \eta) b(\xi, \eta) d\xi d\eta \le \frac{1}{m}, \\ \prod_{(t_0,x_0)<(t_i,x_i)<(t,x)} (1 + m\gamma_i \varphi^m(t_i, x_i)) < \left(1 + \frac{1}{m-1}\right)^{1/1-m}. \end{cases} \tag{7.2.21}$$

Proof As in Theorem 7.2.3, the following inequality can be rewritten as

$$W(t,x) \le 1 + \int_{t_0}^{t} \int_{x_0}^{x} b(\xi, \eta) a^{m-1}(\xi, \eta) W^m(\xi, \eta) d\xi d\eta$$
$$+ \sum_{(t_0,x_0)<(t_i,x_i)<(t,x)} \gamma_i a^{m-1}(t_i, x_i)[W(t_i - 0, x_i - 0)]^m. \tag{7.2.22}$$

By using the generalization of the result given by Bihari for a continuous function of two independent variables, in domain Ω_{11}, an inequality for the function $W(t, x)$ holds

$$W(t,x) \le \left[1 + (1-m) \int_{t_0}^{t} \int_{x_0}^{x} a^{m-1}(\xi, \eta) b(\xi, \eta) d\xi d\eta\right]^{1/(1-m)},$$

only if $0 < m < 1$, for all $(t, x) \in \Omega_{11}$;

$$W(t,x) \le \left[1 - (m-1) \int_{t_0}^{t} \int_{x_0}^{x} a^{m-1}(\xi, \eta) b(\xi, \eta) d\xi d\eta\right]^{-1/(m-1)},$$

only if $m > 1$, for all $(t, x) \in \Omega_1$ satisfying the condition:

$$\int_{t_0}^{t} \int_{x_0}^{x} a^{m-1}(\xi, \eta) b(\xi, \eta) d\xi d\eta < \frac{1}{m-1}. \tag{7.2.23}$$

By using (7.2.21), the inequality (7.2.23) holds for all $(t, x) \in \Omega_{11}$.

Considering $(t, x) \in \Omega_{22}$ and using the relation (7.2.18), (7.2.21) and (7.2.22), we have

$$W(t,x) \le \gamma_1 a^{m-1}(t_1, x_1) \left[1 + (1-m) \int_{t_0}^{t_1} \int_{x_0}^{x_1} a^{m-1}(\xi, \eta) b(\xi, \eta) d\xi d\eta\right]^{m/(m-1)}$$
$$+ \left[1 + (1-m) \int_{t_0}^{t_1} \int_{x_0}^{x_1} a^{m-1}(\xi, \eta) b(\xi, \eta) d\xi d\eta\right]^{1/(1-m)}$$
$$+ \int_{t_1}^{t} \int_{x_1}^{x} a^{m-1}(\xi, \eta) b(\xi, \eta) W^m(\xi, \eta) d\xi d\eta \tag{7.2.24}$$

It is clear (taking into account (7.2.21) and (7.2.24)) that if $0 < m < 1$; for all $(t,x) \in \Omega_{22}$;

$$\varphi(t,x) \le a(t,x)(1+\gamma_1 a^{m-1}(t_1,x_1)) \times \left[1+(1-m)\int_{t_0}^{t}\int_{x_0}^{x} a^{m-1}(\xi,\eta)b(\xi,\eta)d\xi d\eta\right]^{1/(1-m)}; \quad (7.2.25)$$

or if $m > 1$,

$$\varphi(t,x) \le a(t,x)(1+\gamma_1 a^{m-1}(t_1,x_1))\left[1+(m-1)(1+m\gamma_1 a^{m-1}(t_1,x_1))^{m-1} \times \int_{t_0}^{t}\int_{x_0}^{x} a^{m-1}(\xi,\eta)b(\xi,\eta)d\xi d\eta\right]^{-1/(m-1)}.$$

From (7.2.25), it follows that (7.2.19) and (7.2.20) hold for all $(t,x) \in \Omega_{22}$, only if (7.2.21) is fulfilled. The proof is completed by using the procedure of proving Theorem 7.2.3. □

Remark 7.2.2 Theorem 7.2.4 gives us a new analogy of the generalization of Wendroff's results for discontinuous functions which is independent of the result in [104] (see, also Theorem 3.4.1, [585]).

Remark 7.2.3 If $\gamma_i = 0$, the result of the above theorem coincides with the classical result given by Bihari for the functions of two independent variables (see [42, 325]), for continuous functions. In the one-dimensional case ($\varphi(t,x) = \varphi(t), a(t,x) = a(t), b(\xi,\eta) = b(\xi)$), the result of Theorem 7.2.3 coincides with the result in [109].

Theorem 7.2.5 (The Mitropolskiy-Iovane-Borysenko Inequality [399]) *Let a non-negative function $\varphi(t,x)$ satisfy the conditions of Theorem 7.2.4 and the following inequality hold*

$$\varphi(t,x) \le a(t,x) + g(t,x)\int_{t_0}^{t}\int_{x_0}^{x} b(\xi,\eta)\varphi^m(\xi,\eta)d\xi d\eta + \sum_{(t_0,x_0)<(t_i,x_i)<(t,x)} \gamma_i\varphi^m(t_i-0,x_i-0), \quad (7.2.26)$$

where $m > 0$, $g(t,x) \ge 1, a(t,x), b(t,x), \gamma_i$ satisfy the conditions of Theorem 7.2.3.

Then the following estimates hold for any $(t,x) \in \Omega$,

$$\varphi(t,x) \le a(t,x)g(t,x)\prod_{(t_0,x_0)<(t_i,x_i)<(t,x)}(1+\gamma_i a^{m-1}(t_i,x_i)g^m(t_i,x_i)) \times \left[1+(1-m)\int_{t_0}^{t}\int_{x_0}^{x} b(\xi,\eta)a^{m-1}(\xi,\eta)g^m(\xi,\eta)d\xi d\eta\right]^{\frac{1}{m-1}}, \quad (7.2.27)$$

if $0 < m < 1$;

$$\varphi(t,x) \le a(t,x)g(t,x) \prod_{(t_0,x_0)<(t_i,x_i)<(t,x)} (1+\gamma_i g^m(t_i,x_i)) \times \exp\left[\int_{t_0}^{t}\int_{x_0}^{x} b(\xi,\eta)g(\xi,\eta)d\xi d\eta\right], \tag{7.2.28}$$

if $m = 1$;

$$\varphi(t,x) \le a(t,x)g(t,x) \prod_{(t_0,x_0)<(t_i,x_i)<(t,x)} (1+m\gamma_i a^{m-1}(t_i,x_i)g^m(t_i,x_i)) \times\left[1+(m-1)\left(\prod_{(t_0,x_0)<(t_i,x_i)<(t,x)} (1+m\gamma_i a^{m-1}(t_i,x_i)g^m(t_i,x_i))\right)^{m-1} \times \int_{t_0}^{t}\int_{x_0}^{x} b(\xi,\eta)g^m(\xi,\eta)a^{m-1}(\xi,\eta)d\xi d\eta\right]^{-\frac{1}{m-1}}, \tag{7.2.29}$$

if $m > 1$ *with*

$$\begin{cases} \int_{t_0}^{t}\int_{x_0}^{x} b(\xi,\eta)g^m(\xi,\eta)a^{m-1}(\xi,\eta)d\xi d\eta \le 1/m, \\ \prod_{(t_0,x_0)<(t_i,x_i)<(t,x)} \left(1+m\gamma_i a^{m-1}(t_i,x_i)g^m(t_i,x_i)\right) < \left(1+\frac{1}{m-1}\right)^{1/m-1}. \end{cases} \tag{7.2.30}$$

Proof It is obvious that

$$w(t,x) \le g(t,x)\left[1+\int_{t_0}^{t}\int_{x_0}^{x} a^{m-1}(\xi,\eta)b(\xi,\eta)w^m(\xi,\eta)d\xi d\eta + \sum_{(t_0,x_0)<(t_i,x_i)<(t,x)} \gamma_i a^{m-1}(t_i,x_i)w^m(t_i-0,x_i-0)\right]. \tag{7.2.31}$$

Here $w(t,x) = \varphi(t,x)/a(t,x)$. Define

$$w^*(t,x) = 1+\int_{t_0}^{t}\int_{x_0}^{x} a^{m-1}(\xi,\eta)b(\xi,\eta)w^m(\xi,\eta)d\xi d\eta + \sum_{(t_0,x_0)<(t_i,x_i)<(t,x)} (\gamma_i a^{m-1}(t_i,x_i)w^m(t_i-0,x_i-0)). \tag{7.2.32}$$

Then

$$w(t,x) \le g(t,x)w^*(t,x).$$

Thus

$$w^*(t,x) \leq 1 + \int_{t_0}^{t}\int_{x_0}^{x} a^{m-1}(\xi,\eta)b(\xi,\eta)w^{*^m}(\xi,\eta)g^m(\xi,\eta)d\xi d\eta + \sum_{(t_0,x_0)<(t_i,x_i)<(t,x)} \gamma_i a^{m-1}(t_i,x_i)w^{*^m}(t_i-0,x_i-0)g^m(t_i,x_i). \quad (7.2.33)$$

Using the procedure for $w^*(t,x)$, as for $w(t,x)$ in Theorem 7.2.3 (using inequalities (7.2.19)–(7.2.21)), it follows that for all $(t,x) \in \Omega$,

$$w^*(t,x) \leq \prod_{(t_0,x_0)<(t_i,x_i)<(t,x)} (1+\gamma_i a^{m-1}(t_i,x_i)g^m(t_i,x_i)) \times \left[1+(1-m)\int_{t_0}^{t}\int_{x_0}^{x} a^{m-1}(\xi,\eta)b(\xi,\eta)g^m(\xi,\eta)d\xi d\eta\right]^{1/(1-m)}, \quad (7.2.34)$$

if $0 < m < 1$;

$$w^*(t,x) \leq \prod_{(t_0,x_0)<(t_i,x_i)<(t,x)} (1+\gamma_i g(t_i,x_i)) \times \exp\left[\int_{t_0}^{t}\int_{x_0}^{x} b(\xi,\eta)g(\xi,\eta)d\xi d\eta\right], \quad (7.2.35)$$

if $m = 1$;

$$w^*(t,x) \leq \prod_{(t_0,x_0)<(t_i,x_i)<(t,x)} (1+m\gamma_i a^{m-1}(t_i,x_i)g^m(t_i,x_i)) \times \Bigg[1-(m-1)\Big[\prod_{(t_0,x_0)<(t_i,x_i)<(t,x)} (1+m\gamma_i a^{m-1}(t_i,x_i)g^m(t_i,x_i))\Big]^{m-1} \times \int_{t_0}^{t}\int_{x_0}^{x} b(\xi,\eta)g^m(\xi,\eta)a^{m-1}(\xi,\eta)d\xi d\eta\Bigg]^{-1/(m-1)}, \quad (7.2.36)$$

if $m > 1$, which satisfies conditions (7.2.30). From (7.2.22)–(7.2.24), estimates (7.2.14)–(7.2.16) follow immediately. □

Remark 7.2.4 All the remarks for Theorem 7.2.3 coincide with the remarks for Theorem 7.2.4, if $g(t,x) = 1$.

In the next two theorems, we shall consider the class of functions in the E-class of continuous functions $\tau(s) : \mathbb{R} \to \mathbb{R}$, such that $\tau(s) \leq s$, $\lim_{|s|\to+\infty} \tau(s) \leq +\infty$. These inequalities are of retardation. The following results hold.

Theorem 7.2.6 (The Mitropolskiy-Iovane-Borysenko Inequality [399]) *Let $\sigma \in H$ and suppose that the functions φ, a, q, b satisfy the conditions of Theorem 7.2.5, $\gamma_i = const. \geq 0$ and also $\varphi(t,x)$ satisfies the inequality*

$$\varphi(t,x) \leq a(t,x) + q(t,x) \int_{t_0}^{t} b(\xi,\eta)\varphi(\sigma(\xi),\sigma(\eta))d\xi d\eta + \sum_{(t_0,x_0)<(t_i,x_i)<(t,x)} \gamma_i \varphi^m(t_i - 0, x_i - 0), \quad (7.2.37)$$

with $m > 0$.

Then the following estimates hold for $\varphi(t,x)$:

$$\varphi(t,x) \leq a(t,x)q(t,x) \prod_{(t_0,x_0)<(t_i,x_i)<(t,x)} (1 + \gamma_i a^{m-1}(t_i,x_i)q^m(t_i,x_i)) \times \exp\left(\int_{t_0}^{t}\int_{x_0}^{x} \frac{b(\xi,\eta)a(\sigma(\xi),\sigma(\eta))}{a(\xi,\eta)} d\xi d\eta\right), \quad (7.2.38)$$

if $m \in [0, 1)$;

$$\varphi(t,x) \leq a(t,x)q(t,x) \prod_{(t_0,x_0)<(t_i,x_i)<(t,x)} (1 + \gamma_i a^{m-1}(t_i,x_i)q^m(t_i,x_i)) \times \exp\left[m\int_{t_0}^{t}\int_{x_0}^{x} \frac{b(\xi,\eta)a(\sigma(\xi),\sigma(\eta))q(\sigma(\xi),\sigma(\eta))}{a(\xi,\eta)} d\xi d\eta\right], \quad (7.2.39)$$

if $m \geq 1$.

Proof In fact, it follows

$$w(t,x) \leq q(t,x)\left[1 + \int_{t_0}^{t}\int_{x_0}^{x} b(\xi,\eta)\frac{\varphi(\sigma(\xi),\sigma(\eta))}{a(\xi,\eta)} d\xi d\eta + \sum_{(t_0,x_0)<(t_i,x_i)<(t,x)} \gamma_i a^{m-1}(t_i,x_i)W^m(t_i - 0, x_i - 0)\right]. \quad (7.2.40)$$

Here $W = \frac{\varphi}{a}$. Define

$$w^*(t,x) = 1 + \int_{t_0}^{t}\int_{x_0}^{x} b(\xi,\eta)\frac{\varphi(\sigma(\xi),\sigma(\eta))}{a(\xi,\eta)} d\xi d\eta + \sum_{(t_0,x_0)<(t_i,x_i)<(t,x)} \gamma_i a^{m-1}(t_i,x_i)W^m(t_i - 0, x_i - 0). \quad (7.2.41)$$

It is obvious that

$$w^*(t,x) \le 1 + \int_{t_0}^{t}\int_{x_0}^{x} \frac{b(\xi,\eta)a(\sigma(\xi),\sigma(\eta))q(\sigma(\xi),\sigma(\eta))}{a(\xi,\eta)} W^*(\xi,\eta)d\xi d\eta$$
$$+ \sum_{(t_0,x_0)<(t_i,x_i)<(t,x)} \gamma_i a^{m-1}(t_i,x_i)q^m(t_i,x_i)W^{*^m}(t_i-0,x_i-0). \quad (7.2.42)$$

Using Theorem 7.2.2 for inequality (7.2.42), we obtain

$$w^*(t,x) \le \prod_{(t_0,x_0)<(t_i,x_i)<(t,x)} (1+\gamma_i a^{m-1}(t_i,x_i)q^m(t_i,x_i))$$
$$\times \exp\Big(\int_{t_0}^{t}\int_{x_0}^{x} F(\sigma(\xi),\sigma(\eta))d\xi d\eta\Big),$$

if $0 < m < 1$;

$$w^*(t,x) \le \prod_{(t_0,x_0)<(t_i,x_i)<(t,x)} (1+\gamma_i a^{m-1}(t_i,x_i)q^m(t_i,x_i))$$
$$\times \exp\Big(m\int_{t_0}^{t}\int_{x_0}^{x} F(\sigma(\xi),\sigma(\eta))d\xi d\eta\Big), \quad (7.2.43)$$

if $m \ge 1$.

Here

$$F(\sigma(\xi),\sigma(\eta)) = \frac{b(\xi,\eta)a(\sigma(\xi),\sigma(\eta))q(\sigma(\xi),\sigma(\eta))}{a(\xi,\eta)}.$$

Thus from (7.2.43) and the inequality

$$\varphi(t,x) \le a(t,x)q(t,x)W^*(t,x),$$

the conclusion of Theorem 7.2.6 follows readily. □

Remark 7.2.5 From the result of Theorem 7.2.6, the results of the investigations in [589] follow as a particular case. If $m = 1$, the result of Theorem 7.2.6 coincides with Theorem 3.4.3 in [589].

Theorem 7.2.7 (The Mitropolskiy-Iovane-Borysenko Inequality [399]) *Let all the conditions of Theorem 7.2.6 hold and the function $\varphi(t,x)$ satisfy the inequality*

$$\varphi(t,x) \le a(t,x) + q(t,x)\int_{t_0}^{t} b(\xi,\eta)\varphi^m(\sigma(\xi),\sigma(\eta))d\xi d\eta$$
$$+ \sum_{(t_0,x_0)<(t_i,x_i)<(t,x)} \gamma_i\varphi^m(t_i-0,x_i-0), \quad (7.2.44)$$

with $m > 0$.

Then for any $\varphi(t,x) \in \Omega$, *the following estimates hold*

$$\begin{aligned}\varphi(t,x) \le a(t,x)q(t,x) &\prod_{(t_0,x_0)<(t_i,x_i)<(t,x)} (1+\gamma_i a^{m-1}(t_i,x_i)q^m(t_i,x_i)) \\ &\times\left\{1+(1-m)\int_{t_0}^{t}\int_{x_0}^{x} b(\xi,\eta)a^{m-1}(\xi,\eta)q^m(\sigma(\xi),\sigma(\eta))\right. \\ &\left.\times[\frac{a(\sigma(\xi),\sigma(\eta))}{a(\xi,\eta)}]^m d\xi d\eta\right\}^{1/(1-m)} \end{aligned} \tag{7.2.45}$$

if $0 < m < 1$;

$$\begin{aligned}\varphi(t,x) \le a(t,x)q(t,x) &\prod_{(t_0,x_0)<(t_i,x_i)<(t,x)} (1+\gamma_i q(t_i,x_i)) \\ &\times \exp\left[\int_{t_0}^{t}\int_{x_0}^{x} b(\xi,\eta)q(\sigma(\xi),\sigma(\eta))\frac{a(\sigma(\xi),\sigma(\eta))}{a(\xi,\eta)}d\xi d\eta\right] \end{aligned} \tag{7.2.46}$$

if $m = 1$;

$$\begin{aligned}\varphi(t,x) \le a(t,x)q(t,x) &\prod_{(t_0,x_0)<(t_i,x_i)<(t,x)} (1+\gamma_i a^{m-1}(t_i,x_i)q^m(t_i,x_i)) \\ &\times\left\{1-(m-1)[\prod_{(t_0,x_0)<(t_i,x_i)<(t,x)} (1+m\gamma_i a^{m-1}(t_i,x_i)q^m(t_i,x_i))]^{m-1}\right. \\ &\times\int_{t_0}^{t}\int_{x_0}^{x} b(\xi,\eta)a^{m-1}(\xi,\eta)q^m(\sigma(\xi),\sigma(\eta)) \\ &\left.\times[\frac{a(\sigma(\xi),\sigma(\eta))}{a(\xi,\eta)}]^m d\xi d\eta\right\}^{-1/(m-1)}, \end{aligned} \tag{7.2.47}$$

if $m > 1$, *with*

$$\begin{cases} \int_{t_0}^{t}\int_{x_0}^{x} b(\xi,\eta)a^{m-1}(\xi,\eta)q^m(\sigma(\xi),\sigma(\eta))[\frac{a(\sigma(\xi),\sigma(\eta))}{a(\xi,\eta)}]^m d\xi d\eta < \frac{1}{m}, \\ \prod_{(t_0,x_0)<(t_i,x_i)<(t,x)} \left(1+m\gamma_i a^{m-1}(t_i,x_i)q^m(t_i,x_i)\right) < \left(1+\frac{1}{m-1}\right)^{1/(m-1)}. \end{cases} \tag{7.2.48}$$

Proof Define

$$\begin{aligned}\overline{W}(t,x) = 1 &+ \int_{t_0}^{t}\int_{x_0}^{x} a^{m-1}(\xi,\eta)b(\xi,\eta)\times\left(\frac{a(\sigma(\xi),\sigma(\eta))}{a(\xi,\eta)}\right)^m d\xi d\eta \\ &+ \sum_{(t_0,x_0)<(t_i,x_i)<(t,x)} \gamma_i a^{m-1}(t_i,x_i)\left(\frac{\varphi(t_i-0,x_i-0)}{a(t_i,x_i)}\right)^m. \end{aligned}$$

Then

$$\varphi(t,x) \le a(t,x)q(t,x)\overline{W}(t,x). \tag{7.2.49}$$

Noting that

$$\varphi(t_i-0,x_i-0) \le a(t_i,x_i)q(t_i,x_i)\overline{W}(t_i,x_i),$$
$$\varphi(\sigma(\xi),\sigma(\eta)) \le a(\sigma(\xi),\sigma(\eta))q(\sigma(\xi),\sigma(\eta))\overline{W}(\sigma(\xi),\sigma(\eta)),$$

$$\begin{aligned}\overline{W}(\sigma(\xi),\sigma(\eta)) &= 1+\int_{t_0}^{\sigma(t)}\int_{x_0}^{\sigma(x)} a^{m-1}(\xi,\eta)b(\xi,\eta)[\frac{\varphi(\sigma(\xi),\sigma(\eta))}{a(\xi,\eta)}]^m d\xi d\eta \\ &+ \sum_{(t_0,x_0)<(t_i,x_i)<(t,x)} \gamma_i a^{m-1}(t_i,x_i)[\frac{\varphi(t_i-0,x_i-0)}{a(t_i,x_i)}]^m \\ &\le \overline{W}(t,x),\end{aligned}$$

we obtain

$$\begin{aligned}\overline{W}(\sigma(\xi),\sigma(\eta)) &= 1+\int_{t_0}^{t}\int_{x_0}^{x} a^{m-1}(\xi,\eta)b(\xi,\eta) \\ &\times[\frac{a(\sigma(\xi),\sigma(\eta))q(\sigma(\xi),\sigma(\eta))}{a(\xi,\eta)}]^m\overline{W}^m(\xi,\eta)d\xi d\eta \\ &+ \sum_{(t_0,x_0)<(t_i,x_i)<(t,x)} \gamma_i a^{m-1}(t_i,x_i)q^{m-1}(t_i,x_i)\overline{W}^m(t_i,x_i). \end{aligned} \tag{7.2.50}$$

Applying Theorem 7.2.4 to (7.2.50) and using (7.2.49), we obtain the desired estimates (7.2.45)–(7.2.47). □

Remark 7.2.6 For $\varphi(t,x)=\varphi(t), q(t,x)=1, \sigma(\xi)=\xi$, the result of Theorem 7.2.6 coincides with the result in [109]; if $\gamma_i = 0, a(t,x) = \text{const.}, q(t,x) = 1$, estimates (7.2.45)–(7.2.47) are the classical results given by Bellman and Bihari for functions of two independent variables; if $m=1$, from the result of Theorem 7.2.6, the result in Theorem 3.42 in [589] follows.

Theorem 7.2.8 (The Borysenko Inequality [104]) *Let $u(t,x)$ be a non-negative function which is determined in the domain*

$$D=\{\cup_{k,j>1}D_{kj}, D_{kj}=((t,x): t\in[t_{k-1},t_k], x\in[x_{j-1},x_j], k=1,2,\cdots,j=1,2,\cdots)\}.$$

Moreover let it be continuous in D, with the exception of the points $\{t_i, x_i\}$ *of finite jumps:* $u(t_i - 0, x_i - 0) \neq u(t_i + 0, x_i + 0)$ *and satisfy the integro-sum inequality*

$$\begin{aligned} u(t,x) \leq \Psi(t,x) + q(t,x) \int_{t_0}^{t} \int_{x_0}^{x} f(\xi,\eta) u^m(\xi,\eta) d\xi d\eta \\ + \sum_{(t_0,x_0)<(t_i,x_i)<(t,x)} \beta_i u(t_i - 0, x_i - 0), \end{aligned} \quad (7.2.51)$$

$q(t_0, x_0) = 1$, *where* $\Psi(t,x) > 0$, *for all* $(t,x) \in D$ *it is non-decreasing with respect to* (t,x):
for all $p \leq P$, $q \leq Q$, $\Psi(p,q) \leq \Phi(P,Q)$ *at* $(p,q) \in D$, $(P,Q) \in D$; $q(t,x) \leq 1$, *for all* $(t,x) \in D$, *the values* $\beta_i \geq 0$ *for all* $i \in \mathbb{N}$, *the function* f *is non-negative, where* $f(\xi,\eta) = 0$, $(\xi,\eta) \in D_{lp}$, *for* $l \neq p$, *for arbitrary* $l = 1, 2, \cdots, p = 1, 2, \cdots$. *Here* $(t_i, x_i) < (t_{i+1}, x_{i+1})$, *if* $t_i < t_{i+1}, x_i < x_{i+1}$, *for all* $i = 1, 2, \cdots$, *where* $\lim_{i\to+\infty} t_i = +\infty$, $\lim_{i\to+\infty} x_i = +\infty$.

Then the following estimates hold:

(1) if $0 < m < 1$, *then for all* $(t,x) \in D$,

$$u(x,t) \leq \psi(t,x) q(x,t) \prod(t_0,x)[1 + (1-m) \times \int_{t_0}^{t} \int_{x_0}^{x} \psi^{m-1}(\xi,\eta) q^m(\xi,\eta) f(\xi,\eta) d\xi d\eta]^{1/(1-m)};$$

(2) if $m = 1$, *then for all* $(t,x) \in D$,

$$u(x,t) \leq \psi(t,x) q(x,t) \prod(t_0,x) \exp\left[\int_{t_0}^{t} \int_{x_0}^{x} f(\xi,\eta) q(\xi,\eta) d\xi d\eta\right]; \quad (7.2.52)$$

(3) if $m > 1$, *then for all* $(t,x) \in D$,

$$u(x,t) \leq \psi(t,x) q(x,t)[1 - (m-1) \prod^{m-1}(t_0,x) \times [\int_{t_0}^{t} \int_{x_0}^{x} \psi^{m-1}(\xi,\eta) q^m(\xi,\eta) f(\xi,\eta) d\xi d\eta]^{1/(1-m)},$$

with

$$\int_{t_0}^{t} \int_{x_0}^{x} \psi^{m-1}(\xi,\eta) q^m(\xi,\eta) f(\xi,\eta) d\xi d\eta < \left[(m-1) \prod^{m-1}(t_0,x)\right]^{-1}.$$

Here $\prod(t_0,x) := \prod_{(t_0,x_0)<(t_i,x_i)<(t,x)} (1 + \beta_i q(t_i,x_i))$.

Proof Similarly to Theorem 3.2.15, this theorem can be proved, using the inductive method and the methodology of the integral inequalities theory. □

Let us consider the Euclidean space $\mathbb{R}^n$ with points $x = (x^1, x^2, \cdots, x^n), x^0 = (x^{10}, \cdots, x^{n0})$; with the order $x^0 \leq x$ $(x^{i0} \leq x^i)$, $i = 1, \cdots, n$.

We define

$$\int_{x^0}^{x} \cdots du = \int_{x^{n0}}^{x^1} \cdots du_1 \cdots du_n, \quad \sum_{x^0 < x_k < x} \alpha_k = \sum_{x^0 < x_{k1} < x^1, \cdots, x^{n0} < x_{kn} < x^n} \alpha_k.$$

Let us introduce a space of $\mathcal{F}_3$-continuous functions $F : \mathbb{R}^n \to \mathbb{R}^n$, such that:

(A) $F(x) = (F_1(x), F_2(x), \cdots, F_n(x))$, where $F_j : \mathbb{R}^n \to \mathbb{R}^n$ $j = 1, 2, \cdots, n$;
(B) $F(x) \leq x$;
(C) $\lim_{|x| \to +\infty} F_j(x) = +\infty$, $j = 1, 2, \cdots, n$.

We consider the domain $D \subset \mathbb{R}^n$:

$$D = D_{k_1, \cdots, k_n} = \left\{x : x^1 \in [x_{k_1 - 1}, x_{k_1}], \cdots, x^n \in [x_{k_n - 1}, x_{k_n}], k_j = 1, 2, \cdots, n\right\}.$$

We denote by $\{x_k\} = \{x_{k1}, \cdots, x_{kn}\}$ the points of finite jumps of the function $u(x)$: $u(x_i - 0) \neq u(x_i + 0)$, for all $i \in \mathbb{N}$. Let us define as $\mathcal{F}^*$ the space of functions $f(x) : f \geq 0; f = 0$, only if $x \in D_{k_1, \cdots, k_n}$ at $k_i \neq k_j, i, j = 1, 2, \cdots, n$.

Theorem 7.2.9 (The Borysenko Inequality [106]) *Let a non-negative function $u(x)$ be determined in the domain D and satisfy the inequality*

$$u(x) \leq \psi(x) + q(x)\left[\int_{x_0}^{x} f(\tau)u^m(p(\tau))d\tau + \int_{x_0}^{x} f(s)(\int_{x_0}^{x} g(\tau)u^m(\sigma(\tau))d\tau)ds\right]$$
$$+ \sum_{x^0 < x_i < x} \beta_i u(x_i - 0),$$

with $m > 0$, and where $p(t), \sigma(t) \in \mathcal{F}_3$, $\{x_k\}$ are the points of finite jumps of $u(x)$, $\psi(x)$ is a non-decreasing function, $\psi(x) > 0, f \in \mathcal{F}^*, q(x) \geq 1, g(x) \geq 0$, $\beta_i \geq 0$. Then the following estimates hold

(A) for all $x \in D$,

$$u(x) \leq \psi(x)q(x)\prod(x^0, x)\left\{1 + (1 - m)\int_{x_0}^{x} f(t)\right.$$
$$\left.\times\left[\psi^{m-1}(t)q^m(p(t)) + \int_{t_0}^{t} g(\tau)\psi^{m-1}(\tau)q^m(\sigma(\tau))\left(\frac{\psi(\sigma(\tau))}{\psi(\tau)}\right)^m d\tau\right]dt\right\}^{1/(1-m)},$$

if $0 < m < 1$;
(B) if $m = 1$, for all $x \in D$,

$$u(x) \leq \psi(x)q(x)\prod(x^0, x)\exp\left(\int_{x_0}^{x} Q(\tau)d\tau\right);$$

(C) if $m > 1$, for all $(t, x) \in D$,

$$u(x) \le \psi(x)q(x)\prod(x^0, x)\left\{1 + (1-m)\prod^{m-1}(x^0, x)\int_{x_0}^{x} f(t)\right.$$
$$\left.\times\left[\psi^{m-1}(x)q^m(p(t)) + \int_{t_0}^{t} g(\tau)\psi^{m-1}(\tau)q^m(\sigma(\tau))\left(\frac{\psi(\sigma(\tau))}{\psi(\tau)}\right)^m d\tau\right]dt\right\}^{1/(1-m)},$$

$$\int_{x_0}^{x} f(t)\left[\psi^{m-1}(x)q^m(p(t)) + \int_{t_0}^{t} g(\tau)\psi^{m-1}(\tau)q^m(\sigma(\tau))\left[\frac{\psi(\sigma(\tau))}{\psi(\tau)}\right]^m d\tau\right]dt$$
$$< \left((m-1)\prod^{m-1}(t_0, x)\right)^{-1}\frac{\prod^{1-m}(x^0, x)}{m-1}.$$

Here

$$\prod(x^0, x) := \prod_{(x^0 < x_i < x)}(1 + \beta_i q(x_i)),$$

$$Q(t) = \frac{f(t)q(p(t))\psi(p(t)) + g(t)q(\sigma(t))\psi(\sigma(t))}{\psi(p(t))}.$$

7.2.2 Nonlinear Two-Dimensional Discontinuous Integral Inequalities of Bihari Type

We first introduce a definition as follow

Definition 7.2.1 $W(\sigma) \in \mathcal{F}_4$ if and only if

(1) $W(x)$ is non-negative, continuous, non-decreasing for all $x > 0$,
(2) for all $t > 0, u \ge 0, t^{-1}(W(u)) \le W(t^{-1}u)$,
(3) $W(0) = 0$.

Theorem 7.2.10 (The Mitropolskiy-Iovane-Borysenko Inequality [399]) *Assume the function $u(x_1, x_2)$ satisfies the following integro-sum inequality in D^*:*

$$u(x_1, x_2) \le q(x_1, x_2) + g(x_1, x_2)\int\int_{G_n}\overline{\psi}(\tau, s)W[u(\tau, s)]d\tau ds$$
$$+\sum_{j=1}^{n-1}\int_{\Gamma_j \cap G_n}\beta_j(x, y)u(x_1, x_2)d\mu_{\varphi_j}, \tag{7.2.53}$$

where $q(x_1,x_2)$ is positive and non-decreasing, $g(x_1,x_2) \geq 1, \beta_j(x_1,x_2) \geq 0$, $\overline{\psi}(\tau,s) \geq 0$; function W belongs to the class of functions $\overline{\mathcal{F}}$. The function $u(x_1,x_2)$ is a non-negative discontinuous function, which has finite jumps on the curves $\{\Gamma_j\}, j = 1, 2, \cdots$.

Then for arbitrary $(x_1,x_2) \in \mathbb{R}_0 \times \mathbb{R}_0 = (0,+\infty) \times (0,+\infty)$, the following estimate holds

$$u(x_1,x_2) \leq q(x_1,x_2)g(x_1,x_2)\Psi_i^{-1}\left\{\int\int_{D_i} \frac{\overline{\psi}(\tau,s)}{q(\tau,s)} W[q(\tau,s)g(\tau,s)]d\tau ds\right\}, \tag{7.2.54}$$

where for all $x \in D_i$:

$$\int\int_{D_i} \frac{\overline{\psi}(\tau,s)}{q(\tau,s)} W[q(\tau,s)g(\tau,s)]d\tau ds \in Dom\ (\Psi_i^{-1}),$$

$$\Psi_0(V) := \int_1^V \frac{d\sigma}{W(\sigma)}, \quad \Psi_i(V) := \int_{C_i}^V \frac{d\sigma}{W(\sigma)}, \ i = 1, 2,$$

with $V = (V_1, V_2), \sigma = (\sigma_1, \sigma_2)$ and

$$C_i = \left(1 + \int_{\Gamma_j \cap G_n} \beta_j(x,x_2)u(x_1,x_2)d\mu_{\varphi_j}\right) \times \Psi_i^{-1}\left(\int\int_{Gi+1\setminus G_i} \frac{\overline{\psi}(\tau,s)}{q(\tau,s)} W[q(\tau,s)g(\tau,s)]d\tau ds\right).$$

Theorem 7.2.11 (The Mitropolskiy-Iovane-Borysenko Inequality [399]) *Suppose that a non-negative discontinuous function $u(x_1,x_2)$, which has finite jumps on the curves $\{\Gamma_j\}$ satisfies the inequality (7.2.53), where the function W belongs to the class of functions $\mathcal{F}_4$*

All functions $q, g, \overline{\psi}, \beta_j$ satisfy the conditions of Theorem 7.2.10 and $q(x_1,x_2) \geq 1$. Then for arbitrary $0 \leq x_1 \leq x_1^*, 0 \leq x_2 \leq x_2^*$, the following estimate holds

$$u(x_1,x_2) \leq q(x_1,x_2)g(x_1,x_2)\Psi_i^{-1}\left(\int\int_{D_i} \overline{\psi}(\tau,s)g(\tau,s)d\tau ds\right), \ i = 0, 1, \cdots$$

where

$$\overline{\Psi}(V) := \int_1^V \frac{d\sigma}{W(\sigma)}, \quad \overline{\Psi}(V) := \int_{C_i}^V \frac{d\sigma}{W(\sigma)}, \ i = 1, 2$$

$$C_i = \left(1 + \int_{\Gamma_j \cap G_n} \beta_j(x,x_2)u(x_1,x_2)d\mu_{\varphi_j}\right) \overline{\Psi}_i^{-1}\left\{\int\int_{D_i} \overline{\psi}(\tau,s)g(\tau,s)d\tau ds\right\},$$

$$(x_1^*, x_2^*) = x^*$$
$$= \sup_x \left\{ x : \int\int_{G_{i+1}\setminus G_i} \overline{\psi}(\tau, s) g(\tau, s) d\tau ds \in Dom\,(\overline{\Psi}_i^{-1}(V)),\ i = 1, 2, \cdots \right\}.$$

Proof The proof is similar to that of the above theorem. □

However, a common feature of these integral inequalities is that they are based on Lebesgue integrals. The next result, due to Mao [371], is to extend Wendroff's inequality in several variables to the case of Lebesgue-Stieltjes integrals. These results generalize and unify some recent results established by Bondge and Pachpatte [94, 96], Corduneanu [174], Pachpatte [471], Yang [688] and Yeh [701].

Let $X = (X_1, \cdots, X_n) \in \mathbb{R}^n$ be a fixed point with $X \geq 0$. Let $[0, X] = [0, X_1] \times \cdots \times [0, X_n]$. Let $A_i(t) = (A_{i1}(t), \cdots, A_{in}(t))$, $i = 1, 2, \cdots, m$, where $A_{ij}(t) : [0, X_i] \to \mathbb{R}_+$ is right continuous non-decreasing function with $A_{ij}(0) = 0$ for all $i = 1, \cdots, n$ and $j = 1, 2, \cdots, n$. Let $x = (x_1, \cdots, x_n)$, denote $x' = (x_2, \cdots, x_n)$. We also denote

$$\int_0^{x_1} \cdots \int_0^{x_n} \cdots dA_{in}(s_n) \cdots dA_{i1}(s_1) \ or \ \int_0^{x_2} \cdots \int_0^{x_n} \cdots dA_{in}(s_n) \cdots dA_{i2}(s_2)$$

by

$$\int_0^x \cdots dA_i(s) \ or \ \int_{0'}^{x'} \cdots dA_i'(s')$$

respectively.

A function $f : [0, X] \to \mathbb{R}$ is said to be left limit if there exists $\lim_{s\uparrow x} f(s)$ for all $0 < x \leq X$. In this case, we shall denote $\lim_{s\uparrow x} f(s)$ by $f(x_-)$, here we let $f(0_-) = f(0)$. Particularly, we shall use the notation

$$\int_0^{x-} \cdots dA(s) = \lim_{s\uparrow x} f(s) \int_0^y \cdots dA(s).$$

Denote by $M([0, X], \mathbb{R}_+)$ the family of all Borel measurable, bounded, left limit and non-negative functions defined on $[0, X]$. Denote by $B([0, X], \mathbb{R}_+)$ the family of all bounded, Borel measurable and non-negative functions defined on $[0, X]$. We also denote by F the class of all Borel measurable functions $\sigma : [0, X] \to [0, X]$ such that $\sigma(x) \leq x$ for all $0 \leq x \leq X$.

We begin with the following fundamental theorem which claims its origin to Wendroff's inequality [79].

Theorem 7.2.12 (The Mao Inequality [371]) *Let $u(x), n(x) \in M([0, X], \mathbb{R}_+)$ with a positive and non-decreasing $n(x)$. Let $f_i(x)$ $(i = 1, 2, \cdots, m) \in B([0, X], \mathbb{R}_+)$ and $a(x_1) : [0, X_1] \to \mathbb{R}_+$ be right continuous and non-decreasing. Let H be a continuous function defined on $\mathbb{R}_+$ into $\mathbb{R}_+$ such that $H(v)$ is positive and* *non-decreasing for all* $v > 0$. *If for all* $0 \le x \le X$,

$$u(x) \le a(x_1) + n(x) + \sum_{i=1}^{m} \int_0^x f_i(s)H(u(s_-))dA_i(s), \tag{7.2.55}$$

then for all $0 \le x \le \hat{x}$,

$$u(x) \le G^{-1}(G(a(0) + n(x)) + \int_0^{x_1} \frac{da(\tau_1)}{H(a(\tau_{1}-) + n(x))} + \sum_{i=1}^{m} \int_0^x f_i(s)dA_i(s)) \tag{7.2.56}$$

where

$$G(r) = \int_\varepsilon^r (1/H(v))dv, \quad r \ge \varepsilon > 0 \tag{7.2.57}$$

and G^{-1} *is the inverse function of* G, *and* $\hat{x}$ *is chosen so that for all* $0 \le x \le \hat{x}$,

$$G(a(0) + n(x)) + \int_0^{x_1} \frac{da(\tau_1)}{H(a(\tau_{1}-) + n(x))} + \sum_{i=1}^{m} \int_0^x f_i(s)dA_i(s) \in Dom\ (G^{-1}).$$

Proof It follows from (7.2.55) that for all $0 \le \tau \le x \le X$,

$$u(\tau) \le a(x_1) + n(x) + \sum_{i=1}^{m} \int_0^{\tau_1} \left(\int_{0'}^{x'} f_i(s)H(u(s_-))dA_{i'}(s') \right) dA_{i_1}(s_1). \tag{7.2.58}$$

Since $H(\cdot)$ is non-decreasing, from (7.2.58) it follows for all $0 \le \tau \le x \le X$,

$$H(u(\tau_-)) \le H\left(a(\tau_{1}-) + n(x) + \sum_{i=1}^{m} \int_0^{\tau_1 -} \left(\int_{0'}^{x'} f_i(s)H(u(s_-))dA_{i'}(s') \right) dA_{i_1}(s_1) \right).$$

Consequently, we get

$$\int_0^{x_1} \frac{\int_{0'}^{x'} f_i(\tau)H(u(\tau_-))dA_{i'}(\tau')}{H(a(\tau_{1}-) + n(x) + \sum_{i=1}^{m} \int_0^{\tau_1 -} \{\int_{0'}^{x'} f_i(s)H(u(s_-))dA_{i'}(s')\}dA_{i_1}(s_1))} dA_{i_1}(\tau_1)$$
$$\le \int_0^x f_i(\tau)dA_i(\tau). \tag{7.2.59}$$

However, setting

$$Y(\tau_1) = a(\tau_1) + n(x) + \sum_{i=1}^{m} \int_0^{\tau_1} \left\{ \int_{0'}^{x'} f_i(s)H(u(s_-))dA_{i'}(s') \right\} dAi_1(s_1)$$

and using Itô's formula (see [683]), we deduce

$$G(Y(x_1)) - G(a(0)+n(x)) = \int_0^{x_1} (1/H(Y(\tau_{1-})))dY(\tau_1)$$
$$+ \sum_{0\le\tau_1\le x_1} [G(Y(\tau_1)) - G(Y(\tau_{1-})) - \Delta Y(\tau_1)/H(Y(\tau_{1-}))] \quad (7.2.60)$$

where $\Delta Y(\tau_1) := Y(\tau_1) - Y(\tau_{1-})$. Since

$$G(Y(\tau_1)) - G(Y(\tau_{1-})) = \int_{Y(\tau_{1-})}^{Y(\tau_1)} (1/H(v))dv \le \int_{\tau_1}^{Y(\tau_1)} (1/H(Y(\tau_{1-})))dv$$
$$= \Delta Y(\tau_1)/H(Y(\tau_{1-})),$$

it follows from (7.2.60) that

$$G(Y(x_1)) \le G(a(0)+n(x)) + \int_0^{x_1} (1/H(Y(\tau_{1-})))dY(\tau_1)$$
$$\le G(a(0)+n(x)) + \int_0^{x_1} \frac{da(\tau_1)}{H(a(\tau_{1-})+n(x))}$$
$$+ \sum_{i=0}^{m} \int_0^{x_1} \frac{\int_{0'}^{x'} f_i(\tau)H(u(\tau_-))dA_{i'}(\tau')}{H(a(\tau_{1-})+n(x)+\sum_{i=1}^m \int_0^{\tau_1 -}\{\int_{0'}^{x'} f_i(s)H(u(s_-))dA_{i'}(s')\}dA_{i_1}(s_1))}$$
$$\times dA_{i_1}(\tau_1).$$

This, together with (7.2.55) and (7.2.59), yields for all $0 \le x \le X$,

$$G(u(x)) \le G(a(0)+n(x)) + \int_0^{x_1} \frac{da(\tau_1)}{H(a(\tau_{1-})+n(x))} + \sum_{i=1}^m \int_0^x f_i(s)dA_i(s) \quad (7.2.61)$$

which implies the desired result (7.2.56). The proof is thus complete. □

Applying Theorem 7.2.2, we now prove the following theorem which generalizes Theorem 7.2.2.

Theorem 7.2.13 (The Mao Inequality [371]) *Let $u(x)$, $n(x)$, $a(x_1)$ and $H(v)$ be defined as in Theorem 7.2.10. Let $f_i(x,s)$ $(i = 1,2,\cdots,m)$ be bounded Borel measurable non-negative functions defined for all $0 \le s \le x \le X$ with non-decreasing in x. If for all $0 \le x \le X$,*

$$u(x) \le a(x_1) + n(x) + \sum_{i=1}^m \int_0^x f_i(x,s)H(u(s_-))dA_i(s), \quad (7.2.62)$$

then for all $0 \le x \le \hat{x}$,

$$u(x) \le G^{-1}(G(a(0)+n(x)) + \int_0^{x_1} \frac{da(\tau_1)}{H(a(\tau_{1-})+n(x))} + \sum_{i=1}^m \int_0^x f_i(x,s)dA_i(s)) \tag{7.2.63}$$

where G *and* G^{-1} *are defined as in Theorem 7.2.12, and* $\hat{x}$ *is chosen so that for all* $0 \le x \le \hat{x}$,

$$u(x) \le G^{-1}(G(a(0)+n(x)) + \int_0^{x_1} \frac{da(\tau_1)}{H(a(\tau_{1-})+n(x))} + \sum_{i=1}^m \int_0^x f_i(x,s)dA_i(s)).$$

Proof Fixed $x \in [0,\hat{x}]$ arbitrarily. For any $0 \le \overline{x} \le x$, it follows from (7.2.62) that

$$u(\overline{x}) \le a(\overline{x}_1) + n(\overline{x}) + \sum_{i=1}^m \int_0^{\overline{x}} f_i(x,s)H(u(s_-))dA_i(s).$$

By Theorem 7.2.12, we get

$$u(\overline{x}) \le G^{-1}(G(a(0)+n(\overline{x})) + \int_0^{\overline{x_1}} \frac{da(\tau_1)}{H(a(\tau_{1-})+n(\overline{x}))} + \sum_{i=1}^m \int_0^{\overline{x}} f_i(x,s)dA_i(s)).$$

Setting $\overline{x} = x$, we get the desired result (7.2.63) which completes the proof. □

The following corollary follows from Theorem 7.2.13 immediately.

Corollary 7.2.1 (The Mao Inequality [371]) *Let* $u(x),\ n(x),\ a(x_1),\ f_i(x,s)\ (i = 1,2,\cdots,m)$ *be defined as in Theorem 7.2.13 If for all* $0 \le x \le X$,

$$u(x) \le a(x_1) + n(x) + \sum_{i=1}^m \int_0^x f_i(x,s)u(s_-)dA_i(s), \tag{7.2.64}$$

then for all $0 \le x \le X$,

$$u(x) \le (a(0)+n(x)) \exp\left\{ \int_0^{x_1} \frac{da(\tau_1)}{a(\tau_{1-})+n(x)} + \sum_{i=1}^m \int_0^x f_i(x,s)dA_i(s) \right\}. \tag{7.2.65}$$

Remark 7.2.7 Let $n = 2,\ A_1(x) = x,\ n(x) = b(x_2),\ f_1(x,s) = f(s),\ f_i(x,s) = 0$ for all $i = 2,3,\cdots,m$. Suppose $u(x),\ a(x_1),\ b(x_2)$ and $f(s)$ are all continuous. Then inequality (7.2.64) reduces to

$$u(x_1,x_2) \le a(x_1) + b(x_2) + \int_0^{x_1}\int_0^{x_2} f(s_1,s_2)u(x_1,x_2)ds_2ds_1. \tag{7.2.66}$$

Noticing

$$\int_0^{x_1} \frac{da(\tau_1)}{a(\tau_{1-}) + n(x)} \le \int_0^{x_1} \frac{da(\tau_1)}{a(\tau_1) + b(0)} = \log\left(\frac{a(x_1) + b(0)}{a(0) + b(0)}\right),$$

we deduce from (7.2.65) that

$$u(x_1, x_2) \le \frac{[a(0) + b(x_2)][a(x_1) + b(0)]}{a(0) + b(0)} \exp\left(\int_0^{x_1}\int_0^{x_2} f(s_1, s_2)ds_2ds_1\right) \tag{7.2.67}$$

which is exactly (see Corollary 5.1.1, [557]) Wendroff's inequality [47].

If $u(x)$ and $f_1(x, s) = f(s)$ are continuous, $f_i(x, s) = 0$ $(2 \le i \le m)$, $A_1(x) = x$, $a(x_1) = 0$, Corollary 7.2.1 reduces to Theorem 1 of Zahariev and Bainov [714].

Theorem 7.2.14 (The Mao Inequality [371]) *Let* $u(x), n(x)$ *and* $f_i(x, s)$ $(i = 1, 2, \cdots, m)$ *be defined as in Theorem 7.2.10 with* $n(x) \ge 1$. *Let* $r_i \in (0, 1]$, $i = 1, 2, \cdots, m$. *If for all* $0 \le x \le X$,

$$u(x) \le n(x) + \sum_{i=1}^{m} \int_0^x f_i(x, s)(u(s_-))^{r_i} dA_i(s), \tag{7.2.68}$$

then for all $0 \le x \le X$,

$$u(x) \le n(x) \prod_{i=1}^{m} G_i(x), \tag{7.2.69}$$

where

$$G_i(x) = \begin{cases} [1 + (1 - r_i)\{\prod_{k=1}^{i-1} G_k(x)\} \int_0^x f_i(x, s)dA_i(s)]^{1/(1-r_i)}, & \text{if } 0 < r_i < 1, \\ \exp(\{\prod_{k=1}^{i-1} G_k(x) \int_0^x f_i(x, s)dA_i(s)), & \text{if } r_i = 1 \end{cases}$$

here we use notation

$$\prod_{k=1}^{0} G_k(x) = 1.$$

Proof We shall prove the theorem by induction. We first prove the theorem holds for $m = 1$. In fact, in this case, we get from (7.2.68) that for all $0 \le x \le X$,

$$u(x)/n(x) \le 1 + \int_0^x f_1(x, s)[u(s_-)/n(s_-)]^{r_1} dA_1(s). \tag{7.2.70}$$

An application of Theorem 7.2.13 to (7.2.70) implies $u(x) \leq n(x)G_1(x)$ for all $0 \leq x \leq X$, i.e., the theorem holds for $m = 1$. We now suppose it holds for $m = k$. We want to prove it also holds for $m = k + 1$. In this case, we can rewrite inequality (7.2.68) as, for all $0 \leq x \leq X$,

$$u(x) \leq N(x) + \sum_{i=1}^{k} \int_0^x f_i(x,s)(u(s_-))^{r_i} dA_i(s), \tag{7.2.71}$$

where

$$N(x) = n(x) + \int_0^x f_{k+1}(x,s)(u(s_-))^{r_{r+1}} dA_i(s). \tag{7.2.72}$$

Hence, by the inductive assumption, we deduce from (7.2.72) that

$$\begin{aligned} u(x) &\leq N(x) \prod_{i=1}^{k} G_i(x) \\ &= n(x) \prod_{i=1}^{k} G_i(x) + \int_0^x \left(\prod_{i=1}^{k} G_i(x) f_{k+1}(x,s) \right) (u(s_-))^{r_{r+1}} dA_i(s). \end{aligned} \tag{7.2.73}$$

Thus, by the first step of our proof, we derive from (7.2.73) that

$$u(x) \leq \left(n(x) \prod_{i=1}^{k} G_i(x) \right) G_{k+1}(x),$$

which means the theorem holds for $m = k + 1$. The proof is thus complete. □

Remark 7.2.8 If all of the functions are continuous and $A_i(x) = x$ $(i = 1, 2, \cdots, m)$, Theorem 7.2.14 reduces to the result of Corduneanu [174]. For one-dimensional case, Theorem 7.2.14 gives us a generalization of lemma of Yang in [688].

Next, we shall use the results to establish some new Lebesque-Stieltjes integral inequalities with retardation. We begin with the following theorem which generalizes Theorem 4 of Akinyele [28] and Theorem 1 of Dannan [181].

Theorem 7.2.15 (The Mao Inequality [371]) *Let $u(x)$, $n(x)$, $a(x_1)$ and $f_i(x,s)$ $(i = 1, 2, \cdots, m)$ be defined as in Theorem 7.2.13. Let $\sigma(x) \in F$, $q(x)$, $h(x) \in M([0, X], \mathbb{R}_+)$ with $q \geq 1$ and $h(x)$ positive and non-decreasing. Let $H \in \mathcal{H}$ with corresponding multiplier function Φ.*

Assume the inequality holds for all $0 \leq x \leq X$,

$$u(x) \leq h(x)(a(x_1) + n(x))q(x) \sum_{i=1}^{m} \int_0^x f_i(x,s) H(u(\sigma(s)_-)) dA_i(s). \tag{7.2.74}$$

Then we have for all $0 \le x \le \hat{x}$,

$$u(x) \le h(x)q(x)G^{-1}(G(a(0)+n(x)) + \int_0^{x_1} \frac{da(\tau_1)}{H(a(\tau_{1-})+n(x))}$$

$$+\sum_{i=1}^{m}\int_0^x [f_i(x,s)/h(s)]\Phi(h(\sigma(s)_-)q(\sigma(s)_-))dA_i(s)) \quad (7.2.75)$$

where G *and* G^{-1} *are as defined in Theorem 7.2.12 and* $\hat{x}$ *is chosen so that for all* $0 \le x \le \hat{x}$,

$$G(a(0)+n(x)) + \int_0^{x_1} \frac{da(\tau_1)}{H(a(\tau_{1-})+n(x))}$$

$$+\sum_{i=1}^{m}\int_0^x [f_i(x,s)/h(s)]\Phi(h(\sigma(s)_-)q(\sigma(s)_-))dA_i(s) \in Dom\ (G^{-1}).$$

Proof It follows from (7.2.74) that

$$u(x) \le h(x)q(x)\left(a(x_1) + n(x) + \sum_{i=1}^{m}\int_0^x [f_i(x,s)/h(s)]H(u(\sigma(s)_-))dA_i(s)\right).$$

Defining

$$w(x) = a(x_1) + n(x) + \sum_{i=1}^{m}\int_0^x [f_i(x,s)/h(s)]H(u(\sigma(s)_-))dA_i(s), \quad (7.2.76)$$

we have

$$u(x) \le h(x)q(x)w(x). \quad (7.2.77)$$

Consequently,

$$u(\sigma(s)_-) \le h(\sigma(s)_-)q(\sigma(s)_-)w(\sigma(s)_-) \le h(\sigma(s)_-)q(\sigma(s)_-)w(s_-).$$

Therefore,

$$w(x) \le a(x_1) + n(x) + \sum_{i=0}^{m}\int_0^x [f_i(x,s)/h(s)]\Phi(h(\sigma(s)_-)q(\sigma(s)_-))H(w(s_))dA_i(s).$$
$$(7.2.78)$$

In view of Corollary 7.2.1 and (7.2.77), (7.2.78) yields desired result (7.2.75) immediately and the proof is complete. □

Remark 7.2.9 If $A_1(x) = x$, $a(x_1) = 0$, $n(x) = 1$, $f_1(x,s) = f(s)$ $(2 \le i \le m)$ and all functions in Theorem 7.2.15 are continuous, then Theorem 7.2.15 is an improvement of Theorem 4 of Akinyele [27]. For one-dimensional case, if we still have $\sigma(x) = x$, $q(x) = 1$, then Theorem 7.2.15 reduces to Theorem 1 of Dannan [181].

We can similarly prove the following theorem which generalizes Theorem 5 in Akinyele [27], Yeh and Shih Theorem 3 in [706].

Theorem 7.2.16 (The Mao Inequality [371]) *Let $u(x)$, $n(x)$, $q(x)$, $w(x)$, $a(x_1)$ and $f_i(x,s)$ $(1 \le i \le m)$ be defined as in Theorem 7.2.15. Let $P \in p$, $h(x) \in M([0,X],\mathbb{R}_+)$ with $h(x) \ge 1$ and non-decreasing.*

If there holds for all $0 \le x \le X$,

$$u(x) \le h(x)(a(x_1) + n(x)) + q(x)\sum_{i=1}^{m}\int_0^x f_i(x,s)P(u(\sigma(s)_-))dA_i(s)$$

then for all $0 \le x \le \hat{x}$,

$$\begin{aligned} u(x) \le & h(x)q(x)Q^{-1}\Bigg(Q(a(0)+n(x)) + \int_0^{x_1}\frac{da(\tau_1)}{P(a(\tau_{1-})+n(x))} \\ & +\sum_{i=1}^{m}\int_0^x f_i(x,s)q(\sigma(s)_-)dA_i(s)\Bigg) \end{aligned}$$

where

$$Q(r) = \int_\varepsilon^r (1/P(v))dv, \quad r \ge \varepsilon > 0,$$

and Q^{-1} is the inverse function of Q, $\hat{x}$ is chosen so that for all $0 \le x \le \hat{x}$,

$$Q(a(0)+n(x)) + \int_0^{x_1}\frac{da(\tau_1)}{P(a(\tau_{1-})+n(x))} + \sum_{i=1}^{m}\int_0^x f_i(x,s)q(\sigma(s)_-)dA_i(s) \in Dom\,(Q^{-1}).$$

Theorem 7.2.17 (The Mao Inequality [371]) *Let $u(x)$, $n(x)$, $a(x_1)$ and $f_i(x,s)$ $(1 \le i \le m)$ be defined as in Theorem 7.2.13. Let $H : \mathbb{R}_+ \to \mathbb{R}_+$ be a continuous non-negative non-decreasing function. Suppose the following inequality holds for all $0 \le x \le X$,*

$$\begin{aligned} u(x) \le & a(x_1) + n(x) \\ & + \sum_{i=1}^{m}\int_0^{s-} f_i(x,s)\left[u(s_-) + \sum_{i=1}^{m}\int_0^{s-} f_i(s,\tau)H(u(\tau_-))dA_i(\tau)\right]dA_i(s). \end{aligned} \tag{7.2.79}$$

Then for all $0 \leq x \leq \hat{x}$,

$$u(x) \leq F^{-1}\Bigg(F(a(0)+n(x)) + \int_0^{x_1} \frac{da(\tau_1)}{a(\tau_{1-}) + n(x) + H(a(\tau_{1-}) + n(x))} + \sum_{i=1}^{m} \int_0^{x} f_i(x,s) dA_i(s)\Bigg), \tag{7.2.80}$$

and for all $0 \leq x \leq \hat{x}$,

$$u(x) \leq a(x_1) + n(x) + \sum_{i=1}^{m} \int_0^{x} f_i(x,s) F^{-1}(F(a(0)+n(s_-)) + \int_0^{s_{1-}} \frac{da(\tau_1)}{a(\tau_{1-}) + n(x) + H(a(\tau_{1-}) + n(x))} + \sum_{i=1}^{m} \int_0^{s_-} f_i(s,\tau) dA_i(\tau)) dA_i(s) \tag{7.2.81}$$

where

$$F(r) = \int_{\varepsilon}^{r} \frac{dv}{v + H(v)}, \quad r \geq \varepsilon > 0, \tag{7.2.82}$$

and F^{-1} *is the inverse function of* F, *and* x *is chosen so that for all* $0 \leq x \leq \hat{x}$,

$$F(a(0)+n(x)) + \int_0^{x_1} \frac{da(\tau_1)}{a(\tau_{1-}) + n(x) + H(a(\tau_{1-}) + n(x))} + \sum_{i=1}^{m} \int_0^{x} f_i(x,s) dA_i(s) \in Dom\,(F^{-1}).$$

Proof For all $0 \leq x \leq X$, set

$$w(x) = u(x) + \sum_{i=1}^{m} \int_0^{x} f_i(x,s) H(u(s_-)) dA_i(s). \tag{7.2.83}$$

Thus

$$u(x) \leq w(x) \tag{7.2.84}$$

and

$$u(x) \leq a(x_1) + n(x) + \sum_{i=1}^{m} \int_0^{x} f_i(x,s) w(s_-) dA_i(s). \tag{7.2.85}$$

Inserting (7.2.84) and (7.2.85) into (7.2.83), we have

$$w(x) \leq a(x_1) + n(x) + \sum_{i=1}^{m} \int_0^x f_i(x,s)(w(s_-) + H(w(s_-)))dA_i(s). \quad (7.2.86)$$

Applying Theorem 7.2.13 to (7.2.86), we get, for all $0 \leq x \leq \hat{x}$,

$$w(x) \leq F^{-1}\Bigg(F(a(0) + n(x)) + \int_0^{x_1} \frac{da(\tau_1)}{a(\tau_{1-}) + n(x) + H(a(\tau_{1-}) + n(x))}$$
$$+ \sum_{i=1}^{m} \int_0^x f_i(x,s)dA_i(s)\Bigg), \quad (7.2.87)$$

which is the desired result (7.2.80) if we notice (7.2.84). Finally, substituting this bound on $w(x)$ in (7.2.85), we can deduce the required result (7.2.81) and hence complete the proof. □

Remark 7.2.10 If $A_1(x) = x$, $n(x) = \sum_{2 \leq i \leq n} a_i(x_i)$, $f_1(x,s) = f(s)$, $f_i(x,s) = 0$ $(2 \leq i \leq m)$ and all functions are continuous, then Theorem 7.2.17 with estimate (7.2.81) reduces to Theorem 1 of Yeh [701], which in turn, is a generalization Theorem 2 of Yeh and Shih [706].

Theorem 7.2.18 (The Mao Inequality [371]) *Let $u(x)$, $n(x)$, $a(x_1)$ and $f_i(x,s)$ $(1 \leq i \leq m)$ be defined as in Theorem 7.2.13. Let $a(x) \in M([0,X],\mathbb{R}_+)$ with $q(x) \geq 1$, and $\sigma(x)$, $\rho(x) \in F$. Let $g_i(x,s)$ $(i = 1,2,\cdots,m)$ be bounded Borel measurable non-negative functions defined for $0 \leq s \leq x \leq X$ with non-decreasing in x. Suppose the following inequality holds for all $0 \leq x \leq X$,*

$$u(x) \leq a(x_1) + n(x) + q(x)\Bigg\{\sum_{i=0}^{m} \int_0^x f_i(x,s)u(\sigma(s)_-)dA_i(s)$$
$$+ \sum_{i=1}^{m} \int_0^x f_i(x,s)\Bigg[\sum_{i=1}^{m} \int_0^{s-} g_i(s,\tau)u(\rho(\tau)_-)\Bigg]dA_i(s)\Bigg\}. \quad (7.2.88)$$

Then for all $0 \leq x \leq X$, we have

$$u(x) \leq q(x)(a(0) + n(x)) \exp\Bigg[\int_0^{x_1} \frac{da(\tau_1)}{a(\tau_{1-}) + n(x)}$$
$$+ \sum_{i=1}^{m} \int_0^x [f_i(x,\tau)q(\sigma(\tau_-)) + g_i(x,\tau)q(\rho(\tau)_-)]dA_i(s)\Bigg] \quad (7.2.89)$$

and

$$u(x) \leq q(x)\Big(a(x_1) + n(x) + \sum_{i=1}^{m}\int_0^x f_i(x,s)q(\sigma(s)_-)(a(0) + n(s_-))$$
$$\times \exp\Big[\int_0^{s_1-} \frac{da(\tau_1)}{a(\tau_{1-}) + n(s_-)}$$
$$+ \sum_{i=1}^{m}\int_0^{s-} [f_i(s,\tau)q(\sigma(\tau)_-) + g_i(s,\tau)q(\rho(\tau)_-)]dA_i(\tau)\Big]dA_i(s)\Big). \tag{7.2.90}$$

Proof It follows from (7.2.88) that

$$u(x) \leq q(x)(a(x_1) + n(x) + \sum_{i=1}^{m}\int_0^x f_i(x,s)u(\sigma(s)_-)dA_i(s)$$
$$+ \sum_{i=1}^{m}\int_0^x f_i(x,s)\{\sum_{i=1}^{m}\int_0^{s-} g_i(s,\tau)u(\rho(\tau)_-)dA_i(\tau)\}dA_i(s)). \tag{7.2.91}$$

Define

$$w(x) = a(x_1) + n(x) + \sum_{i=1}^{m}\int_0^x f_i(x,s)u(\sigma(s)_-)dA_i(s)$$
$$+ \sum_{i=1}^{m}\int_0^x f_i(x,s)\Big[\sum_{i=1}^{m}\int_0^{s-} g_i(s,\tau)u(\rho(\tau)_-)dA_i(\tau)\Big]dA_i(s). \tag{7.2.92}$$

Then for all $0 \leq x \leq X$,

$$u(x) \leq q(x)w(x). \tag{7.2.93}$$

Therefore

$$\begin{cases} u(\sigma(s)_-) \leq q(\sigma(s)_-)w(\sigma(s)_-) \leq q(\sigma(s)_-)w(s_-), \\ u(\rho(\tau)_-) \leq q(\rho(\tau)_-)w(\rho(\tau)_-) \leq q(\rho(\tau)_-)w(s_-). \end{cases}$$

Hence, we get

$$w(x) \leq a(x_1) + n(x) + \sum_{i=1}^{m} \int_0^x f_i(x,s)q(\sigma(s)_-)$$
$$\times \left(w(s_-) + \sum_{i=1}^{m} \int_0^{s_-} g_i(s,\tau)q(\rho(\tau)_-)w(\tau_-)dA_i(\tau) \right) dA_i(s). \tag{7.2.94}$$

Setting

$$z(x) = w(x) + \sum_{i=1}^{m} \int_0^x g_i(x,\tau)q(\rho(\tau)_-)dA_i(\tau), \tag{7.2.95}$$

we have

$$w(x) \leq z(x) \tag{7.2.96}$$

and

$$w(x) \leq a(x_1) + n(x) + \sum_{i=1}^{m} \int_0^x f_i(x,s)q(\sigma(s)_-)z(s_-)dA_i(s). \tag{7.2.97}$$

Inserting (7.2.96) and (7.2.97) into (7.2.95), we get

$$z(x) \leq a(x_1) + n(x) + \sum_{i=1}^{m} \int_0^x [f_i(x,s)q(\sigma(\tau)_-) + g_i(x,s)q(\rho(\tau)_-)]z(\tau_-)dA_i(s). \tag{7.2.98}$$

Applying Corollary 7.2.1 to (7.2.98) yields

$$z(x) \leq (a(0) + n(x) \times \exp\left[\int_0^{x_1} \frac{da(\tau_1)}{a(\tau_{1-}) + n(x)} \right.$$
$$\left. + \sum_{i=1}^{m} \int_0^x [f_i(x,s)q(\sigma(\tau)_-) + g_i(x,s)q(\rho(\tau)_-)]dA_i(s) \right]. \tag{7.2.99}$$

In view of (7.2.93) and (7.2.96), (7.2.99) implies desired result (7.2.88). Finally, putting (7.2.99) into (7.2.97) and using (7.2.93), we can obtain the required inequality (7.2.90) and hence complete the proof. □

Remark 7.2.11 If $A_1(x) = x$, $\sigma(x) = \rho(x) = x$, $q(x) = 1$, $a(x_1) = 0$, $f_i(x,s) = g_i(x,s) = 0$ $(2 \leq i \leq m)$, and $u(x)$, $n(x)$, $f_i(x,s) = f(x)$, $g_i(x,s) = g(x)$ are all continuous, then Theorem 7.2.18 with estimate (7.2.92) is a little improvement of a result, Theorem 1 of Yeh and Shih [706]. Furthermore, if $g(x) = 0$, Theorem 7.2.18 with estimate (7.2.88) reduces to a result due to Yeh [701], Theorem 1. For $n = 1$, Theorem 7.2.18 is another generalization of the inequality due to Pachpatte [456], and if, in addition, $n(x)$ is a constant, we obtain a generalization of Theorem 1 of Pachpatte [441].

Let $I = [x_0, x_1]$, $J = [y_0, y_1)$, and $\Lambda := I \times J \subset \mathbb{R}^2$. Consider inequality $\psi(u(x,y)) \leq a(x,y) + \int_{b(x_0)}^{b(x)} \int_{c(y_0)}^{c(y)} f(x,y,s,t)[g(s,t,\sigma,\tau)\varphi_2(u(\sigma,\tau))]$, where we suppose that $\phi \in C(\mathbb{R}_+, \mathbb{R}_+)$ is strictly increasing such that $\phi(+\infty) = +\infty$, $b \in C^1(I,I)$, and $c \in C^1(J,J)$ are non-decreasing, such that $b(x) \leq x$ and $c(y) \leq y$, $a \in C^1(\Lambda, \mathbb{R}_+)$, $f \in C^0(\Lambda^2, \mathbb{R}_+)$, and $g(x,y,s,t) \in C(\Lambda^2, \mathbb{R}_+)$ are given, and $\phi_i \in C(\mathbb{R}_+, \mathbb{R}_+)$ $(i = 1,2)$ are functions satisfying $\phi_i(0) = 0$ and $\phi_i(u) > 0$ for all $u > 0$.

Define functions

$$w_1(s) := \max_{\tau\in[0,s]}\{\phi_1(\tau)\},\quad w_2(s) := \max_{\tau\in[0,s]}\{\phi_2(\tau)/w_1(\tau)\}w_1(s),\quad \phi(s) := w_2(s)/w_1(s). \tag{7.2.100}$$

Obviously, w_1, w_2 and ϕ are all non-decreasing and non-negative functions and satisfy $w_i(s) \geq \phi_i(s)$, $i = 1,2$. Let

$$W_1(s) := \int_1^u \frac{ds}{w_1(\psi^{-1}(s))}, \tag{7.2.101}$$

$$W_2(s) := \int_1^u \frac{ds}{w_2(\psi^{-1}(s))}, \tag{7.2.102}$$

$$\Phi(s) := \int_{W_1(1)}^u \frac{ds}{\phi(\psi^{-1}((W_1^{-1}(s))))}. \tag{7.2.103}$$

Obviously W_1, W_2, and Φ are strictly increasing in all $u > 0$, and therefore the inverses W_1^{-1}, W_2^{-1}, and Φ^{-1} are well defined, continuous, and increasing. We note that

$$\begin{aligned}\Phi(s) &:= \int_{W_1(1)}^u \frac{dx}{\phi(\psi^{-1}((W_1^{-1}(x))))} = \int_{W_1(1)}^u \frac{w_1(\psi^{-1}(W_1^{-1}(x)))dx}{w_2(\psi^{-1}((W_1^{-1}(s))))}\\ &= \int_1^{W_1^{-1}(u)} \frac{dx}{w_2(\psi^{-1}(x))} = W_2(W_1^{-1}(u)).\end{aligned} \tag{7.2.104}$$

Furthermore, let $\hat{f}(x,y,s,t) := \max_{\tau\in[x_0,x]} f(\tau,y,s,t)$, which is also non-decreasing in x for each fixed y, s and t and satisfies $\hat{f}(x,y,s,t) \geq f(x,y,s,t) \geq 0$.

Theorem 7.2.19 (The Wang-Shen Inequality [665]) *If the following inequality holds*

$$\psi(u(x,y)) \le a(x,y) + \int_{b(x_0)}^{b(x)} \int_{c(y_0)}^{c(y)} f(x,y,s,t)[\phi_1(u(s,t)) + \int_{b(x_0)}^{s} \int_{c(y_0)}^{t} g(s,t,\sigma,\tau)\phi_2(u(\sigma,\tau))d\tau d\sigma]dtds, \quad (7.2.105)$$

where $\phi \in C(\mathbb{R}_+, \mathbb{R}_+)$, and $u(x,y)$ is a non-negative function, then for all $(x,y) \in [x_0, X_1) \times [y_0, Y_1)$,

$$u(x,y) \le \psi^{-1}\{W_2^{-1}[\Xi(x,y)]\} \quad (7.2.106)$$

where

$$\begin{cases} \Xi(x,y) := W_2[W_1^{-1}(r_2(x,y))] + \int_{b(x_0)}^{b(x)} \int_{c(y_0)}^{c(y)} \hat{f}(x,y,s,t) \\ \qquad [\int_{b(x_0)}^{s} \int_{c(y_0)}^{t} g(s,t,\tau,\sigma)d\tau d\sigma]dtds, \\ r_2(x,y) = W_1(r_1(x,y)) + \int_{b(x_0)}^{b(x)} \int_{c(y_0)}^{c(y)} \hat{f}(x,y,s,t)dtds, \\ r_1(x,y) = a(x_0,y) + \int_{x_0}^{s} |a_x(s,y)|ds, \end{cases} \quad (7.2.107)$$

and $(X_1, Y_1) \in \Lambda$ is arbitrary given on the boundary of the planar region

$$\mathcal{R} := \{(x,y) \in \Lambda : \Xi(x,y) \in Dom\,(W_2^{-1}),\ r_2(x,y) \in Dom\,(W_1^{-1})\}. \quad (7.2.108)$$

Here Dom denotes the domain of a function.

Proof By the definition of functions w_i and $\hat{f}_i$, from (7.2.105) we derive for all $(x,y) \in \Lambda$,

$$\psi(u(x,y)) \le a(x,y) + \int_{b(x_0)}^{b(x)} \int_{c(y_0)}^{c(y)} \hat{f}(x,y,s,t)[w_1(u(s,y)) + \int_{b(x_0)}^{s} \int_{c(y_0)}^{t} g(s,t,\tau,\sigma)w_2(u(\sigma,\tau))d\tau d\sigma]dtds. \quad (7.2.109)$$

First, we discuss the case that $a(x,y) > 0$ for all $(x,y) \in \Lambda$. It means that $r_1(x,y) > 0$ for all $(x,y) \in \Lambda$. In such a case, $r_1(x,y)$ is positive and non-decreasing on Λ and

$$r_1(x,y) \ge a(x_0,y) + \int_{x_0}^{x} a_x(t,y)dt. \quad (7.2.110)$$

Regarding (7.2.105), we consider the auxiliary inequality

$$\psi(u(x,y)) \le r_1(x,y) + \int_{b(x_0)}^{b(x)} \int_{c(y_0)}^{c(y)} \hat{f}(x,y,s,t)[w_1(u(s,y)) + \int_{b(x_0)}^{s} \int_{c(y_0)}^{t} g(s,t,\tau,\sigma) w_2(u(\sigma,\tau)) d\tau d\sigma] dt ds. \quad (7.2.111)$$

For all $(x,y) \in [x_0,X) \times J$, where $x_0 \le X \le X_1$ is chosen arbitrary. We claim that for all $(x,y) \in [x_0,X) \times [y_0,Y_1)$,

$$u(x,y) \le \int_{b(x_0)}^{b(x)} \psi^{-1}\{W_2^{-1}[W_2(W_1^{-1}(W_1(r_1(x,y)) + \int_{b(x_0)}^{b(x)} \int_{c(y_0)}^{c(y)} \hat{f}(X,y,s,t) dt ds)) + \int_{b(x_0)}^{b(x)} \int_{c(y_0)}^{c(y)} \hat{f}_1(x,y,s,t)[\int_{b(x_0)}^{s} \int_{c(y_0)}^{t} g(s,t,\tau,\sigma) w_2(u(\sigma,\tau)) d\tau d\sigma] dt ds]\} \quad (7.2.112)$$

where Y_1 is defined by (7.2.108).

Let $\eta(x,y)$ denote the right-hand side of (7.2.111), which is a non-negative and non-decreasing function on $[x_0,X) \times J$. Then (7.2.111) is equivalent to, for all $(x,y) \in [x_0,Y) \times J$,

$$u(x,y) \le \psi^{-1}(\eta(x,y)). \quad (7.2.113)$$

By the fact that $b(x) \le x$ for all $x \in [x_0,X)$ and the monotonicity of w_i, ψ, η, and $b(x)$, we have for all $(x,y) \in [x_0,X) \times J$,

$$\begin{aligned} \frac{(\partial/\partial x)\eta(x,y)}{w_1(\psi^{-1}(\eta(x,y)))} &\le \frac{(\partial/\partial x) r_1(x,y)}{w_1(\psi^{-1}(r_1(x,y)))} + \frac{b'(x)}{w_1(\psi^{-1}(\eta(x,y)))} \\ &\quad \times \int_{c(y_0)}^{c(y)} \hat{f}_1(X,y,b(x),t)[w_1(u(b(x),y)) \\ &\quad + \int_{b(x_0)}^{b(t)} \int_{c(y_0)}^{t} g(b(x),t,\tau,\sigma) w_2(u(\sigma,\tau)) d\tau d\sigma] dt \\ &\le \frac{(\partial/\partial x) r_1(x,y)}{w_1(\psi^{-1}(r_1(x,y)))} + b'(x) \int_{c(y_0)}^{c(y)} \hat{f}_1(X,y,b(x),t) dt \\ &\quad + b'(t) \int_{c(y_0)}^{c(y)} \hat{f}_1(X,y,b(x),t) \\ &\quad \times [\int_{b(x_0)}^{s} \int_{c(y_0)}^{b(x)} g(b(x),t,\tau,\sigma) w_2(u(\sigma,\tau)) d\tau d\sigma] dt. \end{aligned} \quad (7.2.114)$$

Integrating the above from x_0 to x, we get for all $(x, y) \in [x_0, X) \times J$,

$$W_1(\eta(x,y)) \leq W_1(r_1(x,y)) + \int_{b(x_0)}^{b(x)} \int_{c(y_0)}^{c(y)} \hat{f}_1(X, y, s, t) dt ds$$
$$+ \int_{b(x_0)}^{b(x)} \int_{c(y_0)}^{c(y)} \hat{f}_1(X, y, s, t) [\int_{b(x_0)}^{s} \int_{c(y_0)}^{t} g(s, t, \tau, \sigma) \phi(u(\tau, \sigma)) d\tau\sigma] dt ds. \tag{7.2.115}$$

Let

$$\begin{cases} \psi(\xi(x,y)) := W_1(\eta(x,y)), \\ \hat{r}_2(x,y) := W_1(r_1(x,y)) + \int_{b(x_0)}^{b(x)} \int_{c(y_0)}^{c(y)} \hat{f}_1(X, y, s, t) dt ds. \end{cases} \tag{7.2.116}$$

From (7.2.115), (7.2.116), we derive for any $x_0 \leq x \leq X$, $y_0 \leq y \leq Y_1$,

$$\psi(\xi(x,y)) \leq \hat{r}_2(t) + \int_{b(x_0)}^{b(x)} \int_{c(y_0)}^{c(y)} \hat{f}_1(X, y, s, t)$$
$$\times \left[\int_{b(x_0)}^{s} \int_{c(y_0)}^{t} g(s, t, \tau, \sigma) \phi(u(\tau, \sigma)) d\tau\sigma \right] dt ds. \tag{7.2.117}$$

Let $\beta(x, y)$ denote the right-hand side of (7.2.117), which is a non-negative and non-decreasing function on $[x_0, Y) \times J$. Then (7.2.117) is equivalent to, for all $(x, y) \in [x_0, Y) \times J$,

$$\psi(\xi(x,y)) \leq \beta(x,y). \tag{7.2.118}$$

From (7.2.113), (7.2.116) and (7.2.118), it follows for all $x_0 \leq x \leq X$, $y_0 \leq y \leq Y_1$,

$$u(x,y) \leq \psi^{-1}(\eta(x,y)) = \psi^{-1}(W_1^{-1}(\psi(\xi(x,y)))) \leq \psi^{-1}(W_1^{-1}(\beta(x,y))) \tag{7.2.119}$$

where Y_1 is defined by (7.2.108). By the definitions of ϕ, ψ and W_1, $\phi(\psi^{-1}(W_1^{-1}(s)))$ is continuous and non-decreasing on $[0, +\infty)$ and satisfies $\phi(\psi^{-1}(W_1^{-1}(s))) > 0$ for $s > 0$.

Let $h(s) = \psi^{-1}(W_1^{-1}(s))$. Since $b'(x) \geq 0$ and $b(x) \leq x$ for all $x \in [x_0, X)$, from (7.2.119) we derive for all $(x, y) \in [x_0, X) \times [y_0, Y_1)$,

$$\begin{aligned}\frac{(\partial/\partial x)\beta(x,y)}{w_1(\phi(h(\beta(x,y))))} &\leq \frac{(\partial/\partial x)\hat{r}_2(x,y)}{\psi(h(\hat{r}_2(x,y)))} + \frac{b'(x)}{\phi(h(\beta(x,y)))}\int_{c(y_0)}^{c(y)} \hat{f}_1(X,y,b(x),t)\\ &\quad\times\left[\int_{b(x_0)}^{b(t)}\int_{c(y_0)}^{t} g(b(x),t,\tau,\sigma)\phi(u(\sigma,\tau))d\tau d\sigma\right]dt\\ &\leq \frac{(\partial/\partial x)\hat{r}_2(x,y)}{\phi(h(\hat{r}_2(x,y)))} + b'(t)\int_{c(y_0)}^{c(y)} \hat{f}_1(X,y,b(x),t)\\ &\quad\times\left[\int_{b(x_0)}^{s}\int_{c(y_0)}^{b(x)} g(b(x),t,\tau,\sigma)d\tau d\sigma\right]dt. \end{aligned} \tag{7.2.120}$$

Integrating the above inequality from x_0 to x, by (7.2.103), we get for all $(x, y) \in [x_0, X) \times [y_0, y_1)$,

$$\Phi(\beta(x,y)) \leq \Phi(\hat{r}_2(x,y)) + \int_{b(x_0)}^{b(x)}\int_{c(y_0)}^{c(y)} \hat{f}_1(X,y,s,t)[\int_{b(x_0)}^{s}\int_{c(y_0)}^{t} g(s,t,\tau,\sigma)d\tau d\sigma]dtds. \tag{7.2.121}$$

By (7.2.118) and the above inequality, we obtain for all $(x, y) \in [x_0, X) \times [y_0, Y_1)$,

$$\begin{aligned}u(t) \leq \psi^{-1}\Bigg(W_1^{-1}[\Phi^{-1}(\Phi(\hat{r}_2(x,y)) &+ \int_{b(x_0)}^{b(x)}\int_{c(y_0)}^{c(y)} \hat{f}_1(X,y,s,t)\\ \times[\int_{b(x_0)}^{s}\int_{c(y_0)}^{t} g(s,t,\tau,\sigma)d\tau d\sigma]dtds)]\Bigg)\end{aligned} \tag{7.2.122}$$

where Y_1 is defined by (7.2.108). It follows from (7.2.104) that

$$\begin{aligned}u(x,y) \leq \psi^{-1}\Bigg(W_2^{-1}[W_2(W_1^{-1}(W_1(r_1(x,y)) &+ \int_{b(x_0)}^{b(x)}\int_{c(y_0)}^{c(y)} \hat{f}_1(X,y,s,t)dtds))\\ + \int_{b(x_0)}^{b(x)}\int_{c(y_0)}^{c(y)} \hat{f}_1(X,y,s,t)[\int_{b(x_0)}^{s}\int_{c(y_0)}^{t} g(s,t,\tau,\sigma)d\tau d\sigma]dtds)]\Bigg)\end{aligned} \tag{7.2.123}$$

which proves the claim (7.2.112).

We start from the original inequality (7.2.105) and see that for all $y \in [y_0, Y_1)$,

$$\phi(u(x,y)) \le r_1(x,y) + \int_{b(x_0)}^{b(X)} \int_{c(y_0)}^{c(y)} \hat{f}_1(X,y,s,t)[\psi_1(u(s,t)) + \int_{b(x_0)}^{s} \int_{c(y_0)}^{t} g(s,t,\tau,\sigma)\phi_2(u(\sigma,\tau)d\tau d\sigma]]dtds \quad (7.2.124)$$

that is, the auxiliary inequality (7.2.111) holds for $x = X$, $y \in [y_0, Y_1)$. By (7.2.112), we can derive

$$u(X,y) \le \psi^{-1}\Big(W_2^{-1}[W_2(W_1^{-1}(W_1(r_{1,\varepsilon}(x,y)) + \int_{b(x_0)}^{b(X)} \int_{c(y_0)}^{c(y)} \hat{f}_1(X,y,s,t)dtds)) + \int_{b(x_0)}^{b(X)} \int_{c(y_0)}^{c(y)} \hat{f}_1(X,y,s,t)[\int_{b(x_0)}^{s} \int_{c(y_0)}^{t} g(s,t,\tau,\sigma)d\tau d\sigma]dtds)]\Big). \quad (7.2.125)$$

The remainder case is that $a(x,y) = 0$ for some $(x,y) \in \Lambda$. Let

$$r_{1,\varepsilon}(x,y) := r_1(x,y) + \varepsilon, \quad (7.2.126)$$

where $\varepsilon > 0$ is an arbitrary small number. Obviously, $r_{1,\varepsilon} > 0$ for all $(x,y) \in \Lambda$. Using the same arguments as above, where $r_1(x,y)$ is replaced with $r_{1,\varepsilon}$, we get for all $x_0 \le X \le X_1$, $y_0 \le y \le Y_1$,

$$u(x,y) \le \psi^{-1}\Big(W_2^{-1}[W_2(W_1^{-1}(W_1(r_{1,\varepsilon}(x,y)) + \int_{b(x_0)}^{b(x)} \int_{c(y_0)}^{c(y)} \hat{f}_1(X,y,s,t)dtds)) + \int_{b(x_0)}^{b(x)} \int_{c(y_0)}^{c(y)} \hat{f}_1(X,y,s,t)[\int_{b(x_0)}^{s} \int_{c(y_0)}^{t} g(s,t,\tau,\sigma)d\tau\sigma]dtds)]\Big). \quad (7.2.127)$$

Letting $\varepsilon \to 0^+$, we obtain (7.2.106) because of continuity of $r_{1,\varepsilon}$ in ε and continuity of ψ^{-1}, W_1^{-1}, W, W_2^{-1}. This completes the proof. □

7.3 Nonlinear Multi-Dimensional Discontinuous Integral Inequalities Involving Kernels

7.3.1 Nonlinear Two-Dimensional Discontinuous Integral Inequalities Involving Kernels

In the next theorem, we shall discuss nonlinear integral operators whose kernels are "majorized" by a kernel generating a linear integral operator.

We shall consider complex functions defined in the partially ordered set T, $(T, <\ )$, employing the following notations:

$T_x := \{y : y \in T \text{ and } y < x\}$ for the segment of the element x and

$$\chi(x,y) = \begin{cases} 1, & \text{if} \quad y \leq x \\ 0, & \text{otherwise} \end{cases}$$

for its characteristic function.

Concerning the set T, we shall assume that the following conditions hold:

(C_1) . T is a partially ordered connected topological space with positive measure μ.

(C_2) . For every $x \in T$, the function $\chi(x, \cdot)$ is μ-measurable.

(C_3) . If x_α is a generalized sequence of elements of T, convergent to x, then $||\chi(x_\alpha, \cdot) - \chi(x, \cdot)||_2$ tends to 0.

(C_4) . There exists an element $x_0 \in T$ such that $||\chi(x_0, \cdot)||_2 = 0$, where

$$||\varphi||_2^2 = \int_T |\varphi(x)|^2 d\mu(x).$$

Theorem 7.3.1 (The Ronkov-Bainov Inequality [576]) *Let*

(1) Conditions $(C_1) - (C_4)$ hold for the space T.

(2) Let the kernel $K_1(x, y, z)$ of the integral operator V_1,

$$V_1 h(x) := \int_{T_x} K_1(x, y, h(y)) d\mu(y),$$

be a real function defined in $T \times T \times \mathbb{R}$ satisfying the following conditions:

(2.1) For every two fixed elements x and y from T, $K_1(x, y, z)$ is a monotone increasing and continuous function of z.

(2.2) $|K_1(x, y, z)| \leq K(x, y)|z|$ for every two elements x and y from T and for any real z, $K(x, y)$ being a real function from $L^2(T \times T)$.

(2.3) For every function $h \in L^2(T)$ and for any $x \in T$, the function $\Gamma(y) := K_1(x, y, h(y))$ is μ-measurable.

(3) Two real functions f, g *exist from* $L^2(T)$, *such that the inequality*

$$f \leq g + V_1 f \quad (\textit{respectively, } f \geq g + V_1 f) \tag{7.3.1}$$

holds.
Then the integral equation

$$\psi = g + V_1 \psi \tag{7.3.2}$$

possesses a solution $\varphi \in L^2(T)$ *satisfying the inequality*

$$f \leq \varphi \quad (\textit{respectively, } f \geq \varphi).$$

Proof Assume that the inequality $f \leq g + V_1 f$ holds. Conditions (2.2) and (2.3) imply that

$$V_1 : L^2(T) \to L^2(T)$$

which allows us to consider the sequence $\{f_n\}$ of real functions from $L^2(T)$ defined in the following way:

$$f_0 := f, \; f_{n+1} := g + V_1 f_n.$$

We shall show that for every $x \in T$, the sequence of real numbers $\{f_n(x)\}$ is convergent and if $\varphi(x) := \lim_{n\to+\infty} \{f_n(x)\}$, then $\varphi(x) \in L^2(T)$ satisfies (7.3.2) and inequality $\varphi(x) \leq g(x) \cdot \exp(\int_{T_x} K(x,y)d\mu(y))$. First we shall show that the sequence $\{f_n(x)\}$ is monotonely increasing. Indeed, for $n = 0$, inequality $\chi(x,\cdot)((f(\cdot) - g(\cdot)) - Vf(\cdot)) \leq 0$ yields that $f_0 \leq f_1$.

If we assume that for some natural number k, the inequality $f_k \leq f_{k+1}$ holds, then condition (2.1) implies that $V_1 f_k \leq V_1 f_{k+1}$. But then

$$f_{k+1} = g + V_1 f_k \leq g + V_1 f_{k+1} = f_{k+2}.$$

Now we shall show that the sequence $\{f_n(x)\}$ is bounded. To this end, consider the sequence $\{h_n(x)\}$ defined in the following way:

$$h_0 := |f|, \; h_{n+1} := |g| + Vh_n,$$

where $Vh(x) := \int_{T_s} K(x,y)h(y)d\mu(y)$ (i.e., V is obtained by substituting ϕ by E, the identical operator in $L^2(T)$ in $Vf(x) := \int_T \chi(x,y)K(x,y)\phi f(y)d\mu(y) = \int_{T_x} K(x,y)\phi f(y)d\mu(y)$).

We shall show that $|f_n| \leq h_n$. Indeed, $|f_0| = |f| = h_0$. If we assume that $|f_k| \leq h_k$, then

$$\begin{aligned}|f_{k+1}(x)| &= |g(x) + V_1 f_k(x)| \leq |g(x)| + |V_1 f_k(x)| \\ &= |g(x)| + |\int_{T_s} K_1(x, y, f_k(y)) d\mu(y)| \\ &\leq |g(x)| + \int_{T_s} |K_1(x, y, f_k(y))| d\mu(y) \\ &\leq |g(x)| + \int_{T_s} K_1(x, y)|f_k(y)| d\mu(y) \\ &\leq |g(x)| + \int_{T_s} K(x, y) h_k(y) d\mu(y) \\ &= |g| + Vh_k(x) = h_{k+1}(x).\end{aligned}$$

Thus we have proved that $|f_n| \leq h_n$. On the other hand, the proof of Theorem 3.3.6 in Qin [557] yields that $\lim_{n\to+\infty} h_n$ exists ($\lim_{n\to+\infty} h_n = \Sigma_{n=0}^{+\infty} V^n|g|$). But then $\{h_n(x)\}$, and hence $\{f_n(x)\}$ is also a bounded sequence and since we have already proved that $\{f_n(x)\}$ is monotonely increasing, then it is convergent. By $\varphi(x)$ denote the limit of $\{f_n(x)\}$, i.e.,

$$\varphi(x) = \lim_{n\to+\infty} f_n(x)$$

Obviously, $f_n \leq \varphi \leq \lim_{n\to+\infty} h_n = \Sigma_{n=0}^{+\infty} V^n|g|$.

However, Theorem 3.3.6 in Qin [557] implies that $\Sigma_{n=0}^{+\infty} V^n|g| \in L^2(T)$ and in view of Levi Theorem, $\varphi \in L^2(T)$.

Analogously, since for a fixed x from T, the sequence of functions $\psi_n(y) := K_1(x, y, f_k(y))$ is monotonically increasing (in view of (2.1)) and $\lim_{n\to+\infty} \psi_n(y) = K_1(x, y, \psi(y))$ (in view of (2.1) and since $\psi_n(y) \leq K_1(x, y, \psi(y))$ (in view of (2.1)), and besides, $K_1(x, y, \psi(y)) \in L^2(T)$ (in view of (2.3)), then again in view of Levi Theorem, it follows that

$$\lim_{n\to+\infty} V_1 f_n(x) = V_1\varphi(x)$$

and therefore, the equality

$$f_{n+1} = g + Vf_n$$

yields that

$$\varphi = g + V_1\varphi$$

i.e., φ is a solution of the integral equation (7.3.2). It remains to prove that φ satisfies the inequality $f \leq \varphi$. But this follows from the fact that $f_n \leq \varphi$ for every n, and from the fact that $f = f_0$.

Now consider the case when $K_1(x, y, z)$ for fixed x and y from T is monotonically increasing function of z, and the inequality $f \geq g + V_1 f$ holds.

In this case,

$$-f(x) \leq -g(x) + \int_{T_s} -K_1(x, y, f(y))d\mu(y).$$

If $K_2(x, y, z) = -K_1(x, y, -z)$, then obviously $K_2(x, y, z)$ for fixed x and y from T is monotonically increasing and continuous function of z and $|K_2(x, y, z)| \leq K(x, y)|z|$. Then, as was already proved, there exists a solution $\varphi \in L^2(T)$ of the equation

$$\psi(x) = -g(x) + \int_{T_s} K_2(x, y, \psi(y))d\mu(y),$$

such that $-f \leq \varphi$. Therefore,

$$\begin{aligned} f(x) \geq -\varphi(x) &= g(x) + \int_{T_s} -K_2(x, y, \varphi(y))d\mu(y) \\ &= g(x) + \int_{T_s} K_1(x, y, -\varphi(y))d\mu(y), \end{aligned}$$

i.e., $\varphi_1 = \varphi$ is a solution of (7.3.2), such that $f \geq \varphi_1$. Thus, Theorem 7.3.1 is proved. □

Remark 7.3.1 While proving Theorem 7.3.1, it was actually proved that if for two real functions f and g from $L^2(T)$, the inequality (7.3.1) holds, then

$$|f| \leq h \tag{7.3.3}$$

where h is the unique (Theorem 3.3.6 in Qin [557]) solution of the equation

$$\begin{aligned} \psi(x) &= |g(x)| + \int_{T_s} K_2(x, y, \psi(y))d\mu(y) \\ &= |g(x)| + V\psi(x), \ (h = \Sigma_{n=0}^{+\infty} V^n |g|). \end{aligned}$$

Definition 7.3.1 The solution $\varphi \in L^2(T)$ of the integral (7.3.2) is called maximal, if for every real solution $\psi \in L^2(T)$ of equation (7.3.2), the inequality $\psi \leq \varphi$ is fulfilled. Analogously, $\varphi \in L^2(T)$ is a minimal solution provided that $\psi \geq \varphi$ for any solution $\psi \in L^2(T)$ of the integral equation (7.3.2).

Theorem 7.3.2 ([576]) *If the conditions of Theorem 7.3.1 are fulfilled, then the integral equation (7.3.2) possesses a maximal solution.*

Proof By ζ denote the class of all real solutions of equation (7.3.2) from $L^2(T)$. (Note that Theorem 7.3.1 implies that $\zeta \neq \varnothing$). According to Remark 7.3.1, the inequality $|\psi| \leq |h|$ holds for every element of ζ. But then the function $s(x) := \sup_{\psi\in\zeta}\{\psi(x)\}$ is from $L^2(T)$ (see, e.g., [212], IV 8.22). We shall show that the function $s(\cdot)$ satisfies the inequality (7.3.1). Indeed, let $x \in T$ and ε be an arbitrary positive number. Then there exists an element $\psi \in \zeta$, such that $s(x) \leq \psi(x) + \varepsilon$. Since $\psi(x) = g(x) + V_1\psi(x) \leq g(x) + V_1 s(x)$, then $s(x) \leq g(x) + V_1 s(x) + \varepsilon$. Hence the inequality $s(x) \leq g(x) + V_1 s(x)$ holds and by Theorem 7.3.1, the integral equation (7.3.2) possesses a solution $\varphi \in L^2(T)$ satisfying the inequality $s \leq \varphi$. Thus Theorem 7.3.2 is proved. □

Corollary 7.3.1 ([576]) *If the conditions of Theorem 7.3.1 are fulfilled, then the integral equation (7.3.2) possesses a minimal solution.*

Indeed, then the kernel $K_2(x, y, z) = -K_1(x, y, -z)$ is a monotonely increasing function of z, and hence the integral equation

$$\psi(x) = -g(x) + \int_{T_s} K_2(x, y, \psi(y))d\mu(y)$$

possesses a maximal solution φ. However, then

$$-\varphi(x) = g(x) + \int_{T_s} K_2(x, y, -\varphi(y))d\mu(y)$$

is a minimal solution of equation (7.3.2).

A sufficient condition for uniqueness of the solution of the integral equation (7.3.2) is supplied by the following theorem.

Theorem 7.3.3 ([576]) *Let the conditions of Theorem 7.3.1 be fulfilled. Then, if for the kernel $K_1(x, y, z)$ for all $x, y \in T$ and $z_1, z_2 \in \mathbb{R}$, the following inequality holds,*

$$|K_1(x, y, z_1) - K_1(x, y, z_2)| \leq K(x, y)|z_1 - z_2|, \qquad K(x, y) \in L^2(T \times T)$$

then the integral equation (7.3.2) possesses a unique real solution $\varphi \in L^2(T)$.

Proof By Theorem 7.3.1, we know that (7.3.2) possesses a real solution from $L^2(T)$.

Let the functions φ and $\psi \in L^2(T)$ be two solutions of the integral equation (7.3.2). Then

$$\begin{aligned}|\varphi(x) - \psi(x)| &= \left|\int_{T_s} (K_1(x, y, \varphi(y)) - K_1(x, y, \psi(y)))d\mu(y)\right| \\ &\leq \int_{T_s} |K_1(x, y, \varphi(y)) - K_1(x, y, \psi(y))|d\mu(y) \\ &\leq \int_{T_s} K(x, y)|\varphi(y) - \psi(y)|d\mu(y).\end{aligned}$$

Then, if $f := |\varphi - \psi|$, and $Vh(x) := \int_{T_s} K(x,y)h(y)d\mu(y)$, then $f \leq Vf$, and in view of Corollary 3.3.7 in Qin [557], $f \leq 0$, and hence $\varphi = \psi$. □

Corollary 7.3.2 ([576]) *If under the conditions of Theorem 7.3.3, for two real functions $f_i, f_s \in L^2(T)$, the inequality $f_i - V_1 f_i \leq f_s - V_1 f_s$ holds, then $f_i \leq f_s$.*

Indeed, if $g := f_i - V_1 f_i$, then, in view of Theorem 7.3.3 and Theorem 7.3.1, for the unique solution φ of the integral equation (7.3.2), the inequalities $f_i \leq \varphi$ and $f_s \geq \varphi$, are fulfilled.

To close this subsection, we shall consider Volterra type nonlinear integral operators defined in the space $L^2(T)$, where $T = [t_0, t_1)$ (t_1 may be $+\infty$ as well). The space T is considered with respect to the usual topology, and μ denotes the Lebesgue measure. Let, as before, $T_x = [t_0, t(x))$, where $t(\cdot)$ is a continuous real function defined in T, such that for every $x \in T$, the inequalities $t_0 \leq t(x) \leq x$ hold. It will be assumed that $K(\cdot,\cdot)$ is a non-negative function from $L^2(T \times T)$ and that $h(\cdot)$ is a continuous real function defined in $\mathbb{R}$. By V_1 denote the integral operator defined in $L^2(T)$ in the following way

$$V_1 f(x) := \int_{t_0}^{t(x)} K(x,y)h(f(y))d\mu(y).$$

Lemma 7.3.1 *Let h be a Lipschitzian, monotone increasing real function defined in $\mathbb{R}$. Then, if for two real functions $f, g \in L^2(T)$, the inequality*

$$f \leq g + V_1 f \quad (\textit{respectively, } f \geq g + V_1 f) \tag{7.3.4}$$

holds, then the integral equation

$$\psi = g + V_1 \psi \tag{7.3.5}$$

possesses a unique solution $\varphi \in L^2(T)$, which satisfies the inequality

$$f \leq \varphi \quad (\textit{respectively, } f \geq \varphi) \tag{7.3.6}$$

for the function f.

Proof Obviously, for the space T, conditions (C1)–(C4) in as before hold. Let $h_0(x) := h(x) - h(0)$,

$$g_1(x) := g(x) + \int_{t_0}^{t(x)} K(x,y)d\mu(y) \text{ and } K_1(x,y,z) := K(x,y)h_0(z).$$

It is not difficult to verify that for kernel $K_1(x, y, z)$, the conditions (2) of Theorem 7.3.1 are fulfilled. Inequality (7.3.4) implies that

$$f(x) \leq g_1(x) + \int_{t_0}^{t(x)} K_1(x, y, f(y))d\mu(y)$$

(i.e., condition (3) of Theorem 7.3.1 for the functions f and g_1) holds, and hence the equation

$$\begin{aligned}\psi(x) &= g_1(x) + \int_{t_0}^{t(x)} K_1(x, y, \psi(y))d\mu(y) \\ &= g(x) + \int_{t_0}^{t(x)} K(x, y)h(\psi(y))d\mu(y)\end{aligned}$$

possesses a solution $\varphi \in L^2(T)$ which satisfies inequality (7.3.6).

On the other hand, since for the kernel $K_1(x, y, z)$ the inequality

$$|K_1(x, y, z_1) - K(x, y, z_2)| \leq K(x, y)|h_0(z_1) - h_0(z_2)| < lK(x, y)|z_1 - z_2|$$

holds. (Here l denotes a constant such that for every two reals z_1, and z_2, the inequality $|h(z_1) - h(z_2)| < l|z_1 - z_2|$ holds). But then, in view of Theorem 7.3.3, the integral equation (7.3.5) possesses a unique solution. □

By T^Γ denote the finite sub-interval $[t_0, \Gamma)$ of $T = [t_0, +\infty)$. If f is a function defined in T, then by f_Γ denote the restriction of f on the interval T^Γ .

Theorem 7.3.4 ([576]) *Let the following conditions be fulfilled:*

(1) h is a monotone increasing real function defined in $\mathbb{R}$ which is Lipschitzian in every finite interval.
(2) g is real function defined in $T = [t_0, +\infty)$, for which two real functions f_i and f_s exist, defined in T, whose restrictions g_Γ, $f_{i\Gamma}$ and $f_{s\Gamma}$ on every finite sub-interval $T^\Gamma = [t_0, \Gamma)$ of T belong to $L^2(T^\Gamma)$, and

$$f_{i\Gamma} - V_1 f_{i\Gamma} \leq g_\Gamma \leq f_{s\Gamma} - V_1 f_{s\Gamma}$$

Then

(1) A unique real function φ exists defined in T whose restriction φ_Γ on every finite interval T^Γ is bounded, it belongs to $L^2(T^\Gamma)$ and is a solution of the integral equation $\psi = g_\Gamma + V_1\psi$.
(2) The inequalities $f_i \leq \varphi \leq f_s$ hold.

Proof Let us associate the function

$$h_n(x) = \begin{cases} h(-n), & \text{if} \quad x \le -n \\ h(x), & \text{if} \quad -n \le x \le n \\ h(n), & \text{if} \quad x \ge n \end{cases}$$

to every natural number n.

Then, if $T^\Gamma = [t_0, \Gamma)$ is an arbitrary finite sub-interval of T, then since $f_{i\Gamma}$ and $f_{s\Gamma}$ are bounded, for all sufficiently s large n, $h_n(f_{i\Gamma}) = h(f_{i\Gamma})$ and $h_n(f_{s\Gamma}) = h(f_{s\Gamma})$. Hence it follows that for such n, the inequalities

$$f_{i\Gamma} - \int_{t_0}^{t(x)} K(x,y)h_n(f_{i\Gamma})d\mu(y) \le g_\Gamma(x) \le f_{s\Gamma} - \int_{t_0}^{t(x)} K(x,y)h_n(f_{s\Gamma})d\mu(y)$$

hold, and hence, in view of Lemma 7.3.1, the integral equation

$$\psi(x) = g_\Gamma(x) + \int_{t_0}^{t(x)} K(x,y)h_n(\psi(y))d\mu(y)$$

possesses a unique solution $\varphi_{n\Gamma} \in L^2(T^\Gamma)$ which satisfies the inequalities $f_{i\Gamma} \le \varphi_{n\Gamma} \le f_{s\Gamma}$. This and the fact that h is monotone function imply that $h_n(\varphi_{n\Gamma}) = h(\varphi_{n\Gamma})$ for all sufficiently large n. However, in this case, in view of Lemma 7.3.1, for all sufficiently large n and m, $\varphi_{n\Gamma} = \varphi_{m\Gamma}$ and hence $\lim_{n\to+\infty} \varphi_{n\Gamma}$ exists. Then if $\varphi_\Gamma := \lim_{n\to+\infty} \varphi_{n\Gamma}$, then $\varphi_\Gamma = g_\Gamma + V_1\varphi_\Gamma$ and $f_{i\Gamma} \le \varphi_\Gamma \le f_{s\Gamma}$.

On the other hand, if $T^s \subset T^\Gamma$, then the corresponding solution $\varphi_s \in L^2(T^s)$ of the equation $\varphi_s = g_s + V_1\varphi_s$, will coincide, as is implied by Lemma 7.3.1, with the restriction of φ_Γ on the interval T^s.

Now let ψ be another real function defined in T whose restriction ψ_Γ, on every finite interval T^Γ, is bounded and belongs to $L^2(T^\Gamma)$, and $\psi_\Gamma = g_\Gamma + V_1\psi_\Gamma$. Since ψ_Γ is bounded, then $h_n(\psi_\Gamma) = h(\psi_\Gamma)$ for all sufficiently large n and hence $\psi_\Gamma = g_\Gamma + \int_{t_0}^{t(x)} K(x,y)h_n(\psi_\Gamma)d\mu(y)$.

For all sufficiently large n, φ_Γ also satisfies the same equation and since h_n is a Lipschitzian function, then in view of Lemma 7.3.1, $\varphi_\Gamma = \psi_\Gamma$, whence it follows that $\varphi = \psi$. □

Remark 7.3.2 While proving Theorem 7.3.4, we actually establish that if h is a monotonely real function defined in $\mathbb{R}$, which is Lipschitzian in every finite interval and f_i, f_s are two real functions defined in $[t_0, +\infty)$ whose restrictions $f_{i\Gamma}$, $f_{s\Gamma}$ on every finite interval $[t_0, \Gamma)$ are bounded, they belong to $L^2[t_0, \Gamma)$ and satisfy the inequality $f_{i\Gamma} - V_1 f_{i\Gamma} \le f_{s\Gamma} - V_1 f_{s\Gamma}$, then $f_i \le f_s$.

Theorem 7.3.5 ([576]) *Let h be a monotone increasing, continuous real function defined in $\mathbb{R}$ and let g, f_i, f_s be such real functions defined in $T = [t_0, +\infty)$ that* *for their restrictions* g_Γ, $f_{i\Gamma}$, $f_{s\Gamma}$ *on every finite interval* $T^\Gamma = [t_0, \Gamma)$, *the following conditions hold:*

(1) $g_\Gamma, f_{i\Gamma}, f_{s\Gamma} \in L_2(T^\Gamma)$,
(2) $f_{i\Gamma}$ *and* $f_{s\Gamma}$ *are bounded,*
(3) there exists an $\epsilon > 0$ *such that*

$$f_{i\Gamma} - V_1 f_{i\Gamma} \leq g(x) \leq g(x) + \epsilon \int_{t_0}^{t(x)} K(x,y) d\mu(y)$$
$$\leq f_{s\Gamma} - V_1 f_{s\Gamma}.$$

Then a real function φ *defined in* T *exists, such that*

$$\varphi(x) = g(x) + \int_{t_0}^{t(x)} K(x,y) h(\varphi(y) d\mu(y)$$

and

$$f_i \leq \varphi \leq f_s.$$

Proof First note that a sequence of monotone functions exists, such that it tends to h on every finite interval, the functions from this sequence being Lipschitzian in every similar interval. Indeed, the sequence

$$l_n(x) := n \int_x^{x+1/n} h(y) d\mu(y)$$

consists of monotone functions that are continuously differentiable (since h is monotone and continuous) and hence everyone of them is Lipschitzian in every finite interval. The fact that the sequence $\{l_n\}$ tends uniformly to h on every finite interval is implied by the uniform continuity of h in such intervals. It is clear that if h is monotonically increasing, then all the functions l_n are also increasing. It is not difficult to deduce that we can find a monotonically decreasing sequence $\{h_n\}$ of functions, possessing the listed properties, convergent to the function h.

Now let $T^\Gamma = [t_0, \Gamma)$ be an arbitrary finite sub-interval of T. Then, conditions (2) and (3) imply that for all sufficiently large n, the inequality

$$g_\Gamma(x) \leq f_s(s) + \int_{t_0}^{t(x)} K(x,y) h_n(f_{s\Gamma}(y)) d\mu(y)$$

holds for almost any $x \in T^{\Gamma}$. On the other hand, since $h_n \geq h$ and $K(x, y) \geq 0$, then

$$f_{i\Gamma}(x) - \int_{t_0}^{t(x)} K(x, y)h_n(f_{i\Gamma}(y))d\mu(y)$$

$$\leq f_{i\Gamma}(x) - \int_{t_0}^{t(x)} K(x, y)h(f_{i\Gamma}(y))d\mu(y) \leq g_{\Gamma}(x)$$

for almost every $x \in T^{\Gamma}$. But then, in view of Lemma 7.3.1, a unique solution $\varphi_{n\Gamma} \in L_2(T^{\Gamma})$ of the equation

$$\varphi_{n\Gamma}(x) = g_{\Gamma}(x) + \int_{t_0}^{t(x)} K(x, y)h_n(\varphi_{n\Gamma})d\mu(y) \tag{7.3.7}$$

exists.
Moreover, the inequalities

$$f_{i\Gamma} \leq \varphi_{n\Gamma} \leq f_{s\Gamma}$$

hold, whence, since Γ is arbitrary, the inequality $f_i \leq f_s$ is obtained.

Since the sequence $\{h_n\}$ is monotonically decreasing, then the sequence $\{\varphi_{n\Gamma}\}$ is also monotonically decreasing. Indeed,

$$\varphi_{n+1\Gamma} = g_{\Gamma}(x) + \int_{t_0}^{t(x)} K(x, y)h_{n+1}(\varphi_{n+1}(y))d\mu(y)$$

$$\leq g_{\Gamma}(x) + \int_{t_0}^{t(x)} K(x, y)h_n(\varphi_{n+1}(y))d\mu(y)$$

whence, in view of Lemma 7.3.1, $\varphi_{n+1\Gamma} \leq \varphi_{n\Gamma}$.

Then, since $\{\varphi_{n\Gamma}\}$ is bounded and $\{h_n\}$ tends uniformly to h on T^{Γ}, then in (7.3.7) a boundary transition by n can be carried out. Hence, if $\varphi_{\Gamma} := \lim_{n\to+\infty} \varphi_{n\Gamma}$, then $\varphi_{\Gamma} = g_{\Gamma} + V_1\varphi_{\Gamma}$ and $f_{i\Gamma} \leq \varphi_{\Gamma} \leq f_{s\Gamma}$.

Now, let $T^{\Gamma} \subset T^p = [t_0, p)$, and, besides,

$$\varphi_{np}(x) = g_p(x) + \int_{t_0}^{t(x)} K(x, y)h_n(\varphi_{np})d\mu(y).$$

In view of Lemma 7.3.1, the restriction of φ_{np} on T^{Γ} will coincide with $\varphi_{n\Gamma}$, and therefore the restriction of φ_p on T^{Γ} will coincide with φ_{Γ}. Hence $\varphi = \lim_{p\to+\infty} \varphi_p$

exists and

$$\varphi(x) = g(x) + \int_{t_0}^{t(x)} K(x,y)h(\varphi(y))d\mu(y).$$

The proof is complete. □

7.3.2 Nonlinear Multi-Dimensional Discontinuous Integral Systems of Inequalities

In the sequel, we shall concern the problem of obtaining explicit upper bounds for solutions $u(x)$, of linear or nonlinear Volterra integral equations

$$u(x) = f(x) + \int_a^x K(x,t)u(t)dt + \int_a^x g(x,t,u(t))dt, \tag{7.3.8}$$

or the corresponding inequalities

$$u(x) \leq f(x) + \int_a^x K(x,t)u(t)dt + \int_a^x g(x,t,u(t))dt. \tag{7.3.9}$$

We also prove some global existence theorems for (7.3.8) in both the bounded, measurable case and the L^2-case of f, K, g. Most of these results, deal with the unperturbed cases where either $g \equiv 0$ or $K \equiv 0$.

Here, $u(x) = (u_1(x), \cdots, u_N(x))^T$ and $x = (x_1, \cdots, x_n)$, so we are dealing with systems of N equations or inequalities in n independent variables. If either $n = 1$ or $N = 1$, (7.3.8) and (7.3.9) have been dealt with by many authors. For example, if $n = 1, N \geq 1$, a basic reference is Miller [393]; see also Tricomi [649] and Mikhlin [391]. The case $N = 1, n \geq 1$ of (7.3.9) has been considered as so-called Gronwall-type inequalities or so by Pachpatte and his coworkers [96, 97, 476, 613], by Yeh [701, 703], Young [709–711], and others [57, 59, 286, 643]. Some special cases with $N > 1$ and $n \geq 1$ have been handled by Greene [253], Das [183], and Shinde and Pachpatte [602], and by Chandra and Davis [135], Beesack [60], and Conlan and Wang [161] who considered general N, n.

We consider linear equations, that is, (7.3.8) with $g \equiv 0$. We shall use a lemma proved in [59] to give easy proofs of existence theorems and related matters both for the bounded, measurable case and the Lebesgue square integrable case. These theorems seem to be part of the folklore for linear Volterra equations but, for general $n \geq 1$ at least, complete proofs seem to be hard to come by. The special case having $K(x,t) = G(x)H(t)$ was, however, dealt with by Chandra and Davis [135]. We obtain the Neumann series solution for the solution vector $u(x)$, as well as for the general matrix resolvent kernel $\Gamma(x,t)$. A number of explicit bounds are given for $|\Gamma(x,t)|$ and $|u(x)|$. One of the most interesting of these involves a generalized exponential function,

$$\exp_n(z) = \sum_{r=0}^{+\infty} z^r/(r!)^n,$$

where n is the dimension of the space of $x = (x_1, \ldots, x_n)$. Some estimates of this function which is actually a generalized hypergeometric function-are given in Remark 7.1.3 in Qin [557].

In the sequel, we shall consider general systems of N nonlinear Volterra integral equations in the independent variable $x = (x_1, \ldots, x_n) \in \mathbb{R}^n$, namely,

$$u_i(x) = f_i(x) + \int_a^x g_i(x, t, u_1(t), \ldots, u_N(t))dt, \quad 1 \le i \le N, \tag{7.3.10}$$

or in vector form,

$$u(x) = f(x) + \int_a^x g(x, t, u(t))dt. \tag{7.3.11}$$

We let $J = [a, b]$ and $T = \{(x, t) : a \le t \le x \le b\}$, and suppose that f is either bounded and measurable on J is bounded, or in $L^2(J)$ where J need not be bounded. As for g, we shall always assume that g satisfies the basic hypotheses:

(H) g is measurable on $D = T \times \mathbb{R}^N$, and $g(x, t, u)$ is a continuous function of u on $\mathbb{R}^N$ for each $(x, t) \in T$.

This will, for example, assume that $g(x, t, u(t))$ is measurable on T for each measurable function u on J. In next two existence theorems for (7.3.10), we shall add additional hypotheses for g. In cases where u is non-negative on J, (H) can be modified to (H^+) by replacing $\mathbb{R}^N$ by $\mathbb{R}^N_+$.

The first theorem is, essentially, the n-dimensional version of Theorem 7.1.2 in Qin [557] of the basic monograph [393] by Miller, but obtains a global solution on J rather than just a local solution.

Theorem 7.3.6 ([62]) *Let $J = [a, b]$ be bounded, and f be bounded and measurable on J. Suppose that g satisfies (H_0), that $g(x, t, f(t))$ is bounded on T and that, relative to a vector norm $|\cdot|$ on $\mathbb{R}^N$,*

$$|g(x, t, u) - g(x, t, \bar{u})| \le k(x, t)|u - \bar{u}|, \quad \text{for all } (x, t) \in T, u, \bar{u} \in \mathbb{R}^N, \tag{7.3.12}$$

where k is bounded and measurable on GT. Then equation (7.3.11) has a unique bounded, measurable solution, u on J.

Proof We shall use the Banach fixed-point theorem of the Banach space $(X, \|\cdot\|)$ consisting of all bounded, measurable functions $v = (v_1, \ldots, v_N)$ on J, with norm $\|\cdot\|$ defined by

$$\|v\| = \sup_{x \in J}\{F(x)|v(x)|\}, \quad v \in X, \tag{7.3.13}$$

for an appropriate bounded measurable function $F(x)$ which is bounded away from zero on J. In (7.3.13), the vector norm is that in (7.3.12). To construct F, we proceed as follows: Let $\phi : X \to X$ be the mapping defined by

$$\phi v(x) = f(x) + \int_a^x g(x, t, v(t))dt, \quad x \in J, \quad v \in X. \tag{7.3.14}$$

By setting $u = v(t), \bar{u} = f(t)$ in (7.3.12), we see that $f(x, t, v(t))$ is bounded (as well as measurable) on T, so that $\phi v \in X$ when $v \in X$. If $u, v \in X$, then by (7.3.12)

$$|\phi u(x) - \phi v(x)| \le \int_a^x k(x, t)|u(t) - v(t)|dt,$$

so for any $F > 0$ a. e. on J,

$$\begin{aligned} F(x)|\phi u(x) - \phi v(x)| &\le F(x) \int_a^x \frac{k(x, t)}{F(t)}(F(t)|u(t) - v(t)|)dt \\ &\le \|u - v\| \sup_{x \in J}\Big(F(x) \int_a^x \frac{k(x, t)}{F(t)} dt\Big), \quad x \in J. \end{aligned}$$

Thus we have for any $\rho \in (0, 1)$, for all $u, v, \in X$,

$$\|\phi u - \phi v\| \le \rho\|u - v\|, \tag{7.3.15}$$

provided that we can choose F so that, for all $x \in J$,

$$F(x) \int_a^x (k(x, t)/F(t))dt \le \rho.$$

Set $F(x) = 1/G(x)$, and choose G to be a solution of the linear Volterra equation

$$\int_a^x k(x, t)G(t)dt = \rho(G(x) - 1), \quad x \in J,$$

or

$$G(x) = 1 + \rho^{-1} \int_a^x k(x, t)G(t)dt, \quad x \in J.$$

By Theorem 7.1.2 in Qin [557], this equation has a unique, bounded measurable solution $G = G_\rho(x)$ given by

$$G(x) = 1 + \int_a^x \gamma(x,t)dt\ (\geq 1), \quad x \in J,$$

where γ (≥ 0) is the resolvent kernel of $\rho^{-1}k(x,t)$. Thus $F = 1/G$ is bounded and measurable on J, and also bounded away from zero. With this choice of F in (7.3.13), (7.3.15) holds. Thus ϕ is a contraction mapping on X, and the desired result follows. □

Remark 7.3.3 If $0 \leq k(x,t) \leq M$ on T, so $\rho^{-1}k(x,t) \leq M\rho^{-1}$, it follows from $|u(x)| \leq |f(x)| + \int_a^x |f(t)|B_t(x)\exp(\int_t^x B_t(x)du)dt$ that

$$1 \leq G(x) \leq 1 + M\rho^{-1}\Big\{\exp\Big(M\rho^{-1}\prod_{i=1}^n (x_i - a_i)\Big) - 1\Big\}$$

from which explicit bounds can be obtained for $F(x)$ in (7.3.13). To see that, the space X with norm defined by (7.3.13) is complete, let $\{u_m\}$ be a Cauchy sequence in $(X, \|\cdot\|)$. Since F is bounded away from zero,

$$\sup_{x\in J}|u_n(x) - u_m(x)| \to 0 \quad \text{as} \quad n, m \to +\infty,$$

and so $u(x) = \lim_{m\to+\infty} u_n(x)$ exists on J. Thus u is measurable and bounded on J, so $u \in X$. Also,

$$\sup_{x\in J}|u_n(x) - u(x)| \to 0 \quad \text{as} \quad n \to +\infty,$$

and so, by the boundedness of F, $\|u_n - u\| \to 0$, proving completeness.

Remark 7.3.4 The technique used in (7.3.13) of modifying the usual metric in order that ϕ becomes a contraction mapping on all of J, appears to have been introduced first by Bielecki [80] in 1956 for differential equations. It has been used sporadically since then for differential equations although as noted by Bielecki in [80] without giving any details,

Remark 7.3.5 The usual modification of Theorem 7.1.4 in Qin [557] when $g(x,t,u)$ is defined only for a bounded set, say $(x,t,u) \in T \times D$, where

$$D = \{u = (u_1, \ldots, u_n) : |u - f(x)| \leq M, x \in J\},$$

also applies here. We now only obtain a local theorem. Specifically, we obtain a unique bounded measurable solution of (7.3.11) on a sub-interval $\bar{J} = [a, \bar{b}]$ of J, where $\bar{b}$ is chosen so that

$$\int_a^{\bar{b}} |g(x,t,f(t))|dt + M\int_a^{\bar{b}} k(x,t)dt \le M.$$

Now we need to replace X by the metric space of all bounded, measurable functions v on $\bar{J}$ having $|v(x) - f(x)| \le M$ for all $x \in \bar{J}$, with metric d defined by

$$d(u,v) = \sup_{x\in J}\{F(x)|u(x) - v(x)|\}.$$

We still have $\phi(X) \subset X$ since for all $v \in X$,

$$\begin{aligned} |\phi v(x) - f(x)| &\le \int_a^x |g(x,t,v(t))|dt \\ &\le \int_a^x |g(x,t,f(t))|dt + \int_a^x k(x,t)|v(t) - f(t)|dt \le M \end{aligned}$$

holds for all $x \in \bar{J}$. The rest of the proof proceeds as before.

We now obtain bounds for the unique solution u of equation (7.3.10).

Theorem 7.3.7 ([62]) *Let $|\cdot|$ be any vector norm such that $|v_i| \le |v|$ for $1 \le i \le N$. Under the hypotheses of Theorem 7.3.6, the unique solution u of equation (7.3.10) satisfies*

$$u_i(x) \le f_i(x) + G_i(x) + \int_a^x \gamma(x,s)|G(s)|ds, \quad 1 \le i \le N, \quad x \in J, \quad (7.3.16)$$

where γ is the resolvent kernel of the kernel of k of (7.3.12), and

$$(G_1(x),\ldots,G_N(x)) = G(x) = \int_a^x g(x,t,f(t))dt. \quad (7.3.17)$$

Proof From (7.3.10), for $1 \le i \le N$ and $x \in J$, we get

$$\begin{aligned} u_i(x) &\le f_i(x) + G_i(x) + \int_a^x (g_i(x,t,u(t)) - g_i(x,t,f(t)))dt \\ &\le f_i(x) + G_i(x) + \int_a^x (g(x,t,u(t)) - g(x,t,f(t)))dt \\ &\le f_i(x) + G_i(x) + \int_a^x k(x,t)|u(t) - f(t)|dt. \end{aligned}$$

Also, by (7.3.11),

$$|u(x) - f(x)| \leq G(x) + \int_a^x k(x,t)|u(t) - f(t)|dt, \quad x \in J.$$

Hence by the case $N = 1$ of Theorem 7.1.3 in Qin [557], for all $x \in J$,

$$|u(x) - f(x)| \leq G(x) + \int_a^x \gamma(x,t)|G(t)|dt.$$

The preceding inequality therefore yields

$$\begin{aligned} u_i(x) &\leq f_i(x) + G_i(x) + \int_a^x k(x,t)\Big(|G(t)| + \int_a^t |G(s)|ds\Big)dt \\ &= f_i(x) + G_i(x) + \int_a^x k(x,t)|G(t)|dt + \int_a^x \Big(\int_s^x k(x,t)\gamma(t,s)dt\Big)|G(s)|ds. \end{aligned}$$

This reduces to (7.3.16) on using the resolvent equation $\Gamma(x,t) = K(x,t) + \int_0^x K(x,s)\Gamma(s,t)ds = K(x,t) + \int_0^x \Gamma(s,t)K(x,s)ds$, $(x,t) \in T$ for γ $(N = 1)$. □

Remark 7.3.6 Any of the explicit bounds $|\Gamma(x,t)| \leq B(t)\exp_n[B(t)\prod_1^n(x_i - t_i)]$, $|\Gamma(x,t)| \leq B(t)\exp(\int_t^x B_t(u)du)$, $|\Gamma(x,t)| \leq |K(x,t)| + D(x)E(t)\exp(\int_t^x D(u)du)$, $|\Gamma(x,t)| \leq |K(x,t)| + D_1(x)E_1(t)\exp(\int_t^x E_1(u)du)$, $(x,t) \in T$ given for $|\Gamma(x,t)|$ apply equally to the scalar case $\gamma(x,t)$, provided that $|K(x,t)|$ in these bounds is replaced here by $k(x,t)$. These bounds may be used with (7.3.16) to give explicit upper bounds for the u_i, but we omit these. Similarly, but now by taking $N = 1$ and $K(x,t) = k(x,t)$, the bounds Corollary 7.1.2 in Qin [557] may also be used, as well as the bound

$$\gamma(x,t) \leq g(x)\exp\Big(\int_t^x h(s)g(s)ds\Big)h(t)$$

explicit in Theorem 7.1.4 in Qin [557] when $k(x,t) = g(x,t) = g(x)h(t)$. Since $N = 1$ here, the inequality hypotheses associated with these inequalities are automatically satisfied.

We now turn to the L^2-case of equation (7.3.10). Unfortunately, and perhaps surprisingly, the method of proof used in Theorem 7.3.6 breaks down here, and it may be worthwhile to show why. The present basic assumptions in this case are that $f \in L^2(J)$ and that (7.3.12) holds for some kernel $k \in L^2(T)$. We then have for all u, v *in* $L^2(J)$,

$$|\phi u(x) - \phi v(x)|^2 \leq \Big(\int_a^x k(x,t)|u(t) - v(t)|dt\Big)^2,$$

$$F(x)|\phi u(x) - \phi v(x)|^2 \leq F(x)\int_a^x \frac{k^2(x,t)}{F(t)}dt \cdot \int_a^x F(t)|u(t) - v(t)|^2 dt,$$

provided that F is positive, measurable and bounded. Hence if we use a modified L^2-norm with

$$\|u - v\|^2 = \int_a^b F(t)|u(t) - v(t)|^2 dt,$$

then

$$\|\phi u - \phi v\|^2 \leq \rho^2 \|u - v\|^2$$

will hold provided that $G = 1/F$ can be chosen so that, for all $x \in J$,

$$\int_a^x k^2(x, t)G(t)dt \leq \rho^2 G(x).$$

Here, G must be positive and bounded away form zero. In general, no such G exists. For example, take $n = 1$ and $k^2(x, t) = x^{-3/4}t^{-3/4}, 0 \leq t \leq x \leq 1$. Then $k^2 \in L^1(T)$, but for any measurable G satisfying $G(x) \geq \alpha$ for some $\alpha > 0$, we have

$$\rho^2 G(x) \geq \alpha \int_0^x x^{-3/4}t^{-3/4} dt = 4\alpha x^{-1/2}, \quad \text{or} \quad G(x) \geq 4\alpha\rho^{-2}x^{-1/2},$$

but then $\int_a^x k^2(x, t)G(t)dt$ clearly does not exist. It is now clear a fortiori, that the integral equation

$$G(x) = 1 + \rho^{-2} \int_0^x k^2(x, t)G(t)dt$$

so has no positive solution G for general $k \in L^2(T)$.

In order to obtain global existence theorem for the L^2-case, we shall use the L^2 norm $\|\cdot\|$, and show that for some integer $r \geq 1$, the mapping $\phi^r : X \to X$ is a contraction mapping, where $X = L^2(J)$. This is a common procedure and is used, for example, by Hochstadt [283] for the case of linear equations with $N = n = 1$. It is also essentially the method used here.

Theorem 7.3.8 ([62]) *Let $J = [a, b]$ be bounded or unbounded, let $f \in L^2(J)$ and suppose that $g(x, t, u)$ satisfies the basic assumptions (H), that $\int_a^x |g(x, t, f(t))|dt \in L^2(J)$, and that for all $(x, t) \in T,\ u, \bar{u} \in \mathbb{R}^N$,*

$$|g(x, t, u) - g(x, t, \bar{u})| \leq k(x, t)|u - \bar{u}|, \tag{7.3.18}$$

where $k \in L^2(T)$. The equation (7.3.11) has a unique solution $u \in L^2(J)$.

Proof As in the proof of Theorem 7.3.6, we define the mapping $\phi : X \to X$ (with $X = L^2(J)$) by (7.3.14). From (7.3.12), if $v \in L^2(J)$, then for all $(x,t) \in T$,

$$|g(x,t,v(t))| \leq |g(x,t,f(t))| + k(x,t)|v(t) - f(t)|.$$

It follows that $\int_a^x |g(x,t,v(t))|dt \in L^2(J)$ and hence $\phi v(x)$ is in X for each $v \in X$. Since $k \in L^2(T)$, we have

$$A^2(x) = \int_a^x k^2(x,t)dt < +\infty \quad \text{a. e. on} \quad J,$$

with $\|k\|^2 = \int_J A^2 dx < +\infty$. Now if $u, v \in X$, then

$$|\phi u(x) - \phi v(x)|^2 \leq A^2(x) \int_a^x |u(t) - v(t)|^2 dt.$$

By using the following equality,

$$F(0) + \int_0^x f(t)F'\left(\int_a^t fds\right)dt \leq F\left(\int_a^x fdt\right),$$

we easily prove by induction that holds for all $x \in J, r \geq 1$,

$$|\phi^r u(x) - \phi^r v(x)|^2 \leq A^2(x) \int_a^x \left(\int_t^x A^2(u)du\right)^{r-1} |u(t) - v(t)|dt/(r-1)!.$$

Integrating over J, we obtain

$$\begin{aligned}
\|\phi^r u - \phi^r v\|^2 &\leq \int_a^b \int_a^x A^2(x)\left(\int_t^x A^2(u)du\right)^{r-1} |u(t) - v(t)|dtdx/(r-1)! \\
&= \int_a^b \left(\int_t^b A^2(x)\left(\int_t^x A^2(u)du\right)^{r-1} dx\right)|u(t) - v(t)|dt/(r-1)! \\
&\leq \int_a^b \left(\int_t^b A^2(u)du\right)^r |u(t) - v(t)|dt/(r)!
\end{aligned}$$

Using the following equality, we have

$$F(0) + \int_0^x f(t)F'\left(\int_t^x fds\right)dt \leq F\left(\int_a^x fdt\right).$$

Thus

$$\|\phi^r u - \phi^r v\|^2 \leq \|k\|^{2r} \cdot \|u - v\|^2/r! < \frac{1}{2}\|u - v\|^2$$

holds for large enough r. For such r, ϕ^r is a contraction mapping on X so ϕ^r has a unique fixed point in X, which thus proves the theorem. □

Before considering nonlinear Volterra inequalities, we look briefly at the perturbed equation $u(x) = f(x) + \int_a^x K(x,t)u(t)dt + \int_a^t g(x,t,u(t))dt$ which contains linear and nonlinear terms. In order not to obscure the discussion, we do not state any formal theorems. We are now dealing with the N-vector system

$$u(x) = f(x) + \int_a^x K(x,t)u(t)dt + \int_a^x g(x,t,u(t))dt, \quad x \in J, \tag{7.3.19}$$

where $f, K = (k_{ij})$, and g satisfy either the bounded, measurable hypotheses of Theorems 7.3.2 in Qin [557] and 7.3.6 or the L^2 hypotheses of Theorem 7.3.1 in Qin [557] and Theorem 7.3.8.

Set

$$F(x) = f(x) + \int_a^x g(x,t,u(t))dt,$$

so

$$u(x) = F(x) + \int_a^x K(x,t)u(t)dt.$$

By $u(x) = f(x) + \int_a^x \Gamma(x,s)f(s)ds,\ x \in J$, if Γ denotes the resolvent kernel of K, then

$$u(x) = F(x) + \int_a^x \Gamma(x,t)F(t)dt,$$

which can be rewritten in the form

$$u(x) = \bar{f}(x) + \int_a^x G(x,t,u(t))dt, \quad x \in J, \tag{7.3.20}$$

where

$$\begin{cases} \bar{f}(x) = f(x) + \int_a^x \Gamma(x,t)F(t)dt, & (7.3.21)\\ G(x,t,u) = g(x,t,u) + \int_a^x \Gamma(x,s)g(s,t,u)ds. & (7.3.22)\end{cases}$$

Equation (7.3.20) is of the same type as (7.3.11). Moreover, if g satisfies (7.3.12), then G satisfies a condition of the same kind since

$$|G(x,t,u) - G(x,t,\bar{u})| \leq |g(x,t,u) - g(x,t,\bar{u})| + \int_t^x \Gamma(x,s)|g(s,t,u) - g(s,t,\bar{u})|ds$$
$$\leq \Big(k(x,t) + \int_t^x |\Gamma(x,s)|k(s,t)ds\Big)|u - \bar{u}| \equiv \hat{k}(x,t)|u - \bar{u}|,$$

has precisely the same character as k and $|\Gamma|$. Thus any equation of type (7.3.11) perturbed by the addition of a linear term is actually equivalent to another equation of the same kind.

Suppose now that $u^{(1)}$ and $u^{(2)}$ are the unique solutions of two equations of the form (7.3.19) in which only the free terms, say $f^{(1)}$ and $f^{(2)}$ are different. If $\hat{f}_1, \hat{f}_2$ are the corresponding terms defined by (7.3.21), and if γ denotes the resolvent kernel of $\hat{k}$, then from

$$|u^{(1)}(x) - u^{(2)}(x)| \leq |\hat{f}_1(x) - \hat{f}_2(x)| + \int_a^x \hat{k}(x,t)|u^{(1)}(t) - u^{(2)}(t)|dt,$$

we readily obtain by Theorem 7.1.3 in Qin [557],

$$|u^{(1)}(x) - u^{(2)}(x)| \leq |\hat{f}_1(x) - \hat{f}_2(x)| + \int_a^x \hat{\gamma}(x,t)|\hat{f}_1(t) - \hat{f}_2(t)|dt.$$

If we set $\Delta f(x) = f^{(1)}(x) - f^{(2)}(x), \Delta u(x) = u^{(1)}(x) - u^{(2)}(x)$, we obtain a perturbation estimate

$$|\Delta u(x)| \leq |\Delta f(x)| + \int_a^x \Big(\hat{\gamma}(x,t) + |\Gamma(x,t)| + \int_t^x \hat{\gamma}(x,s)|\Gamma(s,t)|ds\Big)|\Delta f(t)|dt. \tag{7.3.23}$$

Upper bounds can be given for the resolvent kernels $\hat{\gamma}, |\Gamma|$ by using estimates of in Theorem 7.1.2 in Qin [557].

The technique used in (7.3.19)–(7.3.23) has been applied by Corduneanu [171] in the case $N = 1, n = 2$ to obtain a stability theorem.

Now we consider the system of inequalities

$$u_i(x) \leq f_i(x) + \int_a^x g_i(x,t,u_1(t),\ldots,u_N(t))dt, \quad 1 \leq i \leq N, \tag{7.3.24}$$

or

$$u(x) \leq f(x) + \int_a^x g(x,t,u(t))dt, \quad x \in J = [a,b]. \tag{7.3.25}$$

We now always assume that J is bounded, and that f, g satisfy the bounded measurable assumptions of Theorem 7.3.6.

Theorem 7.3.9 (The Beesack Inequality [62]) *Under the hypotheses of Theorem 7.3.6, suppose that u is bounded and measurable on J and satisfies (7.3.25). Let $|\cdot|$ be any vector norm on $\mathbb{R}^N$ such that $|v_i| \le |v|$ for all $1 \le i \le N$ and $|u| \le |v|$ holds provided that $|u_i| \le |v_i|$ for all $1 \le i \le N$. If each component, $g_i(x,t,u)$, is non-negative and non-decreasing in u, then*

$$u_i(x) \le f_i(x) + G_i(x) + \int_a^x \gamma(x,s)|G(s)|ds, \quad 1 \le i \le N, \quad x \in J, \qquad (7.3.26)$$

where γ, G are as in Theorem 7.3.7.

Proof As in the proof of Theorem 7.1.3 in Qin [557], u satisfies an integral equation

$$u(x) = \bar{f}(x) + \int_a^x g(x,t,u(t))dt, \quad x \in J,$$

where $\bar{f}(x) \le f(x)$ on J, and $\bar{f}$ is also bounded and measurable. By Theorem 7.3.7, we get

$$u_i(x) \le \bar{f}_i(x) + \bar{G}_i(x) + \int_a^x \gamma(x,s)|\bar{G}(s)|ds, \quad 1 \le i \le N, \quad x \in J,$$

where

$$0 \le \bar{G}_i(x) = \int_a^x g_i(x,t,\bar{f}(t))dt \le \int_a^x g_i(x,t,f(t))dt = G_i(x).$$

Hence (7.3.22) follows. □

Remark 7.3.7 The comments of Remark 7.3.6 also apply here to give simpler explicit upper bounds for the $u_i(x)$. Observe that there are no results here as precise as Theorem 7.1.3in Qin [557]. In particular, we do not know whether it is true that if $v(x)$ is a solution of (7.3.25) with equality, then $u(x) \le v(x)$ holds, even under the assumptions on g of Theorem 7.3.9. In the case $n = 1$, a comparison theorem of the form $|u(x)| \le V(x)$ is given in Miller [393], but here V is the solution of a related dominating scalar equation.

We also observe that if f and u are non-negative on J, then $g(x,t,u)$ need only be defined for all $u \ge 0$, and the hypotheses (H) can be replaced by (H^+), as noted after (7.3.11).

Next we consider a more strictly component style version of (7.3.24). Suppose that, as in Theorem 7.3.9, each component $g_i(x,t,u)$ is non-decreasing in u, and that for $1 \le i \le N$,

$$|g_i(x,t,u) - g_i(x,t,v)| \le \sum_{j=1}^N k_{ij}(x,t)|u_j - v_j|, \quad (x,t) \in T, \qquad (7.3.27)$$

where each k_{ij} is non-negative, bounded, and measurable on T. Set $K = (k_{ij})$, and suppose $|K|, |v|$, are compatible norms for which $|u_i| \leq |v_i|$ for $1 \leq i \leq N$ implies $|u| \leq |v|$.

Then by (7.3.27), we have

$$|g(x,t,u) - g(x,t,v)| \leq |K(x,t)(|u_1 - v_1|, \cdots, |u_N - v_N|)^t| \leq |K(x,t)||u - v|$$

for all $(x, t) \in T$ and $u, v \in \mathbb{R}^N$.

That is, (7.3.12) holds with $k = |K|$ bounded and measurable on T. In addition, by (7.3.27) and the non-decreasing character of the g_i, we have for $1 \leq i \leq N$,

$$0 \leq g_i(x,t,u) - g_i(x,t,v) \leq \sum_{j=1}^{N} k_{ij}(x,t)(u_j - v_i), \quad \text{if} \quad v \leq u,$$

or

$$0 \leq g(x,t,u) - g(x,t,v) \leq K(x,t)(u - v), \quad (x,t) \in T, \quad \text{if} \quad v \leq u. \tag{7.3.28}$$

In particular, if $f(t) \leq u(t)$ for all $t \in J$, then

$$g(x,t,u(t)) \leq g(x,t,f(t)) + K(x,t)(u(t) - f(t)), \quad (x,t) \in T. \tag{7.3.29}$$

We now assume that $u(x)$ satisfies (7.3.25), that is, for all $x \in J$,

$$u(x) \leq f(x) + \int_a^x g(x,t,u(t))dt. \tag{7.3.30}$$

If we set $U(x) = \max(u(x), f(x))$, then $f(x) \leq U(x)$ and so by (7.3.25) and the non-decreasing character of g, for all $x \in J$,

$$U(x) \leq f(x) + \int_a^x g(x,t,u(t))dt \leq f(x) + \int_a^x g(x,t,U(t))dt.$$

Hence, by (7.3.29) with u replaced by U, we obtain

$$U(x) \leq f(x) + \int_a^x (g(x,t,f(t)) - K(x,t)f(t))dt + \int_a^x K(x,t)U(t)dt,$$

or for all $x \in J$,

$$U(x) \leq F(x) + \int_a^x K(x,t)U(t)dt, \tag{7.3.31}$$

where, for all $x \in J$,

$$F(x) = \int_a^x (g(x,t,f(t)) - K(x,t)f(t))\,dt. \tag{7.3.32}$$

Observe that if $f(t) \geq 0$ on J, then it follows from (7.3.28) with $v = 0$ that, for all $x \in J$,

$$F(x) \leq \int_a^x g(x,t,0)dt. \tag{7.3.33}$$

We summarize some of these results in the following theorem.

Theorem 7.3.10 (The Beesack Inequality [62]) *Let $J = [a,b]$ be bounded, and let f be bounded and measurable on J, while $g(x,t,u)$ satisfies (H), with $g(x,t,f(t))$ bounded on T. Suppose also that $K = (k_{ij})$ is non-negative, measurable and bounded on T and that (7.3.28) holds. If u is any bounded measurable function on J satisfying the inequality (7.3.25), then for all $x \in J$,*

$$u(x) \leq F(x) + \int_a^x \Gamma(x,t)F(t)dt, \tag{7.3.34}$$

where Γ is the resolvent kernel of K and F is given by (7.3.32).

Proof In fact, the inequality (7.3.28) implies that g is non-decreasing in u. With $U(x) = \max\,(u(x), f(x))$ as above, U is a bounded, measurable solution of (7.3.31). It follows from Theorem 7.1.3 in Qin [557] that for all $x \in J$,

$$U(x) \leq F(x) + \int_a^x \Gamma(x,t)U(t)dt$$

and since $u(x) \leq U(x)$, (7.3.34) follows. □

Corollary 7.3.3 (The Beesack Inequality [62]) *If g satisfies the inequalities (7.3.27) and is non-decreasing in u, then (7.3.34) holds if u satisfies (7.3.25).*

For, as noted above, when g is non-decreasing in u and satisfies (7.3.27), then (7.3.28) holds.

Remark 7.3.8 To obtain specific, simpler bounds for $u(x)$ from (7.3.34), we may use the bounds $\Gamma(x,t) \leq B(t)\exp_n[\prod_1^n(x_i - t_i)B(t)]$, $\Gamma(x,t) \leq K(x,t) + D(x)\exp(\int_t^x D(u)du)E(t)$, $\Gamma(x,t) \leq K(x,t) + D_1(x)\exp(\int_t^x D_1(u)du)E_1(t)$, $(x,t) \in T$, for $\Gamma(x,t)$. In case $f(t) \geq 0$, we may also use (7.3.33). If $|\cdot|$ is a vector norm on $\mathbb{R}^N$ such that $|v_i| \leq |v|$ for $1 \leq i \leq N$, and $|u| \leq |v|$ whenever $|u_i| \leq |v_i|$ for $1 \leq i \leq N$, then if $g(x,t,u)$ is non-negative and non-decreasing in u, we may also obtain the bounds

$$u_i(x) \leq f_i(x) + G_i(x) + \int_a^x \gamma_0(x,s)|G(s)|ds, \quad 1 \leq i \leq N, \quad x \in J,$$

with G as in Theorems 7.3.7, 7.3.9 and γ_0 the resolvent kernel of $|K|$. This follows from Theorem 7.3.9, since, as noted above, (7.3.12) holds with $k = |K|$ under the stated conditions.

We conclude introducing with a Bihari-Gronwall type of integral inequality, where we do not assume any condition on g such as (7.3.12) or (7.3.26). In such cases, we have no a priori assurance of the existence of a solution of either the Volterra equation (7.3.10) or of any related dominating linear equation such as

$$U(x) = F(x) + \int_a^x K(x,t)U(t)dt.$$

We assume that u, f are bounded measurable function on a bounded cell $J = [a, b]$ satisfying the system of integral inequalities

$$u_i(x) \leq f_i(x) + \int_a^x g_i(x,t,u_1(t),\ldots,u_N(t))dt, \quad 1 \leq i \leq N,\ x \in J, \tag{7.3.35}$$

where g satisfies the basic hypotheses (H). In addition, we assume that each $g_i(x,t,u)$ is non-negative and non-decreasing in u for each $(x,t) \in T$, and bounded on T for each $u \in \mathbb{R}^N$.

We define functions $\bar{u}, \bar{f}, \bar{g}$ by

$$\bar{u}(x) = \max_{1\leq i\leq N} u_i(x), \quad \bar{f}(x) = \max_{1\leq i\leq N} f_i(x), \quad x \in J, \tag{7.3.36}$$

$$\bar{g}(x,t,v) = \max_{1\leq i\leq N} g_i(x,t,v,v,\ldots v), \quad \text{for all } (x,t,v) \in T \times \mathbb{R}. \tag{7.3.37}$$

Taking maxima in (7.3.24) over $i = 1,\ldots,N$ yields, for all $x \in J$,

$$\bar{u}(x) \leq \bar{f}(x) + \int_a^x \bar{g}(x,t,\bar{u}(t))dt. \tag{7.3.38}$$

Next, for arbitrary $x \in J$, we define

$$\hat{f}(x) = \sup_{a\leq t\leq x} \bar{f}(x), \quad \hat{g}(x,v) = \sup_{a\leq s\leq t\leq x} \bar{g}(t,s,v). \tag{7.3.39}$$

Then $\hat{f}(x)$ and $\hat{g}(x, v)$ are non-decreasing functions of x and, in addition, $\hat{g}(x, v)$ is non-decreasing in v for each $x \in J$. By (7.3.36), we get, for all $x \in J$,

$$\bar{u}(x) \leq \hat{f}(x) + \int_a^x \hat{g}(x, \bar{u}(t))dt. \tag{7.3.40}$$

It follows that for arbitrary fixed $X \in J$, we have, for all $a \leq x \leq X$,

$$\bar{u}(x) \leq \hat{f}(X) + \int_a^x \hat{g}(X, \bar{u}(t))dt. \tag{7.3.41}$$

Now, set

$$U(x) = \int_a^x \hat{g}(X, \bar{u}(t))dt, \quad a \leq x \leq X,$$

and observe that $U(x) = 0$ if any $x_i = a_i$, and that U is non-decreasing in each x_i, we obtain

$$\bar{u}(x) \leq \hat{f}(x) + U(x), \quad a \leq x \leq X,$$
$$\frac{\partial U(x)}{\partial x_1} = \int_a^{\bar{x}} \hat{g}(X, \bar{u}(x_1, \bar{t}))d\bar{t}, \quad a \leq x = (x_1, \bar{x}) \leq X,$$

where $\bar{a} = (a_2, \cdots, a_n), \bar{t} = (t_2, \cdots, t_n), \bar{x} = (x_2, \cdots, x_n)$. Hence

$$\frac{\partial U(x)}{\partial x_1} \leq \int_{\bar{a}}^{\bar{x}} \hat{g}(X, \hat{f}(X) + U(x_1, \bar{t}))d\bar{t},$$

so

$$\frac{\partial U(x)/\partial x_1}{\hat{g}(X, \hat{f}(X) + U(x))} \leq \int_{\bar{a}}^{\bar{x}} 1 \cdot d\bar{t}, \quad a \leq x = (x_1, \bar{x}) \leq X. \tag{7.3.42}$$

For each $x \in J$, we define the function Φ_x by

$$\Phi_x(r) = \int_{r_0}^r ds/\hat{g}(x, s), \quad r \geq r_0 = \hat{f}(a). \tag{7.3.43}$$

Then by (7.3.42),

$$\frac{\partial}{\partial x_1}\Phi_X(\hat{f}(X) + U(X)) \leq \int_{\bar{a}}^{\bar{x}} 1 \cdot d\bar{t}, \quad a \leq x = (x_1, \bar{x}) \leq X,$$

and so, integrating, we obtain for all $a \le x \le X$,

$$\Phi_X(\hat{f}(X) + U(X)) - \Phi_X(\hat{f}(X)) \le \int_a^x 1 \cdot dt = \prod_{i=1}^{n}(x_i - a_i)$$

Hence, if Φ_x^{-1} denotes the inverse function of Φ_x, then by (7.3.41), we have

$$\hat{u}(x) \le \hat{f}(X) + U(x) \le \Phi_X^{-1}\Big(\prod_{i=1}^{n}(x_i - a_i) + \Phi_X(\hat{f}(X))\Big), \quad a \le x \le X,$$

provided that $\prod_{i=1}^{n}(x_i - a_i) + \Phi_x(\hat{f}(X)) \in Dom\ (\Phi_x^{-1})$. In particular, this holds for $x = X$ provided that $\prod_{i=1}^{n}(X_i - a_i) + \Phi_x(\hat{f}(X)) \in Dom\ (\Phi_x^{-1})$.

Thus making this substitution and then replacing X by x, we obtain for all $a \le x \le \bar{x}\ (\le b)$,

$$\bar{u}(x) \le \Phi_x^{-1}\Big(\prod_{i=1}^{n}(x_i - a_i) + \Phi_x(\hat{f}(x))\Big). \tag{7.3.44}$$

We have thus proved the next theorem.

Theorem 7.3.11 (The Beesack Inequality [62]) *Let u, f be bounded measurable function on a bounded cell $J = [a, b] \subset \mathbb{R}^N$, let $g(x, t, v)$ satisfy the basic hypotheses (H) with each g_i non-negative, and non-decreasing in v for each $(x, t) \in T$ as well as bounded on T for each $v \in \mathbb{R}^N$. Let $\hat{f}, \hat{g}(x, v)$ be defined by (7.3.36), (7.3.37), (7.3.39) and suppose that $\hat{g}(a, \hat{f}(a)) > 0$. If $u(x)$ satisfies the inequality (7.3.24), then for each $i = 1, \ldots, N$,*

$$u_i(x) \le \bar{u}(x) \le \Phi_x^{-1}\Big(\prod_{i=1}^{n}(x_i - a_i) + \Phi_x(\hat{f}(x))\Big), \quad a \le x \le \bar{x}\ (\le b), \tag{7.3.45}$$

where Φ_x is defined for all $x \in J$ by (7.3.43), Φ_x^{-1} is the inverse function of Φ_x, and $\bar{x}$ is chosen so that $\prod_{i=1}^{n}(x_i - a_i) + \Phi_x(\hat{f}(x)) \in Dom\ (\Phi_x^{-1})$ for all $a \le x \le \bar{x}$.

The only point requiring clarification is to note that $\hat{g}(x, s) \ge \hat{g}(a, \hat{f}(a)) > 0$ holds for all $a \ge r_0 = \hat{f}(a)$, so that for each $x \in J$, the function Φ_x is well-defined by (7.3.43), and continuous and strictly increasing.

Corollary 7.3.4 ([62]) *Under the above hypotheses, we also have, for all $1 \le i \le N$, and for all $a \le x \le \bar{x}$,*

$$u(x) \le f_i(x) + \int_a^x g_i(x, t, V(t), \ldots, V(t))dt, \tag{7.3.46}$$

where $V(x) = \Phi_x^{-1}\Big(\prod_1^n(x_i - a_i) + \Phi_x(\hat{f}(x))\Big)$.

Remark 7.3.9 The special case $N = n = 1, f(x) \equiv c > 0$, and $g(x,t,u) = g(u)$ positive and non-decreasing, gives us the well-known Bihari inequality [82]. For further special cases, see [710] and [60] and other references in Remark 5 of [60]. Bihari's case shows that, in general, neither the bound nor the restricted domain in (7.3.45) can be improved. Despite this, there are cases where (7.3.46) will give a better bound than (7.3.45), and where the restricted domain is far from best. Such an example is given by

$$u_i(x) \leq c_i + \int_0^x \prod_{j=1}^{i} u_j(t)dt, \quad 1 \leq i \leq N, \quad x > 0, \tag{7.3.47}$$

where we assume all $c_i > 0$ and $c = \max(c_i) \geq 1$. For upper bounds, we may assume all $u_i(x) \geq 0$. Then $g_i(x,t,u_1,\ldots,u_N) = \prod_{j=1}^{i} u_j$ is non-decreasing in each u_j with this restriction. Using the notation of Theorem 7.3.11, we have $\hat{f}(x) = \bar{f}(x) \equiv c, g_i(x,t,v,\ldots,v) = v^i$, and

$$\hat{g}(x,v) = \bar{g}(x,t,v) = \begin{cases} v, & 0 \leq v \leq 1, \\ v^N, & v \geq 1. \end{cases}$$

Thus

$$\Phi_x(r) = \int_c^r \frac{ds}{\hat{g}(x,s)} = \int_c^r s^{-N} ds = \{c^{1-N} - r^{1-N}\}/(N-1), \quad r \geq c,$$

and

$$\Phi_x^{-1}(y) = \{c^{1-N} - (N-1)y\}^{-1/(N-1)}, \quad \text{if} \quad 0 \leq y < c^{1-N}/(N-1).$$

Thus the theorem gives us the bounds, if $0 \leq \prod_{j=1}^{n} x_j < c^{1-N}/(N-1)$,

$$u_i(x) \leq V(x) = \left(c^{1-N} - (N-1)\prod_{j=1}^{n} x_j\right)^{-1/(N-1)}. \tag{7.3.48}$$

If we recall that $c \geq 1$, the domain is thus very restricted, and the bounds become infinite on this domain. On the other hand, we may solve the system (7.3.47) "explicitly", beginning with $i = 1$, as linear inequalities each time using the upper bounds obtain in the succeeding equation $(i+1)$. This shows, among other things, that we may obtain (bounded) for the u_i on every bounded cell $[a,b]$, so that (7.3.48) is rather weak. In this case, although we can not improve the domain in (7.3.48), we do obtain better bounds on this domain by using (7.3.46). In order to obtain explicit result, we assume that $n = 1$ from now on. Then (7.3.46) becomes

$$u_i(x) \le \begin{cases} c_i + \frac{c^{(-N-1-i)}}{N-1-i} - \frac{1}{N-1-i}(c^{1-N} - (N-1)x)^{(N-1-i)/(N-1)}, \text{ if } i \neq N-1, \\ c_{N-1} - \log c - \frac{1}{N-1}\log(c^{1-N} - (N-1)x), \text{ if } i = N-1, \end{cases}$$

provided that $0 \le x < c^{1-N}/(N-1)$. Since all $c_i \le c$, a simple calculation shows that for $1 \le i \le N$,

$$u_i(x) < V(x) = \{c^{1-N} - (N-1)x\}^{-1/(N-1)} \quad \text{if} \quad 0 \le x < c^{1-N}/(N-1);$$

in fact the largest discrepancy occurs for $x = 0$ in all cases.

Remark 7.3.10 The part of the technique used in the proof of Theorem 7.3.9 which consisted of differentiating $U(x)$ with respect to only one variables, followed by a subsequent single integration, seems to have been first used in a similar context by Bainov and his colleagues [286, 287, 714]. This technique is much simpler than the n-variable methods used in [96, 476, 613, 701, 706, 710] and elsewhere.

Remark 7.3.11 We conclude by showing how we may combine the techniques of Theorem 7.3.11 with that used for the perturbed equation (7.3.19) to obtain a corresponding result for the perturbed inequality

$$u(x) \le f(x) + \int_a^x K(x,t)u(t)dt + \int_a^x g(x,t,u(t))dt, \quad x \in J. \tag{7.3.49}$$

If $K \ge 0$ on T, then applying Theorem 7.1.3 in Qin [557] gives us

$$u(x) \le \varphi(x) + \int_a^x \Gamma(x,t)F(t)dt, \quad \left(\phi(x) = f(x) + \int_a^x g(x,t,u(t))dt\right),$$

which, similarly as (7.3.19), reduces to

$$u(x) \le F(x) + \int_a^x G(x,t,u(t))dt, \quad x \in J, \tag{7.3.50}$$

where

$$F(x) = f(x) + \int_a^x \Gamma(x,t)f(t)dt, \tag{7.3.51}$$

and

$$G(x,t,u) = g(x,t,u) + \int_a^x \Gamma(x,t) g(s,t,u)dt, \quad (x,t) \in T. \qquad (7.3.52)$$

Here, Γ is the resolvent kernel of K, so $\Gamma \geq 0$ on T and it follows that if K is also bounded and measurable on T, then F and G satisfy the hypotheses of Theorem 7.3.11 provided that f and g do.

We now indicate briefly how to obtain (dominating functions for) the functions $\bar{F}, \hat{F}, \bar{G}(x,t,v), \hat{G}(x,v)$ which are required for Theorem 7.3.11, in terms of the functions $\bar{f}, \hat{f}, \bar{g}(x,t,v), \hat{g}(x,v)$.

It is easy to verify

$$\bar{F}(x) = \max_{1\leq i\leq N} F_i(x) \leq \bar{f}(x) + \int_a^x \bar{\Gamma}(x,t)\bar{f}(t)dt \quad \text{if} \quad \bar{f}(t) \geq 0 \quad \text{on } J,$$

where

$$\bar{f}(t) = \max_{1\leq j\leq N} f_j(t), \quad \bar{\Gamma}(x,t) = \max_{1\leq j\leq N} \sum_{j=1}^{N} \Gamma_{ij}(x,t).$$

Also

$$\hat{F}(x) = \sup_{a\leq t\leq x} \bar{F}(x) \leq \hat{f}(x) + \int_a^x \hat{\Gamma}(x,t)\hat{f}(t)dt \equiv \hat{F}_0(x),$$

where

$$\hat{f}(x) = \sup_{a\leq t\leq x} \bar{f}(x), \quad \hat{\Gamma}(x,t) = \sup_{t\leq r\leq s\leq x} \bar{\Gamma}(s,r).$$

Observe that $\hat{\Gamma}$ is non-decreasing in x and non-increasing in t. Next,

$$\bar{G}(x,t,v) = \max_{1\leq i\leq N} G_i(x,t,v,\cdots,v) \leq \bar{g}(x,t,v) + \int_a^x \bar{\Gamma}(x,s)\bar{g}(s,t,v)ds,$$

$$\hat{G}(x,v) = \sup_{a\leq s\leq t\leq x} \bar{G}(t,s,v) \leq \hat{g}(x,v) + \int_a^x \bar{\Gamma}(x,r)\bar{g}(r,v)dr \equiv \hat{G}_0(x,v).$$

Observe that if $\hat{g}(a,\bar{f}(a)) > 0$, then also $\hat{G}_0(a,\hat{F}_0(a)) > 0$, where $\hat{F}_0, \hat{G}_0$ are the above dominating functions of $\hat{F}, \hat{G}$.

7.4 Nonlinear Multi-Dimensional Discontinuous Ou-Yang Inequalities

First we introduce some notation. Let $t = (t_1, t_2, \cdots, t_n)$ be a non-negative vector, and denote $[0,t_1]\times[0,t_2]\times\cdots\times[0,t_n]$ by $[0,t]$. The set of all bounded m-vector-valued functions defined on $[0,t]$ which are non-negative will be denoted by $B^{\geq 0}([0,t],m)$. For such a function u and positive numbers q and p, we shall denote by u^q the function obtained by taking the qth power of each component of u, and by $|u|_q$ the pth root of the sum of the component of u^p. For two such functions u_1 and u_2, we shall write $u_1 \geq u_2$ if the inequality $u_1(s) \geq u_2(s)$ holds (componentwise) for all $s \in [0,t]$. Further, a linear operator K on $B^{\geq 0}([0,t],m)$ will be called monotone if $u_1 \geq u_2$ implies that $Ku_1 \geq Ku_2$. The components of the linear operator K will be denoted by k_{ij}, $1 \leq i,j \leq m$.

Theorem 7.4.1 (The Pang-Agarwal Inequality [528]) *Let $q > 1, p > 0, y \in B^{\geq 0}([0,t],m)$, and let K, L be monotone linear operator on the same space of functions, and suppose there exists a constant vector c such that*

$$y^q(t) \leq c + \int_0^{t_1} [qLy^q(\tau, t_2, \cdots, t_n) + Ky(\tau, t_2, \cdots, t_n)]d\tau. \tag{7.4.1}$$

Then the following inequality holds,

$$|y(t)|_p \leq e^{\int_0^{t_1} \bar{L}(\tau,t_2,\cdots,t_n)d\tau} \kappa(q/p,m)^{1/q} \Bigg\{ |c|_1^{(q-1)/q} + (\frac{q-1}{q})\kappa(q,m)^{1/q} \times \int_0^{t_1} e^{-(q-1)\int_0^{\tau} \bar{L}(\xi,t_2,\cdots,t_n)d\xi} \bar{K}(\tau,t_2,\cdots,t_n)d\tau \Bigg\}^{1/(q-1)}, \tag{7.4.2}$$

where

(i) $\bar{K} = \max_{1 \leq j \leq m}\{\sum_{i=1}^m K_{ij}(1)\}, \bar{L} = \max_{1\leq j\leq m}\{\sum_{i=1}^m L_{ij}(1)\}$,
(ii) $\kappa(r,m)$ *is the least constant that satisfies* $(u_1+u_2+\cdots+u_m)^r \leq \kappa(r,m)(u_1^r + u_2^r + \cdots + u_m^r)$, *i.e.,*

$$\kappa(r,m) = \begin{cases} m^{r-1} & \text{if } r \geq 1, \\ 1 & \text{if } 0 < r \leq 1. \end{cases}$$

Proof Let us use the shorthand notation T_1 for $(\tau, t_2, \cdots, t_n)$, and put

$$z(t) = c + \int_0^{t_1} [qLy^q(T_1) + Ky(T_1)]d\tau.$$

First, by the definition of κ and (7.4.1), we have

$$|y|_p \leq \kappa(q/p,m)^{1/q}|y|_q \leq \kappa(q/p,m)^{1/q}|z|_1^{1/q}, \tag{7.4.3}$$

$$|z^{1/q}|_1 \leq \kappa(q,m)^{1/q}|z^{1/q}|_q = \kappa(q,m)^{1/q}|z|_1^{1/q}. \tag{7.4.4}$$

Next, by the monotonicity of L and K and (7.4.4), we obtain

$$\begin{aligned}\frac{\partial |z|_1(t)}{\partial t_1} &= \frac{\partial}{\partial t_1}\left\{|c|_1 + \int_0^{t_1}[q|Ly^q(T_1)|_1 + |Ky^q(T_1)|_1]d\tau\right\}\\ &\leq q|Lz(t)|_1 + |Kz^{1/q}(t)|_1\\ &\leq q\overline{L}(t)|z|_1(t) + \overline{K}(t)|z^{1/q}|_1(t)\\ &\leq q\overline{L}(t)|z|_1(t) + \overline{K}\kappa(q,m)^{1/q}|z|_1^{1/q}(t).\end{aligned} \tag{7.4.5}$$

Thus, by (7.4.1) and (7.4.5), it follows that

$$\begin{aligned}\frac{\partial e^{-q\int_0^{t_1}\overline{L}(T_1)d\tau}|z|_1(t)}{\partial t_1} &= e^{-q\int_0^{t_1}\overline{L}(T_1)d\tau}\frac{\partial |z|_1(t)}{\partial t_1} - e^{-q\int_0^{t_1}\overline{L}(T_1)d\tau}q\overline{L}(t)|z|_1(t)\\ &\leq e^{-q\int_0^{t_1}\overline{L}(T_1)d\tau}\overline{K}(t)(q,m)^{1/q}|z|_1^{1/q}(t)\\ &= e^{-(q-1)\int_0^{t_1}\overline{L}(T_1)d\tau}\overline{K}(t)(q,m)^{1/q}e^{-\int_0^{t_1}\overline{L}(T_1)d\tau}|z|_1^{1/q}(t).\end{aligned} \tag{7.4.6}$$

Now integrating (7.4.6) and combining with (7.4.3), the desired inequality (7.4.2) follows. □

Remark 7.4.1 For $n = m = 1, q = 2, p = 1, L = \alpha, K = 2Ng$, and $c = M^2y^2(0)$, inequality (7.4.2) reduces essentially to $y(t) \leq Me^{\alpha t}y(0) + Ne^{\alpha t}\int_0^t g(\tau)d\tau$.

Remark 7.4.2 We note that all the six results of Pachpatte [252] can be deduced as special cases of (7.4.2) by taking $L = 0, p = 1$, and suitable choices of K. Indeed, to get Theorem 1 in [252], we take $n = m = 1$ and

$$Ky(t) = \int_0^t r_1(\tau_1)\int_0^{\tau_{\upsilon-2}}\cdots r_{\upsilon-1}(\tau_{\upsilon-1})\int_0^{\tau\upsilon-1}h(\tau)y(\tau)d\tau d\tau_{\upsilon-1}\cdots d\tau_1,$$

where $r_1,\cdots,r_{\upsilon-1},h$ are non-negative functions on $[0,t]$.

For Theorem 2 in [252], we take $n = 1, m = 2$, and define the operators K and A by

$$\begin{aligned} K_{ij}y(t) &= \int_0^t r_1(\tau_1)\int_0^{\tau_{v-2}} \cdots r_{v-1}(\tau_{v-1})\int_0^{\tau_{v-1}} \tilde{h}_{2(i-1)+j}(\tau)y_j(\tau)d\tau d\tau_{v-1}\cdots d\tau_1 \\ &= A[t, r_1, \cdots, r_{v-1}, \tilde{h}_{2(i-1)+j}y_j], \quad 1 \le i,j \le 2, \end{aligned}$$

where

$$\tilde{h}_\rho = e^{-(q-1)\mu\tau}h_\rho, \quad \rho = 1, 3, \quad \tilde{h}_\iota = h_\iota, \quad \iota = 2, 4,$$

the functions $r_1, \cdots, r_{v-1}, h_l, 1 \le l \le 4$, are non-negative on $[0, t]$, and μ is a positive number.

Then we make the change of variables

$$y_1(\tau) = e^{-\mu\tau}u(\tau), \quad y_2(\tau) = v(\tau),$$

so that (7.4.1) becomes

$$\begin{cases} e^{-q\mu t}u^q(t) \le c_1 + A[t, r_1, \cdots, r_{v-1}, h_1 e^{-q\mu\tau}u] + A[t, r_1, \cdots, r_{v-1}, h_2 v], \\ v^q(t) \quad \le c_2 + A[t, r_1, \cdots, r_{v-1}, h_3 e^{-q\mu\tau}u] + A[t, r_1, \cdots, r_{v-1}, h_4 v]. \end{cases}$$

These inequalities are equivalent to (34) and (6) of [252].

Hence inequality (7.4.2) gives us

$$e^{-\mu t}u(t) + v(t) \le \left\{[2^{q-1}(c_1 + c_2)]^{(q-1)/q} + 2^{q-1}\left(\frac{q-1}{q}\right)A[t, r_1, \cdots, r_{v-1}, \bar{h}]\right\}^{1/(q-1)},$$

where

$$\bar{h} = \max\{[e^{-(q-1)\mu\tau}(h_1 + h_3)], [h_2 + h_4]\}.$$

It can easily be seen that this inequality is in fact an improvement over Theorem 2 in [252].

For Theorem 3 of [252], we take

$$\begin{aligned} Ky(t) = &\int_0^{t_1}\int_0^{\sigma_{v-1}} \cdots \int_0^{\sigma_1}\int_0^{t_2}\int_0^{\tau_{\mu-1}} \cdots \\ &\times \int_0^{\tau_1} h(\xi_1, \xi_2)y(\xi_1, \xi_2)d\xi_2 d\tau_1 \cdots d\tau_{\mu-1}d\xi_1 d\sigma_1 \cdots d\sigma_{v-1}, \end{aligned}$$

where h is non-negative on $[0, t_1] \times [0, t_2]$; and for Theorem 5 of [252], we take $n = n, m = 1$, and

$$Ky(t) = \int_0^{t_1} \frac{1}{b_1(\xi_1, t_2, \cdots, t_n)} \times \cdots \times \int_0^{t_{n-1}} \frac{1}{b_{n-1}(\xi_1, \cdots, \xi_{n-1}, t_n)} \int_0^{t_n} h(\xi)y(\xi)d\xi,$$

where $b_1, \cdots, b_{n-1}, h$ are non-negative functions defined on $[0, t]$. As in our discussion of Pachpatte's Theorem 2 above, in [252] it is rather easy to show that Theorem 7.2.1 also gives us improvements over Theorems 4 and 6 in [252].

In the next theorem, we shall introduce a generalized retarded integral inequality of Gronwall-like type in two variables, which includes both a nonconstant term outside the integrals and more than one distinct nonlinear integrals without assumption of monotonicity.

Let $I := [x_0, x_1), J := [y_0, y_1)$ and $\Lambda := I \times J \subset \mathbb{R}^2$.

In the next result, we shall introduce a more general form of integral inequality

$$u^p(x, y) \le a(x, y) + \sum_{i=1}^{n} \int_{b_i(x_0)}^{b_i(x)} \int_{c_i(y_0)}^{c_i(y)} f_i(x, y, t, s)\varphi_i(u(t, s))dsdt \tag{7.4.7}$$

for all $(x, y) \in [x_0, x_1) \times [y_0, y_1)$, where $a(x, y)$ is a function and $\varphi_i s$ is may not be monotone. We suppose

(H_1) all φ_i $(i = 1, \dots, n)$ are continuous functions on $\mathbb{R}_+$ and positive on $\mathbb{R}_0 = (0, +\infty)$,

(H_2) $a(x, y) \ge 0$ on Λ,

(H_3) $b_i : I \to I$ $(i = 1, \dots, n)$ and $c_i : J \to J$ $(i = 1, \dots, n)$ are C^1 and non-decreasing such that $b_i(x) \le x$ on I and $c_i(y) \le y$ on J,

(H_4) all f_i $(i = 1, \dots, n)$ are non-negative functions $\Lambda \times \Lambda$.

We now consider a sequence of functions $w_i(s)$, which can be calculated recursively by

$$\begin{cases} w_1(s) := \max_{\tau \in [0,s]} \{\varphi_1(\tau)\}, \\ w_{i+1}(s) := \max_{\tau \in [0,s]} \{\varphi_{i+1}(\tau)/w_i(\tau)\} w_i(s), \quad i = 1, \dots, n. \end{cases} \tag{7.4.8}$$

Then for given constant $u_i > 0$, the function

$$W_{p,i}(u, u_i) := \int_{u_i}^{u} \frac{ds}{w_i(s^{1/p})}$$

is well-defined for all $u > 0$ and strictly increasing. When there is no confusion, we simply let $W_{p,i}(u)$ denote $W_{p,i}(u, u_i)$ and $W_{p,i}^{-1}$ denote its inverse. As explained in Remark 2 in [13], different choices of u_i in $W_{p,i}$ do not affect results here.

Theorem 7.4.2 (The Wang Inequality [664]) *Suppose that* $(H_1 - H_4)$ *hold and* $u(x, y)$ *is a non-negative function on* Λ *satisfying (7.4.7). Then for all* $(x, y) \in [x_0, X_1) \times [y_0, Y_1)$,

$$u(x, y) \leq \left\{W_{p,n}^{-1}(\Xi_n(x, y))\right\}^{1/p} \tag{7.4.9}$$

where

$$\Xi_i(x, y) := W_{p,i}(r_i(x, y)) + \int_{b_i(x_0)}^{b_i(x)} \int_{c_i(y_0)}^{c_i(y)} \max_{(\tau,\xi)\in[x_0,x]\times[y_0,y]} f_i(\tau, \xi, t, s)dsdt,$$

$i = 1, 2, \ldots, n$, *$r_i(x, y)$ is determined recursively by*

$$\begin{cases} r_1(x, y) := a(x_0, y_0) + \int_{x_0}^{x} |a_x(t, y_0)|dt + \int_{y_0}^{y} |a_y(x, s)|ds, \\ r_i(x, y) := W_{p,i-1}^{-1}(\Xi_{i-1}(x, y)), \end{cases} \tag{7.4.10}$$

and $(X_1, Y_1) \in \Lambda$ *is arbitrarily given on the boundary of the planar region*

$$R := \left\{(x, y) \in \Lambda : \Xi_i(x, y) \leq \int_{u_i}^{+\infty} \frac{ds}{w_i(s^{1/p})},\ i = 1, \ldots, n\right\}. \tag{7.4.11}$$

Proof First of all, we monotonize some given functions f_i, φ_i in the integral. Obviously, the sequence $(w_i(s))$ defined by $\varphi_i s$ in (7.4.8) consists of non-decreasing non-negative functions and satisfies $w_i(s) \geq \varphi_i(s),\ i = 1, \ldots, n$. Moreover,

$$w_i \propto w_{i+1}, \quad i = 1, \ldots, n-1, \tag{7.4.12}$$

for comparison of monotonicity of functions, because the ratios $w_{i+1}(s)/w_i(s),\ i = 1, \ldots, n-1$, are all non-decreasing. Furthermore, let

$$\tilde{f}_i(x, y, t, s) := \max_{(\tau,\xi)\in[x_0,x]\times[y_0,y]} f_i(\tau, \xi, t, s),$$

which is also non-decreasing in x and y for each fixed s and t and satisfies $\tilde{f}_i(x, y, t, s) \geq f_i(x, y, t, s) \geq 0$ for all $i = 1, \ldots, n$. With the above defined functions w_i and $\tilde{f}_i$, from (7.4.7) we infer for all $(x, y) \in \Lambda$,

$$u^p(x, y) \leq a(x, y) + \sum_{i=1}^{n} \int_{b_i(x_0)}^{b_i(x)} \int_{c_i(y_0)}^{c_i(y)} \tilde{f}_i(x, y, t, s)w_i(u(t, s))dsdt. \tag{7.4.13}$$

First, we discuss the case that $a(x, y) > 0$ for all $(x, y) \in \Lambda$. It means that $r_1(x, y) > 0$ for all $(x, y) \in \Lambda$. In such a case, $r_1(x, y)$ is positive and non-decreasing on Λ and

$$r_1(x, y) \geq a(x_0, y_0) + \int_{x_0}^{x} a_x(t, y_0)dt + \int_{y_0}^{y} a_y(x, y)ds = a(x, y).$$

Consider the auxiliary inequality to (7.4.13)

$$u^p(x, y) \leq r_1(x, y) + \sum_{i=1}^{n} \int_{b_i(x_0)}^{b_i(x)} \int_{c_i(y_0)}^{c_i(y)} \tilde{f}_i(X, Y, t, s) w_i(u(t, s))dsdt \tag{7.4.14}$$

for all $(x, y) \in [x_0, X) \times [y_0, Y)$, where $x_0 \leq X \leq X_1$ and $y_0 \leq Y \leq Y_1$ are chosen arbitrarily, and claim that

$$u(x, y) \leq \left\{W_{p,n}^{-1}(\Upsilon_n(X, Y, x, y))\right\}^{1/p} \tag{7.4.15}$$

for all $x, X \in [x_0, X_2)$ with $x \leq X$ and y, Y in $[y_0, Y_2)$ with $y \leq Y$, where

$$\Upsilon_i(X, Y, x, y) := W_{p,i}(\tilde{r}_i(X, Y, x, y)) + \int_{b_i(x_0)}^{b_i(x)} \int_{c_i(y_0)}^{c_i(y)} \tilde{f}_i(X, Y, t, s)dsdt,$$

$i = 1, \ldots, n$, $\tilde{r}_i(X, Y, x, y)$ is defined recursively by

$$\begin{cases} \tilde{r}_1(X, Y, x, y) := r_1(x, y), \\ \tilde{r}_i(X, Y, x, y) := W_{p,i-1}^{-1}(\Xi_{i-1}(X, Y, x, y)), \end{cases} \tag{7.4.16}$$

and X_2, Y_2 are both functions of (X, Y) such that $(X_2(X, Y), Y_2(X, Y)) \in \Lambda$ lies on the boundary of the planar region

$$R_1(x, y) := \left\{(x, y) \in \Lambda : \Upsilon_i(X, Y, x, y) \leq \int_{u_i}^{+\infty} \frac{ds}{w_i(s^{1/p})},\ i = 1, \ldots, n\right\}.$$

We can choose X_2, Y_2 appropriately such that for all $(X, Y) \in [x_0, X_1) \times [y_0, Y_1)$,

$$X_2(X, Y) \geq X_1, \quad Y_2(X, Y) \geq Y_1. \tag{7.4.17}$$

In fact, from the fact of (X_1, Y_1) being on the boundary of R, we see that

$$\Upsilon_i(X_1, Y_1, X_1, Y_1) = \Xi_i(X_1, Y_1) = \int_{u_i}^{+\infty} \frac{ds}{w_i(s^{1/p})}. \tag{7.4.18}$$

Moreover, the monotonicity that $\tilde{r}_i(X,Y,x,y)$ and $\tilde{f}_i(X,Y,x,y)$ are both non-decreasing in each variable implies that $\Upsilon_i(X,Y,x,y)$ is also non-decreasing in each variable. Therefore, it follows from (7.4.17) that the rectangles $[x_0,X_1)\times[y_0,Y_1)$, $[x_0,X_2)\times[y_0,Y_2)$ and Λ are nestled one by one, i.e.,

$$[x_0,X_1)\times[y_0,Y_1)\subset[x_0,X_2)\times[y_0,Y_2)\subset\Lambda. \tag{7.4.19}$$

Obviously,

$$[x_0,X_1)\times[y_0,Y_1)\subset R,\ [x_0,X_2(X,Y))\times[y_0,Y_2(X,Y))\subset R_1(X,Y),$$

so that $r_i,\tilde{r}_i$ $(i=1,2,\ldots,n)$ are well-defined.

Now we prove (7.4.15) by induction. Let $\beta_1(x,y)$ denote the function on the right-hand side of (7.4.14), which is a non-negative and non-decreasing function on $[x_0,Y)\times[y_0,Y)$. Then (7.4.14) is equivalent to

$$u^p(x,y)\le\beta_1(x,y),\quad \textit{for all } (x,y)\in[x_0,Y)\times[y_0,Y). \tag{7.4.20}$$

By (H_3), $b_1'\ge 0$ and $b_1(x)$ for all $x\in[x_0,X)$. Moreover, w_1 is non-decreasing. Then

$$\begin{aligned}
&\frac{\frac{\partial}{\partial x}\beta_1(x,y)}{w_1(\beta_1^{1/p}(x,y))}\\
&\le\frac{\frac{\partial}{\partial x}r_1(x,y)}{w_1(r_1^{1/p}(x,y))}+\frac{b_1'(x)}{w_1(\beta_1^{1/p}(x,y))}\int_{c_1(y_0)}^{c_1(y)}\tilde{f}_1(X,Y,b_1(x),s)w_1(u(b_1(x),s))ds\\
&\le\frac{\frac{\partial}{\partial x}r_1(x,y)}{w_1(\beta_1^{1/p}(x,y))}+\frac{b_1'(x)}{w_1(\beta_1^{1/p}(x,y))}\int_{c_1(y_0)}^{c_1(y)}\tilde{f}_1(X,Y,b_1(x),s)w_1(\beta_1^{1/p}(x,s))ds\\
&\le\frac{\frac{\partial}{\partial x}r_1(x,y)}{w_1(\beta_1^{1/p}(x,y))}+b_1'(x)\int_{c_1(y_0)}^{c_1(y)}\tilde{f}_1(X,Y,b_1(x),s)ds.
\end{aligned}$$

Integrating both sides of the above inequality from x_0 to x, we obtain for all $(x,y)\in[x_0,Y)\times[y_0,Y)$,

$$\begin{aligned}
W_{p,1}(\beta_1(x,y))&\le W_{p,1}(r_1(x,y))+\int_{x_0}^{x}b_1'(t)\int_{c_1(y_0)}^{c_1(y)}\tilde{f}_1(X,Y,b_1(t),s)dsdt\\
&=W_{p,1}(r_1(x,y))+\int_{b_1(x_0)}^{b_1(x)}\int_{c_1(y_0)}^{c_1(y)}\tilde{f}_1(X,Y,b_1(t),s)dsdt
\end{aligned}\tag{7.4.21}$$

the right-hand side of which is contained in the domain of $W_{p,1}^{-1}$ by the definition of X_2, Y_2 and (7.4.19). It follows from (7.4.20)–(7.4.21) and the monotonicity of $W_{p,1}^{-1}$ that for all $x_0 \le x \le X < X_2, y_0 \le y \le Y < Y_2$,

$$u(x,y) \le \beta_1^{1/p} \le \left\{ W_{p,1}^{-1}[W_{p,1}(r_1(x,y)) + \int_{b_1(x_0)}^{b_1(x)} \int_{c_1(y_0)}^{c_1(y)} \tilde{f}_1(X,Y,b_1(t),s)dsdt] \right\}^{1/p},$$

implying that (7.4.15) is true for $n = 1$. Next, we make the inductive assumption that (7.4.15) is true for $n = k$.

Consider

$$u^p(x,y) \le r_1(x,y) + \sum_{i=1}^{k+1} \int_{b_i(x_0)}^{b_i(x)} \int_{c_i(y_0)}^{c_i(y)} \tilde{f}_1(X,Y,t,s)w_i(u(t,s))dsdt \qquad (7.4.22)$$

for all $x_0 \le x \le X, y_0 \le y \le Y$. Let $\beta_2(x,y)$ denote the non-negative and non-decreasing function on the right-hand side of (7.4.22) and rewrite (7.4.22) as

$$u^p(x,y) \le \beta_2(x,y), \quad \textit{for all } (x,y) \in [x_0, Y) \times [y_0, Y).$$

Let $\phi_{i+1}(u) := w_{i+1}/w_1(u), i = 1, \dots, k$. Similarly to the above statement for $n = 1$, by the fact that $b_i' \ge 0$ and $b_i(x) \le x$ for all $x \in [x_0, X)$, given by (H_3), and the monotonicity of w_i, we obtain for all $(x,y) \in [x_0, Y) \times [y_0, Y)$,

$$\begin{aligned}
&\frac{\frac{\partial}{\partial x}\beta_2(x,y)}{w_1(\beta_2^{1/p}(x,y))} \\
&\le \frac{\frac{\partial}{\partial x} r_1(x,y)}{w_1(r_2^{1/p}(x,y))} + \sum_{i=1}^{k+1} \frac{b_i'(x)}{w_1(\beta_2^{1/p}(x,y))} \int_{c_i(y_0)}^{c_i(y)} \tilde{f}_i(X,Y,b_i(x),s)w_i(u(b_i(x),s))ds \\
&\le \frac{\frac{\partial}{\partial x} r_1(x,y)}{w_1(r_2^{1/p}(x,y))} + b_1'(x) \int_{c_1(y_0)}^{c_1(y)} \tilde{f}_1(X,Y,b_1(x),s)ds \\
&\quad + \sum_{i=1}^{k} b_{i+1}'(x) \int_{c_{i+1}(y_0)}^{c_{i+1}(y)} \tilde{f}_{i+1}(X,Y,b_{i+1}(x),s)\phi_{i+1}(\beta_2^{1/p}(b_{i+1}(x),s))ds \\
&= \frac{\frac{\partial}{\partial x} r_1(x,y)}{w_1(r_1^{1/p}(x,y))} + b_1'(x) \int_{c_1(y_0)}^{c_1(y)} \tilde{f}_1(X,Y,b_1(x),s)ds \\
&\quad + \sum_{i=1}^{k} b_{i+1}'(x) \int_{c_{i+1}(y_0)}^{c_{i+1}(y)} \tilde{f}_{i+1}(X,Y,b_{i+1}(x),s)\phi_{i+1}(\beta_2^{1/p}(b_{i+1}(x),s))ds.
\end{aligned}$$

(7.4.23)

Integrating the above inequality from x_0 to x, we can derive for all $(x,y) \in [x_0, Y) \times [y_0, Y)$,

$$W_{p,1}(\beta_2(x,y)) \leq W_{p,1}(r_1(x,y)) + \int_{b_1(x_0)}^{b_1(x)} \int_{c_1(y_0)}^{c_1(y)} \tilde{f}_1(X,Y,t,s)dsdt$$
$$+\sum_{i=1}^{k} \int_{b_{i+1}(x_0)}^{b_{i+1}(x)} \int_{c_{i+1}(y_0)}^{c_{i+1}(y)} \tilde{f}_{i+1}(X,Y,t,s)\phi_{i+1}(\beta_2^{1/p}(t,s))ds.$$

Let

$$\xi^p(x,y) := W_{p,1}(\beta_2(x,y)), \tag{7.4.24}$$

$$\theta_1 := W_{p,1}(r_1(x,y)) + \int_{b_1(x_0)}^{b_1(x)} \int_{c_1(y_0)}^{c_1(y)} \tilde{f}_1(X,Y,t,s)dsdt. \tag{7.4.25}$$

It follows that

$$\xi^p \leq \theta_1(x,y) + \sum_{i=1}^{k} \int_{b_{i+1}(x_0)}^{b_{i+1}(x)} \int_{c_{i+1}(y_0)}^{c_{i+1}(y)} \tilde{f}_{i+1}(X,Y,t,s)\phi_{i+1}[(W_{p,1}^{-1}(\xi^p(x,y)))^{1/p}]ds, \tag{7.4.26}$$

the same form as (7.4.14) for $n = k$, for all $(x,y) \in [x_0, Y) \times [y_0, Y)$ and we are ready to use the inductive assumption for (7.4.15). In order to demonstrate the basic condition of monotonicity, let $h(s) := (W_{p,1}^{-1}(sp))^{1/p}$ which is clearly a continuous and non-decreasing function on $\mathbb{R}_+$. Thus each $\phi_i(h(s))$ is continuous and non-decreasing on $[0, +\infty)$ and satisfies $\phi_i(h(s)) > 0$ for all $s > 0$. Moreover,

$$\frac{\phi_{i+1}(h(s))}{\phi_i(h(s))} = \frac{w_{i+1}(h(s))}{w_i(h(s))} = \max_{\tau \in [0,h(s)]} \frac{\varphi_{i+1}(\tau)}{w_i(\tau)},$$

which is also continuous and non-decreasing on $\mathbb{R}_+$ and positive on $\mathbb{R}_0$, implying that $\phi_i(h(s)) \propto \phi_{i+1}(h(s))$, $i = 2, \cdots, k$. Therefore, the inductive assumption for (7.4.15) can be used to (7.4.25) to obtain

$$\xi(x,y) \leq \left(\Phi_{p,k+1}^{-1}(\eta_{k+1}(X,Y,x,y))\right)^{1/p} \tag{7.4.27}$$

for all $x_0 \le x < min(X, X_3)$ and $y_0 \le y < min(Y, Y_3)$, where

$$\Phi_{p,i} := \int_{\varpi(u_i)}^{u} \frac{ds}{\phi_i(h(s))}, \quad u > 0, \ \varpi(u) := (W_{p,1}(u))^{1/p}, \tag{7.4.28}$$

$$\eta_1(X, Y, x, y) := \Phi_{p,i}(r_{i-1}(x, y)) + \int_{b_i(x_0)}^{b_i(x)} \int_{c_i(y_0)}^{c_i(y)} \tilde{f}_1(X, Y, t, s) ds dt, \tag{7.4.29}$$

$$\theta_i := \Phi_{p,i}^{-1}(\eta_i(X, Y, x, y)), \tag{7.4.30}$$

$i = 2, .., , k + 1$, and X_3, Y_3 are functions of (X, Y) such that $(X_3(X, Y), Y_3(X, Y)) \in \Lambda$ lies on the boundary of the planar region

$$R_2(X, Y) := \left\{ (x, y) \in \Lambda : \eta_i(X, Y, x, y) \le \int_{\varpi(u_i)}^{\varpi(+\infty)} \frac{ds}{\phi_i(h(s))}, \ i = 2, \ldots, k + 1 \right\}.$$

Here $\varpi(+\infty)$ denotes either the limit $\lim_{u\to+\infty} \varpi(u)$ if it converges or $+\infty$.

Note that

$$\Phi_{p,i}(u) = \int_{u_i}^{W_{p,i}^{-1}(u^p)} \frac{ds}{w_i(s^{1/p})} = W_{p,i}(W_{p,1}^{-1}(u^p)), \quad i = 2, .., k + 1. \tag{7.4.31}$$

Thus (7.4.26), where we note those functions, defined in (7.4.23)–(7.4.24) and (7.4.29), can be equivalently rewritten as

$$u(x, y) \le \beta_2^{1/p}(x, y) = (W_{p,1}^{-1}(\xi^p(x, y)))^{1/p}$$
$$\le \left\{ W_{p,k+1}^{-1}[W_{p,k+1}(W_{p,1}^{-1}(\theta_k(x, y))) + \int_{b_{k+1}(x_0)}^{b_{k+1}(x)} \int_{c_{k+1}(y_0)}^{c_{k+1}(y)} \tilde{f}_{k+1}(X, Y, t, s) ds dt] \right\}^{1/p} \tag{7.4.32}$$

for all $x_0 \le x < min(X, X_3), y_0 \le y < min(Y, Y_3)$. We further claim that the term $W_{p,1}^{-1}(\theta_i(x, y))$ in the formula (7.4.32) is just the same as $\tilde{r}_{i+1}(X, Y, x, y)$, defined in (7.4.16), for all $i = 1, \ldots, k$. For convenience, let $\tilde{\theta}_i(x, y)$ denote that term. It is trivial to see that $\tilde{\theta}_1(x, y) = \tilde{r}_2(X, Y, x, y)$. Assume that the claimed result is true for some i. Then, using (7.4.31) and noting some definitions of functions in (7.4.29)–(7.4.30), we conclude

$$\tilde{\theta}_{i+1}(x, y) = W_{p,1}^{-1} \left(\Phi_{p,i+1}^{-1}[\Phi_{p,i+1}(\theta_i(x, y)) + \int_{b_{i+1}(x_0)}^{b_{i+1}(x)} \int_{c_{i+1}(y_0)}^{c_{i+1}(y)} \tilde{f}_{k+1}(X, Y, t, s) ds dt] \right)$$
$$= W_{p,i+1}^{-1}[W_{p,i+1}(W_{p,1}^{-1}(\theta_i(x, y))) + \int_{b_{i+1}(x_0)}^{b_{i+1}(x)} \int_{c_{i+1}(y_0)}^{c_{i+1}(y)} \tilde{f}_{k+1}(X, Y, t, s) ds dt]$$

$$= W_{p,i+1}^{-1}[W_{p,i+1}(\tilde{r}_{i+1}(X,Y,x,y)) + \int_{b_{i+1}(x_0)}^{b_{i+1}(x)} \int_{c_{i+1}(y_0)}^{c_{i+1}(y)} \tilde{f}_{k+1}(X,Y,t,s)dsdt]$$
$$= \tilde{r}_{i+2}(X,Y,x,y).$$

Thus the claimed result is proved. Hence (7.4.30) can be equivalently written as

$$u(x,y) \le \left\{ W_{p,k+1}^{-1}[W_{p,k+1}(\tilde{r}_{k+1}(X,Y,x,y)) + \int_{b_{k+1}(x_0)}^{b_{k+1}(x)} \int_{c_{k+1}(y_0)}^{c_{k+1}(y)} \tilde{f}_{k+1}(X,Y,t,s)dsdt] \right\}^{1/p}. \tag{7.4.33}$$

Similarly, from (7.4.29) and (7.4.31) it follows

$$\eta_i(X,Y,x,y) = W_{p,i}(\tilde{r}_{k+1}(X,Y,x,y)) + \int_{b_i(x_0)}^{b_i(x)} \int_{c_i(y_0)}^{c_i(y)} \tilde{f}_i(X,Y,t,s)dsdt$$
$$= \Upsilon_i(X,Y,x,y). \tag{7.4.34}$$

Note that $\int_{\varpi(u_i)}^{\varpi(+\infty)} \frac{ds}{\phi_i(h(s))} = \int_{u_i}^{+\infty} ds/w_i(s^{1/p})$. Then, comparing the definition of R_2 with that of R_1 and noting (7.4.34), we can see that X_3, Y_3 can be chosen appropriately such that for all $(X,Y) \in [x_0, X_1) \times [y_0, Y_1)$,

$$X_3(X,Y) = X_2(X,Y), \quad Y_3(X,Y) = Y_2(X,Y). \tag{7.4.35}$$

It means that (7.4.33) holds for all $x_0 \le x < X \le X_2, y_0 \le y < Y \le Y_2$. It actually proves (7.4.15) by induction. Having (7.4.15), we start from the original inequality (7.4.15) and see that

$$u^p(X,Y) \le r_1(X,Y) + \sum_{i=1}^{n} \int_{b_i(x_0)}^{b_i(X)} \int_{c_i(y_0)}^{c_i(Y)} \tilde{f}_1(X,Y,t,s)w_i(u(t,s))dsdt,$$

i.e., the auxiliary inequality (7.4.14) holds for $x = X, y = Y$. By (7.4.15), we obtain

$$u(X,Y) \le \left\{ W_{p,n}^{-1}[W_{p,n}(\tilde{r}_n(X,Y,X,Y)) + \int_{b_i(x_0)}^{b_i(X)} \int_{c_i(y_0)}^{c_i(Y)} \tilde{f}_i(X,Y,t,s)dsdt] \right\}^{1/p},$$
$$= \left\{ W_{p,n}^{-1}[W_{p,n}(r_n(X,Y)) + \int_{b_i(x_0)}^{b_i(X)} \int_{c_i(y_0)}^{c_i(Y)} \tilde{f}_i(X,Y,t,s)dsdt] \right\}^{1/p}$$

for all $x_0 \le X \le X_1, y_0 \le Y \le Y_1$ since $X_2 \ge X_1, Y_2 \ge Y_1$ and $\tilde{r}_n(X,Y,X,Y) = r_n(X,Y)$. This proves (7.4.17). The remainder case is that $a(x,y) = 0$ for some $(x,y) \in \Lambda$.

Let

$$r_{1,\varepsilon}(x, y) := r_1(x, y) + \varepsilon,$$

where $\varepsilon > 0$ is an arbitrary small number. Obviously, $r_{1,\varepsilon}(x, y) > 0$ for all $(x, y) \in \Lambda$. Using the same arguments as above, where $r_1(x, y)$ is replaced with $r_{1,\varepsilon}$, we get

$$u(x, y) \le \left\{ W_{p,n}^{-1}[W_{p,n}(r_{n,\varepsilon}(x, y)) + \int_{b_n(x_0)}^{b_n(X)} \int_{c_n(y_0)}^{c_n(Y)} \tilde{f}_n(X, Y, t, s)dsdt] \right\}^{1/p},$$

for all $x_0 \le x < X_1, y_0 \le y < Y_1$. Letting $\varepsilon \to 0^+$, we obtain (7.1.11) because of continuity of $r_{i,\varepsilon}$ in ε and continuity of $W_{p,i}$ and $W_{p,i}^{-1}$ for $i = 1, \dots, n$. This completes the proof. □

If we choose $n = 2, \varphi_1(s) := s^q, \varphi_2(s) := s^q\psi(s), f_i(x, y, t, s) := (p/(p - q))g_i(t, s)$, where $i = 1, 2$ and $0 < q < p$, and restrict $a(x, y)$ to be a constant a, then we can give a different estimate from [142] for the unknown function u in the inequality (7.4.14). If we choose $p = 1$ and $u(x, y) := v(x)$, let $a(x, y) := a(x), f_i(x, y, t, s) := g_i(x, t), i = 1, \dots, n$, and restrict all c_is to satisfy that $c_i(y) - c_i(y_0) = 1$ for all $y \in J$, then inequality (7.4.15) reduces to the same form as $v(t) \le a(t) + \sum_{i=1}^n \int_{a_i(t)}^{b_i(t)} g_i(t, s)w_i(u(s))ds, \ t_0 \le t \le t_1$, where we do not require the monotonicity of sequence of functions φ_i. Obviously, Theorem 7.4.2 is applicable to more general form than Theorem 2.1 in [13].

Remark 7.4.3 Note that X_1, Y_1 are defined by (7.4.20). In particular, (7.4.17) is true for all $(x, y) \in \Lambda$ when all $w_i, i = 1, 2, \dots, n$ satisfy $\int_{u_i}^{+\infty} ds/w_i(s^{1/p}) = +\infty$, so we may take $X_1 = x_1, Y_1 = y_1$.

7.5 Nonlinear Multi-Dimensional Discontinuous Bellman-Gronwall Integral Inequalities of Wendorff Type

Walter [658] extended Gronwall's inequality to more than one independent variable using monotone operators. Snow used Riemann's method of integration for the case of two independent variables. Now we give a very simple proof of these results using the methods of recursion. Using this method, we extend the result of Walter and a result of Bondge, Pachpatte and Walter [100] on Wendroff type inequalities.

We first introduce the method used by [658].

Let a, b, s, t and x denote real n-vectors and $u(x)$ a scalar function. If $a = (a_1, \dots, a_n), x = (x_1, \dots, x_n), t = (t_1, \dots., t_n)$ and $b = (b_1, \dots, b_n)$, then we shall denote the integral

$$\int_{a_1}^{x_1} \cdots \int_{a_n}^{x_n} u(t_1, \ldots, t_n)dt_1 \ldots dt_n$$

by $J_{ax}u(t)$ and the region $[0, b_1] \times \cdots \times [0, b_n]$ in $\mathbb{R}^n$ by $D(b)$.

If $T(x, u(t))$ is a real functional, we shall say that the inequality $u(x) \leq T(x, u(t))$ is recursive if the right-hand side can be substituted for $u(t)$ on the right-hand side and the process repeated indefinitely with preservation of the inequality.

Let $f(x)$, $u(x)$ and $v(x) \geq 0$ be bounded integrable functions in $D(b)$. Under these conditions, the inequality

$$u(x) \leq f(x) + J_{0x}(v(t)u(t)) \tag{7.5.1}$$

is recursive in $D(b)$.

Substituting the right-hand side of (7.5.1) for $u(t)$ under the integral sign and interchanging the order of integration, we obtain

$$u(x) \leq f(x) + J_{0x}(v(t)f(t)) + J_{0x}(v(t)g_1(x,t)u(t))$$

where $g_1(x, t) = J_{tx}(v(s))$. Repeating the procedure (assuming we may interchange the order of integration), we obtain

$$u(x) \leq f(x) + J_{0x}\left(v(t)f(t)\sum_{i=0}^{m} g_i(x,t)\right) + J_{0x}(u(t)v(t)g_{m+1}(x,t)), \tag{7.5.2}$$

where

$$g_0(x,t) = 1, g_{k+1} = J_{tx}(v(s)g_k(x,s)), \quad k \geq 0.$$

It is easily seen that

$$|J_{0x}(u(t)v(t)g_{m+1}(x,t))| \leq \frac{M}{((m+1)!)^n}$$

for all $x \in D(b)$ where M is independent of x and letting $m \to +\infty$ in (7.5.2), we obtain

$$u(x) \leq f(x) + J_{0x}(f(t)v(t)N_{\upsilon}(x,t)), \quad x \in D(b), \tag{7.5.3}$$

where $N_{\upsilon}(x,t) = \sum_{i=0}^{+\infty} g_i(x,t)$ is the so-called Neumann series for the function $\upsilon(x)$. This is Walter's solution of (7.5.1). The special case of $f(x) = C = \mathit{constant}$ may be solved in the same way (or with greater effort obtained from (7.5.3)) to give

$$u(x) \leq C \sum_{k=0}^{+\infty} I_k(x) \equiv C\underline{N_{\upsilon}}(x), \quad x \in D(b), \tag{7.5.4}$$

where

$$I_0(x) = 1, I_{k+1}(x) = J_{0x}(\upsilon(t)I_k(t)), \quad k \geq 0.$$

The next result is the Gronwall inequality.

Theorem 7.5.1 (The Abrammovich Inequality [1]) *Let $f(x)$ and $u(x)$ be as in (7.5.1) and $\upsilon(x) \geq 0$ be a bounded, integrable function for all $x \in D(b), t \in D(x)$. Then the inequality*

$$u(x) \leq f(x) + J_{0x}(\upsilon(x,t)u(t)) \tag{7.5.5}$$

is recursive.

Proof This inequality may be solved in exactly the same way as (7.5.1) and we shall omit the details. □

The solution is

$$u(x) \leq f(x) + J_{0x}(f(t)M_{\upsilon}(x,t)), \quad x \in D(b), \tag{7.5.6}$$

where $M_{\upsilon}(x,t) = \sum_{n=0}^{+\infty} \upsilon_n(x,t)$, and

$$\upsilon_0(x,t) = \upsilon(x,t), \quad \upsilon_{k+1}(x,t) = J_{tx}(\upsilon(s,t)\upsilon_k(x,s)), \quad k \geq 0.$$

In the particular case that $\upsilon(x,t) = p(t)w(t)$, (7.5.6) reduces to

$$u(x) \leq f(x) + p(x)J_{0x}(f(t)N_{pw}(x,t)) \tag{7.5.7}$$

where $N_{pw}(x,t)$ is the Neumann series for the function $p(x)w(x)$.

As an application, consider the inequality

$$u(x) \leq f(x) + J_{0x}(p(t)u(t)) + J_{0x}(q(t)J_{0t}(r(s)u(s))) \tag{7.5.8}$$

where $p(t) \geq 0, q(t) \geq 0, r(t) \geq 0$ and $u(t)$ are bounded integrable functions in $D(b)$. Integrating the order of integration in the last term (assuming it is valid) and setting $Q(x,t) = J_{tx}(q(t))$, we obtain

$$u(x) \leq f(x) + J_{0x}(\upsilon(x,t)u(t)) \tag{7.5.9}$$

where $\upsilon(x,t) = p(t) + Q(x,t)r(t)$.

Next, we discuss the Wendroff inequalities.

Theorem 7.5.2 (The Abrammovich Inequality [1]) *Let $f(x)$, $\upsilon(x)$ and $u(x)$ be as in (7.5.1) and $M(x) = \sup_{t \in D(x)} f(t)$. Then the inequality*

$$u(x) \leq f(x) + J_{0x}(\upsilon(t)u(t)) \tag{7.5.10}$$

has the solution

$$u(x) \leq M(x)\underline{N}_{\upsilon}(x) \tag{7.5.11}$$

where $\underline{N}_{\upsilon}(x)$ is as in (7.5.4).

Proof The proof is simple. From (7.5.10), it follows

$$\begin{aligned} u(x) &\leq M(x) + J_{0x}(\upsilon(t)u(t)) \\ &\leq M(x) + J_{0x}(\upsilon(t)(M(t) + J_{0t}(\upsilon(s)u(s)))) \\ &\leq M(x) + M(x)J_{0x}(\upsilon(t)) + J_{0x}(\upsilon(t)J_{0t}(\upsilon(s)u(s))) \end{aligned}$$

and the proof is the same as that for (7.5.4). The inequality $\overline{N_{\upsilon}(x)} \leq \exp(J_{0x}(\upsilon(t)))$ $(\upsilon(t) \geq 0)$ is easily established so that in the case $M(x) \geq 0$, (7.5.11) may be written as

$$u(x) \leq M(x)\exp(J_{0x}(\upsilon(t))) \tag{7.5.12}$$

which, however, is a much 'coarser' inequality than (7.5.11). This generalizes a result due to Bondge, Pachpatte and Walter [100]. □

Note that (7.5.12) need not be valid if $f(x) < 0$ as the example

$$u(x, y) = -1 + \int_0^x \int_0^y u(s,t)dsdt \quad (x, y \textit{ scalars})$$

with the solution $u(x, y) = -\sum_{k=0}^{+\infty}(xy)^k/(k!)^2$ shows, contrary to the assertion in [100].

A result similar to (7.5.11) may be established for the inequality (7.5.5) in exactly the same way

$$u(x) \leq M(x)[1 + J_{0x}(M_{\upsilon}(x, t))]. \tag{7.5.13}$$

As an application of this result, (7.5.1) may be transformed into the inequality

$$u(x) \leq f(x) + J_{0x}(\upsilon(t)f(t)) + J_{0x}(p(x, t)u(t))$$

where $p(x,t) = \upsilon(t)J_{tx}(\upsilon(s))$. From (7.5.13) it follows

$$u(x) \leq M(x)[1 + J_{0x}(M_p(x,t))] \tag{7.5.14}$$

where

$$M(x) = \sup_{s \in D(x)} [f(s) + J_{0s}(\upsilon(t)f(t))].$$

Now we begin with some bounds in the case of two variables.

The solutions of the inequalities in the above are in terms of repeated integrals which may be inconvenient to use. If in (7.5.1) $\upsilon(t) = \upsilon_1(t_1)\cdots\upsilon_n(t_n)$, the function $N_\upsilon(x,t)$ in (7.5.3) simplifies to

$$N_\upsilon(x,t) = E_n(J_{tx}(\upsilon(t))) \tag{7.5.15}$$

where

$$E_n(z) = \sum_{k=0}^{+\infty} \frac{z^k}{(k!)^n}. \tag{7.5.16}$$

Though (7.5.15) holds only in the case when $\upsilon(t)$ is a product of functions of one variable, it is possible to obtain lower and upper bounds for $N_\upsilon(x,t)$ of similar form. This is difficult to do for any number of independent variables and we shall consider the case of two independent variables only.

From now on, we shall abandon vector notation so that x, y, etc, shall denote scalar variables. Partial derivatives shall be denoted by subscripts or by D_x, D_y, D_{xy}, etc. Throughout $\int_{st}^{xy} \upsilon$ shall denote the integral

$$\int_s^x \int_t^y \upsilon(p,q)dpdq$$

and in the case that both lower limits of integration are zero, by $J_{xy}(\upsilon)$. The variables of integration shall be shown explicitly only if necessary for clarity. By $R(a,b)$, we shall denote the rectangle $[0,a] \times [0,b]$ in the Euclidean space $\mathbb{R}^2$.

Lemma 7.5.1 ([1]) *Let $I_n(x,y;s,t)$ denote the integrals*

$$\begin{cases} I_0(x,y;s,t) = 1, \\ I_{n+1}(x,y;s,t) = J_{st}^{xy}(\upsilon(p,q)I_n(x,y;p,q)), \quad n \geq 0, \end{cases}$$

and

$$K_n(x,y;s,t) = \frac{1}{(n!)^2}[J_{st}^{xy}(\upsilon)]^n, \quad n \geq 0.$$

Let

$$Q(x, y; s, t) = \int_s^x \upsilon(p, y)dp \int_t^y \upsilon(x, q)dq/\upsilon(x, y)J_{st}^{xy}(\upsilon),$$

where $0 \le s \le x$ *and* $0 \le t \le y$. *If*

$$\begin{cases} r = \min\left(1, \inf_{(x,y)\in R(a,b)} Q(x, y; s, t)\right), \\ R = \max\left(1, \sup_{(x,y)\in R(a,b)} Q(x, y; s, t)\right), \end{cases} \tag{7.5.17}$$

and if $\upsilon(x, y) \ge 0$, $(x, y) \in R(a, b)$, *then*

$$R^{-n}K_n(x, y; s, t) \le I_n(x, y; s, t) \le r^{-n}K_n(x, y; s, t), \quad n \ge 0. \tag{7.5.18}$$

Furthermore, $r = R = 1$ *if an only if* $\upsilon(x, y) = \upsilon_1(x)\upsilon_2(y)$ *in which case, equality holds in (7.5.18).*

Proof For $n = 0$, (7.5.18) is trivial and by the induction hypothesis,

$$I_{n+1}(x, y; s, t) = J_{st}^{xy}(\upsilon I_n) \le r^{-n}J_{st}^{xy}(\upsilon K_n) \le R^{-n}J_{st}^{xy}(\upsilon K_n)$$

and we need only establish

$$R^{-1}K_{n+1}(x, y; s, t) \le J_{st}^{xy}(\upsilon K_n) \le r^{-1}K_{n+1}(x, y; s, t).$$

A straightforward calculation gives us

$$D_{xy}K_{n+1}(x, y; s, t) = K_n(x, y; s, t)\upsilon(x, y)\left[\frac{nQ(x, y; s, t) + 1}{(n + 1)}\right]$$

which implies

$$\begin{aligned} rK_n(x, y; s, t)\upsilon(x, y) &\le D_{xy}K_{n+1}(x, y; s, t) \\ &\le RK_n(x, y; s, t)\upsilon(x, y). \end{aligned}$$

Integrating above inequality and using the fact that $K_{n+1}(s, y; s, t) = K_{n+1}(x, t; s, t) = K_{n+1}(s, t; s, t) = 0$, we obtain the desired result.

To prove the second assertion, we note that $r = R = 1$ if and only if $Q(x, y; s, t) = 1$. If $\upsilon(x, y) = \upsilon_1(x)\upsilon_2(y)$, then obviously this condition is satisfied. Suppose now that $Q(x, y; s, t) = 1$. This implies that

$$D_xVD_yV - VD_{xy}V = 0, \quad V = J_{st}^{xy}(\upsilon).$$

Since $V > 0$ for all $x > 0, y > 0$, setting $U = \ln(V)$, we get $D_{xy}U = 0$ from which we get $V = V_1(x)V_2(y)$ and $\upsilon(x, y) = D_{xy}V = \upsilon_1(x)\upsilon_2(y)$. This completes the proof of the lemma. □

Theorem 7.5.3 ([1]) *Let*

$$N_\upsilon(x, y; s, t) = \sum_{k=0}^{+\infty} I_k(x, y; s, t)$$

denote the Neumann series for the function $\upsilon(x, y) \geq 0, 0 \leq s \leq x \leq a, 0 \leq t \leq y \leq b$, *and let* r, R *be the numbers defined in (7.5.17). Then for all* $0 \leq s \leq x, 0 \leq t \leq y$,

$$E_2(R^{-1}J_{st}^{xy}(\upsilon)) \leq N_\upsilon(x, y; s, t) \leq E_2(r^{-1}J_{st}^{xy}(\upsilon)) \tag{7.5.19}$$

where $E_2(z)$ *is as in (7.5.16).*

Proof The proof is an immediate consequence of Lemma 7.5.1. □

As an example, if $\upsilon(x, y) = x + y$, then $r = 1, R = 9/8$ and for all $x \geq 0, y \geq 0$,

$$E_2(8/9J_{st}^{xy}(\upsilon)) \leq N_\upsilon(x, y; s, t) \leq E_2(J_{st}^{xy}(\upsilon)).$$

These inequalities are particularly useful if $Q(x, y; s, t)$ has a local maximum and a local minimum in the quadrant $x, y \geq 0$.

Using (7.5.19), we can put (7.5.11) in the simpler form

$$u(x, y) \leq M(x, y)E_2(r^{-1}J_{xy}(\upsilon))$$

provided that $M(x, y) \geq 0$.

It is clear that the method of recursion is applicable to systems of inequalities of the form (7.5.1) or (7.5.5) and to more general functional inequalities. We shall not, however, consider these generalizations here.

An example of the flexibility of the method of recursion is the 'partial integration' of inequalities. Suppose the following inequality holds

$$u(x) \leq f(x, u(x)) + J_{0x}(\upsilon(t)u(t)) \tag{7.5.20}$$

and that $\upsilon(t) \geq 0$. Then we may consider the first term on the right-hand side as a known function of x and using (7.5.3), we obtain

$$u(x) \leq f(x, u(x)) + J_{0x}(\upsilon(t)f(t, u(t))N_\upsilon(x, t)) \tag{7.5.21}$$

which may be more useful than the original. If the inequality (7.5.20) is the result of integrating the differential inequality

$$u(x)_{x_1x_2\cdots x_n} \leq g(x, u(x)) + \upsilon(x)u(x),$$

where the subscripts on the left-hand side denote partial differentiation, then (7.5.21) can be considered as a partial integration of this differential inequality.

As an example of the application of (7.5.21), consider the inequality (7.5.1) in the case when the condition $\upsilon(x) \geq 0$ is not satisfied. Writing $v(x) = p(x)-q(x), p(x) \geq 0, q(x) \geq 0$, and using (7.5.21), we obtain

$$u(x) \leq f(x) + J_{0x}(p(t)f(t)N_p(x,t)) - J_{0x}(J_{0t}(q(s)u(s))N_p(x,t))$$
$$-J_{0x}(q(t)u(t)).$$

If a lower bound for $u(x)$ is known, say $u(x) \geq C$, then a useful upper bound may be obtained by replacing $u(t)$ on the right-hand side by C.

Measure differential equations have been investigated by Das and Sharma [184], Leela [333, 334], Raghavendra and Rao [561] and Schmaedeke [593], among others. These equations provide good models for many physical and biological systems. The fact that their solutions are discontinuous renders the conventional methods of ordinary differential equations unapplicable, and thus their study becomes interesting. In [184, 333, 334, 561], the equation

$$Dx = F(t,x) + G(t,x)Du \tag{7.5.22}$$

was studied as an impulsively perturbed system of the ordinary differential equation

$$x' = F(t,x),$$

with $' = \frac{d}{dt}$. In [593], it was investigated from the view point of optimal control theory, that is, G is assumed to be independent of x. Now we are concerned with the system

$$Dx = f(t,x) + AxDu + g(t,x)Du, \tag{7.5.23}$$

which is treated as a perturbed system of the linear system

$$Dx = AxDu. \tag{7.5.24}$$

This gives us a more clear picture of the effect of impulses on the behavior of solutions. Deviations from the conventional theory, which are obviously expected, are noted in particular.

Let $J = [t_0, +\infty), t_0 \geq 0$ and $\mathbb{R}^n$ denote the n-Euclidean space with any convenient norm $|\cdot|$. The same symbol will be used to denote the norm of an n by n matrix. Consider (7.5.23) where $x \in \mathbb{R}^n$, A is an n by n matrix, $u : J \to \mathbb{R}$ is a right-continuous function of bounded variation on every compact sub-interval of $J, f; J \times \mathbb{R}^n \to \mathbb{R}^n$ is Lebesgue integrable, $g : J \times \mathbb{R}^n \to \mathbb{R}^n$ is integrable with respect to the Lebesgue-Stieltjes measure du and Dx, Du denote the distributional derivatives of x and u respectively.

A function $x(t) = x(t, t_0, x_0)$ is a solution of (7.5.23) on J if and only if it satisfies the integral equation

$$x(t) = x_0 + \int_{t_0}^{t} f(s, x(s))ds + \int_{t_0}^{t} [Ax(s) + g(s, x(s))]du(s). \tag{7.5.25}$$

For the proof of this and for the definition of solution of (7.5.23), along with other relevant details, see, e.g., [184].

Remark 7.5.1 In equation (7.5.23), $f(t, x) + AxDu + g(t, x)Du$ is identified with the derivative (in the sense of distributions) of

$$\int_{t_0}^{t} f(s, x(s))ds + \int_{t_0}^{t} [Ax(s) + g(s, x(s))]du(s).$$

When u is an absolutely continuous function, it has the identification $f(t, x) + [A(x) + g(t, x)]u'$, where u' is the ordinary derivative (which exists a.e. on J) of u. In particular, if $u' \equiv 1$, (7.5.23) reduces to the conventional system $x' = f(t, x) + Ax + g(t, x)$.

Let $t_1 < t_2 < \cdots$ denote the discontinuous of u such that $t_1 > t_0$ and $t_k \to +\infty$ as $k \to +\infty$. Suppose further that these discontinuities are isolated. Throughout, except in Lemma 7.5.2 (in which u may be any function of bounded variation), we assume that u has the form

$$u(t) = t + \sum_{k=1}^{+\infty} a_k H_k(t); \quad H_k(t) = \begin{cases} 0, & \text{for all } t < t_k \\ 1, & \text{for all } t \geq t_k \end{cases} \tag{7.5.26}$$

discontinuities (which are obviously countable) are isolated. The above assumption is true reasonable. Moreover, in this case, the predominant effect of the impulses can be visualized. It follows from (7.5.26) that

$$Du = 1 + \sum_{k=1}^{+\infty} a_k \delta(t_k)$$

where $\delta(t_k)$ is the Dirac measure concentrated at t_k. Note that $u' \equiv 1$ a.e. on J. For any $t \in J$, it is clear that there is a unique integer $k \geq 1$ such that $t \in [t_{k-1}, t_k)$.

Let $B_k = E - a_k A, k = 1, 2, \ldots$ where E is the identity n by n matrix. From the assumption on u, it is easy to establish the following theorem.

Theorem 7.5.4 ([527]) *Let B_k be non-singular for each $k = 1, 2 \ldots$. Then, for all $t \in [t_{k-1}, t_k)$ and any $x_0 \in \mathbb{R}^n$, the (unique) solution $x(t) = x(t, t_0, x_0)$ of (7.5.33) is given by*

$$x(t) = \left(\prod_{i=1}^{k-1} B_{k-i}^{-1}\right) e^{(t-t_0)A} x_0. \tag{7.5.27}$$

Here the product $\prod_{i=1}^{k-1}$ is to be understood as E if $k = 1$.

Remark 7.5.2 If $a_k = 0$ for all k, then $B_k (= E)$ is clearly invertible. In this case, (7.5.27) reduces to $x(t) = e^{(t-t_0)A} x_0$, which obviously solves $x' = Ax$. On the other hand, if $a_k \neq 0$ for some k, then a sufficient condition for B_k to be invertible is that a_k^{-1} is not an eigenvalue of A.

Remark 7.5.3 Suppose that a_k^{-1} is an eigenvalue of A for some k. Then, in general, the solution $x(t)$ of (7.5.24) does not exist at $t = t_k$. If $x_0 = 0$, then $x(t)$ is arbitrarily determined at $t = t_k$.

We need the following lemma which is similar to Lemma 2 in [122] or Lemma 3.6 in [627], when u is an absolutely continuous function.

Lemma 7.5.2 ([527]) *Let u be a scalar function of bounded variation on $[t_0, T]$ and let v denote the total variation function of u. Suppose that r and p are non-negative, scalar functions such that r is integrable and p is dv-integrable on $[t_0, T]$. Then, for any positive constants c and M, the following inequality holds for all $t \in [t_0, T]$,*

$$r(t) \leq c e^{M(t-t_0)} + \int_{t_0}^{t} e^{M(t-s)} p(s) dv(s). \tag{7.5.28}$$

Proof Clearly, $r(t) \leq y(t)$, where $y(t)$ is the maximal solution of the integral equation

$$y(t) = c + \int_{t_0}^{t} My(s) ds + \int_{t_0}^{t} p(s) dv(s), \quad t \in [t_0, T]. \tag{7.5.29}$$

Therefore, it is enough to show that any solution of (7.5.29) satisfies the inequality

$$y(t) < (c + \delta) e^{M(t-t_0)} + \int_{t_0}^{t} e^{M(t-t_0)} p(s) dv(s) = z(t) \tag{7.5.30}$$

for all $t \in [t_0, T]$ and for every $\delta > 0$. This will obviously follow if we show that $z(t)$ in (7.5.30) is a solution of the equation

$$z(t) = (c + \delta) + \int_{t_0}^{t} Mz(s) ds + \int_{t_0}^{t} p(s) dv(s). \tag{7.5.31}$$

Here, for the right-hand side of (7.5.31), we obtain

$$(c+\delta)+\int_{t_0}^{t} Mz(s)ds+\int_{t_0}^{t} p(s)dv(s)=(c+\delta)+\int_{t_0}^{t} M(c+\delta)e^{M(s-t_0)}ds+ \\ \int_{t_0}^{t} Me^{Ms}\left\{\int_{t_0}^{s} e^{-M\tau}p(\tau)dv(\tau)\right\}ds+\int_{t_0}^{t} p(s)dv(s). \tag{7.5.32}$$

Denote the first two integrals on the right-hand side of (7.5.32) by I' and I'' respectively. Then

$$I'=(c+\delta)e^{M(t-t_0)}-(c+\delta), \tag{7.5.33}$$

and by integration by parts,

$$\begin{aligned} I'' &= \int_{t_0}^{t}\int_{t_0}^{s} e^{-M\tau}p(\tau)dv(\tau)d(e^{Ms}) \\ &= \int_{t_0}^{t} e^{M(t-s)}p(s)dv(s)-\int_{t_0}^{t} e^{Ms}\left(\int_{t_0}^{s} e^{-M\tau}p(\tau)dv(\tau)\right) \\ &= \int_{t_0}^{t} e^{M(t-s)}p(s)dv(s)-\int_{t_0}^{t} p(s)dv(s). \end{aligned} \tag{7.5.34}$$

From (7.5.32)–(7.5.34), it follows that the equation (7.5.31) reduces to identity, which, of course, is our objective. This completes the qualitative properties of solutions of linear and nonlinear ordinary differential equations under perturbations, through the use of the variation of parameters formula. The theorem that follows gives us an analytic expression for solutions of (7.5.32) in terms of solution of (7.5.33) and the strength of the impulses a_k. In the absence of the impulses, the result reduces to the well-known formula for ordinary differential equations [156].

Theorem 7.5.5 ([527]) *Let the conditions of Theorem 7.5.4 hold. Then, for all $t \in [t_{k-1}, t_k)$, any solution $y(t)=y(t,t_0,x_0)$ of (7.5.32) is given by*

$$y(t)=x(t)+\int_{t_0}^{t} e^{(t-s)A}f(s,y(s))ds+\int_{t_0}^{t} e^{(t-s)A}g(s,y(s))du(s) \\ +e^{tA}\sum_{i=1}^{k-1} a_i\left(\prod_{j=1}^{k-i} B_{k-j}^{-1}\right)A(I_i+J_i) \tag{7.5.35}$$

where $x(t)$ is given by (7.5.27) and

$$I_i = \int_{t_0}^{t_i} e^{-sA} f(s, y(s))ds; \; J_i = \int_{t_0}^{t_i} e^{-sA} g(s, y(s))du(s), 1 \le i \le k-1.$$

Proof Since $u(t) = t$ for all $t \in [t_0, t_1)$, we have that for all $t \in [t_0, t_1)$,

$$y(t) = e^{(t-t_0)A} x_0 + \int_{t_0}^{t} e^{(t-s)A} f(s, y(s))ds + \int_{t_0}^{t} e^{(t-s)A} g(s, y(s))du(s). \quad (7.5.36)$$

At $t = t_1$, (7.5.25) gives us

$$y(t_1) = y(t_1 - h) + \int_{t_1-A}^{t_1} f(s, y(s))ds + \int_{t_1-A}^{t_1} [Ay(s) + g(s, y(s))]du(s) \quad (7.5.37)$$

where $h > 0$. Letting $h \to 0^+$ and using the fact that

$$\lim_{h \to 0^+} \int_{t_1-A}^{t_1} f(s, y(s))ds = 0,$$

we obtain from (7.5.36) and (7.5.37),

$$y(t_1) = e^{t_1 A} \left[e^{-t_0 A} x_0 + I_1 + \int_{t_0}^{t_1-} e^{-sA} g(s, y(s))du(s) \right] + a_1 Ay(t_1) + a_1 g(t_1, y(t_1)). \quad (7.5.38)$$

Now,

$$\int_{t_0}^{t_1^-} e^{(t_1-s)A} g(s, y(s))du(s) + a_1 g(t_1, y(t_1)) = e^{t_1 A} J_1.$$

Therefore, in view of the facts that B_1 is invertible, and B_1^{-1} and $e^{t_1 A}$ commute with each other, (7.5.38) yields

$$\begin{aligned} y(t_1) &= B_1^{-1} e^{t_1 A} [e^{-t_0 A} x_0 + I_1 + J_1] \\ &= x(t_1) + e^{t_1 A} [I_1 + J_1 + a_1 B_1^{-1} A(I_1 + J_1)]. \end{aligned} \quad (7.5.39)$$

For all $t \in [t_1, t_2)$, we know that

$$y(t) = e^{(t-t_1)}y(t_1) + \int_{t_1}^{t} e^{(t-s)A} f(s, y(s))ds + \int_{t_1}^{t} e^{(t-s)A} g(s, y(s))du(s),$$

where $y(t_1)$ is determined by (7.5.39). Thus for all $t \in [t_1, t_2)$,

$$y(t) = x(t) + \int_{t_0}^{t} e^{(t-s)A} f(s, y(s))ds + \int_{t_0}^{t} e^{(t-s)A} g(s, y(s))du(s)$$
$$+ e^{tA} a_1 B_1^{-1} A(I_1 + J_1).$$

As above, it can be shown that

$$y(t_2) = e^{t_2 A}\left[I_2 + J_2 + \sum_{i=1}^{2} a_i \left(\prod_{j=1}^{3-i} B_{3-j}^{-1} \right) A(I_i + J_i) \right].$$

In general, for all $t \in [t_{k-1}, t_k)$, (7.5.35) follows by induction, completing the proof of the theorem. □

Assume the following hypotheses:

(H_1) given by $\varepsilon > 0$, there exist $\delta(\varepsilon) > 0$ and $T(\varepsilon) > t_0$ such that for all $|x| \leq \delta(\varepsilon)$ *and* $t \geq T(\varepsilon)$,

$$|f(t, x)| \leq \varepsilon |x|;$$

(H_2) g satisfies, for all $t \geq t_0$ *and* $|x| \leq r, r > 0$,

$$|g(t, x)| \leq p(t),$$

where p is a dv-integrable ($v(t)$ is the total variation function of $u(t)$ on $[t_0, t], t \in J$) function such that

$$\int_{t_0}^{+\infty} p(s)dv(s) < +\infty; \tag{7.5.40}$$

(H_3) there exist constants P and Q such that

$$\left(\prod_{i=1}^{k} |B_i^{-1}| \right) \quad \textit{and} \quad \sum_{i=1}^{k} |a_i A| \left(\sum_{j=i}^{k} |B_j^{-1}| \right)$$

are bounded by P and Q respectively, as $k \to +\infty$.

Note that, for each $c > 0$, (7.5.40) implies

$$lim_{t \to +\infty} e^{-ct} \int_{t_0}^{t} e^{cs} p(s) dv(s) = 0. \tag{7.5.41}$$

Theorem 7.5.6 ([527]) *Let $(H_1) - (H_2)$ hold. Suppose that all the characteristic roots of A have negative real parts. Then, under the conditions of Theorem 7.5.5, there exist T_0 and $\delta > 0$ such that for every $t_0 \geq T_0$ and x_0 with $|x_0| < \delta$, any solution $y(t) = y(t, t_0, x_0)$ of (7.5.32) satisfies $|y(t)| \to 0$ as $t \to +\infty$. In particular, if (7.5.32) possesses the null solution, then it is asymptotically stable.*

Proof Let $t \in J$ be arbitrary. Then there is an index k such that $t \in [t_{k-1}, t_k)$. Since all the characteristic roots of A have negative real parts, there are positive constants K and α such that $|e^{tA}| \leq Ke^{-\alpha t}$, for all $t > 0$. Let $0 < \varepsilon < \min(\alpha M^{-1}, r)$, where $M = K(Q + 1)$. By (H_1), choose $T(\varepsilon)$ and $\delta(\varepsilon)$ so that $T(\varepsilon) > t_0$ and $\delta(\varepsilon) \leq \varepsilon$. Select $T_0 \geq T(\varepsilon)$ so large that (by (7.5.41)), for all $t \geq T_0$,

$$\int_{t_0}^{t} e^{-(\alpha - M\varepsilon)(t-s)} p(s) dv(s) < (2M)^{-1} \delta(\varepsilon).$$

Let $\delta = (2PK)^{-1}\delta(\varepsilon)$, and consider any $t_0 \geq T_0$ and x_0 satisfying $|x_0| < \delta$. From (7.5.27), (7.5.35) and the conditions of the theorem, we have

$$|y(t)| \leq PK\delta e^{-\alpha(t-t_0)} + \int_{t_0}^{t} Me^{-\alpha(t-s)} |y(s)| ds + \int_{t_0}^{t} Me^{-\alpha(t-s)} p(s) dv(s),$$

as long as $|y(t)| < \delta(\varepsilon)$. By Lemma 7.5.2, this gives us

$$|y(t)| \leq PK\delta e^{-(\alpha - M\varepsilon)(t-t_0)} + M \int_{t_0}^{t} e^{-(\alpha - M\varepsilon)(t-s)} p(s) dv(s),$$

from which the conclusion of the theorem follows in the usual way (see [122, 627]). □

Theorem 7.5.6 remains valid if the condition (7.5.40) in (H_2) is replaced by a more general condition

$$\int_{t}^{t+1} p(s) dv(s) \to 0 \quad as \quad t \to +\infty. \tag{7.5.42}$$

Example 7.5.1 Let $J = [0, +\infty)$. Consider the system (1.3) where

$$x = \begin{bmatrix} x_1 \\ x_2 \end{bmatrix} \quad \text{and} \quad A = \begin{bmatrix} 2 & 5 \\ 0 & 2 \end{bmatrix}.$$

Let $t_k = k$ and $a_k = k^{-1}$ for $k = 1, 2, \ldots$. Then a_2^{-1} is an eigenvalue of A, but a_1^{-1} is not. Choose $x_0 = \left[\begin{smallmatrix}0\\1\end{smallmatrix}\right]$. It can be verified that the solution of (7.5.24) through on the interval $[0, 2]$ is given by

$$x_1(t) = \begin{cases} 5te^{2t}, & 0 \le t < 1, \\ 5(1-t)e^{2t}, & 1 \le t < 2. \end{cases} \qquad x_2(t) = \begin{cases} e^{2t}, & 0 \le t < 1, \\ -e^{2t}, & 1 \le t < 2. \end{cases}$$

However, $x(t)$ does not exist at $t = 2$. If we choose $x_0 = \left[\begin{smallmatrix}0\\0\end{smallmatrix}\right]$, then $x(t) = \left[\begin{smallmatrix}0\\0\end{smallmatrix}\right]$ for $0 \le t < 2$, whereas $x(2) = \left[\begin{smallmatrix}c\\0\end{smallmatrix}\right]$, c an arbitrary constant.

Example 7.5.2 Let $J = [1, +\infty)$. Choose $A = -1$, $t_k = t$ and $a_k = 2(k^3 - 1)^{-1}$ for $k = 2, 3, \ldots$. Then the hypothesis (H_3) is satisfied. Indeed, we have

$$\prod_{k=2}^{+\infty} |B_k^{-1}| = \prod_{k=2}^{+\infty} \left(1 - 2(k^3 + 1)^{-1}\right) = 2/3$$

and

$$\lim_{k\to+\infty} \sum_{i=2}^{k} |a_i A| \left(\prod_{j=1}^{k} |B_j^{-1}| \right) \sum_{k=2}^{+\infty} 2k^{-2} < +\infty.$$

Chapter 8
Applications of Nonlinear Multi-Dimensional Continuous, Discontinuous Integral Inequalities and Discrete Inequalities

8.1 Applications of Theorems 5.1.19–5.1.21 to Partial Differential and Integral Equations

In this section, we shall give some applications of Theorems 5.1.19–5.1.21 to obtain the bounds on the solutions of some partial differential and integral equations. Consider the following partial integral equation of the form

$$u(x,y) = f(x,y) + \int_0^x \int_0^y F[x,y,s,t,u(s,t)]dsdt, \tag{8.1.1}$$

where all the functions f, F are continuous on their respective domains of their definitions and satisfy for all $x \geq 0,\ y \geq 0$,

$$\begin{cases} |f(x,y)| \leq a(x) + b(y), & (8.1.2) \\ |F[x,y,s,t,u(s,t)]| \leq p(s,t)H(|u|), & (8.1.3) \end{cases}$$

where $a(x),\ b(y),\ p(x,y)$ are as defined in Theorem 5.1.19. Using (8.1.2), (8.1.3) in (8.1.1) and then applying Theorem 5.1.19, we can obtain the bound on the solution $u(x,y)$ of (8.1.1).

We also note that the integral inequalities in Theorems 5.1.19–5.1.21 can be used to obtain the bounds on the solution of

$$u_{xy} = g(x,y,u), \tag{8.1.4}$$

$$u_{xy} = F[x,y,u, \int_0^x \int_0^y k(x,y,s,t,u)dsdt], \tag{8.1.5}$$

$$u_{xy} = h(x,y) + F[x,y,u, \int_0^x \int_0^y k_0(x,y,s,t,u_{st})dsdt] \tag{8.1.6}$$

Y. Qin, *Integral and Discrete Inequalities and Their Applications*,
DOI 10.1007/978-3-319-33304-5_8

respectively, under some suitable conditions on the functions involved in (8.1.4)–(8.1.6) together with the suitable given boundary conditions.

8.2 Applications of Theorem 5.1.23 to Hyperbolic Partial Differential Equation

In this section, we shall present some applications of Theorem 5.1.23 to study certain properties of solutions of the following terminal value problem for the hyperbolic partial differential equation

$$u_{xy}(x,y) = h(x,y,u(x,y)) + r(x,y), \tag{8.2.1}$$

$$u(x,+\infty) = \sigma_{\infty}(x), u(0,y) = \tau(y), u(0,+\infty) = k, \tag{8.2.2}$$

where $h : \mathbb{R}_+^2 \times \mathbb{R} \to \mathbb{R}$, $r : \mathbb{R}_+^2 \to \mathbb{R}$, $\tau(y) : \mathbb{R}_+ \to \mathbb{R}$ are continuous functions and k is a real constant.

The following example deals with the estimate on the solution of the partial differential equation (8.2.1) with the conditions (8.2.2).

Example 8.2.1 Let $c(x,y)$ continuous, non-negative, non-decreasing in x and non-increasing in y for all $x, y \in \mathbb{R}_+$, and let

$$|h(x,y,u)| \le c(x,y)d(x,y)|u|, \tag{8.2.3}$$

$$\left|\sigma_{\infty}(x) + \tau(y) - k - \int_0^x \int_y^{+\infty} r(s,t)dtds\right| \le a(x,y) + \int_{\alpha}^{x} b(x,y)g(|u|)ds, \tag{8.2.4}$$

where $a(x,y), b(x,y), d(x,y), g$ are as defined in Theorem 5.1.23. If $u(x,y)$ is a solution of (8.2.1) with the conditions (8.2.2), then it can be written as (see, e.g., [42])

$$u(x,y) = \sigma_{\infty}(x) + \tau(y) - k - \int_0^x \int_y^{+\infty} r(s,t)dtds, \tag{8.2.5}$$

for all $x, y \in \mathbb{R}_+$. From (8.2.3), (8.2.4), (8.2.5), we get

$$|u(x,y)| \le a(x,y) + \int_{\alpha}^{x} b(s,y)g(|u|)ds + c(x,y)\int_0^x \int_y^{+\infty} d(s,t)|u|dtds. \tag{8.2.6}$$

Now, a suitable application of Theorem 5.1.23 to (8.2.6) yields the required estimate

$$u(x,y) \le p(x,y)\left[a(x,y) + c(x,y)e(x,y)\exp\left(\int_0^x \int_y^{+\infty} d(s,t)p(s,t)c(s,t)dtds\right)\right], \tag{8.2.7}$$

for all $x, y \in \mathbb{R}_+$, where $e(x,y), p(x,y)$ are define in Theorem 5.1.23.

The next result deals with the uniqueness of the solution of the partial differential equation (8.2.1) with the conditions (8.2.2).

Example 8.2.2 Suppose that the function h in (8.2.1) satisfies the condition

$$|h(x,y,u) - h(x,y,v)| \le c(x,y)d(x,y)|u - v|, \tag{8.2.8}$$

where $c(x,y), d(x,y)$ are define in Theorem 5.1.23 with $c(x,y)$ is non-increasing in y. Let $u(x,y), v(x,y)$ be two solutions of equation (8.2.1) with the conditions (8.2.2). From (8.2.5), (8.2.7), we infer

$$|u(x,y) - v(x,y)| \le c(x,y)\int_0^x \int_y^{+\infty} d(x,y)|u(s,t) - v(s,t)|dtds. \tag{8.2.9}$$

Now a suitable application of Theorem 5.1.23 yields $u(x,y) = v(x,y)$, that is, there is at most one solution to the problem (8.2.1) with the conditions (8.2.2).

8.3 An Application of Theorem 5.1.29 to Hyperbolic Partial Differential Equations

In this section, we shall present some applications of the inequalities in Theorem 5.1.29 to obtain the lower bounds on the solutions of a class of hyperbolic partial differential and integro-differential equations.

Example 8.3.1 First, we obtain the lower bound on the solution of a nonlinear hyperbolic partial differential equation of the form

$$u_{xy}(x,y) = F[x,y,u(r,y)], \tag{8.3.1}$$

with the given boundary conditions $u(x,t) = u(s,y) = u(s,t)$, where the functions u and f are real-valued, defined, and continuous on the respective domains of their definitions and

$$|F[x,y,u(x,y)]| \le b(x,y)W(|u(x,y)|), \tag{8.3.2}$$

where b and W are as defined in Theorem 5.1.29. Integrating (8.3.1) first with respect to y from y to t, and then with respect to x from x to s, we have

$$u(x,y) = u(s,t) + \int_x^s \int_y^t F[\,m,\ n,\ u(m,n)]\,dm\,dn. \tag{8.3.3}$$

Using (8.3.2) in (8.3.3), we have

$$|u(x,y)| \leq |\,u(s,t)| + \int_x^s \int_y^t b(m,n)\,W(|u(m,n)|)\,dm\,dn,$$

i.e.,

$$|u(s,t)| \geq |u(x,y)| - \int_x^s \int_y^t b(m,n)\,W(|u(m,n)|)\,dm\,dn.$$

Now applying Theorem 5.1.29 yields

$$|\,u(s,t)| \geq \Omega^{-1}[\Omega(|\,u(x,y)|) - \int_x^s \int_y^t b(m,n)\,dm\,dn], \tag{8.3.4}$$

where Ω and Ω^{-1} are as defined in Theorem 5.1.29.Thus the right-hand side in (8.3.4) gives a lower bound on the solution $u(s,t)$ of (8.3.1).

8.4 Applications of Theorem 5.1.41 to Nonlinear Retarded Differential Equation

In this section, we present some applications of Theorems 5.1.41. First, we obtain an explicit bound on the solution of a retarded partial differential equation of the form

$$D_2(z^{p-1}(x,y)D_1z(x,y)) = F(x,y,z(x-h_1(x),y-g_1(x)),\cdots,z(x-h_n(x),y-g_n(y)))) \tag{8.4.1}$$

for all $(x,y) \in I_1 \times I_2$, with the given initial boundary conditions

$$z(x,y_0) = e_1(x),\ z(x_0,y) = e_2(y),\ e_1(x_0) = e_2(y_0) = 0, \tag{8.4.2}$$

where $p > 1$ is a constant, $F \in C(\Delta \times \mathbb{R}^n, \mathbb{R}), e_1 \in C^1(I_1, \mathbb{R}), e_2 \in C^1(I_2, \mathbb{R})$, and $h_i \in C(I_1, \mathbb{R}_+), g_i \in C(I_2, \mathbb{R}_+)$ are non-increasing and such that $x - h_i(x) \geq 0, x - h_i(x) \in C^1(I_1, I_1), y - g_i(y) \geq 0, y - g_i(y) \in C^1(I_2, I_2), h_i'(t) < 1, g_i'(t) < 1, h_i(x_0) = g_i(y_0) = 0$ for $i = 1, ..., n; x \in I_1, y \in I_2$.

Theorem 8.4.1 ([523]) *Suppose that*

$$\begin{cases} |F(x, y, u_1, \cdots, u_n)| \le \sum_{i=1}^{n} b_i(x, y)|u_i|, & (8.4.3) \\ |e_1^p(x) + e_2^p(y)| \le c, & (8.4.4) \end{cases}$$

where $b_i(x, y)$ and c are as in Theorem 5.1.41. Let

$$M_i = \max_{x \in I_1} \frac{1}{1 - h_i'(x)}, \quad N_i = \max_{y \in I_2} \frac{1}{1 - g_i'(y)}, \quad i = 1, \cdots, n. \tag{8.4.5}$$

If $z(x, y)$ is any solution of the problem (8.4.1)–(8.4.2), then for all $x \in I_1, y \in I_2$,

$$|z(x, y)| \le \left\{ c^{\frac{p-1}{p}} + (p - 1) \sum_{i=1}^{n} \int_{\phi_i(x_0)}^{\phi_i(x)} \int_{\psi_i(x_0)}^{\psi_(y)} \bar{b}(\sigma, \tau) d\tau d\sigma \right\}^{\frac{1}{p-1}} \tag{8.4.6}$$

where $\phi_i(x) = x - h_i(x), x \in I_1, \psi_i(y) = y - g_i(y), y \in I_2, \bar{b}(\sigma, \tau) = M_i N_i(\sigma + h_i(s), \tau + g_i(y))$ for $\sigma, s \in I_1; \tau, t \in I_2$.

Proof It is easy to see that the solution $z(x, y)$ of the problem (8.4.1)–(8.4.2) satisfies the equivalent integral equation

$$z^p(x, y) = e_1^p(x) + s_2^p(y) + p \int_{x_0}^{x} \int_{y_0}^{y} F(s, t, z(s - h_i(s), t - g_i(t)), \cdots, z(s - h_n(s), t - g_n(s))) dt ds. \tag{8.4.7}$$

From (8.4.3)–(8.4.5), (8.4.7) and making the change of variables, we have

$$\begin{aligned} z^p(x, y) &\le c + p \int_{x_0}^{x} \int_{y_0}^{y} \sum_{i=1}^{n} b_i(s, t)|z(s - h_i(s), t - g_i(t))| dt ds \\ &\le c + p \sum_{i=1}^{n} \int_{x_0}^{x} \int_{y_0}^{y} \bar{b}_i(\sigma, \tau)|z(\sigma), \tau)| d\tau d\sigma. \end{aligned} \tag{8.4.8}$$

Now applying Theorem 5.1.41, part (d_1) (when $a_i = 0$) to (8.4.8) yields (8.4.6). □

Next, we obtain an explicit bound on the solution of a retarded partial differential equation of the form

$$\begin{aligned} &D_2(z^{p-1}(x, y) D_1 z(x, y)) \\ &\quad = F(x, y, z(x - h_1(x), y - g_1(y), \cdots, z(x - h_n(x), y - g_n(y)))), \end{aligned} \tag{8.4.9}$$

for $(x, y) \in \Delta$, with the given initial boundary conditions

$$z(x, y_0) = e_1(x), \quad z(x_0, y) = e_2(y), \quad e_1(x_0) = e_2(x_0) = 0, \tag{8.4.10}$$

where $p > 1$ is a constant, $F \in C(\Delta \times \mathbb{R}^n, \mathbb{R}), e_1 \in C^1(I_1, \mathbb{R}), e_2 \in C^1(I_2, \mathbb{R}$, and $h_i \in C(I_1, \mathbb{R}_+), g_i \in C(I_2, \mathbb{R}_+)$ are non-increasing and such that $x - h_i(x) \geq 0$, $x - h_i(x) \in C^1(I_1, I_1), y - g_i(y) \in C^1(I_2, I_2), h_i'(t) < 1, g_i'(t) < 1, h_i(t_0) = g_i(t_0) = 0$ for $i = 1, \cdots, n; x \in I_1, y \in I_2$.

Theorem 8.4.2 ([523]) *Suppose that*

$$|F(x, y, u_1, \cdots, u_n)| \leq \sum_{i=1}^{n} b_i(x, y)|u_i|, \tag{8.4.11}$$

$$|e_1^p(x) + e_2^p(y)| \leq c, \tag{8.4.12}$$

where $b_i(x, y)$ and c are as in Theorem 5.1.41. Let

$$M_i = \max_{x \in I_i} \frac{1}{1 - h_i'(x)}, \quad N_i = \max_{y \in I_2} \frac{1}{1 - g_i'(y)}, \quad i = 1, \cdots, n. \tag{8.4.13}$$

If $z(x, y)$ is any solution of the problem (8.4.9)–(8.4.10), then we have for all $x \in I_1, y \in I_2$,

$$|z(x, y)| \leq \left\{ c^{\frac{p-1}{p}} + (p - 1) \sum_{t=1}^{n} \int_{\phi_i(x_0)}^{\phi_i(x)} \int_{\psi_i(y_0)}^{\psi_i(y)} \bar{b}_i(\sigma, \tau) d\sigma d\tau \right\}^{\frac{1}{p-1}}, \tag{8.4.14}$$

where $\phi_i(x) = x - h_i(x), x \in I_1, \psi_i(y) = y - \psi(y), y \in I_2, \bar{b}_i(\sigma, \tau) = M_i N_i b_i(\sigma + h_i(s), \tau + g_i(t))$ for $\sigma, s \in I_1; \tau, t \in I_2$.

Proof It is easy to see that the solution $z(x, y)$ of the problem (8.4.9)–(8.4.10) satisfies the equivalent integral equation

$$\begin{aligned} z^p(x, y) = e_1^p(x) + e_2^p(y) + p \int_{x_0}^{x} \int_{y_0}^{y} & F(s, t, z(s - h_1(s), t - g_1(t)), \\ & \cdots, z(s - h_n(s), t - g_n(t)) dt ds. \end{aligned} \tag{8.4.15}$$

From (8.4.11)–(8.4.13), (8.4.15) and making the change of variables, we have

$$\begin{aligned} |z(x, y)|^p &\leq c + p \int_{x_0}^{x} \int_{y_0}^{y} \sum_{i=1}^{n} b_i(s, t)|z(s - h_i(s), t - g_i(s))| dt ds, \\ &\leq c + p \sum_{i=1}^{n} \int_{\phi_i(x_0)}^{\phi_i(x)} \int_{\psi_i(y_0)}^{\psi_i(y)} \bar{b}_i(\sigma, \tau)|z(\sigma, \tau)| d\tau d\sigma. \end{aligned} \tag{8.4.16}$$

Now a suitable application of the inequality in Theorem 5.1.41, part (1) (when $a_i = 0$) to (8.4.16) yields (8.4.14). □

8.5 Applications of Theorem 5.1.43 and Corollary 5.1.6 to Partial Differential Equations

In this section, we shall use some results in Theorem 5.1.43 and Corollary 5.1.6 to study the following two problems.

First, we consider the partial differential equation

$$D_1 D_2 u^{\ell}(x,y) = h_1(x,y,u(x,y)) + r(x,y), \tag{8.5.1}$$

$$u^{\ell}(x,\infty) = \sigma_{\infty}(x), \quad u^{\ell}(0,y) = \tau(y), \quad u^{\ell}(0,\infty) = k, \tag{8.5.2}$$

where $h_1 \in C(\mathbb{R}_+^2 \times \mathbb{R}, \mathbb{R})$, $r \in C(\mathbb{R}_+^2, \mathbb{R}_+)$, $\ell \geq 1$ and k are real constants.

Assume that

$$\begin{cases} |h_1(x,y,u)| \leq |u|^{\ell-1}(d(x,y)w(|u|) + e(x,y)), \\ |\sigma_{\infty}(x) + \tau(y) - k| \leq a(x,y), \end{cases} \tag{8.5.3}$$

where $a(x,y)$, $d(x,y)$, $e(x,y)$ and $w(u)$ are defined as in Theorem 5.1.43. If $u(x,y)$ is a solution of (8.5.1) with condition (8.5.2), then it can be written as (see, [42])

$$\begin{aligned} u^{\ell}(x,y) = \sigma_{+\infty}(x) + \tau(y) - k - \int_0^x \int_y^{+\infty} r(s,t)\, dt\, ds \\ - \int_0^x \int_y^{+\infty} h_1(s,t,u(s,t))\, dt\, ds \end{aligned} \tag{8.5.4}$$

for all $x,\ y \in \mathbb{R}_+$. Applying (8.5.3) to (8.5.4), we can get for all $x,\ y \in \mathbb{R}_+$,

$$\begin{aligned} |u(x,y)|^{\ell} &\leq a(x,y) + \int_0^x \int_y^{+\infty} |r(s,t)|\, dt\, ds \\ &+ \int_0^x \int_y^{+\infty} |u(s,t)|^{\ell-1}[d(s,t)w(|u(s,t)|) + e(s,t)]\, dt\, ds. \end{aligned} \tag{8.5.5}$$

Applying Theorem 5.1.43 to (8.5.5) yields for all $0 \leq x \leq \tilde{x}_1,\ \tilde{y}_1 \leq y < +\infty$,

$$\begin{aligned} u(x,y) &\leq G^{-1}\left(G\left[\left(a(x,y) + \int_0^x \int_y^{+\infty} r(s,t)\, dt\, ds\right)^{1/\ell} + E_{\ell}(x,y)\right]\right. \\ &\left. + \frac{1}{\ell}\int_0^x \int_y^{+\infty} d(s,t)\, dt\, ds\right) \end{aligned} \tag{8.5.6}$$

where

$$E_\ell(x,y) = \frac{1}{\ell}\int_0^x \int_y^{+\infty} e(s,t)\,dt\,ds, \tag{8.5.7}$$

and G, G^{-1} are defined as in Theorem 5.1.43, and $\tilde{x}_1,\ \tilde{y}_1 \in \mathbb{R}_+$ are chosen so that the quantity inside the curly brackets in (8.5.6) is in the range of G.

Second, we consider the partial differential equation

$$\begin{cases} D_1D_2u^\ell(x,y) = h_2(x,y,u(x,y),\log u(x,y)) + D_2g(x,y,u(x,y)), & (8.5.8)\\ u^\ell(x,+\infty) = \sigma_\infty(x), \quad u^\ell(0,y) = \tau(y), \quad u^\ell(0,+\infty) = k, & (8.5.9)\end{cases}$$

where $h_2 \in C(\mathbb{R}_+^3 \times \mathbb{R}, \mathbb{R})$, $g \in C(\mathbb{R}_+^3, \mathbb{R})$, σ_∞, $\tau \in C(\mathbb{R}_+, \mathbb{R}_+)$, $\ell, k > 0$ are constants. Assume that for all x, $y \in \mathbb{R}_+$ and all $u > 0$,

$$\begin{cases} |h_2(x,y,u,\log u)| \le u^\ell[f(x,y)w(|\log u|) + e(x,y)],\\ |g(x,y,u)| \le c(x,y)u^\ell,\\ |\sigma_\infty(x) + \tau(y) - k - \int_0^x g(s,\infty,\sigma_\infty(s)\,ds| \le a(x,y), \end{cases} \tag{8.5.10}$$

where $a(x,y)$, $c(x,y)$, $e(x,y)$, $f(x,y)$, and $w(u)$ are as defined in Corollary 5.1.6. If $u(x,y) \in C(\mathbb{R}_+, \mathbb{R}_1)$ is a solution of equation (8.5.8) with condition (8.5.9), then it can be written as (see, [42])

$$\begin{aligned} u^\ell(x,y) &= \sigma_\infty(x) + \tau(y) - k - \int_0^x g(s,+\infty,\sigma_\infty(s))\,ds\\ &\quad + \int_0^x g(s,y,u(s,y))\,ds - \int_0^x\int_y^{+\infty} h_2(s,t,u(s,t),\log u(s,t))\,dt\,ds \end{aligned} \tag{8.5.11}$$

for all x, $y \in \mathbb{R}_+$.

Applying (8.5.10) to (8.5.11), we obtain for all x, $y \in \mathbb{R}_+$,

$$\begin{aligned} u^\ell(x,y) &\le a(x,y) + \int_0^x c(s,t)u^\ell(s,y)\,ds\\ &\quad + \int_0^x\int_y^{+\infty} u^\ell(s,t)[f(s,t)w(\log u(s,t)) + e(s,t)]\,dt\,ds. \end{aligned} \tag{8.5.12}$$

Applying Corollary 5.1.6 to (8.5.12) yields, for all $0 \le x \le \tilde{x}_2,\ \tilde{y}_2 \le y < +\infty$,

$$u(x,y) \le \exp\left(G^{-1}\left(G[\frac{1}{\ell}(p^*(x,y)a(x,y)) + E_\ell(x,y)]\right.\right.$$
$$\left.\left. + \frac{1}{\ell}p^*(x,y)\int_0^x\int_y^\infty f(s,t)\,dt\,ds\right)\right) \tag{8.5.13}$$

where

$$p^*(x,y) = 1 + \int_0^x c(s,y)\exp\left(\int_s^x c(m,y)\,dm\right)ds, \tag{8.5.14}$$

and $G,\ G^{-1}$ are as defined in Corollary 5.1.6, $E_\ell(x,y)$ is as defined above, and $\tilde{x}_2, \tilde{y}_2 \in \mathbb{R}_+$ are chosen so that the quantity inside the curly brackets in (8.5.13) is in the range of G.

8.6 Application of Theorem 5.1.49 to Partial Differential Equations

Consider the partial differential equation

$$\begin{cases} D_1D_2v(x,y) = \dfrac{1}{(x+1)^2(y+1)^2} + \exp(-x)\exp(-y)\sqrt{|v(x,y)|+1} \\ \qquad\qquad + x\exp(-x)\exp(-y)\Im v(x,y), & (8.6.1) \\ v(x,+\infty) = \sigma(x),\ v(0,y) = \tau(y),\ v(0,+\infty) = k & (8.6.2) \end{cases}$$

for all $x, y \in \mathbb{R}_+$, where $\sigma, \tau \in C(\mathbb{R}_+, \mathbb{R})$, $\sigma(x)$ is non-decreasing in x, $\tau(y)$ is non-increasing in y, k is a real constant, and $\Im$ is a continuous operator on $C(\mathbb{R}_+ \times \mathbb{R}_+, \mathbb{R})$ such that $|\Im v| \le c_0|v|$ for a constant $c_0 > 0$. Integrating (8.6.1) with respect to x and y and using the initial conditions (8.6.2), we get

$$v(x,y) = \sigma(x) + \tau(y) - k - \frac{x}{(x+1)(y+1)}$$
$$- \int_0^x\int_y^{+\infty} \exp(-s)\exp(-t)\sqrt{|v(s,t)|+1}\,dt\,ds \tag{8.6.3}$$
$$- \int_0^x\int_y^{+\infty} s\exp(-s)\exp(-t)\Im v(s,t)\,dt\,ds.$$

Thus,

$$
\begin{aligned}
|v(x,y)| \le |\sigma(x)+\tau(y)-k| &+ \frac{x}{(x+1)(y+1)} \\
&+ \int_0^x\int_y^{\infty} \exp(-s)\exp(-t)\sqrt{|v(s,t)|+1}\,dt\,ds \qquad (8.6.4)\\
&+ \int_0^x\int_y^{\infty} s\exp(-s)\exp(-t)c_0|v(s,t)|\,dt\,ds.
\end{aligned}
$$

Letting $u(x,y)=|v(x,y)|$, we have

$$
\begin{aligned}
u(x,y) &\le a(x,y) + \int_0^x\int_y^{+\infty} d_1(x,y,s,t)w_1(u)\,dt\,ds \\
&+ \int_0^x\int_y^{+\infty} d_2(x,y,s,t)w_2(u)\,dt\,ds, \qquad (8.6.5)
\end{aligned}
$$

where $a(x,y)=|\sigma(x)+\tau(y)-k|+x/(x+1)(y+1)$, $w(u)=\sqrt{u+1}$, $w_2(u)=c_0u$, $d_1(x,y,s,t)=\exp(-s)\exp(-t)$, $d_2(x,y,s,t)=s\exp(-s)\exp(-t)$. Clearly, $w_2(u)/w_1(u)=c_0(u\sqrt{u+1})$ is non-decreasing for $u>0$, that is ,$w_1\propto w_2$. Then for all $u_1,\ u_2>0$,

$$b_1(x,y)=a(x,y),\ \tilde{d}_1(x,y,s,t)=d_1(x,y,s,t),\ \tilde{d}_2(x,y,s,t)=d_2(x,y,s,t),$$

$$W_1(u)=\int_{u_1}^{u}\frac{dz}{\sqrt{z+1}}=2(\sqrt{u+1}-\sqrt{u_1+1}),\ W_1^{-1}(u)=(\frac{u}{2}+\sqrt{u_1+1})^2-1,$$

$$W_2=\int_{u_2}^{u}\frac{dz}{c_0z}=\frac{1}{c_0}ln\frac{u}{u_2},\ W_2^{-1}(u)=u_2\exp(c_0u),$$

$$
\begin{aligned}
b_2(x,y) &= W_1^{-1}[W_1(b_1(x,y))+\int_0^x\int_y^{+\infty}\tilde{d}_1(x,y,s,t)\,dt\,ds] \\
&= W_1^{-1}[2(\sqrt{b_1(x,y)+1}-\sqrt{u_1+1})+(1-\exp(-x))\exp(-y)] \\
&= [\sqrt{b_1(x,y)+1}+\frac{1-\exp(-x)}{2}\exp(-y)]^2-1.
\end{aligned}
$$

By Theorem 5.1.49, we conclude

$$
\begin{aligned}
|v(x,y)| &\le W_2^{-1}\left[W_2(b_2(x,y))+\int_0^x\int_y^{+\infty}\tilde{d}_2(x,y,s,t)\,dt\,ds\right] \\
&= W_2^{-1}\left[\frac{1}{c_0}ln\frac{b_2(x,y)}{u_2}+(1-(x+1)\exp(-x))\exp(-y)\right]
\end{aligned}
$$

$$
\begin{aligned}
&= u_2 \exp\left[c_0\left(\frac{1}{c_0}\ln\frac{b_2(x,y)}{u_2} + (1-(x+1)\exp(-x))\exp(-y)\right)\right] \\
&= b_2(x,y)\exp[c_0(1-(x+1)\exp(-x))\exp(-y)] \\
&= \left[\left(\sqrt{|\sigma(x)+\tau(y)-k| + \frac{x}{(x+1)(y+1)} + 1} + \frac{1-\exp(-x)}{2}\exp(-y)\right)^2 - 1\right] \\
&\quad \times [c_0(1-(x+1)\exp(-x))\exp(-y)].
\end{aligned}
$$

This implies that the solution of equation (8.6.1) is bounded for all $x, y \in \mathbb{R}_+$ provided that $\sigma(x)+\tau(y)-k$ is bounded for all $x, y \in \mathbb{R}_+$.

8.7 Applications of Theorem 5.1.55 to Nonlinear Hyperbolic Partial Integro-differential Equations

In this section, we present some applications of Theorem 5.1.55 to study the boundedness and uniqueness of the solutions of some nonlinear hyperbolic partial integro-differential equations. These applications are not stated as theorems so as to obscure the main ideas with technical details. It appears that these inequalities will have as many applications for partial integral and integro-differential equations as the classical integral inequality given in Theorem 1.2.11 in Qin [557] and its various generalizations have had for ordinary integro-differential and integral equations.

Example 8.7.1 As a first application, we obtain the bound on the solution of a nonlinear hyperbolic partial integro-differential equation

$$
u_{xy}(x,y) = f(x,y,u(x,y)) + h[x,y,u(x,y), \textstyle\int_{x_0}^{x}\int_{y_0}^{y} k(x,y,s,t,u(s,t))\mathrm{d}s\mathrm{d}t], \tag{8.7.1}
$$

with the given boundary conditions

$$
u(x,y_0) = a_1(x), \quad u(x_0,y) = a_2(x), \quad a_1(x_0) = a_2(y_0) = 0,
$$

where all the functions are real-valued, continuous and defined on a domain D and such that

$$
|f(x,y,u)| \le c(x,y)|u|, \tag{8.7.2}
$$

$$
|k(x,y,s,t,u)| \le q(s,t)|u|, \tag{8.7.3}
$$

$$
|h[x,y,u,v]| \le p(x,y)[|u|+|v|], \tag{8.7.4}
$$

where $c(x,y), p(x,y)$ and $q(x,y)$ are as in (A_1) (see, Sect. 5.1.4). The equation (8.7.1) is equivalent to the Volterra integral equation

$$u(x,y) = a_1(x) + a_2(y) + \int_{x_0}^{x}\int_{y_0}^{y} f(s,t,u(s,t))\mathrm{d}s\mathrm{d}t + \int_{x_0}^{x}\int_{y_0}^{y} h[s,t,u(s,t), \int_{x_0}^{s}\int_{y_0}^{t} k(s,t,\xi,\eta,u(\xi,\eta)\mathrm{d}\xi\mathrm{d}\eta)]\mathrm{d}s\mathrm{d}t, \tag{8.7.5}$$

where $u(x,y)$ is any solution of equation (8.7.1). Using (8.7.2)–(8.7.4) in (8.7.5) and assuming that $|a_1(x)| + |a_2(y)| \leq a(x,y)$, where $u(x,y)$ is as defined in (H_1), we have

$$|u(x,y)| \leq a(x,y) + \int_{x_0}^{x}\int_{y_0}^{y} c(s,t)|u(s,t)|\mathrm{d}s\mathrm{d}t + \int_{x_0}^{x}\int_{y_0}^{y} p(s,t)(|u(s,t)| + \int_{x_0}^{s}\int_{y_0}^{t} q(\xi,\eta)|u(\xi,\eta)|\mathrm{d}\xi\mathrm{d}\eta)\mathrm{d}s\mathrm{d}t.$$

Now an application of Theorem 5.1.55 with $b(x,y) = 1$ yields

$$|u(x,y)| \leq a(x,y) + \int_{x_0}^{x}\int_{y_0}^{y} \omega(s,t;x,y)(a(s,t)(c(s,t) + p(s,t)) + p(s,t)\int_{x_0}^{s}\int_{y_0}^{t} a(\xi,\eta)[c(\xi,\eta) + p(\xi,\eta) + q(\xi,\eta)]\upsilon(\xi,\eta;s,t)\mathrm{d}\xi\mathrm{d}\eta)\mathrm{d}s\mathrm{d}t, \tag{8.7.6}$$

where $\upsilon(s,t;x,y)$ and $\omega(s,t;x,y)$ are the solutions of the characteristic initial value problems (5.1.462) and (5.1.463) in Theorem 5.1.55 respectively with $b(x,y) = 1$. Thus the right hand side in (8.7.6) gives us the bound on the solution $u(x,y)$ of (8.7.1) in terms of the known functions.

If $|a_1(x)| + |a_2(y)| \leq \varepsilon$, where $\varepsilon \geq 0$ is arbitrary, then the bound obtained in (8.16.6) reduces to

$$|u(x,y)| \leq \varepsilon\Big(1 + \int_{x_0}^{x}\int_{y_0}^{y} \omega(s,t;x,y)(c(s,t) + p(s,t) + p(s,t)\int_{x_0}^{s}\int_{y_0}^{t} [c(\xi,\eta) + p(\xi,\eta) + q(\xi,\eta)]\upsilon(\xi,\eta;s,t)\mathrm{d}\xi\mathrm{d}\eta)\mathrm{d}s\mathrm{d}t\Big). \tag{8.7.7}$$

In this case we note that, **Example 8.7.1** implies not only the boundedness but the stability of the solution $u(x, y)$ of equation (8.7.1), if the bound obtained on the right-hand side in (8.7.7) is small enough.

Example 8.7.2 As a second application, we discuss the uniqueness of the solution of the nonlinear hyperbolic partial integro-differential equation (8.7.1). We assume that the functions f, k and h in (8.7.1) satisfy

$$\begin{cases} |f(x,y,u) - f(x,y,\overline{u})| \leq c(x,y)|u - \overline{u}|, & (8.7.8) \\ |k(x,y,s,t,u) - k(x,y,s,t,\overline{u})| \leq q(s,t)|u - \overline{u}|, & (8.7.9) \\ |h[x,y,u,v] - h[x,y,\overline{u},v]| \leq p(x,y)[|u - \overline{u}| + |r - \overline{r}|], & (8.7.10) \end{cases}$$

where $c(x, y), p(x, y)$ and $q(x, y)$ are as in (A_1). The equation (8.7.1) is equivalent to the Volterra integral equation (8.7.5). Now if $u(x, y)$ and $\overline{u}((x, y)$ be two solutions of the given boundary value problem (8.7.1) with the same boundary conditions, then we have

$$\begin{aligned} u - \overline{u} = & \int_{x_0}^{x}\int_{y_0}^{y} (f(s,t,u(s,t)) - f(s,t,\overline{u}(s,t))\mathrm{d}s\mathrm{d}t \\ & + \int_{x_0}^{x}\int_{y_0}^{y} (h[s,t,u, \int_{x_0}^{s}\int_{y_0}^{t} k(s,t,\xi,\eta,u)\mathrm{d}\xi\mathrm{d}\eta)] \\ & - \int_{x_0}^{x}\int_{y_0}^{y} h[s,t,\overline{u}, \int_{x_0}^{s}\int_{y_0}^{t} k(s,t,\xi,\eta,\overline{u})])\mathrm{d}\xi\mathrm{d}\eta)]\mathrm{d}s\mathrm{d}t \end{aligned} \quad (8.7.11)$$

Using (8.7.8)–(8.7.10) in (8.7.11), we have

$$\begin{aligned} |u - \overline{u}| = & \int_{x_0}^{x}\int_{y_0}^{y} c(s,t)|u - \overline{u}|\mathrm{d}s\mathrm{d}t \\ & + \int_{x_0}^{x}\int_{y_0}^{y} p(s,t)(|u - \overline{u}| + \int_{x_0}^{s}\int_{y_0}^{t} q(\xi,\eta)|u - \overline{u}|\mathrm{d}\xi\mathrm{d}\eta)\mathrm{d}s\mathrm{d}). \end{aligned}$$

Thus a suitable application of Theorem 5.1.55 yields, $|u - \overline{u}| \leq 0$. Therefore $u = \overline{u}$; i.e., there is almost one solution of the problem.

8.8 Applications of Corollary 5.1.6 to Initial Boundary Value Problems for Hyperbolic Partial Differential Equations

We shall in this section apply Corollary 5.1.6 to study the boundedness, uniqueness, and continuous dependence of the solutions of initial boundary value problems for hyperbolic partial differential equations.

Consider the following boundary value problem:

$$z^{p-1}z_{xy} + (p-1)z^{p-2}z_x z_y = F(x, y, z(\rho(x), \lambda(y))), \tag{8.8.1}$$

satisfying

$$z(x, y_0) = f(x), \quad z(x_0, y) = g(y), \; f(x_0) = g(y_0) = 0, \tag{8.8.2}$$

where $p \geq 2,\ F \in C(\Delta \times \mathbb{R}, \mathbb{R}),\ f \in C^1(I, \mathbb{R}),\ g \in C^1(J, \mathbb{R}),\ \rho \in C^1(I, I),\ \lambda \in C^1(J, J),\ 0 < \rho', \lambda' \leq 1,\ \rho(x_0) = x_0, \lambda(y_0) = y_0$.

Remark 8.8.1 Setting $\rho(x) = x - h(x)$ and $\lambda(y) = y - k(y)$, problem (8.8.1)–(8.8.2) becomes an initial boundary value problem with delay.

The first result deals with the boundedness of solutions.

Theorem 8.8.1 ([142]) *If*

$$\begin{cases} |F(x, y, v)| \leq b(x, y)|v|^p, & (8.8.3) \\ |f^p(x) + g^q(y)| \leq k. & (8.8.4) \end{cases}$$

where $b \in C(\Delta, \mathbb{R}_+)$ and $k \geq 0$ is a constant, then all solutions $z(x, y)$ of problem (8.8.1)–(8.8.2) satisfy for all $(x, y) \in \Delta$,

$$|z(x, y)| \leq k^{1/p} \exp(\bar{B}(x, y)),$$

where

$$\begin{cases} \bar{B}(x, y) := MN \displaystyle\int_{\rho(x_0)}^{\rho(x)} \int_{\lambda(y_0)}^{\lambda(y)} \bar{b}(\sigma, \tau) d\tau d\sigma, \;\; \bar{b}(\sigma, \tau) := b(\rho^{-1}(\sigma), \lambda^{-1}(\tau)), \\ M := \max\{\dfrac{1}{\rho'(x)} : x \in I\}, \;\; N := \max\{\dfrac{1}{\lambda'(y)} : y \in J\}. \end{cases}$$

In particular, if $\bar{B}$ is bounded on Δ, then every solution z of problem (8.8.1)–(8.8.2) is bounded on Δ.

Proof First observe that $z = z(x, y)$ solves problem (8.8.1)–(8.8.2) if and only if it satisfies the integral equation

$$z^p(x, y) = f^p(x) + g^q(y) + p \int_{x_0}^{x} \int_{y_0}^{y} F(s, t, z(\rho(s), \lambda(t))) dt ds. \tag{8.8.5}$$

Hence by (8.8.3)–(8.8.4),

$$|z(x, y)|^p \leq k + p \int_{x_0}^{x} \int_{y_0}^{y} b(s, t)|z^p(\rho(s), \lambda(t))| dt ds.$$

By a change of variables $\sigma = \rho(s), \tau = (\lambda(t))$, we have

$$|z(x,y)|^p \leq k + p\int_{\rho(x_0)}^{\rho(x)}\int_{\lambda(y_0)}^{\lambda(y)} b(\rho^{-1}(\sigma), \lambda^{-1}(\tau))|z^p(\sigma,\tau)|(\rho^{-1})'(\sigma)(\lambda^{-1})'(\tau)d\tau d\sigma$$
$$\leq k + pMN\int_{\rho(x_0)}^{\rho(x)}\int_{\lambda(y_0)}^{\lambda(y)} \bar{b}(\sigma,\tau)|z^p(\sigma,\tau)|d\tau d\sigma.$$

Thus, an application of Corollary 5.1.6 to the function $|z(x,y)|$ now gives us the assertion immediately. □

The next result concerns uniqueness of solutions to problem (8.8.1)–(8.8.2).

Theorem 8.8.2 ([142]) *If*

$$|F(x,y,v_1) - F(x,y,v_2)| \leq b(x,y)|v_1^p - v_2^p|,$$

where $b \in C(\Delta, \mathbb{R}_+)$, then problem (8.8.1)–(8.8.2) has at most one solution on Δ.

Proof Let $z(x,y)$ and $\bar{z}(x,y)$ be two solutions of problem. By (8.8.5), we have

$$z^p(x,y) - \bar{z}^p(x,y) = p\int_{x_0}^{x}\int_{y_0}^{y}[F(s,t,\bar{z}(\rho(s),\lambda(t))) - F(s,t,z(\rho(s),\lambda(t)))]dtds.$$

By assumption, we then have

$$|z^p(x,y) - \bar{z}^p(x,y)| \leq p\int_{x_0}^{x}\int_{y_0}^{y} b(s,t)|z^p(\rho(s),\lambda(t)) - \bar{z}^p(\rho(s),\lambda(t))|dtds,$$

which, by a change of variables $\sigma = \rho(s)$, $\tau = \lambda(t)$, yields

$$|z^p(x,y) - \bar{z}^p(x,y)| \leq pMN\int_{\rho(x_0)}^{\rho(x)}\int_{\lambda(y_0)}^{\lambda(y)} \bar{b}(\sigma,\tau)[|z^p(\sigma,\tau) - \bar{z}^p(\sigma,\tau)|]d\tau d\sigma.$$
$$= pMN\int_{\rho(x_0)}^{\rho(x)}\int_{\lambda(y_0)}^{\lambda(y)} \bar{b}(\sigma,\tau)[|z^p(\sigma,\tau) - \bar{z}^p(\sigma,\tau)|^{1/p}]^p d\tau d\sigma.$$

Thus, applying Corollary 5.1.6 to the function $|z^p(x,y) - \bar{z}^p(x,y)|^{1/p}$, we conclude that $|z^p(x,y) - \bar{z}^p(x,y)|^{1/p} \leq 0$ for all $(x,y) \in \Delta$ and hence $z = \bar{z}$ on Δ. □

Finally, we shall investigate the continuous dependence of the solutions of problem on the function F and the boundary data. For this, we consider a variation of problem:

$$z^{p-1}z_{xy} + (p-1)z^{p-2}z_x z_y = \bar{F}(x,y,z(\rho(x),\lambda(y))), \tag{8.8.6}$$

satisfying

$$z(x, y_0) = \bar{f}(x), \quad z(x_0, y) = \bar{g}(y), \bar{f}(x_0) = \bar{g}(y_0) = 0, \tag{8.8.7}$$

where $p \geq 2,\ \bar{F} \in C(\Delta \times \mathbb{R}, \mathbb{R}),\ \bar{f} \in C^1(I, \mathbb{R}),\ \bar{g} \in C^1(J, \mathbb{R}),\ \rho \in C^1(I, I),\ \lambda \in C^1(J, J),\ 0 < \rho', \lambda' \leq 1,\ \rho(x_0) = x_0, \lambda(y_0) = y_0$.

Theorem 8.8.3 ([142]) *Consider problem (8.8.1)–(8.8.2) and problem (8.8.6)–(8.8.7). If*

(i) $|F(x, y, v_1) - F(x, y, v_2)| \leq b(x, y)|v_1^p - v_2^p|$ *for some* $b \in C(\Delta, \mathbb{R}_+)$;
(ii) $|(f(x) - \bar{f}(x)) + (g(y) - \bar{g}(y))| \leq \frac{\varepsilon}{2}$;
(iii) for all solutions $\bar{z}(x, y)$ *of problem (8.8.6)–(8.8.7),*

$$\int_{x_0}^{x} \int_{y_0}^{y} |F(s, t, \bar{z}(\rho(s), \lambda(t))) - \bar{F}(s, t, \bar{z}(\rho(s), \lambda(t)))| dtds \leq \frac{\varepsilon}{2},$$

then

$$|z^p(x, y) - \bar{z}^p(x, y)| \leq \varepsilon \exp\left(p\bar{B}(x, y)\right),$$

where $\bar{B}(x, y)$ *is as defined in Theorem 8.8.1. Hence* $z^p(x, y)$ *depends continuously on* $F, f,$ *and* g*. In particular, if* $z(x, y)$ *does not change sign, it depends continuously on* F, f *and* g.

Proof Let $z = z(x, y)$ and $\bar{z} = \bar{z}(x, y)$ be solutions of problem (8.8.1)–(8.8.2) and problem (8.8.6)–(8.8.7), respectively. Then z satisfies (8.8.5) and $\bar{z}$ satisfies

$$\bar{z}^p(x, y) = \bar{f}^p(x) + \bar{g}^q(y) + p \int_{x_0}^{x} \int_{y_0}^{y} \bar{F}(s, t, \bar{z}(\rho(s), \lambda(t))) dtds.$$

Therefore, by assumption (ii),

$$\begin{aligned}
&|z^p(x, y) - \bar{z}^p(x, y)| \\
&\leq \frac{\varepsilon}{2} + p \int_{x_0}^{x} \int_{y_0}^{y} |F(s, t, z(\rho(s), \lambda(t))) - \bar{F}(s, t, \bar{z}(\rho(s), \lambda(t)))| dtds \\
&\leq \frac{\varepsilon}{2} + p \int_{x_0}^{x} \int_{y_0}^{y} |F(s, t, z(\rho(s), \lambda(t))) - F(s, t, \bar{z}(\rho(s), \lambda(t)))| dtds \\
&\quad + p \int_{x_0}^{x} \int_{y_0}^{y} |F(s, t, z(\rho(s), \lambda(t))) - F(s, t, \bar{z}(\rho(s), \lambda(t)))| dtds.
\end{aligned}$$

Now by assumption (i) and by a change of variables $\sigma = \rho(s), \tau = \lambda(t)$,

$$p\int_{x_0}^{x}\int_{y_0}^{y} |F(s,t,\bar{z}(\rho(s),\lambda(t))) - \bar{F}(s,t,\bar{z}(\rho(s),\lambda(t)))|dtds$$

$$\leq p\int_{x_0}^{x}\int_{y_0}^{y} b(s,t)|z^p(\rho(s),\lambda(t)) - \bar{z}^p(\rho(s),\lambda(t))|dtds$$

$$\leq pMN\int_{\rho(x_0)}^{\rho(x)}\int_{\lambda(y_0)}^{\lambda(y)} \bar{b}(\sigma,\tau)|z^p(\sigma,\tau) - \bar{z}^p(\sigma,\tau)|d\tau d\sigma,$$

while by assumption (iii),

$$p\int_{x_0}^{x}\int_{y_0}^{y} |F(s,t,\bar{z}(\rho(s),\lambda(t))) - \bar{F}(s,t,\bar{z}(\rho(s),\lambda(t)))|dtds \leq \frac{\varepsilon}{2},$$

thus

$$|z^p(x,y) - \bar{z}^p(x,y)| \leq \varepsilon + pMN\int_{\rho(x_0)}^{\rho(x)}\int_{\lambda(y_0)}^{\lambda(y)} \bar{b}(\sigma,\tau)|z^p(\sigma,\tau) - \bar{z}^p(\sigma,\tau)|d\tau d\sigma.$$

Applying Corollary 5.1.6 to the function $|z^p(\sigma,\tau) - \bar{z}^p(\sigma,\tau)|^{1/p}$, we have, for all $(x,y) \in \Delta$,

$$|z^p(\sigma,\tau) - \bar{z}^p(\sigma,\tau)|^{1/p} \leq \varepsilon^{1/p}\exp\left(\bar{B}(x,y)\right),$$

or

$$|z^p(\sigma,\tau) - \bar{z}^p(\sigma,\tau)| \leq \varepsilon\exp(p\bar{B}(x,y)).$$

Now when restricted to any compact set, $\bar{B}(x,y)$ is bounded and so

$$|z^p(\sigma,\tau) - \bar{z}^p(\sigma,\tau)| \leq \varepsilon K$$

for some $K > 0$ for all (x,y) lying in the compact set. Hence z^p depends continuously on F, f and g. □

8.9 An Application of Theorem 5.2.7 to Nonlinear Integral Equation

In this section, we apply Theorem 5.2.7 to study the integral equation

$$u(x) = u(s) + \int_{x}^{s} F(t,u(t))dt \tag{8.9.1}$$

under some suitable conditions on the functions involved in (8.9.1) together with the suitable given boundary conditions.

For (8.9.1), assume

$$|F(x, u(x))| \leq b(x)W(|u(x)|), \tag{8.9.2}$$

where b and W are as defined in Theorem 5.2.7, it follows from (8.9.1) and (8.9.2) that

$$|u(s)| \leq |u(x)| - \int_x^s b(m)W(|u(m)|)dm.$$

Now a suitable application of Theorem 5.2.7 yields

$$|u(s)| \leq Q^{-1}[Q(|u(x)|) - \int_x^s b(m)dm],$$

where Q and Q^{-1} are as defined in Theorem 5.2.7. Thus the right-hand side in the above inequality gives $u(s)$ of (8.9.1) as the lower bound on the solution.

8.10 Applications of Theorem 5.2.25 to Nonlinear Hyperbolic Partial Integro-differential Equation

In this section, we present some applications of results in Theorems 5.2.25 and 5.2.26 to the boundedness and behavioral relationships of the solutions of some nonlinear hyperbolic partial integro-differential equations.

Example 8.10.1 First, we consider a nonlinear hyperbolic partial integro-differential equation of the form

$$D_1 \dots D_n u(x) = A[x, u(x), \int_{x^0}^x B(x, y, u(y))dy] + F(x, u(x)), \tag{8.10.1}$$

with the condition prescribed on $x_i = x_i^0$, $1 \leq i \leq n$, where all the functions are defined and continuous on their respective domains of definitions such that

$$\begin{cases} |B(x, y, u)| \leq c(y)|u|, & (8.10.2) \\ |A[x, u, v]| \leq g(x)[|u| + |v|], & (8.10.3) \\ |F(x, u)| \leq K(x, |u|), & (8.10.4) \end{cases}$$

where $c(y), g(x)$ and $K(x, y, \phi) = K(y, \phi)$ are as in Theorem 5.2.26. Let the boundary conditions be such that the given equation (8.10.1) is equivalent to the integral equation

$$u(x) = h(x) + \int_{x^0}^{x} A[y, u(y), \int_{x^0}^{y} B(y, z, u(z))dz]dy$$
$$+ \int_{x^0}^{x} F(y, u(y))dy, \tag{8.10.5}$$

where $h(x)$ depends on the given boundary conditions. We assume that

$$|h(x)| \leq f(x), \tag{8.10.6}$$

where $f(x)$ is as defined in Theorem 5.2.25. Using (8.10.2)–(8.10.4) and (8.10.6) in (8.10.5), we can get

$$|u(x)| \leq f(x) + \int_{x^0}^{x} g(y)|u(y)|dy + \int_{x^0}^{x} g(y)(\int_{x^0}^{y} c(z)|u(z)|dz)dy$$
$$+ \int_{x^0}^{x} K(y, |u(y)|)dy.$$

Now a suitable application of Theorem 5.2.25 with $q(x) = 1, W(x, \phi) = \phi$ and $K(x, y, \phi) = K(y, \phi)$ yields

$$u(x) \leq E_0^*(x)[f(x) + r(x)] \tag{8.10.7}$$

where $E_0^*(x)$ is obtained by substituting $q(x) = 1$ in (5.2.134) of Theorem 5.2.25 and $r(x)$ is a solution of the equation

$$r(x) = \int_{x^0}^{x} K(y, E_0^*(y)[f(y) + r(y)])dy.$$

If the right-hand side of (8.10.7) is bounded, then we obtain the boundedness of the solution $u(x)$ of (8.10.1).

Example 8.10.2 The second application is an example of behavioral relationships between the solutions of (8.10.1) with the conditions prescribed on $x_i = x_i^0, 1 \leq i \leq n$, and the nonlinear hyperbolic integro-differential equation of the form

$$D_1 \dots D_n u(x) = A_0[x, u(x), \int_{x^0}^{x} B(x, y, u(y))dy], \tag{8.10.8}$$

with conditions prescribed on $x_i = x_i^0$, $1 \le i \le n$, where all the functions are defined and continuous on their domains of definitions and such that

$$\begin{cases} |B(x,y,u) - B_0(x,y,v)| \le c(y)|u-v|, & (8.10.9)\\ |A[x,u,\overline{u}] - A_0[x,v,\overline{v}]| \le g(x)[|u-v| + |\overline{u}-\overline{v}|], & (8.10.10)\\ |F(x,u)| \le K(x,|u|). & (8.10.11) \end{cases}$$

where $c(y), g(x)$ and $K(x,\phi)$ are as in **Example 8.10.1**. Equation (8.10.1) and (8.10.8) are equivalent to the integral equations (8.10.5) and

$$u(x) = \overline{h}(x) + \int_{x^0}^{x} A_0[y, v(y), \int_{x^0}^{y} B_0(y,z,v(z))dz]dy, \tag{8.10.12}$$

where $\overline{h}(x)$ depends on the given boundary conditions. From (8.10.5) and (8.10.12), it follows

$$u - v = h(x) - \overline{h}(x) + \int_{x^0}^{x} \{A[y,u,\int_{x^0}^{y} B(y,z,u)dz]$$
$$-A_0[y,v,\int_{x^0}^{y} B_0(y,z,u)dz]dy\} + \int_{x^0}^{x} F(y,u)dy. \tag{8.10.13}$$

Using (8.10.9)–(8.10.11) and $|u| - |v| \le |u-v|$ and assuming that $|h(x) - \overline{h}(x)| \le f(x)$, and the solution $v(x)$ of (8.10.8) is bounded by a constant M in (8.10.13), where $f(x)$ is as defined in Theorem 5.2.25, we have

$$|u-v| \le f(x) + \int_{x^0}^{x} g(y)(|u-v| + \int_{x^0}^{y} c(z)|u-v|dz)dy$$
$$+ \int_{x^0}^{x} K(y, M + |u-v|)dy.$$

Now applying Theorem 5.2.25 yields

$$|u-v| \le E_0^*(x)[f(x) + r(x)], \tag{8.10.14}$$

where $E_0^*(x)$ is as defined in **Example 8.10.1** and $r(x)$ is a solution of the equation

$$r(x) = \int_{x^0}^{x} K(y, E_0^*(y)[f(y) + r(y)])dy. \tag{8.10.15}$$

If the right-hand side in (8.10.14) is bounded, then we obtain the relative boundedness of the solution $u(x)$ of (8.10.1) and (8.10.8). If $f(x)$ in (8.10.14) is small enough and, say, less than ϵ, where $\epsilon > 0$ is arbitrary, if (8.10.1) admits only identically zero solution, and if $E_0^*(x)$ in (8.10.14) is bounded and $\epsilon \to 0$, then we obtain

$|u(x) - v(x)| \to 0^+$, which gives us the equivalent between the solutions of (8.10.1) and (8.10.8).

8.11 Applications of Theorems 5.2.26 and 5.2.29 to Nonlinear Hyperbolic Partial Integro-differential Equations

In this section, we shall apply Theorems 5.2.26 and5.2.29 by obtaining pointwise bounds on the solutions of a certain class of non-linear equations in n-independent variables. We consider the nonlinear hyperbolic partial integro-differential equation

$$\frac{\partial^n \phi(x)}{\partial x_1 \partial x_2 \ldots \partial x_n} = F(x, \phi(x), \int_{x^0}^{x} K(x, y, \phi(y))dy) + G(x, \phi(x)) \qquad (8.11.1)$$

where $F \in C(\Omega \times \mathbb{R} \times \mathbb{R}, \mathbb{R})$ and $G \in C(\Omega \times \mathbb{R}, \mathbb{R})$ with suitable boundary conditions. The solution of equation (8.11.1) is of the form

$$\phi(x) = h(x) + \int_{x^0}^{x} F(s, \phi(s), \int_{x^0}^{x} K(s, y, \phi(y))dy)ds + \int_{x^0}^{x} G(y, \phi(y))dy. \qquad (8.11.2)$$

We shall assume the following conditions:

(H_6) There exists a continuous function $B : \Omega \times \mathbb{R}_+ \to \mathbb{R}_+$ with B non-decreasing in the second variable such that

$$|G(y, \phi(y))| \leq B(y, |\phi(y)|). \qquad (8.11.3)$$

(H_7) There exists a function $f : \Omega \to \mathbb{R}$ satisfying (H_1) such that $|h(x)| \leq f(x)$, for all $x \in \Omega$.

(H_8) There exists a function $g : \Omega \to \mathbb{R}_+$ satisfying the assumption (H_2) such that for all $s \in \Omega$,

$$|F(s, u, v)| \leq g(s)[|u| + |v|]. \qquad (8.11.4)$$

(H_9) There exist functions $\omega : \Omega \times \Omega \to \mathbb{R}$ and $H : \mathbb{R}_+ \to \mathbb{R}_+$ such that

(i) $\omega(s, y)$ is defined and continuous for all $s \geq y \geq x^0$,

(ii)

$$\omega(s,s) \le h_1(s), D_1\omega(s,s_1,s_2,\ldots,s_{j-1},y_j,\ldots,y_n) = 0,\ \ j = 2,3,\ldots,n,$$
$$D_1D_2\ldots D_i\omega(s,y_1,y_2,\ldots,y_i,s_{i+1},y_{i+2},\ldots,y_n) = 0,\ \ i = 1,2,\ldots,n-1,$$
$$D_1D_2\ldots D_n\omega(s,y) \le p(s)h_2(y),$$

where h_1, p, and h_2 are continuous functions and non-negative on Ω with $D_l = \frac{\partial}{\partial s_l},\ \ 1 \le l \le n$.

(iii) H satisfies assumption (H_5) with $H(1) = 1$.

(iv)

$$|K(s,y,\phi(y))| \le \omega(s,y)H(|\phi(y)|). \tag{8.11.5}$$

Remark 8.11.1 It is easy to see that the function $\omega(s,y) = \prod_{l=1}^{n}(s_l - y_l)p(s)h_2(y) + C$, where C is constant, satisfies (i), (ii), and (iii) if $C \le h_1(s)$ for all $s \in \Omega$.

The following lemma, which is a standard result in calculus of several variables, shall be used to obtain pointwise bounds on the solution of equation (8.11.1).

Lemma 8.11.1 ([27]) *Let $G(s) = \int_{x^0}^{s} \omega(s,y)H(\phi(y))dy$ with $x^0 = (x_1^0,\ldots,x_n^0)$, $y = (y_1,y_2,\ldots,y_n)$ and $s = (s_1,s_2,\ldots,s_n) \in \Omega$ with $x^0 < y < s$, and $D_i = \frac{\partial}{\partial s_i}$, $i = 1,2,\ldots,n$. Assume, also, that for $j = 2,3,\ldots,n$, $D_j\omega(s,s_1,s_2,\ldots,s_{j-1},y_j,\ldots,y_n) = 0$ and $D_1D_2\ldots D_k\omega(s,y_1,y_2,\ldots,y_k,s_{k+1},y_{k+2},\ldots,y_n) = 0$, $k = 1,2,\ldots,n-1$. Then,*

$$D_1D_2\ldots D_nG(s) = \omega(s,s)H(\phi(s)) + \int_{x^0}^{s} D_1D_2\ldots D_n\omega(s,y)H(\phi(y))dy.$$

We now compute the pointwise bounds of the integral equation (8.11.2) taking into account the assumptions (H_6)–(H_9).

Taking the bounds in (8.11.2) and using (8.11.3)–(8.11.5), we obtain

$$\begin{aligned}
|\phi(x)| &\le |h(x)| + \int_{x^0}^{x} |G(y,\phi(y))|dy + \int_{x^0}^{x} |F(s,\phi(s),\int_{x^0}^{s} K(s,y,\phi(y))dy|ds \\
&\le f(x) + \int_{x^0}^{x} B(y,|\phi(y)|)dy + \int_{x^0}^{x} g(s)|\phi(s)|ds + \int_{x^0}^{x} g(s)|\int_{x^0}^{s} K(s,y,\phi(y))dy|ds \\
&\le f(x) + \int_{x^0}^{x} B(y,|\phi(y)|)dy + \int_{x^0}^{x} g(s)|\phi(s)|ds \\
&\quad + \int_{x^0}^{x} g(s)\left(\int_{x^0}^{s} \omega(s,y)H(|\phi(y)|)dy\right)ds.
\end{aligned}$$

In view of hypothesis (H_9) and Lemma 8.11.1, if $R(s) = \int_{x^0}^{s} \omega(s,y)H(|\phi(y)|)dy$, then

$$\begin{aligned}
D_1 \ldots D_n R(s) &= \omega(s,s)H(|\phi(s)|) + \int_{x^0}^{s} D_1 D_2 \ldots D_n \omega(s,y)H(|\phi(y)|)dy \\
&\le h_1(s)H(|\phi(s)|) + \int_{x^0}^{s} p(s)h_2(y)H(|\phi(y)|)dy \\
&= h_1(s)H(|\phi(s)|) + p(s)\int_{x^0}^{s} h_2(y)H(|\phi(y)|)dy.
\end{aligned}$$

Upon integrating from x^0 to s, we obtain

$$R(s) \le \int_{x^0}^{s} h_1(u)H(|\phi(u)|)du + \int_{x^0}^{s} p(u)\left(\int_{x^0}^{u} h_2(y)H(|\phi(y)|)dy\right)du.$$

Hence

$$\begin{aligned}
|\phi(x)| \le & f(x) + \int_{x^0}^{x} g(s)|\phi(s)|ds + \int_{x^0}^{x} g(s)\left(\int_{x^0}^{s} h_1(u)H(|\phi(u)|)du\right)ds \\
&+ \int_{x^0}^{x} g(s)\left(\int_{x^0}^{s} p(u)\left(\int_{x^0}^{u} h_2(y)H(|\phi(y)|)dy\right)du\right)ds \\
&+ \int_{x^0}^{x} B(y,|\phi(y)|)dy.
\end{aligned}$$

We now use Theorem 5.2.29 with $g_i = g, j = 1,2,3, q = 1, H_1(|\phi(s)|) = |\phi(s)|$,

$$H_2(|\phi(s)|) = \int_{x^0}^{s} h_1(u)H(|\phi(u)|)du, \quad H_3(|\phi(s)|) = \int_{x^0}^{s} p(u)\left(\int_{x^0}^{u} h_2(y)H(|\phi(y)|)dy\right)du,$$

$$K(x,y,u) = B(y,u), W(s,z) = z,$$

then $m = 3$ and we have

$$\begin{aligned}
|\phi(x)| &\le \{f(x) + r(x)\}\prod_{l=1}^{3} E_i(x) \qquad (8.11.6) \\
&\le \{f(x) + r(x)\}E_1(x)\cdot E_2(x)\cdot E_3(x)
\end{aligned}$$

where $r(x)$ is a solution of the equation

$$V(x) = \int_{x^0}^{x} B(y, E_1(y)E_2(y)E_3(y)\{f(y) + V(y)\})dy$$

and

$$\begin{cases} E_1 = G_1^{-1}[G_1(1) + \int_{x^0}^{x} g(s)ds], \quad G_1(1) + \int_{x^0}^{x} g(s)ds \in Dom\,(G_1^{-1}), \\ E_2(x) = G_2^{-1}[G_2(1) + \int_{x^0}^{x} g(s)E_1(s)H_2(1)ds], \\ \qquad = G_2^{-1}[G_2(1) + \int_{x^0}^{x} g(s)E_1(s)\left(\int_{x^0}^{x} h_1(u)du\right)ds], \\ E_3(x) = G_3^{-1}[G_3(1) + \int_{x^0}^{x} g(s)E_1(s)E_2(s)\left(\int_{x^0}^{s} p(u)\left(\int_{x^0}^{u} h_2(y)dy\right)du\right)ds]. \end{cases}$$

It is clear that we can compute the pointwise bounds of the solution $\phi(x)$ of the integral equation (8.11.2) as in (8.11.6).

8.12 Applications of Theorems 5.2.40 and 5.2.41 to Retarded Nonlinear Hyperbolic Functional Integro-differential Equations

In this section, we shall apply the results in Theorems 5.2.39 and 5.2.40 to investigate properties of solutions of a certain class of nonlinear hyperbolic functional integro-differential equations of the retarded type. We consider the hyperbolic equation

$$\frac{\partial^n u(x)}{\partial x_1 \partial x_2 \ldots \partial x_n} = G(x, u(\sigma(x)), Tu(x)) \tag{8.12.1}$$

together with the given suitable boundary conditions

$$u(x_1, x_{i-1}, x_i^0, x_{i+1}, \ldots, x_n) = 0,\ 1 \le i \le n$$

where

$$G \in C(\mathbb{R}^n \times \mathbb{R} \times \mathbb{R}, \mathbb{R}) \quad Tu(x) = \int_{x_0}^{x} k(x, y, u(\rho(y)))dy,$$

with $k \in C(\mathbb{R}^n \times \mathbb{R} \times \mathbb{R}, \mathbb{R})$ and $\sigma, \rho \in \mathcal{F}$.

Any solution $u(x)$ of equation (8.12.1) satisfying the boundary conditions is also a solution of the Volterra integral equation

$$u(x) = n(x) + \int_{x_0}^{x} G(s, u(\sigma(s)), Tu(s))ds, \tag{8.12.2}$$

where $n(x)$ subjects to the boundary conditions. The following theorem provides an upper bound on the solutions of equation (8.12.1).

The following theorem gives us another bound on the solutions of equation (8.12.1).

Theorem 8.12.1 ([28]) *Assume that*

(i) $|G(x, u(\sigma(x)), Tu(x))| \le p(x)\Omega(|u(\sigma(x))|)$ *where p is a continuous non-negative real-valued function such that* $\int_{x_1^0}^{+\infty} \cdots \int_{x_n^0}^{+\infty} p(s)ds < +\infty$ *and* $\Omega \in \mathcal{F}_1$ *with* $D_k\Omega(u(x)) \ge 0$ *for* $k = 2, 3, \ldots, n$,

(ii) $n(x)$ *is non-zero, non-decreasing function such that for a* $M > 0$,

$$1 \le |n(x)| \le M.$$

Then solutions of equation (8.12.1) are bounded.

Proof Again using (8.12.2),

$$\begin{aligned}|u(x)| &\le |n(x)| + \int_{x_0}^{x} |G(s, u(\sigma(s)), Tu(s))|ds \\ &\le |n(x)| + \int_{x_0}^{x} p(s)\Omega(|u(\sigma(x))|)ds.\end{aligned}$$

Applying Theorem 5.2.40 to the above inequality, we have

$$|u(x)| \le |n(x)|G^{-1}\Big[G(1) + \int_{x_0}^{x} p(s)ds\Big].$$

Now $G(r) = \int_{r_0}^{r} \frac{ds}{\Omega(s)}$ implies $G'(r) > 0$, so that G^{-1} exists and is an increasing function. Hence by assumption (i),

$$\begin{aligned}|u(x)| &\le M[G(1) + \int_{x_1^0}^{+\infty} \cdots \int_{x_n^0}^{+\infty} p(s)ds] \\ &\le N,\end{aligned}$$

where N is a positive constant. □

Theorem 8.12.2 ([28]) *Assume that*

(i) $|G(x, u(\sigma(x)), Tu(x))| \le f(x)\omega||u(\sigma(x))| + |Tu(x))||$ *where f is a continuous non-negative real-valued function such that* $\int_{x_1^0}^{+\infty} \cdots \int_{x_n^0}^{+\infty} f(s)ds < +\infty$ *and* $\omega \in \mathcal{F}_1$.

(ii) There exist continuous functions g and Ω such that $\Omega : [0, +\infty) \to [0, +\infty)$ *is non-decreasing non-negative sub-multiplicative for all* $u > 0$ *and* $\Omega(0) = 0$, *g is non-negative,* $\int_{x_1^0}^{+\infty} \cdots \int_{x_n^0}^{+\infty} g(s)ds < +\infty$, *and*

$$|Tu(x))| \le g(x)\Omega(|u(\rho(x))|), \quad p \in \mathcal{F},$$

(iii) $n(x)$ *satisfies hypothesis (ii) of Theorem 8.12.1.*

Then solutions of equation (8.12.1) are bounded.

Proof In fact,

$$\begin{aligned}|u(x)| &\leq |n(x)| + \int_{x_0}^{x} f(s)\omega(|u(s)|)ds + \int_{x_0}^{x} g(s)\Omega(|u(\rho(s))|)ds \\ &\leq M + \int_{x_0}^{x} f(s)\omega(|u(\sigma(s))|)ds + \int_{x_0}^{x} g(s)\Omega(|u(\rho(s))|)ds.\end{aligned}$$

Applying Theorem 5.2.41, with $q_1(x) = q_2(x) = 1$, to the above inequality, we obtain

$$\begin{aligned}|u(x)| \leq\ & G^{-1}[G(1) + \int_{x_0}^{x} f(s)ds] \\ &\times F^{-1}\left[F(M) + \int_{x_0}^{x} g(s)ds\Omega\Big(G^{-1}\Big(G(1) + \int_{x_1^0}^{+\infty}\cdots\int_{x_n^0}^{+\infty} f(t)dt\Big)\Big)ds\right]\end{aligned}$$

where

$$G(v) = \int_{v_0}^{v} \frac{ds}{\omega(s)}, \quad v \geq v^0 > 0, \;\; F(r) = \int_{r_0}^{r} \frac{ds}{\Omega(s)}, \quad r \geq r^0 > 0$$

and G^{-1} and F^{-1} are the inverses of G and F, respectively. Clearly G^{-1} and F^{-1} are increasing functions so that

$$\begin{aligned}|u(x)| \leq\ & G^{-1}\left[G(1) + \int_{x_1^0}^{+\infty}\cdots\int_{x_n^0}^{+\infty} f(s)ds\right] \\ &\times F^{-1}\left[F(M) + \int_{x_1^0}^{+\infty}\cdots\int_{x_n^0}^{+\infty} g(s)ds\Omega\right. \\ &\left.\times\Big(G^{-1}\Big(G(1) + \int_{x_1^0}^{+\infty}\cdots\int_{x_n^0}^{+\infty} f(t)dt\Big)\Big)ds\right].\end{aligned}$$

Using hypotheses (i) and (ii), we see clearly that the right-hand side of the last inequality is bounded, which completes the proof. □

8.13 Applications of Theorems 5.2.44, 5.2.46 and 5.2.47 to Integral and Differential Equations

In this section, we shall apply Theorems 5.2.44, 5.2.46 and 5.2.47 to study some integral initial value problems.

Example 8.13.1 Consider the n-variable integral inequality

$$u(x) \leq f(x) + \sum_{i=1}^{N}(T_i u)(x) + g(x)\left\{\int_0^x h(x,s)(u(s))^p ds\right\}^q, \tag{8.13.1}$$

where operators T_i are as Theorem 5.4.30 in Qin [557], p and $q \geq 1$ are positive numbers; $u, f, g : \mathbb{R}_+^n \to \mathbb{R}_+$ and $h, k_{ij} : \mathbb{R}_+^n \times \mathbb{R}_+^n \to \mathbb{R}_+$ are continuous functions. Comparing (8.13.1) with (5.2.262) in Theorem 5.2.46, here we have $G(m) = m^q$, $Q(m) = m^p$, and hence $G \in F(\psi)$ with $\psi(v) = v^{q-1}$ and $Q \in H(\varphi)$ with $\varphi(v) = v^p$ hold. Hence, an upper bound for the solutions of equation (8.13.1) can be obtained by using Theorem 5.2.46 or Theorem 5.2.47. Here we leave this simple computation to the reader.

Remark 8.13.1 The special case of (8.13.1), when $N = 1$, $h(x,s) = h(s)$, and $k_{11}(x,s)$ is directly variable-separable, was discussed in [173] under some additional restrictive assumptions.

Example 8.13.2 Consider the initial value problem

$$\begin{cases} \frac{\partial^2 u(x,y)}{\partial x \partial y} = F\left[x, y, u(x,y), \frac{\partial u(x,y)}{\partial x}, \frac{\partial u(x,y)}{\partial y}\right], \\ u(x,0) = \xi(x),\ u(0,y) = \eta(y), \quad \text{with } \xi(0) = \eta(0),\ x, y \in \mathbb{R}_+, \end{cases} \tag{8.13.2}$$

where functions $\xi, \eta \to \mathbb{R}$ and $F : \mathbb{R}_+ \times \mathbb{R}_+ \times \mathbb{R}^3 \to \mathbb{R}$ are continuous functions. We assume that u and all the involved partial derivatives are continuous. As is well-known, problem (8.13.2) can be reformulated in terms of the following Volterra integral equation

$$u(x,y) = \xi(x) + \eta(y) - \xi(0) + \int_0^x \int_0^y F\left[s, t, u(s,t), \frac{\partial u(s,t)}{\partial s}, \frac{\partial u(s,t)}{\partial t}\right] dtds, \quad x, y \in \mathbb{R}_+. \tag{8.13.3}$$

Suppose that the condition

$$|F[s,t,u,v,w]| \leq j(s,t)Q(|u|) \tag{8.13.4}$$

holds and $Q : \mathbb{R}_+ \to \mathbb{R}_+$ is a known differentiable function with $Q \in H(\varphi)$ for some function φ, and $Q(u) > 0$ for all $u > 0$. Then, using (8.13.4) we derive from (8.13.3) that for all $x, y \in \mathbb{R}_+$,

$$|u(x,y)| \leq |\xi(x) + \eta(y) - \xi(0)| + \int_0^x \int_0^y j(s,t)Q(|u(s,t)|)dtds.$$

Now, applying Theorem 5.2.44 (with $n = 2$, $g(x) = 1$ and $G(m) = m$) to the above inequality yields, for all $x, y \in \mathbb{R}_+$ satisfying $(0,0) \leq (x,y) \leq (X,Y)$,

$$|u(x,y)| \leq K(x,y) \left\{ 1 + \widetilde{H}^{-1} \left[\int_0^x \int_0^y \hat{j}(s,t)\varphi(K(s,t))dtds \right] \right\}, \tag{8.13.5}$$

where $\widetilde{H}^{-1}$ denotes the inverse function of $\widetilde{H}$, and $K(x,y) = \max\{1, |\xi(x) + \eta(y) - \xi(0)|\}$, here in the function $\widetilde{H}$ being defined by, for all $r \geq 0$,

$$\widetilde{H}(r) := \int_0^r \frac{dr}{Q(1+r)},$$

and $X, Y \in \mathbb{R}_+$ are chosen so that

$$\int_0^x \int_0^y \hat{j}(s,t)\varphi(K(s,t))dtds < \widetilde{H}(+\infty) \quad \text{as long as} \quad (0,0) \leq (x,y) \leq (X,Y).$$

In addition, we observe from (8.13.5) that if the functions ξ and η are bounded on $\mathbb{R}_+$ and $\widetilde{H}(+\infty) = +\infty$ holds, then any solution $u(x,y)$ existing on $\mathbb{R}_+$ of problem (8.13.2) is bounded, provided that

$$\int_0^{+\infty} \int_0^{+\infty} \hat{j}(s,t)\varphi(K(s,t))dtds < +\infty.$$

8.14 An Application of Theorem 5.2.52 to Nonlinear Hyperbolic Partial Integro-differential Equation

In this section, we shall apply Theorem 5.2.52 to a nonlinear hyperbolic partial integro-differential equation of n-independent variables.

To this end, we consider the nonlinear hyperbolic partial integro-differential equation

$$\frac{\partial u(x)}{\partial x_1 \partial x_2 \ldots \partial x_n} = F\left(x, u(x), \int_{x_0}^x K(x,s,u(s))ds\right) + G(x, u(x)), \tag{8.14.1}$$

for all $x \in I = [x_0; x_1] \subset \mathbb{R}^n_+$, where $x = (x_1, x_2, \ldots, x_n), x_0 = (x_1^0, x_2^0, \ldots, x_n^0), x^\infty = (x_1^\infty, x_2^\infty, \ldots, x_n^\infty)$ are in $\mathbb{R}^n_+$ and $u \in C(I, \mathbb{R}), F \in C(I \times \mathbb{R} \times \mathbb{R}, \mathbb{R}), K \in C(I \times I \times \mathbb{R}, \mathbb{R})$ and $G \in C(I \times \mathbb{R}, \mathbb{R})$. With suitable boundary conditions, the solution of equation (8.13.1) is of the form

$$u(x) = l(x) + \int_{x_0}^{x} F\left(s, u(s), \int_{x_0}^{s} K(s, t, u(t))dt\right) ds + \int_{x_0}^{x} G(s, u(s))ds. \quad (8.14.2)$$

The following theorem gives us the bound of the solution of equation (8.14.1).

Theorem 8.14.1 ([193]) *Assume that the functions l, F, K and G in equation (8.14.1) satisfy the conditions*

$$\begin{cases} |K(s,t,u(t))| \le k(s,t)\phi(|u(t)|), \ t, s \in I, \ u \in \mathbb{R}, & (8.14.3) \\ |F(t,u,v)| \le \dfrac{1}{2}|u| + |v|, \ u, v \in \mathbb{R}, \ t \in I, & (8.14.4) \\ |G(s,u)| \le \dfrac{1}{2}|u|, \ s \in I, \ u \in \mathbb{R}, & (8.14.5) \\ |l(x)| \le a(x), \ x \in I, & (8.14.6) \end{cases}$$

where a, f, k and ϕ are as defined in Theorem 1.1.41, with $f(x) = b(x) + e(x)$ for all $x \in I$ where $b, e \in C(I, \mathbb{R}_+)$, then we have for all $x_0 \le x \le x^*$,

$$|u(x)| \le \exp\left(\prod_{i=1}^{n}(x_i - x_i^0)\right)\left(a(x) + \int_a^x E(t)dt\right). \quad (8.14.7)$$

Here

$$\begin{cases} E(t) = \psi^{-1}\left(\psi(\eta) + \int_a^t k(x_\infty, s)\phi\Big[\exp\Big(\prod_{i=1}^{n}(s_i - x_i^0)\Big)\int_a^s f(\tau)d\tau\Big]ds\right), & (8.14.8) \\ \eta = \int_{x^0}^{x_\infty} k(x_\infty, s)\phi\Big(a(s)\exp\Big(\prod_{i=1}^{n}(s_i - x_i^0)\Big)\Big)ds, & (8.14.9) \\ \psi(x) = \int_{x^0}^{x} \dfrac{ds}{\phi(s)}, \quad x \ge x^0 > 0, & (8.14.10) \end{cases}$$

where x^* is chosen so that $\psi(\eta) + \int_a^t k(x_\infty, s)\phi\left(\exp(\prod_{i=1}^{n}(s_i - x_i^0))\int_a^s f(\tau)d\tau\right)ds$ is in the domain of ψ^{-1}.

Proof Using the conditions (8.14.3), (8.14.6) in (8.14.2), we have

$$|u(x)| \le a(x) + \int_{x_0}^{x} |G(s,u(s))|ds + \int_{x_0}^{x} f(s)\Big[|u(s)| + \int_{x_0}^{s} |K(s,t,u(t))|dt\Big]ds$$
$$\le a(x) + \int_{x_0}^{x} \Big(|u(s)| + \int_{x^0}^{s} k(s,t)\phi(|u(t)|)dt\Big)ds. \tag{8.14.11}$$

Now applying Theorem 5.2.52 with $f(s) = 1$, $g(u) = u$ and $W(u) = u$ to (8.14.11) yields (8.14.7). □

Remark 8.14.1 If we assume that the functions F and G satisfy the general conditions

$$\begin{cases} |F(t,u,v)| \le f(t)\Big(g(|u|) + W(|v|)\Big), & (8.14.12)\\ |G(t,u)| \le f(t)g(|u|), \ \text{for all } t \in I, \ u \in \mathbb{R}, & (8.14.13)\end{cases}$$

we can obtain an estimation of $u(x)$.

8.15 Applications of Theorem 6.1.3 to Difference Equations

In this section, we shall use Theorem 6.1.3 to study the difference equations.

Example 8.15.1 Consider the difference equation

$$u^{m_1}(m,n) = a(m,n) + \sum_{s=0}^{m-1}\sum_{t=0}^{n-1} k(s,t,u(s,t)). \tag{8.15.1}$$

Let

$$k(s,t,u(s,t)) \le b(s,t)u(s,t), \tag{8.15.2}$$

if we consider $a(s,t) = b(s,t) = t$, it follows from (6.1.29) and (6.1.30) in Theorem 6.1.3,

$$u(m,n) \le n^{\frac{1}{m_1}} \prod_{t=0}^{n-1}\Big[1 + mt^{\frac{1}{m_1}}\Big]^{\frac{1}{m_1}}. \tag{8.15.3}$$

Example 8.15.2 Consider the difference equation

$$u^{m_1}(m,n) = a(m,n) + \sum_{s=0}^{m-1}\sum_{t=0}^{n-1} k(s,t,u(s,t)). \tag{8.15.4}$$

Let

$$k(s,t,u(s,t)) \le b(s,t)u(s,t) + b(s,t)u^{m_2}(s,t), \tag{8.15.5}$$

if we take $m_1 = 3, m_2 = 2, a(s,t) = b(s,t) = c(s,t) = t^3$, then from (8.15.4) we derive from Theorem 6.1.3

$$u(m,n) \le n\prod_{t=0}^{n-1}[1 + mt(t+1)]^{\frac{1}{3}}. \tag{8.15.6}$$

As special cases of (8.15.6), let $m = 2$ and $n = 2$, then $u(2,2) \le 2\sqrt[3]{5}$, if we take $m = 2$ and $n = 3$, then $u(2,3) \le 3\sqrt[3]{45}$, also for $m = 3$ and $n = 2$, then $u(3,2) \le 2\sqrt[3]{7}$.

8.16 Applications of Theorem 6.1.4 and Corollary 6.1.1 to Boundary Value Problems for Difference Equations

In this section, we shall employ Theorem 6.1.4 and Corollary 6.1.1 to study the boundedness, uniqueness, and continuous dependence of the solutions of boundary value problems for difference equations involving two independent variables.

We consider the following boundary value problem

$$\Delta_{12}z^p(m,n) = F(m,n,z(m,n)) \tag{8.16.1}$$

subject to

$$z(m,n_0) = f(m), \quad z(m_0,n) = g(n), \quad f(m_0) = g(n_0) = 0, \tag{8.16.2}$$

where $p > 1,\ F \in F(\Omega \times \mathbb{R}),\ f \in F(J)$ are given.

The first result deals with the boundedness of solutions.

Theorem 8.16.1 ([146]) *If*

$$\begin{cases} |F(m,n,v)| \le b(m,n)|v|^p, & (8.16.3) \\ |f(m)|^p + |g(n)|^p \le k^p, & (8.16.4) \end{cases}$$

for some $k \ge 0$, where $b \in F_+(\Omega)$, then all solutions of problem (8.16.1)–(8.16.2) satisfy for all $(m,n) \in \Omega$,

$$|z(m,n)| \le k \exp(B(m,n)),$$

where $B(m,n)$ is defined as in Theorem 6.1.4. In particular, if $B(m,n)$ is bounded on Ω, then every solution of problem (8.16.1)–(8.16.2) is bounded on Ω.

Proof Observe first that $z = z(m,n)$ solves problem (8.16.1)–(8.16.2) if and only if it satisfies the sumdifference equation

$$z^p(m,n) = f^p(m) + g^p(n) + \sum_{s=m_0}^{m-1}\sum_{t=n_0}^{n-1} F(s,t,z(s,t)). \quad (8.16.5)$$

Hence from (8.16.3) and (8.16.4), we derive for all $(m,n) \in \Omega$,

$$|z(m,n)|^p \le k^p + \sum_{s=m_0}^{m-1}\sum_{t=n_0}^{n-1} b(s,t)|z(s,t)|^p.$$

An application of Theorem 6.1.4 to the function $|z(m,n)|$ gives us the assertion immediately. □

The next result concerns the uniqueness of solutions.

Theorem 8.16.2 ([146]) *If*

$$|F(m,n,v_1) - F(m,n,v_2)| \le b(m,n)|v_1^p - v_2^p|$$

for some $b \in F_+(\Omega)$, then problem (8.16.1)–(8.16.2) has at most one solution on Ω.

Proof Let $z(m,n)$ and $\bar{z}(m,n)$ be two solutions of problem (8.16.1)–(8.16.2) on Ω. By (8.16.5), we have

$$\begin{aligned} |z^p(m,n) - \bar{z}^p(m,n)| &\le \sum_{s=m_0}^{m-1}\sum_{t=n_0}^{n-1} |F(s,t,z(s,t)) - F(s,t,\bar{z}(s,t))| \\ &\le \sum_{s=m_0}^{m-1}\sum_{t=n_0}^{n-1} b(s,t)|z^p(s,t) - \bar{z}^p(s,t)|. \end{aligned}$$

An application of Corollary 6.1.1 to the function $|z^p(s,t) - \bar{z}^p(s,t)|^{1/p}$ shows that for all $(s,t) \in \Omega$,

$$|z^p(s,t) - \bar{z}^p(s,t)|^{1/p} \leq 0,$$

hence $z = \bar{z}$ on Ω. □

Finally, we investigate the continuous dependence of the solutions of problem on the function F and the boundary data f and g. For this we consider the following variation of problem

$$\Delta_{12} z^p(m,n) = \bar{F}(m,n,z(m,n)) \tag{8.16.6}$$

with

$$z(m,n_0) = \bar{f}(m), \quad z(m_0,n) = \bar{g}(n), \quad \bar{f}(m_0) = \bar{g}(n_0) = 0, \tag{8.16.7}$$

where $p > 1, \bar{F} \in F(\Omega \times \mathbb{R}), \bar{f} \in F(I)$, and $\bar{g} \in F(J)$ are given.

Theorem 8.16.3 ([146]) *If*

(i) $|F(m,n,,v_1) - F(m,n,v_2)| \leq b(m,n)|v_1^p - v_2^p|$ *for some* $b \in F_+(\Omega)$;
(ii) $|(f^p(m) - \bar{f}^p(m)) + (g^p(n) - \bar{g}^p(n))| \leq \varepsilon/2$; *and*
(iii) for all solutions $\bar{z}(m,n)$ *of problem (8.16.6)–(8.16.7), for all* $(m,n) \in \Omega$ *and* $v_1, v_2 \in \mathbb{R}$,

$$\sum_{s=m_0}^{m-1}\sum_{t=n_0}^{n-1} |F(s,t,\bar{z}(s,t)) - \bar{F}(s,t,\bar{z}(s,t))| \leq \frac{\varepsilon}{2},$$

then

$$|z^p(m,n) - \bar{z}^p(m,n)| \leq \varepsilon \exp(pB(m,n)),$$

where $B(m,n)$ *is as defined in Theorem 6.1.4. Hence* z^p *depends continuously on* F, f, *and* g. *In particular, if* z *does not change sign, it depends continuously on* F, f *and* g.

Proof Let $z(m,n)$ and $\bar{z}(m,n)$ be solutions of problem (8.16.1)–(8.16.2) and problem (8.16.6)–(8.16.7), respectively, Then z satisfies (8.16.5) and $\bar{z}$ satisfies the corresponding equation

$$\bar{z}(m,n) = \bar{f}^p(m) + \bar{g}^p(n) + \sum_{s=m_0}^{m-1}\sum_{t=n_0}^{n-1} \bar{F}(s,t,\bar{z}(s,t)).$$

Hence

$$
\begin{aligned}
|z^p(m,n) - \bar{z}^p(m,n)| &\le |(f^p(m) - \bar{f}^p(m)) + (g^p(n) - \bar{g}^p(n))| \\
&\quad + \sum_{s=m_0}^{m-1}\sum_{t=n_0}^{n-1} |F(s,t,z(s,t)) - \bar{F}(s,t,\bar{z}(s,t))| \\
&\le \frac{\varepsilon}{2} + \sum_{s=m_0}^{m-1}\sum_{t=n_0}^{n-1} |F(s,t,z(s,t)) - \bar{F}(s,t,\bar{z}(s,t))| \\
&\quad + \sum_{s=m_0}^{m-1}\sum_{t=n_0}^{n-1} |F(s,t,\bar{z}(s,t)) - \bar{F}(s,t,\bar{z}(s,t))| \\
&\le \varepsilon + \sum_{s=m_0}^{m-1}\sum_{t=n_0}^{n-1} |z^p(s,t) - \bar{z}^p(s,t)|
\end{aligned}
$$

by assumptions (i), (ii) and (iii). Now applying Corollary 6.1.1 to the function $|z^p(m,n) - \bar{z}^p(m,n)|^{1/p}$, we have, for all $(m,n) \in \Omega$,

$$|z^p(m,n) - \bar{z}^p(m,n)|^{1/p} \le \varepsilon^{1/p} \exp(B(m,n)),$$

or

$$|z^p(m,n) - \bar{z}(m,n)| \le \varepsilon \exp(pB(m,n)).$$

Now when restricted to any compact sub-lattice, $B(m,n)$ is bounded, so

$$|z^p(m,n) - \bar{z}^p(m,n)| \le \varepsilon K$$

for some constant $K > 0$ for all (m,n) in this compact sublattice. Hence z^p depends continuously on F, f and g. □

8.17 Applications of Theorems 6.1.7–6.1.9 to Sum-difference Equations

In this section we note that the inequalities established in Theorems 6.1.7–6.1.9 can be extended very easily to functions of more than two variables. Next, we shall consider the following sum-difference equation

$$u^2(m,n) = a(m,n) + \sum_{s=m+1}^{+\infty}\sum_{t=n+1}^{+\infty} F[m,n,s,t,u(s,t)] \tag{8.17.1}$$

for all $m, n \in \mathbb{Z}$ where $a(m, n)$ is a function defined for all $m, n \in \mathbb{Z}$ into $\mathbb{R}$ and $F[m, n, s, t, u(s, t)]$ is a function defined for into $\mathbb{R}$ into $\mathbb{R}$. We assume that

$$|a(m,n)| \leq c, \ |F[m,n,s,t,u(s,t)]| \leq f(s,t)|u(s,t)| \tag{8.17.2}$$

where $c \geq 0$ is a constant and $f(m, n)$ is a function defined for all $m, n \in \mathbb{Z}$ into $\mathbb{R}_+$. From (8.17.1) and (8.17.2), we obtain

$$|u(m,n)|^2 \leq c + \sum_{s=m+1}^{+\infty} \sum_{t=n+1}^{+\infty} f(s,t)|u(s,t)|. \tag{8.17.3}$$

Now applying Theorem 6.1.7 yields

$$|u(m,n)| \leq \sqrt{c} + \frac{1}{2} \sum_{s=m+1}^{+\infty} \sum_{t=n+1}^{+\infty} f(s,t). \tag{8.17.4}$$

The right-hand side of (8.17.4) gives us a bound on the solution $u(m, n)$ of the equation (8.17.1).

Now we can also use inequalities in Theorems 6.1.8 and 6.1.9 to obtain bounds on the solutions of the following sum-difference equations

$$u(m,n) = a(m,n) + \sum_{s=m+1}^{+\infty} \sum_{t=n+1}^{+\infty} u(s,t)F[s,t,\log u(s,t)] \tag{8.17.5}$$

and

$$u^2(m,n) = a(m,n) + \sum_{s=m+1}^{+\infty} \sum_{t=n+1}^{+\infty} u(s,t)F[m,n,s,t,u(s,t),Tu(s,t)], \tag{8.17.6}$$

where

$$Tu(s,t) = \sum_{k=s+1}^{+\infty} \sum_{r=t+1}^{+\infty} G(s,t,k,r,u(k,r)) \tag{8.17.7}$$

under some suitable conditions on the functions involved in (8.17.5) and (8.17.6).

8.18 Applications of Theorems 6.1.16 and 6.1.17 to Volterra-Fredholm Sum-difference Equations

In this section, we apply Theorems 6.1.16 and 6.1.17 to study the boundedness, uniqueness, and continuous dependence of the solutions of certain Volterra-Fredholm sum-difference equations of the form

$$u^p(m,n) = l(m,n) + \sum_{s=m_0}^{m-1}\sum_{t=n_0}^{n-1} F(s,t,u(s,t)) + \sum_{s=m_0}^{M-1}\sum_{t=n_0}^{N-1} G(s,t,u(s,t)) \tag{8.18.1}$$

for all $(m,n) \in \Omega$, where $l \in F(\Omega)$, $F, G \in F(\Omega \times \mathbb{R}), p \geq 1$ is a constant.

The following theorem gives the bound on the solutions of (8.18.1).

Theorem 8.18.1 ([359]) *Assume that the functions l, F and G in (8.18.1) satisfy the conditions*

$$\begin{cases} |l(m,n)| \leq k, & (8.18.2)\\ |F(m,n,v)| \leq a(m,n)|v|^{p+q-1}, & (8.18.3)\\ |G(m,n,v)| \leq b(m,n)|v|^{p+q-1}, & (8.18.4)\end{cases}$$

where $a(m,n)$ and $b(m,n)$ are same as in Theorem 6.1.17, $0 < q < 1$ is a constant, then all solutions of (8.18.1) satisfy, for all $(m,n) \in \Omega$,

$$u(m,n) \leq \left\{(c_{11})^{1-q} + (1-q)\sum_{s=m_0}^{m-1}\sum_{t=n_0}^{n-1} a^*(s,t)\right\}^{1/(1-q)} \tag{8.18.5}$$

where c_{11} is the solution of equation

$$\hat{H}_2^*(t) = \frac{1}{q}[((2t-k)^{(1-q)/p} - t^{(1-q)/p} - \sum_{s=m_0}^{M-1}\sum_{t=n_0}^{N-1} a^*(s,t) = 0 \tag{8.18.6}$$

for all $t \geq k$, where $a^*(m,n) \in F_+(\Omega)$ such that both $a(m,n)$ and $b(m,n)$ are less than or equal to $a^*(m,n)$.

Proof Using the conditions (8.18.2)–(8.18.4) in (8.18.1), we have

$$|u(m,n)|^p \leq k + \sum_{s=m_0}^{m-1}\sum_{t=n_0}^{n-1} a^*(s,t)|u(s,t)|^{p+q-1} + \sum_{s=m_0}^{M-1}\sum_{t=n_0}^{N-1} a^*(s,t)|u(s,t)|^{p+q-1}$$

which is a special case of Theorem 6.1.16 when $\varphi(u) = u^p$ and $w(u) = u^q$.

By Theorem 6.1.16, we only need to prove

$$\hat{H}_2(t) = \hat{H}_2^*(t) = G_1 \circ \varphi^{-1}(2t-k) - G_1 \circ \varphi^{-1}(t) - \sum_{s=m_0}^{M-1}\sum_{t=n_0}^{N-1} a^*(s,t)$$

$$= \frac{1}{1-q}[((2t-k)^{(1-q)/p} - t^{(1-q)/p} - \sum_{s=m_0}^{M-1}\sum_{t=n_0}^{N-1} a^*(s,t)$$

is increasing, and equation $H_2^*(t) = 0$ has a solution c_{11}. In fact, taking $r = (1-q)/p$, by computation, we get for all $t \geq k$,

$$\frac{d}{dt}\hat{H}_2^*(t) = \frac{1}{p}\frac{(2^{1/(1-r)}t)^{1-r} - (2t-k)^{1-r}}{[t(2t-k)]^{1-r}} > 0,$$

$$\hat{H}_2^*(t) = -\sum_{s=m_0}^{M-1}\sum_{t=n_0}^{N-1} a^*(s,t) < 0$$

and

$$\lim_{t\to+\infty} \hat{H}_2^*(k) = \lim_{t\to+\infty}\left\{\frac{t^r}{1-q}[2 - \frac{k^r}{l} - 1] - \sum_{s=m_0}^{M-1}\sum_{t=n_0}^{N-1} a^*(s,t)\right\} = +\infty,$$

so $\hat{H}_2^*(t) = 0$ has a unique solution $c_{11} > k$. □

Second, we consider the uniqueness of the solutions of (8.18.1).

Theorem 8.18.2 ([359]) *Assume that the functions F and G in (8.18.1) satisfy the conditions*

$$\begin{cases} |F(m,n,v) - F(m,n,\bar{v})| \leq a(m,n)|v^p - \bar{v}^p|, & (8.18.7) \\ |G(m,n,v) - Gm,n,\bar{v})| \leq b(m,n)|v^p - \bar{v}^p|, & (8.18.8) \end{cases}$$

for some $a, b \in F_+(\Omega)$, and if

$$\Sigma(M,N) = \exp\left(\sum_{s=m_0}^{M-1}\sum_{t=n_0}^{N-1} a^*(s,t)\right) < 2, \tag{8.18.9}$$

where $a^*(m,n)$ is as in Theorem 8.18.1, then (8.18.1) has at most one positive solution on Ω.

Proof Let $u(m,n)$ and $\bar{u}(m,n)$ be two solutions of equation (8.18.1). By (8.18.1) and conditions (8.18.7) and (8.18.8), we derive

$$|u^p(m,n) - \bar{u}^p(m,n)| \leq \sum_{s=m_0}^{m-1}\sum_{t=n_0}^{n-1} a^*(s,t)|u^p(s,t) - \bar{u}^p(s,t)|$$
$$+ \sum_{s=m_0}^{M-1}\sum_{t=n_0}^{N-1} a^*(s,t)|u^p(s,t) - \bar{u}^p(s,t)|.$$

Applying Corollary 6.1.11 to the function $|u^p(m,n) - \bar{u}^p(m,n)|$ yields

$$|u^p(m,n)\bar{u}^p(m,n)| \leq 0, \ for\ all\ (m,n) \in \Omega.$$

Hence, $u(m,n) = \bar{u}(m,n)$ on Ω. □

Finally, we investigate the continuous dependence of the solutions of (8.18.1) on the functions F and G. To this end, we consider the following variation of (8.18.1):

$$u^p(m,n) = \bar{l}(m,n) + \sum_{s=m_0}^{m-1}\sum_{t=n_0}^{n-1} \bar{F}(s,t,u(s,t)) + \sum_{s=m_0}^{M-1}\sum_{t=n_0}^{N-1} \bar{G}(s,t,u(s,t)) \quad (8.18.10)$$

for all $(m,n) \in \Omega$, where $\bar{F}, \bar{G} \in F(\Omega \times \mathbb{R}), p \geq 1$ is a constant as in (8.18.1).

Theorem 8.18.3 ([359]) *Consider equations (8.18.1) and (8.18.10). If*

(i) $|F(s,t,v_1) - F(s,t,v_2)| \leq a(s,t)|v_1^p - v_2^p|$, $|G(s,t,v_1) - G(s,t,v_2)| \leq b(s,t)|v_1^p - v_2^p|$;
(ii) $|l(m,n) - \bar{l}(m,n)| \leq \varepsilon/2$;
(iii) $\Sigma(M,N) = \exp(\sum_{s=m_0}^{M-1}\sum_{t=n_0}^{N-1} a^*(s,t)) < 2$;
(iv) for all solutions $\bar{u}$ *of equation (8.18.10),*

$$\sum_{s=m_0}^{m-1}\sum_{t=n_0}^{n-1} |F(s,t,\bar{u}) - \bar{F}(s,t,\bar{u})| \leq \varepsilon/4,$$

and for all $(m,n) \in \Omega$ *and* $v_1, v_2 \in \mathbb{R}$,

$$\sum_{s=m_0}^{M-1}\sum_{t=n_0}^{N-1} |G(s,t,\bar{u}) - \bar{G}(s,t,\bar{u})| \leq \varepsilon/4,$$

where $\varepsilon > 0$ is an arbitrary constant, $a^*(m,n)$ is defined as in Theorem 8.18.1, then for all $(m,n) \in \Omega$,

$$|u^p(s,t) - \bar{u}^p(s,t)| \leq \frac{\varepsilon}{2 - \Sigma(M,N)} \exp\left(\sum_{s=m_0}^{m-1}\sum_{t=n_0}^{n-1} a^*(s,t)\right). \quad (8.18.11)$$

Hence, u^p depend continuously on F and G. In particular, if u does not change the sign, it depends continuously on F and G.

Proof Let $u(m,n)$ and $\bar{u}(m,n)$ be solutions of equations (8.18.1) and (8.18.10), respectively. Then, $u(m,n)$ satisfies equation (8.18.1) and $\bar{u}(m,n)$ satisfies equation (8.18.10). Hence, by assumptions (i)–(iv) and the definition of a^*,

$$\begin{aligned}
|u^p(m,n) - \bar{u}^p(m,n)| &\leq |l(m,n) - \bar{l}(m,n)| + \sum_{s=m_0}^{m-1}\sum_{t=n_0}^{n-1} |F(s,t,u) - \bar{F}(s,t,\bar{u})| \\
&\quad + \sum_{s=m_0}^{M-1}\sum_{t=n_0}^{N-1} |G(s,t,u) - \bar{G}(s,t,\bar{u})| \\
&\leq \frac{\varepsilon}{2} + \sum_{s=m_0}^{m-1}\sum_{t=n_0}^{n-1} |F(s,t,u) - F(s,t,\bar{u})| \\
&\quad + \sum_{s=m_0}^{m-1}\sum_{t=n_0}^{n-1} |F(s,t,\bar{u}) - \bar{F}(s,t,\bar{u})| \\
&\quad + \sum_{s=m_0}^{M-1}\sum_{t=n_0}^{N-1} |G(s,t,u) - G(s,t,\bar{u})| \\
&\quad + \sum_{s=m_0}^{M-1}\sum_{t=n_0}^{N-1} |G(s,t,\bar{u}) - \bar{G}(s,t,\bar{u})| \\
&\leq \varepsilon + \sum_{s=m_0}^{m-1}\sum_{t=n_0}^{n-1} a(s,t)|u^p(s,t) - \bar{u}^p(s,t)| \\
&\quad + \sum_{s=m_0}^{M-1}\sum_{t=n_0}^{N-1} b(s,t)|u^p(s,t) - \bar{u}^p(s,t)| \\
&\leq \varepsilon + \sum_{s=m_0}^{m-1}\sum_{t=n_0}^{n-1} a^*(s,t)|u^p(s,t) - \bar{u}^p(s,t)| \\
&\quad + \sum_{s=m_0}^{M-1}\sum_{t=n_0}^{N-1} a^*(s,t)|u^p(s,t) - \bar{u}^p(s,t)|.
\end{aligned}$$

Now applying Corollary 6.1.11 to the function $|u^p(m,n) - \bar{u}^p(m,n)|$, we can get

$$|u^p(m,n) - \bar{u}^p(m,n)| \leq \frac{\varepsilon}{2 - \Sigma(M,N)} \exp\left(\sum_{s=m_0}^{m-1} \sum_{t=n_0}^{n-1} a^*(s,t)\right).$$

Evidently, if the function $\sum_{s=m_0}^{m-1} \sum_{t=n_0}^{n-1} a^*(s,t)$ is bounded on Ω, then for some constant $K > 0$ and for all $(m,n) \in \Omega$,

$$|u^p(m,n) - \bar{u}^p(m,n)| \leq \varepsilon K.$$

Hence, u^p depends continuously on F and G. □

8.19 Applications of Theorem 6.1.22 and Corollary 6.1.20 to Discrete Boundary Value Problems

In this section, we use Theorem 6.1.22 and Corollary 6.1.20 to study some properties of positive solutions of the following boundary value problem:

$$\Delta_{12} h(z(m,n)) = F(m,n,z(m,n)) \tag{8.19.1}$$

with

$$z(m,n_0) = f(m),\ z(m_0,n) = g(n),\ f(m_0) = g(n_0) = 0, \tag{8.19.2}$$

where h is defined as in Theorem 6.1.22, $F \in F(\Omega \times \mathbb{R}), f \in F(I)$, and $g \in F(J)$ are given.

The first result deals with the boundedness of solutions.

Theorem 8.19.1 ([147]) *If*

$$|F(m,n,v)| \leq b(m,n)\varphi(|v|), \tag{8.19.3}$$

and

$$|h(f(m)) + h(g(n))| \leq K \tag{8.19.4}$$

for some constant $K \geq 0$, where φ, Φ_h, and Φ_h^{-1} are defined as in Theorem 6.1.18, $b \in F_+(\Omega)$, then all positive solutions of problem (8.19.1)–(8.19.2) satisfy

$$z(m,n) \leq h^{-1}(\Phi_h^{-1}[\Phi_h(K) + B(m,n)]),\ (m,n) \in \Omega, \tag{8.19.5}$$

where $B(m,n)$ is defined as in Theorem 6.1.22. In particular, if $B(m,n)$ is bounded on Ω, then every solution of problem (8.19.1)–(8.19.2) is bounded on Ω.

Proof Observe first that $z = z(m,n)$ solves problem (8.19.1)–(8.19.2) if and only if it satisfies the sum-difference equation

$$h(z(m,n)) = h(f(m)) + h(g(n)) + \sum_{s=m_0}^{m-1}\sum_{t=n_0}^{n-1} F(s,t,z(s,t)). \tag{8.19.6}$$

Hence, by (8.19.3) and (8.19.4), for all $(m,n) \in \Omega$,

$$h(z(m,n)) \leq K + \sum_{s=m_0}^{m-1}\sum_{t=n_0}^{n-1} b(s,t)\varphi(|z(s,t)|).$$

An application of Theorem 6.1.22 to the function $z(m,n)$ yields the assertion immediately. □

The next result concerns the uniqueness of solutions.

Theorem 8.19.2 ([147]) *If*

$$|F(m,n,v_1) - F(m,n,v_2)| \leq b(m,n)|h(v_1) - h(v_2)| \tag{8.19.7}$$

for some $b \in F_+(\Omega)$, then problem (8.19.1)–(8.19.2) has at most one positive solution on Ω.

Proof Let $z(m,n)$ and $\bar{z}(m,n)$ be two solutions of problem (8.19.1)–(8.19.2) on Ω. By (8.19.6), we get

$$\begin{aligned} |h(z(m,n)) - h(\bar{z}(m,n))| &\leq \sum_{s=m_0}^{m-1}\sum_{t=n_0}^{n-1} |F(s,t,z(s,t)) - F(s,t,\bar{z}(s,t))| \\ &\leq \sum_{s=m_0}^{m-1}\sum_{t=n_0}^{n-1} b(s,t)|h(z(s,t)) - h(\bar{z}(s,t))|. \end{aligned} \tag{8.19.8}$$

Applying Corollary 6.1.20 to the function $h^{-1}(|h(z(s,t)) - h(\bar{z}(s,t))|)$ yields for all $(m,n) \in \Omega$,

$$|h(z(m,n)) - h(\bar{z}(m,n))| \leq 0.$$

Hence, $z = \bar{z}$ on Ω. □

Finally, we investigate the continuous dependence of the solutions of problem (8.19.1)–(8.19.2) on the function F and the boundary data f and g. To this end, we consider the following variation of problem:

$$\Delta_{12}h(z(m,n)) = \overline{F}(m,n,z(m,n)) \tag{8.19.9}$$

with

$$z(m,n_0) = \overline{f}(m),\ z(m_0,n) = \overline{g}(n), \overline{f}(m_0) = \overline{g}(n_0) = 0, \tag{8.19.10}$$

where h is defined as in Theorem 6.1.22, $\overline{F} \in F(\Omega \times \mathbb{R}), \overline{f} \in F(I)$, and $g \in F(J)$ are given.

Theorem 8.19.3 ([147]) *Consider problem (8.19.1)–(8.19.2) and problem (8.19.9)–(8.19.10). If*

(i) $|F(m,n,v_1) - F(m,n,v_2)| \le b(m,n)|h(v_1) - h(v_2)|$ *for some* $b \in F_+(\Omega)$;
(ii) $|h(f(m)) - h(\overline{f}(m)) + h(g(n)) - h(\overline{g}(n))| \le \frac{\varepsilon}{2}$;
(iii) $\sum_{s=m_0}^{m-1}\sum_{t=n_0}^{n-1} |F(s,t,u) - \overline{F}(s,t,u)| \le \frac{\varepsilon}{2}$ *for all* $(m,n) \in \Omega$, *and* $v_1, v_2, u \in \mathbb{R}$, *then for all* $(m,n) \in \Omega$,

$$|h(z(m,n)) - h(\overline{z}(m,n))| \le \varepsilon \exp(B(m,n))$$

where $B(m,n)$ *is as defined in Theorem 6.1.22. Hence,* $h(z)$ *depends continuously on* F, f, *and* g.

Proof Let $z(m,n)$ and $\overline{z}(m,n)$ be solutions of problem (8.19.1)–(8.19.2) and problem (8.19.9)–(8.19.10), respectively. Then, z satisfies (8.19.6) and $\overline{z}$ satisfies the corresponding equation

$$h(\overline{z}(m,n)) = h(\overline{f}(m)) + h(\overline{g}(n)) + \sum_{s=m_0}^{m-1}\sum_{t=n_0}^{n-1} \overline{F}(s,t,\overline{z}(s,t)).$$

Hence, from assumptions (i)–(iii) it follows

$$\begin{aligned}
&|h(z(m,n)) - h(\overline{z}(m,n))|\\
&\le |(h(f(m)) - h(\overline{f}(m))) + (h(g(n)) - h(\overline{g}(n)))|\\
&\quad + \sum_{s=m_0}^{m-1}\sum_{t=n_0}^{n-1} |F(s,t,z(s,t)) - \overline{F}(s,t,\overline{z}(s,t))|\\
&\le \frac{\varepsilon}{2} + \sum_{s=m_0}^{m-1}\sum_{t=n_0}^{n-1} |F(s,t,z(s,t)) - F(s,t,\overline{z}(s,t))| \sum_{s=m_0}^{m-1}\sum_{t=n_0}^{n-1} |F(s,t,\overline{z}(s,t)) - \overline{F}(s,t,\overline{z}(s,t))|\\
&\le \varepsilon + \sum_{s=m_0}^{m-1}\sum_{t=n_0}^{n-1} b(s,t)|h(z(s,t)) - h(\overline{z}(s,t))|.
\end{aligned}$$

Now applying Corollary 6.1.20 to the function $h^{-1}(|h(z(m,n)) - h(\bar{z}(m,n))|)$, we have for all $(m,n) \in \Omega$,

$$|h(z(m,n)) - h(\bar{z}(m,n))| \leq \varepsilon \exp(B(m,n)).$$

When restricted to any compact sublattice, $B(m,n)$ is bounded, so

$$|h(z(m,n)) - h(\bar{z}(m,n))| \leq K\varepsilon$$

for some constant $K > 0$ and for all (m,n) in this compact sublattice. Hence, $h(z)$ depends continuously on F, f, and g. □

8.20 An Application of Theorem 6.1.25 to Finite Difference Equations

As an application on the inequality established in Theorem 6.1.25, we consider the following nonlinear partial finite difference equation of the form

$$\Delta_2^m \Delta_1^n u^2(x,y) = u(x,y)F(x,y,u(x,y)) + G(x,y,u(x,y)), \qquad (8.20.1)$$

with the given boundary conditions at $x = 0$ and $y = 0$,

$$\begin{cases} \Delta_2^j u^2(x,0) = \alpha_j(x), & 0 \leq j \leq m-1, \qquad (8.20.2) \\ \Delta_1^i u^2(0,y) = \beta_i(y), & 0 \leq i \leq n-1, \qquad (8.20.3) \end{cases}$$

where F and G are real-valued functions defined on $\mathbb{N}_0 \times \mathbb{R}$ and α_j, β_i are real-valued functions defined on $\mathbb{N}_0$ and $m, n \geq 2$ are integers. Now, by assuming some suitable conditions on the functions involved in problem (8.20.1)–(8.20.3), and following the similar arguments as given above with suitable modifications, we can very easily obtain the bound on the solutions of problem (8.20.1)–(8.20.3) in terms of the known functions.

8.21 An Application of Theorem 6.1.29 to Discrete Partial Integro-differential Equations

In this section, we shall give an application of Theorem 6.1.29 to obtain the bounds on the solutions of discrete versions of partial integro-differential equations involving three independent variables.

To illustrate the application of Theorem 6.1.29, we establish the bound on the solutions of discrete versions of partial integro-differential equations involving three independent variables of the form

$$\Delta^3 u_{xyz} = f(x,y,z,u) + F\left(x,y,z,u,\sum_{s=0}^{x-1}\sum_{t=0}^{y-1}\sum_{r=0}^{z-1} h(x,y,z,s,t,r,u)\right) \quad (8.21.1)$$

with the given boundary conditions at $x = 0, y = 0, z = 0$, where all the functions are defined on their respective domains of definitions and for all $x \geq 0, y \geq 0, z \geq 0$,

$$\begin{cases} |f(x,y,z,u)| \leq p(x,y,z)W(|u|), & (8.21.2) \\ |F(x,y,z,u,v)| \leq b(x,y,z)(|u| + |v|), & (8.21.3) \\ |h(x,y,z,s,t,r,u)| \leq c(s,t,r)|u|, & (8.21.4) \end{cases}$$

where $W, b(x,y,z), c(x,y,z)$, and $p(x,y,z)$ are as defined in Theorem 6.1.29. By using the given boundary conditions, (8.21.1) can be represented by equivalent summary difference equation

$$u(x,y,z) = g(x,y,z) + \sum_{s=0}^{x-1}\sum_{t=0}^{y-1}\sum_{r=0}^{z-1} f(s,t,r,u(s,t,r))$$
$$+\sum_{s=0}^{x-1}\sum_{t=0}^{y-1}\sum_{r=0}^{z-1} F\left(s,t,r,u(s,t,r), \sum_{k=0}^{s-1}\sum_{l=0}^{t-1}\sum_{n=0}^{r-1} h(s,t,r,k,l,n,u(k,l,n))\right)$$
$$(8.21.5)$$

where $g(x,y,z)$ depends on the given boundary conditions. If $|g(x,y,z)| \leq M$, then using (8.21.2)–(8.21.4) in (8.21.5) and then applying Theorem 6.1.29, we can obtain the bound on the solution $u(x,y,z)$ of equation (8.21.1).

8.22 Applications of Theorems 6.2.4–6.2.7 to Difference Equations

In fact, the comparison Theorems 6.2.4–6.2.7 obtained above will be used here to show the dependence of solutions of (6.2.39) on initial values and on parameters. To this end, hereafter we shall denote the term $\sum_{i=1}^{n}(-1)^{i+1}\sum_i u([\bar{x}_i])$ in short by $T(u)$.

Theorem 8.22.1 ([8]) *Assume that $u(x, T(a))$ is the solution of (6.2.39), (6.2.50) and $w(x, 0)$ is the solution of the problem*

$$\Delta_x^n w(x) = F(x, u(x))$$
$$w(((i)x)) = 0,$$

where the function F on A^+ is defined as

$$F(x, c) = \sup_{|u-T(a)|\leq c} |f(x, u)|.$$

Then for all $x, 0 \leq x \leq X$,

$$|u(x, T(a)) - T(a)| \leq w(x, 0).$$

Proof From the definition of $T(a)$ and the summation representation of (6.2.39), (6.2.50) we have

$$u(x, T(a)) = T(a) + S_{s=0}^{x-1} f(s, u(s, T(a)))$$

and hence if we define $y(x) = |u(x, T(a)) - T(a)|$, then it follows that

$$\begin{aligned} y(x) &\leq S_{s=0}^{x-1} |f(s, u(s, T(a)))| \\ &\leq S_{s=0}^{x-1} \sup_{|u-T(a)|\leq y(s)} |f(s, u)| \\ &= S_{s=0}^{x-1} F(s, y(s)). \end{aligned}$$

Next, since $w(x, 0)$ has the summation representation

$$w(x, 0) = S_{s=0}^{x-1} F(s, w(s, 0))$$

and for all fixed x and $1 \leq i \leq m$ the function $F_i(x, u_1, \ldots, u_m)$ is non-decreasing with respect to all $u_1, \ldots, u_m$ as in Theorem 6.2.4 it follows that

$$|u(x, T(a)) - T(a)| = y(x) \leq w(x, 0).$$

□

Theorem 8.22.2 ([8]) *Assume that condition (i) of Theorem 6.2.7 is satisfied, and $u(x, T(a))$ is as in Theorem 8.22.1. Further, assume that $u(x, T(b))$ is the solution of (6.2.39) satisfying*

$$u((i)x) = b([\bar{x}_i]).$$

Then, for all $x, 0 \leq x \leq X$,

$$|u(x, T(a)) - u(x, T(b))| \leq \lambda(x),$$

where $\lambda(x)$ is a solution of

$$\Delta_x^n \lambda(x) = g(x, \lambda(x))$$
$$|T(a) - T(b)| \le T(\lambda). \tag{8.22.1}$$

Proof The result follows by setting $\varepsilon_1(x) = \varepsilon_2(x) \equiv 0$ in Theorem 6.2.7. □

Theorem 8.22.3 ([8]) *Assume that the following conditions hold:*

(i) *$u(x, \mu)$ is the solution of the problem (6.2.47), (6.2.50);*
(ii) $\lim_{\mu \to \mu^0} f(x, u, \mu) = f(x, u, \mu^0)$ *uniformly in $(x, u) \in A$;*
(iii) *for all $(x, u^1, \mu), (x, u^2, \mu)$ in $A \times R^r$,*

$$|f(x, u^1, \mu) - f(x, u^2, \mu)| \le g(x, |u^1 - u^2|),$$

where g is defined on A^+;$g(x, 0) = 0$ for all $x, 0 \le x \le X$; and for all fixed x and $1 \le i \le m$, $g_i(x, u-1, \ldots, u_m)$ is non-decreasing with respect to all $u_1, \ldots, u_m$.

Proof Since $g(x, 0) = 0$ for all $x, 0 \le x \le X$, the solution $\lambda(x, 0)$ of (8.22.1) satisfying $\lambda((i)x) = 0$ is identically zero. Hence, for any $\varepsilon > 0$, there exists an m-dimensional vector $\eta = \eta(\varepsilon)$ such that the solution $\lambda(x, 0, \eta)$ of the difference equation

$$\Delta_x^n \lambda(x) = g(x, \lambda(x)) + \eta$$

satisfying $\lambda((i)x) = 0$ has the property that

$$\lambda(x, 0, \eta) \le \varepsilon.$$

Furthermore, because of (ii), for any given $\eta > 0$, there exists a $\delta = \delta(\eta) > 0$ such that

$$|f(x, u, \mu) - f(x, u, \mu^0)| \le \eta$$

provided that $|u - \mu^0| \le \delta(\eta)$.

Now, let $\varepsilon > 0$ be given, then since

$$\begin{aligned}|u(x, \mu) - u(x, \mu^0)| &\le S_{s=0}^{x-1} |f(s, u(s, \mu), \mu), \mu) - f(s, u(s, \mu^0), u^0)| \\ &\le S_{s=0}^{x-1} [g(s, |u(s, \mu) - u(s, \mu^0)|) + \eta]\end{aligned}$$

as in Theorem 6.2.4 it follows that

$$|u(x, \mu) - u(x, \mu^0)| \le \lambda(x, 0, \eta) \le \varepsilon.$$

Clearly, δ depends on ε since η does. □

In this sequel, we shall state two results showing the boundedness and asymptotic behavior of (6.2.39). The proof of both results is based on the comparison results established.

Theorem 8.22.4 ([8]) *Assume that the followings conditions hold:*

(i) for all (x, u) in A,

$$|f(x,u)| \le g(x,|u|),$$

where the function $g(x,u)$ is defined on A^+, and for all fixed x and $1 \le i \le m$, $g_i(x,u_1,\ldots,u_m)$ is non-decreasing with respect to all $u_1,\ldots,u_m$;

(ii) $u(x)$ is any solution of (6.2.39) and $\lambda(x)$ is any solution of (8.22.1) such that

$$|T(u)| \le T(\lambda).$$

Then the following hold:

(i) if $\lambda(x)$ is bounded, so is $u(x)$;

(ii) if $\lambda(x) \to 0$ as $\|x\| = (x_1^2 + \cdots + x_n^2)^{1/2} \to +\infty$, so is $u(x)$.

Proof It is easy to show that $|u(x)| \le \lambda(x)$ for all $x, 0 \le x \le X$. □

Theorem 8.22.5 ([8]) *Assume that the following conditions hold:*

(i) $u(x)$ is a solution of (6.2.39) and $u^1(x)$ is a solution of the difference equation

$$\Delta_x^n u(x) = f^1(x,u(x));$$

(ii) for all (x,u) and (x,v) in A,

$$|f(x,u) - f^1(x,v)| \le g(x,|u-v|),$$

where the function $g(x,u)$ is defined on A^+, and for all fixed x and $1 \le i \le m$, $g_i(x,u_1,\ldots,u_m)$ is non-decreasing with respect to all $u_1,\ldots,u_m$;

(iii) $\lambda(x)$ is any solution of (8.22.1) such that

$$|T(u) - T(u^1)| \le T(\lambda).$$

Then, if $\lambda(x) \to 0$ as $\|x\| = (x_1^2 + \cdots x_n^2)^{1/2} \to +\infty$, so is $|u(x) - u^1(x)|$.

Proof It is easy to show that $|u(x) - u^1(x)| \le \lambda(x)$ for all $x, 0 \le x \le X$. □

Last, We shall state a comparison result for the summary difference equations

$$u(x) = \theta(x) + S_{s=0}^{x-1} K(x,s)u(s) + S_{s=0}^{x-1} f(x,s,u(s)). \tag{8.22.2}$$

In (8.22.2), $\theta(x) = (\theta_1(x),\ldots,\theta_m(x))^T$; $K(x,s)$ is an $m \times m$ non-negative matrix for all $0 \le s, x \le X$; $f(x,s,u) = (f_1(x,s,u),\ldots,f_m(x,s,u))^T$, and for all fixed

s, x $(0 \leq s, x \leq X)$ and $1 \leq i \leq m$, the function $f_j(x, s, u_1, \ldots, u_m)$ is non-decreasing with respect to all $u_1, \ldots, u_m$.

Theorem 8.22.6 ([8]) *Assume that the following conditions hold:*

(i) $u(x)$ is a solution of (8.22.2), and $\phi(x)$ and $\varphi(x)$ are the solutions of the inequalities

$$\phi(x) \leq \theta(x) + S_{s=0}^{x-1} K(x, s)\phi(s) + S_{s=0}^{x-1} f(x, s, \phi(s))$$

and

$$\varphi(x) \geq \theta(x) + S_{s=0}^{x-1} K(x, s)\varphi(s) + S_{s=0}^{x-1} f(x, s, \varphi(s)).$$

respectively;

(ii) $\phi((i)x) \leq u((i)x) \leq \varphi((i)x)$.

Then, for all $x, 0 \leq x \leq X$,

$$\phi(x) \leq u(x) \leq \varphi(x).$$

Remark 8.22.1 If the non-decreasing nature of f is replaced by mixed monotone property and w consider the corresponding partial inequalities, then the resulting inequalities which follow are in terms of corresponding partial inequalities. The details of these results are similar to the case $n = 1$ discussed in [4].

8.23 Applications of Theorems 7.2.10 and 7.2.11 to Hyperbolic Partial Differential Equations with Impulse Perturbations

In this section, we shall exploit Theorems 7.2.10 and 7.2.11 to investigate some properties of hyperbolic partial differential equations with impulse perturbations concentrated on the surfaces

$$\begin{cases} \dfrac{\partial^2 u(x_1, x_2)}{\partial x_1 \partial x_2} = H(x, u(x)), \quad (x_1, x_2) \in \Gamma_i, \\ u(x_1, 0) = \phi_1(x_1), \\ u(0, x_2) = \phi_2(x_2), \\ \phi_1(0) = \phi_2(0), \\ \Delta u|_{(x_1, x_2) \in \Gamma_i} = \displaystyle\int_{\Gamma_i \cap G_n} \beta_i(x_1, x_2) u((x_1, x_2)) d\mu_{\varphi_i}. \end{cases} \tag{8.23.1}$$

Here $\Delta u|_{(x_1,x_2)\in\Gamma_i}$ are the characterised values of finite jumps $u(x)$ $(x = (x_1, x_2))$, when the solutions of (8.23.1) meet the hypersurfaces $\Gamma_i : u(x) \cap \Gamma_i$.

We investigate (8.23.1) in the domain $D^* \subset \mathbb{R}^2_+$, which was defined as in Sect. 7.2.1.

Denote by $\phi(x_1, x_2)$ the boundary conditions in (8.23.1). Then any solution of equations (8.23.1), satisfying the boundary conditions, is also a solution of the Volterra integro-sum equation

$$u(x_1,x_2) = \phi(x_1,x_2) + \int\int_{G_n} H(\tau,s,u(\tau,s))d\tau ds + \sum_{j=1}^{n-1}\int_{\Gamma_j\cap G_n} \beta_j(x_1,x_2)u(x_1,x_2)d\mu_{\varphi_j}. \tag{8.23.2}$$

Suppose that

$$|H(\tau,s,u(\tau,s))| \le \psi(\tau,s)W\Big[|u(\tau,s)|\Big], \tag{8.23.3}$$

where $\psi(\tau,s) \ge 0$, $W(\sigma) \in \Phi_1$.

Using the result of Theorem 7.2.10, we can obtain the following theorem.

Theorem 8.23.1 ([399]) *Suppose that $H(x,u(x))$ in (8.23.1) satisfies condition (8.23.3). Then for all solutions of problem (8.23.1) the following estimate holds, for all $x \in D_i$,*

$$|u(x_1,x_2)| \le |\phi(x_1,x_2)|\Psi_i^{-1}\left(\int\int_{D_i} \frac{\psi(\tau,s)}{|\phi(\tau,s)|} W[|\phi(\tau,s)|]d\tau ds\right), \tag{8.23.4}$$

with

$$\int\int_{D_i} \frac{\psi(\tau,s)}{|\phi(\tau,s)|} W\Big[|\phi(\tau,s)|\Big]d\tau ds \in Dom\ (\Psi_i^{-1}),$$

where

$$\begin{cases} \Psi_0(V_1) := \int_1^{V_1} \frac{d\sigma}{W(\sigma)}, \quad \Psi_i(V_1) := \int_{C_i}^{V_1} \frac{d\sigma_1}{W(\sigma_1)}, \ i = 1,2, \\ C_i = \left(1 + \int_{\Gamma_i\cap G_n} |\beta_i(x_1,x_2)|d\mu_{\varphi_i}\right)\Psi_{i-1}^{-1} \\ \quad \times\left(\int\int_{Gi+1\setminus G_i} \frac{\psi(\tau,s)}{\phi(\tau,s)} W\Big[|\phi(\tau,s)g(\tau,s)|\Big]d\tau ds\right). \end{cases}$$

By using Theorem 7.2.11, the next theorem can be obtained.

Theorem 8.23.2 ([399]) *In (8.23.1), let the function H satisfy (8.23.3), where W belongs to the class of functions* $\overline{\Phi}_1 : W \in \overline{\Phi}_1$. *Then all solutions of problem (8.23.1) satisfy some estimates:*

$$|u(x_1,x_2)| \le |\phi(x_1,x_2)|\overline{\Psi}_j^{-1}\Big(\int\int_{D_i} \psi(\tau,s)d\tau ds\Big), \quad i = 1,2,\cdots$$

where for all $x : 0 < x < x^*$,

$$\overline{\Psi}_0(V) := \int_1^V \frac{d\sigma}{W(\sigma)}, \quad \overline{\Psi}_i(V) := \int_{C_i}^{V_1} \frac{d\sigma}{W(\sigma)}, \ i = 1,2,$$

$$C_i = \Big(1 + \int_{\Gamma_i \cap G_n} |\beta_i(x_1,x_2)| d\mu_{\varphi_i}\Big)\overline{\Psi}_{i-1}^{-1}\Big(\int\int_{Gi+1\backslash G_i} \psi(\tau,s)d\tau ds\Big),$$

with

$$x^* = \sup_x \Big\{ x : \int\int_{Gi+1\backslash G_i} \overline{\psi}(\tau,s)d\tau ds \in Dom(\overline{\Psi}_i^{-1}), i = 1,2,\cdots \Big\}.$$

From Theorem 7.2.10 and Theorem 8.23.1, the next result follows.

Theorem 8.23.3 ([399]) *Suppose that* $H(x_1,x_2,u(x_1,x_2))$ *satisfies*

(A) $|H(x_1,x_2,u(x_1,x_2))| \le f(x_1,x_2)|u(x_1,x_2)|^{\alpha} = const. > 0$, *where f is a continuous non-negative function in* $\mathbb{R}_+^2$.
(B) Boundary conditions for (8.23.1) are bound: $\exists M = const. > 0 : |\phi(x_1,x_2)| \le M$. *Then for the solutions of problem (8.23.1), the following estimates hold:*

(1) if $\alpha = 1$, *then*

$$|u(x_1,x_2)| \le M \prod_{j=1}^{+\infty} \Big(1 + \int_{\Gamma_i \cap G_{j+1}} |\beta_j(x_1,x_2)| d\mu_{\phi_i}\Big) \times \exp\Big[\int_0^{x_1}\int_0^{x_2} f(\sigma_1,\sigma_2)d\sigma_1 d\sigma_2\Big];$$

(2) if $0 < \alpha < 1$, *then*

$$|u(x_1,x_2)| \le M \prod_{j=1}^{+\infty} \Big(1 + \int_{\Gamma_i \cap G_{j+1}} |\beta_j(x_1,x_2)| d\mu_{\phi_i}\Big) \times \Big[1 + (1-\alpha)M^{\alpha-1}\int_0^{x_1}\int_0^{x_2} f(\sigma_1,\sigma_2)d\sigma_1 d\sigma_2\Big]^{\frac{1}{1-\alpha}};$$

(3) if $\alpha > 1$,

$$|u(x_1, x_2)| \leq M \prod_{j=1}^{+\infty} \left(1 + \int_{\Gamma_i \cap G_{j+1}} |\beta_j(x_1, x_2)| d\mu_{\phi_i}\right)$$

$$\times \left\{ 1 + (\alpha - 1) M^{\alpha-1} \Big[\prod_{j=1}^{+\infty} (1 + \int_{\Gamma_i \cap G_{j+1}} |\beta_j(x_1, x_2)| d\mu_{\phi_i}) \Big]^{\alpha-1} \right.$$

$$\left. \times \int_0^{x_1} \int_0^{x_2} f(\sigma_1, \sigma_2) d\sigma_1 d\sigma_2 \right\}^{-\frac{1}{\alpha-1}},$$

and arbitrary $(x_1, x_2) \in D^*$ *such that*

$$\int_0^{x_1} \int_0^{x_2} f(\sigma_1, \sigma_2) d\sigma_1 d\sigma_2$$

$$\leq \left\{ (\alpha - 1) M^{\alpha-1} \left[\prod_{j=1}^{\infty} (1 + \int_{\Gamma_i \cap G_{j+1}} |\beta_j(x_1, x_2)| d\mu_{\phi_i}) \right]^{\alpha-1} \right\}^{-1}.$$

From Theorem 8.23.3, the next theorem follows immediately.

Theorem 8.23.4 ([399]) *Suppose that for (8.23.1), the following conditions hold:*

(1) $|H(x_1, x_2, u(x_1, x_2))| \leq \psi(x_1, x_2) |u(x_1, x_2)|^\alpha$;
(2) $\exists M = const. > 0$ *: such that* $|\varphi(x_1, x_2)| \leq M$;
(3) $\exists\ \xi, \eta$*: such that*

$$\prod_{j=1}^{+\infty} \left(1 + \int_{\Gamma_i \cap G_{j+1}} |\beta_j(x_1, x_2)| d\mu_{\phi_i}\right) \leq \xi < +\infty,$$

$$\int_0^{x_1} \int_0^{x_2} \psi(\sigma_1, \sigma_2) d\sigma_1 d\sigma_2 \leq \eta < +\infty.$$

Then all solutions $u(x_1, x_2)$ *of problem (8.23.1) are bounded for* $0 < \alpha \leq 1$. *If, additionally,*

$$\prod_{j=1}^{+\infty} \left(1 + \int_{\Gamma_i \cap G_{j+1}} |\beta_j(x_1, x_2)| d\mu_{\phi_i}\right) < \frac{M^{1-\alpha}}{(\alpha - 1)\eta},$$

then all solutions of problem (8.23.1) are also bounded for $\alpha > 1$.

8.24 Applications of Theorem 7.2.19 to Nonlinear Delay Partial Integro-differential Equations

In this section, we shall use Theorem 7.2.19 to study the following nonlinear delay partial integro-differential equation

$$\begin{cases} u_{xy}(x,y) \leq F(x,y,u(x,y), \int_{b(b(x_0))}^{b(x)} \int_{c(c(c_0))}^{c(y)} h(b(x),c(y),\tau,\sigma,u(\tau,\sigma))d\tau d\sigma), \\ u(x,y_0) = \alpha(x), \ u(x_0,y) = \beta(y), \end{cases} \tag{8.24.1}$$

for all $(x,y) \in \Lambda$, where b, c and u are supposed to be as in Theorem 7.2.19, $h : \Lambda^2 \times \mathbb{R} \to \mathbb{R}$, $F : \Lambda \times \mathbb{R}^2 \to \mathbb{R}$, $\alpha : I \to \mathbb{R}$, and $\beta : J \to \mathbb{R}$ are all continuous functions such that $\alpha(0) = \beta(0) = 0$.

We first give an estimate for solutions of (8.24.1) under the conditions

$$\begin{cases} |F(x,y,u,v)| \leq f(x,y)[\varphi_1(|u|) + |v|], \\ |h(x,y,s,t,u(s,t))| \leq g(x,y,s,t)|\varphi_2(u(s,t))|. \end{cases} \tag{8.24.2}$$

The next two corollaries are direct consequences of Theorem 7.2.19, we omit their proofs.

Theorem 8.24.1 ([665]) *If $|\alpha(x) + \beta(y)|$ is non-decreasing in x and y and (8.24.2) holds, then every solution $u(m,n)$ of problem (8.24.1) satisfies, for all $(x,y) \in [x_0, X_1) \times [y_0, Y_1)$,*

$$u(x,y) \leq W_2^{-1}[\Xi(x,y)], \tag{8.24.3}$$

where

$$\begin{aligned} \Xi(x,y) := W_2 \left\{ W_1^{-1}[W_1(|\alpha(x) + \beta(x)|) + \int_{b(x_0)}^{b(X)} \int_{c(y_0)}^{c(y)} \frac{f(b^{-1}(s), c^{-1}(t))}{b'(b^{-1}(s))c'(c^{-1}(t))} dtds] \right\} \\ + \int_{b(x_0)}^{b(X)} \int_{c(y_0)}^{c(y)} \frac{f(b^{-1}(s), c^{-1}(t))}{b'(b^{-1}(s))c'(c^{-1}(t))} \left[\int_{b(x_0)}^{s} \int_{c(y_0)}^{t} g(s,t,\tau,\sigma) d\tau d\sigma \right] dtds, \end{aligned} \tag{8.24.4}$$

and W_1, W_1^{-1}, W_2, W_2^{-1}, and X_1, Y_1 are defined as in Theorem 7.2.19.

Theorem 8.24.1 actually gives a condition of boundedness for solutions. Concretely, if there is a positive constant M such that there holds on $[x_0, X_1) \times [y_0, Y_1)$,

$$\begin{cases} |\alpha(x) + \beta(x)| < M, \quad \int_{b(x_0)}^{b(x)} \int_{c(y_0)}^{c(y)} \frac{f(b^{-1}(s), c^{-1}(t))}{b'(b^{-1}(s))c'(c^{-1}(t))} dtds < M, \\ \int_{b(x_0)}^{b(x)} \int_{c(y_0)}^{c(y)} \frac{f(b^{-1}(s), c^{-1}(t))}{b'(b^{-1}(s))c'(c^{-1}(t))} \left[\int_{b(x_0)}^{s} \int_{c(y_0)}^{t} g(s, t, \tau, \sigma) d\tau d\sigma \right] dtds < M, \end{cases} \tag{8.24.5}$$

then every solution $u(x, y)$ of problem (8.24.1) is bounded on $[x_0, X_1) \times [y_0, Y_1)$.

Next, we give the condition of the uniqueness of solutions for problem (8.24.1).

Corollary 8.24.1 *Suppose*

$$\begin{cases} |F(x, y, u_1, v_1) - F(x, y, u_2, v_2)| \leq f(x, y) \Big[\varphi_1(|u_1 - u_2|) + |v_1 - v_2| \Big], \\ |h(x, y, s, t, u_1) - h(x, y, s, t, u_2)| \leq g(x, y, s, t) \varphi_2(|u_1 - u_2|), \end{cases} \tag{8.24.6}$$

where f, g, ψ_1, ϕ_2 are defined as in Theorem 7.2.19. There is a positive number M such that for all $(x, y) \in [x_0, X_1) \times [y_0, Y_1)$,

$$\begin{cases} \int_{b(x_0)}^{b(x)} \int_{c(y_0)}^{c(y)} \frac{f(b^{-1}(s), c^{-1}(t))}{b'(b^{-1}(s))c'(c^{-1}(t))} dtds < M, \\ \int_{b(x_0)}^{b(x)} \int_{c(y_0)}^{c(y)} \frac{f(b^{-1}(s), c^{-1}(t))}{b'(b^{-1}(s))c'(c^{-1}(t))} [\int_{b(x_0)}^{s} \int_{c(y_0)}^{t} g(s, t, \tau, \sigma) d\tau d\sigma] dtds < M. \end{cases} \tag{8.24.7}$$

Then, (8.24.1) has at most one solution on $[x_0, X_1) \times [y_0, Y_1)$, where X_1, Y_1 are defined as in Theorem 7.2.19.

References

1. J. Abrammovich, On Gronwall and Wendroff type inequalities. Proc. Amer. Math. Soc. **87**, 481–486 (1983)
2. M. Abramowitz, I.A. Stegum, Handbook of Mathematical Functions (New York, 1972)
3. D.R. Adams, Traces of potentials arising from translation invariant operators. Ann. Scoula. Norm. Super. Pisa. **25**, 203–217 (1971)
4. R.P. Agarwal, On finite systems of difference inequalities. J. Math. Phys. Sci. **10**(4), 277–288 (1976)
5. R.P. Agarwal, A note on Grace and Lalli's paper. J. Math. Anal. Appl. **86**, 471–475 (1982)
6. R.P. Agarwal, On an integral inequality in n independent variables. J. Math. Anal. Appl. **85**, 192–196 (1982)
7. R.P. Agarwal, Linear and nonlinear discrete inequalities in n independent variables, in *General Inequalities 5, ISNM 80*, ed. by W. Walter (2nd ed.) (Birkhäuser, Basel, 1987), pp. 303–318
8. R.P. Agarwal, Comparison result for multidimensional difference equations. J. Math. Anal. Appl. **135**, 476–487 (1988)
9. R.P. Agarwal, Systems of multidimensional discrete inequalities. J. Math. Anal. Appl. **140**, 241–250 (1989)
10. R.P. Agarwal, *Difference Equations and Inequalities: Theory, Methods and Applications* (Marcel Dekker, New York, 1991)
11. R.P. Agarwal, M. Bohner, A. Peterson, Inequalities on time scales: a survey. Math. Ineq. Appl. **4**(4), 535–557 (2001)
12. R.P. Agarwal, M. Bohner, D. O'Regan, A. Peterson, Dynamic equations on time scales: A survey. J. Comput. Appl. Math. **141**(1–2), 1–26 (2002)
13. R.P. Agarwal, S. Deng, W. Zhang, Generalization of a retarded Gronwall-like inequality and its applications. Appl. Math. Comput. **165**, 599–612 (2005)
14. R.P. Agarwal, Kim and Sen, New retarded integral inequalities with applications. J. Ineq. Appl. **2008**(Art. ID 908784), 15p. (2008)
15. R.P. Agarwal, S. Leela, A brief biography and survey of collected works of V. Lakshmikantham's legacy: A tribute on his 75th birthday. Nonlinear Anal. TMA **40**, 1–19 (2000)
16. R.P. Agarwal, P.Y.H. Pang, *Opial Inequalities with Applications in Differential and Difference Equations* (Kluwer, Dordrecht, 1995)
17. R.P. Agarwal, C.S. Ryoo, Y.H. Kim, New intergral inequalities for interated integrals, with applications. J. Inequal. Appl. **2007**(Article ID 24385), 18p. (2007). Doi:10.1155/2007/24385
18. R.P. Agarwal, E. Thandapani, On some new discrete inequalities. Appl. Math. Comp. **7**, 205–244 (1980)

Y. Qin, *Integral and Discrete Inequalities and Their Applications*,
DOI 10.1007/978-3-319-33304-5

19. R.P. Agarwal, E. Thandapani, On discrete generalizations of Gronwall's inequality. Bull. Inst. Math. Acad. Sinica **9**, 235–248 (1981)
20. R.P. Agarwal, E. Thandapani, On nonlinear discrete inequalities of Gronwall type. Al. I. Cuza din Tasi Tomul XXVII **5.Ia**, f.1. (1981)
21. R.P. Agarwal, E. Thandapani, Remarks on generalizations of Gronwall-inequality. Chin. J. Math. **9**(2), 1–22 (1981)
22. R.P. Agarwal, E. Thandapani, Some inequalities of Gronwall type. An. st. univ. Iasi **28**, 71–75 (1982)
23. R.P. Agarwal, P.J.Y. Wong, *Advanced Topics in Difference Equations* (Kluwer, Dordrecht, 1997)
24. S. Agmon, L. Nirenberg, Lower bounds and uniqueness theorems for solutions of differential equations in a Hilbert space. Comm. Pure Appl. Math. **20**, 207–229 (1967)
25. N.U. Ahmed, Some remarks on the dynamics of impulsive systems in Banach spaces. Dyn. Continuous Discrete Impulsive Syst. **8**, 261–274 (2001)
26. E. Akin-Bohner, M. Bohner, F. Akin, Pachpatte inequalities on time scales. J. Ineq. Pure Appl. Math. **6**(1), article 6, p. 23. (2005)
27. O. Akinyele, On an integral inequality in n-independent varibales. Inter. J. Math. Math. Sci. **7**, 455–467 (1984)
28. O. Akinyele, On Gronwall-Bellman-Bihari type integral inequalities in several variables with retardation. J. Math. Anal. Appl. **104**, 1–26 (1984)
29. H. Amann, Global existence for semilinear parabolic systems. J. Reine Angew. Math. **360**, 47–83 (1985)
30. H. Amann, Dynamic theory of quasilinear parabolic systems-III, global existence. Math. Z. **202**, 219–250 (1989)
31. H. Amann, Global existence of a class of highly degenerate parabolic systems. Jpn. J. Indust. Appl. Math. **8**(1), 143–159 (1991)
32. K. Ammari, M. Tucsnak, Stabililization of Bernoulli-Euler beams by means of a pointwise feedback force. SIAM J. Control Optim. **39**, 1160–1181 (2000)
33. G. Andrews, On the existence of solutions to the equation $u_{tt} = u_{xxt} + \sigma(u_x)_x$. J. Differ. Equ. **35**, 200–231 (1980)
34. H.A. Antosiewicz, An inequality for approximating solutions of ordinary differential equations. Math. Z. **78**, 44–52 (1962)
35. F.V. Atkinson, *Discrete and Continuous Boundary Problems* (Academic Press, New York/London, 1964)
36. J.P. Aubin, Un théorème de compacité. C. R. Acad. Sc. Paris **256**, 5012–5014 (1963)
37. N.T.J. Bailey, *The Mathematical Theory of Infectious Diseases and Its Applications*, 2nd edn. (Hafner, New York, 1975)
38. D.D. Bainov, S.G. Hristova, Some generalizations of integral inequalities of Bellman-Bihari type fir functions of several variables. Rend. Circ. Mat. Palermo **30**, 378–384 (1981)
39. D.D. Bainov, S.G. Hristova, Integral inequalities of Bihari type for functions of two variables. Riv. Mat. Univ. Parma **8**, 347–354 (1982)
40. D.D. Bainov, S.G. Hristova, Impulsive integral inequalities with a deviation of the argument. Mathematische Nachrichten **171**, 19–27 (1995)
41. D.D. Bainov, P.S. Simeonov, *System with Impulse Effect: Theory and Applications* (Ellis Horwood, Chichister, 1989)
42. D.D. Bainov, P.S. Simeonov, *Integral Inequalities and Applications, Mathematics and Its Applications*, Vol. 57 (Kluwer, Dordrecht, 1992)
43. D.D. Bainov, A.D. Myshkis, A.I. Zahariev, On an abstract analog of the Bellman-Gronwall inequality. Publs Res. Inst. Math. Sci. Kyoto Univ. **20**, 903–911 (1984)
44. V. Barbu, *Differential Equations*, Ed. Junimea. Iasi, 1985 (in Romanian)
45. R. Bartle, *The Elements of Real Analysis* (Wiley, New York, 1964)
46. F. Bauer, Bounds for solutions of ordinary differential equations. Proc. Am. Math. Soc., **14**, 36–43 (1963)
47. E.F. Beckenbach, R. Bellman, *Inequalities* (Springer, Berlin, 1961)

48. E.F. Beckenbach, R. Bellman, *Inequalites*, Second Revised Printing (Springer, New York, 1965)
49. E.F. Beckenbach, R. Bellman, *Inequalities*, 2nd edn. (Springer, Berlin-Heidelberg-New York, 1965); 4th edn. (1983)
50. E. Becker, *Gasdynamik* (Teubner Verlag, Stuttgart, 1966)
51. P.R. Beesack, Comparison theorems and integral inequalities for Volterra integral equations. Proc. Am. Math. Soc. **20**, 61–66 (1969)
52. P.R. Beesack, On certain discrete inequalities involving partial sums. Can. J. Math. **21**, 222–234 (1969)
53. P.R. Beesack, Integral inequalities involving a function and its derivative. Am. Math. Monthly **78**, 705–741 (1971)
54. P.R. Beesack, *Gronwall Inequalities, Carelton Mathematical Lecture Notes*, Vol. 11 (Northfield, Minnesota, 1975)
55. P.R. Beesack, On integral inequalities of Bihari type. Acta Math. Acad. Sci. Hungar. **28**, 81–88 (1976)
56. P.R. Beesack, On Lakshmikantham's comparison method for Gronwall inequalities. Ann. Polon. Math. **35**, 187–222 (1977)
57. P.R. Beesack, Riemann's function has an exponential bound. Proc. Am. Math. Soc. **88**, 313–316 (1983)
58. P.R. Beesack, Asymptotic behavior of solutions of some general nonlinear differential equations and integral inequalities. Proc. Roy. Soc. Edinburgh Ser A **98**, 49–67 (1984)
59. P.R. Beesack, On some Gronwall-type integral inequalities in n independent variables. J. Math. Anal. Appl. **100**, 393–408 (1984)
60. P.R. Beesack, Systems of multidimensional Gronwall integral inequalities. J. Math. Phys. Sci. **18**(3), 215–231 (1984)
61. P.R. Beesack, More generalised discrete Gronwall inequalities. Z. angew. Math. Mech. **65**, 589–595 (1985)
62. P.R. Beesack, Systems of multidimensional Volterra integral equations and inequalities. Nonlinear Anal. TMA **9**(12), 1451–1486 (1985)
63. A. Belleni-Morante, A.C. McBride, *Applied Nonlinear Semigroups, Wiley series in Mathematical Methods in Practice* (Wiley, Chichester, New York, Weinheim, Brisbane, Singapore, Toronto, 1998)
64. R. Bellman, The bounded-ness of solutions of linear differential equations. Duke Math. J. **14**, 83–97 (1943)
65. R. Bellman, The stability of solutions of linear differential equations. Duke Math. J. **10**(4), 643–647 (1943)
66. R. Bellman, The boundedness of solutions of linear differential equations. Duke math. J. **14**, p.84 (1947)
67. R. Bellman, On the existence and boundedness of solutions of nonlinear partial differential equations of parabolic type. Trans. Am. Math. Soc. **64**, 21–44 (1948)
68. R. Bellman, On an inequality of Weinberger. Am. Math. Monthly **60**(402), 402–404 (1953)
69. R. Bellman, *Stability Theory of Differential Equations* (McGraw-Hill, New York, 1953)
70. R. Bellman, *Theory of Solution Stability of Differential Equations* (Isdatelstvo Inostz. Lit, Moscow, 1954)
71. R. Bellman, On an inequality concerning an indefinite form. Am. Math. Monthly **63**, 108–109 (1956)
72. R. Bellman, On inequalities with alternating signs. Proc. Am. Math. Soc. **10**, 807–809 (1959)
73. R. Bellman, *Introduction to Matrix Analysis* (McGraw-Hill, New York, 1960)
74. R. Bellman, *Adaptive Control Processes: A Guided Tour* (Princeton University Press, Princeton, New Jersey, 1961)
75. R. Bellman, A note on asymptotic behavior of diferential equations. Boll. UMI **18**, 16–18 (1963)
76. R. Bellman, *Introduction to the Mathematical Theory of Control Processes*, Vol. I (Academic Press, New York, 1967)

77. R. Bellman, Upper and lower bounds for solution of the matrix Riccati equation. J. Math. Anal. Appl. **17**, 373–379 (1967)
78. R. Bellman, Scattering processes and invariant imbedding. J. Math. Anal. Appl. **23**, 254–268 (1968)
79. R. Bellman, K.L. Cooke, *Differential-Difference Equations* (Academic Press, New York, 1963)
80. A. Bielecki, Une remarque sur la methode de Banach-Cacciopoli-Tikhonov dans la theórie des eq́uations differentielles ordinaires. Bull. Acad. Polon. Sci. **4**, 261–264 (1951)
81. A. Bielecki, Une remarque sur l'application de la méthod de Banach-Cacciopoli-Tikhonov dans la théorie de léquation $s = f(x, y, z, p, q)$. Bull. Acad. Polon. Sci. Sér. Sci. Math. Phys. Astr. **4**, 265–268 (1956)
82. I. Bihari, A generalization of a lemma of Bellman and its applications to uniqueness problems of differential equations. Acta Math. Acad. Sci. Hungar. **7**, 81–94 (1956)
83. I. Bihari, The asymptotic behavior of a system of nonlinear differental equations. Publ. Math. Sci. Hungar Acad. Sci. **8**, 475–488 (1964)
84. G. Birkhoff, *Lattice Theory*, Vol. 25 (Am. Math. Soc. Colloq. Publ., New York, 1948)
85. A. Blaquire, Differential games with piecewise continuous trajectories. Lect. Notes Control Inform. Sci. **3**, 34–69 (1977)
86. H.D. Block, Discrete analogues of certain integral inequalities. Proc. Am. Math. Soc. **8**, 852–859 (1957)
87. R.P. Boas, More about quotients of monotone functions. Am. Math. Monthly. **72**, 59–60 (1965)
88. M.L. Boas, R.P. Boas, N. Levinson, The growth of solutions of a differential equation. Duke Math. J. **9**, 847–853 (1942)
89. L.E. Bobisud, Existence of solutions to some nonlinear diffusion problems. J. Math. Anal. Appl. **168**, 413–424 (1992)
90. M. Bohner, A. Peterson, *Dynamic Equations on Time Scales: An Introduction with Applications* (Birkhäuser, Boston, 2001)
91. M. Bohner, A. Peterson, *Advances in Dynamic Equations on Time Scales* (Birkhäuser, Boston, 2003)
92. E. Bompiani, Un teorema di confronto ed un teorema di unicita per l'equazione differenziale $y' = f(x, y)$, Rendiconti dell'Accad. dei Lincei, Classe di Scienze Fisiche **1**(6), 298–302 (1925)
93. B.K. Bondge, Some Contributions to Integral Inequalities, Ph.D. thesis, Marathwada Univ., India, 1979
94. B.K. Bondge, B.G. Pachpatte, On nonlinear integral inequalitis of the Wendroff's type. J. Math. Anal. Appl. **70**, 161–169 (1979)
95. B.K. Bondge, B.G. Pachpatte, On some fundamental integral inequalities in two independent variables. J. Math. Anal. Appl. **72**, 533–544 (1979)
96. B.K. Bondge, B.G. Pachpatte, On Wenndroff's type integral inequalities in n independent variables. Chin. J. Math. **7**, 37–46 (1979)
97. B.K. Bondge, B.G. Pachpatte, On some fundamental integral inequalities in n independent variables. Bull. Inst. Math. Acad. Sinica **8**, 553–560 (1980)
98. B.K. Bondge, B.G. Pachpatte, On some partial integral inequalities in two independent variables. Funkcialaj Ekvacioj **23**, 327–334 (1980)
99. B.K. Bondge, B.G. Pachpatte, On two independent variable generalization of certain integral inequalities. Tamkang J. Math. **11**, 37–47 (1980)
100. B.K. Bondge, B.G. Pachpatte, W. Walter, On generalized Wendroff type inequalities and their applications. Nonlinear Anal. TMA **4**, 491–495 (1979)
101. K.B. Bopaev, On some discrete inequalities. Differentsial'nye Uraveniya(Alma-Ata), 35–44 (1981) (In Russian)
102. S.D. Borysenko, About asymptotical stability on linear approximation of the systems with impulse influence. Ukr. Mat. Zh. **35**(2), 144–150 (1983)

103. S.D. Borysenko, On solution stability on linear approximation of the systems with impulse. Differ. Equ. **22**(5), 884–886 (1986)
104. S.D. Borysenko, Integro-sum inequalities for functions of many independent variables. Differ. Equ. **25**(9), 1638–1641 (1989)
105. S.D. Borysenko, *Construction of Mathematical Models* (Vipol, Kiev, 1995)
106. S.D. Borysenko, Integro-sum inequalities and their use in the study of impulse systems, in *Proceeding of VI Internat.* M. Kravchuk Conf., 1997
107. S.D. Borysenko, Multidimensional integro-sum inequalities. Ukr. Mat. Zh. **50**(2), 172–177 (1998)
108. S.D. Borysenko, On integro-sum functional inequalities. Differ. Equ. **34**(6), 850 (1998)
109. S.D. Borysenko, About one integral inequality for piece-wise continuous functions, in *Proc. X Int. Kravchuk Conf.*, Kyiv, 2004
110. S.D. Borysenko, G. Iovane, About estimates of solutions for nonlinear hyperbolic equations with impulsive perturbations on some hypersurfaces, University of Salerno, Preprint DIIMA no. 7, March 2006
111. S.D. Borysenko, G. Iovane, Integro-sum Inequalities and Qualitative Analysis Dynamical Systems with Perturbations, Tipografia-Legatoria Elda, Universit'a of Salerno, Salerno, 2006
112. S.D. Borysenko, G. Iovane, About some new integral inequalities of Wendroff type for discontinuous functions. Nonlinear Anal. **66**, 2190–2203 (2007)
113. S.D. Borysenko, M. Ciarletta, G. Iovane, Integro-sum inequalities and motion stability of systems with impulse perturbations. Nonlinear Anal. TMA **62**, 417–428 (2005)
114. S.D. Borysenko, M. Ciarletta, G. Iovane, About stability in nonlinear approximation of systems with impulse influence. Rep. Acad. Sci. Ukr. **1**, 8–12 (2006)
115. S.D. Borysenko, A. Gallo, R. Toscano, Integral inequalities Gronwall-Bellman type for discontinuous functions and estimates of solutions impulsive systems. Visnik Kyiv University (4), 60–66 (2005)
116. S.D. Borysenko, G. Iovane, P. Giordano, About Some Hyperbolic Impulsive Equations and Estimate Solutions, in *Proc. DE&CAS*, Brest, 2005
117. S.D. Borysenko, G. Iovane, P. Giordano, Investigations of the properties motion for essential nonlinear systems perturbed by impulses on some hypersurfaces. Nonlinear Anal. TMA **62**, 345–363 (2005)
118. S.D. Borysenko, V.I. Kosolapov, A.Y. Obolenskiy, *Stability of Processes under Continuous and Discrete Disturbances* (Naukova Dumka, Kiev, 1988)
119. J. Bourgain, *Global Solutions of Nonlinear Schrödinger Equations.* Colloquium Publications, vol. 46 (Amer. Math. Soc., Providence, RI, 1999)
120. D.W. Boyd, Y.S. Wong, On nonlinear contractions. Proc. Am. Math. Soc. **20**(2), 458–464 (1969)
121. F. Brauer, Bounds for solutions of ordinary differential equations. Proc. Am. Math. Soc. **14**, 36–43 (1963)
122. F. Brauer, Nonlinear differential equations with forcing terms. Proc. Am. Math. Soc. **15**, 758–765 (1964)
123. F. Brauer, A nonlinear variation of constants formula for Volterra equations. Mat. Syst. Th. **6**, 226–234 (1972)
124. H. Brill, A semilinear Sobolev evolution equation in a Banach space. J. Differ. Equ. **24**, 412–425 (1977)
125. H. Brunner, P.J. van der Houwen, *The Numerical Solution of Volterra Equation* (North-Holland, Amsterdam, 1986)
126. G. Bulter, T. Rogers, A generalization of a lemma of Bihari and applications to pointwise estimates for integral equations. J. Math. Anal. Appl. **33**(1), 77–81 (1971)
127. P.J. Bushell, W. Okrasifski, Uniqueness of solutions for a class of nonlinear Volterra integral with convolution kernel. Math. Proc. Camb. Philos. **106**, 547–552 (1989)
128. P.J. Bushell, W. Okrasinski, On the maximal interval of existence for solutions to some nonlinear Volterra integral equations with convolution kernel. Bull. Lond. Math. Soc. **28**, 59–65 (1995)

129. G. Butler, T. Rogers, A generalization of a lemma of Bihari and applications to pointwise estimates for integral equations. J. Math. Anal. Appl. **33**(1), 77–81 (1971)
130. F. Cafiero, Su un problem ai limiti relativo all'equazione $y' = f(x, y, \lambda)$. Giornale Matematici Battaglini **77**, 145–165 (1947)
131. F. Cafiero, Sui teoremi di unicità relativi al un'equazione differenziale ordinaria del primo ordine. Ibidem **78**, 10–41 (1948)
132. T. Caraballo, P.M. Rubin, J. Valero, Autonomous and non-autonomous attractors for differential equations with delays. J. Differ. Equ. **208**, 9–41 (2005)
133. J. Caristi, Fixed point theorems for mappings satisfying inwardness conditions. Trans. Am. Math. Soc. **215**, 241–251 (1976)
134. T.K. Caughey, R.T. Shield, Instability and the energy criteria for continuous systems. Z. Angew. Math. Phys. **19**, 485–492 (1968)
135. J. Chandra, P.W. Davis, Linear generalization of Gronwall's inequality. Proc. Am. Math. Soc. **60**, 157–160 (1976)
136. J. Chandra, B.A. Fleishman, Bounds and Maxima solutions of nonlinear functional equations. Bull. Am. Math. Soc. **74**, 512–516 (1968)
137. J. Chandra, B.A. Fleishman, On a generalization of the Gronwall-Bellman Lemma in partially ordered Banach spaces. J. Math. Anal. Appl. **31**, 668–681 (1970)
138. J. Chandra, B.A. Fleishman, Existence and comparison results for a class of nonlinear boundary value problems. Ann. Math. Pura. Appl., **150**, 247–261 (1974)
139. S. Chapman, T.G. Cowling, *The Mathematical Theory of Nonuniform Gases*, 3rd edn. (Cambridge University Press, Cambridge, 1970)
140. W.S. Cheung, On some new integro-differential inequalities of the Gronwall and Wendroff type. J. Math. Anal. Appl. **178**, 438–449 (1993)
141. W.S. Cheung, Some discrete nonlinear inequalities and applications to boundary value problems for difference equations. J. Differ. Equ. Appl. **10**, 213–223 (2004)
142. W.S. Cheung, Some new nonlinear inequalities and applications to boundary value problems. Nonlinear Anal. **64**, 2112–2128 (2006)
143. W.S. Cheung, Some retarded Gronwall-Bellman-Ou-Yang-type inequalities and applications to initial boundary value problems, Update
144. W.S. Cheung, Q.H. Ma, Nonlinear retarded integral inequalities for functions in two variables. J. Concrete Appl. Math. **2**, 119–134 (2004)
145. W.S. Cheung, Q.H. Ma, On certain new Gronwall-Ou-Yang type integral inequalities in two variables and their applications. J. Ineq. Appl. **2005**(4), 347–361 (2005)
146. W.S. Cheung, Q.H. Ma, Josip, Some discrete nonlinear inequalities and applications to difference equations. Acta Math. Sinica **2008**(B), 417–430 (2008)
147. W.S. Cheung, J.L. Ren, Discrete nonlinear inequalities and applications to boundary value problems. J. Math. Anal. Appl. **319**, 708–724 (2006)
148. C. Chicone, Y. Latushkin, *Evolution Semigroups in Dynamical Systems and Differential Equations, Mathematical Surveys and Monographs*, Vol. 70 (Amer. Math. Soc., Providence, RI, 1999)
149. T.S. Chihara, P.G. Neval, Orthogonal polynomials and measures with finitely many point masses. J. Approx. Theory **35**, 370–380 (1982)
150. Y.J. Cho, S.S. Dragomir, Y.H. Kim, On some integral inequations with iterated integrals. J. Korean Math. Soc. **43**(3), 563–578 (2006)
151. S.K. Choi, S. Deng, N.J. Koo, W. Zhang, Nonlinear integral inequalities of Bihari-type without class H. Math. Ineq. Appl. **8**(4), 643–654 (2005)
152. Y.C. Chow, On inequalities of Hilbert and Widder. J. Lond. Math. Soc. **14**, 151–154 (1939)
153. S.C. Chu, F.T. Metcalf, On Gronwall's inequality. Proc. Am. Math. Soc. **18**, 439–440 (1967)
154. M. Ciarletta, S.D. Borysenko, G. Iovane, About stability systems with impulse influence on nonlinear approximation, in *Proc. X Int. Kravchuk Conf.*, Kiev, 2004, pp. 275–276
155. Z. Ciesielski, A note on some inequalities of Jensen's type. Ann. Polon. Math. **4**, 269–274 (1958)

156. E.A. Coddington, N. Levinson, *Theory of Ordinary Differential Equations* (McGraw-Hill, New York, 1955)
157. D.S. Cohen, The asymptotic behavior of a class of nonlinear differential equations. Proc. Am. Math. Soc. **18**, 607–609 (1967)
158. J. Conlan, A generalization of Gronwall's inequality. Contrib. Differ. Equ. **3**, 37–42 (1964)
159. J. Conlan, J.B. Diaz, Existence of solutions of an nth order hyperbolic partial differential equations. Contrib. Differ. Equ. **2**, 277–289 (1963)
160. J. Conlan, C.L. Wang, Higher dimensional Gronwall-Bellman type inequalities. Appl. Anal. **18**, 1–12 (1984)
161. J. Conlan, C.L. Wang, Gronwall-Bellman type inequalitities for systems. J. Math. Anal. Appl. **120**, 622–630 (1986)
162. J. Conlan, C.L. Wang, A unified approach to Gronwall-Bellman inequalities. Math. Appl. **4**, 47–55 (1991)
163. J. Conlan, C.L. Wang, A unified approach to continuous and discrete Gronwall-Bellman inequalities. Appl. Anal. **44**, 243–252 (1992)
164. A. Constantin, Solutions globales déquations différentielles perturbées. C. R. Acad. Sci. Paris **7**, 81–94 (1965)
165. A. Constantin, Global existence of solutions for perturbed differential equations. Ann. Mat. Pura Appl. **168**, 237–299 (1995)
166. A. Constantin, Solutions globales des équations différentielles perturbeés. C. R. Acad. Sci. Paris **320**, 1319–1322 (1995)
167. A. Constantin, S. Peszat, Global existence of solutions of semilinear parabolic evolution equations. Differ. Integral Equ. **13**, 99–114 (2000)
168. P. Constantin, C. Foias, R. Temam, Connexion entre la théorie mathématique des équations de Navier-Stokes et la théorie conventionnelle de la tuebulence. C. R. Acad. Sci. Paris Série I **297**, 599–602 (1983)
169. W.A. Coppel, *Stability and Asymptotic Behaviour of Differential Equations* (Heath, Boston, 1965)
170. E.T. Copson, Some integral inequalities. Proc. Roy. Soc. Edinburg Sect. A **75**, 157–164 (1975–1976)
171. A. Corduneanu, A note on Gronwall inequality in two independent variables. J. Integral Equ. **4**, 271–276 (1982)
172. A. Corduneanu, Bielecki's method in the theory of integral equations. Ann. Univ. Mariae Curie-Sklodowska Sec. A **38**(2), 23–40 (1984)
173. A. Corduneanu, Nonlinear integral inequalities in n-independent variables. An. Stin. Univ. A1 I Cuza, Iasi Ia Mat. **31**, 281–288 (1985)
174. A. Corduneanu, Nonlinear integral inequalities in n-independent variables. An. Strint. Univ. Al. I. Giza Iasi Sect.Ia Mat.(N.S) **31**(3), 281–288 (1988)
175. A. Corduneanu, *Integral Equations and Applications* (Cambridge Univ. Press, Cambridge, 1991)
176. R. Courant, K.O. Friedrichs, H. Lewy, Über die partiellen Differentialgleichungen der mathematischen Physik. Math. Ann. **100**, 32–74 (1928)
177. R. Courant, D. Hilbert, *Partial Differential Equations, Meth. Math. Phys.*, vol. II (Wiley, New York, 1962)
178. J. Cronin, *Differential Equations, Introduction and Qualitative Theory* (Dekker, New York, 1980)
179. R.F. Curtain, *Foundations of Functional Analysis Theory, Control Theory and Topics in Functional Analysis*, vol. I (International Atomic Energy Agency, Vienna, 1976), pp. 101–178
180. C.M. Dafermos, The second law of thermodynamics and stabiity. Arch. Rat. Mech. Anal. **70**(2), 167–179 (1979)
181. F.M. Dannan, Integral inequalities of Gronwall-Bellman-Bihari type and asymptotic behavior of certain second order nonlinear differential equations. J. Math. Anal. Appl. **108**(1), 151–164 (1985)

182. F. M. Dannan, Sub-multiplicative and sub-additive functions and integral inequalities of Bellmann-Bihari type. J. Math. Anal. Appl. **120**, 631–646 (1986)
183. J.N. Das, A note on an inequality due to Greene. Proc. Am. Math. Soc. **77**, 424–425 (1979)
184. P.G. Das, R.R. Sharma, Some stieltjes integral inequalities. J. Math. Anal. Appl. **73**, 423–433 (1980)
185. G.S. Davis, G.M. Peterson, On an inequality of Hardy's (II). Quart. J. Math. (Oxford) **15**, 35–40 (1964)
186. P.J. Davis, P. Rabinnowitz, *Methods of Numeric Integration* (Academic Press, New York, 1984)
187. D.E. Daykin, C.J. Eliezer, Generalization of Höder's and Minkowski's inequalities. Math. Proc. Camb. Phil. Soc. **64**, 1023–1027 (1968)
188. F.S. De Blasi, J. Schinas, On the stability of difference equations in Banach spaces. An. Sti. Univ. Al. I. Cuza Iasi, Sec. I **20**, 65–80 (1974)
189. K. Deckelnick, L^2 Decay for the compressible Navier-Stokes equations in unbounded domains. Comm. Partial Differ. Equ. **18**, 1445–1476 (1993)
190. R.J. Defranco, Gronwall's inequality for systems of multiple Volterra ingegral equations. Funkcial. Ekva. **19**, 1–9 (1976)
191. D. Delbosco, Sur une inégalite de la norme. Univ. Beograd Publ. Elek. Fak. Ser. Mat. Fiz. No. 678-No. **715**, 206–208 (1980)
192. D. Delbosco, L. Rodino, Existence and uniqueness for a nonlinear fractional differential equation. J. Math. Anal. Appl. **204**(2), 609–625 (1996)
193. M. Denche, H. Khellaf, Integral inequalities similar to Gronwall inequality. Electron. J. Differ. Equ. **2007**(176), 1–14 (2007)
194. S.G. Deo, U.D. Dhongade, Some generalizations of Bellman-Bihari integral inequalities. J. Math. Anal. Appl. **44**, 218–226 (1973)
195. S.G. Deo, D.U. Dhongade, Ponitwise estimates of solutions of some Volterra integral equation. J. Math. Anal. Appl. **45**, 615–628 (1974)
196. S.G. Deo, M.G. Murdeshwar, A note on Gronwall's inequality. Bull. Lond. Math. Soc. **3**, 34–36 (1972)
197. U.D. Dhongade, S.G. Deo, Some generalization of Bellman-Bihari integral inequalities. J. Math. Anal. Appl. **44**, 218–226 (1973)
198. U.D. Dhongade, S.G. Deo, Pointwise estimates of solutions of some Volterra integral equations. J. Math. Anal. Appl. **45**, 615–628 (1974)
199. U.D. Dhongade, S.G. Deo, Some generalizations of Bellman-Bihari integral inequalities. J. Math. Anal. Appl. **44**, 218–226 (1974)
200. U.D. Dhongade, S.G. Deo, Nonlinear generalization of Bihari's inequality. Proc. Am. Math. Soc. **54**, 211–216 (1976)
201. J.B. Diaz, On an analogue of Euler-Cauchy polygon method for the numerical solution of $u_{xy} = f(x, y, u, u_x, u_y)$. Arch. Rat. Mech. Anal. **1**, 357–390 (1958)
202. G. Dong, *Nonlinear Partial Differential Equations of Second Order, Applied Mathematics Series* (Tsinghua University Press, Beijing, 1988)
203. S.S. Dragomir, On Gronwall type lemmas and application. Monografii Matematice, Univ. Timisoara **29** (1987)
204. S.S. Dragomir, On discrete generalization of Pachpatte's inequality and applications. Bull. Math. Soc. Sci. Math. Roumanie **36**, 45–58 (1992)
205. S.S. Dragomir, Some refinements of Jensen's inequality. J. Math. Anal. Appl. **168**(2), 518–522 (1992)
206. S.S. Dragomir, M.M. Ionescu, On nonlinear integral inequalities in two independent variables. Studia Univ. Babes-Bolyai Math. **34**, 11–17 (1989)
207. S.S. Dragomir, Y.H. Kim, On certain new integral inequalities and their applications. JIPAM J. Inequal. Pure Appl. Math. **3**, artical 65 (2002)
208. S.S. Dragomir, Y.H. Kim, Some integral inequalities for functions of two variables. Electron. J. Differ. Equ. **2003**(10), 1–13 (2003)

209. S.S. Dragomir, B. Mond, Some inequalities of Aczĺęl type for gramians in inner product spaces. Nonlinear Funct. Anal. Appl. **6**, 411–424 (2001)
210. R. Driver, Existence and continuous dependence of solutions of neutral functional differential equations. Arch. Rat. Mech. Anal. **19**, 149–166 (1965)
211. O. Dtekmann, A note on the asymptotic speed of propagation of an epidemic. J. Differ. Equ. **33**, 58–73 (1979)
212. N. Dunford, J. Schwartz, *Linear Operators, Part I* (Interscience, New York, 1958)
213. N. Dunford, J. Schwartz, *Linear operators, Part II, Spectral Theory, Self-Adjoint Operators in Hilbert Space* (Interscience, New York/London, 1963)
214. Sh.T. Dzabbarov, G.K. Mamyedov, An integral inequality for functions in two variables and its application. Izv. Akad. Nauk Azerb. SSR **4**, 32–37 (1984) (in Russian)
215. M.S.P. Eastham, Asymptotic formula of Liouville-Green type for higher-order differential equations. J. Lond. Math. Soc. **28**(2), 507–518 (1983)
216. D.E. Edmunds, W.D. Evans, *Spectral Theory and Differential Operators* (Oxford University Press, Oxford, 1987)
217. R.E. Edwards, Functional Analysis (Holt Rinehart and Winston, New York, 1965)
218. H. El-Owaidy, A.A. Ragab, A. Abdeldaim, Some new integral inequalities of Gronwall-Bellman type. Appl. Math. Comput. **106**, 289–303 (1999)
219. H. Engler, Global regular solutions for the dynamic antiplane shear problem in nonlinear viscoelasticity. Math. Z. **202**, 251–259 (1989)
220. H. Erbe, QingKai Kong, Stieltjes integral inequalities of Gronwall type and applications. Ann. Math. Pura Appl. **157**, 77–97 (1990)
221. A. Erdelyi, *A Result on Nonlinear Volterra Integral Equations*. Studies in Mathematical Analysis and Related Topics (Stanford University Press, Stanford, 1962)
222. L.C. Evans, *Weak Convergence Methods for Nonlinear Partial Differential Equations, CBMS Regional Conference Series in Mathematics*, vol. 74 (Amer. Math. Soc., Providence, 1990)
223. W.N. Everitt, On the Höder inequality. J. Lond. Math. Soc. **36**, 145–158 (1961)
224. M. Fabrizio, B. Lazzari, On the existence and asymptotic stability of solutions for linearly viscoelastic solids. Arch. Rat. Mech. Anal. **116**, 139–152 (1991)
225. S. Faedo, su un teorema de esistenza di calcolo delle variazioni e una proposizione generale di cacolo funzionale. Ann Scuola Norm. Super. Pisa. **12**, 119–133 (1947)
226. K. Fan, O. Taussky, J. Tood, Discrete analogues of inequalities of Wirtinger. Monatsh. Math. **59**, 73–90 (1955)
227. L.H. Fatori, J.E.M. Rivera, Energy decay for hyperbolic thermoelastic systems of memory type. Quart. Appl. Math. **59**(3), 441–458 (2001)
228. J.M. Feinberg, Some Wirtinger-like inequalities. SIAM J. Math. Anal. **10**, 1258–1271 (1979)
229. R.A.C. Ferreira, D.F.M. Torres, Generalizations of Gronwall-Bihari inequalities on time scales. J. Differ. Equ. Appl. **15**(6), 529–539 (2009)
230. A.N. Filatov, L.A. Sharova, *Integral Inequalities and Theory of Nonlinear Oscillations* (Nauka, Moscow, 1976)
231. M. Fila, Boundedness of global solutions of nonlinear differential equations. J. Differ. Equ. **28**(2), 226–240 (1992)
232. M. Fila, H.A. Levine, On the boundedness of global solutions of abstract semilinear parabolic equations. J. Math. Anal. Appl. **216**, 654–666 (1997)
233. A.M. Fink, Discrete inequalities of generalized Wirtinger type. Aequationes Math. **11**, 31–39 (1974)
234. C. Foias, G. Prodi, Sue le comporetment global des solutions non-stationnaires des équations de Navier-Stokes en dimension 2. Rend. Sem. Mat. Univ. Padova **39**, 1–34 (1967)
235. H. Freedman, An explicity estimate on the norm of the inverse of a matrix. SIAM Rev. **11**, 254–256 (1967)
236. A. Friedman, *Foundations of Modern Analysis* (Holt. Rinehart and Winston, New York, 1970)
237. K.O. Friedrichs, H. Lewy, Über die Eindeutigkeit und das Abhängigkeitsgebiet der Lsosungen beim Anfangöwertproblem linear hyperbolischer Döifferentialgleichungen. Math. Ann. **98**, 192–204 (1927)

238. K.M. Furati, M. Kirane, N.E. Tatar, Existence and asymptotic behavior for a convection problem. Nonlinear Anal. TMA **59**, 407–424 (2004)
239. K.M. Furati, N.E. Tatar, An existence result for a nonlocal nonlinear fractional differential problem. J. Fract. Calc. **26**, 43–51 (2004)
240. K.M. Furati, N.E. Tatar, Behavior of solutions for a weighted Cauchy-type fractional differential problem. J. Fract. Calc. **28**, 23–42 (2005)
241. K.M. Furati, N.E. Tatar, Power type estimates for a nonlinear fractional differential equation. Nonlinear Anal. **62**, 1025–1036 (2005)
242. H. Fujita-Yashima, R. Benabidallah, Equation á symétrie sphérique d'un gaz visqueux et calorifére avec la surface libre. Annali Mat. pura ed applicata **CLXVIII**, 75–117 (1995)
243. A. Gallo, A.M. Piccirillo, About new analogies of Gronwall-Bellman Bihari type inequalities for discountinous functions and estimated solutions for impulsive differential systems. Nonlinear Anal. **67**, 1550–1559 (2007)
244. C. Gao, K. Li, E. Feng, Z. Xiu, Nonlinear impulsive system of fed-batch culture in fermentative production and its properties. Chaos Solutions Fractals **28**, 271–277 (2006)
245. J.S. Geronimo, K.M. Case, Scattering theory and polynomials orthogonal on the real line. Trans. Am. Math. Soc. **258**, 467–494 (1980)
246. S.K. Ghoshal, M.A. Masood, Gronwall's vector inequality and its application to a class of non-self-adjoint linear and nonlinear hyperbolic partial differential equations. J. Indian Math. Soc. **38**, 383–394 (1974)
247. S. Ghoshal, A. Ghoshal, M.A. Massod, Generalized n-dimensional Gronwall's inequality and its applications to non-linear hyperbolic equations. Ann-Polon. Math. **33**, 223–233 (1977)
248. J.A. Goldstein, *Semigroups of Linear Operators and Applications* (Oxford Univ. Press, New York & Clarendon Press, Oxford, 1985)
249. H.E. Gollwitzer, *A Note on a Functional Inequality, Summer ODE Colloquium* (University of Alberta, Edmonton, 1968)
250. H.E. Gollwizer, A note on a functional inequality. Proc. Am. Math. Soc. **23**, 642–647 (1969)
251. S.R. Grace, B.S. Lalli, Asymptotic behaviour of certain second order integro-differential equations. J. Math. Anal. Appl. **76**, 84–90 (1980)
252. J.R. Graff, P.W. Spikes, On the nonoscillation, converge to zero and integrability of solutions of a second order nonlinear differential equation. Math. Nachr. **130**, 139–149 (1987)
253. D.E. Greene, An inequality for a class of integral systems. Proc. Am. Math. Soc. **62**, 101–104 (1977)
254. G. Gripenberg, Periodic solutions of an epidemic model. J. Math. Oil. **10**, 217–280 (1980)
255. G. Gripenberg, On some epidemic model. Quart. Appl. Math. **39**, 317–327 (1981)
256. G. Gripenberg, Unique solutions of some Volterra integral equations. Math. Scand. **48**, 59–67 (1981)
257. J. Groh, A nonlinear Volterra-Stieltjes integral equation and Gronwall inequality in one dimension. Ill. J. Math. **24**, 244–263 (1980)
258. H.E. Gronwall, A note on a function inequality. Proc. Am. Math. Sci. **43**, 642–647 (1969)
259. T.H. Gronwall, Note on the derivatives with respect to a parameter of the solutions of a system of diferential equations. Ann. Math. **20**, 292–296 (1918–1919)
260. I. Györi, A Generalization of Bellman's inequality for stieltjes integrals and a uniqueness theorem. Stud. Sci. Math. Hungar. **6**, 137–145 (1971)
261. I. Györi, I. Stavroulakis, Positive solutions of functional differential equations. Boll. Un. Mat. Ital. **3**(B), 185–198 (1989)
262. L. Hącia, On solving Volterra-Fredholm integral equation. Fasc. Math. **13**, 21–30 (1981)
263. L. Hącia, On integral inequalities in two variables and their applications. J. Math. Anal. Appl. **206**(2), 611–622 (1997)
264. L. Hącia, Note on some integral inequalities and their applications. J. Math. Anal. Appl. **206**, 611–622 (1997)
265. A. Halanay, D. Wexler, *Qualitative Theory of Impulsive Systems* (Acad. Romania, Bucuresti, 1968)
266. J.K. Hale, *Theory of Functional Diferential Equations* (Springer, New York, 1977)

267. J.K. Hale, *Asymptotic Behavior of Dissipative Systems, Mathematical Surveys and Monographs*, vol. 25 (Amer. Math. Soc., Providence, 1988)
268. P.R. Halmos, *A Hilbert Space Problem Book* (Van Nostrand Princeton, New Jersey, 1967)
269. D.L. Hanson, P. Waltman, A note on a functional equation. J. Math. Anal. Appl. **10**, 330–333 (1965)
270. T. Hara, T. Yoneyama, J. Sugie, Continuability of solutions of perturbed differential equations. Nonlinear Anal. TMA **8**, 963–975 (1984)
271. A. Haraux, *Nonlinear Evolution Equation: Global Behavior of Solutions*, Lecture Notes in Math., vol. 841 (Springer, Berlin, 1981)
272. G.H. Hardy, J.E. Littlewood, G. Pólya, *Inequalities*, 2nd edn. (Cambridge University Press, Cambridge, 1952)
273. S.P. Hastings, Bounds on the growth of a nonlinear system. SIAM J. Appl. Math. **14**, 1412–1417 (1966)
274. S.P. Hastings, Errata to On the asymptotic growth of solutions to a non-linear equation. Proc. Am. Math. Soc. **17**, 1473 (1966)
275. S.P. Hastings, On the asymptotic growth of solutions to a non-linear equation. Proc. Am. Math. Soc., **17**, 40–47 (1966)
276. O. Haupt, Über das asymptotische verhalten der Lösungen gewisser linearer gewöhnlicher differentialgleichungen. Math. Z., **48**, 289–292 (1942–1943)
277. V.B. Headley, A multidimensional nonlinear Gronwall inequality. J. Math. Anal. Appl. **47**, 250–255 (1974)
278. P. Henrici, *Discrete Variable Methods in Ordinary Differential Equations* (Wiley, New York, 1962)
279. D. Henry, *Geometric Theory of Semilinear Parabolic Equations*, Lecture Notes in Math., vol. 840 (Springer, New York, 1981)
280. T.H. Hildebrandt, On systems of linear differentio-Stieljes integral equations. Ill. J. Math. **3**, 352–373 (1959)
281. E. Hille, Non-oscillation theorems. Trans. Am. Math. Soc. **64**, 234–252 (1948)
282. T. Hishida, Existence and regularizing properties of solutions for the nonstationary convection problem. Funkcial. Ekvac. **34**, 1449–1474 (1991)
283. H. Hochstadt, *Integral Equations* (Wiley, New York 1973)
284. S.G. Hristova, Nonlinear delay integral inequalities for piecewise continuous functions and applications. J. Ineq. Pure Appl. Math. **5**(4), article 88, 1–14 (2004)
285. S.G. Hristova, D.D. Bainov, On some generalization of Bihari's inequality. Izvestija AN Kaz SSR, Ser. Fiz. Mat. **5**, 88–89 (1978)
286. S.G. Hristova, D.D. Bainov, Integral inequalities for functions of many variables. Math. Cluj 23 **46**, 205–210 (1981)
287. S.G. Hristova, D.D. Bainov, On some integral inequalities for a scalar function of a vector argument. Godishnik Vissh. Uchebn. Zaved. Prilozhna Math. **17**, 67–70 (1981)
288. S.G. Hristova, D.D. Bainov, Some generalizations of integral inequalities of Bellman-Bihari type for functions of several variables. Rendiconti Del Cricdo Matematico Di Palermo Serie II Tomo **XXX**, 378–384 (1981)
289. F. Huang, Characteristic conditions for exponential stability of linear dynamical systems in Hilbert spaces. Ann. Differ. Equ. **1**, 43–56 (1985)
290. T.E. Hull, W.A.J. Luxemburg, Numerical methods and existence theorems for ordinary differential equations. Numer. Math. **2**, 30–41 (1960)
291. R. Ikehata, N. Okazawa, Yosida approximations and nonlinear hyperbolic equation. Nonlinear Anal. TMA **15**, 479–495 (1990)
292. G. Iovane, S.D. Borysenko, Boundedness, stability, practical stability of motion impulsive systems, in *Proc. DECAS*, Brest, 2005, 15–21
293. A. Iserles, On the generalised pantograph functional differential equation. Eur. J. Appl. Math. **4**, 1–38 (1993)
294. M. Izumi, S. Izumi, On some inequalities for Fourier series. J. Anal. Math. **21**(1), 277–291 (1968)

295. R. James, A non-reflexive Banach spaces isometric with its second conjugate space. Proc. Natl. Acad. Sci. **37**, 174–177 (1951)
296. J.L.W.V. Jensen, Sur les fonctions convexes et les inéqalitiés entire les valeurs moyennes. Acta Math. **30**, 175–193 (1906)
297. G.S. Jones, Fundamental inequalities for discrete and discontinuous functional equations. SIAM J. Appl. Math. **12**, 43–47 (1964)
298. D.D. Joseph, E.M. Sparrow, Nonlinear diffusion induced by nonlinear sources. Quart. Appl. Math. **28**, 327–342 (1970)
299. C.S. Kahane, On the asymptotic behavior of solutions of nonlinear parabolic systems under Robin type boundary conditions. Funkcial. Ekvac. **26**, 51–78 (1983)
300. I.A. Kaliev, M.S. Podkuiko, On a boundary value problem for the equations of a viscous heat-conducting gas in noncylindrical domains shrinking in time. Differ. Equ. **42**, 1426–1446 (2006)
301. E. Kamke, *Differentialgleichungen Reeler Funktionen* (Chelsea, New York, 1947)
302. A.G. Kartsatos, M.E. Parrott, On a class of nonlinear functional pseudoparabolic problems. Funkcial. Ekac. **25**, 207–221 (1982)
303. D.Y. Kasture, S.G. Deo, Inequalities of Gronwall type in two independent variables. J. Math. Anal. Appl. **58**, 361–372 (1977)
304. S. Kawashima, M. Nakao, K. Ono, On the decay property of solutions to the Cauchy problem of the semilinear wave equatioin with a dissiaptive term. J. Math. Soc. Jpn. **47**, 617–653 (1995)
305. H. Khellaf, On integral inequalities for functions of several independent variables. Electron. J. Differ. Equ. **2003**(123), 1–12 (2003)
306. B.I. Kim, On some Gronwall type inequalities for a system integral equation. Bull. Korean Math. Soc. **42**(4), 789–805 (2005)
307. Y.H. Kim, On some new integral inequalities for functions in one and two variables. Acta Math. Sinica **21**, 423–434 (2005)
308. M. Kirane, N. Tatar, Global existence and stability of some semilinear problems. Arch. Math. Brno **36**, 33–44 (2000)
309. M. Kirane, N. Tatar, Rates of convergence for a reaction-diffusion system. Z. Anal. Anw. **20**(2), 347–357 (2001)
310. A.N. Kolmogorov, S.V. Fomin, *Elemnents of the Theory of Fnctions and Functional Analysis*, vol. 1 (Graylock Press, Rochester, 1957)
311. Q. Kong, B. Zhang, Some generalization of Gronwall-Bihari integral inequalities and their applications. Chin. Ann. Math. **10B**(3), 371–385 (1989)
312. M.A. Krasnoseiskii, *Topological Methods in the Theory of Nonlinear Integral Equations* (Macmillan, New York, 1964)
313. N.V. Krylov, *Lectures on Elliptic and Parabolic Equations in Höder Spaces, Graduate Studies in Mathematics*, vol. 12 (Amer. Math. Soc., Providence, RI, 1996)
314. N.N. Krylov, N.N. Bogolyubov, *Introduction to Nonlinear Mechanics* (Izd. Acad. Sci. Ukr. SSR, Kiev, 1937)
315. J. Kuang, *Applied Inequalities*, 3nd edn. (Shandong Science and Technology Press, Jinan, Shangdong Provincea, 2004), (in Chinese)
316. M. Kuczma, *An Introduction to the Theory of Functional Equations and Inequationes: Cauchy's Equation and Jensen's Inequlity* (University of Katowice, Katowice, 1985)
317. G. Ladas, V. Lakshmikantham, Lower bounds and uniqueness for solutions of evolution inequalities in a Hilbert space. Arch. Rational Mech. Anal. **43**, 293–303 (1971)
318. O.A. Ladyzhenskaya, *Attractors for Semigroups and Evolution Equations* (Cambridge Univ. Press, Cambridge, 1991)
319. O.A. Ladyzhenskaya, V. Solonnikov, N. Uralceva, *Linear and Quasilinear Equations of Parabolic Type, Translations of Mathematical Monographs*, vol. 23 (Amer. Math. Soc., Rhode Island, 1968)
320. V. Lakshmikantham, On the boundedness of solutions of non-linear differential equations. Proc. Am. Math. Soc. **8**, 1044–1048 (1957)

321. V. Lakshmikantham, Upper and lower bounds of the norm of solutions of differential equations. Proc. Am. Math. Soc., **13**, 615–616 (1962)
322. V. Lakshmikantham, Upper and lower bounds of the norm of solutions of differential equations. Proc. Am. Math. Soc. **14**, 509–513 (1963)
323. V. Lakshmikantham, A variation of constants formula and Bellman-Gronwall-Reid inequalities. J. Math. Anal. Appl. **41**, 199–204 (1973)
324. V. Lakshmikantham, A. McRae Farzana, Basic results for fuzzy impulsive differential equations. Math. Inequal. Appl. **4**(2), 239–246 (2001)
325. V. Lakshmikantham, S. Leela, *Differential and Integral Inequalities, Theory and Applications*, vol. 1,2 (Academic Press, New York, 1969)
326. V. Lakshmikantham, M.R.M. Rao, *Theory of Integro-Diferential Equations* (Gordon and Breach Science Publ., Switzerland, 1995)
327. V. Lakshmikantham, D.D. Bainov, P.S. Simeonov, *Theory of Impulsive Differential Equations, Ser. Modern Appl. Math.*, vol. 6 (World Scientific, New Jersey, 1989)
328. C.E. Langenhop, Bounds on the norm of a solution of a general differential equation. Proc. Am. Math. Soc. **11**, 795–799 (1960)
329. A. Lasota, A discrete boundary value problem. Ann. Polon. Math. **20**, 183–190 (1968)
330. J.P. LaSalle, Uniqueness theorems and successive approximations. Ann. Math. **50**, 722–730 (1949)
331. M. Lavrentiev, Sur une équation différentielle du premier ordre. Math. Zeitschrift **23**, 197–198 (1925)
332. M. Lech, Why Hölder's inequality should be called Rogers' inequality. Math. Inequal. Appl., **1**(1), 69–83 (1998)
333. S. Leela, Stability of measure differential equations. Pacific J. Math. **55**, 489–498 (1974)
334. S. Leela, Stability of differential systems with impulsive perturbations in terms of two measures. Nonlinear Anal. Theory Methods Appl. **1**(6), 667–677 (1977)
335. M. Lees, Approximate solutions of parabolic equations. J. Soc. Ind. Appl. Math. **7**, 167–183 (1959)
336. M. Lees, Asymptotic behavior of solutions of parabolic differential inequalities. Canad. J. Math. **14**, 626–631 (1962)
337. J.J. Levin, J.A. Nohel, On a system of integro-differential equations occurring in reactor dynamics. J. Math. Mech. **9**, 347–368 (1960)
338. V. Levin, On the two-parameter extension and analogue of Hilbert's inequality. J. Lond. Math. Soc. **11**, 119–124 (1936)
339. H.A. Levine, Logarithmic convexity, first order differential inequalities and some applications. Trans. Am. Math. Soc. **152**, 299–320 (1970)
340. H.A. Levine, The role of critical exponents in blowup theorems. SIAM Rev. **32**(2), 262–288 (1990)
341. J.D. Li, Opial-type integral inequalities involving several higher derivatives. J. Math. Anal. Appl. **167**, 98–110 (1982)
342. T. Li, Y. Chen, *Global Classical Solutions for Nonlinear Evolution Equations, Pitman Monographs and Surveys in Pure and Applied Mathematics*, vol. 45 (Longman Scientific and Technical, London, 1992)
343. T. Li, T. Qin, *Physics and Partial Differential Equations* (Higher Education Press, Beijing, 1997), (in Chinese)
344. W.N. Li, Some new dynamic inequalities on time scales. J. Math. Anal. Appl. **319**(2), 802–814 (2006)
345. W.N. Li, M.A. Han, F.W. Meng, Some new delay integral inequalities and their applications. J. Comput. Appl. Math. **180**, 191–200 (2005)
346. W.N. Li, W.H. Sheng, Some nonlinear integral inequalities on time scales. J. Inequal. Appl. **2007**, Article ID 70465, 15p. (2007)
347. X. Li, H. Wu, S. Zheng, The equivalence between two types of exponential stability, Preprint
348. Y. Li, The bound, stability and error estimates for the solution of nonlinear differential equations. Chin. Math. Acta **3**, 34–41 (1963)

349. A.-C. Liang (C.-C. Liang), The boundedness of solutions of certain nonlinear differential equations. Acta Math. Sinica **12**, 156–169 (1962); English translation, Chin. Math. **3**, 169–183 (1963)
350. G.M. Lieberman, *Second Order Parabolic Differential Equations* (World Scitific Publishing, Singapore, 1996)
351. J.H. Lightbourne III, S.M. Rankin III, A partial functional differential equation of Sobolev type. J. Math. Anal. Appl. **93**, 328–337 (1983)
352. P. Linz, *Analytical and Numerical Methods for Volterra Equations* (SIAM Studies in Applied Mathematics, Philadelphia, 1985)
353. J.L. Lions, *Quelques Methods de Resolution des Problemes aux Limites Non Lineares* (Dunod, Paris, 1969)
354. J.L. Lions, E. Magenes, *Probémes aux limites non homogénes et applications*, vol. 1, 2 (Dunod, Paris, 1968); vol. 3, 1970, 1972. English Translation: *Nonhomogeous Boundary Value Problems and Applications* (Springer, New York, 1972)
355. O. Lipovan, A retarded Gronwall-like inequality and its applications. J. Math. Anal. Appl. **252**, 389–401 (2000)
356. O. Lipovan, A retarded integral inequality and its applications. J. Math. Anal. Appl. **285**, 436–443 (2003)
357. O. Lipovan, Integral inequalities for retarded Volterra equations. J. Math. Anal. Appl. **322**(1), 349–358 (2006)
358. Q.H. Ma, N-independent-variable discrete inequalities of Gronwall-Ou-Yang type. Ann. Differ. Equ. **16**, 813–820 (2000)
359. Q.H. Ma, Some new nonlinear Volterra-Fredholm-type discrete inequalities and their applications. J. Comupt. Appl. Math. **216**, 451–466 (2008)
360. Q.H. Ma, W.S. Cheung, Some new nonlinear difference inequalities and their applications. J. Comupt. Appl. Math. **202**, 339–351 (2007)
361. Q.H. Ma, L. Debnath, A more generalized Gronwall-like integral inequality with applications. Inter. J. Math. Math. Sci. **2003**, 15 (2003)
362. Q.H. Ma, J. Pecaric, On certain new nonlinear retarded integral inequalities for functions in two variables and their applications. J. Korean Math. Soc. **45**(1), 121–136 (2008)
363. Q.H. Ma, J. Pecaric, On some new nonlinear discrete inequalities and their applications, Preprint
364. Q.H. Ma, E.H. Yang, Generalizations of Pachpatte's two-variable nonlinear integral inequalities. Acta Math. Sinica **43**(5), 813–820 (2000)
365. Q.M. Ma, E.H. Yang, On some new nonlinear delay integral inequalities. J. Math. Anal. Appl. **252**, 864–878 (2000)
366. Q.M. Ma, E.H. Yang, Some new Gronwall-Bellman-Bihari type integral inequalities with delay. Period. Math. Hungar. **44**, 225–238 (2002)
367. Q. Ma, S. Wang, C. Zhong, Necessary and sufficient conditions for the existence of global attractors for semigroups and applications. Indiana Univ. Math. J. **51**(6), 1541–1559 (2002)
368. E. Magnucka-Blandzi, J. Popenda, R.P. Agarwal, Best possible Gronwall inequalities. Math. Comput. Model. **26**(3), 1–8 (1997)
369. F.O. Mamekonyan, Estimation for solutions of certain class of nonlinear integral inequalities in several variables. Izv. Akad. Nauk Armjan. SSR **XXII**, 133–151 (1987), (in Russian)
370. B. Mandelbrot, *Fractals: Form, Chance and Dimension* (Freeman, San Francisco, 1977)
371. X. Mao, Lebesgue-stieltjes integral inequalities of the Wendroff type in n-independent variables. Chin. J. Math. **17**(1), 29–50 (1989)
372. G.I. Marchuk, V.V. Shaidurov, *Difference Methods and Their Extrapolations* (Springer, New York, 1983)
373. A.H. Martin, Global existence questions for reactioníCdiffusion systems. Pitman Res. Notes Math. **1**, 169–177 (1986)
374. A.A. Martyniuk, R. Gutowski, *Integral Inequalities and Stability of Motion* (Naukova Dumka, Kiev, 1979), (in Russian)

375. A.A. Martyniuk, V. Lakshmikanthan, S. Leela, *Motion Stability: The Method of Integral Inequalities* (Naukova Dumka, Kiev, 1977), (in Russian)
376. L.V. Masolockaja, Stability of difference inequalites. *Differentsialnye Uraveniya* **34**, 147–156 (1967), (in Russian)
377. E.V. Massalitina, On the Perow integro-summable inequality for functions of two variables. Ukrainian Math. J. **56**, 1864–1872 (2004)
378. A. Mate, P. Nevai, Sublinear perturbations of the differential equation $y^{(n)} = 0$ and of the analogous difference equation. J. Differ. Equ. **53**, 234–257 (1984)
379. S. Mazouzi, N.E. Tatar, Global existence for some integro-differential equations with delay subject to nonlocal conditions. Z. Anal. Anw. **21**(1), 249–256 (2002)
380. S. Mazouzi, N.E. Tatar, An improved exponential decay result for some semilinear integro-differential equations. Arch. Math. Brno **39**(3), 163–171 (2003)
381. S. Mckee, Generalized discrete Gronwall lemmas. Z. angew. Math. Mech. **62**(9), 429–434 (1982)
382. M. Medved', A generalization of Bihari's lemma and their application. Mat. Časopis. **20**, 225–232 (1970), (In Slovak)
383. M. Medved', Bihari type inequalities with multiple integral and delay. Period. Math. Hungar. **27**, 207–212 (1993)
384. M. Medved', A new approach to an analysis of Henry type integral inequalities and their Bihari type versions. J. Math. Anal. Appl. **214**, 349–366 (1997)
385. M. Medved', Singular integral inequalities and stability of semilinear parabolic equations. Arch. Math. Brno **34**, 183–190 (1998)
386. M. Medved', Nonlinear singular difference inequalities suitable for discretizations of parabolic equations. Demonstrat. Math. **XXXIII**(3), 517–529 (2000)
387. M. Medved', Nonlinear singular integral inequalities for functions in two and *n* independent variables. J. Ineq. Appl. **5**, 287–308 (2000)
388. F.W. Meng, W.N. Li, On some new nonlinear discrete inequalities and their applications. J. Comupt. Appl. Math. **158**, 407–417 (2003)
389. F.W. Meng, W.N. Li, On some new integral inequalities and their applications. Appl. Math. Comput. **148**, 381–392 (2004)
390. M.W. Michalski, Derivatives of noninteger order and their applications, Dissertationes Mathematicane, Polska Akademia Nauk., Instytut Mathematyczny, Warszawa, 1993
391. S.G. Mikhlin, *Linear Integral Equations* (Hindustan, Delhi, 1960)
392. M. Miklavčič, Stability for semilinear parabolic equations with noninvertible linear operator. Pacific J. Math. **118**(1), 199–214 (1985)
393. R.K. Miller, *Nonlinear Volterra Integral Equations* (Benjamin, Menlo Park, 1971)
394. G. Mingze, On Hilbert's inequality and it's application. J. Math. Anal. Appl. **212**, 316–323 (1997)
395. D.S. Mitrinović, *Analytic Inequalities* (Springer, Berlin, New York, 1970)
396. D.S. Mitrinovicć, J.E. Pečarić, *Differential and Integral Inequalities* (Naucna Knjiga, Belgrade, 1988)
397. D.S. Mitrinović, J.E. Pečarić, A.M. Fink, *Inequalities Involving Functions and Their Integrals and Derivatives, Mathematics and Its Applications* (East European Series), vol. 53 (Kluwer Academic, Dordrecht, 1991)
398. D.S. Mitrinović, J.E. Pecaric, On inequalities of Hilbert and Widder. Proc. Edinburgh. Math. Soc. **34**, 411–414 (1991)
399. Y.A. Mitropolskiy, G. Iovane, S.D. Borysenko, About a generalization of Bellman-Bihari type inequalities for discontinuous functions and their applications. Nonlinear Anal. **66**, 2140–2165 (2007)
400. M. Mizoguchi, E. Yanagida, Blow-up of solutions with sign-changes for semilinear diffusion equation. J. Math. Anal. Appl. **204**, 283–290 (1996)
401. A. Morro, A Gronwall-like inequality and its application to continuum thermodynamics. Boll. Un. Mat. Ital. **6**(B), 553–562 (1982)

402. J.S. Muldowney, J.S.W. Wong, Bounds for solutions of non-linear integro-differential equation. J. Math. Anal. Appl. **23**, 487–499 (1968)
403. W. Mydlarczyk, The existence of nontrivial solutions of Volterra equations. Math. Scand. **68**, 83–88 (1991)
404. W. Mydlarczyk, The existence of nontrivial solutions of Volterra equations. Math. Scand. **78**, 286–292 (1996)
405. W. Mydlarczyk, The existence of solutions to a Volterra integral equation. Ann. Polon. Math. **64**, 175–182 (1996)
406. W. Mydlarczyk, A Volterra inequality with the power type nonlinear kernel. J. Inequal. Appl. **6**, 625–631 (2001)
407. M. Nagumo, Eine hinreichende bedingung für die unität der lösung von differentialgleichungen erster ordnung. Japanese J. Math. **3**, 107–112 (1926)
408. M. Nagumo, Note in Kansü-Höteisiki, 1939, No. 15 (in Japanese)
409. M. Nagumo, S. Simoda, Note sur l'inégalité differential concernant les équations du type parabolique. Proc. Jpn. Acad. **27**, 536–539 (1951)
410. M. Nakao, emphMemoirs of the Faculty of Science, Ser. A, vol. **30**(2) (Kyushu Univ., Fukuoka, Kyushu, 1976)
411. M. Nakao, Asymptotic stability of the bounded or almost periodic solution of the wave equation with nonlinear dissipative term. J. Math. Anal. Appl. **56**, 336–343 (1977)
412. M. Nakao, Decay of solutions of some nonlinear evolution equations. J. Math. Anal. Appl. **60**, 542–549 (1977)
413. M. Nakao, A difference inequality and its applications to nonlinear evolution equations. J. Math. Soc. Jpn. **30**, 747–762 (1978)
414. M. Nakao, Decay of solutions of the wave equation woith a local time-dependent nonlinear dissipation. Adv. Math. Sci. Appl. **7**, 317–331 (1997)
415. J. Navot, A further extension of EulerĺCMaclaurin summation formula. J. Math. Phys. **41**, 155–184 (1962)
416. E. Nelson, An existence theorem for second order parabolic equations. Trans. Am. Math. Soc. **88**, 414–429 (1958)
417. J. Németh, Generalization of the Hardy-Littlewood inequality. Acta. Sci. Math. (Szeged) **32**, 295–299 (1971)
418. V.V. Nemytskii, V.V. Stepanov, *Qualitative Theory of Differential Equations* (Princeton Univ. Press, Princeton, New Jersey, 1960)
419. M. Niezgódka, J. Sprekels, Existence of solutions for a mathematical model of structural phase transitions in shape memory alloys. Math. Meth. Appl. Sci. **10**, 197–223 (1988)
420. M. Niezgódka, S. Zheng, J. Sprekels, Global solutions to a model of structural phase transitions in shape memory alloys. J. Math. Anal. Appl. **130**, 39–54 (1988)
421. V.B. Nikolaev, On the solvability of mixed problem for one-dimensional axisymmetrical viscous gas flow. Dinamicheskie zadachi Mekhaniki sploshnoj sredy, **63**, Sibirsk. Otd. Acad. Nauk SSSR, Inst. Gidrodinamiki, 1983 (in Russian)
422. S.M. Nikol'skii, An embedding theorem for functions with partial derivaives considered in different matrics. Izv. Akad. Nauk. SSSR Ser. Mat. **22**, 321–336 (1958), (in Russian)
423. J.A. Nohel, Some problems in non-linear Volterra integral equations. Bull. Am. Math. Soc. **68**, 323–329 (1962)
424. J.A. Nohel, Qualitative behavior of solutions of non-linear Volterra equations, In *Stability Problems of Solutions of differential Equations* (*Proceedings of NATO Advanced Study Institute*, Padua, Italy, 1965, Gubbio (Italy), 1966, 177–210
425. H. Ogawa, Lower bounds for solutions of parabolic differential inequalities. Canad. J. Math. **19**, 667–672 (1967)
426. H. Ogawa, On the maximum rate of decay of solutions of parabolic differential inequalities. Arch. Rat. Mech. Anal. **38**, 173–177 (1970)
427. J.A. Oguntuase, On integral inequalities of Gronwall-Bellman-Bihari type in several variables. J. Ineq. Pure Appl. Math. **1**(2), 1–7 (2000)

428. J.A. Oguntuase, On an inequality of Gronwall. J. Ineq. Pure Appl. Math. **2**(1), article 9, 1–6 (2001)
429. W. Okrasinski, On a linear convolution equation occurring in the theory of water percolation. Ann. Polon. Math. **33**, 223–229 (1980)
430. W. Okrasiriski, Nontrivial solutions for a class of nonlinear Volterra equations with convolution kernels. J. Integral Equ. Appl. **3**, 399–409 (1991)
431. W. Okrasirski, Nontrivial solutions to nonlinear Volterra integral equations. SIAM J. Math. Anal. **22**, 1007–1015 (1991)
432. C. Olech, On a system of integral inequalities. Colloq. Math. **16**, 137–139 (1967)
433. C. Olech, Z. Opial, Sur une inégalité différentielle. Ann. Polon. Math. **7**, 247–254 (1960)
434. S.N. Olekhnik, Boundedness and unboundedness of solutions of some systems of ordinary differential equations. Vestnik Moskou Univ. Mat. **27**, 34–44 (1972)
435. Z. Opial, On a system of integral inequalities. Ann. Polon. Math. **3**, 200–209 (1957)
436. W.F. Osgood, Beweis der existenz einer Lösung der differentialgleichung $y' = f(x, y)$ ohne hinzuname der Cauchy-Lipschitz bedingung. Monatshefte f. Math. u. Phys. **9**, 331–345 (1898)
437. H. El-Owaidy, A. Ragab, A. Abdeldaim, On some new integral inequalities of Gronwall-Bellman type. Appl. Math. Comp. **106**(2), 289–303 (1999)
438. L. Ou-Yang, The boundedness of solutions of linear differential equation $y'' + A(t)y = 0$. Adv. Math. **3**, 409–415 (1957), (In Chinese)
439. B.G. Pachpatte, On the existence and uniqueness of solutions of Volterra-Fredholm integral equations. SIAM J. Appl. Math. **12**, 43–47 (1964)
440. B.G. Pachpatte, Finite difference inequalities and an extension of Lyapunov's method. Michigan Math. J. **18**, 385–391 (1971)
441. B.G. Pachpatte, A note on Gronwall-Bellman inequality. J. Math. Anal. Appl. **44**, 758–762 (1973)
442. B.G. Pachpatte, Finite difference inequalities and their applications. Proc. Natl. Acad. Sci. India **43A**, 348–356 (1973)
443. B.G. Pachpatte, Integral perturbations of nonlinear system of differential equations. Bull. Soc. Math. Grece **14**, 92–97 (1973)
444. B.G. Pachpatte, On the discrete generalizations of Gronwall's inequality. J. Indian Math. Soc. **37**, 147–156 (1973)
445. B.G. Pachpatte, An integral inequality similar to Bellman-Bihari inequality. Bull. Soc. Math. Grece **15**, 7–12 (1974)
446. B.G. Pachpatte, Integral inequalities of the Gronwall-Bellman type and their applications. J. Math. Phys. Sci. **8**, 309–318 (1974)
447. B.G. Pachpatte, On perturbed stochastic discrete systems. Bull. Aust. Math. Soc. **11**, 385–393 (1974)
448. B.G. Pachpatte, On some integral inequalities similar to Bellman-Bihari inequalities. J. Math. Anal. Appl. **49**, 250–255 (1974)
449. B.G. Pachpatte, Stability and asymptotic behavior of perturbed nonlinear systems. J. Differ. Equ. **16**, 14–25 (1974)
450. B.G. Pachpatte, A note on integral inequality. J. Math. Phys. Sci. **9**, 11–14 (1975)
451. B.G. Pachpatte, A note on integral inequalities of Bellman-Bihari type. J. Math. Anal. Appl. **49**, 295–301 (1975)
452. G.B. Pachpatte, On an integral inequality of Gronwall-Bellman. J. Math. Phys. Sci. **9**(5), 405–416 (1975)
453. B.G. Pachpatte, On some discrete inequalities of the Bellman-Bihari type. Indian J. Pure. Appl. Math. **6**, 1479–1487 (1975)
454. B.G. Pachpatte, On some discrete inequalities of the Gronwall-Bellman type. Indian J. Pure. Appl. Math. **6**, 769–772 (1975)
455. G.B. Pachpatte, On some generalizations of Bellman's lemma. J. Math. Anal. Appl. **5**, 141–150 (1975)

456. B.G. Pachpatte, On some integral inequalities similar to Bellman-Bihari inequalities. J. Math. Anal. Appl. **49**, 794–802 (1975)
457. B.G. Pachpatte, Perturbations of nonlinear systems of differential equations. J. Math. Anal. Appl. **51**, 550–556 (1975)
458. B.G. Pachpatte, A new integral inequality for differential and integral equations. Proc. Natl. Acad. Sci. India **46A**, 21–26 (1976)
459. B.G. Pachpatte, On some new discrete inequalities and their applications. Proc. Natl. Acad. Sci. India **46A**, 255–262 (1976)
460. B.G. Pachpatte, A note on some fundamental discrete inequalities of the Gronwall-Bell man type. Bull. Inst. Math. Acad. Sinica **5**, 121–128 (1977)
461. B.G. Pachpatte, On discrete inequalities related to Gronwall's inequality. Proc. Indian Acad. Sci. Math. Sci. Part A **85**, 26–40 (1977)
462. B.G. Pachpatte, On some fundamental integro-differential and integral. Ann. Sti. Univ. Al. I.Cuza'din Iasi **XXIII**, 77–86 (1977)
463. B.G. Pachpatte, On some fundamental finite difference inequalities. Univ. Beograd Publ. Elek-trotekhn. Fak. Ser. Mat. Fiz. **577–598**, 65–73 (1977)
464. B.G. Pachpatte, On some new integral inequalities and their discrete analogues. Indian J. Pure Appl. Math. **8**, 1093–1107 (1977)
465. B.G. Pachpatte, On some nonlinear discrete inequalities of the Gronwall type. Bull. Inst. Math. Acad. Sinica **5**, 305–315 (1977)
466. B.G. Pachpatte, Finite difference inequalities and discrete time control systems. Indian J. Pure. Appl. Math. **9**, 1282–1290 (1978)
467. B.G. Pachpatte, On some fundamental integrodifferential inequalities and their discrete analogue. Proc. Indian Acad. Sci. Sect. A **87A**, 201–207 (1978)
468. B.G. Pachpatte, On some new discrete inequalities and their applications to a class of sum]difference equations. Ann. Sti. Univ. Al. I. Cuza Iasi Sec. I Mat. **24**, 315–326 (1978)
469. B.G. Pachpatte, On some integro-differential inequalities similar to Wendroff's inequality. Ann. Sti. Univ. Al. I. Cuza din Iasi, Sec. I Math. **25**, 127–136 (1979)
470. B.G. Pachpatte, On some new integral and integro-differential inequalities in two independent variables and their applications. J. Differ. Equ. **33**, 249–272 (1979)
471. B.G. Pachpatte, On Wendroff type partial integral inequalities. Tamkang J. Math. **10**, 141–150 (1979)
472. B.G. Pachpatte, On some fundamental partial integral inequalities. J. Math. Anal. Appl. **73**, 238–251 (1980)
473. B.G. Pachpatte, On some new integral inequalities for nonself-adjoint hyperbolic partial integro-differential equations. J. Math. Anal. Appl. **76**, 58–71 (1980)
474. B.G. Pachpatte, On some new integro-differential inequalities of the Wendroff type. J. Math. Anal. Appl. **73**, 491–500 (1980)
475. B.G. Pachpatte, On some fundamental discrete inequalities in two independent variables. Tamkang J. Math. **12**, 21–33 (1981)
476. B.G. Pachpatte, On some partial integral inequalities in n independent variables. J. Math. Anal. Appl. **79**, 256–272 (1981)
477. B.G. Pachpatte, J. Math. Phys. Sci. **17**, 545–550 (1983)
478. B.G. Pachpatte, A note on Greene's inequality. Tamkang J. Math. **15**, 49–54 (1984)
479. B.G. Pachpatte, Discrete analogues of some inequalities ascribed to Wirtinger. Utilitas Math. **28**, 137–143 (1985)
480. B.G. Pachpatte, On Opial type inequalities in two independent variables. Proc. Roy. Soc. Edinburgh **100A**, 263–270 (1985)
481. B.G. Pachpatte, A note on an inequality ascribed to Wirtinger. Tamkang J. Math. **17**, 69–73 (1986)
482. B.G. Pachpatte, On certain multidimensional discret inequalities. Chin. J. Math. **14**, 185–195 (1986)
483. B.G. Pachpatte, On Poincare type integral inequalities. J. Math. Anal. Appl. **114**, 111–115 (1986)

484. B.G. Pachpatte, On some new multidimensional discrete inequalities. Tamkang J. Math. **17**, 21–29 (1986)
485. B.G. Pachpatte, On some new integral inequalities in two independent variables. J. Math. Anal. Appl. **129**, 375–382 (1988)
486. B.G. Pachpatte, On a class of new discrete inequalities. Tamkang J. Math. **20**, 19–28 (1989)
487. B.G. Pachpatte, On some new discrete inequalities in two independent variables. Indian J. Pure Appl. Math. **20**, 1197–1212 (1989)
488. B.G. Pachpatte, Discrete inequalities in two variables and their applications. Radovi Mat. **6**, 235–247 (1990)
489. B.G. Pachpatte, On multidimensional discrete inequalities and their applications. Tamkang J. Math. **21**, 111–122 (1990)
490. B.G. Pachpatte, On Bihari like integral and discrete inequalities. Soochow J. Math. **17**, 213–232 (1991)
491. B.G. Pachpatte, On Gronwall-like integral inequalites. Indian J. Pure Appl. Math. **23**(2), 131–140 (1992)
492. B.G. Pachpatte, A note on certain multivariable integral inequalities. Tamkang J. Math. **24**, 105–112 (1993)
493. B.G. Pachpatte, On certain new finite difference inequalities. Indian J. Pure Appl. Math. **24**, 373–384 (1993)
494. B.G. Pachpatte, On some new discrete inequalities related to a certain integral. Libertas. Math. **13**, 85–97 (1993)
495. B.G. Pachpatte, Some new finite difference inequalities. Comput. Math. Appl. **28**, 227–241 (1994)
496. B.G. Pachpatte, On a certain inequality arising in the theory of differential equations. J. Math. Anal. Appl. **182**, 143–157 (1994)
497. B.G. Pachpatte, Inequalities related to a certain inequality arising in the theory of differential equations. Studia Univ. Babes-Bolyai Math. **39**, 33–50 (1994)
498. B.G. Pachpatte, A note on certain integral inequalities with delay. Period. Math. Hungar. **31**, 229–234 (1995)
499. B.G. Pachpatte, On a new inequality suggested by the study of certain epidemic models. J. Math. Anal. Appl. **195**, 638–644 (1995)
500. B.G. Pachpatte, On some new inequalities related to certain inequalities in the theory of differential equations. J. Math. Anal. Appl. **189**, 128–144 (1995)
501. B.G. Pachpatte, A note on some series inequalities. Tamkang J. Math. **27**, 77–79 (1996)
502. B.G. Pachpatte, Comparsion theorems related to a certain inequality used in the theory of differential equations. Soochow J. Math. **22**, 383–394 (1996)
503. B.G. Pachpatte, Inequalities similar to a certain inequality used in the theory of differential equations. Chin. J. Math. **24**, 55–68 (1996)
504. B.G. Pachpatte, On some new discrete inequalities useful in the theory of certain partial finite difference equations. Ann. Differ. Equ. **12**, 1–12 (1996)
505. B.G. Pachpatte, A note on Hilbert type inequality. Tamkang J. Math. **29**, 293–298 (1998)
506. B.G. Pachpatte, Inequalities applicable in the theory of finite difference equations. J. Math. Anal. Appl. **222**, 438–459 (1998)
507. B.G. Pachpatte, *Inequalities for Differential and Integral Equations, Mathematics in Science and Engineering*, vol. 197 (Academic Press, California, 1998)
508. B.G. Pachpatte, On some new inequalities similar to Hilbert's inequality. J. Math. Anal. Appl. **226**, 166–179 (1998)
509. B.G. Pachpatte, Integral inequalities of Wirtinger and Opial type in three independent variables. Fasc. Math. **30**, 113–129 (1999)
510. B.G. Pachpatte, A note on inequality of Hilbert type. Demonstratio Math. **32** (2000)
511. B.G. Pachpatte, On an inequality similar to Hilbert's inequality. Bull. Inst. Polytech. Iasi. Sect. I. Mat. Mec. Teor. Fiz **46**(50), 31–36 (2000)
512. B.G. Pachpatte, On some new inequalities related to a certain inequality arising in the theory of differential equations. J. Math. Anal. Appl. **251**, 736–751 (2000)

513. B.G. Pachpatte, *Inequalities for Finite Difference Equations* (Marcel Dekker, New York, 2001)
514. B.G. Pachpatte, On some finite difference inequalities in two independent variables. J. Math. Anal. Appl. **254**, 587–598 (2001)
515. B.G. Pachpatte, On some fundamental finite difference inequalities. Tamkang J. Math. **32**(3), 217–223 (2001)
516. B.G. Pachpatte, On some fundamental integral inequalities and their discrete analogue. J. Ineq. Pure Appl. Math. **2**, article 15 (2001)
517. B.G. Pachpatte, Bounds on certain integral inequalities. J. Ineq. Pure Appl. Math. **3**(3), article 47 (2002)
518. B.G. Pachpatte, Explicit bounds on certain integral inequalities. J. Math. Anal. Appl. **267**, 48–61 (2002)
519. B.G. Pachpatte, Integral inequalities of the Bihari type. Math. Ineq. Appl. **5**(4), 649–657 (2002)
520. B.G. Pachpatte, On some retarded integral inequalities and applications. J. Ineq. Pure Appl. Math. **3**(2), article 18 (2002)
521. B.G. Pachpatte, Inequalities applicable to certain partial differential equations. J. Ineq. Pure Appl. **5**(2), article 27 (2004)
522. B.G. Pachpatte, On a certain retarded integral inequality and applications. J. Ineq. Pure Appl. Math. **5**(3), article 19 (2004)
523. B.G. Pachpatte, On some new nonlinear retarded integral inequalities. J. Ineq. Pure Appl. Math. **5**(3), article 80 (2004)
524. B.G. Pachpatte, Explicit estimates on integral inequalities with time scale. J. Ineq. Pure Appl. Math. **7**(4), article 143 (2006)
525. S.B. Pachpatte, B.G. Pachpatte, Inequalities for terminal value problems for differential equations. Tamkang J. Math. **33**(3), 199–208 (2002)
526. B.G. Pachpatte, S.M. Singare, Discrete generalized Grownwall inequalities in three independent variables. P. J. A. **82**(1), 197–211 (1979)
527. S.G. Pandit, Differential systems with impulse perturbations. Pacific J. Math. **86**(2), 553–560 (1980)
528. P.Y.H. Pang, R.P. Agarwal, On an integral inequality and its discrete analogue. J. Math. Anal. Appl. **194**, 569–577 (1995)
529. A. Pazy, *Semigroups of linear Operators and Applications to Paratial Differential Equations*, Appl. Math. Sci., vol. 44 (Springer, New York, 1983)
530. A. Pelczar, On some inequalities. Zeszyty Nauk. Univ. Jagiello. Prace Mat. **9**, 77–80 (1963)
531. A. Pelczar, On invariant points of monotone transformations in partially ordered spaces. Ann. Polon. Math. **17**, 49–53 (1965)
532. O. Perron, Über ein- und mehrdeutigkeit des integrals eines systems von differentialgleichungen. Math. Ann. **95**, 98–101 (1926)
533. A.M. Pfeffer, On certain discrete inequalities and their continuous analogues. J. Res. Nat. Bur. Standards **70B**, 221–231 (1967)
534. C. Philos, Oscillations of first order linear retarded differential equations. J. Math. Anal. Appl. **157**, 17–33 (1991)
535. A.M. Piccirillo, Estimates of the solutions hyperbolic equations with impulse perturbations, in *Proc. XI Int. Kravchuk Conf. Kyiv*, 2006, pp. 77–80
536. A.M. Piccirillo, I. Verigina, About some generalization of Bihari result for integro-sum inequalities, in *Proc. XI Int. Kravchuk Conf. Kyiv*, 2006, 187p.
537. M. Pinto, Perturbations of asymptotically stable differential systems. Analysis **4**, 161–175 (1984)
538. M. Pinto, Asymptotic integration of second order linear differential equations. J. Math. Anal. Appl. **111**, 388–405 (1985)
539. M. Pinto, Nonlinear integral inequalities and their application to the study of asymptotic behavior of nonlinear defferential equations, in *Symp. on Nonlinear Analysis*, 1986, Publ. University of Concepcion (1987), pp. 109–123

540. M. Pinto, Des Inégalités Fonctionnelles et Leurs Applications, Publ. Inst. Rech. I Av. Strasbourg (July 1988),364/TS-06, 142 pp.
541. M. Pinto, Integral inequalities of Bihari-type and applications. Funkcial. Ekvac. **33**, 387–430 (1990)
542. H. Poincaré, Sur les equations lineaires aux differential et aux differences finies. Am. J. Math. Soc. **7**, 203–258 (1885)
543. C.J.V. Poussin, Sur l'integrale de Lebesgue. Trans. Am. Math. Soc. **16**, 435–501 (1915)
544. J. Popenda, Finite difference inequalities. Fasc. Math. **13**, 79–87 (1981)
545. J. Popenda, On the discrete analogy of Gronwall lemma. Demonstr. Math. **16**, 11–15 (1983)
546. J. Popenda, On some discrete Gronwall type inequalities. Fasc. Math. **14**, 109–114 (1985)
547. J. Popenda, On the discrete analogy of Gronwall-Wendroff inequality. Demonstr. Math. **18**, 1083–1103 (1985)
548. J. Popenda, A note on Gronwall-Bellman inequality. Fax. Math. **16**, 24–42 (1986)
549. J. Popenda, Remark on the paper of Turinici. Demonstr. Math. **22**, 203–212 (1989)
550. J. Popenda, Gronwall type inequalities. Z. angew. Math. Mech. **75**, 669–677 (1995)
551. J. Popenda, Discrete Gronwall inequalities in many variables. Inter. Comput. Math. Appl. **38**, 63–70 (1999)
552. J.E. Potter, Some statistical properties of the motion of nonlinear oscillator driven by white noise, Ph.D. Thesis Massachusetts Institute of Technology, Cambridge, 1962
553. M.H. Protter, Properties of parabolic equations and inequalities. Can. J. Math. **13**, 331–345 (1961)
554. E.J. Putzer, The rate of growth of solutions of a second order differential equations. SIAM J. Appl. Math. **10**, 454–468 (1962)
555. Y. Qin, *Nonlinear Parabolic-Hyperbolic Coupled Systems and Their Attractors, Operator Theory*, vol. 184, Adv. PDE (Springer/Birkhäuser, Basel/Boston/Berlin 2008)
556. Y. Qin, *Analytic Inequalities and Applications in PDEs, Operator Theory*, Adv. PDE (Springer/Birkhäuser, Basel/Boston/Berlin, 2016) (to appear)
557. Y. Qin, *Integral and Discrete inequalities and Their Applications* (Springer International Publishing AG, Birkhauser, 2016)
558. Y. Qin, L. Huang, *Global Wellposedness of Nonlinear Parabolic-Hyperbolic Coupled Systems, Frontier in Mathematics* (Springer/Birkhäuser, Basel/Boston/Berlin 2012)
559. Y. Qin, X. Liu, T. Wang, *Global Existence and Uniqueness of Nonlinear Evolutionary Fluid Equations-Frontier in Mathematics* (Springer/Birkhäuser, Basel/Boston/Berlin, 2016)
560. M. Rab, Linear integral inequality. Arch. Math. Scripta Fac. Sci. Natur. UJEP Brunensis-Math. **15**, 37–46 (1979)
561. V. Raghavenda, M. Rama Mohana Rao, Stability of impulsively perturbed differential equations of Lyapunov's method. Indian J. Pure Appl. Math. **13**(8), 890–897 (1982)
562. D.R.K. Rao, Unicity of solutions of a differential equation. Monatsh. Math. **68**, 150–153 (1964)
563. M.R.M. Rao, System of ordinary difference equations and stability with respect to manifolds. J. Math. Phys. Sci. **8**, 285–296 (1973)
564. V.S.H. Rao, Asymptotically of measure differential equations. Nonlinear Anal. TMA **27**(2), 483–489 (1978)
565. V.S.H. Rao, Integral inequalities of Gronwall type for distributions. J. Math. Anal. Appl. **72**, 545–550 (1979)
566. M.R.M. Rao, V.S.H. Rao, Stability of impulse by perturbed systems. Bull. Aust. Math. Soc. **16**(1), 99–110 (1976)
567. M.R.M. Rao, C.P. Tsokos, Integro-differential equations of Volterra type. Bull. Aust. Math. Soc. **3**, 9–22 (1970)
568. D. Rasmussen, Gronwall's inequality and its generalizations for functions of one independent variable, M. S. thesis, Brigham young University, 1972
569. D.L. Rasmussen, Gronwall's inequalities for functions of two independent variables. J. Math. Anal. Appl. **55**, 407–417 (1976)

570. J. Rauch, X. Zhang, E. Zuazua, Polynomial decay for a hyperbolic-parabolic coupled system. J. Math. Pure Appl. **84**, 407–470 (2005)
571. R. Redheffer, W. Walter, A comparison theorem for difference inequalities. J. Differ. Equ. **44**, 111–117 (1982)
572. R.M. Redheffer, Differential and integral inequalities. Proc. Am. Math. Soc. **15**, 715–716 (1964)
573. M. Renardy, W.J. Hrusa, J.A. Nohel, Mathematical problems in viscoelasticity. π *Pitman Monographs and Surveys in Pure and Applied Mathematics*, vol. 35 (Longman Scientific & Technical, New York, 1987)
574. J.C. Robinson, *Infinite-Dimensional Dynamical Systems* (Cambridge Univ. Press, Cambridge, 2001)
575. J. Rauch, X. Zhang, E. Zuazua, Polynomial decay for a hyperbolic-parabolic coupled system. J. Math. Pures Appl. **84**, 407–470 (2005)
576. A.B. Ronkov, D.D. Bainov, Integral equations and inequalities of Volterra type for functions defined in partially ordered spaces. J. Math. Anal. Appl. **125**(2), 483–507 (1987)
577. A.B. Ronkov, D.D. Bainov, Nonlinear integral inequalities for functions defined in partially ordered topological spaces. Nonlinear Anal. TMA **11**(3), 297–304 (1987)
578. T. Runst, W. Sickel, *Sobolev Spaces of Fractional Order, Nemytskij Operators, and Nonlinear Partial Differential Equations* (Waltwer de Gruyter, Berlin, New York, 1996)
579. S. Salem, On some system of two discrete inequalities of Gronwall Type. J. Math. Anal. Appl. **208**, 553–566 (1997)
580. S. Salem, K. Raslan, Some new discrete inequalities and their applications. J. Inequ. Pure Appl. Math. **5**(1), article 2, 1–9 (2004)
581. S.G. Samko, A.A. Kilbas, O.I. Marichev, *Fractional Integrals and Derivatives, Theory and Applications* (Gordon and Breach, Yverdon, 1993)
582. A.M. Samoilenko, S.D. Borysenko, Integro-sum inequalities and stability of processes with discrete disturbances, in *Proc. Third Int. Conf. Differential Equations and Applications*, Russe, Bulgaria, 1987, pp. 377–380
583. A.M. Samoilenko, S.D. Borysenko, On functional inequalities of Bihari type for discontinuous functions. Uspekhi Mat. Nauk **53**(4), 147–148 (1998)
584. A.M. Samoilenko, N.A. Preestyuk, Stability of solutions of differential equations with impluse effect. Differ. Equ. **13**(11), 1981–1992 (1977) (in Russian)
585. A.M. Samoilenko, N. Perestyuk, *Differential Equations with Impulse Effect* (Visha Shkola, Kyiv, 1987)
586. A.M. Samoilenko, N.A. Preestyuk, *Implusive Differential Equations, world Scientific Series on Nonlinear Science, Series A: Monographs and Treatises*, vol. 14 (World Scientific, New Jersey, 1995)
587. A.M. Samoilenko, S.D. Borysenko, E. Laserra, G. Matarazzo, Wendroff type integro-sum inequalities and applications. Math. Notes **3**(2), 123–132 (2002)
588. A.M. Samoilenko, S.D. Borysenko, C. Cattanl, V. Yasinsky, *Differential Models: Construction, Irepresentions and Applications* (Naukova Dumka, Kyiv, 2001)
589. A.M. Samoilenko, S.D. Borysenko, C. Cattani, G. Matarazzo, V. Yasinsky, *Differential Models: Stability, Inequalities and Estimates* (Naukova Dumka, Kyiv, 2001)
590. H. Sano, N. Kunimatsu, Modified Gronwall's inequality and it's application to stabilization problem for semilinear parabolic systems. Syst. Control Lett. **22**, 145–156 (1994)
591. G. Sansone, R. Conti, *Nonlinear Differential Equations* (Pergaman, London/New York, 1964)
592. T. Sato, A. Iwasaki, Sur'équation intégral de Volterra. Proc. Jpn. Acad. **31**, 395–398 (1955)
593. W.W. Schmaedeke, Optimal control theory for nonlinear vector differenial equations containing mesures. J. Siam Control **3**, 231–280 (1965)
594. W.W. Schmaedeke, G.R. Sell, The Gronwall inequality for modified Stieljes integrals. Proc. Am. Math. Soc. **19**, 1217–1222 (1986)
595. L. Schwartz, *Théorie des distributions, I, II*. Act. Sci. Ind., vol. 1091, 1122 (Hermann et Cie., Paris, 1951)
596. I. Segal, Nonlinear semigroups. Ann. Math. **78**(2), 339–364 (1963)

597. T. Sengadir, Existence and stability of nonlinear functional differential equations. J. Math. Anal. Appl. **197**, 890–907 (1996)
598. R.P. Shastri, D.Y. Kasture, Wendroff type inequalities. Proc. Am. Math. Soc. **72**, 248–250 (1987)
599. R.P. Shastri, D.Y. Kasture, A note on Bihari type inequalities in two independent variables. Indian J. Pure Appl. Math. **19**, 659–663 (1988)
600. M.H. Shih, C.C. Yeh, Some integral inequalities in n-independent variables. J. Math. Anal. Appl. **84**, 569–583 (1981)
601. M.H. Shih, C.C. Yeh, The Gronwall-Bellman's inequality in several variables. J. Math. Anal. Appl. **86**, 157–167 (1982)
602. B.B. Shinde, B.G. Pachpatte, On multidimensional simultaneous integral inequalities. J. Math. Phys. Sci. **17**, 219–243 (1983)
603. J.A. Shohat, J.D. Tamarkin, *The problem of Moments, Mathematical Surveys, Number 1* (Amer. Math. Soc., Providence, 1943)
604. S. Shoshal, M.A. Massod, Gronwall's vector inequality and its applications to a class of non-self adjoint linear and nonlinear hyperbolic partial differential equations. J. Indian Math. Soc. **38**, 383–394 (1974)
605. R.E. Showalter, Existence and representation theorems for a semilinear Sobolev equation in Banach space. SIAM J. Math. Anal. **3**, 527–543 (1972)
606. R.E. Showalter, *Hilbert Space Methods for Partial Differential Equations* (Pitman, London, 1977)
607. D.T. Shum, On a class of new inequailities. Trans. Am. Math. Soc. **204**, 299–341 (1975)
608. B.A. Shuvar, M.N. Kopach, Multidimensional Analogues of Gronwall-Bellman inequality. Dokl. Akad. Nauk Ukr. SSS. **4**(A), 32–35 (1983) (in Russian)
609. P.S. Simeonov, D.D. Bainov, On some generalizations of the Gronwall-Bellman integral inequality for scalar functions of many variables. Circ. Mat. Palermo **32**(2), 27–38 (1983)
610. S.M. Singare, B.G. Pachpatte, On some discrete generalizations of an inequality of Gronwall. Chin. J. Math. **6**, 121–135 (1978)
611. S.M. Singare, B.G. Pachpatte, Discrete generalized Gronwall inequalities in three independent variables. Pacific J. Math. Vol. **82**, 197–210 (1979)
612. S.M. Singare, B.G. Pachpatte, On certain discrete inequalities of the Wendroff type. Indian J. Pure Appl. Math. **11**, 727–736 (1980)
613. V.M. Singare, B.G. Pachpatte, Lower bounds on some integral inequalities in n-independent variables. Indian J. Pure Appl. Math. **12**(3), 318–331 (1981)
614. L.J. Slater, *Generalized Hypergeometric Functions* (Cambridge Univ. Press, Cambridge, 1966)
615. M. Slater, H.S. Wilf, A class of linear differential-difference equations. Pacific J. Math. **10**, 1419–1427 (1960)
616. M. Slemrod, E. Infante, *An Invariance Principle for Dynamical Systems on Banach Space, Instability of Continuous Systems*, ed. by H. Leipholz (Springer, Berlin, 1971), pp. 215–221
617. J. Smoller, *Shock Waves and Reaction-Diffusion Equations* (Springer, New York, 1983)
618. I.N. Sneddon, *Elements of Partial Differential Equations* (McGraw-Hill, New York, 1957)
619. D.R. Snow, *A Two Independent Variables Gronwall Type Inequality, in Inequalities III* (Proceedings of a sysmposium held at the Univ. California, Los Angeles, 1969), ed. by O. Shisha (Academic Press, New York, 1972)
620. D.R. Snow, Gronwall's inequality for systems of partial differential equations in two independent variables. Proc. Am. Math. Soc. **33**, 46–54 (1972)
621. V.A. Solonnikov, Estimates of solutions of a non-stationary linearized system of Navier-Stokes equations. Trudy Mat. Inst. Steklov **70**, 213–317 (1964) (in Russian)
622. P. Souplet, étude des solutions globales de certaines équations différentielles ordinaires du second ordre non linéaires. C. R. Acad. Sci. Paris **313**, 365–370 (1991)
623. R.P. Sperb, *Maximum Principles and Their Applications* (Academic Press, New York, London, Toronto, Sydney San Francisco, 1981)

624. O.J. Staffans, On a nonlinear hyperbolic Volterra equation. SIAM J. Math. Anal. **11**, 793–812 (1980)
625. A. Stokes, The application of a fixed point theorem to a variety of non-linear stability problems. Contrs. Theory Nonl. Oscills. **5**, 173–184 (1960)
626. M.H. Stone, *Linear Transformation in Hilbert Space and Their Applicatons to Analysis*. Amer. Math. Soc., Colloquium Publications, vol. XV (Amer. Math. Soc., Probidence, 1932)
627. A. Strauss, J.A. Yorke, Perturbation theorem for ordinary differential equations. J. Differ. Equ. **3**, 15–30 (1967)
628. S. Sugiyama, On the stability problems of difference equations. Bull. Sci. Enger. Res. Lab. Waseda Univ. **35**, 140–144 (1969)
629. S. Sugiyama, Comparsion theorems on difference equations. Bull. Sci. Eng. Res. Lab. Waseda Univ. **47**, 77–82 (1970)
630. S. Sugiyama, On the asymptotic behaviors of solutions of difference equations. Proc. Jpn. Acad. (2nd edn.) **47**, 477–480 (1971)
631. J. Sugie, Continuability of solutions of the generalized Liénard system with time lag. Proc. Jpn. Acad. **60**, 357–360 (1984)
632. R. Szwarc, Nonlinear integral inequalities of Volterra type. Math. Proc. Camb. Phil. Soc. **111**, 599–608 (1992)
633. G. Szegö, *Orthogonal Polynomials*, 4th edn. Amer. Math. Soc. Colloquium Publications, vol. XXIII (Amer. Math. Soc., Providence, Rhode Island, 1975)
634. T. Takahashi, Remarks on some inequalities. Tôhoku Math. J. **36**, 99–108 (1932)
635. J. Tamarkine, Sur. le théorème d'unicité des solutions des équations différentielles ordinaires. Math. Zeitschrift **16**, 207–212 (1923)
636. H. Tanabe, *Functional Analytic Methods for Partial Differential Equations*, Monographs and Textbooks in Pure and Appl. Math., vol. 204 (Marcel Dekker, New York/Basel/Hong Kong 1997)
637. N.E. Tatar, Exponential decay for a semilinear problem with memory. Arab J. Math. Sci. **7**(1), 29–45 (2001)
638. N.E. Tatar, The role of an integral inequality in the study of certain differential equations. J. Ineq. Pure Appl. Math. **6**(5), article 136 (2005)
639. N.E. Tatar, An impulsive nonlinear singular version of the Gronwall-Bihari inequality. J. Ineq. Appl. **2006**, 1–12 (2006)
640. O. Taussky, A recurring theorem on determinants. Am. Math. Monthly **55**, 672–676 (1949)
641. C.C. Teh, Wendroff's inequality in n independent variables. J. Math. Anal. Appl. **78**, 78–87 (1980)
642. E. Thandapani, A note on Bellman-Bihari type integral inequalities. J. Math. Phy. Sci. **21**, 1–3 (1987)
643. E. Thandapani, R.P. Agarwal, On some new inequalities in n indepednet variables. J. Math. Anal. Appl. **86**, 542–561 (1982)
644. E.C. Titchmarsh, *Eigenfunction Expansions Associated with Second Order Differential Equations, Part I* (Oxford Univ. Press, London/New York, 1946)
645. J. Tong, The asymptotic behavior of a class of nonlinear differential equations of second order. Proc. Am. Math. Soc. **84**(2), 235–236 (1982)
646. P.Ch. Tsamatos, S.K. Ntouyas, On a Bellman-Bihari type inequality with delay. Period. Math. Hungar. **23**, 91–94 (1991)
647. J. Traple, On a boundary value oproblem for systems of ordinary differential equations of second order. Zeszyty Nauk. Uniw. Jagiellon Prace Mat. **15**, 159–168 (1971)
648. W.F. Trench, On the asymptotic behavior of solutions of second order linear differential equations. Proc. Am. Math. Soc. **14**, 12–14 (1963)
649. F.G. Tricomi, *Integral Equations* (Interscience Publishers, New York, 1957)
650. W.J. Trjitzinsky, Singular non-linear integral equations. Duke Math. J. **11**, 517–564 (1944)
651. Z.B. Tsalyuk, On some multidimensional integral inequalities. Differentsialnyie Uravnenija **19**, 1828–1830 (1983) (in Russian)

652. P.C. Tsamatos, S.K. Ntouyas, On a Bellman-Bihari type inequality with delay. Period. Math. Hungar. **23**, 91–94 (1991)
653. M. Tsutsumi, I. Fukunda, On solutions of the derivative nonlinear Schrödinger equations: existence and uniqueness theorem. Funkcial. Ekvac. **23**, 259–277 (1980)
654. V.A. Vaigant, Problem of existence of global solutions of Navier-Stokes equations of compressible, Continuum, Doctoral (Phys.-Math.) Dissertation, Barnaul, 1998
655. P.T. Vaz, Ph. D. Thesis, University of Bombay, 1985, pp. 36–40
656. M. Vidyasagar, Nonlinear functional inequalities of Gronwall-Bihari type (to appear)
657. B. Viswanatham, A generalization of Bellman's lemma. Proc. Am. Math. Soc. **14**, 15–18 (1963)
658. W. Walter, *Differential and Integral Inequalities* (Springer, Berlin, 1970). German edition 1964
659. P. Waltman, Some properties of solutions of $u'' + a(t)f(u) = 0$. Monatsh. Math. **67**, 50–54 (1963)
660. P. Waltman, On the asymptotic behavior of solutions of a nonlinear equation. Proc. Am. Math. Soc. **15**, 919–923 (1964)
661. P. Waltman, A note on a functional equation. J. Math. Anal. Appl. **10**, 330–333 (1965)
662. P. Waltman, *Deterministic Threshold Models in the Theory of Epidemics*, Lecture Notes in Biomathematics, vol. 1 (Springer, New York, 1974)
663. C.L. Wang, A short proof of a Greene theorem. Proc. Am. Math. Soc. **69**, 357–358 (1978)
664. W.S. Wang, A generalized retarded Gronwall-like inequality in two variables and applications to BVP. Appl. Math. Comput. **191**, 144–154 (2007)
665. W.S. Wang, C.X. Shen, On a generalized retarded integral inequality with two variables. J. Ineq. Appl. **2008**, Article ID 518646, 9 p. (2008)
666. T. Ważewski, Systèmes des équations et des inégalités différentielles ordinaires aux deuxièmes membres monotones et leurs applications. Ann. Soc. Polon. Math. **23**, 112–166 (1950)
667. T. Ważewski, Remarque sur un système d'inégalités intégrales. Ann. Polon. Math. **3**, 210–212 (1957)
668. T. Ważewski, Sur un système des inéqualities integrales ordinaires non linéares. Bull. Acad. Polon. Sci. Ser. Sci. Math. Astronom. Phys. **17**, 226–229 (1969)
669. D.V. Widder, An inequality related to one of Hilbert's. J. Lond. Math. Soc. **4**, 194–198 (1929)
670. J.E. Wilkins, On the growth of solutions of linear differential equations. Bull. Am. Math. Soc. **50**, 388-394 (1944)
671. D. Willett, Nonlinear vector integral equations as contraction mappings. Arch. Rat. Mech. Anal. **15**, 79–86 (1964)
672. D. Willett, A linear generalization of GronWall's inequality. Proc. Am. Math. Soc. **16**, 774–778 (1965)
673. D. Willett, J.S.W. Wong, On the discrete analogues of some generalizations of GronWall's inequality. Monatsh. Math. **69**, 362–367 (1965)
674. D. Willett, J.S.W. Wong, On the discrete analogues of some generalization of GronWall's inequality. Monatsh **69**, 362–367 (1965). H. El-OWaidy et al., Appl. Math. Comput. **106**, 289–303 (1999)
675. F.H. Wong, C.C. Yeh, C.H. Hong, GronWall inequalities on time scales. Math. Ineq. Appl. **9**(1), 75–86 (2006)
676. J.S.W. Wong, On two theorems of Waltman. SIAM J. Appl. Math. **14**, 724–728 (1966)
677. J.S.W. Wong, A discrete analogue of Opial's inequality. Cand. Math. Bull. **10**, 115–118 (1967)
678. J.S.W. Wong, Boundedness theorems for solutions of $u''(t) + a(t)f(u)g(u') = 0$ (IV). Enseignement Math. **13**, 157–165 (1967)
679. J.S.W. Wong, On an integral inequality of Gronwall. Rev. Roum. Math. Pures Appl. **12**, 1519–1522 (1967)
680. J.S.W. Wong, Bounds for solutions of nonlinear integro-differential equations. J. Math. Anal. Appl. **23**, 487–499 (1968)
681. V.A. Yakubovič, On the asymptotic behavior of the solutions of systems of differential equations. Dokl. Akad. Nauk. SSSR **63**, 363–366 (1948)

682. Y. Yamada, Some remarks on the equation $y_{tt} - \sigma(y_x)y_{xx} - y_{xtx} = f$. Osaka J. Math. **17**, 303–323 (1980)
683. J.A. Yan, *Introduction to Martingales and Stochastic Integrals* (Shanghai Sci. Tech. Press, Shanghai, 1981)
684. J. Yan, A. Zhao, Oscillation and stability of linear impulsive delay differential equations. J. Math. Anal. Appl. **227**, 187–194 (1998)
685. B. Yang, On Hilbert's integral inequality. J. Math. Anal. Appl. **220**, 778–785 (1998)
686. E. Yang, A generalization of Bihari's inequality and its applications to nonlinear Volterra integral equations. Chin. Ann. Math. **3**(2), 209–216 (1982)
687. E.H. Yang, On some new discrete inequalities of the Bellman-Bihari type. Nonlinear Anal. **7**, 1237–1246 (1983)
688. E.H. Yang, Boundedness conditions for solutions of the differential equation $(a(t)x')' + f(t,x) = 0$. Nonlinear Anal. TMA **8**(5), 541–547 (1984)
689. E.H. Yang, On the most general form of Bellman-type linear inequalities involving multiple-fold integral functionals. J. Math. Anal. Appl. **101**(1), 184–197 (1984)
690. E.H. Yang, On some new integral inequalities in N independent variables. J. Math. Anal. Appl. **109**, 171–181 (1985)
691. E.H. Yang, On some new n-independent-variable discrete inequalities of the Gronwall type. Int. J. Math. Math. Sci. **9**, 485–495 (1986)
692. E.H. Yang, On some new discrete generalizations of Gronwall's inequality. J. Math. Anal. Appl. **129**, 505–516 (1988)
693. E.H. Yang, Some integral inequalities in N independent variables. Acta Math. Appl. Sinica **10**(2), 158–168 (1994)
694. E.H. Yang, Generalizations of Pachpatte's integral and discrete inequalities. Ann. Differ. Equ. **13**, 180–188 (1997)
695. E.H. Yang, On some nonlinear integral and discrete inequalities related to Ou-Yang's inequality. Acta Math. Sinica (New Series) **14**(3), 353–360 (1998)
696. E.H. Yang, A new integral inequality with power nonlinearity and its discrete analogue. Acta Math. Appl. Sinica English Ser. **17**(2), 233–239 (2001)
697. E.H. Yang, L.R. Gan, Boundedness and global uniform lipschitz stability for solutions of two-dimensional differential systems. Acta Math. Appl. Sinica **17**(3), 364–373 (1994)
698. G.S. Yang, Inequalites of opial type in two variables. Tamkang J. Math. **13**, 255–259 (1982)
699. Q. Ye, Z. Li, *An Introduction to Reaction Diffusion Equations* (Science Press, Beijing, 1994), (in Chinese)
700. H. Ye, J. Gao, Y. Ding, A generalized Gronwall inequality and its application to a fractional differential equation. J. Math. Anal. Appl. **328**, 1075–1081 (2007)
701. C.C. Yeh, Wendroff's inequality in n-independent variables. J. Math. Anal. Appl. **78**, 78–87 (1980)
702. C.C. Yeh, Bellman-Bihari integral inequalities in several variables. J. Math. Anal. Appl. **87**, 311–332 (1982)
703. C.C. Yeh, On some integral inequalities in n-independent variables and their applications. J. Math. Anal. Appl. **86**, 387–410 (1982)
704. C.C. Yeh, Discrete inequalities of the Gronwall-Bellman type in n-independent variables. J. Math. Anal. Appl. **105**, 322–332 (1985)
705. C.C. Yeh, Discrete inequalities of the Gronwall-Bellman type in n-independent variables. J. Math. Anal. Appl. II **106**, 282–285 (1985)
706. C.C. Yeh, M.H. Shih, Some integral inequalities in n independent variables. J. Math. Anal. Appl. **84**, 569–583 (1981)
707. C.C. Yeh, M.H. Shih, The Gronwall-Bellman inequality in several variables. J. Math. Anal. Appl. **86**(1), 157–167 (1982)
708. K. Yosida, *Functional Analysis*, 6th edn. (Springer, New York, 1980)
709. E.C. Young, Gronwall's inequality in n independent variables. Proc. Am. Math. Soc. **41**, 241–244 (1973)
710. E.C. Young, Functional integral inequalities in N variables. Chin. J. Math. **10**, 1–7 (1982)

711. E.C. Young, On Bellman-Bihari integral inequalities. Int. J. Math. Math. Sci. **5**, 97–103 (1982)
712. E.C. Young, On integral inequalities of Gronwall-Bellman type. Proc. Am. Math. Soc. **94**(4), 636–640 (1985)
713. A.C. Zaanen, *Linear Analysis* (North-Holland, Amsterdam, 1953)
714. A.I. Zahariev, D.D. Bainov, A note on Bellman-Gronwall's inequality. J. Math. Anal. Appl. **80**(1), 147–149 (1981)
715. A.D. Ziebur, On the Gronwall-Bellman lemma. J. Math. Anal. Appl. **22**, 92–95 (1968)
716. W.P. Ziemer, *Weakly Differentiable Functions*, Graduate Texts in Mathematics, vol. 120 (Springer, New York, 1989)
717. B.-G. Zhang (Ping-ken Chang), Boundedness of solutions of ordinary differential equations of the second order. Acta Math. Sinica **14**, 128–136 (1964). English translation, Chin. Math. **4**, 139–148 (1964)
718. W. Zhang, Generalized exponential dichotomies and invariant manifolds for diferential equations. Adv. Math. Chin. **22**, 1–45 (1993)
719. B. Zhang, Y. Shen, A generalization of Bellman-Gronwall integral inequality. J. Math. Res. Expo. **2**, 83–88 (1985)
720. W. Zhang, S. Deng, Projected Gronwall-Bellman's inequality for integral functions. Math. Comput. Model. **34**(3–4), 393–402 (2001)
721. W. Zhang, R.P. Agarwal, E. Akin-Bóhner, On well-posedness of impulsive problems for nonlinear parabolic equations. Nonlinear Stud. **9**(2), 145–153 (2002)
722. X. Zhao, F. Meng, On some advanced integral inequalities and their applications. J. Ineq. Pure Appl. Math. **6**(3), article 60, pp. 8 (2005) pages.
723. K. Zheng, Y. Wu, S. Deng, Nonlinear integral inequalities in two independent variables and their applications. J. Ineq. Appl. **2007**, article no. 032949 (2007)

Index

Y. Qin, *Integral and Discrete Inequalities and Their Applications*,
DOI 10.1007/978-3-319-33304-5

Printed by Printforce, the Netherlands